CONCEPTS IN ABSTRACT ALGEBRA

The Brooks/Cole Series in Advanced Mathematics
Paul J. Sally, Jr., Editor

Series editor Paul J. Sally, Jr. and Brooks/Cole have developed this prestigious list of books for classroom use. Written for post-calculus to first year graduate courses, these books maintain the highest standards of scholarship from authors who are leaders in their mathematical fields.

Titles in this prestigious series include:

Probability: The Science of Uncertainty with Applications to Investments, Insurance, and Engineering (0-534-36603-1)
Michael A. Bean

A Course in Approximation Theory (0-534-36224-9)
Ward Cheney and Will Light

Advanced Calculus: A Course in Mathematical Analysis (0-534-92612-6)
Patrick M. Fitzpatrick

Fourier Analysis and Its Applications (0-534-17094-3)
Gerald B. Folland

Introduction to Analysis, Fifth Edition (0-534-35177-8)
Edward Gaughan

The Mathematics of Finance: Modeling and Hedging (0-534-37776-9)
Joseph Stampli and Victor Goodman

Algebra: A Graduate Course (0-534-19002-2) and
Geometry for College Students (0-534-35179-4)
I. Martin Isaacs

Numerical Analysis: Mathematics of Scientific Computing, Second Edition (0-534-33892-5)
David Kincaid and Ward Cheney

Ordinary Differential Equations (0-534-36552-3)
Norman Lebovitz

Introduction to Mathematical Modeling (0-534-38478-1)
Daniel Maki and Maynard Thompson

Introduction to Fourier Analysis and Wavelets (0-534-37660-6)
Mark A. Pinsky

Real Analysis (0-534-35181-6)
Paul Sally, Jr.

Beginning Topology (0-534-42426-0)
Sue Goodman

CONCEPTS IN ABSTRACT ALGEBRA

Charles Lanski
University of Southern California

THOMSON
BROOKS/COLE

Australia • Canada • Mexico • Singapore • Spain
United Kingdom • United States

THOMSON
BROOKS/COLE

Publisher: *Bob Pirtle*
Assistant Editor: *Stacy Green*
Editorial Assistant: *Katherine Cook*
Marketing Manager: *Tom Ziolkowski*
Marketing Assistant: *Erin Mitchell*
Advertising Project Manager: *Bryan Vann*
Project Manager, Editorial Production: *Cheryll Linthicum*
Art Director: *Vernon Boes*
Print/Media Buyer: *Doreen Suruki*
Permissions Editor: *Kiely Sexton*
Production Service: *Matrix Productions*
Text Designer: *John Edeen*
Copy Editor: *Connie Day*
Illustrator: *International Typesetting and Composition*
Cover Designer: *Vernon Boes and Denise Davidson*
Compositor: *International Typesetting and Composition*
Text and Cover Printer: *Quebecor World Color/Kingsport*

Printed in the United States of America

1 2 3 4 5 6 7 08 07 06 05 04

Library of Congress Control Number: 2004105799

ISBN 0-534-42323-X

Thomson Brooks/Cole
10 Davis Drive
Belmont, CA 94002
USA

Asia
Thomson Learning
5 Shenton Way #01-01
UIC Building
Singapore 068808

Australia/New Zealand
Thomson Learning
102 Dodds Street
Southbank, Victoria 3006
Australia

Canada
Nelson
1120 Birchmount Road
Toronto, Ontario M1K 5G4
Canada

Europe/Middle East/Africa
Thomson Learning
High Holborn House
50/51 Bedford Row
London WC1R 4LR
United Kingdom

Latin America
Thomson Learning
Seneca, 53
Colonia Polanco
11569 Mexico D.F.
Mexico

Spain/Portugal
Paraninfo
Calle Magallanes, 25
28015 Madrid, Spain

CONTENTS

0

REVIEW 1

1

PRELIMINARIES 51

2

GROUPS 95

3

SPECIAL GROUPS 128

4

SUBGROUPS 157

5

NORMAL SUBGROUPS AND QUOTIENTS 191

6

MORPHISMS 216

7

STRUCTURE THEOREMS 246

8

CONJUGATION 274

9

GROUP ACTIONS 294

10

RINGS 321

11

IDEALS, QUOTIENTS, AND HOMOMORPHISMS 355

12

FACTORIZATION IN INTEGRAL DOMAINS 390

13

COMMUTATIVE RINGS 412

14

FIELDS 456

15

GALOIS THEORY 488

PREFACE

What prompted me to add to the collection of upper-division-level abstract algebra texts? Those I have used have not been a great help to most students in mastering the concepts of abstract algebra. In addition, these texts usually do not have enough exercises on the theorems and techniques. Some students learn the lecture material well, but the time constraints of lectures typically allow only a minimal discussion of the material. Unfortunately, the treatment of material in most texts is not much different from the presentation in a lecture. A text ought to be a reference for students and should provide a fuller and richer account of material than a lecture can. A text ought to present interesting examples of sufficient complexity that a student can see the concepts and results used in a nontrivial setting. A text ought to provide many exercises that require the use and synthesis of the techniques and results. Finally, a text must be readable and mathematically interesting enough for a typical student to want to read it. *Concepts in Abstract Algebra* is a text that does all of this.

The style and structure of *Concepts in Abstract Algebra* are intended to help students learn the core concepts and associated techniques in algebra deeply and well. In addition, the text will help students learn the art of constructing mathematical arguments. My experience in using the text has been gratifying; the typical student has done noticeably better (still far from perfect!) in learning the statements of the definitions and theorems and in constructing arguments using them. The better students actually read sections I tell the classes to skip—they have found typos in them and cited results from them on exams. I find in this experience justification for thinking that students really can and will read and learn from this text.

Much in the text, even in early sections of chapters, may not be appropriate for a given course. A bit later I will discuss the content of the chapters and indicate the core material for groups, for rings, and for fields, and I will mention those theorems that are cited in many subsequent proofs. First I will comment on pedagogical features of the exposition that distinguish this text from most others. Little specific prerequisite material is assumed, but prior experience with college-level mathematics is important. Thus students should already have had courses in calculus. A course in linear algebra is also helpful since it is a bit more abstract than calculus, and some matrix calculation is used here for examples, although the relevant definitions are given in Chapter 2.

Deep and thorough learning of mathematical ideas requires an accurate recall of definitions and results. This same precision is needed for constructing arguments; the justification for each step in a proof is usually some definition or result, so the failure to have an extensive and accurate recall of these leads to an inability to complete, and often to begin, a proof. *Concepts in Abstract Algebra* helps students to learn the definitions and results and gives meaning to them via extensive discussion, example, development in nontrivial contexts, repetition, and use in exercises. These principles pervade the exposition. Besides the many examples discussed, additional numbered examples give a nontrivial context for thinking about the ideas in a richer setting. For example, students will understand cosets better by seeing the cosets of $SL(n,\boldsymbol{R})$ in $GL(n,\boldsymbol{R})$, or of $\langle(2, 3)\rangle$ in $\boldsymbol{Z} \oplus \boldsymbol{Z}$, than they will if the only examples are variants of the cosets of $\langle[3]_{12}\rangle$ in $\boldsymbol{Z}_{12}$.

In proofs the text references both core notions and associated results, reinforcing the learning of these. Complete proofs give students a template for writing their own arguments. It is important for students to see how elementary material is developed, more for the reinforcement of the core ideas than for the result of the development; consequences of definitions and elementary results are more interesting than the basic notions. For example, Cauchy's Theorem for Abelian groups is useful and important, and it is proved in three places to illustrate different techniques. One proof is an application of results on products of subgroups, another uses quotient groups, and the third uses group actions.

Discussion of an idea and then its development are useful for understanding the idea, but repetition of it is important for real learning. The text consciously tries to use earlier results in the development of new material. This approach takes precedence over the simplest or slickest arguments. One example is to introduce equivalence relations whenever feasible, another is to use congruences frequently, and yet another is to choose examples and results that require the correspondence theorem.

No presentation of mathematics, by itself, will lead to learning material thoroughly and well. Real understanding requires independent work and thought about the concepts, results, and techniques. Thus a vital part of any text is its collection of exercises. Computational exercises are necessary for students to gain familiarity with some ideas, but often texts do not provide enough exercises on the theorems themselves. *Concepts in Abstract Algebra* contains a wealth of exercises focused on the results and associated techniques. Many of these exercises are quite challenging, and I hope interesting, for the typical student. The point, of course, is to have students think seriously about the results and how they are used.

The exposition provides the extended discussion, the nuance, and the examples that time does not permit in a standard course of lectures, so it is best used in conjunction with instruction to the students on what parts to study, what parts to skim, and what parts to skip. Dependence on earlier material is mostly obvious, so sections and parts of sections can be omitted easily as appropriate for different courses. Since references are specific, any not covered in class can be found, examined, and understood quite readily.

In the next few paragraphs, I discuss briefly the material in each chapter and indicate the essential results for later references. It is important to remember that often much of the material in a given section illustrates the core notions and could be skipped. Chapter 0 is a reference on topics that students should know, but often do not, and can be omitted, assigned, or covered in part as needed. I have found it useful to have students skim this chapter and review the material on partitions in §0.2. There is extensive material on induction for those students who need it. It is important for the students to read §0.6 on the connection between bijections and inverses: Theorems 0.11 and 0.13 are cited fairly often, and the symmetric groups are previewed in Theorem 0.14. This latter result can be deferred harmlessly until symmetric groups are introduced.

Chapter 1 contains core material on divisibility in $\boldsymbol{Z}$, equivalence relations, and the integers modulo n ($\boldsymbol{Z}_n$). The division theorem in $\boldsymbol{Z}$ (Theorem 1.1) is crucial, and

Theorem 1.2 is important for $\mathbf{Z}_n$. §1.2 and §1.3 through Theorem 1.12 give the basic results on divisibility, GCDs, the Euclidean algorithm, and relative primeness. These are all required later, particularly for groups. Theorem 1.21 through Theorem 1.24, and Theorem 1.28 are the basic results on equivalence relations, congruence computations, and $\mathbf{Z}_n$. The rest of the material—on squares, sums of squares, prime factorization in $\mathbf{Z}$, solutions of linear congruences, divisibility tests by arbitrary primes, and localizations of $\mathbf{Z}$—is not particularly relevant later, and some of it may already be familiar to students.

Chapter 2 introduces groups with the definition and examples constituting §2.1, and uniqueness properties and powers discussed in §2.2. §2.3 on groups of symmetries, characterizing such finite groups, is not used later (the dihedral groups are defined again, algebraically, in §3.4). The material in §2.4 is used often, particularly Theorems 2.7 through 2.9. The last two sections contain the subgroup theorem (Theorem 2.12), the introduction of cyclic subgroups, the often-cited Theorem 2.16 relating the order of an element to the order of the cyclic subgroup it generates, and less important material on the subgroup generated by a subset, centralizers, and the commutator subgroup.

In Chapter 3, the important Theorem 3.1 characterizes subgroups of cyclic groups. §3.2 introduces the unit groups U_n of $\mathbf{Z}_n$, used for examples, and gives results on the noncyclicity of U_n and the multiplicative property of the Euler phi-function. §3.3 treats S_n, even and odd permutations, and A_n. Theorem 3.16 on the generation of A_n is used only for the simplicity of A_n in Chapter 5. §3.4 introduces the dihedral groups D_n algebraically via congruences to make computation in these groups trivial. Theorem 3.19 and Theorem 3.20 on finite direct sums are useful for examples, but the rest of §3.5 on the cyclicity of direct sums, semi-direct products, and infinite direct sums is more peripheral.

The crucial material in Chapter 4 is contained in the first two or three sections. These introduce cosets (Poincaré's Theorem on finite intersections is not used later), Lagrange's Theorem and its consequences, including Euler's Theorem and Fermat's Theorem, and, in §4.3, results on the product of subgroups. The rest of the chapter explores and develops these ideas. Results in §4.4 are needed mainly for the Fundamental Theorem of Finite Abelian groups in Chapter 7. The remaining sections prove Cauchy's Theorem for Abelian groups, that finite multiplicative subgroups in a field are cyclic, and that U_n is cyclic for $n = p^t$ or $2p^t$, $p > 2$ prime. Carmichael numbers are characterized, and the chapter ends with applications to the RSA algorithm, linear codes, syndrome decoding, and Hamming codes.

§5.1 covers normal subgroups and leads to quotient groups in §5.2, which consists mostly of examples. Although it would be easier to identify the quotient groups in these examples using isomorphisms, our goal is to acquaint students with computation with cosets, to show the importance of nice coset representatives, and to motivate the need for a formal way to identify different groups that are essentially the same. §5.3 uses quotients to prove Cauchy's Theorem for Abelian groups and the standard result on the commutator. The last section proves the simplicity of A_n for $n \geq 5$ and essentially proves the simplicity of $PSL(n,F)$ for $n > 2$ or $|F| \geq 4$.

Chapter 6 starts with homomorphisms (Theorem 6.1 characterizes groups of order $2p$, p an odd prime) and gives the basic computational results on them. Theorem 6.2 and Theorem 6.4 are the most important ones, leading to the First Isomorphism Theorem (Theorem 6.6 in §6.3). §6.3 gives applications of this result, including the Chinese Remainder Theorem, but the most referenced is Theorem 6.8 on cyclic groups. §6.4 gives more sophisticated results: the regular representation, producing normal subgroups, the butterfly lemma, and a hint of Jordan-Holder. The last section defines automorphisms. The key results here, Theorem 6.19 and the description of $Aut(\mathbf{Z}_n^k)$, are needed later for the classification of groups via Sylow's theorems in §8.4 or Chapter 9.

Chapter 7 is the heart of a more substantial treatment of groups. It begins with the correspondence theorem for quotients, covers the Second and Third Isomorphism Theorems, and discusses direct sum decompositions. With the aim of exploring the techniques developed so far and exhibiting their limitations, §7.4 classifies all groups of order at most 15, except for those of order 12, done later in Chapter 9 (Example 9.22). §7.5 proves and applies the Fundamental Theorem of Finite Abelian Groups.

Chapter 8 introduces conjugacy and derives the class equation that is used to explore conjugates and centralizers in S_n, and the structure of p-groups. The last section discusses Sylow subgroups, proves their existence using induction, and applies them to obtain some structure results. Much of this chapter can be viewed as special cases of the material in Chapter 9 and could be skipped and then sampled while covering Chapter 9.

§9.1 introduces group actions. The orbit-stabilizer result is proved, leading to a proof of Cauchy's Theorem, and is applied in §9.2 to discuss orbits and Polya-Burnside counting. In §9.3 the three Sylow theorems are proved using group actions, and a number of nontrivial applications of these follow to end the chapter and the material on group structures.

Chapter 10 begins the study of rings with definitions and examples, including endomorphism rings, group rings, and rings of functions. §10.2 introduces subrings, intersections and unions, and centralizers. §10.3 defines polynomial and power series rings in several indeterminates and defines Laurent series rings; Theorem 10.10 identifies the units in these rings. §10.4 introduces zero divisors and domains: Theorems 10.11 and 10.12 concern zero divisors in polynomial rings. Fields are defined, quotient fields of integral domains are discussed, and characteristic is introduced. The last section shows how indeterminates and polynomial rings in arbitrary sets of indeterminates can be defined formally using rings of functions.

Chapter 11 introduces techniques for studying the structure of rings. The notion of ideals (also one-sided ideals) is the important one in §11.1 that characterizes ideals in matrix rings and in $F[x]$ for F a field and proves results on intersections, unions, sums, and products of ideals. Quotient rings, using the construction for groups, appear in §11.2. Homomorphisms are defined in §11.3, and the three usual isomorphism theorems are in §11.4 along with consequences, including direct sum decompositions for $\mathbf{Z}_n$ and for the groups U_n. §11.5 treats the correspondence theorem and describes the fields $F[x]/(m(x))$ in Theorem 11.18, which is used to introduce finite fields other than $\mathbf{Z}_p$ and is essential for §14.4 on splitting fields. §11.6 defines and explores the notions of (right) Artinian and Noetherian rings.

§12.1 covers divisibility, primes, and irreducibles in integral domains. In §12.2 PIDs are studied and shown to be UFDs, the subject of §12.3, which covers primitive polynomials, treats Gauss's Lemma, shows that R a UFD implies $R[X]$ is a UFD, and ends with Eisenstein's Criterion for irreducibility. The last section, on Euclidean domains, shows that $\mathbf{Z}[(-m)^{1/2}]$ is not Euclidean when $m \geq 3$ and is Euclidean when $m = 2, 1, -2, -3, -6$, or -7. The Gaussian integers $\mathbf{Z}[i]$ also appear, along with the characterization of primes and prime factorization in this ring.

Chapter 13 consists of more advanced material on commutative rings for more sophisticated courses or for independent study. §13.1 covers Zorn's Lemma, treats prime and maximal ideals, characterizes the Jacobson radical, presents a version of Nakayama's Lemma, and characterizes the nil radical and units in $R[x]$. §13.2 introduces the important notion of localization, including the contraction and extension of ideals and the relations between these. §13.3 discusses Noetherian rings, proves the Hilbert Basis Theorem, shows that the radical of an ideal is a finite intersection of primes, and characterizes the zero divisors in Noetherian rings as the union of the finitely many maximal annihilator ideals. §13.4 covers integrality, including integral closures, lying over, and going up. A consequence is Zariski's Theorem, that a field finitely generated over a subfield is algebraic, and this yields the maximal ideals in $F[x_1, \dots, x_n]$ for F algebraically closed. §13.5 continues with the basic notions of algebraic geometry, the Hilbert Nullstellensatz, and irreducible varieties. §13.6 shows the equivalence of Zorn's Lemma, the Axiom of Choice, and the existence of a well-ordering of sets. Zorn's Lemma is used to prove cardinality results for infinite sets, needed for the discussion of algebraic closures and transcendence bases in Chapter 14.

Chapter 14 addresses field extensions and starts with a review of vector space concepts using Zorn's Lemma to prove the existence of bases and the uniqueness of dimension for arbitrary vector spaces. This generality is not required if only finite dimensional field extensions are discussed in the course. §14.2 describes simple extensions, relative dimension, and joins. §14.3 proves the impossibility of the three classical problems on straightedge and compass constructions. The existence of splitting fields and of finite fields, and the important isomorphism theorems (Theorems 14.16 through 14.18) for Galois theory make up §14.4. These lead to the existence and uniqueness of algebraic closures in §14.5. Finally, §14.6 covers transcendental extensions and transcendence dimension. The elementary symmetric functions are shown to be a transcendence basis of $F(x_1, \dots, x_n)$, and the chapter ends with a proof of Lüroth's Theorem.

Chapter 15 is on Galois theory. §15.1 covers the Galois group, priming operations, separability, and equivalences for a Galois extension (Theorem 15.2). §15.2 contains the characterization of simple extensions and the Fundamental Theorem. Applications in §15.3 are to finite field extensions and to symmetric polynomials, and those in §15.4 are to cyclotomic polynomials and extensions. Some remarks on solvable groups and on solvable Galois groups in §15.5 prepare the way for §15.6, which proves the main results relating solvable Galois groups to radical extensions.

The descriptions above should give a good indication of which sections might be appropriate to cover for a given course. By a judicious choice of material, one could use the text for an array of courses from an introductory one on the integers, groups and rings, to a more sophisticated, almost graduate-level course. It is impractical to suggest what sections to cover for each of many possible courses, but I will make some comments that should be helpful. The sections mentioned need not be covered completely, and the results needed for reference are mentioned above.

It is useful for typical students to have at least a quick review of partitions (pp. 11–13), induction as needed, and §0.6 on bijections and inverses. Chapter 1 through Theorem 1.15, and §1.5 through Theorem 1.24 in §1.7 should be covered or assigned in part. The basic material on groups is Chapter 2, except for §2.3, then §3.1, §3.3, and pp. 148–151 in §3.5 (§3.2 and §3.4 are useful for examples), §4.1–§4.3, §5.1, §5.2, and §6.1–§6.3 through p. 228. Material on groups that could be added include any of the following: §5.3 to Cauchy's theorem, then perhaps Theorems 4.16 and 4.17; the beginning of §6.4 and/or §6.5; §7.1

and/or §7.3; §8.1, perhaps adding §8.2 or §8.3; §9.1 and §9.2. For courses that do more on groups, all of Chapter 8 or most of Chapter 9 could be covered. To get through all of Chapter 7 and Chapter 9 would come close to a graduate-level course on groups.

The basic material on rings is §10.1, most of §10.4, §11.1 through Theorem 11.3 (and perhaps Theorem 11.5), the first half of §11.2, §11.3, the first half of §11.4, and Theorem 11.18 in §11.5. Additional material includes any of §10.2, §10.3, the missing parts of §11.1, §11.2, or §11.4, and §12.1, then any of §12.2, §12.3, or §12.4 (§12.3 and §12.4 are mostly independent of §12.2, but the definition of PID is needed in §12.4). For more advanced students or courses, §10.5 is worth a peek, §11.5 is important, and §11.6 could be added. To cover Chapter 13 would give an almost graduate-level introduction to commutative rings. §13.1 is essential for this chapter, and the remaining sections are more or less independent, except that §13.3 is needed for §13.5.

Some very introductory material on fields appears in the last half of §10.4 and in §11.1 but is not required in Chapter 14 on field extensions. The basic material here is §14.2–§14.4 and §15.1. §14.1 is not needed for finite dimensional extensions if students know about dimension in vector spaces. Additional material might be any of §15.2 with perhaps either §15.3 or §15.4. More advanced material would be what is left in Chapter 14 and Chapter 15: §14.5 and §14.6 require §13.6, and §15.6 requires §15.5.

These descriptions and suggestions ought to make it clear that the text is adaptable to a wide range of courses, semester- or year-long, from an introductory overview to a deeper and more sophisticated treatment. Whatever the case, students will find a readable reference that will help them to learn results and techniques deeply and well.

There are a few people I would like to thank for their part in this project. I was fortunate that there was a textbook series whose pedagogical goals fit well with my intentions for the text and I am very grateful to Paul Sally for recognizing this and accepting the text for his series with Brooks/Cole. I want to thank Bob Pirtle for his interest in the text and his willingness to put up, in good humor, with my concerns and anxieties at the outset of the whole publication process. There are many people at Brooks/Cole and at Matrix Productions whose hard work was essential in the production of this text and who remain unknown to me personally, but I thank them nonetheless. The two people that I have worked with most are Cheryll Linthicum at Brooks/Cole and Merrill Peterson at Matrix Productions, and I want to express my gratitude to them for their professionalism, their kindness, and their efforts to help me through the production process.

Finally to my wife Elizabeth, whom I dearly love, who solves the important problems, and who makes my life as good as one has a right to expect, I thank profusely for her love, understanding, and humor over the last few years during all of those evenings and weekends that I spent working on the text; it is not clear which of us will be happier that it is now finished.

CHAPTER 0

REVIEW

This chapter reviews some basic but important notions about sets, induction, and functions and provides a substantial reference for the idea of partition, methods of induction, and the notions of inverse image and bijection. These topics arise in important ways in the material to follow and are not often discussed in sufficient detail in other texts.

0.1 SETS

We assume that the concept of set and the standard operations with sets are familiar, so we will review these quickly but carefully. A **set** is a collection of objects called the **elements** of the set. We write $v \in D$ to mean that v is an element of the set D and write $v \notin D$ otherwise. Sets A and B are *equal*, written $A = B$, when they have precisely the same elements.

We can describe a set by specifying its elements inside brackets { }, so $\{a, e, i, o, u\}$ is the set of vowels. The particular manner or order of listing the elements of a set is not relevant, and listing some elements multiple times yields the same set as listing each element once. For example, $\{1, 4, 9\} = \{4, 9, 1\} = \{4, 1, 4, 9\} = \{(-1)^2, (-2)^2, (-3)^2, 1, (10 - 1)\}$.

It is not always convenient or possible to list explicitly all elements of a set. Typically $\{a, b, \ldots, z\}$ denotes the set of letters in the English alphabet, but the "..." introduces ambiguity since it assumes that a clear pattern is described. It is important that "..." convey exactly what we mean. Is $\{2, 4, \ldots, 128\}$ the set of all even whole numbers from 2 to 128, or is it the first seven powers of 2? The meaning of $\{2, 4, 6, 8, \ldots, 128\}$ is clearer.

Often a set is defined as those objects satisfying some condition. Two sets whose elements are described by different conditions are equal if the elements satisfying either condition satisfy the other. For example, let $A =$ {all nonnegative real numbers} and $B =$ {all squares of all real numbers}. If $r \geq 0$ is a real number, then r has $r^{1/2}$ as a real square root and $(r^{1/2})^2 = r$, so r is the square of some real number. Thus all numbers in A satisfy the condition describing numbers in B. But if $r = a^2$ for some real number a, then $r \geq 0$. Since the conditions describing the elements in A and in B are satisfied by the same real numbers, $A = B$. Let us look at another example like this.

Example 0.1 If $T = \{\text{all real roots of } (x^4 - 2x^3 - x^2 + 2x)^{10}\}$ and $S = \{\text{all integers } z \text{ whose square is at most 5 more than } z\}$, then $S = T$.

The integers in S are those satisfying $x^2 - x \leq 5$, and a quick check shows that $S = \{-1, 0, 1, 2\}$. The roots of $(x^4 - 2x^3 - x^2 + 2x)^{10}$ are the same as the roots of the polynomial $(x^4 - 2x^3 - x^2 + 2x) = (x^3 - x)(x - 2)$, so we have $T = \{-1, 0, 1, 2\}$ also, and $S = T$.

We give the standard notation for a few important sets:

$\boldsymbol{N} = \{1, 2, 3, 4, \ldots\}$, called the *natural numbers*;

$\boldsymbol{Z} = \{\ldots, -3, -2, -1, 0, 1, 2, \ldots\}$, called the *integers*;

$\boldsymbol{Q} = \{\text{all quotients of integers}\}$, called the *rational numbers*;

$\boldsymbol{R} = \{\text{all real numbers}\}$.

Every element in $\boldsymbol{N}$ is in $\boldsymbol{Z}$, and the elements in either of these sets are in $\boldsymbol{R}$. The general situation is described next.

DEFINITION 0.1 A set S is a **subset** of a set T, written $S \subseteq T$, if every element of S is also an element of T. When $S \subseteq T$ and $S \neq T$, we say that S is a **proper subset** of T and may write $S \subset T$.

When S and T are sets but S is not a subset of T, we write $S \not\subseteq T$. The definition of subset is quite useful and tells us how to show a set S is a subset of a set T: take an arbitrary element of S and show that it must be in T.

Subsets of a set S are usually defined by some property $P(x)$ that each element $x \in S$ either has or fails to have. The collection of all elements in S satisfying $P(x)$ is a subset of S written $\{x \in S \mid P(x)\}$ and read as "the set of x in S such that $P(x)$." For example, $\boldsymbol{N} = \{x \in \boldsymbol{Z} \mid x > 0\} \subseteq \boldsymbol{Z}$, and in $\boldsymbol{R}$ the interval $[0, 1] = \{x \in \boldsymbol{R} \mid 0 \leq x \leq 1\} \subseteq \boldsymbol{R}$. Note that *any* $A \subseteq S$ can be written as $A = \{x \in S \mid x \in A\}$. For any set S, letting $P(x)$ be "$x \notin S$" yields the subset $\{x \in S \mid x \notin S\} \subseteq S$ containing all elements in S that are not in S! That is, $\{x \in S \mid x \notin S\}$ contains no elements. This set without elements is called the **empty set** and is denoted by $\varnothing$, so $\{x \in S \mid x \notin S\} = \varnothing \subseteq S$ for *any set* S. If we think of a subset as a choice of elements from a set, then the choice of no elements at all is a (mathematically!) legitimate one. Note that $\varnothing \notin \varnothing$ because $\varnothing$ has no elements, but that $\varnothing \subseteq \varnothing$. To prove this inclusion using Definition 0.1, we must verify the implication: if $x \in \varnothing$ then $x \in \varnothing$. Logically this is true because the hypothesis is always false.

DEFINITION 0.2 The implication "P implies Q," written $P \Rightarrow Q$, is true if either its hypothesis P is false or its conclusion Q is true. The statement $P \Leftrightarrow Q$, which is read "P if and only if Q," abbreviates $P \Rightarrow Q$ *and* $Q \Rightarrow P$.

The definition shows why the proof of a statement $P \Rightarrow Q$ often proceeds by assuming that P is true and using this to prove that Q is true: when P is false the implication is automatically true. Also the definition shows what is logically behind many proofs by contradiction that assume Q is false and argue that P must be false. From Definition 0.2, the implication $P \Rightarrow Q$ is true when Q is true, and if Q is false forces P to be false, then again Definition 0.2 shows that the implication $P \Rightarrow Q$ is true.

The concept of subset is useful in showing that two sets are equal. Since $A \subseteq B$ means that the elements of A are in B, and since $A = B$ exactly when they have the same elements, it follows that $A = B$ exactly when both $A \subseteq B$ and $B \subseteq A$. This observation is important and will be used frequently.

Example 0.2 If $T = \{$rightmost (units) digits of all positive integer multiples of 7$\}$ and $S = \{$all digits$\} = \{0, 1, 2, \ldots, 9\}$, then $S = T$.

Every element of T is a digit and so is in S; thus, $T \subseteq S$. To see that $S \subseteq T$ we need each digit to be the end (rightmost digit) of some multiple of 7. This is easy, since $7\underline{0} = 10 \cdot 7$, $2\underline{1} = 3 \cdot 7$, $4\underline{2} = 6 \cdot 7$, $6\underline{3} = 9 \cdot 7$, $1\underline{4} = 2 \cdot 7$, and the others are left to the reader.

Before doing other examples we introduce some useful notation.

DEFINITION 0.3 For $n, a \in \mathbf{Z}$, $n\mathbf{Z} + a = \{x \in \mathbf{Z} \mid x = nt + a \text{ for } t \in \mathbf{Z}\}$ and $nN + a = \{x \in \mathbf{Z} \mid x = nt + a \text{ for } t \in N\}$.

Thus $2\mathbf{Z} + 1$ is the set of odd integers, $7\boldsymbol{N}$ is the set of positive integer multiples of 7, and $5\mathbf{Z} + 3 = \{\ldots, -12, -7, -2, 3, 8, 13, 18, \ldots\}$ is the set of integers that are three more than a multiple of 5. Note that $n\mathbf{Z} = (-n)\mathbf{Z}$.

Example 0.3 $7\mathbf{Z} + 2 = 7\mathbf{Z} + 16$.

We must show that each set contains the other. Let $x \in 7\mathbf{Z} + 16$, so by Definition 0.3 $x = 7t + 16$ for some $t \in \mathbf{Z}$. We need $x \in 7\mathbf{Z} + 2$, and clearly $x = 7t + 16 = 7t + 14 + 2 = 7(t + 2) + 2 \in 7\mathbf{Z} + 2$. Thus $7\mathbf{Z} + 16 \subseteq 7\mathbf{Z} + 2$. Now take $x \in 7\mathbf{Z} + 2$ and write $x = 7m + 2$ for some $m \in \mathbf{Z}$. But $7m + 2 = 7(m - 2) + 16 \in 7\mathbf{Z} + 16$, so $7\mathbf{Z} + 2 \subseteq 7\mathbf{Z} + 16$ and the two inclusions show that $7\mathbf{Z} + 16 = 7\mathbf{Z} + 2$.

The approach in the example can show that two sets are not equal. Take $2 = 5 \cdot 0 + 2 \in 5\mathbf{Z} + 2$. If $2 \in 7\mathbf{Z} + 4$, then $2 = 7k + 4$ for some $k \in \mathbf{Z}$, and $7k = -2$ for $k \in \mathbf{Z}$ would follow, which is impossible. Thus not every element in $5\mathbf{Z} + 2$ is in $7\mathbf{Z} + 4$, so $5\mathbf{Z} + 2 \not\subseteq 7\mathbf{Z} + 4$ and these sets cannot be equal. To prove set equality in Example 0.3, we had to show that every element in each set was contained in the other, but to see that sets are not the same, it suffices to find a single element that is in one set but is not in the other. *A single example can disprove a general statement but usually cannot prove it.*

Next we give some ways of combining given sets to get new ones.

DEFINITION 0.4 The **difference** of sets S and T is $S - T = \{s \in S \mid s \notin T\}$; the **intersection** of S and T is $S \cap T = \{x \in S \text{ and } x \in T\}$; and the **union** of S and T is $S \cup T = \{x \in S \text{ or } x \in T\}$.

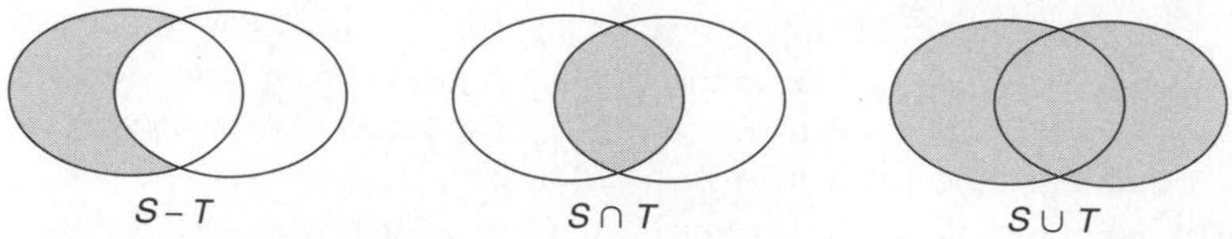

FIGURE 0.1

Pictures (called Venn diagrams) representing these relations are indicated in Fig. 0.1 by the shaded regions of the two ovals, where S is the oval on the left and T is the oval on the right.

Observe that $S - T \subseteq S$. When $T \subseteq S$, the difference $S - T$ is called the **complement** of T in S. Some easy illustrations are: $\mathbf{Z} - 2\mathbf{Z} = 2\mathbf{Z} + 1$, since eliminating the even integers from the set of all integers leaves the odd integers; $\{1, 3, 4, 5, 7\} - \{2, 4, 5, 8, 9\} = \{1, 3, 7\}$; and, using the usual interval notation in $\boldsymbol{R}$, $(-2, 5) - (1, +\infty) = (-2, 1]$.

Example 0.4 $2\mathbf{Z} - 4\mathbf{Z} = 4\mathbf{Z} + 2$ is the complement of $4\mathbf{Z}$ in $2\mathbf{Z}$.

Note that $2\mathbf{Z} - 4\mathbf{Z} \neq \{2t - 4s \mid t, s \in \mathbf{Z}\}$: $2\mathbf{Z} - 4\mathbf{Z}$ consists of even integers that are not multiples of 4. Now $4\mathbf{Z} + 2 \subseteq 2\mathbf{Z} - 4\mathbf{Z}$ if any $x \in 4\mathbf{Z} + 2$ is even but is not a multiple of 4. But $x = 4y + 2 = 2(2y + 1) \in 2\mathbf{Z}$ for some $y \in \mathbf{Z}$. Suppose that $x = 4k \in 4\mathbf{Z}$ also. Then $4k = 4y + 2$ forcing $2 = 4(k - y)$, a contradiction since $k, y \in \mathbf{Z}$ imply $k - y \in \mathbf{Z}$ but 2 is not an integer multiple of 4. Thus we must have $x \notin 4\mathbf{Z}$, and so $x \in 2\mathbf{Z} - 4\mathbf{Z}$ by Definition 0.4, proving the inclusion $4\mathbf{Z} + 2 \subseteq 2\mathbf{Z} - 4\mathbf{Z}$. On the other hand, if $x \in 2\mathbf{Z} - 4\mathbf{Z}$, then $x = 2m$ for some $m \in \mathbf{Z}$ but $x \neq 4s$ for any $s \in \mathbf{Z}$. Now m is even or odd. If $m = 2t$, then $x = 2m = 4t$, a contradiction. Hence m is odd so $m = 2t + 1$ for $t \in \mathbf{Z}$ and $x = 2m = 2(2t + 1) = 4t + 2 \in 4\mathbf{Z} + 2$. Therefore $2\mathbf{Z} - 4\mathbf{Z} \subseteq 4\mathbf{Z} + 2$, and the two containments prove that $2\mathbf{Z} - 4\mathbf{Z} = 4\mathbf{Z} + 2$.

We have defined similar-looking expressions that represent very different things, so we must be careful to use correct notation. For example, $2\mathbf{Z} - 1 = \{2t - 1 \in \mathbf{Z} \mid t \in \mathbf{Z}\}$ is the set of all integers that are one less than an even integer: the set of all odd integers, by Definition 0.3. However, from Definition 0.4 the set difference $2\mathbf{Z} - \{1\}$ is the set of even integers except for 1. Since 1 is not even, no $t \in 2\mathbf{Z}$ is removed, so $2\mathbf{Z} - \{1\} = 2\mathbf{Z}$. Thus using brackets { } can change the whole meaning of the expression.

The union of sets combines all the elements in either set into a new one, and the intersection consists of all elements common to both sets. The use of "or" in the definition of union is the "mathematical or"; *to say that an element is in one set or another means it is in* ***at least one*** *of the sets, not in exactly one of them.* Thus $S \cap T \subseteq S \cup T$. Also note that $S \cap T \subseteq S \subseteq S \cup T$. If we let $S = \{1, 2, 3, 4, 5\}$ and $T = \{4, 5, 6, 7\}$, then $S \cup T = \{1, 2, 3, 4, 5, 6, 7\}$ and $S \cap T = \{4, 5\}$. For intervals, $(-5, 4) \cup [-4, 4] = (-5, 4]$, $[1, 4] \cap (0, 3) = [1, 3)$, and $\boldsymbol{R} - (-1, 2) = (-\infty, -1] \cup [2, +\infty)$.

Since the definition of either union or intersection is symmetric in the two sets, it is clear that $A \cap B = B \cap A$ and $A \cup B = B \cup A$. There are a number of relations among intersection, union, and set difference. Some of these are given in the exercises, and we discuss them more generally in the next section. Two easy and useful facts for later reference come next.

THEOREM 0.1 Let S be a set and $A \subseteq S$. Then: a) $A \cup (S - A) = S$; and b) $A \cap (S - A) = \emptyset$.

Proof a) If $x \in A \cup (S - A)$, then $x \in A$ or $x \in S - A$ by the definition of union, and in either case $x \in S$, so $A \cup (S - A) \subseteq S$. But for $x \in S$, either $x \in A$ or $x \notin A$, so either $x \in A$ or $x \in S - A$ from the definition of set difference. Again the definition of union shows $S \subseteq A \cup (S - A)$. These inclusions prove $A \cup (S - A) = S$. b) Should $x \in A \cap (S - A)$, then $x \in A$ and $x \in S - A$ from the definition of intersection, so both $x \in A$ and $x \notin A$, a contradiction. Thus it is impossible for $x \in A \cap (S - A)$, which means that $A \cap (S - A) = \emptyset$. ■

Two sets are called **disjoint** when their intersection is empty, as in part b) of the theorem. For example, $2\boldsymbol{Z}, 2\boldsymbol{Z} + 1 \subseteq \boldsymbol{Z}$ are disjoint, and the intervals $(1, 2)$ and $[3, 4)$ in $\boldsymbol{R}$ are disjoint. We end this section by introducing two other constructions on sets. The first will be important later when we consider relations, and the second will be useful for examples.

DEFINITION 0.5 The **Cartesian product** of two sets A and B, denoted by $A \times B$, is $\{(a, b) \mid a \in A \text{ and } b \in B\}$. For elements $(a, b), (a', b') \in A \times B$, $(a, b) = (a', b')$ if and only if $a = a'$ and $b = b'$.

If either $A = \emptyset$ or $B = \emptyset$, then $A \times B = \emptyset$. For example, $\emptyset \times B = \emptyset$ since there can be no ordered pair whose first entry is an element of $\emptyset$.

Example 0.5 If $A \neq B$ and neither is empty, then $A \times B \neq B \times A$.

Since $A \neq B$ one of these sets has an element not in the other. If $a \in A - B$, then for $b \in B (B \neq \emptyset!)$, $(a, b) \in A \times B$ but $(a, b) \notin B \times A$. The reason is that if $(a, b) = (b', a') \in B \times A$, then by Definition 0.5 $a = b' \in B$, a contradiction to the choice of $a \in A - B$. Thus $A \times B \neq B \times A$ as claimed. If $B - A$ is not empty a corresponding argument gives the desired conclusion.

The example shows that the order of the sets in the Cartesian product is crucial. For a specific illustration, if $A = \{1, 2\}$ and $B = \{1, 2, 3\}$, then $A \times B = \{(1, 1), (1, 2), (1, 3), (2, 1), (2, 2), (2, 3)\}$ and $B \times A = \{(1, 1), (1, 2), (2, 1), (2, 2), (3, 1), (3, 2)\}$. Also $A \times A = \{(1, 1), (1, 2), (2, 1), (2, 2)\}$ consists of *all* possible ordered pairs chosen from $A : A \times A \neq \{(a, a) \in A \times A \mid a \in A\} = \{(1, 1), (2, 2)\}$. A familiar example of the use of a Cartesian product is the usual coordinate system in the plane, identifying points in the plane with ordered pairs in $\boldsymbol{R} \times \boldsymbol{R}$. We can use Cartesian products to represent the elements of other sets. For example, each $q = a/b \in \boldsymbol{Q}$ can be identified with some element of $\boldsymbol{Z} \times \boldsymbol{N}$ when a and b have no common factor and $b > 0$, by associating $q = a/b$ with (a, b).

DEFINITION 0.6 For any set S the collection of all subsets of S is a set $P(S)$ called the **power set** of S.

For any set S, $A \subseteq S \Leftrightarrow A \in P(S)$, so $\emptyset \in P(S)$ and $S \in P(S)$. Thus $P(\emptyset) = \{\emptyset\} \neq \emptyset$ since $\emptyset \subseteq \emptyset$ and $\emptyset$ is the only subset of $\emptyset$ since $\emptyset$ has no elements. When $S = \{a\}$ any subset of S other than $\emptyset$ must contain a and so be S itself: that is, $P(\{a\}) = \{\emptyset, \{a\}\}$. What is $P(A)$ if $A = \{a, b, c\}$? Any subset of A has either 0, 1, 2, or 3 elements, so $P(A) = \{\emptyset, \{a\}, \{b\}, \{c\}, \{a, b\}, \{a, c\}, \{b, c\}, \{a, b, c\}\}$. It is difficult to visualize $P(S)$ for S an infinite set. Even $S = \boldsymbol{N}$ has a huge collection of infinite subsets, besides its infinite collection of finite subsets.

If $s \in S$ then $\{s\} = \{x \in S \mid x = s\} \subseteq S$, so $\{s\} \in P(S)$. These single element subsets are called **singletons.** It is important to distinguish between $s \in S$ and the singleton $\{s\} \subseteq S$. The idea is to look at an element $s \in S$ in two ways: as itself and as the subset of S whose only element is s. Thus $\{s \in S\} = S \neq \{\{s\} \subseteq S \mid s \in S\}$, the set of all single-element *subsets* of S.

As we mentioned, our interest in power sets is mainly for examples. Using the notion of power set we can construct a set T with both $t \in T$ and $t \subseteq T$! Take $S \neq \emptyset$, let $T = S \cup P(S)$, and note that for any $s \in S$, $s \in T$ also. Hence $t_s = \{s\} \subseteq T$. But $\{s\} \subseteq S$ so $t_s \in P(S)$ and $P(S) \subseteq T$, so $t_s \in T$. Consequently both $t_s \in T$ and $t_s \subseteq T$.

EXERCISES

1. Write the elements of the following sets explicitly:
   a) {all integers less than 30 and a sum of two squares of integers}
   b) {all integers less than 30 and a sum of three integer squares}
   c) {all three-digit positive integers whose digits are all odd and strictly increasing from left to right}
   d) {all real numbers r whose square equals an integer power of r}
   e) {all real numbers r satisfying $r^2 - 3r < -2$}
   f) {all subsets of $\{a, b, c\}$ that contain a}
2. Write each of the following sets as $\{x \in S \mid P(x)\}$ and specify $P(x)$:
   a) {all integer multiples of 3 between -100 and 1,000}
   b) {all points in the first quadrant of the plane, with the usual Cartesian coordinate system, but not on the x-axis or the y-axis}
   c) {all fractions with odd denominators}
   d) {all points in $\boldsymbol{R} \times \boldsymbol{R}$ whose distance from (0, 0) is rational}
   e) {all *pairs* of points in $\boldsymbol{Z} \times \boldsymbol{Z} \subseteq \boldsymbol{R} \times \boldsymbol{R}$ so that the midpoint in $\boldsymbol{R} \times \boldsymbol{R}$ of the segment joining the two points is in $\boldsymbol{Z} \times \boldsymbol{Z}$}
3. Determine whether the two sets in each given pair are equal.
   a) $\{\sum_{i=0} a_i x^i \in \boldsymbol{R}[x] \mid \sum_{i=0} a_i = 0\}$ and $\{(x-1)g(x) \in \boldsymbol{R}[x] \mid g(x) \in \boldsymbol{R}[x]\}$
   b) $[-1, 1] \subseteq \boldsymbol{R}$ and $\{r \in \boldsymbol{R} \mid r = \sin t - \cos t$ for some $t \in \boldsymbol{R}\}$
   c) $\{n \in \boldsymbol{N} \mid n$ is the units digit of 3^k for some $k \in \boldsymbol{N}\}$, and $\{n \in \boldsymbol{N} \mid n$ is the units digit of 7^k for some $k \in \boldsymbol{N}\}$.
4. If $S = 3\boldsymbol{N}$ and $T = \{n \in \boldsymbol{N} \mid$ the digits in n sum to $m \in 3\boldsymbol{N}\}$, show that $S = T$. (Write $n = a_k 10^k + \cdots + a_1 10 + a_0$ for digits $0 \le a_j \le 9$ and use $10^i - 1 = 3z_i$.)
5. Let $\mathrm{I}(50) = \{1, 2, 3, \ldots, 50\} \subseteq \boldsymbol{N}$, $S = \{x \in \mathrm{I}(50) \mid x = z^2$ for some $z \in \boldsymbol{Z}\}$, $E = \{x \in \mathrm{I}(50) \mid x \in 2\boldsymbol{Z}\}$, and $A_i = \{i, i+1, i+2, i+3\}$ for $1 \le i \le 47$.

Explicitly find the integers in each of the following:

a) $(\mathrm{I}(50) - E) \bigcap (A_1 \bigcup A_{47})$
b) $S \bigcup (A_5 \bigcup A_{47})$
c) $S \bigcap (A_4 \bigcup E)$
d) $(A_7 \bigcap A_8) \bigcap A_9$
e) $S \bigcap (E - ((A_1 \bigcup A_8) \bigcup A_{15}))$

6. Let $\mathrm{I}(30) = \{1, 2, 3, \ldots, 30\} \subseteq \boldsymbol{N}$. Explicitly find:
a) $\mathrm{I}(30) \bigcap (5\mathbf{Z} - 3\mathbf{Z})$
b) $(10\mathbf{Z} + 2) - ((5\mathbf{Z} + 2) \bigcap 2\mathbf{Z})$
c) $(\mathrm{I}(30) - 2\mathbf{Z}) \bigcap ((\mathrm{I}(30) - 3\mathbf{Z}) \bigcap (\mathrm{I}(30) - 5\mathbf{Z}))$

7. If A and B are sets, show that:
i) $A \bigcap B = A \Leftrightarrow A \subseteq B$
ii) $A \bigcup B = A \Leftrightarrow B \subseteq A$

8. For any sets A and B, show that $A = B \Leftrightarrow P(A) = P(B)$.

9. For any sets A, B, and C, use the definitions to show that:
a) $A \bigcap (B \bigcap C) = (A \bigcap B) \bigcap C$
b) $A \bigcup (B \bigcup C) = (A \bigcup B) \bigcup C$
c) $A \bigcap (B \bigcup C) = (A \bigcap B) \bigcup (A \bigcap C)$
d) $A - (A - B) = A \bigcap B$
e) $A \bigcup (B \bigcap C) = (A \bigcup B) \bigcap (A \bigcup C)$
f) $(A - B) - C = A - (B \bigcup C)$
g) $(A \bigcap B) - C = (A - C) \bigcap (B - C)$
h) $A \bigcap (B - C) = (A \bigcap B) - (A \bigcap C)$
i) $(A \bigcup B) - C = (A - C) \bigcup (B - C)$
j) $A - (B - C) = (A - B) \bigcup (A \bigcap B \bigcap C)$

10. In each case find sets A, B, C, and D so that the statement is false.
a) $A = B \Leftrightarrow A - C = B - C$
b) $A = B \Leftrightarrow A \bigcup C = B \bigcup C$
c) $A = B \Leftrightarrow A \bigcap C = B \bigcap C$
d) $(A - B) \times (C - D) = (A \times C) - (B \times D)$

11. If $C \times D = C \times F$, show that either $C = \varnothing$ or $D = F$.

12. For sets A, B, and C show that:
i) $(A \bigcup B) \times C = (A \times C) \bigcup (B \times C)$
ii) $(A \bigcap B) \times C = (A \times C) \bigcap (B \times C)$
iii) $(A - B) \times C = (A \times C) - (B \times C)$

13. For sets A and B, prove or disprove:
i) $P(A \bigcup B) = P(A) \bigcup P(B)$
ii) $P(A \bigcap B) = P(A) \bigcap P(B)$
iii) $P(A - B) = P(A) - P(B)$

0.2 INDEX SETS AND PARTITIONS

To distinguish one set from another in a small collection of sets, we can use different letters, but this is impractical in general. Integer subscripts such as $A_1, A_2, \ldots$ will do for finite and some infinite collections, but we need arbitrary sets of labels for general collections of sets (see §7).

DEFINITION 0.7 For any set $\Lambda \neq \emptyset$, $\{S_\lambda\}_\Lambda$ denotes a collection of sets labeled by Λ; that is, for each $i \in \Lambda$, there is one set $S_i \in \{S_\lambda\}_\Lambda$, and every set $A \in \{S_\lambda\}_\Lambda$ satisfies $A = S_i$ for some $i \in \Lambda$. Λ is called the **index** set for $\{S_\lambda\}_\Lambda$.

Finite collections of sets can be considered indexed. For example, given the sets A, B, and C, setting $A = A_1$, $B = A_2$, and $C = A_3$ shows that $\{A, B, C\} = \{A_i\}_{\{1,2,3\}}$ is indexed by $\Lambda = \{1, 2, 3\}$. The indexing labels for large collections of sets are usually naturally related to the sets.

Example 0.6 For $n \in \boldsymbol{N}$ let $H_n = \{-n, -n+1, \ldots, 0, 1, 2, \ldots, n^2\}$.

Here we have $\{H_n\}_{\boldsymbol{N}}$ with $H_n \subseteq \boldsymbol{Z}$ finite; for example, $H_1 = \{-1, 0, 1\}$, $H_3 = \{-3, -2, \ldots, 0, 1, 2, \ldots, 9\}$, and $H_{20} = \{-20, -19, \ldots, 400\}$. When $m \leq n$, $H_m \subseteq H_n$ and $H_m \neq H_n$ if $m \neq n$. No two H_j are disjoint since $H_1 \subseteq H_i \cap H_j$.

Other infinite sets arise naturally as labels. Recall from algebra or calculus that $\boldsymbol{Q}[x] = \{a_n x^n + \cdots + a_1 x + a_0 \mid \text{all } a_j \in \boldsymbol{Q} \text{ and } n \geq 0\}$ is the set of polynomials with coefficients in $\boldsymbol{Q}$. Similarly $\boldsymbol{R}[x]$ is the set of all polynomials having all coefficients $a_i \in \boldsymbol{R}$.

Example 0.7 For $p(x) \in \boldsymbol{Q}[x] - \{0\}$, set $A_{p(x)} = \{r \in \boldsymbol{R} \mid p(r) = 0\}$, so $A_{p(x)}$ is the set of roots of $p(x)$ in $\boldsymbol{R}$.

Each $A_{p(x)} \subseteq \boldsymbol{R}$ is finite since nonzero polynomials in $\boldsymbol{Q}[x]$ (or in $\boldsymbol{R}[x]$) of degree n have at most n roots in $\boldsymbol{R}$. For $\{r_1, \ldots, r_k\} \subseteq \boldsymbol{Q}$, $\{r_1, \ldots, r_k\} = A_{p(x)}$ where $p(x) = (x - r_1)(x - r_2) \cdots (x - r_k)$. Many $A_{p(x)}$ are empty, for example $A_{x^2+k} = \emptyset$ when $k > 0$, and many of the $A_{p(x)}$ are equal; indeed $A_{p(x)} = A_{p(x)^m}$ for any $m \in \boldsymbol{N}$, and also $A_{x^4-1} = A_{x^2-1} = \{-1, 1\}$.

Next we consider sets of intervals in $\boldsymbol{R}$.

Example 0.8 For $i \in \boldsymbol{Z}$, let $B_i = [i, i+1] \subseteq \boldsymbol{R}$ and $D_i = (-\infty, i) \subseteq \boldsymbol{R}$.

This gives two collections of subsets of $\boldsymbol{R}$ indexed by $\boldsymbol{Z}$: $\{B_i\}_{\boldsymbol{Z}}$ and $\{D_i\}_{\boldsymbol{Z}}$. In the first case $B_i \cap B_{i+1} = \{i+1\}$ but when $j \neq i \pm 1$, $B_i \cap B_j = \emptyset$. However, when $i \leq j$ then $D_i \subseteq D_j$ so $D_i \cap D_j \neq \emptyset$ for any $i, j \in \boldsymbol{Z}$.

Index sets allow us to extend the notions of intersection and union from two sets to many. The ideas remain the same: the intersection consists of elements common to all sets in the collection, and the union consists of those elements in at least one set in the collection.

DEFINITION 0.8 The **union** of the sets S_λ in $\{S_\lambda\}_\Lambda$, is $\bigcup_\Lambda S_\lambda = \{x \text{ in at least one } S_\lambda\}$ and the **intersection** of the sets S_λ is $\bigcap_\Lambda S_\lambda = \{x \text{ in every } S_\lambda\}$.

Equivalently, $\bigcup_\Lambda S_\lambda = \{x \mid \text{there is } \lambda \in \Lambda \text{ with } x \in S_\lambda\}$ and $\bigcap_\Lambda S_\lambda = \{x \mid \text{for all } \lambda \in \Lambda, x \in S_\lambda\}$. When Λ is understood, we may write $\bigcup_\lambda S_\lambda$ or $\bigcup S_\lambda$ for $\bigcup_\Lambda S_\lambda$ and $\bigcap_\lambda S_\lambda$ or

$\bigcap S_\lambda$ for $\bigcap_\Lambda S_\lambda$. The definitions of union and intersection are symmetric in all the constituent sets, so the order in which the sets are given is irrelevant. Two useful consequences of the definitions are that $\bigcap_\lambda S_\lambda \subseteq S_i$ for all $i \in \Lambda$ and that if all $S_\lambda \subseteq T$, then $\bigcup_\lambda S_\lambda \subseteq T$. The proofs of these use Definition 0.8 (verify!).

Let us look again at $\{B_i\}_Z$ and $\{D_i\}_Z$ in Example 0.8. It is fairly easy to see that $\bigcup B_i = \boldsymbol{R}$ since every real number is in some interval $[i, i+1]$ and to see that $\bigcap B_i = \emptyset$. Similarly $\bigcup D_i = \boldsymbol{R}$ and $\bigcap D_i = \emptyset$ (why?). Note that the intersection of any *finite* number of the D_j is not empty! A more interesting example in $\boldsymbol{R}$ follows.

Example 0.9 Find $\bigcup_N A_j$ and $\bigcap_N A_j$ for $A_n = [1/2^n, 1 - 1/2^{n+1}] \subseteq \boldsymbol{R}$.

To see what the A_i are, write out the first few: $A_1 = [1/2, 3/4]$, $A_2 = [1/4, 7/8]$, $A_3 = [1/8, 15/16]$, and $A_4 = [1/16, 31/32]$. As n increases it appears that the endpoints of A_n approach 0 and 1, so we guess that $\bigcup A_j = (0, 1)$. To prove this we need to show two containments. For any $n \in \boldsymbol{N}$, $0 < 1/2^n < 1 - 1/2^{n+1} < 1$ so each $A_n \subseteq (0, 1)$, and from the definition of union, $\bigcup A_j \subseteq (0, 1)$. For the other inclusion, if $x \in (0, 1)$ choose $m \in \boldsymbol{N}$ with both $1/2^m < x$ and $1/2^m < 1 - x$. This is possible since the elements in $\{1/2^j \mid j \in \boldsymbol{N}\}$ get arbitrarily small and $x, 1 - x > 0$. But $1/2^m < x < 1 - 1/2^m < 1 - 1/2^{m+1}$ so $x \in A_m$, and it follows that $x \in \bigcup A_j$. Hence $(0, 1) \subseteq \bigcup A_j$, proving that $\bigcup A_j = (0, 1)$. Next observe that $A_i \subseteq A_{i+1}$ for all $i \in \boldsymbol{N}$, so $A_1 \subseteq \bigcap A_j \subseteq A_1$ since the intersection is a subset of each constituent set. Thus $\bigcap A_j = A_1 = [1/2, 3/4]$.

The next result gives general distributive laws for intersections and unions and gives some relations between intersection, union, and difference.

THEOREM 0.2 Let A be a set and $\{S_\lambda\}_\Lambda$ a collection of sets. Then

a) $A \cap \bigcup_\lambda S_\lambda = \bigcup_\lambda (A \cap S_\lambda)$;
b) $A \cup \bigcap_\lambda S_\lambda = \bigcap_\lambda (A \cup S_\lambda)$;
c) $A - \bigcup_\lambda S_\lambda = \bigcap_\lambda (A - S_\lambda)$;
d) $A - \bigcap_\lambda S_\lambda = \bigcup_\lambda (A - S_\lambda)$.

Proof We prove these using the definitions to show that the sets appearing in each equality are subsets of each other. The definitions translate the symbols into their English equivalents. Throughout we let $\bigcup B_\lambda = \bigcup_\lambda B_\lambda$ and $\bigcap B_\lambda = \bigcap_\lambda B_\lambda$. In a), $x \in A \cap \bigcup S_\lambda \Leftrightarrow x \in A$ and $x \in \bigcup S_\lambda$ from the definition of intersection. But $x \in A$ and $x \in \bigcup S_\lambda \Leftrightarrow x \in A$ and $x \in S_\lambda$ for some $\lambda \in \Lambda$ (definition of union), so $x \in A \cap \bigcup S_\lambda \Leftrightarrow$ for some $\lambda \in \Lambda$, $x \in S_\lambda$, and $x \in A$. The definition of intersection shows that this last condition is equivalent to $x \in A \cap S_\lambda$ for some $\lambda \in \Lambda$, which itself is equivalent to $x \in \bigcup (A \cap S_\lambda)$ by the definition of union. Therefore $x \in A \cap \bigcup S_\lambda \Leftrightarrow x \in \bigcup (A \cap S_\lambda)$, proving a). The proof of b) is similar and is left as an exercise. To see that c) holds, let $x \in A - \bigcup S_\lambda$. The definition of difference shows this is equivalent to $x \in A$ and $x \notin \bigcup S_\lambda$. But $x \notin \bigcup S_\lambda$ forces $x \notin S_\lambda$ for every $\lambda \in \Lambda$, so $x \in A - \bigcup S_\lambda \Leftrightarrow$ for all $\lambda \in \Lambda$, $x \in A$ and $x \notin S_\lambda \Leftrightarrow$ for all $\lambda \in \Lambda$, $x \in A - S_\lambda$. Finally $x \in A - \bigcup S_\lambda \Leftrightarrow x \in \bigcap (A - S_\lambda)$ from the definition of intersection, proving c). The proof of d) is left as an exercise. ■

Parts a) and b) of the theorem are the distributive laws relating unions and intersections. Since $A \cap B = B \cap A$, part a) can be written as $(\bigcup S_\lambda) \cap A = \bigcup A \cap S_\lambda$ and part b) can be changed similarly. Part d) is illustrated by $D - (A \cap B \cap C) = (D - A) \cup (D - B) \cup (D - C)$. When all $S_\lambda \subseteq A$, parts c) and d) of Theorem 0.2 are known as DeMorgan's Laws. Their statements and proofs are the logical negation of union and intersection: an element is not in the union of $\{S_\lambda\}$ exactly when it fails to be in *every* S_λ, and an element is not in their intersection exactly when it fails to be in some S_λ.

Example 0.10 Characterize $S = \bigcap_{i \in \mathbf{Z}}((-\infty, i] \cup [i+1, \infty))$.

Write $(-\infty, i] \cup [i+1, \infty) = \mathbf{R} - (i, i+1)$ and use Theorem 0.2 to see that $S = \bigcap_{\mathbf{Z}}(\mathbf{R} - (i, i+1)) = \mathbf{R} - \bigcup_{\mathbf{Z}}(i, i+1) = \mathbf{Z}$, since this set difference represents all real numbers that are not strictly between two consecutive integers.

Example 0.11 $\bigcup_{\mathbf{N}}(\mathbf{Q} \cap [1/2^n, 1 - 1/2^{n+1}]) = \mathbf{Q} \cap (0, 1)$.

If $A_n = [1/2^n, 1 - 1/2^{n+1}] \subseteq \mathbf{R}$, then Example 0.9 shows that $\bigcup A_j = (0, 1)$, so $\mathbf{Q} \cap (0, 1) = \mathbf{Q} \cap (\bigcup A_j) = \bigcup(\mathbf{Q} \cap A_j)$ using Theorem 0.2.

Let us quickly discuss one more interesting consequence of Theorem 0.2. Suppose that $A_1, \ldots, A_k \subseteq S$ with the property that each $S - A_j$ is finite. Then $\bigcap A_i$ also has finite complement in S since by Theorem 0.2, $S - \bigcap A_i = \bigcup (S - A_i)$, and a finite union of finite sets clearly is finite.

We have considered unions and intersections for arbitrary collections of sets. What about Cartesian products of arbitrary collections of sets? An appropriate definition must await a discussion of functions, but we will briefly consider Cartesian products of finite collections of sets.

DEFINITION 0.9 The **Cartesian product** of the sets $A_1, \ldots, A_n$ is the set $A_1 \times A_2 \times \cdots \times A_n = \{(a_1, a_2, \ldots, a_n) \mid a_j \in A_j \text{ for each } 1 \le j \le n\}$. If all $A_j = A$, we write A^n for $A_1 \times A_2 \times \cdots \times A_n$. Finally $(a_1, a_2, \ldots, a_n) = (b_1, b_2, \ldots, b_n)$ in $A_1 \times A_2 \times \cdots \times A_n$ when $a_j = b_j$ for all $1 \le j \le n$.

The Cartesian product of $A_1, \ldots, A_n$ agrees with Definition 0.5 for two sets. As before, the Cartesian product here depends on the order in which the sets are listed. We could use Definition 0.5 to define the Cartesian product of $A_1, \ldots, A_n$, but we would need to specify how to place parentheses. For three sets A, B, and C, we could write either $(A \times B) \times C$ or $A \times (B \times C)$, or even $(C \times A) \times B$, as the Cartesian product. These sets are different: the elements of the first look like $((a, b), c)$, but those of the second look like $(a, (b, c))$. Definition 0.9 simplifies notation by using a "parentheses-free" approach. Observe that there is a fairly clear way of identifying the elements in $(A \times B) \times C$ with those in $A \times (B \times C)$ or in $A \times B \times C$ by associating $((a, b), c)$ with $(a, (b, c))$ or with (a, b, c).

Cartesian products are often useful for representing the elements in a set. The usual Cartesian coordinate system in the plane associates points in the plane with the elements of $\boldsymbol{R} \times \boldsymbol{R} = \boldsymbol{R}^2$. Also every circle of positive radius in the plane is uniquely determined by its center (p, q) and radius $r > 0$, so the set of all circles corresponds to all $(p, q, r) \in \boldsymbol{R} \times \boldsymbol{R} \times (0, +\infty)$ (how?). Finally, the possible sequences resulting from flipping a standard coin n times correspond naturally to the elements in $\{H, T\}^n$.

Many problems in mathematics are attacked by breaking the problem into smaller pieces or special cases and then examining these. This simple but very important approach is formalized next.

DEFINITION 0.10 A **partition** of a set S is a collection of subsets $\{A_\lambda\}_\Lambda$ of S satisfying:

i) $\bigcup_\Lambda A_\lambda = S$; and
ii) if $\lambda, \mu \in \Lambda$ with $\lambda \neq \mu$, then $A_\lambda \bigcap A_\mu = \varnothing$.

A partition breaks the set S into nonoverlapping subsets. Some of the subsets in the partition may be $\varnothing$, but any such may be omitted to give a new partition consisting only of nonempty sets, at least when $S \neq \varnothing$. Thus a partition of S is a collection of pairwise disjoint subsets whose union is the whole set S. A simple picture representing this situation for five subsets is given in Fig. 0.2.

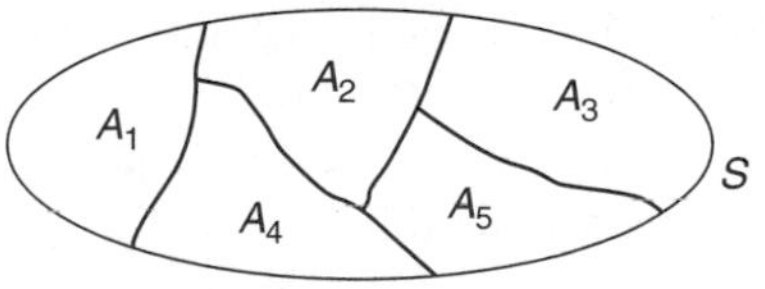

FIGURE 0.2

By Theorem 0.1, any $A \subseteq S$ gives rise to a partition of S into two pieces: A and its complement $S - A$. In $\boldsymbol{Z}$ we can take the even integers $2\boldsymbol{Z}$ and the odd integers $\boldsymbol{Z} - 2\boldsymbol{Z} = 2\boldsymbol{Z} + 1$, or $3\boldsymbol{Z}$ and $\boldsymbol{Z} - 3\boldsymbol{Z}$, the nonmultiples of 3. For $\boldsymbol{R}$, one partition into three subsets is to take $(-\infty, -1)$, $[-1, 2]$, and $(2, +\infty)$. If we take the subset $S \subseteq S$, then $\{S\}$ is a partition of S, as is $\{S, \varnothing\}$. It is important to remember that $\{S\} \neq S$, but rather $\{S\}$ means the set whose only element is S. Another partition of S is $\{\{a\} \subseteq S \mid a \in S\}$, the collection of all single-element subsets of S. Again, this collection of subsets of S should not be confused with S itself.

There are many partitions for most sets, although a set $S = \{a\}$ of one element has only the partition $\{S\}$ into nonempty subsets. If $S = \{a, b\}$ with $a \neq b$, there are two partitions of S into nonempty subsets, namely $P_1 = \{S\}$ and $P_2 = \{\{a\}, \{b\}\}$. For $T = \{a, b, c\}$ with three elements, there are five partitions into nonempty subsets: $\{T\}$, $\{\{a, b\}, \{c\}\}$, $\{\{a, c\}, \{b\}\}$, $\{\{b, c\}, \{a\}\}$, and $\{\{a\}, \{b\}, \{c\}\}$. If P is a partition of $W = \{a, b, c, d\}$ into nonempty subsets, then $d \in W$ must be in some element of P. When $\{d\}$ itself is one subset of P, the other subsets are those of a partition of $\{a, b, c\}$ and there are five possibilities, as we have just seen. There are six possibilities for P when $d \in W$ is in a two-element subset of P, namely each of the three subsets $\{a, d\}$, $\{b, d\}$, and $\{c, d\}$, together with either of the two partitions of the remaining two elements. Using this idea that $d \in W$ must be in a subset of P containing one, two, three, or all four elements of W shows that the number of possible such P is 15 (verify!).

These computations lead to an interesting problem: if S has n elements, how many partitions of S use only nonempty subsets? Although we have no use for this result, its solution provides a nice example of how to divide a problem into smaller, manageable pieces. We defer the argument until the end of this section after we have stated a result on counting. Now we look at additional examples of partitions.

It should be clear that $\{[i, i+1) \subseteq \boldsymbol{R} \mid i \in \boldsymbol{Z}\}$ is a partition of $\boldsymbol{R}$ into nonoverlapping half closed and half open intervals: every real number is in one of the sets $[j, j+1)$ and any two different intervals of this form are disjoint. The following examples may not be quite so obvious.

Example 0.12 For any $r \in [0, 1) \subseteq \boldsymbol{R}$, let $r + \boldsymbol{Z} = \{t \in \boldsymbol{R} \mid t = r + z \text{ for some } z \in \boldsymbol{Z}\}$. Then $\{r + \boldsymbol{Z} \mid r \in [0, 1)\}$ is a partition of $\boldsymbol{R}$.

Each $r + \boldsymbol{Z}$ is the set of translates of r by all integers so $0 + \boldsymbol{Z} = \boldsymbol{Z}$ and $0.1 + \boldsymbol{Z} = \{\ldots, -1.9, -0.9, 0.1, 1.1, 2.1, \ldots\}$. As for the verification, $r + \boldsymbol{Z} \subseteq \boldsymbol{R}$ so the union over $r \in [0, 1)$ is also (why?). If $y \in \boldsymbol{R}$ then $j \le y < j + 1$ for some $j \in \boldsymbol{Z}$, hence $r = y - j \in [0, 1)$ and $y = r + j \in r + \boldsymbol{Z}$. Thus every $y \in \boldsymbol{R}$ is in some $r + \boldsymbol{Z}$. By definition of union, $\boldsymbol{R} \subseteq \bigcup_{[0,1)}(r + \boldsymbol{Z})$, and with the opposite inclusion proves $\boldsymbol{R} = \bigcup_{[0,1)}(r + \boldsymbol{Z})$. If $x \in (s + \boldsymbol{Z}) \bigcap (r + \boldsymbol{Z})$ then $x \in (s + \boldsymbol{Z})$ and $x \in (r + \boldsymbol{Z})$ from the definition of intersection. The definition of $a + \boldsymbol{Z}$ gives $x = s + z = r + w$ for some $z, w \in \boldsymbol{Z}$ forcing $|s - r| \in \boldsymbol{Z}$. But $s, r \in [0, 1)$ so $|s - r| < 1$, and putting this together with $|s - r| \in \boldsymbol{Z}$, we conclude that $s = r$. Consequently $(s + \boldsymbol{Z}) \bigcap (r + \boldsymbol{Z}) = \emptyset$ when $s \neq r$, and the definition of a partition is satisfied for the sets $\{r + \boldsymbol{Z}\}_{(0,1)}$.

Example 0.13 For each $r \in \boldsymbol{R}$, let $B(r) = \{(a, b) \in \boldsymbol{R}^2 \mid b = a^2 + a + r\}$. Then $\{B(r)\}_{\boldsymbol{R}}$ is a partition of $\boldsymbol{R}^2$.

Geometrically, $B(r)$ is the graph of the parabola $y = x^2 + x + r$, and the different $B(r)$ represent all the vertical translations of any one of them, so we expect $\{B(r)\}_{\boldsymbol{R}}$ to be a partition of the plane. Let us show this formally. First, for $(a, b) \in \boldsymbol{R}^2$, if $r = b - a^2 - a$ then $b = a^2 + a + r$ and $(a, b) \in B(r)$. Thus every element of $\boldsymbol{R}^2$ in some $B(r)$, or $\boldsymbol{R}^2 \subseteq \bigcup B(r)$, so $\boldsymbol{R}^2 = \bigcup B(r)$ since each $B(r) \subseteq \boldsymbol{R}^2$. When $(x, y) \in B(r) \cap B(s)$ then (x, y) is in each set so both $y = x^2 + x + r$ and $y = x^2 + x + s$. Subtraction shows that $r = s$, proving $B(r) \cap B(s) = \emptyset$ when $r \neq s$. Therefore $\{B(r)\}_{\boldsymbol{R}}$ is a partition of $\boldsymbol{R}^2$.

If each subset in a partition of a set S is itself partitioned, then the collection of these smaller sets is also a partition of S. Roughly, if S decomposes into nonoverlapping pieces, and each of these pieces itself decomposes into nonoverlapping smaller pieces, then all the smaller pieces decompose S into nonoverlapping pieces. To write this formally is a little complicated, the motivation for the next example. It shows that understanding the meaning behind the symbols often leads to the correct approach and a fairly easy argument. Suppose that $\{A_i\}_I$ is a partition of S. We need to denote the sets in the partition of each A_i. In general the various A_i might have vastly different numbers of subsets used for their partitions, so the partition for each A_i needs its own index set. We choose to label the sets in the partition of A_i with elements from a set we denote by $J(i)$.

Example 0.14 Let $\{A_i\}_I$ be a partition of S, and for each $i \in I$ let $\{B(i, j)\}_{J(i)}$ be a partition of A_i. The collection of all $B(i, j)$ is a partition of S.

$J(i)$ is the index set for the partition of A_i and depends on $i \in I$. For $j \in J(i)$, $B(i, j) \subseteq A_i$. The labels (i, j) are in the set $I \times \bigcup_I J(i)$. Rather than try to specify an index set to represent $\{B(i, j)\} = \{B_t\}_T$ as in Definition 0.10 on partitions, it is less confusing to forgo this formalism and concentrate on the meaning of Definition 0.10. We must show that S is the union of all $B(i, j)$ and use that $\{A_i\}_I$ is a partition of S. Since $S = \bigcup A_i$, for any $s \in S$ there is $i \in I$ with $s \in A_i$. But $\{B(i, j)\}_{J(i)}$ is a partition of A_i, so A_i is the union of the sets in $\{B(i, j)\}_{J(i)}$ and it follows that $s \in B(i, j)$ for some $j \in J(i)$. Thus by definition, S is the union of all the $B(i, j)$. For the disjointness property of partitions, since $B(i, j) \subseteq A_i$, $B(i, j) \bigcap B(u, v) \subseteq A_i \bigcap A_u = \varnothing$ if $i \neq u$ because $\{A_i\}_I$ is a partition of S. When $i = u$ and $j \neq v$, then $B(i, j) \bigcap B(i, v) = \varnothing$ since $\{B(i, j)\}_{J(i)}$ is a partition of A_i. Thus $B(i, j) \bigcap B(u, v) = \varnothing$ when $(i, j) \neq (u, v)$, the collection of all $B(i, j)$ is pairwise disjoint, and it is a partition of S as claimed.

DEFINITION 0.11 If S is a finite set, the number of its elements is denoted $|S|$, or card S, and is called the **cardinality** of S.

Of course $|\varnothing| = 0$ and $\varnothing$ is the only set with cardinality zero. Two fundamental facts about the cardinality of finite sets related to Cartesian products and to partitions will be stated but not proved.

THEOREM 0.3 Let $A_1, \ldots, A_n$ be finite sets.

i) $|A_1 \times A_2 \times \cdots \times A_n| = |A_1||A_2| \cdots |A_n|$.
ii) If $\{A_1, \ldots, A_n\}$ is a partition of S, then $|S| = |A_1| + |A_2| + \cdots + |A_n|$.

For i), there are $|A_1|$ ways of filling the first component of an n-tuple in $A_1 \times A_2 \times \cdots \times A_n$, and for each such choice there are $|A_2|$ ways of filling the second component, and for any two initial choices, $|A_3|$ ways of filling the third component, etc. Part ii) follows from the definition of partition: each element of S is counted in $\Sigma|A_i|$ since each element is in some A_i, and no double counting occurs because the different A_j are disjoint. Theorem 0.3 gives a powerful technique to use on finite sets.

The four digit positive integers whose two left digits are multiples of 3 and whose two right digits are even correspond to the elements of $\{3, 6, 9\} \times \{0, 3, 6, 9\} \times \{0, 2, 4, 6, 8\} \times \{0, 2, 4, 6, 8\}$. We conclude from Theorem 0.3 that there are $3 \cdot 4 \cdot 5 \cdot 5 = 300$ such integers. A couple of other illustrations showing how Theorem 0.3 can be used are based on the observation that any k-digit element of $\mathbf{N}$, say $a_1 a_2 \cdots a_k$ with $a_1 \neq 0$, corresponds to the k-tuple $(a_1, a_2, \ldots, a_k) \in \{0, 1, \ldots, 8, 9\}^k$, and conversely if $a_1 \neq 0$. Let us find $|S|$ for

$S = \{n \in N \mid 1 \leq n < 1{,}000{,}000$ and all digits of n are either 1, 2, or 3$\}$. It should be clear that if $S(i) = \{n \in S \mid n$ has exactly i digits$\}$, then $\{S(1), S(2), \ldots, S(6)\}$ is a partition of S, so to find $|S|$ it suffices by Theorem 0.3 to find $|S(i)|$ for each i. Now $\{1, 2, 3\}^i$ corresponds to $S(i)$, so by the theorem, $|S(i)| = 3^i$ and $|S| = |S(1)| + \cdots + |S(6)| = 3 + \cdots + 3^6 = 1092$.

Let us find $|A|$ for $A = \{n \in N \mid 10^5 \leq n < 10^6$ with the digits of n alternating between even and odd$\}$. If A_e are those integers in A whose first digit is even and A_o are those whose first digit is odd then $\{A_e, A_o\}$ is a partition of A and $|A| = |A_e| + |A_o|$ by Theorem 0.3. Now A_o corresponds to 6-tuples $(a_1, \ldots, a_6)$ satisfying $a_1, a_3, a_5 \in \{1, 3, 5, 7, 9\}$ and $a_2, a_4, a_6 \in \{0, 2, 4, 6, 8\}$, and it follows that $|A_o| = 5^6$ from Theorem 0.3. Since the integers in A have six digits, the first is not zero so those in A_e must have first digit in $\{2, 4, 6, 8\}$, $a_2, a_4, a_6 \in \{1, 3, 5, 7, 9\}$, and $a_3, a_5 \in \{0, 2, 4, 6, 8\}$. In this case Theorem 0.3 yields $|A_e| = 4 \cdot 5^5$, so $|A| = 4 \cdot 5^5 + 5^6 = 9 \cdot 5^5 = 28{,}125$.

An interesting consequence of our counting involves infinite series introduced in calculus. If $T(k) = \{n \in N \mid n = a_1 a_2 \cdots a_k$ with all $a_j \neq 0\}$, then as above $|T(k)| = |\{1, \ldots, 9\}^k| = 9^k$ using Theorem 0.3. This implies the convergence of the infinite series of reciprocals of all these integers! Thus if $T = \bigcup_N T(i)$ then $\sum_{n \in T} 1/n$ converges although the harmonic series $\sum_N 1/n$ diverges. The idea is to observe that for each $n \in T(k)$, $1/n < 1/10^{k-1}$ and to use the fact that $|T(k)| = 9^k$ to see that $\sum_{n \in T(k)} 1/n < 9(9/10)^{k-1}$. Using this we can compare $\sum_{n \in T} 1/n$ with the convergent geometric series $\sum_N 9(9/10)^{k-1}$.

Finally, for a set S with $|S| = n$, we look at the number of partitions $P(n)$ of S that use only nonempty subsets. It is straightforward to compute that $P(1) = 1$, $P(2) = 2$, and $P(3) = 5$, and we claimed above that $P(4) = 15$. Suppose that $|S| = n \geq 3$ and fix $x \in S$. Let $V(k)$ be the number of k-element subsets of S that contain x, so $P(n - k)$ is the number of partitions of the remaining $n - k$ elements into nonempty subsets. Then $V(k)P(n - k)$ is the number of partitions of S for which x is contained in some k-element subset. But for every partition of S, x is in a j-element subset for some $1 \leq j \leq n$. If $S(j)$ are those partitions of S with a j-element subset containing x, then $\{S(1), \ldots, S(n)\}$ is a partition of the set of partitions of S, and $P(n) = |S(1)| + \cdots + |S(n)|$ by Theorem 0.3. From above, $|S(j)| = V(j)P(n - j)$, $V(1) = 1$ since $\{x\}$ is the only singleton containing x, $V(n) = 1$ since S is the only n-element subset of S, and $P(n) = V(1)P(n - 1) + \cdots + V(n - 1)P(1) + V(n)$. From results in the next section, $V(i) = (n - 1)!/(i - 1)!(n - i)!$. Although it is not feasible to write $P(n)$ as a simple expression in n, we can compute $P(n)$ from smaller values. Specifically, $P(4) = 5 + 3 \cdot 2 + 3 \cdot 1 + 1 = 15$, then $P(5) = 15 + 4 \cdot 5 + 6 \cdot 2 + 4 \cdot 1 + 1 = 52$, and $P(6) = 52 + 5 \cdot 15 + 10 \cdot 5 + 10 \cdot 2 + 5 \cdot 1 + 1 = 203$.

EXERCISES

1. a) If A is a set and $B \subseteq A$, find $\bigcap_B (A - \{b\})$ in terms of A and B.

b) If $C \subseteq A$, $|A| = m$, and $0 \leq |C| \leq m - 1$, show that C is an intersection of subsets of A each having exactly $m - 1$ elements.

2. For each collection $\{A(i)\}_I$ below, find $\bigcup_I A(i)$ and $\bigcap_I A(i)$:

i) $I = N - \{1\}$ and for $n \in I$, $A(n) = n\boldsymbol{Z} \subseteq \boldsymbol{Z}$

ii) $I = (0, 1) \subseteq \boldsymbol{R}$ and for $s \in I$, $A(s) = (0, 1 - s) \subseteq \boldsymbol{R}$

iii) $I = (0, +\infty) \subseteq \boldsymbol{R}$ and for $r \in I$, $A(r) = (-1/r, 1 + 1/r) \subseteq \boldsymbol{R}$

iv) $I = \{(a, b) \in \boldsymbol{R}^2 \mid a \geq 0 \text{ and } b \geq 0\}$, and for $(a, b) \in I$, $A((a, b)) = \{(x, y) \in \boldsymbol{R}^2 \mid x = a \text{ or } y = b\} \subseteq \boldsymbol{R}^2$
v) $I = \{(a, b) \in \boldsymbol{R}^2 \mid a = 0\}$ and for $(a, b) \in I$, $A((a, b))$ as in iv)
vi) $I = \boldsymbol{R}$ and for $r \in \boldsymbol{R}$, $A(r) = \{(a, b) \in \boldsymbol{R}^2 \mid b = ra^2\} \subseteq \boldsymbol{R}^2$

3. Let S be a set and let $\{A_i\}_I$ and $\{B_j\}_J$ be collections of sets indexed by the sets I and J, respectively. Prove each equality below.
a) $S \cup (\bigcup_J B_j) = \bigcup_J (S \cup B_j)$ and $S \cap (\bigcap_J B_j) = \bigcap_J (S \cap B_j)$
b) $(\bigcup_I A_i) - S = \bigcup_I (A_i - S)$ and $(\bigcap_I A_i) - S = \bigcap_I (A_i - S)$
c) $(\bigcup_I A_i) \cap (\bigcup_J B_j) = \bigcup_J (\bigcup_I (A_i \cap B_j)) = \bigcup_I (\bigcup_J (A_i \cap B_j))$
d) $(\bigcap_I A_i) \cup (\bigcap_J B_j) = \bigcap_I (\bigcap_J (A_i \cup B_j)) = \bigcap_J (\bigcap_I (A_i \cup B_j))$
e) $(\bigcup_I A_i) - (\bigcup_J B_j) = \bigcup_I (\bigcap_J (A_i - B_j))$
f) $(\bigcap_I A_i) - (\bigcap_J B_j) = \bigcup_J (\bigcap_I (A_i - B_j))$

4. Let $S = \{A \subseteq \boldsymbol{N} \mid \boldsymbol{N} - A \text{ is finite}\}$.
i) If $A \in S$ and $A \subseteq B \subseteq \boldsymbol{N}$, show that $B \in S$.
ii) If $A_1, A_2, \ldots, A_k \in S$, show that $A_1 \cap \cdots \cap A_k \in S$.
iii) If $\{A_\lambda\}_I \subseteq S$, show that $\bigcup_I A_\lambda \in S$ and also that $\bigcap_I A_\lambda = \varnothing \Leftrightarrow \bigcup_I (\boldsymbol{N} - A_\lambda) = \boldsymbol{N}$.

5. Find all partitions of:
a) $S = \{a, b, c, d\}$ into two nonempty subsets
b) $S = \{a, b, c, d\}$ into three nonempty subsets
c) $T = \{a, b, c, d, e\}$ into four nonempty subsets

6. Fix $m \in \boldsymbol{R} - \{0\}$, and for any $a \in \boldsymbol{R}$, set $L(a) = \{(r, mr + a) \in \boldsymbol{R}^2 \mid r \in \boldsymbol{R}\}$. Show that $\{L(a)\}_{\boldsymbol{R}}$ is a partition of $\boldsymbol{R}^2$. (Geometry!)

7. Set $W = \{(x, y) \in \boldsymbol{R}^2 \mid xy \neq 0\}$, and for $r \in I = \boldsymbol{R} - \{0\}$ let $B(r) = \{(a, b) \in W \mid b = ra^2\}$. Show that $\{B(r)\}_I$ is a partition of W.

8. If $\{A_i\}_I$ partitions the set S and if $T \subseteq S$, show that $\{A_i \cap T\}_I$ partitions T.

9. If $\{A_i\}_I$ and $\{B_j\}_J$ are partitions of the set S, set $C_{(i,j)} = A_i \cap B_j$ and show that $\{C_{(i,j)}\}_{I \times J}$ is a partition of S.

10. If $\{A_i\}_I$ is a partition of the set S and $\{B_j\}_J$ is a partition of the set T, let $D_{(i,j)} = A_i \times B_j \subseteq S \times T$. Is $\{D_{(i,j)}\}_{I \times J}$ a partition of $S \times T$?

11. a) Show that $\{2\boldsymbol{Z} + 1, 4\boldsymbol{Z}, 4\boldsymbol{Z} + 2\}$ is a partition of $\boldsymbol{Z}$.
b) For $k \in \boldsymbol{N} - \{1\}$, show that $\{2\boldsymbol{Z} + 1, 4\boldsymbol{Z} + 2, \ldots, 2^{k-1}\boldsymbol{Z} + 2^{k-2}, 2^{k-1}\boldsymbol{Z}\}$ is a partition of $\boldsymbol{Z}$ into k infinite subsets. (Use that every $z \in \boldsymbol{Z} - \{0\}$ may be written uniquely as $z = 2^s m$ for m odd, $s \geq 0$.)
c) Show that $\{2^k\boldsymbol{Z} + 2^{k-1} \mid k \in \boldsymbol{N}\}$ is a partition of $\boldsymbol{Z}$ into infinitely many infinite sets.

12. For any set S and subsets $A_1, \ldots, A_k$, let $A_j^1 = A_j$ and $A_j^{-1} = S - A_j$. Show that the set of intersections $\{A_1^{e(1)} \cap A_2^{e(2)} \cap \cdots \cap A_k^{e(k)} \mid \text{all } e(j) \in \{\pm 1\}\}$ is a partition of S. When $k = 2$ and the subsets are A and B, the intersections are $A \cap B$, $A \cap (S - B)$, $(S - A) \cap B$, and $(S - A) \cap (S - B)$. If there are three subsets, A, B, and C, two intersections are $A^1 \cap B^1 \cap C^{-1} = A \cap B \cap (S - C)$ and $A^{-1} \cap B^{-1} \cap C^{-1} = (S - A) \cap (S - B) \cap (S - C)$.

13. For a set S and any subsets $A_1, \ldots A_k$, show $\{S - A_1, A_1 - A_2, (A_1 \cap A_2) - A_3, \ldots, (A_1 \cap A_2 \cap \cdots \cap A_{k-1}) - A_k, A_1 \cap A_2 \cap \cdots \cap A_k\}$ is a partition of S.

14. If the sequence of heads and tails resulting from flipping a standard coin 10 times is recorded, how many sequences are possible?

15. How may sequences of five rolls of a die have no rolls of 3 appearing?
16. How many 9-digit positive integers $a_1 \cdots a_9$ have $a_i \in i\mathbf{Z}$? Note that $0 \in i\mathbf{Z}$.
17. A positive integer $x \in \mathbf{N}$ is a **palindrome** if the sequences of its digits from left to right and from right to left are the same, such as 33133 and 3001881003. How many palindromes are less than one billion?
18. A bag of coins contains nine quarters, one dime, one nickel, and four pennies. If some of the coins are grabbed randomly from the bag and the values of these are added, how many different amounts can result? Include the possibilities that no coins are removed and that all coins are removed.
19. Let $S = \{n \in \mathbf{N} \mid n < 1{,}000{,}000\}$ and $S(6) = \{s \in S \mid s \geq 100{,}000\}$.
a) How may elements of $S(6)$ have all their digits integer squares?
b) How many elements of $S(6)$ have some odd digit?
c) How many elements of $S(6)$ have exactly one even digit?
d) Answer each of a), b), and c) for S.

0.3 INDUCTION I

Mathematical Induction is an axiom for $\mathbf{N}$ and an important proof technique. It shows that $S \subseteq \mathbf{N}$ is all of $\mathbf{N}$ if $1 \in S$ and if $k \in S \Rightarrow k+1 \in S$.

IND-I (The First Principle of Mathematical Induction). Let $t \in S \subseteq \mathbf{N}$. If $k \geq t$ and $k \in S$ imply that $k+1 \in S$, then $n \in S$ for all $n \geq t$.

The usual statement of Induction takes $t = 1$, but it will be useful to have the more general form. IND-I is an implication, so to use it we must verify both conditions of its hypothesis: we must *show* that some $t \in S$; then, *assuming* $k \in S$ for an arbitrary $k \geq t$, we must *prove* that $k+1 \in S$. If all this is done, *then* the conclusion of IND-I is $n \in S$ for all $n \geq t$. IND-I is the formal, mathematically correct way of saying, "$k \in S$ implies $k+1 \in S$ and as $t \in S$, then $t+1 \in S$, then $t+2 \in S$, etc., so all $m \geq t$ are in S."

Often Induction is stated in terms of a **proposition** on $\mathbf{N}$. A proposition P on a set S means that for each $s \in S$, $P(s)$ is a statement about s and $P(s)$ is either true or false. IND-I is equivalent to: *Let P be a proposition on* $\mathbf{N}$ *with P(t) true for some* $t \in \mathbf{N}$. *If for any* $k \in \mathbf{N}$ *with* $k \geq t$, *the truth of P(k) implies the truth of* $P(k+1)$, *then P(n) is true for all* $n \in \mathbf{N}$ *with* $n \geq t$.

This statement follows from IND-I by taking $S = \{n \in \mathbf{N} \mid P(n) \text{ is true}\}$ and also implies IND-I. To see this, let $t \in S \subseteq \mathbf{N}$ so that $k \geq t$ and $k \in S$ imply $k+1 \in S$. Let the proposition $P(n)$ be "$n \in S$." Now $P(t)$ is true and when $k \geq t$, $P(k) \Rightarrow P(k+1)$. Hence $P(n)$ is true when $n \geq t$, so $n \in S$ for all $n \geq t$ by the statement, proving IND-I. Therefore IND-I is equivalent to the formulation for propositions. We illustrate the use of IND-I in various contexts and show that proving "$k \in S \Rightarrow k+1 \in S$" can require some care.

Example 0.15 For all $n \in \mathbf{N}$, $1 + 2 + \cdots + n = n(n+1)/2$.

Let the equality be $P(n)$ and $S = \{k \in \mathbf{N} \mid P(k) \text{ is true}\}$. Now $1 \in S$ since the sum of the integers up to 1 is $1 = 1 \cdot (1+1)/2$. Next let $k \in S$ for some $k \geq 1$, so $1 + 2 + \cdots + k = k(k+1)/2$. Does this force $k+1 \in S$? In trying to use the truth of

$P(k)$ to show that $P(k+1)$ holds, note that the left-hand sides of these two equalities are simply related. Namely,

$$1+2+\cdots+(k+1)=(1+2+\cdots+k)+k+1$$

so by using the truth of $P(k)$, we may write

$$1+2+\cdots+k+(k+1)=k(k+1)/2+(k+1)=(k+1)(k+2)/2$$

Thus $P(k+1)$ is true, $k+1 \in S$, and IND-I yields $n \in S$ whenever $n \geq 1$.

Example 0.16 For all $n \in \boldsymbol{N}$, $1 \cdot 1! + 2 \cdot 2! + \cdots + n \cdot n! = (n+1)! - 1$.

Let $P(n)$ be the equality and $S = \{n \in \boldsymbol{N} \mid P(n) \text{ is true}\}$. When $n = 1$, $1 \cdot 1! = 1 = 2! - 1$, so $1 \in S$. Next assume $k \in S$ for some $k \geq 1$, so we have $1 \cdot 1! + 2 \cdot 2! + \cdots + k \cdot k! = (k+1)! - 1$. To prove $P(k+1)$ from $P(k)$, note that $1 \cdot 1! + 2 \cdot 2! + \cdots + (k+1) \cdot (k+1)! = (1 \cdot 1! + 2 \cdot 2! + \cdots + k \cdot k!) + (k+1) \cdot (k+1)!$ and use $P(k)$ to replace $1 \cdot 1! + 2 \cdot 2! + \cdots + k \cdot k!$ with $(k+1)! - 1$. Then

$$\begin{aligned}
&(1 \cdot 1! + 2 \cdot 2! + \cdots + k \cdot k!) + (k+1) \cdot (k+1)! \\
&= (k+1)! - 1 + (k+1)(k+1)! \\
&= (1 + (k+1))(k+1)! - 1 \\
&= (k+2)! - 1
\end{aligned}$$

which is just $P(k+1)$. Therefore $k \in S$ does indeed imply that $k+1 \in S$, and applying IND-I shows that $n \in S$ for all $n \geq 1$.

The next example shows why we want a more general statement of Induction than the special case with $t = 1$.

Example 0.17 Find all $n \in \boldsymbol{N}$ so that $3^n < n!$.

Let $S = \{n \in \boldsymbol{N} \mid 3^n < n!\}$ and note that $3 > 1!$ and $9 > 2!$. A direct computation shows that the minimal $t \in S$ is $t = 7$, that $3^8 < 8!$ and $3^9 < 9!$. This suggests that $S = \{n \in \boldsymbol{N} \mid n \geq 7\}$. Assume that $k \in S$ and $k \geq 7$. We need to use $3^k < k!$ to prove $3^{k+1} < (k+1)!$. Now $3^{k+1} = 3 \cdot 3^k < 3 \cdot k!$ since $k \in S$, and we can conclude that $k+1 \in S$ *if* $3k! \leq (k+1)!$. But this holds exactly when $3 \leq k+1$, which is true since $k \geq 7$. Therefore the hypothesis of IND-I is satisfied, so IND-I, with $t = 7$, yields $3^n < n!$ whenever $n \geq 7$.

Example 0.18 For all $n \geq 2$, $x^n - 1 = (x-1)(x^{n-1} + x^{n-2} + \cdots + x + 1)$ as polynomials in $\boldsymbol{Z}[x]$. In particular $r^n + r^{n-1} + \cdots + 1 = (r^{n+1} - 1)/(r-1)$ for any $r \in \boldsymbol{R} - \{1\}$, and so $2^n + 2^{n-1} + \cdots + 2 + 1 = 2^{n+1} - 1$.

This equality $P(n)$ holds when $n = 2$ since $x^2 - 1 = (x-1)(x+1)$. Assume that for some $k \geq 2$, $x^k - 1 = (x-1)(x^{k-1} + \cdots + x + 1)$. Now use the distributive law for multiplication on the right side of $P(k+1)$:

$$(x-1)(x^k+x^{k-1}+\cdots+x+1)=(x-1)x^k+(x-1)(x^{k-1}+\cdots+x+1)$$

Our induction assumption that $P(k)$ is true shows that

$$(x-1)(x^k+x^{k-1}+\cdots+x+1)=(x-1)x^k+x^k-1=x^{k+1}-1$$

which proves that $P(k+1)$ is true. Hence IND-I implies that $P(n)$ holds for all $n \geq 2$. The remaining statements are clear by substituting $r \in \boldsymbol{R}$ for x.

Recall that $P(S)$ is the collection of all subsets of S (Definition 0.6) and $|S|$ is the number of elements in S (Definition 0.11).

Example 0.19 If S is a set with $|S|=n \in \boldsymbol{N}$, then $|P(S)|=2^n$.

We have seen $P(\varnothing)=\{\varnothing\}$ so the result holds when $S=\varnothing$ also. If $S=\{a\}$ then $P(S)=\{\varnothing, S\}$, so $|S|=1$ implies $|P(S)|=2$. To apply IND-I let $k \geq 1$, assume that $|S|=k$ implies $|P(S)|=2^k$, and take $T=\{a_1, a_2, \ldots, a_{k+1}\}$. For $A \in P(T)$ either $a_{k+1} \in A$ or not, and this dichotomy gives a partition of $P(T)$. Let $U=\{B \subseteq T \mid a_{k+1} \in B\}$ and $V=P(T)-U=\{B \subseteq T \mid a_{k+1} \notin B\}=P(\{a_1, a_2, \ldots, a_k\})$. Now $\{U, V\}$ partitions $P(T)$ by Theorem 0.1, $|P(T)|=|U|+|V|$ by Theorem 0.3, and by assumption $|V|=2^k$. How are elements of U and V related? For $A, A' \in V$, $A \cup \{a_{k+1}\} \in U$ and if $A' \neq A$, $A \cup \{a_{k+1}\} \neq A' \cup \{a_{k+1}\}$ results. Also each $B \in U$ arises in this way since $a_{k+1} \in B$ means that $B-\{a_{k+1}\} \in V$ and $B=(B-\{a_{k+1}\}) \cup \{a_{k+1}\}$. Hence $U=\{A \cup \{a_{k+1}\} \mid A \in V\}$ and it follows that $|U|=|V|=2^k$ so $|P(T)|=2^k+2^k=2^{k+1}$. The result for sets of k elements implies the result for sets with $k+1$ elements. Applying IND-I shows that $|P(S)|=2^n$ whenever $|S|=n$, for all $n \geq 1$.

Recall that for $n \in \boldsymbol{N}$ and integer $0 \leq k \leq n$, the **binomial coefficient n choose k** is $\binom{n}{k}=n!/k!(n-k)!=n(n-1)\cdots(n-k+1)/k!$, where $0!=1$. Observe that $\binom{n}{0}=\binom{n}{n}=1$, that $\binom{n}{1}=n$, and that $\binom{n}{k}=\binom{n}{n-k}$. A useful result following from the definitions and known as **Pascal's identity** is

$$\binom{n}{k}+\binom{n}{k+1}=\binom{n+1}{k+1}$$

The important Binomial Theorem is proved by applying IND-I. Its proof is an exercise in manipulating summations and using Pascal's identity.

Example 0.20 (Binomial Theorem) $(x+y)^n=\sum_{k=0}^n \binom{n}{k} x^{n-k}y^k$ for all $n \in \boldsymbol{N}$.

To compute $(x+y)^n$ take all products formed by choosing either x or y in each of its n factors. For example, multiplying y from the first, second, and last factor with an x in all the other factors yields $x^{n-3}y^3$. This same term arises in many other ways depending on which three of the n factors $(x+y)$ we choose the y term from.

Example 0.20 shows that there are exactly $\binom{n}{3}$ of these products $x^{n-3}y^3$. Similarly the terms $x^{n-k}y^k$ arise by choosing y from exactly k of the n factors in $(x+y)^n$, and the Binomial Theorem shows there are $\binom{n}{k}$ of these products containing y exactly k times. This is the number of choices of k factors from the n available and is clearly the same as the number of choices of k elements from a set with n elements. Hence $\binom{n}{k}$ represents the number of k-element subsets in a set with n elements. Using this interpretation and setting $x = y = 1$ in the Binomial Theorem shows again that if $|S| = n$, then $|P(S)| = 2^n$ since every subset has between 0 and n elements, and the number of subsets with exactly k elements is $\binom{n}{k}$.

We turn to some examples that require formulating the statement to be shown by induction.

Example 0.21 Let $\varnothing \neq T_1 = \{r_1, \ldots, r_k\} \subseteq \boldsymbol{R}$ with $r_1 < \cdots < r_k$. Let $T_1^{\#} \subseteq \boldsymbol{R}$ be any $T_1 \cup \{s_1, \ldots, s_{k+1}\}$ with $s_j < r_j < s_{j+1}$ for all j. Set $T_2 = T_1^{\#}$, $T_3 = T_2^{\#}$, and in general $T_{n+1} = T_n^{\#}$. Find $|T_n|$ in terms of $|T_1|$.

By construction $|T_1| = k$, $|T_2| = k + (k+1) = 2k+1$, $|T_3| = 2k+1+(2k+2) = 4k+3$, and $|T_4| = 4k+3+(4k+4) = 8k+7$. A natural guess is that $|T_n| = 2^{n-1}k + 2^{n-1} - 1$ and we try to prove this by induction. When $n = 1$, $|T_1| = k = 2^0k + 2^0 - 1$, proving the $n = 1$ case. Assume $|T_m| = 2^{m-1}k + 2^{m-1} - 1$ for some $m \geq 1$. As above, T_{m+1} adjoins $|T_m| + 1 = 2^{m-1}k + 2^{m-1}$ additional numbers to T_m. Thus $|T_{m+1}| = 2^{m-1}k + 2^{m-1} - 1 + 2^{m-1}k + 2^{m-1} = 2^m k + 2^m - 1$, proving that the expression for $|T_{m+1}|$ follows from the one for $|T_m|$. Applying IND-I shows that $|T_n| = 2^{n-1}|T_1| + 2^{n-1} - 1$ is correct for all $n \geq 1$.

Example 0.22 Find a simple expression for $s_n = 4/3 + 8/9 + \cdots + 4n/3^n$.

Computing shows $s_1 = 4/3$, $s_2 = 20/9$, $s_3 = 72/27$, $s_4 = 232/81$, and $s_5 = 716/243$, so we can write the denominator of s_n as 3^n. The numerators have no recognizable pattern but for s_n seem to lie between 3^n and 3^{n+1}. Write the numerator of s_n as $3^{n+1} - r_n$ to see what results. In order these are $4 = 9 - 5$; $20 = 27 - 7$; $72 = 81 - 9$; $232 = 243 - 11$; and $716 = (729 - 13)$. This reveals the pattern $s_n = (3^{n+1} - (2n+3))/3^n = 3 - (2n+3)/3^n$. Let us prove this equality $P(n)$ by induction. For $n = 1$, $s_1 = 4/3 = 3 - 5/3$, so $P(1)$ is true. If $P(n)$ is true for some $n \geq 1$ then $s_n = 3 - (2n+3)/3^n$. Therefore $s_{n+1} = s_n + 4(n+1)/3^{n+1} = 3 - (2n+3)/3^n + 4(n+1)/3^{n+1} = 3 + (4n+4-6n-9)/3^{n+1}$. Continuing, $s_{n+1} = 3 - (2n+5)/3^{n+1} = 3 - (2(n+1)+3)/3^{n+1}$, and $P(n+1)$ is true if $P(n)$ is, so using IND-I shows that $s_n = 3 - (2n+3)/3^n$ for all $n \in \boldsymbol{N}$.

A rich source of examples using induction is the **Fibonacci sequence** defined by $F_1 = F_2 = 1$ and $F_n = F_{n-1} + F_{n-2}$ for $n > 2$. Its first few terms are (1, 1, 2, 3, 5, 8, 13,

21, 34, ...). Many relations among the terms of (F_n) can be proved by IND-I. An example is that $F_{4n} \in 3N$ for all $n \in N$. When $n = 1$, $F_4 = 3 \in 3N$ by direct calculation. Assume $F_{4k} = 3m \in 3N$ for some $k \geq 1$. Using $F_n = F_{n-1} + F_{n-2}$, $F_{4(k+1)} = F_{4k+2} + F_{4k+3} = 2F_{4k+2} + F_{4k+1} = 3F_{4k+1} + 2F_{4k} = 3(F_{4k+1} + 2m) \in 3N$. Since the truth of the $n = k$ case implies the truth of the $n = k + 1$ case, IND-I shows that $F_{4n} \in 3N$ for all $n \in N$. Other examples of induction arguments using (F_n) appear in the exercises.

The last example of this section is curious and surprising. Consider towers of exponents of $r \in R$ with $r > 1$ defined by $E(r, 0) = 1$, $E(r, 1) = r$, $E(r, 2) = r^r$, $E(r, 3) = r^{E(r,2)} = r^{r^r}$, and in general for $k \in N$, $E(r, k + 1) = r^{E(r,k)}$. It is important to compute these towers from the top down rather than from the bottom up. Specifically, $E(3, 3) = 3^{27}$, not 27^3, and $E(2, 4) = 2^{E(2,3)} = 2^{16}$. $E(r, k)$ gets huge as k increases and grows so rapidly that it is impractical to compute it except for very small values of r and k. For example, $E(2, 4) = 2^{16} = 65{,}536$, but $E(2, 5) = 2^{65{,}536}$ is too large to compute in any practical sense. The reader should also consider $E(10, 4)$. Now let $r = 10^9 = 1{,}000{,}000{,}000$ and take some large $n \in N$. Although $E(10^9, n)$ is huge, there will be some $s \in N$ with $E(10, s)$ even larger (see the exercises). For example, replacing each 10^9 in $E(10^9, n)$ with 10^{10} will give a bigger number but less than $E(10, 2n)$. It is surprising that in fact $E(10^9, n) < E(10, n + 1)$ for any $n \in N$! A general criterion from which this follows is given next.

Example 0.23 If $s \in N - \{1\}$ then for all $n \in N$, $E(s^{s-1}, n) < E(s, n + 1)$.

We use IND-I. For $n = 1$, $E(s^{s-1}, 1) = s^{s-1} < s^s = E(s, 2)$ so the statement for $n = 1$ is true. Assume $k \geq 1$ with $E(s^{s-1}, k) < E(s, k + 1)$. Then $(s^{s-1})^{E(s^{s-1}, k-1)} = E(s^{s-1}, k) < E(s, k + 1) = s^{E(s,k)}$ so comparing exponents gives $(s - 1)E(s^{s-1}, k - 1) < E(s, k)$. But since these are integers, $(s - 1)E(s^{s-1}, k - 1) + 1 \leq E(s, k)$. Using this, $(s - 1)E(s^{s-1}, k) = (s - 1)s^{(s-1)E(s^{s-1}, k-1)} < s \cdot s^{(s-1)E(s^{s-1}, k-1)} = s^{(s-1)E(s^{s-1}, k-1)+1} \leq s^{E(s,k)} = E(s, k + 1)$. Therefore $E(s^{s-1}, k + 1) = (s^{s-1})^{E(s^{s-1}, k)} < s^{E(s, k+1)} = E(s, k + 2)$, proving that the $n = k + 1$ case follows from the $n = k$ case. Applying IND-I shows that $E(s^{s-1}, n) < E(s, n + 1)$ for all $n \in N$.

The exercises show that whenever $1 < t \leq s^{s-1}$, then for all $n \in N$, $E(t, n) < E(s, n + 1)$. Other special cases of Example 0.23 are that $E(9, 5) < E(3, 6)$, $E(625, 5) < E(5, 6)$, $E(64, 10) < E(4, 11)$, and $E(10^9, 3) < E(10, 4)$.

We want to stress the importance of verifying both parts of the hypothesis of IND-I before applying it. Consider $p(x) = x^2 + x + 41$ and note that $p(1) = 43$, $p(2) = 47$, $p(3) = 53$, and $p(4) = 61$ are all primes (i.e., have no proper factorization in N). Is $p(n)$ always prime? It is known that $p(n)$ is prime for all $0 \leq n \leq 39$ but that $p(40)$ is not. *Verifying a statement for a large number of consecutive integers does not make the statement true in general.* Now consider the equality $P(n): 2 \cdot 3 + 2 \cdot 3^2 + \cdots + 2 \cdot 3^n = 3^{n+1} - 1$. Assume $P(k)$ holds for some $k \in N$ and add $2 \cdot 3^{k+1}$ to both sides to obtain

$$2 \cdot 3 + 2 \cdot 3^2 + \cdots + 2 \cdot 3^k + 2 \cdot 3^{k+1} = (3^{k+1} - 1) + 2 \cdot 3^{k+1} = 3 \cdot 3^{k+1} - 1 = 3^{k+2} - 1.$$

Thus $P(k)$ implies $P(k + 1)$ but $P(n)$ is false for *all* $n \in N$, since $3^{n+1} - 1 = 2(1 + 3 + 3^2 + \cdots + 3^n)$ by Example 0.18. The point here is that IND-I cannot be applied without knowing the truth of some specific case.

Finally consider the (false!) statement: for every $\emptyset \neq S \subseteq N$ and finite, either $S \subseteq 2N$ or $S \subseteq 2N+1$. Clearly, if $|S| = 1$ then the statement is true for S. Assume that for some $k \geq 1$ the statement is true for any $S \subseteq N$ with $|S| = k$, and let $T \subseteq N$ with $|T| = k+1$. If $a \in T$ and $W = T - \{a\}$, then $|W| = k$, so by the induction assumption, $W \subseteq 2N$ or $W \subseteq 2N+1$. Let $w \in W$, $t \in W - \{w\}$ and set $V = T - \{t\}$. Again $|V| = k$ so $V \subseteq 2N$ or $V \subseteq 2N+1$. But $w \in W \cap V$ so if w is even, $W \cup V = T \subseteq 2N$, and if w is odd, $T \subseteq 2N+1$. Applying IND-I appears to prove the statement. The error here is the tacit assumption that $k \geq 2$, allowing $t \in W - \{w\}$ so $w \in W \cap V$ can transfer the odd or even condition to all elements of T. Our "argument" holds if $k \geq 2$ but fails when $|T| = 2$—that is, when $k = 1$. This illustrates the importance of making arguments that are valid for the smallest possible case covered by the induction assumption.

EXERCISES

1. Prove the following for all $n \in N$:
  a) $1 + 3 + \cdots + (2n-1) = n^2$
  b) $1 + 4 + \cdots + n^2 = n(n+1)(2n+1)/6$
  c) $1 + 8 + \cdots + n^3 = (n(n+1)/2)^2$
  d) $1/1 \cdot 2 + \cdots + 1/n(n+1) = n/(n+1)$
  e) $1 \cdot 2 + \cdots + n(n+1) = n(n+1)(n+2)/3$

2. If a set has $n \geq 2$ elements, show by induction that it has exactly $n(n-1)/2$ subsets with exactly two elements.

3. Find all $n \in N$ satisfying: a) $2n + 1 < 2^{n-1}$; b) $n^2 < (n-1)!$; and c) $n^2 < 2^{n-1}$.

4. Given $i \in N$, find all $i \leq n \in N$ satisfying:
  a) $\binom{n}{i} \leq 2^{n-1}$; and
  b) $\binom{n}{i} \leq 2^{n-2}$.

5. Recall that $E(r, n)$ is the number that is n iterated powers of r: $E(r, 0) = 1$, $E(r, 1) = r$, and $E(r, n+1) = r^{E(r,n)}$. Prove the following:
  a) If $1 < b < a$, then for all $n \in N$, $E(b, n) < E(a, n)$.
  b) If $r > 1$, then $E(r, n) < E(r, n+1)$ for all $n \in N$.
  c) For $r \geq 2$, $r^n < E(r, n)$ unless $n = 1$, or $r = n = 2$.

6. Let $(F_n) = (1, 1, 2, 3, 5, 8, 13, 21, 34, 55, \ldots)$ be the Fibonacci sequence, so $F_{n+2} = F_{n+1} + F_n$ with $F_1 = F_2 = 1$. Show that the following hold for each $n \in N$:
  a) $F_1 + F_2 + \cdots + F_n = F_{n+2} - 1$
  b) $F_2 + F_4 + \cdots + F_{2n} = F_{2n+1} - 1$
  c) $F_1 + F_3 + \cdots + F_{2n-1} = F_{2n}$
  d) $F_1^2 + \cdots + F_n^2 = F_n F_{n+1}$
  e) $F_n F_{n+2} = F_{n+1}^2 + (-1)^{n+1}$

7. If (F_n) is the Fibonacci sequence, for each $n \in N$ and $m \in N$, show that:
  a) $\binom{n}{0} F_m + \binom{n}{1} F_{m+1} + \cdots + \binom{n}{n} F_{m+n} = F_{2n+m}$
  b) $\binom{n}{1} F_1 + \binom{n}{2} F_2 + \cdots + \binom{n}{n} F_n = F_{2n}$ (use part a))

8. For any $1 \le k \le n \in \boldsymbol{N}$, show that $\binom{k}{k} + \binom{k+1}{k} + \cdots + \binom{n}{k} = \binom{n+1}{k+1}$.

9. If $p > 2$ is a prime, argue that for all $2 \le i \le p-1$, the binomial coefficient $\binom{p}{i} \in p\boldsymbol{Z}$. Use IND-I on $a \in \boldsymbol{N}$ to show that for all $a \in \boldsymbol{Z}$, $a^p - a \in p\boldsymbol{Z}$.

10. Prove the Binomial Theorem: $(x+y)^n = \sum_{k=0}^{n} \binom{n}{k} x^{n-k} y^k$. Use this to show that:

a) $\sum_{k=0}^{n} \binom{n}{k} 2^k = 3^n$ for all $n \in \boldsymbol{N}$

b) For all $a, b \in \boldsymbol{N}$, $(a+b)^n \in a\boldsymbol{N} + b^n$

c) $\binom{n}{0} + \binom{n}{2} + \binom{n}{4} + \cdots = \binom{n}{1} + \binom{n}{3} + \binom{n}{5} + \cdots$

d) If p is a prime, show that $(x_1 + \cdots + x_m)^p = x_1^p + \cdots + x_m^p + p \cdot f(x_1, \ldots, x_m)$, where $f(x_1, \ldots, x_m)$ is a polynomial in $\boldsymbol{Z}[x_1, \ldots, x_m]$.

e) For p a prime, $(x_1 + \cdots + x_m)^{p^n} = x_1^{p^n} + \cdots + x_m^{p^n} + p \cdot f(x_1, \ldots, x_m)$, where $f(x_1, \ldots, x_m)$ is a polynomial in $\boldsymbol{Z}[x_1, \ldots, x_m]$.

11. Define a function on $\boldsymbol{N}$ by $f(1) = 1$, $f(2) = 2 - 1$, and $f(n) = n - f(n-1)$ when $n > 2$. Find $f(n)$ as a simple expression in n and prove that it holds.

12. a) For $n \in \boldsymbol{N}$ find and verify a simple expression $f(n)$ giving the number of ways of writing n as a sum using only 1 and 2. Ignore the order of the summands, so consider $2+2+1$ and $2+1+2$ to be the same sums.

b) Find and verify a simple expression $g(n)$, giving the number of ways of writing n as a sum using only 1, 2, and 4. As in a), ignore the order of the summands. (The answer varies with $n \in 4\boldsymbol{Z} + i$ for $0 \le i \le 3$.)

13. Define a sequence (a_k) by setting $a_1 = 2$ and, for $k \ge 1$, $a_{k+1} = 1/(1 - a_k)$. Find an explicit expression for a_m and prove it is correct.

14. Set $a_1 = \sqrt{2}$, $a_2 = \sqrt{2 + \sqrt{\sqrt{2}}}$, and in general $a_{n+1} = \sqrt{2 + \sqrt{a_n}}$. Prove that $a_k < 2$ for all $k \ge 1$ and that $a_1 < a_2 < \cdots < a_k < a_{k+1} < \cdots$.

15. Note that $1^3 + 2^3 = 9 = 2 + 3 + 4$, $2^3 + 3^3 = 35 = 5 + 6 + 7 + 8 + 9$, and $3^3 + 4^3 = 91 = 10 + 11 + \cdots + 16$. Is every sum of two consecutive cubes a sum of consecutive integers? Formulate a general statement and prove it.

16. a) Let $A_n = 1/2 + 2/2^2 + 3/2^3 + \cdots + n/2^n$. Compute the first several values for A_n, find a simple expression for A_n in terms of n, and prove it holds.

b) Let $B_n = 16/5 + 32/5^2 + \cdots + 16n/5^n$. Compute the first several values for B_n, find a simple expression for B_n in terms of n, and prove it holds.

17. Define sets in $\boldsymbol{R}$ by $A_1 = \{0, 1\}$, $A_2 = \{0, a, b, 1\}$ for $0 < a < b < 1$, and if $A_n = \{a_1, \ldots, a_k\}$ with $0 = a_1 < a_2 < \cdots < a_k = 1$ then A_{n+1} adjoins to A_n two numbers strictly between each pair a_i and a_{i+1}. Note $|A_3| = 10$ and $|A_4| = 28$. Find an expression for $|A_n|$ in terms of n and prove that it is correct.

18. Define $B_j \subseteq \boldsymbol{R}$ by $B_1 = \{0, 1\}$, $B_2 = \{0, a, 1\}$ with $0 < a < 1$, $B_3 = \{0, b, c, a, d, e, 1\}$ with $0 < b < c < a < d < e < 1$, and if $B_n = \{0 = b_1, b_2, \ldots, b_k = 1\}$ with the $\{b_j\}$ strictly increasing, then B_{n+1} adjoins to B_n n different numbers strictly between each pair b_j and b_{j+1}. Thus $|B_4| = 25$ and $|B_5| = 121$. Find an expression for $|B_n|$ in terms of n and prove that it is correct.

19. Let $T(1)$ be an equilateral triangle in the plane, and $T(2)$ the result of connecting the midpoints of the sides of $T(1)$. This decomposes $T(1)$ into four smaller equilateral triangles. If each small triangle in $T(2)$ is similarly decomposed by connecting the midpoints of its sides, we obtain $T(3)$. Continuing in this fashion, let $T(n+1)$ be the result of decomposing each small equilateral triangle in $T(n)$ by connecting the midpoints of its sides.

a) Find an expression in n that gives the number of (smallest) equilateral triangles in $T(n)$ and prove by induction that it holds.

b) If $B(n)$ is the number of segments in $T(n)$, each a side of one of the smallest triangles in $T(n)$, prove that $B(n) = 3(2^{2n-3} + 2^{n-2})$ for all $n \geq 2$.

20. Find the number $G(n)$ of rearrangements of $I(n) = \{1, 2, \ldots, n\}$ that from left to right strictly decrease to some point and then strictly increase. Prove that your expression is correct. Some allowable rearrangements of $I(4)$ are 1234 and 4213, and of $I(5)$ are 12345, 54321, 42135, and 41235.

21. Let D denote the usual derivative and D^n the nth derivative. Find and prove a general expression for: a) $D^n(1/(2x+1))$; and b) $D^n(xe^{2x})$. c) Prove that

$D^n(fg) = \sum_{k=0}^{n} \binom{n}{k} D^k(f) D^{n-k}(g)$, where $D^0(f) = f$.

22. **a)** Let $A = \begin{bmatrix} 7 & -4 \\ 9 & -5 \end{bmatrix} \in M_2(\boldsymbol{Q})$. Find A^n in terms of A and I_2, the 2×2 identity matrix (see how A^2 compares to A).

b) Do the same for $B = \begin{bmatrix} -7 & -5 \\ 11 & 8 \end{bmatrix}$.

23. For $A, B \in M_2(\boldsymbol{R})$, let $f(B) = AB - BA$, so $f^{k+1}(B) = Af^k(B) - f^k(B)A$ if $k \geq 1$. Show $f^n(B) = \sum_{k=0}^{n} \binom{n}{k} (-1)^k A^{n-k} B A^k$. If $A^2 = A$, show that $f^{2m+1}(B) = f(B)$.

24. Show that any set of n straight lines in the plane divides the plane into at most $(n^2 + n + 2)/2$ regions.

25. For all $n \in \boldsymbol{N}$ show: a) $2^{5n+1} + 5^{n+2} \in 9\boldsymbol{Z}$; and b) $4^{n+2} + 5^{2n+1} \in 7\boldsymbol{Z}$.

26. If $n \in \boldsymbol{N}$ with $n \geq 2$, show that for any $0 \leq k < (n-1)/2$, $\binom{n}{k} < \binom{n}{k+1}$.

27. (Towers of Hanoi) A board has three vertical sticks, A, B, and C, and n rings stacked on Stick A. The rings strictly increase in size from top to bottom. Rings can be moved from one stick to another, one at a time, but *cannot* be placed on a smaller ring. Show that the minimal number of moves required to transfer the stack of rings to stick B is $2^n - 1$.

0.4 WELL ORDERING AND INDUCTION II

In this section we discuss two less familiar forms of induction that are closely related. Often when a proof "by induction" is given, the particular version used is not specified, so recognizing the different forms of induction ensures that the structure of the proof is understandable. The clearest of the three induction axioms for $\boldsymbol{N}$ is called Well Ordering.

Well Ordering (WO) If $\emptyset \neq S \subseteq \boldsymbol{N}$, then S contains a smallest element.

Well Ordering is an axiom for $\boldsymbol{N}$ that we often refer to as WO. It should seem completely reasonable since if $m \in S \subseteq \boldsymbol{N}$ then we could check all $t < m$ in $\boldsymbol{N}$ to find the smallest in S. WO ensures a smallest element *in* each nonempty subset, not just a lower bound in $\boldsymbol{N}$. Note that in $\boldsymbol{Z}$, neither $3\boldsymbol{Z}$ nor $S = \{x \in \boldsymbol{Z} \mid x < 0\}$ has a smallest element. Even in $\boldsymbol{Q}^+ = \{x \in \boldsymbol{Q} \mid x > 0\}$ with no negatives, $\boldsymbol{Q}^+$ itself has no smallest element. Also $T = \{x \in \boldsymbol{Q} \mid 1 < x < 2\}$ contains no smallest element and has lower bound 1, but $1 \notin T$. Thus Well Ordering is a rather special property of $\boldsymbol{N}$. There are other subsets of $\boldsymbol{R}$ whose nonempty subsets have minimal elements—for example, any finite $T \subseteq \boldsymbol{R}$, any $T \subseteq \boldsymbol{N}$, and the union of any finite subset of $\boldsymbol{R}$ with $\boldsymbol{N}$.

Next we use Well Ordering to prove that every $n \in \boldsymbol{N} - \{1\}$ is a product of primes, an important result needed later. A **prime** is $p \in \boldsymbol{N} - \{1\}$ so if $p = ab$ with $a, b \in \boldsymbol{N}$, then $a = p$ or $b = p$, or, equivalently, $a = 1$ or $b = 1$. Note that $p > 1$ is not prime, or is **composite,** when there are $a, b \in \boldsymbol{N}$ with $1 < a, b < p$, and $p = ab$. The first few primes are 2, 3, 5, 7, 11, and 13.

THEOREM 0.4 Every $n \in \boldsymbol{N} - \{1\}$ is prime or a product of primes.

Proof The idea is that if $n \in \boldsymbol{N}$ is not a product of primes then WO yields a smallest such, leading to a contradiction. Let $S = \{n \in \boldsymbol{N} \mid n > 1$ with n neither prime nor a product of primes$\}$. If $m \in S$ is minimal then $m > 1$ is not a prime so $m = ab$ for $a, b \in \boldsymbol{N}$ with $1 < a < m$ and $1 < b < m$. The minimality of m forces each of a and b to be prime or a product of primes. Thus $m = ab$ is a product of primes, contradicting $m \in S$. Since S cannot have a smallest element, $S = \emptyset$ by WO, proving the theorem. ∎

We present another example using WO.

Example 0.24 The set $D = 6\boldsymbol{N} + 17\boldsymbol{N} = \{n \in \boldsymbol{N} \mid n = 6 \cdot a + 17 \cdot b$ for $a, b \in \boldsymbol{N} \cup \{0\}\}$ contains all $n \geq 80$.

Computation shows $79 \notin D$ and any given $n \geq 80$ seems to be in D. For example, $80 = 6 \cdot 2 + 17 \cdot 4$ and $89 = 6 \cdot 12 + 17 \cdot 1$. Verifying the claim for examples is not a proof since we cannot verify it for all $n \geq 80$. We use WO on $S = \{n \in \boldsymbol{N} \mid n \geq 80$ and $n \neq 6 \cdot a + 17 \cdot b$ for $a, b \geq 0\}$ and want $S = \emptyset$. If not, then S contains a minimal integer m and we claim $m - 1 \in D$. We saw that $80 \in D$ so $m > 80$, forcing $m - 1 \geq 80$. But $m - 1 < m$, so from the minimality of $m \in S$, $m - 1 \notin S$. Using $m - 1 \geq 80$ we have $m - 1 \in D$ and can write $m - 1 = 6 \cdot a + 17 \cdot b$ for some $a, b \geq 0$. Now $m = 6 \cdot a + 17 \cdot b + 1 = 6 \cdot (a + 3) + 17 \cdot (b - 1) = 6 \cdot (a - 14) + 17 \cdot (b + 5)$. It follows that $m \in D$, contradicting $m \in S$, *if* either $b \geq 1$ or $a \geq 14$. But both $a < 14$ and $b = 0$ imply $m - 1 = 6 \cdot a + 17 \cdot b = 6 \cdot a \leq 6 \cdot 13 = 78 < 80$. We know that $m - 1 \geq 80$ so in fact either $a \geq 14$ or $b \geq 1$, forcing $m \in D$, and this contradiction to the existence of m shows that $S = \emptyset$.

The key computation in Example 0.24 writes $1 = 6 \cdot u + 17 \cdot v$ in two ways, once with u negative and once with v negative. Whenever this happens for $1 = au + bv$ with $a, b \in \boldsymbol{N}$, a statement like that in the example for $aN + bN$ holds. Specifically, $1 = 10(-3) + 31 \cdot 1 = 10 \cdot 28 + 31(-9)$ so the argument in Example 0.24 will show that $10N + 31N$ contains all $n \geq 270$.

Sometimes IND-I does not work since we need to know the truth of one or more smaller cases and not necessarily the immediately preceding one. For example, in proving Theorem 0.4 we saw that when $n + 1$ is not a prime, then $n + 1 = ab$ for both $1 < a, b < n + 1$. Here IND-I is useless since knowing only that n is a prime or a product of primes tells us nothing about a or b. We need to know about the integers *arbitrarily* smaller than $n + 1$. To handle arguments like this, we need another form of induction.

IND-II (The Second Principle of Mathematical Induction). Let $t \in S \subseteq N$. If $k \geq t$ and $t, t + 1, \ldots, k \in S$ imply that $k + 1 \in S$, then $n \in S$ for all $n \geq t$.

Like IND-I, IND-II is an implication whose hypothesis has two parts: $t \in S$; and $t, t + 1, \ldots, k \in S$ imply $k + 1 \in S$. If we *show* that $t \in S$ for some t and then use $t, t + 1, \ldots, k \in S$ to *prove* $k + 1 \in S$, *then* IND-II tells us that $n \in S$ for all $n \geq t$. As for IND-I there is a formulation of IND-II for propositions: *If P is a proposition on* **N** *with P(t) true for some $t \in N$, and if for $k \geq t$ the truth of $P(t)$, $P(t + 1)$, ..., and $P(k)$ imply the truth of $P(k + 1)$, then P(n) is true if $n \geq t$.*

Relations among the Fibonacci numbers are good examples for using IND-II. Specifically, $F_n F_m + F_{n+1} F_{m+1} = F_{n+m+1}$ seems to be true. When $n = 3$ and $m = 8$, $2 \cdot 21 + 3 \cdot 34 = 144 = F_{12}$, and trying $n = 4$ and $m = 6$ gives $3 \cdot 8 + 5 \cdot 13 = 89 = F_{11}$. Since $F_n F_m + F_{n+1} F_{m+1} = F_{n+m+1}$ has two variables, it is not clear that induction is applicable. In the next example we see how to finesse this problem and why IND-I is not always sufficient.

Example 0.25 If (F_n) is the Fibonacci sequence, then for all $n, m \in N$, $F_n F_m + F_{n+1} F_{m+1} = F_{n+m+1}$.

Let $P(n)$ be the proposition: $F_n F_m + F_{n+1} F_{m+1} = F_{n+m+1}$ for all $m \in N$. When $n = 1$, since $F_1 = F_2 = 1$ and $F_m + F_{m+1} = F_{m+2} = F_{1+m+1}$ for all $m \in N$, $P(1)$ holds. If $n = 2$ then $F_2 F_m + F_3 F_{m+1} = F_m + 2F_{m+1} = F_m + F_{m+1} + F_{m+1} = F_{m+2} + F_{m+1} = F_{m+3} = F_{2+m+1}$, so $P(2)$ holds. Assume that $k \geq 2$ and that all of $P(2)$, $P(3)$, ..., $P(k)$ are true. To prove that $P(k + 1)$ is true, use the definition of (F_n) to write $F_{k+1} F_m + F_{k+2} F_{m+1} = (F_{k-1} + F_k) F_m + (F_k + F_{k+1}) F_{m+1} = (F_{k-1} F_m + F_k F_{m+1}) + (F_k F_m + F_{k+1} F_{m+1}) = F_{k-1+m+1} + F_{k+m+1} = F_{k+1+m+1}$, proving $P(k + 1)$. The next to last equality uses the truth of both $P(k)$ *and* $P(k - 1)$ so IND-I cannot be used here. If $k = 2$ then the case $k - 1 = 1$ was verified directly. Therefore IND-II shows that $P(n)$ is true for all $n \geq 2$, and so for all $n, m \in N$ by our computation for the $n = 1$ case. We emphasize that we need to "start" the induction with $k \geq 2$ since if we assumed that $k \geq 1$, then the $k - 1$ case would make no sense when $k = 1$.

Example 0.26 Each Fibonacci number F_n satisfies $F_n < (5/3)^n$.

As in the last example, the argument reduces the $k + 1$ case to two previous cases so we will need IND-II. We must also verify the inequality for $n = 1$ and

$n = 2$ so we can begin the induction with $k \geq 2$ and make sense of the $k - 1$ case. When $n = 1$, $F_1 = 1 < 5/3$ is true, and when $n = 2$, $F_2 = 1 < 25/9$ is true. Let $k \geq 2$ and assume the inequality holds for $n = 2, 3, \ldots, k$. Now $F_{k+1} = F_k + F_{k-1}$ so by our induction assumption and the truth of the case $n = 1$ when $k = 2$, $F_{k+1} < (5/3)^k + (5/3)^{k-1} = (5/3)^{k-1}(5/3 + 1) = (5/3)^{k-1}(8/3) < (5/3)^{k-1}(25/9) = (5/3)^{k+1}$, proving the $k + 1$ case. By IND-II, $F_n < (5/3)^n$ for all $n \geq 2$, so for all $n \in \boldsymbol{N}$ since $F_1 < 5/3$ also.

We present one more example where IND-I seems insufficient. It demonstrates how the need for IND-II arises naturally. The result is a generalization of the usual decimal representation of integers and gives what is called the ***m*-adic** or **base *m*** representation of the positive integers.

Example 0.27 If $n, m \in \boldsymbol{N}$ and $m > 1$, then n has a unique expression as $n = a_k m^k + \cdots + a_1 m + a_0$, for $k \geq 0$, all $a_j \in \{0, 1, \ldots, m - 1\}$, and $a_k > 0$.

Results claiming the uniqueness of a particular expression are usually proved in two independent parts: the existence of the expression and the uniqueness of it. We prove first that each n is a sum of powers of m as required. Any $i \in \{1, 2, \ldots, m - 1\}$ has such a representation with $k = 0$ and $a_0 = i$. Take $s \geq 1$ and assume that $1, 2, \ldots, s$ have required representations. We know this for $1 \leq i < m$ so we may assume that $s + 1 \geq m$. If $s + 1 = bm^v$ for $1 \leq b \leq m - 1$ and some $v \geq 1$, then $s + 1$ has a required representation. If not, $bm^v < s + 1 < (b + 1)m^v$ for some $1 \leq b \leq m - 1$ and $v \geq 1$. Subtracting yields $0 < (s + 1) - bm^v < (b + 1)m^v - bm^v = m^v$. But $1 \leq (s + 1) - bm^v \leq s$, so the induction assumption for all these integers means that $(s + 1) - bm^v = a_k m^k + \cdots + a_1 m + a_0$ for appropriate a_j. Using $(s + 1) - bm^v < m^v$ it follows that $k < v$ so $s + 1 = bm^v + a_k m^k + \cdots + a_1 m + a_0$ as required with $m^v > m^k$. Thus n is a sum of powers of m as described if all integers $1 \leq t < n$ are also, so IND-II shows that each $n \in \boldsymbol{N}$ has a required representation.

To prove that the representation for each $n \in \boldsymbol{N}$ is unique, suppose that there are two representations for some integer $w \in \boldsymbol{N}$. By WO there is a minimal $n \in \boldsymbol{N}$ having two different representations, and for it we can write

$$n = a_k m^k + a_{k-1} m^{k-1} + \cdots + a_1 m + a_0 = c_j m^j + c_{j-1} m^{j-1} + \cdots + c_1 m + c_0$$

with $0 \leq a_i, c_i \leq m - 1$ for each i. If $a_0 = c_0$ then subtracting and dividing by m gives two different representations for $(n - a_0)/m < n$, contradicting the minimality of n unless $n = a_0$. In this case $a_0 \leq m - 1$ and $c_0 = a_0$ forces all $c_i = 0$ when $i \geq 1$, contradicting the assumption that the representations are different. When $a_0 \neq c_0$, say $a_0 < c_0$, then $c_0 - a_0 = ms$ for some $s \in \boldsymbol{N}$. This is impossible since $1 \leq c_0 - a_0 \leq m - 1$. Thus if $w \in \boldsymbol{N}$ has two different representations we get a contradiction, so there are no such integers and the m-adic representation must be unique.

We shall see that IND-II is very useful. A rather surprising application of IND-II is to a game played with $n > 1$ marbles. Divide the marbles into two subsets of a and b marbles each, so $n = a + b$, and earn ab points. Keep dividing each set of marbles into two smaller subsets, earning the product of the number of marbles in each of the two new sets. The game ends when all subsets contain one marble, and the total score of the game is the sum of all points received at each division. Fig. 0.3 illustrates one possible game starting with six marbles. There are other ways of dividing the original six marbles to end with six sets of one marble each, and as n gets large there are a huge number of ways of getting down to n sets of one marble each. It is remarkable, but fairly easy to prove using IND-II (Exercise 15), that for any fixed n there is only one possible total score for the game! To see what this is for n marbles, divide every set into one marble and the remainder, earning $(n-1)+(n-2)+\cdots+1$ total points: this is $n(n-1)/2$ by Example 0.15, the only score for n marbles.

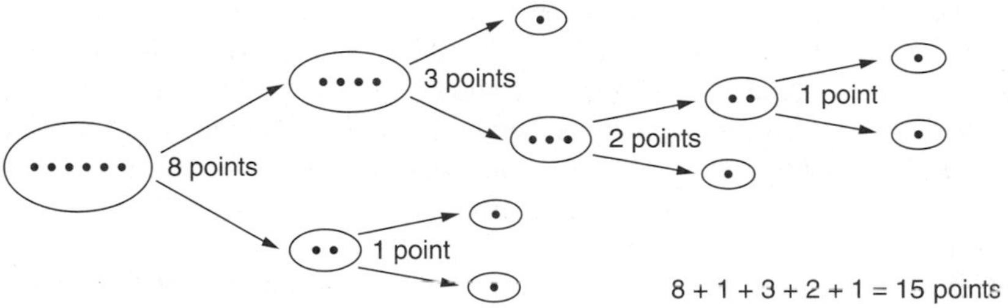

FIGURE 0.3

We briefly discuss the relation between the forms of induction above and their more standard versions. This is not crucial for using induction later but is interesting mathematically and is helpful in understanding induction arguments. Our proofs are not complete and leave details to the reader.

THEOREM 0.5 IND-I is equivalent to its special case when $t=1$: Let $1 \in S \subseteq \boldsymbol{N}$. If $k \in S$ and $k \geq 1$ imply that $k+1 \in S$, then $S=\boldsymbol{N}$. Also IND-II is equivalent to its special case when $t=1$: Let $1 \in S \subseteq \boldsymbol{N}$. If $k \geq 1$ and all of $1, \ldots, k \in S$ imply $k+1 \in S$, then $S=\boldsymbol{N}$.

Proof Certainly the special case of IND-I when $t=1$ holds if we assume IND-I. Now we assume the special case and prove IND-I. Let $t \in S \subseteq \boldsymbol{N}$ so that whenever $k \geq t$ and $k \in S$, then $k+1 \in S$. To prove IND-I we must show that $n \in S$ when $n \geq t$. Define $W \subseteq \boldsymbol{N}$ by $W=\{n \in \boldsymbol{N} \mid t-1+n \in S\}$. Now $1 \in W$ since $t \in S$ and applying the special case to W shows $W=\boldsymbol{N}$ (why?), and so for all $m \in \boldsymbol{N}$, $t-1+m \in S$. Therefore $n \in S$ for all $n \geq t$, proving that IND-I holds. The special case of IND-II when $t=1$ is equivalent to IND-II by the approach above and using the same $W \subseteq \boldsymbol{N}$. ■

We mentioned earlier that the three forms of induction that we have discussed are equivalent. We now indicate why this is true.

THEOREM 0.6 WO, IND-I, and IND-II are equivalent statements.

Proof To prove the equivalence of several statements, normally we prove that each implies the next and that the last implies the first. To see how WO implies IND-I, let $t \in S \subseteq \boldsymbol{N}$ so that $k \geq t$ and $k \in S$ force $k+1 \in S$. Set $E=\{k \in \boldsymbol{N} \mid k \geq t \text{ and } k \notin S\}$ with $m \in E$ minimal if $E \neq \emptyset$. Now $t \in S$ means $m > t$, so $m - 1 \geq t$ and $m - 1 \notin E$ by the choice of m. Thus $m - 1 \in S$ and $m - 1 \geq t$ so $m = m - 1 + 1 \in S$, contradicting $m \in E$. Since E cannot have a minimal element it follows from WO that $E = \emptyset$, proving IND-I: $n \in S$ when $n \geq t$.

Assume that IND-I holds, that $t \in S \subseteq N$, and that $t, t+1, \ldots, k \in S$ imply $k + 1 \in S$. Let $H = \{k \in N \mid k \geq t \text{ and } t, t+1, \ldots, k \in S\}$. Using our assumption on S and IND-I shows that $n \in H$ for all $n \geq t$, so in particular $n \in S$ for all $n \geq t$, proving that IND-II holds.

Assuming IND-II, let $S \subseteq N$ and set $T = N - S$. If S has no minimal element then $1 \notin S$ and so $1 \in T$. Use IND-II on T to obtain $T = N$, forcing $S = \emptyset$ if S has no smallest element and so proving WO. ∎

Although the three forms of induction are equivalent, in any specific problem one of them usually seems more natural to use than the others. It is not clear how Theorem 0.4 on prime factorization or the examples in this section could be proved using IND-I. Of course, it is possible to use IND-II in all cases so why bother with IND-I at all? In mathematics it is customary to use as little as possible to prove results. IND-II appears to assume more than IND-I so perhaps for this and also historical reasons it is common to use IND-I when the proof of the $n + 1$ case follows from the preceding one.

At times we need to use induction on finite sets. The formal result is given in our next theorem.

THEOREM 0.7 Let $S \subseteq N$ have maximal element M and let $t \in S$. If $t \leq k < M$ and $k \in S$ imply that $k + 1 \in S$, then $n \in S$ for all $t \leq n \leq M$. Similarly, if $t \leq k < M$ and $t, \ldots, k \in S$ imply that $k + 1 \in S$, then $n \in S$ for all $t \leq n \leq M$.

Proof Let $W = \{w \in N \mid w \in S \text{ or } w > M\} \subseteq N$. We claim that the first assumption on S implies that $n \in W$ for all $n \geq t$. Since $t \in S$, $t \in W$ follows. Suppose that $t \leq k$ and $k \in W$. If $k \geq M$ then $k + 1 > M$ so $k + 1 \in W$ by definition. When $k < M$ then the definition of W forces $k \in S$, and our assumption on S gives $k + 1 \in S$. Thus $k + 1 \in W$ and applying IND-I to W shows that W contains all $n \geq t$, so S must contain all $t \leq n \leq M$. Using the second assumption leads to the same conclusion by applying IND-II to W. ∎

Example 0.28 $S = 100N + 101N = \{100a + 101b \in N \mid a, b \in N \cup \{0\}\}$ contains all $n \in N$ that satisfy $9800 \leq n \leq 9898$.

Let $T = \{s \in S \mid 9800 \leq s \leq 9898\}$. Now $9800 = 100 \cdot 98 + 101 \cdot 0 \in T$ and also $9898 = 100 \cdot 0 + 101 \cdot 98 \in T$, so 9898 is maximal in T. Assume that $9800 \leq k < 9898$ and that $k \in T$, and write $k = a \cdot 100 + b \cdot 101$. If $a = 0$, then since $9797 < k = b \cdot 101 < 9898$ we have $97 < b < 98$, a contradiction. Hence $a \geq 1$ so $k + 1 = (a - 1) \cdot 100 + (b + 1) \cdot 101 \in T$, and applying Theorem 0.7 shows that $n \in T$ for all $9800 \leq n \leq 9898$. Consequently S does contain these integers. Note that $9799 \notin S$ and $9899 \notin S$ (why?).

A result about the binomial coefficients suggests another example using Theorem 0.7. A complete argument requires knowing that when a prime p divides a product of integers, it must divide one of the factors. This fact will be proved in Chapter 1 but may be familiar to the reader.

Example 0.29 If p is an odd prime, then $\binom{p-1}{i} - (-1)^i \in p\mathbf{Z}$.

For example, if $p = 11$, then $\binom{10}{4} - (-1)^4 = 210 - 1 = 209 = 19 \cdot 11$, and if $p = 13$, then $\binom{12}{3} - (-1)^3 = 220 + 1 = 221 = 17 \cdot 13$. Define $S = \left\{ j \in \mathbf{Z} \mid \binom{p-1}{j} - (-1)^j \in p\mathbf{Z} \right\}$. Since $\binom{p-1}{1} - (-1)^1 = (p-1) - (-1) = p \in p\mathbf{Z}$ and $\binom{p-1}{p-1} - (-1)^{p-1} = 1 - 1 = 0 \in p\mathbf{Z}$, using p is odd, we have $1 \in S$ and $p - 1 \in S$ is maximal. It is easy to see that $0 \in S$. For $1 \le k < p - 1$, $\binom{p-1}{k} = (p-1)!/(k!(p-1-k)!)$ and $(k+1)\binom{p-1}{k+1} = (p-k-1)\binom{p-1}{k}$ follows easily. If $k \in S$, $\binom{p-1}{k} - (-1)^k = pt \in p\mathbf{Z}$, so $(k+1)\binom{p-1}{k+1} = (p-k-1)(pt + (-1)^k) = ps - (k+1)(-1)^k$. Hence $(k+1)\left(\binom{p-1}{k+1} + (-1)^k\right) = ps$, and since p is a prime, $k + 1 \in p\mathbf{Z}$ or $\binom{p-1}{k+1} - (-1)^{k+1} \in p\mathbf{Z}$. Now $1 \le k \le p - 2$ shows $k + 1 \le p - 1$, so p cannot divide $k + 1$ and must divide $\binom{p-1}{k+1} - (-1)^{k+1}$. Thus $k + 1 \in S$, so $S = \{1, 2, \ldots, p - 1\}$ by Theorem 0.7, completing the argument.

An interesting consequence of Example 0.29 is that $2^{p-1} - 1 \in p\mathbf{Z}$ when p is an odd prime, a special case of what is known as Euler's Theorem. Write $2^{p-1} = (1+1)^{p-1} = \sum_{i=0}^{p-1} \binom{p-1}{i}$ (Binomial Theorem) and use Example 0.29 ro replace $\binom{p-1}{i}$ with $ps_i + (-1)^i$ for $s_i \in \mathbf{N}$. Since p is odd, $ps + \sum_{i=0}^{p-1} (-1)^i = 2^{p-1} = ps + 1$ so $2^{p-1} - 1 \in p\mathbf{Z}$. Examples are $2^4 - 1 = 15 = 5 \cdot 3$, $2^6 - 1 = 63 = 7 \cdot 9$, $2^{10} - 1 = 1023 = 11 \cdot 93$, and $2^{16} - 1 = 65{,}535 = 3855 \cdot 17$.

EXERCISES

1. If $T \subseteq \mathbf{N}$, show that each nonempty subset of T has a smallest element.
2. **a)** Use Well Ordering to prove that the binomial coefficients are integers.
   b) Show that the product of any k consecutive $m \in \mathbf{N}$ is a multiple of $k!$.
3. Show that for a fixed $z \in \mathbf{Z}$, in $\{m \in \mathbf{Z} \mid m \ge z\}$ every nonempty subset contains a minimal element.
4. Let $S = \mathbf{N}^k = \mathbf{N} \times \cdots \times \mathbf{N} = \{(a_1, \ldots, a_k) \mid a_i \in \mathbf{N}\}$. The **lexicographic ordering** on S is $(a_1, \ldots, a_k) \le (b_1, \ldots, b_k)$ when either $a_i = b_i$ for all i, $a_1 < b_1$, or for some

$j < k$, $a_i = b_i$ when $1 \le i \le j$ and $a_{j+1} < b_{j+1}$. Show that each nonempty subset of S has a smallest element with respect to this $\le$.

5. Let $T = \bigcup_N N^k$. For $x, y \in T$ with $x \in N^c$ and $y \in N^d$, set $x \le y$ if either $c < d$ or $c = d$ and $x \le y$ in the lexicographic ordering in Exercise 4. Show that each nonempty subset of T has a smallest element with respect to $\le$.

6. Show that each set $S(a, b) = \{n \in N \mid n = a \cdot u + b \cdot v$, with $u, v \in N \cup \{0\}\}$ contains all $n \ge m$ and that m is minimal with this property when:

i) $a = 5, b = 17$, and $m = 64$

ii) $a = 3, b = 71$, and $m = 140$

iii) $a = 17, b = 75$, and $m = 1184$

7. For (F_n) the Fibonacci sequence, show that $F_n F_{n+k} = F_{n+1} F_{n+k-1} + (-1)^{n+1} F_{k-1}$ for any $k, n \in N$ with $k \ge 2$.

8. For (F_n) the Fibonacci sequence, show that for any $k, n \in N$, $F_{kn} \in F_n N$; that is, show that F_{kn} is a multiple of F_n. (Use Example 0.25.)

9. For (F_n) the Fibonacci sequence, find all $n \in N$ such that:

a) $F_n > (7/5)^n$; b) $F_n < (6/7)(7/4)^{n-2}$; c) $F_n > (9/8)(4/3)^{n-1}$.

10. For (F_n) the Fibonacci sequence, show that every $n \in N$ can be expressed as $n = F_u + F_v + \cdots + F_w$ for some $2 \le u < v < \cdots < w$.

11. a) If $z \in Z$, satisfies $-(3^{k+1} - 1)/2 \le z \le (3^{k+1} - 1)/2$, show that there is a unique representation $z = c_k 3^k + \cdots + c_1 3 + c_0$ with each $c_j \in \{-1, 0, 1\}$.

b) What integral gram weights can be found using a two-pan balance with a set of 1, 3, 9, 27, and 81 gram weights? For example, an object weighing 32 grams could be determined since it, together with the 1- and 3-gram weights, in one pan balances the 9- and 27-gram weights in the other pan.

12. Prove that any $n \in N$ has a unique expression $n = c_1 1! + c_2 2! + \cdots + c_k k!$ for some $k \ge 1$, each $0 \le c_j \le j$ and $c_k \ne 0$. (See Theorem 0.3 and Example 0.16.)

13. An ordered partition of $n \in N$ is some $(a_1, \ldots, a_k)$ with $a_j \in N$ satisfying $n = a_1 + \cdots + a_k$. Note that (1, 2) and (2, 1) are different ordered partitions of 3. Find the number of ordered partitions $g(n)$ of $n \in N$ and prove your answer.

14. Find $G(n)$, the number of n-tuples of zeros and ones with no three consecutive equal digits, and prove your answer. For example, $G(1) = 2$, $G(2) = 4$, $G(3) = 6$, $G(4) = 10$, $G(5) = 16$, $G(6) = 26$, and $G(7) = 42$.

15. Recall the game described in the text: $n > 1$ marbles are divided into two subsets of a and b marbles each $(a + b = n)$ and ab points are awarded. If either $a > 1$ or $b > 1$, then that subset is further divided into sets of c and d elements (say $a = c + d$) and cd additional points are awarded. Continue to divide the remaining groups of more than one marble, adding the corresponding points to the previous total, until all subsets have one marble each. Show that the only possible point total is $(n - 1)n/2$.

16. The Morse sequence (a_i) is defined by $a_1 = 1$, $a_2 = 0$, and for $k \ge 1$ with $2^k < j \le 2^{k+1}$, $a_j = 1 - a_{j-2^k}$. In effect, each initial block of length 2^k gives the next 2^k elements by exchanging 0 and 1. The first 32 terms are 1, 0, 0, 1, 0, 1, 1, 0, 0, 1, 1, 0, 1, 0, 0, 1, 0, 1, 1, 0, 1, 0, 0, 1, 1, 0, 0, 1, 0, 1, 1, 0, ...

a) For all $k \ge 1$, show that $a_{2^k - 1} \ne a_{2^k}$.

b) For every $k \ge 2$, show that $a_{2^k + 2} = a_{2^k + 3} = 1$ and $a_{2^k - 2} = a_{2^k - 1}$.

c) Show that neither 0, 1, 0, 1, 0 nor 1, 0, 1, 0, 1 occurs as five consecutive terms.

d) Show that neither 1, 1, 0, 1, 1 nor 0, 0, 1, 0, 0 occurs as five consecutive terms.
e) Show that (a_i) does not contain either 0, 0, 0, or 1, 1, 1 as three consecutive terms.
f) Show that no two or three consecutive terms in (a_i) occur three consecutive times.

0.5 FUNCTIONS

We review the basic notion of function usually given as some rule or procedure that assigns each element in one set to an element in another set. A more precise notion of function will be useful to have.

DEFINITION 0.12 Given nonempty sets S and T, $f \subseteq S \times T$ is a **function from S to T**, denoted $f : S \to T$, if for each $s \in S$ there is a *unique* $(s, t) \in f$. S is the **domain** of f, written Dom f, and T is the **codomain** of f.

Other terms used for function are *map, mapping,* and *transformation.* A function involves three sets: a domain S, a codomain T, and a subset of $S \times T$ that consists of precisely one ordered pair starting with each element of S. Therefore $\{(1, 2), (2, 2), (3, 1)\}$ could be either $f : \{1, 2, 3\} \to \{1, 2\}$ or $h : \{1, 2, 3\} \to \{1, 2, 3\}$. These are different functions since they have different codomains. Now $g = \{(1, 2), (1, 3), (2, 2), (3, 1)\}$ is not a function since $(1, 2), (1, 3) \in g$ have the same first coordinate. In practice, given $f : S \to T$ as a set of pairs, we can consider its codomain to be any subset of T containing all second coordinates of pairs in f, although each choice formally gives a different function.

The definition of function as a collection of pairs is precise but is often awkward to work with. For $f : S \to T$ we write $f(s) = t$ when $(s, t) \in f$ and call $f(s) = t$ the **image** or **value** of s under f. We think of f *as a rule* that associates $s \in S$ to $t \in T$ when $(s, t) \in f$, so f "acts" on S. This is illustrated by Fig. 0.4.

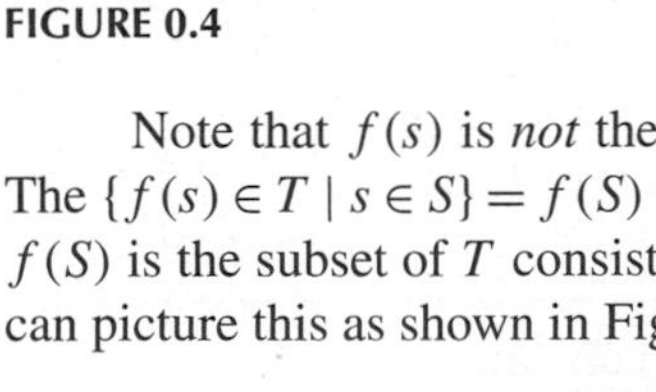

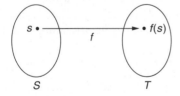

FIGURE 0.4

Note that $f(s)$ is *not* the function f but the element of T associated to $s \in S$ by f. The $\{f(s) \in T \mid s \in S\} = f(S) = \text{Im } f$ is the **image of f**, also called the range of f. Thus $f(S)$ is the subset of T consisting of all second coordinates of the pairs that define f. We can picture this as shown in Fig. 0.5.

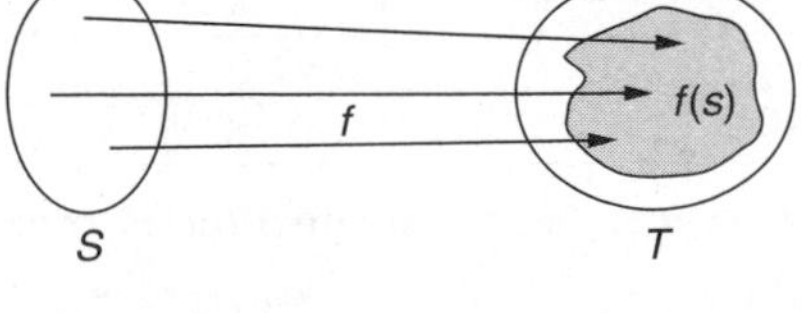

FIGURE 0.5

In practice, we can list the ordered pairs of a function only when the domain has very few elements. We usually define f with domain S by specifying $f(s)$ for all $s \in S$, such as $f(r) = r^3 - 4r$ for $r \in \boldsymbol{R}$. Here $\{(s, f(s)) = (s, s^3 - 4s) \mid s \in \boldsymbol{R}\}$ determines $f : \boldsymbol{R} \to \boldsymbol{R}$. It is essential that the expression $f(s)$ be *unambiguous* in associating each element of the domain with some element of the codomain, in which case $f = \{(s, f(s)) \in S \times T \mid s \in S\}$. For example, the usual squaring function on $\boldsymbol{R}$ is $sq = \{(r, r^2) \in \boldsymbol{R}^2 \mid r \in \boldsymbol{R}\} \subseteq \boldsymbol{R}^2$, and the greatest integer function $[\,] \subseteq \boldsymbol{R} \times \boldsymbol{R}$ is $[\,] = \{(r, n) \in \boldsymbol{R} \times \boldsymbol{R} \mid n \in \boldsymbol{Z}$ and $n \leq r < n + 1\}$. We could consider $sq' : \boldsymbol{R} \to [0, +\infty)$, which still gives $sq'(r) = r^2$, but this function is different from sq since it has a different codomain. Similarly, we might consider $[\,] \subseteq \boldsymbol{R} \times \boldsymbol{Z}$, again technically different from $[\,] \subseteq \boldsymbol{R} \times \boldsymbol{R}$ (why?).

Example 0.30 Is $g = \{(r, [r]) \mid r < 1\} \bigcup \{(r, 0) \mid r \geq 0\} \subseteq \boldsymbol{R} \times \boldsymbol{R}$ a function?

Since for all $0 \leq r < 1$ both $(r, [r])$, $(r, 0) \in g$, it may seem that g is not a function. However, for all of these, $[r] = 0$, so g contains exactly one pair beginning with each $r \in \boldsymbol{R}$; thus $g : \boldsymbol{R} \to \boldsymbol{R}$. Now $g \subseteq \boldsymbol{R} \times (-\infty, 0]$ and also $g \subseteq \boldsymbol{R} \times ((-\infty, 0] \bigcap \boldsymbol{Z})$ so we could take the codomain of g to be $(-\infty, 0] \bigcap \boldsymbol{Z}$, $\boldsymbol{Z}$, or $(-\infty, 0]$, although technically all of these are different functions.

Example 0.31 Is $f = \{(a/b, a - b) \in \boldsymbol{Q} \times \boldsymbol{Z}\}$ a function?

Here $f(a/b) = a - b$ certainly looks like a function from $\boldsymbol{Q}$ to $\boldsymbol{Z}$, but $(2/3, -1) \in f$ and $(2/3, -5) = (10/15, -5) \in f$, so f is *not* a function.

For any $f : S \to T$, $f(s)$ must be uniquely determined by $s \in S$ but can be given "abstractly." For example $P = \{(n, n^{\text{th}}$ decimal digit in $\pi)\} \subseteq \boldsymbol{N} \times \boldsymbol{Z}$ defines a function. The pairs $(1, 1)$, $(2, 4)$, $(3, 1)$, $(4, 5)$, and $(5, 9)$ are in P since $\pi = 3.14159\ldots$. The second coordinate of $(10^{1000}!, P(10^{1000}!)) \in P$ will probably never be found explicitly, although it is a uniquely defined digit.

Let us consider a few general and useful examples of functions.

Example 0.32 For any nonempty set S, $I_S = \{(s, s) \in S \times S \mid s \in S\}$ is the **identity** function $I_S : S \to S$ on S.

We can write $I_S(x) = x$ for all $x \in S$. Here $\text{Dom}\, I_S = S = \text{Im}\, I_S$.

Example 0.33 For $y \in T$, $C_y = S \times \{y\} \subseteq S \times T$ is a **constant** function.

$C_y(s) = y$ for all $s \in S$, $\text{Im}\, C_y = \{y\}$, and the codomain of C_y is T.

Example 0.34 If $S_1, S_2, \ldots, S_n$ are nonempty sets, the jth **projection** is $\pi_j : S_1 \times S_2 \times \cdots \times S_n \to S_j$ given by $\pi_j((x_1, \ldots, x_n)) = x_j$.

As ordered pairs $\pi_j = \{((x_1, \ldots, x_n), x_j)\} \subseteq (S_1 \times S_2 \times \cdots \times S_n) \times S_j$, $\operatorname{Im} \pi_j = S_j$, and the codomain of $\pi_j = S_j$.

For nonempty sets $S_1, \ldots, S_n$ we can define $\iota_j : S_j \to (S_1 \times S_2 \times \cdots \times S_n)$ by $\iota_j = \{(s_j, (y_1, \ldots, y_{j-1}, s_j, y_{j+1}, \ldots, y_n)) \mid s_j \in S_j\}$ for fixed $y_i \in S_i$ when $i \neq j$, so $\iota_j(s_j) = (y_1, \ldots, y_{j-1}, s_j, y_{j+1}, \ldots, y_n)$. Here $\operatorname{Dom} \iota_j = S_j$, and the image $\iota_j(S_i) = \{y_1\} \times \cdots \times \{y_{j-1}\} \times S_j \times \{y_{j+1}\} \times \cdots \times \{y_n\}$ depends on the choice of $\{y_i\}$. The codomain of ι_j is $S_1 \times S_2 \times \cdots \times S_n$.

Common arithmetic operations can be viewed as functions. For examples on $\boldsymbol{R}$, addition is the function $+ : \boldsymbol{R}^2 \to \boldsymbol{R}$ given by $+(r, s) = r + s$, or $+ = \{((r, s), r + s) \mid r, s \in \boldsymbol{R}\} \subseteq \boldsymbol{R}^2 \times \boldsymbol{R}$, and multiplication is $\cdot : \boldsymbol{R}^2 \to \boldsymbol{R}$ given by $\cdot(r, s) = rs$, or $\cdot = \{((r, s), rs) \mid r, s \in \boldsymbol{R}\} \subseteq \boldsymbol{R}^2 \times \boldsymbol{R}$. Clearly, $\operatorname{Im}(+) = \boldsymbol{R} = \operatorname{Im}(\cdot)$.

Defining a function as a set gives meaning to the notion of unions of functions. Thus when a function is defined by different "rules" on different parts of its domain, like the greatest integer function, we can regard it as unions of functions on the pieces. We give the general example of this next.

Example 0.35 Let $\{A_j\}_J$ be a partition of S into nonempty subsets and let $f_i : A_j \to T$ for each $j \in J$. Then $f = \bigcup f_j$ defines $f : S \to T$ by $f(s) = f_j(s)$ if $s \in A_j$.

For each $s \in S$ there is a unique A_j with $s \in A_j$ since $\{A_j\}$ is a partition of S. But $f_j : A_j \to T$ means that for $s \in A_j$ there is a unique pair $(s, f_j(s)) \in f_j$. Now $f_i \cap f_j = \varnothing$ when $i \neq j$ because $\{A_j\}$ is pairwise disjoint, so $(s, f_j(s))$ is the only pair in f starting with $s \in A_j$ and f is a function.

To illustrate Example 0.35 we can consider $[\,] : \boldsymbol{R} \to \boldsymbol{Z}$ as the union over $j \in \boldsymbol{Z}$ of the constant functions $C_j : [j, j + 1) \to \boldsymbol{Z}$ given by $C_j(r) = j$, since $\{[j, j + 1) \subseteq \boldsymbol{R} \mid j \in \boldsymbol{Z}\}$ is a partition of $\boldsymbol{R}$.

Example 0.36 For $1 \leq i \leq k$ and $f_i : S \to T_i$, there is $f : S \to (T_1 \times \cdots \times T_k)$ given by $f = \{(s, (f_1(s), \ldots, f_k(s))) \in S \times (T_1 \times \cdots \times T_k) \mid s \in S\}$.

Suppose that both $(s, (t_1, \ldots, t_k)), (s, (t'_1, \ldots, t'_k)) \in f$. Since f_i is a function, $f_i(s)$ is uniquely determined by $s \in S$ so $t_i = f_i(s) = t'_i$ for $1 \leq i \leq k$. Hence $(t_1, \ldots, t_k) = (t'_i, \ldots, t'_k)$ so f is a function with $f(s) = (f_1(s), \ldots, f_k(s))$.

We saw in Example 0.31 that a rule that looks like a function may not be. A related example is $f(2\boldsymbol{Z} + i) = i$ from $\{2\boldsymbol{Z} + i \mid i \in \boldsymbol{Z}\}$ to $\boldsymbol{Z}$. Observe that $2\boldsymbol{Z} + 2 = 2\boldsymbol{Z} = 2\boldsymbol{Z} + 4$ so $f(2\boldsymbol{Z} + 2) = 2$ and also $f(2\boldsymbol{Z} + 2) = f(2\boldsymbol{Z} + 4) = 4$. Thus f cannot be a function. This example and Example 0.31 run into the same problem: there are different ways of representing each element in the domain, and the rule of the function depends on the particular representation of the element rather than on the element itself.

Related to these last examples but not usually recognized as a problem is the frequent need to pick an element from each of several different sets. For example, given a set S, we will often want to choose one element from each nonempty subset in a partition of S, or we might want to choose an element from every nonempty subset of S. In the latter case this amounts to specifying a function $g : P(S)^* \to S$ satisfying $g(A) \in A$, for $P(S)$ the power set of S (Definition 0.6) and $P(S)^* = P(S) - \{\varnothing\}$. There are two problems with this. First, it is impossible in general even to describe all the nonempty subsets. Think, for example, of the subsets of $\boldsymbol{R}$, or even of $\boldsymbol{N}$. Next, if $A \in P(S)$ how can we actually specify $g(A) \in A$? When $S = \boldsymbol{N}$ we can use Well Ordering and let $g(A)$ be the minimal element in A. In general there is no obvious way to specify an element in an arbitrary nonempty subset, although intuitively we ought to be able to do so. The accepted way around this problem is to take the existence of such choices as an axiom.

THE AXIOM OF CHOICE For any nonempty set S there is a **choice function** $h : P(S)^* \to S$ satisfying $h(A) \in A$ for every $A \in P(S)^* = P(S) - \{\varnothing\}$.

It is not obvious, but the Axiom of Choice is equivalent to the existence of an ordering $\leq$ on any $S \neq \varnothing$ (see Chapter 1) so that every nonempty subset of S contains a smallest element with respect to $\leq$. In particular there must be such as ordering on $\boldsymbol{R}$ although one has never been described explicitly. Our use of the Axiom of Choice will be restricted to a few proofs and examples.

The definition of function shows that two functions are equal exactly when they are the same set of pairs and have the same domain and the same codomain. We record an observation about equality of functions.

THEOREM 0.8 Given two functions $f : S \to T$ and $g : S \to T$, then $f = g$ precisely when $f(s) = g(s)$ for all $s \in S$.

Proof If $f = g$ these are the same subset of $S \times T$ by Definition 0.12. Thus for each $s \in S$, each of f and g has a unique pair starting with s, say $(s, f(s))$ and $(s, g(s))$, respectively, so $(s, f(s)) = (s, g(s))$ and $f(s) = g(s)$ follows. Of course if this equality holds, then the sets $f = \{(s, f(s)) \in S \times T \mid s \in S\} = \{(s, g(s)) \in S \times T \mid s \in S\} = g$, so $f = g$. ∎

Suppose that $f : S \to T$ and $g : T \to U$. The image $f(s) \in T$ of any $s \in S$ under f is uniquely determined, and the image $g(f(s)) \in U$ of $f(s)$ under g is uniquely determined. Therefore applying the functions consecutively associates with each element of S a unique element in U, and this defines a function from S to U, the composition of f with g.

DEFINITION 0.13 Given $f : S \to T$ and $g : T \to U$, the **composition** of f with g is $g \circ f : S \to U$ given by $g \circ f(s) = g(f(s))$.

A standard picture of composition as given in the definition is shown in Fig. 0.6.

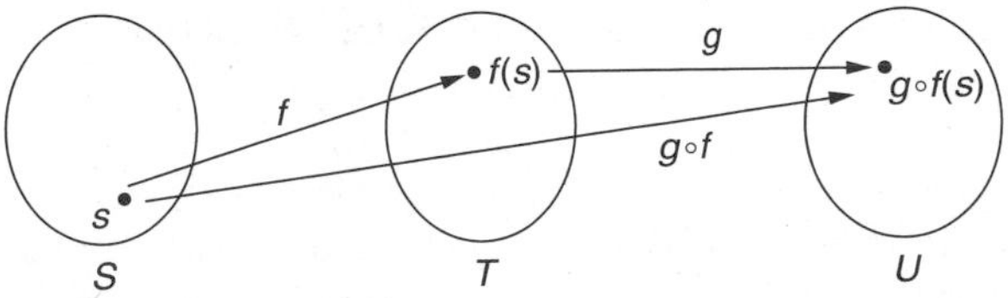

FIGURE 0.6

Example 0.37 If $f: S \to T$, then $f \circ I_S = f$ and $I_T \circ f = f$.

Since $I_S: S \to S$ and $I_T: T \to T$, $I_T \circ f$, $f \circ I_S: S \to T$ so we use Theorem 0.8 to verify the required equalities. If $s \in S$, $I_T \circ f(s) = I_T(f(s)) = f(s)$ because $f(s) \in T$, so $I_T \circ f = f$. Similarly, $f \circ I_S(s) = f(I_S(s)) = f(s)$ for all $s \in S$ shows that $f \circ I_S = f$.

When $f: S \to T$ and $g: T \to U$, we can compose $g \circ f: S \to U$ with $h: U \to V$ to get $h \circ (g \circ f): S \to V$. In this situation we could also compose g with h first and then f with $h \circ g$ to get $(h \circ g) \circ f$. It is easy to see that these are the same, so composition of functions is associative when defined.

THEOREM 0.9 If $f: S \to T$, $g: T \to U$, and $h: U \to V$, then $h \circ (g \circ f) = (h \circ g) \circ f$.

Proof Each of these compositions has domain S and codomain V, so by Theorem 0.8 they are the same if each gives the same image for every $s \in S$. But by Definition 0.13, $h \circ (g \circ f)(s) = h(g \circ f(s)) = h(g(f(s))) = h \circ g(f(s)) = ((h \circ g) \circ f)(s)$, so the functions are equal as claimed. ∎

To follow the action of $f: S \to T$ with $g: U \to V$, we need Im $f \subseteq U$ but the definition of $g \circ f$ requires $T = U$. For example, if $[\,]: \mathbf{R} \to \mathbf{R}$ is the greatest integer function and $g: \mathbf{Z} \to \mathbf{Z}$ is given by $g(z) = z + 1$, then $g([r]) = [r] + 1$ makes sense for all $r \in \mathbf{R}$, and $h = \{(r, [r] + 1) \in \mathbf{R} \times \mathbf{Z} \mid r \in \mathbf{R}\}$ is a function. However, $h \neq g \circ [\,]$ since $g \circ [\,]$ is not defined: the codomain of $[\,] = \mathbf{R} \neq \mathbf{Z} = \text{Dom}\, g$. For the sake of expediency we deal with this problem as follows: *in any composition replace $f: S \to T$ with $f: S \to U$ as necessary, where Dom $f \subseteq U$, ignoring the fact that these are different functions when $T \neq U$*. In the example just given we *will* write $h = g \circ [\,]$ by simply considering $[\,] \subseteq \mathbf{R} \times \mathbf{Z}$.

A different problem arises if some $f(s) \notin \text{Dom}\, g$. Let $f: \mathbf{R} \to \mathbf{R}$ be given by $f(x) = x^2$ or $g: \mathbf{R} \to \mathbf{R}$ be given by $g(x) = [x]$. Then neither can be composed with $h: (\mathbf{R} - \{0, 1\}) \to \mathbf{R}$ given by $h(r) = 1/(r^2 - r)$ since $\{0, 1\} \subseteq \text{Im}\, f \cap \text{Im}\, g$ but $0, 1 \notin \text{Dom}\, h$. It does no good to replace the codomain of f or g with its image; to compose we must avoid the $x \in \text{Dom}\, f$, or $y \in \text{Dom}\, g$, with $f(x), g(y) \in \{0, 1\}$. Now $x^2 \in \{0, 1\} \Leftrightarrow x \in \{-1, 0, 1\}$, so to form $h \circ f$, $\{-1, 0, 1\}$ must be excised from Dom f, and to form $h \circ g$, the interval $[0, 2)$ must be excised from Dom g since $[y] \in \{0, 1\} \Leftrightarrow y \in [0, 2)$. In either case the change gives ordered pairs that are a proper subset of the original function. The formal way to do this is given next.

DEFINITION 0.14 Given $f : S \to T$ and a nonempty subset $A \subseteq S$, the **restriction** $f\,|_A : A \to T$ is defined by $f\,|_A = f \bigcap (A \times T) = \{(a, f(a)) \in f \mid a \in A\}$.

In other words, for $a \in A$, $f\,|_A\,(a) = f(a)$ and $f\,|_A$ is not defined on $S - A$. For the functions g and h of the last paragraph, if $g^{\#} = g\,|_{(\boldsymbol{R}-[0,2))}$ then $h \circ g^{\#} : (\boldsymbol{R} - [0, 2)) \to \boldsymbol{R}$ with $h \circ g^{\#}(r) = 1/([r]^2 - [r])$ for all $r \notin [0, 2)$.

The simple idea of restriction will be useful to have. With the notation in Example 0.35, when $f = \bigcup f_j : S \to T$ for $f_j : A_j \to T$ and $\{A_j\}$ a partition of S, then $f_j = f\,|_{A_j}$.

EXERCISES

1. In each case show why the example is a function or why it is not:
   i) $f = \{(n, m) \in \boldsymbol{N} \times \boldsymbol{N} \mid m$ is the integer obtained from n by reversing its digits$\}$, e.g., $f(123) = 321$ and $f(11271) = 17211$
   ii) $F = \{(a/b, 2^a 3^b) \in \boldsymbol{Q}^+ \times \boldsymbol{N}\}$, where $\boldsymbol{Q}^+ = \{q \in \boldsymbol{Q} \mid q > 0\}$
   iii) $h = \{(p(x), x^n p(1/x)) \in (\boldsymbol{R}[x] - \{0\}, \boldsymbol{R}[x] - \{0\}) \mid n = \deg p(x)\}$
   iv) $k = \{(r, s) \in ([0, +\infty) \times \boldsymbol{R} \mid s^2 = r\}$
   v) $g = \{(re^{it}, (r, t)) \in \boldsymbol{C} \times ([0, +\infty) \times \boldsymbol{R})\}$, where $e^{ib} = \cos b + i \cdot \sin b$
2. Determine whether each of the following is a function with domain $T = \{$triangles in the plane $\boldsymbol{R}^2\}$:
   i) $h = \{(\Delta, P) \in T \times \boldsymbol{R}^2 \mid P \in \Delta$ has minimal distance from the line $y = x\}$
   ii) $f = \{(\Delta, d) \in T \times \boldsymbol{R} \mid d$ is the minimal distance from Δ to the y-axis$\}$
   iii) $g = \{(\Delta, (r_1, r_2, r_3)) \in T \times \boldsymbol{R}^3 \mid \Delta$ has angle measures r_1, r_2, and $r_3\}$
   iv) $k = \{(\Delta, p) \in T \times \boldsymbol{R} \mid p$ is the perimeter of $\Delta\}$
3. If $f : S \to T$ and $g : T \to U$, define $g \circ f \subseteq S \times U$ using $f \subseteq S \times T$ and $g \subseteq T \times U$.
4. If $f : S \to T$ and $\varnothing \neq A \subseteq B \subseteq S$, show that $(f|_B)|_A = f|_A$.
5. Let $\varnothing \neq A \subseteq S$ and $g : A \to T$. Find $f : S \to T$ with $f|_A = g$. Is this f unique?
6. Let $f : S \to T$, $g : T \to U$, and $\varnothing \neq A \subseteq S$. Show that $(g \circ f)|_A = g \circ f|_A$.
7. Assume that $f_1, f_2, \ldots, f_m$ are functions so that $((\cdots((f_1 \circ f_2) \circ f_3)\cdots) f_{m-1}) \circ f_m$ is defined. Show that $((\cdots((f_1 \circ f_2) \circ f_3)\cdots) f_{m-1}) \circ f_m =$ $f_1 \circ (f_2 \circ (f_3 \circ \cdots (f_{m-1} \circ f_m)\cdots))$.
8. For a nonempty set S, let $f \subseteq S \times S^k$ and let $\pi_j : S^k \to S$ be the jth projection map of Example 0.34. Let $\pi_j \circ f = \{(s, s_j) \in S \times S \mid (s, (s_1, \ldots, s_k)) \in f\} \subseteq S \times S$. Show that f is a function—that is $f : S \to S^k$—if and only if each $\pi_j \circ f : S \to S$.
9. Let $f_1 \subseteq S \times T_1, \ldots, f_m \subseteq S \times T_m$ and $f = \{(s, (t_1, \ldots, t_m)) \in S \times (T_1 \times \cdots \times T_m) \mid (s, t_j) \in f_j\}$. Show that f is a function if and only if each f_j is a function.
10. Given $f : \boldsymbol{R} \to \boldsymbol{R}$ and $T \subseteq \boldsymbol{R}$, find the largest $A \subseteq \boldsymbol{R}$ so that $f|_A : A \to T$ is defined:
    i) $f(r) = r^2 - 4r + 1$, $T = [0, +\infty)$
    ii) $f(r) = r - [r]$, $T = \boldsymbol{R} - \{0\}$
    iii) $f(r) = r - [r] - 1/2$, $T = (0, +\infty)$
    iv) $f(r) = r^2$, $T = [4, 16]$

0.6 BIJECTIONS AND INVERSES

Given $f : S \to T$ and $t \in T$, it turns out to be useful to consider those elements in S that have t as their image. More generally, we have

DEFINITION 0.15 If $f : S \to T$ and $B \subseteq T$, then $f^{-1}(B) = \{s \in S \mid f(s) \in B\}$ is the **inverse image** of B under f. When $B = \{t\}$, $f^{-1}(\{t\})$ is also called the **fiber** of f over t.

Thus $f^{-1}(B)$ consists of all the elements in S, if any, whose images under f lie in B. Illustrations for general $B \subseteq T$, and $B = \{t\}$ are given in Fig. 0.7.

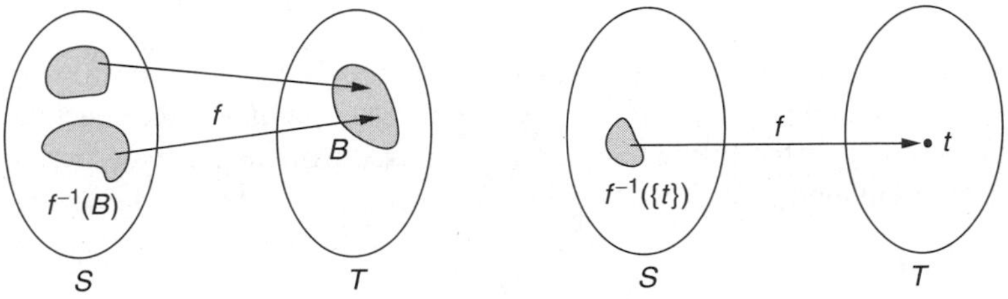

FIGURE 0.7

Taking inverse images looks like we "apply" f^{-1}, but f^{-1} is *not* a function from T to S since inverse images are defined for *subsets* of T, not elements of T. The inverse image does give rise to a function $F : P(T) \to P(S)$ defined by $F(B) = f^{-1}(B)$. For $t \in T$, the inverse image or fiber $f^{-1}(\{t\})$ is the set of all elements in S with image t. Observe that $f^{-1}(\{t\}) = \emptyset$ exactly when $t \notin \text{Im } f$. It is important to remember that $s \in f^{-1}(B)$ exactly when $f(s) \in B$.

Example 0.38 If $f : S \to T$, then $\{f^{-1}(\{t\}) \mid t \in T\}$ is a partition of S.

Since $f(s) \in T$ for each $s \in S$, $s \in f^{-1}(\{f(s)\})$ by Definition 0.15. Thus S is the union of the fibers of f. If $s \in f^{-1}(\{t\}) \cap f^{-1}(\{t'\})$, then $f(s) = t$ and $f(s) = t'$, but f is a function so $t = t'$. Hence $\{f^{-1}(\{t\}) \mid t \in T\}$ is pairwise disjoint so $\{f^{-1}(\{t\}) \mid t \in T\}$ is a partition of S by Definition 0.10.

If $\pi_1 : \mathbf{R}^2 \to \mathbf{R}$ is the projection $\pi_1((a, b)) = a$, then $\pi_1^{-1}(\{-2\}) = \{(x, y) \in \mathbf{R}^2 \mid \pi_1((x, y)) = -2\}$ is the line $x = -2$. Also $\pi_1^{-1}((-\infty, 0])$ is the left half-plane, including the y-axis. For the general projection $\pi_j : (S_1 \times \cdots \times S_n) \to S_j$ of Example 0.34 and $B \subseteq S_j$, $\pi_j^{-1}(B) = S_1 \times \cdots S_{j-1} \times B \times S_{j+1} \times \cdots \times S_n$.

For the identity function $I_S : S \to S$, $I_S^{-1}(B) = B$ for any $B \subseteq S$ (why?). If $C_y : S \to T$ is the constant function in Example 0.33 then since $C_y(s) = y$ for all $s \in S$, we have $C_y^{-1}(B) = S$ when $y \in B$ and $C_y^{-1}(B) = \emptyset$ when $y \notin B$.

If $f : \mathbf{R} \to \mathbf{R}$ is $f(r) = r^3 - 3r$ then f increases on $(-\infty, -1]$ to a local maximum $f(-1) = 2$, then decreases to its local minimum $f(1) = -2$, then increases on $[1, +\infty)$ (calculus!). These facts, $f(-2) = -2$, and $f(2) = 2$, show that when $|x| > 2$, $f^{-1}(\{x\})$ consists of one element; $f^{-1}(\{-2\}) = \{-2, 1\}$; $f^{-1}(\{2\}) = \{-1, 2\}$; and when $|x| < 2$, $f^{-1}(\{x\})$

has exactly three elements. Also $f^{-1}(\{0\}) = \{0, \pm 3^{1/2}\}$, $f^{-1}([0, +\infty)) = [-3^{1/2}, 0] \cup [3^{1/2}, +\infty)$ and $f^{-1}([-2, 2]) = [-2, 2]$.

For the usual sine function $\sin : \boldsymbol{R} \to \boldsymbol{R}$, $\sin(\boldsymbol{R}) = [-1, 1]$. We have $\sin^{-1}([2, +\infty)) = \emptyset$ since no $r \in \boldsymbol{R}$ satisfies $\sin r > 1$. If $r \in [-1, 1]$, $\sin^{-1}(\{r\})$ is infinite with $\sin^{-1}(\{-1\}) = \{2\pi z + 3\pi/2 \mid z \in \boldsymbol{Z}\}$ and $\sin^{-1}(\{0\}) = \{z\pi \mid z \in \boldsymbol{Z}\}$. More complicated is $\sin^{-1}([0, 1]) = \sin^{-1}([0, +\infty)) = \bigcup_{z \in \boldsymbol{Z}}[2z\pi, (2z+1)\pi]$.

DEFINITION 0.16 $f : S \to T$ is **injective,** or an injection, if $f(s) = f(s')$ implies $s = s'$, and is **surjective,** or a surjection, if $\text{Im } f = f(S) = T$. When f is both injective and surjective it is called **bijective,** or a bijection.

Other terms for these concepts are *one-to-one* or *monomorphism* for injective, *onto* or *epimorphism* for surjective, and *equivalence* or *one-to-one correspondence* for bijective. Injectivity is equivalent to any of the following: different ordered pairs in f have different second coordinates; different elements in S have different images; $|f^{-1}(\{s\})| \leq 1$. Fig. 0.8 illustrates this condition.

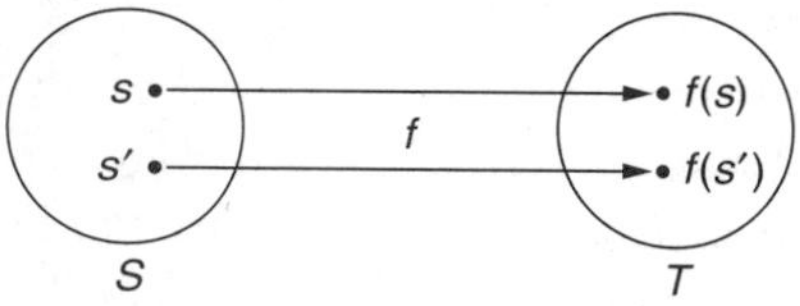

FIGURE 0.8

The surjectivity of f is equivalent to $|f^{-1}(\{t\})| \geq 1$ since any $t \in T$ is the image of some $s \in S$. Combining this with the condition for injectivity shows that f is a bijection exactly when $|f^{-1}(\{t\})| = 1$ for each $t \in T$.

The existence of a bijection $\varphi : S \to T$ implies that the elements in S and in T can be matched. Since φ is injective, $s \neq s' \Rightarrow \varphi(s) \neq \varphi(s')$, and since $\varphi(S) = T$, every element of T is associated with a unique element in S. Therefore it is natural to interpret the existence of a bijection between S and T to mean that S and T have the "same size." We will explore this idea in the next section and see that not all infinite sets have the same size!

If $f : S \to T$ is not surjective we may consider $f : S \to \text{Im } f$, which is surjective, although technically a different function. The greatest integer function $[\,] : \boldsymbol{R} \to \boldsymbol{R}$ is not surjective since $[\boldsymbol{R}] = \boldsymbol{Z} \neq \boldsymbol{R}$. If instead we take $[\,] : \boldsymbol{R} \to \boldsymbol{Z}$, then it is surjective but not injective. Similarly, the squaring function $sq : \boldsymbol{R} \to \boldsymbol{R}$ is neither injective, since $(-r)^2 = r^2$, nor surjective, since $\text{Im } sq = [0, +\infty)$. But $sq : \boldsymbol{R} \to [0, +\infty)$ is surjective, and $sq|_{[0,+\infty)} : [0, +\infty) \to [0, +\infty)$ is a bijection. Suitably restricting the domain of a given function will produce an injective function: for the sine function, radical surgery on its domain is needed. One of infinitely many ways to do this is to take $\sin|_{[-\pi/2,\pi/2]} : [-\pi/2, \pi/2] \to \boldsymbol{R}$. Finally, $\sin|_{[-\pi/2,\pi/2]} : [-\pi/2, \pi/2] \to [-1, 1]$ is bijective.

We have just seen that a function can be injective or surjective without being both, but our examples use infinite sets. Injectivity and surjectivity are equivalent for a function between finite sets.

THEOREM 0.10 Let S and T be sets with $|S| = |T| = n \in \boldsymbol{N}$. If $f : S \to T$, then f is injective $\Leftrightarrow$ f is surjective, so either implies that f is a bijection.

Proof If f is injective, $f(s) = f(s')$ implies $s = s'$ so $\mathrm{Im}\, f = \{f(s) \mid s \in S\} \subseteq T$ must have n distinct elements. Thus $|T| = n$ forces $\mathrm{Im}\, f = T$, and f is surjective. By Example 0.38, $\{f^{-1}(\{t\}) \mid t \in T\}$ is a partition of S so from Theorem 0.3, $n = |S| = \sum_{t \in T} |f^{-1}(\{t\})|$. When f is surjective $|f^{-1}(\{t\})| \geq 1$ if $t \in T$ and it follows that $n = \sum_{t \in T} |f^{-1}(\{t\})| \geq |T| = n$. Hence $|f^{-1}(\{t\})| = 1$ for any $t \in T$, and f is injective. Since f injective or surjective implies both, f is a bijection. ■

Some of the exercises use Theorem 0.10 but we want to give a nontrivial illustration of its use here. We need the well-known fact that each integer $n > 1$ has a unique factorization into primes, which we prove in the next chapter.

Example 0.39 Let $n, m \in \boldsymbol{N} - \{1\}$ with no common factor except 1, and let $I(n+m) = \{1, 2, \ldots, n+m\}$. There is a bijection $f : I(n+m) \to I(n+m)$ so that $f(n+m) = n+m$ and $f(k+1) - f(k) \in \{n, -m\}$ for all $1 \leq k < n+m$.

Consecutive values of f must differ by n or $-m$, so $f(t+1) = f(t) + n$ or $f(t+1) = f(t) - m$. Define f on $I(n+m)$ by $f(1) = n$ and for $k > 1$, $f(k) = f(k-1) + n$ if $f(k-1) \leq m$ but $f(k) = f(k-1) - m$ if $f(k-1) > m$. We keep adding n until the sum is larger than m and then subtract m as many times as we can. Thus when $n = 3$ and $m = 7$, the values of f in order are 3, 6, 9, 2, 5, 8, 1, 4, 7, 10. If $n = 7$ and $m = 4$, we get the values 7, 3, 10, 6, 2, 9, 5, 1, 8, 4, 11. It is not immediately clear why f is a bijection.

Now $f(1) \in I(n+m)$, and if $f(k-1) \in I(n+m)$ then $f(k) \in I(n+m)$ by the definition of f, so $f : I(n+m) \to I(n+m)$ by induction (Theorem 0.7). Since $f(k)$ either adds n to or subtracts m from $f(k-1)$, if $1 \leq s < t \leq n+m$ then $f(t) = f(s) + an - bm$ with $a, b \geq 0$ and $a + b = t - s$, so $f(s) = f(t)$ yields $an = bm$. But n and m have no common factor so every prime power factor of n must appear as a factor of b. Thus $b = nc$ and now $an = ncm$ yields $a = cm$. Hence $a + b = c(n+m)$ and $a + b = t - s < n + m$ force $c = 0$, so $a = 0 = b$. This means that $s = t$, proving that f is injective. By Theorem 0.10, f is a bijection, and the definition of f shows that $f(k+1) - f(k) \in \{n, -m\}$. Finally, since f is surjective, $f(s) = n+m$ for some $s \in I(n+m)$. Should $s < n+m$ then $f(s+1) = f(s) - m = n = f(1)$, contradicting the injectivity of f, so indeed $f(n+m) = n+m$ as required.

When $f : S \to T$, $|S| = n$, and $|T| < n$, the proof of Theorem 0.10 yields another interesting consequence. Since $\{f^{-1}(\{t\}) \subseteq S \mid t \in T\}$ is a partition of S into at most $|T| < |S|$ sets, $f^{-1}(\{t\})$ must have more than one element for some $t \in T$. That is, when $|S| > |T|$ then $f(s_1) = f(s_2)$ for $s_1 \neq s_2$ and f is not injective. This is often called the **pigeonhole principle:** *If n objects (pigeons) are put into fewer than n boxes (holes) then at least two objects wind up in the same box.* We use this and let $I(n) = \{1, 2, \ldots, n\}$.

Example 0.40 If $n > 1$ people are at a party, then two different people have the same number of acquaintances at the party.

We assume that "acquaintance" is always mutual and that each person is an acquaintance of himself. Let $W = \{w_1, \ldots, w_n\}$ represent the people and $f(w_j)$ the number of acquaintances of w_j. Then $f : W \to I(n)$ and if some $f(w_j) = n$ then w_j is an acquaintance of everyone so all $f(w_i) \geq 2$. Thus either $f : W \to I(n-1)$ or $f : W \to \{2, 3, \ldots, n\}$. Since $|\{2, 3, \ldots, n\}| = n - 1$ and $|W| = n$, the pigeonhole principle implies that $f(w_i) = f(w_j)$ for some $i \neq j$.

Example 0.41 Let $P_1, \ldots, P_9 \in \mathbf{Z}^3 \subseteq \boldsymbol{R}^3$ be 9 different points. Then for some $i \neq j$ the midpoint in $\boldsymbol{R}^3$ of the segment joining P_i to P_j is in $\mathbf{Z}^3$.

It is an exercise to see that for $a \in \mathbf{Z}$, $2\mathbf{Z} + a = 2\mathbf{Z}$ when a is even and $2\mathbf{Z} + a = 2\mathbf{Z} + 1$ when a is odd. Define $f : \{P_1, \ldots, P_9\} \to \{2\mathbf{Z}, 2\mathbf{Z} + 1\}^3$ by $f((a_1, a_2, a_3)) = (2\mathbf{Z} + a_1, 2\mathbf{Z} + a_2, 2\mathbf{Z} + a_3)$. Now $|\{P_1, \ldots, P_9\}| = 9$ and by Theorem 0.3, $|\{2\mathbf{Z}, 2\mathbf{Z} + 1\}^3| = 8$ so the pigeonhole principle implies that $f(P_i) = f(P_j)$ for some $i \neq j$. Thus if $P_i = (u_1, u_2, u_3)$ and $P_j = (v_1, v_2, v_3)$ then for each i, u_i and v_i are both even or both odd so $u_i + v_i \in 2\mathbf{Z}$, $(u_i + v_i)/2 \in \mathbf{Z}$, and the midpoint of the segment joining P_i and P_j has integer coordinates.

We turn now to inverses and their connection with bijections. Recall from Example 0.32 that $I_S(x) = x$ is the identity function on S.

DEFINITION 0.17 For $f : S \to T$ and $g : T \to S$, g is a **left inverse** for f when $g \circ f = I_S$, g is a **right inverse** for f when $f \circ g = I_T$, and g is an **inverse** for f when both $f \circ g = I_T$ and $g \circ f = I_S$.

Familiar examples of functions with inverses are $f(x) = x + 1$ with inverse $g(x) = x - 1$ on $\boldsymbol{R}$; $f(x) = x^3$ with inverse $g(x) = x^{1/3}$ on $\boldsymbol{R}$; $f(x) = 1/x$ is its own inverse on $\boldsymbol{R} - \{0\}$; and $f(x) = e^x$ from $\boldsymbol{R}$ to $(0, +\infty)$ with inverse $g(x) = ln(x)$. A left inverse g for $f : S \to T$ must undo the action of f since $g(f(s)) = s$, and if g is an inverse then each function must undo the action of the other. One way to picture this is shown in Fig. 0.9.

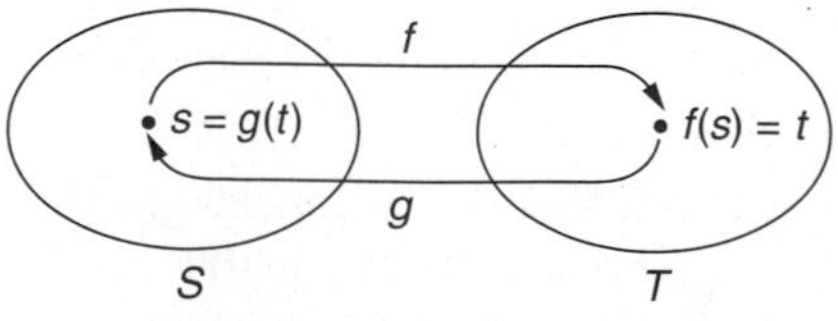

FIGURE 0.9

It will be helpful to have another characterization of inverses using the notions of injection and surjection.

THEOREM 0.11 If $f : S \to T$, $g : T \to S$, and $h : T \to S$ then:

a) If g is a left inverse for f and h is a right inverse for f, then $g = h$.
b) f is injective $\Leftrightarrow$ some $g' : T \to S$ is a left inverse for f.
c) f is surjective $\Leftrightarrow$ some $h' : T \to S$ is a right inverse for f.
d) f is bijective $\Leftrightarrow$ some $g'' : T \to S$ is an inverse for f.

Proof For a) use Example 0.37 on compositions with I_S and Theorem 0.9 on the associativity of "$\circ$" to write $g = g \circ I_T = g \circ (f \circ h) = (g \circ f) \circ h = I_S \circ h = h$. In b) when $g' \circ f = I_S$, if $f(s) = f(s')$ for $s, s' \in S$ apply g' to obtain $g'(f(s)) = g'(f(s'))$. Hence $s = I_S(s) = g' \circ f(s) = g' \circ f(s') = I_S(s') = s'$, so f is injective. Assuming f is injective we must define $g' : T \to S$ satisfying $g'(f(s)) = s$ for all $s \in S$. Now $|f^{-1}(\{t\})| \le 1 (f(s') = f(s) \Rightarrow s' = s)$ so if $t \in \text{Im } f$ then $f^{-1}(\{t\}) = \{s_t\}$ for $f(s_t) = t$ and we set $g'(t) = s_t$ unambiguously. Fix $s_0 \in S$ and let $g'(t) = s_0$ if $t \notin \text{Im } f$. Since $T = \text{Im } f \cup (T - \text{Im } f)$, $g' : T \to S$ by Example 0.35. For $x \in S$, $f(x) \in \text{Im } f$, so $g'(f(x)) = x$. Therefore $g' \circ f = I_S$ by Theorem 0.8, and g' is a left inverse for f.

For c), if $f \circ h' = I_T$ then $t = I_T(t) = f \circ h'(t) = f(h'(t))$ for any $t \in T$, and since $h'(t) \in S$, $t \in \text{Im } f$. Thus $T = \text{Im } f$, and f is surjective. When f is surjective, $f^{-1}(\{t\}) \neq \varnothing$ for each $t \in T$. We need $h' : T \to S$ so that $f(h'(t)) = t$; that is, $h'(t) \in f^{-1}(\{t\})$. To choose $h'(t) \in f^{-1}(\{t\})$ use the Axiom of Choice to get $H : P(S)^* \to S$ satisfying $H(A) \in A$. For $t \in T$, set $h'(t) = H(f^{-1}(\{t\})) \in f^{-1}(\{t\})$. This defines $h' : T \to S$ and by its definition $f(h'(t)) = t$ so $f \circ h' = I_T$ (Theorem 0.8 again!) and h' is a right inverse for f.

Finally, in d) g'' is both a left and a right inverse for f so by b) and c), f is both injective and surjective. Therefore f is a bijection. When f is a bijection it is both injective and surjective, so again by b) and c), f has both a left and a right inverse. Part a) forces these to be equal, thus an inverse for f. ∎

The identity function I_S on S is its own inverse and so is a bijection. Let us consider some functions with inverses on one side only. The greatest integer function $[\,] : \boldsymbol{R} \to \boldsymbol{Z}$ is not injective but is surjective so will have a right inverse but no left inverse. Two right inverses are $u : \boldsymbol{Z} \to \boldsymbol{R}$ given by $u(z) = z$, since $[\,] \circ u(z) = [z] = z$, and $v : \boldsymbol{Z} \to \boldsymbol{R}$ given by $v(z) = z + 1/2$, since $[\,] \circ v(z) = [z + 1/2] = z$. By Theorem 0.11 each of u and v is injective, but clearly neither is surjective. Another example is $G : \boldsymbol{N} \to \boldsymbol{N}$ given by $G(n) = n - [n/2]$. Now G is not injective since $G(1) = 1 = G(2)$ but is surjective because for any $n \in \boldsymbol{N}$, $G(2n) = n$; that is, $H(n) = 2n$ is a right inverse $H : \boldsymbol{N} \to \boldsymbol{N}$ for G.

We can use Theorem 0.11 to conclude that a bijection has an inverse without explicitly finding the inverse. In calculus this procedure is often used to define e^x as the inverse of $ln(x)$ and is used to define the inverse trigonometric functions. If $f(r) = r^7 + 3r^5 + 2r - 41$ defines $f : \boldsymbol{R} \to \boldsymbol{R}$ then f has an inverse that cannot be given explicitly. To see that f is a bijection, note its derivative $f'(r) > 0$, so f is strictly increasing, forcing f to be injective (verify!). Since f is continuous and $f(\boldsymbol{R})$ is not bounded above or below (why?), the Intermediate Value Theorem from calculus implies that $\text{Im } f = \boldsymbol{R}$, and f is a surjection.

Our examples showing that there may be many left inverses or right inverses for a function cannot hold for bijections by Thoerem 0.11.

THEOREM 0.12 If $f : S \to T$ then any left inverse for f is equal to any right inverse for f. In particular, if f has an inverse, the inverse is unique.

Proof The first statement is just Theorem 0.11. If g and h are two inverses for f, then since g is a left inverse and h is a right inverse, they are equal. ■

DEFINITION 0.18 If $f : S \to T$ has an inverse, the inverse is denoted by f^{-1}.

The notation for inverse and "inverse image" are similar, but there is no confusion in their use. The inverse image of $f : S \to T$ is defined for any f and acts only on subsets of T. When f has an inverse, $f^{-1} : T \to S$, then f must be a bijection, so for any $t \in T, t = f(f^{-1}(t))$. Thus $f^{-1}(t) \in f^{-1}(\{t\})$ and in fact $f^{-1}(\{t\}) = \{f^{-1}(t)\}$ since f is injective. Therefore for $B \subseteq T$, $f^{-1}(B) = \{f^{-1}(t) \mid t \in B\}$. We give some important facts about inverses and compositions.

THEOREM 0.13 Let $h : S \to T$ and $g : T \to U$.

a) If h has an inverse $h^{-1} : T \to S$, then h^{-1} has an inverse and $(h^{-1})^{-1} = h$.
b) If h has an inverse $h^{-1} : T \to S$, then h^{-1} is a bijection.
c) If h and g have inverses, then $g \circ h$ has inverse $h^{-1} \circ g^{-1}$.
d) If h and g are bijections, so is $g \circ h$.

Proof If $h^{-1} : T \to S$ then $h^{-1} \circ h = I_S$ and $h \circ h^{-1} = I_T$. By Definition 0.17, h is an inverse for h^{-1} and hence is its unique inverse by Theorem 0.12 and $(h^{-1})^{-1} = h$. From Theorem 0.11, h^{-1} is a bijection. For c) compute the compositions of $g \circ h$ and $h^{-1} \circ g^{-1}$ using Theorem 0.9 on associativity of composition and Example 0.37. Thus $(g \circ h) \circ (h^{-1} \circ g^{-1}) = ((g \circ h) \circ h^{-1}) \circ g^{-1} = (g \circ (h \circ h^{-1})) \circ g^{-1} = (g \circ I_T) \circ g^{-1} = g \circ g^{-1} = I_U$. Similarly $(h^{-1} \circ g^{-1}) \circ (g \circ h) = I_S$ so by Definition 0.17 $h^{-1} \circ g^{-1}$ is an inverse for $g \circ h$ and hence is its unique inverse $(g \circ h)^{-1}$. Finally, if h and g are bijections, each has an inverse by Theorem 0.11. Therefore $g \circ h$ has an inverse by part c), and Theorem 0.11 shows that $g \circ h$ is a bijection. ■

The converses of parts c) and d) are not true. Our example above, $G, H : N \to N$ via $G(n) = n - [n/2]$ with right inverse $H(n) = 2n$, shows that $G \circ H = I_N$ is a bijection and so has an inverse, but neither G nor H is a bijection or has an inverse. Also, although the composition is a bijection, and thus is both injective and surjective, G is not injective and H is not surjective.

DEFINITION 0.19 Let $\mathrm{Bij}(S, T) = \{f : S \to T \mid f \text{ is a bijection}\}$. We write $\mathrm{Bij}(S)$ for $\mathrm{Bij}(S, S)$ and, when $S = I(n) = \{1, 2, \ldots, n\}$, set $\mathrm{Bij}(I(n)) = S_n$.

Note that $\mathrm{Bij}((N, \{1, 2\}) = \varnothing$ and $\mathrm{Bij}(I(5), I(6)) = \varnothing$. In the latter case if $f \in \mathrm{Bij}(I(5), I(6))$, then $f(I(5))$ has at most five elements so f cannot be surjective. The basic properties of $\mathrm{Bij}(S)$ come next.

THEOREM 0.14 If S is a nonempty set, then $\mathrm{Bij}(S)$ satisfies:

i) $I_S \in \mathrm{Bij}(S)$.
ii) If $f, g, h \in \mathrm{Bij}(S)$, then $f \circ g \in \mathrm{Bij}(S)$ and $h \circ (g \circ f) = (h \circ g) \circ f$.
iii) If $h \in \mathrm{Bij}(S)$, then h has an inverse and $h^{-1} \in \mathrm{Bij}(S)$.

iv) If $|S| > 2$, then there are $f, g \in \text{Bij}(S)$ with $f \circ g \neq g \circ f$.
v) $|S_n| = n!$.

Proof Part i) is clear from Theorem 0.11 since $I_S \circ I_S = I_S$. Part ii) is just Theorem 0.13 d) and Theorem 0.9, and for iii) apply Theorem 0.11 and Theorem 0.13. In iv) if $|S| > 2$ let $a, b, c \in S$ and distinct, and write $S = \{a, b, c\} \cup T$ for $T = S - \{a, b, c\}$. Define $f, g : S \to S$ by $f(a) = b$, $f(b) = c$, $f(c) = a$, and $f(t) = t$ for all $t \in T$, and $g(a) = b$, $g(b) = a$, and $g(s) = s$ when $s \notin \{a, b\}$. It is easy to see that $f, g \in \text{Bij}(S)$ directly, or by noting that $g \circ g = I_S$ and $f \circ (f \circ f) = (f \circ f) \circ f = I_S$, so both f and g have inverses and are bijections by Theorem 0.11. But $g(f(a)) = g(b) = a$ and $f(g(a)) = f(b) = c$, so $f \circ g \neq g \circ f$ as claimed, using Theorem 0.8. Finally, we argue informally that $|S_n| = n!$, using that $\varphi \in S_n$ exactly when $\varphi : I(n) \to I(n)$ is an injection (Theorem 0.10). To define $\varphi \in S_n$ let $\varphi(1) \in I(n)$: there are n such choices. For each choice of $\varphi(1)$, $\varphi(2) \neq \varphi(1)$ or φ will not be injective so there are $n - 1$ choices for $\varphi(2) \in I(n) - \{\varphi(1)\}$. Thus there are $n(n-1)$ choices for $\varphi(1)$ and $\varphi(2)$. Continuing, and using that φ must be injective, after $\varphi(1), \ldots, \varphi(k)$ are chosen, there are exactly $n - k$ choices for $\varphi(k+1)$ in $I(n) - \{\varphi(1), \ldots, \varphi(k)\}$. This results in $n(n-1) \cdots 2 \cdot 1 = n!$ total possibilities in defining $\varphi \in S_n$. But any $f \in S_n$ can be viewed as arising by such a sequence of choices so $|S_n| = n!$. ■

EXERCISES

1. Find the indicated inverse images. [ ] is the greatest integer function.
   i) For $f : \boldsymbol{R} \to \boldsymbol{R}$ given by $f(r) = r^2$, find $f^{-1}([-1, 9))$; $f^{-1}((-\infty, 0])$; and $f^{-1}(\boldsymbol{Z})$.
   ii) For $[\ ] : \boldsymbol{R} \to \boldsymbol{R}$, find $[\ ]^{-1}(\boldsymbol{Q})$; $[\ ]^{-1}((-1, \pi])$; $[\ ]^{-1}([3/2, 2])$; and $[\ ]^{-1}([-1/2, -1/3])$.
   iii) For $g : \boldsymbol{R} \to \boldsymbol{R}$ given by $g(r) = r - [r]$, find $g^{-1}([1, 9])$; $g^{-1}(\{0\})$; $g^{-1}((0, 1))$; and $g^{-1}([-1/2, 1/2])$.
   iv) For $+ : \boldsymbol{R}^2 \to \boldsymbol{R}$, the usual addition, describe $+^{-1}(\{r\})$ geometrically;
   v) For $\cdot : \boldsymbol{R}^2 \to \boldsymbol{R}$ the usual multiplication, describe $\cdot^{-1}(\{r\})$ geometrically.
   vi) For $\cdot : \boldsymbol{R}^2 \to \boldsymbol{R}$, the usual multiplication, what is $\cdot^{-1}(\boldsymbol{R} - \{0\})$ (geometry!)?

2. For each function $f : \boldsymbol{R} \to \boldsymbol{R}$ and $r \in \boldsymbol{R}$, how many elements are in $f^{-1}(\{r\})$?
   i) $f(x) = 1/(x^2 + 1)$
   ii) $f(x) = x + [x]$
   iii) $f(x) = x^3 - x^2$
   iv) $f(x) = x^4 - 2x^2$
   v) $f(x) = |x^4 - x^2|$
   vi) $f(x) = x^2[x]^2$

3. Let $f : S \to T$, $A \subseteq S$, $B \subseteq T$, and $\{B_i\}_I$ a collection of subsets of T. Show that:
   i) $A \subseteq f^{-1}(f(A))$ and give an example with $A \neq f^{-1}(f(A))$.
   ii) $B \supseteq f(f^{-1}(B))$, and $B = f(f^{-1}(B)) \Leftrightarrow B \subseteq \text{Im}\, f$
   iii) $f^{-1}(\bigcap_I B_i) = \bigcap_I f^{-1}(B_i)$
   iv) $f^{-1}(\bigcup_I B_i) = \bigcup_I f^{-1}(B_i)$
   v) If $\{B_i\}_I$ is a partition of T, then $\{f^{-1}(B_i)\}_I$ is a partition of S.

4. Let $f : S \to T$, $A \subseteq S$, $A \neq \emptyset$, and $B \subseteq T$. Show that:
   i) $f(f^{-1}(f(A))) = f(A)$
   ii) $f^{-1}(f(f^{-1}(B))) = f^{-1}(B)$

5. If $f : S \to T$, $\varnothing \neq A \subseteq S$, $B \subseteq T$, and $f|_A : A \to T$, show that $(f|_A)^{-1}(B) = f^{-1}(B) \cap A$.

6. Let $f : S \to T$, $A \subseteq S$ with $A \neq \varnothing$.

a) If f is injective show that:

(i) $f|_A$ is also.

(ii) $F : A \to f(A)$ given by $F(a) = f|_A(a)$ is a bijection.

b) If $f|_A$ is injective show by example that f need not be injective.

7. For S and T nonempty sets, argue that $\text{Bij}(S, T) \in P(P(S \times T))$.

8. If $f \in \text{Bij}(S, T)$, $\varnothing \neq A \subseteq S$ with $A \neq S$, show that $f|_{S-A} \in \text{Bij}(S - A, T - f(A))$.

9. Let $f_1, \ldots, f_m$ be functions so that $F = (\cdots((f_1 \circ f_2) \circ f_3) \cdots) \circ f_{m-1}) \circ f_m$ is defined, and assume each f_i has an inverse. Show that $F^{-1} = ((\cdots(f_m^{-1} \circ f_{m-1}^{-1}) \circ \cdots) \circ f_2^{-1}) \circ f_1^{-1}$.

10. Use only the definitions to show that if $f : S \to T$ and $g : T \to U$ are both injections or surjections, then so is $g \circ f$.

11. Can a constant function $C_y : S \to T$ via $C_y(s) = y \in T$ be injective? surjective?

12. Let $f : S \to T$ and $g : T \to U$. Show that:

a) $g \circ f$ injective $\Rightarrow f$ is injective

b) $g \circ f$ surjective $\Rightarrow g$ is surjective

c) What can be said if $g \circ f$ is a bijection?

13. Let $f : S \to T$ and $g : T \to S$ so that $g \circ f$ is injective and $f \circ g$ is surjective. Show that both compositions are bijections.

14. If $f : S \to T$ has a unique left or right inverse, show that f is a bijection.

15. For $1 \leq i \leq n$ let $f_i : S_i \to T_i$ and define $F \subseteq (S_1 \times S_2 \times \cdots \times S_n) \times (T_1 \times \cdots \times T_n)$ by $F = \{((s_1, \ldots, s_n), (f_1(s_1), \ldots, f_n(s_n)) \mid (s_1, \ldots, s_n) \in S_1 \times \cdots \times S_n)\}$. a) Show that F is a function. b) Show that F is injective, surjective, or bijective $\Leftrightarrow$ each f_i is.

16. Let $\{S_i\}_I$ and $\{T_i\}_I$ be collections of nonempty sets, and for each $i \in I$ let $f_i : S_i \to T_i$. If for all $x \in S_i \bigcap S_j$, $f_i(x) = f_j(x) \in T_i \bigcap T_j$ then define $f = \{(s, t) \in \bigcup S_i \times \bigcup T_i \mid t = f_i(s) \text{ if } s \in S_i\}$. Show that:

i) $f : \bigcup S_i \to \bigcup T_i$ (a function)

ii) all f_i surjective $\Rightarrow f$ is surjective

iii) f surjective and $\{T_i\}_I$ pairwise disjoint $\Rightarrow$ each f_i is surjective

iv) f injective $\Rightarrow$ each f_i is injective

v) each f_i injective and $\{T_i\}_I$ pairwise disjoint $\Rightarrow f$ is injective

vi) If $\{T_i\}_I$ is pairwise disjoint then f has an inverse $\Leftrightarrow$ each f_i has an inverse.

17. Let $n, m \in \mathbf{N}$ with no common factor except 1. Find $\varphi \in \text{Bij}(\mathbf{Z})$ satisfying $\varphi(k + 1) - \varphi(k) \in \{n, -m\}$ for all $k \in \mathbf{Z}$ (use Example 0.39).

18. For $f : S \to T$ define $F : P(T) \to P(S)$ (see Definition 0.6) by $F(B) = f^{-1}(B)$. Show that:

a) F is surjective $\Leftrightarrow f$ is injective

b) F is injective $\Leftrightarrow f$ is surjective

19. Let $f : P(P(S)) \to P(S)$ be $f(C) = \bigcup_{c \in C} c$. Find two different right inverses for f.

20. Set $T^S = \{f : S \to T\}$. If $\varnothing \neq A \subseteq S$ show that:

i) $\sigma(f) = f|_A : A \to T$ defines a surjection $\sigma : T^S \to T^A$.

ii) If $|T| > 1$ then σ is a bijection $\Leftrightarrow A = S$.

21. Define $f : \mathbf{N} \times \mathbf{N} \to \mathbf{N}$ by $f(s, t) = 2^{s-1}(2t - 1)$. Show that f is a bijection and find f^{-1}.

22. Let $S=\{-1, 1\}\subseteq \boldsymbol{R}$, $f: S^n \to S^n$ with $n\geq 2$, and $f((s_1, \ldots, s_n))=(b_1, \ldots, b_n)$ as defined below. Is $f\in \mathrm{Bij}(S^n)$?
   i) $b_j = s_1 s_2 \cdots s_j$
   ii) $b_j = -s_1 s_j s_n$
   iii) $b_j = s_j s_{j+1}$ when $j<n$ and $b_n = s_n s_1$
   iv) $b_j = s_1 \cdots s_{j-1} s_{j+1} \cdots s_n$
   v) $b_j = s_j s_{j+1} s_{j+2}$ when $j<n-1$, $b_{n-1} = s_{n-1} s_n s_1$, and $b_n = s_n s_1 s_2$ (The answer here depends on whether n is a multiple of 3.)

23. Let $S\subseteq \boldsymbol{R}^2$ be a set of $m>1$ points in the plane, no three of which lie on the same straight line. Connect some of the points by straight line segments. Show that two different points in S must be connected to the same number (including zero) of other points in S.

24. If $S\subseteq \boldsymbol{Z}$ and $|S|=2k>0$, show that $S=\{x_1, \ldots, x_k, y_1, \ldots, y_k\}$ so that $x_j - y_j$ is a multiple of $2j-1$. For example, $\{2, 4, 6, 8, 10, 20, 21, 23, 26, 27\}$ can be written as $\{23, 10, 26, 27, 20, 8, 4, 21, 6, 2\}$ with $10-4=3\cdot 2$, $26-21=5\cdot 1$, $27-6=7\cdot 3$, and $20-2=9\cdot 2$. (Look at remainders on division by $2k-1$.)

25. If A, B, C, and A_j are nonempty sets, show that:
   i) $\mathrm{Bij}(A\times B, B\times A)\neq\varnothing$
   ii) $\mathrm{Bij}((A\times B)\times C, A\times(B\times C))\neq\varnothing$
   iii) $\mathrm{Bij}(A_1\times\cdots\times A_n, (A_1\times\cdots\times A_{n-1})\times A_n)\neq\varnothing$

26. Let $2^S=\{f: S\to\{0, 1\}\}$. For $A\subseteq S$, define $\chi_A\in 2^S$ by
$\chi_A(s)=\begin{cases}0 & \text{if } s\notin A\\ 1 & \text{if } s\in A\end{cases}$. Show that $\mu: P(S)\to 2^S$ given by $\mu(A)=\chi_A$ is a bijection.

27. Let $S_\infty=\{f\in \mathrm{Bij}(\boldsymbol{N}) \mid \text{for some } n_f\in \boldsymbol{N},\ f(k)=k \text{ if } k\geq n_f\}$.
   i) Show that if $f, g\in S_\infty$ then $g\circ f\in S_\infty$ and $f^{-1}\in S_\infty$.
   ii) Describe $gS_\infty=\{g\circ f\in \mathrm{Bij}(\boldsymbol{N}) \mid f\in S_\infty\}$ for a fixed $g\in \mathrm{Bij}(\boldsymbol{N})$.
   iii) Let $f, g\in \mathrm{Bij}(\boldsymbol{N})$ be defined by $f(2m-1)=2m$ and $f(2m)=2m-1$, and $g(1)=1$, $g(2m)=2m+1$, and $g(2m+1)=2m$. Show that $f\circ f=g\circ g=I_{\boldsymbol{N}}$. Describe $f\circ g$ and $g\circ f$, and argue that repeated compositions of either of these two with itself never result in $I_{\boldsymbol{N}}$.

0.7 CARDINALITY AND INFINITE SETS

We interpret $\mathrm{Bij}(S, T)\neq\varnothing$ to mean that S and T have the "same number" of elements because a bijection gives a matching of their elements. We explore this idea for infinite sets. Although peripheral to our needs, the material is quite interesting and is needed for more advanced topics. Since the results are not central for us, our arguments may be somewhat abridged.

DEFINITION 0.20 If $\mathrm{Bij}(S, T)\neq\varnothing$ then S and T have the same **cardinality,** or the same **cardinal number,** and we write $|S|=|T|$. When $|S|=|I(n)|$ for some $I(n)=\{1, 2, \ldots, n\}$ then S is finite and $|S|=n$. $\varnothing$ is finite and $|\varnothing|=0$.

From Theorem 0.14, $I_S\in \mathrm{Bij}(S)$, so $|S|=|S|$, and Theorem 0.13 shows that if $f\in \mathrm{Bij}(S, T)$ and $g\in \mathrm{Bij}(T, U)$, then $f^{-1}\in \mathrm{Bij}(T, S)$ and $g\circ f\in \mathrm{Bij}(S, U)$. Thus $|S|=|T|$ implies that $|T|=|S|$, and both $|S|=|T|$ and $|T|=|U|$ imply that $|S|=|U|$. These observations on cardinality will be useful.

It not obvious whether there are bijections between sets like $\boldsymbol{N}$, Bij$(\boldsymbol{N})$, $\boldsymbol{Z}$, $\boldsymbol{R}$, $2\boldsymbol{N}$, or $(0, 1) \subseteq \boldsymbol{R}$. We can see that $g : \boldsymbol{N} \to 2\boldsymbol{N}$ via $g(n) = 2n$ is a bijection even though $2\boldsymbol{N}$ appears to have only half the elements of $\boldsymbol{N}$. Perhaps all infinite sets have the same cardinality. A simple result of Cantor shows that there are different infinite cardinals (i.e., cardinal numbers). Recall from Definition 0.6 that $P(S)$ is the set of all subsets of S.

THEOREM 0.15 For any set $S \neq \varnothing$ there is no surjection $f : S \to P(S)$.

Proof If $f : S \to P(S)$ then for $s \in S$, $f(s) = T \subseteq S$, so $s \in f(s)$ or $s \notin f(s)$. Let $A = \{s \in S \mid s \notin f(s)\} \in P(S)$. If f is surjective write $A = f(y)$ for $y \in S$. Since $A \subseteq S$ either $y \in A$ or $y \notin A$. If $y \in A$ then the definition of A shows that $y \notin f(y) = A$, a contradiction. But if $y \notin A = f(y)$ the definition of A yields $y \in A$, again a contradiction. Thus there can be no surjection $f : S \to P(S)$. ∎

A bijection must be a surjection so a consequence of the theorem is $|S| \neq |P(S)|$ and also $|P(S)| \neq |P(P(S))|$. Are $|S|, |P(S)|, |P(P(S))|, |P(P(P(S)))|, \ldots$ all different? The answer will follow by resolving another interesting question that we illustrate next. Different infinite cardinals exist by Theorem 0.15. It might appear that $|\boldsymbol{N}|$ and $|\boldsymbol{Z}|$ are different since we may identify $\boldsymbol{N}$ in $\boldsymbol{Z}$ using the injection f given by $f(n) = n$, leaving all $z \leq 0$ in $\boldsymbol{Z}$ left over. However, $g : \boldsymbol{Z} \to \boldsymbol{N}$ given by $g(z) = 4z + 1$ when $z \geq 0$ and $g(z) = -4z$ when $z < 0$ is injective, but $g(\boldsymbol{Z})$ is only about half of $\boldsymbol{N}$. Do injections from each of two sets to a proper subset of the other force there to be a bijection between the sets? The answer is not obvious and is given by our next result, which is known as the Schroeder-Bernstein Theorem.

THEOREM 0.16 If $f : S \to T$ and $g : T \to S$ are injective, then $|S| = |T|$.

Proof We must find $\varphi \in$ Bij(S, T). Since g is injective, if $s_0 \in \text{Im } g$ then $g^{-1}(\{s_0\}) = \{t_1\}$, and similarly, if $t_1 \in \text{Im } f$ then $f^{-1}(\{t_1\}) = \{s_2\}$. Given $s_0 \in S$, let its chain of pre-images be $\{s_0, t_1, s_2, t_3, \ldots\}$, alternately using g and f. Either this chain is infinite, it stops with some $s_{2k} \notin \text{Im } g$ including s_0 itself, or it stops with some $t_{2k+1} \notin \text{Im } f$. These conditions define $S^{\#}$, S_S, $S_T \subseteq S$ respectively, so for example $s \in S_S$ has its chain of pre-images end in S. Now $\{S^{\#}, S_S, S_T\}$ is a partition of S, so to define $\varphi : S \to T$ it suffices by Example 0.35 to define φ on each of $S^{\#}$, S_S, and S_T. Similarly, $\{T^{\#}, T_S, T_T\}$ partitions T, where $t \in T^{\#}$ has an infinite chain of pre-images, $t \in T_S$ has its chain of pre-images end in S, and $t \in T_T$ has its chain of pre-images end in T.

If $x \in S^{\#}$ then $f(x) \in T^{\#}$ since x is the pre-image of $f(x)$ so $f(x)$ has an infinite chain of pre-images if x does. For $y \in T^{\#}$, $y = f(s)$ for $s \in S$, and if the chain of pre-images for y is $\{y, s, t_1, s_2, t_2, \ldots\}$, then the chain of pre-images for s is $\{s, t_1, s_2, t_2, \ldots\}$ and $s \in S^{\#}$. Thus $f(S^{\#}) = T^{\#}$. Similarly $f(S_S) = T_S$ and $g(T_T) = S_T$. Now g is injective so $g|_{T_T} : T_T \to S_T$ is a bijection, it has an inverse $h : S_T \to T_T$ by Theorem 0.11, and h is a bijection by Theorem 0.13. Define the bijection $\varphi : S \to T$ by setting $\varphi(s) = f(s)$ if $s \in S - S_T$ and $\varphi(s) = h(s)$ if $s \in S_T$. By Example 0.35, φ is a function and is injective since f and h are, and also $f(S - S_T) \cap h(S_T) = f(S^{\#} \cup S_S) \cap T_T = (T^{\#} \cup T_S) \cap T_T = \varnothing$. Finally, $\varphi(S) \supseteq \varphi(S^{\#}) \cup \varphi(S_S) \cup \varphi(S_T) = f(S^{\#}) \cup f(S_S) \cup h(S_T) = T$ (why?) so φ is surjective. ∎

COROLLARY Let $f : S \to T$ and $g : T \to U$ be injective.

a) If $|S| = |U|$, then $|T| = |U|$.
b) If $|S| = |T|$ and $f(S) \subseteq B \subseteq T$, then $|B| = |T|$.
c) If $S \subseteq B \subseteq T$ and $|S| = |T|$, then $|B| = |T|$.

Proof For a) $|S| = |U|$ implies there is $h \in \mathrm{Bij}(U, S)$ so $f \circ h : U \to T$ is injective (verify!), and now $g : T \to U$ injective gives $|T| = |U|$ by the theorem. Part b) follows from a) since $F : S \to B$ given by $F(s) = f(s)$ is an injection and the restriction $I_T|_B : B \to T$ is an injection. Lastly, c) follows from a) as well by using the injections $I_B|_S : S \to B$ and $I_T|_B : B \to T$. ■

Consider again the sets $S, P(S), P(P(S)), \ldots$. For $T \neq \emptyset$ there is an injection $f : T \to P(T)$ via $f(t) = \{t\}$. Compositions of injections are injections (exercise!) so there is an injection from any set U in this sequence of power sets to any set V appearing later. If $|U| = |V|$ then there are injections from U to $P(U)$ and from $P(U)$ to V. The corollary shows that $|P(U)| = |V| = |U|$, contradicting Theorem 0.15. Thus $|S|, |P(S)|, |P(P(S))|, \ldots$ are all different, so there are infinitely many infinite cardinals!

The "smallest" infinite set is $\boldsymbol{N}$, and $|S| = |\boldsymbol{N}|$ is often denoted by $\aleph_0$, "aleph" zero. We can use Theorem 0.16 to show that $|\boldsymbol{N}^k| = \aleph_0$. Clearly $f : \boldsymbol{N} \to \boldsymbol{N}^k$ given by $f(n) = (n, \ldots, n)$ is injective. To define an injection $g : \boldsymbol{N}^k \to \boldsymbol{N}$ we use prime factorization in $\boldsymbol{N}$, proved in the next chapter. We mean the familiar statement that any $n \in \boldsymbol{N} - \{1\}$ is a product of primes, uniquely except for order. Using it, $g((n_1, \ldots, n_k)) = p_1^{n_1} \cdots p_k^{n_k}$ (p_j is the jth prime) defines an injection $g : \boldsymbol{N}^k \to \boldsymbol{N}$ so Theorem 0.16 shows $|\boldsymbol{N}^k| = \aleph_0$. However, we prefer to use other important ideas and avoid Theorem 0.16.

DEFINITION 0.21 A set S is **denumerable** if $|S| = |\boldsymbol{N}| = \aleph_0$ and is **countable** if it is either finite or denumerable.

Clearly, every infinite set ought to have a denumerable subset, but this is not easy to show and requires a statement equivalent to the Axiom of Choice. We will not do this here (see §13.6). Our remaining comments mostly concern countable sets, although we will see that $|\boldsymbol{R}| \neq \aleph_0$.

THEOREM 0.17 If S is a countable set and if $A \subseteq S$, then A is countable, and if $f : S \to B$ is a surjection, then B is countable.

Proof First take $S = \boldsymbol{N}$. If $A \subseteq \boldsymbol{N}$ is finite it is countable. When A is infinite we must find a bijection $g : \boldsymbol{N} \to A$. Define g recursively, or by induction, by $g(1) = \min(A)$, and given $g(1), \ldots, g(k) \in A$ set $g(k+1) = \min\{A - \{g(1), \ldots, g(k)\}\}$. Each $g(j)$ exists by Well Ordering and the assumption that A is infinite. From the definition of g, if $i < j$ then $g(j) > g(i)$, so g must be injective. If $m \in A - g(\boldsymbol{N})$, the definition of g forces $m > g(i)$ for all $i \in \boldsymbol{N}$, a contradiction (why?). Consequently $g(\boldsymbol{N}) = A$, so g is a bijection, $|A| = |\boldsymbol{N}|$, and A is denumerable.

When $f : \boldsymbol{N} \to B$ is surjective, then for $b \in B$, $\emptyset \neq f^{-1}(\{b\}) \subseteq \boldsymbol{N}$, so by Well Ordering $h(b) = \min\{f^{-1}(\{b\})\}$ makes sense. This defines $h : B \to \boldsymbol{N}$ with $f(h(b)) = b$ so f is a left inverse for h, and h is injective by Theorem 0.11. The function $h: B \to h(B)$ is a bijection; hence $|B| = |h(B)|$. The first part of the proof shows that $h(B) \subseteq \boldsymbol{N}$ is countable, so B is also.

Returning to our general S, Definition 0.21 yields an injection $g: S \to \boldsymbol{N}$ and a surjection $h: \boldsymbol{N} \to S$. Now $|A| = |g(A)|$ (why?) and $g(A) \subseteq \boldsymbol{N}$. Hence $g(A)$, so A is countable by the special case when $S = \boldsymbol{N}$. Next observe that $f \circ h: \boldsymbol{N} \to B$ is surjective so B is countable by the special case when $S = \boldsymbol{N}$. ∎

Next we prove that some other sets are countable but avoid using Theorem 0.16, which is often required for nondenumerable infinite sets.

THEOREM 0.18 For any $m \in \boldsymbol{N}$, if S is countable so is S^m. In particular, $\boldsymbol{N}^m$, $\boldsymbol{Z}^m$, and $\boldsymbol{Q}^m$ are denumerable.

Proof Use prime factorization to write $n \in \boldsymbol{N}$ as $n = 2^{n_1}3^{n_2} \cdots p_k^{n_k}$, where p_k is the kth prime, $k \geq m$, and each $n_j \geq 0$. Then $f_m: \boldsymbol{N} \to \boldsymbol{N}^m$ defined by $f_m(2^{n_1}3^{n_2} \cdots p_k^{n_k}) = (n_1 + 1, \ldots, n_m + 1)$ is surjective so by Theorem 0.17, $|\boldsymbol{N}^m| = \aleph_0$. Let $h: \boldsymbol{N} \to S$ be a surjection and $H((n_1, \ldots, n_m)) = (h(n_1), \ldots, h(n_m))$. From Exercise 15 in §6, $H: \boldsymbol{N}^m \to S^m$ is surjective so S^m is countable by Theorem 0.17 since $|\boldsymbol{N}^m| = \aleph_0$. Similarly $g((a, b, c, d)) = a/b - c/d$ defines a surjection $g: \boldsymbol{N}^4 \to \boldsymbol{Q}$ (verify!) and $\boldsymbol{Q}$ is denumerable, as is $\boldsymbol{Z} \subseteq \boldsymbol{Q}$, by Theorem 0.17, so $\boldsymbol{Q}^m$ and $\boldsymbol{Z}^m$ are denumerable by the proof above for S^m. ∎

Another application of Theorem 0.17 shows $\boldsymbol{R}$ is *not* denumerable.

THEOREM 0.19 $\boldsymbol{R}$ is not denumerable.

Proof Define $g: P(\boldsymbol{N}) \to [0, 1) \subseteq \boldsymbol{R}$ by $g(S) = .a_1a_2a_3\ldots$, where $a_j = 1$ if $j \in S$ and $a_j = 0$ if $j \notin S$. For example $g(\varnothing) = .000000\ldots$, $g(\{2, 3, 5, 7, 8\}) = .01101011000\ldots$, $g(2\boldsymbol{N}) = .01010101\ldots$, and $g(\boldsymbol{N}) = .1111111\ldots$. If $g(S) = g(T)$ then $S = T$, so g is injective and $|P(\boldsymbol{N})| = |g(P(\boldsymbol{N}))|$. Should $|\boldsymbol{N}| = |\boldsymbol{R}|$ then Theorem 0.17 and $g(P(\boldsymbol{N})) \subseteq \boldsymbol{R}$ force $|g(P(\boldsymbol{N}))| = \aleph_0$. Thus $|\boldsymbol{N}| = |g(P(\boldsymbol{N}))| = |P(\boldsymbol{N})|$, contradicting Theorem 0.15. Hence $|\boldsymbol{N}| \neq |\boldsymbol{R}|$, proving the theorem. ∎

Our next result turns out to be a very useful one for countable sets.

THEOREM 0.20 Any countable union of countable sets is countable.

Proof Write the collection of sets as $\{B_i\}_I$ for a countable set I. There is a surjection $h: \boldsymbol{N} \to I$ and define $A_i = B_{h(i)}$. We replace $\{B_i\}_I$ with $\{A_i\}_N$ since $\bigcup B_j = \bigcup A_i$. Each A_i is countable so there is a surjection $f_i: \boldsymbol{N} \to A_i$. Define $g: \boldsymbol{N}^2 \to \bigcup A_i$ by $g((n, m)) = f_m(n)$. If $y \in \bigcup A_i$ then for some k, $y \in A_k$ so $y = f_k(n)$ for some $n \in \boldsymbol{N}$. Thus $g((n, k)) = y$, and g is surjective. But by Theorem 0.18, $\boldsymbol{N}^2$ is denumerable, so $\bigcup A_i$ is countable as required by Theorem 0.17. ∎

The final two theorems are consequences of Theorem 0.20. The first shows that two interesting sets related to $\boldsymbol{N}$ are denumerable.

THEOREM 0.21 $|\bigcup_{k \in \boldsymbol{N}} \boldsymbol{N}^k| = \aleph_0$ and $|\text{Fin}(\boldsymbol{N}) = \{A \subseteq \boldsymbol{N} \mid A \text{ is finite}\}| = \aleph_0$.

Proof Each N^k is denumerable by Theorem 0.18, so $\bigcup N^k$ is denumerable directly from Theorem 0.20. Now $G((a_1, \ldots, a_k)) = \{a_1, \ldots, a_k\}$ defines a surjection $G : \bigcup_{k \in N} N^k \to \text{Fin}(N)$, and by Theorem 0.17, Fin(N) is denumerable. ■

It is left for the exercises to show that the results in the last theorem extend to any denumerable set S in place of N.

THEOREM 0.22 $Z[x]$, $Q[x]$, and $Q[x_1, \ldots, x_n]$ are denumerable. Also $\{c \in R \mid p(c) = 0 \text{ for some nonzero } p(x) \in Q[x]\} \subseteq R$ is denumerable.

Proof Observe that $f_k((q_1, \ldots, q_{k+1})) = q_1 + q_2 x + \cdots + q_{k+1} x^k$ gives a bijection $f_k : Q^{k+1} \to \{f(x) \in Q[x] \mid \deg f \leq k \text{ or } f = 0\}$ whose restriction to Z^{k+1} is a bijection with $\{f(x) \in Z[x] \mid \deg f \leq k \text{ or } f = 0\}$. From Example 0.35 setting $f((q_1, \ldots, q_{m+1})) = f_m((q_1, \ldots, q_{m+1}))$ defines $f : \bigcup Q^n \to Q[x]$, or $f : \bigcup Z^n \to Z[x]$. It is an exercise to see that f is surjective, so Theorems 0.18, 0.20, and 0.17 show that $Z[x]$ and $Q[x]$ are denumerable. To show that $|Q[x_1, \ldots, x_n]| = |N|$ also, use induction. The case $n = 1$ is just $Q[x]$. Assume $|Q[x_1, \ldots, x_k]| = |N|$ for some $k \geq 1$. Since $|Q| = |N|$ from Theorem 0.18, $|Q| = |N| = |Q[x_1, \ldots, x_k]|$ so there is a bijection $g : Q \to Q[x_1, \ldots, x_k]$. But we have $Q[x_1, \ldots, x_{k+1}] = Q[x_1, \ldots, x_k][x_{k+1}]$ so $\varphi(q_0 + q_1 x + \cdots + q_m x^m) = g(q_0) + \cdots + g(q_m) x_{k+1}{}^m$ gives $\varphi : Q[x] \to Q[x_1, \ldots, x_{k+1}]$, which is a bijection (why?). Using $|N| = |Q[x]| = |Q[x_1, \ldots, x_{k+1}]|$, $Q[x_1, \ldots, x_{k+1}]$ is denumerable, and applying IND-I proves that $Q[x_1, \ldots, x_n]$ is denumerable for any $n \in N$. Finally, for $p(x) \in Q[x] - \{0\}$, set $V_{p(x)} = \{c \in R \mid p(c) = 0\}$, and when $p(x) = 0$, set $V_0 = \varnothing$. Each $V_{p(x)}$ is a finite subset of R and so is countable, and we showed that $Q[x]$ is countable. The last statement to be proved is that $\bigcup V_{p(x)}$ is countable, and this now follows directly from Theorem 0.20. ■

Of course $Z[x_1, \ldots, x_n] \subseteq Q[x_1, \ldots, x_n]$ is also countable using both Theorem 0.17 and Theorem 0.22.

EXERCISES

1. Show directly (without using Theorem 0.16 or its corollary) that $|S| = |P(P(S))|$ implies that there is a surjection $f : S \to P(S)$.
2. If $Q^+ = \{q \in Q \mid q > 0\}$, show that for any $m \in N$, $(Q^+)^m$ is denumerable.
3. For sets A and B let $A^B = \{f : B \to A\}$. If $|A| = |C|$ and $|B| = |D|$, show that:
   i) $|A^B| = |C^D|$
   ii) $|\text{Bij}(B, A)| = |\text{Bij}(D, C)|$
4. Let $S^T = \{f : T \to S\}$ and show $|(A \times B)^C| = |A^C \times B^C|$ for nonempty A, B, C.
5. If $|S| = |T|$, show that:
   i) $|P(S)| = |P(T)|$
   ii) $|\text{Bij}(S)| = |\text{Bij}(T)|$
6. Describe a denumerable collection S of circles in the plane so that every circle with positive radius properly contains some circle in S.
7. A finite set $P_1, \ldots, P_k \in R^2$, together with straight line segments joining some P_i with some $P_j (i \neq j)$, is a finite graph. Show that the set of all finite graphs that have all vertices with both coordinates in Q is countable.

8. a) Show that $|(a, b)| = |(c, d)|$ for any two open intervals $(a, b), (c, d) \subseteq \boldsymbol{R}$.
   b) Show that $|(a, b)| = |\boldsymbol{R}|$ for any open interval $(a, b) \subseteq \boldsymbol{R}$ (arctangent!).
9. Show that $|I| = |\boldsymbol{R}|$ for any interval $I \subseteq \boldsymbol{R}$. I can have finite length or not, and I can be open, closed, or half open and half closed. (If $S = \{0, 1/2, 1/3, \ldots, 1/n, \ldots\}$, write $[0, 1) = S \cup ([0, 1) - S)$. Now show that $|[0, 1)| = |(0, 1)|$.)
10. If $|S| = |\boldsymbol{N}|$, show that:
    i) $|\bigcup S^n| = |\boldsymbol{N}|$
    ii) $|\{B \subseteq S \mid B \text{ is finite}\}| = |\boldsymbol{N}|$
11. Let $\prod_\Lambda A_\lambda = \{f : \Lambda \to \bigcup_\Lambda A_\lambda \mid f(\lambda) \in A_\lambda\}$ be the **Cartesian product** of the nonempty sets $\{A_\lambda\}_\Lambda$. For the same Λ, let $\{B_\lambda\}_\Lambda$ be another collection of nonempty sets with with $g_\lambda : A_\lambda \to B_\lambda$ for each $\lambda \in \Lambda$. Construct some $F : \prod_\Lambda A_\lambda \to \prod_\Lambda B_\lambda$ satisfying: F is injective when all g_λ are, F is surjective when all g_λ are, and F is a bijection when all g_λ are.
12. If $\{A_\lambda\}_\Lambda$ is a partition of a set S into nonempty subsets and $T \neq \varnothing$, show that $|T^S| = |\prod_\Lambda T^{A_\lambda}|$. (See Exercises 3 and 11. If $f_\lambda : A_\lambda \to T$, show there is $f : S \to T$ satisfying $f|_{A_\lambda} = f_\lambda$.)
13. Let T be a nonempty set and $\{A_\lambda\}_\Lambda$ a collection of nonempty sets.
    a) Show that $\left|\left(\prod_\Lambda A_\lambda\right)^T\right| = \left|\prod_\Lambda \left(A_\lambda^T\right)\right|$.
    b) If all $A_\lambda = A$ show that $\left|\left(\prod_\Lambda A_\lambda\right)^T\right| = |A^{\Lambda \times T}|$. (See Exercise 11.)
14. Let $X = \{x_i \mid i \in \boldsymbol{N}\}$ be indeterminates over $\boldsymbol{Q}$. Show that $\boldsymbol{Q}[X]$, the set of polynomials in X with coefficients in $\boldsymbol{Q}$, is denumerable. (Each $f \in \boldsymbol{Q}[X]$ satisfies $f \in \boldsymbol{Q}[y_1, \ldots, y_s]$ for a finite subset $\{y_j\} \subseteq X$.)
15. Show that $\text{Bij}(\boldsymbol{N})$ is not countable. (Let $\{A_i\}_{\boldsymbol{N}}$ be a partition of $\boldsymbol{N}$ with each $|A_i| = |\boldsymbol{N}|$ (see §2 Exercise 11) and fix $f_i \in \text{Bij}(A_i) - \{I_{A_i}\}$. For each $S \subseteq \boldsymbol{N}$, define $g_S \in \text{Bij}(\boldsymbol{N})$ by $g_S|_{A_i} = f_i$ when $i \in S$ and $g_S|_{A_i} = I_{A_i}$ when $i \notin S$.)
16. a) Show that $|\boldsymbol{R}| = |\boldsymbol{R}^2|$. (Write $r \in \boldsymbol{R}$ as $r = a_{2k}a_{2k-1} \cdots a_1.b_1b_2 \cdots b_j \cdots$. Define $f(a_{2k}a_{2k-1} \cdots a_1.b_1b_2 \cdots b_j \cdots) =$ $(a_{2k}a_{2k-2} \cdots a_2.b_2b_4 \cdots b_{2j} \cdots, a_{2k-1}a_{2k-3} \cdots a_1.b_1b_3 \cdots b_{2j-1} \cdots)$, where all a_i and b_j are digits and $a_{2k} = 0$ is allowed, but r does not have an infinite repetition of consecutive nines.)
    b) Show that $|\boldsymbol{R}| = |\boldsymbol{C}|$.
    c) Show that for any $m \in \boldsymbol{N}$, $|\boldsymbol{R}| = |\boldsymbol{R}^m| = |\boldsymbol{C}^m|$.
17. a) Show that there is an injection $G : \boldsymbol{R} \to \text{Bij}(\boldsymbol{R})$.
    b) Show that $\text{Bij}(\boldsymbol{R}, \text{Bij}(\boldsymbol{R})) = \varnothing$. (Let $f : \boldsymbol{R} \to \boldsymbol{R}^2$ be a bijection (Exercise 16) and let $A_r = f^{-1}(\{r\} \times \boldsymbol{R})$. Prove that $|A_r| = |\boldsymbol{R}|$ and that $\{A_r\}_{\boldsymbol{R}}$ is a partition of $\boldsymbol{R}$ into $|\boldsymbol{R}|$ subsets. Let $g_r \in \text{Bij}(A_r)$ but not the identity map on A_r and proceed as in Exercise 15 to show that $|\boldsymbol{R}| \neq |\text{Bij}(\boldsymbol{R})|$.)
18. Assume that for any infinite set S there is a partition $\{A_i\}_I$ of S with $|I| = |S|$ and A_i denumerable for each i. When $S = \boldsymbol{R}$, this follows from Example 0.12 and Exercise 9 above. Show that:
    i) $|S| = |S^2|$
    ii) $|\boldsymbol{N} \times S| = |S|$
    iii) For any $m \in \boldsymbol{N}$, $|S| = |S^m|$.
    iv) Show that there is no surjection $g : S \to \text{Bij}(S)$, using the argument in Exercise 15 or Exercise 17.

CHAPTER 1

PRELIMINARIES

This chapter covers elementary but important properties of $\boldsymbol{Z}$, namely divisibility, the Euclidean Algorithm for finding and representing greatest common divisors, and prime factorization. It also introduces the crucial notions of equivalence relation and congruence modulo n.

1.1 REMAINDERS

We begin with a slight generalization of usual long division of integers: dividing by n produces a remainder smaller than n.

THEOREM 1.1 Let $a, n \in \boldsymbol{Z}$ with $n \neq 0$. There are $q, r \in \boldsymbol{Z}$ with $a = qn + r$ and $0 \leq r < |n|$, and both q and r are unique with these properties.

Proof We first show that there are $q, r \in \boldsymbol{Z}$ with $a = qn + r$ and $0 \leq r < |n|$ and afterward show that q and r are unique. It suffices to assume that $n > 0$ because if $a = qn + r$ in this case, then when $n < 0$ we have $a = q(-n) + r = (-q)n + r$. Thus assume that $n > 0$. To find some $r \geq 0$ with $r = a - qn$ we consider $S = \{a - zn \mid z \in \boldsymbol{Z}\}$ and show that $S \cap \boldsymbol{N} \neq \varnothing$. If $z = -(|a| + 1)$, then $a - zn = a + (|a| + 1)n \geq a + |a| + n \geq n > 0$. Consequently $S \cap \boldsymbol{N} \neq \varnothing$ so by Well Ordering there is a minimal $m \in S \cap \boldsymbol{N}$. But $m \in S$ implies $m = a - tn$ for $t \in \boldsymbol{Z}$, or $a = tn + m$. If $m \geq n$, then $a = (t + 1)n + (m - n)$. When $m > n$ then $m - n = a - (t + 1)n \in S \cap \boldsymbol{N}$, contradicting the minimality of m, and when $m = n$ then $a = (t + 1)n$ so take $q = t + 1$ and $r = 0$. If $m < n$ just take $q = t$ and $r = m$ to obtain the desired representation. Finally, if $a = qn + r = q'n + r'$ with $0 \leq r, r' < n$, subtracting yields $(q - q')n = r' - r$. Hence $|r - r'| = |q - q'||n|$ and $r, r' \in [0, n) \subseteq \boldsymbol{R}$ imply that the distance between r and r' must be less than n. Thus $n > |r - r'| = |q - q'||n|$, which forces $|q - q'| = 0$ and shows first that $q = q'$ and then that $r = r'$, completing the proof. ■

We may refer to Theorem 1.1 as the *Division Theorem*. Note that the remainder on division by n or $-n$ is the same since $qn + r = (-q)(-n) + r$. For example,

$24 = 3 \cdot 7 + 3$ and also $24 = (-3) \cdot (-7) + 3$, so the remainder on dividing 24 by either 7 or -7 is 3. How do the remainders on division by n of a and $-a$ compare? Suppose that $n, a \in \mathbf{N}$ and $a = qn + r$ as in Theorem 1.1. Then $-a = -qn - r = -(1+q)n + (n-r)$ with $1 \le n - r \le n$, so $r = 0$ or $n - r$ is the remainder when $-a$ is divided by n. Specifically, the remainder on dividing 2 by 27 is 2 itself, so the remainder on dividing -2 by 27 is $27 - 2 = 25$ and $-2 = (-1) \cdot 27 + 25$. Also $122 = 2 \cdot 57 + 8$, so the remainder on dividing -122 by 57 is $57 - 8 = 49$ and $-122 = -3 \cdot 57 + 49$.

Recall Definition 0.3 that for $n, j \in \mathbf{Z}$, $n\mathbf{Z} + j = \{t \in \mathbf{Z} \mid t = nz + j \text{ for some } z \in \mathbf{Z}\}$. Using Theorem 1.1 we get a partition of $\mathbf{Z}$ from these sets.

THEOREM 1.2 For $n \in \mathbf{N}$ the following hold:

i) For $a, b \in \mathbf{Z}$, $n\mathbf{Z} + a = n\mathbf{Z} + b \Leftrightarrow a \in n\mathbf{Z} + b \Leftrightarrow a - b \in n\mathbf{Z}$.
ii) $\{n\mathbf{Z}, n\mathbf{Z}+1, n\mathbf{Z}+2, \ldots, n\mathbf{Z}+(n-1)\}$ is a partition of $\mathbf{Z}$.
iii) For $k \in \mathbf{Z}$, $\{n\mathbf{Z} + i \mid k \le i \le k + n - 1\}$ is a partition of $\mathbf{Z}$.

Proof i) We show that $n\mathbf{Z} + a = n\mathbf{Z} + b \Rightarrow a \in n\mathbf{Z} + b$, then that $a \in n\mathbf{Z} + b \Rightarrow a - b \in n\mathbf{Z}$, and finally that $a - b \in n\mathbf{Z} \Rightarrow n\mathbf{Z} + a = n\mathbf{Z} + b$. If $n\mathbf{Z} + a = n\mathbf{Z} + b$ then $a = 0 + a \in n\mathbf{Z} + a = n\mathbf{Z} + b$. When $a \in n\mathbf{Z} + b$ then $a = nt + b$ for some $t \in \mathbf{Z}$, forcing $a - b = nt \in n\mathbf{Z}$. Lastly, if $a - b = zn$ for some $z \in \mathbf{Z}$ we may write $a = zn + b$, so $nq + a = n(z+q) + b \in n\mathbf{Z} + b$, showing $n\mathbf{Z} + a \subseteq n\mathbf{Z} + b$. But also $b = (-z)n + a$, so repeating the last computation shows that $n\mathbf{Z} + b \subseteq n\mathbf{Z} + a$, and these two inclusions prove that $n\mathbf{Z} + a = n\mathbf{Z} + b$.

ii) We need the union of the n sets in question to be $\mathbf{Z}$ and pairwise disjoint. Now $\bigcup (n\mathbf{Z} + i) \subseteq \mathbf{Z}$ follows since each $n\mathbf{Z} + i \subseteq \mathbf{Z}$. If $t \in \mathbf{Z}$ then $t = qn + r \in n\mathbf{Z} + r$ with $0 \le r \le n - 1$ by Theorem 1.1, so $\mathbf{Z} \subseteq \bigcup (n\mathbf{Z} + i)$ for $0 \le i < n$ by the definition of union, proving that $\mathbf{Z} = \bigcup (n\mathbf{Z} + i)$. Suppose next that for $0 \le i, j \le n - 1$, some $m \in (n\mathbf{Z} + i) \cap (n\mathbf{Z} + j)$. The definitions of intersection and the $n\mathbf{Z} + a$ imply that $m = nt + i = ns + j$ for some $t, s \in \mathbf{Z}$. Consequently, $i = j$ follows from the uniqueness of remainders in Theorem 1.1. Hence $\{n\mathbf{Z}, n\mathbf{Z}+1, n\mathbf{Z}+2, \ldots, n\mathbf{Z}+(n-1)\}$ is pairwise disjoint so Definition 0.10 on partitions shows that this set is a partition of $\mathbf{Z}$.

iii) For each $0 \le i \le n - 1$, $k + i = q(i)n + r(i)$ for some $q(i), r(i) \in \mathbf{Z}$ with $0 \le r(i) \le n - 1$ by Theorem 1.1. Part i) yields $n\mathbf{Z} + k + i = n\mathbf{Z} + r(i)$. If for some $0 \le i, j \le n - 1$, $n\mathbf{Z} + k + i = n\mathbf{Z} + k + j$, then part i) gives $j - i \in n\mathbf{Z}$, and now $0 \le |j - i| < n$ forces $i = j$. Therefore $\{n\mathbf{Z} + k + i \mid 0 \le i \le n - 1\} = \{n\mathbf{Z} + i \mid 0 \le i \le n - 1\}$ is a partition of $\mathbf{Z}$ by ii). ∎

COROLLARY Let $n \in \mathbf{N}$ and $k, k+1, \ldots, k+n-1$ be n consecutive integers. For each $i \in \mathbf{Z}$ with $0 \le i < n$, exactly one of the $k + j$ is in $n\mathbf{Z} + i$.

Theorem 1.2 enables us to write an arbitrary integer in one of n specific forms that can be used to reduce an apparently infinite number of calculations to a finite number. We illustrate this idea with a few examples.

Example 1.1 For every $t \in \mathbf{Z}$, $t^5 - t \in 5\mathbf{Z}$.

We cannot actually compute $t^5 - t$ for all $t \in \mathbf{Z}$ so another attack is needed. Induction could be used here (see Exercise 9 in §0.3) but we want to employ

Theorem 1.2. By that result, for any $t \in \mathbf{Z}$, $t = 5z + i$ with $0 \le i \le 4$ and some $z \in \mathbf{Z}$. Thus $t^5 - t = (5z+i)^5 - (5z+i) = 5s + i^5 - 5z - i$ for $s \in \mathbf{Z}$ by the Binomial Theorem. Hence $t^5 - t = 5(s - z) + (i^5 - i)$, and $t^5 - t \in 5\mathbf{Z}$ is equivalent to $i^5 - i \in 5\mathbf{Z}$ (why?). That $i^5 - i \in 5\mathbf{Z}$ is easy to verify if $0 \le i \le 4$.

Example 1.2 $3\mathbf{Z} + 2$ contains no integer square.

Since $\{3\mathbf{Z}, 3\mathbf{Z}+1, 3\mathbf{Z}+2\}$ is a partition of $\mathbf{Z}$ from Theorem 1.2, it suffices to show that $x^2 \in 3\mathbf{Z} \cup (3\mathbf{Z}+1)$ for $x \in \mathbf{Z}$. But $x = 3s + i$ for $0 \le i \le 2$ by Theorem 1.1, so $(3s)^2 \in 3\mathbf{Z}$, $(3s+1)^2 = 9s^2 + 6s + 1 \in 3\mathbf{Z} + 1$, and $(3s+2)^2 = 3t + 4 = 3(t+1) + 1 \in 3\mathbf{Z} + 1$. Hence $x^2 \in 3\mathbf{Z} \cup (3\mathbf{Z}+1)$.

Next is a useful result about squares.

THEOREM 1.3 If $t \in \mathbf{Z}$ is even, $t^2 \in 8\mathbf{Z} \cup (8\mathbf{Z}+4) = 4\mathbf{Z}$; if t is odd, $t^2 \in 8\mathbf{Z} + 1$.

Proof It is an exercise that $8\mathbf{Z} \cup (8\mathbf{Z}+4) = 4\mathbf{Z}$. By Theorem 1.2 every $t \in \mathbf{Z}$ is in some $8\mathbf{Z} + i$ for $-3 \le i \le 4$; taking $0 \le i \le 7$ leads to more computation. If $t = 8z + i$ then t is odd (even) exactly when i is and $t^2 = (8z+i)^2 = 8q + i^2$ so $t^2 \in 8\mathbf{Z} + j$ exactly when i^2 is. Clearly, $0^2, 4^2 \in 8\mathbf{Z}$, $(\pm 1)^2 \in 8\mathbf{Z}+1$, $(\pm 2)^2 \in 8\mathbf{Z}+4$, and $(\pm 3)^2 = 9 = 8 + 1 \in 8\mathbf{Z} + 1$. ∎

Information about sums of squares in $\mathbf{Z}$ follows from Thereom 1.3.

Example 1.3 For any $x, y, z \in \mathbf{Z}$: i) If $t = x^2 + y^2$ is odd then $t \in 4\mathbf{Z} + 1$; ii) $x^2 + y^2 + z^2 \notin 8\mathbf{Z} + 7$.

If $t = x^2 + y^2$ is odd then exactly one of x or y is odd, so assume x is odd and y is even. By Theorem 1.3, $x^2 = 8u + 1$ and $y^2 = 8v$ or $y^2 = 8v + 4$, and if $s = u + v$ either $t = 8s + 1$ or $t = 8s + 5 = 4(2s+1) + 1$ so $t \in 4\mathbf{Z} + 1$ as claimed. Now use Theorem 1.3 to write $x^2 = 8a + i$, $y^2 = 8b + j$, and $z^2 = 8c + k$ for $i, j, k \in \{0, 1, 4\}$. A quick check shows that $i + j + k \notin 8\mathbf{Z} + 7$ so $x^2 + y^2 + z^2 = 8(a+b+c) + i + j + k \notin 8\mathbf{Z} + 7$.

Using the example may reduce computation. Thus $1{,}000{,}000{,}003 \in 4\mathbf{Z} + 3$ is not the sum of two integer squares since it is not in $4\mathbf{Z} + 1$ (Theorem 1.2), and since $8{,}168{,}432{,}807 \in 8\mathbf{Z} + 7$, it cannot be a sum of three integer squares. These conclusions do not require computing sums of squares. Which integers can be written as a sum of two, three, or more squares? We will not study this question here, but a famous theorem of Lagrange proves that every $n \in \mathbf{N}$ is the sum of four integer squares and those that are sums of two or of three squares can be determined precisely.

EXERCISES

1. If $a \in \mathbf{Z}$ and $n \in \mathbf{N}$, show $a = qn + r$ with $-n/2 \le r < n/2$ for some $q, r \in \mathbf{Z}$.
2. Find $q, r \in \mathbf{Z}$ as in Theorem 1.1 with $0 \le r < |n|$, when $a = \pm 85$ and $n = \pm 16$.

3. Show carefully that $\mathbf{Z} - 2\mathbf{Z} = (4\mathbf{Z} + 1) \bigcup (4\mathbf{Z} + 3)$.

4. Which of the following pairs of subsets of $\mathbf{Z}$ are equal?

i) $7\mathbf{Z} - 3$ and $7\mathbf{Z} + 66$

ii) $13\mathbf{Z} + 81$ and $13\mathbf{Z} + 146$

iii) $11\mathbf{Z} - 61$ and $11\mathbf{Z} - 1161$

iv) $79\mathbf{Z} + 30000$ and $79\mathbf{Z} + 40000$

5. Use Theorem 1.2 to determine whether the set $\{p\mathbf{Z}\} \bigcup \{p\mathbf{Z} + 2^k \mid 1 \le k < p\}$ is a partition of $\mathbf{Z}$ when p is: a) 5; b) 7; c) 11; d) 13; or e) 17.

6. In each case determine whether the set $\{n\mathbf{Z} + ka \mid 0 \le k < n\}$ is a partition of $\mathbf{Z}$:

i) $n = 7$ and $a = 3$

ii) $n = 8$ and $a = 3$

iii) $n = 8$ and $a = 10$

iv) $n = 99$ and $a = 6$

v) $n = 159$ and $a = 2$

7. Which pairs of the following sets have nonempty intersection?

i) $17\mathbf{Z} + 2601$; $17\mathbf{Z} + 2700$; $17\mathbf{Z} + 2802$; and $17\mathbf{Z} + 2907$

ii) $41\mathbf{Z} + 10^{21} + 11{,}215$; $41\mathbf{Z} + 10^{21} + 11{,}245$; and $41\mathbf{Z} + 10^{21} + 11{,}297$

8. If $n \in \mathbf{N}$ and $a, b \in \mathbf{Z}$, show there is $i \in \mathbf{Z}$ with $a, b \in n\mathbf{Z} + i \Leftrightarrow a - b \in n\mathbf{Z}$.

9. Given $n \in \mathbf{N}$, if $x \in n\mathbf{Z} + i$ and $y \in n\mathbf{Z} + j$, show that $x + y \in n\mathbf{Z} + (i + j)$ and $xy \in n\mathbf{Z} + ij$.

10. Use the corollary of Theorem 1.2 to show that for all $x \in \mathbf{Z}$, $f(x) = x^2(x^2 - 1)(x^2 - 4) \in 8\mathbf{Z} \bigcap 9\mathbf{Z} \bigcap 5\mathbf{Z}$.

11. If $a \in \mathbf{Z}$ is both a square and a cube, show that $a \in 7\mathbf{Z} \bigcup (7\mathbf{Z} + 1)$.

12. If $a \in \mathbf{Z}$, show that $a^2 \in 4\mathbf{Z} \bigcup (4\mathbf{Z} + 1)$. If $a, b \in \mathbf{Z}$ are *odd* show that there is no $c \in \mathbf{Z}$ satisfying $a^2 + b^2 = c^2$.

13. If $a, b, c \in \mathbf{Z}$ are all odd, show that $a^2 + b^2 + c^2$ is not the sum of two squares in $\mathbf{Z}$. (Note that $1^2 + 3^2 + 8^2 = 74 = 5^2 + 7^2$.)

1.2 DIVISIBILITY

Next we prove a very useful result that combines Theorem 1.1 with induction to characterize those $S \subseteq \mathbf{Z}$ closed under subtraction: $a, b \in S$ imply $a - b \in S$. Clearly $\mathbf{N}$ is not closed under subtraction since $2 - 5 \notin \mathbf{N}$. Also $2\mathbf{Z} + 1$ is not closed under subtraction since $5 - 17$ is even. A little thought shows that $2\mathbf{Z}$, $3\mathbf{Z}$, and more generally any $n\mathbf{Z}$ is closed under subtraction. In fact, these are the only subsets of $\mathbf{Z}$ that are closed under subtraction.

THEOREM 1.4 If $\emptyset \ne S \subseteq \mathbf{Z}$ then S satisfies $x, y \in S$ imply $x - y \in S$ precisely when either $S = \{0\}$ or $S = n\mathbf{Z}$ for n the minimal integer in $S \cap \mathbf{N}$.

Proof If $S = n\mathbf{Z}$ for some $n \in \mathbf{N}$, or $n = 0$, then $nz_1 - nz_2 = n(z_1 - z_2) \in n\mathbf{Z}$ so $n\mathbf{Z}$ is closed under subtraction. Now let $S \ne \emptyset$ be closed under subtraction. We are finished if $S = \{0\}$, so assume $S \ne \{0\}$. Since the theorem mentions a minimal element in $S \cap \mathbf{N}$, we begin by showing that $S \bigcap \mathbf{N} \ne \emptyset$. If $a \in S$ then $a, a \in S$ result in $0 = a - a \in S$. But now both $0, a \in S$ so $-a = 0 - a \in S$ also. Thus $-a \in S$ when $a \in S$, and $S \ne \{0\}$, imply $S \bigcap \mathbf{N} \ne \emptyset$. By Well Ordering there is a minimal $m \in S \bigcap \mathbf{N}$ and we need to show $S = m\mathbf{Z}$. First we see that $mN \subseteq S$ using induction. The first step, that $m \cdot 1 = m \in S$, is true, and so $-m \in S$ by our computation above. Assuming that $mk \in S$ for some $k \in N$ gives both $mk, -m \in S$, so $m(k+1) = mk - (-m) \in S$. By IND-I, $mN \subseteq S$ as claimed. We saw that $0 \in S$ and that $y = zm \in S$ implies $-y = (-z)m \in S$, so $mZ \subseteq S$. To finish the proof it suffices to show that $S \subseteq mZ$. For $s \in S$ use Theorem 1.1 to write $s = qm + r$ for $q, r \in Z$ with $0 \leq r < m$. It follows that $r = s - qm \in S$ since we have $qm \in mZ \subseteq S$ and know that $s \in S$. But $r \in S$, $r < m$, and the choice of m force $r = 0$, so $s = qm \in mZ$ and $S \subseteq mZ$. ■

DEFINITION 1.1 If $a, b \in Z$ and $a \neq 0$ then a **divides** b, written $a \mid b$, when $b = ac$ for some $c \in Z$, or, equivalently, when $b \in aZ$. When it is false that a divides b, we write $a \nmid b$.

Writing "$a \mid b$" requires $a \neq 0$. Examples are that $-4 \mid 12$ since $12 = (-4)(-3)$, $17 \mid 0$ since $0 = 0 \cdot 17$, and $2 \nmid 5$ since $5 = 2(5/2)$, but $5/2 \in Q$ and not in Z. Easy consequences we will use without reference are: $1 \mid a$ for all $a \in Z$ since $a = 1 \cdot a$; if $a \mid 1$ then $a = \pm 1$ since $1 = ac$ forces $a = c = \pm 1$; and $a \mid ab$ for all $a \neq 0$ since $ab = a \cdot b$, so in particular $a \mid \pm a$ and $a \mid 0$. Also, divisibility is independent of sign in that all of $a \mid b$, $-a \mid b$, $a \mid -b$, and $-a \mid -b$ are equivalent since $b = ac$, $b = (-a)(-c)$, $-b = a(-c)$, and $-b = (-a)c$ are equivalent. A final observation is that when $a \mid b$, then either $b = 0$ or $|a| \leq |b|$. We collect together some other very easy consequences of Definition 1.1.

THEOREM 1.5 For $a, b, c, d \in Z$, the following hold:

i) $a \mid b$ and $b \mid a \Rightarrow b = \pm a$
ii) $a \mid b \Rightarrow a \mid bc$
iii) $ab \mid c \Rightarrow a \mid c$
iv) If $c \neq 0$, then $a \mid b \Leftrightarrow ca \mid cb$
v) $a \mid b$ and $b \mid c \Rightarrow a \mid c$
vi) $a \mid b$ and $c \mid d \Rightarrow ac \mid bd$.

Proof We prove i) and v) and leave the other parts as exercises. For i) the assumptions and definition of division give $b = ax$ and $a = by$ with $x, y \in Z$. Substituting yields $a = (ax)y = a(xy)$ and since $a \neq 0$, $1 = xy$ results. Thus $x = y = \pm 1$ and $b = \pm a$. In v) the assumptions mean $b = ax$ and $c = by$ for some $x, y \in Z$, so $c = (ax)y = a(xy)$, proving that $a \mid c$ by Definition 1.1. ■

Another statement about divisibility is easy, but we will use this fact so often that we state it for reference.

THEOREM 1.6 For all $a, b, c, s, t \in Z$, if $a \mid b$ and $a \mid c$, then $a \mid (sb + tc)$.

Proof Since $a \mid b$ and $a \mid c$, we have $b = ax$ and $c = ay$ for some $x, y \in Z$. Now write $sb + tc = s(ax) + t(ay) = a(sx + ty)$, so by Definition 1.1, $a \mid (sb + tc)$. ■

From Theorem 1.6, if $a \mid n$ and $a \mid (n+k)$ then $a \mid k$ since $k = 1 \cdot (n+k) + (-1) \cdot n$. Thus for any $n \in N$, ± 1 are the only divisors of both n and $n+1$, and the only possible divisors of both $n-1$ and $n+1$ are ± 1 or ± 2. A less direct application of the theorem is that the only possible divisors in $\mathbf{N}$ of $3n+2$ and $8n+3$ are 1 and 7 since $7 = 8(3n+2) - 3(8n+3)$. It is an exercise to show that 7 *can* be a common divisor. A useful observation is that if $b, c, s, t \in \mathbf{Z}$ with $sb + tc = 1$ then by Theorem 1.6 the only common divisors of b and c are ± 1. For example, $n^4 - 2$ and $n^2 + 1$ have no common divisors except for ± 1 since $(n^2 - 1)(n^2 + 1) - (n^4 - 2) = 1$. In fact when $a, b \in \mathbf{Z}$ have no common divisor except ± 1, some integer linear combination of them is 1, but this is not obvious; we prove it shortly. Next we consider least common multiples.

DEFINITION 1.2 $\mathbf{Z}^*$ denotes the set of **nonzero integers.**

DEFINITION 1.3 For $S = \{a_1, \ldots, a_k\} \subseteq \mathbf{Z}^*$, $c \in \mathbf{Z}$ is a **common multiple** for S if $a_j \mid c$ for each $a_j \in S$. The **least common multiple** of S, written LCM(S), is the smallest, positive, common multiple of S.

Whenever $S \subseteq \mathbf{Z}^*$ is finite, LCM(S) exists. By Theorem 1.5 ii, the product of all the elements in S is a common multiple for S so its negative is also. Thus there is a common multiple for S in $\mathbf{N}$, and Well Ordering guarantees a minimal such multiple. For sets of a few small integers in $\mathbf{N}$, looking at multiples of the largest will usually produce the LCM quickly. For example, $\text{LCM}\{6, 9, 12\} = 36$, and $\text{LCM}\{2, 4, 8, 16, 32\} = 32$. We will see later how to compute LCM(S), but we now turn to a somewhat surprising fact about it.

THEOREM 1.7 Let $S \subseteq \mathbf{Z}^*$ be a finite set with $\text{LCM}(S) = m$. If c is any common multiple for S then $m \mid c$.

Proof If $T = \{c \in \mathbf{Z} \mid c \text{ is a common multiple for } S\}$ then m is minimal in $T \cap \mathbf{N}$ by Definition 1.3. For $c, d \in T$ and $s \in S$, $s \mid c$ and $s \mid d$, so $s \mid (c - d)$ from Theorem 1.6, forcing $c - d \in T$. Now Theorem 1.4 shows that $T = m\mathbf{Z}$. Therefore every common multiple for S is a multiple of LCM(S). ∎

This notion of least common multiple will be needed and used, but more important is the companion idea of greatest common divisor.

DEFINITION 1.4 A **common divisor** for $\emptyset \neq S \subseteq \mathbf{Z}^*$ is $d \in \mathbf{Z}$ satisfying $d \mid s$ for all $s \in S$. The **greatest common divisor** of S, denoted GCD(S), is the largest common divisor for S.

Every $\emptyset \neq S \subseteq \mathbf{Z}^*$ has a greatest common divisor. Since $1 \mid s$ if $s \in S$, there are common divisors for S, and since any common divisor d satisfies $d \mid s$ for all $s \in S$, $|d| \le |s|$. Thus the set of common divisors is bounded, so there is a largest one. Trial and error shows that $\text{GCD}(\{12, 18, 45\}) = 3$ and $\text{GCD}(\{6, 10, 15\}) = 1$, although $\text{GCD}(\{6, 10\}) = 2$, $\text{GCD}(\{6, 15\}) = 3$, and $\text{GCD}(\{10, 15\}) = 5$. A useful fact is that if $a \mid b$ then $\text{GCD}(\{a, b\}) = |a|$.

The next result gives useful, important, and surprising properties of GCD(S) when S is a finite set.

THEOREM 1.8 Let $S=\{a_1,\ldots,a_k\}\subseteq \mathbf{Z}^*$ with GCD(S) $=d$, and define $T=\{z_1a_1+\cdots+z_ka_k\in \mathbf{Z}\mid z_i\in \mathbf{Z}\}$. Then the following hold:

i) $d=y_1a_1+\cdots+y_ka_k\in T$ for some $y_j\in \mathbf{Z}$
ii) $T=d\mathbf{Z}$ and d is minimal in $T\cap \mathbf{N}$
iii) If $c\mid a_j$ for all $a_j\in S$, then $c\mid d$.

Proof We proceed by showing that the minimal element in $T\cap \mathbf{N}$ satisfies i), ii), and iii), and is a common divisor for S. Now $a_i=0\cdot a_1+\cdots+1\cdot a_i+\cdots+0\cdot a_k\in T$, and if $t=\Sigma z_ia_i\in T$ then $-t=\Sigma(-z_i)a_i\in T$, so $T\cap \mathbf{N}\neq\emptyset$. By Well Ordering there is a minimal $m\in T\cap \mathbf{N}$. Since $\Sigma x_ia_i-\Sigma y_ia_i=\Sigma(x_i-y_i)a_i$, Theorem 1.4 applies to show $T=m\mathbf{Z}$. All $a_j\in m\mathbf{Z}$ show m is a common divisor for S, and $m\in m\mathbf{Z}=T$ implies that $m=\Sigma b_ia_i$ for $b_i\in \mathbf{Z}$. If c is a common divisor for S, set $a_i=w_ic$ and replace a_j in the representation of m to get $m=\Sigma b_iw_ic=(\Sigma b_iw_i)c$. Therefore $c\mid m$ so $c\le m$. Now d is a common divisor for S so $d\mid m$, and m is a common divisor for S so $m\le d$ must hold. Thus $d=m$ satisfies the three statements of the theorem since m does. ∎

A common error in applying Theorem 1.8 is to conclude that $y=\text{GCD}(\{a_i\})$ if $\{a_i\}\subseteq \mathbf{Z}^*$ with $y=c_1a_1+\cdots+c_ka_k\in \mathbf{N}$ without showing y to be the minimal such combination. Consider: $\text{GCD}(\{\text{GCD}(\{a,b\}),c\})=\text{GCD}(\{a,b,c\})$. If we write $\text{GCD}(\{\text{GCD}(\{a,b\}),c\})=u\text{GCD}(\{a,b\})+vc$ and $\text{GCD}(\{a,b\})=sa+tb$, for $u,v,s,t\in \mathbf{Z}$, using Theorem 1.8, then $\text{GCD}(\{\text{GCD}(\{a,b\}),c\})=(us)a+(ut)b+vc$. This does *not* show that GCD $(\{\text{GCD}(\{a,b\}),c\})=\text{GCD}(\{a,b,c\})$ since this integer combination of a, b, and c may not be the minimal positive one. We do have that $\text{GCD}(\{\text{GCD}(\{a,b\}),c\})$ is a *multiple* of $\text{GCD}(\{a,b,c\})$ from Theorem 1.8. Similarly, $\text{GCD}(\{a,b,c\})$ is a multiple of $\text{GCD}(\{\text{GCD}(\{a,b\}),c\})$ (how?), and since each of these positive integers divides the other, they are equal by Theorem 1.5. This result about greatest common divisors is a special case of one of the exercises. A straightforward way to show it is simply to use the definition of GCD together with the part of Theorem 1.8 that shows that GCD(S) is a multiple of every common divisor for S.

For $\{a_i\}\subseteq \mathbf{Z}^*$ if $c_1a_1+\cdots+c_ka_k=1$ then $\text{GCD}(\{a_i\})=1$: clearly $1\in \mathbf{N}$ is the minimal integer combination of $\{a_i\}$. Even here Theorem 1.8 can be misused. We prove a bit later that $\text{GCD}(\{a,b\})=1\Leftrightarrow\text{GCD}(\{a^m,b\})=1$ for any $m\in \mathbf{N}$. It is tempting here to use Theorem 1.8 to write $1=sa+tb$, let $s=a^{m-1}$, write $1=a^m+tb$, and conclude that $\text{GCD}(\{a^m,b\})=1$. The point is that Theorem 1.8 shows the existence of *some* $s,t\in \mathbf{Z}$ satisfying $1=sa+tb$, and although there may be many such choices, the theorem neither specifies which choices work *nor allows arbitrary choices*.

If $\text{GCD}(S)=z_1a_1+\cdots+z_ka_k$, how do we find the z_j? This assumes we can find GCD(S), not always easy to do. For small numbers, trial and error may produce the required integers: $\text{GCD}(\{6,10,15\})=1=6+10-15$ and $\text{GCD}(\{12,20\})=4=(2)12-20=(-3)12+2\cdot 20$. GCD $(\{17,19\})=1$ since these are primes, but it is not obvious how to write $1=17s+19t$. Also, it is not obvious what integer $\text{GCD}(\{2747,3649\})$ is.

A nice algorithm finds both GCD($\{a, b\}$) and the integers representing it as $sa + tb$. We need a preliminary result but first give another notation for GCD($\{a, b\}$).

DEFINITION 1.5 For $a, b \in \mathbf{Z}^*$, set $(a, b) = \text{GCD}(\{a, b\})$.

THEOREM 1.9 If $a, b, c, d \in \mathbf{Z}^*$ with $a = bc + d$, then $(a, b) = (b, d)$.

Proof From the definition of GCD($\{a, b\}$) we conclude that $(a, b) \mid a$ and $(a, b) \mid b$, so applying Theorem 1.6 to $1 \cdot a + (-c)b = d$ shows that $(a, b) \mid d$. But now $(a, b) \mid b$ and $(a, b) \mid d$ so $(a, b) \mid (b, d)$ by Theorem 1.8. Similarly, starting with $a = bc + d$ shows that $(b, d) \mid (a, b)$, so $(b, d) = \pm(a, b)$ by Theorem 1.5. Since both of these are positive, $(a, b) = (b, d)$. ■

The **Euclidean Algorithm** that finds GCD($\{a, b\}$) $= sa + tb$ is extremely useful, very important, and generally easy to use. We state it for $a, b \in \mathbf{N}$ with $0 < b < a$ and when b does not divide a. If $b \mid a$ then $(a, b) = b = 0 \cdot a + 1 \cdot b$. In other cases, use GCD($\{a, b\}$) $=$ GCD($\{\pm a, \pm b\}$) and the algorithm when both are positive; that is, if $b < 0 < a$ then by the algorithm, $(a, b) = (a, -b) = sa + t(-b) = sa + (-t)b$. Proceed similarly if $b < a < 0$.

EUCLIDEAN ALGORITHM Let $a, b \in \mathbf{N}$ with $b < a$ and b not a divisor of a. Use Theorem 1.1, repeatedly as necessary, to write

$$\begin{aligned} a &= q_1 b + r_1 \\ b &= q_2 r_1 + r_2 \\ &\vdots \\ r_{n-3} &= q_{n-1} r_{n-2} + r_{n-1} \\ r_{n-2} &= q_n r_{n-1} + r_n \end{aligned}$$

where r_n is the last nonzero remainder. Then $(a, b) = r_n = sa + tb$ for $s, t \in \mathbf{Z}$.

Since $b > r_1 > \cdots > r_j$, there will be a last nonzero remainder r_n in at most b steps, and $r_n \mid r_{n-1}$ since the next remainder is zero. First we use Theorem 1.9 to argue that $(a, b) = r_n$. From $a = q_1 b + r_1$ the theorem gives $(a, b) = (b, r_1)$. When $n = 1$, $r_1 \mid b$ so $(a, b) = r_1$. When $n > 1$ use $b = q_2 r_1 + r_2$ to get $(b, r_1) = (r_1, r_2)$, and from the other equations $(a, b) = (b, r_1) = (r_1, r_2) = \cdots = (r_{n-1}, r_n)$. Since r_n is the last nonzero remainder, $r_n \mid r_{n-1}$ so $(a, b) = (r_{n-1}, r_n) = r_n$. Find $(a, b) = r_n = sa + tb$ from the equations above, using the last equation to write r_n in terms of r_{n-1} and r_{n-2}, then replacing r_{n-1} with its representation as a sum of r_{n-2} and r_{n-3} in the next to last equation, and continuing to obtain s and t. Examples will make this procedure clear. Using $(17, 19) = 1$ as above, write the equations $19 = 1 \cdot 17 + 2$ and $17 = 8 \cdot 2 + 1$. The second equation shows $1 = 17 - 8 \cdot 2$, and using the first equation to express 2 in terms of 17 and 19 yields $1 = 17 - 8(19 - 17)$, so $1 = 9 \cdot 17 - 8 \cdot 19$.

Example 1.4 Find $s, t \in \mathbf{Z}$ so that $(2747, 3649) = s \cdot 2747 + t \cdot 3649$.

Write $3649 = 1 \cdot 2747 + 902$; $2747 = 3 \cdot 902 + 41$; and $902 = 22 \cdot 41$. Thus $(2747, 3649) = 41 = 2747 - 3 \cdot 902 = 2747 - 3 \cdot (3649 - 2747)$: $s = 4$, $t = -3$.

Example 1.5 Find $s, t \in \mathbf{Z}$ so that $(969, 1273) = s \cdot 969 + t \cdot 1273$.

Again write $1273 = 969 + 304$; $969 = 3 \cdot 304 + 57$; $304 = 5 \cdot 57 + 19$; and $57 = 3 \cdot 19$. Thus $(969, 1273) = 19 = 304 - 5 \cdot 57 = 304 - 5(969 - 3 \cdot 304) = 16 \cdot 304 - 5 \cdot 969 = 16 \cdot (1273 - 969) - 5 \cdot 969$, so $19 = 16 \cdot 1273 - 21 \cdot 969$.

The integers $s, t \in \mathbf{Z}$ satisfying $(a, b) = sa + tb$ are never unique since also $(a, b) = (s + zb)a + (t - za)b$ for any $z \in \mathbf{Z}$. By results in the next section, $(a, b) = sa + tb = s'a + t'b$ forces $s' = s + zb'$ and $t' = t - za'$ where $a = (a, b)a'$ and $b = (a, b)b'$, showing how different solutions are related. The Euclidean Algorithm can be used to find $\mathrm{GCD}(\{a_1, \ldots, a_k\})$ as a combination of the $\{a_i\}$. This needs $\mathrm{GCD}(\{a_1, \ldots, a_k\}) = \mathrm{GCD}(\{\mathrm{GCD}(\{a_1, \ldots, a_{k-1}\}), a_k\})$, one of the exercises. We illustrate this for $k = 3$ since the computations are tedious if $k > 3$. If some $\mathrm{GCD}(\{a_i, a_j\}) = 1$, then $1 = sa_i + ta_j$ for $s, t \in \mathbf{Z}$.

Example 1.6 Find $\mathrm{GCD}(\{385, 455, 1001\}) = a \cdot 385 + b \cdot 455 + c \cdot 1001$.

The equations of the Euclidean Algorithm for (385, 455) are $455 = 385 + 70$ and $385 = 5 \cdot 70 + 35$. Thus $(385, 455) = 35 = 385 - 5 \cdot 70 = 385 - 5(455 - 385)$, so $(385, 455) = 35 = 6 \cdot 385 - 5 \cdot 455$. Now find (35, 1001) by writing $1001 = 28 \cdot 35 + 21$, $35 = 21 + 14$, and $21 = 14 + 7$. By the exercise mentioned above, $\mathrm{GCD}(\{385, 455, 1001\}) = (35, 1001) = 7$ and $7 = 21 - 14 = 21 - (35 - 21) = 2 \cdot (1001 - 28 \cdot 35) - 35 = 2 \cdot 1001 - 57 \cdot 35$. Our computation $35 = 6 \cdot 385 - 5 \cdot 455$ above gives $7 = 2 \cdot 1001 - 57 \cdot (6 \cdot 385 - 5 \cdot 455)$, so $\mathrm{GCD}(\{385, 455, 1001\}) = 7 = 2 \cdot 1001 + 285 \cdot 455 - 342 \cdot 385$.

EXERCISES

1. Describe $S \subseteq \mathbf{Z}$ consisting of all sums of cubes, not necessarily distinct, of elements from $5\mathbf{Z}$. For example, this set contains $(-5)^3 + (-5)^3 + 10^3 = 750$, $(-15)^3 = -3375$, $10^3 + 10^3 + 10^3 + 10^3 = 4000$, and $0^3 = 0$.
2. Show that for each $z \in \mathbf{Z}$ there are odd integers $a_1, a_2, \ldots, a_k \in \mathbf{Z} - \{\pm 1\}$, not necessarily distinct, such that $z = a_1^5 + \cdots + a_k^5$. (Consider the set of all possible sums of such fifth powers.) Is this still true if seventh powers are used instead of fifth powers? What if ninth powers are used?
3. Complete the proof of Theorem 1.5.
4. If $a, c_1, \ldots, c_k \in \mathbf{N}$ and $a \mid c_i$ for all i, show $a \mid (y_1 c_1 + \cdots + y_k c_k)$ where all $y_j \in \mathbf{Z}$.

5. For any given $n \in \boldsymbol{N}$, show that:
i) $(n^2 + n + 1, n + 1) = 1$
ii) $(n^4 + 1, n^2 + 1) \mid 2$
iii) $(5n + 4, 6n + 1) \mid 19$
iv) $(n^2 + 1, n^3 - 1) \mid 2$
v) $(n^5 + 1, n^7 - 1) \mid 2$

6. Find LCM(S) for each S below:
i) $S = \{1, 2, 3, 4, 5, 6, 7\}$
ii) $S = \{1, 3, 6, 10, 15, 21, 28\}$
iii) $S = \{1, 2, 3, \ldots, 9, 10, 11\}$
iv) $S = \{2 \cdot 3^{10}, 2^2 \cdot 3^9, 2^3 \cdot 3^8, \ldots, 2^9 \cdot 3^2, 2^{10} \cdot 3\}$

7. Let $S = \{s_i\}$ and $T = \{t_j\}$ be finite subsets of $\boldsymbol{N}$. Show that $\text{GCD}(S) = \text{GCD}(T)$ if and only if any common divisor for S is a common divisor for T, *and* any common divisor for T is a common divisor for S.

8. If $S = \{s_1, \ldots, s_n\} \subset \boldsymbol{N}$, show that $\text{GCD}(S) = \text{GCD}(\{\text{GCD}(\{s_1, \ldots, s_{n-1}\}), s_n\})$.

9. For $n_1, n_2, \ldots, n_k \in \boldsymbol{N}$, show: a) $n_1\boldsymbol{Z} + \cdots + n_k\boldsymbol{Z} = \text{GCD}(\{n_1, \ldots, n_k\})\boldsymbol{Z}$ where $n_1\boldsymbol{Z} + \cdots + n_k\boldsymbol{Z} = \{c_1n_1 + \cdots + c_kn_k \mid c_j \in \boldsymbol{Z}\}$; and b) $n_1\boldsymbol{Z} \bigcap n_2\boldsymbol{Z} \bigcap \cdots \bigcap n_k\boldsymbol{Z} = \text{LCM}(\{n_1, \ldots, n_k\})\boldsymbol{Z}$.

10. For each pair of integers a and b below, find $s, t \in \boldsymbol{Z}$ such that $(a, b) = sa + tb$.
i) $a = 111 \quad b = 126$
ii) $a = 89 \quad b = 144$
iii) $a = 153 \quad b = 225$
iv) $a = 1331 \quad b = 3113$
v) $a = 12449 \quad b = 30031$

11. As in Example 1.6, find the GCD as a integer linear combination of the given integers; i) $\{30, 42, 105\}$; ii) $\{330, 462, 1155\}$.

12. For $d, n \in \boldsymbol{N}$, show $d \mid n \Leftrightarrow (2^d - 1) \mid (2^n - 1)$. (Hint: Use Example 0.18 and also Theorem 1.1.) For example this result implies that $17 \mid (2^{2048} - 1)$ since $17 \mid (2^8 - 1)$ and $8 \mid 2048$, and also $13 \mid (2^{3000} - 1)$ since $13 \mid (2^{12} - 1)$ and $12 \mid 3000$.

1.3 RELATIVE PRIMENESS

The situation when $\text{GCD}(\{a_1, \ldots, a_k)\} = 1$ is of special importance.

DEFINITION 1.6 $\{a_k\} \subseteq \boldsymbol{Z}^*$ is **relatively prime** when $\text{GCD}(\{a_k\}) = 1$ and is **pairwise relatively prime** when $(a_i, a_j) = 1$ for $i \neq j$.

Clearly $(10, 21) = 1$ although neither is a prime, and $\{6, 10, 15\}$ is relatively prime even though any two of its elements are not. Any set that is pairwise relatively prime must be relatively prime by definition. Also, if p is a prime and $a \in \boldsymbol{Z}$, then since $(p, a) \mid p$ either $(p, a) = p$ and $p \mid a$, or $(p, a) = 1$.

The special case of Theorem 1.8 when $(a, b) = 1$ occurs frequently and so we isolate it as a theorem and observe that its converse is true as well.

THEOREM 1.10 For $a, b \in \boldsymbol{Z}^* (a, b) = 1 \Leftrightarrow sa + tb = 1$ for some $s, t \in \boldsymbol{Z}$. Thus in this case, $(s, t) = 1$ as well.

Proof By Theorem 1.8, $\text{GCD}\{a, b\} = 1$ implies $sa + tb = 1$ for some $s, t \in \mathbf{Z}$. If $sa + tb = 1$ for some $s, t \in \mathbf{Z}$, then since 1 is the smallest integer combination of a and b in $\mathbf{N}$, again Theorem 1.8 shows that $1 = \text{GCD}\{a, b\}$. ■

Theorem 1.10 is extremely useful in dealing with relatively prime integers. We record a few consequences needed later both in proofs and for examples. Our next result shows that the property of being relatively prime to a fixed integer is preserved under multiplication.

THEOREM 1.11 If $a, b \in \mathbf{Z}^*$ and $n, m \in \mathbf{N}$, then: i) $(a, n) = 1 = (b, n) \Leftrightarrow (ab, n) = 1$; and ii) $(a, b) = 1 \Leftrightarrow (a^m, b) = 1$.

Proof For i) if $(ab, n) = 1$ then by Theorem 1.10 there are $u, v \in \mathbf{Z}$ with $u(ab) + vn = 1$. In particular $(ub)a + vn = (ua)b + vn = 1$, so both $(a, n) = 1$ and $(b, n) = 1$ using Theorem 1.10 again. Now assume that $(a, n) = 1 = (b, n)$ and use the same theorem to write $1 = sa + tn = qb + wn$. Multiplying these yields $1 = (sa + tn)(qb + wn) = (sq)ab + zn$ so $(ab, n) = 1$, which follows from a final application of Theorem, 1.10, proving i). When $(a^m, b) = 1$ for $m > 1$, then $(a, b) = 1$ follows from part i) since $a^m = a \cdot a^{m-1}$. If $(a, b) = 1$, an induction argument shows that $(a^m, b) = 1$ for all $m \in \mathbf{N}$. Specifically, assuming that $(a^k, b) = 1$ for $k \geq 1$ and using both $(a, b) = 1$ and i) shows that $(a^{k+1}, b) = 1$, so IND-I proves that $(a^m, b) = 1$ for all $m \in \mathbf{N}$, and ii) is proved. ■

The results in Theorem 1.11 are usually more useful for proving other results than for computational problems. Other results needed about relatively prime integers are grouped into two theorems. The first deals with divisibility itself, and will be cited often. The second theorem concerns other multiplicative properties.

THEOREM 1.12 If $a, b \in \mathbf{Z}^*$ and $c \in \mathbf{Z}$, then the following hold:

i) If $(a, b) = 1$ and $a \mid bc$, then $a \mid c$.
ii) If $(a, b) = 1$, $a \mid c$ and $b \mid c$, then $ab \mid c$.
iii) If $a = a'(a, b)$ and $b = b'(a, b)$, then $(a', b') = 1$.

Proof For i), $(a, b) = 1$ implies $1 = sa + tb$ for $s, t \in \mathbf{Z}$ from Theorem 1.10. To try to use $a \mid bc$, multiply the last equality by c to obtain $c = (cs)a + t(bc)$. Now $a \mid bc$ and also $a \mid a$ so $a \mid c$ by Theorem 1.6 since c is an integer combination of bc and a. In ii), from the definition of division, $b \mid c$ implies $c = bx$ for some $x \in \mathbf{Z}$, so now $a \mid bx$. But since $(a, b) = 1$, part i) shows that $a \mid x$, or, equivalently, $x = ay$ for some $y \in \mathbf{Z}$. Thus $c = bx = bay = (ab)y$ and $ab \mid c$ results. For iii), write $(a, b) = sa + tb = sa'(a, b) + tb'(a, b)$ for some $s, t \in \mathbf{Z}$ using Theorem 1.8. Cancellation of (a, b) yields $1 = sa' + tb'$, and now Theorem 1.10 shows that $(a', b') = 1$. ■

To see why we need $(a, b) = 1$ in the first two parts of the theorem, note that $8 \mid 40$, that $40 = 4 \cdot 10$, but $8 \nmid 4$ and $8 \nmid 10$. Similarly, $8 \mid 40$ and $10 \mid 40$ but $80 \nmid 40$. Part ii) is also useful to show that products of distinct primes divide some expression. By Example 1.1, $5 \mid (t^5 - t)$ for all $t \in \mathbf{Z}$, and the same approach shows that $2 \mid (t^5 - t)$ and that

$3 \mid (t^5 - t)$. Since $(2, 5) = 1$ it follows from Theorem 1.12 that $10 \mid (t^5 - t)$, and now, since $(10, 3) = 1$, that $30 \mid (t^5 - t)$. Similarly, by Exercise 10 in §1, $a \mid x^2(x^2 - 1)(x^2 - 4)$ for all $x \in \mathbf{Z}$ if $a \in \{5, 8, 9\}$ so we may conclude that $360 \mid x^2(x^2 - 1)(x^2 - 4)$ since $360 = 5 \cdot 8 \cdot 9$, $(8, 9) = 1$, and $(72, 5) = 1$. Our final result on relative primeness will be useful after we discuss congruences in the next chapter.

THEOREM 1.13 Let $n \in N - \{1\}$ and $a \in \mathbf{Z}$ with $(a, n) = 1$.

i) For any $k \in \mathbf{Z}$, $(a + kn, n) = 1$.
ii) There is a unique $r \in N$ with $1 \le r < n$ and $(r, n) = 1$, so that $n \mid (ar - 1)$.

Proof Since $(a, n) = 1$ use Theorem 1.10 to write $1 = sa + tn$ for $s, t \in \mathbf{Z}$. Clearly this can be rewritten as $1 = s(a + kn) + (t - sk)n$, so $(a + kn, n) = 1$ by Theorem 1.10 again, proving i). Theorem 1.1 and Theorem 1.2 show that for each $1 \le i < n, n\mathbf{Z} + ia = n\mathbf{Z} + r$ for $0 \le r < n$, the remainder of ia upon division by n. Recall from Theorem 1.2 that $\{n\mathbf{Z} + k \mid 0 \le k < n\}$ are n different subsets of $\mathbf{Z}$. Should $n\mathbf{Z} + ia = n\mathbf{Z} + ja$ for $1 \le i, j < n$, then Theorem 1.2 forces $ia - ja \in n\mathbf{Z}$, or $n \mid (i - j)a$. Since $(a, n) = 1$, Theorem 1.12 shows that $n \mid (i - j)$, but now $1 \le i, j < n$ force $i = j$. Therefore the $n - 1$ subsets $n\mathbf{Z} + ia$ are distinct, and each is some $n\mathbf{Z} + k$ for $0 \le k < n$. If some $n\mathbf{Z} + ia = n\mathbf{Z}$, then as above $n \mid ia$, then $n \mid i$, a contradiction, so $n\mathbf{Z} + ia \ne n\mathbf{Z}$. Hence $\{n\mathbf{Z} + ia \mid 1 \le i \le n - 1\} = \{n\mathbf{Z} + k \mid 1 \le k \le n - 1\}$ and for exactly one $1 \le i < n, n\mathbf{Z} + ia = n\mathbf{Z} + 1$. Once again Theorem 1.2 gives $ia - 1 \in n\mathbf{Z}$, so $n \mid (ia - 1)$. Finally, $ia - 1 = nz$, or $1 = ai + (-z)n$, so $(i, n) = 1$ by Theorem 1.10. ∎

An important implication of Theorem 1.13 i) is that if $(a, n) = 1$ then every $a' \in n\mathbf{Z} + a$ is also relatively prime to n. But $n\mathbf{Z} + a = n\mathbf{Z} + a'$ exactly when $a' \in n\mathbf{Z} + a$ by Theorem 1.2, so whenever $n\mathbf{Z} + a' = n\mathbf{Z} + a$, $(a, n) = 1 \Leftrightarrow (a', n) = 1$. To find the $r \in N$ in the second part of the theorem, use the Euclidean Algorithm to write $ra + tn = 1$; when $r < 0$ rewrite this as $(r + kn)a + (t - ka)n = 1$ with $0 \le r + kn < n$. A special case is when n is odd and $a = 2$, in which case $2((n + 1)/2) - n = 1$, so $n \mid (2((n + 1)/2) - 1)$. Let us look at a couple of other examples. For $n = 15$ and $a = 4$ it is easy to see that $1 = 4 \cdot 4 - 15$, so $15 \mid (4 \cdot 4 - 1)$. When $n = 13$ and $a = 8$ the Euclidean Algorithm uses the divisions $13 = 8 + 5$, $8 = 5 + 3$, $5 = 3 + 2$, and $3 = 2 + 1$ to write $1 = 5 \cdot 8 - 3 \cdot 13$, so $13 \mid (5 \cdot 8 - 1)$. When $n = 55$ and $a = 16$, a complication arises. The successive divisions $55 = 3 \cdot 16 + 7$, $16 = 2 \cdot 7 + 2$, and $7 = 3 \cdot 2 + 1$ give $1 = 7 \cdot 55 - 24 \cdot 16 = (7 - 16) \cdot 55 + (-24 + 55) \cdot 16$. Hence $1 = -9 \cdot 55 + 31 \cdot 16$ and $55 \mid (31 \cdot 16 - 1)$.

EXERCISES

1. If $a, b, i, j \in N$, show that $(a, b) = 1 \Leftrightarrow (a^i, b^j) = 1$.
2. Let $a, b, c \in N$.
  i) If $a \mid b$ then $(b, c) = 1 \Rightarrow (a, c) = 1$.
  ii) $(ab, ac) = a(b, c)$
  iii) If $(a, b) = 1$ then $(a, bc) = (a, c)$.

3. If $n, k \in \boldsymbol{N}$ with $1 \leq k < n$ and $(n, k) = 1$, show that $n \mid \binom{n}{k}$.
4. Given $a, b \in \boldsymbol{N}$, use only the properties of division and the definition of $\mathrm{LCM}(\{a, b\})$ to show that $\mathrm{LCM}(\{a, b\}) = ab/(a, b)$.
5. Let $a, b \in \boldsymbol{Z}^*$ and suppose that $sa + tb = ua + vb$ for $s, t, u, v \in \boldsymbol{Z}$. Show $u = s - zb'$ and $v = t + za'$ for some $z \in \boldsymbol{Z}$, with $a = (a, b)a'$ and $b = (a, b)b'$.
6. Let $\{a_1, \ldots, a_k\} \subseteq \boldsymbol{N}$ be pairwise relatively prime. If $c \in \boldsymbol{N}$ and each $a_j \mid c$, show carefully that $a_1 a_2 \cdots a_k \mid c$.
7. **i)** Let $a, b \in \boldsymbol{N}$. If $d \in \boldsymbol{N}$ and $d \mid ab$, show that $d = ef$ where $e \mid a$ and $f \mid b$.
   ii) If $(a, b) = 1$, show that $e \mid a$, $e' \mid a$, $f \mid b$, $f' \mid b$, and $ef = e'f' \Rightarrow e = e'$ and $f = f'$.
   iii) If $\tau(n)$ denotes the number of positive divisors of $n \in \boldsymbol{N}$, use i) and ii) to show that for $a, b \in \boldsymbol{N}$ with $(a, b) = 1$, $\tau(ab) = \tau(a)\tau(b)$.
8. Let $(F_n) = (1, 1, 2, 3, 5, 8, 13, \ldots)$ be the Fibonacci sequence as in §0.3.
   i) Show that $(F_n, F_{n+1}) = 1$ for all $n \in \boldsymbol{N}$.
   ii) Using Example 0.25, Theorem 1.9 part i), and Exercise 2, show that $(F_k, F_m) = (F_k, F_{m-k})$ if $1 \leq k < m$.
   iii) Using part ii) and the idea behind the Euclidean Algorithm, show that for all $k, m \in \boldsymbol{N}$, $(F_k, F_m) = F_{(k,m)}$.
9. In each case, find $1 \leq r < b$ so that $b \mid (ar - 1)$:
   i) $a = 19$ and $b = 23$
   ii) $a = 113$ and $b = 126$
   iii) $a = 89$ and $b = 144$
   iv) $a = 16$ and $b = 101$
10. Note that $\{2, 3, 4, 5\} = \{2, 4\} \cup \{3, 5\}$ with both $7 \mid (2 \cdot 4 - 1)$ and $7 \mid (3 \cdot 5 - 1)$.
    i) Find a partition of $\{2, 3, \ldots, 11\}$ into five sets of two elements each so that for each subset $\{a, b\}$, $13 \mid (ab - 1)$: $\{2, 7\}$ is one such subset.
    ii) Find a partition of $\{2, 3, \ldots, 15\}$ into seven sets of two elements each so that for each subset $\{a, b\}$, $17 \mid (ab - 1)$.
    iii) If $p > 3$ is prime, show there is a partition of $\{2, 3, \ldots, p - 2\}$ into $(p - 3)/2$ sets of two elements each so that for each subset $\{a, b\}$, $p \mid (ab - 1)$.
    iv) Use iii) to show (Wilson's Theorem): if p is a prime, then $p \mid ((p - 1)! + 1)$.
11. If $n \in \boldsymbol{N} - \{1\}$, $a \in \boldsymbol{Z}$, and $(a, n) = 1$, show there is $1 \leq i < n$ with $n \mid (a^i - 1)$.
12. In each case find the minimal $m \in \boldsymbol{N}$ satisfying:
    i) $5 \mid (2^m - 1)$
    ii) $7 \mid (3^m - 1)$
    iii) $9 \mid (5^m - 1)$
    iv) $11 \mid (2^m - 1)$
    v) $21 \mid (2^m - 1)$
13. Let $R(n) \in \boldsymbol{N}$ have exactly n digits, all equal to 1. If $p > 5$ is a prime, show that $p \mid R(n)$ for infinitely many $n \in \boldsymbol{N}$. (Use Theorem 1.11 and Exercise 11.)
14. Let $5 < q_1 < \cdots < q_k$ be primes. Show that $q_1 q_2 \cdots q_k \mid R(n)$ (Exercise 13) for infinitely many $n \in \boldsymbol{N}$.
15. For $R(n)$ as in Exercise 13 and $m \in \boldsymbol{N}$, show that there are infinitely many $n \in \boldsymbol{N}$ so that $m \mid R(n)$ if: a) $(m, 30) = 1$; and b) $(m, 10) = 1$.

1.4 PRIME FACTORIZATION

We prove that every ingeter $n \geq 2$ is a product of primes, uniquely except for order, and observe some consequences of this. By Theorem 0.4 we need only show the uniqueness of the representation, and the key result needed is sometimes called Euclid's Lemma.

THEOREM 1.14 If p is a prime and $a_1, \ldots, a_n \in \mathbf{Z}$ so that $p \mid a_1 \cdots a_n$, then $p \mid a_i$ for some $1 \leq i \leq n$.

Proof We use induction on n. If $n = 1$ there is nothing to prove since $p \mid a_1$ implies that $p \mid a_1$. Assume the theorem holds when $p \mid a_1 \cdots a_k$ and let $p \mid a_1 \cdots a_{k+1}$. Set $a_1 \cdots a_k = x$. We have $p \mid xa_{k+1}$, and if $p \mid a_{k+1}$ then p divides one of $a_1, \ldots, a_{k+1}$. If p does not divide a_{k+1} then since p is a prime $(p, a_{k+1}) = 1$, so $p \mid xa_{k+1}$ and Theorem 1.12 force $p \mid x$. Therefore the induction assumption yields $p \mid a_i$ for some $1 \leq i \leq k$, proving the $k + 1$ case, and the theorem holds by IND-I. ■

COROLLARY If p is a prime, $a \in \mathbf{Z}$, and for some $n \in \mathbf{N}$, $p \mid a^n$, then $p \mid a$.

The key to proving the uniqueness of prime factorization in Theorem 0.4 is the division property for primes in Theorem 1.14. This characterizes primes in $\mathbf{N}$: if $p \neq 1$ and $p \mid ab$ for $a, b \in \mathbf{Z}$ forces $p \mid a$ or $p \mid b$, then p *must* be a prime. If not, then $p = ab$ with $a, b > 1$ and $p \mid ab$, but p fails to divide either factor. In general this divisibility condition for p is not the same as p having no proper factorization, as we see in the next example.

Example 1.7 If $R = \{a + b\sqrt{-19} \mid a, b \in \mathbf{Z}\} \subseteq \mathbf{C}$, the complex numbers, there is $\alpha \in R$ whose only factorizations in R are $\alpha = 1 \cdot \alpha$ or $\alpha = (-1)(-\alpha)$, but α fails to satisfy the divisibility property in Theorem 1.14.

The sum and product of elements in R are again in R. Define the norm $N(a + b\sqrt{-19}) = a^2 + 19b^2$. The following hold: $N(a + b\sqrt{-19}) = 1$ implies $b = 0$ and $a = \pm 1$; $N(a + b\sqrt{-19}) \neq 2, 5, 10$; and for any $x, y \in R$, $N(xy) = N(x)N(y)$ (why?). Now $N(1 + \sqrt{-19}) = 20$, so if $\alpha = 1 + \sqrt{-19} = xy$ in R, then $20 = N(x)N(y)$. We just observed that $N(x) \neq 2, 5, 10$, so one of $N(x)$ or $N(y)$ has to be 1, which means that α does not factor in R except trivially as $\alpha = (-1)(-\alpha)$. In R, $\alpha \mid 20$ since $N(\alpha) = \alpha(a - b\sqrt{-19}) = 20 = 4 \cdot 5$, but both $\alpha \mid 4$ and $\alpha \mid 5$ are impossible. If $4 = \alpha u$ for $u \in R$ then $16 = N(4) = N(\alpha)N(u) = 20N(u)$ for $N(u) \in \mathbf{Z}$, and similarly $5 \neq \alpha v$ for $v \in R$.

We turn to the uniqueness of prime factorization in $\mathbf{Z}$, called the *Fundamental Theorem of Arithmetic.*

THEOREM 1.15 For $n \in \mathbf{N} - \{1\}$, $n = p_1^{a_1} \cdots p_k^{a_k}$ for $k \geq 1$, $p_1 < \cdots < p_k$ primes, and $a_i \in \mathbf{N}$ for $1 \leq i \leq k$. Also k, $\{p_j\}$, and $(a_1, \ldots, a_k)$ are unique.

Proof Now $n = p_1^{a_1} \cdots p_k^{a_k}$ by Theorem 0.4 so we need uniqueness: if $n = p_1^{a_1} \cdots p_k^{a_k} = q_1^{b_1} \cdots q_t^{b_t}$ where $q_1 < \cdots < q_t$ are primes, we must show $k = t$ and for each i both $p_i = q_i$

and $a_i = b_i$. Use IND-II starting with $n = 2$. Since $p \geq 2$ for each prime p, 2 is neither another prime nor a product of primes so $2 = 2^1$ is its unique representation. Assume the theorem for all $2 \leq d < n$ but $n = p_1^{a_1} \cdots p_k^{a_k} = q_1^{b_1} \cdots q_t^{b_t}$. Now $p_1 \mid n$ so $p_1 \mid q_1^{b_1} \cdots q_t^{b_t}$ and Theorem 1.14 forces $p_1 \mid q_j$ for some $1 \leq j \leq t$. Thus $q_j = p_1 m$ and q_j prime force $p_1 = q_j$. Similarly $q_1 = p_i$ and comparing these yields $p_1 = q_j \geq q_1 = p_i$, so $i = 1$ and $p_1 = q_1$. Let $n = p_1 d$ so $d = p_1^{a_1 - 1} \cdots p_k^{a_k} = q_1^{b_1 - 1} \cdots q_t^{b_t}$. When all exponents appearing are positive, the induction assumption implies $k = t$, all $p_i = q_i$, and all $a_i = b_i$ as required. If $d = 1$ then $k = t = 1$ and $a_1 = b_1 = 1$ so we are finished again. When $a_1 - 1 = 0$ and p_2 is the smallest prime in $d = p_2^{a_2} \cdots p_k^{a_k}$ our argument shows that p_2 is the smallest prime in $q_1^{b_1 - 1} \cdots q_t^{b_t}$. Thus $b_1 - 1 = 0$ also, and again the induction assumption applies to d. Finally $b_1 = 1$ also forces $a_1 = 1$, and yet again the induction assumption shows the uniqueness of the representations for n. Thus the theorem holds by IND-II. ■

The previous two results give a useful approach to some problems by observing that if $(a, b) = d > 1$ then some prime p divides d and therefore $p \mid a$ and $p \mid b$. In Theorem 1.11 if $(a, n) = 1 = (b, n)$ but $(ab, n) = d > 1$, then for any prime divisor p of d, $p \mid n$ and $p \mid ab$. By Theorem 1.14 either $p \mid a$ or $p \mid b$ so from Theorem 1.8 on GCDs we obtain the contradiction $p \mid (a, n)$ or $p \mid (b, n)$.

A standard argument for the famous result of Euclid on the existence of infinitely many primes also depends on prime factorization.

THEOREM 1.16 There are infinitely many primes in $\boldsymbol{N}$.

Proof Assume that $p_1 < p_2 < \cdots < p_n$ are all of the primes, let $m = p_1 p_2 \cdots p_n + 1$, and use Theorem 1.15 to write m as a product of primes. Thus some $p_j \mid m$. But $m - 1 = p_1 p_2 \cdots p_n$ so $p_j \mid (m - 1)$ also. Since $1 = m - (m - 1)$, Theorem 1.6 forces $p_j \mid 1$, a contradiction. Therefore there must be infinitely many primes. ■

Next we use Theorem 1.15 to characterize powers in terms of prime factorization and obtain as a special case that $\sqrt{2}$ is not a rational number.

THEOREM 1.17 Let $n \in \boldsymbol{N} - \{1\}$ with prime factorization $n = p_1^{a_1} \cdots p_t^{a_t}$.

i) For $k \in \boldsymbol{N}$, $n = d^k$ for some $d \in \boldsymbol{N} \Leftrightarrow k \mid a_j$ for all $1 \leq j \leq t$.
ii) For $k \in \boldsymbol{N}$, $n^{1/k} \in \boldsymbol{Q} \Leftrightarrow n = d^k$ for some $d \in \boldsymbol{N}$.

Proof If $n = p_1^{a_1} \cdots p_t^{a_t}$ then $n^k = p_1^{ka_1} \cdots p_t^{ka_t}$, which must be the unique prime factorization of n^k by Theorem 1.15. Thus each exponent in the prime factorization of every k^{th} power in $\boldsymbol{N}$ is a multiple of k. On the other hand, if each $a_j = kb_j$ then $n = d^k$ for $d = p_1^{b_1} \cdots p_t^{b_t}$, proving i). Now if $n = d^k$ for $d \in \boldsymbol{N}$ then $n^{1/k} = d \in \boldsymbol{N} \subseteq \boldsymbol{Q}$. Finally if $n^{1/k} \in \boldsymbol{Q}$ write $n^{1/k} = c/b \in \boldsymbol{Q}$ with $c, b \in \boldsymbol{N}$ and $(c, b) = 1$. Note that, $bn^{1/k} = c$ so $b^k n = c^k$. Clearly $c = 1$ is not possible, and if $b = 1$ then $n = c^k$. Otherwise use Theorem 1.15 to write $c = q_1^{c_1} \cdots q_s^{c_s}$ and $b = w_1^{b_1} \cdots w_v^{b_v}$ where the sets of primes $\{q_j\}$ and $\{w_i\}$ are disjoint since $(c, b) = 1$. Thus $b^k n = c^k$ shows that $w_1^{kb_1} \cdots w_v^{kb_v} p_1^{a_1} \cdots p_t^{a_t} = q_1^{kc_1} \cdots q_s^{kc_s}$.

By Theorem 1.15 $\{w_i, p_i\} = \{q_j\}$, but no w_i appears on the right side of the equality, a contradiction. Hence $b = 1$, $n = c^k$, and ii) holds. ∎

COROLLARY If $n \in N$ is not a square, so $n \neq d^2$ for all $d \in N$, $n^{1/2} \notin Q$.

Two specific cases of Theorem 1.17 are that $n \in N$ is a square exactly if all exponents of its prime factors are even, and n is a cube exactly when all these exponents are multiples of 3. The idea in the proof of Theorem 1.17 leads to a characterization of the divisors of $n \in N$. First it will be helpful to change the way we denote prime factorization by allowing zero to be an exponent. The proof of Theorem 1.17 uses different symbols for the primes in each of n, b, and c since different sets of primes may appear in each integer. If we use zero as an exponent, then for any finite subset $S \subseteq N$ we can find a set of primes to represent all the integers in S: use all those primes dividing some element of S. For example, we may write $15 = 2^0 \cdot 3 \cdot 5$, $18 = 2 \cdot 3^2 \cdot 5^0$, $30 = 2 \cdot 3 \cdot 5$, $40 = 2^3 \cdot 3^0 \cdot 5$, and $600 = 2^3 \cdot 3 \cdot 5^2$.

THEOREM 1.18 Let $n = p_1^{a_1} \cdots p_k^{a_k}$ for $p_1 < \cdots < p_k$ primes and all $a_i \in N$.

i) If $d \in N$ then $d \mid n \Leftrightarrow d = p_1^{b_1} \cdots p_k^{b_k}$ with all $0 \leq b_i \leq a_i$ for all $i \leq k$.
ii) n has exactly $(a_1 + 1)(a_2 + 1) \cdots (a_k + 1)$ distinct divisors in N.

Proof When $d = p_1^{b_1} \cdots p_k^{b_k}$ with all $0 \leq b_i \leq a_i$ then if $c = p_1^{a_1 - b_1} \cdots p_k^{a_k - b_k}$, $n = dc$ and it follows that $d \mid n$. On the other hand if $d \mid n$, so $n = cd$, then by the uniqueness in Theorem 1.15, each prime factor of c and of d must be one of the p_i. Consequently $d = p_1^{b_1} \cdots p_k^{b_k}$ and $c = p_1^{c_1} \cdots p_k^{c_k}$ with $0 \leq b_i, c_i$ for each i, so $cd = p_1^{c_1 + b_1} \cdots p_k^{c_k + b_k} = n$, and using Theorem 1.15 again yields $b_j + c_j = a_j$ for each j. But each $c_j \geq 0$ so $b_j \leq a_j$ as claimed. Since the divisors of n in N are the $p_1^{b_1} \cdots p_k^{b_k}$ with all $0 \leq b_i \leq a_i$ for all $i \leq k$, clearly they correspond to choices of exponents between 0 and a_i for each p_i. Each choice gives a different divisor by the uniqueness of prime factorization in Theorem 1.15, so the number of divisors in N is just $|\{0, 1, \ldots, a_1\} \times \cdots \times \{0, 1, \ldots, a_k\}|$. This is the product $(a_1 + 1) \cdots (a_k + 1)$ by Theorem 0.3. ∎

Using Theorem 1.18 we can study integers with a fixed and special number of divisors. For example, if $n = p_1^{a_1} \cdots p_k^{a_k}$ has a prime number q of divisors then $q = (a_1 + 1) \cdots (a_k + 1)$, so $k = 1$ and $a_1 = q - 1$; thus $n = p^{q-1}$ for some prime p. Also, if n has exactly six divisors then $6 = (a_1 + 1) \cdots (a_k + 1)$ so either $k = 1$ and $a_1 = 5$, or $k = 2$ and $\{a_1, a_2\} = \{1, 2\}$. Hence either $n = p^5$ for some prime p, or there are different primes p and q such that $n = pq^2$.

The product $(a_1 + 1) \cdots (a_k + 1)$ is odd only if every a_i is even. Therefore $n = p_1^{a_1} \cdots p_k^{a_k}$ has an odd number of divisors in N if and only if n is a square. Another way to see this is to observe that the divisors of n come in pairs, namely d and n/d, and $d = n/d$ only if $d = \sqrt{n}$. Hence if n is a square, its divisors come in pairs except for $\sqrt{n}$. This observation is used in the exercises to solve a standard and interesting problem about switches; its solution is not otherwise very obvious.

Suppose that $p_1 < \cdots < p_k$ are primes, $S \subseteq N$ is finite, and each $s \in S$ can be written $s = p_1^{s_1} \cdots p_k^{s_k}$ with each $s_i \geq 0$. By Theorem 1.18 a common divisor for S must be expressed as $c = p_1^{c_1} \cdots p_k^{c_k}$ with each $c_i \leq s_i$ for every $s \in S$, and any such $c \in \boldsymbol{N}$ will be a common divisor for S. Since GCD(S) is the largest common divisor, it must have each exponent c_i as large as possible and satisfying $c_i \leq s_i$ for every $s \in S$. Hence for each i, c_i is the smallest exponent of p_i appearing in the prime factorizations of the elements of S. A similar analysis shows that to obtain LCM(S) we take the largest exponent of each p_i appearing in all elements of S. This is stated as our next theorem, whose formal proof is left as an exercise.

THEOREM 1.19 Let $S = \{s_1, \ldots, s_n\} \subseteq \boldsymbol{N}$, $p_1 < \cdots < p_k$ primes, and assume that for each $s_j \in S$, $s_j = p_1^{a_{j1}} \cdots p_k^{a_{jk}}$ with each $a_{ji} \geq 0$. For each $1 \leq i \leq k$, set $m_i = \min\{a_{1i}, \ldots, a_{ni}\}$ and $M_i = \max\{a_{1i}, \ldots, a_{ni}\}$. Then $\text{GCD}(S) = p_1^{m_1} \cdots p_k^{m_k}$ and $\text{LCM}(S) = p_1^{M_1} \cdots p_k^{M_k}$.

COROLLARY For $a, b \in \boldsymbol{N}$, $\text{GCD}(\{a, b\})\text{LCM}(\{a, b\}) = ab$, so $\text{LCM}(\{a, b\}) = ab/(a, b)$ and $\text{LCM}(\{a, b\}) = ab$ exactly when $(a, b) = 1$.

Proof Let $a = p_1^{a_1} \cdots p_k^{a_k}$ and $b = p_1^{b_1} \cdots p_k^{b_k}$ so that for each i, $\{\max\{a_i, b_i\}, \min\{a_i, b_i\}\} = \{M_i, m_i\} = \{a_i, b_i\}$. Then $ab = p_1^{a_1+b_1} \cdots p_k^{a_k+b_k} = p_1^{m_1+M_1} \cdots p_k^{m_k+M_k} = p_1^{m_1} \cdots p_k^{m_k} p_1^{M_1} \cdots p_k^{M_k} = \text{GCD}(\{a, b\})\text{LCM}(\{a, b\})$. ∎

To illustrate Theorem 1.19 let $S = \{20, 32, 45\}$. Write $20 = 2^2 \cdot 3^0 \cdot 5, 32 = 2^5 \cdot 3^0 \cdot 5^0$, and $45 = 2^0 \cdot 3^2 \cdot 5$, so $\text{GCD}(S) = 2^0 \cdot 3^0 \cdot 5^0 = 1$ uses the minimal power of each prime appearing in the elements of S and $\text{LCM}(S) = 2^5 \cdot 3^2 \cdot 5 = 1440$ uses the maximal power. For $T = \{3^3 \cdot 7, 3 \cdot 5^2 \cdot 7^2, 2^5 \cdot 5^3 \cdot 7^2, 2 \cdot 3^4 \cdot 5 \cdot 7\}$, using the minimal and maximal exponent of each prime appearing in the elements of T yields $\text{GCD}(T) = 2^0 \cdot 3^0 \cdot 5^0 \cdot 7 = 7$ and $\text{LCM}(T) = 2^5 \cdot 3^4 \cdot 5^3 \cdot 7^2$.

It might seem unnecessary now to use the Euclidean Algorithm to find the greatest common divisor of two integers a and b since GCD($\{a, b\}$) can be read off from the prime factorizations of a and of b. However, the prime factorizations do not give the useful expression of (a, b) as an integer combination of a and b. A more fundamental problem is that finding a prime factorization is in general extraordinarily difficult. We are used to looking at integers that are really quite small, on the order of a few thousand at the most, and often it is not difficult to factor these. It is easy to factor 992 by taking out factors of 2 or to factor 945 by dividing by 9, but it requires a little work to see that 317 is prime or to factor 2867. Of course a computer can factor such numbers very easily, but even the most powerful computers likely to be developed will probably not be able to factor randomly given integers with millions, trillions, or more digits. Even these numbers are not really large compared to others in $\boldsymbol{N}$. By using the Euclidean Algorithm, a sophisticated computer could easily find the greatest common divisor of two integers each having a huge number of digits.

As a final application of factorization, consider the following question: if $n \in \boldsymbol{N}$ and $p < n$ is prime what power of p divides $n!$? Suppose that $m, n \in \boldsymbol{N}$ with $m \leq n$ and use Theorem 1.1 to write $n = tm + r$ for $t, r \in \boldsymbol{Z}$ with $0 \leq r < m$. Since $m \leq n$, t counts the number of multiples of m not exceeding n. Also $n/m = t + r/m$ and $0 \leq r/m < 1$ so $t = [n/m]$, the greatest integer less than or equal to n/m. In particular, for a prime p,

$[n/p^a]$ is the number of multiples of p^a that are at most n. The factors of p in $n!$ come from multiples $p^a \mid k$ for some $1 \le k \le n$. Using these ideas gives:

THEOREM 1.20 If $n \in N$, p is a prime, and $n! = p^s m$ with $(p, m) = 1$, then $s = [n/p] + [n/p^2] + \cdots$.

Proof The sum $[n/p] + [n/p^2] + \cdots$ is finite since eventually $n < p^k$ so then $[n/p^k] = 0$. For $1 \le v \le n$ and $p \mid v$, write $v = p^{a(v)} t_v$ with $(p, t_v) = 1$. Then $s = \Sigma a(v)$ by Theorem 1.15. As we saw above, $[n/p^j]$ counts the number of multiples of p^j that do not exceed n. Hence if $v = p^{a(v)} t_v \le n$ then v is counted $a(v)$ times in $[n/p] + [n/p^2] + \cdots$: once as a multiple of p, once as a multiple of $p^2, \ldots,$ and once as a multiple of $p^{a(v)}$. Therefore $[n/p] + [n/p^2] + \cdots$ is the sum over all $a(v)$, which is just s. ∎

Let us find the number of times that 7 divides 4000!. We just add $[4000/7] + [4000/7^2] + \cdots = 571 + 81 + 11 + 1 = 664$ since $7^5 > 4000$, so $4000! = 7^{664} m$ with $(7, m) = 1$. It follows that $21^{664} \mid 4000!$ since 3 divides 4000! at least as many times as 7 does. To see this from the theorem, observe that $[4000/7^k] \le [4000/3^k]$. Since $n!$ ends with k zeros when 10^k is the largest power of 10 dividing it, to find k it suffices to find the largest power of 5 dividing $n!$ (why?). For example, $[135/5] + [135/25] + \cdots = 27 + 5 + 1 = 33$, so 135! ends with 33 zeros.

EXERCISES

1. For $n \in N$, $n = p_1^{a_1} \cdots p_k^{a_k}$, $p_1 < \cdots < p_k$ primes, $k > 1$, and all $a_j \ge 1$, if $n_j = n/p_j^{a_j} = \prod_{i \ne j} p_i^{a_i}$ show that $1 = y_1 n_1 + \cdots + y_k n_k$ for $y_1, \ldots, y_k \in Z$. Thus if $n = 90 = 2 \cdot 3^2 \cdot 5$, then $n_1 = 45$, $n_2 = 10$, $n_3 = 18$, and $1 = (-1)45 + 1 \cdot 10 + 2 \cdot 18$; and if $n = 2 \cdot 3 \cdot 5 \cdot 7$, then $n_1 = 105$, $n_2 = 70$, $n_3 = 42$, $n_4 = 30$, and we can write $1 = (-5)105 + 4 \cdot 70 + 3 \cdot 42 + 4 \cdot 30$.

2. **i)** Show that if $a, b \in nZ + 1$ for some $n \in N$, then $ab \in nZ + 1$.
ii) If $a \in 4N - 1$ show that a must have a prime factor in $4N - 1$.
iii) Show there are infinitely many primes in $4N + 3$ (see Theorem 1.16).

3. **i)** If $p > 3$ is a prime, show that $p \in (6N + 1) \cup (6N + 5)$.
ii) If $a \in 6N - 1$, show that a must have a prime factor in $6N - 1$.
iii) Show that $6N + 5$ must contain an infinite number of primes.

4. If $n \in N - \{1\}$ show that n is prime or there is a prime factor p of n with $p \le \sqrt{n}$.

5. If $n, a, b \in N$, $n^2 = ab$, and $(a, b) = 1$, show that $a = s^2$ and $b = t^2$ for $s, t \in N$.

6. Given $a, b \in N$ with $a^2 = b^3$, show that for some $y \in N$, $a = y^3$ and $b = y^2$.

7. Find all $n \in N$ so that whenever $a, b \in N$ with $a \mid n$ and $b \mid n$ then either $a \mid b$ or $b \mid a$; that is, given any two positive divisors of n, one divides the other.

8. Any $n \in N$ is called **square free** if $d \in N$ with $d^2 \mid n$ implies $d = 1$. Show that $n > 1$ is square free if and only if $n = p_1 p_2 \cdots p_k$ for primes $p_1 < \cdots < p_k$.

9. How many positive divisors does a square free $n \in N$ (Exercise 8) have?

10. If $n \in N$ is written $n = \prod_{i=1}^{k} p_i^{a_i}$ for $p_1 < \cdots < p_k$ primes, let $\mathrm{rad}(n) = p_1 p_2 \cdots p_k$. Show that $\mathrm{rad}(a) = \mathrm{rad}(b) \Leftrightarrow$ there are $s, t \in N$ so that both $a \mid b^s$ and $b \mid a^t$.

11. For $n \in N$ and $n > 1$, show that $n = p_1 p_2 \cdots p_k$ for primes $p_1 < \cdots < p_k$ if and only if whenever $n \mid a^t$ for $a, t \in N$ then $n \mid a$.

12. For $a_1, \ldots, a_k \in N$, $c = \mathrm{LCM}(\{a_j \mid 1 \le i \le k\})$, and $d = \mathrm{GCD}(\{a_j \mid 1 \le i \le k\})$, show that $c = d \cdot \mathrm{LCM}(\{c/a_j \mid 1 \le i \le k\})$. (Write the a_j as products of primes).

13. Find the number of positive integer divisors of $n \in N$:
 i) when $n = 900$;
 ii) when $n = 1{,}000{,}000$;
 iii) when $n = 10!$.

14. Which positive integers $n < 100$ have the greatest number of divisors in N?

15. For p a prime and $n, i \in N$ show that $[n/p^{i+1}] = [[n/p^i] \mid p]$.

16. **i)** How many zeros end 1111!?
 ii) What power of 30 divides 1111!?

17. **a)** Show by example that Theorem 1.20 is false if p is not prime.
 b) When p is a prime and $k \in N$, how many times does p divide $p^k!$?

18. A museum has a model control room with panels containing switches numbered in order from 1 to n. Over time, n children visit the room and each in turn flips some of the switches. All switches start in the off position, and the first child flips all n switches to the "on" position. The second child flips every other switch starting with the second and so flips the second, fourth, sixth, etc., to the "off" position. The third child flips every third switch starting with switch 3, and in general the jth child flips every j^{th} switch, starting with switch j. After the nth child visits, which switches are in the "on" position?

19. **i)** If $n = \prod_{i=1}^{k} p_i^{a_i}$ for $p_1 < \cdots < p_k$ primes, show that the sum of all the positive divisors of n is $\sigma(n) = \prod_{i=1}^{k} \dfrac{p_i^{a_i+1} - 1}{p_i - 1}$.
 ii) Show that $\sigma : N \to N$ is not injective.
 iii) If $a, b \in N$ with $(a, b) = 1$, show that $\sigma(ab) = \sigma(a)\sigma(b)$.

1.5 RELATIONS

We use Cartesian products to formalize the fundamental and crucial idea of how elements in one set are related to elements in another. A special case of this in the last chapter was the notion of function. Although there are rather complicated relations, the general notion is rather simple.

DEFINITION 1.7 Any $W \subseteq S \times T$ is a **relation** from the set S to the set T. When $S = T$ we say that W is a **relation on** S.

If $W \subseteq S \times T$ then $s \in S$ is related to $t \in T$ by W exactly when $(s, t) \in W$ and we write sWt: related elements are those that form a pair in W. Often other symbols such as $\le$, $\sim$, or $*$ are used for a relation, and in these cases when $s \in S$ is related to $t \in T$ we write $s \le t$, $s \sim t$, or $s * t$, respectively. At the extremes, $\varnothing \subseteq S \times T$ relates no element of S to any element of T, and $S \times T$ itself relates every element of S to every element of T. These are not particularly interesting relations. If we take $W = \{(1, 4), (1, 5), (3, 5), (4, 6)\} \subseteq \{1, 2, 3, 4\} \times \{4, 5, 6, 7\}$ then W relates 1 to 4 and to 5, 3 to 5, 4 to 6, and 2 to nothing in $\{4, 5, 6, 7\}$. No element in $\{1, 2, 3, 4\}$ is related to 7 by W.

We can define $W \subseteq S \times T$ relating any particular $s \in S$ to each element in any $B \subseteq T$ by taking W to be the corresponding pairs in $S \times T$. However, the relations of interest arise naturally and are not usually given as ordered pairs. Viewing relations as ordered pairs makes this notion precise but the idea behind the relation is *usually* more important than the formalism. Some familiar relations are "less than" on $\boldsymbol{R}$, $<=\{(x, y) \in \boldsymbol{R}^2 \mid x < y\}$; set inclusion on $P(S)$, $\subseteq = \{(A, B) \in P(S)^2 \mid A \subseteq B\}$; and relating a subset of S to its elements, $\ni = \{(A, y) \in P(S) \times S \mid y \in A\}$. Most elements in the first of these sets are related to many elements in the second. In the last example, $\varnothing \in P(S)$ is not related to anything in S, and $S \in P(S)$ is related to every element in S. The solutions $(x, y) \in \boldsymbol{R}^2$ of an equation or inequality in two variables naturally define a relation. Some examples are: $\{(x, y) \in \boldsymbol{R}^2 \mid x^2 + y^2 = 1\}$, the unit circle about the origin; $\{(x, y) \in \boldsymbol{R}^2 \mid x^2 - y = 2\}$, a parabola; and $\{(x, y) \in \boldsymbol{R}^2 \mid xy \geq 0\}$, the first and third quadrants of the plane.

Equality relates each element to itself only, so is written $\{(a, b) \in S \times S \mid a = b\}$, or as $\{(s, s) \in S \times S \mid s \in S\} = D_S$, the **diagonal** of S^2.

DEFINITION 1.8 If $W \subseteq S \times S$ then:

i) W is **reflexive** if for every $s \in S$, $(s, s) \in W$: that is, if sWs for all $s \in S$.
ii) W is **symmetric** if $(s, s') \in W \Rightarrow (s', s) \in W$: that is, $sWs' \Rightarrow s'Ws$.
iii) W is **anti-symmetric** if both $(s, s'), (s', s) \in W \Rightarrow s = s'$: that is, if both sWs' and $s'Ws \Rightarrow s = s'$.
iv) W is **transitive** if both $(s, s'), (s', s'') \in W \Rightarrow (s, s'') \in W$: that is, sWs' and $s'Ws'' \Rightarrow sWs''$.

The reflexive property requires each element to be related to something, namely itself. Thus W on S is reflexive when the diagonal $D_S \subseteq W$. The other properties are implications so Definition 0.2 comes into play. For example, when $S \neq \varnothing$, $\varnothing \subseteq S \times S$ is symmetric, anti-symmetric, and transitive, because the hypothesis of each implication defining these terms is false. Specifically $(s, s') \in \varnothing$ is false so all of the implications $(s, s') \in \varnothing \Rightarrow (s', s) \in \varnothing$; $(s, s'), (s', s) \in \varnothing \Rightarrow s = s'$; and $(s, s'), (s', s'') \in \varnothing \Rightarrow (s, s'') \in \varnothing$ are true. However $\varnothing$ cannot be reflexive on S since $D_S \not\subseteq \varnothing$. The relation D_S of equality on $S \neq \varnothing$ satisfies all four properties above. Now $D_S \subseteq D_S$ is reflexive. If $(s, s') \in D_S$ then $s = s'$, so $(s', s) \in D_S$ and D_S is symmetric. If $(s, s'), (s', s) \in D_S$ then $s = s'$ and D_S is anti-symmetric. Finally, if both $(s, s'), (s', s'') \in D_S$ then $s = s'$ so $(s, s'') = (s', s'') \in D_S$ and D_S is transitive.

Next consider "divides" on $\boldsymbol{Z}$ given by $Div = \{(a, b) \in \boldsymbol{Z}^2 \mid a \mid b\}$. Div is nearly reflexive but $(0, 0) \notin Div$ by Definition 1.1. Although $2 \mid 4$, $4 \nmid 2$ so Div is not symmetric, and both $2 \mid -2$ and $-2 \mid 2$ so Div is not anti-symmetric. Finally, if $a \mid b$ and $b \mid c$, then $a \mid c$ does hold by Theorem 1.5, so Div is transitive.

A relation on a set S that is reflexive, anti-symmetric, and transitive is called a **partial order** on S and we say that S is a partially ordered set, or a **poset**. This notion is quite important in mathematics but not one that will concern us much here. The natural ordering $\leq$ on the real numbers is easily seen to be a partial order. Two other standard examples left to the exercises are set inclusion on $P(S)$ and divisibility on $\boldsymbol{N}$. A partial order # on $\boldsymbol{R}^2$ is given by $(a, b) \# (c, d)$ if $(a, b) = (c, d)$ or if $a^2 + b^2 < c^2 + d^2$. Thus a $P \in \boldsymbol{R}^2$ is related to $P' \neq P$ if P is closer to the origin than P', and different points at the same distance from $(0, 0)$ are not related.

DEFINITION 1.9 If $\leq$ is a partial order on S then $T \subseteq S$ is a **chain** if for any $t, t' \in T$ either $t \leq t'$ or $t' \leq t$. S is **linearly ordered** if S itself is a chain.

In a chain every two elements are comparable in the ordering. $\boldsymbol{R}$ with the usual $\leq$ is linearly ordered. For "$\subseteq$" on a set S, nonempty disjoint subsets of S are noncomparable, and for division on $\boldsymbol{N}$, different primes are not comparable. It is an exercise to see that every subset of a poset is a poset using the same ordering, and every subset of a chain is a chain. Thus any collection of subsets of S is partially ordered by inclusion, and any subset of $\boldsymbol{R}$ is linearly ordered by the usual $\leq$. To get a chain in $\boldsymbol{R}^2$ using $(a, b) \# (c, d)$ if $(a, b) = (c, d)$ or $a^2 + b^2 < c^2 + d^2$, we need a set with each point on a different circle centered at the origin. Examples are points on any straight line segment lying inside the first quadrant. Finally, the set of divisors in $\boldsymbol{N}$ of a fixed $n > 1$ will be a chain with respect to division only if $n = p^k$ for a prime p (why?). Other chains in $\boldsymbol{N}$ with division as the partial order would be $\{n! \mid n \in \boldsymbol{N}\}$, or, for some fixed $m \in \boldsymbol{N}$, $\{m^k \mid k \in \boldsymbol{N}\}$.

EXERCISES

1. Describe the following relations on $\boldsymbol{R}$ geometrically as subsets of $\boldsymbol{R}^2$:
   i) $\{(x, y) \in \boldsymbol{R}^2 \mid |y| = x^2\}$
   ii) $\{(x, y) \in \boldsymbol{R}^2 \mid |x| + |y| \leq 1\}$
   iii) $\{(x, y) \in \boldsymbol{R}^2 \mid |x| \neq |y|\}$
   iv) $\{(x, y) \in \boldsymbol{R}^2 \mid |y| < |x|^{1/3}\}$
   v) $\{(x, y) \in \boldsymbol{R}^2 \mid [x] < [y]\}$.
2. Describe algebraically the relations on $\boldsymbol{R}$ given by:
   i) All points inside the circle of radius 3 centered at $(0, 0)$ and on or outside the circle of radius 1 with center $(0, 0)$.
   ii) The union of all straight lines with integer slopes passing through $(0, 0)$.
   iii) All points in the band bordered by the lines $x - y = 1$ and $x - y = 3$.
   iv) The union of all circles with center at $(0, 0)$ and having a rational radius.
3. Give three examples showing that a relation can satisfy any one of reflexivity, symmetry, or transitivity without satisfying the other two.
4. Verify the partial orderings: i) $\mid$ (divides) on $\boldsymbol{N}$; and ii) $\subseteq$ on $P(S)$.
5. Define $\leq_k$ on $\boldsymbol{R}^k$, the Cartesian product of $\boldsymbol{R}$ with itself k times, by setting $(a_1, \ldots, a_k) \leq_k (b_1, \ldots, b_k)$ if for each i, $a_i \leq b_i$. Show that $\leq_k$ is a partial order.
6. Define # on $\boldsymbol{N} \times \boldsymbol{N}$ by $(a, b) \# (c, d)$ if $a + b = c + d$ and $a \leq c$.
   i) Show # is a partial order.
   ii) Find an infinite subset of noncomparable elements in $\boldsymbol{N} \times \boldsymbol{N}$ with respect to #.
   iii) Show that using #, every chain in $\boldsymbol{N} \times \boldsymbol{N}$ is finite.
7. Define $*$ on $\boldsymbol{N} \times \boldsymbol{N}$ by $(a, b) * (c, d)$ if $a + b < c + d$, or $a + b = c + d$ and $a \leq c$.
   a) Show $*$ is a partial order and $\boldsymbol{N} \times \boldsymbol{N}$ is linearly ordered with respect to $*$.
   b) Using $*$, does every nonempty subset of $\boldsymbol{N} \times \boldsymbol{N}$ have a minimal element?
8. Define $\sim$ on $\boldsymbol{N}^k$ by $(a_1, \ldots, a_k) \sim (b_1, \ldots, b_k)$ if for each i, $a_i \mid b_i$. Show that $\sim$ is a partial order. For each $(b_1, \ldots, b_k) \in \boldsymbol{N}^k$, show that there are only finitely many $(a_1, \ldots, a_k) \in \boldsymbol{N}^k$ satisfying $(a_1, \ldots, a_k) \sim (b_1, \ldots, b_k)$.

9. Let the **lexicographic order** on N^n be $(a_1, \ldots, a_n) \leq (b_1, \ldots, b_n)$ if either $a_1 < b_1$; if for $1 \leq k < n$, $a_i = b_i$ for $i \leq k$ and $a_{k+1} < b_{k+1}$; or if $a_i = b_i$ for all i.
Show that:
  i) $\leq$ is a partial order
  ii) N^n with partial order $\leq$ is a chain
  iii) With respect to $\leq$, every nonempty subset of N^n has a minimal element.

10. For $q \in Q^+ = \{q \in Q \mid q > 0\}$ write $q = a/b$ for $a, b \in N$ and use prime factorization to write $q = p_1^{q(1)} p_2^{q(2)} \cdots p_k^{q(k)}$ where p_i is the ith prime and each $q(i) \in Z$. For example, $14/9 = 2^1 \cdot 3^{-2} \cdot 5^0 \cdot 7^1 = p_1^1 p_2^{-2} p_3^0 p_4^1 \cdots p_k^0$ for any $k > 4$. Define $\leq'$ on Q^+ by $u \leq' v$ when either $u = v$ or $u(2) < v(2)$.
  a) Show that $\leq'$ is partial order on Q^+.
  b) Find an infinite set of noncomparable elements in Q^+ with respect to $\leq'$.
  c) Show that any chain $C \subseteq Q^+$ is contained in a maximal chain M: $C \subseteq M$, a chain, and $M \subseteq W \subseteq Q^+$ with W a chain $\Rightarrow M = W$.

11. Let S be a nonempty set with partial order θ and $T \subseteq S$. Show $(T \times T) \cap \theta$ is a partial order on T. Further, if S is a chain, show that T is a chain.

12. Describe a linear order on; a) R^2; b) R^k for any $k > 2$; and c) $R[x]$.

13. Let $\leq_d$ be the partial order on N via $a \leq_d b$ when $a \mid b$. Find $S \subseteq N$ so that all chains in S are finite and for *any* $k \in N$ there is a chain C_k in S having exactly k elements and not properly contained in any other chain in S.

14. Let W be the set of all partial orders on a nonempty set S. The elements of W are subsets of $S \times S$ so W itself is partially ordered by set inclusion.
  a) If C is a chain in W under set inclusion, show $\bigcup_{c \in C} c$ is a partial order on S.
  b) If $w \in W$ is not a chain, show there is $w' \in W$ with $w \subseteq w'$ and $w \neq w'$.

1.6 EQUIVALENCE RELATIONS

Equivalence relations are very important and arise naturally when identifying elements that share some aspect of a property under study. A simple property for $P \in R^2$ is its distance to the origin; related points are those at the same distance as P from the origin. All mutually related points form a circle about the origin, and the totality of these circles is a partition of R^2. For any equivalence relation on any set S, the subsets of related elements form a partition V of S. When S is finite, $|V| \leq |S|$ so induction can sometimes be used to draw conclusions about V that can help in studying S. This is a vague description of a powerful technique. We will use it often but some preliminaries are needed, including the definition of an equivalence relation!

DEFINITION 1.10 A relation $\sim$ on a nonempty set S is an **equivalence relation** if $\sim$ is reflexive, symmetric, and transitive.

We saw above that $D_s \subseteq S^2$, the relation of equality, satisfies these properties so is an equivalence relation. Every equivalence relation $\sim$ on S is reflexive so $D_s \subseteq \sim$, and $\sim \subseteq S \times S$, the largest equivalence relation on S, relating every two elements and describing the property of belonging to S. For T the set of all triangles in the plane, it is easy to see that the usual relation of similarity is an equivalence relation on T. Similarity identifies triangles having the same set of angle measures. Other equivalence relations on T are congruence, identifying triangles that have both the same set of angle measures and the same set of side lengths; and "perimeter," identifying two triangles when the sums of the lengths of their sides are equal. We proceed with several more concrete examples.

Example 1.8 For $(a, b), (c, d) \in \boldsymbol{R}^2$, set $(a, b) \sim (c, d)$ if $a^2 + b^2 = c^2 + d^2$. Then $\sim$ is an equivalence relation on $\boldsymbol{R}^2$.

To see that $\sim$ is reflexive we need $(a, b) \sim (a, b)$ for any $(a, b) \in \boldsymbol{R}^2$ so take any $(a, b) \in \boldsymbol{R}^2$ and note that $a^2 + b^2 = a^2 + b^2$. Thus $(a, b) \sim (a, b)$ is true. If $(a, b) \sim (c, d)$ then $a^2 + b^2 = c^2 + d^2$, so $c^2 + d^2 = a^2 + b^2$ implying $(c, d) \sim (a, b)$ and $\sim$ is symmetric. For transitivity assume $(a, b) \sim (c, d)$ and $(c, d) \sim (e, f)$, so $a^2 + b^2 = c^2 + d^2$ and $c^2 + d^2 = e^2 + f^2$. It follows that $a^2 + b^2 = e^2 + f^2$ resulting in $(a, b) \sim (e, f)$. This proves that $\sim$ is an equivalence relation on $\boldsymbol{R}^2$. Elements of $\boldsymbol{R}^2$ are related by $\sim$ if their distances from (0, 0) are equal, so this distance property is what the relation identifies.

Example 1.9 $\# = \{(a, b) \in \boldsymbol{C}^2 \mid a^4 = b^4\}$ is an equivalence relation on $\boldsymbol{C}$.

Recall that $\boldsymbol{C} = \{u + vi \mid u, v \in \boldsymbol{R}\}$ with $i^2 = -1$. For any $a \in \boldsymbol{C}$, $a^4 = a^4$ so $a \# a$ and # is reflexive. If $a \# b$ then $a^4 = b^4$. Clearly $b^4 = a^4$ so $b \# a$, proving # is symmetric. Lastly, if $a \# b$ and $b \# c$ we need $a \# c$. But $a^4 = b^4$ and $b^4 = c^4$ yield $a^4 = c^4$, so $a \# c$, # is transitive, and # is an equivalence relation. When $a \# b$ then $a = b = 0$ or $(a/b)^4 = 1$. The solutions of $x^4 = 1$ in $\boldsymbol{C}$ are the powers of i, namely $i, -1, -i$, and 1, since these *are* four roots in $\boldsymbol{C}$ of $x^4 - 1$, so the only roots. Thus $a \# b \Rightarrow a = i^k b \in \{b, ib, -b, -ib\} = \{a, ia, -a, -ia\}$ (why?) and # identifies elements that are some i^k times each other.

Example 1.10 $\approx = \{(a, b) \in \boldsymbol{R}^2 \mid a - b \in \boldsymbol{Q}\}$ is an equivalence relation.

For all $a \in \boldsymbol{R}$, $a - a = 0 \in \boldsymbol{Q}$, so $a \approx a$ and $\approx$ is reflexive. When $a \approx b$ then $a - b \in \boldsymbol{Q}$, and to show $b \approx a$, note that $b - a = -(a - b) \in \boldsymbol{Q}$. Thus $\approx$ is symmetric. If $a \approx b$ and $b \approx c$ then $a - b, b - c \in \boldsymbol{Q}$, adding yields $a - c \in \boldsymbol{Q}$, so $a \approx c$ results. Thus $\approx$ is transitive and therefore an equivalence relation on $\boldsymbol{R}$. All rational numbers are related but it is not easy to tell if two arbitrary numbers are related. Since $a \approx b \Leftrightarrow a - b \in \boldsymbol{Q}$, for any fixed $r \in \boldsymbol{R}$, $\{r + q \mid q \in \boldsymbol{Q}\}$ are all related. For example, $\pi + 3 \approx \pi - 3/4$, although it is difficult to visualize all of $\{\pi + q \mid q \in \boldsymbol{Q}\} \subseteq \boldsymbol{R}$.

Example 1.11 For $a, b \in \boldsymbol{Z}^* = \boldsymbol{Z} - \{0\}$, the relation $a \sim b$ when $b = 2^i a$ for some $i \in \boldsymbol{Z}$ is an equivalence relation on $\boldsymbol{Z}^*$.

This condition is a bit strange as we may be multiplying integers by fractions. As examples of $\sim$, $32 \sim 1$ since $1 = 2^{-5} \cdot 32$, $-6 \sim -48$ since $-48 = 2^3 \cdot (-6)$, and

$60 \sim 15$ since $15 = 2^{-2} \cdot 60$. To see that $\sim$ is reflexive let $a \in \boldsymbol{Z}^*$ and write $a = 2^0 \cdot a$, so $a \sim a$. If $a \sim b$ then by definition of $\sim$, $b = 2^i a$ for some $i \in \boldsymbol{Z}$ so $a = 2^{-i} b$ and $b \sim a$, proving that $\sim$ is symmetric. This shows why we need the exponents of 2 to be in $\boldsymbol{Z}$ rather than in $\boldsymbol{N}$. Finally, if $a \sim b$ and if $b \sim c$ then $b = 2^i a$ and $c = 2^j b$ for some $i, j \in \boldsymbol{Z}$, and substitution shows that $c = 2^{j+i} a$. Therefore $a \sim c$, $\sim$ is transitive, and # is an equivalence relation. By the unique factorization in Theorem 1.15 we see that $z \sim z'$ identifies those integers with the same sign whose prime factorizations differ only in their powers of 2: that is, having the same "odd" part.

Example 1.12 The relation on $S = \boldsymbol{Z} \times \boldsymbol{Z}^*$ given by $(a, b) \# (c, d)$ when $ad = bc$ is an equivalence relation.

For reflexivity, $(a, b) \# (a, b)$ for any $(a, b) \in S$ because $ab = ba$. For symmetry, suppose $(a, b) \# (c, d)$, so $ad = bc$. We need $(c, d) \# (a, b)$, requiring $cb = da$, which is equivalent to $ad = bc$, so $(c, d) \# (a, b)$. Assume $(a, b) \# (c, d)$ and $(c, d) \# (e, f)$, so $ad = bc$ and $cf = de$. We must use these to show $(a, b) \# (e, f)$, or equivalently that $af = be$. Multiply $ad = bc$ by f and $cf = de$ by b to get $adf = bcf = bde$. Thus $adf = bde$ so $af = be$ results ($d \neq 0$). Consequently $(a, b) \# (e, f)$ so # is transitive and an equivalence relation. A little thought shows that # identifies all (a, b) that represent the same fraction in $\boldsymbol{Q}$.

Since the collection of elements related to each other by an equivalence relation is important, we give a name to it now.

DEFINITION 1.11 For $\sim$ an equivalence relation on S and $x \in S$, the **equivalence class** of x is $[x]_\sim = \{y \in S \mid x \sim y\}$. If $X \subseteq S$ is an equivalence class expressed as $X = [x]_\sim$, then x is called a **representative** for X.

When the equivalence relation $\sim$ is understood, we may denote the equivalence class of $x \in S$ simply by $[x]$. Since $\sim$ is reflexive, $x \in [x]_\sim \neq \varnothing$ for every $x \in S$. The symmetry of $\sim$ means $a \sim b$ exactly when $b \sim a$ so we can write $[x]_\sim = \{y \in S \mid y \sim x\}$ and may do so without further reference. We turn to some illustrations of the notion of equivalence class.

The relation of equality on S has $[x]_= = \{x\}$ for all $x \in S$, since $x \in S$ is related to itself only. For the relation $S \times S$, $[x] = S$ for each $x \in S$ since any $x \in S$ is related to every $y \in S$ and the single equivalence class S has any element in S as a representative. In Example 1.8, $[(0, 0)] = \{(0, 0)\}$, and for every other $(a, b) \in \boldsymbol{R}^2$, $[(a, b)]$ is the set of all points on the same circle about the origin as (a, b), so $[(a, b)] = [P]$ if $P \in [(a, b)]$. As we saw for # on $\boldsymbol{C}$ in Example 1.9, $[x]_\# = \{x, ix, -x, -ix\} = [ix]_\# = [-x]_\# = [-ix]_\#$ (verify!).

In Example 1.10 for real numbers, $a \approx b \Leftrightarrow a - b \in \boldsymbol{Q} \Leftrightarrow b - a \in \boldsymbol{Q} \Leftrightarrow b = a + q$ for $q \in \boldsymbol{Q}$. Thus $[a]_\approx = \{a + q \mid q \in \boldsymbol{Q}\}$. A routine exercise shows that for any $q' \in \boldsymbol{Q}$, $[a + q']_\approx = [a]_\approx$ so any element in $[a]_\approx$ is a representative for this class. Finally, consider $^\wedge \subseteq \boldsymbol{R} \times \boldsymbol{R}$ given by $x^\wedge y$ when either $x = y$ or $xy > 0$. This equivalence relation (verify!) has exactly three classes: $[0]_\wedge = \{0\}$, $\{r \in \boldsymbol{R} \mid r > 0\} = [a]_\wedge$ for any $a > 0$, and $\{r \in \boldsymbol{R} \mid r < 0\} = [b]_\wedge$ for any $b < 0$. The next result gives information in general about the relation between the equivalence classes of different elements.

THEOREM 1.21 Let $\sim$ be an equivalence relation on the set S and $a, b \in S$ with equivalence classes $[a]$ and $[b]$, respectively, Then:

i) The following are equivalent: $[a] = [b]$, $b \in [a]$, and $a \sim b$.
ii) Either $[a] = [b]$ or $[a] \cap [b] = \varnothing$.

Proof i) We show each condition in turn implies the next. If $[a] = [b]$ then by the reflexivity of $\sim$, $b \in [b] = [a]$. If $b \in [a]$ then $a \sim b$ by the definition of equivalence class. Finally, assume $a \sim b$ and let $x \in [b]$, so $b \sim x$ from the definition of equivalence class. But now $a \sim b$ and $b \sim x$ so the transitivity of $\sim$ yields $a \sim x$, or, equivalently, $x \in [a]$. Thus $a \sim b$ implies $[b] \subseteq [a]$. But $\sim$ is symmetric so $a \sim b$ implies that $b \sim a$, and now our argument forces $[a] \subseteq [b]$. These two inclusions show that $[a] = [b]$, completing the proof of i).

ii) Assume that $c \in [a] \cap [b]$ so $c \in [a]$ and $c \in [b]$. By part i), $[a] = [c] = [b]$ and $[a] = [b]$ unless $[a] \cap [b] = \varnothing$, proving ii). ■

The first part of Theorem 1.21 will be very useful and tells us that the representatives for any class are just the elements in the class. That two classes are equal precisely when their representatives are related, and otherwise are disjoint, essentially means that the equivalence classes form a partition of S. These observations imply that each equivalence class consists of mutually related elements, unrelated to elements in any other class.

Often the collection of all equivalence classes of $\sim$ on S is written as $S/\!\sim\, = \{[x]_\sim \mid x \in S\}$. Remember that each of the subsets $[x]_\sim \subseteq S$ appears only once in $S/\!\sim$ even though it may be that $[x]_\sim = [y]_\sim$ for many $y \in S$. As an extreme case, $S/(S \times S) = \{[x]_{S\times S} \mid x \in S\} = \{S\}$ has only one element.

The next result shows that the study of equivalence relations and the study of partitions are really the same thing.

THEOREM 1.22 Let $\sim$ be an equivalence relation on the set S. Then $S/\!\sim\, = \{[s]_\sim \mid s \in S\}$ is a partition of S into nonempty subsets. Conversely, if $\{A_\lambda\}_\Lambda$ is a partition of S into nonempty subsets then for some equivalence relation $\approx$ on S, each $A_\lambda = [s]_\approx$ for some $s \in S$, so $\{A_\lambda\}_\Lambda = S/\!\approx$.

Proof To see that $S/\!\sim$ is a partition we use Definition 0.10. First, $s \in [s]_\sim$ for each $s \in S$ so by the definition of union, $\bigcup [s]_\sim = S$. Next, Theorem 1.21 proves that different $[s]_\sim$ are disjoint so $\{[s]_\sim \mid s \in S\}$ is a partition of S into nonempty subsets. If $\{A_\lambda\}_\Lambda$ is a partition of S with all $A_\lambda \neq \varnothing$ we must define $\approx$ on S so the A_λ are its equivalence classes. An equivalence class consists of related elements so it is natural to define $\approx$ by $s \approx t$ when $s, t \in A_\lambda$ for some $\lambda \in \Lambda$. Since $\{A_\lambda\}_\Lambda$ is a partition of S, the union of the A_λ is S so for any $s \in S$, $s \in A_\lambda$ for some $\lambda \in \Lambda$. Clearly $s, s \in A_\lambda$ so $s \approx s$ and $\approx$ is reflexive. If $s \approx t$ then $s, t \in A_\lambda$ for some $\lambda \in \Lambda$. Now $t, s \in A_\lambda$ as well so $t \approx s$ and $\approx$ is symmetric. Next, if $s \approx t$ and $t \approx w$ then $s, t \in A_\lambda$ and $t, w \in A_\mu$ for some $\lambda, \mu \in \Lambda$. Note that once we designate λ and A_λ for s and t we must use a different symbol A_μ for t and w

and not tacitly *assume* that A_λ contains all three elements. Since $\{A_\lambda\}_\Lambda$ is a partition of S, $t \in A_\lambda \cap A_\mu$ forces $\lambda = \mu$ and proves that $s, t, w \in A_\lambda$. In particular, $s, w \in A_\lambda$ so $s \approx w$ and $\approx$ is transitive and so an equivalence relation. To show $\{A_\lambda\}_\Lambda = S/\approx$ consider any A_λ and any $x \in A_\lambda$, which is possible since $A_\lambda \neq \emptyset$. Using again that the different A_λ are pairwise disjoint, the definition of $\approx$ shows that $x \approx y$ precisely when $y \in A_\lambda$. Thus $A_\lambda = [x]_\approx$ and $\{A_\lambda\} \subseteq S/\approx$. But any $s \in S$ is in some A_λ, as we saw in proving reflexivity, so as just above, $[s]_\approx = A_\lambda$. Hence $S/\approx \subseteq \{A_\lambda\}_\Lambda$, completing the proof of the theorem. ∎

The theorem shows that equivalence relations and partitions are equivalent concepts. Normally the more useful fact is that equivalence relations give partitions. However, let us look at two examples of partitions and determine the underlying equivalence relations. By Example 0.12, $\{r + \boldsymbol{Z} \mid 0 \leq r < 1\}$ is a partition of $\boldsymbol{R}$ where $r + \boldsymbol{Z} = \{t \in \boldsymbol{R} \mid t = r + z \text{ for some } z \in \boldsymbol{Z}\}$. The $r + \boldsymbol{Z}$ are the classes of the underlying relation and it is an exercise to prove that $a, b \in r + \boldsymbol{Z}$ for some $r \in \boldsymbol{R}$ exactly when $a - b \in \boldsymbol{Z}$: $a \sim b \Leftrightarrow a - b \in \boldsymbol{Z}$. Finally, Theorem 1.2 shows that for any $n \in \boldsymbol{N}$, $\{n\boldsymbol{Z}, n\boldsymbol{Z} + 1, \ldots, n\boldsymbol{Z} + (n-1)\}$ is a partition of $\boldsymbol{Z}$, and $a, b \in n\boldsymbol{Z} + j$ for some $0 \leq j \leq n - 1$ exactly when $n \mid (a - b)$, or $a \approx b \Leftrightarrow a - b \in n\boldsymbol{Z}$. This relation is very interesting and important, is called **congruence modulo *n***, and will be discussed in the next section.

If $\sim$ is an equivalence relation on S, then by Theorem 1.21, $[s]_\sim = [t]_\sim$ when $t \in [s]_\sim$. It is often helpful to distinguish the different classes by fixing a representative for each class. In practice it is not always easy to do this.

DEFINITION 1.12 A **transversal** for an equivalence relation $\sim$ on a set S is any subset $T \subseteq S$ consisting of exactly one element from each equivalence class of $\sim$ on S: T is a set of representatives for the equivalence classes.

The existence of a transversal technically depends on the Axiom of Choice, at least when $S/\sim$ is infinite, since we need to pick an element from each class. A transversal T enables us to write any class in a specified way since $\{[x]_\sim \mid x \in S\} = \{[t]_\sim \mid t \in T\}$ and for $t, t' \in T$, $[t]_\sim = [t']_\sim$ forces $t = t'$ by Theorem 1.21. Later we will see the advantage of specifying a transversal in some uniform way rather than by making a more random choice from each class. For equality on S, each class $[x]_= = \{x\}$ so the only transversal is $T = S$. If $\sim = S \times S$, $S = [x]_\sim$ for each $x \in S$ so any singleton $\{s\}$ is a transversal. For the relation $x \sim y$ in $\boldsymbol{R}$ when $x - y \in \boldsymbol{Z}$ from Example 0.12, as mentioned above for $r \in \boldsymbol{R}$, $[r]_\sim = r + \boldsymbol{Z}$. It is impossible to specify a representative from each of these infinitely many sets but by Example 0.12 a transversal is the interval $[0, 1)$. Another transversal would be $[0, 1/2] \cup (3/2, 2)$ (why?). There are only n classes, $\{n\boldsymbol{Z} + j \mid 0 \leq j < n\}$, for congruence modulo n, also mentioned above, so if n is small we can pick one element from each class in many ways, although no choice is nicer than $\{0, 1, 2, \ldots, n-1\}$.

What are transversals for equivalence relations presented above? In Example 1.8 two points of $\boldsymbol{R}^2$ are related if they lie on the same circle centered at the origin and the equivalence classes are the different circles. Except for $[(0, 0)] = \{(0, 0)\}$, each circle has infinitely many points so a transversal must be a choice of one point on each circle, together with the origin. One uniform way of describing such a choice is to take all points on a straight half-line starting at the origin, since any such line contains exactly one point on each circle. The nonnegative x-axis gives the transversal $T = \{(a, 0) \mid a \geq 0\}$, and the "negative" half-line $y = x$ gives $T' = \{(-a, -a) \mid a \geq 0\}$. The parabola $y = x^2$ also yields a transversal, namely $T'' = \{(a, a^2) \mid a \geq 0\}$.

In Example 1.9, $a \# b$ on $\boldsymbol{C}$ when $a^4 = b^4$ has the equivalence classes $[c]_\# = \{c, ic, -c, -ic\}$ if $c \neq 0$ and $[0]_\# = \{0\}$. It is difficult to describe any transversal for #. We proceed geometrically by viewing $\boldsymbol{C}$ as points in $\boldsymbol{R}^2$, associating $a + bi \in \boldsymbol{C}$ with $(a, b) \in \boldsymbol{R}^2$. The *polar form* for $a + bi \in \boldsymbol{C}$ is $re^{i\theta} = r(\cos\theta + i \cdot \sin\theta)$ where $r = (a^2 + b^2)^{1/2}$, $\theta = \arctan(b/a)$ if $a \neq 0$, and when $a = 0$: $\theta = \pi/2$ if $b > 0$, $\theta = 3\pi/2$ if $b < 0$, and $\theta = 0$ if $b = 0$. Using trigonometric identities, it follows that multiplication by $i = i \cdot \sin(\pi/2)$ is rotation counterclockwise about (0, 0) by $\pi/2$ radians. Thus when $c \neq 0$ the four complex numbers in $[c]_\#$ lie on the same circle about (0, 0) and $\pi/2$ radians apart. Hence one transversal is the first quadrant, omitting the positive y-axis, or $\{a + bi \mid a > 0 \text{ and } b \geq 0\} \cup \{0\}$. Any similar "wedge" of one quarter of the plane starting at the origin will do but may be difficult to describe in terms of complex numbers without using polar form.

In Example 1.10 where $a, b \in \boldsymbol{R}$ are related when $a - b \in \boldsymbol{Q}$, it seems impossible to describe all the different classes $[a] = \{a + q \mid q \in \boldsymbol{Q}\}$, so finding any transversal explicitly appears to be hopeless! Example 1.11 is much easier. Here the nonzero elements of $\boldsymbol{Z}$ are related if they have the same "odd part" and sign, so one explicit transversal is $(2\boldsymbol{Z} + 1) \cup \{0\}$, and another would be $\{2^{|m|}(2m - 1) \mid m \in \boldsymbol{Z}\} \cup \{0\}$.

Finally, in Example 1.12, an equivalence class of # contains those (a, b) that represent the same fraction a/b. For $p/q \in \boldsymbol{Q} - \{0\}$ we claim there is a unique $p'/q' = p/q$ with $\text{GCD}\{p', q'\} = 1$, and $q' > 0$. If two fractions satisfy these conditions, say $a/b = c/d$, then $ad = bc$. Thus $d \mid bc$ so $(d, c) = 1$ and Theorem 1.12 show $d \mid b$. Similarly, $b \mid d$ and $b = d$ follows since both are positive, so also $a = c$. Therefore $[(p, q)]_\#$ has exactly one element (p', q') of this special form in each "nonzero" class. Since $(0, s) \# (0, t)$ for all $s, t \in \boldsymbol{Z}^*$, $[(0, s)]_\# = \{(0, t) \mid t \in \boldsymbol{Z}^*\}$ (proof!) so a transversal for # is $\{(a, b) \in \boldsymbol{Z}^* \times \boldsymbol{Z}^* \mid \text{GCD}\{a, b\} = 1 \text{ and } b > 0\} \cup \{(0, 1)\}$.

EXERCISES

1. In each case determine which properties of an equivalence relation hold:

i) $\sim$ on $\boldsymbol{N} - \{0\}$ given by $a \sim b \Leftrightarrow (a, b) \neq 1$

ii) $\approx$ on $\{f : \boldsymbol{R} \to \boldsymbol{R}\}$ given by $f \approx g \Leftrightarrow$ for some $t \in \boldsymbol{R}$, $f(t) = g(t)$

iii) $*$ on $\boldsymbol{R}^2$ via $(s, t) * (u, v) \Leftrightarrow s^3 + t^3 = u^3 + v^3$

iv) # on $\boldsymbol{R}$ via $a \# b \Leftrightarrow ab \in \boldsymbol{Q}$

v) #′ on $\boldsymbol{R} - \{0\}$ given by $a \#' b \Leftrightarrow ab^{-1} \in \boldsymbol{Q}^+$

vi) $\sim$ on $P(S)$ for S any set via $A \sim B \Leftrightarrow (A - B) \cup (B - A)$ is finite

vii) $*$ on $M_n(\boldsymbol{R})$, the $n \times n$ matrices over $\boldsymbol{R}$, given by $A * B \Leftrightarrow$ for some invertible $P \in M_n(\boldsymbol{R})$, $B = P^{-1}AP$

viii) $\triangle$ on $M_n(\boldsymbol{R})$ given by $A \triangle B \Leftrightarrow$ for some $P, Q \in M_n(\boldsymbol{R})$, $B = PAQ$

2. If $f : S \to T$ show that $\sim_f$ defined on S by $s_1 \sim_f s_2 \Leftrightarrow f(s_1) = f(s_2)$ is an equivalence relation, and that $|S/{\sim_f}| = |\text{Im } f|$ (Definition 0.11).

3. If $S \neq \emptyset$ and $\sim, \approx \subseteq S \times S$ are equivalence relations, show that $\sim \cap \approx$ is also.

4. If $\emptyset \neq T \subseteq S$ and $\sim$ is an equivalence relation on S, show that $\sim \cap (T \times T)$ is an equivalence relation on T.

5. Show that the given relation is an equivalence relation and find a transversal:

i) $\sim$ on $\boldsymbol{Q}$ given by $x \sim y \Leftrightarrow x - y \in \boldsymbol{Z}$

ii) $*$ on $\boldsymbol{R}^2$ via $(a, b) * (c, d) \Leftrightarrow b^2 - a = d^2 - c$ (think geometrically)

iii) $//$ on $M_n(\boldsymbol{R})$ given by $A//B \Leftrightarrow \det(A) = \det(B)$

iv) $\approx$ on $\boldsymbol{N} \times \boldsymbol{N}$ given by $(a, b) \approx (c, d) \Leftrightarrow a - b = c - d$

v) $\#$ on $\boldsymbol{N}$ given by $s \# t \Leftrightarrow$ there is $k \in \boldsymbol{N}$ with $2^{k-1} \leq s, t < 2^k$

vi) $\triangle$ on $\boldsymbol{N}$ given by $a \triangle b \Leftrightarrow$ there are $n, m \in \boldsymbol{N}$ with $a \mid b^n$ and $b \mid a^m$

vii) $\sim$ on $\boldsymbol{N} \times \boldsymbol{N}$ given by $(a, b) \sim (c, d) \Leftrightarrow$ there is $i \in \boldsymbol{Z}$ with $ad = 2^i bc$ ($\{(a, b) \in \boldsymbol{N} \times \boldsymbol{N} \mid a, b \in 2\boldsymbol{Z} + 1$ and $(a, b) = 1\}$ is a transversal.)

viii) $\#$ on the set $P(\{1, 2, \ldots, 1000\})$ given by $A \# B \Leftrightarrow |A| = |B|$

ix) $\approx$ on $\boldsymbol{N}$ via $a \approx b \Leftrightarrow a$ and b have the same number of divisors in $\boldsymbol{N}$

x) $\sim$ on $\boldsymbol{N}$ given by $a \sim b \Leftrightarrow a$ and b have the same number of distinct prime divisors

6. Let $\leq$ be a partial order on S, and define $\sim$ on S by $a \sim b$ if for some $k \in \boldsymbol{N}$ there are $a = c_1, \ldots, c_k = b \in S$ so that $c_j \leq c_{j+1}$ or $c_{j+1} \leq c_j$ for each $1 \leq j < k$.

a) Prove $\sim$ is an equivalence relation—its classes are the **components** of S.

b) Show that the partial order $\#$ on $\boldsymbol{N}^k$ via $(a_1, \ldots, a_k) \# (b_1, \ldots, b_k)$ when $a_i \mid b_i$ for each $1 \leq i \leq k$ has only the one component, $\boldsymbol{N}^k$.

c) Define $\approx$ on $\boldsymbol{R}^2$ by $(a, b) \approx (c, d)$ when $a \leq c$ and $b = d$. Show that $\approx$ is a partial order and describe its components geometrically.

7. If $\sim$ is an equivalence relation on the set S, show that $T \subseteq S$ is a transversal for $\sim$ if and only if the elements of T are mutually incomparable and T is a maximal subset (with respect to set inclusion) with this property.

1.7 CONGRUENCE MODULO *n*

We explore an important and useful relation on $\boldsymbol{Z}$. Given $n \in \boldsymbol{N}$, Theorem 1.2 shows that $\{n\boldsymbol{Z}, n\boldsymbol{Z} + 1, \ldots, n\boldsymbol{Z} + (n-1)\}$ is a partition of $\boldsymbol{Z}$ and $a, b \in n\boldsymbol{Z} + i \Leftrightarrow n \mid (a - b)$, and Theorem 1.22 shows that this defines an equivalence relation. The standard notation for it comes next.

DEFINITION 1.13 If $n \in \boldsymbol{N}$, the relation **congruence modulo *n*** on $\boldsymbol{Z}$ is given by $a \equiv b \pmod{n}$ when $n \mid (a - b)$.

We often say congruence "mod n" and write $a \equiv_n b$, or just $a \equiv b$ when n is understood. Note that $a \equiv 0 \pmod{n}$ exactly when $n \mid a$. Also the case $n = 1$ is quite uninteresting since every two integers are related. Some examples are $17 \equiv -5 \pmod{11}$ since $11 \mid 22$, $100 \equiv 2 \pmod{14}$ since $14 \mid 98$, and $12 \equiv 61 \pmod{7}$ since $7 \mid (-49)$. That congruence modulo n is an equivalence relation on $\boldsymbol{Z}$ follows from the definition and properties of division without reference to Theorem 1.22.

THEOREM 1.23 Congruence modulo n is an equivalence relation. The equivalence class of $a \in \boldsymbol{Z}$ is denoted $[a]_n$, $[a]_n = n\boldsymbol{Z} + a$, and the following are equivalent: $a \equiv b \pmod{n}$; $[a]_n = [b]_n$; $a \in [b]_n$; and $a \in n\boldsymbol{Z} + b$.

Proof We need to verify the three properties of an equivalence relation. For reflexivity let $m \in \boldsymbol{Z}$ and note that $m \equiv_n m$ holds since $m - m = 0$ and $n \mid 0$. Next, if $a \equiv_n b$ then $n \mid (a - b)$

and we need $b \equiv_n a$, or that $n \mid (b-a)$. But $b-a=-(a-b)$, so $n \mid (b-a)$ and $\equiv_n$ is symmetric. Finally, if $a \equiv_n b$ and $b \equiv_n c$ then transitivity holds if $a \equiv_n c$, or if $n \mid (a-c)$. By our assumption $n \mid (a-b)$ and $n \mid (b-c)$. Since $a-c=(a-b)+(b-c)$, $n \mid (a-c)$ results from Theorem 1.6 and congruence modulo n is an equivalence relation. If $b \in [a]_n$ then $a \equiv_n b$ from Definition 1.11 of equivalence classes so $b \equiv_n a$ by symmetry. Hence $n \mid (b-a)$ so $b-a \in n\mathbf{Z}$, and $b \in n\mathbf{Z}+a$ follows. Thus $[a]_n \subseteq n\mathbf{Z}+a$. But for any $y = nz+a \in n\mathbf{Z}+a$, $a-y \in n\mathbf{Z}$ and $a \equiv_n y$ results. Definition 1.11 shows that $y \in [a]_n$ so $n\mathbf{Z}+a \subseteq [a]_n$, forcing $[a]_n = n\mathbf{Z}+a$. That $a \equiv_n b$ is equivalent to $[a]_n=[b]_n$ follows from Theorem 1.21, and the equivalence of the other conditions follows from $[a]_n = n\mathbf{Z}+a$ and Theorem 1.2. ■

Any integer is congruent mod n to its remainder on division by n. This follows from Theorem 1.1 on division and Theorem 1.23 since $a = qn + r$ exactly when $a \equiv r \pmod{n}$. Thus $[19]_6 = [61]_6$ since the remainder when either 19 or 61 is divided by 6 is 1, or also since $6 \mid 42$ and $42 = 61 - 19$. Similarly, $[100]_7 = [2]_7$ and $[15]_9 = [6]_9$. The language of congruences is particularly useful for describing results about division or remainders. Thus Example 1.1 is now $t^5 \equiv_5 t$ for every $t \in \mathbf{Z}$; Theorem 1.3 is $t^2 \equiv_8 k$ for $k = 0, 1$, or 4 for any $t \in \mathbf{Z}$, and $t^2 \equiv_8 1$ when t is odd; and from Example 1.3, if $t \equiv_8 7$ then t is not the sum of three squares.

Next we examine the extent to which congruence works like equality with respect to the usual arithmetic operations.

THEOREM 1.24 Let $n \in N - \{1\}$ and $a, b, c, d \in \mathbf{Z}$. The following hold:

i) If $a \equiv b \pmod{n}$ and $c \equiv d \pmod{n}$, then $(a+c) \equiv (b+d) \pmod{n}$.
ii) If $a \equiv b \pmod{n}$, then $ac \equiv bc \pmod{n}$.
iii) If $a \equiv b \pmod{n}$ and $c \equiv d \pmod{n}$, then $ac \equiv bd \pmod{n}$.
iv) If $ac \equiv bc \pmod{n}$ and $(c, n) = 1$, then $a \equiv b \pmod{n}$.
v) If $a \equiv b \pmod{n}$, then for any $m \in N$, $a^m \equiv b^m \pmod{n}$.

Proof Write $a \equiv b$ for $a \equiv b \pmod{n}$. The conclusion of i), $(a+c) \equiv (b+d)$ will follow from the definition of congruence if $n \mid ((a+c)-(b+d))$. By assumption, $n \mid (a-b)$ and $n \mid (c-d)$, so adding $a-b$ to $c-d$ and using Theorem 1.6 give $n \mid ((a+c)-(b+d))$. From Theorem 1.5 we conclude that $n \mid (a-b)c$ when $n \mid (a-b)$, proving ii). Using $a \equiv b$ and ii) imply that $ac \equiv bc$, and also $c \equiv d$ implies $bc \equiv bd$ so the transitive property of congruence yields $ac \equiv bd$, proving iii). The assumption $ac \equiv bc$ of iv) means that $n \mid c(a-b)$ and we conclude that $n \mid (a-b)$ from Theorem 1.12 since $(c, n) = 1$. Therefore $a \equiv b$. Part v) follows by induction on m. The case $m = 1$ is just the assumption $a \equiv b$, and this with $a^k \equiv b^k$ for any $k \geq 1$ yields $a^{k+1} \equiv b^{k+1}$ by an application of iii). Therefore v) holds by IND-I. ■

COROLLARY Let $n \in N - \{1\}$ and $a, b, a_1, \ldots, a_k, b_1, \ldots, b_k \in \mathbf{Z}$.

i) If all $a_i \equiv_n b_i$ then $\Sigma a_i \equiv_n \Sigma b_i$ and $a_1 \cdots a_k \equiv_n b_1 \cdots b_k$.
ii) If $a \equiv_n b$ and $p(x) \in \mathbf{Z}[x]$, then $p(a) \equiv_n p(b)$.

Proof Part i) is an induction on k using parts i) and iii) of the theorem; its proof is left as an exercise. For ii) write $p(x) = c_m x^m + \cdots + c_1 x + c_0$ and note that for $1 \leq j \leq m$, if $a \equiv b$

(mod n) then $a^j \equiv b^j$ and then $c_j a^j \equiv c_j b^j$ by the theorem. Using part i) together with $c_0 \equiv c_0$ and adding all these together shows that $c_m a^m + \cdots + c_1 a + c_0 \equiv c_m b^m + \cdots + c_1 b + c_0$, so $p(a) \equiv p(b)$. ∎

Do not regard the last part of Theorem 1.24 as an equivalence or forget that $m \in \boldsymbol{N}$! For example, $5^2 \equiv_{15} 10^2$ (why?) but clearly $5 \equiv_{15} 10$ is false. Although for $a \in \boldsymbol{N}$ and $q \in \boldsymbol{Q}$, $a^q \in \boldsymbol{R}$ is defined, if $a \equiv_n b$, $a^q \equiv_n b^q$ generally makes no sense since usually $a^q \notin \boldsymbol{Z}$ or $b^q \notin \boldsymbol{Z}$. Even if $a^q, b^q \in \boldsymbol{N}$ for $q \in \boldsymbol{Q} - \boldsymbol{N}$, $a \equiv_n b$ may not force $a^q \equiv_n b^q$: $64 \equiv_7 1$ but $64^{1/3} \equiv_7 1^{1/3}$ is false since $4 \equiv_7 1$ is false.

Before looking at mathematical consequences of Theorem 1.24 we consider it informally. Essentially it shows that we can compute mod n by using integers congruent to the given ones, and in particular their remainders on division by n. Thus computations resulting in integers larger than n may be reduced by subtracting multiples of n. This is commonly used with units of time, although we do not consciously think in terms of congruences. For example, to see what time is fourteen hours from the present we would automatically add two hours to the present time (and change between P.M. and A.M.) since the addition of twelve hours will not change the "hour": we compute mod 12. Or we know quickly that twenty hours after 7 A.M. will be 3 A.M. the next day, since it will be four hours less than twenty-four, computing mod 24. A similar computation mod 60 using minutes is that 55 minutes after 3:20 we recognize as 4:15 by subtracting 5 minutes from 60 (one hour). Even for computations with months, it is often easier to think "mod 12." Rather than counting out nine months from July, the seventh month, we add 9 to 7 and subtract 12 to get the fourth month, April. A final amusing calendar "application" of congruences is the following example, which shows that there is at least one Friday the 13th in every year.

Example 1.13 Given any date from 1 to 31 and any day of the week, the given date falls on the given day in some month at least once each year with a single exception: the 31st of no month is on the same day as July 1st.

The days on which different dates occur in any month depend solely on what day the first of the month is. Hence it suffices to check that every day of the week except the day starting June starts some month with 31 days. Let the day that starts January be day 1. For example, if January 1 is Thursday then day 2 of the week is Friday, day 3 is Saturday, etc. Any multiple of seven days after day k will be the same day of the week, and in general m days after day k will be the day $1 \leq x \leq 7$ with $k + m \equiv x$ (mod 7). January has 31 days so February starts on day 4, since $1 + 31 \equiv_7 4$. In a nonleap year, February has 28 days and $4 + 28 \equiv_7 4$ so March starts on day 4 also, and April starts on day 7 since March has 31 days and $4 + 31 \equiv_7 7$. The sequence of first days of each month for a nonleap year is 1, 4, 4, 7, 2, 5, 7, 3, 6, 1, 4, 6. Note that the first of June is on day 5 of the week, starting no other month, and that every other day of the week starts a month with 31 days. A similar computation for leap years gives the corresponding result, with day 6 of the week starting June, but no other month.

We return to things more formally mathematical. By Theorem 1.24, when $x \equiv_n y$ then either may be replaced by the other in a congruence mod n. Let us use this idea to solve linear congruences $ax \equiv_n b$ whenever $(a, n) = 1$.

THEOREM 1.25 If $n \in \boldsymbol{N}$, $a, b \in \boldsymbol{Z}$, and $(a, n) = 1$, then there is a unique solution $r' \in \{0, 1, \ldots, n-1\}$ of $ax \equiv b \pmod{n}$, and the set of all integer solutions is $[r']_n = n\boldsymbol{Z} + r'$.

Proof For brevity write $p \equiv q$ for $p \equiv q \pmod{n}$. By Theorem 1.13, $ar \equiv 1$ for $1 \le r < n$, so r acts like a^{-1}, and we claim the set of solutions of $ax \equiv b$ is $[rb]_n$, in analogy with solving $ax = b$ in $\boldsymbol{R}$. If $m \in [rb]_n$ then by the definition of equivalence class, $m \equiv rb$. Multiply $m \equiv rb$ by a and use Theorem 1.24 first to get $am \equiv arb$ and then to conclude that $arb \equiv b$ from $ar \equiv 1$. The transitivity of $\equiv$ forces $am \equiv b$ so any $m \in [rb]_n$ is a solution of $ax \equiv b$. Now if $y \in \boldsymbol{Z}$ satisfies $ay \equiv b$ then $ay \equiv arb$, and $y \equiv rb$ follows from Theorem 1.24 using $(a, n) = 1$. Thus $y \in [rb]_n$ by Theorem 1.23, every solution of $ax \equiv b$ is in $[rb]_n$, and $[rb]_n$ is the set of all solutions of $ax \equiv b$. The remainder r' of rb on division by n satisfies $rb = qn + r'$ by Theorem 1.1 so Theorem 1.23 shows that $[rb]_n = [r']_n = n\boldsymbol{Z} + r'$ is the set of solutions of $ax \equiv b$. Thus $r' \in [r']_n$ is a solution, $0 \le r' < n$, and if $0 \le j < n$ satisfies $j \in [r']_n$ then $j \equiv r'$ by Theorem 1.21 so $j = r'$ (why?) and r' is the only solution in $\{0, 1, \ldots, n-1\}$. ∎

The proof of the theorem shows how to get the solutions of $ax \equiv_n b$: multiply by the "inverse" r of a, reduce rb modulo n, and take all integers congruent to that result. Note that in the proof of Theorem 1.25 we can replace r there with any $s \in \boldsymbol{Z}$ satisfying $as \equiv_n 1$. The reason is that $as \equiv_n ar$, so using $(a, n) = 1$ and Theorem 1.24 shows that $s \equiv_n r$, and then $sb \equiv_n rb$. A couple of examples should clarify this procedure.

Example 1.14 Solve $33x \equiv 8 \pmod{19}$.

Since $(33, 19) = 1$ there is a solution. Use the Euclidean Algorithm to find an "inverse" for 33. The equations are $33 = 19 + 14$, $19 = 14 + 5$, $14 = 2 \cdot 5 + 4$, and $5 = 4 + 1$, yielding $1 = 5 - (14 - 2 \cdot 5) = 3 \cdot (19 - 14) - 14 = -4(33 - 19) + 3 \cdot 19 = 7 \cdot 19 - 4 \cdot 33$. Thus $-4 \cdot 33 \equiv_{19} 1$ so by the proof of Theorem 1.25, $-4 \cdot 8 = -32$ is a solution of $33x \equiv 8 \pmod{19}$, and the set of solutions is $[-32]_{19} = [6]_{19} = 19\boldsymbol{Z} + 6$, or all $z \in \boldsymbol{Z}$ with $z \equiv_{19} 6$. As $14 \equiv_{19} 33$ it follows by Theorem 1.24 that $14x \equiv 33x \pmod{19}$ for any $x \in \boldsymbol{Z}$, so we could have solved $14x \equiv 8 \pmod{19}$ instead.

Example 1.15 Solve $89x \equiv 230 \pmod{75}$.

It is easier working with small integers and $89 \equiv_{75} 14$ so from Theorem 1.24, $89x \equiv_{75} 14x$ for any $x \in \boldsymbol{Z}$. The symmetry and transitivity of $\equiv$ show that $89x \equiv_{75} 230$ is equivalent to $14x \equiv_{75} 230$. This in turn is equivalent to $14x \equiv_{75} 5$ using $230 \equiv_{75} 5$. We find the "inverse" for 14 (mod 75) by the Euclidean Algorithm. The equations for the algorithm are $75 = 5 \cdot 14 + 5$, $14 = 2 \cdot 5 + 4$, and $5 = 4 + 1$, resulting in $1 = 3 \cdot 75 - 16 \cdot 14$. It follows that $-16 \cdot 14 \equiv_{75} 1$ so our discussion above yields the solution $-16 \cdot 5$ for $89x \equiv 230 \pmod{75}$. However $-80 \equiv_{75} -5$ and $-5 \equiv_{75} 70$, so the solutions are all $z \equiv_{75} 70$, or $[70]_{75} = 75\mathbf{Z} + 70$, by Theorem 1.25.

This method of solving linear congruences extends to $ax \equiv_n b$ when $(a, n) \mid b$. Write $n = n'(a, n)$, $a = a'(a, n)$, $b = b'(a, n)$ and use Theorem 1.5 to see that $n \mid (ax - b)$ exactly when $n' \mid (a'x - b')$. Therefore, the solutions in $\mathbf{Z}$ of $a'x \equiv_{n'} b'$ and of $ax \equiv_n b$ are the same. Since $(a', n') = 1$ by Theorem 1.12, applying Theorem 1.25 gives a unique solution $0 \le r' < n'$ of $a'x \equiv_{n'} b'$ and any solution of $a'x \equiv_{n'} b'$ must be in $n'\mathbf{Z} + r' = [r']_{n'}$ by Theorem 1.25. Now $r', r' + n', \ldots, r' + ((a, n) - 1)n' \in n'\mathbf{Z} + r'$ are (a, n) solutions of both $a'x \equiv_{n'} b'$ and $ax \equiv_n b$ in $\{0, 1, 2, \ldots, n - 1\}$, and no two are congruent modulo n (why?). Any $y \in n'\mathbf{Z} + r'$ is of the form $y = r' + mn'$ and dividing m by (a, n) gives $y = r' + (q(a, n) + s)n' = r' + sn' + qn \in [r' + sn']_n$ where $0 \le s < (a, n)$. Since any $r' + sn' + cn \in [r']_{n'}$, the solutions $[r']_{n'}$ of $ax \equiv_n b$ are just the union of the (a, n) classes $[r']_n, [r' + n']_n, \ldots,$ and $[r' + ((a, n) - 1)n']_n$. Thus there are exactly (a, n) noncongruent solutions of $ax \equiv_n b$ in the set $\{0, 1, \ldots, n - 1\}$ and these are $r', r' + n', \ldots,$ and $r' + ((a, n) - 1)n'$.

Example 1.16 Find all solutions in $\{0, 1, \ldots, 79\}$ of $55x \equiv 65 \pmod{80}$.

Since $(55, 80) = 5$ and $5 \mid 65$ we eliminate the common factor of 5 and consider $11x \equiv_{16} 13$. The Euclidean Algorithm or a little thought gives $3 \cdot 11 \equiv_{16} 1$ so the solutions of $11x \equiv_{16} 13$ are all $z \equiv_{16} 3 \cdot 13$. This reduces to $z \equiv_{16} 7$ since $39 \equiv_{16} 7$. In the notation of the last paragraph, $(a, n) = 5$ and $n' = 16$, so the solutions of $55x \equiv_{80} 65$ in $\{0, 1, 2, \ldots, 79\}$ are 7, 23, 39, 55, and 71, and the solution in $\mathbf{Z}$ is $[7]_{16} = [7]_{80} \cup [23]_{80} \cup [39]_{80} \cup [55]_{80} \cup [71]_{80}$.

Example 1.17 Find all solutions in $\mathbf{Z}$ of $6x \equiv 15 \pmod{33}$.

Here $(6, 33) = 3$ and $3 \mid 15$ so we consider the congruence $2x \equiv_{11} 5$. Again the Euclidean Algorithm or judicious guessing and Theorem 1.25 show that the solutions are all $z \equiv_{11} 6 \cdot 5 \equiv_{11} 8$. Now $33 = 3 \cdot 11$, so $n' = 11$ and the three solutions of $6x \equiv 15 \pmod{33}$ in $\{0, 1, 2, \ldots, 32\}$ are 8, 19, and 30. The totality of solutions in $\mathbf{Z}$ is $[8]_{11} = [8]_{33} \cup [19]_{33} \cup [30]_{33}$.

Note that $ax \equiv b \pmod{n}$ cannot have a solution *unless* $(a, n) \mid b$. If for some $z \in \mathbf{Z}$, $az \equiv b \pmod{n}$, then $az = b + kn$ for some $k \in \mathbf{Z}$, and $(a, n) \mid b$ by Theorem 1.6, since $(a, n) \mid a$ and $(a, n) \mid n$. We turn to other examples that show the utility of computing powers mod n (see Exercise 11 in §1.3).

Example 1.18 What is the remainder when 3^{10000} is divided by 7?

We must find $0 \le x < 7$ with $3^{10000} \equiv_7 x$. Now $3^2 = 9 \equiv_7 2$ so by Theorem 1.24 we have $(3^2)^3 \equiv_7 2^3$, or, equivalently, $3^6 \equiv_7 1$ since $8 \equiv_7 1$. Write

$10000 = 6 \cdot 1666 + 4$, so $3^{10000} = (3^6)^{1666} \cdot 3^4$ and using Theorem 1.24 again gives $3^{10000} \equiv_7 1^{1666} \cdot 3^4$. Since $3^4 = 81 \equiv_7 4$, by transitivity $3^{10000} \equiv 4 \pmod 7$.

Example 1.19 Find the last two digits of $7^{1234567}$.

For $n \in \boldsymbol{N}$, $n = q \cdot 100 + r$ with $0 \leq r < 100$ by Theorem 1.1. Thus $n \equiv_{100} r$ by definition, the last two digits of n equals r, and we want to find $7^{1234567} \equiv_{100} x$ for $0 \leq x < 100$. Now $7^2 = 49 = 50 - 1$, so $7^4 = 2500 - 100 + 1$ and $7^4 \equiv_{100} 1$. Dividing 1234567 by 4 gives $1234567 = 4 \cdot 308641 + 3$. As in Example 1.18, $7^{1234567} \equiv_{100} 7^3 = 343$, and it follows that $7^{1234567} \equiv_{100} 43$.

By Exercise 11 in §1.3 if $n \in \boldsymbol{N}$, $a \in \boldsymbol{Z}$, and $(a, n) = 1$ then $a^m \equiv_n 1$ for some $m \in \boldsymbol{N}$. Later we will see the importance of finding the minimal m.

Example 1.20 Find the minimal $m \in \boldsymbol{N}$ with $2^m \equiv 1 \pmod{25}$.

Now $2^{10} = 1024$ (everyone should know this!) so $2^{10} \equiv_{25} -1$ and squaring gives $2^{20} \equiv_{25} 1$ by Theorem 1.24. If $m < 20$ then $20 = qm + r$ for $r < m$ (Theorem 1.1) and we would have $(2^m)^q \cdot 2^r \equiv_{25} 1$. Hence Theorem 1.24 and $2^m \equiv_{25} 1$ yield $2^r \equiv_{25} 1$, contradicting the minimality of m unless $r = 0$: that is, $m \mid 20$. But $2^{10} \equiv_{25} -1$, and $-1 \not\equiv_{25} 1$ so $m \neq 10$, and it is clear that none of $2^2 = 4$, $2^4 = 16$, or $2^5 = 32$ is congruent to 1 (mod 25), so indeed $m = 20$.

Since $2^{10} = 1024$ it follows that $2^{10} \equiv_{1023} 1$. Thus $1023 \mid (2^{10} - 1)$ and Theorem 1.5 implies that $k \mid (2^{10} - 1)$ for any divisor k of 1023. But $1023 = 3 \cdot 11 \cdot 31$, so $31 \mid (2^{10} - 1)$ and $2^{10} \equiv 1 \pmod{31}$. In this case 10 is not the minimal m in $\boldsymbol{N}$ satisfying $2^m \equiv 1 \pmod{31}$ since $2^5 = 32 \equiv 1 \pmod{31}$.

We end this section by developing tests for the divisibility of integers by various primes. These tests are easy for small primes but not very practical for large ones. For any $n \in \boldsymbol{N}$ write $n = a_k a_{k-1} \cdots a_1 a_0 = a_k 10^k + \cdots + a_1 10 + a_0 = p_n(10)$, where the a_i are the digits of n and $p_n(x) = a_k x^k + \cdots + a_1 x + a_0$. By the corollary of Theorem 1.24, $10 \equiv_m b$ implies that $n = p_n(10) \equiv_m p_n(b)$. This has the following simple consequence.

THEOREM 1.26 If $n \in \boldsymbol{N}$ write $n = a_k a_{k-1} \cdots a_1 a_0 = a_k 10^k + \cdots + a_1 10 + a_0 = p_n(10)$ for $p_n(x) = a_k x^k + \cdots + a_1 x + a_0$. Then $n \equiv_3 \Sigma a_i$, $n \equiv_9 \Sigma a_i$, and $n \equiv_{11} ((a_0 + a_2 + \cdots) - (a_1 + a_3 + \cdots))$. Therefore $3 \mid n \Leftrightarrow 3 \mid \Sigma a_i$, $9 \mid n \Leftrightarrow 9 \mid \Sigma a_i$, and $11 \mid n \Leftrightarrow 11 \mid ((a_0 + a_2 + \cdots) - (a_1 + a_3 + \cdots))$.

Proof Since $10 \equiv_3 1$ and $10 \equiv_9 1$, the corollary of Theorem 1.24 shows $n = p_n(10) \equiv_3 p_n(1)$ and $n = p_n(10) \equiv_9 p_n(1)$. But $p_n(1) = \Sigma a_i$ so the first two statements follow. Also $10 \equiv_{11} -1$ so $n \equiv_{11} p_n(-1)$ and the result for 11 holds since $p_n(-1) = a_k(-1)^k + \cdots + a_1(-1) + a_0 = ((a_0 + a_2 + \cdots) - (a_1 + a_3 + \cdots))$. ■

Some examples using the theorem are that $3 \mid 56472$ but $9 \nmid 56472$ since the sum of the digits of 56472 is $24 \in 3\mathbf{Z} - 9\mathbf{Z}$. Note that $11 \nmid 56472$ since $(2+4+5)-(7+6) = -2 \notin 11\mathbf{Z}$. The digits of 30789 add to 27, so $9 \mid 30789$. Also $(9+7+3)-(8+0) = 11$ so $11 \mid 30789$, and since $(9, 11) = 1$ we conclude from Theorem 1.12 that $99 \mid 30789$. A curious consequence of Theorem 1.26 is that a positive integer is divisible by 3 (or 9) exactly when any rearrangement of its digits is also divisible by 3 (or 9), so 37089 and 93780 are divisible by 9. To maintain divisibility by 11, we can rearrange the sets of alternating digits, so $11 \mid 78309$ and $11 \mid 90387$.

Theorem 1.26 and Theorem 1.24 explain an old procedure for checking multiplication called "casting out nines." By Theorem 1.24, in computing mod n we can replace any integer by what it is congruent to. Using Theorem 1.26 provides a quick check for integer multiplication. Specifically, $157 \cdot 42 = 6594$, $157 \equiv_9 4$, $42 \equiv_9 6$, $6594 \equiv_9 6$, and $4 \cdot 6 \equiv_9 6$. Is $(521162)(387411) = 201{,}913{,}891{,}582 = y$? By adding digits, $521162 \equiv_9 8$ and $387411 \equiv_9 6$, so if the correct product is y, then $y \equiv_9 3$ since $8 \cdot 6 = 48 \equiv_9 3$. But adding the digits of y gives $y \equiv_9 4$, forcing $(521162)(387411) \neq y$. In fact the correct product replaces the "middle" 1 in y with a 0. Note that $201{,}813{,}891{,}582 \equiv_9 3$ but is not the product of the original two integers. Thus this procedure cannot guarantee a correct result.

Are there simple divisibility tests for other primes? Of course it is easy to tell if an integer is divisible by either 2 or 5. Let $p > 5$ be a prime and write $n \in \mathbf{N}$ as $n = a_k a_{k-1} \cdots a_1 a_0 = a_k a_{k-1} \cdots a_1 \cdot 10 + a_0$. Since $p > 5$, clearly $(p, 10) = 1$ so $10 \cdot m(p) \equiv_p 1$ for some integer $1 \le m(p) < p$ by Theorem 1.13. From Theorem 1.24, $nm(p) \equiv_p a_k a_{k-1} \cdots a_1 \cdot 10m(p) + a_0 m(p)$, and with $10 \cdot m(p) \equiv_p 1$ yields $nm(p) \equiv_p a_k a_{k-1} \cdots a_1 + a_0 m(p)$. We have seen that $p \mid t \Leftrightarrow t \equiv_p 0$, so $p \mid (a_k a_{k-1} \cdots a_1 + a_0 m(p)) \Leftrightarrow p \mid nm(p)$. But $(p, m(p)) = 1$, so $p \mid nm(p) \Leftrightarrow p \mid n$ by Theorem 1.12. The result is that $p \mid n \Leftrightarrow p \mid (a_k a_{k-1} \cdots a_1 + a_0 m(p))$. Now if $k \equiv_p m(p)$ then $ka \equiv_p m(p)a$ and $(a_k a_{k-1} \cdots a_1 + a_0 m(p)) \equiv_p (a_k a_{k-1} \cdots a_1 + a_0 k)$ so $p \mid n \Leftrightarrow p \mid (a_k a_{k-1} \cdots a_1 + a_0 k)$. This change allows us to add or subtract a smaller multiple of a_0 to the truncation $a_k a_{k-1} \cdots a_1$ of n and proves the next result.

THEOREM 1.27 Let $n = k \cdot 10 + a \in \mathbf{N}$, $0 \le a \le 9$, and let $p > 5$ be prime. If $r \in \mathbf{Z}$ satisfies $10r \equiv_p 1$ and if $s \equiv_p r$, then $p \mid n$ exactly when $p \mid (k + sa)$.

Proof Our comments above show that there is $r = m(p)$ satisfying the theorem. For any $s \equiv_p r$ Theorem 1.24 yields $(k + ra) \equiv_p (k + sa)$, so $p \mid n \Leftrightarrow p \mid (k + ra) \Leftrightarrow (k + ra) \equiv_p 0 \Leftrightarrow (k + sa) \equiv_p 0 \Leftrightarrow p \mid (k + sa)$. ■

Now $10 \cdot 5 \equiv_7 1$ so $m(7) = 5$ but $5 \equiv_7 -2$ and it is easier to subtract twice a_0 rather than adding five times it. Thus $7 \mid 623 \Leftrightarrow 7 \mid 56$ since $62 - 2 \cdot 3 = 56$, so indeed $7 \mid 623$. To see that $7 \mid 6223$ we consider first $622 - 6 = 616$, and to see if $7 \mid 616$ we consider $61 - 12 = 49$. Since $7 \mid 49$ we must have $7 \mid 6223$. It is immediate from Theorem 1.26 that 6223 is divisible by neither 3 nor 11. Factoring 6223 gives $6223 = 7 \cdot 889$ and since neither 3 nor 11 divides 889 its smallest possible prime divisor is 7. Now $88 - 18 = 70$ is divisible by 7 so $7 \mid 889$ also and $6223 = 7^2 \cdot 127$. Neither 2, 3, 5, nor 11 divides 127, and $12 - 14 = -2$, so $7 \nmid 127$. Thus any prime divisor of 127 is at least 13. But $127 < 169 = 13^2$ so 127 must be prime. This illustrates how these tests can be used to factor small integers by hand.

A different test for 11 follows from Theorem 1.27. Now $10 \cdot 10 \equiv_{11} 1$ so $m(11) = 10$ and $10 \equiv_{11} -1$, so $11 \mid (10k + a)$ when $11 \mid (k - a)$. It is usually easier and requires fewer steps to apply the test in Theorem 1.26 for divisibility by 11. By that test $11 \mid 123321$ since $(1+3+2)-(2+3+1)=0$, but it is not obvious that $12331 = 12332 - 1$ is divisible by 11. Just as for 7, we can continue this process on 12331 to obtain 1232, then $121 = 11^2$.

For 13, $10 \cdot 4 = 40 \equiv_{13} 1$ so $m(13) = 4$. To see if $13 \mid 1194$ we look at $119 + 4 \cdot 4 = 135$, not divisible by 13 since clearly 130 is. Trying 17407 leads to $1740 + 4 \cdot 7 = 1768$, $176 + 4 \cdot 8 = 208$, and finally $20 + 4 \cdot 8 = 52 = 13 \cdot 4$, so in turn $13 \mid 208$, $13 \mid 1768$, and $13 \mid 17407$. To factor $17407 = 13 \cdot 1339$, note that $13 \mid 1339$ and $17407 = 13^2 \cdot 103$ with 103 prime since it is not divisible by 2, 3, 5, or 7 $(10 - 6 = 4)$.

Next is 17 and $10 \cdot 12 = 120 \equiv_{17} 1$, so $m(17) = 12 \equiv_{17} -5$. Note that $17 \mid 85$ and $8 - 5 \cdot 5 = 8 - 25 = -17$ is divisible by 17. To see that $17 \mid 3587$, consider first $358 - 5 \cdot 7 = 323$, then $32 - 15 = 17$. Since $17 \mid 17$ we have that $17 \mid 323$, so $17 \mid 3587$. Clearly $10 \cdot 2 \equiv_{19} 1$ so the test for divisibility by 19 adds twice the last digit to the truncation. Thus $19 \mid 114$ since $11 + 2 \cdot 4 = 19$, and $19 \mid 6859$ since $685 + 2 \cdot 9 = 703$ and $70 + 2 \cdot 3 = 76 = 19 \cdot 4$. As a final observation, the Euclidean Algorithm shows that $(53, 10) = 1 = 16 \cdot 10 - 3 \cdot 53$, so $m(53) = 16$, and the test for divisibility by 53 is to add 16 times the last digit to the truncation, not a pleasant procedure!

EXERCISES

1. Show that any transversal for congruence mod n consists of exactly n integers, no two of which are congruent modulo n. Use this to show that the following are transversals:
  i) $\{0, 3, 3^2, \ldots, 3^6\}$ for congruence mod 7
  ii) $\{0, 2, 2^2, \ldots, 2^{10}\}$ for congruence mod 11
  iii) $\{2, 2^2, \ldots, 2^{20}\} \cup \{0, 5, 10, 15, 20\}$ for congruence mod 25
  iv) $\{2, 4, 6, \ldots, 2n\}$ for congruence mod n if n is odd
  v) $\{a, 2a, \ldots, na\}$ for congruence mod n, if $(a, n) = 1$

2. Are the following congruences valid?
  a) $131942 \equiv 132036 \pmod{47}$
  b) $3141596 \equiv 3141696 \pmod{19}$
  c) $10^n \equiv 0 \pmod{2^n}$
  d) $384^2 \equiv 593^2 \pmod{977}$
  e) $513^3 + 484^3 \equiv 0 \pmod{997}$

3. Find all $x \in \mathbf{Z}$ so that:
  a) $100x \equiv 165 \pmod{77}$
  b) $177x \equiv 10 \pmod{77}$
  c) $25x \equiv 80 \pmod{77}$
  d) $22x \equiv 0 \pmod{77}$

4. a) Find all $x \in \{0, 1, \ldots, 71\}$ so that $x^k \equiv 0 \pmod{72}$ for some $k \in \mathbf{N}$.
  b) Describe all $x \in \{0, 1, \ldots, 80999\}$ so $x^k \equiv 0 \pmod{81000}$ for some $k \in \mathbf{N}$.

5. Characterize all $x \in \mathbf{Z}$ that satisfy:
  a) $7x \equiv 14 \pmod{21}$
  b) $20x \equiv 28 \pmod{64}$
  c) $36x \equiv 78 \pmod{96}$
  d) $30x \equiv 36 \pmod{72}$
  e) $98x \equiv 175 \pmod{245}$
  f) $35x \equiv 63 \pmod{77}$

6. Find $x \in \{0, 1, \ldots, 31\}$ with:
a) $10^4 \equiv x \pmod{32}$
b) $5^{1000} \cdot 10^4 \equiv x \pmod{32}$
c) For $n \in \mathbf{N}$ show that $32 \mid n \Leftrightarrow 32$ divides the last five digits of n (e.g., $32 \mid 70410016$ because $32 \mid 10016$).

7. Show that $(n^2 + 3n)/2 \equiv [n/2] \pmod{2}$ for every $n \in \mathbf{N}$, where [ ] is the greatest integer function.

8. **a)** If k is an odd integer, show that $k^2 \equiv 1 \pmod{8}$.
b) Use induction to prove that $k^{2^n} \equiv 1 \pmod{2^{n+2}}$ for any odd integer k.

9. Find:
a) the last two digits of 41^{8192}
b) the last three digits of 251^{893}

10. Find $m > 1$ and minimal so that:
a) $3^m \equiv_{17} 1$
b) $13^m \equiv_{17} 1$
c) $5^m \equiv_{19} 1$
d) $3^m \equiv_{32} 1$
e) $13^m \equiv_{32} 1$
f) $2^m \equiv_{41} 1$

11. In each case characterize all $n \in \mathbf{N}$ that satisfy the given condition:
a) $1 + 2 + \cdots + (n-1) \equiv 0 \pmod{n}$
b) $1 + 2^2 + \cdots + (n-1)^2 \equiv 0 \pmod{n}$
c) $1 + 2^3 + \cdots + (n-1)^3 \equiv 0 \pmod{n}$

12. Let $k, n \in \mathbf{N}$ with $1 < k < n$. A game has n playing pieces, and two players alternately remove from one to k pieces at each turn. The player removing the last piece *loses*. If $n = 4$ and $k = 2$ the second player wins: she takes one piece if the first player takes two, two pieces if the first player takes one.
a) If $k = 2$, what condition on n guarantees a win for the first player? What condition on n guarantees a win for the second player?
b) Do part a) when $k = 3$.
c) What relation between k and n can guarantee the first player a win? Under what conditions can the second player be assured of a win?

13. If $n \in \mathbf{N}$, show that $3 \mid n \Rightarrow 3 \mid m$ for m any rearrangement of the digits of n.

14. **a)** If $0 \le a_1, a_2, \ldots, a_k \le 9$ and $a_1 \ne 0$, show that $11 \mid a_1a_2 \cdots a_ka_ka_{k-1} \cdots a_1$.
b) For which $k \in \mathbf{N}$ does $11 \mid (10^k + 1)$?

15. Find all digits x for which $12x{,}527{,}846{,}531$ is divisible by 3; by 9; or by 11.

16. For which digits x is $12x{,}527{,}846{,}531$ divisible by 7?

17. For all $k \in \mathbf{N}$, show that $10^{3^k} \equiv 1 \pmod{3^{k+2}}$ and $3^{k+3} \nmid (10^{3^k} - 1)$.

18. Set $M(k) = 100^{k-1} + \cdots + 100 + 1 = 1010\ldots01$ having exactly k ones.
a) Show that each of 7, 9, and 11 divides infinitely many $M(k)$. Note that $7 \mid 10101$.
b) For which $k > 1$ is $M(k)$ prime? (If $R(k) = 111\ldots1$ has exactly k ones, note that $11 \cdot M(k) = R(2k) = R(k) \cdot (10^k + 1)$.)

19. Find the minimal $m > 0$ with $9^m \equiv 1 \pmod{19}$. Show that $19 \mid (18^{9^{18}} + 9^{18^9})$.

20. For $n, m \in \mathbf{N}$ with $(n, m) = 1$, show the following:
a) Both $a \equiv b \pmod{n}$ and $a \equiv b \pmod{m} \Leftrightarrow a \equiv b \pmod{nm}$.
b) If c and d are each solutions of both $x \equiv_n a$ and $x \equiv_m b$, then $c \equiv_{nm} d$.

c) Given $a, b \in \mathbf{Z}$, there is a solution of both $x \equiv_n a$ and $x \equiv_m b$. (For each $0 \le x \le nm - 1$, $x \equiv_n u$ and $x \equiv_m v$ for some $u, v \in \mathbf{Z}$. Apply part b).)

21. If $\{n_1, \ldots, n_k\} \subseteq \mathbf{N}$ is pairwise relatively prime and if $a, b \in \mathbf{Z}$, then $a \equiv b \pmod{n_j}$ for each $1 \le j \le k$ if and only if $a \equiv b \pmod{n_1 \cdots n_k}$.

22. The Chinese Remainder Theorem. If $\{n_1, \ldots, n_k\} \subseteq \mathbf{N}$ is pairwise relatively prime, and $\{a_1, \ldots, a_k\} \subseteq \mathbf{Z}$, then there is a common solution to all of the conguences: $x \equiv a_1 \pmod{n_1}, \ldots, x \equiv a_k \pmod{n_k}$. (Set $s_i = \prod_{j \ne i} n_j$ so that $(n_i, s_i) = 1$ and write $1 = n_i z_i + s_i y_i$. Show that $t = \Sigma a_i s_i y_i$ is a solution.) Show that any two solutions are congruent mod $n_1 \cdots n_k$ (use Exercise 21).

23. Use Exercise 22 to find a solution of the systems of congruences:

a) $x \equiv 3 \pmod 8$, $x \equiv 4 \pmod 9$, $x \equiv 5 \pmod{11}$

b) $x \equiv 1 \pmod 7$, $x \equiv 7 \pmod{11}$, $x \equiv 10 \pmod{13}$

c) $x \equiv 2 \pmod 3$, $x \equiv 1 \pmod 4$, $x \equiv 3 \pmod 5$, $x \equiv 1 \pmod 7$

24. For $n, m \in \mathbf{N}$, show there are n consecutive integers each divisible by an mth power $t^m > 1$. For example: $4^2 \mid 48$, $7^2 \mid 49$, and $5^2 \mid 50$; $7^2 \mid 5047$, $2^2 \mid 5048$, $3^2 \mid 5049$, and $5^2 \mid 5050$; and $5^3 \mid 1375$, $2^3 \mid 1376$, and $3^3 \mid 1377$. (Use Exercise 22.)

25. **a)** Show that for $p > 3$ a prime, $\{2, 3, \ldots, p-2\}$ can be partitioned into two element subsets $\{a, b\}$ so that $ab \equiv 1 \pmod p$. (See Exercise 10 in §1.3.)

b) Find the partition in part a) explicitly when $p = 11$, $p = 17$, and $p = 19$.

c) Use a) to prove Wilson's Theorem: $(p-1)! \equiv -1 \pmod p$.

26. Use Exercise 25 part c) to solve:

a) $27! \equiv x \pmod{31}$ with $0 \le x \le 30$

b) $55! \equiv x \pmod{61}$ with $0 \le x \le 60$

c) $(14!)^2 \equiv x \pmod{29}$ with $0 \le x \le 28$

d) Show that $(18!)^2 \equiv 1 \pmod{437}$

e) $(27!)^6 \equiv 1 \pmod{899}$

27. If $p = 4k + 1$ is a prime, show that $(2k)!$ is a solution of $x^2 \equiv -1 \pmod p$. (Use Exercise 25 part c) and replace half the terms in $(p-1)! \equiv -1 \pmod p$ with their negatives.)

28. Use the tests for divisibility by small primes to find the prime factorization for all integers n satisfying $700 < n < 725$.

1.8 THE RING OF INTEGERS MOD *n*

For a fixed $n > 1$ set $J_n = \{[a]_n \mid a \in \mathbf{Z}\}$, the equivalence classes for congruence modulo n. It is natural to use the addition and multiplication in $\mathbf{Z}$ to define similar operations in J_n as $[a]_n + [b]_n = [a+b]_n$ and $[a]_n \cdot [b]_n = [ab]_n$. We must show that these definitions do not depend on the many possible representatives for the classes $[a]_n$ and $[b]_n$, but only on the classes themselves. For example, $[3]_{61} \cdot [7]_{61} = [21]_{61}$ and since $[3]_{61} = [125]_{61}$ and $[7]_{61} = [68]_{61}$, $[21]_{61} = [125 \cdot 68]_{61}$ should be true. In general if $X, Y \in J_n$, $X = [a]_n = [a']_n$, and $Y = [b]_n = [b']_n$, then $X + Y = [a+b]_n$ and $X + Y = [a' + b']_n$. To see that $[a+b]_n = [a'+b']_n$ use Theorems 1.23 and 1.24: $[a]_n = [a']_n$ and $[b]_n = [b']_n$ imply $a \equiv_n a'$ and $b \equiv_n b'$ so $a + b \equiv_n a' + b'$. Thus $[a+b]_n = [a'+b']_n$ so the addition in J_n depends only on $X, Y \in J_n$ and not on their representatives. Multiplication is also independent of the representatives since $a \equiv_n a'$ and $b \equiv_n b'$ yield $ab \equiv_n a'b'$ so $[ab]_n = X \cdot Y = [a'b']_n$. Since now $+, \cdot : J_n \times J_n \to J_n$ we can try to verify that the usual arithmetic properties of $\mathbf{Z}$ hold in J_n.

DEFINITION 1.14 For $n \in N$ the **ring of integers mod** $\boldsymbol{n}$ is $\mathbf{Z}_n$, the set J_n of equivalence classes of congruence mod n together with the multiplication and addition of classes: $[a]_n \cdot [b]_n = [ab]_n$ and $[a]_n + [b]_n = [a+b]_n$.

We will write products in $\mathbf{Z}_n$ by juxtaposition without a "$\cdot$". When computing sums or products in $\mathbf{Z}_n$ we are really computing mod n since by Theorem 1.23, $[a]_n = [b]_n \Leftrightarrow a \equiv_n b$. Thus $[8]_{11} + [7]_{11} = [4]_{11}$ since $15 \equiv_{11} 4$, and $[8]_{11}[7]_{11} = [1]_{11}$ since $56 \equiv_{11} 1$. Also $[6]_{15}[6]_{15} = [6]_{15}$ since $36 \equiv_{15} 6$, and $[4]_8 + [4]_8 = [0]_8$ since $8 \equiv_8 0$. The reason $\mathbf{Z}_n$ is called the ring of integers mod n is that it satisfies arithmetic properties like those of $\mathbf{Z}$. We state these next and note that any set with an addition and multiplication satisfying the first eight properties listed below is called a **commutative ring** with 1.

THEOREM 1.28 For all $[a]_n, [b]_n, [c]_n \in \mathbf{Z}_n$ the following hold:

i) $([a]_n + [b]_n) + [c]_n = [a]_n + ([b]_n + [c]_n)$
ii) $[a]_n + [b]_n = [b]_n + [a]_n$
iii) $[0]_n + [a]_n = [a]_n$
iv) $-[a]_n = [-a]_n$ satisfies $[a]_n + (-[a]_n) = [0]_n$
v) $([a]_n[b]_n)[c]_n = [a]_n([b]_n[c]_n)$
vi) $[a]_n[b]_n = [b]_n[a]_n$
vii) $([a]_n + [b]_n)[c]_n = [a]_n[c]_n + [b]_n[c]_n$
viii) $[a]_n[1]_n = [a]_n$
ix) If $[a]_n[d]_n = [b]_n[d]_n$ and $(d, n) = 1$ then $[a]_n = [b]_n$.

Proof All except ix) follow from the definitions of $+$ and $\cdot$ in $\mathbf{Z}_n$ together with the corresponding property in $\mathbf{Z}$. For the associativity of addition in i), $([a]_n + [b]_n) + [c]_n = ([a+b]_n) + [c]_n = [(a+b)+c]_n$ using the definition of addition in $\mathbf{Z}_n$. But addition in $\mathbf{Z}$ is associative so $(a+b)+c = a+(b+c)$. Thus $[(a+b)+c]_n = [a+(b+c)]_n = [a]_n + ([b+c]_n) = [a]_n + ([b]_n + [c]_n)$, proving i). We leave the other properties through viii) as exercises. For ix), the assumption $[ad]_n = [bd]_n$ implies $ad \equiv bd \pmod{n}$ by Theorem 1.23. Now $(d, n) = 1$ shows $a \equiv b \pmod{n}$ by Theorem 1.24, and $[a]_n = [b]_n$ follows. ∎

By iii), $[0]_n$ is a zero for addition in $\mathbf{Z}_n$. When $0 \le a < n$ the additive inverse $-[a]_n$ of $[a]_n$ is usually written $[n-a]_n$, so $-[5]_{11} = [6]_{11}$, $-[17]_{25} = [8]_{25}$, and $-[93]_{100} = [7]_{100}$. Although $\mathbf{Z}_n$ satisfies most of the same arithmetic properties as $\mathbf{Z}$ there are some interesting differences. Adding an element of $\mathbf{Z}_n$ to itself doubles the representative: that is, $[a]_n + [a]_n = [a+a]_n = [2a]_n$, and an easy induction shows that adding $[a]_n$ to itself k times yields $[ka]_n$. Hence adding $[a]_n$ to itself n times results in $[na]_n = [0]_n$, but adding $t \in \mathbf{Z}$ to itself is zero only for $t = 0$. Since $\mathbf{Z}_n$ is finite, not all powers of an element can be distinct so for any $[a]_n \in \mathbf{Z}_n ([a]_n)^m = ([a]_n)^s$ for some $1 \le m < s$ depending on $[a]_n$.

For example in $\boldsymbol{Z}_{12}$, $([2]_{12})^2 = ([2]_{12})^4$, $([3]_{12})^3 = [3]_{12}$, and $([6]_{12})^2 = ([6]_{12})^3 = [0]_{12}$. Thus the power of a nonzero element in $\boldsymbol{Z}_n$ can be zero, and powers of various nonidentity elements can be themselves. Also the computation $[3]_{12}[4]_{12} = [0]_{12}$ shows that the product of distinct nonzero elements can be zero. This occurs only when n is not a prime, as we see next. When n is a prime, $\boldsymbol{Z}_n$ has another interesting property.

THEOREM 1.29 For all $[a]_n, [b]_n \in \boldsymbol{Z}_n - \{[0]_n\}$, $[a]_n[b]_n \neq [0]_n$ exactly when n is a prime. When n is a prime, if $[a]_n \in \boldsymbol{Z}_n - \{[0]_n\}$ there is $[c]_n \in \boldsymbol{Z}_n$ with $[a]_n[c]_n = [1]_n$.

Proof From Theorem 1.23, $[y]_n = [0]_n$ exactly when $y \equiv_n 0$. If n is not prime there are integers $1 < c, d < n$ with $n = cd$. Thus $[c]_n \neq [0]_n$, $[d]_n \neq [0]_n$, but $[c]_n[d]_n = [cd]_n = [n]_n = [0]_n$. When n is a prime, if $[b]_n \in \boldsymbol{Z}_n - \{[0]_n\}$ then $b \notin n\boldsymbol{Z}$, and $(b, n) = 1$ results (why?). By Theorem 1.10, $1 = sb + tn$ for some $s, t \in \boldsymbol{Z}$, leading first to $bs \equiv_n 1$ and then to $br \equiv_n 1$ for $0 < r < n$, the remainder when n divides s (see Theorem 1.13). Hence $[br]_n = [1]_n$, so $[b]_n[r]_n = [1]_n$, proving the second statement. Note that $0 < r < n$ means that $[r]_n \neq [0]_n$. Finally, if n is a prime and $[a]_n[b]_n = [0]_n$ for $[b]_n \neq [0]_n$, then using the "inverse" $[r]_n$ just found for $[b]_n$, $[0]_n = [0r]_n = [0]_n[r]_n = ([a]_n[b]_n)[r]_n = [a]_n([b]_n[r]_n) = [a]_n[1]_n = [a]_n$ by the various parts of Theorem 1.28. This finishes the proof of the theorem. ∎

The theorem shows that when p is a prime, all nonzero classes in $\boldsymbol{Z}_p$ have inverses under multiplication so arithmetically $\boldsymbol{Z}_p$ behaves much like $\boldsymbol{Q}$. In particular, linear equations $[a]_p x = [b]_p$ can be solved in $\boldsymbol{Z}_p$ when $[a]_p \neq [0]_p$, and as we will see later, polynomials of degree k with coefficients in $\boldsymbol{Z}_p$ have at most k roots in $\boldsymbol{Z}_p$, just as for $\boldsymbol{Q}$. This latter statement fails in $\boldsymbol{Z}_n$ when n is not prime. An example is that $[0]_{10}, [1]_{10}, [5]_{10}, [6]_{10} \in \boldsymbol{Z}_{10}$ are all roots of $[1]_{10}x^2 - [1]_{10}x \in \boldsymbol{Z}_{10}[x]$ since for each of them $[a]_{10}^2 - [a]_{10} = [0]_{10}$.

We often study similar mathematical objects by looking at functions between them. If we try this for $\boldsymbol{Z}_n$ it is natural to consider $f : \boldsymbol{Z}_n \to \boldsymbol{Z}_m$ given by $f([a]_n) = [a]_m$. Although this certainly appears to be a function, if $n = 4$ and $m = 6$ then $f([1]_4) = [1]_6$ and $f([1]_4) = f([5]_4) = [5]_6$! The difficulty here is that $f([a]_4)$ is described by using the representative a of $[a]_4$, but different representatives may have different "images" under f. The condition required for f to be a function is given next and will prove useful to have.

THEOREM 1.30 $f = \{([a]_n, [a]_m) \in \boldsymbol{Z}_n \times \boldsymbol{Z}_m \mid a \in \boldsymbol{Z}\}$ is a function $\Leftrightarrow m \mid n$.

Proof If f is a function then $([0]_n, [0]_m), ([n]_n, [n]_m) \in f$ and $[0]_n = [n]_n$ force these pairs to be equal. Thus $[0]_m = [n]_m$ so $0 \equiv_m n$ by Theorem 1.23 and $m \mid n$. Now assume that $m \mid n$, and $([a]_n, [b]_m), ([a]_n, [c]_m) \in f$. The definition of f yields $([a]_n, [b]_m) = ([d]_n, [d]_m)$ and $([a]_n, [c]_m) = ([e]_n, [e]_m)$ for some $d, e \in \boldsymbol{Z}$, so $[d]_n = [a]_n = [e]_n$, $[b]_m = [d]_m$, and $[c]_m = [e]_m$. It follows that $n \mid (d - e)$, $m \mid (b - d)$, and $m \mid (e - c)$ (why?). But $m \mid n$ so $m \mid (d - e)$ and by Theorem 1.6, m divides $b - c = (b - d) + (d - e) + (e - c)$, or $b \equiv_m c$. Thus $[b]_m = [c]_m$ by Theorem 1.23 and f is a function. ∎

Of course when $n = m$, the function f in the theorem is just the identity function on $\boldsymbol{Z}_n$.

EXERCISES

1. Prove parts ii) through viii) of Theorem 1.28.
2. Find all $X \in \mathbf{Z}_n$ satisfying $X^2 = X$ when: a) $n = 10$; b) $n = 12$; c) $n = 30$.
3. Let $n \in N - \{1\}$. Characterize $\{[a]_n \in \mathbf{Z}_n \mid [a]_n^k = [0]_n$ for some $k \in N$ depending on $a\}$ in terms of the prime factorization of n.
4. **a)** If p is a prime, show that the solutions of $X^2 = [1]_p$ in $\mathbf{Z}_p$ are precisely $[1]_p$ and $[-1]_p$.
   b) If $p > 2$, show that the solutions of $X^2 = [1]_{p^n}$ are $[1]_{p^n}$ and $[-1]_{p^n}$.
5. **a)** If $p_1 < \cdots < p_k$ are primes and $n = p_1 \cdots p_k$, use Exercise 4, and Exercise 20 or Exercise 22 in §1.7, to determine the number of solutions of $X^2 = [1]_n$.
   b) Determine the number of solutions of $X^2 = [1]_m$ for any given odd $m \in N$.
6. Determine when $\{([a]_n, ([a]_m, [a]_k)) \in \mathbf{Z}_n \times (\mathbf{Z}_m \times \mathbf{Z}_k)\}$ is a function.
7. For $m, n, k \in N$ with $m \mid n$, show that $\{([a]_n, [ka]_m) \in \mathbf{Z}_n \times \mathbf{Z}_m\}$ is a function.
8. Show why the example is a function or demonstrate explicitly that it is not:
   i) $g = \{([a]_{63}, [a]_{28}) \in \mathbf{Z}_{63} \times \mathbf{Z}_{28} \mid a \in \mathbf{Z}\}$
   ii) $h = \{([a]_{60}, [a]_{15}) \in \mathbf{Z}_{60} \times \mathbf{Z}_{15} \mid a \in \mathbf{Z}\}$
   iii) $\varphi = \{(([a]_n, [b]_m), [ab]_{nm}) \in (\mathbf{Z}_n \times \mathbf{Z}_m) \times \mathbf{Z}_{nm} \mid a, b \in \mathbf{Z}\}$
   iv) $\theta = \{(([a]_n, [b]_m), [a+b]_{nm}) \in (\mathbf{Z}_n \times \mathbf{Z}_m) \times \mathbf{Z}_{nm} \mid a, b \in \mathbf{Z}\}$
   v) $\eta = \{(([a]_3, [b]_2), [3b - 2a]_6) \in (\mathbf{Z}_3 \times \mathbf{Z}_2) \times \mathbf{Z}_6 \mid a, b \in \mathbf{Z}\}$
9. a) Use Example 1.1 to see that $f, g : \mathbf{Z}_5 \to \mathbf{Z}_5$ given by $f(X) = X^5$ and $g(X) = X^{25}$ are the same function. b) Show that $h(X) = X^5 - X$ is constant on $\mathbf{Z}_5$.
10. For $n, m \in N$ with $(n, m) = 1$, show that $F(([a]_n, [b]_m)) = [ma + nb]_{nm}$ defines a function from $\mathbf{Z}_n \times \mathbf{Z}_m$ to $\mathbf{Z}_{nm}$ that is a bijection.
11. Let $n, m \in N$ with $(n, m) = 1$ and set $g = \{([a]_{nm}, ([a]_n, [a]_m)) \mid a \in \mathbf{Z}\}$. Show that: i) $g : \mathbf{Z}_{nm} \to \mathbf{Z}_n \times \mathbf{Z}_m$; ii) g is an injection; iii) g is a bijection.
12. Fix $n \in N$, $k \in \mathbf{Z}$, and let $\varphi_k = \{([a]_n, [ka]_n) \in \mathbf{Z}_n \times \mathbf{Z}_n \mid a \in \mathbf{Z}\}$.
    i) Show that $\varphi_k : \mathbf{Z}_n \to \mathbf{Z}_n$.
    ii) For any $t \in \mathbf{Z}$ show $\varphi_k = \varphi_t \Leftrightarrow k \equiv t \pmod{n}$.
    iii) Show that φ_k is a bijection $\Leftrightarrow (k, n) = 1$. (Hint: Is φ_k injective?)

1.9 LOCALIZATION

The construction of $\mathbf{Z}_n$ from $\mathbf{Z}$ shows how equivalence relations can be used to construct new objects. We discuss another important construction that uses equivalence relations to get fractions. This useful technique is called **localization,** a special case of which is the construction of the rational numbers from the integers. A simple motivation for this is to be able to solve $Ax = B$. If $A, B \in \mathbf{Z}$ with no solution in $\mathbf{Z}$, there is $B/A \in \mathbf{Q}$ when $A \neq 0$. Can we solve $(y^3 - 2yz)x = y + z$ by taking $x = (y + z)/(y^3 - 2yz)$? If A and B are matrices does B/A make sense? Our approach will focus on $\mathbf{Z}$ and construct part of $\mathbf{Q}$, but it is the same approach used in more general situations so we want to avoid using special properties of $\mathbf{Z}$.

There are problems using "formal fractions," or symbols b/a, aside from taking $a \neq 0$. For example, $1, 2, 3, 6 \in \mathbf{Z}$ are different, so $1/2$ and $3/6$ are formally different fractions but are equal in $\mathbf{Q}$. For polynomials we would want the quotient $(x^3 + x)/(x^2 + 1)^2$ to be the same as $x/(x^2 + 1)$. We could insist that common factors of a and of b be removed before writing b/a, and for $\mathbf{Z}$ or $\mathbf{Q}[x]$ this gives each fraction a unique representation, but the definition and computation of sums and products would be more complicated. Also, as in Example 1.7, unique factorization into primes does not always hold. The construction ought to be applicable to a wide range of examples so the use of "common factors" is not desirable. Thus we need some abstract way of identifying different fractions as the same element. Our experience computing in $\mathbf{Q}$ indicates that to form fractions that can be added and multiplied, we should expect the set of denominators to be closed under multiplication. This is the first step in the construction.

DEFINITION 1.15 A nonempty subset $S \subseteq \mathbf{Z}$ is called **multiplicatively closed** if $0 \notin S$ and if whenever $s, s' \in S$ then $ss' \in S$.

Examples of multiplicatively closed sets in $\mathbf{Z}$ are $\mathbf{Z}^* = \mathbf{Z} - \{0\}$, or for a fixed $n > 1$ either $\{n^k \mid k \in \mathbf{N}\}$ or $\{z \in \mathbf{Z} \mid \mathrm{GCD}\{z, n\} = 1\}$. The first two examples are straightforward to verify, and the last follows since the product of integers relatively prime to n is also relatively prime to n (Theorem 1.11). If $S \subseteq \mathbf{Z}$ is multiplicatively closed we construct a set $\mathbf{Z}_S$ in which we can identify a copy of $\mathbf{Z}$ via an injection $f : \mathbf{Z} \to \mathbf{Z}_S$ and in which division by elements of $f(S)$ is possible; $\mathbf{Z}_S$ are fractions with denominators in S. Remember that $S \subseteq \mathbf{Z}$ is a model for more general situations.

Let $M = \mathbf{Z} \times S$ for $S \subseteq \mathbf{Z}$ and multiplicatively closed. If $(a, s), (b, t) \in M$, define $(a, s) \sim (b, t)$ when $at = sb$. That $\sim$ is an equivalence relation on M is the argument of Example 1.12 when $S = \mathbf{Z}^*$. Denote the equivalence class of $(a, s) \in M$ by $[(a, s)] = a/s$, and the set of all these classes by $\mathbf{Z}_S$. It is vital to realize that here $a/s \notin \mathbf{Q}$ but is simply our notation for the equivalence class $[(a, s)] \in \mathbf{Z}_S$. We are essentially constructing $\mathbf{Q}$, when $S = \mathbf{Z}^*$, so must not use it. The notation a/s suggests a natural identification with rational numbers having denominators in S but our fractions are actually equivalence classes. Theorem 1.21 shows that $[(a, s)] = a/s = b/t = [(b, t)] \Leftrightarrow (a, s) \sim (b, t) \Leftrightarrow at = bs$.

We want to define an addition and multiplication on $\mathbf{Z}_S$. For $X, Y \in \mathbf{Z}_S$, a natural way to define $X + Y$ is to choose representatives, say $X = a/s$ and $Y = b/t$, and set $X + Y = (at + bs)/st$, which looks like addition in $\mathbf{Q}$. Since S is multiplicatively closed, $st \in S$ so $[(at + bs, st)] \in \mathbf{Z}_S$. The problem is that $X + Y$ appears to depend on the choice of representatives in M for X and Y, and there are many representatives for each class. The sum $X + Y$ should depend only on X and on Y, not on their representatives. This uncertainty arises whenever one tries to define a function on equivalence classes by using representatives for the classes. We digress to illustrate this problem with a simple example. Let # be the equivalence relation on $\mathbf{Z}$ given by $a \# b$ when $b = \pm a$, with $[a]_\# = \{-a, a\}$. Suppose we try to define a sum on the set of classes by $[a]_\# + [b]_\# = [a + b]_\#$, mimicking addition in $\mathbf{Z}$. Then $[7]_\# = [2]_\# + [5]_\# = [2]_\# + [-5]_\# = [-3]_\#$ but $[7]_\# \neq [-3]_\#$. Here our "addition" *does depend* on the representatives we choose for these classes, not simply on the classes themselves. Thus mimicking addition in $\mathbf{Z}$ does not work as an addition of classes for all equivalence relations on $\mathbf{Z}$.

Returning to $\mathbf{Z}_S$ and $X + Y = (at + bs)/st$ when $X = a/s$ and $Y = b/t$, we show that this sum depends only on the classes themselves. If we take other representatives, say $X = a/s = a'/s'$ and $Y = b/t = b'/t'$, then $X + Y = (at + bs)/st$ and $X + Y = (a't' + b's')/s't'$. Now $[(at + bs, st)] = [(a't' + b's', s't')] \Leftrightarrow (at + bs, st) \sim (a't' + b's', s't') \Leftrightarrow (at + bs)s't' = (a't' + b's')st$. Since $a/s = a'/s'$ and $b/t = b'/t'$, we know that $as' = sa'$ and $bt' = tb'$. Consequently $as'tt' = sa'tt'$ and $bt'ss' = tb'ss'$. Adding these yields $(at + bs)s't' = (a't' + b's')st$, proving that $(at + bs)/st = (a't' + b's')/s't'$. Therefore $X + Y$ depends only on X and Y and not on their representatives. A similar computation shows that for $X = a/s$ and $Y = b/t \in \boldsymbol{Z}_S$, $XY = ab/st$ defines a multiplication on $\boldsymbol{Z}_S$ that depends only on the classes being multiplied and not on the elements of M that represent those classes (verify!).

Arithmetic properties of $\boldsymbol{Z}$ like those in Theorem 1.28 are easily verified for $\boldsymbol{Z}_S$, so it is a commutative ring. To prove $a/s + b/t = b/t + a/s$ use the definition of addition in $\boldsymbol{Z}_S$ and the commutativity of both addition and multiplication in $\boldsymbol{Z}$ to write $a/s + b/t = (at + bs)/st = (bs + at)/ts = b/t + a/s$. In verifying other properties of $\boldsymbol{Z}_S$ it helps to make two observations. We know that $[(a, s)] = a/s = b/t = [(b, t)] \Leftrightarrow at = sb$ so for s, $t \in S$, $a/s = at/st$ and $0/s = 0/t$. Now we can see that $0/s$ is an identity element for addition, like $0 \in \boldsymbol{Z}$, since $0/s + b/t = (0t + bs)/st = bs/ts = b/t$. Also $0/s \cdot b/t = 0b/st = 0/st = 0/s$, so $0/s$ "acts" like $0 \in \boldsymbol{Z}$ under multiplication.

The additive inverse, or negative, of a/s is $(-a)/s$, since $a/s + (-a)/s = (as - as)/s^2 = 0/s^2 = 0/s$, using the computation above once again. As usual we set $(-a)/s = -(a/s)$. Finally, like $\boldsymbol{Z}$, $\boldsymbol{Z}_S$ has as identity element under multiplication. It is tempting to say it must be 1/1 but there is no guarantee that $1 \in S$. Instead, for any $s \in S$ use the element s/s, which of course is the same as t/t for $s, t \in S$ since $st = ts$. To see that this works we compute that $a/s \cdot t/t = at/st = a/s$. All of these arithmetic properties hold for $\boldsymbol{Z}$. The new property of $\boldsymbol{Z}_S$ is that certain elements other than the identity and its negative have reciprocals. If $s, t \in S$, then $s/t, t/s \in \boldsymbol{Z}_S$ and $s/t \cdot t/s = st/ts = st/st = s/s$, so $t/s = (s/t)^{-1}$ acts like a reciprocal of s/t. Note that when $S = \boldsymbol{Z}^*$ every element of $\boldsymbol{Z}_S - \{0/s\}$ has a reciprocal.

We want to argue somewhat informally that for any multiplicatively closed S we may think of $\boldsymbol{Z} \subseteq \boldsymbol{Z}_S$ via $f_S = f : \boldsymbol{Z} \to \boldsymbol{Z}_S$ given by $f(m) = ms/s$ for any $s \in S$. Note that f really is a function since for $s, t \in S$, $ms/s = mt/t$ (why?). Next, if $f(m) = ms/s = f(n)$ then $ms/s = nt/t$ so $mst = nst$ and $m = n$ results from $st \neq 0$. Consequently f is an injection and we can identify $\boldsymbol{Z}$ in $\boldsymbol{Z}_S$ as $\{ns/s \mid n \in \boldsymbol{Z} \text{ and } s \in S\}$. Observe that $f(n)f(m) = ns/s \cdot mt/t = mnst/st = (mn)s/s = f(nm)$ and $f(n) + f(m) = ns/s + mt/t = (nst + mst)/st = (m + n)st/st = (m + n)s/s = f(n + m)$. Thus the addition and multiplication of corresponding elements in $\boldsymbol{Z}$ and in $\boldsymbol{Z}_S$ correspond, so we can consider $\boldsymbol{Z} \subseteq \boldsymbol{Z}_S$ via f as commutative rings rather than simply as sets. If $s, t \in S$ then $st \in S$, so $f(s) = st/t$ has reciprocal $t/st \in \boldsymbol{Z}_S$: the elements in $\boldsymbol{Z}_S$ corresponding to S have inverses in $\boldsymbol{Z}_S$. Finally, for any $a/s \in \boldsymbol{Z}_S$, we write $a/s = at^2/st^2$ for any $t \in S$ (why?), so $a/s = at/t \cdot t/st = at/t(st/t)^{-1} = f(a)f(s)^{-1}$. This implies that $\boldsymbol{Z}_S$ is the smallest commutative ring containing a copy of $\boldsymbol{Z}$ and the inverses of all $s \in S$. When $S = \boldsymbol{Z}^*$ we may naturally identify $\boldsymbol{Z}_S$ with $\boldsymbol{Q}$, constructing $\boldsymbol{Q}$.

Our main purpose in this section has been to illustrate an important use of equivalence relations and to observe some of the complications that arise in working with equivalence classes. Before ending this section we take a very informal look at some other examples of our construction. When $S = \{2^i \mid i \in \boldsymbol{N}\}$ then $\boldsymbol{Z}_S$ is essentially the set of those rational numbers with denominator a power of 2, and when $S = \{n \in \boldsymbol{Z} \mid \text{GCD}\{n, 2\} = 1\}$, then $\boldsymbol{Z}_S$ is identified with the set of fractions with odd denominators.

Recall that a commutative ring satisfies i)–vii) of Theorem 1.28. Multiplicatively closed subsets S can be defined for commutative rings R other than $\boldsymbol{Z}$, such as the polynomials $\boldsymbol{Q}[x]$ with rational coefficients, or $\mathbf{Z}_n[x]$ with coefficients in $\mathbf{Z}_n$. The properties we want in general are those in Definition 1.11: $0 \notin S$ and closure of S under multiplication. Our construction above works in general to give $R \subseteq R_S$ as long as $a, b \in R$ with $a, b \neq 0 \Rightarrow ab \neq 0$. Some multiplicative sets in $\boldsymbol{Q}[x]$ are: $S = \{x^k \mid k \geq 0\}$, $T = \boldsymbol{Q}[x] - \{0\}$, and $W = \{p(x) \in \boldsymbol{Q}[x] \mid p(0) \neq 0\}$. $\boldsymbol{Q}[x]_S$ can be identified with $\boldsymbol{Q}[x, x^{-1}]$, the set of polynomials over $\boldsymbol{Q}$ in both x and x^{-1} (how?), called the *Laurent polynomials*. For $\boldsymbol{Q}[x]_T$ we get the set of quotients of all polynomials with nonzero denominators, called the *ring of rational functions* and denoted $\boldsymbol{Q}(x)$. It is hard to describe $\boldsymbol{Q}[x]_W$ directly as quotients of polynomials other than as the set of quotients of polynomials with denominators having a nonzero constant term. However, any nonzero $p(x) \in \boldsymbol{Q}[x]$ may be written $p(x) = x^k q(x)$ where $q(x)$ has a nonzero constant term and may be a constant itself. Hence $q(x) \in W$ and has an inverse in $\boldsymbol{Q}[x]_W$, so any $\alpha \in \boldsymbol{Q}[x]_W$ can be written $\alpha = x^k w$ for $k \geq 0$ and some invertible $w \in \boldsymbol{Q}[x]_W$. Thus in $\boldsymbol{Q}[x]_W$, we invert all polynomials except the powers of x, just the opposite of what happened in $\boldsymbol{Q}[x]_S$.

When $R = 2\mathbf{Z}$ and $T = (2\mathbf{Z})^*$ it turns out that $(2\mathbf{Z})_T$ is essentially $\boldsymbol{Q}$! To see why, use our construction above when $S = \mathbf{Z}^* \subseteq \mathbf{Z}$ to identify $\mathbf{Z}_S$ with $\boldsymbol{Q}$. For any $a \in \mathbf{Z}$ and $b \in \mathbf{Z}^*$ set $g(a/b) = 2a/2b$. It follows that $g : \mathbf{Z}_S \to (2\mathbf{Z})_T$, that g is a bijection, and that $g(\alpha\beta) = g(\alpha)g(\beta)$ and $g(\alpha + \beta) = g(\alpha) + g(\beta)$ (verify!). Thus the elements in $\mathbf{Z}_S$ and in $(2\mathbf{Z})_T$ are identified, and the sums and products of corresponding elements correspond.

Starting with $\mathbf{Z}_{\{2^i\}}$, take $S = \{3^j/2^i \in \mathbf{Z}_{\{2^i\}} \mid i, j \in \boldsymbol{N}\}$. It is an exercise to see that S is multiplicatively closed so we may consider $(\mathbf{Z}_{\{2^i\}})_S$. By using the usual computations with fractions, we may appropriately identify elements of $(\mathbf{Z}_{\{2^i\}})_S$ with $\mathbf{Z}_T$ for the multiplicatively closed set $T = \{2^i, 3^j, 2^i3^j \mid i, j \in \boldsymbol{N}\} \subseteq \mathbf{Z}$. Specifically, $(m/2^i)/(3^j/2^k) \in (\mathbf{Z}_{\{2^i\}})_S$ can be identified with $m/(2^{i-k}3^j) \in \mathbf{Z}_T$ when $i \geq k$ and with $2^{k-i}m/3^j \in \mathbf{Z}_T$ otherwise.

EXERCISES

1. If R is a commutative ring (or is $\mathbf{Z}$) and $S, T \subseteq R$ are multiplicatively closed, show that $S \cup T \cup ST$ is the smallest multiplicatively closed subset of R containing both S and T, where $ST = \{st \in R \mid s \in S \text{ and } t \in T\}$.
2. Let $\{S_i\}_I$ be a set of multiplicatively closed subsets of the commutative ring R.
   a) Show that $\bigcap S_i$ is multiplicatively closed.
   b) If $\{S_i\}_I$ is a chain (see Definition 1.9), show that $\bigcup S_i$ is multiplicatively closed.
3. If $m \in \mathbf{Z}$, find a maximal multiplicatively closed set $T \subseteq \mathbf{Z}$ that satisfies $T \cap \{m^k \mid k > 0\} = \emptyset$ and identify $\mathbf{Z}_T$ in $\boldsymbol{Q}$ if:
   i) $m = p$ a prime
   ii) $m = p^n$ a prime power
   iii) $m = 6$
   iv) $m = 12$
4. If R is a commutative ring with 1 and if $S \subseteq R$ is multiplicatively closed and consists of units ($s \in S \Rightarrow$ there is $t \in R$ with $st = 1$) show that $R_S = f_S(R)$.
5. If $M^* = \mathbf{Z} \times \mathbf{Z}$ with $\sim$ on M^* defined by $(a, b) \sim (c, d)$ when $ad = bc$, as above, show that $\sim$ is not an equivalence relation.

6. Construct $\boldsymbol{Z}$ from $\boldsymbol{N}$ in analogy with $\boldsymbol{Z}_S$ above as the classes of some equivalence relation on $\boldsymbol{N} \times \boldsymbol{N}$.

7. For $S = \boldsymbol{Q}[x] - \{0\}$, let $\boldsymbol{Q}(x) = \boldsymbol{Q}[x]_S = \{p(x)/q(x) \mid p(x) \in \boldsymbol{Q}[x], q(x) \in S\}$. Show informally how to identify $\boldsymbol{Q}(x)$ with:

i) $\boldsymbol{Z}[x]_T$ for $T = S \cap \boldsymbol{Z}[x]$

ii) $\boldsymbol{Z}[x]_{2T}$ for $2T = \{2t \in \boldsymbol{Z}[x] \mid t \in T\}$

8. Let $S \subseteq \boldsymbol{Z}$ be multiplicatively closed and take $W \subseteq \boldsymbol{Z}_S$ multiplicatively closed. Show there is a multiplicatively closed $T \subseteq \boldsymbol{Z}$ so that the elements of $\boldsymbol{Z}_T$ and of $(\boldsymbol{Z}_S)_W$ can be identified.

CHAPTER 2

GROUPS

2.1 BASIC NOTIONS AND EXAMPLES

Much of the power and utility of mathematics arises from its abstraction of important features common to different situations. Many sets we encounter have a natural multiplication defined on their elements. These multiplications often satisfy a few common properties that we want to isolate and study. Besides the obvious examples in various number systems, we can add and multiply polynomials, add and multiply appropriate matrices, or sometimes add, multiply, and compose functions. The properties of interest define a group, a fundamental algebraic structure.

DEFINITION 2.1 A **group** is a pair $(G, \cdot)$ for G a nonempty set and $\cdot : G \times G \to G$ satisfying: i) $(a \cdot b) \cdot c = a \cdot (b \cdot c)$ for all $a, b, c \in G$; ii) there is $e_G \in G$ with $a \cdot e_G = a = e_G \cdot a$ for all $a \in G$; and iii) for each $a \in G$ there is $b \in G$ with $a \cdot b = b \cdot a = e_G$. If $g \cdot h = h \cdot g$ for all $g, h \in G$, then G is called **Abelian.**

The three group properties are called associativity, the existence of an identity, and the existence of an inverse for each element, respectively. If the multiplication "$\cdot$" is clear, we write G instead of $(G, \cdot)$ and gh instead of $g \cdot h$. We say that G is **finite** when $|G|$ is finite and call $|G|$ the **order** of G. Otherwise G is called infinite and has infinite order. Consider the set of real numbers $\boldsymbol{R}$ that has both an addition and a multiplication. Addition in $\boldsymbol{R}$ is associative and commutative, zero is an identity for addition, and the inverse of $r \in \boldsymbol{R}$ is $-r$, so $(\boldsymbol{R}, +)$ is an Abelian group. Multiplication in $\boldsymbol{R}$ is associative and commutative, and 1 is an identity for it. The inverse for $r \in \boldsymbol{R}$ using multiplication is $1/r$, requiring $r \neq 0$. Hence if $\boldsymbol{R}^* = \boldsymbol{R} - \{0\}$ then $(\boldsymbol{R}^*, \cdot)$ is an Abelian group. Other infinite Abelian groups are $(\boldsymbol{Q}, +)$, $(\boldsymbol{Q}^*, \cdot)$, $(\boldsymbol{Q}^+, \cdot)$, where $\boldsymbol{Q}^+ = \{q \in \boldsymbol{Q} \mid q > 0\}$, $(\boldsymbol{Z}, +)$, and $(n\boldsymbol{Z}, +)$. Although $\boldsymbol{Z}$ has a multiplication, the only integers with integer reciprocals are ± 1, so the only group we could define using the multiplication in $\boldsymbol{Z}$ is $\{\pm 1\}$, or $\{0\}$. In the latter case $0 \cdot 0 = 0$, so 0 is the identity element and its own inverse! This also holds for addition since $0 + 0 = 0$ so $(\{0\}, +)$ is a group.

If $n \in N$ then $(\boldsymbol{Z}_n, +)$ is a finite Abelian group by Theorem 1.28. Just as for $\boldsymbol{R}$, $(\boldsymbol{Z}_n, \cdot)$ is not a group since $[1]_n$ is the identity but $[0]_n$ has no inverse. For $[a]_n \in \boldsymbol{Z}_n$, $[a]_n[b]_n = [1]_n \Leftrightarrow [ab]_n = [1]_n \Leftrightarrow ab - 1 = nt$ for some $t \in \boldsymbol{Z} \Leftrightarrow (a, n) = 1$ by Theorem 1.10. Thus when p is a prime, $[a]_p \in \boldsymbol{Z}_p^* = \boldsymbol{Z}_p - \{[0]_p\}$ has an inverse under multiplication. Since $\boldsymbol{Z}_p^*$ is closed under multiplication by Theorem 1.29, since $[1]_p$ is an identity for multiplication, and since the multiplication in $\boldsymbol{Z}_p$ is associative and commutative by Theorem 1.28, $(\boldsymbol{Z}_p^*, \cdot)$ is an Abelian group.

DEFINITION 2.2 For p a prime, $U_p = \boldsymbol{Z}_p - \{[0]_p\}$ is an Abelian group using the multiplication in $\boldsymbol{Z}_p$, and $|U_p| = p - 1$.

For example, $U_{11} = \{[1], [2], \ldots, [10]\} \subseteq \boldsymbol{Z}_{11}$ has order 10 with inverses identified by: $[2][6] = [3][4] = [5][9] = [7][8] = [10][10] = [1]$. Other computations are $[5][8] = [40] = [7]$, $[3][9] = [27] = [5]$, and $[9][9] = [81] = [4]$. Let us turn to some collections of non-Abelian groups.

By Theorem 0.14, when $|S| > 2$, Bij(S) is a non-Abelian group using composition as the operation. Further, if $I(n) = \{1, 2, \ldots, n\}$, Bij$(I(n)) = S_n$ is finite and $|S_n| = n!$. We will examine these groups in detail in Chapter 3.

Other non-Abelian groups arise from matrices. To describe them, we review briefly the basic definitions starting with the notion of a **ring.** A ring is a set R with an addition and multiplication satisfying properties i)–vii) in Theorem 1.28 for $\boldsymbol{Z}_n$. We will study rings in later chapters but here it will suffice to think of a ring as one of $\boldsymbol{Z}, n\boldsymbol{Z}, \boldsymbol{Q}, \boldsymbol{R}$, or $\boldsymbol{Z}_n$, or polynomials over them. A ring R need not be commutative ($ab = ba$ for all $a, b \in R$). For $n \in N$ and R a ring, $M_n(R)$ is the set of all $n \times n$ matrices with entries in R. Each $A \in M_n(R)$ is a square array of n^2 elements of R arranged in n rows and n columns with the element in the ith row and the jth column denoted by A_{ij}. The **diagonal** of A is the set of entries $A_{11}, A_{22}, \ldots, A_{nn}$.

Addition and multiplication in $M_n(R)$ are defined using "+" and "$\cdot$" in R. For $A, B \in M_n(R)$, $A + B \in M_n(R)$ is defined by $(A + B)_{ij} = A_{ij} + B_{ij}$ and the product $AB \in M_n(R)$ by $(AB)_{ij} = A_{i1}B_{1j} + A_{i2}B_{2j} + \cdots + A_{in}B_{nj}$. Then $M_n(R)$ is a ring but its multiplication is not commutative, with a few exceptions. The associativity of matrix multiplication is not obvious but is a straightforward, but tedious, computation. Many of our examples will use $M_2(R)$ so for this case we write out these operations explicitly. Specifically, $\begin{bmatrix} a & c \\ b & d \end{bmatrix} + \begin{bmatrix} a' & c' \\ b' & d' \end{bmatrix} = \begin{bmatrix} a + a' & c + c' \\ b + b' & d + d' \end{bmatrix}$, and $\begin{bmatrix} a & c \\ b & d \end{bmatrix}\begin{bmatrix} a' & c' \\ b' & d' \end{bmatrix} = \begin{bmatrix} aa' + cb' & ac' + cd' \\ ba' + db' & bc' + dd' \end{bmatrix}$. Note that the matrix I_n, where $(I_n)_{jj} = 1$ and $(I_n)_{ij} = 0$ if $i \neq j$, is an identity element for multiplication since $(I_nB)_{ij} = (I_n)_{ii}B_{ij} = B_{ij} = B_{ij}(I_n)_{jj} = (BI_n)_{ij}$. For example, $I_2 = \begin{bmatrix} 1 & 0 \\ 0 & 1 \end{bmatrix}$ and $I_3 = \begin{bmatrix} 1 & 0 & 0 \\ 0 & 1 & 0 \\ 0 & 0 & 1 \end{bmatrix}$. We call I_n the $n \times n$ **identity matrix.** There is also a **scalar multiplication** in $M_n(R)$, $(rA)_{ij} = rA_{ij}$ for $r \in R$ and $A \in M_n(R)$: each entry of A is multiplied by $r \in R$.

DEFINITION 2.3 For any ring R and $n > 1$ let $GL(n, R) = \{A \in M_n(R) \mid AB = BA = I_n$ for some $B \in M_n(R)\}$, that is all $A \in M_n(R)$ with an inverse.

It is difficult to determine whether a matrix has an inverse and, if so, to find it. One approach to the first problem when R is a commutative ring is to use the **determinant,** $\det : M_n(R) \to R$, studied in linear algebra. We give a very brief introduction to it that should suffice for the limited use we will make of this idea. For $A = \begin{bmatrix} a & c \\ b & d \end{bmatrix} \in M_2(R)$, $\det(A) = ad - bc$. For $A \in M_n(R)$ with $n > 2$, computing $\det(A)$ is not easy and by induction turns out to be:

$$\det(A) = A_{11} \det(A(1 \mid 1)) - A_{12} \det(A(1 \mid 2)) + \cdots + (-1)^{n+1} A_{1n} \det(A(1 \mid n))$$

where $A(1 \mid j) \in M_{n-1}(R)$ is obtained from A by deleting its first row and j^{th} column. When $A_{1j} = 0$ that term in $\det(A)$ may be omitted. For example:

$$\begin{aligned} \det\begin{bmatrix} 1 & 2 & -1 \\ 2 & -1 & 0 \\ 1 & -2 & 3 \end{bmatrix} &= \det\begin{bmatrix} -1 & 0 \\ -2 & 3 \end{bmatrix} - 2\det\begin{bmatrix} 2 & 0 \\ 1 & 3 \end{bmatrix} + (-1)\det\begin{bmatrix} 2 & -1 \\ 1 & -2 \end{bmatrix} \\ &= -3 - 12 + 3 = -12, \end{aligned}$$

$$\det\begin{bmatrix} 1 & 2 & 3 \\ 4 & 5 & 6 \\ 7 & 8 & 9 \end{bmatrix} = \det\begin{bmatrix} 5 & 6 \\ 8 & 9 \end{bmatrix} - 2\det\begin{bmatrix} 4 & 6 \\ 7 & 9 \end{bmatrix} + 3\det\begin{bmatrix} 4 & 5 \\ 7 & 8 \end{bmatrix} = -3 + 12 - 9 = 0,$$

and

$$\begin{aligned} \det\begin{bmatrix} 0 & 1 & 0 & 2 \\ 2 & 0 & 1 & 0 \\ 0 & 2 & 0 & 1 \\ 1 & 0 & 2 & 0 \end{bmatrix} &= -\det\begin{bmatrix} 2 & 1 & 0 \\ 0 & 0 & 1 \\ 1 & 2 & 0 \end{bmatrix} - 2\det\begin{bmatrix} 2 & 0 & 1 \\ 0 & 2 & 0 \\ 1 & 0 & 2 \end{bmatrix} \\ &= -2\det\begin{bmatrix} 0 & 1 \\ 2 & 0 \end{bmatrix} + \det\begin{bmatrix} 0 & 1 \\ 1 & 0 \end{bmatrix} - 4\det\begin{bmatrix} 2 & 0 \\ 0 & 2 \end{bmatrix} - 2\det\begin{bmatrix} 0 & 2 \\ 1 & 0 \end{bmatrix} \\ &= -2(-2) + (-1) - 4(4) - 2(-2) = -9 \end{aligned}$$

We state some useful facts about determinants. The most important is $\det(AB) = \det(A)\det(B)$. Also $\det(A) = 0$ if A has a zero row or column or has two equal rows or columns. A matrix $A \in M_n(R)$ is **upper triangular** when $A_{ij} = 0$ for all $i > j$: all entries below the diagonal are zero. **Lower triangular** is defined similarly: $A_{ij} = 0$ when $i < j$. The determinant of an upper (or lower) triangular matrix is the product of its diagonal entries. This exercise follows from the definition and facts above, using induction. Hence $\det(I_n) = 1$ so if A has an inverse B then $1 = \det(I_n) = \det(A)\det(B)$ and $\det(A)$ has an inverse in R under multiplication. The converse is also true when R is commutative, but is not obvious: $A \in M_n(R)$ has an inverse under multiplication $\Leftrightarrow$ $\det(A)$ has an inverse in R under multiplication. Thus for $R = \boldsymbol{Q}$, $\boldsymbol{R}$, or $\boldsymbol{Z}_p$ (p a prime), $A \in M_n(R)$ has an inverse $\Leftrightarrow \det(A) \neq 0$, and, if $R = \boldsymbol{Z}$, exactly when $\det(A) = \pm 1$.

The situation for $M_2(R)$ is easy. If $A = \begin{bmatrix} a & c \\ b & d \end{bmatrix}$ and $\det(A) = ad - bc$ has an inverse in R then the inverse of A is $A^{-1} = (1/\det(A))\begin{bmatrix} d & -c \\ -b & a \end{bmatrix}$. Hence since $\det\begin{bmatrix} 2 & 3 \\ 3 & 4 \end{bmatrix} = -1$,

$\begin{bmatrix} 2 & 3 \\ 3 & 4 \end{bmatrix}^{-1} = \begin{bmatrix} -4 & 3 \\ 3 & -2 \end{bmatrix}$ so $\begin{bmatrix} 2 & 3 \\ 3 & 4 \end{bmatrix} \in GL(2, \boldsymbol{Z})$, and as $\det\begin{bmatrix} 2 & 3 \\ 1 & 4 \end{bmatrix} = 5$, $\begin{bmatrix} 2 & 3 \\ 1 & 4 \end{bmatrix}^{-1} = \begin{bmatrix} 4/5 & -3/5 \\ -1/5 & 2/5 \end{bmatrix}$ so $\begin{bmatrix} 2 & 3 \\ 1 & 4 \end{bmatrix} \in GL(2, \boldsymbol{Q})$, but $\begin{bmatrix} 2 & 3 \\ 1 & 4 \end{bmatrix} \notin GL(2, \boldsymbol{Z})$. Similarly, if $A = \begin{bmatrix} [2]_5 & [2]_5 \\ [1]_5 & [2]_5 \end{bmatrix} \in M_2(\boldsymbol{Z}_5)$, $\det(A) = [2]_5$ and $[2]_5[3]_5 = [1]_5$, so $A^{-1} = [3]_5\begin{bmatrix} [2]_5 & [3]_5 \\ [4]_5 & [2]_5 \end{bmatrix} = \begin{bmatrix} [1]_5 & [4]_5 \\ [2]_5 & [1]_5 \end{bmatrix}$. When they exist, inverses for larger matrices are much more complicated to find.

Example 2.1 Using matrix multiplication, $(GL(n, R), \cdot)$ is a group.

As we remarked above, matrix multiplication is associative. The matrix I_n is an identity for this multiplication, and by definition each element in $GL(n, R)$ has an inverse under multiplication. This does not quite show that $GL(n, R)$ is a group since we must verify that matrix multiplication is a multiplication for $GL(n, R)$: $A, B \in GL(n, R) \Rightarrow AB \in M_n(R)$ but is $AB \in GL(n, R)$? This does hold using $(AB)^{-1} = B^{-1}A^{-1}$ (verify!) or from the relation between inverses and determinants and the fact that $\det(AB) = \det(A)\det(B)$.

From Example 2.1, $GL(n, \boldsymbol{Z}_m)$ is a finite group and is not Abelian when $n > 1$ and $m > 1$ (exercise). Using linear algebra to see that the rows of an invertible matrix are linearly independent, it follows that when p is prime $|GL(n, \boldsymbol{Z}_p)| = (p^n - 1)(p^n - p) \cdot (p^n - p^2) \cdots (p^n - p^{n-1})$. We will use this mostly when $n = 2$ so $|GL(2, \boldsymbol{Z}_p)| = (p^2 - 1)(p^2 - p)$. Thus $|GL(2, \boldsymbol{Z}_2)| = 6$, $|GL(2, \boldsymbol{Z}_3)| = 48$, $|GL(2, \boldsymbol{Z}_5)| = 480$, and $|GL(2, \boldsymbol{Z}_7)| = 2016$. The orders grow quickly if $n > 2$: $|GL(3, \boldsymbol{Z}_2)| = 168$, $|GL(3, \boldsymbol{Z}_3)| = 11{,}232$, and $|GL(5, \boldsymbol{Z}_2)| = 9{,}999{,}360$.

Another group is the usual Euclidean n-dimensional space $\boldsymbol{R}^n$, the Cartesian product of $\boldsymbol{R}$ with itself n times. There is a natural coordinatewise addition on $\boldsymbol{R}^n$: If $\alpha = (a_1, a_2, \ldots, a_n)$, $\beta = (b_1, b_2, \ldots, b_n) \in \boldsymbol{R}^n$, that $\alpha + \beta = (a_1 + b_1, a_2 + b_2, \ldots, a_n + b_n)$. With this definition, $\boldsymbol{R}^n$ is a group.

Example 2.2 $(\boldsymbol{R}^n, +)$ is an Abelian group.

It is straightforward to verify that the addition in $\boldsymbol{R}^n$ is both associative and commutative and to see that an identity element is $(0, \ldots, 0)$. Finally, the inverse for $\alpha = (a_1, a_2, \ldots, a_n) \in \boldsymbol{R}^n$ is $-\alpha = (-a_1, -a_2, \ldots, -a_n)$.

The addition for $\boldsymbol{R}^n$ is the usual vector addition just as for the plane $\boldsymbol{R}^2$ or 3-space $\boldsymbol{R}^3$. Using the same coordinate addition, we can get similar groups like $(\boldsymbol{Z}^n, +)$ or $((\boldsymbol{Z}_m)^n, +)$, and using multiplication in each coordinate gives the groups $((\boldsymbol{R}^*)^n, \cdot)$ and $((U_p)^n, \cdot)$. In the latter case when $p = 7$ and $n = 3$, $([2]_7, [3]_7, [5]_7) \cdot ([4]_7, [6]_7, [5]_7) = ([1]_7, [4]_7, [4]_7)$, and $(([2]_7, [3]_7, [5]_7))^{-1} = ([4]_7, [5]_7, [3]_7)$. We will discuss this construction in more detail in Chapter 3.

EXERCISES

1. Show that each of the following sets with the indicated multiplication is a group and determine whether the group is Abelian.
   a) $(\boldsymbol{R}-\{-1\}, *)$ where $a*b=a+b+ab$ (show that $*:(\boldsymbol{R}-\{-1\})^2 \to \boldsymbol{R}-\{-1\})$
   b) $(\boldsymbol{R}, \#)$ where $r \# s=(r^3+s^3)^{1/3}$
   c) $(S, *)$ for $S=\{\text{all circles } C_r(a, b) \subseteq \boldsymbol{R}^2 \text{ with radius } r>0 \text{ and center } (a, b)\}$, and $C_r(a, b) * C_s(a', b')=C_{rs}(a+a', b+b')$
   d) $(P(S), \#)$ where $P(S)=\{A \subseteq S\}$ and $A \# B=(A-B) \cup (B-A)$
   e) $([0, 1), +)$ where $[0, 1) \subset \boldsymbol{R}$ and $a+b=\begin{cases} a+b & \text{when } a+b<1 \\ a+b-1 & \text{when } a+b \geq 1 \end{cases}$
   f) $(\{[3k]_{15} \in \boldsymbol{Z}_{15} \mid 1 \leq k \leq 4\}, \cdot)$ where $\cdot$ is the usual multiplication in $\boldsymbol{Z}_{15}$
   g) $(\{[2k]_{14} \in \boldsymbol{Z}_{14} \mid 1 \leq k \leq 6\}, \cdot)$ where $\cdot$ is the usual multiplication in $\boldsymbol{Z}_{14}$
   h) $(\{[3k]_{27} \in \boldsymbol{Z}_{27} \mid k \in \boldsymbol{N}\}, *)$ where $[a]_{27} * [b]_{27}=[a]_{27}+[b]_{27}-[ab]_{27}$
   i) $(\boldsymbol{Z}_3 \times \boldsymbol{Z}_7, \oplus)$ where $([a]_3, [b]_7) \oplus ([c]_3, [d]_7)=([a]_3+[c]_3, [b]_7+[d]_7)$
   j) $(\boldsymbol{Z}_3 \times \boldsymbol{Z}_7, \cdot)$ where we take $[s]_3$ always with $0 \leq s \leq 2$ and $[t]_7$ always with $0 \leq t \leq 6$ and then $([a]_3, [b]_7) \cdot ([c]_3, [d]_7)=([a+c]_3, [4^c b+d]_7)$
   k) $(\boldsymbol{Z}_2 \times \boldsymbol{Z}, \cdot)$ where $([a]_2, s) \cdot ([b]_2, t)=([a+b]_2, (-1)^b s+t)$
   l) $(\{f: S \to \boldsymbol{C}^*\}, \cdot)$ where S is a nonempty set, $\boldsymbol{C}^*$ is the set of nonzero complex numbers, and $f \cdot g(s)=f(s)g(s)$
2. Let S be a nonempty set and $\cdot: S \times S \to S$ an associative multiplication on S. If there is $e_L \in S$ satisfying $e_L \cdot s=s$ for all $s \in S$, and if for each $s \in S$ there is $t \in S$ with $t \cdot s=e_L$, show that $(S, \cdot)$ is a group.
3. Let $T=\left\{\begin{bmatrix} 1 & q \\ 0 & 0 \end{bmatrix} \in M_2(\boldsymbol{Q})\right\}$ and use matrix multiplication to consider $(T, \cdot)$. Show that $\begin{bmatrix} 1 & 0 \\ 0 & 0 \end{bmatrix}=e_L$ is a "left" identity element for T ($e_L \cdot t=t$ if $t \in T$) and that for each $t \in T$ there is $w \in T$ with $t \cdot w=e_L$. Show that $(T, \cdot)$ is *not* a group.
4. Show that in general $(\{[a]_n \in \boldsymbol{Z}_n-\{[0]_n\} \mid [a]_n^2=[a]_n\}, \cdot)$, with $\cdot$ the regular multiplication in $\boldsymbol{Z}_n$, is not a group (consider $pq \mid n$ for $p \neq q$ primes).
5. Show carefully that if $A \in M_n(R)$ for R a commutative ring, and A is upper triangular, then $\det(A)$ is the product of the diagonal entries of A.
6. If R is a commutative ring show carefully that the inverse of an invertible upper triangular matrix in $GL(n, R)$ is also upper triangular.
7. Show that $GL(n, \boldsymbol{Z}_m)$ is not an Abelian group if $n>1$ and $m>1$.

2.2 UNIQUENESS PROPERTIES

We look at results that will make computing in groups easier, and we begin by showing that the identity and inverses in a group are unique.

THEOREM 2.1 For any group G the following hold:

i) e_G is the only element $x \in G$ satisfying $gx=g=xg$ for all $g \in G$.
ii) For each $g \in G$ there is a unique $g^{-1} \in G$ satisfying $gg^{-1}=e_G=g^{-1}g$.
iii) For each $g \in G$, $(g^{-1})^{-1}=g$.
iv) For any $g, h \in G$, $(gh)^{-1}=h^{-1}g^{-1}$.

Proof If both $e_G, f \in G$ are identity elements then as f is an identity, $e_G = e_G f$, and since e_G is an identity, $e_G f = f$, so $e_G = f$ proving i). For ii), if $a, c \in G$ are inverses for g then $a = ae_G = a(gc) = (ag)c = e_G c = c$. Thus the inverse for g is unique, denoted g^{-1}. But $gg^{-1} = e_G = g^{-1}g$ so g is an inverse for g^{-1}. By ii), g must be the unique inverse: $g = (g^{-1})^{-1}$. Using associativity, $(gh)(h^{-1}g^{-1}) = g(h(h^{-1}g^{-1})) = g((hh^{-1})g^{-1}) = g(e_G g^{-1}) = gg^{-1} = e_G$. Similarly, $(h^{-1}g^{-1})(gh) = e_G$, so by ii), $h^{-1}g^{-1}$ must be the unique inverse for gh: that is, $(gh)^{-1} = h^{-1}g^{-1}$. ∎

It is a nontrivial exercise using induction that part iv) extends to any $\{g_1, \ldots, g_n\} \subseteq G: (\cdots(((g_1g_2)g_3)\cdots g_n))^{-1} = ((\cdots(g_n^{-1}g_{n-1}^{-1})\cdots))g_1^{-1}$. Since elements in a group G have inverses, equations like $gx = h$ or $xg = h$ have unique solutions in G obtained by multiplying by g^{-1}. Similarly, cancellation holds in a group. We state this together with facts about squares, $g^2 = gg$.

THEOREM 2.2 In any group G the following hold:

i) For $a, b, c, \in G$, if either $ab = ac$ or $ba = ca$, then $b = c$.
ii) For $g \in G$, $g^2 = e_G \Leftrightarrow g^{-1} = g$.
iii) e_G is the only solution in G of $x^2 = x$.

Proof If $ab = ac$ then $a^{-1}(ab) = a^{-1}(ac)$, so $(a^{-1}a)b = (a^{-1}a)c$ and $e_G b = e_G c$, forcing $b = c$. Similarly $ba = ca \Rightarrow b = c$. If $gg = e_G$ then g is an inverse for itself so $g = g^{-1}$ by the uniqueness of inverses in Theorem 2.1. When $g = g^{-1}$, clearly $gg = gg^{-1} = e_G$. Finally, e_G is a solution of $x^2 = x$, and if $g \in G$ is any solution we have $gg = g = ge_G$ and conclude from i) that $g = e_G$. ∎

We may want to compute the product of several elements in a group G, but must define what this means since we can multiply only two elements at a time. Any product involving many elements must be obtained from a sequence of multiplications. The product of $a, b, c \in G$, written in order, is either $a(bc)$ or $(ab)c$ and these are equal by the associative property. For $a, b, c, d \in G$, two products in the given order are $(ab)(cd)$ and $(a(bc))d$. For many elements there are many possible products with the elements written in a fixed order. Some examples for five elements would be $(ab)(c(df))$, $((ab)(cd))f$, $a(b(c(df)))$, and $(a(bc))(df)$. For the remainder of this section denote any possible prouduct of $g_1, g_2, \ldots, g_k \in G$ written with increasing subscripts as $[[g_1, g_2, \ldots, g_k]]$. We emphasize that this symbol is not a fixed product but represents any of many possibilities, as above. The next result shows that all of these products are equal. Hence it is not necessary to use parentheses to define the product and we can write $g_1g_2\cdots g_k$ unambiguously as the product of these elements in the given order, no matter what sequence of multiplications is used to obtain it.

THEOREM 2.3 For G a group, all products of $g_1, g_2, \ldots, g_k \in G$ written in this same order are equal, so any such product is $(\cdots((g_1g_2)g_3)\cdots)g_k$, written $g_1g_2\cdots g_k$.

Proof We use induction on $n \geq 2$. For $n = 2$ the only product with increasing subscripts is g_1g_2, and for $n = 3$ there are only two such products $g_1(g_2g_3) = (g_1g_2)g_3$, by associativity. Assume that for some $n \geq 2$ the theorem holds for all $2 \leq k \leq n$, and consider any product $[[g_1, g_2, \ldots, g_{n+1}]]$ of $n + 1$ elements written with increasing subscripts. This product arises from a sequence of products, the last of which is $[[g_1, g_2, \ldots, g_k]][[g_{k+1}, \ldots, g_{n+1}]]$: some product of the first k elements with increasing subscripts multiplied by some product of the last $n+1-k$ elements with increasing subscripts. Should $k=n$, then by our induction hypothesis $[[g_1, g_2, \ldots, g_n]]g_{n+1} = (\cdots((g_1g_2)g_3)\cdots g_n)g_{n+1}$, which would prove the $n+1$ case of the theorem. When $k<n$, then $n+1-k \geq 2$, and using the induction hypothesis again, together with the associative property, shows that

$$\begin{aligned}
[[g_1, g_2, \ldots, g_k]][[g_{k+1}, \ldots, g_{n+1}]] &= [[g_1, g_2, \ldots, g_k]]((((g_{k+1}g_{k+2})\cdots)g_n)g_{n+1}) \\
&= ([[g_1, g_2, \ldots, g_k]](((g_{k+1}g_{k+2})\cdots)g_n))g_{n+1} \\
&= [[g_1, g_2, \ldots, g_n]]g_{n+1} \\
&= (\cdots((g_1g_2)\cdots)g_n)g_{n+1}
\end{aligned}$$

proving the $n+1$ case again. Applying IND-II proves the theorem. ■

An important consequence of Theorem 2.3 is that we need not parenthesize a product of elements in G or consciously apply associativity. For example, we may write $(x(yg^{-1}))(gh) = xyh$ since by Theorem 2.3 we could compute this product by starting with $g^{-1}g = e_G$ and next multiplying by h on the right (or y on the left), etc. Similarly $(x(yg))(g^{-1}(y^{-1}z)) = xz$. Another use of Theorem 2.3 is to define powers unambiguously. For $n \in \boldsymbol{N}$ and $g \in G$, $g^n = [[g, g, \ldots, g]]$, a product of g with itself n times.

DEFINITION 2.4 For G any group and $g \in G$, set $g^0 = e_G$, and for $n \in \boldsymbol{N}$, set $g^n = g^{n-1}g$ and $g^{-n} = (g^n)^{-1} = (g^{-1})^n$.

Note that $g^1 = g$, and for $n > 0$, $g^n = g^{n-1}g = (\cdots((gg)g)\cdots g)g$ is just the product of g with itself n times. That $(g^n)^{-1} = (g^{-1})^n$ requires justification. By Theorem 2.3 if $n \geq 1$, $g^n = g^{n-1}g = gg^{n-1}$ and now $g^n(g^{-1})^n = g^{n-1}(g(g^{-1})^n) = g^{n-1}((gg^{-1})(g^{-1})^{n-1}) = g^{n-1}(e_G(g^{-1})^{n-1}) = g^{n-1}(g^{-1})^{n-1}$. It follows from IND-I that $g^n(g^{-1})^n = e_G$ (verify!). Similarly $(g^{-1})^n g^n = e_G$, so $(g^{-1})^n$ is an inverse for g^n and by Theorem 2.1 is the unique inverse $(g^n)^{-1}$, so $(g^n)^{-1} = (g^{-1})^n$.

We emphasize that g^k means the product of g with itself *using the multiplication in G*. If $f \in (\text{Bij}(\boldsymbol{R}), \circ)$ is defined by $f(r) = r+1$ then $f^2(r) = f(f(r)) = r+2$, not $(r+1)^2$, since the multiplication in the group $\text{Bij}(\boldsymbol{R})$ is composition. Also for $5 \in (\boldsymbol{Z}, +)$, $5^3 = 5+5+5 = 15 \in \boldsymbol{Z}$, not 125. The danger of miscomputing powers is even greater for $\boldsymbol{Q}$ since $(\boldsymbol{Q}, +)$ and $(\boldsymbol{Q}^*, \cdot)$ are groups. As a group element $3^4 \in \boldsymbol{Q}$ is ambiguous: if $3 \in (\boldsymbol{Q}, +)$ then $3^4 = 3+3+3+3 = 12$ and $3^{-4} = -12$, but if $3 \in (\boldsymbol{Q}^*, \cdot)$ then $3^4 = 3\cdot 3\cdot 3\cdot 3 = 81$ and $3^{-4} = 81^{-1} = 1/81$. For $g \in (\boldsymbol{Z}, +)$ and $n \in \boldsymbol{N}$ $g^n = g^{n-1} + g = (n-1)g + g = ng$, using induction. Also $g^{-n} = -ng$ so $g\boldsymbol{Z} = \{g \cdot z \mid z \in \boldsymbol{Z}\} = \{g^z \mid z \in \boldsymbol{Z}\}$ computed in $(\boldsymbol{Z}, +)$. Next we see that the usual rules for exponents hold in groups.

THEOREM 2.4 Let G be a group, $g, h \in G$, and $i, j \in \boldsymbol{Z}$.

i) $g^i = (g^{-1})^{-i} = (g^{-i})^{-1}$
ii) $g^i g^j = g^{i+j}$
iii) $(g^i)^j = g^{ij}$

iv) $(g^{-1}hg)^j = g^{-1}h^j g$
v) If $gh = hg$, then $g^j h^i = h^i g^j$
vi) If $gh = hg$, then $(gh)^j = g^j h^j$.

Proof The arguments are not difficult, but we must be careful not to use facts about powers that we have not yet proved.

Part i). If $i = 0$ all three terms are e_G, so equal. If $i < 0$ then $-i > 0$ and Definition 2.4 gives $g^i = g^{-(-i)} = (g^{-i})^{-1} = (g^{-1})^{-i}$. When $i > 0$, $g^{-i} = (g^{-1})^i = (g^i)^{-1}$ (Definition 2.4), so taking inverses yields $(g^{-i})^{-1} = ((g^{-1})^i)^{-1} = g^i$ by Theorem 2.1. Apply Definition 2.4 to g^{-1} to get $((g^{-1})^i)^{-1} = (g^{-1})^{-i}$, proving i).

Part ii). Since $g^0 = e_G$, ii) is clear if $i = 0$ or $j = 0$. If $j = 1$, then ii) is just the definition of powers when $i \geq 1$. If $i < 0$, then i) and Definition 2.4 show $g^i g = (g^{-1})^{-i} g = (g^{-1})^{-i-1}(g^{-1})g = (g^{-1})^{-i-1} = g^{i+1}$. Assume for some $n \geq 1$ that $g^i g^n = g^{i+n}$ for any $g \in G$ and $i \in \mathbf{Z}$. This assumption, the $n = 1$ case, and Definition 2.4 show that $g^i g^{n+1} = g^i(g^n g) = (g^i g^n)g = g^{i+n}g = g^{i+n+1}$, proving ii) when $j \geq 1$ by IND-I. If $j < 0$ then use i) to write $g^i g^j = (g^{-1})^{-i}(g^{-1})^{-j} = (g^{-1})^{-i-j}$ since $-j > 0$, so applying i) again results in $g^i g^j = g^{i+j}$.

Part iii). The cases $j = 0$ and $j = 1$ are clear. The truth of $(g^i)^n = g^{in}$ for $n \geq 1$, along with ii) and Definition 2.4, gives $(g^i)^{n+1} = (g^i)^n(g^i) = g^{in}g^i = g^{in+i} = g^{i(n+1)}$. Thus IND-I proves $(g^i)^j = g^{ij}$ when $j \geq 1$. When $j < 0$, then iii) for $-j > 0$ and i) give $(g^i)^j = ((g^i)^{-1})^{-j} = ((g^{-1})^i)^{-j} = (g^{-1})^{-ij} = g^{ij}$, proving iii).

Part iv). The result is clear when $j = 0$ or $j = 1$, and if true for $m \geq 1$ then from Definition 2.4, $(g^{-1}hg)^{m+1} = (g^{-1}hg)^m(g^{-1}hg) = (g^{-1}h^m g)g^{-1}hg$. Hence $(g^{-1}hg)^{m+1} = (g^{-1}h^m)hg = g^{-1}h^{m+1}g$ by Theorem 2.3, and IND-I shows iv) holds for $j \geq 1$. By Theorem 2.3, $(g^{-1}hg)(g^{-1}h^{-1}g) = e_G = (g^{-1}h^{-1}g)(g^{-1}hg)$ so $(g^{-1}hg)^{-1} = g^{-1}h^{-1}g$. When $j < 0$ then $-j > 0$ so using i) and Definition 2.4 leads again to $(g^{-1}hg)^j = ((g^{-1}hg)^{-1})^{-j} = (g^{-1}h^{-1}g)^{-j} = g^{-1}(h^{-1})^{-j}g = g^{-1}h^j g$, proving iv).

Part v). Since $g^0 = e_G$ the case $j = 0$ holds. Assume $gh = hg$ and for some $m \geq 1$ that $g^m h = hg^m$. Using Theorem 2.3 we see that $g^{m+1}h = gg^m h = ghg^m = hgg^m = hg^{m+1}$ so IND-I implies that $g^j h = hg^j$ for $j \geq 1$. Multiplying on both sides by h^{-1} yields $h^{-1}g^j = g^j h^{-1}$ and taking inverses gives $(g^j)^{-1}h = h(g^j)^{-1}$, or $g^{-j}h = hg^{-j}$ by iii). Hence $g^j h = hg^j$ for $j < 0$, so for all $j \in \mathbf{Z}$. But now for any $i, j \in \mathbf{Z}$ if $x = g^j$ then $hx = xh$ so $h^i x = xh^i$, or $g^j h^i = h^i g^j$, proving v).

Part vi). This clearly holds when $j = 0$ or $j = 1$. If $(gh)^m = g^m h^m$ for some $m \geq 1$, then $(gh)^{m+1} = (gh)^m gh = (g^m h^m)gh = g^m(h^m g)h$ by Definition 2.4 and Theorem 2.3, so using v), $(gh)^{m+1} = g^m gh^m h = g^{m+1}h^{m+1}$. Thus vi) holds if $j \geq 1$ by IND-I. Taking inverses of $gh = hg$ shows that $g^{-1}h^{-1} = h^{-1}g^{-1}$ as well. For $j < 0$ use i) to observe that $(gh)^j = ((gh)^{-1})^{-j} = (h^{-1}g^{-1})^{-j} = (g^{-1}h^{-1})^{-j} = (g^{-1})^{-j}(h^{-1})^{-j} = g^j h^j$, finishing the proof of the theorem.

■

An informal way of looking at Theorem 2.4 when $i, j > 0$ would be to say in ii) that the product of g with itself i times multiplied with the product j times is the product $i + j$ times, and for iii) the product of g with itself i times multiplied by *itself* j times is just the product of g with itself ij times. For iv), $(g^{-1}hg)^j = (g^{-1}hg)(g^{-1}hg)\cdots(g^{-1}hg)$ and all the interior products $gg^{-1} = e_G$ so these cancel, resulting in $g^{-1}h^j g$. Similarly, $g^j h = gg\cdots gh$ and since $gh = hg$, h moves past each g until it is on the left, leaving hg^j. Finally, writing

$(gh)^j = ghgh \cdots gh$, use $gh = hg$ to move each h to the right to get $g^j h^j$. When $i < 0$ or $j < 0$ the same approach works using Definition 2.4 to write $g^k = (g^{-1})^{-k}$.

EXERCISES

1. If G is a group and $(ab)^2 = a^2b^2$ for all $a, b \in G$, show that G is Abelian.
2. If G is a group so any $g, h \in G - \{e_G\}$ with $h \neq g \Rightarrow gh \neq hg$, show $|G| \leq 2$.
3. If G is a group and $a, x \in G$ show that $xa = ax \Leftrightarrow x^{-1}ax = a \Leftrightarrow xax^{-1} = a$.
4. For $g \in G$ a group, show that $g^n = e_G$ for some $n \in \boldsymbol{N} \Leftrightarrow g^{-1} = g^k$, for some $k \in \boldsymbol{N}$.
5. If G is a finite group and $g \in G$, show that $g^k = e_G$ for some $k \in \boldsymbol{N}$.
6. Let G be a group and $g_1, g_2, \ldots, g_k, g, h \in G$. Show carefully that:
   a) for $n \in \boldsymbol{Z}$, $e_G^n = e_G$
   b) $(g_1g_2 \cdots g_k)^{-1} = g_k^{-1}g_{k-1}^{-1} \cdots g_1^{-1}$
   c) if $g^{-1}hg = h^j$ for some $j \in \boldsymbol{Z}$, then for all $m \in \boldsymbol{N}$, $g^{-m}hg^m = h^{j^m}$
7. If G is a group with $|G| < 6$, show that G is Abelian.
8. For G any group, show that the following functions on G are in Bij(G):
   a) $f : G \to G$ given by $f(g) = g^{-1}$
   b) for each $x \in G$, $\tau_x : G \to G$ given by $\tau_x(g) = gx$
   c) for each $x \in G$, $\sigma_x : G \to G$ given by $\sigma_x(g) = xg$
   d) for each $x \in G$, $T_x : G \to G$ given by $T_x(g) = x^{-1}gx$
9. If $G = \{g_1, g_2, \ldots, g_n\}$ is an Abelian group of order n, show that $(g_1g_2 \cdots g_n)^2 = e_G$.
10. If G is a finite group of even order, show that there is $g \in G - \{e_G\}$ with $g^2 = e_G$.

2.3 GROUPS OF SYMMETRIES

We consider some groups arising from the geometry in $\boldsymbol{R}^n$. Just as in $\boldsymbol{R}^2$ or $\boldsymbol{R}^3$ we define length and distance in $\boldsymbol{R}^n$. If $\alpha = (a_1, a_2, \ldots, a_n) \in \boldsymbol{R}^n$, the **length** of α is $\|\alpha\| = (a_1^2 + \cdots + a_n^2)^{1/2}$ and the distance between $\alpha, \beta \in \boldsymbol{R}^n$ is $\text{dist}(\alpha, \beta) = \|\alpha - \beta\|$. These reduce to the usual definitions when $n = 2$ or $n = 3$. The basic properties of distance are satisfied by $\text{dist} : \boldsymbol{R}^n \times \boldsymbol{R}^n \to \boldsymbol{R}$: $\text{dist}(\alpha, \beta) \geq 0$ and $\text{dist}(\alpha, \beta) = 0 \Leftrightarrow \alpha = \beta$; $\text{dist}(\alpha, \beta) = \text{dist}(\beta, \alpha)$; and $\text{dist}(\alpha, \beta) + \text{dist}(\beta, \gamma) \geq \text{dist}(\alpha, \gamma)$, the triangle inequality. The last fact is not obvious but we will accept it since its proof would entail a major digression. Our study of groups will be mostly abstract, and the examples here provide something more concrete, although we will use them rarely after this section.

DEFINITION 2.5 $E(\boldsymbol{R}^n) = \{f : \boldsymbol{R}^n \to \boldsymbol{R}^n \mid \text{dist}(f(\alpha), f(\beta)) = \text{dist}(\alpha, \beta), \alpha, \beta \in \boldsymbol{R}^n\}$.

Any $f \in E(\boldsymbol{R}^n)$ is called a **Euclidean transformation** of $\boldsymbol{R}^n$. Since f preserves distances it is easy to see that it is an injection but it is not clear if f is a bijection. We restrict our attention to $E(\boldsymbol{R}^2)$, characterize its elements, and show that it is a group. Since $f \in E(\boldsymbol{R}^2)$ preserves distances, if $\mathrm{C}_r(\alpha)$ is the circle of radius r about α, then $f(\mathrm{C}_r(\alpha)) \subseteq \mathrm{C}_r(f(\alpha))$. Also, for any straight line L, $f(L)$ is contained in another straight line. To see this let $P, Q \in L$ be distinct and take any third $S \in L$. Rename these H, I, and J with I between H and J, so $\text{dist}(H, J) = \text{dist}(H, I) + \text{dist}(I, J)$. Since $f \in E(\boldsymbol{R}^2)$, $\text{dist}(f(H), f(J)) = \text{dist}(f(H), f(I)) + \text{dist}(f(I), f(J))$ and this forces $f(H)$, $f(I)$, and $f(J)$ to be collinear (why?). Since two of these points are $f(P)$ and $f(Q)$, $f(S)$ must be on the line passing through $f(P)$ and $f(Q)$, for any S on L.

We claim that any $f \in E(\boldsymbol{R}^2)$ is determined by $f(U)$, $f(V)$, and $f(W)$ for any noncollinear points U, V, and W. For any $P \in \boldsymbol{R}^2$ let $d_1 = \text{dist}(P, U)$, $d_2 = \text{dist}(P, V)$, and $d_3 = \text{dist}(P, W)$. These distances uniquely determine P. Note that $P \in C_{d_1}(U) \cap C_{d_2}(V)$ and since $U \neq V$ there are at most two such points of intersection, say P and P'. When $P = P'$, $\text{dist}(U, V) = d_1 + d_2$ and P is uniquely determined by d_1 and d_2 on the line through U and V, so by (d_1, d_2, d_3). If $P \neq P'$ and $P, P' \in C_{d_3}(W)$ then W lies on the perpendicular bisector of the segment joining P and P' and this is also the line through U and V. But U, V, and W are not collinear, so only P can lie on all three circles and is uniquely determined by (d_1, d_2, d_3). If $f, g \in E(\boldsymbol{R}^2)$ and agree on U, V, and W then $f(P)$ and $g(P)$ lie on the circles of radius d_1 about $f(U) = g(U)$, of radius d_2 about $f(V) = g(V)$, and of radius d_3 about $f(W) = g(W)$. Since there is only one such point, $f(P) = g(P)$. Thus f is determined by its action on any three noncollinear points. For example, if $f \in E(\boldsymbol{R}^2)$ fixes each of three noncollinear points then f must be the identity map, and if f fixes the origin and interchanges $(1, 2)$ and $(-1, 2)$ then f must be the reflection about the y-axis since this reflection acts the same way on these three points.

Note next that if $f, g \in E(\boldsymbol{R}^2)$, then $f \circ g \in E(\boldsymbol{R}^2)$, and if $\gamma \in \boldsymbol{R}^2$, the translation $A_\gamma(\alpha) = \alpha + \gamma$ is in $E(\boldsymbol{R}^2)$: $\text{dist}(A_\gamma(\alpha), A_\gamma(\beta)) = \text{dist}(\alpha + \gamma, \beta + \gamma) = \|((\alpha + \gamma) - (\beta + \gamma)\| = \|\alpha - \beta\| = \text{dist}(\alpha, \beta)$. Also $A_\gamma^{-1} = A_{-\gamma}$ so A_γ is a bijection by Theorem 0.11. Clearly the counter-clockwise rotation $T_{\theta,\alpha}$ of $\boldsymbol{R}^2$ about the point α by θ radians for $0 \leq \theta < 2\pi$ preserves distances so $T_{\theta,\alpha} \in E(\boldsymbol{R}^2)$. If $T_\theta = T_{\theta,(0,0)}$ then $T_0 = I$, the identity map on $\boldsymbol{R}^2$, and for $\theta > 0$, $T_\theta^{-1} = T_{2\pi-\theta}$.

Take any $f \in E(\boldsymbol{R}^2)$, let $f((0, 0)) = \gamma$, and set $g = A_{-\gamma} \circ f \in E(\boldsymbol{R}^2)$, so $g((0, 0)) = (0, 0)$. If $g((1, 0)) = \alpha \in C_1((0, 0))$ then some $T_\theta(\alpha) = (1, 0)$ and $H = T_\theta \circ g$ fixes both $(0, 0)$ and $(1, 0)$. Now $H \in E(\boldsymbol{R}^2)$ so $\text{dist}(H((0, 1)), (0, 0)) = \text{dist}(H((0, 1)), H((0, 0))) = \text{dist}((0, 1), (0, 0)) = 1$. Similarly, $\text{dist}(H((0, 1)), (1, 0)) = \text{dist}((0, 1), (1, 0)) = 2^{1/2}$ so $H((0, 1)) = (0, 1)$ or $H((0, 1)) = (0, -1)$. Since H is uniquely determined by its action on $(0, 0)$, $(1, 0)$, and $(0, 1)$, in the first case $H = I$, the identity, and $f = A_\gamma \circ T_{2\pi-\theta}$. When $H((0, 1)) = (0, -1)$ then H agrees with the reflection about the x-axis on three noncollinear points so must *be* the reflection about the x-axis. In this case $f = A_\gamma \circ (T_{2\pi-\theta} \circ H)$. Therefore every $f \in E(\boldsymbol{R}^2)$ is a composition of some A_γ, some T_θ, and perhaps H, and since each of these is a bijection with an inverse of the same type, $E(\boldsymbol{R}^2) \subseteq \text{Bij}(\boldsymbol{R}^2)$ and contains the inverse of each of its elements. Furthermore, if $f \in E(\boldsymbol{R}^2)$ fixes the origin then no A_γ is required and either $f = T_\theta$ or $f = T_\theta \circ H$.

Example 2.3 $(E(\boldsymbol{R}^2), \circ)$ is a group.

We verify the properties in Definition 2.1. We have observed that $f \circ g \in E(\boldsymbol{R}^2)$ when $f, g \in E(\boldsymbol{R}^2)$, so "$\circ$" is a multiplication on $E(\boldsymbol{R}^2)$ and is associative by Theorem 0.9. Clearly the identity function is in $E(\boldsymbol{R}^2)$, and we showed above that $f \in E(\boldsymbol{R}^2) \Rightarrow f^{-1} \in E(\boldsymbol{R}^2)$ so $(E(\boldsymbol{R}^2), \circ)$ is a group by Definition 2.1.

To get some additional examples, we look at the following:

DEFINITION 2.6 If $S \subseteq \boldsymbol{R}^2$ is not empty let $E(S) = \{f \in E(\boldsymbol{R}^2) \mid f(S) = S\}$ and call $E(S)$ the set of **symmetries** of S.

Example 2.4 If $S \subseteq \boldsymbol{R}^2$ is not empty then $(E(S), \circ)$ is group.

Clearly $E(S)$ contains the identity function. If $f, g \in E(S)$ then $g(f(S)) = g(S) = S$, so $f \circ g \in E(S)$, and from Theorem 0.9, composition is associative. Finally, $f \in E(S) \subseteq E(\boldsymbol{R}^2) \Rightarrow f^{-1} \in E(\boldsymbol{R}^2)$ and $f^{-1}(S) = f^{-1}(f(S)) = f^{-1} \circ f(S) = I(S) = S$ so $f^{-1} \in E(S)$ and $E(S)$ is a group by Definition 2.1.

An important special case of Example 2.4 comes next. Let C_1 denote the circle of radius 1 about the origin, $H_{L(\theta)}$ the reflection about the straight line $L(\theta) = T_\theta(x\text{-axis})$, and $H = H_{L(0)}$, the reflection about the x-axis.

Example 2.5 $E(\{(0, 0)\}) = E(C_1) = \{T_\theta, T_\theta \circ H = H_{L(\theta/2)} \mid \theta \in [0, 2\pi)\}$ is not Abelian. Furthermore, $H \circ T_\theta \circ H = T_\theta^{-1} = T_{2\pi-\theta}$ and $H \circ T_\theta = T_{2\pi-\theta} \circ H = H_{L(\pi-\theta/2)}$.

Any $f \in E(\{(0, 0)\})$ preserves distances so $f((0, 0)) = (0, 0) \Rightarrow f(C_1) \subseteq C_1$, and $f^{-1} \in E(\boldsymbol{R}^2)$ satisfies the same property. The restrictions of f and f^{-1} to C_1 are still inverses, so bijections on C_1 by Theorem 0.11 and $f(C_1) = C_1$ results, proving $E(\{(0, 0)\}) \subseteq E(C_1)$. If $f \in E(C_1)$ then $f((0, 0))$ must remain one unit from every point in $f(C_1) = C_1$. Thus $f((0, 0)) = (0, 0)$ and $E(C_1) \subseteq E(\{(0, 0)\})$, proving that $E(\{(0, 0)\}) = E(C_1)$. From our discussion above, $E(\{(0, 0)\}) = \{T_\theta, T_\theta \circ H \mid \theta \in [0, 2\pi)\}$. Clearly $T_\theta \circ H$ fixes $(0, 0)$, $T_\theta \circ H((1, 0)) = T_\theta((1, 0)) = (\cos(\theta), \sin(\theta))$, and $T_\theta \circ H((\cos(\theta), \sin(\theta))) = T_\theta((\cos(\theta), -\sin(\theta))) = T_\theta((\cos(-\theta), \sin(-\theta))) = (\cos(\theta - \theta), \sin(\theta - \theta)) = (1, 0)$. Therefore $T_\theta \circ H$ and the reflection about $L(\theta/2) = T_{\theta/2}(x\text{-axis})$ agree on the noncollinear points $(0, 0)$, $(1, 0)$, and $(\cos(\theta), \sin(\theta))$ so $T_\theta \circ H = H_{L(\theta/2)}$. Since each $T_\theta \circ H$ is a reflection, $(T_\theta \circ H) \circ (T_\theta \circ H) = I$, and from Theorem 2.3, $H \circ T_\theta \circ H = T_\theta^{-1} \circ I = T_\theta^{-1} = T_{2\pi-\theta}$ follows. Hence $H \circ T_\theta = T_{2\pi-\theta} \circ H = H_{L((2\pi-\theta)/2)} = H_{L(\pi-\theta/2)}$. In general $T_\theta \circ H = H_{L(\theta/2)} \neq H_{L(\pi-\theta/2)} = H \circ T_\theta$ (example?) so $E(\{(0, 0)\})$ is not Abelian.

The specific examples we will make use of later come from the following theorem characterizing finite groups in $E(\{(0, 0)\})$. It will enable us to describe all finite groups in $E(\boldsymbol{R}^2)$ and all $E(S)$ for $S \subseteq \boldsymbol{R}^2$ and finite.

THEOREM 2.5 If $G \subseteq E(\{0, 0\})$ is a finite group under composition then there is $n \in \boldsymbol{N}$ so that either $G = \langle T_{2\pi/n} \rangle = \{T_{2\pi j/n} \mid 0 \leq j < n\}$ or $G = \{T_{2\pi j/n}, T_{2\pi j/n} \circ (T_\theta \circ H) = T_{\theta+2\pi j/n} \circ H \mid 0 \leq j < n$ and $\theta \in [0, \pi)$ is fixed$\}$. In the first case, $|G| = n$ and G is Abelian, and in the second, $|G| = 2n$, G is not Abelian when $n \geq 3$, and $(T_{\theta+2\pi j/n} \circ H) \circ (T_{\theta+2\pi i/n} \circ H) = T_{2\pi(j-i)/n}$.

Proof From Example 2.5, if $g \in G$ then $g = T_\eta$ is a rotation or $g = T_\eta \circ H$ is a reflection. We write T_α for *any* $\alpha \in \boldsymbol{R}$, using $T_\alpha = T_\beta \Leftrightarrow \alpha = \beta + 2z\pi$ for some $z \in \boldsymbol{Z}$. Since $T_\eta \circ T_\rho = T_{\eta+\rho}$, for $k \in \boldsymbol{N}$, $T_\eta^k = T_{k\eta}$ and $T_\eta^{-1} = T_{-\eta}$ so $T_\eta^{-k} = T_{-k\eta}$. If G contains any T_η then $\{T_\eta^m \mid m \in \boldsymbol{Z}\} = \{T_{m\eta} \mid m \in \boldsymbol{Z}\} \subseteq G$, and G finite forces $T_{k\eta} = T_{m\eta}$ for $k < m$ so $k\eta = m\eta + 2z\pi$ for $z \in \boldsymbol{Z}$. Hence $\eta = 2\pi j/n$ for some $j \in \boldsymbol{Z} - \{0\}$, $n \in \boldsymbol{N}$, and $\text{GCD}\{j, n\} = 1$. Write $sj + tn = 1$ from Theorem 1.10 and observe from above that $(T_\eta)^s = T_{2\pi sj/n} = T_{2\pi(1-tn)/n} = T_{2\pi/n} \in G$.

Thus $\langle T_{2\pi/n}\rangle \subseteq G$, $(T_{2\pi/n})^{-1} = T_{-2\pi/n} = T_{2\pi(n-1)/n} \in \langle T_{2\pi/n}\rangle$, and $T_{2\pi j/n} = T_{2\pi k/n} \Leftrightarrow j \equiv_n k$ so $\langle T_{2\pi/n}\rangle = \{(T_{2\pi/n})^z \mid z \in \mathbf{Z}\}$. Choose $n \in \boldsymbol{N}$ with $T_{2\pi/n} \in G$ and n maximal. If $T_\upsilon = T_{2\pi i/m} \in G$ with $(i, m) = 1$ then from above, $T_{2\pi/m} \in G$. When $m \mid n$, say $n = sm$, then $T_{2\pi/m} = T_{2\pi s/n} = (T_{2\pi/n})^s$ so $T_{2\pi/m}, T_\upsilon \in \langle T_{2\pi/n}\rangle$. Otherwise let $(m, n) = d < m$ and use Theorem 1.8 to write $sm + tn = d$. Now as we have seen, $(T_{2\pi/n})^s = T_{2\pi s/n} \in G$ and $(T_{2\pi/m})^t = T_{2\pi t/m} \in G$. Hence $T_{2\pi d/mn} = T_{2\pi(sm+tn)/mn} = T_{2\pi s/n} \circ T_{2\pi t/m} \in G$ but if $m = dm'$ we have $T_{2\pi/nm'} \in G$ and $nm' > n$, contradicting the choice of n. Consequently $\langle T_{2\pi/n}\rangle$ are all the rotations in G.

If G contains a reflection $T_\theta \circ H$ then G contains all $T_{2\pi j/m} \circ (T_\theta \circ H) = (T_{2\pi j/n} \circ T_\theta) \circ H = T_{\theta+2\pi j/n} \circ H$ (note that $n = 1$ is possible!). Any $T_\theta \circ H, T_\upsilon \circ H \in G$ give $(T_\upsilon \circ H) \circ (T_\theta \circ H) = T_\upsilon \circ (H \circ T_\theta \circ H) = T_\upsilon \circ T_{-\theta} = T_{\upsilon-\theta} \in G$ using Example 2.5. This forces $T_{\upsilon-\theta} = T_{2\pi j/n}$ and $\upsilon = \theta + 2\pi j/n + 2\pi z$. In conclusion, $T_\upsilon \circ H = T_{2\pi j/n} \circ (T_\theta \circ H)$ so $\{T_{2\pi j/n}, T_{\theta+2\pi j/n} \circ H\} = G$. Finally, note that the rotations in $\langle T_{2\pi/n}\rangle$ commute so if $G = \langle T_{2\pi/n}\rangle$ it is an Abelian group of order n. In the other case, $|G| = 2n$ and a straightforward calculation using Example 2.5 proves the other statements in the theorem. Here when $n = 1$, $G = \{I, g = T_\theta \circ H\}$ and $g \circ g = I$, and when $n = 2$, $G = \{I, T_\pi, T_\theta \circ H, T_{\pi+\theta} \circ H\}$, each element is its own inverse, and the composition of any two different nonidentity elements is the third, so G is Abelian of order 4. ■

The most important examples for us using Theorem 2.5 are the groups $E(S)$ when $S = P_n$, the regular n-sided polygon inscribed in C_1 with a vertex at $(1, 0)$. These $D_n = E(P_n)$ are the *groups of symmetries* of the regular n-gons. Note that P_n is symmetric about the x-axis and has n equal sides: P_3 is an equilateral triangle, P_4 a square, P_5 a regular pentagon, P_8 a regular octagon, etc. It should be clear that the maximal distance between two points on P_n occurs between appropriate vertices, so any $f \in E(P_n)$ must send vertices to vertices. The bisector of the angle at each vertex passes through the opposite vertex when n is even, and it is the perpendicular bisector of the opposite side when n is odd. Therefore all of these pass through (0, 0) and $E(P_n) \subseteq E((0, 0))$. Using Theorem 2.5 shows the following:

Example 2.6 $D_n = E(P_n) = \{T_{2\pi j/n}, T_{2\pi j/n} \circ H\}$ is non-Abelian of order $2n$.

The $T_{2\pi j/n} \in D_n$ are n rotations and the $T_{2\pi j/n} \circ H$ are n reflections. When n is odd there is a reflection about the perpendicular bisector of each of the n sides, and when $n = 2m$ there are just m of these, but an additional m reflections about the diameters, or segments through opposite vertices. In Fig. 2.1, these reflections are illustrated for $n = 5$ and for $n = 6$

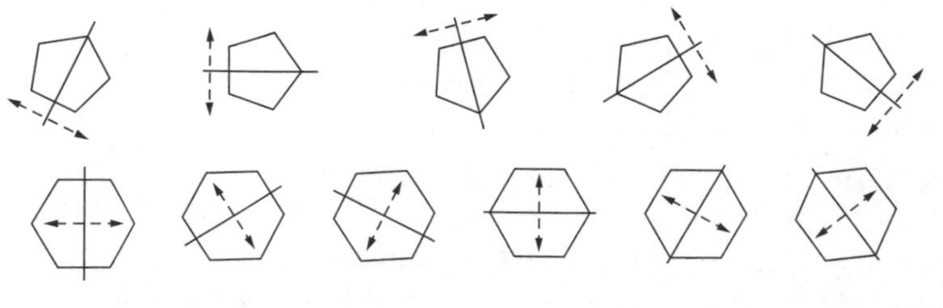

FIGURE 2.1

as representative cases; the rotations about the center moving vertices to vertices are clear and need not be displayed.

We will introduce a different representation for D_n in the next chapter based on congruences. In any case having the $\{D_n\}$ for examples will be valuable since they are the easiest non-Abelian groups to deal with.

To see that $\langle T_{2\pi/n}\rangle = E(S)$ in Theorem 2.5 can occur, take $S = P_n$ but with each edge of P_n an arrow. In Fig. 2.2, we illustrate this for $n = 4$ but the general case is similar. Also illustrated are an isosceles triangle S_1 with $E(S_1) = \{I, H\}$, a rectangle S_2 with $E(S_2) = \{I, T_\pi, H, T_\pi H\}$, and a square S_3 with $T_{2\theta} \circ H \in E(S_3)$ but $H \notin E(S_3)$.

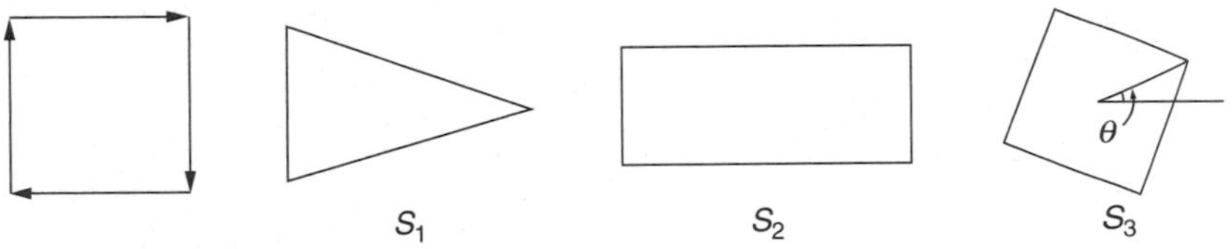

FIGURE 2.2

We want to use Theorem 2.5 to describe $E(S)$ when S is finite and also whenever $E(S)$ itself is finite, but we need two observations. First, if $S \subseteq \boldsymbol{R}^2$, $\gamma \in S$, and $W = \{\beta - \gamma \mid \beta \in S\}$, then $E(S) = A_\gamma E(W) A_{-\gamma}$ where $A_\gamma E(W) A_{-\gamma} = \{A_\gamma \varphi A_{-\gamma} \mid \varphi \in E(W)\}$. The details are left as an exercise (note that $(A_\gamma \varphi A_{-\gamma})^{-1} = A_\gamma \varphi^{-1} A_{-\gamma}$). Second, if also $g(\gamma) = \gamma$ for all $g \in E(S)$, then $E(W) \subseteq E((0, 0))$. Indeed, here $E(S)$ and $E(W)$ are abstractly the same group, one centered at γ and the other at $(0, 0)$, since $A_\gamma T_\theta A_{-\gamma}$ is the rotation $T_{\theta,\gamma}$ about γ through θ radians and $A_\gamma H A_{-\gamma}$ is the reflection about the line $y = b$ through γ where $\gamma = (a, b)$ (why?).

THEOREM 2.6 If $S \subseteq \boldsymbol{R}^2$ then S finite and $|S| \geq 2 \Rightarrow E(S)$ is finite. If $E(S)$ is finite then $E(S) = A_\gamma G A_{-\gamma}$ for some $\gamma \in S$ and $G \subseteq E((0, 0))$ a finite group.

Proof If $S = \{\alpha, \beta\}$ then each $f \in E(S)$ either fixes both α and β or interchanges them. In either case the midpoint γ of the segment with end points α and β is fixed by all $f \in E(S)$. Thus $A_{-\gamma} f A_\gamma \in E((0, 0))$ and $A_{-\gamma} f A_\gamma \in E(\{\alpha - \gamma, \beta - \gamma\})$ so, by Example 2.5, must be either I, T_π, the reflection K about the line through $(0, 0)$ and perpendicular to the line through α and β, or $T_\pi \circ K$. Thus there are only four possibilities for f. The same conclusion holds if S is finite and $S \subseteq L$ a straight line: consider the two points in S farthest apart. If S is finite and contains three noncollinear points then every $f \in E(\boldsymbol{R}^2)$ is determined by the images of these points. But $g \in E(S) \Rightarrow g(S) = S$, and S finite means there are only finitely many possibilities for g.

Now assume $E(S)$ is finite, although S may not be. By our comments preceding the theorem it suffices to see that there is $\gamma \in \boldsymbol{R}^2$ fixed by all $f \in E(S)$ since then $A_{-\gamma} E(S) A_\gamma$ is a finite group in $E((0, 0))$. Let $E(S) = \{I = f_1, \ldots, f_n\}$, take $\alpha \in \boldsymbol{R}^2$, and set $W = \{f_j(\alpha) = \alpha_j\}$. Since $E(S)$ is a group, $\{g f_j\} = E(S)$ for any $g \in E(S)$ because for any f_j, some $f_k = g^{-1} f_j$ and $f_j = g f_k$. This implies that $g(W) = W$ for all $g \in E(S)$. Let $V \subseteq \boldsymbol{R}^2$ be the solid region obtained as follows: connect every two points in W by a straight line segment, giving a finite set $\{B_i\} = M$ of segments, and take all points on all segments with endpoints on some B_i, $B_k \in M$, or, equivalently, $V = \{c_1\alpha_1 + \cdots + c_n\alpha_n \mid \text{all } c_j \geq 0 \text{ and } c_1 + \cdots + c_n = 1\}$. Let γ be the centroid of V. If we had a thin lamina of uniform material in the shape of V it would balance on the point γ, so on any straight line through γ. It follows that half the area of V falls on either side of any line through γ. Further, any line through V and bisecting its area must contain γ. But it is clear that all A_v, T_θ, $H \in E(\boldsymbol{R}^2)$ send any region $U \subseteq \boldsymbol{R}^2$ to a congruent version of itself, so any element of $E(\boldsymbol{R}^2)$ acting on a region preserves its area. Thus if $g \in E(S)$ and L is a line through γ, then since $g(W) = W$ and g sends lines to lines, $g(V) = V$ and $g(L)$ must be a line through V dividing it into equal areas. Thus V must balance on all the $f_j(L)$, and either each of these is L or γ is their unique point of intersection. In the first case, as for $|S| = 2$ above, the midpoint of L is fixed by all $f_j \in E(S)$, and in the second case, $f_j(\gamma) = \gamma$ for all f_j, and there is a fixed point for $E(S)$ in $\boldsymbol{R}^2$ as required. ∎

EXERCISES

1. **a)** Show $H \in E((0, 0))$ satisfies $H(\alpha + \beta) = H(\alpha) + H(\beta)$ for any $a, \beta \in \boldsymbol{R}^2$.
   b) Do the same for $T_\theta \in E((0, 0))$ (write $\alpha = (a, b) = (\|\alpha\| \cos(\eta), \|\alpha\| \sin(\eta))$.
2. For $A_\alpha, T_\theta, H \in E((0, 0))$ show that: i) $T_\theta \circ A_\alpha = A_{T_\theta(\alpha)} \circ T_\theta$;
   ii) $H \circ A_\alpha = A_{H(\alpha)} \circ H$.
3. **a)** Show that $H_{L(\rho)}(\cos(\theta), \sin(\theta)) = (\cos(2\rho - \theta), \sin(2\rho - \theta))$.
   b) Show that $H_{L(\rho)} \circ T_\theta \circ H_{L(\rho)} = T_\theta^{-1}$.
   c) Show that $T_\theta^{-1} \circ H_{L(\rho)} \circ T_\theta = H_{L(\rho - \theta)}$.
4. If $T_\theta \circ H, T_\eta \circ H \in E((0, 0))$, show that their composition is a rotation.
5. Let D_n be given as in Example 2.6 and let $T = T_{2\pi/n}$. Show that $H \circ T \circ H = T^{-1}$ and $H \circ T^k \circ H = (T^k)^{-1}$ where $T^k = T \circ \cdots \circ T$ k-times.
6. Find all $g \in D_n$ satisfying $g \circ h = h \circ g$ for all $h \in D_n$ (this depends on n).
7. If C_r is the circle of radius r about $(0, 0)$, show that $E(\mathrm{C}_1) = E(\mathrm{C}_r)$.
8. If W is the union of the x-axis and the y-axis in $\boldsymbol{R}^2$, describe $E(W)$.
9. If $S \subseteq \boldsymbol{R}^2$ is proper and not empty and if $E(S)$ contains all rotations about $(0, 0)$, show that $E(S) = E((0, 0))$.
10. If $\varnothing \neq S \subseteq \boldsymbol{R}^2$ so that $E(S)$ contains all reflections about all straight lines, show that $E(S) = E(\boldsymbol{R}^2)$.
11. If $S \subseteq \boldsymbol{R}^2$, $\gamma \in S$, and $W = \{\beta - \gamma \mid \beta \in S\}$, show that $E(S) = A_\gamma E(W) A_{-\gamma}$.

2.4 ORDERS OF ELEMENTS

An important approach in studying groups and understanding their structure is to look at the set of all powers of a fixed element. A relevant question is, given $g \in G$, is there $n \in \boldsymbol{N}$ with $g^n = e_G$? We ignore negative powers since $g^{-n} = (g^n)^{-1}$ by Definition 2.4. For any $[a] \in (\boldsymbol{Z}_n, +)$ and $m \in \boldsymbol{N}$, $[a]^m = [a] + \cdots + [a] = [ma]$, so $[a]^n = [na] = [0]$, the identity of $(\boldsymbol{Z}_n, +)$. If p is a prime then in $U_p = (\boldsymbol{Z}_p^*, \cdot)$ (Definition 2.2), $[1] \in U_p$ is the identity and elements $[a]_p \in U_p$ can satisfy $[a]_p^m = [1]_p$. For example, $[-1]_p^2 = [(-1)^2]_p = [1]_p$, $[2]_7^3 = [8]_7 = [1]_7$, $[3]_5^4 = [81]_5 = [1]_5$, and $[2]_{17}^4 = [16]_{17} = [-1]_{17}$, so $[2]_{17}^8 = [(-1)^2]_{17} = [1]_{17}$. It is not at all clear if $[2]_{977}^n = [1]_{977}$ for some $n \in \boldsymbol{N}$.

The only $g \in G$ with $g^n = e_G$ for $n \in \boldsymbol{N}$ may be e_G. If $G = (\boldsymbol{R}, +)$, then $e_G = 0$ and for $r \in (\boldsymbol{R}, +)$ and $n \in \boldsymbol{N}$, $r^n = r + \cdots + r = nr$, so $nr = 0$ forces $r = 0$. When $r \in (\boldsymbol{R}^*, \cdot)$, which has 1 as its identity, then $r^n = 1$ for $n > 0$ implies $r = \pm 1$. In $(\boldsymbol{C}^* = \boldsymbol{C} - \{0\}, \cdot)$ infinitely many g can have some power equal to the identity. Recall that $\boldsymbol{C} = \{a + bi \mid a, b \in \boldsymbol{R}\}$ is the set of **complex numbers** with addition $(a + bi) + (c + di) = (a + c) + (b + d)i$ and multiplication $(a + bi) \cdot (c + di) = (ac - bd) + (ad + bc)i$. Then $(\boldsymbol{C}^*, \cdot)$ is an Abelian group with identity 1, $i^2 = -1$, and $(a + bi)^{-1} = a/r - (b/r)i$ for $r = a^2 + b^2$. If $e^{ri} = \cos(r) + \sin(r)i$ then $e^{ri}e^{si} = e^{(r+s)i}$, so for any $k, n \in \boldsymbol{N}$, the element $z = e^{2\pi ik/n} = \cos(2\pi k/n) + \sin(2\pi k/n)i \in (\boldsymbol{C}^*, \cdot)$ satisfies $z^n = 1$. If $G = GL(2, \boldsymbol{Q})$, non-Abelian with identity $I_2 = \begin{bmatrix} 1 & 0 \\ 0 & 1 \end{bmatrix}$, then many $A \in G$ satisfy $A^2 = I_2$, for example, $\begin{bmatrix} 1 & k \\ 0 & -1 \end{bmatrix}$ for any $k \in \boldsymbol{Q}$. Also $\begin{bmatrix} 1 & 1 \\ -1 & 0 \end{bmatrix}^6 = \begin{bmatrix} 0 & 1 \\ -1 & 0 \end{bmatrix}^4 = \begin{bmatrix} 0 & -1 \\ 1 & -1 \end{bmatrix}^3 = I_2$. Observe that $\begin{bmatrix} 1 & 1 \\ 0 & 1 \end{bmatrix}^n = \begin{bmatrix} 1 & n \\ 0 & 1 \end{bmatrix}$ for any $n \in \boldsymbol{N}$, so $\begin{bmatrix} 1 & 1 \\ 0 & 1 \end{bmatrix}^n \neq I_2$ for all $n \in \boldsymbol{N}$. It is time for a definition to name this property.

DEFINITION 2.7 If G is a group, $g \in G$, and $g^n = e_G$ for some $n \in \boldsymbol{N}$, then the **order** of g, written $o(g)$, is the smallest $m \in \boldsymbol{N}$ so that $g^m = e_G$. If $g^n \neq e_G$ for all $n \in \boldsymbol{N}$ then we say that g has **infinite order.**

If for some $n \in \boldsymbol{N}$, $g^n = e_G$, then by Well Ordering, $\{n \in \boldsymbol{N} \mid g^n = e_G\}$ has a minimal element so $o(g)$ makes sense. Since $e_G^1 = e_G$ and $1 \in \boldsymbol{N}$ is surely minimal, $o(e_G) = 1$. The converse of this is true and will arise many times: if $g \in G$ and $o(g) = 1$ then $g^1 = e_G$, so $g = e_G$. The underlying group must be clear when we consider orders of elements. If $1 \in (\boldsymbol{R}^*, \cdot)$ then $o(1) = 1$, but 1 has infinite order when $1 \in (\boldsymbol{R}, +)$ (why?). Also $[2]_5 \in (\boldsymbol{Z}_5, +)$ has order 5 but $[2]_5 \in U_5$ has order 4 (verify!). It is very important to realize that if $g^m = e_G$ for some $m \in \boldsymbol{N}$, we *cannot* conclude that $o(g) = m$, but only that $o(g) \leq m$: as above, $\begin{bmatrix} 0 & -1 \\ 1 & -1 \end{bmatrix}^9 = I_2$ but $o\left(\begin{bmatrix} 0 & -1 \\ 1 & -1 \end{bmatrix}\right) = 3$. Our next result will be very useful, characterizing $\{m \in \boldsymbol{Z} \mid g^m = e_G\}$ for $g \in G$.

THEOREM 2.7 Let G be a group and $g \in G$ with finite order m. Then $S = \{z \in \boldsymbol{Z} \mid g^z = e_G\} = m\boldsymbol{Z}$. In particular, $g^k = e_G \Leftrightarrow m \mid k$.

Proof The only means of obtaining the conclusion is from Theorem 1.4 so we need $S \neq \emptyset$ and closed under subtraction. Since $o(g) = m \in \boldsymbol{N}$, $m \in S$ by Definition 2.7. If $u, v \in S$ then the definition of S shows $g^u = e_G = g^v$. We need $u - v \in S$ so consider $g^{u-v} = g^u g^{-v} = g^u(g^v)^{-1} = e_G(e_G)^{-1} = e_G$, using Theorem 2.4. Therefore $u - v \in S$, and Theorem 1.4 yields $S = n\boldsymbol{Z}$ for n minimal in $S \cap \boldsymbol{N}$. But then $n = o(g) = m$ from Definition 2.7. ∎

To know that $g^k = e_G$ *exactly* when $o(g) \mid k$ is helpful in finding orders of elements since it restricts the powers that need to be considered. Given $[a]_n \in (\boldsymbol{Z}_n, +)$, recall that $[a]_n^n = [a]_n + \cdots + [a]_n = [na]_n = [0]_n$ so $o([a]_n) \mid n$. Specifically, when $n = 200$, $[120]_{200}^{10} = [1200]_{200} = [0]_{200}$ and $o([120]_{200}) \mid 10$. Now $o([120]_{200}) \neq 1, 2$ (why?), but $[120]_{200}^5 = [600]_{200} = [0]_{200}$, hence $o([120]_{200}) = 5$. It is more interesting to find orders in U_{13}. The successive powers of $[3] \in U_{13}$ are $[3]$, $[9]$, and $[27] = [1]$, so $o([3]) = 3$ in U_{13}. For $[2] \in U_{13}$ $[2]^2 = [4]$, $[2]^4 = [16] = [3]$, and $[2]^8 = [2]^4[2]^4 = [3]^2 = [9]$. We just showed $[3]^3 = [9][3] = [1]$, so $[2]^{12} = [2]^8[2]^4 = [9][3] = [1]$ and $o([2]) \mid 12$ in U_{13}. Clearly $o([2]) \neq 2, 3$, or 4, and $[2]^6 = [64] = [-1] \in U_{13}$; thus $o([2]) = 12$. Still in U_{13}, $[10]^2 = [9]$ and $[10]^4 = [81] = [3]$, so $[10]^6 = [10]^4[10]^2 = [3][9] = [1]$. Is $o([10]) = 6$? Now $[10]^2 = [9]$ and $[10]^3 = [9][10] = [90] = [-1]$, so $o([10]) \neq 1, 2, 3$ forcing $o([10]) = 6$.

We can use Theorem 2.7 to describe $o(g^k)$ when $o(g)$ is finite, so we can find $o([a]_n)$ in $(\mathbf{Z}_n, +)$ since $[a]_n = ([1]_n)^a$. Recall that $(k, n) = \mathrm{GCD}\{k, n\}$.

THEOREM 2.8 If G is a group, $g \in G$, and $o(g) = n \in \mathbf{N}$, then for $k \in \mathbf{Z}^*$, $o(g^k) = n/(k, n)$. In particular, $o(g^k) = n \Leftrightarrow (k, n) = 1$.

Proof Since $(k, n) \mid k$ and $(k, n) \mid n$ write $k = (k, n)k'$ and $n = (k, n)n'$. If $o(g^k) = m$ we show $m = n' = n/(k, n)$ by using Theorem 2.7 to prove $m \mid n'$ and $n' \mid m$. Now $(g^k)^{n'} = g^{kn'} = g^{k'(k,n)n'} = (g^n)^{k'} = e_G^{k'} = e_G$, using Theorem 2.4. Thus $m \mid n'$ by Theorem 2.7. But by definition of order, $(g^k)^m = e_G$, so Theorem 2.4 shows $g^{km} = e_G$ and we conclude that $n \mid km$ by Theorem 2.7. Thus $n'(k, n) \mid k'(k, n)m$, implying $n' \mid k'm$ (Theorem 1.5), and Theorem 1.12 shows first that $(k', n') = 1$ and then that $n' \mid m$. Therefore $m \mid n'$ and $n' \mid m$, so $m = n'$ since $m, n' > 0$. Finally, it is clear that $n/(k, n) = n \Leftrightarrow (k, n) = 1$, finishing the proof. ∎

We saw above in $(\mathbf{Z}_n, +)$ that $[a]_n^k = [ka]_n$ so $o([1]_n) = n$ and if $m > 0$ then $[m]_n = [1]_n^m$. Thus by Theorem 2.8, $o([m])_n = o([1]_n^m) = n/(m, n)$. For example, in $(\mathbf{Z}_{36}, +)$, $o([22]_{36}) = 36/(22, 36) = 18$, $o([30]_{36}) = 36/(30, 36) = 6$, and $o([27]_{36}) = 36/(27, 36) = 4$. Since $o([m]_n) = n/(m, n)$ in $(\mathbf{Z}_n, +)$, for $[a]_n \in (\mathbf{Z}_n, +)$, $o([a]_n) = n \Leftrightarrow (a, n) = 1$. In $(\mathbf{Z}_{10}, +)$, the classes of order 10 are $[1]_{10}$, $[3]_{10}$, $[7]_{10}$, and $[9]_{10}$, and when $n = 2^k$ the classes in $(\mathbf{Z}_n, +)$ of order 2^k are those represented by $\{a \in \mathbf{N} \mid (a, 2^k) = 1\}$, namely the odd integers. As we saw in U_{17}, $[2]_{17}^4 = [-1]_{17}$ so $[2]_{17}^8 = [1]_{17}$ and $o([2]_{17}) = 8$ (why?). By Theorem 2.8 other $[a]_{17} \in U_{17}$ of order 8 are $[2]_{17}^k$ with $(k, 8) = 1$: $[2]_{17}^3 = [8]_{17}$, $[2]_{17}^5 = [15]_{17}$, and $[2]_{17}^7 = [9]_{17}$. Also since $[4]_{17} = [2]_{17}^2$, $o([4]_{17}) = 8/(2, 8) = 4$, and $[4]_{17}^3 = [13]_{17}$ implies that $o([13]_{17}) = 4/(3, 4) = 4$ as well.

We saw that $GL(2, \mathbf{Z})$ contains elements of infinite order and others of finite order. In some infinite groups every element has finite order.

Example 2.7 $G = \{e^{iq\pi} = \cos(q\pi) + \sin(q\pi)i \mid q \in \mathbf{Q}\} \subseteq (\mathbf{C}^*, \cdot)$ is a group using the multiplication in $\mathbf{C}$. For $g \in G$, $o(g) \in \mathbf{N}$ and all orders occur.

Using multiplication of complex numbers in polar form or just the trigonometric identities, $e^{iq\pi}e^{iq'\pi} = e^{i(q+q')\pi}$ so the associative multiplication in $\mathbf{C}$ restricts to one on G. Now $e^{i0\pi} = e^0 = 1 \in G$ is the identity and $(e^{iq\pi})^{-1} = e^{i(-q)\pi}$, so G is a group and is infinite (why?). Since $(e^{iq\pi})^n = e^{iqn\pi}$, if $q = m/n \in \mathbf{Q}$ with $n > 0$, then $(e^{iq\pi})^{2n} = 1$. Hence every element in G has finite order. For any $n \in \mathbf{N}$, $o(e^{i2\pi/n}) = n$, so all finite orders occur.

It turns out that every element in a finite group must have finite order. Our next result shows this and obtains some other facts about powers.

THEOREM 2.9 Let G be a group and $g \in G$.

i) If G is finite then $o(g) \in \boldsymbol{N}$ and $o(g) \leq |G|$.
ii) If $o(g) = m$, then $g^{-1} = g^{m-1}$.
iii) If $o(g) = m$, and if $s, t \in \boldsymbol{Z}$, then $g^s = g^t \Leftrightarrow s \equiv t \pmod{m}$.

Proof Suppose G is finite and let $S = \{g, g^2, \ldots, g^{|G|+1}\} \subseteq G$. Since G cannot contain $|G| + 1$ elements, for some $1 \leq d < n \leq |G| + 1$, $g^d = g^n = g^d g^{n-d}$ by Theorem 2.4. Canceling g^d (Theorem 2.2) yields $g^{n-d} = e_G$ so $o(g) \in \boldsymbol{N}$ and $o(g) \leq n - d \leq |G|$. When $o(g) = m$, the definitions of order and powers and Theorem 2.4 yield $e_G = g^m = g^{m-1}g = gg^{m-1}$, so g^{m-1} is an inverse for g, forcing $g^{m-1} = g^{-1}$ by Theorem 2.1. Using Theorem 2.4 on powers again, $g^s = g^t \Leftrightarrow e_G = g^s g^{-t} = g^{s-t}$ (why?). If $o(g) = m, e_G = g^{s-t} \Leftrightarrow m \mid (s - t)$ from Theorem 2.7, so $g^s = g^t \Leftrightarrow s \equiv t \pmod{m}$ by definition of congruence. ■

Theorem 2.9 will be used often, but to illustrate now what it says, consider the group Bij(S) of bijections, or permutations, of S and recall from Theorem 0.14 that when S is finite, $|\text{Bij}(S)| = |S|!$. Therefore if $|S| = n$, every $\theta \in \text{Bij}(S)$ has finite order, so applying θ enough times gives I_S, the identity. Imagine we could duplicate a particular shuffle of a deck of cards as many times in a row as desired. The shuffle is a permutation of the 52 cards, so sufficient repetition of it will return the deck to its original order. Another illustration is to note that since $GL(n, \boldsymbol{Z}_m)$ is a finite group, if $A \in GL(n, \boldsymbol{Z}_m)$ then $A^k = I_n$ for some $k \in \boldsymbol{N}$. It is certainly not true that every element in $GL(2, \boldsymbol{R})$, or even in $GL(2, \boldsymbol{Z})$, has finite order, as we have seen.

In the next chapter we will see how to compute the order of permutations of finite sets. Let us look at a special case of this in Bij$(\boldsymbol{Z}_n)$. By Exercise 12 in §1.8, if for $k \in \boldsymbol{Z}$, $\varphi_k : \boldsymbol{Z}_n \to \boldsymbol{Z}_n$ is given by $\varphi_k([a]_n) = [ka]_n$ then $\varphi_k \in \text{Bij}(\boldsymbol{Z}_n) \Leftrightarrow (k, n) = 1$. To see this when $(k, n) = 1$, write $1 = sk + tn$ using Theorem 1.10 and compute directly that $(\varphi_k)^{-1} = \varphi_s$. Thus $\varphi_k \in \text{Bij}(\boldsymbol{Z}_n)$ by Theorem 0.11 and must have finite order by Theorem 2.9 since Bij$(\boldsymbol{Z}_n)$ is finite. Note that $(\varphi_k)^m([a]_n) = [k^m a]_n$, so $o(\varphi_k) = m \Leftrightarrow k^m a \equiv_n a$ for all $a \in \boldsymbol{Z}$ with $m \in \boldsymbol{N}$ minimal. In particular, $k^m = k^m \cdot 1 \equiv_n 1$. Of course if $k^m \cdot 1 \equiv_n 1$ then $k^m a \equiv_n a$ for all $a \in \boldsymbol{Z}$ by Theorem 1.24, so $o(\varphi_k)$ is the minimal m with $k^m \equiv_n 1$. To find $o(\varphi_5) \in \text{Bij}(\boldsymbol{Z}_{32})$ compute: $5^2 = 25 \equiv_{32} -7$, $5^4 \equiv_{32} (-7)^2 \equiv_{32} 17$, and $17^2 = (16 + 1)^2 \equiv_{32} 1$. It follows from Theorem 2.8 that $o(\varphi_5) = 8$ in Bij$(\boldsymbol{Z}_{32})$. As an illustration of Theorem 2.9, using $o(\varphi_5) = 8$ in Bij$(\boldsymbol{Z}_{32})$ we have $(\varphi_5)^{-1} = (\varphi_5)^7$ and $(\varphi_5)^3 = (\varphi_5)^{19} = (\varphi_5)^{-13}$ since $-1 \equiv_8 7$ and $3 \equiv_8 19 \equiv_8 -13$.

There is a simple but useful situation when the orders of the elements put a strong constraint on the structure of the group.

THEOREM 2.10 If G is a group and $g \in G \Rightarrow g^2 = e_G$, then G is Abelian.

Proof We need to show that $xy = yx$ for all $x, y \in G$ using that each square in G is the identity. But $g^2 = e_G$ means that every element in G is its own inverse by Theorem 2.2. Thus from Theorem 2.1 $xy = (xy)^{-1} = y^{-1}x^{-1} = yx$. ■

Let $W \subseteq E(\boldsymbol{R}^2)$ be the set of all reflections of the plane about straight lines through the origin, together with the identity map I of $\boldsymbol{R}^2$. For any $w \in W$, $w^2 = I$ but by

Theorem 2.10, W cannot be a group using composition since the reflections about the x-axis and about the line $y=x$ do not commute (try a few points!). There *are* groups G satisfying $g^2=e_G$ for all $g\in G$, for example $(\{\pm 1\},\ \cdot)\subseteq(\mathbf{Q}^*,\ \cdot)$. For any $n>1$, another example is the group $(\{Diag(\pm 1,\ldots,\pm 1)\},\ \cdot)\subseteq GL(n,\mathbf{R})$ where $A=Diag(a_1,\ldots,a_n)$ when $A_{jj}=a_j$ and $A_{ij}=0$ for $i\neq j$, and "$\cdot$" is matrix multiplication. An example of an infinite group with this property from Exercise 1 in §1 is the collection of subsets of an infinite set $S: G=(P(S),\#)$ with $A\# B=(A-B)\cup(B-A)$. The empty set $\varnothing=e_G$ and $A\# A=\varnothing$ for all $A\in P(S)$.

EXERCISES

1. Find $A, B\in GL(2,\mathbf{Z})$ so that $o(A), o(B)\in \mathbf{N}$ but AB has infinite order.
2. In any group G and $g, h\in G$, show that the two given elements have infinite order or the same finite order:
   a) g and g^{-1}
   b) $h^{-1}gh$ and g
   c) gh and hg
3. In a group G, if $g\in G$, find the possibilities for $o(g)$:
   a) $g^9=e_G=g^{15}$
   b) $g^3=e_G=g^{20}$
   c) $g^{24}=g^{54}$ and $g^{18}=e_G$
   d) $g^2=g^6$ and $g^3=g^{12}$
4. Compute $o([2])$ in:
   a) U_7
   b) U_{11}
   c) U_{17}
   d) U_{41}
5. **a)** Show that $U_{13}=\{([2]_{13})^k \mid k\in \mathbf{N}\}$.
   In U_{13} find:
   b) $o([2])$
   c) $o([5])$ and $o([2][5])$
   d) $o([8])$ and $o([5][8])$
   e) $o([10])$ and $o([5][10])$
6. Given $o([2]_{37})=36$ in U_{37} and that $U_{37}=\{([2]_{37})^k \mid k\in \mathbf{N}\}$, find all elements in U_{37} of order:
   a) 3
   b) 4
   c) 6
   d) 12
   e) 36
7. **a)** Show that $U_{61}=\{([2]_{61})^k \mid k\in \mathbf{N}\}$.
   b) Find all $[a]\in U_{61}$ of order 10.
   Describe all classes in U_{61} of order:
   c) 12
   d) 20
   e) 60

8. Define $f, g \in \text{Bij}(\boldsymbol{N})$ by $f(2i-1)=2i$ and $f(2i)=2i-1$ for all $i \in \boldsymbol{N}$, and by $g(1)=1$, $g(2i)=2i+1$, and $g((2i+1))=2i$ for all $i \in \boldsymbol{N}$. Show that $o(f)=o(g)=2$ but that the compositions $f \circ g$ and $g \circ f$ have infinite order.

9. a) If $f \in \text{Bij}(\boldsymbol{R}) - \{I_{\boldsymbol{R}}\}$ and is increasing, can f have finite order?

b) Let $c, d \in \boldsymbol{C}$, the complex numbers, with $c \neq 0$, and set $f_{(c,d)}(x) = cx + d$. Show that $f_{(c,d)} \in \text{Bij}(\boldsymbol{C})$ and determine when $f_{(c,d)}$ can have finite order.

c) For any prime $p > 2$, $g : \boldsymbol{Z}_p^2 \to \boldsymbol{Z}_p^2$ given by $g([a], [b]) = ([a+b], [a-b])$ is a bijection because for some $k > 1$, $g^k = I_{\boldsymbol{Z}_p^2}$. Why is this so? Describe $o(g)$ in $\text{Bij}(\boldsymbol{Z}_p^2)$ in terms of $o([2])$ in U_p.

10. Define g: $\boldsymbol{Z}_{49} \times \boldsymbol{Z}_{49} \to \boldsymbol{Z}_{49} \times \boldsymbol{Z}_{49}$ by $g([a], [b]) = ([b], [2a])$.

a) Show that $g \in \text{Bij}(\boldsymbol{Z}_{49} \times \boldsymbol{Z}_{49})$.

b) Find $o(g)$ in $\text{Bij}(\boldsymbol{Z}_{49} \times \boldsymbol{Z}_{49})$.

c) Do parts a) and b) for the function $h([a], [b]) = ([b], [47a])$.

d) Show that $f([a], [b]) = ([a+b], [47a])$ satisfies $f^k = I_{\boldsymbol{Z}_{49} \times \boldsymbol{Z}_{49}}$, some $k \in \boldsymbol{N}$.

11. Let $F : \boldsymbol{Z}_{100} \times \boldsymbol{Z}_{100} \to \boldsymbol{Z}_{100} \times \boldsymbol{Z}_{100}$ be given by $F([i], [j]) = ([j+2], [7i+3j])$. For any $Y \in \boldsymbol{Z}_{100} \times \boldsymbol{Z}_{100}$ show some composition $F^m(Y) = Y$. Note $F([56], [54]) = ([56], [54])$ if $Y = ([56], [54])$. Also $F([6], [4]) = ([6], [54])$, $F([6], [54]) = ([56], [4])$, and $F([56], [4]) = ([6], [4])$ so $m = 3$ when $Y = ([6], [4])$.

2.5 SUBGROUPS

A strategy in studying any mathematical structure is to consider subsets with the same structure. The idea is that small objects should be easier to study than large ones and that by understanding enough parts of the whole, we can answer questions about it more easily. A group G is a set with its multiplication $\cdot_G$ as its structure. A subset $H \subseteq G$ will have a structure similar to that of G if H is a group itself. Since we are interested in understanding G, we want H to be a group using $\cdot_G$ so that the structure of H as a group is closely related to the structure of G. For $H \subseteq G$ to be a group using $\cdot_G$, we need $\cdot_G$ to restrict to a multiplication on H: elements in H must have products back in H.

DEFINITION 2.8 Let $(G, \cdot)$ be a group and $H \subseteq G$. Then H is a **subgroup** of G,written $H \leq G$, if $H \neq \varnothing$ and $(H, \cdot\,|_{H \times H})$ is a group.

The definition requires what is called **closure under multiplication;** the statement that $\cdot\,|_{H \times H}$ is a multiplication on H means that $x, y \in H \Rightarrow xy \in H$ also. With this multiplication, $H \leq G$ when the group properties in Definition 2.1 hold. It is easy to see that $G \leq G$ and $\{e_G\} \leq G$ since $e_G e_G = e_G$. It is also clear that for $n \in \boldsymbol{N}$, $(n\boldsymbol{Z}, +) \leq (\boldsymbol{Z}, +) \leq (\boldsymbol{Q}, +) \leq (\boldsymbol{R}, +)$ since the addition is the same in each set and we know that the other properties defining a group hold. Other examples from §3 are $(E(\boldsymbol{R}^2), \circ) \leq (\text{Bij}(\boldsymbol{R}^2), \circ)$, and $(E(S), \circ) \leq (E(\boldsymbol{R}^2), \circ)$ for $\varnothing \neq S \subseteq \boldsymbol{R}^2$. On the other hand, $\boldsymbol{N} \cup \{0\} \subseteq (\boldsymbol{Z}, +)$ is closed under addition in $\boldsymbol{Z}$ and has 0 as an identity, but for $n \in \boldsymbol{N}$ the inverse under addition is $-n \notin \boldsymbol{N} \cup \{0\}$, so $\boldsymbol{N} \cup \{0\}$ is not a subgroup of $\boldsymbol{Z}$. If $B = \{0, \pm 2, \pm 4, \pm 8, \ldots, \pm 2^k, \ldots\} \subseteq (\boldsymbol{Z}, +)$ then $0 \in B$ is an identity for addition and B contains inverses of its elements under addition, but $(B, +)$ is not a group since $2 + 4 \notin B$. Note that B is closed under multiplication in $\boldsymbol{Z}$.

We need not verify all the properties defining a group to show that a subset closed under multiplication is a subgroup. In fact, closure under multiplication and the existence of inverses are sufficient. First we need to deal with a technical problem. When $H \le G$ then H must have an identity and an inverse for each of its elements. Are these the same identity and inverses as those in G? Fortunately they are.

THEOREM 2.11 Let G be a group and $H \le G$. Then $e_H = e_G$, and if $h \in H$ with inverse $h^\wedge \in H$, then $h^\wedge = h^{-1} \in G$.

Proof If e_H is the identity for H then $e_H e_H = e_H$. But $e_H \in G$ and is a solution of $x^2 = x$ so by Theorem 2.2, $e_H = e_G$. Using this we have $h^\wedge \in G$ satisfies $hh^\wedge = e_G = hh^{-1}$, so $h^\wedge = h^{-1}$ by cancellation in Theorem 2.1. ■

THEOREM 2.12 If G is a group and $\varnothing \neq H \subseteq G$, then the following are equivalent:

i) $H \le G$
ii) $x, y \in H \Rightarrow xy, x^{-1} \in H$
iii) $x, y \in H \Rightarrow xy^{-1} \in H$
iv) $x, y \in H \Rightarrow x^{-1}y \in H$

If H is finite, then $H \le G \Leftrightarrow (x, y \in H \Rightarrow xy \in H)$.

Proof Instead of proving that each statement implies the next, because of the similarity of iii) and iv), we show that i), ii), and iii) are equivalent and leave the similar arguments that ii) implies iv) and iv) implies i) as exercises. If $H \le G$ the multiplication in G restricts to one on H so $x, y \in H \Rightarrow xy \in H$, and Theorem 2.11 shows $x^{-1} \in H$ for $x \in H$, proving ii). Given ii), if $y \in H$ then $y^{-1} \in H$, so $x, y \in H \Rightarrow x, y^{-1} \in H \Rightarrow xy^{-1} \in H$, proving iii) If $H \neq \varnothing$ some $x \in H$, so $x, x \in H$ and iii) yield $e_G = xx^{-1} \in H$, and then both $e_G, x \in H$ force $x^{-1} = e_G x^{-1} \in H$. Thus H contains e_G and the inverse of each of its elements. Now for $x, y \in H$ we have just seen $x, y^{-1} \in H$ so $xy = x(y^{-1})^{-1} \in H$, proving that the restriction of multiplication on G is a multiplication on H. The associative property holds for any three elements in G so it holds if the three elements are in H, proving $H \le G$ by Definition 2.1. Finally, if $H \le G$ then it is closed under multiplication by the definition of subgroup. When $H \subseteq G$ is finite with $x, y \in H \Rightarrow xy \in H$ then by the equivalence of i) and ii), $H \le G$ when $x \in H \Rightarrow x^{-1} \in H$. By Theorem 2.9 it suffices to show that $o(x) \in \boldsymbol{N}$ for all $x \in H$ and the argument in Theorem 2.9 shows this. Namely, H closed under multiplication and $x \in H \Rightarrow \{x, x^2, \ldots, x^{|H|+1}\} \subseteq H$ so $x^i = x^j$ for $1 \le i < j \le |H| + 1$. Thus $x^{j-i} = e_G$ with $j - i > 0$, resulting in $o(x)$ finite. ■

Theorem 2.12 will be used many times to show that subsets of groups are subgroups. For example, the set of rotations about the origin, $\{T_\theta \in E(\boldsymbol{R}^2) \mid 0 \le \theta < 2\pi\} \le E(\boldsymbol{R}^2)$. This follows from Theorem 2.12 since $T_\theta T_\rho^{-1} = T_\theta T_{2\pi-\rho} = T_\alpha$ where $\alpha = \theta - \rho$ if $\theta \ge \rho$ and $\alpha = 2\pi + \theta - \rho$ otherwise.

It is very important to realize that if the group operation is written as addition, then the conditions $xy, x^{-1}, xy^{-1} \in H$ become $x + y, -x, x - y \in H$. In this case $H \le G$ when $x, y \in H \Rightarrow x - y \in H$. For $k, n \in \boldsymbol{N}$ consider $H = \{[ka]_n \in \boldsymbol{Z}_n \mid a \in \boldsymbol{Z}\}$. Certainly $H \neq \varnothing$ and $[ka]_n + [kb]_n = [ka + kb]_n = [k(a + b)]_n$. Therefore H is closed under the addition in

$\mathbf{Z}_n$, so $\mathbf{Z}_n$ finite and Theorem 2.12 show $H \le (\mathbf{Z}_n, +)$. Although we do not yet know that these are all of the subgroups of $\mathbf{Z}_n$, a characterization of the subgroups of $(\mathbf{Z}, +)$ does follow from Theorem 2.12 and is really just Theorem 1.4.

THEOREM 2.13 $H \le (\mathbf{Z}, +) \Leftrightarrow H = \{0\}$ or $H = n\mathbf{Z}$ for some $n \in \mathbf{N}$.

Proof If $H = k\mathbf{Z}$ for $k \ge 0$ then $ka, kb \in k\mathbf{Z} \Rightarrow ka - kb = k(a - b) \in k\mathbf{Z}$, and since $ka - kb$ is the additive version of $ka(kb)^{-1}$, $H \le (\mathbf{Z}, +)$ by Theorem 2.12. Assume $H \le \mathbf{Z}$. By Theorem 2.12 $x, y \in H \Rightarrow x - y \in H$ so H is closed under subtraction and Theorem 1.4 shows $H = \{0\}$ or $H = n\mathbf{Z}$ for $n \in \mathbf{N}$. ∎

We use Theorem 2.12 on $Cube(G) = \{g^3 \mid g \in G\}$ for any group G. Since $e_G = e_G^3 \in Cube(G)$, $Cube(G) \ne \emptyset$, and from Definition 2.4 on powers, $(g^3)^{-1} = (g^{-1})^3$ so $x \in Cube(G) \Rightarrow x^{-1} \in Cube(G)$. Hence by Theorem 2.12 $Cube(G) \le G$ if it is closed under multiplication. When G is Abelian, $g^3h^3 = (gh)^3$ by Theorem 2.4 and $Cube(G) \le G$. The commutativity in G is crucial for this. As in Example 2.6, $D_3 = \{I, T, T^2, H, TH, T^2H\}$ for $T = T_{2\pi/3}$ the rotation about (0, 0) of $2\pi/3$ radians, and H, TH, and T^2H the reflections about the altitudes. Clearly each rotation in D_3 has order 3, and each reflection has order 2 so is its own cube. Thus $Cube(D_3) = \{I, H, TH, T^2H\}$ is *not* a subgroup of D_3 since $(TH)H = T \notin Cube(D_3)$. However, for D_5, the symmetries of the regular pentagon, $Cube(D_5) = D_5$, so is a subgroup. Again each reflection is its own cube and now each rotation $T^k = T^k_{2\pi/5}$ is a cube as well (exercise!), for example, $T = (T^2)^3 \in Cube(D_5)$.

Let us find $Cube(G)$ for some Abelian groups. In $(\mathbf{Z}_n, +)$, $[a]_n^3 = [3a]_n$ so $Cube((\mathbf{Z}_n, +)) = \{[3a]_n \mid a \in \mathbf{Z}\}$ and direct computation shows that $Cube((\mathbf{Z}_5, +)) = (\mathbf{Z}_5, +)$, $Cube((\mathbf{Z}_{12}, +)) = \{[0]_{12}, [3]_{12}, [6]_{12}, [9]_{12}\}$, and $Cube((\mathbf{Z}_{13}, +)) = (\mathbf{Z}_{13}, +)$. For $U_p = (\mathbf{Z}_p^*, \cdot)$ in Definition 2.2, direct calculation (verify!) shows $Cube(U_5) = U_5$, $Cube(U_7) = \{[1]_7, [6]_7\}$, which is essentially $\{\pm 1\}$ under multiplication, $Cube(U_{11}) = U_{11}$, $Cube(U_{13}) = \{[1]_{13}, [5]_{13}, [8]_{13}, [12]_{13}\}$, and $Cube(U_{19}) = \{[1]_{19}, [7]_{19}, [8]_{19}, [11]_{19}, [12]_{19}, [18]_{19}\}$.

There are some useful examples of subgroups of the group of invertible matrices, $GL(n, R)$ when $n > 1$.

DEFINITION 2.9 If $n > 1$ and R is a ring with identity 1, then $Diag(n, R) = \{A \in GL(n, R) \mid A_{ij} = 0 \text{ when } i \ne j\}$ is the set of **diagonal** invertible matrices, $UT(n, R) = \{A \in GL(n, R) \mid A_{ij} = 0 \text{ if } i > j\}$ is the set of **upper triangular** invertible matrices, and when R is commutative $SL(n, R) = \{A \in GL(n, R) \mid \det(A) = 1\}$.

Example 2.8 $Diag(n, R) \le GL(n, R)$.

Now $I_n = Diag(1, \ldots, 1) \in Diag(n, R) \ne \emptyset$. If $A \in Diag(n, R)$ has inverse $B \in GL(n, R)$, then using $AB = I_n = BA$ and the definition of matrix multiplication force A to have invertible elements of R on its diagonal and $B \in Diag(n, R)$ so $A^{-1} = Diag(A_{11}^{-1}, \ldots, A_{nn}^{-1})$ (proof?). It is straightforward to verify that $Diag(a_1, \ldots, a_n)Diag(b_1, \ldots, b_n) = Diag(a_1b_1, \ldots, a_nb_n)$. Thus Theorem 2.12 shows that $Diag(n, R) \le GL(n, R)$. Also note that $Diag(n, R)$ is Abelian if R is a commutative ring.

Since ± 1 are the only invertible elements in $\boldsymbol{Z}$, $\{Diag(\pm 1, \ldots, \pm 1)\} = Diag(n, \boldsymbol{Z})$ is an Abelian group of order 2^n (Theorem 0.3), and $A^2 = I_n$ for each $A \in Diag(n, \boldsymbol{Z})$. This was mentioned after Theorem 2.10 and shows that there are groups of arbitrarily large order having all elements of small order. Another group like this is $Diag(n, \boldsymbol{Z}_p)$, whose elements have diagonal entries in U_p, so $|Diag(n, \boldsymbol{Z}_p)| = (p-1)^n$.

Example 2.9 For $n \geq 2$ and R a commutative ring, $UT(n, R) \leq GL(n, R)$.

For $A \in UT(n, R)$, $A_{ij} = 0$ if $i > j$ so using the definition of matrix multiplication shows $UT(n, R)$ is closed under matrix multiplication. Also $UT(n, R)$ is not empty since the identity matrix $I_n \in UT(n, R)$. If $A \in UT(n, R)$ with inverse $B \in GL(n, R)$ then using $A_{nj} = 0$ when $j < n$ and $AB = I_n$ force $1 = (AB)_{nn} = A_{nn}B_{nn}$. Thus $A_{nn}, B_{nn} \in R$ are inverses since R is commutative, and for $j < n$, $0 = (AB)_{nj} = A_{nn}B_{nj}$, so $B_{nj} = 0$. From this $(AB)_{n-1\,n-1} = 1$ shows that $A_{n-1\,n-1} \in R$ has an inverse, and $(AB)_{n-1\,j} = 0$ for $j < n-1$ yields $B_{n-1\,j} = 0$. Continuing this argument leads to $B \in UT(n, R)$. Therefore $A^{-1} \in UT(n, R)$ so $UT(n, R) \leq GL(n, R)$ by Theorem 2.12. Similarly, the invertible lower triangular matrices form a subgroup of $GL(n, R)$.

Example 2.10 For $n \geq 2$ and R a commutative ring, $SL(n, R) \leq GL(n, R)$.

If $A, B \in SL(n, R)$ then $\det(A) = \det(B) = 1$ and $\det(AB^{-1}) = \det(A)\det(B^{-1}) = 1 \cdot \det(B)^{-1} = 1 \cdot 1$, so by definition $AB^{-1} \in SL(n, R)$. Since $I_n \in SL(n, R)$, $\varnothing \neq SL(n, R) \leq GL(n, R)$ by Theorem 2.12.

Although $SL(n, R)$ is quite important, other subgroups can be defined by determinants when R is commutative, using $\det(AB) = \det(A)\det(B)$ and $\det(A^{-1}) = \det(A)^{-1}$. As in Example 2.10, using Theorem 2.12 shows $\{A \in M_n(R) \mid \det(A) = \pm 1\} \leq GL(n, R)$, $\{A \in M_n(\boldsymbol{Q}) \mid \det(A) > 0\} \leq GL(n, \boldsymbol{Q})$, and $\{A \in GL(n, \boldsymbol{R}) \mid \det(A) \in \boldsymbol{Q}\} \leq GL(n, \boldsymbol{R})$. But $\{A \in GL(n, \boldsymbol{R}) \mid \det(A) < 0\}$ and $\{A \in GL(n, \boldsymbol{R}) \mid \det(A) \in \boldsymbol{N}\}$ are not subgroups of $GL(n, \boldsymbol{R})$ (why?).

EXERCISES

1. In each case determine whether the subset of the given group is a subgroup, and if so, determine whether it is an Abelian group.
   a) $\{(a_1, a_2, \ldots, a_n) \in \boldsymbol{R}^n \mid \Sigma a_i = 0\} \subseteq (\boldsymbol{R}^n, +)$
   b) $\{\text{all reflections}\} \subseteq D_n$
   c) $\{A \in GL(2, \boldsymbol{R}) \mid A^2 = I_2\} \subseteq GL(2, \boldsymbol{R})$
   d) $\left\{\begin{bmatrix} a & b \\ c & d \end{bmatrix} \in GL(2, \boldsymbol{Z}) \mid b, c \in 2\boldsymbol{Z}\right\} \subseteq GL(2, \boldsymbol{Z})$
   e) for p a prime, $\{a/p^k \in \boldsymbol{Q} \mid a, k \in \boldsymbol{Z}\} \subseteq (\boldsymbol{Q}^*, \cdot)$
   f) for p a prime, $\{a/b \in \boldsymbol{Q} \mid a, b \in \boldsymbol{Z} \text{ and } (p, b) = 1\} \subseteq (\boldsymbol{Q}, +)$
   g) for $m \in \boldsymbol{N}$ fixed, $\{a/m^k \in \boldsymbol{Q} \mid a, k \in \boldsymbol{Z}\} \subseteq (\boldsymbol{Q}, +)$
   h) $\{f \in \text{Bij}(\boldsymbol{N}) \mid f(k) = k \text{ for infinitely many } k \in \boldsymbol{N}\} \subseteq (\text{Bij}(\boldsymbol{N}), \circ)$

2. Show that the following are subgroups of $(\boldsymbol{C}^*, \cdot)$
a) $\{\rho \in \boldsymbol{C}^* \mid \rho^k = 1 \text{ for some } k \in \boldsymbol{N} \text{ depending on } \rho\}$
b) $\{\rho \in \boldsymbol{C}^* \mid |\rho| \in \boldsymbol{Q}\}$ (for $\rho = a + bi$, $|\rho| = (a^2 + b^2)^{1/2}$)
c) $\{\rho \in \boldsymbol{C}^* \mid |\rho| = 1\}$
d) For c), does every element in the subgroup have finite order?

3. Determine whether the following sets are subgroups of $\{\text{Bij}(\boldsymbol{R}), \circ)$.
a) $\{f \in \text{Bij}(\boldsymbol{R}) \mid f(x) = ax + b \text{ with } a \in \boldsymbol{Z}^* \text{ and } b \in \boldsymbol{R}\}$
b) $\{g \in \text{Bij}(\boldsymbol{R}) \mid g(x) = ax + b \text{ with } a \in \boldsymbol{Q}^* \text{ and } b \in \boldsymbol{R}\}$
c) $\{h \in \text{Bij}(\boldsymbol{R}) \mid \text{for some } c(h) \geq 0, |h(r) - r| \leq c(h) \text{ for all } r \in \boldsymbol{R}\}$
d) $\{f \in \text{Bij}(\boldsymbol{R}) \mid \text{for some } j \in \boldsymbol{Z} \text{ depending on } f, f(j) = j\}$
e) $\{g \in \text{Bij}(\boldsymbol{R}) \mid g(\boldsymbol{Z}) \subseteq \boldsymbol{Z}\}$
f) $\{h \in \text{Bij}(\boldsymbol{R}) \mid h \text{ is increasing on } \boldsymbol{R}\}$
g) $\{f \in \text{Bij}(\boldsymbol{R}) \mid f \text{ is increasing on } \boldsymbol{R} \text{ or decreasing on } \boldsymbol{R}\}$

4. If G is a finite Abelian group and $m \in \boldsymbol{N}$, show $\{g \in G \mid (o(g), m) = 1\} \leq G$.

5. If G is a group, use Exercise 8 in §2 to show $S \leq (\text{Bij}(G), \circ)$:
a) $S = \{\tau_g \mid g \in G\}$.
b) $S = \{\sigma_g \mid g \in G\}$.
c) $S = \{T_g \mid g \in G\}$.
d) Show that $\{\tau_g \mid g \in G\} = \{\sigma_g \mid g \in G\} \Leftrightarrow G$ is Abelian.
e) Show that $\{T_g \mid g \in G\} = \{I_G\} \Leftrightarrow G$ is Abelian.

6. If $n \in \boldsymbol{N} - \{1\}$ let $\text{Nil}(\boldsymbol{Z}_n) = \{[a] \in \boldsymbol{Z}_n \mid [a]^k = [0], k \in \boldsymbol{N}$ *depending* on $[a]\}$. Here $[a]^k = [a^k]$, not $[ka]$ so $[6]_{72}^3 = [0]_{72}$ and $[12]_{72}^2 = [0]_{72}$, but $[1]_{72}^k \neq [0]_{72}$ for any $k \geq 0$, and $[8]_{72}^3 = [8]_{72}$, so also $[8]_{72}^k \neq [0]_{72}$ for any $k \geq 0$.
a) Show that $\text{Nil}(\boldsymbol{Z}_n) \leq (\boldsymbol{Z}_n, +)$.
b) Describe $\text{Nil}(\boldsymbol{Z}_n)$ explicitly when $n = 6$, $n = 8$, $n = 24$, and $n = 80$.
c) For any n, describe $\text{Nil}(\boldsymbol{Z}_n)$ in terms of the prime factorization of n.

7. Let G be a group and $H \leq G$.
a) If $g \in G$, show that $g^{-1}Hg = \{g^{-1}hg \in G \mid h \in H\} \leq G$.
b) Show that $N(H) = \{g \in G \mid g^{-1}Hg = H\} \leq G$ and that $H \leq N(H)$.

8. Let G be a group, fix $m \in \boldsymbol{N}$, and set $G(m) = \{g^m \in G \mid g \in G\}$.
a) Show that G Abelian $\Rightarrow G(m) \leq G$ but that $G(m)$ is not always a subgroup.
b) If G is finite and Abelian, show $G(m) = G \Leftrightarrow (o(g), m) = 1$ for all $g \in G$.

9. **a)** If G is an Abelian group and $m \in \boldsymbol{N}$, show that $\{g \in G \mid g^m = e_G\} \leq G$.
b) Find $\{g \in G \mid g^m = e_G\}$ explicitly when $G = \boldsymbol{Z}_{36}$ and $m = 8$; when $G = Diag(2, \boldsymbol{Z}_{13})$ and $m = 3$; and when $G = U_{31}$ and $m = 6$.
c) Is $\{g \in D_n \mid g^2 = e_G\} \leq D_n$? What about $\{g \in D_n \mid g^n = e_G\}$?

10. Let $\emptyset \neq B \subseteq S$ and show $U, V \leq \text{Bij}(S)$:
a) $U = \{f \in \text{Bij}(S) \mid f(B) = B\}$
b) $V = \{g \in \text{Bij}(S) \mid g|_B = I_B\}$

11. If G is an Abelian group, show that $T(G) = \{g \in G \mid o(g) \in \boldsymbol{N}\} \leq G$.

2.6 SPECIAL SUBGROUPS

Here we consider subgroups arising naturally in any group. Recall from Chapter 0 that $\{H_j\}_J$ denotes a nonempty collection of sets indexed by the set J. Also recall from Definition 1.9 that $\{H_j\}_J$ is a *chain* with respect to inclusion when for all $i, j \in J$ either $H_i \subseteq H_j$ or $H_j \subseteq H_i$.

THEOREM 2.14 Let $\{H_j\}_J$ be a collection of subgroups of the group G. Then $\bigcap_j H_j \le G$ and if $\{H_j\}_J$ is a chain using inclusion then $\bigcup_j H_j \le G$.

Proof By the definition of intersection, $g, h \in \bigcap H_j \Rightarrow g, h \in H_j$ for all $j \in J$. But $H_j \le G$ so from Theorem 2.12, $gh^{-1} \in H_j$ for all $j \in J$, and $gh^{-1} \in \bigcap H_j$ follows. Hence $\bigcap H_j \le G$ using Theorem 2.12 again. When $\{H_j\}_J$ is a chain under inclusion let $g, h \in \bigcup H_j$. From the definition of union $g \in H_i$ and $h \in H_j$ for some $i, j \in J$, and since $\{H_j\}_J$ is a chain we may assume $H_i \le H_j$. Thus $g, h \in H_j$ and from Theorem 2.12, $gh^{-1} \in H_j$, so $gh^{-1} \in \bigcup H_j$. Consequently, $\bigcup H_j \le G$ by a final application of Theorem 2.12. ∎

By the theorem $\bigcap H_j \le H_i$ for each H_i, and when the union is a subgroup each $H_i \le \bigcup H_j$. In $(\mathbf{Z}, +)$ Theorem 2.13 shows $n\mathbf{Z}, m\mathbf{Z} \le \mathbf{Z}$ for $n, m \in \mathbf{N}$ so by Theorem 2.14, $n\mathbf{Z} \bigcap m\mathbf{Z} \le \mathbf{Z}$. Thus $n\mathbf{Z} \bigcap m\mathbf{Z} = k\mathbf{Z}$ for some $k \in \mathbf{N}$ by Theorem 2.13. What is k?

Example 2.11 For $n, m \in \mathbf{N}$, $n\mathbf{Z} \bigcap m\mathbf{Z} = \mathrm{LCM}\{n, m\}\mathbf{Z}$.

We know $n\mathbf{Z} \bigcap m\mathbf{Z} = k\mathbf{Z}$, want $k = \mathrm{LCM}\{n, m\}$, and use $z \in n\mathbf{Z} \Leftrightarrow n \mid z$. Now $n \mid \mathrm{LCM}\{n, m\}$ and $m \mid \mathrm{LCM}\{n, m\}$ so $\mathrm{LCM}\{n, m\} \in n\mathbf{Z} \bigcap m\mathbf{Z} = k\mathbf{Z}$ and $k \mid \mathrm{LCM}\{n, m\}$. But $k \in n\mathbf{Z} \bigcap m\mathbf{Z}$ so $n \mid k$ and $m \mid k$, implying $\mathrm{LCM}\{n, m\} \mid k$ by Theorem 1.7. Thus $k = \mathrm{LCM}\{n, m\}$ since both are positive.

The union of subgroups is not always a subgroup. For $2\mathbf{Z}, 3\mathbf{Z} \le (\mathbf{Z}, +)$, $2, 3 \in 2\mathbf{Z} \bigcup 3\mathbf{Z}$ but $5 = 2 + 3 \notin 2\mathbf{Z} \bigcup 3\mathbf{Z}$. It can happen that a union of subgroups, not a chain, is a subgroup. For example, the group $Diag(2, \mathbf{Z})$ is the union of the three subgroups (verify that these *are* subgroups!) $\{I_2, Diag(-1, 1)\}$, $\{I_2, Diag(1, -1)\}$, and $\{I_2, Diag(-1, -1)\}$. Next we consider a somewhat less obvious example that will prove useful to have.

Example 2.12 $S_\infty = \{f \in \mathrm{Bij}(\mathbf{N}) \mid f(m) = m, \text{ all } m \ge n(f) \in \mathbf{N}\} \le \mathrm{Bij}(\mathbf{N})$.

Let us be clear about S_∞. Each $f \in S_\infty$ is a bijection of $\mathbf{N}$ and acts like the identity map for all large enough m, where "large" means $m \ge n(f)$ and $n(f)$ depends on f. We could use Theorem 2.12 alone to show $S_\infty \le \mathrm{Bij}(\mathbf{N})$ but prefer to illustrate Theorem 2.14. For $i \in \mathbf{N}$ let $H_i = \{f \in \mathrm{Bij}(\mathbf{N}) \mid f(m) = m, \text{ all } m \ge i\} \subseteq \mathrm{Bij}(\mathbf{N})$. Now $I_{\mathbf{N}} \in H_i$ and H_i is finite since it can be identified with $\mathrm{Bij}(\{1, 2, \ldots, i-1\})$ if $i > 1$. For $f, g \in H_i$ and $m \ge i$, $f \circ g(m) = f(g(m)) = f(m) = m$, so $f \circ g \in H_i$ and $H_i \le \mathrm{Bij}(\mathbf{N})$ by the second statement in Theorem 2.12. It is clear that $H_i \le H_j$ whenever $i \le j$, so $\{H_i\}_{\mathbf{N}}$ is a chain and Theorem 2.14 shows that $\bigcup H_i \le \mathrm{Bij}(\mathbf{N})$. But $\bigcup H_i = S_\infty$ from its definition: if $f(m) = m$ for $m \ge n(f)$ then $f \in H_{n(f)}$.

An extremely important and easy way to generate subgroups in a group G is to take the set of all powers of a fixed element.

DEFINITION 2.10 For $g \in G$ a group, $\langle g \rangle = \{g^j \mid j \in \mathbf{Z}\}$ is the **cyclic subgroup generated** by g. If $\langle g \rangle = G$, we say G is **cyclic** with **generator** g.

It is important to remember that $\langle g\rangle$ contains *all* integer powers of g, not just the positive ones. Since by Theorem 2.4, $g^i(g^j)^{-1} = g^{i-j} \in \langle g\rangle$, it is really true that $\langle g\rangle \leq G$ using Theorem 2.12. In any group, $\langle e_G\rangle = \{e_G\}$ and also each $\langle g\rangle$ is Abelian since powers of g commute by Theorem 2.4. If $g^j \in \langle g\rangle$ and $k \in \mathbf{Z}$ then $(g^j)^k = g^{jk} \in \langle g\rangle$ by Theorem 2.4 on powers, so $\langle g^j\rangle \leq \langle g\rangle$. It follows that $\langle g\rangle = \langle g^{-1}\rangle$ since $\langle g^{-1}\rangle \leq \langle g\rangle$ and $g = (g^{-1})^{-1}$ forces $\langle g\rangle \leq \langle g^{-1}\rangle$.

Since $k^z = zk$ in $(\mathbf{Z}, +)$, here $\langle k\rangle = k\mathbf{Z}$. Similarly, if $2 \in (\mathbf{Q}, +)$, $\langle 2\rangle = 2\mathbf{Z}$, but if $2 \in (\mathbf{Q}^*, \cdot)$, then $\langle 2\rangle = \{\ldots, 1/8, 1/4, 1/2, 1, 2, 4, 8, \ldots\}$. In $(\mathbf{Z}_n, +)$, *Cube* $(\mathbf{Z}_n, +) = \{[a]_n^3 = [3a]_n = [3]_n^a \mid a \in \mathbf{Z}\} = \langle [3]_n\rangle$ is cyclic. In the group $GL(2, \mathbf{Z})$, $\left\langle \begin{bmatrix} 1 & 1 \\ 0 & 1 \end{bmatrix} \right\rangle = \left\{ \begin{bmatrix} 1 & z \\ 0 & 1 \end{bmatrix} \mid z \in \mathbf{Z} \right\}$ using that $\begin{bmatrix} 1 & 1 \\ 0 & 1 \end{bmatrix}^{-1} = \begin{bmatrix} 1 & -1 \\ 0 & 1 \end{bmatrix}$, and finally $\left\langle \begin{bmatrix} 0 & 1 \\ -1 & 0 \end{bmatrix} \right\rangle = \left\{ I_2, -I_2, \begin{bmatrix} 0 & 1 \\ -1 & 0 \end{bmatrix}, \begin{bmatrix} 0 & -1 \\ 1 & 0 \end{bmatrix} \right\}$ since $\begin{bmatrix} 0 & 1 \\ -1 & 0 \end{bmatrix}^4 = I_2$.

Some Abelian groups are not cyclic, for example $Diag(2, \mathbf{Z})$ since for $A \in Diag(2, \mathbf{Z})$, $\langle A\rangle = \{I_2, A\}$ (why?) but $|Diag(2, \mathbf{Z})| = 4$. Consider $(\mathbf{Z}, +)$. Since $zn = n^z = (-n)^{-z}$ in $(\mathbf{Z}, +)$ by Theorem 2.4, $n\mathbf{Z} \leq (\mathbf{Z}, +)$ is the cyclic subgroup $\langle n\rangle = \langle -n\rangle$ with generators n and $-n$. If also $\langle n\rangle = \langle m\rangle$ then $n \in m\mathbf{Z}$ implies $m \mid n$ and similarly $m \in n\mathbf{Z}$ implies $n \mid m$, forcing $m = \pm n$ so $\langle n\rangle$ has exactly two generators. In particular, $(\mathbf{Z}, +)$ is cyclic with ± 1 as generators.

Clearly $(\mathbf{Z}_n, +) = \langle [1]_n\rangle$ since for $k > 0$, $[1]_n^k = [k]_n$. Now $\mathbf{Z}_n$ is finite so by Theorem 2.9, for $[a]_n \in \mathbf{Z}_n$, $o([a]_n) = m \in \mathbf{N}$ and for $z \in \mathbf{Z}$, $[a]_n^z = [a]_n^k \Leftrightarrow z \equiv k \pmod{m}$. Therefore $\langle [a]_n\rangle = \{[a]_n, [a]_n^2, \ldots, [a]_n^m\}$ and $|\langle [a]_n\rangle| = m$. It follows that $\langle [a]_n\rangle = \mathbf{Z}_n$ when $|\langle [a]_n\rangle| = n$, so when $o([a]_n) = n$. Since $[a]_n = [1]_n^a$ and $o([1]_n) = n$ we have from Theorem 2.8 that $\langle [a]_n\rangle = \mathbf{Z}_n$, and $[a]_n$ is a generator for $\mathbf{Z}_n$ exactly when $(a, n) = 1$. We state this result for reference.

THEOREM 2.15 $(\mathbf{Z}, +)$ is cyclic with only ± 1 as generators, and $(\mathbf{Z}_n, +) = \langle [a]_n\rangle$ exactly when $(a, n) = 1$.

For examle, $(\mathbf{Z}_4, +)$ has the two generators $[1]_4$ and $[3]_4$, $(\mathbf{Z}_7, +)$ is generated by each of its six nonzero classes, $(\mathbf{Z}_9, +)$ is generated by the classes of 1, 2, 4, 5, 7, and 8, and $(\mathbf{Z}_{10}, +)$ is generated by the classes of 1, 3, 7, and 9. Computing powers of elements shows that $U_7 = \langle [3]_7\rangle = \langle [5]_7\rangle$, since for example $5 \equiv_7 5$, $5^2 \equiv_7 4$, $5^3 \equiv_7 6$, $5^4 \equiv_7 2$, $5^5 \equiv_7 3$, and $5^6 \equiv_7 1$. Similar computations show that $U_{11} = \langle [2]_{11}\rangle = \langle [6]_{11}\rangle = \langle [7]_{11}\rangle = \langle [8]_{11}\rangle$ and that these are the only four generators of U_{11}. It is true and important that for p a prime, U_p is a cyclic group, but this fact is neither obvious nor easy.

When $o(g)$ is finite there is a simple relation among $|\langle g\rangle|$, $o(g)$, and the set of powers of g. The result will be used often as we continue and enables us to find all generators of a finite cyclic group from a given one.

THEOREM 2.16 If G is a group and $g \in G$ then $o(g) = m \in \mathbf{N} \Leftrightarrow |\langle g\rangle| = m$ and $\langle g\rangle = \{e_G, g, g^2, \ldots, g^{m-1}\}$. Thus if $o(g) = m$ then $\langle g^k\rangle = \langle g\rangle \Leftrightarrow (k, m) = 1$.

Proof Assume first that $o(g)=m$. Since every $z\in \mathbf{Z}$ is congruent to its remainder mod m, the statement in Theorem 2.9 that $g^s=g^t \Leftrightarrow s\equiv t \pmod m$ proves that $\langle g\rangle = \{g^m=e_G, g, g^2, \ldots, g^{m-1}\}$ and $|\langle g\rangle|=m$. On the other hand, when $|\langle g\rangle|=m$ then since $g\in\langle g\rangle$, $o(g)=d\le m$ by Theorem 2.9. But the first part of the proof now shows that $|\langle g\rangle|=d$, so $m=d$. When $o(g)=m$, a generator of $\langle g\rangle$ is any g^k such that $\langle g^k\rangle=\langle g\rangle$, and this holds if and only if $\langle g^k\rangle$ has $|\langle g\rangle|=m$ elements (why?). Thus the first part of the theorem shows that $\langle g^k\rangle=\langle g\rangle \Leftrightarrow o(g^k)=m$. By Theorem 2.8 this is equivalent to $(k,m)=1$. ∎

Since $(\mathbf{Z}_n,+)=\langle [1]_n\rangle$ we see that Theorem 2.15 is a special case of Theorem 2.16. Also for our claim just above about U_{11}, once we verify that $U_{11}=\langle [2]_{11}\rangle$ then since $|U_{11}|=o([2]_{11})=10$, the other generators for U_{11} must be $[2]_{11}^k$ for $(k,10)=1$. These are $[2]_{11}^3=[8]_{11}$, $[2]_{11}^7=[128]_{11}=[7]_{11}$, and $[2]_{11}^9=[2]_{11}^{-1}=[6]_{11}$ as claimed. We illustrate these ideas using U_{31}.

Example 2.13 The group $U_{31}=\langle [3]_{31}\rangle$ and has exactly eight generators.

For simplicity write $[a]$ for $[a]_{31}$ and $a\equiv b$ for $a\equiv_{31} b$. Now $|U_{31}|=30$ so by Theorem 2.16, we need to show $o([3])=30$. Since $3^4=81\equiv 19$, Theorem 1.24 yields $3^5\equiv 57\equiv -5$, so $3^{10}\equiv 25$, and $3^{15}\equiv -125\equiv -1$. Hence $3^{30}\equiv(-1)^2=1$, or, equivalently, $[3]^{30}=[1]$, and Theorem 2.7 shows that $o([3])\mid 30$. By our computations $o([3])$ cannot be 1, 5, 10, or 15, and it is clear that it is not 2 or 3. Finally, $3^6\equiv 3^5\cdot 3\equiv -15$, so we are forced to conclude that $o([3])=30$ and $U_{31}=\langle[3]\rangle$. From Theorem 2.16 the other generators of U_{31} are those $[3]^k=[3^k]$ with $(k,o([3]))=(k,30)=1$: namely $k=1, 7, 11, 13, 17, 19, 23$, and 29. We must do some additional computation with congruences to see which specific classes generate U_{31}. Since, as we have seen, $3^5\equiv -5$, we have $3^7\equiv -45\equiv 17$, and then $3^{11}=3^7\cdot 3^4\equiv 17\cdot 81\equiv 17\cdot 19=323$ so $3^{11}\equiv 13$. Continuing in this way, the eight generators of U_{31} are $\{[3], [11], [12], [13], [17], [21], [22], [24]\}$.

Another general way of obtaining subgroups is to consider elements that commute with each other.

DEFINITION 2.11 If G is a group and $\emptyset\ne S\subseteq G$, the **centralizer** of S in G is $C_G(S)=\{g\in G\mid gs=sg \text{ for all } s\in S\}$. $C_G(G)=Z(G)$ is the **center** of G.

When G is understood we usually write $C(S)$ for $C_G(S)$. Note that $C(S)\ne\emptyset$ since $e_G\in C(S)$. If G is Abelian then $C(S)=G$ for any $S\subseteq G$ since every $x, y\in G$ commute, and then $G=Z(G)$. Also $C(\{e_G\})=G$ for any group G (why?). We prove that $C(S)\le G$ using Theorem 2.14, although a minor variant of the argument can avoid reference to this result.

THEOREM 2.17 Given a group G and nonempty $S\subseteq G$, $C(S)\le G$.

Proof Consider when $S=\{h\}$ and use Theorem 2.12. If $x, y\in C(\{h\})$ then $xh=hx$ and $yh=hy$, so $x^{-1}xhx^{-1}=x^{-1}hxx^{-1}\Rightarrow hx^{-1}=x^{-1}h$, proving $x^{-1}\in C(\{h\})$. Now $(xy)h=x(yh)=x(hy)=(xh)y=(hx)y=h(xy)$ shows $xy\in C(\{h\})$, and $C(\{h\})\le G$ by applying

Theorem 2.12. Finally, the definitions of $C(S)$ and of intersection, together with Theorem 2.14, imply that $C(S) = \bigcap_{s \in S} C(\{s\}) \le G$. ∎

In any group, $\langle g \rangle \le C(\{g\}) = C(\langle g \rangle)$ and $\langle g \rangle \le C(\{g^j\})$ for any $j \in \mathbf{Z}$, by Theorem 2.4. Centralizers are uninteresting in Abelian groups, and at present we have few non-Abelian groups in which computation is fairly easy. We claim that in D_3 (Example 2.6), $C(\{g\}) = \langle g \rangle$ for $g \neq I$, which implies that $Z(D_3) = \{I\}$. The reason is that different reflections do not commute and no reflection commutes with either nonidentity rotation. This is verified using Example 2.5 or evaluating the functions in D_3 on the vertices of P_3 (Do it!).

A little work is needed to see that $Z(GL(n, \boldsymbol{R})) = \boldsymbol{R}^* \cdot I_n$, the nonzero scalar matrices. For $i \neq j$ the matrices $I_n + rE_{ij}$ with r in the i-j entry, 1's on the diagonal, and zeros elsewhere are in $GL(n, \boldsymbol{R})$ since $(I_n + rE_{ij})^{-1} = I_n - rE_{ij}$. Thus $A \in Z(GL(n, \boldsymbol{R})) \Rightarrow A(I_n + E_{ij}) = (I_n + E_{ij})A \Rightarrow AE_{ij} = E_{ij}A$. It is an exercise that any matrix A commuting with all the E_{ij} with $i \neq j$ must be scalar and that scalar matrices over $\boldsymbol{R}$ are central, so $Z(GL(n, \boldsymbol{R})) = \boldsymbol{R}^* \cdot I_n$. Since the determinant of a triangular matrix is the product of its diagonal entries, $\det(I_n + rE_{ij}) = 1$ and $I_n + rE_{ij} \in SL(n, \boldsymbol{R})$. As above $Z(SL(n, \boldsymbol{R})) \subseteq \boldsymbol{R}^* \cdot I_n$ but now these must have determinant 1: that is, $A = rI_n \subseteq Z(SL(n, \boldsymbol{R})) \Leftrightarrow r^n = 1$. Hence $Z(SL(n, \boldsymbol{R})) = \{I_n\}$ when n is odd and $Z(SL(n, \boldsymbol{R})) = \{\pm I_n\}$ when n is even. These same computations hold in any commutative ring R, so in general $Z(GL(n, R)) = U \cdot I_n$ for $U = \{r \in R \mid rs = 1$, some $s \in R\}$ and $Z(SL(n, R)) = \{uI_n \in U \cdot I_n \mid u^n = 1\}$. For $R = \boldsymbol{C}$, $Z(GL(n, \boldsymbol{C})) = \boldsymbol{C}^* \cdot I_n$, and $Z(SL(n, \boldsymbol{C})) = \{\rho I_n \mid \rho^n = 1\} = \{e^{2\pi ik/n} I_n \mid 0 \le k < n\}$, where $e^{\pi ik/n} = \cos(2\pi k/n) + \sin(2\pi k/n)i$. Also $Z(GL(n, \boldsymbol{Z}_p)) = U_p \cdot I_n$ and $Z(SL(n, \boldsymbol{Z}_p))$ depends on both n and p: $Z(SL(2, \boldsymbol{Z}_5)) = \langle [4]_5 \rangle \cdot I_2$, $Z(SL(3, \boldsymbol{Z}_5)) = \{I_3\}$, $Z(SL(4, \boldsymbol{Z}_5)) = U_5 \cdot I_4$, $Z(SL(3, \boldsymbol{Z}_7)) = \langle [2]_7 \rangle I_3$, $Z(SL(5, \boldsymbol{Z}_7)) = \{I_5\}$, and $Z(SL(6, \boldsymbol{Z}_7)) = U_7 \cdot I_6$.

Other subgroups arising in G are the "smallest" containing a given $\varnothing \neq S \subseteq G$. We mean that $S \subseteq H \le G$ and if $S \subseteq K \le G$ then $H \le K$. Is there always such an H? Certainly $S \subseteq G$ so $\{K \le G \mid S \subseteq K\} \neq \varnothing$. If $\{H_j\}_J$ is the set of *all* subgroups of G containing S then $L = \bigcap H_j \le G$ by Theorem 2.14 and $S \subseteq \bigcap H_j$ from the definition of intersection. As we observed, $\bigcap H_j \le H_i$ so indeed $\bigcap H_j$ must be the "smallest" subgroup of G containing S.

DEFINITION 2.12 If G is a group, $\varnothing \neq S \subseteq G$, and $L = \{H \le G \mid S \subseteq H\}$, then the **subgroup generated** by S is $\langle S \rangle = \bigcap_{H \in L} H$.

Although $\langle S \rangle$ really does exist as an intersection of subgroups, this definition is not very helpful in identifying and using $\langle S \rangle$. A little analysis leads to a more concrete description of $\langle S \rangle$. Suppose that $\varnothing \neq S \subseteq H \le G$. Then $S^{-1} = \{s^{-1} \mid s \in S\} \subseteq H$ by Theorem 2.12, and $H \le G$ so must contain all products of elements from both S and S^{-1}. This set of products is just $\langle S \rangle$.

THEOREM 2.18 If G is a group, $\varnothing \neq S \subseteq G$, and $S^{-1} = \{s^{-1} \in G \mid s \in S\}$, then $\langle S \rangle = \{s_1 s_2 \cdots s_k \mid k \in \boldsymbol{N}$, and all $s_j \in S \cup S^{-1}\}$.

Proof Set $K = \{s_1 s_2 \cdots s_m \mid m \in \boldsymbol{N}$, and all $s_j \in S \cup S^{-1}\}$. To show $K \le G$ we use Theorem 2.12. Given $s_1 s_2 \cdots s_m, t_1 t_2 \cdots t_q \in K$, the product $s_1 s_2 \cdots s_m t_1 t_2 \cdots t_q$ still has each factor s_i or t_j in $S \cup S^{-1}$, so is in K. Also $(s_1 s_2 \cdots s_m)^{-1} = s_m^{-1} s_{m-1}^{-1} \cdots s_1^{-1}$ is in K since $s_j \in S \Rightarrow s_j^{-1} \in S^{-1}$ and $s_j \in S^{-1} \Rightarrow s_j^{-1} \in S$ (why?). Thus $K \leq G$ by Theorem 2.12. Now $S \subseteq K$ by definition of K and we observed above that $S \subseteq H \leq G \Rightarrow K \leq H$. Therefore $K \leq \bigcap_{H \in L} H = \langle S \rangle$ as in Definition 2.12. But since $K \in L$, the collection of all subgroups of G containing S, $\langle S \rangle = \bigcap_{H \in L} H \leq K$, proving that $K = \langle S \rangle$. ∎

In the characterization of $\langle S \rangle$ in Theorem 2.18, the $s_j \in S$ appearing in a given product need not be different, so we get $e_G = gg^{-1} \in \langle S \rangle$ and $\langle g \rangle \leq \langle S \rangle$ when $g \in S$. If $S = \{g\}$, then $\langle S \rangle = \langle g \rangle$. When G is Abelian its elements commute so any $x \in \langle S \rangle$ can be rewritten as $x = s_1^a \cdots s_k^b$ for $s_j \in S$. In $(\mathbf{Z}, +)$, if $S = \{4, 6\}$, then $\langle S \rangle = \{4^s + 6^t = 4s + 6t \mid s, t \in \mathbf{Z}\}$ in $(\mathbf{Z}, +)$. We know from Theorem 2.13 that $\langle \{4, 6\} \rangle = k\mathbf{Z}$ and Theorem 1.8 on GCDs shows that $\langle \{4, 6\} \rangle = (4, 6)\mathbf{Z} = 2\mathbf{Z}$. Similarly in $(\mathbf{Q}^+, \cdot)$, the positive rationals under multiplication, if $S = \{2, 3, 5\}$ (or $\{2, 1/3, 1/5\}$) then $\langle S \rangle = \{2^i 3^j 5^k \mid i, j, k \in \mathbf{Z}\}$. By unique factorization in $\mathbf{N}$, any $q \in \mathbf{Q}^+$ can be written $q = p_1^{a_1} \cdots p_k^{a_k}$ for p_i primes and each $a_i \in \mathbf{Z}$, so it follows that $(\mathbf{Q}^+, \cdot) = \langle S \rangle$ for $S = \{p \in \mathbf{N} \mid p \text{ is prime}\}$. Note that $(\mathbf{Q}^+, \cdot) \neq \langle T \rangle$ for any finite subset $T \subseteq \mathbf{Q}^+$.

Clearly $G = \langle G \rangle$ for group G, but it is preferable to have a small subset that generates G. Using Example 2.6, it is easy to see that $D_n = \langle \{H, T_{2\pi/n}\} \rangle$. The elements in the Abelian group $Diag(n, \mathbf{Z})$ have ± 1 on the diagonal so $|Diag(n, \mathbf{Z})| = 2^n$. What is a small generating set? For any $A \in Diag(n, \mathbf{Z})$, $A^2 = I_n$ so $A = A^{-1}$. Thus if $S = \{s_1, \ldots, s_k\} \subseteq Diag(n, \mathbf{Z})$ then each element of $\langle S \rangle$ is some $s_{j_1} \cdots s_{j_t}$ for $j_1 < \cdots < j_t$ (why?) so can be identified with $\{s_{j_1}, \ldots, s_{j_t}\} \subseteq S$, where we identify I_n with $\emptyset \subseteq S$. By Example 0.19, S has $2^{|S|}$ subsets, so $\langle S \rangle$ can have at most $2^{|S|}$ elements. Therefore a necessary condition for $\langle S \rangle = Diag(n, \mathbf{Z})$ is that $|S| \geq n$. If $|S| = n$ then $\langle S \rangle = Diag(n, \mathbf{Z})$ when the products identified with different subsets of S are different, or, equivalently (why?), when $s_{j_1} \cdots s_{j_t} = I_n$ is impossible with $t > 0$. Finding a "small" or nicely described set that generates a given group may be quite difficult, especially for an infinite group. One of the exercises shows that $UT(2, \mathbf{Z})$ can be generated by a three-element set, although no finite set can generate $UT(2, \mathbf{Q})$, or even $(\mathbf{Q}, +)$ (exercises!).

Our discussion above about $\langle S \rangle \subseteq Diag(n, \mathbf{Z})$ will be useful for our next example, which at first glance has nothing to do with groups.

Example 2.14 Each of 50 switches is to control one of 50 lighting fixtures in a circular arrangement in the ceiling of a concert hall. By mistake an electrician makes switch s_i control fixture i *and* the adjacent fixture on either side of it as well. He claims that any collection of the fixtures can still be turned on with the others off. Is he correct?

How do we analyze this problem and how do groups enter? The analysis will not depend on the number 50 so we generalize to n fixtures and n switches for n large. The circular arrangement makes it convenient to label a fixture as F_1, label the others in order (clockwise) by F_i, and then for any $i, j \in \mathbf{Z}$ set $F_i = F_j \Leftrightarrow i \equiv_n j$ since n fixtures past F_j is F_{j+n}, which is just F_j. Identify the effect of switch s_j with $g_j \in Diag(n, \mathbf{Z})$ where $(g_j)_{ii} = 1$ if s_j does not change F_i and $(g_j)_{ii} = -1$ if it does. Writing $Diag(a, b, c, \ldots)$ as $(a, b, c, \ldots)$, s_2 corresponds to $g_2 = (-1, -1, -1, 1, \ldots, 1)$, s_3 to $g_3 = (1, -1, -1, -1, 1, \ldots, 1)$, etc. Since the arrangement is circular, $g_n = (-1, 1, 1, \ldots 1, -1, -1)$ and $g_1 = (-1, -1, 1, \ldots, 1, -1)$. *The effect of throwing any collection of switches is represented by multiplying the corresponding g_i in $Diag(n, \mathbf{Z})$ in any order.* For example throwing switches s_2 and s_4 has the effect of turning on exactly F_1, F_2, F_4, and F_5 and in $Diag(n, \mathbf{Z})$ the corresponding product is $g_4g_2 = g_2g_4 = (-1, -1, -1, 1 \ldots) \cdot (1, 1, -1, -1, -1, 1, \ldots) = (-1, -1, 1, -1, -1, 1, \ldots, 1)$. Also $g_3g_2 = g_2g_3 = (-1, -1, -1, 1, 1, \ldots) \cdot (1, -1, -1, -1, 1, 1, \ldots) = (-1, 1, 1, -1, 1, 1 \ldots)$, so throwing switches s_2 and s_3 leaves on F_1 and F_4 only, which can be verified directly. Thus the possibilities of turning on subsets of fixtures correspond to the elements in $\langle\{g_1, \ldots, g_n\}\rangle \le Diag(n, \mathbf{Z})$ and the fixture F_i is left on by a product of switches $s_{m(1)} \cdots s_{m(k)}$ exactly when -1 is the ith coordinate of the corresponding n-tuple $g_{m(1)} \cdots g_{m(k)}$. To be able to turn on any collection of the $\{F_j\}$ we need $\langle\{g_1, \ldots, g_n\}\rangle = Diag(n, \mathbf{Z})$.

Since $g_j^2 = I_n$, $g_j^{-1} = g_j$, and the differrent g_i commute, Theorem 2.18 shows that $\langle\{g_1, \ldots, g_n\}\rangle = \{g_{m(1)} \cdots g_{m(k)} \mid m(1) < m(2) < \cdots < m(k)\} \cup \{I_n\}$. If two *different* products of this kind are equal, say $g_{v(1)} \cdots g_{v(d)} = g_{t(1)} \cdots g_{t(q)}$, then using $\{v(i)\} \ne \{t(j)\}$ and $g_j^2 = I_n$, we have $I_n = g_{v(1)} \cdots g_{v(d)} g_{t(1)} \cdots g_{t(q)} = g_{m(1)} \cdots g_{m(k)}$ for $m(1) < \cdots < m(k)$ (why?). Can this happen? Note that F_i is changed by exactly s_{i-1}, s_i, and s_{i+1}, so there is -1 in position i for exactly three g_j. Hence if $I_n = g_{m(1)} \cdots g_{m(k)}$ and $k \ge 1$ then two or none of the g_j appearing have -1 in any given position. Suppose some F_j is not changed by any of the $g_{m(i)}$ but F_{j+1} is. Since $I_n = g_{m(1)} \cdots g_{m(k)}$, exactly two of g_j, g_{j+1}, and g_{j+2} must appear among $\{g_{m(i)}\}$. But g_j and g_{j+1} change F_j so cannot appear. Consequently if F_j is not changed by any $g_{m(i)}$ then neither is F_{j+1}. This argument applies to F_{j+1} as well, and continuing shows that no F_s is changed by any $g_{m(i)}$, contradicting $k \ge 1$. An inductive argument can be constructed to prove this more formally. Therefore every F_j is changed by some $g_{m(i)}$ so the total number of -1's appearing in all the $g_{m(i)}$ is $2n$ but is also $3k$, three for each $g_{m(i)}$. Thus $2n = 3k$ and $3 \mid n$ by Theorem 1.12. The upshot is that $I_n = g_{m(1)} \cdots g_{m(k)}$ with $k \ge 1$ forces $3 \mid n$ and otherwise the products $g_{m(1)} \cdots g_{m(k)}$ are different. That is, if $(3, n) = 1$ and $n > 3$ then $\langle\{g_1, \ldots, g_n\}\rangle$ has 2^n elements, one for each subset of $\{g_1, \ldots, g_n\}$, where I_n corresponds to $\varnothing$. Consequently for $n = 50$ the electrician is correct! Since for each coordinate, -1 occurs in exactly three of the g_j, $g_1g_2 \cdots g_n = (-1, -1, \ldots, -1)$. When $n = 3t$ we also have that $(-1, -1, \ldots, -1) = g_2g_5 \cdots g_{3t-1}$ implying that $I_{3t} = g_1g_3g_4g_6 \cdots g_{3t-2}g_{3t}$. Thus $g_{3t} \in \langle\{g_1, \ldots, g_{3t-1}\}\rangle$ and it follows that $\langle\{g_1, \ldots, g_{3t}\}\rangle \le \langle\{g_1, \ldots, g_{3t-1}\}\rangle$ so the total number of different products in $\langle\{g_1, \ldots, g_{3t}\}\rangle$ is at most 2^{3t-1}. In conclusion, whenever $(3, n) = 1$ and $n > 3$ then $\langle\{g_1, \ldots, g_n\}\rangle = Diag(n, \mathbf{Z})$ and every subset of fixtures can be turned on with the others off by flipping some set of switches, but this is not possible if $3 \mid n$.

We observed in Example 2.14 that flipping all the switches turns on all the fixtures: each fixture is changed three times so left on. As in the example, when $n = 6$, $g_6 = g_1g_3g_4$, and some experimenting shows that also $g_5 = g_1g_2g_4$. Thus $\langle\{g_i\}\rangle = \langle\{g_1, g_2, g_3, g_4\}\rangle$ (why?) has only 16 elements.

Other examples of $\langle S\rangle$ are for $S = \{g^m \mid g \in G\}$, or $S = \{g \in G \mid g^m = e_G\}$, in each case for a fixed $m \in N$. For $D_n = \{T^i, T^iH\}$ where $T = T_{2\pi/n}$ as in Example 2.6, when m is odd

$(T^iH)^m = T^iH$ so $\langle\{g^m \mid g \in D_n\}\rangle = D_n$, and when m is even $\langle\{g^m \mid g \in D_n\}\rangle \le \langle T\rangle$ (verify!). Consider the group $S_\infty = \{f \in \text{Bij}(\boldsymbol{N}) \mid f(m) = m$ for all $m \ge n(f)\}$ in Example 2.12 for a result involving $\langle\{g \in S_\infty \mid g^2 = I\}\rangle$. Call $f \in S_\infty$ a **transposition** if it interchanges two integers and leaves the others fixed. Denote the one interchanging $i, j \in \boldsymbol{N}$ by (i, j). Although this notation has various meanings, its interpretation is usually clear from the context. Writing $(i, j) \in \boldsymbol{N}^2$ means (i, j) is an ordered pair; writing $(i, j) = 1$ means $\text{GCD}\{i, j\} = 1$; and writing $(i, j) \in S_\infty$ means that (i, j) is a function on $\boldsymbol{N}$ with $(i, j)(i) = j$, $(i, j)(j) = i$, and $(i, j)(m) = m$ if $m \ne i, j$.

Example 2.15 $S_\infty = \langle\{(i, j) \in S_\infty \mid i < j\}\rangle$.

Let $T = \{(i, j) \in S_\infty \mid i < j\}$, the set of transpositions, and observe that $T \subseteq \{g \in S_\infty \mid g^2 = I\}$ so $(i, j)^{-1} = (i, j)$. As in Example 2.12, let $H_i = \{f \in \text{Bij}(\boldsymbol{N}) \mid f(m) = m$ for all $m \ge i\}$ and recall $H_i \le H_j$ when $i \le j$ and that S_∞ is their union. Thus it suffices to show that each $f \in H_n$ is a product of transpositions and we do so by induction, starting with $n = 3$. By definition $H_3 = \{I_N, (1, 2)\}$ and $I_N = (1, 2)(1, 2)$ so the claim holds for H_3. Assume that for some $m \ge 3$, $H_m = \langle T \cap H_m\rangle$ and let $g \in H_{m+1}$. If $g \in H_m$ then by the induction assumption $g \in \langle T \cap H_m\rangle \le \langle T \cap H_{m+1}\rangle$, and otherwise $g(m) = j < m$. But now $(j, m)g(m) = m$, and $(j, m), g \in H_{m+1}$ imply $(j, m)g \in H_{m+1}$. Thus $(j, m)g(s) = s$ for all $s \ge m$, forcing $(j, m)g \in H_m$. By assumption there are $d_s = (i_s, k_s) \in H_m$ with $(j, m)g = d_1d_2 \cdots d_q$ and it follows that $g = (j, m)d_1d_2 \cdots d_q \in \langle T \cap H_{m+1}\rangle$. Thus $H_{m+1} = \langle T \cap H_{m+1}\rangle$ so $H_n = \langle T \cap H_n\rangle$ for all $n \ge 3$ by IND-I, and $S_\infty = \langle T\rangle$ as claimed. Our argument shows that for $n \ge 3$, $H_n = \langle T \cap H_n\rangle$ and an easy modification proves that $S_n = \text{Bij}(\{1, 2, \ldots, n\})$ is generated by its transpositions.

Our last example of the chapter is given as a theorem because the subgroup described is quite important and will be used later.

THEOREM 2.19 Let G be a group and $S = \{x^{-1}y^{-1}xy \mid x, y \in G\}$.

i) If $w \in \langle S\rangle \le G$ and $g \in G$, then $g^{-1}wg \in \langle S\rangle$.
ii) G is Abelian $\Leftrightarrow \langle S\rangle = \langle e_G\rangle$.

Proof By definition $\langle S\rangle \le G$. To be clear, the elements of $\langle S\rangle$ are not simply $x^{-1}y^{-1}xy$ for $x, y \in G$, but products of these and their inverses. Set $c(x, y) = x^{-1}y^{-1}xy$ and use Theorem 2.1 to see that $c(x, y)^{-1} = (x^{-1}y^{-1}xy)^{-1} = y^{-1}x^{-1}yx = c(y, x)$. Therefore any element of $\langle S\rangle$ is some $c(x_1, y_1)c(x_2, y_2) \cdots c(x_k, y_k)$, and

$$g^{-1}c(x_1, y_1)c(x_2, y_2) \cdots c(x_k, y_k)g = g^{-1}c(x_1, y_1)gg^{-1}c(x_2, y_2)gg^{-1} \cdots gg^{-1}c(x_k, y_k)g.$$

But $g^{-1}c(x, y)g = g^{-1}x^{-1}gg^{-1}y^{-1}gg^{-1}xgg^{-1}yg = (g^{-1}xg)^{-1}(g^{-1}yg)^{-1}(g^{-1}xg) \cdot (g^{-1}yg) = c(g^{-1}xg, g^{-1}yg)$, so replacing each $g^{-1}c(x_i, y_i)g$ with $c(g^{-1}x_ig, g^{-1}y_ig)$ shows that $g^{-1}c(x_1, y_1)c(x_2, y_2) \cdots c(x_k, y_k)g \in \langle S\rangle$ as claimed. When G is Abelian then each $x^{-1}y^{-1}xy = x^{-1}xy^{-1}y = e_G$, so $\langle S\rangle = \langle e_G\rangle$. On the other hand, if $\langle S\rangle = \langle e_G\rangle$, then since each $x^{-1}y^{-1}xy \in \langle S\rangle$, $x^{-1}y^{-1}xy = e_G$, and multiplying by yx on the left gives $xy = yx$. Consequently, G is Abelian when $\langle S\rangle = \langle e_G\rangle$, finishing the proof. ∎

DEFINITION 2.13 In any group G, if $S = \{x^{-1}y^{-1}xy \mid x, y \in G\}$ then $G' = \langle S \rangle$ is called the **commutator subgroup** of G.

The commutator subgroup is a measure of how close G is to being Abelian: $G' = \langle e_G \rangle \Leftrightarrow G$ is Abelian by Theorem 2.19. The chain of subgroups that arises from taking successive commutator subgroups is quite useful. Briefly, since $G' = G^{(1)}$ is a group we may consider $(G')' = G^{(2)}$, and then $(G^{(2)})' = G^{(3)}$, etc. to obtain the chain $G \geq G' \geq G^{(2)} \geq \cdots$. When G is finite this chain must either terminate in $\langle e_G \rangle$, with say $G^{(k)} = \langle e_G \rangle$, or become stable with some $G^{(k+1)} = G^{(k)}$, and $G^{(k)} \neq \langle e_G \rangle$. In the former case, G is called **solvable** and the smaller k is, the closer G is to being Abelian. This notion is useful for doing arguments by induction, although we will not need it here.

For the groups D_n, $D_n{}^{(2)} = \{I\}$. This follows from Example 2.6 by writing $D_n = \{T^i, T^iH \mid 0 \leq i < n\}$, where $T = T_{2\pi/n}$. By Example 2.5, $HT^iH = T^{-i}$, which implies that all $T^{-i}H$ have order 2. The simple commutators $x^{-1}y^{-1}xy$ are $HT^iHT^jHT^iHT^j = (T^{-i+j})^2 = T^{2(j-i)}$ and $HT^iT^jHT^iT^{-j} = T^{-i-j}T^{i-j} = T^{-2j}$. Note that powers of T commute and that $(x^{-1}y^{-1}xy)^{-1} = y^{-1}x^{-1}yx$. It follows that $D_n' = \langle T^2 \rangle$, which is Abelian, so $D_n{}^{(2)} = \{I\}$ by Theorem 2.19.

It turns out that $(S_n)' = S_n^{(2)}$ when $n \geq 5$, which is interpreted to mean that these S_n are "very non-Abelian." This fact requires an analysis of S_n that will come in the next chapter and in Chapter 5. As an aside, we mention that this result about S_n underlies the fact that for a polynomial $p(x)$ of degree at least five, there is no general algebraic expression for a root of $p(x)$ in terms of its coefficients. This is proved in Chapter 15.

Examples with $G^{(k-1)} \neq \langle e_G \rangle$ and $G^{(k)} = \langle e_G \rangle$ for larger k are the $UT(n, \boldsymbol{Q})$ (Example 2.9). Matrix calculation shows that $UT(n, \boldsymbol{Q})' = \{A \in UT(n, \boldsymbol{Q}) \mid \text{all } A_{jj} = 1\}$, and $UT(n, \boldsymbol{Q})^{(2)} = \{A \in UT(n, \boldsymbol{Q})' \mid \text{all } A_{i,i+1} = 0\}$. For larger k, $A \in UT(n, \boldsymbol{Q})^{(k)} \Leftrightarrow$ for all i $A_{ii} = 1$ and $A_{ij} = 0$ whenever $i + 1 \leq j < i + 2^{k-1}$. Therefore, eventually $G^{(k)} = \{I_n\}$ and if n is sufficiently large, k must be large before $G^{(k)} = \{I_n\}$.

EXERCISES

1. Let G be an Abelian group and $H_1, H_2, \ldots, H_k \leq G$.
   a) Show that $H_1H_2 \cdots H_k = \{x_1x_2 \cdots x_k \in G \mid x_j \in H_j\} \leq G$.
   b) If $G = (\boldsymbol{Z}, +)$ and for $1 \leq i \leq k$, $H_i = a_i\boldsymbol{Z}$ with $a_i \in \boldsymbol{N}$, find $H_1 + \cdots + H_k$.
   c) If $G = (\boldsymbol{Q}^+, \cdot)$, $H = \langle 6 \rangle$, $K = \langle 7 \rangle$, and $L = \langle 10 \rangle$, describe HKL.
2. Let G be an Abelian group and $\{H_m\}_{\boldsymbol{N}}$ a collection of subgroups of G. Set $\sum H_m = \{x_1x_2 \cdots x_k \in G \mid k \in \boldsymbol{N} \text{ and each } x_j \in H_j\}$.
   a) Show that $\sum H_m \leq G$.
   b) Show that $\sum H_m = \bigcup_{m \in \boldsymbol{N}} H_1H_2 \cdots H_m$ (see Exercise 1).
   c) If $G = (\boldsymbol{Q}^+, \cdot)$ and $H_i = \langle p_i \rangle$ for p_i the ith prime, find $\sum H_m$ and $\sum_{i>1} H_i$.
3. a) If $H, K, L \leq G$ a group with $L \subseteq H \cup K$, as sets, show $L \leq H$ or $L \leq K$.
   b) Show that as a set, a group cannot be the union of two proper subgroups.
4. Show that the groups in §1 Exercise 1 parts f), g), h) and i) are cyclic.
5. For $n \in \boldsymbol{N}$, set $\boldsymbol{Q}_n = \{a/n \in \boldsymbol{Q} \mid a \in \boldsymbol{Z}\}$.
   a) Show that $\boldsymbol{Q}_n \leq (\boldsymbol{Q}, +)$.
   b) Find $\bigcap_{\boldsymbol{N}} \boldsymbol{Q}_n$.

c) Show that $\{\boldsymbol{Q}_{n!} \mid n \in \boldsymbol{N}\}$ is a chain and find $\bigcup_{\boldsymbol{N}} \boldsymbol{Q}_{n!} = W$.

d) Find $H \leq W$ with $H \neq W$ and H not a subgroup of any $\boldsymbol{Q}_{n!}$.

6. For a fixed prime p and any $m \in \boldsymbol{N}$, set $\boldsymbol{Q}_{[m]} = \{a/p^m \in \boldsymbol{Q} \mid a \in \boldsymbol{Z}\}$.

a) Show that $\boldsymbol{Q}_{[m]} \leq (\boldsymbol{Q}, +)$.

b) Show $\boldsymbol{Q}_{[m]}(p) = \{pq \in \boldsymbol{Q} \mid q \in \boldsymbol{Q}_{[m]}\} \neq \boldsymbol{Q}_{[m]}$.

c) If $m < m'$ show $\boldsymbol{Q}_{[m]} \leq \boldsymbol{Q}_{[m']}$.

d) If $\boldsymbol{Q}_{[\infty]} = \bigcup_{\boldsymbol{N}} \boldsymbol{Q}_{[m]}$ show that $\boldsymbol{Q}_{[\infty]}(p) = \boldsymbol{Q}_{[\infty]}$.

7. a) For p a prime, show $G_p = \{c \in \boldsymbol{C} \mid c^{p^t} = 1, \text{ some } t = t(c) \geq 0\} \leq (\boldsymbol{C}^*, \cdot)$.

b) If $H \leq G_p$ with $H \neq G_p$, show H is finite and cyclic.

8. a) If G is a group and $g \in G$ has infinite order, show $\langle g^n \rangle = \langle g^m \rangle \Leftrightarrow m = \pm n$.

b) Show $G = \langle g^2 \rangle$ for some $g \in G \Leftrightarrow G$ is a finite cyclic group of odd order.

c) If $m > 1$, the group $G = \langle g^m \rangle$ for some $g \in G \Leftrightarrow G$ is a finite cyclic group with $(|G|, m) = 1$.

9. Describe the cyclic subgroups generated by the following elements:

a) $\begin{bmatrix} -1 & 1 \\ 0 & 1 \end{bmatrix} \in GL(2, \boldsymbol{Z})$

b) $\begin{bmatrix} 0 & 1 \\ -1 & 0 \end{bmatrix} \in GL(2, \boldsymbol{Z})$

c) $\begin{bmatrix} 0 & 1 \\ -1 & -1 \end{bmatrix} \in GL(2, \boldsymbol{Z})$

d) $x^2 \in \text{Bij}((0, +\infty))$

e) $[2]_{31} \in U_{31}$

f) $[6]_{31} \in U_{31}$

g) $(x-1)/x \in \text{Bij}(\boldsymbol{R} - \{0, 1\})$

h) $2x + 1 \in \text{Bij}(\boldsymbol{R})$

i) $ix + \pi \in \text{Bij}(\boldsymbol{C})$

10. For any group G and $h, g \in G$, show that:

i) $C_G(\{h\}) = C_G(\{h^{-1}\})$

ii) $C_G(\{h\}) = C_G(\langle h \rangle)$

iii) $g^{-1}C_G(\{h\})g = C_G(\{g^{-1}hg\})$ (§5 Exercise 7!)

11. Find the centralizer of each $T_{2\pi j/n} \in D_n$. (This depends on n.)

12. In $GL(2, \boldsymbol{Z})$ find the centralizer of:

a) $\begin{bmatrix} 1 & 0 \\ 0 & -1 \end{bmatrix}$

b) $\begin{bmatrix} 0 & 1 \\ -1 & 0 \end{bmatrix}$

c) $\begin{bmatrix} 1 & 1 \\ 0 & 1 \end{bmatrix}$.

13. Find $Z(G)$ for:

a) D_n

b) $UT(n, \boldsymbol{R})$

c) $UT(n, \boldsymbol{R})' = \{A \in UT(n, \boldsymbol{R}) \mid \text{all } A_{jj} = 1\}$.

14. In any group G and for any $x, y \in G$, show that $xy \in Z(G) \Leftrightarrow yx \in Z(G)$.

15. a) For any *odd* $m \in \boldsymbol{N}$ show that D_n is generated by $\{g^m \mid g \in D_n\}$.

b) For any $m \in \boldsymbol{N}$ determine $\langle \{g^{2m} \mid g \in D_n\} \rangle$.

16. Show that $Diag(n, \boldsymbol{Z})$ is generated by $S = \{D(i) = I_n - 2E_{ii} \mid 1 \leq i \leq n\}$ but not by $T = \{D(i)D(j) \mid D(i), D(j) \in S\}$.

17. **a)** Show that $UT(2, \mathbf{Z})$ is generated by $\left\{Diag(1, -1), Diag(-1, 1), \begin{bmatrix} 1 & 1 \\ 0 & 1 \end{bmatrix}\right\}$.

b) Show that $UT(2, \mathbf{Z})$ cannot be generated by any two of its matrices.

c) Show that $GL(2, \mathbf{Q})$ cannot be generated by any finite subset.

18. Show that $GL(2, \mathbf{Z}) = \left\langle S = \left\{ \begin{bmatrix} 1 & 1 \\ 0 & 1 \end{bmatrix}, \begin{bmatrix} 1 & 0 \\ 1 & 1 \end{bmatrix}, \begin{bmatrix} 0 & 1 \\ 1 & 0 \end{bmatrix}, \begin{bmatrix} 1 & 0 \\ 0 & -1 \end{bmatrix}, \begin{bmatrix} -1 & 0 \\ 0 & 1 \end{bmatrix} \right\} \right\rangle$.

$\Bigg($If $A = \begin{bmatrix} a & b \\ c & d \end{bmatrix} \in GL(2, \mathbf{Z})$ then $(a, b) = 1$, $A \begin{bmatrix} 1 & -q \\ 0 & 1 \end{bmatrix} = \begin{bmatrix} a & b - aq \\ c & * \end{bmatrix}$, and $A \begin{bmatrix} 0 & 1 \\ 1 & 0 \end{bmatrix} = \begin{bmatrix} b & a \\ d & c \end{bmatrix}$. Using the Euclidean Algorithm, some product $Ag_1g_2 \cdots g_k = \begin{bmatrix} 1 & 0 \\ * & 1 \end{bmatrix}$ with all $g_j \in S$ or $g_j^{-1} \in S$, so $A \in \langle S \rangle$.$\Bigg)$

19. Show that the group $(\mathbf{Z}[x], +)$ is generated by each of the sets:

a) $\{1, x, x^2, \ldots, x^n, \ldots\}$

b) $\{1, x - 1, x^2 - 1, \ldots, x^n - 1, \ldots\}$

c) $\{1, x - 1, \ldots, x^n - x^{n-1}, \ldots\}$

20. If G is a group and $S \subseteq G$ with S countable and nonempty, use the last section of Chapter 0 to show that $\langle S \rangle$ is countable.

21. Referring to Example 2.14, involving the circular arrangement of n fixtures, if the electrician wires the switches so that switch s_j controls fixtures F_{j-2}, F_j, and F_{j+2}, what condition on n enables any collection of fixtures to be turned on with the others off?

22. A rectangular array of n lights denoted by $\{L_j\}$ in some order is controlled by n switches $\{s_j\}$ so that when switch s_j is thrown, all the lights *except* light L_j are changed from on to off or from off to on. What condition on n, if any, enables an arbitrary pattern of lights to be turned on with the others off?

CHAPTER

3

SPECIAL GROUPS

This chapter contains important standard examples of groups. Having a collection of familiar groups will be useful for illustrating various concepts we will discuss, for recognizing some new groups we encounter as "old friends" in disguise, and for identifying a given group as simply constructed from known groups.

3.1 CYCLIC GROUPS

We characterize the subgroups of finite cyclic groups. For $(\mathbf{Z}, +)$, $H \le \mathbf{Z} \Rightarrow H = n\mathbf{Z}$, as we saw in Theorem 2.13. The situation for finite cyclic groups is a little more involved, but it is just as decisive and quite useful.

THEOREM 3.1 Let $G = \langle g \rangle$ with $|G| = n$. If $k \in \mathbf{N}$ and $k \mid n$ then $H_k = \langle g^{n/k} \rangle$ is the only subgroup of G of order k, and $\{H \le G\} = \{H_k \mid k \mid n\}$. Thus any $H \le G$ is cyclic, $|H| \mid |G|$, and there is exactly one subgroup of G for each $k \mid n$.

Proof If $k > 0$ and $k \mid n$ then $\langle g^{n/k} \rangle \le G$ by Definition 2.10 of cyclic subgroup, and $|H_k| = o(g^{n/k})$ from Theorem 2.16. In particular, $H_n = G$ and $o(g) = n$ so Theorem 2.8 shows that $o(g^{n/k}) = n/(n/k, n) = n/(n/k) = k$ and $|H_k| = k$. Next if $H \le G$ with $|H| = m$ we want $H = H_m$. The elements of H are powers of g so if $S = \{z \in \mathbf{Z} \mid g^z \in H\}$ then $H = \{g^z \mid z \in S\}$. If we show $S = d\mathbf{Z}$ then $H = \{g^{dz} \mid z \in \mathbf{Z}\} = \{(g^d)^z \mid z \in \mathbf{Z}\} = \langle g^d \rangle$. But $S = d\mathbf{Z}$ follows from Theorem 1.4 (or Theorems 2.12 and 2.13). To see this take $i, j \in S$, so $g^i, g^j \in H$, and use $H \le G$ to conclude that $g^i(g^j)^{-1} \in H$. By Theorem 2.4 on exponents, $g^{i-j} \in H$, so $i - j \in S$ and $S = d\mathbf{Z}$ for d minimal in $S \cap \mathbf{N}$, by Theorem 1.4. Therefore $H = \langle g^d \rangle$, but we cannot conclude that $H = H_{n/d}$ until we know that $d \mid n$. Now $o(g) = n \Rightarrow g^n = e \in H$ so $n \in S$, which forces $d \mid n$. Consequently, $H = \langle g^d \rangle = H_{n/d}$ so $m = |H| = |H_{n/d}| = n/d$, and $H = H_m$. ■

One consequence of Theorem 3.1 is that a finite Abelian group G cannot be cyclic if it has different subgroups of the same order. Thus for any prime p, there can be at most $p - 1$ elements of order p in a cyclic group. To see this, suppose $x, y \in G = \langle g \rangle$

with $o(x) = o(y) = p$, and note that G must be finite (exercise!). Then $|\langle x\rangle| = |\langle y\rangle| = p$ by Theorem 2.16 so $\langle x\rangle = \langle y\rangle$ by Theorem 3.1 and $y \in \langle x\rangle - \langle e_G\rangle$ follows. In particular a cyclic group contains at most one element of order 2. For example, when $n > 1$, $Diag(n, \mathbf{Z}_3)$ and $Diag(n, \mathbf{Z})$ are not cyclic since every element in these groups has order at most 2. In Chapter 7 we will prove the interesting fact that a finite Abelian group is cyclic if and only if there are exactly $p - 1$ elements of order p for each prime divisor p of the order of the group.

It is an exercise to show that a cyclic group G with only finitely many subgroups must be finite. By Theorem 3.1 the subgroups of G and the divisors of $|G|$ correspond. Thus Theorem 1.18 on the divisors of $n \in \mathbf{N}$ allows us to characterize $|G|$ given the number of subgroups of G. When a cyclic group G has only two subgroups, $|G| = p$ a prime, with exactly three subgroups, $|G| = p^2$ for a prime p, and with exactly four subgroups, $|G| = p^3$ or pq for primes $p < q$. Let us look at a couple of examples illustrating the content of Theorem 3.1.

Example 3.1 Describe the subgroups of $(\mathbf{Z}_{15}, +) = \langle [1]\rangle$.

Since $|\mathbf{Z}_{15}| = 15$ with divisors $1, 3, 5$, and 15, $\mathbf{Z}_{15}$ has exactly four corresponding subgroups, namely: $\langle [1]^{15/1}\rangle = \{[0]\}$; $\langle [1]^{15/3}\rangle = \langle [5]\rangle = \{[0], [5], [10]\}$; $\langle [1]^{15/5}\rangle = \langle [3]\rangle = \{[0], [3], [6], [9], [12]\}$; and $\langle [1]^{15/15}\rangle = \mathbf{Z}_{15}$.

In general the subgroups of $\mathbf{Z}_n$ will be the cyclic groups $\langle [1]_n^m\rangle = \langle [m]_n\rangle = \{[0]_n, [m]_n, [2m]_n, \ldots, [n-m]_n\}$ for the divisors m of n, so as in $(\mathbf{Z}, +)$ look like the "multiples of m." When k is not a divisor of n but $o([k]_n) = o([m]_n)$ for $m \mid n$, then $\langle [k]_n\rangle = \langle [m]_n\rangle$ by Theorem 2.16 and Theorem 3.1. For example, in $\mathbf{Z}_{16}$, $o([12]_{16}) = 4 = o([4]_{16})$ and $\langle [12]_{16}\rangle = \{[0]_{16}, [12]_{16}, [8]_{16}, [4]_{16}\} = \langle [4]_{16}\rangle$. Multiplicative examples can look quite different.

Example 3.2 Describe the subgroups of $U_{13} = \langle [2]\rangle$.

Since $|U_{13}| = 12$ there are exactly six subgroups of U_{13}, and their orders are the divisors of 12: 1, 2, 3, 4, 6, and 12. These subgroups are $\langle [2]^{12/1}\rangle = \{[1]\}$; $\langle [2]^{12/2}\rangle = \langle [2]^6\rangle = \{[1], [12]\}$; $\langle [2]^{12/3}\rangle = \langle [2]^4\rangle = \{[1], [3], [9]\}$; $\langle [2]^{12/4}\rangle = \{[1], [8], [12], [5]\}$; $\langle [2]^{12/6}\rangle = \{[1], [4], [3], [12], [9], [10]\}$; and $\langle [2]^{12/12}\rangle = U_{13}$. From Theorem 2.8 the generators of U_{13} are the powers of $[2]$ relatively prime to $o([2])$: $[2]$, $[2]^5 = [6]$, $[2]^7 = [11]$, and $[2]^{11} = [7]$.

EXERCISES

1. If G is a cyclic group and $o(x) \in \mathbf{N}$, some $x \in G - \{e_G\}$, show that G is finite.
2. If a cyclic group G has only finitely many subgroups, show that G is finite.
3. If $G \neq \langle e_G\rangle$ are the only subgroups of G, show that $G = \langle g\rangle$ with $|G|$ prime.
4. In each case find all the subgroups and the inclusions between them:
   i) $(\mathbf{Z}_{77}, +)$
   ii) $(\mathbf{Z}_{12}, +)$
   iii) $(\mathbf{Z}_{27}, +)$

iv) $(\mathbf{Z}_{24}, +)$
v) $(\mathbf{Z}_{30}, +)$
vi) $(\mathbf{Z}_{80}, +)$

5. If G is a finite group, and some $m \in N$ satisfies $m > 1$, $m \mid |G|$, and $|\{g \in G \mid g^m = e_G\}| > m$, show that G is not a cyclic group.

6. Let G be a group with $a, b \in G$, $o(a), o(b) \in N$, and $ab = ba$.
i) Show that $|\langle a\rangle \cap \langle b\rangle| \mid \text{GCD}\{o(a), o(b)\}$.
ii) If $o(a) = o(b)$ is prime, show that $\langle a\rangle = \langle b\rangle$ or that $\langle a\rangle \cap \langle b\rangle = \langle e_G\rangle$.
iii) If $\text{GCD}\{o(a), o(b)\} = 1$ show that $o(ab) = o(a)o(b)$.

7. For $G = \langle g\rangle$ and finite, show that the following two statements are equivalent:
a) $|G| = p^n$ for p a prime; and b) For any $H, K \leq G$, either $H \leq K$ or $K \leq H$.
c) Does b) $\Rightarrow$ a) without assuming initially that $G = \langle g\rangle$ is finite?

8. Let $x, y \in G$, a finite cyclic group, so that $o(x) = o(y)$. Show that there are $s, m \in N$ with both $x^s = y$ and $y^m = x$.

9. Let G be a finite cyclic group, $H, K \leq G$, and $HK = \{hk \mid h \in H \text{ and } k \in K\}$.
i) Show that $HK \leq G$, $H \leq HK$, and $K \leq HK$.
ii) Show that $|HK| = \text{LCM}\{|H|, |K|\}$.
iii) For which $|G|$ are there $H, K \leq G$, $H \neq G$, $K \neq G$, but $HK = G$?
iv) Find a pair $H, K \leq (\mathbf{Z}_{12}, +)$ with $H, K \neq \mathbf{Z}_{12}$ but $HK = \mathbf{Z}_{12}$.
v) Find pairs of proper subgroups $\{H, K\} \neq \{H_1, K_1\}$ with $HK = H_1K_1 = \mathbf{Z}_{30}$.

3.2 THE GROUPS U_n

There is a natural generalization from the groups U_p for p a prime to $\mathbf{Z}_n$ by taking the classes $[k]_n \in \mathbf{Z}_n$ with $(k, n) = 1$. This set $\{[k]_n \mid (k, n) = 1\}$ is ambiguously defined *unless* $[k]_n = [s]_n$ with $(k, n) = 1$ implies $(s, n) = 1$. Fortunately this is true. If $[k]_n = [s]_n$ then $s = nt + k$ for some $t \in \mathbf{Z}$ by Theorem 1.23, so Theorem 1.9 shows that $(s, n) = (k, n) = 1$.

DEFINITION 3.1 For any integer $n > 1$ set $U_n = \{[k]_n \in \mathbf{Z}_n \mid (k, n) = 1\}$.

THEOREM 3.2 $(U_n, \cdot)$ is an Abelian group using the multiplication in $\mathbf{Z}_n$.

Proof Is the multiplication in $\mathbf{Z}_n$ really a multiplication on U_n? If $[k], [s] \in U_n$ then $(k, n) = 1$ and $(s, n) = 1$, so $(ks, n) = 1$ follows from Theorem 1.11. Thus $[k][s] = [ks] \in U_n$. Multiplication in $\mathbf{Z}_n$ is associative, is commutative, and has identity $[1]$ by Theorem 1.28 or directly, so we need only show that each $[k] \in U_n$ has an inverse. Since $(k, n) = 1$, $rk + tn = 1$ for some $r, t \in \mathbf{Z}$ by Theorem 1.10, which also shows that $(r, n) = 1$; hence $[k][r] = [kr] = [1]$, and $[r] = [k]^{-1}$ in U_n. ∎

The groups U_n provide examples of Abelian groups in which computation is fairly easy. As a bonus, not all of the U_n are cyclic. We saw above that a cyclic group has at most one element of order 2, so U_8 is not cyclic since $o([3]_8) = 2 = o([5]_8)$, U_{15} is not cyclic since $o([4]_{15}) = o([11]_{15}) = 2$, and in either U_{16} or U_{20}, $o([9]) = o([-1]) = 2$. However, $U_5 = \langle [2]_5\rangle$, $U_7 = \langle [3]_7\rangle$, and we have seen that $U_{13} = \langle [2]_{13}\rangle$ and $U_{31} = \langle [3]_{31}\rangle$ (Example 2.13). Which U_n are cyclic?

THEOREM 3.3 The group U_n is not cyclic if $n = 2^m$ with $m \geq 3$, if $4p \mid n$ for p an odd prime, or if $pq \mid n$ for $p < q$ odd primes.

Proof In such U_n $o([-1]) = 2$ so it suffices to find another $[k] \in U_n$ of order 2; then U_n will have different subgroups of order 2 and so cannot be cyclic by Theorem 3.1. When $n = 2^m$ and $m \geq 3$ then $o([2^{m-1} - 1]) = 2$ since $(2^{m-1} - 1)^2 = 2^{2m-2} - 2 \cdot 2^{m-1} + 1 = (2^{m-2} - 1)2^m + 1$, and $[2^{m-1} - 1]_n \neq [\pm 1]_n$. In the other cases, $n = st$ with $(s, t) = 1$ and $s, t > 2$. Use Theorem 1.10 to write $cs + dt = 1$ for some $c, d \in \mathbf{Z}$. The definition of congruence gives $dt \equiv_s 1$ and $cs \equiv_t 1$. Thus $cs - dt \equiv_s -1$ and $cs - dt \equiv_t 1$, so $(cs - dt)^2 \equiv_s 1$ and $(cs - dt)^2 \equiv_t 1$ using Theorem 1.24. Hence $s \mid ((cs - dt)^2 - 1)$ and $t \mid ((cs - dt)^2 - 1)$. But $(s, t) = 1$ so $st \mid ((cs - dt)^2 - 1)$ from Theorem 1.12, which yields $[cs - dt]_n^2 = [1]_n$. Should $[cs - dt]_n = [-1]_n$ then $cs - dt \equiv_{st} -1$, resulting in $cs - dt \equiv_t -1$ (why?). Since $cs - dt \equiv_t 1$ and $t > 2$ we get the contradiction $-1 \equiv_t 1$. Similarly, if $[cs - dt]_n = [1]_n$ then $cs - dt \equiv_s 1$ and again we get $-1 \equiv_s 1$ contradicting $s > 2$. Therefore $[cs - dt]_n \neq [-1]_n$ and both have order 2 in U_n, completing the proof. ■

In the other cases, when $n = 1, 2, 4, p^k$, or $2p^k$ for p an odd prime, U_n is cyclic. We prove this in the next chapter although it is not usually easy to find a generator. Often $U_p = \langle [2]_p \rangle$ for small primes, but the minimal $k \in \mathbf{N}$ with $U_p = \langle [k]_p \rangle$ is $k = 5$ for U_{23}, $k = 7$ for U_{71}, and $k = 17$ for U_{311}. What is $|U_n|$? Since $U_n = \{[k]_n \in \mathbf{Z}_n \mid 1 \leq k \leq n$ with $(k, n) = 1\}$, $|U_n|$ is the number of positive integers not greater than and relatively prime to n.

DEFINITION 3.2 For $n \in \mathbf{N}$, the **Euler phi-function** $\varphi(n)$ is the number of positive integers not exceeding n and relatively prime to n.

We may write $|U_n| = \varphi(n)$, but this is not particularly enlightening. Clearly $\varphi(1) = 1$, and for p a prime $\varphi(p) = p - 1$ since $1, 2, \ldots, p - 1$ are all relatively prime to p. For a prime power p^k, $1 \leq s \leq p^k$ is *not* relatively prime to $p \Leftrightarrow p \mid s$, or when $s \in \{p, 2p, 3p, \ldots, p^{k-1}p\}$. There are p^{k-1} such integers so $\varphi(p^k) = p^k - p^{k-1} = p^{k-1}(p - 1)$. Thus $\varphi(243) = \varphi(3^5) = 3^4 \cdot 2 = 162$ and $\varphi(256) = \varphi(2^8) = 2^7 \cdot 1 = 128$. Evaluating $\varphi(n)$ uses the fact that $(s, m) = 1 \Rightarrow \varphi(sm) = \varphi(s)\varphi(m)$. Functions on $\mathbf{N}$ satisfying this property are called *multiplicative*.

THEOREM 3.4 If $n > 1$ has prime factorization $n = p_1^{a_1} \cdots p_k^{a_k}$, then $\varphi(n) = \prod \varphi(p_j^{a_j}) = \prod p_j^{a_j - 1}(p_j - 1)$.

Proof We have seen that $\varphi(p^m) = p^{m-1}(p - 1)$. If we prove $\varphi(st) = \varphi(s)\varphi(t)$ when $(s, t) = 1$, then induction on k will give the result (exercise!). As $\varphi(1) = 1$, we may assume $s, t \geq 2$. We show that $\varphi(st) = |U_{st}| = |U_s \times U_t| = \varphi(s)\varphi(t)$. If $[k]_{st} \in U_{st}$ then $(k, st) = 1$. so $(k, s) = 1 = (k, t)$ by Theorem 1.13: $[k]_s \in U_s$ and $[k]_t \in U_t$. Therefore by Theorem 1.30, $f([k]_{st}) = ([k]_s, [k]_t)$ defines $f : U_{st} \to U_s \times U_t$. Since $(s, t) = 1$, $cs + dt = 1$ for $c, d \in \mathbf{Z}$ by Theorem 1.10. We use this to define $g(([a]_s, [b]_t)) = [adt + bcs]_{st} \in \mathbf{Z}_{st}$ for $([a]_s, [b]_t) \in U_s \times U_t$. Is g a function? Note that $adt + bcs \equiv_s adt \equiv_s a \cdot 1 = a$, and $adt + bcs \equiv_t b$. If $([a]_s, [b]_t) = ([a']_s, [b']_t)$ then $adt + bcs \equiv_s a \equiv_s a' \equiv_s a'dt + b'cs$, and similarly, $adt + bcs \equiv_t a'dt + b'cs$. Thus s and t divide $x = (adt + bcs - (a'dt + b'cs))$ so $st \mid x$ by Theorem 1.12 since $(s, t) = 1$. Hence $[adt + bcs]_{st} = [a'dt + b'cs]_{st}$ and

$g: U_s \times U_t \to \mathbf{Z}_{st}$. Further, if $(st, adt + bcs) = m > 1$ and $p \mid m$, p a prime, then $p \mid st$ implies that $p \mid s$ or $p \mid t$ by Theorem 1.14. Say $p \mid s$. Now $(a, s) = 1$ by definition of U_s and $(d, s) = (t, s) = 1$ using Theorem 1.10 and $cs + dt = 1$, so $p \mid adt$ is impossible by Theorem 1.14. But $p \mid m$ so $p \mid (adt + bcs)$, forcing $p \mid adt$ by Theorem 1.6. A similar contradiction arises if $p \mid t$ so $(st, adt + bcs) = 1$, $[adt + bcs]_{st} \in U_{st}$, and $g: U_s \times U_t \to U_{st}$. Our computations above show that $f \circ g(([a]_s[b]_t)) = f([adt + bcs]_{st}) = ([adt + bcs]_s, [adt + bcs]_t) = ([a]_s, [b]_t)$ and $g \circ f([k]_{st}) = g([k]_s, [k]_t) = [kdt + kcs]_{st} = [k(dt + cs)]_{st} = [k]_{st}$. Thus $g = f^{-1}$ so f is a bijection (Theorem 0.11), forcing $\varphi(st) = |U_{st}| = |U_s \times U_t| = \varphi(s)\varphi(t)$. ■

A consequence of Theorem 3.4 is that $|U_n|$ is even for $n > 2$. Some computations of $\varphi(n)$ are: $\varphi(72) = \varphi(2^3)\varphi(3^2) = 4 \cdot 1 \cdot 3 \cdot 2 = 24$, $\varphi(105) = \varphi(3)\varphi(5)\varphi(7) = 2 \cdot 4 \cdot 6 = 48$, and $\varphi(2700) = \varphi(2^2)\varphi(3^3)\varphi(5^2) = 2 \cdot 9 \cdot 2 \cdot 5 \cdot 4 = 720$. Different U_n can have the same order, such as $|U_{10}| = 4 = |U_{12}|$ or $|U_{35}| = 24 = |U_{56}| = |U_{72}|$. We will see later that U_{56} and U_{72} are abstractly the same group but U_{35} and U_{56} have different group structures (Exercise 3). Similarly, U_{10} and U_{12} have different structures since for $[k] \in U_{12} = \{[1], [5], [7], [11]\}$, $[k]^2 = [1]$, so U_{12} is not cyclic, but $U_{10} = \langle [3] \rangle = \{[1], [3], [9], [7]\}$. In Chapter 7 we will determine when U_n and U_m have the same group structure. As a final illustration of computation in U_n we consider U_{25}.

Example 3.3 $U_{25} = \langle [2] \rangle$ has eight generators and six subgroups.

Using $|U_{25}| = \varphi(25) = 5 \cdot 4 = 20$, by Theorem 2.16 we need $o\langle [2] \rangle = 20$. Now $2^{10} = 1024 \equiv_{25} -1$ so $[2]^{10} = [-1]$ and then $[2]^{20} = [-1]^2 = [1]$. Thus $o([2]) \mid 20$ from Theorem 2.7, and $o([2])$ is not a divisor of 10 (why?), so is either 4 or 20. Clearly $[2]^4 \neq [1]$, so $o([2]) = 20$ and $U_{25} = \langle [2] \rangle$ as claimed. Theorem 2.8 shows that $U_{25} = \langle [2]^k \rangle$ for $1 \leq k < 20$ with $(k, 20) = 1$ so there are $\varphi(20) = \varphi(4)\varphi(5) = 2 \cdot 4 = 8$ of them: $[2]^1 = [2]$, $[2]^3 = [8]$, $[2]^7 = [3]$, $[2]^9 = [12]$, $[2]^{11} = [23]$, $[2]^{13} = [17]$, $[2]^{17} = [22]$, and $[2]^{19} = [13]$. From Theorem 3.1, $U_{25} = \langle [2] \rangle$ has exactly six subgroups, one for each of the divisors of 20, i.e., 1, 2, 4, 5, 10, and 20. These subgroups, H_k of order k, are $H_1 = \langle [2]^{20} \rangle = \langle [1] \rangle$, $H_2 = \langle [2]^{10} \rangle = \langle [24] \rangle$, $H_4 = \langle [2]^5 \rangle = \langle [7] \rangle$, $H_5 = \langle [2]^4 \rangle$, $H_{10} = \langle [2]^2 \rangle$ and $H_{20} = U_{25} = \langle [2] \rangle$. By §1 Exercise 8, two elements in U_{25} of the same order lie in the same cyclic subgroup. Since any element of order m generates a cyclic subgroup of the same order by Theorem 2.16, and since $o(g^k) = o(g) \Leftrightarrow (k, o(g)) = 1$ by Theorem 2.8, it follows that U_{25} has exactly $\varphi(2) = 1$ element of order 2, $\varphi(4) = 2$ elements of order 4, $\varphi(5) = 4$ elements of order 5, and $\varphi(10) = 4$ elements of order 10.

The observation in the example about the number of generators of each cyclic subgroup is useful to record.

THEOREM 3.5 A cyclic group of order n has $\varphi(n)$ generators, and if U_n is cyclic, it has $\varphi(\varphi(n))$ generators.

Proof Suppose that $G = \langle g \rangle$ and apply Theorem 2.8 and Theorem 2.16 to conclude that $G = \langle g^k \rangle$ exactly when $1 \leq k < n$ and $(k, n) = 1$. Thus G has $\varphi(n)$ generators. Since $|U_n| = \varphi(n)$, if it is cyclic then it has $\varphi(|U_n|) = \varphi(\varphi(n))$ generators. ■

EXERCISES

1. By direct calculation: a) show that the group is cyclic and find all its generators; and b) find a generator for each subgroup and its order:
   i) U_9
   ii) U_{14}
   iii) U_{18}
   iv) U_{22}
   v) U_{49}
   vi) U_{50}
   vii) U_{98}
2. Match each element with its inverse (Note that $ab = 1 \Rightarrow (-a)(-b) = 1$.) in:
   i) U_{12}
   ii) U_{13}
   iii) U_{16}
   iv) U_{28}
   v) U_{35}
   vi) U_{40}
   vii) U_{44}
3. Show that $[8]_{35} \in U_{35}$ has order 4 but that no class in U_{56} has order 4.
4. Use §1 Exercise 8 to show that for any $n \in \boldsymbol{N}$, $n = \sum_{d|n} \varphi(d)$. (That is, $9 = \varphi(1) + \varphi(3) + \varphi(9)$, and $12 = \varphi(1) + \varphi(2) + \varphi(3) + \varphi(4) + \varphi(6) + \varphi(12)$.)
5. Find all cyclic subgroups in:
   i) U_{15}
   ii) U_{32}
   iii) U_{45}
   iv) U_{52}
   v) U_{65}
6. For $m \geq 3$, prove that $k^{2^m} \equiv 1 (\text{mod } 2^{m+2})$ for all $k \in 2\boldsymbol{N} + 1$ by Induction. (See Theorem 1.3 and §1.7 Exercise 8.) Show that U_{2^m} is not cyclic.
7. In each case, find a bijection of groups so that products in the first group correspond to the product of their images in the second group:
   i) U_8 and U_{12}
   ii) $(\boldsymbol{Z}_6, +)$ and U_9
   iii) U_{15} and U_{16}
   iv) $(\boldsymbol{Z}_{20}, +)$ and U_{25}
   (In iii) write each group $G = HK = \{hk \mid h \in H \text{ and } k \in K\}$ for $H, K \leq G$.)

3.3 THE SYMMETRIC GROUPS S_n

Here we study the groups $\text{Bij}(I(n)) = S_n$, for $I(n) = \{1, 2, \ldots, n\}$. From Theorem 0.14, S_n is a group with composition of functions as the multiplication, S_n is not Abelian when $n > 2$, and $|S_n| = n!$. Of course everything we say about S_n has an analogue for $\text{Bij}(S)$ for any $S = \{s_1, \ldots, s_n\}$ of cardinality n; we could refer to s_i and s_j instead of to i and j. We want a notation for the **permutations** $\theta \in S_n$ that will make computation in S_n fairly easy. A standard way to do this *initially* is to write $\theta \in S_n$ as a double row with the image $\theta(i)$ of $i \in I(n)$ underneath i. A particular $\theta \in S_8$ is $\theta = \begin{pmatrix} 1 & 2 & 3 & 4 & 5 & 6 & 7 & 8 \\ 5 & 3 & 7 & 1 & 8 & 6 & 2 & 4 \end{pmatrix}$ representing the permutation with $\theta(1)=5$, $\theta(2)=3$, $\theta(3)=7$, $\theta(4)=1$, etc. The multiplication of two permutations is composition so if $\rho=\begin{pmatrix}1&2&3&4&5&6\\3&6&1&2&5&4\end{pmatrix}$, $\sigma=\begin{pmatrix}1&2&3&4&5&6\\2&3&4&6&1&5\end{pmatrix}\in S_6$ then $\rho\sigma=\begin{pmatrix}1&2&3&4&5&6\\3&6&1&2&5&4\end{pmatrix}\cdot\begin{pmatrix}1&2&3&4&5&6\\2&3&4&6&1&5\end{pmatrix}=\begin{pmatrix}1&2&3&4&5&6\\6&1&2&4&3&5\end{pmatrix}$, since, for example, $\rho\sigma(1)=\rho(\sigma(1))=\rho(2)=6$ and $\rho\sigma(2)=\rho(\sigma(2))=\rho(3)=1$. This notation is not bad but seems inefficient for the identity or the permutation that just interchanges $i, j\in I(n)$. A better representation, giving additional information, arises from an equivalence relation on $I(n)$ defined for each element in S_n.

DEFINITION 3.3 For $\theta\in S_n$ and $i, j\in I(n)$, i and j are **congruent modulo θ**, written $i\equiv_\theta j$, if for some $k\in\mathbf{Z}$, $j=\theta^k(i)$.

Since S_n is a group, θ^k is defined by Definition 2.4 as the product (composition!) of θ or of θ^{-1} with itself $|k|$ times. For $\theta\in S_8$ given above, $8\equiv_\theta 1$ since $1=\theta(4)=\theta^2(8)$, $1\equiv_\theta 4$ since $4=\theta^3(1)$, $3\equiv_\theta 7$ since $7=\theta(3)$, and $2\equiv_\theta 2$ since $2=\theta^3(2)$. If $\eta\in S_n$ interchange 1 and 2 and $\eta(i)=i$ for $i\neq 1, 2$ then $\eta(1)=2$ and $\eta(2)=1$ so $2\equiv_\eta 1$ and $1\equiv_\eta 2$. For every other $j\in I(n)$, $\eta(j)=j$ so $j\equiv_\eta j$, and finally $1\equiv_\eta 1$ and $2\equiv_\eta 2$ since $\eta^2(1)=1$ and $\eta^2(2)=2$. These are the only relations of congruence modulo η.

Because S_n is finite, any $\theta\in S_n$ has finite order by Theorem 2.9, say $o(\theta)=m$, so $j=\theta^m(j)$ and $j\equiv_\theta j$ for each $j\in I(n)$: $\equiv_\theta$ is reflexive. Also since θ^{-1} is a positive power of θ (Theorem 2.9 again), we may take the integers k appearing in Definition 3.3 to be in $\boldsymbol{N}$. The relation $\equiv_\theta$ connects those $i, j\in I(n)$ that can be obtained from one another as images under consecutive applications of θ and is always an equivalence relation.

THEOREM 3.6 For each $\theta\in S_n$, $\equiv_\theta$ is an equivalence relation on $I(n)$.

Proof We must show $\equiv_\theta$ is reflexive, symmetric, and transitive. The comments above show that $\equiv_\theta$ is reflexive. Now suppose $i\equiv_\theta j$ with $j=\theta^m(i)$ for $m\in\mathbf{Z}$. Applying θ^{-m} yields $\theta^{-m}(j)=\theta^{-m}(\theta^m(i))=I_{I(n)}(i)=i$ so $j\equiv_\theta i$ and $\equiv_\theta$ is symmetric. Finally if $i\equiv_\theta j$ and $j\equiv_\theta k$ then for some $s, t\in\mathbf{Z}$, $j=\theta^s(i)$ and $k=\theta^t(j)$ so $k=\theta^t(\theta^s(i))=\theta^{t+s}(i)$ using Theorem 2.4 on powers, and $i\equiv_\theta k$ results. Therefore $\equiv_\theta$ is transitive and so is an equivalence relation.

■

Since $\equiv_\theta$ is an equivalence relation on $I(n)$, by Theorem 1.22 its equivalence classes partition $I(n)$. From Definition 1.11 the equivalence class of $i\in I(n)$ is just $[i]_\theta=\{\theta^k(i)\mid k\in\mathbf{Z}\}$, all elements of $I(n)$ that can be obtained from i by application of θ some number of times.

DEFINITION 3.4 For $\theta\in S_n$ and $i\in I(n)$, by the **orbit of *i* under θ** is the equivalence class $[i]_\theta$. The **length** of $[i]_\theta$ is its cardinality. If $[i]_\theta=\{i\}$ then i is called a **fixed point** of θ.

Since S_n is finite, to find $[i]_\theta$ it suffices to consider $\theta^k(i)$ for $k\in\boldsymbol{N}$ and note that some $\theta^m(i)=I_{I(n)}(i)=i$. For $\theta=\begin{pmatrix}1&2&3&4&5&6&7&8\\5&3&7&1&8&6&2&4\end{pmatrix}\in S_8$ from above:

$[1]_\theta = \{\theta^0(1)=1, \theta^1(1)=5, \theta^2(1)=\theta(5)=8, \theta^3(1)=\theta(8)=4, \theta^4(1)=1\} = \{1, 5, 8, 4\}$; $[2]_\theta = \{\theta^0(2)=2, \theta^1(2)=3, \theta^2(2)=\theta(3)=7, \theta^3(2)=2\} = \{2, 3, 7\}$; and $[6]_\theta = \{\theta^0(6) = 6, \theta^1(6)=6\} = \{6\}$, so 6 is the only fixed point of θ. Thus this θ has three different orbits, of lengths 1, 3, and 4. Recall from Theorem 1.21 that $[i]_\theta = [j]_\theta \Leftrightarrow i \equiv_\theta j$, so $[1]_\theta = [5]_\theta = [8]_\theta = [4]_\theta$, and $[2]_\theta = [3]_\theta = [7]_\theta$. The permutation $\sigma = \begin{pmatrix} 1 & 2 & 3 & 4 & 5 & 6 \\ 2 & 3 & 4 & 6 & 1 & 5 \end{pmatrix} \in S_6$ has only one orbit of length 6 since $[1]_\sigma = \{1, \sigma(1)=2, \sigma^2(1)=3, \sigma^3(1)=4, \sigma^4(1) = 6, \sigma^5(1)=5\}$, so it has no fixed points. Figure 3.1 is a picture with arrows representing the action of these permutations and their orbits.

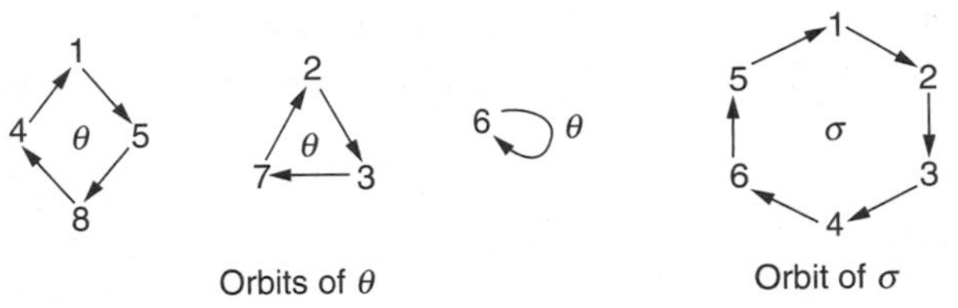

FIGURE 3.1

The identity $I_{I(n)}$ will have n orbits, all $\{j\}$ in $I(n)$ each of length 1, so all elements of $I(n)$ are fixed points. In general, as in our examples, the orbits are the consecutive powers of θ with the initial element as the first repetition.

THEOREM 3.7 For any $\theta \in S_n$ and $i \in I(n)$, if $m \in \boldsymbol{N}$ is minimal with $\theta^m(i) = i$, then $[i]_\theta$ is of length m and $[i]_\theta = \{i, \theta(i), \theta^2(i), \ldots, \theta^{m-1}(i)\}$.

Proof If $o(\theta) = s$ then $\theta^s(i) = i$ so Well Ordering yields a minimal $m \in \boldsymbol{N}$ satisfying $\theta^m(i) = i$. Note that $\theta^{-m}(i) = \theta^{-m}(\theta^m(i)) = i$. For any $k \in \boldsymbol{Z}$ use the division theorem (Theorem 1.1) to write $k = qm + r$ with $0 \le r < m$. Thus by Theorem 2.4 on powers, $\theta^k(i) = \theta^{qm+r}(i) = \theta^r(\theta^{qm}(i)) = \theta^r(i)$, since $\theta^{qm}(i) = (\theta^m)^q(i) = (\theta^{-m})^{-q}(i) = i$. It follows from Definition 3.4 that $[i]_\theta \subseteq \{i, \theta(i), \theta^2(i), \ldots, \theta^{m-1}(i)\}$, and since the opposite inclusion is clear, these sets are equal. Finally, if $\theta^t(i) = \theta^j(i)$ for some $0 \le t < j < m$, then $i = \theta^{-t}(\theta^t(i)) = \theta^{-t}(\theta^j(i)) = \theta^{j-t}(i)$, contradicting the minimality of m since $0 < j - t < m$. Therefore $\{i, \theta(i), \theta^2(i), \ldots, \theta^{m-1}(i)\} = [i]_\theta$ must have cardinality m. ■

The action of θ on any $[i]_\theta$ is particularly simple and pleasing: each element is moved to its neighbor, and the last is sent to the first. This is the key to writing a general $\eta \in S_n$ in a simple way, and it leads to a definition for the special case when there is only one orbit of length more than 1.

DEFINITION 3.5 For $k \ge 2$, a ***k*-cycle** is any $\rho \in S_n$ with a single orbit of lenght k containing all the nonfixed points under ρ, and all other orbits of ρ have length 1. If $[i]_\rho = \{i, \rho(i), \ldots, \rho^{k-1}(i)\}$ with $k > 1$ then we write $\rho = (i, \rho(i), \ldots, \rho^{k-1}(i))$. A 1-cycle is the identity map $I_{I(n)} = (1) = (2) = \cdots = (n)$.

The notation for a cycle is ambiguous. Now $\rho = (1, 2, 3)$ sends 1 to 2, 2 to 3, and 3 to 1, but $\rho \in S_n$ for any $n \ge 3$, so unless we know n we cannot tell how many fixed points ρ has.

Also $(1, 2, 3), (2, 3, 1), (3, 1, 2) \in S_n$ all represent the same 3-cycle. Observe that the action of a cycle is read from left to right. We must be careful not to proceed from the far left to the right when evaluating a product of cycles since the product is a composition of functions applied in order from right to left. Specifically, if $\sigma = (1, 2, 5)$ and $\rho = (2, 3, 4)$ then $\sigma\rho = (1, 2, 5)(2, 3, 4) = (1, 2, 3, 4, 5)$ since, for example, $\sigma\rho(1) = \sigma(\rho(1)) = \sigma(1) = 2$ and $\sigma\rho(2) = \sigma(\rho(2)) = \sigma(3) = 3$. It is incorrect to start from the far left thinking that the effect of $\sigma\rho$ on 1 is to send 1 to 2, which $\rho = (2, 3, 4)$ sends to 3. This would be the computation for $\rho\sigma$.

In general the product of cycles is not a cycle. For example, take $\alpha = (1, 3, 4)$ and $\beta = (1, 5, 4)$ with $\alpha\beta = (1, 3, 4)(1, 5, 4)$. Then $\alpha\beta(1) = \alpha(\beta(1)) = \alpha(5) = 5$, $\alpha\beta(5) = \alpha(\beta(5)) = \alpha(4) = 1$, $\alpha\beta(2) = 2$, $\alpha\beta(3) = 4$, $\alpha\beta(4) = 3$, and $\alpha\beta(k) = k$ for $k > 5$, so $\alpha\beta$ has two orbits of length 2, $[1] = \{1, 5\}$ and $[3] = \{3, 4\}$, and all others of length 1. The next result gives facts about cycles.

THEOREM 3.8 For $k \geq 2$, let $\rho = (i_1, i_2, \ldots, i_k)$ be a k-cycle in S_n.

a) $\rho = (i_2, i_3, \ldots, i_k, i_1) = (i_3, i_4, \ldots, i_k, i_1, i_2) = \cdots = (i_k, i_1, \ldots, i_{k-1})$
b) $o(\rho) = k$
c) $\rho^{-1} = (i_k, i_{k-1}, \ldots, i_1) = (i_1, i_k, i_{k-1}, \ldots, i_2)$
d) S_n contains $n!/(k \cdot (n-k)!)$ k-cycles.

Proof By the definition of k-cycle, $\rho(x) = x$ if $x \notin \{i_1, \ldots, i_k\}$ and $\rho(i_j) = i_{j+1}$, where the subscript $j + 1$ is understood to be 1 when $j = k$; the action of ρ on $\{i_j\}$ adds one to the subscript *computed modulo k*. Hence a) is clear since all these k-cycles have the same effect on every $y \in I(n)$. Since $\rho^m(i_j) = i_{j+m}$, computed modulo k, the smallest $m \in \mathbf{N}$ satisfying $\rho^m(i_1) = i_1$ is $m = k$ and also $\rho^k(i_j) = i_j$ for all $1 \leq j \leq k$. If $\rho(x) = x$ then $\rho^k(x) = x$, so $\rho^k = I_{I(n)}$ and k minimal implies $o(\rho) = k$. Now $\sigma = (i_k, i_{k-1}, \ldots, i_1)$ fixes all $x \notin \{i_1, \ldots, i_k\}$ and $\sigma(i_j) = i_{j-1}$ with the subscripts computed modulo k. Then $\sigma\rho$ and $\rho\sigma$ both fix all $x \notin \{i_1, \ldots, i_k\}$ and send i_j to i_j (why?) so are inverses. Finally, there are $n(n-1)\cdots(n-k+1) = n!/(n-k)!$ lists of k distinct elements from $I(n)$ (why?) and each written as a k-tuple represents a k-cycle. By part a) each k-cycle has k representations and grouping these together partitions the $n!/(n-k)!$ k-tuples into k-element subsets corresponding to the different k-cycles. Thus the number of k-cycles is $(1/k)(n!/(n-k)!) = n!/(k \cdot (n-k)!)$. ■

As in the theorem, $(2, 3, 5, 6) = (3, 5, 6, 2) = (5, 6, 2, 3) = (6, 2, 3, 5) \in S_6$ and $(2, 3, 5, 6)^{-1} = (6, 5, 3, 2) = (2, 6, 5, 3)$. The number of distinct 4-cycles in S_6 is $6!/4 \cdot 2! = 90$, and the number in S_4 is $4!/4 \cdot 0! = 6$ (find them!). Using cycles will give a good way to represent an arbitrary $\theta \in S_n$. From Theorem 3.7 on orbits, $[i]_\theta = \{i, \theta(i), \theta^2(i), \ldots, \theta^{k-1}(i)\}$ so the restriction of θ to $[i]_\theta$ acts like the k-cycle $(i, \theta(i), \theta^2(i), \ldots, \theta^{k-1}(i))$ on $[i]_\theta$. We claim that θ is the product of such cycles, one for each orbit of θ. In general, products in S_n are not commutative so the order in which we write these cycles may matter.

DEFINITION 3.6 If $k > 1$ and ρ is a k-cycle $(i_1, i_2, \ldots, i_k) \in S_n$, the **orbit** of ρ is $\{i_1, i_2, \ldots, i_k\} = \mathrm{orb}(\rho)$. Cycles $\rho, \sigma \neq I_{I(n)}$ are **disjoint** if $\mathrm{orb}(\rho) \bigcap \mathrm{orb}(\sigma) = \varnothing$. The identity is considered to be disjoint from every cycle.

Disjointness of ρ and σ means no element in $I(n)$ is moved by both cycles, so if some integer is moved by one of the disjoint cycles it is fixed by the other. Thus $I_{I(n)}$ fixing all $j \in I(n)$ ought to be disjoint from all $\theta \in S_n$.

THEOREM 3.9 Disjoint cycles in S_n commute.

Proof Let $\rho = (i_1, i_2, \ldots, i_k)$ and $\sigma = (j_1, j_2, \ldots, j_t)$ be disjoint cycles. We may assume that $k, t \geq 2$ since $I_{I(n)}$ commutes with any permutation. We prove that $\rho\sigma = \sigma\rho$ by showing $\rho\sigma(j) = \sigma\rho(j)$ for all $j \in I(n)$ (Theorem 0.8). By definition of disjoint cycles, {orb(ρ), orb(σ), $I(n) - (\text{orb}(\rho) \cup \text{orb}(\sigma))$} is a partition of $I(n)$ so it suffices to show that $\rho\sigma$ and $\sigma\rho$ agree on each of these three subsets. If $m \in I(n) - (\text{orb}(\rho) \cup \text{orb}(\sigma))$, then $\rho(m) = m = \sigma(m)$, so $\rho\sigma(m) = m = \sigma\rho(m)$. If $m \in \text{orb}(\rho)$ then also $\rho(m) \in \text{orb}(\rho)$, and ρ, σ disjoint implies $\sigma(m) = m$ and $\sigma(\rho(m)) = \rho(m) = \rho(\sigma(m))$. A similar computation when $m \in \text{orb}(\sigma)$ shows $\sigma(\rho(m)) = \sigma(m) = \rho(\sigma(m))$, completing the proof. ■

The idea behind the proof of our next theorem is illustrated in Fig. 3.2 with an example showing how the orbits of θ give cycles whose product is θ.

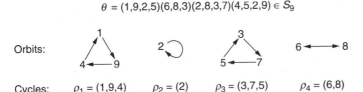

$\theta = (1,9,2,5)(6,8,3)(2,8,3,7)(4,5,2,9) \in S_9$

Orbits:

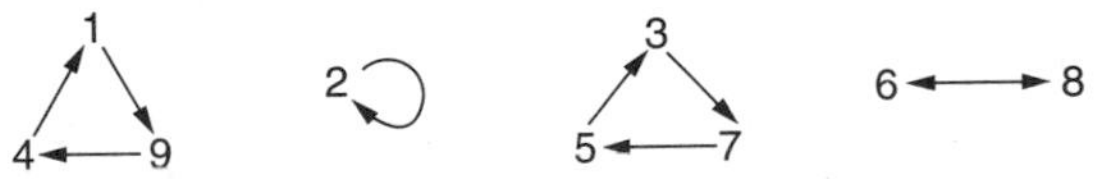

Cycles: $\rho_1 = (1,9,4)$ $\rho_2 = (2)$ $\rho_3 = (3,7,5)$ $\rho_4 = (6,8)$

$\theta = \rho_1\rho_2\rho_3\rho_4 = (1,9,4)(2)(3,7,5)(6,8) = (1,9,4)(3,7,5)(6,8) = \rho_1\rho_3\rho_4$

FIGURE 3.2

THEOREM 3.10 If $\theta \in S_n$ has orbits $\{[x_1], \ldots, [x_t]\}$ on $I(n)$ of lengths $k_1, k_2, \ldots, k_t$ respectively, set $\rho_j = (x_j, \theta(x_j), \ldots, \theta^{k_j-1}(x_j))$. Then $\theta = \rho_1\rho_2\cdots\rho_t$, and except for order and the number of 1-cycles (identity map) this is the unique representation of θ as a product of pairwise disjoint cycles.

Proof Let $\sigma_1, \sigma_2, \ldots, \sigma_w$ be pairwise disjoint cycles in S_n with $\eta = \sigma_1\sigma_2\cdots\sigma_w$. If $m \in \text{orb}(\sigma_j)$, then $\sigma_j(m) \in \text{orb}(\sigma_j)$ also, so $\sigma_i(m) = m$ and $\sigma_i(\sigma_j(m)) = \sigma_j(m)$ for $i \neq j$, from the definition of disjoint cycles. Hence $\eta(m) = \sigma_1\sigma_2\cdots\sigma_w(m) = \sigma_1\sigma_2\cdots\sigma_{j-1}(\sigma_j(m)) = \sigma_j(m)$, η and σ_j agree on orb(σ_j), and each orb(σ_j) is an orbit of η acting on $I(n)$. Any $j \in I(n)$ is moved by at most one σ_i so $[j]_\eta = \text{orb}(\sigma_i)$ or j is fixed by all σ_i. The latter case is equivalent to $\eta(j) = j$. Thus the orbits of a product of pairwise disjoint cycles $\{\sigma_i\}$ are $\{\text{orb}(\sigma_i)\}$ and any common fixed points.

If $\theta(x_i) = x_i$ then $\rho_i = I_{I(n)}$ and $\text{orb}(\rho_i) = \{x_i\}$. Now ρ_j agrees with θ on $\text{orb}(\rho_j) = [x_j]_\theta$ and $\{[x_j]_\theta\}$ is a partition of $I(n)$, so the $\{\rho_i\}$ is pairwise disjoint. From above, $\sigma = \rho_1\rho_2\cdots\rho_t$ has orbits $\{\text{orb}(\rho_i)\}$, and $\sigma(m) = \rho_j(m)$ for all $m \in \text{orb}(\rho_j)$. But $\theta(m) = \rho_j(m)$ for all $m \in [x_j]_\theta = \text{orb}(\rho_j)$, for all j, forcing $\theta = \sigma = \rho_1\rho_2\cdots\rho_t$. If $\theta = \sigma_1\sigma_2\cdots\sigma_s$ for a collection $\{\sigma_i\}$ of pairwise disjoint cycles, then σ_i and θ agree on orb(σ_i), which we have seen is some $[x_j]_\theta$. Thus $\text{orb}(\sigma_i) = [x_j]_\theta = \text{orb}(\rho_j)$ for some j, and σ_i, θ, and ρ_j all agree on this set, so $\sigma_i = \rho_j$, by the definition of cycle. From Theorem 3.9, σ_i and σ_j commute, and ρ_i and ρ_j commute, so in $\theta = \rho_1\rho_2\cdots\rho_t = \sigma_1\sigma_2\cdots\sigma_s$ we may rearrange terms so that $\rho_j = \sigma_i$ appears on the right and multiplication by σ_i^{-1} eliminates it from the products. Using induction on s, cancel each σ_i and some $\rho_d = \sigma_i$, leaving $I_{I(n)}$ equal to a product of some of the ρ_v. Now each such ρ_v and $I_{I(n)}$ must agree on $\text{orb}(\rho_v) = [x_v]_\theta$, an orbit of $I_{I(n)}$. Since each orbit of $I_{I(n)}$ is a singleton, $\rho_v = I_{I(n)}$, completing the proof of the theorem. ∎

DEFINITION 3.7 Let $\theta \in S_n$ have orbits on $I(n)$ of lengths $k_1 \geq k_2 \geq \cdots \geq k_t$. Then the **cycle structure** of θ is $(k_1, k_2, \ldots, k_t)$.

For any cycle structure $(k_1, k_2, \ldots, k_t)$, $\Sigma k_i = n$ and the number of $k_j = 1$ is the number of fixed points. An n-cycle in S_n has cycle structure (n), the identity in S_n has cycle structure $(1, 1, \ldots, 1)$ with n ones, a 3-cycle in S_n has cycle structure $(3, 1, 1, \ldots, 1)$ where 1 appears $n-3$ times, and any product of pairwise disjoint cycles of lengths $k_1 \geq k_2 \geq \cdots \geq k_s$ has cycle structure $(k_1, k_2, \ldots, k_s, 1, 1, \ldots, 1)$, with as many ones as necessary to have the sum equal n. For example, $(5, 10)(7, 9)(2, 4, 8) \in S_{10}$ has cycle structure $(3, 2, 2, 1, 1, 1)$.

By Theorem 3.10, to write $\theta \in S_n$ as a product of disjoint cycles, find its orbits. For $\rho\eta \in S_7$ with $\rho = (1, 3, 6, 2, 4)$ and $\eta = (3, 5, 4, 7, 6)$, compute $\rho\eta(1) = \rho(1) = 3$, $\rho\eta(3) = \rho(5) = 5$, and $\rho\eta(5) = \rho(4) = 1$, so $\{1, 3, 5\} = [1]_{\rho\eta}$ with corresponding cycle $(1, 3, 5)$. Similarly, $\rho\eta(2) = \rho(2) = 4$, $\rho\eta(4) = \rho(7) = 7$, and $\rho\eta(7) = \rho(6) = 2$, so $\{2, 4, 7\} = [2]_{\rho\eta}$ with cycle $(2, 4, 7)$. Also $\rho\eta(6) = \rho(3) = 6$, so $\{6\} = [6]_{\rho\eta}$, and $\rho\eta = (1, 3, 5)(2, 4, 7)(6) = (1, 3, 5)(2, 4, 7)$ has three orbits (including one fixed point) and has cycle structure $(3, 3, 1)$. A similar computation gives $\eta\rho = (1, 5, 4)(2, 7, 6) \neq \rho\eta$. This does not contradict Theorem 3.9 since ρ and η are not disjoint. It is straightforward to check that $o(\rho\eta) = o(\eta\rho) = 3$, although $o(\rho) = o(\eta) = 5$.

Now let $\alpha = (1, 2, 3)$, $\beta = (2, 3, 4)$, and $\gamma = (3, 4, 5)$ in S_5 and compute $\alpha\beta\gamma(1) = \alpha\beta(1) = \alpha(1) = 2$ and $\alpha\beta\gamma(2) = 1$, $\alpha\beta\gamma(3) = \alpha\beta(4) = \alpha(2) = 3$, and finally $\alpha\beta\gamma(4) = \alpha\beta(5) = 5$ and $\alpha\beta\gamma(5) = 4$. Thus $\alpha\beta\gamma$ has three orbits, $\{1, 2\}$, $\{3\}$, and $\{4, 5\}$, so cycle structure $(2, 2, 1)$, and $\alpha\beta\gamma = (1, 2)(4, 5)$. Note here that each of α, β, and γ has order 3, but $o(\alpha\beta\gamma) = 2$. A slight variation shows that a product of three 3-cycles can be a 5-cycle. Specifically, $(1, 2, 4)(2, 5, 4)(3, 4, 5) = (1, 2, 5, 3, 4)$ in S_7 (verify!) has cycle structure $(5, 1, 1)$.

Let us consider some computations with 7-cycles in S_7 (verify these!). If $\rho = (1, 2, 3, 4, 5, 6, 7)$ and $\theta = (1, 7, 6, 5, 4, 2, 3)$ then $\rho\theta = (2, 4, 3)$, a 3-cycle, so has cycle structure $(3, 1, 1, 1, 1)$. For $\lambda = (1, 7, 5, 2, 4, 3, 6)$, $\rho\lambda = (2, 5, 3, 7, 6)$ with cycle structure $(5, 1, 1)$, and the product $\rho(1, 4, 7, 3, 6, 2, 5) = (1, 5, 2, 6, 3, 7, 4)$, another 7-cycle. Another simple example using 4-cycles is that $(1, 3, 2, 7)(2, 6, 4, 8)(1, 3, 6, 2)(5, 4, 3, 8)(1, 8, 2, 6) = (1, 5, 8, 3, 7)(2, 6)$ (why?), so this product of five 4-cycles has cycle structure $(5, 2, 1)$ in S_8 and turns out to have order 10, as we will see shortly.

The examples above indicate that for $\theta = \sigma_1 \cdots \sigma_k$ there is no obvious relation between $o(\theta)$ and $\{o(\sigma_j)\}$. For a product of disjoint cycles there is.

THEOREM 3.11 If the cycle structure of $\theta \in S_n$ is $(k_1, k_2, \ldots, k_t)$ then $o(\theta) = \text{LCM}\{k_1, k_2, \ldots, k_t\}$.

Proof Use Theorem 3.10 to write $\theta = \rho_1\rho_2\cdots\rho_t$ for ρ_j a k_j-cycle with $\{\rho_j\}$ pairwise disjoint. Different ρ_j commute by Theorem 3.9, so it follows from Theorem 2.4 that $\theta^m = (\rho_1\rho_2\cdots\rho_t)^m = \rho_1^m\rho_2^m\cdots\rho_t^m$ for any $m \in \boldsymbol{N}$. If all $k_j \mid m$, then since $o(\rho_j) = k_j$ (Theorem 3.8), $\rho_j^m = I_{I(n)}$ by Theorem 2.7, so $\theta^m = I_{I(n)}$, and by Theorem 2.7 again, $o(\theta) \mid m$. But LCM$\{k_j\}$ is a common multiple for $\{k_j\}$ so if $m = \text{LCM}\{k_j\}$ then $o(\theta) \mid \text{LCM}\{k_j\}$. Say $o(\theta) = d$. Then, as we just saw, $I_{I(n)} = \theta^d = (\rho_1\rho_2\cdots\rho_t)^d = \rho_1^d\rho_2^d\cdots\rho_t^d$. For $x \in \text{orb}(\rho_j)$, $\rho_i(x) = x$ when $i \neq j$, so commuting the ρ_w to have ρ_j applied last, $x = I_{I(n)}(x) = \rho_j^d\rho_1^d\cdots\rho_t^d(x) = \rho_j^d(x)$. Clearly $\rho_j^d(y) = y$ for $y \notin \text{orb}(\rho_j)$, so $\rho_j^d = I_{I(n)}$ and $k_j \mid d$ by Theorem 3.8 and Theorem 2.7. Thus LCM$\{k_j\} \mid d(= o(\theta))$ using the basic property of least common multiples in Theorem 1.7. Since $o(\theta)$, LCM$\{k_j\} > 0$, they are equal. ∎

From computations above, $\theta = (1, 3, 6, 2, 4)(3, 5, 4, 7, 6) = (1, 3, 5)(2, 4, 7)$ in S_7 has order 3 since its cycle structure is (3, 3, 1) and LCM$\{3, 3, 1\} = 3$, the product $(1, 2, 3)(2, 3, 4)(3, 4, 5) = (1, 2)(4, 5)$ has cycle structure (2, 2, 1) in S_5 so order $2 = \text{LCM}\{2, 2, 1\}$, and $(1, 3, 2, 7)(2, 6, 4, 8)(1, 3, 6, 2)(5, 4, 3, 8)(1, 8, 2, 6) = (1, 5, 8, 3, 7)(2, 6) \in S_8$ has order LCM$\{5, 2, 1\} = 10$.

The cycle structures in S_n are the different ways of expressing n as a decreasing sum of positive integers, called **partitions** of n. For any partition of n there is $\theta \in S_n$ with the corresponding cycle structure. For example, $15 = 4 + 3 + 3 + 2 + 1 + 1 + 1$ and $(1, 2, 3, 4)(5, 6, 7)(8, 9, 10)(11, 12) \in S_{15}$, among others, has cycle structure (4, 3, 3, 2, 1, 1, 1) and order 12. In this way, using partitions we can (in theory!) determine all possible orders of elements in S_n.

Example 3.4 Cycle structures and orders of elements in S_5.

The cycle structures are (5), (4, 1), (3, 2), (3, 1, 1), (2, 2, 1), (2, 1, 1, 1), and (1, 1, 1, 1, 1), corresponding to the seven partitions of 5, and with orders 5, 4, 6, 3, 2, 2, and 1, respectively. Therefore, although $|S_5| = 5! = 120$, there are only seven possible orders for its elements, and the largest order is only 6.

Computing the largest possible order for $\theta \in S_n$ becomes unwieldy for large n, but some general observations can be useful. For example, if all orbits of θ have lengths $k_j \leq 3$, $o(\theta) \leq 6$ since LCM$\{2, 1\} = 2$, LCM$\{3, 1\} = 3$, and LCM$\{3, 2, 1\} = 6$. Similarly, if all $k_j \leq 4$, $o(\theta) \leq 12$: LCM$\{4, 3\} = \text{LCM}\{4, 3, 2\} = 12$. If $\theta \in S_8$ has cycle structure (5, 3) then $o(\theta) = 15$, so to find $\eta \in S_8$ of larger order, we need to consider partitions of 8 containing an integer at least as large as 5. In fact, 15 is the largest order for elements in S_8.

Given $\theta \in S_n$ with $o(\theta) = m$, we can describe its possible cycle structures by determining which k_j can satisfy LCM$\{k_j\} = m$. Any $\theta \in S_7$ of order $6 = \text{LCM}\{k_j\}$ forces all $k_j = 6, 3, 2$, or 1 since each $k_i \mid 6$. The cycle structures are (6, 1), (3, 2, 1, 1) and (3, 2, 2) so $\theta = (a, b, c, d, e, f)$, $\theta = (a, b, c)(d, e)$, or $\theta = (a, b, c)(d, e)(f, g)$. For $\theta \in S_8$ we also have the cycle structures (6, 2) and (3, 3, 2). The next theorem describes elements of prime order.

THEOREM 3.12 If $\theta \in S_n$ and $o(\theta) = p$, a prime, then θ is a product of pairwise disjoint p-cycles. If in addition θ has no fixed points, then $p \mid n$.

Proof From Theorem 3.11, if θ has cycle structure $(k_1, k_2, \ldots, k_s)$ then $p = \mathrm{LCM}\{k_j\}$. But each $k_j \mid p$ and p is prime so either $k_j = 1$ or $k_j = p$ and the cycle structure of θ is $(p, p, \ldots, p, 1, \ldots, 1)$. Thus θ is a product of pairwise disjoint p-cycles. If θ has no fixed points then it has no orbit of length one so its cycle structure is $(p, p, \ldots, p)$ and since these add to n, it follows that $p \mid n$. ■

Our next example illustrates a nontrivial way to use the relation between order and cycle structures.

Example 3.5 For $n \geq 5$ but $n \neq 2^k + 2$ there is $\theta \in S_n$ with $o(\theta) = 2(n-2)$.

Note that if $n = 4 = 2^1 + 2$ then S_4 contains 4-cycles of order 4. If $n = 2^k + 2$ with $k \geq 2$, then $2(n-2) = 2^{k+1}$. Should $\theta \in S_n$ have order 2^{k+1} then Theorem 3.11 implies that an orbit of θ must have 2^{k+1} elements (why?). This is impossible when $k \geq 2$ because $2^{k+1} > n = 2^k + 2$. Thus if $n = 2^k + 2 \geq 5$ then S_n cannot have an element of order $2(n-2)$, so assume that $n \neq 2^k + 2$. When n is odd then $n-2$ is also, so $\mathrm{GCD}\{2, n-2\} = 1$. Therefore $\theta \in S_n$ with cycle structure $(n-2, 2)$ has order $\mathrm{LCM}\{2, n-2\} = 2(n-2)$, and since $2 + (n-2) = n$, such elements exist. Finally, when n is even, $n-2$ is even and we have $n - 2 = 2^k m$ with m odd. If $m = 1$ then $n = 2^k + 2$, a contradiction, so $m \geq 3$. Now $2(n-2) = 2^{k+1}m$ and $\mathrm{GCD}\{2^{k+1}, m\} = 1$, so if $\theta \in S_n$ has cycle structure $(2^{k+1}, m, 1, \ldots, 1)$ or $(m, 2^{k+1}, 1, \ldots, 1)$, then $o(\theta) = 2^{k+1}m = 2(n-2)$, by Theorem 3.11 again. Is such a cycle structure possible in S_n? We need that $2^{k+1} + m \leq n = 2^k m + 2$. But $2^{k+1} + m \leq 2^k m + 2$ is equivalent to $2^{k+1} - 2 \leq m(2^k - 1)$, which is just $2(2^k - 1) \leq m(2^k - 1)$, and this is true since $3 \leq m$.

The next idea we introduce depends on writing permutations as products of 2-cycles. A 2-cycle (a, b) interchanges or transposes a and b and leaves the other integers fixed.

DEFINITION 3.8 A **transposition** is a 2-cycle in S_n.

We claim that any $\theta \in S_n$ is a product of transpositions, but not uniquely. For example $(1,2,3) = (1,2)(2,3) = (2,3)(1,3) = (1,2)(2,4)(2,3)(3,4)$ and $I_{I(n)} = (1,2)(1,2) = (1,2)(2,3)(3,4)(3,4)(2,3)(1,2)$. Using Theorem 3.10, it suffices to show that every cycle is a product of transpositions.

THEOREM 3.13 For $k \geq 2$, any k-cycle in S_n is a product of $k-1$ transpositions and every $\theta \in S_n$ is a product of transpositions; that is, $S_n = \langle\{\sigma \in S_n \mid \sigma \text{ is a transposition}\}\rangle$.

Proof If $\rho = (a_1, a_2, \ldots, a_k)$ is a k-cycle then a straightforward computation shows that $\rho = (a_1, a_2)(a_2, a_3) \cdots (a_{k-1}, a_k)$. Using Theorem 3.10, any $\theta \in S_n$ can be written as $\theta = \rho_1 \rho_2 \cdots \rho_s$ where each ρ_j is a k_j cycle. Since ρ_j is a product of transpositions when $k_j > 2$, ρ_j is itself a transposition when $k_j = 2$, and $I_{I(n)} = (1,2)(1,2)$, the representation of each ρ_j in this way shows that θ is a product of transpositions. Clearly this means that S_n is generated as a group by its set of transpositions. ■

By the proof of the theorem, $(3, 4, 2, 6, 5) = (3, 4)(4, 2)(2, 6)(6, 5)$, which is also $(3, 5)(3, 6)(3, 2)(3, 4)$. By writing its disjoint cycles as products of 2-cycles, $(2, 4, 6, 3)(8, 10)(1, 7, 5, 9) = (2, 4)(4, 6)(6, 3)(8, 10)(1, 7)(5, 7)(5, 9) = (4, 6)(3, 6)(2, 3)(8, 10) \cdot (1, 9)(1, 7)(5, 7)$. The different representations for a permutation as a product of transpositions in our examples always use an even number of transpositions or always use an odd number; they have the same **parity.** This is true generally but not obvious, and, it lets us distinguish permutations by this property. Of the several standard ways to prove this uniqueness of parity, we choose to use congruences and show that for $\theta \in S_n$, the parity of the number of transpositions for θ is the parity of the sum of n and the number of orbits of θ on $I(n)$.

THEOREM 3.14 Let $\theta \in S_n$ with $\theta = \tau_1\tau_2 \cdots \tau_t = \sigma_1\sigma_2 \cdots \sigma_m$ with all τ_i and σ_j transpositions. Then $t \equiv m \pmod 2$, so t and m are both even or both odd.

Proof Let $[z_1], \ldots, [z_k]$ be the orbits of θ on $I(n)$ with corresponding cycles $\rho_1, \ldots, \rho_k$, so by Theorem 3.10, $\theta = \rho_1 \cdots \rho_k$. Let $\upsilon(\theta) = k$, the number of orbits of θ on $I(n)$. For example, $\upsilon(I_{I(n)}) = n$ since each $j \in I(n)$ is its own orbit, $\upsilon(\rho) = 1$ if ρ is an n-cycle, and $\upsilon(\sigma) = n - 1$ if σ is a transposition (why?). It is essential to compare $\upsilon(\theta)$ and $\upsilon(\theta\sigma)$ for $\sigma = (a, b)$. Now $a, b \in [a]_\theta = [z_j]$ or else $[a]_\theta \neq [b]_\theta$. In the first case, $\rho_j = (a, x_1, \ldots, x_m, b, y_1, \ldots, y_s)$ with $m, s \geq 0$, where $m = 0$ means that $\rho_j = (a, b, y_1, \ldots, y_s)$, $s = 0$ means $\rho_j = (a, x_1, \ldots, x_m, b)$, and $m = s = 0$ means that $\rho_j = (a, b)$. Write $\theta(a, b) = \beta\rho_j(a, b)$ for β the product of all the ρ_i except for ρ_j, using that the $\{\rho_w\}$ commute by Theorem 3.9. Computing yields $\rho_j(a, b) = (a, x_1, \ldots, x_m, b, y_1, \ldots, y_s) \cdot (a, b) = (a, y_1, \ldots, y_s)(b, x_1, \ldots, x_m)$. Thus the orbit $\{a, x_1, \ldots, x_m, b, y_1, \ldots, y_s\}$ of θ breaks up into two orbits of $\theta(a, b)$, and the other orbits of θ remain orbits of $\theta(a, b)$. That is, if $q \in [z_i]$ for $[z_i] \neq [a]_\theta$ then since $a, b \notin [z_i]$, $(a, b)(q) = q$ so $(\theta(a, b))(q) = \theta(q)$. If $[z_u] = [a]_\theta \neq [b]_\theta = [z_v]$, then $\rho_u\rho_v = (a, y_1, \ldots, y_s)(b, x_1, \ldots, x_m)$, as above. As in the previous case, $\theta(a, b) = \eta\rho_u\rho_v(a, b)$ for η the product of all the ρ_i except for ρ_u and ρ_v. Also $(a, y_1, \ldots, y_s)(b, x_1, \ldots, x_m)(a, b) = (a, x_1, \ldots, x_m, b, y_1, \ldots, y_s)$ as above since $(a, b)^2 = I_{I(n)}$, so now two orbits of θ combine to give a single orbit of $\theta(a, b)$. Again the other orbits of θ remain orbits of $\theta(a, b)$. In either case, $\upsilon(\theta\sigma) = \upsilon(\theta) \pm 1$.

Define $T(\theta) = [\upsilon(\theta)]_2 \in \mathbf{Z}_2$. Since $[-1]_2 = [1]_2$, we have $T(\theta(a, b)) = [\upsilon(\theta(a, b))]_2 = [\upsilon(\theta) \pm 1]_2 = [\upsilon(\theta)]_2 + [\pm 1]_2 = T(\theta) + [1]_2$ using our argument above. Since $\theta = \tau_1\tau_2 \cdots \tau_t = \sigma_1\sigma_2 \cdots \sigma_m$ with all τ_i and σ_j transpositions, $I_{I(n)} = \theta(\sigma_1\sigma_2 \cdots \sigma_m)^{-1} = \theta\sigma_m^{-1}\sigma_{m-1}^{-1} \cdots \sigma_1^{-1} = \theta\sigma_m\sigma_{m-1} \cdots \sigma_1$. Therefore $[n]_2 = T(I_{I(n)}) = T(\theta\sigma_m\sigma_{m-1} \cdots \sigma_1) = T(\theta\sigma_m\sigma_{m-1} \cdots \sigma_2) + [1]_2$, and by induction on m, $[n]_2 = T(\theta\sigma_m\sigma_{m-1} \cdots \sigma_1) = T(\theta) + [m - 1]_2 + [1]_2 = T(\theta) + [m]_2$. Similarly, writing $I_{I(n)} = \theta(\tau_1\tau_2 \cdots \tau_t)^{-1}$ leads to $[n]_2 = T(\theta) + [t]_2$. Hence $[m]_2 = T(\theta) + [n]_2 = [t]_2$, which is equivalent to $t \equiv m \pmod 2$ by Theorem 1.21. ■

Theorem 3.14 ensures that no matter how $\theta \in S_n$ is written as a product of transpositions, an even number must always be used or an odd number must always be used. Therefore the next definition is unambiguous.

DEFINITION 3.9 Any $\theta \in S_n$ is **even** if it is a product of an even number of transpositions and is **odd** if it is a product of an odd number of transpositions. The set of all even permutations is denoted by A_n.

In view of Theorem 3.14, $S_n - A_n$ is the set of odd permutations so S_n is partitioned by $S_n - A_n$ and A_n. Do not confuse the terms *odd* and *even* with the order of the permutation! For example $(1, 2)$ is odd with even order and $(1, 2)(3, 4)$ is an even permutation with even order. It is an exercise using Theorem 3.11 and Theorem 3.13 that $\theta \in S_n$ of odd order is in A_n.

We can tell whether $\theta \in S_n$ is even or odd by its cycle structure. From Theorem 3.13, a k-cycle is a product of $k-1$ transpositions so a k-cycle is odd when k is even and is even when k is odd. Since $I_{I(n)} = (1, 2)^2$, it is in A_n. Any nonidentity $\theta \in S_n$ is a product of cycles, say $\theta = \rho_1\rho_2 \cdots \rho_s$ for ρ_i a k_i-cycle (Theorem 3.10) so θ is a product of $(k_1 - 1) + (k_2 - 1) + \cdots + (k_s - 1)$ transpositions. Thus a product of cycles all of odd lengths is even, but a product of all even-length cycles may not be odd, such as $(1, 2, 3, 4)(5, 6) = (1, 2)(2, 3)(3, 4)(5, 6)$. If $\theta \in S_{13}$ has cycle structure $(5, 4, 4)$ then $\theta \in A_{13}$ since it is a product of $4 + 3 + 3 = 10$ transpositions. However, a permutation with cycle structure $(5, 5, 4, 3)$ would be odd since $4 + 4 + 3 + 2 = 13$.

Odd and even permutations arise in writing a general expression for $\det(A)$ when $A \in M_n(R)$ with R a commutative ring. For $\theta \in S_n$, let $(-1)^\theta = 1$ if $\theta \in A_n$, and $(-1)^\theta = -1$ if $\theta \notin A_n$. Then $\det(A) = \Sigma(-1)^\sigma A_{1\sigma(1)} \cdots A_{n\sigma(n)}$ over all $\sigma \in S_n$ follows from linearity properties of determinants. Using this we show that $\det(A) = 0$ when A has two equal rows. If $A_{ik} = A_{jk}$ for all k and some $i \neq j$ then each term $(-1)^\theta A_{1\theta(1)} \cdots A_{i\theta(i)} \cdots A_{j\theta(j)} \cdots A_{n\theta(n)}$ has the companion term $(-1)^{\sigma\theta} A_{1\sigma\theta(1)} \cdots A_{i\sigma\theta(i)} \cdots A_{j\sigma\theta(j)} \cdots A_{n\sigma\theta(n)}$ where $\sigma = (\theta(i), \theta(j))$. Since σ is a transposition $(-1)^{\sigma\theta} = -(-1)^\theta$, and from the definition of σ, $\sigma\theta(i) = \theta(j)$, $\sigma\theta(j) = \theta(i)$, and $\sigma\theta(m) = \theta(m)$ for all $m \neq i, j$. But $A_{i\sigma\theta(i)} = A_{i\theta(j)} = A_{j\theta(j)}$ and $A_{j\sigma\theta(j)} = A_{j\theta(i)} = A_{i\theta(i)}$ so the companion terms are negatives and cancel in the sum representing $\det(A)$. Thus all terms cancel and $\det(A) = 0$. Similarly, $\det(A) = \det(A^T)$, where A^T is the *transpose* of A (recall that $(A^T)_{ij} = A_{ji}$). Write $\det(A^T) = \Sigma(-1)^\sigma \cdot (A^T)_{1\sigma(1)} \cdots (A^T)_{n\sigma(n)} = \Sigma(-1)^\sigma A_{\sigma(1)1} \cdots A_{\sigma(n)n}$. Now $\sigma^{-1}(\sigma(i)) = i$, so each term $(-1)^\sigma A_{\sigma(1)1} \cdots A_{\sigma(n)n} = (-1)^\sigma A_{1\sigma^{-1}(1)} \cdots A_{n\sigma^{-1}(n)}$ since $A_{\sigma(i)i} = A_{\sigma(i)\sigma^{-1}(\sigma(i))}$ and $\{\sigma(i) \mid i \in I(n)\} = I(n)$. Furthermore, it is an exercise to show that $(-1)^\sigma = (-1)^{\sigma^{-1}}$, so indeed $\det(A^T) = \det(A)$.

The set A_n is an important mathematical object. It is easy to see that $A_n \leq S_n$, verified shortly. When $n \geq 5$ the group properties of A_n are essential for proving that for polynomials in $\boldsymbol{Q}[x]$ of degree at least 5 there is no general and fixed expression in its coefficients that always gives a root of the polynomial in $\boldsymbol{C}$. This situation contrasts with that for polynomials of degree 2, 3, or 4, where such general expressions do exist. For quadratic polynomials there is the well-known quadratic formula. The expressions in degree 3 or 4 are quite complicated. The reason for all of this, and how the structure of A_n comes into play is a deep result proved in Chapter 15. We will be content for now to see that $A_n \leq S_n$ and in Chapter 5 will prove the group theoretic property of A_n related to the impossibility of a formula solving equations of degree five or higher.

THEOREM 3.15 For $n > 1$, $A_n \leq S_n$; $|A_n| = n!/2$; and for any $\theta \in S_n$, $A_n = \{\theta^{-1}\eta\theta \mid \eta \in A_n\} = \theta^{-1}A_n\theta$.

Proof Theorem 2.12 shows that $A_n \leq S_n$. $I_{I(n)} = (1, 2)^2 \in A_n$ so $A_n \neq \varnothing$, and S_n is finite so it suffices to see that A_n is closed under the multiplication in S_n. For $\alpha, \beta \in A_n$, write $\alpha = \sigma_1 \cdots \sigma_{2k}$ and $\beta = \tau_1 \cdots \tau_{2m}$ where all σ_i and τ_j are transpositions. Clearly $\alpha\beta = \sigma_1 \cdots \sigma_{2k}\tau_1 \cdots \tau_{2m} \in A_n$ since it is a product of $2(k+m)$ transpositions, so $A_n \leq S_n$ by Theorem 2.12. For $\alpha = \sigma_1 \cdots \sigma_{2k} \in A_n$, $(1,2)\alpha$ is an odd permutation. Thus $f(\alpha) = (1,2)\alpha$ defines $f : A_n \to (S_n - A_n)$. Similarly, if $\beta \in S_n - A_n$ is odd then $g(\beta) = (1,2)\beta \in A_n$, so $g : (S_n - A_n) \to A_n$ and since $(1,2)^2 = I_{I(n)}$ it is immediate that f and g are inverses. Consequently, f is a bijection by Theorem 0.11 and $|A_n| = |S_n - A_n|$. But A_n and $S_n - A_n$ partition S_n, so Theorem 0.3 shows $n! = |S_n| = |A_n| + |S_n - A_n| = 2|A_n|$ and $|A_n| = n!/2$ follows. Finally, if $\alpha = \sigma_1\sigma_2 \cdots \sigma_{2k} \in A_n$ for transpositions σ_i, and $\theta \in S_n$ is the product of transpositions $\rho_1\rho_2 \cdots \rho_m$ as in Theorem 3.13, then $\theta^{-1}\alpha\theta = (\rho_1\rho_2 \cdots \rho_m)^{-1}\sigma_1\sigma_2 \cdots \sigma_{2k}\rho_1\rho_2 \cdots \rho_m$. But $(\rho_1\rho_2 \cdots \rho_m)^{-1} = \rho_m \cdots \rho_2\rho_1$ so $\theta^{-1}\alpha\theta$ is a product of $2(k+m)$ transpositions, forcing $\theta^{-1}A_n\theta \subseteq A_n$. A similar argument shows $\theta A_n \theta^{-1} \subseteq A_n$ so $A_n = \theta^{-1}(\theta A_n \theta^{-1})\theta \subseteq \theta^{-1}A_n\theta \subseteq A_n$, proving $\theta^{-1}A_n\theta = A_n$. ■

Let us look at A_n for small n. Now $S_2 = \{I_{I(2)}, (1,2)\}$ so $A_2 = \{I_{I(2)}\}$. Elements in S_3 have cycle structures (3), (2, 1), or (1, 1, 1) so by Theorem 3.13 only the two 3-cycles and the identity are even. In this case $A_3 = \langle (1,2,3) \rangle$ is a cyclic group, even though S_3 is not Abelian. The cycle structures of the nonidentity elements in A_4 are (3, 1) and (2, 2), since the other possible cycle structures are (4) and (2, 1, 1) giving 4-cycles (products of three transpositions) and the transpositions. Write $A_4 = \{I_{I(4)}, (a,b,c), (a,b)(c,d)\}$ where "(a,b,c)" represents all eight 3-cycles (see Theorem 3.8) and "$(a,b)(c,d)$" represents all three products of two disjoint transpositions. In A_5 the cycle structures are (5), (3, 1, 1),)(2, 2, 1) and (1, 1, 1, 1, 1) and $A_5 = \{I_{I(5)}, (a,b,c,d,e), (a,b,c), (a,b)(c,d)\}$ so each nonidentity element has prime order. When $n > 5$, A_n has elements with cycle structure (4, 2) so composite order 4 by Theorem. 3.11.

From Theorem 3.13, S_n is generated by its transpositions, and by definition, A_n is generated by all products of two transpositions. There are other important sets that generate A_n, in particular the 3-cycles.

THEOREM 3.16 For $n > 2$, $A_n = \langle \{(a,b,c) \in S_n\} \rangle = \langle \{(a_1, a_2, \ldots, a_{2m+1}) \in S_n\} \rangle$ if $m \geq 1$ and fixed. If $n \geq 5$, $A_n = \langle \{(a,b)(c,d) \in S_n \mid (a,b), (c,d) \text{ are disjoint}\} \rangle$.

Proof All of the sets described are in A_n. Since $\theta \in A_n \Rightarrow \theta = \tau_1 \cdots \tau_{2m}$ for the τ_j transpositions, $A_n = \langle\{\text{3-cycles}\}\rangle$ if any product of two transpositions is a product of 3-cycles. Now $I_{I(n)} = (a,b)^2 = (a,b,c)^3 \in \langle\{\text{3-cycles}\}\rangle$, and also $(a,b)(b,c) = (a,b,c) \in \langle\{\text{3-cycles}\}\rangle$. If $n \geq 4$ and a, b, c, and d are distinct, $(a,b)(c,d) = (a,b,d)(b,d,c)$, so $A_n = \langle\{\text{3-cycles}\}\rangle$ when $n > 2$. For an odd $s > 3$, $(a,c,b,d_1,\ldots,d_{s-3})(a,c,d_{s-3},d_{s-2},\ldots,d_2,d_1,b) = (a,b,c)$ so $A_n = \langle\{\text{3-cycles}\}\rangle \subseteq \langle\{s\text{-cycles in } S_n\}\rangle \subseteq A_n$, proving that the set of s-cycles generates A_n when $n \geq 5$. Finally, observe that $(a,b,c) = (a,b)(d,e) \cdot (b,c)(d,e)$ if $n \geq 5$, so once again $A_n = \langle\{\text{3-cycles}\}\rangle \subseteq \langle\{(a,b)(c,d) \in S_n \mid (a,b) \text{ and } (c,d) \text{ are disjoint}\}\rangle \subseteq A_n$, and A_n is generated by all products of two disjoint transpositions. ■

Now $H = \{I, (1,2)(3,4), (1,3)(2,4), (1,4)(2,3)\} \leq A_4$ and is proper (exercise!) so $n \geq 5$ is required for the last part of Theorem 3.16. This $H \leq A_4$ will be useful for examples in later chapters.

EXERCISES

1. For each $\theta \in S_8$ below: a) find its orbits; b) write θ as a product of pairwise disjoint cycles; c) find $o(\theta)$; d) write θ^{-1} as a product of pairwise disjoint cycles; e) determine whether $\theta \in A_n$; f) find the cycle structures of θ^2 and θ^3.

- **i)** $(5,8,4,6)(3,5,7,8)(1,3,5,7)$
- **ii)** $(2,3,4)(2,4,5)(1,2,3)$
- **iii)** $(1,3,8)(1,2,5)(1,7,3)(2,4,3)(2,6,8)$
- **iv)** $(1,6,4,8)(3,5,7)(2,6,8)(3,7)$
- **v)** $(1,2)(1,5,6,8)(5,8)(1,7)(1,2,4,5)(3,6,4)$
- **vi)** $(1,5,4,8)(2,5,7)(2,6,8)(3,7)$
- **vii)** $(1,2,8,6,7,5,3)(1,8,3)(1,5)$
- **viii)** $(1,3,6,5,4)(2,4,6)(1,2,6,5,3)$

2. Find the cycle structures of all elements of:

- **i)** order 2 in S_7
- **ii)** order 3 in S_9
- **iii)** order 4 in S_{10}
- **iv)** order 4 in S_{12}
- **v)** order 6 in S_8
- **vi)** order 8 in S_{15}
- **vii)** order 9 in S_{18}
- **viii)** order 12 in S_{20}

3. Find an element of maximal order in a) S_n and in b) A_n when:

- **i)** $n=7$
- **ii)** $n=8$
- **iii)** $n=9$
- **iv)** $n=10$
- **v)** $n=12$
- **vi)** $n=14$

4. If $n>2$ show that $Z(S_n)=\langle I_{I(n)}\rangle$ (see Definition 2.11).

5. Let $\theta, \rho \in S_n$, $\rho=(x_1,\ldots,x_k)$ a k-cycle. Show that $\theta\rho\theta^{-1}=(\theta(x_1),\ldots,\theta(x_k))$.

6. Show $\theta, \eta \in S_n \Rightarrow \theta$ and $\eta\theta\eta^{-1}$ have the same cycle structure (Exercise 5!).

7. For $\eta, \sigma \in S_n$ show $\eta\sigma$ and $\sigma\eta$ have the same cycle structure (Exercise 6!).

8. Show that θ and θ^{-1} have the same cycle structure for any $\theta \in S_n$.

9. For $\theta \in S_n$ show that θ and θ^2 have the same cycle structure $\Leftrightarrow o(\theta)$ is odd.

10. If $\theta \in S_{2n+1}$ with $o(\theta)=2^k$, show that θ must have a fixed point.

11. For $n \geq 2$ show that:

- **a)** $(1,2)$ is not a square in S_n ($(1,2) \neq \theta^2$ for any $\theta \in S_n$)
- **b)** for $2 \leq k \leq n$, $(1,2,\ldots,k)$ is not a kth power in S_n

12. When $n \geq 2$ show that S_n is generated by each of the following subsets:

- **i)** $\{(1,2),(1,3),\ldots,(1,n)\}$ (Show that this set generates all 2-cycles.)
- **ii)** $\{(1,2),(2,3),\ldots,(n-1,n)\}$ (Use Exercise 5 and part i).)
- **iii)** $H_n \cup \{\rho_n\}$ where $H_n=\{\alpha \in S_n \mid \alpha(n)=n\}$ and $\rho_n=(1,2,3,\ldots,n)$ (Show that $H_n \leq S_n$ and that for $\theta \in S_n$, $\rho_n^k(\theta(n))=n$ for some $k \in \boldsymbol{N}$.)
- **iv)** $\{(1,2),(1,2,3),\ldots,(1,2,3,\ldots,n)\}$ when $n \geq 3$ (Induction and part iii).)
- **v)** $\{(1,2),(1,2,3,\ldots,n)\}$ (Use Exercise 5 and part ii).)

13. Let T be a tetrahedron in $\boldsymbol{R}^3$: T is a solid with four faces, each congruent to a fixed equilateral triangle. Let G be the subgroup of symmetries $E(T)$ of T generated by all rotations of T about its altitudes. The elements of G permute the faces of T and so correspond to elements in S_4. Show that the subgroup of S_4 generated by these is A_4.

14. **i)** Find $H \leq S_6$, $|H| = 6$, H Abelian, no 6-cycle in H.
ii) Find a non-Abelian $K \leq S_6$ with $|K| = 6$.
iii) Find $L \leq S_6$ with $|L| = 9$.
iv) Show that $\{(1,3)^i(1,2,3,4)^j \mid 1 \leq i \leq 2 \text{ and } 1 \leq j \leq 4\} \leq S_6$ of order 8.
v) Find $M \leq S_6$, $|M| = 16$ (use iv)).

15. Let $\{a, b, c, d, e\} = \{1, 2, 3, 4, 5\} = I(5)$ and $H = \langle (a, b)(c, d), (a, b)(d, e) \rangle \leq S_5$.
i) Show $|H| = 6$.
ii) Show $\theta \in H \Rightarrow \theta(j) = j$, some $j \in I(5)$ but $\eta(j) \neq j$ some $\eta \in H$.

16. Find a subgroup in A_6 of order 8 (use Exercise 14).

17. If $\theta \in S_n$ is an m-cycle, find all $k \in \boldsymbol{N}$ so that θ^k are also m-cycles when:
i) $m = 3$
ii) $m = 4$
iii) $m = 8$
iv) $m = 9$
v) $m = 10$
vi) any fixed $m \in \boldsymbol{N}$

18. Let $\theta \in S_n$ satisfy: for any $k \in \boldsymbol{N}$, either $\theta^k = I_{I(n)}$ or θ^k has no fixed point. Show that $o(\theta) \mid n$.

19. Let $\rho \in S_n$ be a k-cycle.
i) If $t \in \boldsymbol{N}$ and $t \mid k$, find the cycle structure of ρ^t.
ii) If $m \in \boldsymbol{N}$ and $(m, k) = 1$, show that ρ^m is a k-cycle.
iii) For any $m \in \boldsymbol{N}$, find the cycle structure of ρ^m.

20. For $(1, 2, \ldots, m)(2, 3, \ldots, m+1) \cdots (n-m+1, n-m+2, \ldots, n) \in S_n$, find a general expression for the following cases, and prove it holds by induction:
i) $n \geq 4$ and $m = 3$
ii) $n \geq 4$ and $m = 4$ (compute these for at least $n = 4, 5, 6, 7, 8$, and 9 to guess what the result should be)
iii) $n \geq 5$ and $m = 5$ (compute for $5 \leq n \leq 10$)
iv) $n \geq 7$ and $m = 6$ (the answer depends on $n = 3k$, $n = 3k+1$, or $n = 3k+2$).

21. **i)** For $n \geq 4$, show that some $\theta \in S_n$ has $o(\theta) = 3(n-3)$ unless $n = 3^k + 3$ or $n = 2 \cdot 3^k + 3$. (Follow the argument of Example 3.5.)
ii) Find a statement like i) describing when some $\theta \in S_n$ has $o(\theta) = 5(n-5)$.
iii) Find a statement like i) describing when some $\theta \in S_n$ has $o(\theta) = p(n-p)$ for p a prime with $p < n$.
iv) If $n \geq 11$ some $\theta \in S_n$ has $o(\theta) = 4(n-4)$ unless $n = 2^k + 4$ or $n = 3 \cdot 2^k + 4$.

22. If $k \geq 2$, ρ is a k-cycle in S_n, $\rho = \tau_1\tau_2 \cdots \tau_m$ for τ_j transpositions, then $m \geq k-1$. (If $\rho = (a_1, \ldots, a_k)$ then a_1 must apear in some τ_j. Show that when m is minimal, using $(a, c)(a, b) = (a, b, c) = (a, b)(b, c)$ reduces to the case when $\tau_1 = (a_1, a_2)$.)

23. For any $\theta, \eta \in S_n$, show that $\theta^{-1}\eta^{-1}\theta\eta \in A_n$.

24. If $\theta \in S_n$ has odd order, show that $\theta \in A_n$.

25. Let $\theta \in S_n - A_n$. Show $\theta A_n = \{\theta\eta \mid \eta \in A_n\} = S_n - A_n = A_n\theta = \{\eta\theta \mid \eta \in A_n\}$.

26. Let $n \geq 4$ and $F_n = \{\theta \in S_n \mid \theta$ has no fixed points$\}$, so F_n consists of all permutations whose orbits all have length greater than one.

i) Show that $S_n = \langle F_n \rangle$: that is, F_n generates S_n.

ii) Show that $\langle F_n \bigcap A_n \rangle = A_n$ for $n > 4$. (Consider separately when n is odd and when n is even, and note that $2m = 3 + (2m - 3)$.)

27. Let α be the product of all $n!$ $\theta \in S_n$ in some order, using each θ once. For which n is $\alpha \in A_n$? The particular order of multiplication is not relevant.

28. Let β be the product of all cycles in S_n in some order, using each cycle once. For which n is $\beta \in A_n$? The order of multiplication is not relevant.

29. Let $S_\infty = \{f \in \text{Bij}(\boldsymbol{N}) \mid f(m) = m$ for $m \geq n(f)$ depending on $f\} \leq \text{Bij}(\boldsymbol{N})$ (see Example 2.12). Set $A_\infty = \{g \in S_\infty \mid g|_{I(k)} \in A_k$ if $g(j) \neq j \Rightarrow j \leq k\}$.

a) Show that $A_\infty \leq S_\infty$.

b) For any $f \in S_\infty$, show that $f^{-1}A_\infty f = \{f^{-1}gf \in S_\infty \mid g \in A_\infty\} = A_\infty$.

c) For $f \in S_\infty$ if $fA_\infty = \{fg \in S_\infty \mid g \in A_\infty\}$, show $\{fA_\infty \mid f \in S_\infty\} = \{A_\infty, S_\infty - A_\infty\}$.

d) If $f = (a, b, c) \in S_\infty$ is definied by $f(a) = b$, $f(b) = c$, $f(c) = a$, and $f(k) = k$ for $k \notin \{a, b, c\}$ then show that A_∞ is generated by $\{(a, b, c) \mid a, b, c \in \boldsymbol{N}\}$.

3.4 THE DIHEDRAL GROUPS D_n

Recall from §2.3 that P_n is the regular n-gon inscribed in the unit circle, with vertices $\{v_k = (\cos(2\pi k/n), \sin(2\pi k/n)) \mid k \in \boldsymbol{Z}\}$, and $D_n = E(P_n)$ is the set of Euclidean transformations of $\boldsymbol{R}^2$ sending P_n back to itself and consisting of n rotations about (0, 0) through the angles $2\pi t/n$, and n reflections. We want a different view of D_n that will make it easy to compute in this group. We saw that since $F \in D_n$ preserves distances, it must permute $\{v_j\}$ so we can identify F with some $\theta \in S_n$. Now $T_{2\pi/n} \in D_n$ sends v_k to v_{k+1} so corresponds to the n-cycle $(1, 2, \ldots, n)$. H fixes $v_n = v_0$, interchanges v_1 and v_{n-1}, v_2 and v_{n-2}, and, in general, v_i and v_{n-i}. It is hard to identify a general reflection of P_n in S_n but $H \in D_5$ corresponds to $(1, 4)(2, 3)$ in S_5, and $H \in D_6$ corresponds to $(1, 5)(2, 4)$ in S_6 since both v_6 and v_3 lie on the x-axis. Because it is not easy to identify reflections in S_n and so to compute with them, we give another definition of D_n based on congruences and the formulas $T_{2\pi i/n}(v_k) = v_{i+k}$ and $H(v_k) = v_{n-k}$. This approach is indepenent of §2.3 and makes computation in D_n very easy.

DEFINITION 3.10 If $n \geq 3$, the **dihedral group** $D_n = \{W_j, T_j : \boldsymbol{Z}_n \to \boldsymbol{Z}_n \mid j \in \boldsymbol{Z}, W_j([k]) = [j - k]$ and $T_j([k]) = [j + k]\}$.

Several questions arise immediately. Are W_j and T_j functions? Is $D_n \leq \text{Bij}(\boldsymbol{Z}_n)$? Is $|D_n| = 2n$? Can D_n be identified naturally with $E(P_n)$?

THEOREM 3.17 $D_n \leq \text{Bij}(\boldsymbol{Z}_n)$ with identity $T_0 = I$, D_n is not Abelian, $|D_n| = 2n$, $W_j^2 = I$, $T_iT_j = T_{i+j}$, $W_iW_j = T_{i-j}$, $T_jW_i = W_{j+i}$, and $W_iT_j = W_{i-j}$.

Proof Note first that $[x]_n = [y]_n \Leftrightarrow x \equiv_n y \Leftrightarrow j + x \equiv_n j + y$ and $j - x \equiv_n j - y$ for any $j \in \boldsymbol{Z}$, using Theorem 1.24. In this case $T_j([x]) = [j + x] = [j + y] = T_j([y])$ and similarly,

$W_j([x]) = [j - x] = [j - y] = W_j([y])$, so indeed T_j and W_j are functions. Now $T_0 = I$ since $T_0([k]) = [0 + k] = [k]$, and also $T_i T_j([k]) = T_i([j + k]) = [i + j + k] = T_{i+j}([k])$, so $T_i T_j = T_{i+j}$. It follows that $T_j^{-1} = T_{-j}$ so $T_j \in \mathrm{Bij}(\mathbf{Z}_n)$ by Theorem 0.11, and $W_j^2 = I$ (verify) forces $W_j \in \mathrm{Bij}(\mathbf{Z}_n)$. Next, $T_i = T_j \Leftrightarrow T_i([k]) = T_j([k])$ for all $[k] \in \mathbf{Z}_n \Leftrightarrow$ $[i + k] = [j + k]$ for all $k \in \mathbf{Z} \Leftrightarrow i \equiv_n j$ (why?) and similarly, $W_i = W_j \Leftrightarrow i \equiv_n j$. It is an exercise to see that $W_i \neq T_j$ for any i and j, so $|D_n| = 2n$. Using Theorem 2.12, $D_n \leq \mathrm{Bij}(\mathbf{Z}_n)$ if it is closed under multiplication. We have observed $T_i T_j = T_{i+j}$ and $W_j^2 = I = T_0$. In addition, $T_j W_i([k]) = T_j([i - k]) = [j + i - k] = W_{j+i}([k])$, so $T_j W_i = W_{j+i}$, $W_i T_j([k]) = W_i([j + k]) = [i - j - k] = W_{i-j}([k])$, so $W_i T_j = W_{i-j}$, and $W_i W_j([k]) = W_i([j - k]) = [i - j + k] = T_{i-j}([k])$, so $W_i W_j = T_{i-j}$. Therefore D_n is closed under multiplication and $D_n \leq \mathrm{Bij}(\mathbf{Z}_n)$. Since $W_1 W_2 = T_{-1}$, $W_2 W_1 = T_1$, and $T_{-1} = T_{n-1} \neq T_1$ because $n \geq 3$, D_n cannot be Abelian. ■

We have a new definition of D_n but does it correspond to $E(P_n)$? By identifying $[k]_n$ with $v_k \in P_n$ it is clear that T_1 can be identified with $T_{2\pi/n} \in E(P_n)$ and similarly that T_j corresponds to $T_{2\pi j/n}$. Does W_j corresond to a reflection of P_n? When n is odd, each reflection of P_n is about a diameter through a vertex, which is fixed. When n is even, P_n has reflections about the diameters fixing opposite vertices and reflections about the bisectors of sides, interchanging two pairs of adjacent vertices. Note that $W_j([k]) = [k] \Leftrightarrow j - k \equiv_n k \Leftrightarrow j \equiv_n 2k$. Using §1.7, when n is odd this has the unique solution $k \equiv_n (n + 1)j/2$, and when n and j are even the solutions are $k \equiv_n j/2$ and $k \equiv_n (n + j)/2$. Thus $W_3 \in D_5$ fixes $[3 \cdot 3]_5 = [4]_5$ so corresponds to the reflection of P_5 about the diameter through v_4, $W_3 \in D_7$ fixes $[4 \cdot 3]_7 = [5]_7$ so corresponds to the reflection of P_7 about the diameter through v_5, and $W_4 \in D_6$ fixes $[2]_6$ and $[5]_6$ so corresponds to the reflection of P_6 about the diameter through v_2 and v_5. When n is even and j odd, $W_j \in D_n$ has no fixed points but $W_j([k]) = [k + 1] \Leftrightarrow j - k \equiv_n k + 1 \Leftrightarrow j - 1 \equiv_n 2k$. Here by §1.7, the solutions are $k \equiv_n (j - 1)/2$ and $k \equiv_n (n + j - 1)/2$. Thus $W_5 \in D_8$ interchanges $[2]_8$ and $[3]_8$, as well as $[6]_8$ and $[7]_8$, so corresponds to the reflection of P_8 about the bisector of the side joining v_2 and v_3, and $W_7 \in D_8$ interchanges $[3]_8$ and $[4]_8$, and also $[7]_8$ and $[8]_8$, so corresponds to the reflection of P_8 about the bisector of the side joining v_3 and v_4.

We use D_n primarily to illustrate concepts in non-Abelian groups so its identification with $E(P_n)$ will not matter much. It is important to remember that W_j and T_j are somewhat ambiguous since they depend on n for their evaluation: we must know n before computing with them. For example, W_1 and W_7 and different in D_n if $n > 6$ but are the same when $n = 3$ or 6. Also $W_1 T_2 = W_{n-1}$ for any n, but in D_5, $W_1 T_2 = W_4$ and in D_8, $W_1 T_2 = W_7$.

We end this section with a result demonstrating the utility of Definition 3.10 for D_n by characterizing $Z(D_n)$ (recall Definition 2.11).

THEOREM 3.18 $Z(D_n) = \langle T_0 \rangle$ if n is odd and $Z(D_n) = \langle T_{n/2} \rangle$ if n is even.

Proof From Definition 2.11, $Z(D_n) = \{F \in D_n \mid FL = LF$ for all $L \in D_n\}$. Using Theorem 3.17, $W_i T_1 = W_{i-1}$ and $T_1 W_i = W_{1+i}$ so if W_i and T_1 commute then $W_{i-1} = W_{i+1}$, implying that $-1 \equiv_n 1$ (why?) and contradicting $n \geq 3$. Thus $W_i \notin Z(D_n)$ and $Z(D_n) \leq \langle T_1 \rangle$. Since $\langle T_1 \rangle$ is an Abelian group, $T_j \in Z(D_n) \Leftrightarrow T_j W_i = W_i T_j$ for all W_i. This condition is equivalent to $W_{i+j} = W_{i-j}$ for all $i \in \mathbf{Z}$ by Theorem 3.17, which in turn is equivalent to $i + j \equiv_n i - j$ for all $i \in \mathbf{Z}$. Hence $T_j \in Z(D_n) \Leftrightarrow 2j \equiv_n 0$. When n is odd, $2j \equiv_n 0 \Leftrightarrow j \equiv_n 0$

(Theorem 1.24) and $Z(D_n) = \langle T_0 \rangle = \langle I \rangle$ follows. When n is even, by the results in §1.7, $2j \equiv_n 0$ has the solutions $j \equiv_n 0$ and $j \equiv_n n/2$, resulting in $T_j \in Z(D_n) \Leftrightarrow j = 0$ or $j = n/2$. This means that $Z(D_n) = \{T_0, T_{n/2}\} = \langle T_{n/2} \rangle$, completing the proof. ■

EXERCISES

1. Compute the indicated products in D_n explicitly to show the following:
 i) $W_i W_j \in \langle T_1 \rangle$
 ii) $W_i T_j, T_j W_i \in \{W_k\}$
 iii) $W_i T_j W_i = T_j^{-1}$
 iv) $T_j^{-1} W_i T_j \in \{W_k\}$

2. If $m > 2$ show that D_n contains an element of order $m \Leftrightarrow m \mid n$.

3. Find all elements:
 i) of order 4 in D_8
 ii) of order 3 in D_{12}
 iii) of order 10 in D_{10}
 iv) of order 5 in D_{15}
 v) of order 6 in D_{18}
 vi) of order 6 in D_{30}

4. a) Show $D_n = \langle \{W_1, W_2\} \rangle$, e.g., $W_3 = W_2 W_1 W_2$ and $T_2 = W_2 W_1 W_2 W_1$.
 b) For $i, j \in \boldsymbol{N}$ show that $D_n = \langle \{W_i, W_j\} \rangle \Leftrightarrow (i - j, n) = 1$.

5. a) Let $E_n = \{W_{2j}, T_{2j} \in D_n \mid j \in \boldsymbol{Z}\}$. Show that $E_n \le D_n$ and find $|E_n|$.
 b) If $m, n \in \boldsymbol{N}$ with $m \mid n$ show $K_m = \{T_{mj}, W_{mj} \in D_n \mid j \in \boldsymbol{Z}\} \le D_n$. Find $|K_m|$.

6. In D_{2m}, for each $0 \le i < 2m$, show that $H_i = \{T_0, T_m, W_i, W_{m+i}\} \le D_{2m}$, that H_i is Abelian, but that H_i is not a cyclic group.

7. Given $i \in \boldsymbol{Z}$, show that $\{W_j \mid j \in \boldsymbol{Z}\} = \{W_j T_i \mid j \in \boldsymbol{Z}\} = \{T_i W_j \mid j \in \boldsymbol{Z}\}$.

3.5 DIRECT SUMS

In this section we study a simple construction of a new group from a finite set of given groups. A model is $(\boldsymbol{R}^n, +)$ in Example 2.2, and like it, the underlying set for the new group will be the Cartesian product of the given groups. Recall from Definition 0.9 that the Cartesian product of the nonempty sets $A_1, \ldots, A_n$ is $\prod_{i=1}^{n} A_i = A_1 \times \cdots \times A_n = \{(a_1, \ldots, a_n) \mid a_j \in A_j\}$. When each A_j is a group, the simplest way to define a multiplication on the Cartesian product is to multiply elements in corresponding coordinates.

THEOREM 3.19 Given groups $G_1, \ldots, G_k$, and $G = \prod G_i = G_1 \times \cdots \times G_k$, $(G, *)$ is a group, where $(g_1, \ldots, g_k) * (h_1, \ldots, h_k) = (g_1 h_1, \ldots, g_k h_k)$.

Proof The multiplication in each coordinate is the multiplication in the appropriate group. That $*$ is associative is an exercise using the associativity of multiplication in each G_j. For $k = 2$, we have $((g_1, g_2) * (g_1', g_2')) * (g_1'', g_2'') = (g_1 g_1', g_2 g_2') * (g_1'', g_2'') = ((g_1 g_1')g_1'', (g_2 g_2')g_2'')$. The associativity in G_1 and G_2 shows $((g_1 g_1')g_1'', (g_2 g_2')g_2'') = (g_1(g_1' g_1''), g_2(g_2' g_2'')) = (g_1, g_2) * (g_1' g_1'', g_2' g_2'') = (g_1, g_2) * ((g_1', g_2')) * (g_1'', g_2''))$. If e_j is the identity of G_j, then it follows easily that $e_G = (e_1, \ldots, e_k)$ and $(g_1, \ldots, g_k)^{-1} = (g_1^{-1}, \ldots, g_k^{-1})$, so indeed $(G, *)$ is a group by Definition 2.1. ■

DEFINITION 3.11 The **direct sum** of the groups $G_1, \ldots, G_k$ is the group $\oplus G_i = \oplus_{i=1}^k G_i = (\prod_{i=1}^k G_i, *)$ where $(g_1, \ldots, g_k) * (h_1, \ldots, h_k) = (g_1h_1, \ldots, g_kh_k)$.

We know that $\oplus G_i$ is a group by Theorem 3.19. At times we write $G_1 \oplus \cdots \oplus G_k$ for $\oplus G_i$. The direct sum of finitely many groups is also the **direct product,** denoted by $\prod G_i$ or by $G_1 \times \cdots \times G_k$. We will reserve this term for an infinite collection of groups and will discuss both the direct product and the direct sum for infinitely many groups at the end of this section.

We illustrate direct sums with a few examples. From Example 2.2, $(\boldsymbol{R}^n, +)$ is just $\oplus (\boldsymbol{R}, +)$ using n copies of $(\boldsymbol{R}, +)$. In general there will be a different group in each coordinate of a direct sum. If we take $(2, 3), (3, 4) \in (G, *) = (\boldsymbol{R}^*, \cdot) \oplus (\boldsymbol{R}, +)$, then $e_G = (1, 0)$, $(2, 3) * (3, 4) = (6, 7)$, $(3, 4)^2 = (9, 8)$, and $(2, 3)^{-1} = (1/2, -3)$. In $\boldsymbol{Z}_3 \oplus S_5 \oplus D_4$, $([2], (1, 3, 4), W_2) * ([2], (1, 5)(2, 4), W_3) = ([2] + [2], (1, 3, 4)(1, 5)(2, 4), W_2W_3) = ([1], (1, 5, 3, 4, 2), T_3)$, and the product in the other order is $([1], (1, 3, 2, 4, 5), T_1)$. Sometimes a direct sum can be recognized as a simpler group.

Example 3.6 The group $H = \boldsymbol{Z}_5 \oplus U_5$ is cyclic of order 20.

Now $|H| = |\boldsymbol{Z}_5||U_5| = 5 \cdot 4 = 20$ (Theorem 0.3). As $[2] \in (\boldsymbol{Z}_5, +) \cap U_5$ and both $\boldsymbol{Z}_5 = \langle [2]_5 \rangle$ and $U_5 = \langle [2] \rangle$, it is natural to see if $H = \langle ([2], [2]) \rangle$. Computing powers in H, $([2], [2])^{10} = ([20], [2^{10}]) = ([0], [1024]) = ([0], [4])$ and so $([2], [2])^{20} = ([0], [4])^2 = ([0], [16]) = ([0], [1]) = e_H$. Therefore from Theorem 2.7, $o(([2], [2])) \mid 20$ but does not divide 10. Since $([2], [2])^4 = ([3], [1]) \neq e_H$, we conclude that $o(([2], [2])) = 20$ so $|\langle ([2], [2]) \rangle| = 20$ from Theorem 2.16, forcing $H = \langle ([2], [2]) \rangle$.

The utility of direct sums is not so much in constructing new groups but rather in recognizing a group as "essentially" the direct sum of a collection of easily understood groups. In this case, knowledge of the smaller constituent groups and the simple way they are combined yield answers to many questions about the given group. This approach of putting together small and understandable objects to study a larger one is a recurring theme in mathematics. The details of it for groups require some further development in the next few chapters. We look at some examples of how groups, not direct sums, can be identified with direct sums of other groups.

Consider $Diag(n, \boldsymbol{R})$, the invertible diagonal matrices with entries in $\boldsymbol{R}$ (Definition 2.9). Identify $A \in Diag(n, \boldsymbol{R})$ with $(A_{11}, \ldots, A_{nn}) \in ((\boldsymbol{R}^*)^n, \cdot) = \oplus (\boldsymbol{R}^*, \cdot)$, and for $A, B \in Diag(n, \boldsymbol{R})$, AB is identified with $(A_{11}B_{11}, \ldots, A_{nn}B_{nn})$, the product in $\oplus (\boldsymbol{R}^*, \cdot)$ of the tuples identified with A and B. Thus $Diag(n, \boldsymbol{R})$ is "essentially" $\oplus (\boldsymbol{R}^*, \cdot)$. Let us consider some less obvious examples.

Example 3.7 $(\boldsymbol{R}^*, \cdot)$ can be identified with $\langle -1 \rangle \oplus (\boldsymbol{R}^+, \cdot) \leq (\boldsymbol{R}^*, \cdot) \oplus (\boldsymbol{R}^+, \cdot)$.

Recall $\boldsymbol{R}^+ = \{r \in \boldsymbol{R} \mid r > 0\} \leq (\boldsymbol{R}^*, \cdot)$. For $r \in \boldsymbol{R}^*$, $f(r) = (r/|r|, |r|)$ defines $f \in \text{Bij}(\boldsymbol{R}^*, \langle -1 \rangle \times (\boldsymbol{R}^+, \cdot))$; f is a bijection by Theorem 0.11 since $g((\varepsilon, a)) = \varepsilon a$ is an inverse for it. In $\langle -1 \rangle \oplus \boldsymbol{R}^+$, $(r/|r|, |r|) * (s/|s|, |s|) = (sr/|rs|, |rs|)$ so the multiplication in each group is consistent with the identification of their elements:

$f(r)f(s) = f(rs)$. Thus we may consider $(\mathbf{R}^*, \cdot)$ and $\langle -1\rangle \oplus \mathbf{R}^+$ to be "essentially" the same. They are not actually the same since the elements of $(\mathbf{R}^*, \cdot)$ are not ordered pairs!

Example 3.8 The group U_{15} can be identified with $\mathbf{Z}_4 \oplus \mathbf{Z}_2$.

We need to associate each element in $U_{15} = \{[1], [2], [4], [7], [8], [11], [13], [14]\}$ with some $([a]_4, [b]_2) \in \mathbf{Z}_4 \oplus \mathbf{Z}_2$ so that corresponding elements have corresponding product in each group. It is not obvious how to do this at present. In U_{15}, $\langle[2]\rangle = \{[1], [2], [4], [8]\}$, $o([2]) = 4$, and $U_{15} = \{[2^j], [-2^j] \mid 0 \le j \le 3\}$. But $[-1]_{15} = [14]_{15}$ so $U_{15} = \{[2^j 14^i] \mid 0 \le j \le 3 \text{ and } 0 \le i \le 1\}$. Define $\varphi : U_{15} \to \mathbf{Z}_4 \oplus \mathbf{Z}_2$ by $\varphi([2^j 14^i]) = ([j]_4, [i]_2)$, and note that φ is a bijection. Furthermore $[2^j 14^i][2^s 14^t] = [2^u 14^v]$ where $j + s \equiv_4 u$ and $i + t \equiv_2 v$, using $o([2]) = 4$ and $o([14]) = 2$. Hence $([u]_4, [v]_2) = ([j+s]_4, [i+t]_2) = ([j]_4, [i]_2) + ([s]_4, [t]_2)$ in $\mathbf{Z}_4 \oplus \mathbf{Z}_2$, so $\varphi([2^j 14^i][2^s 14^t]) = \varphi([2^j 14^i]) + \varphi([2^s 14^t])$ and corresponding elements have corresponding multiplication.

Example 3.8 is a special case of a theorem in Chapter 7 showing that any finite Abelian group can be identified with a direct sum of cyclic groups. Consequences of this result will be: for any positive $k \mid |G|$, G a finite Abelian group, there is $H \le G$ with $|H| = k$; and a description of the maximal order of elements in G. Thus representing a group as a direct sum can be very useful. We present one last example that uses non-Abelian groups.

Example 3.9 $GL(n, \mathbf{C})$ can be identified with the direct sum of the invertible scalar matrices $\mathbf{C}^* \cdot I_n$ and $SL(n, \mathbf{C})$, the matrices of determinant one.

We need to identify $A \in GL(n, \mathbf{C})$ with something in $\mathbf{C}^* \cdot I_n \times SL(n, \mathbf{C})$ and will assume some standard facts about determinants. Note that the center $Z(GL(n, \mathbf{C})) = \mathbf{C}^* \cdot I_n$ and any $A \in GL(n, \mathbf{C})$ may be written as $A = (\det(A)^{1/n}) \cdot I_n \cdot B$ where $B \in SL(n, \mathbf{C})$ is defined by $\det(A)^{1/n} B_{ij} = A_{ij}$. If $A, A' \in GL(n, \mathbf{C})$ and $A' = (\det(A')^{1/n}) I_n \cdot B'$ then using $\mathbf{C}^* \cdot I_n = Z(GL(n, \mathbf{C}))$ we compute

$$\begin{aligned} AA' &= \left(\det(A)^{1/n}\right) I_n \cdot B \left(\det(A')^{1/n}\right) I_n \cdot B' \\ &= \left(\det(A)^{1/n}\right)\left(\det(A')^{1/n}\right) I_n \cdot BB' \\ &= \left(\det(AA')^{1/n}\right) I_n \cdot BB'. \end{aligned}$$

Thus if we identify A with $((\det(A)^{1/n}) I_n, B) \in \mathbf{C}^* \cdot I_n \times SL(n, \mathbf{C})$, then AA' is identified with $((\det(AA')^{1/n}) \cdot I_n, BB') = ((\det(A)^{1/n}) \cdot I_n, B) * ((\det(A)^{1/n}) \cdot I_n, B')$, so the multiplication of identified elements in the two groups corresponds. In this way we consider $GL(n, \mathbf{C})$ to be essentially the same as $\mathbf{C}^* \cdot I_n \oplus SL(n, \mathbf{C})$.

THEOREM 3.20 For $G_1, \ldots, G_k$ groups, e_j the identity of G_j, and $G = \oplus G_j$:

i) $Z(\oplus G_j) = \oplus Z(G_j)$.

ii) If each G_j is finite, then $|G| = |G_1| \cdot |G_2| \cdots |G_k|$.

iii) If $g=(g_1,\ldots,g_k)\in G$, then $o(g)=n\in \mathbf{N} \Leftrightarrow o(g_i)\in \mathbf{N}$ for each $1\le i\le k$, and then $n=\mathrm{LCM}\{o(g_i)\}$.

Proof Part i) follows from the definitions: for $(x_1,\ldots,x_k)\in \bigoplus G_j$ and all $(g_1,\ldots,g_k)\in \bigoplus G_j$, $(x_1,\ldots,x_k)(g_1,\ldots,g_k)=(g_1,\ldots,g_k)(x_1,\ldots,x_k) \Leftrightarrow x_jg_j=g_jx_j$ for all $j \Leftrightarrow$ each $x_j\in Z(G_j)$ (why?). Thus $Z(\bigoplus G_j)=\bigoplus Z(G_j)$. The definition of $G=G_1\times\cdots\times G_k$ and Theorem 0.3 imply ii). As for iii), the multiplication in G shows that for $m\in \mathbf{N}$, $g^m=(g_1^m,\ldots,g_k^m)$, so $g^m=e_G=(e_1,\ldots,e_k)\Leftrightarrow$ each $g_j^m=e_j$. If $o(g)=n\in \mathbf{N}$ then $o(g_i)\mid n$ by Theorem 2.7, so $\mathrm{LCM}\{o(g_j)\}\mid n$ by Theorem 1.7. If each $o(g_i)$ is finite then $g_i^M=e_i$ for $M=\mathrm{LCM}\{o(g_j)\}$, again by Theorem 2.7, so $g^M=e_G$ and $o(g)\mid M$ using Theorem 2.7 yet again. Consequently, $o(g)\in \mathbf{N}\Leftrightarrow$ each $o(g_i)\in \mathbf{N}$ and in this case $o(g)\mid \mathrm{LCM}\{o(g_j)\}$ and $\mathrm{LCM}\{o(g_j)\}\mid o(g)$, so these are equal, proving iii). ■

We use Theorem 3.20 to characterize when the direct sum of finite cyclic groups is itself cyclic.

THEOREM 3.21 Let $k\ge 2$ and $G_1,\ldots,G_k$ be finite groups with identities $e_i\in G_i$. Then $\bigoplus G_j$ is cyclic $\Leftrightarrow$ each G_j is cyclic and $\{|G_j|\}$ is pairwise relatively prime. In particular, $\bigoplus \mathbf{Z}_{n_i}$ is cyclic $\Leftrightarrow \{n_i\}$ is pairwise relatively prime.

Proof By Theorem 3.20, if $G=\bigoplus G_j$ then $|G|=|G_1|\cdots|G_k|$. If each $G_i=\langle g_i\rangle$ then $o(g_i)=|G_i|$ by Theorem 2.16, and if $g=(g_1,\ldots,g_k)$ in G, $o(g)=\mathrm{LCM}\{o(g_i)\}$ by Theorem 3.20. When $\{|G_j|\}$ is pairwise relatively prime then Theorem 1.19 shows $o(g)=|G_1|\cdots|G_k|=|G|$. But $|\langle g\rangle|=o(g)$ (Theorem 2.16) so $G=\langle g\rangle$. Now assume $G=\langle g\rangle$ with $g=(x_1,\ldots,x_k)$. Clearly $\langle x_i\rangle\le G_i$. By definition of cyclic group, $G=\{g^m=(x_1^m,\ldots,x_k^m)\mid m\in \mathbf{Z}\}$ so for $y\in G_1$, $(y,e_2,\ldots,e_k)\in G=\langle g\rangle$ implies $y=x_1^m\in\langle x_1\rangle$ for $m\in \mathbf{Z}$. Therefore $G_1=\langle x_1\rangle$ and similarly, $G_j=\langle x_j\rangle$ is cyclic. Finally, suppose $p\mid |G_1|$ and $p\mid |G_2|$ for some prime p. Since the G_j are cyclic, by Theorem 3.1 there are $\langle z_1\rangle\le G_1$, $\langle z_2\rangle\le G_2$, each of order p and $o(z_1)=o(z_2)=p$ (Theorem 2.16 again). It follows that $\alpha=(z_1,e_2,\ldots,e_k)$, $\beta=(e_1,z_2,e_3,\ldots,e_k)\in G$ have order p. But then $|\langle\alpha\rangle|=p=|\langle\beta\rangle|$, and clearly $\langle\alpha\rangle\cap\langle\beta\rangle=e_G$. Thus G contains different subgroups of order p, contradicting Theorem 3.1; when G is cyclic, $\mathrm{GCD}\{|G_1|,|G_2|\}=1$. We used G_1 and G_2 for convenience. Our argument shows $\mathrm{GCD}\{|G_i|,|G_j|\}=1$ for $i\ne j$ so $\{|G_j|\}$ is pairwise relatively prime. The second result of the theorem follows from the first since each $\mathbf{Z}_m$ is cyclic. ■

The theorem shows why in Example 3.6, $\mathbf{Z}_5\oplus U_5$ is cyclic: $\mathbf{Z}_5=\langle[1]\rangle$, $U_5=\langle[2]\rangle$, and $\mathrm{GCD}\{|\mathbf{Z}_5|,|U_5|\}=(5,4)=1$. Similarly, $\mathbf{Z}_{14}\oplus\mathbf{Z}_{15}$ is cyclic of order $14\cdot 15=210$ with generator $([1]_{14},[1]_{15})$, and $\varphi(210)=48$ generators in all by Theorem 2.8. Also $\mathbf{Z}_5\oplus\mathbf{Z}_9\oplus\mathbf{Z}_{77}$ is cyclic of order 3465 with generator $([1]_5,[1]_9,[1]_{77})$ (why?). However, $\mathbf{Z}_6\oplus\mathbf{Z}_{10}$ is not cyclic since $\langle([3]_6,[0]_{10})\rangle\ne\langle([0]_6,[5]_{10})\rangle$ each have order 2. The identification of $\mathbf{Z}_4\oplus\mathbf{Z}_2$ with U_{15} in Example 3.8 shows that U_{15} cannot be cyclic.

It is easy to see that the direct sum of groups is Abelian if and only if each constituent group is Abelian. There are useful constructions similar to the direct sum that start with Abelian groups but result in non-Abelian groups. We briefly outline such a construction called the **semi-direct product,** leaving most of the details to the exercises. This construction turns out to be very useful in describing certain abstract groups.

Let H and K be groups and set $G = H \times K$. We want to change the multiplication in the direct sum to get a non-Abelian group with identity (e_H, e_K). Set $(h, k) \cdot (h', k') = (hh^{\#}, kk')$, where we consider $h^{\#} \in H$ to be $h^{\#} = f_k(h')$ for $f_k : H \to H$ so we write $(h, k) \cdot (h', k') = (hf_k(h'), kk')$. What conditions on the f_k are required by the group axioms to make $(G, \cdot)$ into a group? Since $e_G = (e_H, e_K)$, we have $(h, k) = (h, k) \cdot (e_H, e_K) = (hf_k(e_H), ke_K)$, so $h = hf_k(e_H)$, forcing $f_k(e_H) = e_H$ for all $k \in K$. Similarly, using $(e_H, e_K) \cdot (h, k) = (h, k)$ results in $f_{e_K}(h) = h$ for all $h \in H$, so $f_{e_K} = I_H$. Since $\cdot$ must be associative, by computing $((h, k) \cdot (h', k')) \cdot (h'', k'') = (h, k) \cdot ((h', k') \cdot (h'', k''))$ we get additional information. The first coordinates of the two products yield $hf_k(h')f_{kk'}(h'') = hf_k(h'f_{k'}(h''))$ so $f_k(h')f_{kk'}(h'') = f_k(h'f_{k'}(h''))$ for all $k, k' \in K$ and $h', h'' \in H$. When $k' = e_K$ this gives $f_k(h')f_k(h'') = f_k(h'h'')$ since $f_{e_K} = I_H$; when $h' = e_H$ using $f_k(e_H) = e_H$ results in $f_{kk'}(h'') = f_k(f_{k'}(h''))$. Thus $f_k \circ f_{k'} = f_{kk'}$ and when $k' = k^{-1}$, $I_H = f_{e_K} = f_k \circ f_{k^{-1}}$. Similarly, $f_{k^{-1}} \circ f_k = I_H$, so $f_k^{-1} = f_{k^{-1}}$ and all $f_k \in \text{Bij}(H)$ by Theorem 0.11. Finally, since each element of G needs an inverse, say $(e_H, e_K) = (h, k) \cdot (h^\wedge, k^\wedge) = (hf_k(h^\wedge), kk^\wedge)$, we have $k^\wedge = k^{-1}$ and $f_k(h^\wedge) = h^{-1}$, so $h^\wedge = f_k^{-1}(h^{-1})$, and an easy computation shows that $(f_k^{-1}(h^{-1}), k^{-1}) = (h, k)^{-1}$.

The properties that $\{f_k : H \to H \mid k \in K\}$ *must* satisfy when $(H \times K, \cdot)$ is a group are also sufficient to guarantee that it *is* a group. We state this as a theorem but leave the verification as an exercise.

THEOREM 3.22 Let H and K be groups, $G = H \times K$, and $F = \{f_k : H \to H \mid k \in K\}$. Then $(G, \cdot)$ with $(h, k) \cdot (h', k') = (hf_k(h'), kk')$ is a group with identity $(e_H, e_K) \Leftrightarrow$ each $f_k \in \text{Bij}(H)$ and: i) $f_{e_K} = I_H$; ii) $f_k(hh') = f_k(h)f_k(h')$ for all $k \in K$, $h, h' \in H$; and iii) $f_k \circ f_{k'} = f_{kk'}$ for all $k, k' \in K$.

It is not at all clear how to obtain a collection of functions F as in the theorem. In fact such sets F arise naturally, but to see how, we must wait for a discussion of normal subgroups, homomorphisms, and automorphisms. One choice of functions satisfying the conditions in Theorem 3.22 is to take all $f_k = I_H$. This is not particularly interesting since then $(H \times K, \cdot) = H \oplus K$. To get a non-Abelian group when H and K are Abelian, *some* $f_k \neq I_H$ suffices, for if $f_k(h) \neq h$, then $(h, e_K) \cdot (e_H, k) = (h, k)$, but $(e_H, k) \cdot (h, e_K) = (f_k(h), k)$. Let us present a specific example of this construction.

Example 3.10 $G = (\mathbf{Z}_n \times \mathbf{Z}_2, \cdot)$ with $n > 2$, $f_{[0]} = I_{\mathbf{Z}_n}$ and $f_{[1]}([a]_n) = [-a]_n = [n - a]_n$ is a non-Abelian group that is essentially D_n.

Since $f_{[1]}^2 = I_{\mathbf{Z}_n}$, $f_{[1]} \in \text{Bij}(\mathbf{Z}_n)$ follows from Theorem 0.11, and as $F = \{f_{[0]}, f_{[1]}\}$, property iii) in Theorem 3.22 holds. To see that property ii) is satisfied and so that G is a group, compute $f_{[1]}([a]_n + [b]_n) = f_{[1]}([a + b]_n) = [-a - b]_n = [-a]_n + [-b]_n = f_{[1]}([a]_n) + f_{[1]}([b]_n)$. This same property surely holds for $f_{[0]} = I_{\mathbf{Z}_n}$. Now G is non-Abelian since $f_{[1]}([1]_n) = [-1]_n \neq [1]_n$ using $n > 2$. Clearly $|G| = 2n$. Also $([1]_n, [0]_2)^m = ([m]_n, [0]_2)$ so $o(([1]_n, [0]_2)) = n$ (verify!); $([0]_n, [1]_2)^2 = e_G = ([0]_n, [0]_2)$; and $([0]_n, [1]_2)([1]_n, [0]_2)([0]_n, [1]_2) = ([n - 1]_n, [0]_2) = ([1]_n, [0]_2)^{-1}$. By identifying T_1^k in D_n with $([1]_n, [0]_2)^k$, any W_i with $([0]_n, [1]_2)$, and then W_j with $([j - i]_n, [1]_2)$, we get a bijection between D_n and G for which products of corresponding elements correspond. For example, $W_j W_i = T_{j-i}$ and $([j-i]_n, [1]_2)([0]_n, [1]_2) = ([j-i]_n, [0]_2) = ([1]_n, [0]_2)^{j-i}$.

Another example of this type is to let $H = \mathbf{Z}_{13}$, $K = \mathbf{Z}_3$, $f_{[0]} = I_{\mathbf{Z}_{13}}$, $f_{[1]}([a]_{13}) = [3a]_{13}$, and $f_{[2]}([a]_{13}) = [9a]_{13}$. It is an exercise to verify that the properties of the $f_{[c]}$ required by Theorem 3.22 are satisfied, resulting in a non-Abelian group of order 39. Also, in the resulting group every nonidentity element has order 3 or order 13.

We end with a brief description of direct products and direct sums for an infinite collection of groups. Actually the collection of groups may be finite, but then either object is the direct sum in Definition 3.11. These notions require the definition of Cartesian product for an infinite collection of sets that appeared in Exercise 11 in §0.7.

DEFINITION 3.12 If $\{S_\lambda\}_\Lambda$ is a nonempty collection of nonempty sets, then $\prod_\lambda S_\lambda = \{f : \Lambda \to \bigcup_\lambda S_\lambda \mid f(\lambda) \in S_\lambda\}$ is the **Cartesian product** of the $\{S_\lambda\}_\Lambda$.

We will generally write $\prod S_\lambda$ for $\prod_\lambda S_\lambda$. There is a logical problem in Definition 3.12 as to whether $\prod S_\lambda \neq \varnothing$ when Λ is infinite. That $\prod S_\lambda \neq \varnothing$, in all cases, is actually equivalent to the Axiom of Choice (§0.5). The Axiom of Choice guarantees $\prod S_\lambda \neq \varnothing$, since for any choice function F on $\bigcup S_\lambda$ we obtain $f \in \prod S_\lambda$ via $f(\lambda) = F(S_\lambda)$. When S is any nonempty set and $\{S_\lambda\}_\Lambda$ are all its nonempty subsets, then any $h \in \prod S_\lambda$ defines a choice function H on S by $H(S_\lambda) = h(\lambda)$. When all the S_λ are groups and $e_\lambda \in S_\lambda$ is the identity, this formalism is not necessary since we can use the distinguished identity element in each S_λ to define $f(\lambda) = e_\lambda$.

We think of each element in the direct product as a "tuple" with the λth coordinate in S_λ, in analogy with the finite case. When $\Lambda = I(n)$, any $(s_1, s_2, \ldots, s_n) \in \prod S_\lambda$ corresponds to the function sending $j \in I(n)$ to s_j. Clearly any function $f : I(n) \to \bigcup S_\lambda$ satisfying $f(i) \in S_i$ can be represented by the element $(f(1), \ldots, f(n)) \in \prod S_\lambda$. When Λ is infinite we cannot easily write out the "tuple," especially when Λ is not countable, so we use the formalism of functions to specify the elements (tuples!) in the direct product. When each S_λ is a group, we can define a group structure on $\prod S_\lambda$ just as for direct sums of finitely many groups: multiply coordinate by coordinate.

DEFINITION 3.13 The **direct product** of the groups $\{G_\lambda\}_\Lambda$ is the group $(\prod_\lambda G_\lambda, *) = \prod_\lambda G_\lambda$ where $f * g(\lambda) = f(\lambda)g(\lambda)$.

We often write $\prod G_\lambda$ for $\prod_\lambda G_\lambda$. It is easy to see that $G = \prod G_\lambda$ really is a group. The element $e \in \prod G_\lambda$ defined by $e(\lambda) = e_\lambda$, the identity of G_λ, will be e_G, and for $f \in \prod G_\lambda$, f^{-1} is defined by $f^{-1}(\lambda) = f(\lambda)^{-1}$. The associativity of the multiplication in each G_λ implies that $*$ is associative on $\prod G_\lambda$ (verify!). The multiplication should be thought of in the same way as when Λ is finite: coordinatewise multiplication. Aside from the fact that the direct product produces very large groups from finite ones, subgroups of it serve as a source of examples. It is useful to point out that each constituent group can be identified with a subgroup of the direct product.

Example 3.11 Given groups $\{G_\lambda\}_\Lambda$ with e_λ the identity of G_λ, $H_\lambda = \{f \in \prod G_\lambda | f(\alpha) = e_\alpha \text{ for all } \alpha \notin \lambda\} \le \prod G_\lambda$ is identified with G_λ.

Clearly the identity of $\prod G_\lambda$ is in each H_λ and if $f, g \in H_\lambda$ then for all $\alpha \neq \lambda$, $fg^{-1}(\alpha) = f(\alpha)g^{-1}(\alpha) = e_\alpha$ so $H_\lambda \le \prod G_\lambda$ by Theorem 2.12. The identification of $h \in H_\lambda$ with $h(\lambda) \in G_\lambda$ is a bijection (why?) and $h, h' \in H_\lambda$ implies $hh'(\lambda) = h(\lambda)h'(\lambda)$, so the products of corresponding elements correspond. Thus we may identify H_λ with G_λ.

The subgroup generated by all the H_λ in Example 3.11 is the *direct sum* of the G_λ and generalizes direct sums to an infinite collection of groups.

DEFINITION 3.14 The **direct sum** of the groups $\{G_\lambda\}_\Lambda$, with e_λ the identity of G_λ, is the group $\oplus_\lambda G_\lambda \le \prod G_\lambda$ given by $\oplus_\lambda G_\lambda = \{f \in \prod G_\lambda \mid f(\alpha) = e_\alpha$ for all but finitely many $\alpha \in \Lambda\}$. We may also write $\oplus G_\lambda$ for $\oplus_\lambda G_\lambda$.

Note that the "finitely many" mentioned in the definition includes zero, so $\oplus G_\lambda$ contains the identity of $\prod G_\lambda$: the function $e(\lambda) = e_\lambda$. By Theorem 2.12, $\oplus G_\lambda$ is a subgroup since $f, g \in \oplus G_\lambda \Rightarrow fg^{-1}(\lambda) = f(\lambda)g^{-1}(\lambda) = e_\lambda$ whenever both $f(\lambda) = g(\lambda) = e_\lambda$. This can fail to happen for at most finitely many λ. Hence $fg^{-1} \in \oplus G_\lambda$, so $\oplus G_\lambda \le \prod G_\lambda$.

Let us look at a few examples. First we find a huge group with each nonidentity element of the same order. For each $n \in \boldsymbol{N}$ and a fixed prime p, let $G_n = \boldsymbol{Z}_p$. Then $\prod G_n$ is an uncountable (Abelian) group and every nonidentity element has order p. That $G = \prod G_n$ is uncountable follows from the proof of Theorem 0.19 that $\boldsymbol{R}$ is not countable: we can produce an injection from $P(\boldsymbol{N})$ to $\prod G_n$, using "tuples" of $[0]_p$ and $[1]_p$ rather than decimals. On the other hand, by Theorems 0.21 and 0.20, $\oplus G_n$ is countable (verify!), and every element in it satisfies $x^p = e_G$. Next take $H_n = \boldsymbol{Z}_n$. Then again $\oplus H_n$ is a countable group, every element in it has finite order, but there is no bound on these orders since $f \in \oplus H_n$ via $f(m) = [1]_m$ and $f(k) = [0]_k$ for $k \neq m$ has order m. Note that $\prod H_n$ has elements of infinite order, for example, the function given by $F(n) = [1]_n$ $(([1]_1, \ldots, [1]_n, \ldots))$. Also, although each $\alpha \in H_n$ has finite order, the cyclic subgroup $\langle F \rangle \le \prod H_n$ is infinite (why?). Finally, for an example that does not lend itself to visualization as "sequences" let $G_r = (\boldsymbol{Q}, +)$ for each $r \in \boldsymbol{R}$ and consider $\prod G_r$. It is an exercise that $H = \{f \in \prod G_r \mid f(r) \cdot n \in \boldsymbol{Z}$ for all $r \in \boldsymbol{R}$ and some fixed $n = n_f \in \boldsymbol{N}$ depending on $f\} \le \prod G_r$. This H consists of all $f \in \prod G_r$ with a fixed bound for the denominators of all $f(r)$. A little thought shows that $\oplus G_r \le H$, that these are not equal, and that $H \neq \prod G_r$.

EXERCISES

1. If $G_1, \ldots, G_k$ are groups, show that:
   i) $\oplus G_i$ is Abelian $\Leftrightarrow$ each G_j is Abelian
   ii) if $H_i \le G_i$ then $\oplus H_j \le \oplus G_j$.
2. If G is a group, show that i) $\{(g, g) \in G \oplus G \mid g \in G\} \le G \oplus G$ but
   ii) $\{(g, g^{-1}) \in G \oplus G \mid g \in G\} \le G \oplus G \Leftrightarrow G$ is Abelian.
3. For groups H and K, show that $H_1 = \{(h, e_K) \in H \oplus K \mid h \in H\} \le H \oplus K$.

4. Are the following cyclic? If so find a generator and determine how many generators there are:

i) $U_7 \oplus \mathbf{Z}_5$
ii) $U_7 \oplus \mathbf{Z}_{15}$
iii) $\mathbf{Z}_4 \oplus \mathbf{Z}_5 \oplus \mathbf{Z}_6$
iv) $U_6 \oplus \mathbf{Z}_3 \oplus \mathbf{Z}_5$
v) $\mathbf{Z}_3 \oplus \mathbf{Z}_5 \oplus \mathbf{Z}_7 \oplus \mathbf{Z}_{11}$
vi) $U_n \oplus U_m$ with $n, m > 2$

5. If $G_1, \ldots, G_n$ are groups, $n > 1$, $|G_j| > 1$ all j, and $G = G_1 \oplus \cdots \oplus G_n$ is cyclic, show that G is finite.

6. Find the number of elements:

i) of order 2 in $U_3 \oplus U_4 \oplus \cdots \oplus U_9$
ii) of order 3 in $U_3 \oplus U_4 \oplus \cdots \oplus U_9$
iii) of order 2 in $D_4 \oplus D_5 \oplus D_6$
iv) of order 3 in $S_4 \oplus U_7 \oplus D_9$
v) of order 5 in $D_{15} \oplus \mathbf{Z}_{20} \oplus U_{25}$

7. Find all subgroups of: a) $\mathbf{Z}_2 \oplus \mathbf{Z}_2$; and b) $\mathbf{Z}_3 \oplus \mathbf{Z}_3$.

8. If $G = \mathbf{Z}_2 \oplus (\mathbf{Z}, +)$ show that $\mathbf{Z}_2 \oplus n\mathbf{Z} \leq G$ for any $n \in \mathbf{N}$. Find $\langle e_G \rangle \neq H \leq G$ with $H \neq G$ with H neither $\mathbf{Z}_2 \oplus n\mathbf{Z}$, $\mathbf{Z}_2 \oplus \langle 0 \rangle$, nor $\langle [0]_2 \rangle \oplus n\mathbf{Z}$.

9. If $p < q$ are primes, show that the only subgroups of $G = \mathbf{Z}_p \oplus \mathbf{Z}_q$ are $\langle e_G \rangle$, $\mathbf{Z}_p \oplus \langle [0]_q \rangle$, $\langle [0]_p \rangle \oplus \mathbf{Z}_q$, and G.

10. Find all subgroups of: a) $\mathbf{Z}_6 \oplus \mathbf{Z}_{11}$; b) $\mathbf{Z}_9 \oplus \mathbf{Z}_{10}$; and c) $\mathbf{Z}_3 \oplus \mathbf{Z}_7 \oplus \mathbf{Z}_{10}$.

11. Let $G = (\mathbf{Z}, +) \oplus (\mathbf{Z}, +)$.

i) Show that $n, m \in \mathbf{N} \Rightarrow n\mathbf{Z} \oplus m\mathbf{Z} \leq G$.
ii) Show that $H = \{(a, b) \in G \mid 2a + 3b = 0\} \leq G$ and is cyclic.
iii) If $n, m \in \mathbf{N}$, show that $K = \{(a, b) \in G \mid na + mb = 0\} \leq G$ and is cyclic.
iv) For $m \in \mathbf{N}$, show that $L = \{(a, b) \in G \mid a + b \equiv_m 0\} \leq G$ but is not cyclic.

12. Find the centers of the following groups:

a) $\mathbf{Z}_8 \oplus D_3$
b) $U_7 \oplus A_4 \oplus D_4$
c) $A_2 \oplus A_3 \oplus A_4$
d) $A_3 \oplus D_4 \oplus D_6 \oplus \mathbf{Z}_7$
e) $D_4 \oplus D_5 \oplus \cdots \oplus D_{10}$

13. Verify Theorem 3.22.

14. If $G = (\mathbf{Z}_7 \times \mathbf{Z}_3, \cdot)$ with $([a]_7, [b]_3) \cdot ([c]_7, [d]_3) = ([a + 2^b c]_7, [b + d]_3)$, with $b \in \{0, 1, 2\}$, verify directly that G is a non-Abelian group, $o(G) = 21$, and if $g \in G$ with $g \neq e_G$ then $o(g) = p$, some prime. What are the $f_{[b]}$ in Theorem 3.22?

15. If $G = (\mathbf{Z}_{13} \times \mathbf{Z}_3, \cdot)$ with $([a]_{13}, [b]_3) \cdot ([c]_{13}, [d]_3) = ([a + 3^b c]_{13}, [b + d]_3)$, where $b \in \{0, 1, 2\}$, verify directly that G is a non-Abelian group with each nonidentity element of prime order. What are the $f_{[b]}$ in Theorem 3.22?

16. Let $\{G_\lambda\}_\Lambda$ be a collection of groups, e_λ the identity of G_λ, and $H_\lambda = \{f \in \prod G_\lambda \mid f(\alpha) = e_\alpha \text{ for } \alpha \neq \lambda\}$. Show that in $\prod G_\lambda$ the subgroup $\oplus G_\lambda$ is generated by all the H_λ; that is, show that $\oplus G_\lambda = \langle \bigcup H_\lambda \rangle$.

17. For each $i \in \mathbf{N}$, let $G_i = (\mathbf{Z}_2, +)$ and set $G = \oplus G_i$. If $H = \{g \in G \mid g(j) = [1]_2 \text{ for an even number of } j\}$, show that $H \leq G$.

18. For each $i \in \boldsymbol{N}$, let $G_i = (\boldsymbol{Z}, +)$ and set $G = \prod G_i$. Show that the following are subgroups of G:

i) $A = \{f \in G \mid f(j) \in m_j\boldsymbol{Z}\}$ where $m_j \in \boldsymbol{N} \cup \{0\}$ depends on j

ii) $T = \{f \in G \mid \text{for } m \in \boldsymbol{N} \text{ depending on } f,\ f(i+1) = f(i) \text{ for all } i \geq m\}$

iii) $H = \{f \in G \mid \text{for } m \in \boldsymbol{N} \text{ depending on } f,\ f(i+1) = 2f(i) \text{ for all } i \geq m\}$

iv) $K = \{f \in G \mid \text{for all } i, j \in \boldsymbol{N},\ f(i) \equiv f(j) \pmod 2\}$

v) $L = \{f \in G \mid \text{for } k \in \boldsymbol{N} \text{ depending on } f,\ f(i) = f(j) \text{ when } i \equiv_k j\}$

19. Recall Theorem 1.30: $F_{n,m}([a]_n) = [a]_m$ defines $F_{n,m} : \boldsymbol{Z}_n \to \boldsymbol{Z}_m \Leftrightarrow m \mid n$. When p is a prime and $j \geq i$, for convenience set $g_{j,i} = F_{p^j, p^i}$, consider the group $G = \prod \boldsymbol{Z}_{p^i}$ over $i \in \boldsymbol{N}$, and set $H = \{f \in G \mid \text{whenever } j \geq i,\ g_{j,i}(f(j)) = f(i)\}$. H is called an **inverse limit** of $\{\boldsymbol{Z}_{p^i}\}$. It may be easier to consider $f \in G$ to be given by $f(i) = [f_i] \in \boldsymbol{Z}_{p^i}$ for representative $f_i \in \{0, 1, 2, \ldots, p^i - 1\}$.
Thus $H = \{f \in G \mid \text{whenever } j \geq i,\ f_j \equiv f_i \pmod{p^i}\}$.

i) Show that H makes sense: that is, it unambiguously defines elements of G. (The problem is that H is defined by conditions on all pairs of its coordinates, but it has infinitely many coordinates, so one must check that the conditions are consistent.)

ii) Find elements of H such that the set $\{f_i\}$ of representatives is not bounded.

iii) Given $f, g \in H$ with $f \neq g$, show that $f(j) = g(j)$ is possible only for finitely many $j \in \boldsymbol{N}$.

iv) If $f \in H$ is given and if for some $g \in H$, $f(k) = g(k)$ for a given $k \in \boldsymbol{N}$, how many different values for $g(k+3)$ are possible?

v) Show that $H \leq G$.

vi) Show that $f \in H$ has finite order $\Leftrightarrow f = e_H$.

CHAPTER 4

SUBGROUPS

We study elementary but important ideas based on counting that lead to a number of results on finite groups. Our approach illustrates how simple ideas can be developed into results of greater complexity and sophistication. This gradual accretion of knowledge and understanding is a common and powerful method used in mathematics. The ideas we introduce are quite general, but their main applications here are useful results about Abelian groups and some interesting and nontrivial facts about integers.

4.1 COSETS

We can get useful results from certain equivalence relations on a finite group G if we can count the elements in the equivalence classes. The sum of these numbers is $|G|$ and interesting results can arise from this observation. Our first example of the power of this simple idea is based on a generalization to arbitrary groups of the relation congruence modulo n in $\mathbf{Z}$.

DEFINITION 4.1 For G any group, $H \leq G$, and $x, y \in G$, we call x **congruent to y modulo H**, written $x \equiv_H y$, when $xy^{-1} \in H$.

This notion really does generalize $\equiv_n$ in $\mathbf{Z}$. If $G = (\mathbf{Z}, +)$ and $H = n\mathbf{Z}$ then using $+$ in $\mathbf{Z}$, $xy^{-1} \in H$ is just $x - y \in n\mathbf{Z}$ so $x \equiv_H y$ is the same as $x \equiv_n y$. From Theorem 1.2 the integers congruent to one another mod n are the sets $n\mathbf{Z} + i$ for $0 \leq i < n$. In general nothing interesting results from $\equiv_H$ if $H = G$ or $H = \langle e_G \rangle$, since in the first case $x \equiv_G y$ for all x and y, and in the second case $x \equiv_H y$ only when $x = y$. We mention that $h, h' \in H$ and $H \leq G$ imply $h \equiv_H h'$. When $G = S_3$ and $H = \langle (1, 2, 3) \rangle$, any two transpositions are congruent modulo H (why?). In general for $G = S_n$ and $H = A_n$, any two odd permutations are congruent modulo A_n and also any two even permutations are congruent modulo A_n (why?). In these examples the sets of elements related to each other give a partition of G and so an equivalence relation on G.

THEOREM 4.1 If G is a group and $H \leq G$, then $\equiv_H$ is an equivalence relation on G. The equivalence class of $g \in G$ is $[g]_H = Hg = \{hg \mid h \in H\}$.

Proof We need $\equiv_H$ to be reflexive, symmetric, and transitive. For $x \in G$, $xx^{-1} = e_G \in H$ by Theorem 2.11, so $x \equiv_H x$ and $\equiv_H$ is reflexive. If $x \equiv_H y$ then by definition $xy^{-1} \in H \le G$, so $yx^{-1} = (xy^{-1})^{-1} \in H$, $y \equiv_H x$ and $\equiv_H$ is symmetric. When both $x \equiv_H y$ and $y \equiv_H z$, then $xy^{-1}, yz^{-1} \in H$ and multiplying these yields $xz^{-1} = xy^{-1}yz^{-1} \in H$, so $x \equiv_H z$ and $\equiv_H$ is transitive, proving that it is an equivalence relation. For any $g \in G$ the definitions of equivalence class and symmetry yield $[g]_H = \{y \in G \mid y \equiv_H g\} = \{y \in G \mid yg^{-1} \in H\}$. Now $yg^{-1} \in H \Leftrightarrow y = hg$ for some $h \in H$, so $[g]_H = Hg = \{hg \mid h \in H\}$. ∎

DEFINITION 4.2 For $H \le G$ and $g \in G$, the **right coset** of H **generated** by g is $Hg = \{hg \mid h \in H\}$. The **left coset generated** by g is $gH = \{gh \mid h \in H\}$.

Since $\equiv_H$ is an equivalence relation on G, from Theorem 1.22 the set of its classes, $\{Hg \mid g \in G\}$, is a partition of G. The left cosets of H give a partition of G as well and are the equivalence classes of the relation $x \equiv'_H y$ when $x^{-1}y \in H$. The argument for this follows that given in Theorem 4.1. We work mostly with right cosets, which we often refer to just as "cosets," and properties of them are true as well for left cosets. There is no difference at all between left and right cosets when G is Abelian. Before computing some examples, we record the special case of Theorem 1.21 for the relation $\equiv_H$.

THEOREM 4.2 If G is a group, $H \le G$, and $x, y \in G$, then:

i) $xy^{-1} \in H \Leftrightarrow Hx = Hy \Leftrightarrow y \in Hx$
ii) $Hx = Hy$ or $Hx \cap Hy = \emptyset$

We can write each equivalence class for $\equiv_H$ as the "translate" of H by a fixed $g \in G$ using Theorem 4.1, and Theorem 4.2 tells us that given a coset Hg, a new coset must start with some $x \notin Hg$. Now H is always a right (or left) coset of itself since $He = H$ so by Theorem 4.2, $Hx = H \Leftrightarrow x \in H$. When $H = G$, the only coset is G, and when $H = \langle e_G \rangle$, $Hx = \{x\}$, so each singleton in G is a coset. When the group operation is addition the coset of H generated by g is $H + g = \{h + g \mid h \in H\}$. Recall that the subgroups of the cyclic groups $(\mathbf{Z}_n, +)$ are just the $\langle [1]_n^m \rangle = \langle [m]_n \rangle$ for $m \mid n$, by Theorem 3.1.

Example 4.1 The cosets of $H = \langle [3] \rangle$ in $G = \mathbf{Z}_{12}$.

A coset of H in G is $H = H + [0] = \{[0], [3], [6], [9]\}$. Another is $H + [1] = \{[1], [4], [7], [10]\}$ obtained by adding $[1] \notin H$ to each element in H. From Theorem 4.2, or directly, $H + [1] = H + [4] = H + [7] = H + [10]$. Another coset is $H + [2] = \{[0] + [2], [3] + [2], [6] + [2], [9] + [2]\} = \{[2], [5], [8], [11]\} = H + [5] = H + [8] = H + [11]$. Every element of $\mathbf{Z}_{12}$ is in one of H, $H + [1]$, or $H + [2]$, so these three cosets are all cosets of H in G.

Example 4.2 The cosets of $H = \langle [4] \rangle$ in $G = U_{21}$.

Since G is Abelian, $Hg = gH$. As $H = \{[4], [4]^2 = [16], [4]^3 = [64] = [1]\} = \{[1], [4], [16]\}$ and $[2] \in U_{21} - H$, another coset is $H[2] = \{[1][2], [4][2],$

$[16][2]\} = \{[2], [8], [11]\}$, which is also $H[8]$ and $H[11]$ (but $H[2]$ is easier to compute!). Since $[5] \notin H \cup H[2]$, $H[5] = \{[5], [20], [17]\}$ will be a third coset. Finally, $U_{21} - (H \cup H[2] \cup H[5]) = \{[10], [13], [19]\} = H[10] = H[13] = H[19]$, computed directly as $\{[1][10], [4][10], [16][10]\}$. Therefore $\langle [4] \rangle = H$ has four cosets in $U_{21} : H$, $H[2] = \{[2], [8], [11]\}$, $H[5] = \{[5], [17], [20]\}$, and $H[10] = \{[10], [13], [19]\}$.

Example 4.3 The cosets of $H = \langle (1, 2) \rangle$ in S_3.

Since S_3 is not Abelian we compute both the right and left cosets. Right cosets are $H = \{I, (1, 2)\}$ and also $H(1, 3) = \{(1, 3), (1, 2)(1, 3) = (1, 3, 2)\}$. Now $S_3 - (H \cup H(1, 3)) = \{(2, 3), (1, 2, 3)\} = H(2, 3) = \{(2, 3), (1, 2)(2, 3)\}$ and these three cosets partition S_3. Both $H = \{I, (1, 2)\}$ and $(1, 3)H = \{(1, 3), (1, 3)(1, 2) = (1, 2, 3)\}$ are left cosets. As above, $(2, 3)H = \{(2, 3), (1, 3, 2)\}$ is the third left coset. Except for H, the left and the right cosets are different, although in this case we may use the same transversal for the two partitions.

In non-Abelian groups the left and right cosets of a subgroup often give different partitions, but not always. In S_n the odd permutations form one right coset of A_n, and A_n is another, and similarly the odd permutations form one left coset, so the right cosets and the left cosets give the same partition of S_n, namely $\{S_n - A_n, A_n\}$. Another example of this phenomenon occurs in D_6 (Definition 3.10).

Example 4.4 The cosets of $H = \langle T_2 \rangle = \{I = T_6, T_2, T_4\}$ in D_6.

One right coset is H itself and another is $H \cdot T_1 = \{T_1, T_2T_1, T_4T_1\} = \{T_1, T_3, T_5\}$. Since $(H \cup H \cdot T_1) \cap \{W_i\} = \emptyset$, we consider $H \cdot W_0 = \{W_0, T_2W_0, T_4W_0\}$ and use $T_iW_j = W_{j+i}$ from Theorem 3.17 to write $H \cdot W_0 = \{W_0, W_2, W_4\}$. Also $H \cdot W_1 = \{W_1, W_3, W_5\}$ (why?). These four cosets partition D_6 and are all of the right cosets of H in D_6. Similarly, left cosets of H in D_6 are H and $T_1 \cdot H = \{T_1, T_3, T_5\}$. Now using $W_jT_i = W_{j-i}$ (Theorem 3.17) yields $W_0 \cdot H = \{W_0, W_{-2}, W_{-4}\} = \{W_0, W_4, W_2\}$ and $W_1 \cdot H = \{W_1, W_{-1}, W_{-3}\} = \{W_1, W_5, W_3\}$, so $\langle T_2 \rangle g = g\langle T_2 \rangle$ for all $g \in D_6$ although D_6 is not Abelian.

In the next chapter we study subgroups of G having the same left and right cosets. These subgroups are quite important because their set of cosets is a group with multiplication closely related to the multiplication in G. This construction of a group of cosets provides a powerful technique for proving things about finite groups by using induction on the group order.

In large groups or in infinite groups it is usually impractical to write out the cosets of most subgroups, but we can hope to describe them anyway. When $H \leq G$, $\{Hg \mid g \in G\}$ is a partition of G and the cosets Hg are determined by H, so we can hope to describe them by some property of elements of G related to the properties of H. Specifically, in $GL(n, R)$, the invertible $n \times n$ matrices over a commutative ring R, we can easily describe the cosets of $SL(n, R)$, the subgroup of matrices of determinant one.

Example 4.5 For R any commutative ring and $A \in GL(n, R)$, $SL(n, R) \cdot A = A \cdot SL(n, R) = \{B \in GL(n, R) \mid \det(B) = \det(A)\}$.

By Theorem 4.2, $B \in SL(n, R) \cdot A \Leftrightarrow AB^{-1} \in SL(n, R)$. But this means that $1 = \det(AB^{-1}) = \det(A)\det(B)^{-1}$, or, equivalently, that $\det(A) = \det(B)$. Similarly, $B \in A \cdot SL(n, R)$ leads to $A^{-1}B \in SL(n, R)$ (exercise!), equivalent to $\det(A)^{-1}\det(B) = \det(A^{-1}B) = 1$, and so to $\det(A) = \det(B)$.

For $A \in GL(n, \mathbf{Z})$, $\det(A) = \pm 1$ so by Example 4.5 there are only two cosets of $SL(n, \mathbf{Z})$ in $GL(n, \mathbf{Z})$ but we cannot explicitly write out either of them. There are $\varphi(m)$ cosets of $SL(n, \mathbf{Z}_m)$ in $GL(n, \mathbf{Z}_m)$: U_m are the invertible elements in $\mathbf{Z}_m$. It is impractical to write out $SL(n, \mathbf{Z}_m)$ or its cosets in $GL(n, \mathbf{Z}_m)$, but $\{Diag([a]_m, [1]_m, \ldots, [1]_m) \mid [a]_m \in U_m\}$ is a transversal for its cosets. This follows from Example 4.5 since $\det(Diag([a]_m, [1]_m, \ldots, [1]_m)) = [a]_m$. Similarly, a nice transversal for the cosets of $SL(n, \mathbf{C}) \leq GL(n, \mathbf{C})$ is $\{Diag(c, 1, \ldots, 1) \mid c \in \mathbf{C}^*\}$.

Another example that identifies cosets by a common property of their elements is $H = S_\infty = \{f \in \mathrm{Bij}(\mathbf{N}) \mid f(k) = k \text{ if } k \geq n(f)\} \leq \mathrm{Bij}(\mathbf{N})$ (see Example 2.12). Consider the *left* cosets of H in G. Now $g \in fH \Leftrightarrow f^{-1}g \in H$ (why?) $\Leftrightarrow f^{-1}g(k) = k$ for all sufficiently large $k \Leftrightarrow g(k) = f(k)$ for all large enough k. Thus $fH = \{\varphi \in \mathrm{Bij}(\mathbf{N}) \mid \varphi(k) = f(k)$ for all large enough $k\}$. It follows that if $f, g \in \mathrm{Bij}(\mathbf{N})$ agree on all but finitely many $k \in \mathbf{N}$, then $g \in f\mathbf{N}$ and $f \in g\mathbf{N}$ so $f\mathbf{N} = g\mathbf{N}$ by Theorem 4.2 (for left cosets!). Consequently, two elements of $\mathrm{Bij}(\mathbf{N})$ are in the same left coset of H when they agree on all but finitely many integers. This gives a description for the left cosets of S_∞ in $\mathrm{Bij}(\mathbf{N})$, but it is not clear how to specify a transversal for these cosets. Also it turns out to be true, but not obvious, that this same condition of agreement on all but finitely many $k \in \mathbf{N}$ describes the right cosets of S_∞ in $\mathrm{Bij}(\mathbf{N})$, so the set of right cosets of S_∞ in $\mathrm{Bij}(\mathbf{N})$ is the same as the set of left cosets. This is left as an exercise.

We end this section with a sometimes useful observation illustrating how the properties of cosets in Theorem 4.2 are used.

THEOREM 4.3 If G is a group and $H, K \leq G$, the cosets of $H \cap K$ are the nonempty intersections of the cosets of H with the cosets of K. Thus if $|\{Hg \mid g \in G\}| = n$ and $|\{Kg \mid g \in G\}| = m$, then $|\{(H \cap K)g | g \in G\}| \leq nm$.

Proof When $\{Hg \mid g \in G\}$ and $\{Kg \mid g \in G\}$ are finite there are only finitely many intersections $Hx \cap Ky$, so the second statement follows from the first. Now $H \cap K \subseteq H$ and $H \cap K \subseteq K$ so $(H \cap K)g \subseteq Hg \cap Kg$ if $g \in G$. If $x = hg = kg \in Hg \cap Kg$, then $h = k \in H \cap K$ and $x \in (H \cap K)g$ so $(H \cap K)g = Hg \cap Kg$. This shows $\{(H \cap K)g \mid g \in G\} \subseteq \{Hx \cap Ky \mid x, y \in G\}$. Next, if $x \in Hg \cap Kg'$, then by Theorem 4.2, $Hg = Hx$ and $Kg' = Kx$, so $Hg \cap Kg' = Hx \cap Kx = (H \cap K)x$ by the observation above. Therefore $\{Hx \cap Ky \neq \emptyset \mid x, y \in G\} \subseteq \{(H \cap K)g \mid g \in G\}$, proving the theorem. ∎

It is an exercise to extend Theorem 4.3 to $H_1 \cap \cdots \cap H_k$ when all $H_j \le G$ have only finitely many cosets in G. To illustrate the theorem, take $G = (\mathbf{Z}, +)$, $H = 4\mathbf{Z}$ with cosets $4\mathbf{Z} + i$ for $0 \le i < 4$, and $K = 6\mathbf{Z}$ with cosets $6\mathbf{Z} + i$ for $0 \le i \le 6$. Now $4\mathbf{Z} \cap 6\mathbf{Z} = 12\mathbf{Z}$ (see Example 2.11) has twelve cosets. Some of these are $12\mathbf{Z} + 5 = (4\mathbf{Z} + 1) \cap (6\mathbf{Z} + 5)$ and $12\mathbf{Z} + 6 = (4\mathbf{Z} + 2) \cap 6\mathbf{Z}$. However, $4\mathbf{Z} \cap (6\mathbf{Z} + 1) \subseteq 2\mathbf{Z} \cap (2\mathbf{Z} + 1) = \varnothing$ and also $(4\mathbf{Z} + 1) \cap (6\mathbf{Z} + 2) \subseteq (2\mathbf{Z} + 1) \cap 2\mathbf{Z} = \varnothing$.

EXERCISES

1. In each case for $H \le G$, find the right cosets of H in G and the left cosets of H in G. Determine whether $\{gH \mid g \in G\} = \{Hg \mid g \in G\}$.
  a) $H = \langle [5]_{20} \rangle \le \mathbf{Z}_{20}$
  b) $H = \langle g^3 \rangle \le \langle g \rangle$ with $o(g) = 15$
  c) $H = \langle [11]_{16} \rangle \le U_{16}$
  d) $H = \langle [4]_{17} \rangle \le U_{17}$
  e) $H = \{\theta \in S_4 \mid \theta(4) = 4\} \le S_4$
  f) $H = \langle W_1 \rangle \le D_4$
  g) $H =$ "D_4" $= \{(2, 4)^i (1, 2, 3, 4)^j\} \le S_4$
  h) $H = \{I, (a, b)(c, d)\} \le S_4$ ($|H| = 4$)
  i) $H = \langle T_2 \rangle \le D_4$
  j) $H = \langle W_0 \rangle \le D_5$
  k) $H = \{I, W_1, T_3, W_4\} \le D_6$
  l) $H = \{I, T_4, W_0, W_4\} \le D_8$
  m) $H = \{W_{2j}, T_{2j} \mid j \in \mathbf{Z}\} \le D_{2m}$

2. If $H \le Z(G)$, show that each left coset of H in G is a right coset of H in G.

3. For $H \le G$, show that $x \approx y \Leftrightarrow x^{-1}y \in H$ is an equivalence relation on G and that $[g]_{\approx} = gH = \{gh \mid h \in H\}$.

4. If $H \le G$, show carefully that $|\{gH \mid g \in G\}| = |\{Hg \mid g \in G\}|$.

5. Let $G = \left\{ \begin{bmatrix} a & b \\ 0 & c \end{bmatrix} \in GL(2, \mathbf{Z}) \mid ac = \pm 1 \right\}$, $H = \left\{ \begin{bmatrix} a & 0 \\ 0 & c \end{bmatrix} \in G \right\}$, and $A = \begin{bmatrix} 1 & 1 \\ 0 & 1 \end{bmatrix}$.
  a) Show $G \le GL(2, \mathbf{Z})$.
  b) Explicitly describe the matrices in $\langle A \rangle$.
  c) Show that $H \le G$ and that $|H| = 4$.
  d) Show that H can be used as a transversal for the right cosets of $\langle A \rangle$ in G.
  e) Show that $\{\langle A \rangle g \mid g \in G\} = \{g\langle A \rangle \mid g \in G\}$.

6. If G and H are groups, show that either $G \oplus \langle e_H \rangle$ or $\langle e_G \rangle \oplus H \le G \oplus H$ can be used as a transversal for the cosets of the other in $G \oplus H$.

7. If $H = \langle (1, 1) \rangle \le \mathbf{Z} \oplus \mathbf{Z}$, show that the elements in $\mathbf{Z} \oplus \langle 0 \rangle$ can be used as a transversal for the cosets of H in $\mathbf{Z} \oplus \mathbf{Z}$.

8. Show that $[0, 1) \subseteq \boldsymbol{R}$ is a transversal for the cosets of $(\mathbf{Z}, +)$ in $(\boldsymbol{R}, +)$.

9. If $G = (\boldsymbol{R}^n, +)$, set $H = \{(a_1, \ldots, a_n) \in G \mid a_1 + a_2 + \cdots + a_n = 0\}$. Show that $H \le G$ and that $\{(0, \ldots, 0, r) \in G \mid r \in \boldsymbol{R}\}$ is a transversal for H in G.

10. If $H, K \leq G$ with $H \cap K = \langle e_G \rangle$, show that there is a transversal T for the cosets of H in G such that $K \subseteq T$.

11. Let $G = D_{2m}$, $k \mid m$ for some $k \in \mathbf{N}$, fix $0 \leq i < k$, and set $H = \{T_{kz}, W_{i+kz} \mid z \in \mathbf{Z}\}$. Show $H \leq D_{2m}$. Find a transversal for the right cosets of H in D_{2m}.

For Exercises 12–13, recall that $I(n) = \{1, \ldots, n\}$, and $S_\infty = \{f \in \mathrm{Bij}(\mathbf{N}) \mid f(k) = k$ for all $k \geq n(f)\}$.

12. Show that for $f \in \mathrm{Bij}(\mathbf{N})$, $S_\infty f = \{g \in \mathrm{Bij}(\mathbf{N}) \mid f(k) = g(k)$ for all $k \geq m(g)\}$.

13. Set $A_\infty = \{f \in S_\infty \mid$ if $f(k) = k$ for all $k \geq n$, then $f|_{I(n)} \in A_n\}$. Show that $A_\infty \leq S_\infty$ and find a transversal for its cosets. Show that the left and the right cosets of A_∞ in S_∞ give the same partition of S_∞.

14. If $g_1, \ldots, g_k \in G$ so that $H_1 g_1 \cup \cdots \cup H_k g_k = G$, where each $H_i \leq G$ but the H_j are not necessarily different, show that some H_i has only finitely many cosets in G. (Reduce to $H_1, \ldots, H_s$ distinct with s minimal and show that if all H_j have infinitely many cosets in G, then H_s is in the union of finitely many cosets of $H_1, \ldots, H_{s-1}$. Use this to contradict the choice of s.)

15. If $H_1, \ldots, H_k \leq G$ so that $H_1 \cup \cdots \cup H_k = G$ show that some H_j has only finitely many cosets in G (see Exercise 14).

16. Write $(\mathbf{Z} \oplus \mathbf{Z}, +)$ as the union of three proper subgroups (use even and odd!).

17. Write $(\mathbf{Z} \oplus \mathbf{Z}_3, +)$ as a union of proper subgroups.

18. Write $(\mathbf{Z}, +)$ as the union of cosets of two different $H, K \leq \mathbf{Z}$ so that not every coset of either H or K is used in the union.

4.2 LAGRANGE'S THEOREM AND CONSEQUENCES

Our first major result about abstract finite groups is Lagrange's Theorem that the order of a subgroup divides the order of the group. Its proof is quite simple: the group is the disjoint union of the cosets of the subgroup, and each coset has the same number of elements as the subgroup.

DEFINITION 4.3 If G is a group and $H \leq G$ then the **index** of H in G is $|\{Hg \mid g \in G\}|$, the number of cosets of H in G, and is denoted $[G : H]$.

When G is finite, $[G : H]$ is the number of distinct right cosets of H in G so should really be called the "right" index, but in fact $|\{gH \mid g \in G\}| = [G : H]$ as well. To see this it is tempting to try associating right cosets of H to left cosets by $Hg \to gH$, that is, to define $F : \{Hg \mid g \in G\} \to \{gH \mid g \in G\}$ by $F(Hg) = gH$. Unfortunately, F is not in general a function! The problem is that we are trying to define $F(X)$ by using a representative for X. Consider "F" in Example 4.3 with $G = S_3$ and $H = \langle (1, 2) \rangle$. Here $H(1, 3) = \{(1, 3), (1, 3, 2)\} = H(1, 3, 2)$, $F(H(1, 3)) = (1, 3)H = \{(1, 3), (1, 2, 3)\}$, but $F(H(1, 3, 2)) = (1, 3, 2)H = \{(2, 3), (1, 3, 2)\}$, so $F(H(1, 3)) \neq F(H(1, 3, 2))$. Instead observe that $f : G \to G$ given by $f(g) = g^{-1}$ is a bijection and so $f^\wedge(Hg) = \{f(hg) \mid h \in H\} = \{g^{-1}h^{-1} \mid h \in H\} = g^{-1}H$ (why?). This gives $f^\wedge : \{Hg \mid g \in G\} \to \{gH \mid g \in G\}$, a bijection. The details are an exercise but we will henceforth use the fact that $|\{Hg \mid g \in G\}| = [G : H] = |\{gH \mid g \in G\}|$.

With this notation, for $H, K \leq G$, Theorem 4.3 says that $[G : H]$ and $[G : K]$ finite $\Rightarrow [G : H \cap K]$ is finite and in fact $[G : H \cap K] \leq [G : H][G : K]$. We proceed to the last piece needed to prove Lagrange's Theorem.

THEOREM 4.4 If $H \leq G$ and $x, y \in G$, then $|Hx| = |H| = |yH|$.

Proof We show $|Hx| = |H|$, since the proof of $|H| = |yH|$ is similar, and to do so we must produce $f \in \text{Bij}(H, Hx)$. The most natural guess is $f : H \to Hx$ given by $f(h) = hx$. We can verify directly that f is a bijection or let $F : Hx \to H$ be $F(hx) = (hx)x^{-1} = h$ and show easily that f and F are inverses. Then f is a bijection by Theorem 0.11. ■

We can now prove Lagrange's Theorem.

THEOREM 4.5 (Lagrange) If G is a finite group and $H \leq G$, then $|G| = |H|[G : H]$, so in particular, $|H| \mid |G|$.

Proof From Theorem 4.1 and Theorem 1.22, $\{Hg \mid g \in G\}$ is a partition of G. Since G is finite, it is the union of finitely many of the Hg, in fact $[G : H]$ of them. Let $\{g_1, g_2, \ldots, g_{[G:H]}\}$ be a transversal for the partition of G by cosets of H, so G is the disjoint union of $\{Hg_1, \ldots, Hg_{[G:H]}\}$. It follows from Theorem 0.3 that $|G| = |Hg_1| + \cdots + |Hg_{[G:H]}|$. But $|Hg_j| = |H|$ by Theorem 4.4, so in fact $|G| = |H| + \cdots + |H| = |H|[G : H]$, proving the theorem. ■

The proof of Lagrange's theorem is illustrated in Fig. 4.1.

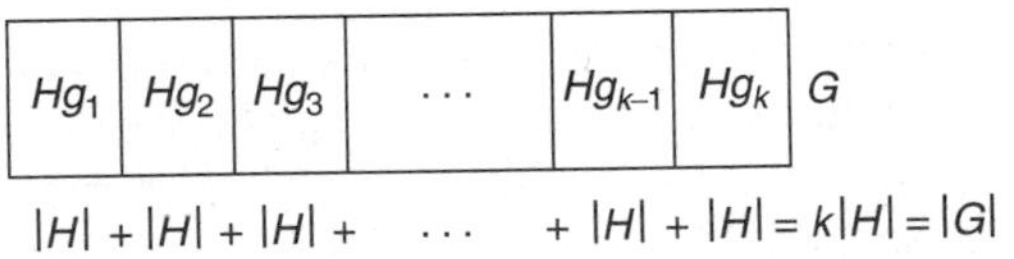

FIGURE 4.1

The value of Lagrange's Theorem cannot be overemphasized, and it will be used throughout our development of the theory of finite groups. We note that the converse of Lagrange's Theorem is false; that is, if $m \in N$ and $m \mid |G|$ there need not be $H \leq G$ with $|H| = m$. This converse is true for finite cyclic groups (Theorem 3.1) and for D_n by a nontrivial exercise. A group of smallest order for which the converse of Lagrange's Theorem is false is A_4. It would be easier to show this later with the help of some other concepts, but it is not too hard to do it now using Lagrange's Theorem.

Example 4.6 A_4 has no subgroup of order 6.

Recall from §3.3 that $A_4 = \{I, (a, b, c), (a, b)(c, d)\}$ consists of the identity, the eight 3-cycles in S_4, and the three products of two disjoint transpositions in S_4. Suppose $K \leq A_4$ with $|K| = 6$. It is easy to verify that $H = \{I, (a, b)(c, d)\} \leq A_4$ and $|H| = 4$. Since $H \cap K \leq H$ and $H \cap K \leq K$, by Lagrange's Theorem $|H \cap K| \mid 4$ and $|H \cap K| \mid 6$, forcing $|H \cap K| = 1$ or $|H \cap K| = 2$. Thus K contains at most one product of two disjoint transpositions, so contains at least four 3-cycles. Let $\alpha, \beta \in K$ be 3-cycles with $\beta \in K - \langle \alpha \rangle$. Since $\alpha^2 = \alpha^{-1}$ and $\beta^2 = \beta^{-1}$ we must have $K = \{I, \alpha, \alpha^2, \beta, \beta^2, \gamma\}$. Now $\alpha\beta, \beta\alpha \in K$ and it follows that $\alpha\beta, \beta\alpha \notin \{I, \alpha, \alpha^2, \beta, \beta^2\}$, so $\alpha\beta = \gamma = \beta\alpha$. Because $\alpha\beta = \beta\alpha$, $\gamma^3 = (\alpha\beta)^3 = \alpha^3\beta^3 = I$ so $o(\gamma) = 3$ by

Theorem 2.7. Now $\gamma^2 \notin \{I, \gamma\}$ and if $\gamma^2 = \alpha$ then $\gamma = (\gamma^2)^2 = \alpha^2$. Similarly, $\gamma^2 \neq \alpha^2, \beta, \beta^2$, and this contradiction shows that A_4 can have no subgroup of order 6.

Note that $L = \{\varphi \in S_4 \mid \varphi(4) = 4\} \leq S_4$ is a "copy" of S_3 and has order 6. We continue with some consequences of Lagrange's Theorem.

THEOREM 4.6 Let G be a finite group with $H, K \leq G$. Then:

i) $|H \cap K|$ divides each of $|H|$, $|K|$, and GCD$\{|H|, |K|\}$.
ii) If $(|H|, |K|) = 1$, then $H \cap K = \langle e_G \rangle$.
iii) If $|H| = |K| = p$, a prime, then either $H = K$ or $H \cap K = \langle e_G \rangle$.

Proof Since $H \cap K \leq H$ and $H \cap K \leq K$, $|H \cap K|$ divides $|H|$ and $|K|$ by Theorem 4.5. Hence $|H \cap K| \mid GCD\{|H|, |K|\}$ using Theorem 1.8 on greatest common divisors. Thus $(|H|, |K|) = 1$ forces $|H \cap K| = 1$ so $H \cap K = \langle e_G \rangle$. If $|H| = |K| = p$, a prime, then $(|H|, |K|) = p$, so $|H \cap K| = 1$ or $|H \cap K| = p$. In the first case $H \cap K = \langle e_G \rangle$, and in the second $H \cap K$, H, and K have the same order, so $H = H \cap K = K$. ■

COROLLARY In a finite group G, if p is prime then there are $(p-1)k$ elements in G of order p, where k is the number of cyclic subgroups of G of order p.

Proof If $x \in G$ has order p then $|\langle x \rangle| = p$ by Theorem 2.16, and Theorem 2.8 shows that each $g \in \langle x \rangle - \langle e \rangle$ has order p and generates $\langle x \rangle$, again by Theorem 2.16. Using Theorem 4.6, the intersection of different subgroups of order p is the identity. Hence $\{g \in G \mid o(g) = p\}$ is the set of nonidentity elements in all cyclic subgroups of order p, and these subgroups have no intersections except for $\langle e \rangle$. Each of these subgroups has $p-1$ elements of order p appearing in no other such subgroup, so $|\{g \in G \mid o(g) = p\}| = (p-1)k$, where k is the number of distinct cyclic subgroups of order p. ■

In S_5 the elements of order 5 are the 5-cycles, $\{(a, b, c, d, e)\}$. There are 5! ways of writing such expressions but each cycle can be written with any integer first, so the number of different 5-cycles is $5!/5 = 24$. Since $24 = 4 \cdot 6$, there are exactly 6 different cyclic subgroups of order 5 in S_5. The computation in S_6 shows that there are $6 \cdot 5 \cdot 4 \cdot 3 \cdot 2/5 = 144$ different elements of order 5, so $144/4 = 36$ different cyclic subgroups of order 5.

Example 4.7 $\mathbf{Z}_5 \oplus \mathbf{Z}_5 \oplus \mathbf{Z}_5$ has 31 different cyclic subgroups of order 5.

The definition of direct sum, or Theorem 3.20, shows that $x^5 = e$ for each $x \in G = \mathbf{Z}_5 \oplus \mathbf{Z}_5 \oplus \mathbf{Z}_5$. Since $|G| = 5^3 = 125$, by the Corollary above G must have exactly $124/4 = 31$ cyclic subgroups of order 5.

Example 4.8 $U_7 \oplus U_7 \oplus U_7$ has 13 different cyclic subgroups of order 3.

Now $|U_7| = 6$ and $U_7 = \langle [3]_7 \rangle$, so from Theorem 3.1 on cyclic groups, $\langle [2]_3 \rangle$ is the unique subgroup of order 3 in U_7. Thus U_7 has exactly two elements of order

3 by the corollary above. To count the elements of order 3 in $G = U_7^3$ use Theorem 3.20: $g = (g_1, g_2, g_3)$ has order 3 when $o(g_i) \mid 3$ and some $g_i \neq [1]_7$. There are $2^3 = 8$ such g with no $g_i = [1]$, $3 \cdot 2^2 = 12$ with exactly one $g_i = [1]$, and $3 \cdot 2 = 6$ with two of the g_i equal to $[1]$. Thus G has 26 elements of order 3, so $26/2 = 13$ different cyclic subgroups of order 3.

THEOREM 4.7 For G a finite group, $g \in G \Rightarrow o(g) \mid |G|$, so $g^{|G|} = e_G$.

Proof From Theorem 4.5 $|\langle g \rangle| \mid |G|$, but $|\langle g \rangle| = o(g)$ by Theorem 2.16 so indeed $o(g) \mid |G|$. But now $g^{|G|} = e_G$ by Theorem 2.7 since $|G| \in o(g)\mathbf{Z}$. ■

Theorem 4.7 is useful for many arguments and also is helpful in finding orders of elements since it restricts the possibilities. To find $o([5])$ in U_{36}, observe that $|U_{36}| = \varphi(36) = \varphi(4)\varphi(9) = 2 \cdot 6 = 12$, so $o([5]) | 12$. Now $[5]^2 = [25]$, $[5]^3 = [125] = [17]$, and $[5]^4 = [85] = [13]$, so $o([5]) = 6$ or 12. But $[5]^6 = [17]^2 = [289] = [1] \in U_{36}$, so $o([5]) = 6$ in U_{36}.

Nice applications of Theorem 4.7 to number theory include a group theory approach to a famous theorem of Euler on congruences and its corollary, the "Little Theorem" of Fermat.

THEOREM 4.8 (Euler) If $k \in \mathbf{Z}$, $n \in \mathbf{N}$, and $(k, n) = 1$, then $k^{\varphi(n)} \equiv_n 1$.

Proof Since $(k, n) = 1$, we have $[k]_n \in U_n$. Using $|U_n| = \varphi(n)$, a direct application of Theorem 4.7 yields $[k]_n^{\varphi(n)} = [1]_n$, which is equivalent to $[k^{\varphi(n)}]_n = [1]_n$. The elements of U_n are equivalence classes of congruence modulo n, so $k^{\varphi(n)} \equiv_n 1$ follows from Theorem 1.23. ■

COROLLARY (Fermat) If p is a prime and $a \in \mathbf{Z}$, then $a^p \equiv_p a$.

Proof When $p \mid a$ then clearly $a \equiv_p 0$ and $a^p \equiv_p 0$, so $a^p \equiv_p a$ since $\equiv_p$ is an equivalence relation. If $(a, p) = 1$, Euler's Theorem applies to yield $a^{p-1} \equiv_p 1$. Now multiply by a and use Theorem 1.24 on congruences to see that $a^p \equiv_p a$. ■

Immediate consequences of these results are that $19 \mid (5^{19} - 5)$ since $5^{19} \equiv_{19} 5$, the last two digits of $p^{40} - 1$ are zeros for any prime $p > 5$ since $(p, 100) = 1$ and $\varphi(100) = \varphi(2^2)\varphi(5^2) = 2 \cdot 5 \cdot 4 = 40$ so $p^{40} \equiv_{100} 1$, and $2^{48} \equiv_{105} 1$ since $\varphi(105) = \varphi(3)\varphi(5)\varphi(7) = 2 \cdot 4 \cdot 6 = 48$. We continue with additional examples illustrating the results of Euler and Fermat. In the last section of this chapter we will discuss briefly how Euler's Theorem is related to modern cryptography.

Example 4.9 If p is a prime, $x \in \mathbf{Z}$, and $m, k \in \mathbf{N}$, then $x^m \equiv_p x^{m+k(p-1)}$.

If $p \mid x$ then $p \mid x^m$ and $p \mid x^{m+k(p-1)}$ so $x^m \equiv_p 0 \equiv_p x^{m+k(p-1)}$. When $(x, p) = 1$ then $x^{m+k(p-1)} = x^m (x^{p-1})^k \equiv_p x^m$ by Euler's Theorem and Theorem 1.24.

Specific instances of Example 4.9 are $x^7 \equiv_5 x^{31}$, $x^2 \equiv_3 x^{102}$, and finally $x^{11} \equiv_7 x^{71}$. Fermat's Theorem can be useful even when n is composite.

Example 4.10 $x^{13} \equiv_{35} x$ for all $x \in \mathbf{Z}$.

Verify the congruence for each prime factor of 35 and then apply Theorem 1.12. Since $13 = 1 + 3 \cdot 4 = 1 + 2 \cdot 6$, we have $x^{13} \equiv_5 x$ and $x^{13} \equiv_7 x$ by Example 4.9. A direct computation is also feasible because the numbers here are small. That is, $x^{13} = x^8 x^5 \equiv_5 x^8 x = x^9$ by Fermat's Theorem, then similarly $x^9 = x^4 x^5 \equiv_5 x^4 x \equiv_5 x^5 \equiv_5 x$. In any event, $5 \mid (x^{13} - x), 7 \mid (x^{13} - x)$, and $(5, 7) = 1$ result in $35 \mid (x^{13} - x)$ by Theorem 1.12. Thus $x^{13} \equiv_{35} x$ for all $x \in \mathbf{Z}$.

Example 4.11 Find the last three digits of $13^{3605603}$.

This problem is a complicated version of Example 1.19. Any $n \in \mathbf{N}$ is congruent to its last three digits modulo 1000 (why?), so we really want to solve $13^{3605603} \equiv_{1000} x$ with $0 \le x \le 999$. In Chapter 1 we would have tried to find $m \in \mathbf{N}$ with $13^m \equiv_{1000} 1$, but now we use Euler's Theorem. Since $(13, 1000) = 1$ and $\varphi(1000) = \varphi(2^3)\varphi(5^3) = 4 \cdot 5^2 \cdot 4 = 400$, we have $13^{400} \equiv_{1000} 1$ from Theorem 4.8. Thus Theorem 1.24 yields $13^{400k} = (13^{400})^k \equiv_{1000} 1$. Writing $3605603 = 9014 \cdot 400 + 3$, $13^{3605603} = (13^{400})^{9014} 13^3 \equiv_{1000} 13^3$ follows. A direct computation gives $13^3 = 2197$, so $13^{3605603} \equiv_{1000} 197$ and the last three digits of $13^{3605603}$ are 197. Similarly, since $13^3 = 2197$ and $\varphi(10000) = 4000$, the last four digits of $13^{12000003}$ would be 2197.

Example 4.12 For $p < q$ primes, $pq \mid (p^{q-1} + q^{p-1} - 1)$.

Now $(p, q) = 1$ since these are different primes, so Euler's Theorem shows that $p \mid (q^{p-1} - 1)$. But $p \mid p^{q-1}$ so $p \mid (p^{q-1} + q^{p-1} - 1)$ follows (Theorem 1.6). Similarly, $q \mid (p^{q-1} + q^{p-1} - 1)$ and using again that $(p, q) = 1$, Theorem 1.12 forces $pq \mid (p^{q-1} + q^{p-1} - 1)$.

A numerical illustration is $55 \mid (5^{10} + 11^4 - 1)$ or $143 \mid (11^{12} + 13^{10} - 1)$. Also for any odd prime p, $2p \mid (2^{p-1} + p - 1)$. For the next example let $R(n) = (10^n - 1)/9$, the integer with n digits, all of which are 1.

Example 4.13 Let $k \in \mathbf{N}$ with $(k, 10) = 1$. There are infinitely many $n \in \mathbf{N}$ with $k \mid R(n)$, and if also $(k, 30) = 1$ then the minimal such n divides $\varphi(k)$.

When $(10, k) = 1$ then $10^{\varphi(k)} \equiv_k 1$ by Euler's Theorem. The properties of congruences in Theorem 1.24 show $10^{\varphi(k)s} \equiv_k 1$ for all $s \in \mathbf{N}$ so for infinitely many $n \in \mathbf{N}$, $10^n \equiv_k 1$. Thus $k \mid (10^n - 1)$, or $k \mid 9 \cdot R(n)$. Now assume $(k, 30) = 1$, so $(k, 3) = 1$ and $k \mid R(n)$ by Theorem 1.12. But if $k \mid R(t)$ then $k \mid 9R(t)$, so $10^t \equiv_k 1$. Hence when $(k, 30) = 1$ then $k \mid R(t) \Leftrightarrow 10^t \equiv_k 1$, and this holds for infinitely many $t \in \mathbf{N}$. Also the minimal $m \in \mathbf{N}$ with $k \mid R(m)$ is the minimal $m \in \mathbf{N}$ so that $10^m \equiv_k 1$, which by definition is $o([10]_k)$ in U_k. Applying Theorem 4.7 shows $m \mid \varphi(k)$. Return to $(k, 10) = 1$ and write $k = 3^b \cdot m$ with $(3, m) = 1$. Clearly $(m, 30) = 1$ so $m \mid R(n)$ for infinitely many $n \in \mathbf{N}$. We claim that $3^s \mid R(3^s n)$, all $s, n \in \mathbf{N}$. The digits in $R(3n)$ sum to $3n$ so

$3 \mid R(3n)$ by Theorem 1.26. Using induction, let $3^{t-1} \mid R(3^{t-1}n)$ with $t > 1$ and write $R(3^t n) = R(3^{t-1}n)(10^{2\cdot 3^{t-1}n} + 10^{3^{t-1}n} + 1)$. The digits of the second factor sum to 3 so it is divisible by 3 using Theorem 1.26 again, and $3^{t-1} \mid R(3^{t-1}n)$ by the induction assumption. Thus $3^t \mid R(3^t n)$ so $3^s \mid R(3^s n)$ for all $s \in N$ by IND-I. But we may also write $R(3^t n) = R(n)((10^n)^{3^t-1} + \cdots + (10^n)^2 + 10^n + 1)$ so if $m \mid R(n)$, then $m \mid R(3^t n)$ as well. Therefore when $t = b$ both $3^b \mid R(3^b n)$ and $m \mid R(3^b n)$, forcing $k \mid R(3^b n)$ by Theorem 1.12 since $k = 3^b m$ and $(3^b, m) = 1$.

There is one more consequence of Lagrange's Theorem to record.

THEOREM 4.9 For any group $G \neq \langle e_G \rangle$ the following are equivalent:

i) If $H \leq G$ then either $H = G$ or $H = \langle e_G \rangle$.
ii) $G = \langle g \rangle$ for any $g \in G - \langle e_G \rangle$.
iii) $|G| = p$, a prime.

Proof If i) holds and $g \in G - \langle e_G \rangle$ then $\langle g \rangle \leq G$ and $\langle g \rangle \neq \langle e_G \rangle$, so i) forces $\langle g \rangle = G$. Next assume $G = \langle g \rangle$ for any $g \neq e_G$. If G is infinite and $G = \langle x \rangle$ then $x^2 \neq e_G$ so $G = \langle x^2 \rangle$. Since every $h \in G$ is power of x^2, in particular $x = (x^2)^k$ and it follows that $e = x^{2k-1}$. Hence $o(x) \in N$ and by Theorem 2.16, G must be finite. But now Theorem 3.1 on cyclic groups, together with ii) and $|G| > 1$, shows that $|G|$ is a prime. Finally, assume that iii) holds. If $H \leq G$ then Lagrange's Theorem yields $|H| \mid p$, forcing either $|H| = 1$ or $|H| = p$. In the first case $H = \langle e_G \rangle$, and in the second $H = G$, proving i) and the theorem. ■

EXERCISES

1. If $H \leq (Q, +)$, show that $H = Q$ or $[(Q, +) : H]$ is infinite. (If $q \notin H$ then $q/n \notin H$ for all $n \in N$, and also if $[(Q, +) : H] = m$, then $m! \cdot q \in H$ for all $q \in Q$.)
2. Show $(Q^+, \cdot)$ has proper subgroups of finite index. (Write $q = 2^{q(2)}3^{q(3)} \cdots p^{q(p)}$ with exponents in Z and consider $\{q \in Q \mid q(2) \in mZ$, some fixed $m \in N\}$.)
3. Let G be a cyclic group and $\langle e_G \rangle \neq H \leq G$. Show that $[G : H]$ is finite. Use this observation to show that the following groups are not cyclic:
   i) $Z \oplus Z$
   ii) $(Q, +)$
   iii) $(Q^+, \cdot)$
   iv) $\{c \in C \mid c^n = 1$, some $n \in N\}$
   v) $\bigcup_{k=0}^{\infty} \langle 2^{1/2^k} \rangle \leq (R^+, \cdot)$
   vi) $\{(a, b, c) \in (Z, +)^3 \mid a + b + c = 0\}$
4. If G is any group with an infinite strictly decreasing chain of subgroups $G \geq H_1 \geq H_2 \geq \cdots \geq H_n \geq \cdots$, $H_i \neq H_j$ if $i \neq j$, then $[G : \bigcap H_i]$ is infinite.
5. If G is a group with $H_1, \ldots, H_k \leq G$ and each $[G : H_j]$ is finite, show that $[G : \bigcap H_i]$ is at most the product of the $[G : H_j]$.
6. Let $H \leq K \leq G$. Show $[G : H]$ is finite $\Leftrightarrow$ both $[G : K]$ and $[K : H]$ are finite, in which case $[G : H] = [G : K][K : H]$, so $[G : K] \mid [G : H]$.(G may be infinite!)
7. If $H, K \leq G$ with $[G : K]$ finite, then $[H : H \bigcap K]$ is also finite.

8. Let G be an Abelian group so that for all $g \in G - \langle e \rangle$, $o(g)$ is infinite. Show that $x, y \in G - \langle e \rangle \Rightarrow \langle x \rangle \bigcap \langle y \rangle = \langle e \rangle$ or $\langle x \rangle \langle y \rangle = \langle w \rangle$. (Hint: If $y^k \in \langle x \rangle$ and $k = 1$ then $\langle x \rangle \langle y \rangle = \langle x \rangle$; if $k > 1$ and minimal with $y^k = x^m$, show that $(k, m) = 1$ and use $1 = sk + tm$ to find a generator for $\langle x \rangle \langle y \rangle$.)

9. If G is an infinite Abelian group so that for any $H \leq G$ with $H \neq \langle e \rangle$, $[G : H]$ is finite, show that G must be cyclic. (Use Exercise 8 and Exercise 6.)

10. If $H, K \leq G$ and $[G : H \bigcap K] = p$, a prime, then $H = K$, $H = G$, or $K = G$.

11. If $H \leq G$ and $g \in G$, then $g^{-1}Hg \leq G$ (Exercise 7 in §2.5).

i) Show that $[G : g^{-1}Hg] = [G : H]$. (See Exercise 8d in §2.2.)

ii) Let $H^{\#} = \bigcap_{g \in G} g^{-1}Hg$. Show that for all $x \in G$, $xH^{\#} = H^{\#}x$.

12. Let $H_1, \ldots, H_k, H \leq G$. If $H \subseteq \bigcup_{i=1}^{k} H_i g_i$ for $g_i \in G$, show for some $1 \leq j \leq k$, $[H : H \bigcap H_j]$ is finite. (For $h \in H$, $h \in H_i g_i$ and it follows that $H_i g_i$ may be replaced by $(H \bigcap H_i)h$ or is not needed in the union; use Exercise 14, §4.1).

13. If $k \in \mathbf{N}$ with $k \mid 24$, show that S_4 has a subgroup of order k.

14. For $n \geq 3$, if $k \mid 2n$ for $k \in \mathbf{N}$, show that D_n has a subgroup of order k.

15. For G a finite Abelian group, $n \in \mathbf{N}$, $A = \{g \in G \mid g^n = e_G\}$, and $K = \{g^n \mid g \in G\}$, show: a) $|A| \mid |G|$; b) $|K| \mid |G|$; and c) $|A||K| = |G|$ (relate K to $\{Ag \mid g \in G\}$).

16. Determine the index of each subgroup in the given group:

i) $\langle [3]_{61} \rangle$ in U_{61}

ii) $\langle [7]_{100} \rangle$ in U_{100}

iii) $\langle [11]_{100} \rangle$ in U_{100}

iv) $\langle W_1 \rangle$ in D_n

v) $\langle T_6 \rangle$ in D_{24}

vi) $L = \{\theta \in S_9 \mid \theta(\{2, 5, 7\}) = \{2, 5, 7\}\}$ in S_9

vii) $H = \left\{ \begin{bmatrix} a & b \\ 0 & c \end{bmatrix} \in GL(2, \mathbf{Z}_5) \mid a, c \in \langle [4]_5 \rangle \right\} \leq UT(2, \mathbf{Z}_5)$ (Definition 2.9)

viii) $K = \left\{ \begin{bmatrix} a & b \\ 0 & c \end{bmatrix} \in GL(2, \mathbf{Z}_7) \mid c = [1]_7 \right\}$ in $UT(2, \mathbf{Z}_7)$

ix) $3\mathbf{Z} \oplus 3\mathbf{Z}$ in $\mathbf{Z} \oplus \mathbf{Z}$

x) $SL(n, \mathbf{Z}_m)$ in $GL(n, \mathbf{Z}_m)$

xi) $SL(n, \mathbf{Z})$ in $GL(n, \mathbf{Z})$

xii) $SL(n, \mathbf{Q})$ in $GL(n, \mathbf{Q})$

xiii) $\{f(x) \in \mathbf{Z}_n[x] \mid f([1]_n) = [0]_n\}$ in $(\mathbf{Z}_n[x], +)$

xiv) $x^3\mathbf{Z}_n[x] = \{f(x) \in \mathbf{Z}_n[x] \mid f(x) = a_k x^k + \cdots + a_3 x^3 \text{ some } k \geq 3\}$ in $(\mathbf{Z}_n[x], +)$

17. For all $x \in \mathbf{Z}$, show each of the following:

i) $x^{19} \equiv x \pmod 7$

ii) $x^{17} \equiv x \pmod 5$

iii) $x^{34} \equiv x^4 \pmod{77}$

iv) $x^{185} \equiv x^{65} \pmod{62062}$

18. Find $m \in \mathbf{N}$ minimal so that $x^{21+m} \equiv x^{21} \pmod n$ holds for all $x \in \mathbf{Z}$ if $n =$:

i) 22

ii) 23

iii) 91

iv) 187

v) 1430

vi) $77531 = 31 \cdot 41 \cdot 61$

19. If $n, m \in N$ with $(n, m) = 1$, show that $nm \mid (n^{\varphi(m)} + m^{\varphi(n)} - 1)$.
20. For $p \neq q$ primes, show: a) $pq \mid (p^q + q^p - p - q)$; and b) $p^{q^3+q} \equiv p^2 \pmod{q}$.
21. Find the remainder when:
  i) 3^{95} is divided by 31
  ii) 15^{68} is divided by 67
  iii) 19^{578} is divided by 70
  iv) 6^{1719} is divided by 169
  v) 22^{53235} is divided by 105
  vi) $11^{40000000000}$ is divided by 37
22. Find the last three digits of: a) $2^{96,421}$; b) 7^{1608}; c) $17^{124,803}$.
23. If G is a finite group and $H_1, \ldots, H_k \le G$, show that LCM $\{|H_1|, \ldots, |H_k|\} \mid |G|$.
24. If $H \le G$ with G finite and $|G - H| \mid |G|$, show that $|G - H| = |H| = |G|/2$.
25. If G is a finite Abelian group, $H \le G$, $k \in N$, and $L = \{g \in G \mid g^k \in H\}$, show that $|L| \mid |G|$.
26. If G is a finite group, $H \le G$, $x \in G$, and $A = \{g \in G | gx \in H\}$, show $|A| \mid |G|$.

4.3 PRODUCTS OF SUBGROUPS

It is often useful to consider products of elements from different subgroups. We begin with a slightly more general situation.

DEFINITION 4.4 If G is a group and $S, T \subseteq G$ are not empty, then the **product** of S and T is $ST = \{st \in G \mid s \in S \text{ and } t \in T\}$.

We are interested mostly in products of cosets or of subgroups. Note that for $H \le G$ and $g \in G$, $Hg = H\{g\}$ is a product, and if $S, T \subseteq (G, +)$ the product $ST = S + T = \{s + t \mid s \in S \text{ and } t \in T\}$. Also observe that $ST \neq \langle\{st \in G \mid s \in S \text{ and } t \in T\}\rangle$ but is just the set of products. If $S_1, \ldots, S_k \subseteq G$ are nonempty then from Theorem 2.3, $S_1 S_2 \cdots S_k = \{s_1 s_2 \cdots s_k \mid s_i \in S_i\} = ((\cdots (S_1 S_2) \cdots S_{k-1}) S_k)$ is unambiguous: if $A, B, C \subseteq G$, $ABC = (AB)C = A(BC)$.

Can products of cosets be a coset? If $G = S_3$ and $H = \langle(1, 2)\rangle$, then $H(1, 3) = \{(1, 3), (1, 3, 2)\}$ and $H(2, 3) = \{(2, 3), (1, 2, 3)\}$, so $H(1, 3)H(2, 3) = \{(1, 3)(2, 3), (1, 3)(1, 2, 3), (1, 3, 2)(2, 3), (1, 3, 2)(1, 2, 3)\} = \{(1, 3, 2), (1, 2), (1, 3), I\}$, not another coset of $\langle(1, 2)\rangle$. When G is an Abelian group and $H \le G$, then using Theorem 2.3, $HaHb = H(\{a\}H)\{b\} = H(H\{a\})\{b\} = HHab = Hab$. The last equality follows since $H \le G \Rightarrow HH \subseteq H = H\{e\} \subseteq HH$. Hence the product of cosets of a subgroup of an Abelian group is always another coset! In U_{13}, if $H = \langle[12]\rangle = \{[1], [12]\}$, then $H[2]H[3] = H[6]$ from above, and computing directly shows $H[2]H[3] = \{[2][3], [2][10], [11][3], [11][10]\} = \{[6], [7]\} = H[6]$.

Some $H \le G$ non-Abelian may have products of cosets be a coset. The cosets of A_n in S_n are A_n and $S_n - A_n$, the even and the odd permutations, so $(S_n - A_n)(S_n - A_n) = A_n$, and $(S_n - A_n)A_n = A_n(S_n - A_n) = S_n - A_n$. For $H = \langle T_2\rangle$ in D_6, $HW_1HW_0 = \{W_1, W_3, W_5\}\{W_0, W_2, W_4\} = \{T_1, T_3, T_5\} = HT_1$, where we have used computations given in Theorem 3.17.

In general the product of subgroups is not a subgroup. For example, in any D_n, $\langle W_1\rangle\langle W_0\rangle = \{I, W_1\}\{I, W_0\} = \{I, W_1, W_0, T_1\}$ is not a subgroup since $T_1^2 = T_2 \notin \langle W_1\rangle\langle W_0\rangle$.

But in S_n, $\langle(1,2,3)\rangle\langle(1,2)\rangle = \{I, (1,2), (1,3), (2,3), (1,2,3), (1,3,2)\} =$ "S_3" is a subgroup. There is a useful technical condition that tells when the product of subgroups is a subgroup.

THEOREM 4.10 If $H, K \le G$ then:

i) $HH = H$
ii) $H, K \subseteq HK$
iii) $HK \le G \Leftrightarrow HK = KH$

Proof For i) $e_G \in H$ (Theorem 2.11) $\Rightarrow H = \{e_G\}H \subseteq HH$ and $HH \subseteq H$ since $H \le G$ so $H = HH$. Also $H = H\{e_G\} \subseteq HK$ and $K = \{e_G\}K \subseteq HK$ so ii) holds. In iii) $HK = KH \Rightarrow \{hk \mid h \in H \text{ and } k \in K\} = \{kh \mid h \in H \text{ and } k \in K\}$, *not that* $hk = kh$ for each choice of $h \in H$ and $k \in K$. First assume $HK = KH$. To show that $HK \le G$ we use Theorem 2.12. Clearly $e = ee \in HK \ne \emptyset$. To see that HK is closed under multiplication, let $hk, h'k' \in HK$, so $hkh'k' = h(kh')k'$. Now $kh' \in KH = HK$ so $kh' = h''k''$ for some $h'' \in H$ and $k'' \in K$. Thus $hkh'k' = hh''k''k' \in HK$ since $H, K \le G$. For inverses, if $hk \in HK$ then $(hk)^{-1} = k^{-1}h^{-1} \in KH = HK$, so HK contains the inverse of each of its elements and $HK \le G$ by Theorem 2.12. Finally, assume that $HK \le G$ so for $hk \in HK$, $(hk)^{-1} \in HK$ also, say $(hk)^{-1} = h'k'$. Now $hk = ((hk)^{-1})^{-1} = (h'k')^{-1} = (k')^{-1}(h')^{-1} \in KH$, so $HK \subseteq KH$. But $H, K \subseteq HK$ by ii) and $HK \le G$ so $KH \subseteq (HK)(HK) = HK$ using i). The two inclusions prove that $HK = KH$, finishing the proof. ∎

The condition $HK = KH$ certainly holds for Abelian groups since $xy = yx$ for all x and y. We record this observation and a slight generalization of it. Recall that $Z(G) = \{x \in G \mid xg = gx \text{ for all } g \in G\}$ is the center of G.

THEOREM 4.11 For a group G:

i) $H \le Z(G)$ and $K \le G \Rightarrow HK \le G$
ii) G Abelian and $H_1, \ldots, H_m \le G \Rightarrow H_1 \cdots H_m \le G$

Proof If $h \in Z(G)$ then $hg = gh$ for all $g \in G$ so $HK = KH$ when $H \le Z(G)$ by definition of product, and $HK \le G$ by Theorem 4.10. When G is Abelian, $k \ge 1$, and $H_j \le G$, assume $H_1 \cdots H_k \le G$. Given $H_1, \ldots, H_{k+1} \le G$, by our assumption $K = H_1 \cdots H_k \le G = Z(G)$, so $H_1 \cdots H_{k+1} = KH_{k+1} \le G$ by i). Thus any finite product of subgroups of G is a subgroup by IND-I. ∎

When $H, K \le G$ are finite and $HK \le G$, what is the relation among the orders of H, K, and HK? Our next result answers this important question.

THEOREM 4.12 If $H, K \le G$ are finite, then $|HK| = |H||K|/|H \cap K|$. In particular, if either $H \cap K = \langle e_G\rangle$ or $(|H|, |K|) = 1$ then $|HK| = |H||K|$.

Proof We need not assume that G is finite or $HK \le G$. Write $HK = \bigcup Hk$ over $k \in K$. Since K is finite, $\{Hk \mid k \in K\}$ has $s \in N$ distinct cosets so choose $Hk_1, \ldots, Hk_s$ with

$\{Hk_1, \ldots, Hk_s\} = \{Hk \mid k \in K\}$. Thus $HK = \bigcup Hk_i$ and since $\{Hk_j\}$ is pairwise disjoint by Theorem 4.2, $|HK| = |Hk_1| + \cdots + |Hk_s| = s|H|$ using Theorem 1.3 and Theorem 4.4, so we need $|K|/|H \cap K| = s$. Now $H \cap K \le K$ by Theorem 2.14 so Lagrange's Theorem yields $[K : H \cap K| = |K|/|H \cap K|$. Hence we want $s = [K : H \cap K]$. Certainly $(H \cap K)k_i$ is a coset of $H \cap K$ in K, and if $(H \cap K)k_i = (H \cap K)k_j$ then $k_i(k_j)^{-1} \in H \cap K$ (why?). But $k_i(k_j)^{-1} \in H \Leftrightarrow Hk_i = Hk_j$ (Theorem 4.2) and $\{Hk_t\}$ distinct forces $i = j$. Therefore $\{(H \cap K)k_1, \ldots, (H \cap K)k_s\}$ contains s distinct cosets of $H \cap K$ in K. For $k \in K \subseteq HK$, using that $\{Hk_t\}$ partitions HK, $k \in Hk_j$ for some k_j, so $kk_j^{-1} \in H$ (why?). But now $kk_j^{-1} \in H \cap K$ and $k \in (H \cap K)k_j$. This proves $\{(H \cap K)k \mid k \in K\} = \{(H \cap K)k_1, \ldots, (H \cap K)k_s\}$ so $s = [K : H \cap K]$. For the final statement, if $H \cap K = \langle e_G \rangle$ then $|H \cap K| = 1$ so $|HK| = |H||K|$ by the first statement of the theorem. The condition $(|H|, |K|) = 1$ forces $H \cap K = \langle e_G \rangle$ by Theorem 4.6 and so results in $|HK| = |H||K|$. ∎

Let us see how the last two theorems can be used. Computing in D_6, $H = \langle T_3 \rangle \langle W_1 \rangle = \{I, T_3, W_1, W_4\} = \langle W_1 \rangle \langle T_3 \rangle$, so $H \le D_6$. Now $H \cap \langle W_2 \rangle = \langle I \rangle$ so $|H\langle W_2 \rangle| = |H||\langle W_2 \rangle| = 8$ (why?). But $|D_6| = 12$ and $8 \nmid 12$ so $H\langle W_2 \rangle$ cannot be a subgroup of D_6 by Lagrange's Theorem. In particular $H\langle W_2 \rangle \ne \langle W_2 \rangle H$. For $L = \langle (1, 2)(3, 4, 5) \rangle$, $K = \langle (1, 2, 3)(4, 5) \rangle \le S_5$, L and K are both cyclic of order 6 and intersect in I (compute their powers!). Hence $|LK| = 36$, not dividing $120 = |S_5|$, so LK is not a subgroup of S_5, and $LK \ne KL$. Finally, let $U = \langle (1, 2, 3, 4, 5, 6) \rangle$, $V = \langle (1, 4, 3, 6, 5, 2) \rangle \le S_6$, cyclic of order 6. Computing powers of the two generators shows $U \cap V = \{I, (1, 3, 5)(2, 4, 6), (1, 5, 3)(2, 6, 4)\}$, so $|U \cap V| = 3$. Theorem 4.12 gives $|UV| = 6 \cdot 6/3 = 12$, and $U \subseteq UV$, so $UV = U \cup U(1, 4, 3, 6, 5, 2)$ containing 12 elements (why?). A quick check shows that the latter coset contains $(1, 5, 3)$, but its inverse $(1, 3, 5)$ is in neither coset, so UV cannot be a subgroup of S_6.

Example 4.14 If $H, K \le G$, a finite cyclic group, $|HK| = \text{LCM}\{|H|, |K|\}$.

From Theorem 3.1, every subgroup of a cyclic group is cyclic and there is a unique subgroup for every divisor of the order of the group. Certainly $d = \text{GCD}\{|H|, |K|\}$ divides both $|H|$ and $|K|$ and since $d \mid |G|$, there is $L \le G$ with $|L| = d$. Each of H and K must have a subgroup of this same order and there is only one such subgroup in G so $L \le H \cap K$. But by Theorem 4.6, $|H \cap K| \mid \text{GCD}\{|H|, |K|\}$, so $|H \cap K| \mid d$ and it follows that $L = H \cap K$. Applying Theorem 4.12 yields $|HK| = |H||K|/\text{GCD}\{|H|, |K|\} = \text{LCM}\{|H|, |K|\}$ by the corollary to Theorem 1.19. Of course $HK \le G$ by Theorem 4.11.

For a nontrivial illustration let $\langle [54] \rangle, \langle [84] \rangle \le \mathbf{Z}_{216}$. By Theorem 2.8, in $\mathbf{Z}_{216} o([54]) = 216/(54{,}216) = 4$ and $o([84]) = 216/(84{,}216) = 18$ so $|\langle [54] \rangle| = 4$ and $|\langle [84] \rangle| = 18$ using Theorem 2.16. Since $\text{LCM}\{4, 18\} = 36$, the "product" $\langle [54] \rangle + \langle [84] \rangle = \langle [6] \rangle$, the unique subgroup of order 36 in $\mathbf{Z}_{216} (216 = 6^3)$. This says that in $\mathbf{Z}_{216}$ every "multiple of 6" is the sum of a multiple of 54 and a multiple of 84. Note that $(84, 54) = 6$, so using the Euclidean Algorithm we can write $6 = 2 \cdot 84 + (-3) \cdot 54$ and in $\mathbf{Z}_{216}$ get $[6] = [2 \cdot 84] + [-3 \cdot 54] = [2 \cdot 84] + [54]$ $(o([54]) = 4)$. This representation is not unique since also $[6] = [11 \cdot 84] + [3 \cdot 54]$.

EXERCISES

1. In each case, show that the product of any two cosets is again a coset:
   i) $H \le G$ for $H \le Z(G)$
   ii) $\langle T_2\rangle \le D_{10}$
   iii) $H = \{I, (a, b)(c, d)\} \le S_4$ $(|H| = 4)$

2. Are the following products subgroups?
   i) $\langle T_2\rangle\langle W_1\rangle$ in D_8
   ii) $\langle T_m\rangle\langle W_0\rangle$ in D_{nm}
   iii) $\langle \alpha\rangle\langle \beta\rangle$ in S_{12} for $\alpha, \beta \in S_{12}$ any 7-cycles

3. Let $H, K, L \le G$ with G Abelian. If $H \cap K = \langle e\rangle$ and $HK \cap L = \langle e\rangle$, show that $H \cap (KL) = \langle e\rangle$.

4. If $H, K \le G$ with G finite and $|H||K| > |G|$, show that $H \cap K \ne \langle e\rangle$.

5. If G is a group of order 315 and if $H, K \le G$ with $|H| = 45$ and $|K| = 63$, show that there is $L \le G$ with $|L| = 9$.

6. If $H, K \le G$ with $|G| = 300$, find the possibilities below for $|H \cap K|$:
   a) $|H| = 75$, $|K| = 60$
   b) $|H| = 75$, $|K| = 100$
   c) $|H| = 100$, $|K| = 60$
   d) $|H| = 100 = |K|$
   e) $|H| = 60 = |K|$
   f) $|H| = 12$, $|K| = 60$

7. If $H \le G$ and $N(H) = \{g \in G \mid g^{-1}Hg = H\} \le G$ (Exercise 7 in §2.5), show that for any $K \le N(H)$, $HK \le G$.

8. For $n > 2$ in S_n, fix $1 \le i \le n$ and set $H_i = \{\theta \in S_n \mid \theta(i) = i\}$. Show $H_i \le S_n$. If $1 \le i < j \le n$, find $|H_iH_j|$. Can $H_iH_j = H_jH_i$?

9. In D_{2m}, if $0 \le i < m$ show that $K_i = \langle W_i\rangle\langle T_m\rangle \le D_{2m}$. When is $K_iK_j \le D_{2m}$?

10. In S_8 find the number of elements in the following products:
   a) $\langle(1, 2, 3, 4, 5)(6, 7, 8)\rangle\langle(1, 2, 3, 4)(5, 6, 7, 8)\rangle$
   b) $\langle(1, 2, 3)(4, 5, 6, 7, 8)\rangle\langle(1, 2, 3, 4, 5)(6, 7, 8)\rangle$
   c) $\langle(1, 2, 3, 4)(5, 6)(7, 8)\rangle\langle(1, 4, 3, 2)(5, 6)\rangle$
   d) $\langle(1, 2, 3, 4, 5, 6, 7, 8)\rangle\langle(1, 4, 3, 2, 5, 8, 7, 6)\rangle$
   and for $K = \{\theta \in S_8 \mid \theta(5) = 5, \theta(6) = 6, \theta(7) = 7, \theta(8) = 8\} \le S_8$
   e) $K\langle(1, 2, 3, 4)(5, 6, 7, 8)\rangle$
   f) $K\langle(1, 2, 3)(4, 5)(6, 7)\rangle$

11. Let G be a group, $|G| = p^t m$ for p a prime and $(p, m) = 1$, and $H \le G$ with $|H| = p^t$. If $K \le G$ with $|K| = p^s$ and $HK = KH$, then show that $K \le H$.

12. If G is an Abelian group, $|G| = p^n$ for p a prime, and $1 \le s \le n$, show there is $H \le G$ with $|H| = p^s$. (Use induction on n. If $|G| = p^{n+1}$ but G has no subgroup of order p^n, let $L \le G$ have maximal order $p^m < p^n$. Show that for $g \in G$ and $o(g) = p$, $g \in L$, and then show that $L = G$.)

4.4 PRODUCTS IN ABELIAN GROUPS

We prove two technical theorems on cardinality that build on the results developed above. Eventually we want to represent some groups as direct sums of subgroups. The initial step will look at products of subgroups, and then it will be very helpful to have the information developed here.

THEOREM 4.13 Let $H_1, H_2, \ldots, H_m \leq G$ and finite so that the products $H_1 H_2 \cdots H_k \leq G$ for any $1 \leq k \leq m$. Then the following hold:

i) If $\{|H_j|\}$ is pairwise relatively prime, then $|H_1 H_2 \cdots H_m| = |H_1||H_2| \cdots |H_m|$.
ii) If for some $n \in N$, $(|H_j|, n) = 1$ for all $1 \leq j \leq m$, then $(|H_1 H_2 \cdots H_m|, n) = 1$.
iii) If for all $1 \leq j \leq m$, $|H_j| = p^{a(j)}$ for a fixed prime p, then $|H_1 H_2 \cdots H_m| = p^t$.

Proof We use IND-I on ii) and iii) and Theorem 0.7 on induction in finite sets for part i). Certainly $|H_1| = |H_1|$, so assume $\{|H_j|\}$ is pairwise relatively prime and $|H_1 H_2 \cdots H_k| = |H_1||H_2| \cdots |H_k|$ for some $1 \leq k < m$. Since $(|H_j|, |H_{k+1}|) = 1$ for $j \leq k$, our induction assumption and Theorem 1.11 (and induction!) imply that $(|H_1 H_2 \cdots H_k|, |H_{k+1}|) = (|H_1||H_2| \cdots |H_k|, |H_{k+1}|) = 1$. Now $H_1 H_2 \cdots H_k \leq G$ by assumption, so Theorem 4.12 applies to show that $|H_1 H_2 \cdots H_k H_{k+1}| = |H_1 H_2 \cdots H_k||H_{k+1}| = |H_1||H_2| \cdots |H_k||H_{k+1}|$, proving i) by Theorem 0.7.

For ii) there is nothing to prove when $m = 1$. Assume for any $K_1, \ldots, K_s \leq G$ satisfying $K_1 K_2 \cdots K_t \leq G$ for any $t \leq s$ and with all $(|K_j|, n) = 1$, that $(|K_1 K_2 \cdots K_t|, n) = 1$. Let $H_1, \ldots, H_{s+1} \leq G$ satisfy $H_1 H_2 \cdots H_k \leq G$ for any $k \leq s+1$ and $(|H_j|, n) = 1$ for all $j \leq s+1$. Certainly these same conditions hold for $\{H_1, \ldots, H_s\}$ so by the induction assumption, $(|H_1 H_2 \cdots H_s|, n) = 1$. Using $H_1 H_2 \cdots H_s \leq G$ and Theorem 4.12, $|H_1 H_2 \cdots H_s H_{s+1}||(H_1 H_2 \cdots H_s) \cap H_{s+1}| = |H_1 H_2 \cdots H_s||H_{s+1}|$. As we saw above, the assumptions $(|H_1 H_2 \cdots H_s|, n) = 1$ and $(|H_{s+1}|, n) = 1$ show that $(|H_1 H_2 \cdots H_s||H_{s+1}|, n) = 1$ using Theorem 1.11. It follows that $(|H_1 H_2 \cdots H_s H_{s+1}||(H_1 H_2 \cdots H_s) \cap H_{s+1}|, n) = 1$, and Theorem 1.11 implies that $(|H_1 H_2 \cdots H_s H_{s+1}|, n) = 1$, proving ii) by IND-I.

The case $m = 1$ is trivial for iii) so assume iii) holds for some $m \geq 1$ and any m subgroups with the properties described. Given $H_1, \ldots, H_{m+1} \leq G$ with $H_1 H_2 \cdots H_k \leq G$ for any $k \leq m+1$ and $|H_j| = p^{a(j)}$ for each $j \leq m+1$, the induction assumption applied to $\{H_1, \ldots, H_m\}$ yields $|H_1 H_2 \cdots H_m| = p^s$. Since $H_1 H_2 \cdots H_m \leq G$ by assumption, by using Theorem 4.12 we can write $|H_1 H_2 \cdots H_m H_{m+1}||(H_1 H_2 \cdots H_m) \cap H_{m+1}| = |H_1 H_2 \cdots H_m||H_{m+1}| = p^s p^{a(m+1)} = p^r$. Unique factorization in N forces $|H_1 H_2 \cdots H_{m+1}| = p^c$ and proves iii) by IND-I. ■

The basic hypothesis that any $H_1 H_2 \cdots H_k \leq G$ in Theorem 4.13 will hold when G is Abelian by Theorem 4.11. We use this result and our next one to obtain some interesting consequences for Abelian and cyclic groups. More sophisticated results will be given in Chapter 7. First we introduce some notation for subsets of elements of prime power order in a group.

DEFINITION 4.5 For G a group and p a prime, set $G_p = \{g \in G \mid o(g) = p^t, \text{ some } t \geq 0\}$.

Note that $e_G \in G_p$ since $o(e_G) = p^0$. In general G_p need not be a subgroup, but it is when G is Abelian, as we will see. For example, $(S_5)_3$ in S_5 consists of all twenty 3-cycles and the identity. This is not a subgroup as $(1, 2, 3)(2, 3, 4) = (1, 2)(3, 4) \notin (S_5)_3$, or by Lagrange's Theorem since $21 \nmid 120$ $(= o(S_5))$. Now $\langle (D_n)_2 \rangle = D_n$, (see Exercise 4 in §3.4) but $(D_n)_2$ cannot be a subgroup unless $n = 2^k$ (why?). On the other hand, $(D_{10})_5 = \langle T_2 \rangle \leq D_{10}$. Since for $g \in G$, $o(g) \mid |G|$ by Theorem 4.7, if $(|G|, p) = 1$ then $G_p = \langle e_G \rangle$. Some examples in Abelian groups are; $(\mathbf{Z}_{24})_3 = \langle [8]_{24} \rangle$ and $(\mathbf{Z}_{24})_2 = \langle [3]_{24} \rangle$ using Theorem 3.1;

$(U_8)_2 = U_8$; $(U_{13})_2 = \{[1], [5], [8], [12]\}$; and $(\mathbf{Z}_9 \oplus \mathbf{Z}_{12})_3 = \langle [1]_9 \rangle \oplus \langle [4]_{12} \rangle$ using Theorem 3.20 on orders of elements in direct sums.

Next we describe the relation between $\{G_p\}$ and G for Abelian groups. The proof is a good illustration of how a number of earlier results can be combined to give new and nontrivial information. The argument could be simplified by using results (e.g., Cauchy's Theorem) we will prove later, but it is interesting that these results can be avoided here. We do not yet know that if a prime p divides $|G|$ then there is $g \in G$ with $o(g) = p$ (Cauchy's Theorem), so it might be that $G_p = \langle e \rangle$ when $p \mid |G|$. The next theorem shows that this is impossible for Abelian groups.

THEOREM 4.14 Let G be a finite Abelian group with $|G| = p_1^{a_1} \cdots p_k^{a_k}$ for $p_1 < \cdots < p_k$ primes, and set $G_i = G_{p_i}$. Then $G_i \leq G$; $|G_i| = p_i^{a_i}$; for $s < k$ both $|G_1 \cdots G_s| = p_1^{a_1} \cdots p_s^{a_s}$ and $(G_1 \cdots G_s) \cap G_{s+1} = \langle e_G \rangle$; and $G = G_1 \cdots G_k$.

Proof If $x, y \in G_i$ then by definition $o(x) = p^a$ and $o(y) = p^b$ where we write p for p_i. If $c = \max\{a, b\}$ then $p^a \mid p^c$ and $p^b \mid p^c$ so it follows from Theorem 2.7 that $x^{p^c} = e = y^{p^c}$. Use Theorem 2.4 and $xy = yx$ to write $(xy)^{p^c} = x^{p^c} y^{p^c} = e$, forcing $xy \in G_i$, again by Theorem 2.7; G_i is closed under multiplication and is finite so $G_i \leq G$ by Theorem 2.12. Let $G_i = \{g_1, \ldots, g_m\}$. Since G is Abelian and $\langle g_j \rangle \leq G_i$, $\langle g_1 \rangle \cdots \langle g_m \rangle \leq G_i$ by Theorem 4.11, and Theorem 4.10 implies that each $g_j \in \langle g_j \rangle \subseteq \langle g_1 \rangle \cdots \langle g_m \rangle$ so $G_i = \langle g_1 \rangle \cdots \langle g_m \rangle$. But $g_i \in G_i$ means $o(g_i) = p^{v(i)}$ so $|\langle g_j \rangle| = p^{v(i)}$ using Theorem 2.16 and $|G_i| = |\langle g_1 \rangle \cdots \langle g_m \rangle| = p_i^{d_i}$ follows from Theorem 4.13. Note that $d_i \leq a_i$ since $|G_i| \mid |G|$ by Lagrange's Theorem. Clearly $\{|G_j|\} = \{p_j^{d_j}\}$ is pairwise relatively prime. When $s \leq k$, $|G_1 G_2 \cdots G_s| = p_1^{d_1} \cdots p_s^{d_s}$ by Theorem 4.13, and if $s < k$, $(G_1 \cdots G_s) \cap G_{s+1} = \langle e_G \rangle$ by Theorem 4.6. It remains to show that $G = G_1 \cdots G_k$ and that $d_i = a_i$. Since $|G_1 G_2 \cdots G_k| = p_1^{d_1} \cdots p_k^{d_k}$, $d_i = a_i$ will follow from $G = G_1 \cdots G_k$.

To show $G = G_1 \cdots G_k$ take $h \in G$ and observe from Theorem 4.7 that $o(h) \mid |G|$, so $o(h) = p_1^{b_1} \cdots p_k^{b_k}$ with each $b_j \geq 0$. If $g \in G$ and $o(g) = p_1^{b_1}$, then $g \in G_1 \subseteq G_1 \cdots G_k$. Proceed by induction, assuming that if $g \in G$ with $o(g) = p_1^{b_1} \cdots p_m^{b_m}$ for $1 \leq m < k$, then $g \in G_1 \cdots G_k$, and take $g \in G$ with $o(g) = p_1^{b_1} \cdots p_m^{b_m} p_{m+1}^{b}$. If $b = 0$ our induction assumption shows that $g \in G_1 \cdots G_k$, so assume $b \geq 1$. Let $q = p_1^{b_1} \cdots p_m^{b_m}$ and note that $(q, p_{m+1}^b) = 1$. It follows from Theorem 1.10 that $up_{m+1}^b + vq = 1$ for $u, v \in \mathbf{Z}$, so $g = g^1 = g^{qv} g^{up_{m+1}^b}$. Now Theorem 2.8 implies first that $o(g^q) = p_{m+1}^b$ and $o(g^{p_{m+1}^b}) = q$ and then that $o(g^{vq}) = o((g^q)^v) = p_{m+1}^r$ and $o(g^{up_{m+1}^b}) \mid q$. By Definition 4.5, $g^{vq} \in G_{m+1}$, and by the induction assumption, $g^{up_{m+1}^b} \in G_1 \cdots G_k$ so $g \in (G_1 \cdots G_k) G_{m+1} \subseteq G_1 \cdots G_k$. Thus $g \in G \Rightarrow g \in G_1 \cdots G_k$ by IND-I, so $G = G_1 \cdots G_k$, completing the proof. We sketch another approach showing that $G = G_1 \cdots G_k$ (see §1.4 Exercise 1). Write $o(g) = p_1^{b_1} \cdots p_k^{b_k}$ and set $h_j = o(g)/p_j^{b_j}$. Then show that $\{h_j\}$ is relatively prime and use Theorem 1.8 to write $1 = u_1 h_1 + \cdots + u_k h_k$. Next observe that $g^{h_i u_i} \in G_i$ and $g = g^{h_1 u_1} \cdots g^{h_k u_k}$, so it follows that $g \in G_1 \cdots G_k$. ■

A consequence of Theorem 4.14 is that if G is an Abelian group with $|G| = p^t m$, for p a prime and $(p, m) = 1$, then $G_p = \{g \in G \mid o(g) = p^k\}$ is the unique subgroup of G of order p^t. Note that S_5 has order $5 \cdot 24$ but has many subgroups of order 5, each some $\langle (a, b, c, d, e) \rangle$. On the other hand $\langle T_1 \rangle \leq D_5$ is the unique subgroup of order 5. A major result called Sylow's Theorem states that any group G with $|G| = p^t m$ as above must have some subgroup of order p^t. We prove this in a later chapter.

EXERCISES

1. In each case find G_p explicitly for each prime divisor of $|G|$:
  i) $G = \mathbf{Z}_{30}$
  ii) $G = \mathbf{Z}_{36}$
  iii) $G = \mathbf{Z}_{81}$
  iv) $G = U_{21}$
  v) $G = U_{25}$
  vi) $G = U_{30}$
  vii) $G = U_{27}$

2. (Cauchy's Theorem for Abelian groups) If G is a finite Abelian group and $p \mid |G|$ for a prime p, show that there is $g \in G$ with $o(g) = p$.

3. If G is a group, $|G| = pq$ for $p < q$ primes, and if $x \in G$ with $o(x) = q$, show that $G_q = \langle x \rangle$. Thus $\langle x \rangle$ is the only subgroup of G of order q.

4. Let G be a group with $|G| = pq$ for $p < q$ primes.
  i) Show that $G \neq G_q$.
  ii) If $(p-1) \nmid (q-1)$, show $G_q \neq \langle e_G \rangle$ (Note that $pq - 1 = q(p-1) + (q-1)$.)

5. Let G be an Abelian group with $|G| = p_1^{a(1)} p_2^{a(2)} \cdots p_k^{a(k)}$ for $p_1 < \cdots < p_k$ primes. For any $H \leq G$, show that $H = (H \cap G_{p_1}) \cdots (H \cap G_{p_k})$.

6. If G is an Abelian group, $|G| = n$, and $m \mid n$ for $m \in N$, show that there is $H \leq G$ with $|H| = m$. (Use Exercise 12 in §4.3.)

4.5 CAUCHY'S THEOREM AND CYCLIC GROUPS

The material above leads to nontrivial, interesting, and useful results on when Abelian groups must be cyclic. Our first observation is a pretty theorem of Cauchy that we mentioned above, but for now can prove only for Abelian groups. Other proofs using different techniques will be given in later chapters. That Cauchy's Theorem for Abelian groups follows just from Theorem 4.14 is interesting in itself and illustrates how the slow accretion of elementary ideas can lead to nice and nontrivial results.

THEOREM 4.15 (Cauchy for Abelian groups) If G is a finite Abelian group and $p \mid |G|$ for p prime, then there is $x \in G$ with $o(x) = p$, so $|\langle x \rangle| = p$ also.

Proof Let $|G| = p^t m$ with $(p, m) = 1$ and note that $t \geq 1$ since $p \mid |G|$. By Theorem 4.14, $|G_p| = p^t$ so there is $y \in G$ with $o(y) = p^s$ for some $1 \leq s \leq t$. But now Theorem 2.8 yields $o(y^{p^{s-1}}) = p$, then $|\langle y^{p^{s-1}} \rangle| = p$ by Theorem 2.16. ∎

Suppose that G is Abelian and $|G|$ is a product of *distinct* primes. From Cauchy's Theorem there are $x_p \in G$ with $o(x_p) = p$ for each prime $p \mid |G|$ and we would like to say something about the order of their product. Our next result shows that this product has order equal to $|G|$, so G must be cyclic.

THEOREM 4.16 Let G be a group and $x_1, x_2, \ldots, x_k \in G$ with all $x_i x_j = x_j x_i$. If $\{o(x_i)\}$ is pairwise relatively prime, then $o(x_1 x_2 \cdots x_k) = o(x_1)o(x_2) \cdots o(x_k)$.

Proof We use induction on $k \geq 2$. When $k = 2$ denote the two elements by x and y. Since $xy = yx$, by Theorem 2.4 $(xy)^{o(x)o(y)} = (x^{o(x)})^{o(y)}(y^{o(y)})^{o(x)} = ee = e$. It follows from Theorem 2.7 that $o(xy) \mid o(x)o(y)$. The definition of order shows that $e = (xy)^{o(xy)} = x^{o(xy)}y^{o(xy)}$, using $xy = yx$ and Theorem 2.4 again, and this equality yields $x^{o(xy)} = y^{-o(xy)} \in \langle x \rangle \cap \langle y \rangle$. Now $|\langle x \rangle| = o(x)$ and $|\langle y \rangle| = o(y)$ by Theorem 2.16 so GCD$\{|\langle x \rangle|, |\langle y \rangle|\} = 1$, and Theorem 4.6 forces $\langle x \rangle \cap \langle y \rangle = \langle e \rangle$. Thus $x^{o(xy)} = y^{-o(xy)} = e = y^{o(xy)}$, implying $o(x)|o(xy)$ and $o(y) \mid o(xy)$ using Theorem 2.7 again. Finally, since GCD$\{o(x), o(y)\} = 1$, Theorem 1.12 yields $o(x)o(y) \mid o(xy)$, resulting in $o(xy) = o(x)o(y)$ (why?). Now assume the theorem holds for $k \geq 2$ elements satisfying the hypotheses and suppose that $x_1, \ldots, x_{k+1} \in G$ with all $x_i x_j = x_j x_i$ and $\{o(x_i)\}$ pairwise relatively prime. By the induction assumption, $o(x_1 x_2 \cdots x_k) = o(x_1)o(x_2) \cdots o(x_k)$, so it follows from Theorem 1.11 (and induction) that GCD$\{o(x_1 x_2 \cdots x_k), o(x_{k+1})\} = 1$. Also $x_i x_j = x_j x_i$ implies $(x_1 x_2 \cdots x_k)x_{k+1} = x_{k+1}(x_1 x_2 \cdots x_k)$ (exercise!) and we can apply the case $k = 2$. Therefore $o(x_1 x_2 \cdots x_{k+1}) = o((x_1 x_2 \cdots x_k)x_{k+1}) = o(x_1 x_2 \cdots x_k)o(x_{k+1}) = o(x_1)o(x_2) \cdots o(x_k)o(x_{k+1})$ by our assumption, so the theorem holds by IND-I. ■

The assumptions in Theorem 4.16 are crucial for its conclusion. When $xy \neq yx$ in G nothing much can be said about $o(xy)$. For example, in S_5, $\alpha\beta = (1, 2, 3, 4, 5)(1, 4, 3, 2) = (1, 5)$, $o(\alpha) = 5$, $o(\beta) = 4$, but $o(\alpha\beta) = 2$. More extreme in $GL(2, \mathbf{Z})$ is to take $A = \begin{bmatrix} 0 & -1 \\ 1 & 0 \end{bmatrix}$ and $B = \begin{bmatrix} 0 & 1 \\ -1 & -1 \end{bmatrix}$. Then $AB = \begin{bmatrix} 1 & 1 \\ 0 & 1 \end{bmatrix}$ has infinite order even though $o(A) = 4$ and $o(B) = 3$ (verify!). When $xy = yx$ but $(o(x), o(y)) \neq 1$, xy might have various orders, although the first part of the proof of Theorem 4.16 shows that $o(xy) \mid o(x)o(y)$. Of course $gg^{-1} = e$ has order 1 but $o(g) = o(g^{-1})$ could be anything, and if $xy = yx$ with $(o(x), o(y)) = 1$ then for $z = xy$, $zy^{-1} = x$, so $o(zy^{-1}) = o(x) = o(z)/o(y)$. For a specific example in $\mathbf{Z}_{24}$, $o([2]) = 12$, $o([3]) = 8$, and $o([2] + [3]) = o([5]) = 24$. Finally, using $U_{37} = \langle [2]_{37} \rangle$ (show that $2^{18} \equiv -1 \pmod{37}$), we have $o([2]_{37}) = 36$. Now by Theorem 2.8, $o([2]^6) = 6$, $o([2]^4) = 9$, and $o([2]^{10}) = 36/(10, 36) = 18$.

We come to a condition on $|G|$ that forces G to be cyclic when it is Abelian. It is a simple condition, but the result is far from obvious by just considering the definitions of group and of "cyclic." Recall Definition 4.5 that for a prime p, $G_p = \{g \in G \mid o(g) = p^t$ for some $t \geq 0\}$.

THEOREM 4.17 If G is a finite Abelian group, then G is cyclic $\Leftrightarrow G_p$ is cyclic for each prime divisor p of $|G|$. Thus if $|G| = p_1 p_2 \cdots p_k$ for $p_1 < \cdots < p_k$ primes, then G is a cyclic group.

Proof If G is cyclic then since each $G_p \leq G$ (Theorem 4.14), G_p is cyclic by Theorem 3.1. Next assume $|G| = p_1^{a_1} \cdots p_k^{a_k}$ for $p_1 < \cdots < p_k$ primes and that each $G_i = G_{p_i} = \langle x_i \rangle$. From Theorem 4.14, $|G_i| = p_i^{a_i}$ so $o(x_i) = p_i^{a_i}$ as well by Theorem 2.16. Clearly $\{o(x_i) \mid 1 \leq i \leq k\}$ is pairwise relatively prime and because G is Abelian, Theorem 4.16 yields $o(x_1 x_2 \cdots x_k) = p_1^{a_1} \cdots p_k^{a_k}$. But again $|\langle x_1 x_2 \cdots x_k \rangle| = o(x_1 x_2 \cdots x_k) = |G|$, and $\langle x_1 x_2 \cdots x_k \rangle \leq G$, so $G = \langle x_1 x_2 \cdots x_k \rangle$ is cyclic. When all the exponents $a_i = 1$, each

$|G_i| = p_i$ so each $G_i = \langle x_i \rangle$ for any $x_i \in G_i - \langle e_G \rangle$ by Theorem 4.9 and G is cyclic by the first statement. ■

Theorem 4.17 gives another proof of Theorem 3.21 and helps in showing that certain direct sums are cyclic. In $G = \mathbf{Z}_5 \oplus \mathbf{Z}_7 \oplus \mathbf{Z}_9$, using the multiplication in direct sums or Theorem 3.20, $G_3 = \langle([0]_5, [0]_7, [1]_9)\rangle$, $G_5 = \langle([1]_5, [0]_7, [0]_9)\rangle$, and $G_7 = \langle([0]_5, [1]_7, [0]_9)\rangle$, so G is cyclic by Theorem 4.17 and has generator the sum $([1]_5, [1]_7, [1]_9)$ of the generators of the G_p. The same conclusion holds for direct sums like $H = \mathbf{Z}_{18} \oplus \mathbf{Z}_{25}$ since $H_2 = \langle([9]_{18}, [0]_{25})\rangle$, $H_3 = \langle([2]_{18}, [0]_{25})\rangle$, and $H_5 = \langle([0]_{18}, [1]_{25})\rangle$. Here the sum of the generators is $([11]_{18}, [1]_{25})$ but a simpler generator is $([1]_{18}, [1]_{25})$.

Some of the U_n groups that could be cyclic by Theorem 3.3 are cyclic by Theorem 4.17. For example, U_{49} is cyclic since $|U_{49}| = 2 \cdot 3 \cdot 7$, a product of distinct primes; U_{103} is cyclic since $|U_{103}| = 2 \cdot 3 \cdot 17$; and U_{62} is cyclic since $|U_{62}| = \varphi(2 \cdot 31) = 2 \cdot 3 \cdot 5$. It does not follow directly that $G = U_{61}$ is cyclic since $|G| = 60 = 4 \cdot 3 \cdot 5$, but G_3 and G_5 are cyclic since each has prime order (Theorem 4.9). Here $|G_2| = 4$ and since $[11]_{61}^2 = [-1]_{61}$ it follows that $o([11]_{61}) = 4$, so $G_2 = \langle [11]_{61} \rangle$. Now Theorem 4.17 shows that U_{61} is cyclic.

EXERCISES

1. If G is a group with $|G| = 2p$, $p > 2$ a prime, show $G_p \neq \langle e_G \rangle$. (Theorem 2.10)

2. Use Theorem 4.17 to show that each of the following is cyclic:

- **i)** U_{86}
- **ii)** U_{206}
- **iii)** U_{242}
- **iv)** U_{422}
- **v)** U_{961}
- **vi)** U_{3698} $(1849 = 43^2)$

3. For each group G below and each prime divisor p of $|G|$, describe G_p (Definition 4.5), find $|G_p|$, and then determine whether G is cyclic.

- **i)** $U_{19} \oplus \mathbf{Z}_{49} \oplus \mathbf{Z}_{55}$
- **ii)** $U_{17} \oplus U_{19} \oplus \mathbf{Z}_5$
- **iii)** $U_{14} \oplus \mathbf{Z}_7 \oplus \mathbf{Z}_{25}$
- **iv)** $\mathbf{Z}_{36} \oplus U_{23}$
- **v)** $\mathbf{Z}_{49} \oplus \mathbf{Z}_{91} \oplus \mathbf{Z}_{170}$
- **vi)** $\mathbf{Z}_{81} \oplus \mathbf{Z}_{400}$
- **vii)** $U_{47} \oplus \mathbf{Z}_{81} \oplus \mathbf{Z}_{400}$

4. Let G be an Abelian group with $p_1 < \cdots < p_k$ all the prime divisors of $|G|$. Show there is $g \in G$ with $o(g) = p_1 p_2 \cdots p_k$.

5. If G is a finite Abelian group and $g \in G$ has maximal order m among all elements of G, then for any $x \in G$ show that $o(x) \mid o(g)$.

In Exercises 6–7, for any finite group G, $\mathrm{Exp}(G) = \min\{m \in \mathbf{N} \mid g^m = e \text{ for all } g \in G\}$.

6. Find Exp(G) for the following groups:

- **i)** $G = S_3$
- **ii)** $G = S_4$
- **iii)** $G = D_5$
- **iv)** $G = D_{12}$

v) $G = U_{15}$
vi) $G = D_{15} \oplus \mathbf{Z}_{15}$
vii) $G = \mathbf{Z}_{10} \oplus \mathbf{Z}_{15} \oplus \mathbf{Z}_{24}$

7. If G is a finite Abelian group, show that $\mathrm{Exp}(G) = |G| \Leftrightarrow G$ is cyclic. (Hint: consider $\mathrm{Exp}(G_p)$ for $p \mid |G|$.)

8. Let G be an Abelian group, p a prime, and $g^p = e$ for all $g \in G$.
i) If $\langle e \rangle \neq H \leq G$ with $H \neq G$, show that there is $\langle e \rangle \neq K \leq G$ with $H \cap K = \langle e \rangle$.
ii) If G is finite and $H \leq G$, show that there is $L \leq G$ with $H \bigcap L = \langle e \rangle$ and $HL = G$. (Take L of maximal order with $H \bigcap L = \langle e \rangle$; use Exercise 3 in §4.3.)

9. Use Exercise 8 and Theorem 4.14 to show that if G is a finite Abelian group with $\mathrm{Exp}(G) = p_1 \cdots p_k$ for primes $p_1 < \cdots < p_k$, then given $H \leq G$ there is $K \leq G$ satisfying $H \bigcap K = \langle e \rangle$ and $HK = G$.

4.6 THE GROUPS U_{p^n} ARE CYCLIC

One consequence of the results in the last section is that for any odd prime p, U_p is cyclic, and this implies that U_{p^n} is cyclic as well. This fact is a special case of a more general result for fields, a notion we will discuss more fully in a later chapter. A **field,** $(F, \cdot, +)$, is a set with an addition and multiplication satisfying the usual properties for these operations: specifically those given in Theorem 1.28 for $\mathbf{Z}_n$ and with each nonzero element having an inverse under multiplication. For our purposes it suffices to take "field" to mean either $\boldsymbol{C}$ or $\boldsymbol{Z}_p$ for p a prime. These two examples satisfy the required properties, $\boldsymbol{Z}_p$ does by Theorem 1.29, and there are many other examples also, such as $\boldsymbol{Q}$ and $\boldsymbol{R}$. In any field F the elements commute under multiplication so the nonzero elements F^* form an Abelian group using the multiplication in F. We will show that any finite subgroup of $(F^*, \cdot)$ is cyclic. Note that $(\mathbf{Z}_p^*, \cdot) = U_p$.

We need to see that for any field F the usual relation holds between the degree and roots of a polynomial in $F[x] = \{a_n x^n + \cdots + a_1 x + a_0 \mid a_j \in F\}$.

THEOREM 4.18 Let F be a field and $f(x) = a_n x^n + \cdots + a_1 x + a_0 \in F[x]$ with $a_n \neq 0$, so $\deg f = n$. Then $f(c) = 0$ for at most n distinct $c \in F$.

Proof When $n = 0$, $f(x) = a \neq 0$ so $f(c) = a \neq 0$ for all $c \in F$. If $n = 1$, $f(x) = ax + b$, so $f(c) = 0$ for $c \in F \Leftrightarrow c = -ba^{-1}$, and $f(x)$ has only one root in F. Assume now that any polynomial of degree $n \geq 1$ has at most n distinct roots in F and let $g(x) = a_{n+1}x^{n+1} + \cdots + a_1 x + a_0$ with $a_{n+1} \neq 0$. If $g(c) = 0$ for $c \in F$, then $g(x) = g(x) - g(c) = \sum_{i=0}^{n+1} a_i x^i - \sum_{i=0}^{n+1} a_i c^i = \sum_{i=1}^{n+1} a_i (x^i - c^i)$. Now $(x - c)$ is a factor of each $x^i - c^i$ (why?), so $g(x) = (x - c)h(x)$ with $deg\, h = n$. Let $c \neq d$ in F with $0 = g(d) = (d - c)h(d)$. Since $d \neq c$ and F is a field, $(d - c)^{-1}$ exists and $h(d) = 0$ follows. By our induction assumption, $h(x)$ has at most n different roots, and each root of $g(x)$ other than c is a root of $h(x)$, so $g(x)$ can have at most $n + 1$ different roots in F. Hence the theorem holds by IND-I. ■

Our main use of Theorem 4.18 is for $x^q - 1$ or $[1]_p x^q - [1]_p$ when $F = \mathbf{Z}_p$, for q a prime. The theorem tells us that the group $(F^*, \cdot)$ has at most $q - 1$ elements of order $q (1^q = 1!)$, so at most one subgroup of order q. The idea behind proving that any finite

$G \leq (F^*, \cdot)$ is cyclic is to show that each G_q (Definition 4.5) is cyclic and apply Theorem 4.17. The proof that G_q is cyclic uses the fact that these have a unique subgroup of order q.

THEOREM 4.19 If G is an Abelian group with $|G| = p^n > 1$ for p a prime, then G is a cyclic group $\Leftrightarrow$ G has a unique subgroup of order p.

Proof If G is a cyclic group then it has a unique subgroup of order p by Theorem 3.1. Now assume $|G| = p^n$ and that G has a unique subgroup of order p. For $e \neq g \in G$, $o(g) = p^a$ for $a \leq n$ by Theorem 4.7, and so $|\langle g \rangle| = p^a$ by Theorem 2.16. Let $g \in G$ have maximal order p^t. Using $|\langle g \rangle| = p^t$ and Theorem 3.1, we know that $\langle g \rangle$ contains a subgroup of order p, so the unique subgroup of order p in G. But if $x \in G$ with $o(x) = p$, then $|\langle x \rangle| = p$ forces $x \in \langle g \rangle$; $\langle g \rangle$ contains all elements of order p in G. If $y \in G - \langle g \rangle$ with $o(y) = p^s$ then $s > 1$. By Theorem 2.8, $o(y^{p^{s-1}}) = p$ so, as we have seen, $y^{p^{s-1}} \in \langle g \rangle$. Hence, because $y \notin \langle g \rangle$ there is some minimal $m(y) > 0$ with $y^{p^{m(y)}} \in \langle g \rangle$. Among all $y \notin \langle g \rangle$ choose one with minimal $m \geq 1$ satisfying $y^{p^m} \in \langle g \rangle$. Since $(y^p)^{p^{m-1}} \in \langle g \rangle$ and $m - 1 < m$, the minimality of m forces $m = 1$, so $y^p \in \langle g \rangle$. If $o(y) = p^d$, then as we have seen, $o(y^p) = p^{d-1}$. Notice that $d \leq t$ since p^t is the maximal order of any element of G. Now Theorem 3.1 shows $\langle g^{p^{t-d+1}} \rangle$ is the unique subgroup of $\langle g \rangle$ of order p^{d-1} and so is the subgroup generated by any element of order p^{d-1} in $\langle g \rangle$ (why?). Thus $y^p \in \langle g^{p^{t-d+1}} \rangle$ so $y^p = (g^{p^{t-d+1}})^v = g^{vp^{t-d+1}}$. Using $t \geq d$ by the choice of $g \in G$ and that G is Abelian, it follows that $(g^{-vp^{t-d}} y)^p = (g^{vp^{t-d+1}})^{-1} y^p = e_G$. The assumption $y \notin \langle g \rangle$ means that $x = g^{-vp^{t-d}} y \notin \langle g \rangle$. But now $o(x) = p$ forces $x \in \langle g \rangle$. This contradiction implies that $y \notin \langle g \rangle$ is impossible so $G = \langle g \rangle$ must be cyclic, proving the theorem. ■

We now have the pieces needed to prove the very pretty and surprising result mentioned above about finite $G \leq (F^*, \cdot)$ for F a field.

THEOREM 4.20 If F is a field and $G \leq (F^*, \cdot)$ is finite, then G is cyclic. In particular, U_p is a cyclic group when p is a prime.

Proof The outline of the proof was given just before Theorem 4.19. Since G is Abelian, it will be a cyclic group $\Leftrightarrow$ $G_q = \{g \in G \mid o(g) = q^s$, some $s \geq 0\}$ is a cyclic group for each prime $q \mid |G|$, by Theorem 4.17. Now $|G_q| = q^m$ by Theorem 4.14 so there is $h \in G_q$ with $o(h) = q$ by Theorem 4.15 or Theorem 2.8. Thus $(h^i)^q = (h^q)^i = 1$ and Theorem 4.18 shows that $\langle h \rangle$ contains the only q solutions of $x^q - 1$ in F. This means that $\langle h \rangle$ is the only subgroup of order q in G_q. Therefore Theorem 4.19 forces G_q to be cyclic, so G is cyclic, proving the theorem. The statement about U_p follows since $U_p = (\mathbf{Z}_p^*, \cdot)$ and $\mathbf{Z}_p$ is a field. ■

Although it is useful to know that U_p is cyclic when p is a prime, there is no easy way to find a generator. Before proving that U_{p^n} is cyclic when p is odd, we illustrate some consequences of the fact that U_p is cyclic. The first is called *Euler's Criterion* for when $[a] \in U_p$ is a square.

Example 4.15 If $p > 2$ is prime and $b \in \mathbf{Z}$ with $(b, p) = 1$ then there is a solution in $\mathbf{Z}$ of the congruence $x^2 \equiv b \pmod{p} \Leftrightarrow b^{(p-1)/2} \equiv 1 \pmod{p}$.

By Theorem 4.20, $U_p = \langle [a] \rangle$ is cyclic so $[c] \in U_p \Rightarrow [c] = [a]^j$ and $\{[c]^2 \in U_p \mid [c] \in U_p\} = \{[a]^{2j} = [a^{2j}] \in U_p \mid j \in \mathbf{Z}\}$. Note that $c^2 \equiv_p b$ forces $(c, p) = 1$, so showing $c^2 \equiv_p b \Leftrightarrow b^{(p-1)/2} \equiv_p 1$ is equivalent to $[c]^2 = [b] \Leftrightarrow [1] = [b]^{(p-1)/2}$ in U_p by Theorem 1.23. If $[b] = [c]^2 = [a^{2j}]$ then $[b]^{(p-1)/2} = [a^{(p-1)}]^j = [1]^j = [1]$ from Theorem 4.8. When $[b] = [a]^i$, then $[1] = [b]^{(p-1)/2}$ implies that $[1] = [a]^{i(p-1)/2}$. Now $o([a]) = p - 1$ and Theorem 2.7 force $(p-1) \mid i(p-1)/2$. Thus for some $k \in \mathbf{Z}$, $2k(p-1) = i(p-1)$ so $i = 2k$ and indeed $[b] = ([a^k])^2$.

The example shows when $[b] \in U_p$ is a square in U_p. We could instead compute all the squares, but unless p is small this is more difficult than using the congruence. Is $[5] \in U_{23}$ a square? Here $(p-1)/2 = 11$. Now $5^2 \equiv_{23} 2$ so $5^{10} \equiv_{23} 2^5 = 32 \equiv_{23} 9$ and then $5^{11} \equiv_{23} 45 \equiv_{23} -1$. Thus $[5]$ is not a square in U_{23}. For $[3] \in U_{37}$ with $(p-1)/2 = 18$, $3^3 \equiv_{37} -10$, so $3^6 \equiv_{37} 100 \equiv_{37} -11$, and then $3^{12} \equiv_{37} 121 \equiv_{37} 10$. Finally, $3^{18} \equiv_{37} -110 \equiv_{37} 1$ so $[3] \in U_{37}$ is a square (of what?). A special case is of independent interest.

Example 4.16 For p an odd prime, $x^2 \equiv_p -1$ has a solution $\Leftrightarrow p \equiv_4 1$.

From the last example $x^2 \equiv_p -1$ has a solution $\Leftrightarrow (-1)^{(p-1)/2} \equiv_p 1 \Leftrightarrow (p-1)/2 = 2k \Leftrightarrow p = 4k + 1 \Leftrightarrow p \equiv_4 1$ by definition of congruence.

Some illustrations are: $5^2 \equiv_{13} -1$, $4^2 \equiv_{17} -1$, $12^2 \equiv_{29} -1$, $6^2 \equiv_{37} -1$, and $9^2 \equiv_{41} -1$. Although there is a solution of $x^2 \equiv -1 \pmod{541}$, it is not easy to see what a solution is (52 works!).

We return to the study of U_{p^n}. The key is to look at U_{p^2} first.

THEOREM 4.21 If $p > 2$ is a prime, then U_{p^2} is a cyclic group. Furthermore, if $U_p = \langle [a]_p \rangle$, then either $U_{p^2} = \langle [a]_{p^2} \rangle$ or $U_{p^2} = \langle [a+p]_{p^2} \rangle$.

Proof Using $U_p = \langle [a]_p \rangle$ from Theorem 4.20, $o([a]_p) = \varphi(p) = p - 1$. Note that for $[b]_p \in U_p$, $(b, p) = 1$ so $[b]_{p^2} \in U_{p^2}$. What is $m = o([a]_{p^2})$ in U_{p^2}? Since $|U_{p^2}| = \varphi(p^2) = p(p-1)$, $m \mid p(p-1)$ by Theorem 4.7. But $a^m \equiv_{p^2} 1$ so $a^m \equiv_p 1$ and $(p-1) \mid m$ by Theorem 2.7. Hence $m = p - 1$ or $m = p(p-1)$. In the latter case $U_{p^2} = \langle [a]_{p^2} \rangle$ by Theorem 2.16. When $m = p - 1$ consider $o([a+p]_{p^2})$. Since $[a]_p = [a+p]_p$, $o([a+p]_p) = p - 1$ so again $o([a+p]_{p^2})$ is $p - 1$, or is $p(p-1)$ and $\langle [a+p]_{p^2} \rangle = U_{p^2}$. Assuming $(a+p)^{p-1} \equiv_{p^2} 1$, expanding by the Binomial Theorem gives $a^{p-1} + (p-1)pa^{p-2} + p^2 t \equiv_{p^2} 1$. Now $o([a]_{p^2}) = p - 1$ yields $a^{p-1} \equiv_{p^2} 1$ and it follows that $(p-1)pa^{p-2} + p^2 t \equiv_{p^2} 0$, or that $-pa^{p-2} \equiv_{p^2} 0$. But $p^2 \mid pa^{p-2} \Leftrightarrow p \mid a^{p-2}$ forcing $p \mid a$ by Theorem 1.14, a contradiction. Therefore, if $o([a]_p) = p - 1 = o([a]_{p^2})$, then $o([a+p]_{p^2}) = \varphi(p^2) = p(p-1)$ and $U_{p^2} = \langle [a+p]_{p^2} \rangle$ is cyclic, completing the proof. ∎

Theorem 4.21 shows that for p an odd prime and some $b \in \mathbf{Z}$, $U_p = \langle [b]_p \rangle$ and $U_{p^2} = \langle [b]_{p^2} \rangle$. Using this and an argument much like that for Theorem 4.21, we can show in general that U_{p^n} is cyclic.

THEOREM 4.22 If $p > 2$ is a prime, then U_{p^n} is a cyclic group. Furthermore, $a \in \mathbf{Z}$ with $U_p = \langle [a]_p \rangle$ and $U_{p^2} = \langle [a]_{p^2} \rangle \Rightarrow U_{p^n} = \langle [a]_{p^n} \rangle$.

Proof Apply Theorem 4.21 to get $a \in \mathbf{Z}$ with $U_p = \langle [a]_p \rangle$ and $U_{p^2} = \langle [a]_{p^2} \rangle$. To show $U_{p^n} = \langle [a]_{p^n} \rangle$, it suffices to prove that $o([a]_{p^n}) = |U_{p^n}| = \varphi(p^n) = p^{n-1}(p-1)$. Now if $o([a]_{p^n}) = m$ then Theorem 4.7 shows that $m \mid p^{n-1}(p-1)$. But $a^m \equiv 1 \pmod{p^n}$ so $a^m \equiv 1 \pmod{p^2}$ and $p(p-1) \mid m$ by Theorem 2.7 using $o([a]_{p^2}) = p(p-1)$. Thus $m = p^s(p-1)$ for $1 \le s < n$ and we use induction on $n \ge 3$ to prove $s = n-1$. Let $n = 3$ but $a^{p(p-1)} \equiv 1 \pmod{p^3}$. Because $o([a]_p) = p-1$, $a^{p-1} \equiv 1 \pmod{p}$ and so $a^{p-1} = 1 + pt$. If $p \mid t$ then $a^{p-1} \equiv 1 \pmod{p^2}$, contradicting $o([a]_{p^2}) = p(p-1)$, so $(p, t) = 1$. Thus $a^{p(p-1)} \equiv 1 \pmod{p^3}$ yields $(1+pt)^p \equiv 1 \pmod{p^3}$ and by expanding, $1 + p^2t + p^3u \equiv 1 \pmod{p^3}$. It follows that $p^2t \equiv 0 \pmod{p^3}$, or that $p \mid t$, a contradiction: $a^{p(p-1)} \equiv 1 \pmod{p^3}$ is impossible. Hence $o([a]_{p^3}) = p^2(p-1)$ and $U_{p^3} = \langle [a]_{p^3} \rangle$. Take $n \ge 3$, assume that for $3 \le k \le n$, $U_{p^k} = \langle [a]_{p^k} \rangle$, and let $o([a]_{p^{n+1}}) = m$ in $U_{p^{n+1}}$. As above, $m \mid p^n(p-1)$ and $a^m \equiv 1 \pmod{p^{n+1}}$ implies $a^m \equiv 1 \pmod{p^n}$ so together with $o([a]_{p^n}) = p^{n-1}(p-1)$ yields $p^{n-1}(p-1) \mid m$. Thus $m = p^{n-1}(p-1)$, or $m = p^n(p-1)$ and $\langle [a]_{p^{n+1}} \rangle = U_{p^{n+1}}$ as required. We show that $m \ne p^{n-1}(p-1)$. Now $o([a]_{p^{n-1}}) = p^{n-2}(p-1)$ by assumption so $a^{p^{n-2}(p-1)} = 1 + p^{n-1}t$, and $(p, t) = 1$ since $a^{p^{n-2}(p-1)} \equiv 1 \pmod{p^n}$ contradicts $o([a]_{p^n}) = p^{n-1}(p-1)$. Should $a^{p^{n-1}(p-1)} \equiv 1 \pmod{p^{n+1}}$, then $a^{p^{n-1}(p-1)} = (1 + p^{n-1}t)^p \equiv 1 \pmod{p^{n+1}}$, or, equivalently, $1 + p^n t + \dfrac{p(p-1)}{2} p^{2n-2}t^2 + p^{n+1}u \equiv 1 \pmod{p^{n+1}}$, (why?). But $p^{n+1} \mid p^{2n-1}$ since $n \ge 3$ so in fact $p^n t \equiv 0 \pmod{p^{n+1}}$, forcing the contradiction $p \mid t$. Thus $\langle [a]_{p^{n+1}} \rangle = U_{p^{n+1}}$ and IND-II proves the theorem. ■

The last result and Theorem 3.3 tell us when U_n is cyclic, unless $n = 2p^m$ for p an odd prime. That these are cyclic follows from Theorem 4.22, requiring only a common odd generator for U_p and U_{p^2}, and is left as an exercise. We can use Theorem 4.22 to find generators for U_{p^n} when p is not too large by finding a common generator for U_p and U_{p^2}. For example, $U_5 = \langle [2]_5 \rangle$ and $U_{25} = \langle [2]_{25} \rangle$ (Example 3.3), so $U_{5^n} = \langle [2]_{5^n} \rangle$. Hence the minimal $m \in \mathbf{N}$ with $2^m \equiv_{3125} 1$ is $m = \varphi(5^5) = 2500$. For many small primes the smallest generator of U_p also generates U_{p^2}. Can we find $U_p = \langle [a]_p \rangle$ but $\langle [a]_{p^2} \rangle \ne U_{p^2}$? Since $[2]_5 = [7]_5$, $U_5 = \langle [7]_5 \rangle$, but $[7]_{25} = [32]_{25} = [2^5]_{25}$ and $(2^5)^4 = 2^{\varphi(25)} \equiv_{25} 1$ by Euler's Theorem, so $\langle [7]_{25} \rangle \ne U_{25}$. By Theorem 4.21, $\langle [12]_5 \rangle = U_5$ and $\langle [12]_{25} \rangle = U_{25}$, so $U_{5^n} = \langle [12]_{5^n} \rangle$ as well.

EXERCISES

1. If G is an Abelian group with $|G| = p^n$ for p a prime so that all $H, K \le G$ with $H \ne \langle e \rangle$ and $K \ne \langle e \rangle$ satisfy $H \cap K \ne \langle e \rangle$, show that G must be cyclic.
2. If $p > 2$ is a prime, argue carefully that for all $k \in \mathbf{N}$, U_{2p^k} is cyclic.
3. Show $o([2]_{23}) = 11$ in U_{23}. Find a generator for U_{23}. Does it generate U_{529}?
4. In U_{71} compute $o([5]_{71})$ and $o([20]_{71})$. Find a generator for U_{71}.
5. Find a generator for U_{41}. (Use $2^{10} = 1024$, $1025 = 25 \cdot 41$, and $2 \cdot 41 = 81 + 1$ to show that $o([2]_{41}) = 20$ and $o([3]_{41}) = 8$.)
6. Find a generator for U_{343} (note that $7|343$).
7. For $k \ge 2$ find: a) a generator for U_{11^k}; and b) a generator for $U_{2 \cdot 11^k}$.
8. **a)** If $p > 2$ is a prime, show that $a^{(p-1)/2} \equiv \pm 1 \pmod{p}$ for $a \in \mathbf{Z}$ with $(a, p) = 1$.
   b) If $(a, p) = 1 = (b, p)$ and if neither $a \equiv x^2 \pmod{p}$ nor $b \equiv x^2 \pmod{p}$ has a solution in $\mathbf{Z}$, then there *is* a solution in $\mathbf{Z}$ of $ab \equiv x^2 \pmod{p}$.
9. For p a prime with $p \equiv_4 1$, what is $\text{card}(\{[j]_p^{(p-1)/4} \in U_p \mid 1 \le j < p\})$?
10. Use U_p cyclic for p a prime to prove Wilson's Theorem: $(p-1)! \equiv_p -1$.

11. Determine whether the given class is a square in U_p:
i) $[2] \in U_{37}$
ii) $[5] \in U_{41}$
iii) $[7] \in U_{41}$
iv) $[2] \in U_{43}$
v) $[11] \in U_{43}$
vi) $[2] \in U_{73}$
vii) $[3] \in U_{73}$

12. Use $U_{101} = \langle [2]_{101} \rangle$ to find a solution in U_{101} for each of:
i) $x^{11} = [64]$
ii) $x^{13} = [16]$
iii) $x^{67} = [2]$
iv) $x^{78} = [4]$

13. If p is prime and $p \equiv_4 1$ show $a^2 + b^2 = p$ for some $a, b \in \mathbf{N}$. (Show $1 \le c < p$ with $c^2 \equiv_p -1$. Let $D = \{ac - b \mid a, b \in \mathbf{Z}$ and $0 \le a, b < p^{1/2}\}$, observe that there are more than p such pairs (a, b), so two different elements of D are congruent mod p. Thus $uc \equiv_p v$ with $1 \le |u|, |v| < p^{1/2}$. Show $p = u^2 + v^2$.)

4.7 CARMICHAEL NUMBERS

We know from Fermat's Theorem (corollary to Theorem 4.8) that $a^p \equiv_p a$ for any $a \in \mathbf{Z}$ when p is prime. It would be nice if this characterized primes since using computers it is easy to take large powers of numbers but very difficult to factor them. Experiments with small odd n show $2^n \equiv_n 2$ only when n is a prime. For example, $2^4 \equiv_{15} 1$ so $2^{15} \equiv_{15} 2^3 = 8$, not 2. Consider $2^{10} = 1024$ and $1023 = 3 \cdot 341 = 3 \cdot 11 \cdot 31$. Thus 341 is not prime, $2^{10} \equiv_{341} 1$, so $2^{340} \equiv_{341} 1$ and finally $2^{341} \equiv_{341} 2$. In fact 341 is the smallest composite with this property. But $7^3 = 343 \equiv_{341} 2$ yields $7^{30} \equiv_{341} 2^{10} \equiv_{341} 1$ and $7^{330} \equiv_{341} 1$. Now $7^9 \equiv_{341} 2^3 = 8$, so $7^{341} \equiv_{341} 1 \cdot 8 \cdot 7^2 = 7^2 + 7^3 \equiv_{341} 7^2 + 2 = 51 \not\equiv_{341} 7$, proving 341 is not prime. Could there be $n \in \mathbf{N}$ composite and satisfying $a^n \equiv_n a$ for all $a \in \mathbf{Z}$? If not, then this property in Fermat's Theorem will characterize primes. Unfortunately, there are such composites.

DEFINITION 4.6 A **Carmichael number,** or **absolute pseudoprime,** is any *composite* $n > 1$ satisfying $a^n \equiv a \pmod{n}$ for all $a \in \mathbf{Z}$.

When a new term is defined it is a good idea to make sure that there really are examples of it! We present an example of a Carmichael number.

Example 4.17 1105 is a Carmichael number.

Note first that $1105 = 5 \cdot 221 = 5 \cdot 13 \cdot 17$, so to show $a^{1105} \equiv_{1105} a$ for all $a \in \mathbf{Z}$, or, equivalently, that $1105 \mid (a^{1105} - a)$, it suffices by Theorem 1.12 to show that each of 5, 13, and 17 divides $(a^{1105} - a)$ (verify!). Take $a \in \mathbf{Z}$. If $5 \mid a$ then $a^{1105} \equiv_5 0 \equiv_5 a$. When $(a, 5) = 1$, Euler's Theorem (Theorem 4.8) show that $a^4 \equiv_5 1$, and since $4 \mid 1104$, $a^{1104} \equiv_5 1$ follows from Theorem 1.24 on properties of congruences. This same result then gives $a^{1105} \equiv_5 a$, so this congruence holds for all $a \in \mathbf{Z}$ and $5 \mid (a^{1105} - a)$. Using that $12 \mid 1104$ and $16 \mid 1104$, similar computations show $a^{1105} \equiv_{13} a$ and $a^{1105} \equiv_{17} a$. Therefore $13 \mid (a^{1105} - a)$ and $17 \mid (a^{1105} - a)$ as required.

Are there other Carmichael numbers? Suppose $n > 1$ and $n = p^2 s$ for some prime p. Then $(ps)^2 = ns \equiv_n 0$, so $(ps)^n \equiv_n 0$ and n is not a Carmichael number. Thus a Carmichael number cannot have a repeated prime divisor. Also no Carmichael number is even since $(-1)^{2n} \equiv_{2n} 1$. Surprisingly, we can obtain a simple characterization of these numbers using that U_p is cyclic.

THEOREM 4.23 If $n \in \mathbf{N}$ with $n > 1$, then n is a Carmichael number $\Leftrightarrow$ $n = p_1 p_2 \cdots p_k$ for $2 < p_1 < \cdots < p_k$ primes, $k \geq 3$, and each $(p_j - 1) \mid (n - 1)$.

Proof Let n be a Carmichael number. We mentioned that n must factor as $n = p_1 p_2 \cdots p_k$ for $2 < p_1 < \cdots < p_k$ primes with $k \geq 2$ since n is composite. From Theorem 4.20, for each $1 \leq j \leq k$ there is $a_j \in \mathbf{Z}$ with $o([a_j]_{p_j}) = p_j - 1$ in U_{p_j}. We claim there is $b_j \in \mathbf{Z}$ with $[a_j]_{p_j} = [b_j]_{p_j}$ and $(b_j, n) = 1$. Set $q_j = n/p_j$, so q_j is the product of all the p_i except for p_j. Clearly $(q_j, p_j) = 1$ so by Theorem 1.10, $tq_j + sp_j = 1$ for $t, s \in \mathbf{Z}$. Let $b_j = tq_j a_j + p_j$ and observe that for $p_i \neq p_j$, $p_i \mid q_j$, so p_i cannot divide b_j or else it would divide p_j. Similarly, $p_j \mid b_j$ implies $p_j \mid tq_j a_j$, but $(t, p_j) = 1$ by Theorem 1.10 and we have seen $(q_j, p_j) = 1 = (a_j, p_j)$ so this contradicts Theorem 1.14, forcing $(p_j, b_j) = 1$. Now that no p_i divides b_j we have $(b_j, n) = 1$, and $b_j \equiv a_j \pmod{p_j}$ implies $[a_j]_{p_j} = [b_j]_{p_j}$. Hence we may assume $(a_j, n) = 1$ and so $[a_j]_n \in U_n$. But n is a Carmichael number so $a_j^n \equiv_n a_j$. Thus $(a_j, n) = 1$ implies $a_j^{n-1} \equiv_n 1$ by Theorem 1.24, then $a_j^{n-1} \equiv 1 \pmod{p_j}$ so using $o([a_j]_{p_j}) = p_j - 1$ and Theorem 2.7, $(p_j - 1) \mid (n - 1)$. Finally, if $k = 2$, so $n = pq$ with $p < q$ primes, then we have just shown that $(q - 1) \mid (pq - 1)$, or $u(q - 1) = pq - 1 = pq - p + p - 1 = p(q - 1) + (p - 1)$. This forces the contradiction $(q - 1) \mid (p - 1)$ and proves that $k \geq 3$.

Let $n = p_1 \cdots p_k$ be as described, let $p \mid n$ be prime and take $a \in \mathbf{Z}$. If $p \mid a$ then $a^n \equiv_p 0 \equiv_p a$, and if $(a, p) = 1$ then by Euler's Theorem $a^{p-1} \equiv_p 1$. But $(p - 1) \mid (n - 1)$ so $n - 1 = s(p - 1)$ and $a^{n-1} \equiv_p (a^{p-1})^s \equiv_p 1$. It follows from Theorem 1.24 that $a^n \equiv_p a$, or $p \mid (a^n - a)$ so $n \mid (a^n - a)$ by Theorem 1.12 (and induction). Therefore n is a Carmichael number. ■

Finding Carmichael numbers from this description is not easy. The smallest Carmichael number is $561 = 3 \cdot 11 \cdot 17$ and the second smallest is 1105 in Example 4.17. It is known that there are infinitely many Carmichael numbers but they are quite scarce. The smallest with four prime factors is $41041 = 7 \cdot 11 \cdot 13 \cdot 41$.

EXERCISES

1. If for $m \in \mathbf{N}$, all of $6m + 1$, $12m + 1$, and $18m + 1$ are prime, show that their product is a Carmichael number. Using this, find two Carmichael numbers.

2. a) If $n, m \in \mathbf{N}$, show that $(n - 1) \mid (nm - 1) \Leftrightarrow (n - 1) \mid (m - 1)$. In particular, for $p_1 < \cdots < p_k$ primes, $(p_j - 1) \mid (p_1 p_2 \cdots p_k - 1) \Leftrightarrow$ $(p_j - 1) \mid (p_1 \cdots p_{j-1} p_{j+1} \cdots p_k - 1)$.

b) Use a) to find Carmichael numbers of the forms $p_1 < p_2 < p$:

i) $5 \cdot 17 \cdot p$	**iv)** $7 \cdot 19 \cdot p$
ii) $7 \cdot 13 \cdot p$	**v)** $41 \cdot 61 \cdot p$
iii) $7 \cdot 23 \cdot p$	

4.8 ENCRYPTION AND CODES

We want to use some ideas and results in this chapter to address the problems of how to send a message so that only the intended receiver can understand it, and how to ensure that a message received is the one sent. To send a secret message that cannot be read if intercepted is an age-old problem. A simple way to encode or encrypt a message is to permute letters. For example, interchanging the ith and $27 - i^{\text{th}}$ letters encodes "arrive at ten" as "ziirev zg gvm." A third party intercepting the message might have trouble decoding it. However, with enough intercepted coded material this type of code can be broken easily using known letter combination frequency distributions. The permutation can be changed frequently to stymie such analysis, but this both is inconvenient and introduces other difficulties.

One of the first practical methods (using computers) for sending a message that a third party cannot hope to decode was described by R. Rivest, A. Shamir, and L. Adleman. It is called the RSA algorithm and is based on Euler's Theorem. It is a *public key code cryptograph:* the means of *encoding* messages to us can be made public (or not), but only we can decode an encrypted message. Its security depends on the fact that (currently!) huge integers cannot in general be factored in any reasonable time, even using the most advanced computing technology.

A simple way to encode a message as a number is to assign 01 to a, 02 to b, and continue with 26 assigned to z, and use other two-digit expressions for punctuation. Let us use 29 to begin a message, 00 for a space, and 28 for a period. The expression "Math is good." becomes 29130120080009190007151504 28 since 29 starts the message, 13 corresponds to the letter m, 01 to a, 20 to the twentieth letter t, etc. Similarly 2629090012091105000301201928 0102 encodes "I like cats." If we simply send such a message it could be read by anyone who intercepts it.

The idea of RSA takes $n = pq$ for $p < q$ very large primes, say having tens or hundreds of digits each, then chooses $k < n$ with $(k, \varphi(n)) = (k, (p-1)(q-1)) = 1$. The public is given the pair (n, k). A message is encoded into a large number as above, say M. If $M \geq n$ then cut M into shorter pieces, making several messages each less than n. Thus assume that $M < n$. Now compute $X \equiv_n M^k$ with $0 < X < n$ and send X. Computers can easily compute X from M. Since $(k, \varphi(n)) = 1$, $[k] \in U_{\varphi(n)}$ so has an inverse, say $[s] \in U_{\varphi(n)}$, and we may write $1 + r\varphi(n) = ks$. Even for large integers, computers make it easy to find s via the Euclidean Algorithm. The message is decoded by computing X^s modulo n. Note that $X^s \equiv_n M^{ks} \equiv_n M(M^{\varphi(n)})^r$ so if $(M, n) = 1$ Euler's Theorem gives $(M^{\varphi(n)})^r \equiv_n 1$ so $X^s \equiv_n M$, recovering M. If $p \mid M$, then $X^s \equiv_p M^{ks} \equiv_p 0 \equiv_p M$, and now $M < n$ forces $(q, M) = 1$, so $X^s \equiv_q M^{ks} \equiv_q M(M^{q-1})^{(p-1)r}$ and again Euler's Theorem yields $X^s \equiv_q M$. Since $p \mid (X^s - M)$ and $q \mid (X^s - M)$ we have by Theorem 1.12 that $n \mid (X^s - M)$, so $X^s \equiv_n M$ and again the message M is recovered. A similar computation works when $q \mid M$. The point to remember is that the "inverse" s needed to decode the message can be computed only from $\varphi(n)$, which cannot be found from n without knowing its factorizations. But n cannot in practice be factored because it is too large.

If the means of encoding is public, how can we tell if a message is from the expected source? As above, if $(k, \varphi(n)) = 1$ and $[k]^{-1} = [s]$ in $U_{\varphi(n)}$, then also $[s]^{-1} = [k]$ so s can be used to encode and (n, k) to decode. The expected source of the message makes known its own $n' > n$ and k' and knows $[k']^{-1} = [s']$ in $U_{\varphi(n')}$. As above, the source computes

$X \equiv M^k \pmod{n}$ and then sends $Y = X^{s'} \pmod{n'}$. We compute $Y^{k'} = X \pmod{n'}$. Since $n' > n$ we have $X \equiv M^k \pmod{n}$ and decode this as above. Although anyone can decode Y to get $X = M^k$, they cannot decode X and cannot encode a message since s' is unknown.

Example 4.18 RSA encoding using $n = 4661$.

Although $4661 = 59 \cdot 79$ is unrealistically small for coding, it will serve to illustrate the approach. Now $\varphi(4661) = 58 \cdot 78 = 4524$ and $4525 = 4524 + 1 = 5 \cdot 905$. We publish (4661, 5) and use 905 to decode received messages. Let us encode "Get it for me." in the manner described above: 29 starts the message, 01 represents A, etc., 00 is a space, and 28 is a period. Then $M = 29070520000920000615180013\text{-}0528$ encodes the message. Since we need $M < 4661$, we have to break up M into the three-digit submessages: 290, 705, 200, 009, 200, 006, 151, 800, 130, 528, then send the fifth power of each mod 4661. This can be done with a hand calculator. The coded parts to send are (in order): 3542, 1704, 0505, 3117, 0505, 3115, 1047, 4410, 2389, and 3120. To be clear, $290^5 \equiv_{4661} 3542$, $705^5 \equiv_{4661} 1704$, etc. Decoding would compute $3542^{905} \equiv_{4661} 290$, $1704^{905} \equiv_{4661} 705$, and so on.

It is fair to view any electronically transmitted message as a string of zeros and ones, so we assume the message is encoded as such a string. Random errors changing 0 to 1 or 1 to 0 do occur in the transmission of these digits. In a long message a few errors may not matter since the context will enable us to understand the true message. Nevertheless it is important to detect errors and best to correct them. Directions sent electronically to give position or to target a powerful weapon must be received accurately!

Sophisticated methods are used to detect and correct errors in transmitted messages. Our aim is to introduce general ideas concerning this problem and to see how groups and cosets enter the picture. Our assumption is that error is random and relatively infrequent, so the goal is to detect and correct a small number of errors efficiently. Now messages are strings of 0 and 1 like $\alpha = (1, 0, 0, 1, 1, \ldots) \in \{0, 1\}^n$. We will identify any such α with the corresponding $([1], [0], [0], [1], [1], \ldots) \in \mathbf{Z}_2^n = \mathbf{Z}_2 \oplus \cdots \oplus \mathbf{Z}_2$ and conversely. Since $\mathbf{Z}_2^n$ is a group under addition, we will add elements in $\{0, 1\}^n$ as if they are in $\mathbf{Z}_2^n$ so $(1, 1, 0, 0) + (1, 0, 1, 0) = (0, 1, 1, 0)$. A **code** is any $C \subseteq \{0, 1\}^n$ and each $\alpha \in C$ is a *code word* representing some bit of information to be sent, say a digit, a letter, or a specific instruction. If $C = \{0, 1\}^n$ then any received message is some code word so we cannot detect an error in its transmission. Similarly, we do not want a code word to differ from another in just one place since an error in one digit might be another code word.

How can we choose a code $C \subseteq \mathbf{Z}_2^n$ so that errors can be detected and corrected? A simple example is $C = \{O = (0, 0, 0, 0, 0), I = (1, 1, 1, 1, 1\} \subseteq \mathbf{Z}_2^5$. If $(0, 1, 1, 0, 0) \notin C$ is received, there is clearly an error. Assuming that errors are random and infrequent it is more likely that two errors were made in transmission rather than three, so we decode $(0, 1, 1, 0, 0)$ as $O \in \mathbf{Z}_2^5$ since it is "closer" to $O \in C$ than to $I \in C$. Similarly, in general C can correct up to two errors. If $(0, 0, 0, 0), (1, 1, 1, 1) \in \mathbf{Z}_2^4$ are the code words and we receive (0, 1, 1, 0) we can tell an error has occurred but cannot correct it since (0, 1, 1, 0) is equally "close" to each code word.

DEFINITION 4.7 The **weight** of $\alpha = (a_1, \ldots, a_n) \in \{0, 1\}^n$ is $wt(\alpha) = \sum a_i$, the number of ones in α. The **Hamming distance** between $\alpha, \beta \in \{0, 1\}^n$ is $D(\alpha, \beta) = wt(\alpha + \beta)$.

Now $\alpha, \beta \in \{0, 1\}^n$ differ in m positions $\Leftrightarrow D(\alpha, \beta) = wt(\alpha + \beta) = m$, since an entry of $\alpha + \beta$ is $1 \Leftrightarrow \alpha$ and β differ in that position. Also it is an exercise that D in Definition 4.7 satisfies the usual properties of distance: $D(\alpha, \beta) \geq 0$ and is zero $\Leftrightarrow \alpha = \beta$; $D(\alpha, \beta) = D(\beta, \alpha)$; and $D(\alpha, \beta) + D(\beta, \gamma) \geq D(\alpha, \gamma)$. This last inequality holds since if α and γ differ in a position, β must differ from one of them in that position. Just as for our simple example above, we decode a received n-tuple as the closest code word to it, using the assumption that errors are random and infrequent. If more than one code word is closest, we either do not decode the message or replace it with one of these closest code words. This standard approach is called **maximum-likelihood decoding.** When this method can correct up to k errors on transmission we say that the code C is ***k*-error correcting.** For this to work, code words cannot be too close together. Since distance is measured by adding, it is convenient to have $C \leq \mathbf{Z}_2^n$, although there are other advantages of using subgroups. In this case C is called a **binary linear code** and we focus our attention on it.

DEFINITION 4.8 If $\langle(0, 0, \ldots, 0)\rangle \neq C \leq \mathbf{Z}_2^n$, the **minimum weight** of C is $wt(C) = \min\{wt(\alpha) \mid \alpha \in C, \alpha \neq (0, 0, \ldots, 0)\}$.

Using Definition 4.7, when $C \leq \mathbf{Z}_2^n$ then $\alpha, \beta \in C \Rightarrow \alpha + \beta \in C$ so if $\alpha \neq \beta$, $D(\alpha, \beta) \geq wt(C)$. For binary linear codes C, $wt(C)$ tells us how many errors in a received n-tuple we can detect and correct.

THEOREM 4.24 If $C \leq \mathbf{Z}_2^n$ is a binary linear code, then $\alpha \in \mathbf{Z}_2^n$ received with $d - 1$ errors can be detected as incorrect $\Leftrightarrow wt(C) \geq d$, and any α with k errors can be corrected $\Leftrightarrow wt(C) \geq 2k + 1$.

Proof Let $\alpha \in C$ be sent and β received. Suppose $wt(C) \geq d$ and β contains $h \leq d - 1$ errors. Then $wt(\alpha + \beta) = h < d \leq wt(C)$ so $\beta \notin C$, since $\beta \in C$ and $C \leq \mathbf{Z}_2^n$ would force $\alpha + \beta \in C$ and then $wt(\alpha + \beta) \geq d$. Thus C can detect up to $d - 1$ errors. Note that for $\upsilon, \eta \in C$, $\upsilon = \eta + (\upsilon + \eta)$. Assume C can detect $d - 1$ errors but $wt(C) \leq d - 1$ and let $\gamma, \eta \in C$ with $wt(\gamma + \eta) = wt(C) \leq d - 1$. Then γ can be sent and $\eta \in C$ received so not detected as an error, although η and γ differ in at most $d - 1$ places. Hence, when C detects $d - 1$ errors, $wt(C) \geq d$. Assume now that β has at most k errors and that $wt(C) \geq 2k + 1$ We claim that β is closest to α in C and no $\gamma \in C$ is as close to β as α is. By assumption, $D(\alpha, \beta) \leq k$. If for some other $\gamma \in C$, $D(\beta, \gamma) \leq k$ also, then $D(\alpha, \gamma) \leq D(\alpha, \beta) + D(\beta, \gamma) \leq 2k$, contradicting $wt(C) \geq 2k + 1$. Thus maximum-likelihood decoding replaces β with α, the correct message, and C corrects up to k errors. Finally, assume that any received β with at most $k \geq 1$ errors can be corrected but $wt(C) \leq 2k$. Let $\gamma, \eta \in C$ with $1 \leq wt(\gamma + \eta) = wt(C) = m \leq 2k$ and let $i_1 < \cdots < i_m$ be the positions where γ and η differ. If $m = 2s$, change γ at positions $i_1, \ldots, i_s$ and call this n-tuple υ. Then υ agrees at positions $i_1, \ldots, i_s$ with η and differs from η exactly at positions $i_{s+1}, \ldots, i_{2m}$

so $wt(\gamma + \upsilon) = wt(\eta + \upsilon) = s \leq k$, but υ cannot be corrected if received for γ since it has equal distances from both γ and $\eta \in C$. When $m = 2s + 1 \leq 2k$, letting ρ differ from γ in positions $i_1, \ldots, i_{s+1}$ yields $wt(\gamma + \rho) = s + 1 \leq k$ and $wt(\eta + \rho) = s < k$ so by maximum-likelihood decoding ρ will not be decoded as γ. Therefore if C is k-error correcting then $wt(C) \geq 2k + 1$. ■

The error detecting and correcting properties of C depend on $wt(C)$. Can we find C with a given $d = wt(C)$? To correct a message with at most $(d - 1)/2$ errors seems to require storing all code words and adding each $\gamma \in C$ to the received β until we find $wt(\beta + \gamma) \leq (wt(C) - 1)/2$. For large C this needs a lot of storage.

Now $\mathbf{Z}_2^n$ is a vector space with $\mathbf{Z}_2$ as scalars and $C \leq \mathbf{Z}_2^n$ is a subspace. If the dimension of C is k then $|C| = 2^k$ and C is called an *$[n, k]$ code*. From a group theory point of view, a basis of C is just a minimal generating set. If $\{\alpha_1, \ldots, \alpha_k\}$ generates C with k minimal, then every $\gamma \in C$ is a sum of α_j. But $\alpha_i + \alpha_j = \alpha_j + \alpha_i$ and $\alpha_i + \alpha_i = e \in \mathbf{Z}_2^n$, so each $\alpha \in C$ is a sum of different elements in $\{\alpha_j\}$. When k is minimal, different subsets of $\{\alpha_j\}$ yield different sums (exercise!) so if we regard $e \in C$ as the sum of the empty subset, it follows that $|C| = 2^k$ (Example 0.19). The relation between n, $|C| = 2^k$, and the number d of errors that C can correct is next.

THEOREM 4.25 If the $[n, k]$ binary linear code $C \leq \mathbf{Z}_2^n$ with 2^k elements is d-error correcting, then $\left(1 + \binom{n}{1} + \cdots + \binom{n}{d}\right) \leq 2^{n-k}$.

Proof If C is d-error correcting via maximum-likelihood decoding, then $\gamma \in \{0, 1\}^n$ with $D(\gamma, \alpha) \leq d$ for some $\alpha \in C$ satisfies $D(\gamma, \beta) > d$ for any other $\beta \in C$. Given $\alpha \in C$, how many γ satisfy $D(\gamma, \alpha) \leq d$? Now γ must be α with at most d coordinates of α changed. But there are $\binom{n}{s}$ ways of changing s coordinates, so a total of $1 + \binom{n}{1} + \cdots + \binom{n}{d}$ n-tuples have distance at most d from α. The fact that C is d-error correcting implies that these sets are disjoint from the n-tuples at a distance at most d from any other element of C. There are $|C| = 2^k$ such sets so the number of received n-tuples that can be corrected by C is $2^k\left(1 + \binom{n}{1} + \cdots + \binom{n}{d}\right) \leq 2^n = |\mathbf{Z}_2^n|$, proving the theorem. ■

The theorem shows that a 1-error correcting code can be quite large since the inequality is $1 + n \leq 2^{n-k}$. If $k = 5$ we must expect $n \geq 9$, if $k = 20$ then $n \geq 25$, and if $k = 100$ then $n \geq 107$. When $d > 1$, n must be quite large relative to k. For example, a 2-error correcting code with 32 code words satisfies $(1 + n + n(n - 1)/2) \leq 2^{n-5}$, or $(n^2 + n + 2) \leq 2^{n-4}$. A little computation shows that $n \geq 12$ and for $k = 20$, $n \geq 29$. For a 3-error correcting code, if $k = 5$ then we need $n \geq 14$, when $k = 20$, $n \geq 32$, and for $k = 100$, $n \geq 118$. These calculations do not guarantee C exists but just describe the possible n.

A nice method for obtaining linear codes and efficiently detecting and correcting errors depends on solving linear equations. A system of m homogeneous linear equations in n variables corresponds to the matrix equation $HX = 0$ for H the $m \times n$ matrix of coefficients of the system. If H has rank k, or, equivalently, if k is the maximal number of independent rows (or columns) of H, then there are $n - k$ independent solutions of

$HX = 0$ that generate all solutions. Now these results hold for equations with coefficients in $\mathbf{Z}_2$ (or any field) as well as in $\mathbf{R}$. In our case, if the rows of H are independent then $C = \{\alpha \in \mathbf{Z}_2^n \mid H\alpha^T = 0\} \leq \mathbf{Z}_2^n$ (α^T is the transpose of α) has a minimal generating set of $n - k$ elements, so $|C| = 2^{n-k}$.

Let H be an $(n-k) \times n$ matrix over $\mathbf{Z}_2$ of rank $n-k$, so the transposes of the solutions of $HX = 0$ generate $C \leq \mathbf{Z}_2^n$ with $|C| = 2^k$. For $\alpha \in \mathbf{Z}_2^n$, $H\alpha^T$ adds the columns of H corresponding to the positions where a 1 appears in α, so if any d columns of H are independent then no nonzero solution of $HX = 0$ can have fewer than $d + 1$ ones: that is, $wt(C) \geq d + 1$. Thus C can detect up to d errors in a received n-tuple β simply by computing that $H\beta \neq 0$. When the columns of H are distinct and not zero, no two of them are dependent so $d \geq 2$ and $wt(C) \geq 3 : C$ is 1-error correcting by Theorem 4.24.

In the situation above, when $C \leq \mathbf{Z}_2^n$ is defined as the (transposes) of solutions of $HX = 0$ with the rank of $H = n - k$ and with any d columns of H independent there is an easy decoding process called **syndrome decoding.** If $\gamma \in \mathbf{Z}_2^n$ then for $\alpha \in C$, $H\gamma^T = H(\alpha^T + \gamma^T)$. Thus $\rho \in C + \gamma \Rightarrow H\rho^T = H\gamma^T$ and this common value $H\gamma^T$ is called the **syndrome** of γ, written syn(γ). If $\upsilon \in \mathbf{Z}_2^n$ satisfies syn(υ) = syn(γ), then $H\upsilon^T = H\gamma^T \Rightarrow H(\upsilon - \gamma)^T = 0$ so $\nu - \gamma \in C$ and $\upsilon \in C + \gamma$. We claim there is at most one element of weight at most $d/2$ in any coset $C + \gamma$. If $\alpha, \beta, \in C + \gamma$ with $wt(\alpha), wt(\beta) \leq d/2$, then we have $d \geq wt(\alpha + \beta)$ (why?) so $\alpha + \beta \in C$ contradicts $wt(C) \geq d + 1$. Suppose there is $\mu(\gamma) \in C + \gamma$ with minimal weight at most $d/2$. We have just shown that $\mu(\gamma)$ is unique in $C + \gamma$, so there is a unique $\alpha \in C$ with $\mu(\gamma) = \alpha + \gamma$. But $wt(\mu(\gamma)) = \min\{wt(\beta + \gamma) \mid \beta \in C\} = \min\{D(\beta, \gamma) \mid \beta \in C\} = D(\alpha, \gamma)$ so γ decodes (maximum-likelihood) as $\alpha = \mu(\gamma) + \gamma$.

Since $|C| = 2^k$, C has 2^{n-k} cosets by Lagrange's Theorem. For each of these that has an element of weight at most $d/2$, its **coset leader** is the unique element of minimal weight. List each coset leader with its syndrome. If γ is received and has at most $d/2$ errors, then $C + \gamma$ contains a coset leader so we simply compute $H\gamma^T$, find the coset leader ρ with syn(γ) = syn(ρ), and decode γ as $\rho + \gamma$. This method of syndrome decoding requires storing at most 2^{n-k} coset leaders and their syndromes, which can be a substantial savings for single-error correcting codes. For example, if we start with H a 4×14 matrix of rank 4 with distinct columns then the code $C \leq \mathbf{Z}_2^{14}$ arising from the solutions of $HX = 0$ will be 1-error correcting and $|C| = 2^{10}$. This code has only $2^4 = 16$ cosets so rather than store all 2^{10} elements of C to use for maximum-likelihood decoding we need only store H and at most 16 coset leaders and their syndromes. The savings in storage become huge if $n = 200$ and $k = 150$. Let us look at an example of syndrome decoding.

Example 4.19 A [6, 3] single-error correcting code.

Let $H = \begin{bmatrix} 1 & 0 & 0 & 1 & 0 & 1 \\ 0 & 1 & 0 & 1 & 1 & 0 \\ 0 & 0 & 1 & 0 & 1 & 1 \end{bmatrix}$ and C the code of solutions of $HX = 0$ in $\mathbf{Z}_2^6$.

Now H has rank 3 and no two columns are equal so, as we saw above, $wt(C) \geq 3$ and C is at least single-error correcting. It is not necessary here to compute C explicitly to correct single errors! Observe that since the rank of H is 3, C is generated by $6 - 3 = 3$ independent solutions so $|C| = 8$ and C has $2^6/8 = 8$

cosets. If $\varepsilon_i \in \mathbf{Z}_2^6$ with $wt(\varepsilon_i) = 1$ and the 1 in position i, then each of these is in a different coset of C since $wt(C) \geq 3$ and $wt(\varepsilon_i + \varepsilon_j) = 2$ if $i \neq j$. Clearly $\text{syn}(\varepsilon_i) = H\varepsilon_i^T = \text{Col}_i(H)$. If $\alpha = (0, 1, 1, 1, 1, 1)$ is received then since $\text{syn}(\alpha) = (0, 1, 1)^T = \text{Col}_5(H) = \text{syn}(\varepsilon_5)$, the correct message is $\alpha + \varepsilon_5 = (0, 1, 1, 1, 0, 1) \in C$. If we receive $\beta = (1, 1, 1, 0, 0, 0)$, then $\text{syn}(\beta) = H\beta^T = (1, 1, 1)^T \neq \text{syn}(\varepsilon_j)$. Computing the elements of C shows that β has three errors and $C + \beta$ contains three elements of minimal weight 2 : $(0, 0, 1, 1, 0, 0)$, $(1, 0, 0, 0, 1, 0)$, and $(0, 1, 0, 0, 0, 1)$. Thus β cannot be corrected. Finally, note that $C \cup \{C + \varepsilon_j\}$ contains 56 of the 64 elements in $\{0, 1\}^6$, so 7/8 of all 6-tuples can be decoded.

A nice collection of specific examples of single-error correcting codes that arise in the manner just described are the *Hamming codes*. For any k write the first $2^k - 1$ positive integers in their 2-adic representation as powers of 2 and use these as the columns of a $k \times (2^k - 1)$ matrix. For $k = 3$ we get the matrix $H_3 = \begin{bmatrix} 0 & 0 & 0 & 1 & 1 & 1 & 1 \\ 0 & 1 & 1 & 0 & 0 & 1 & 1 \\ 1 & 0 & 1 & 0 & 1 & 0 & 1 \end{bmatrix}$. Note that H_3 has rank 3 and no two rows are equal. The code arising as the solutions of $H_3X = 0$ is the [7, 4] Hamming code with 2^4 elements. In this case there are $2^{7-4} = 8$ cosets of C and other than C itself, the cosets are $C + \varepsilon_i$ where ε_i is the 7-tuple with exactly one 1 which occurs in the ith position. These ε_j are in different cosets since otherwise the sum of two of them, of weight 2, is in C but $wt(C) \geq 3$. Further, $\{\varepsilon_i\}$ are clearly coset leaders of minimal weight. Since $\text{syn}(\varepsilon_j) = \text{Col}_i(H_3)$, if $\gamma \in \mathbf{Z}_2^7$ is received with a single error then decode γ as $\gamma + \varepsilon_j$ if $\text{syn}(\gamma)$ is the 2-adic representation of j, or, equivalently, $\text{Col}_j(H_3)$.

The analysis above holds for the $[2^k - 1, 2^k - k - 1]$ Hamming code. Here H_k is a $k \times (2^k - 1)$ matrix of rank k with distinct rows and the solutions of $H_kX = 0$ give C with 2^s elements for $s = 2^k - k - 1$. It follows from Lagrange's Theorem that there are 2^k cosets of C in $\mathbf{Z}_2^t$ where $t = 2^k - 1$. Once again $wt(C) \geq 3$ so the $2^k - 1$ different ε_i are the coset leaders for the cosets other than C itself. Decoding γ with at most one error proceeds by evaluating $\text{syn}(\gamma) = H_3\gamma^T$ and correcting γ to $\gamma + \varepsilon_i$ if $\text{syn}(\gamma)$ is the 2-adic representation of i. Thus for the Hamming codes only the matrix H_k needs to be stored!

EXERCISES

1. Is the given information legitimate for using the RSA algorithm? If so, find the decoding integer s:
   i) (671, 11);
   ii) (6161, 33);
   iii) (6161, 77).

2. **a)** Can (25060027, 142) be used for the RSA algorithm (n here is OK!)?
   b) Using RSA with (25060027, 101), how many digits can be encoded as a single transmission?
   c) Find primes $p < q$ so that a four-digit message can be encoded using RSA with (pq, k).

3. If $\langle e \rangle \neq C \leq \mathbf{Z}_2^n$ show for any $\gamma \in C$ there is $\eta \in C$ with $wt(\gamma + \eta) = wt(C)$.

4. If $\langle e\rangle \neq C \leq \mathbf{Z}_2^n$, $C = \langle\{\alpha_1, \ldots, \alpha_k\}\rangle$ with k minimal, $\beta_i, \gamma_j \in \{\alpha_m\}$ with $\beta_1 + \cdots + \beta_s = \gamma_1 + \cdots + \gamma_t$, $\beta_i \neq \beta_j$ and $\gamma_i \neq \gamma_j$ when $i \neq j$, show $s = t$ and $\{\beta_j\} = \{\gamma_j\}$.

5. Which of the following describe $C \leq \mathbf{Z}_2^n$?

i) $\{\alpha \mid wt(\alpha) \in 2\mathbf{Z}\}$

ii) $\{\alpha \mid wt(\alpha) \in 3\mathbf{Z}\}$

iii) $\{\alpha = (a_1, \ldots, a_n) \mid a_j + a_{n-j+1} = 0\}$

6. How many errors could the following $[n, k]$ binary linear codes correct?

i) [20, 10]

ii) [20, 8]

iii) [20, 7]

iv) [15, 10]

v) [30, 20]

vi) [30, 10]

7. For which n might C be a 4-error correcting binary linear $[n, 6]$ code?

8. **a)** Describe a 2-error correcting [15, 3] code (find C with $wt(C) = 5$).

b) Describe how to obtain a d-error correcting linear $[(2d+1)k, k]$ code.

9. Let C be a binary linear $[n, n-k]$ code defined by the solutions of $HX = 0$ for H a $k \times n$ matrix with independent rows. If α is received and if $H\alpha = \mathrm{Col}_j(H)$, show that α can be corrected and specify to what.

10. Let C be the [9, 5] code obtained as the solutions of $HX = 0$ for the matrix

$$H = \begin{bmatrix} 0 & 0 & 1 & 1 & 0 & 0 & 1 & 1 & 0 \\ 1 & 0 & 1 & 1 & 1 & 0 & 0 & 1 & 0 \\ 0 & 1 & 0 & 1 & 0 & 0 & 0 & 1 & 1 \\ 1 & 1 & 1 & 0 & 0 & 1 & 0 & 1 & 0 \end{bmatrix}.$$

i) How many errors can C correct?

ii) If we receive (1, 1, 1, 1, 1, 1, 1, 1, 1) with one error, what is the correct message?

iii) Can the message (1, 1, 1, 1, 1, 0, 0, 1, 0) be corrected?

iv) Can (0, 1, 1, 1, 0, 0, 1, 1, 0) be corrected?

11. For the code C in Exercise 10, if for $\alpha \in C$, $\alpha + \varepsilon_2 + \varepsilon_3$ is received, show that it can be corrected. Can $\alpha + \varepsilon_4 + \varepsilon_6 + \varepsilon_7$ and $\alpha + \varepsilon_1 + \varepsilon_5$ be corrected?

CHAPTER 5

NORMAL SUBGROUPS AND QUOTIENTS

There are $H \leq G$ whose set of cosets has a natural group structure closely related to the structure of G. In this case induction often can be used on the group of cosets to study G when H is finite and $H \neq \langle e \rangle$ since there are fewer cosets of H in G than elements of G. The hope is to translate inductive information back to G. This approach of using the sets in a partition to form a new object turns out to be very useful also in other mathematical contexts. The construction of the group of cosets of $H \leq G$ is essentially the same as the construction of $\mathbf{Z}_n$ from $n\mathbf{Z} \leq \mathbf{Z}$ in §1.8. We study those subgroups whose cosets have a natural group structure, describe the new group, discuss a number of examples, and give some consequences.

5.1 NORMAL SUBGROUPS

Suppose $H \leq G$ and we want to make $\{Hg \mid g \in G\}$ into a group. We are not interested in *some* multiplication that does this, but rather in one closely related to "$\cdot$" in G. It is natural to use the multiplication in G and try the product of subsets of G given in Definition 4.4 as the multiplication of cosets: $Hx \cdot Hy = HxHy = \{hxky \mid h, k \in H\}$. Of course $HxHy$ may not be another coset of H, as we have seen when $G = S_3$ and $H = \langle (1, 2) \rangle$. Thus we need H to satisfy $HxHy = Hz$ and could take this as the defining property of the subgroups we want to consider. A more standard definition is often easier to verify and is equivalent to the property that $HxHy = Hz$.

DEFINITION 5.1 $H \leq G$ is **normal,** denoted $H \triangleleft G$, if for all $x \in H$ and all $g \in G$, $g^{-1}xg \in H$, or, equivalently, if $g^{-1}Hg = \{g^{-1}\}H\{g\} \subseteq H$ for all $g \in G$.

The equivalence in the definition holds using Definition 4.4 on products of subsets since $g^{-1}Hg = \{g^{-1}\}H\{g\} = \{g^{-1}xg \mid x \in H\}$. Clearly $\langle e \rangle \triangleleft G$ and $G \triangleleft G$, the first because $g^{-1}eg = g^{-1}g = e$ if $g \in G$. The similar computation with $x \in Z(G)$

(Definition 2.11) and $g \in G$ implies that $g^{-1}xg = g^{-1}gx = ex = x \in Z(G)$ and shows that if $H \le Z(G)$ then $H \lhd G$. But $G = Z(G)$ when G is Abelian so it follows that every subgroup of an Abelian group is normal. These observations will be used frequently, so for convenience of reference we state them as a theorem.

THEOREM 5.1 For any group G, if $H \le Z(G)$ then $H \lhd G$. In particular, every subgroup of an Abelian group is a normal subgroup.

Non-Abelian groups G can have normal subgroups other than $\langle e \rangle$ and G itself. It is important to note that to show $H \lhd G$, $g^{-1}hg = h$ for each $h \in H$ and $g \in G$ is *not required;* we need only the set $g^{-1}Hg \subseteq H$. In D_n (Definition 3.10), $W_j^{-1}W_iW_j = W_jW_iW_j = W_{2j-i}$. Thus if $\langle W_i \rangle = \{I, W_i\} \lhd D_n$ then using $j = i + 1$ we must have $W_{i+2} = W_i$, since $W_{i+2} \ne I$. But $W_s = W_t \Leftrightarrow n \mid (s - t)$ so $W_{i+2} = W_i$ forces $n \mid 2$, a contradiction. Therefore $\langle W_i \rangle$ cannot be normal in D_n. On the other hand, we have the following example.

Example 5.1 For $n, m \in \mathbf{N}$ with $m \mid n$, $\langle T_m \rangle \lhd D_n$.

Since $D_n = \{T_i, W_i \mid 0 \le i < n\}$ it suffices to show that $T_j^{-1}\langle T_m \rangle T_j \subseteq \langle T_m \rangle$ and $W_j^{-1}\langle T_m \rangle W_j \subseteq \langle T_m \rangle$. Now $\langle T_m \rangle = \{T_m^k = T_{km} \mid k \in \mathbf{Z}\}$ and elements in $\{T_i\}$ commute, so $T_j^{-1}T_{km}T_j = T_{km}$ for all j, resulting in $T_j^{-1}\langle T_m \rangle T_j = \langle T_m \rangle$. From Theorem 3.17 (see Exercise 1 in §3.4), $W_jT_{km}W_j = T_{-km}$ for any j; therefore $W_j^{-1}\langle T_m \rangle W_j \subseteq \langle T_m \rangle$, and $\langle T_m \rangle \lhd D_n$ follows.

It is hard to describe general normal subgroups for arbitrary groups. One example is $A_n \lhd S_n$ by Theorem 3.15. In any group G, $\langle \{x^{-1}y^{-1}xy \in G \mid x, y \in G\} \rangle \lhd G$, using Theorem 2.19. For another important example, recall (Example 2.10) that $SL(n, R) = \{A \in GL(n, R) \mid \det(A) = 1\} \le GL(n, R)$.

Example 5.2 If R is a commutative ring, then $SL(n, R) \lhd GL(n, R)$.

Given any $B \in SL(n, R)$ and $A \in GL(n, R)$, then $A^{-1}BA \in SL(n, R)$ if $\det(A^{-1}BA) = 1$. By the basic property of determinants we have $\det(A^{-1}BA) = \det(A^{-1})\det(B)\det(A) = \det(A^{-1}) \cdot 1 \cdot \det(A) = \det(A^{-1}A) = \det(I_n) = 1$, so indeed $A^{-1}BA \in SL(n, R)$ and it follows that $SL(n, R) \lhd GL(n, R)$ by Definition 5.1.

Most other $H \le GL(n, R)$ will not be normal. For example $Diag(n, \mathbf{Z})$ (Example 2.8) is not normal in $GL(n, \mathbf{Z})$ since if $x = Diag(-1, 1, \ldots, 1)$ and $g = I_n + E_{12}$ is the matrix in $GL(n, \mathbf{Z})$ with $g_{12} = 1$, all $g_{jj} = 1$, and zeros elsewhere, then $(g^{-1}xg)_{12} = -2$, so $g^{-1}xg \notin Diag(n, \mathbf{Z})$. We give some conditions on $H \le G$ equivalent to normality.

THEOREM 5.2 If $H \le G$, then the following are equivalent:

i) $H \lhd G$

ii) $gHg^{-1} \subseteq H$ for all $g \in G$

iii) $g^{-1}Hg = H = gHg^{-1}$ for all $g \in G$
iv) $gH = Hg$ for all $g \in G$
v) $HxHy = Hxy$ for all $x, y \in G$

Proof We show that each statement implies the next. Condition i) means $g^{-1}Hg \subseteq H$ for all $g \in G$, so replacing g by g^{-1} yields $gHg^{-1} = (g^{-1})^{-1}Hg^{-1} \subseteq H$ for all $g \in G$ and ii) holds. But ii) and Theorem 2.3 yield $H = g^{-1}(gHg^{-1})g \subseteq g^{-1}Hg = g^{-1}H(g^{-1})^{-1} \subseteq H$ by ii) with g^{-1} replacing g. This proves $H = g^{-1}Hg$ for all $g \in G$, so again replacing g with g^{-1} gives iii). Now iii) implies iv) since right-multiplying the sets $H = gHg^{-1}$ by g shows $Hg = \{gHg^{-1}\}g = gH$ for any $g \in G$. Given iv) and $x, y \in G$, $HxHy = \{Hx\}\{Hy\} = \{(hx)(ky) \mid h, k \in H\} = H\{xH\}y = H\{Hx\}y = HHxy$, using Theorem 2.3, so $HxHy = Hxy$ since $HH = H$ by Theorem 4.10. Finally, assuming v), for any $g \in G$, $Hg^{-1}Hg = Hg^{-1}g = He = H$, so $g^{-1}Hg = eg^{-1}Hg \subseteq Hg^{-1}Hg \subseteq H$ and $H \lhd G$ from Definition 5.1. ■

Parts of Theorem 5.2 show that in Definition 5.1 we could have used $g^{-1}Hg = H$ instead of containment and could also have put the "inverse" on the other side, taking $H \lhd G$ when $gHg^{-1} \subseteq H$. That $Hg = gH$ in part iv) is useful, but the most important part of the theorem is that the set product of two cosets of a normal subgroup is again a coset and that a representative for the product is any product of representatives of the two cosets. In effect, this says we have a natural multiplication on the collection of cosets that resembles the multiplication in the group. Next we give some useful consequences of Theorem 5.2, convenient to have for future reference.

THEOREM 5.3 If $H \le G$, then $H \lhd G$ when any of the following holds:

i) H is finite and is the only subgroup in G of order $|H|$.
ii) $[G : H] = k \in N$ and H is the only subgroup of G with index k.
iii) $[G : H] = 2$.

Proof For i) observe that $g^{-1}Hg \le G$ (Exercise 7 in §2.5) and $|H| = |g^{-1}Hg|$ since $T(h) = g^{-1}hg$ is a bijection from H to $g^{-1}Hg$ whose inverse is $F(g^{-1}hg) = g(g^{-1}hg)g^{-1} = h$ (Exercise 8 in §2.2). Thus by i), $H = g^{-1}Hg$ so Theorem 5.2 implies that $H \lhd G$. When G is a finite group, Lagrange's Theorem tells us that $|G| = [G : H]|H|$. In this case for $K \le G$, $[G : K] = [G : H] \Leftrightarrow |K| = |H|$ so ii) implies i), forcing $H \lhd G$. For general groups, if ii) holds we claim that $[G : H] = [G : g^{-1}Hg]$ for all $g \in G$ (Exercise 11 in §4.2). Let $\{x_1, \ldots, x_k\}$ be a transversal for the right cosets of H in G so $\{Hx_j \mid 1 \le j \le k\}$ is a partition of G. As above, if $T(x) = g^{-1}xg$ then $T \in \text{Bij}(G)$ so G is partitioned by the k subsets $\{T(Hx_j) = g^{-1}Hx_jg = g^{-1}Hg(g^{-1}x_jg) \mid 1 \le j \le k\}$ (verify!). This proves that $\{g^{-1}x_1g, \ldots, g^{-1}x_kg\}$ is a transversal for the right cosets of $g^{-1}Hg$ in G and so $g^{-1}Hg = H$ by ii). Hence $H \lhd G$ by Theorem 5.2. Finally, if $[G : H] = 2$ then for $y \in G - H$, H and Hy are the two cosets of H in G. Certainly $HH = H$ and $HHy = Hy$ by Theorem 4.10. For $h \in H$, $yh \notin H$ so $Hyh = Hy$ results (why?). It follows that $HyH = \bigcup\{Hyh \mid h \in H\} = Hy$ and then that $HyHy = Hy^2$. Therefore the product of two cosets of H is again a coset, so by Theorem 5.2, $H \lhd G$. Another approach for iii) uses $|\{Hg\}| = |\{gH\}|$ (Exercise 4 §4.1) so $\{H, G - H\}$ are the right cosets as well as the left cosets. Any right coset is a left coset and again H is normal in G by Theorem 5.2. ■

We mention a few applications of the theorem that do not really require it but illustrate that its hypotheses do actually arise. Theorem 3.15 and Theorem 5.3 show $A_n \lhd S_n$ since $|A_n| = |S_n|/2$, although we can see this directly from Theorem 3.15. Similarly, $\langle T_1 \rangle \lhd D_n$ from Example 5.1 follows also from Theorem 5.3 since $|\langle T_1 \rangle| = |D_n|/2$. Finally, suppose that G is a group with $|G| = pq$ for $p < q$ primes and that $H \le G$ with $|H| = q$. If $K \le G$ and $|K| = q$ then by Theorem 4.12, $|HK| = q^2/|H \cap K| \le |G| = pq$. This forces $|H \cap K| = q$ (why?) so $K = H \cap K = H$ and H is the only subgroup of G with order q. By Theorem 5.3, $H \lhd G$. This is an important first step in determining the structure of these groups. We do this in a later chapter.

Next we consider products and intersections of normal subgroups.

THEOREM 5.4

i) If for $\{N_\lambda\}_\Lambda$ each $N_\lambda \lhd G$, then $\bigcap_\lambda N_\lambda \lhd G$.
ii) If $N \lhd G$ and $H \le G$, then $NH \le G$ and $N \cap H \lhd H$.
iii) If $N_1, \ldots, N_k \lhd G$, then $N_1 N_2 \cdots N_k \lhd G$.

Proof We use Definiton 5.1 to prove i). If $x \in \bigcap N_\lambda$ and $g \in G$, then from the definition of intersection $x \in N_\lambda$ for each $\lambda \in \Lambda$, and since $N_\lambda \lhd G$, $g^{-1}xg \in N_\lambda$. Thus $g^{-1}xg \in \bigcap N_\lambda$ so $\bigcap N_\lambda \lhd G$, using Theorem 2.14. For ii), Theorem 5.2 and $h \in H$ give $Nh = hN$ so $NH = \bigcup \{Nh \mid h \in H\} = \bigcup \{hN \mid h \in H\} = HN$. Therefore $NH \le G$ by Theorem 4.10. If $x \in N \cap H$ and $h \in H$, then $h^{-1}xh \in N$ because $x \in N$ and $N \lhd G$. Also $h^{-1}xh \in H$ since $x, h \in H$, so $h^{-1}xh \in N \cap H$, showing that $N \cap H \lhd H$ by definition. We use induction for iii). When $k = 2$, $N_1 N_2 \le G$ by ii) and for any $g \in G$, $(N_1 N_2)g = N_1(N_2 g) = N_1(gN_2) = (N_1 g)N_2 = (gN_1)N_2 = gN_1N_2$, so $N_1N_2 \lhd G$ from Theorem 5.2. Assume iii) holds for some $k \ge 2$ and let $\{N_1, \ldots, N_{k+1}\}$ be normal subgroups of G. Then $N_1N_2 \lhd G$ so the induction assumption for k normal subgroups yields $N_1 N_2 \cdots N_{k+1} = (N_1N_2)N_3 \cdots N_{k+1} \lhd G$ and iii) is true by IND-I.

■

Results from Chapter 4 together with Theorem 5.4 yield a technical result used later to decompose certain groups into simpler ones.

THEOREM 5.5 If $N_1, \ldots, N_k \lhd G$, each N_j is finite, and $\{|N_j|\}$ is pairwise relatively prime, then $|N_1 \cdots N_k| = |N_1| \cdots |N_k|$ and if $j < k$, $N_1 \cdots N_j \cap N_{j+1} = \langle e_G \rangle$.

Proof By Theorem 5.4, $N_1 \cdots N_j \le G$ for each $j \le k$, so $|N_1 \cdots N_k| = |N_1| \cdots |N_k|$ by Theorem 4.13, and it now follows from Theorem 4.6 and Theorem 1.11 that $(N_1 \cdots N_j) \cap N_{j+1} = \langle e_G \rangle$ when $j < k$.

■

EXERCISES

1. In each case determine whether the given subset N is a normal subgroup:
  i) for G and H groups, $N = G \oplus \langle e_H \rangle \subseteq G \oplus H$
  ii) for G a group, $N = \{(g, g) \in G \oplus G \mid g \in G\} \subseteq G \oplus G$
  iii) for $G = \{f \in \mathrm{Bij}(\boldsymbol{R}) \mid f(x) = ax + b$, with $a, b \in \boldsymbol{R}$ and $a \ne 0\}$, and $\boldsymbol{Q}^+ = \{q \in \boldsymbol{Q} \mid q > 0\}$, $N = \{g \in G \mid g(x) = ax + b$ for $a \in \boldsymbol{Q}^+\} \subseteq (G, \circ)$

iv) $N = \{f \in \text{Bij}(\mathbf{N}) \mid f(2\mathbf{N}) = 2\mathbf{N}\} \subseteq \text{Bij}(\mathbf{N})$
v) $N = \{I, W_0, W_n, T_n\} \subseteq D_{2n}$
vi) $N = \{W_{2k+1}, T_{2k} \in D_{2n} \mid k \in \mathbf{Z}\} \subseteq D_{2n}$
vii) if $n, m \in \mathbf{N}$ with $m \mid n$, $N = \{W_{mk}, T_{mk} \in D_n \mid k \in \mathbf{Z}\} \subseteq D_n$
viii) if p is a prime, $N = \left\{\begin{bmatrix} a & b \\ 0 & a^{-1} \end{bmatrix} \in UT(2, \mathbf{Z}_p) \mid a \in U_p\right\} \subseteq GL(2, \mathbf{Z}_p)$
ix) if p is a prime, $N = \left\{\begin{bmatrix} a & b \\ 0 & a^{-1} \end{bmatrix} \in UT(2, \mathbf{Z}_p) \mid a \in U_p\right\} \subseteq UT(2, \mathbf{Z}_p)$
x) $N = UT(n, \mathbf{R}) \subseteq GL(n, \mathbf{R})$
xi) $N = \{A \in GL(n, \mathbf{Z}) \mid A_{ij} \in 3\mathbf{Z} \text{ if } i \neq j\} \subseteq GL(n, \mathbf{Z})$
xii) for $p > 2$ prime, $N = \{A \in UT(3, \mathbf{Z}_p) \mid \text{all } A_{ii} \in \langle [-1]_p \rangle\} \subseteq UT(3, \mathbf{Z}_p)$

2. For groups $G_1, \ldots, G_k$ and $N_i \lhd G_i$, show that $N_1 \oplus \cdots \oplus N_k \lhd G_1 \oplus \cdots \oplus G_k$.

3. Show for a group G and $m > 1$ that $N = \{(g, \ldots, g) \in G^m = G \oplus \cdots \oplus G (m \text{ times}) \mid g \in G\} \lhd G^m$ if and only if G is Abelian.

4. For $n \in \mathbf{N}$ and G any group, when $H = \{g \in G \mid g^n = e_G\} \leq G$, then $H \lhd G$.

5. If $N \lhd G$ and N is a cycle group, show that for any $H \leq N$, $H \lhd G$.

6. If $N, M \leq G$, $H \lhd N$, and $K \lhd M$, show that:
a) $H \bigcap K \lhd N \bigcap M$
b) if in addition $M \lhd G$, then $HM \lhd NM$.

7. If $N \lhd G$ and $H \lhd N$, show that H need not be normal in G. (Try $G = A_4$.)

8. Let $A \lhd B \leq G$ and $C \lhd D \leq G$. Show that:
i) $A(B \bigcap C), A(B \bigcap D) \leq B$
ii) $B \bigcap C \lhd B \bigcap D$
iii) $A(B \bigcap C) \lhd A(B \bigcap D)$

For Exercises 9–10, recall from Example 2.12 that $S_\infty = \{f \in \text{Bij}(\mathbf{N}) \mid \text{for some } n(f) \in \mathbf{N}, f(m) = m \text{ if } m \geq n(f)\} \leq \text{Bij}(\mathbf{N})$, and also that for $m \in \mathbf{N}$, $I(m) = \{1, 2, \ldots, m\}$.

9. Show that $S_\infty \lhd \text{Bij}(\mathbf{N})$.

10. Let $A_\infty = \{f \in S_\infty \mid \text{when } f(n) = n \text{ for all } n \geq M, \text{ then } f|_{I(M)} \in A_M\}$.
a) Show that $A_\infty \lhd S_\infty$.
b) Is $A_\infty \lhd \text{Bij}(\mathbf{N})$?

11. If $3 \leq m < n$ in $\mathbf{N}$, regard $A_m \leq S_n$ by extension of each $f \in A_m$ to $I(n)$, setting $f(k) = k$ for all $k > m$. Show that A_m is not a normal subgroup of S_n.

12. Find all normal subgroups in:
i) D_4
ii) D_5
iii) A_4
iv) S_4

13. Define the *quaternion group* to be $Q = \{\pm 1, \pm i, \pm j, \pm k\}$ with an associative multiplication satisfying $1 = e$, $(-1)^2 = 1$, $-1 \cdot y = y \cdot (-1) = -y$ and $-1 \cdot -y = -y \cdot -1 = y$ for $y \in \{i, j, k\}$, $x^2 = -1$ if $x \neq \pm 1$, $ij = k$, $jk = i$, and $ki = j$.
i) If $a, b \in Q - \{\pm 1\}$ with $a \neq \pm b$, show that $ab = -1 \cdot ba$.
ii) If $H \leq Q$ and is proper, show that H is cyclic and $H \lhd Q$.

14. Let p be a prime and G be a finite group. If $A = \{\langle e \rangle \neq H \leq G \mid |H| = p^s, \text{ some } s \geq 1\}$, then if $A \neq \varnothing$, show that $\bigcap_A H \lhd G$.

15. Let G be a finite group and p a prime. Set $A = \{H \lhd G \mid |H| = p^s \text{ some } s \geq 0\}$ and let $B = \{H \lhd G \mid (|H|, p) = 1\}$. Show that with respect to set inclusion:
   i) A contains a unique maximal subgroup.
   ii) B contains a unique maximal subgroup.

16. Let $H, K \lhd G$ with $H \cap K = \langle e \rangle$. Show that for all $h \in H$ and $k \in K$, $hk = kh$. (Hint: consider $h^{-1}k^{-1}hk$.)

17. Let $H \leq G$, $S = \{gH \mid g \in G\}$, and assume $[G : H] = |S| = n \in N$.
   a) For each $x \in G$, show that $F_x(gH) = xgH$ defines $F_x : S \to S$.
   b) Show that each $F_x \in \mathrm{Bij}(S)$.
   c) If $F : G \to \mathrm{Bij}(S)$ is $F(x) = F_x$, show $F(xy) = F(x)F(y)$ and $F(x)^{-1} = F(x^{-1})$.
   d) Show that $F(G) \leq \mathrm{Bij}(S)$.
   e) If $K(F) = \{x \in G \mid F(x) = I_S\}$, show that $K(F) \lhd G$ with $K(F) \leq H$.
   f) Show that $K(F) \neq \langle e_G \rangle$ (part e)) $\Leftrightarrow$ F is not injective.

18. Let $\langle e_G \rangle \neq H \leq G$, G finite with $[G : H] = p$ the smallest prime divisor of $|G|$. Use Exercise 17 to show that there is $\langle e \rangle \neq K \leq H$ and normal in G.

5.2 QUOTIENT GROUPS

Normal subgroups of a group G are important primarily because their cosets naturally become a new group with multiplication closely related to that in G. We mentioned this earlier and it follows from Theorem 5.2. If $H \leq G$ and $X = \{Hg \mid g \in G\}$ we can define a product of the cosets $Hx, Hy \in X$ as the set product $HxHy \subseteq G$. Examples in §4.3 show that this set product is not always another coset of H, but Theorem 5.2 proves that when $N \lhd G$ then not only is $NxNy = Nz \in X$, it is the coset Nxy. In this case the theorem ensures that the natural set multiplication $NxNy = Nxy$ really is a function $\cdot : X \times X \to X$, so there is a possibility that $(X, \cdot)$ is a group.

THEOREM 5.6 Let $N \lhd G$, $G/N = (\{Ng \mid g \in G\}, \cdot)$, and $Nx \cdot Ny = NxNy = Nxy$. Then G/N is a group and if G is finite, $|G/N| = [G : N] = |G|/|N|$.

Proof From Theorem 5.2 and our comments above, if $X = \{Ng \mid g \in G\}$ the coset multiplication $Nx \cdot Ny = \{Nx\}\{Ny\} = Nxy$ is a function from X^2 to X, the first step in showing that G/N is a group. Another way to see that $Nx \cdot Ny = Nxy$ is a multiplication is to show that this product depends only on the cosets as sets and not on the particular elements used to represent them. If $Nx = Nx'$ and $Ny = Ny'$, then using Theorem 5.2, $Nx \cdot Ny = Nxy = NxNy$, and $Nx' \cdot Ny' = Nx'y' = Nx'Ny'$. In G the set products $NxNy = Nx'Ny'$ so $Nx \cdot Ny = Nx' \cdot Ny'$ and "$\cdot$" *is* a function from X^2 to X. To prove G/N is a group we must use Definition 2.1. If $Nx, Ny, Nz \in G/N$, the definiton of "$\cdot$" shows $(Nx \cdot Ny) \cdot Nz = Nxy \cdot Nz = N((xy)z)$. Now "$\cdot$" in G is associative so $((xy)z) = (x(yz))$ and $(Nx \cdot Ny) \cdot Nz = N(x(yz)) = Nx \cdot Nyz = Nx \cdot (Ny \cdot Nz)$ follows, proving that "$\cdot$" is associative. Since $ge_G = g = e_G g$ for all $g \in G$, clearly $e_{G/N} = Ne_G = N$. Lastly, $Ng \cdot Ng^{-1} = Ngg^{-1} = Ne_G = N = Ng^{-1} \cdot Ng$, so each $Ng \in G/N$ has inverse $(Ng)^{-1} = Ng^{-1}$. Thus G/N is a group. The elements of G/N are just the cosets of N in G, so when G is finite, Lagrange's Theorem (Theorem 4.5) tells us that $|G/N| = [G : N] = |G|/|N|$, completing the proof.

■

We write $Nx \cdot Ny$ as $NxNy$ in G/N to remind us that the multiplication in G/N is just set multiplication of cosets in G. The proof of Theorem 5.6 shows that Ne_G is the identity of G/N and that $Ng^{-1} = (Ng)^{-1}$, indicating again how the structures of G/N and of G are closely related.

DEFINITION 5.2 If $N \lhd G$, then the group G/N with coset multiplication is called the **quotient group** or **factor group** of G by N.

We will use quotient groups, often just called quotients, to derive new results but will first look at a number of examples. It is important to become familiar with the quotient construction and also to see how to identify the structure of G/N in terms of groups we know. The simplest examples are when $N = G$ or when $N = \langle e \rangle$. In the first case the only coset of G in G is G itself, so $G/G = \{G\} = \{e_{G/G}\}$. When $N = \langle e_G \rangle$ its cosets are $\langle e_G \rangle g = \{g\}$, so there is an obvious bijection between G and $G/\langle e_G \rangle = \{\{g\} \mid g \in G\}$. Since $\{g\}\{h\} = \{gh\}$, multiplication in $G/\langle e_G \rangle$ looks like that in G, and in a sense we have yet to make precise, $G/\langle e_G \rangle$ and G are *essentially* the same group. These cannot be the same group since their underlying sets differ. Figure 5.1 gives a picture of the quotient.

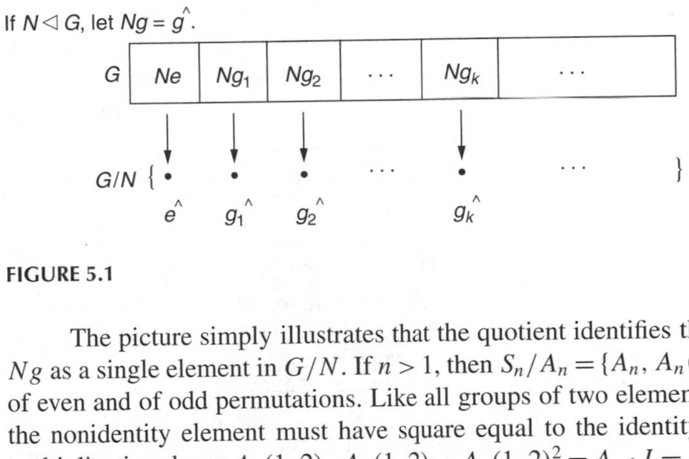

FIGURE 5.1

The picture simply illustrates that the quotient identifies the set of elements in each Ng as a single element in G/N. If $n > 1$, then $S_n/A_n = \{A_n, A_n(1,2) = S_n - A_n\}$, the sets of even and of odd permutations. Like all groups of two elements, S_n/A_n must by cyclic: the nonidentity element must have square equal to the identity. The definition of coset multiplication shows $A_n(1,2) \cdot A_n(1,2) = A_n(1,2)^2 = A_n \cdot I = A_n$, so $S_n/A_n = \langle A_n(1,2) \rangle$ even though S_n is not Abelian when $n > 2$. Next we identify all quotients of $(\mathbf{Z}, +)$.

Example 5.3 For $n > 1$, $(\mathbf{Z}, +)/n\mathbf{Z} = (\mathbf{Z}_n, +)$.

A typical coset is $n\mathbf{Z} + a$ for $a \in \mathbf{Z}$. We know from Theorem 1.2 that $\{n\mathbf{Z} + a \mid a \in \mathbf{Z}\} = \{n\mathbf{Z} + i \mid 0 \le i < n\} = \mathbf{Z}_n$. The group operation in $\mathbf{Z}/n\mathbf{Z}$ is $(n\mathbf{Z} + a) + (n\mathbf{Z} + b) = n\mathbf{Z} + (a + b)$, which is exactly the addition in $\mathbf{Z}_n$. Consequently, the groups $(\mathbf{Z}, +)/n\mathbf{Z}$ and $(\mathbf{Z}_n, +)$ are the same.

The example shows that infinite groups can have nontrivial finite quotients. We turn to a number of examples of quotient groups. In each case we will try to identify the quotient with a known group. This identification would be easier using Theorem 6.6 in the next chapter, but to become familiar with the quotient construction it is important to work directly with cosets and their multiplication.

Example 5.4 $G = \mathbf{Z}_{12}/\langle[4]_{12}\rangle$ is cyclic of order 4.

Since $\mathbf{Z}_{12}$ is Abelian each of its subgroups is normal by Theorem 5.1 so $G = \mathbf{Z}_{12}/\langle[4]_{12}\rangle$ is a group using set multiplication of cosets. Now $\langle[4]_{12}\rangle = \{[0]_{12}, [4]_{12}, [8]_{12}\}$ and $\langle[4]_{12}\rangle + [11]_{12} = \{[0+11]_{12}, [4+11]_{12}, [8+11]_{12}\} = \{[11]_{12}, [3]_{12}, [7]_{12}\}$ are in G. The other cosets are $\langle[4]_{12}\rangle + [2]_{12} = \{[2]_{12}, [6]_{12}, [10]_{12}\}$ and $\langle[4]_{12}\rangle + [5]_{12} = \{[5]_{12}, [9]_{12}, [1]_{12}\}$. Write $\langle[4]_{12}\rangle + [a]_{12} = a'$ so $G = \{0', 2', 5', 11'\}$. The group operation is addition so $2'^2 = 2' + 2' = (\langle[4]_{12}\rangle + [2]_{12}) + (\langle[4]_{12}\rangle + [2]_{12}) = \langle[4]_{12}\rangle + [4]_{12} = \langle[4]_{12}\rangle = 0' = e_G$ and $5'^2 = 5' + 5' = (\langle[4]_{12}\rangle + [5]_{12}) + (\langle[4]_{12}\rangle + [5]_{12}) = \langle[4]_{12}\rangle + [10]_{12} = \langle[4]_{12}\rangle + [2]_{12} = 2'$ by Theorem 4.2 since $[10]_{12} \in \langle[4]_{12}\rangle + [2]_{12}$. But $o(5') \mid 4$ by Theorem 4.7 so $G = \langle 5'\rangle$. We can choose more "natural" representatives for the cosets using Theorem 4.2 and write $G = \{0', 1', 2', 3'\}$ since $5' = \langle[4]_{12}\rangle + [1]_{12}$ and $11' = \langle[4]_{12}\rangle + [3]_{12}$. Now $1'^2 = 2'$, $1'^3 = 3'$, and $1'^4 = 0'$ (verify!), so G looks like the cyclic group $\mathbf{Z}_4$.

Example 5.4 is a guide for arbitrary finite cyclic groups.

Example 5.5 If $G = \langle g\rangle$, $|G| = n$, and $m \mid n$, then $G/\langle g^m\rangle = \langle\langle g^m\rangle g\rangle$ of order m.

Since G is Abelian, $\langle g^m\rangle \lhd G$ (Theorem 5.1) so $G/\langle g^m\rangle$ is a group. Now $\langle g^m\rangle = \{g^m, \ldots, g^{m(n/m)} = e\}$ by Theorem 2.16 and we want "nice" representatives for its cosets. From Theorem 2.9, $g^i = g^j \Leftrightarrow n \mid (i-j)$, so if $g^{mi} = g^k$ then $m \mid n$ and Theorem 1.6 force $m \mid k$. Thus m is the smallest positive power of g in $\langle g^m\rangle$. If $0 \le i < j < m$ and $\langle g^m\rangle g^i = \langle g^m\rangle g^j$ then by Theorem 4.2, $g^{j-i} = g^j g^{-i} \in \langle g^m\rangle$, contradicting $0 < j - i < m$. Hence $\langle g^m\rangle, \langle g^m\rangle g, \ldots, \langle g^m\rangle g^{m-1}$ are distinct. But $o(g^m) = n/m$ by Theorem 2.8 so $|\langle g^m\rangle| = n/m$ from Theorem 2.16, and Theorem 5.6 shows $|G/\langle g^m\rangle| = m$, forcing $G/\langle g^m\rangle = \{\langle g^m\rangle, \langle g^m\rangle g, \ldots, \langle g^m\rangle g^{m-1}\}$. Using coset multiplication, $\langle g^m\rangle g^j = (\langle g^m\rangle g)^j$, so $G/\langle g^m\rangle = \langle\langle g^m\rangle g\rangle$.

In particular, since $[k]_n = [1]_n^k$ in $\mathbf{Z}_n$, we have $\mathbf{Z}_{16}/\langle[8]_{16}\rangle = \langle\langle[8]_{16}\rangle + [1]_{16}\rangle$ has order 8 and $(\langle[8]_{16}\rangle + [1]_{16})^m = \langle[8]_{16}\rangle + [m]_{16}$ (why?). Also $U_7 = \langle[3]_7\rangle$ and $[3]_7^3 = [27]_7 = [6]_7$ so $U_7/\langle[6]_7\rangle = \langle\langle[6]_7\rangle[3]_7\rangle$ of order 3. Here $(\langle[6]_7\rangle[3]_7)^2 = \langle[6]_7\rangle[3]_7^2 = \langle[6]_7\rangle[2]_7$ and $(\langle[6]_7\rangle[3]_7)^3 = \langle[6]_7\rangle$ since $[2]_7[3]_7 = [6]_7 \in \langle[6]_7\rangle$.

Not all quotient groups are cyclic, as we see next! It is hard to find normal subgroups in non-Abelian groups so for now we continue with examples of Abelian groups. The importance of finding good coset representatives becomes more evident in the examples that follow.

Example 5.6 Let $G = \mathbf{Z}_5^3 = \mathbf{Z}_5 \oplus \mathbf{Z}_5 \oplus \mathbf{Z}_5$ and $H = \{(a, a, a) \in G \mid a \in \mathbf{Z}_5\}$. Then G/H is Abelian, is not cyclic, is of order 25, and is essentially $\mathbf{Z}_5 \oplus \mathbf{Z}_5$.

Now $H \le G$ by Theorem 2.12, using that G is finite and $(a, a, a) + (b, b, b) = (a+b, a+b, a+b)$. Since G is Abelian, $H \lhd G$ by Theorem 5.1 so G/H is a group and $|H| = 5$ yields $|G/H| = |G|/|H| = 25$ by Theorem 5.6. To describe G/H we want to find nice coset representatives. Now $H + (a, b, c) = \{(x + a, x + b, x + c) \mid x \in \mathbf{Z}_5\}$, so for $x = -c$, $(a - c, b - c, [0]_5) \in H + (a, b, c)$ follows and $H + (a, b, c) = H + (a - c, b - c, [0]_5)$ by Theorem 4.2. Thus each $H + (a, b, c)$ is represented by some $(u, v, [0]_5)$. If $(u, v, [0]_5), (u', v', [0]_5) \in H + (a, b, c)$ then using Theorem 4.2 forces their difference, $(u - u', v - v', [0]_5) \in H$. By definition of H, $u - u' = v - v' = [0]_5$, so $u = u'$ and $v = v'$. Consequently, different elements in $\{(u, v, [0]_5) \mid u, v \in \mathbf{Z}_5\}$ are in different cosets so $G/H = \{H + (a, b, [0]_5) = (a, b)^\# \mid a, b \in \mathbf{Z}_5\}$. Now $(\ ,\ )^\#: \mathbf{Z}_5 \oplus \mathbf{Z}_5 \to G/H$ is a bijection and also $(a, b)^\# + (a', b')^\# = (H + (a, b, [0]_5)) + (H + (a', b', [0]_5)) = H + ((a, b, [0]_5) + (a', b', [0]_5)) = H + (a + a', b + b', [0]_5) = (a + a', b + b')^\#$. It follows that G/H is Abelian and is "essentially" $(\mathbf{Z}_5 \oplus \mathbf{Z}_5, +)$. Since $((a, b)^\#)^5 = (H + (a, b, [0]_5))^5 = H + (5a, 5b, [0]_5) = H + ([0]_5, [0]_5, [0]_5) = H$, each $X \in G/H$ satisfies $X^5 = e_{G/H}$ and G/H cannot be cyclic (why?). Observe that finding some $K \leq G$ to represent the cosets of H makes G/H look like K.

Example 5.7 If $G = U_{16}$, $H = \langle [9]_{16} \rangle$, and $K = \langle [15]_{16} \rangle$, then H and K are cyclic of order 2, $|G/H| = |G/K| = 4$, and G/K is cyclic but G/H is not cyclic.

$U_{16} = \{[1], [3], [5], [7], [9], [11], [13], [15]\}$ is Abelian, $H, K \lhd U_{16}$ by Theorem 5.1, so G/H and G/K are defined. Clearly $[9]^2 = [81] = [1]$ shows $|H| = 2$ so $|G/H| = 4$ by Theorem 5.6. Its cosets are $H = \{[1], [9]\}$, $H[3] = \{[3], [11]\}$, $H[5] = \{[5], [13]\}$, and $H[7] = \{[7], [15]\}$, denoted by $1'$, $3'$, $5'$, and $7'$, respectively, so $G/H = \{1', 3', 5', 7'\}$. Using Theorem 4.2, $3'^2 = H[3]H[3] = H[9] = H = 1'$ and $5'^2 = H[5]H[5] = H[25] = H[9] = H = 1'$. Similarly, $7'^2 = 1'$, so G/H is not cyclic. It is straightforward to see that the product of two different elements in $\{3', 5', 7'\}$ is the third, so G/H is Abelian, e.g., $3'7' = H[3]H[7] = H[21] = H[5] = 5'$. This quotient is like U_8 in that we can identify $[a]_8 \in U_8$ with a' for $a \in \{1, 3, 5, 7\}$ and products of corresponding elements correspond. For $K = \{[1], [15]\}$, $[15]^2 = [-1]^2 = [1]$, so $|K| = 2$ and $|G/K| = 4$. The cosets in G/K are K, $K[3] = \{[3], [13]\}$, $K[5] = \{[5], [11]\}$, and $k[7] = \{[7], [9]\}$ represented by $1''$, $3''$, $5''$, and $7''$. Now Theorem 4.2 shows $3''^2 = K[3]K[3] = K[9] = K[7] = 7''$, $3''^3 = 7''3'' = K[7]K[3] = K[21] = K[5] = 5''$, and $3'^4 = 5''\,3'' = K[5]K[3] = K[15] = K = 1''$. Thus $G/K = \langle 3'' \rangle$, so G/K and $G/H = \{1', 3', 5', 7'\}$ have different structures.

The next examples for infinite groups reinforce the importance of finding nice representatives for the cosets in G/N.

Example 5.8 If $G = \mathbf{Z} \oplus \mathbf{Z}$ and $N = \{(2a, 3b) \in G \mid a, b \in \mathbf{Z}\}$, then G/N is cyclic of order 6.

Since $(2a, 3b) - (2a', 3b') = (2(a - a'), 3(b - b')) \in N$, Theorem 2.12 implies $N \leq G$. From Theorem 5.1, $N \lhd G$ because G is Abelian and so G/N is defined. Let us try to find nice representatives for the cosets of N in G. Using Theorem 4.2, $(u, v) \in N + (x, y) \Leftrightarrow (x, y) - (u, v) = (x - u, y - v) \in N \Leftrightarrow 2 \mid (x - u)$

and $3 \mid (y - v)$ from the definition of N. The division theorem (Theorem 1.1) yields $x = 2s + i$ and $y = 3t + j$ with $0 \le i \le 1$ and $0 \le j \le 2$, so $(x, y) - (i, j) = (2s, 3t) \in N$. Therefore $(i, j) \in N + (x, y)$ and each coset of N in G contains an element in $S = \{(i, j) \in G \mid 0 \le i \le 1 \text{ and } 0 \le j \le 2\}$. If two such, say (i, j) and (i', j'), were in the same coset then $(i - i', j - j') \in N$ so $2 \mid (i - i')$ and $3 \mid (j - j')$, forcing $i = i'$ and $j = j'$. Hence $G/N \subseteq \{s' = N + s \mid s \in S\} \subseteq G/N$ so $|G/N| = 6$. We compute that $(1, 1)'^2 = (N + (1, 1)) + (N + (1, 1)) = N + ((1, 1) + (1, 1)) = N + (2, 2) = N + (0, 2) = (0, 2)'$, since $(2, 2) - (0, 2) \in N$. Now $(1, 1)'^3 = (1, 1)'^2 + (1, 1)' = (N + (0, 2)) + ((N + (1, 1)) = N + (1, 3) = N + (1, 0)$ (why?). But using Theorem 4.7, $o((1, 1)') \mid 6$, forcing $o((1, 1)') = 6$ and $G/N = \langle (1, 1)' \rangle$ is cyclic of order 6 from Theorem 2.16.

A slightly different example gives a much different result.

Example 5.9 If $N = \langle (2, 3) \rangle \le G = \mathbf{Z} \oplus \mathbf{Z}$, G/N is an infinite cyclic group.

The group operation in $(\mathbf{Z}, +)$ is addition so using Definition 2.4, powers become $k^t = tk$ for $t, k \in \mathbf{Z}$. The group operation in G is also written as addition and is given coordinatewise, so if $t \in \mathbf{Z}$ and $(x, y) \in G$, it follows that $t(x, y) = (x, y)^t = (x^t, y^t) = (tx, ty)$. These yield $\langle (x, y) \rangle = \mathbf{Z} \cdot (x, y) = \{z(x, y) = (zx, zy) \in G \mid z \in \mathbf{Z}\}$. Thus $N = \{(2z, 3z) \in \mathbf{Z}^2 \mid z \in \mathbf{Z}\}$ and from Theorem 5.1, $N \lhd G$ because G is Abelian. It is not easy to exhibit a good choice of coset representatives for N in G. For any $(a, b) \in G$, the two equations $2x + y = a$ and $3x + y = b$ have solution $x = -a + b$ and $y = 3a - 2b$. Therefore $(a, b) = x(2, 3) + y(1, 1) \in N + y(1, 1)$ so $N + (a, b) = N + y(1, 1)$ by Theorem 4.2 and we can use elements from $\langle (1, 1) \rangle$ to represent the cosets in G/N. Of course, every $N + (t, t) \in G/N$. Note that $\langle (2, 3) \rangle \cap \langle (1, 1) \rangle = \langle (0, 0) \rangle$ since $s(2, 3) = t(1, 1)$ forces $2s = t = 3s$, so $s = 0 = t$. Thus no coset of N contains different $\alpha, \beta \in \langle (1, 1) \rangle$ since otherwise $\alpha - \beta \in \langle (2, 3) \rangle \cap \langle (1, 1) \rangle$, using Theorem 4.2 again, so $G/N = \{N + (z, z) \mid z \in \mathbf{Z}\}$. It is an exercise that $\{N + (1, 1)\}^k = N + (k, k)$: if $k > 0$ then $\{N + (1, 1)\}^k = (N + (1, 1)) + \cdots + (N + (1, 1)) = N + (1, 1)^k = N + (k, k)$. Hence $G/N = \langle N + (1, 1) \rangle$ is an infinite cyclic group.

Finding normal subgroups and computing are harder in non-Abelian groups than in Abelian groups, but we should look at a few such examples.

Example 5.10 If $H = \{I, (a, b)(c, d) \in S_4 \mid \{a, b, c, d\} = I(4)\}$ then $H \lhd S_4$ and G/H is essentially S_3.

It is easy to verify that $H \le S_4$. Later it will be easy to see that $H \lhd S_4$, but for now we just compute that left and right cosets of H in G are the same. These sets are: $H = \{I, (1, 2)(3, 4), (1, 3)(2, 4), (1, 4)(2, 3)\}$; $H(1, 2) = (1, 2)H = \{(1, 2), (3, 4), (1, 4, 2, 3), (1, 3, 2, 4)\}$; $H(1, 3) = (1, 3)H = \{(1, 3), (1, 4, 3, 2), (2, 4), (1, 2, 3, 4)\}$; $H(2, 3) = (2, 3)H = \{(2, 3), (1, 2, 4, 3), (1, 3, 4, 2), (1, 4)\}$; $H(1, 2, 3) = (1, 2, 3)H = \{(1, 2, 3), (2, 4, 3), (1, 4, 2), (1, 3, 4)\}$; $H(1, 3, 2) = (1, 3, 2)H = \{(1, 3, 2), (1, 4, 3), (2, 3, 4), (1, 2, 4)\}$. Using Theorem 4.2, $H\theta = \theta H$ for any $\theta \in S_4$

so $H \triangleleft S_4$ by Theorem 5.2. By our computation we can write $S_4/H = \{Hs = s' \mid s \in S_4 \text{ and } s(4) = 4\}$. Clearly there is a bijection from S_3 to S_4/H and matching elements multiply consistently: $s't' = HsHt = Hst$ and $st(4) = 4$, so $s't' = (st)'$. For example, $(1, 2)'(1, 3)' = H(1, 2)H(1, 3) = H(1, 2)(1, 3) = H(1, 3, 2) = (1, 3, 2)' = ((1, 2)(1, 3))'$. Therefore S_4/H is "essentially" S_3.

Example 5.11 If $N = \langle T_2 \rangle \leq D_6$, then $N \triangleleft D_6$ and D_6/N is Abelian of order 4 with every nonidentity element having order 2.

From Theorem 3.17, $D_6 = \{W_i, T_i \mid 0 \leq i \leq 5\}$ and $\langle T_2 \rangle = \{T_0, T_2, T_4\}$. Now $N = \langle T_2 \rangle \triangleleft D_6$ by Example 5.1 so Theorem 5.6 shows D_6/N is defined with $|D_6/N| = 4$. Using $T_i W_j = W_{i+j}$ the cosets of N are $N = \{I, T_2, T_4\}$, $NT_1 = \{T_1, T_3, T_5\}$, $NW_0 = \{W_0, W_2, W_4\}$, and $NW_1 = \{W_1, W_3, W_5\}$, so $G = D_6/N = \{N, NT_1, NW_0, NW_1\}$. Squares are $NT_1NT_1 = NT_1T_1 = NT_2 = N$, $NW_0NW_0 = NW_0^2 = NI = N$, and $NW_1NW_1 = NW_1^2 = NI = N$; Theorem 2.10 now implies that G is Abelian. This follows directly using Theorem 4.2; for example, $NT_1NW_0 = NT_1W_0 = NW_1$ and $NW_0NT_1 = NW_0T_1 = NW_{-1} = NW_5 = NW_1$.

Example 5.12 $G = GL(n, \mathbf{R})/SL(n, \mathbf{R})$ is essentially $(\mathbf{R}^*, \cdot)$.

From Example 5.2, $SL(n, \mathbf{R}) \triangleleft GL(n, \mathbf{R})$ and from Example 4.5, each coset consists of all matrices with a fixed determinant in $\mathbf{R}^*$. For any $r \in \mathbf{R}^*$, $r'' = Diag(r, 1, \ldots, 1) \in GL(n, \mathbf{R})$ satisfies $\det(r'') = r$ so we may consider $G = \{r^\# = SL(n, \mathbf{R})r'' \mid r \in \mathbf{R}^*\}$. Computing the product of $r^\#, s^\# \in G$ yields $r^\# s^\# = SL(n, \mathbf{R})r''SL(n, \mathbf{R})s'' = SL(n, \mathbf{R})r''s'' = SL(n, \mathbf{R})Diag(rs, 1, \ldots, 1) = (rs)^\#$. Consequently, identifying $r^\# \in G$ with $r \in \mathbf{R}^*$ shows that G is structurally the same as the group $(\mathbf{R}^*, \cdot)$. The key to recognizing G as $(\mathbf{R}^*, \cdot)$ is that we can use $\{r''\}$, an obvious copy of $\mathbf{R}^*$ in $GL(n, \mathbf{R})$, as the coset representatives.

Example 5.13 If $H = \{A \in UT(n, \mathbf{R}) \mid \text{all } A_{ii} = 1\}$ then $UT(n, \mathbf{R})/H$ is essentially $Diag(n, \mathbf{R}) \leq GL(n, \mathbf{R})$.

Recall that $UT(n, \mathbf{R}) \leq GL(n, \mathbf{R})$ is the group of all invertible upper triangular matrices (Example 2.9). Hence $A \in UT(n, \mathbf{R}) \Rightarrow A^{-1} \in UT(n, \mathbf{R})$ and from the definition of matrix multiplication, $(A^{-1})_{ii} = (A_{ii})^{-1}$. Using $A_{ij} = 0$ when $j < i$ and that for $h \in H$ all $h_{jj} = 1$, it follows that $(A^{-1}hA)_{ii} = (A^{-1})_{ii}h_{ii}A_{ii} = (A_{ii})^{-1} \cdot 1 \cdot A_{ii} = 1$, so $A^{-1}hA \in H$ and $H \triangleleft UT(n, \mathbf{R})$. To understand $UT(n, \mathbf{R})/H$ we want to exhibit simple representatives for its cosets. For $A \in UT(n, \mathbf{R})$, $A\,Diag(A_{11}, \ldots, A_{nn})^{-1} = A\,Diag(A_{11}^{-1}, \ldots, A_{nn}^{-1}) \in H$, so by Theorem 4.2, $HA = HDiag(A_{11}, \ldots, A_{nn})$. If $\alpha = Diag(a_1, \ldots, a_n)$, $\beta = Diag(b_1, \ldots, b_n)$, and $H\alpha = H\beta$ then $Diag(a_1b_1^{-1}, \ldots, a_nb_n^{-1}) = \alpha\beta^{-1} \in H$, forcing all $a_j = b_j$ and so $\alpha = \beta$. Hence each coset of H in $UT(n, \mathbf{R})$ contains exactly one invertible diagonal matrix so $UT(n, \mathbf{R})/H = \{HD \mid D \in Diag(n, \mathbf{R})\}$, where $Diag(n, \mathbf{R}) \leq GL(n, \mathbf{R})$ is the group of invertible diagonal matrices (Example 2.8). Now $UT(n, \mathbf{R})/H$ looks like $(Diag(n, \mathbf{R}), \cdot)$

by computing $HDiag(a_1, \ldots, a_n)HDiag(b_1, \ldots, b_n) = HDiag(a_1, \ldots, a_n)Diag(b_1, \ldots, b_n) = HDiag(a_1b_1, \ldots, a_nb_n)$.

One last example of the quotient construction shows that G and G/N can have surprisingly different properties. The example is more complex than the previous ones but is good for practicing with coset multiplication. We start with a group G so that $o(g)$ is infinite for all $g \in G - \langle e \rangle$ and construct an infinite G/N in which every proper subgroup is finite and cyclic.

Example 5.14 Let p be a prime and $\boldsymbol{Z}_{(p)} = \{a/b \in \boldsymbol{Q} \mid a, b \in \boldsymbol{Z}, b = p^m$ for some $m \geq 0\}$, so $\boldsymbol{Z}_{(p)} \leq (\boldsymbol{Q}, +)$ and has no nonzero element of finite order. Then $\boldsymbol{Z}_{p^\infty} = \boldsymbol{Z}_{(p)}/\boldsymbol{Z}$ is an infinite Abelian group, each of its elements has order p^k for some $k \geq 0$, and every proper subgroup of it is finite and cyclic.

It follows from Theorem 2.12 that $\boldsymbol{Z}_{(p)} \leq (\boldsymbol{Q}, +)$ since $(a/p^m) - (b/p^s) = (ap^s - bp^m)/p^{m+s}$ with $m + s \geq 0$ when $m, s \geq 0$, and certainly $\boldsymbol{Z} \leq \boldsymbol{Z}_{(p)}$. Since $(\boldsymbol{Q}, +)$ is Abelian, $\boldsymbol{Z}_{(p)}$ is also so $\boldsymbol{Z}_{p^\infty} = \boldsymbol{Z}_{(p)}/\boldsymbol{Z}$ exists by Theorem 5.1 and Theorem 5.6. It is an exercise that $\boldsymbol{Z}_{(p)}$ Abelian forces $\boldsymbol{Z}_{p^\infty}$ to be Abelian. For $a/b \in \boldsymbol{Q}$, $o(a/b) = n$ implies $na/b = 0$, forcing $a = 0$, so $\boldsymbol{Z}_{(p)} \leq (\boldsymbol{Q}, +)$ has no nonzero element of finite order. Once again, to understand $\boldsymbol{Z}_{p^\infty}$ we need to exhibit representatives for the cosets of $\boldsymbol{Z}$ in $\boldsymbol{Z}_{(p)}$. Given $\boldsymbol{Z} + a/p^m$ write $a = qp^m + r$ with $0 \leq r < p^m$, using Theorem 1.1. As $a/p^m = q + r/p^m$, $\boldsymbol{Z} + a/p^m = \boldsymbol{Z} + r/p^m$ by Theorem 4.2, so each coset is $\boldsymbol{Z} + r/p^m$ with $0 \leq r < p^m$ and $r = 0 \Leftrightarrow \boldsymbol{Z} + r/p^m = \boldsymbol{Z}$. Thus if $\boldsymbol{Z} + r/p^m \neq \boldsymbol{Z}$ we may assume $r, m \geq 1$ and $(r, p) = 1$ (why?). Next we show that r/p^m is the only such fraction in $\boldsymbol{Z} + r/p^m \neq \boldsymbol{Z}$. Suppose $s/p^k \in \boldsymbol{Z} + r/p^m$ with $1 \leq s < p^k, k \geq 1$, and $(s, p) = 1$. Using Theorem 4.2, $(r/p^m) - (s/p^k) = (rp^k - sp^m)/p^{m+k} \in \boldsymbol{Z}$ so $rp^k - sp^m = zp^{k+m}$ for $z \in \boldsymbol{Z}$. If $k < m$ then $r = sp^{m-k} + zp^m$ forces $p \mid r$, contradicting $(r, p) = 1$. The contradiction $p \mid s$ results if $k > m$. When $k = m, r - s = zp^m$, which contradicts $1 \leq r, s < p^m$ unless $z = 0$. But $z = 0$ forces $r/p^m = s/p^m$. This shows that each $\boldsymbol{Z} + r/p^m \neq \boldsymbol{Z}$ contains exactly one element r/p^m, satisfying $1 \leq r < p^m$, and $(r, p) = 1$. Thus $\boldsymbol{Z}_{p^\infty} = \{[r/p^m] = \boldsymbol{Z} + r/p^m \mid m \geq 1, 0 \leq r < p^m$, and when $r > 0, (r, p) = 1\}$. In this quotient $[r/p^m] + [s/p^k] = (\boldsymbol{Z} + r/p^m) + (\boldsymbol{Z} + s/p^k) = \boldsymbol{Z} + ((r/p^m) + (s/p^k)) = [c/p^s]$ where $r/p^m + s/p^k = z + c/p^s$ for $Z \in \{0, 1\}$. Hence we add the $[r/p^m]$ as if they are the fractions r/p^m, then subtract 1 if the sum is at least 1, and write the result in lowest terms. For example, with $p = 5 : [3/5] + [4/5] = [2/5]$ since $\boldsymbol{Z} + (3/5 + 4/5) = \boldsymbol{Z} + 7/5 = \boldsymbol{Z} + 2/5$; $[7/25] + [13/25] = [4/5]$ since $\boldsymbol{Z} + (7/25 + 13/25) = \boldsymbol{Z} + 20/25 = \boldsymbol{Z} + 4/5$; and $[19/25] + [4/5] = [14/25]$ since $\boldsymbol{Z} + (19/25 + 4/5) = \boldsymbol{Z} + 39/25 = \boldsymbol{Z} + 14/25$.

Note that in $\boldsymbol{Z}_{p^\infty}$, $[a/p^m]^{-1} = [(p^m - a)/p^m]$, and $[a/p^m]^n = \boldsymbol{Z} + na/p^m$ for $n \in \boldsymbol{N}$. Now $o([a/p^m]) = n \Leftrightarrow n$ is minimal, satisfying $p^m \mid na$, so $(a, p) = 1$ implies $o([a/p^m]) = p^m$ (why?). Thus for every $[a/p^m] \in \boldsymbol{Z}_{p^\infty}$, $o([a/p^m]) = p^m$. Also $(a, p) = 1$ and Theorem 1.10 lead to $sa + tp^m = 1$ for some $s, t \in \boldsymbol{Z}$, so $1/p^m = t + sa/p^m \in \boldsymbol{Z} + sa/p^m = (\boldsymbol{Z} + a/p^m)^s$. Therefore $\langle [a/p^m] \rangle = \langle [1/p^m] \rangle$. Let $\langle [0] \rangle \neq H \leq \boldsymbol{Z}_{p^\infty}$ and $T = \{m \in \boldsymbol{N} \mid [a/p^m] \in H\}$. If $k \in T$ is maximal, some $[b/p^k] \in H$ so $\langle [1/p^k] \rangle = \langle [b/p^k] \rangle \leq H$ and also $|H| \leq p^k$ by the definition of T.

But $|\langle[1/p^k]\rangle| = p^k$ by Theorem 2.16, forcing $H = \langle[1/p^k]\rangle$. If T has no maximal element and $[a/p^m] \in \mathbf{Z}_{p^\infty}$ then some $[b/p^k] \in H - \langle[0]\rangle$ with $k \geq m$, so $[a/p^m] = [1/p^k]^{ap^{k-m}} \in \langle[1/p^k]\rangle = \langle[b/p^k]\rangle \leq H$. This shows that $H = \mathbf{Z}_{p^\infty}$.

By Exercise 7 in §2.6, $G_p = \{c \in \mathbf{C} \mid o(c) = p^m$ for some $m \geq 0\} \leq (\mathbf{C}^*, \cdot)$ satisfies the same properties as $\mathbf{Z}_{p^\infty}$. Identifying $[a/p^m] \in \mathbf{Z}_{p^\infty}$ with $e^{2\pi i a/p^m} \in G_p$ shows that these two groups are essentially the same since products of corresponding elements correspond.

EXERCISES

1. If $N_1, \ldots, N_k \triangleleft G$ and $\bigcap N_i = \langle e\rangle$, show that there is an injective mapping $f : G \to G/N_1 \oplus \cdots \oplus G/N_k$ satisfying $f(xy) = f(x)f(y)$ for all $x, y \in G$.
2. Construct the following quotients explicitly and determine their structure:
   i) $U_{10}/\langle[9]_{10}\rangle$
   ii) $U_{20}/\langle[19]_{20}\rangle$
   iii) $U_{20}/\langle[11]_{20}\rangle$
   iv) $U_{20}/\langle[9]_{20}\rangle$
   v) $U_{32}/\langle[9]_{32}\rangle$
   vi) $U_{32}/\langle[7]_{32}\rangle$
   vii) $2\mathbf{Z}/6\mathbf{Z}$
   viii) $D_n/\langle T_1\rangle$
   ix) $D_6/\langle T_3\rangle$
   x) $(\mathbf{Z} \oplus \mathbf{Z})/(\mathbf{Z} \oplus \langle 0\rangle)$
   xi) $(S_n \oplus S_n)/(A_n \oplus A_n)$
   xii) S_∞/A_∞
3. If $N \triangleleft G$, $g \in G$, and $o(g)$ is finite, show that in G/N, $o(Ng) \mid o(g)$.
4. If $K \triangleleft G$, $g \in G$, $o(g) \in \mathbf{N}$, $|G/K| \in \mathbf{N}$, and $(o(g), |G/K|) = 1$, then $g \in K$.
5. If G is a finite group, $N \triangleleft G$, $K \leq G$, and $([G:N], |K|) = 1$, then $K \leq N$.
6. Identify the structure of the quotient with a known group:
   i) $m\mathbf{Z}/n\mathbf{Z}$ for $m, n \in \mathbf{N}$ with $m \mid n$
   ii) $D_8/\langle T_2\rangle$
   iii) $D_8/\langle T_4\rangle$
   iv) $D_{15}/\langle T_3\rangle$
7. For G any group, set $T(G) = \{g \in G \mid o(g)$ is finite$\}$.
   i) If G is Abelian, show that $T(G) \leq G$. (Exercise 11 § 2.5)
   ii) If $T(G) \leq G$ (some G), show that $T(G) \triangleleft G$.
   iii) If $T(G) \triangleleft G$, show that $T(G/T(G)) = \langle e_{G/T(G)}\rangle$. That is, if $X \in G/T(G)$, $X \neq T(G)e$, then X has infinite order in $G/T(G)$.
8. a) For groups H and K, identify $(H \oplus K)/(H \oplus \langle e_k\rangle)$ with K.
   b) If $G = \langle g\rangle$ and $\langle e\rangle \neq H \leq G$ show G/H is a finite group. (Exercise 3 §4.2)
   c) For $n, m > 1$ show the following are not cyclic: i) $n\mathbf{Z} \oplus m\mathbf{Z}$; ii) $\mathbf{Z}_n \oplus \mathbf{Z}$.
9. If G is a group, $N \triangleleft G$, and every element in N and in G/N has finite order, show that every element in G has finite order.
10. For $N \triangleleft G$ and $H \leq G$, set $H^\# = \{Nx \in G/N \mid x \in H\}$.
    a) Show that $H^\# \leq G/N$.
    b) If $H \triangleleft G$, then show that $H^\# \triangleleft G/N$.

11. If $N \lhd G$ and $K \leq G/N$, then there is $S \subseteq G$ with $K = \{Nx \in G/N \mid x \in S\}$. Set $K^{\sim} = \bigcup_S Nx \subseteq G$. Show that:
  a) $K^{\sim} \leq G$
  b) $N \leq K^{\sim}$
  c) If $K \lhd G/N$ then $K^{\sim} \lhd G$
  d) If N and K are finite then $|K^{\sim}| = |N| \cdot |K|$.

12. Let $N \lhd G$, $H \leq G$, and $K \leq G/N$. Use Exercises 10 and 11 to show:
  a) $(K^{\sim})^{\#} = K$
  b) If $N \leq H$ then $(H^{\#})^{\sim} = H$.

13. If $N \lhd G$, show that $H = \{y \in G \mid Ny \in Z(G/N)\} \lhd G$.

14. If $m \in \mathbf{N}$, G is a group, and $S = \{g^m \in G \mid g \in G\}$, then $H = \langle S \rangle = \{s_1^{e(1)} \cdots s_k^{e(k)} \in G \mid k \in \mathbf{N}, s_i \in S, \text{ and each } e(i) = \pm m\} \leq G$ by Theorem 2.18.
  i) Show that $H \lhd G$.
  ii) If $X \in G/H$, show that $X^m = e_{G/H} = H$.
  iii) If $m = 2$, show that G/H is Abelian (see Theorem 2.10).
  iv) When $m = 2$, find H for: a) $G = \mathbf{Z}_n$; b) $G = D_n$; and c) $G = S_n$.

15. Let $N, K \leq G$ with $N \lhd G$, $N \bigcap K = \langle e \rangle$, and $NK = G$. Show that G/N can be identified with the group K.

16. Let $N \leq \mathbf{Z}(G)$ be finite with $|N|$ odd. If for each $X \in G/N$ there is $Y \in G/N$ so that $X = Y^2$, show that for any $g \in G$ there is $h \in G$ with $g = h^2$.

17. Consider G, N and H below as groups under addition and identify G/N with H.
  i) $G = \mathbf{Z}^3 = (\mathbf{Z} \oplus \mathbf{Z} \oplus \mathbf{Z})$, $N = (2\mathbf{Z})^3$, $H = \mathbf{Z}_2^3$
  ii) $G = \mathbf{Z}^2$, $N = n\mathbf{Z} \oplus m\mathbf{Z}$, $H = \mathbf{Z}_n \oplus \mathbf{Z}_m$ for $n, m > 0$
  iii) $G = \mathbf{Z}^2$, $N = \langle (2,2) \rangle$, $H = \mathbf{Z} \oplus \mathbf{Z}_2$

5.3 SOME RESULTS USING QUOTIENT GROUPS

We want to see how the quotient construction is used. First we state a result giving some group properties inherited by quotients. Recall that $Z(G) = \{x \in G \mid xg = gx \text{ for all } g \in G\}$ is the center of G.

THEOREM 5.7 Let G be a group and $N \lhd G$.

i) If $|G| = p^t$ for p a prime, then $|G/N| = p^k$, $k \leq t$ and $k < t$ unless $N = \langle e \rangle$.
ii) If $y \in Z(G)$, then $Ny \in Z(G/N)$.
iii) If G is Abelian, then G/N is Abelian.
iv) If $g \in G$ and $k \in \mathbf{Z}$, then $Ng^k = (Ng)^k$.
v) If $G = \langle x \rangle$ is cyclic, then $G/N = \langle Nx \rangle$ is cyclic.

Proof We note that i) is just a special case of Theorem 5.6 since $|G/N| = [G:N]$ and $[G:N] \mid |G|$. For ii) we need $Ny \in G/N$ to commute with all $X \in G/N$. By definition, $G/N = \{Ng \mid g \in G\}$ so $X = Ng$ for $g \in G$ and the definition of multiplication in G/N yields $XNy = NgNy = Ngy = Nyg$, since $y \in Z(G)$. Continuing gives $Nyg = NyNg = NyX$, showing $XNy = NyX$, and $Ny \in Z(G/N)$ as claimed. If G is Abelian, $G = Z(G)$, so for every $X \in G/N$, $X = Ng \in Z(G/N)$ by ii), which means that $G/N = Z(G/N)$ is

Abelian. If $g \in G$ and $k > 0$, $Ng^k = NgNg \cdots Ng = (Ng)^k$ by the definition of powers, the definition of multiplication in G/N, and induction, and clearly $N = Ng^0 = (Ng)^0$. Using this and Theorem 2.4 on powers, when $h = -m$ for $m > 0$, $Ng^h = N(g^{-1})^m = (Ng^{-1})^m = ((Ng)^{-1})^m = (Ng)^{-m} = (Ng)^h$, proving iv). Finally, when $G = \langle x \rangle$ and $Y \in G/N$, then $Y = Ng$ for $g = x^k \in \langle x \rangle$. Thus using iv), $Y = Nx^k = (Nx)^k \in \langle Nx \rangle$ and it follows that $G/N = \langle Nx \rangle$ is cyclic. ■

The two generators $1, -1 \in \mathbf{Z}$ give generators $n\mathbf{Z} + 1 \in \mathbf{Z}/n\mathbf{Z}$ and $n\mathbf{Z} - 1 = n\mathbf{Z} + (n-1) \in \mathbf{Z}/n\mathbf{Z}$. However, $\mathbf{Z}/n\mathbf{Z}$ may have other generators: if $p > 3$ is prime then all nonzero elements of $\mathbf{Z}_p$ are generators by Theorem 2.15. Let us see how information about G can be obtained from a quotient.

THEOREM 5.8 If G is a group and $G/Z(G)$ is cyclic, then G is Abelian.

Proof By assumption, $G/Z(G) = \langle X \rangle$ for some $X \in G/Z(G)$. The definition of the quotient group shows that $X = Z(G)x$ for some $x \in G$, and so for each $g \in G$ and some $k \in \mathbf{Z}$, $Z(G)g = (Z(G)x)^k = Z(G)x^k$, using Theorem 5.7. Thus $g = zx^k$ for some $z \in Z(G)$. Similarly, for any other $g' \in G$ write $g' = z'x^s$ for $z' \in Z(G)$ and $s \in \mathbf{Z}$. Then from $z, z' \in Z(G)$ we have $gg' = zx^kz'x^s = zz'x^kx^s = z'zx^sx^k = z'x^szx^k = g'g$, and so G must be Abelian. ■

One application of Theorem 5.8 is that the center of a non-Abelian group cannot have prime index since in this case $|G/Z(G)| = p$, a prime, so $G/Z(G)$ cyclic by Theorem 4.9. Next we use quotients to prove Cauchy's Theorem for Abelian groups somewhat more easily than in Theorem 4.15. Our purpose is not simply to give an easier argument but rather to illustrate a typical use of the quotient construction together with induction.

THEOREM 5.9 (Cauchy) If G is a finite Abelian group and $p \mid |G|$ for p a prime, then there is $x \in G$ with $o(x) = p$. Also $H = \langle x \rangle \leq G$ and $|H| = p$.

Proof The second statement follows from the first by Theorem 2.16. For the first statement, fix p, let $|G| = pm$, and proceed by induction on m. When $m = 1$, $|G| = p$ so for any $x \in G$, $o(x) \mid p$ by Theorem 4.7, resulting in $o(x) = p$ for any $x \in G - \langle e \rangle$. Assume next that G is an Abelian group, $|G| = pm$ for $m > 1$, and that any Abelian group H with $|H| = pk$ and $1 \leq k < m$ contains an element of order p. Since $m > 1$, G has proper nonidentity subgroups by Theorem 4.9. If $L \leq G$ is any such and if $p \mid |L|$ then $|L| = pk$ for $k < m$ and our induction assumption yields $x \in L$ with $o(x) = p$. Since also $x \in G$ we are finished in this case. Suppose instead that $(p, |L|) = 1$ for any proper $L \leq G$ with $L \neq \langle e \rangle$. Now Lagrange's Theorem shows that $|G| = pm = |L|[G:L]$, so $p \mid [G:L]$ by Theorem 1.12 or unique factorization in N, giving $[G:L] = pk$. Because G is Abelian, L is normal (Theorem 5.1), so from Theorem 5.6, G/L exists and $|G/L| = [G:L] = pk$, and $k < m$ since $|L| > 1$. Furthermore, G/L is an Abelian group by Theorem 5.7 so the induction assumption yields $X \in G/L$ with $o(X) = p$. We need to use this to get an element in G having order p.

From the definition of G/L, $X = Lx$ for some $x \in G$. Now $o(X) = p$ means $Lx^p = (Lx)^p = e_{G/L} = L$ using Theorem 5.7, so $x^p \in L$ and Theorem 4.7 shows that $(x^p)^{|L|} = e$. Thus $(x^{|L|})^p = e$ (Theorem 2.4) and now $o(x^{|L|}) \mid p$ by Theorem 2.7. Clearly we want

$o(x^{|L|}) = p$ but perhaps $x^{|L|} = e$. If this happens then applying Theorem 5.7 again yields $X^{|L|} = (Lx)^{|L|} = Lx^{|L|} = L$, and since $o(X) = p$, $p \mid |L|$ follows from Theorem 2.7, contradicting $(p, |L|) = 1$. Therefore $o(x^{|L|}) = p$, proving the theorem when $|G| = pm$ from its truth for $|H| = pk$ for $k < m$, so the theorem holds by IND-II. ∎

Recall from Definition 2.13 that $G' = \langle\{x^{-1}y^{-1}xy \mid x, y \in G\}\rangle$, the subgroup generated by all $x^{-1}y^{-1}xy$, and that $G' \lhd G$ from Theorem 2.19. There is a close connection between G' and Abelian quotients of G.

THEOREM 5.10 If G' is the commutator subgroup of the group G, then:

i) G/G' is Abelian.
ii) If $N \lhd G$, then G/N is Abelian $\Leftrightarrow G' \le N$.
iii) If $H \le G$ with $G' \le H$, then $H \lhd G$.

Proof Since i) follows from ii), we begin by proving ii). The definition of the quotient group and Theorem 4.2 show that if $N \lhd G$ then G/N is Abelian $\Leftrightarrow NxNy = NyNx$, all $x, y \in G \Leftrightarrow Nxy = Nyx$, all $x, y \in G \Leftrightarrow xy(yx)^{-1} \in N$, all $x, y \in G \Leftrightarrow xyx^{-1}y^{-1} \in N$, all $x, y \in G$. Replacing x with x^{-1} and y with y^{-1} gives G/N is Abelian $\Leftrightarrow$ every $x^{-1}y^{-1}xy \in N \Leftrightarrow G' \le N$ from the definition of G', proving ii). Now suppose $G' \le H \le G$ and take $g \in G$. To see that $H \lhd G$ we need $g^{-1}Hg \subseteq H$. By definition of G', $h^{-1}g^{-1}hg \in G' \subseteq H$ for any $h \in H$ and it follows that $g^{-1}hg = h(h^{-1}g^{-1}hg) \in H$, which proves that $H \lhd G$. ∎

The theorem shows that G' is the smallest normal subgroup of G whose quotient is Abelian. Also G' has the pleasant property that every larger subgroup is normal in G and produces an Abelian qotient group. This characterization can be helpful in finding G'. Of course $G' = \langle e\rangle \Leftrightarrow G$ is Abelian. Let us describe G' for D_n and for S_n.

Example 5.15 $D_n' = \langle T_2\rangle$, so when n is odd, $D_n' = \langle T_1\rangle$.

Using Theorem 3.17, $W_2^{-1}W_1^{-1}W_2W_1 = (W_2W_1)^2 = T_1^2 = T_2 \in D_n'$. Thus $\langle T_2\rangle \le D_n'$ and by Example 5.1, $\langle T_2\rangle \lhd D_n$. If n is odd, $o(T_2) = o(T_1^2) = n/(2, n) = n$ by Theorem 2.8, so from Theorem 2.16, $|\langle T_2\rangle| = n$, and $\langle T_2\rangle = \langle T_1\rangle$, since $\langle T_2\rangle \le \langle T_1\rangle$ with the same order. In any case $|\langle T_2\rangle| = o(T_2) = n/(2, n)$ so Theorem 5.6 yields $|D_n/\langle T_2\rangle| = 2(2, n)$. When $2(2, n) = 2$, $D_n/\langle T_2\rangle$ is cyclic, so Abelian. If $2(2, n) = 4$ then $G = D_n/\langle T_2\rangle$ must be Abelian. One way to see this is by direct computation with cosets as in Example 5.11, and another is to observe that for $X \in G$, $X^4 = e_G$ by Theorem 4.7. Thus either G is cyclic or for each $X \in G$, $X^2 = e_G$ so Theorem 2.10 forces G to be Abelian. Now Theorem 5.10 yields $D_n' \le \langle T_2\rangle$ and together with $\langle T_2\rangle \le D_n'$ proves $D_n' = \langle T_2\rangle$.

Example 5.16 $S_n' = A_n$ and for $n \ge 5$, $A_n' = A_n$.

From Theorem 3.15 $A_n \lhd S_n$ and $|S_n/A_n| = 2$, so S_n/A_n is cyclic and $S_n' \le A_n$ by Theorem 5.10. Now $A_2 = \langle I\rangle = S_2'$ and if $n > 2$, Theorem 3.16 shows

$A_n = \langle\{\rho \in S_n \mid \rho \text{ is a 3-cycle}\}\rangle$ so $S_n' = A_n$ follows if each 3-cycle is in S_n'. But $(a, b, c) = (a, b)^{-1}(a, c)^{-1}(a, b)(a, c) \in S_n'$. Finally, let $n \geq 5$ and take distinct $a, b, c, d, e \in \{1, \ldots, n\}$. For $x = (a, e, b)$ and $y = (a, d, c)$, $(a, b, c) = x^{-1}y^{-1}xy \in A_n'$. Hence every 3-cycle is in A_n', which shows that $A_n = A_n'$.

We indicate how the commutator subgroup and quotients can be used to prove things about a large class of groups called **solvable,** mentioned at the end of Chapter 2. To review the definition, set $G^{(1)} = G'$, $G^{(2)} = (G^{(1)})'$, and in general, $G^{(k+1)} = (G^{(k)})'$. Then G is *solvable of length* k if $G^{(k)} = \langle e\rangle$ for some $k \geq 1$ with k minimal. Now G is solvable of length $1 \Leftrightarrow G' = \langle e\rangle \Leftrightarrow G$ is Abelian. From Example 5.15, D_n is solvable of length 2 since $D_n' = \langle T_2\rangle$ is Abelian. However, S_n is not solvable when $n \geq 5$ since then $S_n' = A_n$ by Example 5.16, and $S_n^{(2)} = A_n' = A_n$. It follows that $S_n^{(k)} = A_n$ for any $k \geq 1$.

Sometimes results for Abelian groups hold for solvable groups by an induction argument on the solvable length. A specific example is Cauchy's Theorem (Theorem 5.9): If G is a finite solvable group and $p \mid |G|$ for p a prime, then for some $x \in G$, $o(x) = p$. We will not state this as a theorem since we will prove the result later for any finite group. Our purpose here is merely to illustrate informally a very important and useful approach for dealing with solvable groups. The case of solvable groups of length one is Theorem 5.9. Assume G is solvable of length $k > 1$ with $p \mid |G|$. Now G' is solvable of length $k - 1$ so if $p \mid |G'|$ then G', so G, has an element of order p by induction. Otherwise $(p, |G'|) = 1$ and G/G' is an Abelian group with $p \mid |G/G'|$ so G/G' has an element of order p. Using $(p, |G'|) = 1$, the proof in Theorem 5.9 produces an element of order p in G.

EXERCISES

1. If $N \triangleleft G$, $X = Ng \in G/N$, and $o(g) \in \mathbf{N}$, show that $o(X) \mid o(g)$.

2. If $N \triangleleft G$ is finite, $X \in G/N$, and $o(X) = k \in \mathbf{N}$, show $o(g) = k$ for some $g \in G$.

3. Let p be a prime and $H, K \triangleleft G$. Show $o(V) = p^{a(V)}$, all $V \in G/(H \cap K) \Leftrightarrow o(X) = p^{b(X)}, o(Y) = p^{c(Y)}$, all $X \in G/H$, $Y \in G/K$.

4. Let G be a group, p a prime, and $H = \{x \in G \mid o(x) = p^s, \text{ some } s \geq 0\}$. If $H \leq G$, show $H \triangleleft G$ and for $Y \in G/H$ either $o(Y)$ is infinite or $(p, o(Y)) = 1$.

5. Let $N \triangleleft G$, G finite $p \neq q$ primes, $pq \mid |G|$, and $|N| = q$. If G/N is Abelian, show that there is $H \leq G$ with $|H| = pq$.

6. If G is a finite Abelian group, $H, K \leq G$, and if $x \in HK$ with $o(x) = p$ a prime, show that H or K must contain an element of order p.

7. Let G be a group and $\{H_i\}_I$ a set of normal subgroups of G so that all G/H_i are Abelian and $\bigcap H_i = \langle e_G\rangle$. Show that G is Abelian.

8. For p a prime, let $G = \{A \in UT(3, \mathbf{Z}_p) \mid \text{all } A_{jj} = [1]_p\} = \{I_3 + aE_{12} + bE_{13} + cE_{23} \mid a, b, c \in \mathbf{Z}_p\} \leq GL(3, \mathbf{Z}_p)$. a) If $N = \{A \in G \mid A_{12} = A_{23} = [0]_p\}$, show that $Z(G) = N$ and identify G/N as a known group. b) Determine G'.

9. For G' the commutator subgroup of G, if $K \leq G$, show that $K' \leq G'$.

10. If $H \triangleleft G$, show that $H' \triangleleft G$ and then show that $H^{(j)} \triangleleft G$ for all $j \geq 1$.

11. **a)** For groups G and H, show that $(G \oplus H)' = G' \oplus H'$.
b) Show that $(G \oplus H)^{(j)} = G^{(j)} \oplus H^{(j)}$.

c) Find all $G^{(j)}$ for:

i) $G = \mathbf{Z} \oplus D_5 \oplus S_3$

ii) $G = D_8 \oplus S_8$

iii) $G = S_5 \oplus A_5$

12. For $N \lhd G$, show that $NG'/N = \{Nx \in G/N \mid x \in G'\} = (G/N)'$.

13. a) If $H \lhd G$ and $|H| = |G/H| = p$ a prime, show G is Abelian. (If $G \neq \langle g \rangle$ find $x, y \in G$ with $o(x) = o(y) = p$, $H = \langle x \rangle$, $y \notin H$, note that $y^{-1}xy = x^j$, and show that $xy = yx$.)

b) Let p be a prime, G solvable of length k, set $G^{(0)} = G$, and assume $|G^{(j)}/G^{(j+1)}| = p$ for all $0 \leq j < k$. Show that G is Abelian.

14. If $|G/G'| = q_1 \cdots q_k$ for $q_1 < \cdots < q_k$ primes and $k > 1$, then for $1 \leq j < k$ show that there is $H_j \lhd G$ with $|G/H_j| = q_{j+1} \cdots q_k$ (use Exercise 11 §5.2).

5.4 SIMPLE GROUPS

If G is a finite group and $N \lhd G$ then in general $|N|, |G/N| < |G|$ so these smaller groups should be easier to understand than G. Further, an induction argument may be used to draw conclusions about G/N, and perhaps N, to use in studying G. This a useful technique for analyzing G, but it requires that G contain nontrivial normal subgroups. When G does not have any nontrivial normal subgroup, we give G a special name.

DEFINITION 5.3 If $N \lhd G \Rightarrow N = \langle e_G \rangle$ or $N = G$, then G is called **simple.**

Every subgroup of an Abelian group is normal, so by Theorem 4.9 a finite Abelian group is simple precisely when it is of prime order or has order one. Are there finite, non-Abelian, simple groups? The commutator subgroup in any group is normal and we saw in Example 5.16 that for $n \geq 5$, $A_n' = A_n$. Perhaps A_n is simple since it is not clear how else we can produce a nontrivial normal subgroup in it. Unfortunately, when G is not Abelian $G' = G$ does *not* imply that G is a simple group, for example, take $G = A_5 \oplus A_5$. Using Exercise 11 in §5.3, $G' = A_5' \oplus A_5' = A_5 \oplus A_5 = G$ but $A_5 \oplus \langle I \rangle \lhd G$. We will show that A_n is in fact simple when $n \geq 5$ and also that $SL(n, \mathbf{Z}_p)$ is "essentially" simple when p is a prime and $n \geq 3$ or $p \geq 5$.

We will have little to say in general about finite simple groups, but they are important as basic building blocks for all finite groups. The idea is that to understand a group G, it suffices to understand $N \lhd G$ and G/N. Call N a *maximal normal subgroup* when $N \lhd G$, $N \neq G$, and if $N \leq M \lhd G$ then $M = N$ or $M = G$. Let G be a finite group, N_1 a maximal normal subgroup of G, N_2 a maximal normal subgroup of the group N_1, N_3 a maximal normal subgroup of the group N_2, etc. Since G is finite, the chain $G \geq N_1 \geq N_2 \geq \cdots$ must end with some $N_k = \langle e_G \rangle$. As we will see in Chapter 7, each N_j/N_{j+1} is a simple group. Thus to understand G we want to understand the simple group G/N_1 and the group N_1. To understand N_1 we want to understand the simple group N_1/N_2 and the group N_2. Continuing, we see that to understand G we should aim to understand the simple groups $G/N_1, N_1/N_2, \ldots$ and also N_{k-1}/N_k. This is all quite vague but in practice we can tell a great deal about a group G if we have a good understanding of these "simple factors."

Starting in the 1960s, much of the research in finite group theory tried to describe and understand all finite simple groups. This feat was accomplished after many years of hard work by many mathematicians and represents a stunning achievement in modern mathematics. As of this writing, it takes a thousand or more printed pages to verify the complete list of the finite simple groups! We will be content to describe two general classes of finite simple groups that have been known for a very long time.

THEOREM 5.11 The group A_n is simple when $n \geq 5$.

Proof Now $A_2 = \langle I \rangle$ is simple, $A_3 = \langle (1, 2, 3) \rangle$ is simple of order 3, but in A_4 $H = \{I, (a, b)(c, d)\} \lhd A_4$ by Example 5.10. We assume now that $n \geq 5$, $\langle I \rangle \neq N \lhd A_n$, and show that $N = A_n$. Although we will clearly indicate the necessary computations, we will not always write them out in detail.

By Theorem 3.16, A_n is generated by its 3-cycles so it suffices to see that N contains all 3-cycles. Suppose N contains *some* 3-cycle, say (a, b, c). Since $N \lhd A_n$ and A_n contains all 3-cycles, if $1 \leq a, b, c, d, e \leq n$ are distinct, then $(c, d, e)(a, b, c)(c, d, e)^{-1} = (a, b, d) \in N$. Also $(a, b, d)^2 = (a, d, b) \in N$ and $(b, c, a) = (a, b, c) \in N$. It follows that by a succession of computations like these, we can change one symbol at a time in (a, b, c) to prove that any 3-cycle $(x, y, z) \in N$, and so $N = A_n$. Therefore, we need only show that N contains some 3-cycle.

Let $\theta \in N - \langle I \rangle$ and use Theorem 3.10 to write $\theta = \rho_1\rho_2 \cdots \rho_t$, a product of disjoint cycles of lengths $k_1 \geq \cdots \geq k_t \geq 2$. Assume for some θ that $k_1 = s \geq 4$ and set $\rho_1 = (a_1, a_2, \ldots, a_s)$. Let $\sigma = (a_1, a_2, a_3) \in A_n$, observe that σ and ρ_j are disjoint when $j > 1$, and consider $(\sigma\theta\sigma^{-1})\theta^{-1} \in N$. If $\theta = \rho_1\tau$ for $\tau = \rho_2 \cdots \rho_t$ then τ commutes with ρ_1 and σ by Theorem 3.9 so $\sigma\theta\sigma^{-1}\theta^{-1} = \sigma\rho_1\sigma^{-1}\tau\rho_1^{-1}\tau^{-1} = \sigma\rho_1\sigma^{-1}\rho_1^{-1}\tau\tau^{-1} = (a_1, a_2, a_3)(a_1, a_2, \ldots, a_s)(a_1, a_3, a_2)(a_1, a_s, \ldots, a_2) = (a_1, a_2, a_4)$. Consequently, N contains a 3-cycle if $k_1 \geq 4$ for some $\theta \in N$. Assume now that $k_1 \leq 3$. If $k_1 = k_2 = 3$, write $\theta = (a, b, c)(d, e, f)\tau$, where $\tau = \rho_3 \cdots \rho_t$ commutes with (a, b, c), (d, e, f), and $\sigma = (a, b, d) \in A_n$ by Theorem 3.9. Again $(\sigma\theta\sigma^{-1})\theta^{-1} \in N$ and so $\sigma\theta\sigma^{-1}\theta^{-1} = \sigma(a, b, c)(d, e, f)\sigma^{-1}(a, c, b) \cdot (d, f, e)\tau\tau^{-1} = (a, b, e, c, d) \in N$ with its $k_1 = 5$, so N contains a 3-cycle by the first case above. If there is $\theta \in N$ with $k_1 = 3$ and $k_2 < 3$ then $\theta^2 \in N$ is a 3-cycle.

Finally, assume that some $\theta \in N$ is a product of pairwise disjoint transpositions; that is, all $k_j = 2$, and $t = 2q$ since $N \leq A_n$. When $q > 1$ for some $\theta \in N$ then $\theta = (a, b)(c, d) \cdot (e, f)\rho_4 \cdots \rho_{2q} = (a, b)(c, d)(e, f)\tau$. If $\sigma = (a, c, e) \in A_n$, note that τ commutes with all of (a, b), (c, d), (e, f), and σ, and consider $(\sigma\theta\sigma^{-1})\theta^{-1} \in N$. Now $\sigma(a, b)(c, d)(e, f) \cdot \sigma^{-1}(a, b)(c, d)(e, f)\tau\tau^{-1} = (a, c, e)(b, f, d) \in N$, and again N contains a 3-cycle by the cases above. Finally, suppose that $\theta \in N$ has cycle structure $(2, 2, 1, \ldots, 1)$, so $\theta = (a, b)(c, d)$. Using $n \geq 5$, again set $\sigma = (a, c, e) \in A_n$ so $\eta = (\sigma\theta\sigma^{-1})\theta^{-1} \in N$ and $\eta = \sigma(a, b)(c, d)\sigma^{-1}(a, b)(c, d) = (a, c, e, d, b)$ is a 5-cycle with $k_1 = 5$ and N contains a 3-cycle from above. This exhausts the possibilities for cycle structures of elements of N, so indeed N must contain a 3-cycle, forcing $N = A_n$. ■

We turn to another collection of simple groups. Recall from §4.6 that a *field* is a nonempty set F with addition and multiplication satisfying the properties in $\boldsymbol{Z}$ or $\boldsymbol{Z}_n$ as in Theorem 1.28, and in addition, if $a \in F - \{0\}$ then $ab = 1$ for some $b \in F$. Just as in §4.6, there is no harm here in thinking of a field as either $\boldsymbol{Z}_p$ for p a prime, $\boldsymbol{Q}$, $\boldsymbol{R}$, or $\boldsymbol{C}$. From Example 2.10, for any field F, $\{A \in GL(n, F) \mid \det(A) = 1\} = SL(n, F) \leq GL(n, F)$. For the rest of this chapter *assume F is a field* and set $F^* = F - \{0\}$.

That $F^* \cdot I_n = Z(GL(n, F))$ is an exercise so by Theorem 5.1, $SL(n, F) \cap F^* \cdot I_n \lhd SL(n, F)$, and $Z(SL(n, F)) = SL(n, F) \cap F^* \cdot I_n$, which we prove in a bit. Therefore $Z(SL(n, F)) = \{aI_n = Diag(a, a, \ldots, a) \mid \det(aI_n) = a^n = 1\}$. When $F = \boldsymbol{R}$ then $a = 1$, or $a = -1$ when n is even: $Z(SL(2n, \boldsymbol{R})) = \langle -I_{2n} \rangle = \{\pm I_{2n}\}$ and $Z(SL(2n+1, \boldsymbol{R})) = \langle I_{2n+1} \rangle$. When $\boldsymbol{F}$ is a finite field, $aI_n \in Z(SL(n, F)) \Leftrightarrow o(a) \mid (n, |F| - 1)$ (why?). Now $\boldsymbol{Z}_7^* = \langle [3]_7 \rangle$ has order 6, so $Z(SL(n, \boldsymbol{Z}_7)) \neq \langle I_n \rangle$ if $n \in 2\boldsymbol{N} \cup 3\boldsymbol{N}$. When $n \in 6\boldsymbol{N}$ then $Z(SL(n, \boldsymbol{Z}_7)) = \boldsymbol{Z}_7^* \cdot I_n$, when $n \in 2\boldsymbol{N} - 3\boldsymbol{N}$ then $Z(SL(n, \boldsymbol{Z}_7)) = \langle [6]_7 \rangle \cdot I_n$ has two elements, and when $n \in 3\boldsymbol{N} - 2\boldsymbol{N}$ then $Z(SL(n, \boldsymbol{Z}_7)) = \langle [2]_7 \rangle \cdot I_n$ has three elements.

We now know that $SL(n, F)$ may have nontrivial normal subgroups. We will prove that $N \lhd SL(n, F)$ implies N is central or $N = SL(n, F)$ except when $|F| < 4$ and $n = 2$. This is equivalent to the statement that the group $PSL(n, F) = SL(n, F)/Z(SL(n, F))$ is simple, but demonstrating the equivalence requires results from Chapter 7. We will not need this result on normal subgroups of $SL(n, F)$ but present it to have another important class of simple groups. The argument uses matrix computations that are easily checked but sometimes messy. There are more elegant proofs that exploit pretty geometric ideas, but we prefer not to make such a digression.

Our approach to proving that proper normal subgroups of $SL(n, F)$ are central is in the spirit of the argument showing that A_n is a simple group when $n \geq 5$. The role of 3-cycles there will be played here by $\{I_n + aE_{ij}\}$, which have the entry "a" in the i-j position for $i \neq j$, ones on the diagonal, and zeros elsewhere. We show that these generate $SL(n, F)$ and that every noncentral normal subgroup of $SL(n, F)$ contains all of them. We begin with the formal definition of these matrices and then make a few simple computations with them. *Assume throughout that* $n \leq 2$.

DEFINITION 5.4 For $a \in F$ and $i \neq j$, $T_{ij}(a) = I_n + aE_{ij} \in M_n(F)$ is a **transvection,** where $E_{ij} \in M_n(F)$ with $(E_{ij})_{ij} = 1$ and $(E_{ij})_{st} = 0$ if $(s, t) \neq (i, j)$.

Now $T_{ij}(a) \in SL(n, F)$ since it is a triangular matrix with $\det(T_{ij}(a)) = T_{ij}(a)_{11} \cdots T_{ij}(a)_{nn} = 1 \cdots 1 = 1$. Note also $T_{ij}(0) = I_n$, $T_{ij}(a) + T_{ij}(b) = T_{ij}(a+b)$, $T_{ij}(a)^{-1} = T_{ij}(-a)$, $T_{ij}(a)T_{ik}(b) = T_{ik}(b)T_{ij}(a)$ and $T_{ji}(a)T_{ki}(b) = T_{ki}(b)T_{ji}(a)$ for i, j, and k distinct, $\prod_{j \neq i} T_{ij}(a_j) = I_n + a_1E_{i1} + \cdots + a_{i-1}E_{ii-1} + a_{i+1}E_{ii+1} + \cdots + a_nE_{in}$, and similarly, $\prod_{j \neq i} T_{ji}(a_j) = I_n + a_1E_{1i} + \cdots + a_{i-1}E_{i-1\,i} + a_{i+1}E_{i+1\,i} + \cdots + a_nE_{ni}$.

THEOREM 5.12 If $i \neq j$ and $v \in F^*$, then $D_{ij}(v) = I_n - E_{ii} - E_{jj} + vE_{ii} + v^{-1}E_{jj} \in \langle \{T_{sk}(a) \mid a \in F^* \text{ and } s \neq k\} \rangle$ and $SL(n, F) = \langle \{T_{sk}(a) \mid a \in F^* \text{ and } s \neq k\} \rangle$

Proof $D_{ij}(v)$ is the diagonal matrix with $D_{ij}(v)_{ii} = v$, $D_{ij}(v)_{jj} = v^{-1}$, and $D_{ij}(v)_{kk} = 1$ for $k \neq i, j$. That $D_{ij}(v) \in G = \langle \{T_{sk}(a) \mid a \in F\} \rangle$, generated by all transvections, follows from $D_{ij}(v) = T_{ji}(v^{-1})T_{ij}(v-1)T_{ji}(1)T_{ij}(v^{-1}-1)T_{ji}(-2v)$. Since all $T_{ij}(a) \in SL(n, F)$ we need any $A \in SL(n, F)$ to be a product of $T_{ij}(a)$.

For $A \in SL(n, F)$ we show that a product of various $T_{ij}(a)$ on the left of A results in a diagonal matrix. Since A is invertible, its first column is not zero so some $A_{j1} \neq 0$. If $V = T_{1j}(x)A$, $V_{11} = A_{11} + xA_{j1}$ so $V_{11} \neq 0$ for some x and $V \in G \Leftrightarrow A \in G$. Therefore we may assume that $A_{11} \neq 0$. But now $B = (\prod_{j \neq 1} T_{j1}(-A_{11}^{-1}A_{j1}))A$ satisfies $B_{k1} = -A_{11}^{-1}A_{k1}A_{11} + A_{k1} = 0$ for all $k > 1$, and $B \in G \Leftrightarrow A \in G$. Since B is invertible and $B_{k1} = 0$ for $k > 1$, we must have $B_{j2} \neq 0$ for some $j \geq 2$ and so as above some $T_{2j}(y)B = C$

satisfies $C_{22} \neq 0$. We still have $C_{k1} = B_{k1} = 0$ for $k > 1$ and also $C \in G \Leftrightarrow B \in G$. Consider $D = (\prod_{j \neq 2} T_{j2}(-C_{22}^{-1}C_{j2}))C : D_{k1} = C_{k1} = 0$ if $k > 1$, $D_{k2} = -C_{22}^{-1}C_{k2}C_{22} + C_{k2} = 0$ for $k \neq 2$, and $D \in G \Leftrightarrow C \in G$. Either D is diagonal or $n > 2$, and we may continue this procedure to obtain $Diag(Y_{11}, \ldots, Y_{nn}) = PA$ with $P \in G$, a product of various $T_{ij}(a)$. Now $D_{12}(Y_{11}^{-1}) \in G$ from above, so first we obtain $D_{12}(Y_{11}^{-1})PA = Diag(1, Y_{11}Y_{22}, Y_{33}, \ldots, Y_{nn})$, then $D_{23}(Y_{11}^{-1}Y_{22}^{-1})D_{12}(Y_{11}^{-1})PA = Diag(1, 1, Y_{11}Y_{22}Y_{33}, Y_{44}, \ldots, Y_{nn})$. In general we get

$$\begin{aligned} H &= D_{n-1\,n}\left(Y_{11}^{-1} \cdots Y_{n-1\,n-1}^{-1}\right) \cdots D_{23}\left(Y_{11}^{-1}Y_{22}{}^{-1}\right) D_{12}\left(Y_{11}^{-1}\right) PA \\ &= Diag(1, \ldots, 1, Y_{11} \cdots Y_{nn}) \end{aligned}$$

Since $PA \in SL(n, F)$, $1 = \det(PA) = Y_{11} \cdots Y_{nn}$, so $H = I_n$. We have seen $D_{ij}(v)$, P, $D_{ij}(v)^{-1}$, $P^{-1} \in G$ so $A = (D_{n-1\,n}(Y_{11}^{-1} \cdots Y_{n-1\,n-1}^{-1}) \cdots D_{23}(Y_{11}^{-1}Y_{22}^{-1})D_{12}(Y_{11}^{-1})P)^{-1}$ is in G, completing the proof of the theorem. ■

Using Theorem 5.12, each $A \in Z(SL(n, F))$ commutes with all $T_{ij}(a)$, so with each E_{ij} for $i \neq j$. It is an easy exercise to see that the only $Y \in M_n(F)$ that commute with all the E_{ij} for $i \neq j$ are the scalar matrices $F \cdot I_n$. Thus $Z(GL(n, F)) = F^* \cdot I_n$, and as stated above, $Z(SL(n, F)) = SL(n, F) \bigcap F^* \cdot I_n$.

THEOREM 5.13 Proper normal subgroups of $SL(2, F)$ are central if $|F| \geq 4$.

Proof Let $N \lhd SL(2, F)$. If $N \not\subset F \cdot I_2$ we will use Theorem 5.12 to prove that $N = SL(2, F)$ by showing that N contains all $T_{12}(a)$ and $T_{21}(a)$.

Case I. Some $A = \begin{bmatrix} a & 0 \\ b & c \end{bmatrix} \in N - F \cdot I_2$ with $a \neq c$.

Since $\det(A) = ac = 1, c = a^{-1} \neq a$. For $x \in F$, $\begin{bmatrix} a & 0 \\ b + (a-c)x & c \end{bmatrix} = T_{21}(x)AT_{21}(x)^{-1}$ is in N, and if $x = (a-c)^{-1}(y-b)$ for $y \in F$ then $\begin{bmatrix} a & 0 \\ y & c \end{bmatrix} \in N$. Hence $y, t \in F$ imply $\begin{bmatrix} a & 0 \\ y & c \end{bmatrix}\begin{bmatrix} a & 0 \\ t & c \end{bmatrix}^{-1} = T_{21}(c(y-t)) \in N$, so $\{T_{21}(f) \mid f \in F\} \subseteq N$. But $E = \begin{bmatrix} 0 & 1 \\ -1 & 0 \end{bmatrix} \in SL(2, F)$, so $T_{12}(f) = ET_{21}(-f)E^{-1} \in N$ for all $f \in F$, and N contains all transvections. By Theorem 5.12, $SL(2, F) \leq N$ so $N = SL(2, F)$.

Case II. Some $A = \begin{bmatrix} a & b \\ c & d \end{bmatrix} \in N - F \cdot I_2$ with $bc \neq 0$ and $|F| \neq 5$.

Set $B = T_{12}(c^{-1}a)^{-1}AT_{12}(c^{-1}a) = \begin{bmatrix} 0 & -c^{-1} \\ c & a+d \end{bmatrix} \in N$, using $ad - bc = 1$, and observe that for any $v \in F^*$ $Diag(v, v^{-1}) \in SL(2, F)$ by Theorem 5.12, so $B(v) = Diag(v, v^{-1})B\,Diag(v, v^{-1})^{-1}B^{-1} \in N$. Computing, $B(v) = \begin{bmatrix} v^2 & 0 \\ c(a+d)(v^{-2}-1) & v^{-2} \end{bmatrix}$ so we have reduced to Case I when $v^2 \neq v^{-2}$, or, equivalently, when $v^4 \neq 1$. There is some $v \in F^*$ with $v^4 \neq 1$ whenever $|F| \neq 5$. If $|F| = 4$

then $(F^*, \cdot)$ has three elements so is cyclic (Theorem 4.9), say $(F^*, \cdot) = \langle v \rangle$. Thus $o(v) = 3$ and v cannot satisfy $v^2 = v^{-2}$ (why?). When $|F| > 5$ the polynomial $x^4 - 1 \in F[x]$ has at most four roots in F by Theorem 4.18 so there is $v \in F^*$ with $v^4 \neq 1$. Consequently, when $|F| = 4$ or $|F| > 5$ there is some $B(v)$ satisfying the condition in Case I, which shows that $N = SL(2, F)$.

Case III. $|F| = 5$ and some $A = \begin{bmatrix} a & b \\ c & d \end{bmatrix} \in N - F \cdot I_2$ with $bc \neq 0$.

Proceed as in Case II to get the same B, $B(v) \in N$. Now $|F^*| = 4$ so $v^4 = 1$ if $v \in F^*$ (why?), $v^2 = v^{-2}$, and the elements of F^* are distinct roots of $x^4 - 1$ in F. But $x^4 - 1 = (x - 1)(x + 1)(x^2 + 1)$, so $-1 \neq 1$ in F and some $v \in F$ satisfies $v^2 = -1$. If $a + d \neq 0$, for $v^2 = -1$, $B(v) = \begin{bmatrix} u & 0 \\ w & u \end{bmatrix} \in N$ with $u = -1$ and $w \neq 0$: note $-2 \neq 0$ since $-2 = 0 \Rightarrow 2 = 0 \Rightarrow -1 = 1$. Thus $B(v)^2 = T_{21}(-2w) \in N$ with $-2w \neq 0 \in F$ as $w \neq 0$, so using $(F, +) = \langle 1 \rangle$ (Theorem 4.9), $\{T_{21}(f) \mid f \in F\} = \{T_{21}(k(-2w)) \mid k \in \mathbf{N}\} = \langle T_{21}(-2w) \rangle \leq N$ and for $E = \begin{bmatrix} 0 & 1 \\ -1 & 0 \end{bmatrix}$ as in Case I, $T_{12}(f) = ET_{21}(-f)E^{-1} \in N$. This shows that N contains all transvections, and Theorem 5.12 yields $N = SL(2, F)$. If $a + d = 0$, from above $B = \begin{bmatrix} 0 & h \\ c & 0 \end{bmatrix} \in N$ with $hc = -1$ since $B \in SL(2, F)$, and $T_{12}(x)^{-1} B T_{12}(x) = \begin{bmatrix} -cx & h - cx^2 \\ c & cx \end{bmatrix} \in N - F \cdot I_2$. Thus if some $h - cx^2 = 0$ we are in Case I. But for $x = 2c^{-1}$, it follows that $h - c(4c^{-2}) = h + c^{-1} = c^{-1}(hc + 1) = 0$ (exercise!) as $hc = -1$. Thus for $x = 2c^{-1}$ we are in the situation of Case I so $N = SL(2, F)$.

Case IV. Some $A = \begin{bmatrix} a & b \\ c & d \end{bmatrix} \in N - F \cdot I_2$ with $bc = 0$.

If $b = 0 = c$ then $a \neq d$ so $N = SL(2, F)$ by Case I. Assume $b = 0$ but $c \neq 0$ by replacing A with EAE^{-1} for $E = \begin{bmatrix} 0 & 1 \\ -1 & 0 \end{bmatrix}$ with $b \neq 0$. When $a \neq d$ Case I applies so assume $A = \begin{bmatrix} a & 0 \\ c & a \end{bmatrix}$ with $\det(A) = a^2 = 1$, so $a = \pm 1$. Now $EAE^{-1} = \begin{bmatrix} a & -c \\ 0 & a \end{bmatrix} \in N$ so $AEAE^{-1} = \begin{bmatrix} 1 & -ca \\ ca & 1 - c^2 \end{bmatrix} \in N$ and $(-ca)(ca) = -c^2 \neq 0$ so Cases II and III show that $N = SL(2, F)$. The four cases exhaust all the possibilities when N is not central, proving the theorem. ∎

The result for normal subgroups of $SL(n, F)$ when $n > 2$, does not depend on $|F|$. The argument is different but still depends on showing that a noncentral normal subgroup contains a transvection.

THEOREM 5.14 Proper normal subgroups of $SL(n, F)$ are central if $n \geq 3$.

Proof Let $N \lhd SL(n, F)$ with N not central. We want to find a transvection in N and begin by taking $A \in N - F \cdot I_n$. First assume that for some $1 \leq j \leq n$, $AE_{jj} = aE_{jj}$ for some

$a \in F^*$, so $A_{ij} = 0$ for all $i \neq j$. Now $E_{jj} = A^{-1}AE_{jj} = aA^{-1}E_{jj}$, or, equivalently, $A^{-1}E_{jj} = a^{-1}E_{jj}$. It follows that for $s \neq j$, $E_{js}A^{-1}E_{js} = E_{js}A^{-1}E_{jj}E_{js} = a^{-1}E_{js}E_{jj}E_{js} = 0$. Using this computation for $k \neq j$ and that $T_{jk}(f) \in SL(n, F)$, $B(k) = T_{jk}(1)^{-1}A^{-1}T_{jk}(1)A = I_n - E_{jk} + a^{-1}E_{jk}A \in N$. If $B(k) \in F \cdot I_n$ then $0 = B(k)_{jk} = -1 + a^{-1}A_{kk}$ which implies that $A_{kk} = a = A_{jj}$. Also for $s \neq j, k$, we have $0 = B(k)_{js} = a^{-1}A_{ks}$ forcing $A_{ks} = 0$. Now $A_{kj} = 0$ since $AE_{jj} = aE_{jj}$ so we have shown that when $B(k)$ is central then $A_{kk} = A_{jj}$ and $A_{kt} = 0$ when $t \neq k$. When $B(k) \in F \cdot I_n$ for all $k \neq j$ then $A = aI_n + \sum_{t \neq j} A_{jt}E_{jt} \in N - F \cdot I_n$, and $B(k) \notin F \cdot I_n$ has the same form since $B(k)_{jj} = 1 + a^{-1}A_{kj} = 1$ because $A_{kj} = 0$ follows from $AE_{jj} = aE_{jj}$. Thus $AE_{jj} = aE_{jj}$ gives some $B = bI_n + \sum_{s \neq j} B_{js}E_{js} \in N - F \cdot I_n$ and $B^{-1} = b^{-1}I_n - \sum_{s \neq j} b^{-2}B_{js}E_{js}$ (exercise!). Since $B \notin F \cdot I_n$ there is $k \neq j$ with $B_{jk} \neq 0$. Choose $s \neq j, k$, and set $C = T_{ks}(1)^{-1}B^{-1}T_{ks}(1)B \in N$. Using $E_{ks}B^{-1} = b^{-1}E_{ks}$ results in $C = (B^{-1} - b^{-1}E_{ks})(B + bE_{ks}) = I_n + bB^{-1}E_{ks} - b^{-1}E_{ks}B = I_n + b(b^{-1}E_{ks} - b^{-2}B_{jk}E_{js}) - E_{ks} = I_n + E_{ks} - b^{-1}B_{jk}E_{js} - E_{ks} = T_{js}(-b^{-1}B_{jk}) \in N - F \cdot I_n$.

Assume some $A \in N - F \cdot I_n$ satisfies $E_{ii}A = A_{ii}E_{ii}$ for some i. Multiplying by A^{-1} on the right yields $E_{ii}A^{-1} = A_{ii}^{-1}E_{ii}$. For any $k \neq i$, set $B = AT_{ki}(1)A^{-1}T_{ki}(1)^{-1} \in N$, so $B = (A + AE_{ki})(A^{-1} - A^{-1}E_{ki}) = I_n - E_{ki} + AE_{ki}A^{-1}$ since $E_{ii}A^{-1} = A_{ii}^{-1}E_{ii}$ implies $E_{ki}A^{-1}E_{ki} = 0$. Also $B = I_n - E_{ki} + AA_{ii}^{-1}E_{ki}$ so if $B \notin F \cdot I_n$ then $BE_{kk} = E_{kk}$ and by the first part of the proof, N contains a nonidentity transvection. When $B \in F \cdot I_n$ it is diagonal so $0 = B_{ki} = -1 + A_{kk}A_{ii}^{-1}$ forcing $A_{kk} = A_{ii}$ and for $t \neq k, i$, we get $0 = B_{ti} = A_{tk}A_{ii}^{-1}$ so $A_{tk} = 0$. But $E_{ii}A = A_{ii}E_{ii}$ so also $A_{ik} = 0$ and $AE_{kk} = A_{kk}E_{kk}$. Again by the first paragraph, N contains some $T_{st}(f) \neq I_n$. Therefore N contains a nonidentity transvection if for some $A \in N - F \cdot I_n$ either $AE_{jj} = A_{jj}E_{jj}$ for some j or $E_{ii}A = A_{ii}E_{ii}$ for some i.

Next we show N must contain $A \notin F \cdot I_n$ with $AE_{jj} = A_{jj}E_{jj}$ or $E_{ii}A = A_{ii}E_{ii}$. Let $A \in N - F \cdot I_n$ and assume $A_{jk} \neq 0$ for some $j \neq k$. Let $i \neq j, k$, set $B = T_{ij}(x)^{-1}AT_{ij}(x) = T_{ij}(-x)AT_{ij}(x) \in N$, and note that $B \notin F \cdot I_n$. Expanding shows $B = A - xE_{ij}A + AxE_{ij} - x^2A_{ji}E_{ij}$, so $B_{ik} = A_{ik} - xA_{jk}$ and $B_{ik} = 0$ for $x = A_{jk}^{-1}A_{ik}$. Consequently, there is $A \in N - F \cdot I_n$ with $A_{ij} = 0$, some $i \neq j$. If $A_{ik} = 0$ for all $k \neq i$ then $E_{ii}A = A_{ii}E_{ii}$ and we have seen this leads to a transvection in $N - F \cdot I_n$. Thus assume $A_{ik} \neq 0$ for some $k \neq i$. For $s \neq i, k$, consider $B = A^{-1}T_{si}(1)^{-1}AT_{si}(1) \in N$. Then $B = I_n - A^{-1}E_{si}A + E_{si} - A^{-1}E_{si}AE_{si}$, and using $A_{ij} = 0$ results in $BE_{jj} = E_{jj}$. Again N will contain a nonidentity transvection unless $B \in F \cdot I_n$. In this case $B_{tk} = 0$ for $t \neq k$, so $0 = B_{tk} = -A_{ts}^{-1}A_{ik}$ and $A_{ts}^{-1} = 0$ for each $t \neq k$. Also $B_{jj} = 1 - A_{js}^{-1}A_{ij} = 1$ since $A_{ij} = 0$, so $B \in F \cdot I_n$ means $1 = B_{kk} = 1 + -A_{ks}^{-1}A_{ik}$ forcing $A_{ks}^{-1} = 0$, so $A^{-1}E_{ss} = 0$. But A^{-1} is invertible so this is impossible, $B \in N - F \cdot I_n$, and N must contain some nonidentity transvection.

The argument has shown that if $N \not\subset F \cdot I_n$ then some $T_{ij}(x) \in N - F \cdot I_n$. Now use the identity $T_{st}(y)^{-1}T_{vs}(x)T_{st}(y)T_{vs}(x)^{-1} = T_{vt}(xy)$ for v, s, and t distinct to get $T_{ij}(f) \in N$ for *all* $i \neq j$ and all $f \in F$ (verify!). Therefore N contains all transvections so $SL(n, F) = N$ by Theorem 5.12, proving the theorem. ∎

The special cases of Theorem 5.14 that yield simple groups are those with $Z(SL(n, F)) = \langle I_n \rangle$. As we saw above, when $m \in 2N + 1$ then $SL(m, \mathbf{R})$ is simple, and when F is finite and $(n, |F| - 1) = 1$ then $SL(n, F)$ is simple. In particular, all the $SL(n, \mathbf{Z}_2)$ are simple when $n > 2$; $SL(m, \mathbf{Z}_3)$, $SL(m, \mathbf{Z}_5)$, and $SL(m, \mathbf{Z}_{17})$ are simple when $m > 1$ is odd;

and $SL(m, \mathbf{Z}_7)$ is simple when $(m, 6) = 1$. We can compute the orders of these groups by using both $[GL(n, \mathbf{Z}_p) : SL(n, \mathbf{Z}_p)] = p - 1$ from Example 4.5 and that $|GL(n, \mathbf{Z}_p)| = (p^n - 1)(p^n - p) \cdots (p^n - p^{n-1})$ mentioned in Chapter 2. Even for small n and p the orders get large quickly. Examples are $|SL(3, \mathbf{Z}_2)| = |GL(3, \mathbf{Z}_2)| = 7 \cdot 6 \cdot 4 = 168$, $|SL(4, \mathbf{Z}_2)| = |GL(4, \mathbf{Z}_2)| = 15 \cdot 14 \cdot 12 \cdot 8 = 20{,}160$, $|SL(5, \mathbf{Z}_2)| = 9{,}999{,}360$ (verify!) and $|SL(3, \mathbf{Z}_5)| = |GL(3, \mathbf{Z}_5)|/4 = 124 \cdot 120 \cdot 100/4 = 372{,}000$.

Using that Theorem 5.13 and Theorem 5.14 imply that $PSL(n, F) = SL(n, F)/Z(SL(n, F))$ is a simple group, which will follow from results in Chapter 7, we compute orders of other simple groups. Thus $|PSL(3, \mathbf{Z}_7)| = |SL(3, \mathbf{Z}_7)|/3$ since $Z(SL(3, \mathbf{Z}_7)) = \langle [2] \rangle \cdot I_3$, and $|SL(3, \mathbf{Z}_7) = |GL(3, \mathbf{Z}_7)|/6$, so $|PSL(3, \mathbf{Z}_7)| = |GL(3, \mathbf{Z}_7)|/18 = 342 \cdot 336 \cdot 294/18 = 1{,}876{,}896$. Also $|PSL(4, \mathbf{Z}_5)|$ is $|SL(4, \mathbf{Z}_5)|/4$ since $Z(SL(4, \mathbf{Z}_5) = U_5 \cdot I_4$, and $|SL(4, \mathbf{Z}_5)| = |GL(4, \mathbf{Z}_5)|/4$, so $|PSL(4, \mathbf{Z}_5)| = |GL(4, \mathbf{Z}_5)|/16 = 624 \cdot 620 \cdot 600 \cdot 500/16 = 7{,}254{,}000{,}000$.

EXERCISES

1. Find a chain $G = N_0 \geq N_1 \geq \cdots \geq N_{k+1} = \langle e \rangle$ with each $N_{i+1} \lhd N_i$ and N_i/N_{i+1} a simple group, for G the group:
   i) $\mathbf{Z}_9$
   ii) $\mathbf{Z}_{18}$
   iii) $\mathbf{Z}_6 \oplus \mathbf{Z}_6$
   iv) S_4
   v) D_{12}
   vi) S_5
   vii) $\mathbf{Z}_6 \oplus S_6$
   viii) $GL(7, \mathbf{Z}_3)$

2. Let F be a field with $|F| = 5$ and identity $1 = 1_F$. Show that:
   i) $(F, +) = \langle 1_F \rangle$
   ii) if $n \in \mathbf{N}$, $a \in F$, and $n \cdot a = a + \cdots + a$ (n times), $F = \{0, 1, 2 \cdot 1, 3 \cdot 1, 4 \cdot 1\}$ and $1 + 4 \cdot 1 = 0 = 2 \cdot 1 + 3 \cdot 1$

3. If $H \neq \langle e \rangle$ is a simple and solvable group, show that H is cyclic of prime order.

4. If $n \neq 4$ and $H \lhd S_n$, show that either $H = \langle I \rangle$, $H = A_n$, or $H = S_n$.

5. If $H \leq S_n$ for $n \geq 5$, use Exercise 17 in §5.1 to show that $A_n \leq H$ or $[S_n : H] \geq n$.

6. Fix $n, m \in \mathbf{N}$ with $n \geq 2$. If $A = \{f \in S_n \mid f^m = I\} \neq \langle I \rangle$, identify the subgroup of S_n generated by A, depending on m. (See Exercise 4.)

7. If $n \geq 3$, show that $GL(n, \mathbf{Z}_2)$ is a simple group.

8. **a)** For $n \geq 2$, show that each $A \in GL(n, \mathbf{Z}_2)$ is a product of matrices of order 2.
   b) For $p > 3$ a prime and $n \geq 2$, show that every $A \in SL(n, \mathbf{Z}_p)$ is a product of matrices in $SL(n, \mathbf{Z}_p)$, each of order p.
   c) For $n \geq 3$, let $A = E_{12} - E_{21} + E_{33} + \cdots + E_{nn} \in SL(n, \mathbf{Q})$: note that $o(A) = 4$. Show any $Y \in SL(n, \mathbf{Q})$ is some product of $g_i \in SL(n, \mathbf{Q})$, each of order 4.

9. If $2 < p < q$ are primes, argue that $SL(q, \mathbf{Z}_p)$ is a simple group.

10. Show that A_∞ (Exercise 10 in §5.1) is a simple group.

11. For which $n > 1$ are the following groups simple? i) $SL(n, \mathbf{Z}_{11})$; ii) $SL(n, \mathbf{Z}_{13})$; and iii) $SL(n, \mathbf{Z}_{23})$.

12. Let $p \geq 5$ be a prime and $n > 1$. Find: a) $SL(n, \mathbf{Z}_p)'$; b) $GL(n, \mathbf{Z}_p)'$.

13. As a function of $n > 1$, find the number of proper normal subgroups of:

i) $SL(n, \boldsymbol{Q})$

ii) $SL(n, \boldsymbol{R})$

iii) $SL(n, \boldsymbol{C})$

14. For p a prime, find the number of proper normal subgroups in $SL(n, \mathbf{Z}_p)$:

i) $n = 2$ and $p > 3$

ii) $n = 8$ and:

(a) $p = 7$

(b) $p = 13$

(c) $p = 67$

(d) $p = 73$

iii) $n = 15$ and:

(a) $p = 5$

(b) $p = 11$

(c) $p = 13$

(d) $p = 31$

CHAPTER 6

MORPHISMS

It is often useful to compare sets with related properties or structures by considering functions between them that preserve these properties or structures. Such functions are called *morphisms*. The set of real numbers $\boldsymbol{R}$ has a structure given by the distance between points, so the functions $f: \boldsymbol{R} \to \boldsymbol{R}$ of greatest interest are continuous, which, loosely speaking, preserve distances. We concentrate on algebraic structures. The structure of a group is determined by its multiplication, and the structure of a ring (Theorem 1.28) by its addition and multiplication. The morphisms of these sets should relate the products, and sums, of elements to those of their images. We give these maps between groups a special name.

6.1 HOMOMORPHISMS

DEFINITION 6.1 For $(G, \cdot)$ and $(B, \#)$ groups, $\varphi : G \to B$ is a **homomorphism** if $x, y \in G \Rightarrow \varphi(x \cdot y) = \varphi(x) \# \varphi(y)$. If in addition φ is a bijection, then φ is called an **isomorphism** and G is **isomorphic** to B, written $G \cong B$.

When the group operations are suppressed, a homomorphism satisfies $\varphi(xy) = \varphi(x)\varphi(y)$. We must remember that the product xy is in G and the product $\varphi(x)\varphi(y)$ is in B. These group products may be written as $+$, so the equality defining a homomorphism might be written in particular cases as $\varphi(xy) = \varphi(x) + \varphi(y)$, $\varphi(x + y) = \varphi(x)\varphi(y)$, or $\varphi(x + y) = \varphi(x) + \varphi(y)$.

When φ is a homomorphism the product of images under φ is the image of the product. A simple picture of this is shown in Fig. 6.1.

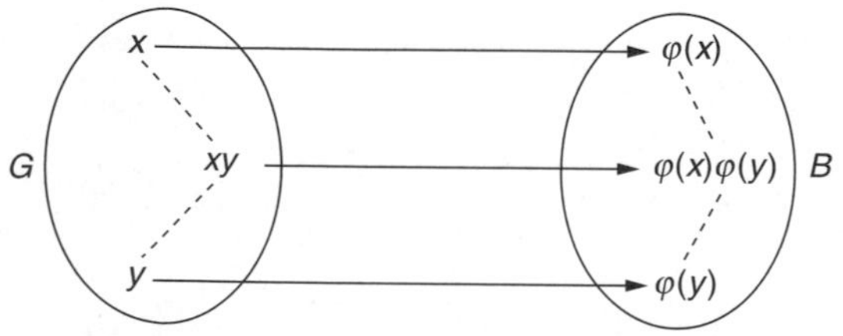

FIGURE 6.1

Thus if we regard φ as identifying elements in G with elements in B, then the multiplications of the identified elements in each group are themselves identified. In this way we think of φ as consistent with, or preserving, the multiplication in each group. When φ is an isomorphism from G to B it is a matching, or correspondence, between all the elements of G and all those of B. Since the products of corresponding elements in each group correspond, we think of $G \cong B$ as meaning that these groups are abstractly the same; the isomorphism φ identifies elements in the two groups, and up to this identification, the group multiplications are the same.

Before starting a general discussion of homomorphisms and their properties, we present examples that illustrate the idea of isomorphism described above. First we observe that $(\boldsymbol{Z}_2, +) \cong \langle -1 \rangle \leq (\boldsymbol{R}^*, \cdot)$ via $\varphi = \{([0]_2, 1), ([1]_2, -1)\} \subseteq \boldsymbol{Z}_2 \times \boldsymbol{R}^*$. Now $\varphi([0]_2) = 1$ and $\varphi([1]_2) = -1$ so φ is a bijection and it is easy to see that φ is a homomorphism; for example, $\varphi([1]_2 + [1]_2) = \varphi([0]_2) = 1 = -1 \cdot -1 = \varphi([1]_2) \cdot \varphi([1]_2)$. More generally we have:

Example 6.1 If $G = \langle g \rangle$ has order n then $G \cong \boldsymbol{Z}_n$.

There is a natural identification $\varphi : \boldsymbol{Z}_n \to G$ as $\varphi([j]_n) = g^j$. Note that since $o(g) = n$ by Theorem 2.16, φ is a function using Theorem 2.9 since $[i]_n = [j]_n \Leftrightarrow i \equiv_n j \Leftrightarrow g^i = g^j \Leftrightarrow \varphi([i]_n) = \varphi([j]_n)$. Clearly φ is a bijection and is a homomorphism: $\varphi([i]_n + [j]_n) = \varphi([i + j]_n) = g^{i+j} = g^i g^j = \varphi([i]_n)\varphi([j]_n)$.

Example 6.1 shows that for each $n \in \boldsymbol{N}$ there is abstractly only one cyclic group of order n. Let us next characterize groups of order 4.

Example 6.2 If G is a group and $|G| = 4$ then $G \cong \boldsymbol{Z}_4$ or $G \cong \boldsymbol{Z}_2 \oplus \boldsymbol{Z}_2$.

Recall from Definition 3.11, in $\boldsymbol{Z}_2, \oplus \boldsymbol{Z}_2$, $([a]_2, [b]_2) + ([c]_2, [d]_2) = ([a + c]_2, [b + d]_2)$. If G is cyclic then $G \cong \boldsymbol{Z}_4$ by Example 6.1 and otherwise $G = \{e, a, b, c\}$ with $g^2 = e$ for all $g \in G$ by Theorem 4.7. It follows that $ab, ba \notin \{e, a, b\}$ so $ab = c = ba$, and similarly $ac = b = ca$ and $bc = a = cb$. Using this, any bijection between G and $\boldsymbol{Z}_2 \oplus \boldsymbol{Z}_2$ that sends e to $([0]_2, [0]_2)$ is an isomorphism, since the nonidentity elements of $\boldsymbol{Z}_2 \oplus \boldsymbol{Z}_2$ multiply (add here!) just as $\{a, b, c\}$ do (verify this!). That is, each square is the identity and the product of two different nonidentity elements is the third.

The noncyclic group $\boldsymbol{Z}_2 \oplus \boldsymbol{Z}_2$ of order 4 is often called the **Klein four group.** Other groups that are isomorphic to $\boldsymbol{Z}_2 \oplus \boldsymbol{Z}_2$ by the example are U_8, U_{12}, and $Diag(2, \boldsymbol{Z})$ (Example 2.8).

Example 6.3 $(\boldsymbol{R}^+, \cdot) \cong (\boldsymbol{R}, +)$, where $\boldsymbol{R}^+ = \{r \in \boldsymbol{R} \mid r > 0\}$.

We must find a bijection between the set of positive reals and the set of all reals so that multiplication in $\boldsymbol{R}^+$ corresponds to addition of the images. Any logarithm function satisfies this property and is a bijection since its inverse is an exponential function. The arithmetic property of logarithms is the group homomorphism condition we need. Specifically, if $log = log_{10}$ with base 10 then for any $a, b \in \mathbf{R}^+$, $log(ab) = log(a) + log(b)$.

This example shows that groups may be isomorphic, so abstractly the same, even when they appear quite different. Next we see how groups given abstractly can be recognized as a known group.

Example 6.4 For $n > 2$ and G a group with $|G| = 2n$, if there are $x, y \in G$ with $o(x) = n$, $o(y) = 2$, and $y^{-1}xy = yxy = x^{-1}$ then $G \cong D_n$.

Note that D_n has the required elements: $x = T_1$ and $y = W_0$. Now suppose $x, y \in G$ are as described. Since $xy = yx^{-1} \neq yx$ (why?), $y \notin \langle x \rangle$, so $\langle x \rangle \cap \langle y \rangle = \langle e \rangle$ because $\langle y \rangle = \{e, y\}$, and finally $|\langle x \rangle \langle y \rangle| = n \cdot 2/1 = 2n$ by Theorem 4.12. Thus $G = \langle x \rangle \langle y \rangle = \{x^i, x^i y \mid 0 \le i \le n-1\}$, and $yxy = x^{-1}$ implies that $yx^j y = (yxy)^j = x^{-j}$ (Theorem 2.4) so $yx^j = x^{-j}y$ using $y^2 = e$. How do elements of G and D_n correspond? Now $x^i y = x^j y \Leftrightarrow x^i = x^j \Leftrightarrow i \equiv_n j$ by Theorem 2.9. In D_n, $W_i = W_j$ and $T_i = T_j \Leftrightarrow i \equiv_n j$ so $\varphi(x^i) = T_i$ and $\varphi(x^j y) = W_j$ defines a function and bijection $\varphi : G \to D_n$. Note that $(x^i y)(x^j y) = x^i x^{-j} y^2 = x^{i-j}$, $(x^i y)x^j = x^i x^{-j} y = x^{i-j} y$, and also $x^i(x^j y) = x^i x^j y = x^{i+j} y$. Using Theorem 3.17 we have $W_i W_j = T_{i-j}$, $W_i T_j = W_{i-j}$, and $T_i W_j = W_{i+j}$. These computations imply that φ is an isomorphism, e.g., $\varphi(x^i x^j) = \varphi(x^{i+j}) = T_{i+j} = T_i T_j = \varphi(x^i)\varphi(x^j)$ and $\varphi((x^i y)(x^j y)) = \varphi(x^{i-j}) = T_{i-j} = W_i W_j = \varphi(x^i y)\varphi(x^j y)$.

The example identifies an abstract group with D_n but it is not so clear when we can find appropriate x and $y \in G$ if $|G| = 2n$. The following theorem gives one case when we can do this. It also demonstrates some standard arguments used in determining the structure of abstract groups.

THEOREM 6.1 If $p > 2$ is prime and $|G| = 2p$, then $G \cong \mathbf{Z}_{2p}$ or $G \cong D_p$.

Proof From Theorem 4.7 if $g \in G$ then $o(g) \mid 2p$ so $o(g) = 1, 2, p$, or $2p$. If $o(g) = 2p$ for some $g \in G$, then $|\langle g \rangle| = 2p$ by Theorem 2.16, so $G = \langle g \rangle$, and $G \cong \mathbf{Z}_{2p}$ by Example 6.1. Now assume that no $g \in G$ has order $2p$. Should $g \in G \Rightarrow g^2 = e$ then G is Abelian by Theorem 2.10 and some $x \in G$ has order p by Cauchy's Theorem (Theorem 5.9), a contradiction. Hence for some $x \in G$, $o(x) = p$, and $|\langle x \rangle| = p$ using Theorem 2.16 again. If $H \le G$ with $|H| = p$, then Theorem 4.6 shows that $H = \langle x \rangle$ or $H \cap \langle x \rangle = \langle e \rangle$. In the latter case, by using Theorem 4.12, $2p = |G| \ge |H\langle x \rangle| = |H| \cdot |\langle x \rangle| / |H \cap \langle x \rangle| = p^2$. This contradiction forces $H = \langle x \rangle$. One consequence using Theorem 2.16 is that if $y \in G$ with $o(y) = p$ then $y \in \langle x \rangle$, and another from Theorem 5.3 is that $\langle x \rangle \lhd G$. Thus if $y \notin \langle x \rangle$, then $o(y) = 2$ and so $|G| = 2p$ implies that $G = \langle x \rangle \cup \langle x \rangle y$. Using $\langle x \rangle \lhd G$, $y^{-1}xy \in \langle x \rangle - \langle e \rangle$ (why?), so $y^{-1}xy = x^k$ for some $1 \le k < p$. But $y^2 = e$ now yields $x = y^{-2}xy^2 = y^{-1}(y^{-1}xy)y = y^{-1}x^k y = (y^{-1}xy)^k = (x^k)^k$, and Theorem 2.9 shows that $1 \equiv_p k^2$. But p is a prime so $k \equiv_p 1$ or $k \equiv_p -1$ (verify!) and we conclude that $y^{-1}xy = x$ or $y^{-1}xy = x^{-1}$. The second possibility gives $G \cong D_p$ from Example 6.4, and the first implies that $xy = yx$ so G is

Abelian and $o(xy) = 2p$ by Theorem 4.16, or directly. This contradicts our assumption that no element of G has order $2p$, completing the proof. ■

The theorem shows that *any* group of order 38 is essentially D_{19} or $\mathbf{Z}_{38}$. This is not immediate from the definition of group and shows the advantage of some of the abstract ideas we have been developing.

Homomorphisms are important for two reasons. Isomorphisms provide a way to identify two groups as abstractly the same, as above, and as we did informally when discussing quotient groups. Many examples of this will arise as we proceed. The notion of isomorphism also will give a clearer picture of how some groups can be represented as a direct sum of simpler groups (see Chapter 3). In this case we can deduce properties of a given group from properties of smaller subgroups. The second and perhaps more important use of homomorphisms is to uncover properties of a group by examining its images under homomorphisms. These images turn out to be isomorphic to the quotients G/N of G. Properties of homomorphisms will help us to use information about the quotients to deduce properties of G.

Before the effectiveness of the approaches just mentioned can become apparent there are many facts that need to be developed. First we give several standard and important examples of homomorphisms.

Example 6.5 For any groups $G$and B, $f : G \to B$ given by $f(g) = e_B$ for all $g \in G$ is a homomorphism, called the **trivial homomorphism** of G into B.

Simply observe that for $x, y \in G$, $f(xy) = e_B = e_B e_B = f(x)f(y)$.

Example 6.6 The identity function $I_G : G \to G$ is an isomorphism.

We know that I_G is a bijection by Theorem 0.14, and that for any $x, y \in G$, $I_G(xy) = xy = I_G(x)I_G(y)$, so I_G is also a homomorphism.

Next is perhaps the most important single example of the notion of homomorphism. The example illustrates the primary method of construction of homomorphisms used in the study of groups.

Example 6.7 If G is a group and $N \lhd G$, then $\rho : G \to G/N$ given by $\rho(g) = Ng$ is a surjective homomorphism, and ρ is an isomorphism $\Leftrightarrow N = \langle e_G \rangle$.

By definition, G/N is the set of cosets of N, so ρ is a function, and from the definition of coset multiplication in G/N, for $x, y \in G$, $\rho(xy) = Nxy = NxNy = \rho(x)\rho(y)$. Hence ρ is a homomorphism. Any $X \in G/N$ is $X = Ng$ for some $g \in G$, so $\rho(g) = X$ and ρ is surjective. If $y \in N$ then $\rho(y) = Ny = N = \rho(e)$, so ρ injective would force $y = e$ and $N = \langle e_G \rangle$. Finally, if $N = \langle e_G \rangle$ and $\rho(x) = \rho(y)$, then $Nx = Ny$ so $xy^{-1} \in N$ by Theorem 4.2. Therefore $xy^{-1} = e$, or $x = y$, showing that ρ is injective.

From the example, $G \cong G/\langle e\rangle$ by identifying $g \in G$ with the singleton $\langle e\rangle g = \{g\} \in G/\langle e\rangle$. An important special case of Example 6.7 is for $G = \mathbf{Z}$ and $N = n\mathbf{Z}$. From Example 5.3 $(\mathbf{Z}, +)/n\mathbf{Z} = \mathbf{Z}_n$ so here $\rho(k) = n\mathbf{Z} + k = [k]_n$.

Example 6.8 For G a group and $x \in G$, $T_x : G \to G$ given by $T_x(g) = xgx^{-1}$ is an isomorphism of G with itself.

To see that T_x is a homomorphism, let $g, h \in G$ and compute $T_x(gh) = xghx^{-1} = xgx^{-1}xhx^{-1} = T_x(g)T_x(h)$. We show that T_x is a bijection by finding an inverse for it and then using Theorem 0.11. It is straightforward to compute that an inverse for T_x is $T_{x^{-1}}$ given by $T_{x^{-1}}(g) = (x^{-1})g(x^{-1})^{-1} = x^{-1}gx$.

Example 6.9 $\det : GL(n, \mathbf{R}) \to (\mathbf{R}^*, \cdot)$ is a surjective homomorphism.

The verification is just the multiplicative property of determinants: $\det(AB) = \det(A)\det(B)$. This homomorphism is surjective since if $r \in \mathbf{R}^*$ then $r = \det(r^\#)$ for $r^\# = Diag(r, 1, 1, \ldots, 1)$ and $r^\# \in GL(n, \mathbf{R})$ since clearly $Diag(r, 1, 1, \ldots, 1)^{-1} = Diag(r^{-1}, 1, 1, \ldots, 1)$.

Example 6.10 For groups $G_1, \ldots, G_k$, the projections $\pi_j : G_1 \oplus \cdots \oplus G_k \to G_j$ are surjective homomorphisms.

By Definition 3.11, $G_1 \oplus \cdots \oplus G_k = \{(g_1, \ldots, g_k) \in G_1 \times \cdots \times G_k \mid g_i \in G_i\}$ with coordinate multiplication, and by Example 0.34, $\pi_j((g_1, \ldots, g_k)) = g_j$. If e_i is the identity of G_i and $g \in G_j$ then $\pi_j((e_1, \ldots, e_{j-1}, g, e_{j+1}, \ldots, e_k)) = g$ and π_j is a surjective map. Now let $(g_1, \ldots, g_k), (h_1, \ldots, h_k) \in G_1 \oplus \cdots \oplus G_k$ and use the multiplication in $G_1 \oplus \cdots \oplus G_k$ to see that $\pi_j((g_1, \ldots, g_k)(h_1, \ldots, h_k)) = \pi_j((g_1h_1, \ldots, g_kh_k)) = g_jh_j = \pi_j((g_1, \ldots, g_k))\pi_j((h_1, \ldots, h_k))$, showing that π_j is a homomorphism. We remark that if $G = G_1 \oplus \cdots \oplus G_k$ and we define f_j by $f_j((g_1, \ldots, g_k)) = (e_1, \ldots, g_j, \ldots, e_k)$, then f_j is a homomorphism from G to G, neither injective nor surjective if $k > 1$.

EXERCISES

1. Determine whether the following are group homomorphisms:

i) $f(r) = 2r$ for $f : (\mathbf{R}^*, \cdot) \to (\mathbf{R}^*, \cdot)$
ii) $g(r) = 2r$ for $g : (\mathbf{R}, +) \to (\mathbf{R}, +)$
iii) $\varphi(r) = r - 1$ for $\varphi : (\mathbf{R}^*, \cdot) \to (\mathbf{R} - \{-1\}, \circ)$ where $a \circ b = ab + a + b$
iv) $\sigma(g)(x) = g^{-1}x$ for $\sigma : G \to \mathrm{Bij}(G)$
v) $\beta(g)(x) = xg^{-1}$ for $\beta : G \to \mathrm{Bij}(G)$
vi) $\eta(x, y)(g) = xgy^{-1}$ for $\eta : G \oplus G \to \mathrm{Bij}(G)$
vii) $\delta(A) = \det(A) \cdot A$ for $\delta : GL(n, \mathbf{R}) \to GL(n, \mathbf{R})$
viii) $\alpha(a, r)(x) = ax + r$ for $\alpha : (\mathbf{R}^*, \cdot) \oplus (\mathbf{R}, +) \to \mathrm{Bij}(\mathbf{R})$

ix) $\varphi(r) = 2^r r$ for $\varphi : (\boldsymbol{R}^*, \cdot) \to (\boldsymbol{R}^*, \cdot)$
x) $\rho(f(x)) = x^n f(1/x)$ if $deg f = n$ and $\rho(0) = 0$ for $\rho : (\boldsymbol{R}[x], +) \to (\boldsymbol{R}[x], +)$
xi) $\eta(A) = (A^{-1})^{\mathrm{T}} (B_{ij}^{\mathrm{T}} = B_{ji})$ for $\eta : GL(n, \boldsymbol{R}) \to GL(n, \boldsymbol{R})$

2. For $1 \le j \le k$ let $\varphi_j : G \to B_j$ be a group homomorphism and let $\varphi : G \to \oplus B_i$ be given by $\varphi(g) = (\varphi_1(g), \ldots, \varphi_k(g))$. Show that:
i) φ is a group homomorphism
ii) φ is injective if some φ_j is injective
iii) φ can be injective even if no φ_j is injective
iv) if φ is surjective then all φ_j are surjective
v) if all φ_j are surjective, φ need not be surjective

3. Given an Abelian group A and $\gamma_i : G_i \to A$ group homomorphisms, show that $\gamma(g_1, \ldots, g_m) = \gamma_1(g_1)\gamma_2(g_2) \cdots \gamma_m(g_m)$ is a group homomorphism $\gamma : \oplus G_i \to A$.

4. Suppose that for $1 \le i \le k$ there are group homomorphisms $\varphi_i : G_i \to B_i$. Define $\varphi : \oplus G_i \to \oplus B_i$ by $\varphi(g_1, \ldots, g_k) = (\varphi_1(g_1), \ldots, \varphi_k(g_k))$. Prove the following:
a) φ is a homomorphism of groups
b) φ is injective $\Leftrightarrow$ each φ_j is injective
c) φ is surjective $\Leftrightarrow$ each φ_j is surjective
d) φ is an isomorphism $\Leftrightarrow$ each φ_j is an isomorphism

5. **a)** For any groups A and B, show that $A \oplus B \cong B \oplus A$.
b) If $A \cong B$ and $C \cong D$ are group isomorphisms, show that $A \oplus C \cong B \oplus D$.
c) If $k > 1$ and $G_j \cong H_j$ are groups, show that $G_1 \oplus \cdots \oplus G_k \cong H_1 \oplus \cdots \oplus H_k$.
d) For $k > 1$, G_j groups, and $\sigma \in S_k$, prove $G_1 \oplus \cdots \oplus G_k \cong G_{\sigma(1)} \oplus \cdots \oplus G_{\sigma(k)}$.

6. For each pair of groups find an explicit isomorphism:
i) $U_8 \cong U_4 \oplus U_6$
ii) $U_8 \cong \boldsymbol{Z}_2 \oplus \boldsymbol{Z}_2$
iii) $U_{14} \cong \boldsymbol{Z}_6$
iv) $U_{16} \cong \boldsymbol{Z}_2 \oplus \boldsymbol{Z}_4$
v) $U_{15} \cong \boldsymbol{Z}_2 \oplus \boldsymbol{Z}_4$
vi) $\boldsymbol{Z}_3 \oplus \boldsymbol{Z}_5 \cong \boldsymbol{Z}_{15}$
vii) $D_3 \cong S_3$
viii) $GL(2, \boldsymbol{Z}_2) \cong S_3$
ix) $UT(3, \boldsymbol{Z}_2) \cong D_4$

7. If $G = \langle\{2, 1/3\}\rangle \le (\boldsymbol{Q}^+, \cdot)$ is the subgroup generated by $2, 1/3 \in \boldsymbol{Q}^+$ (see Theorem 2.18), show that $G \cong \boldsymbol{Z}^2 = \boldsymbol{Z} \oplus \boldsymbol{Z}$.

8. Show that $(\boldsymbol{Q}^+, \cdot) \cong (\boldsymbol{Z}[x], +)$. (Use unique factorization in $\boldsymbol{N}$ to see that $q \in \boldsymbol{Q}^+$ can be written $q = p_1^{a(1)} \cdots p_k^{a(k)}$ for $p_1 < \cdots < p_k$ primes and all $a(j) \in \boldsymbol{Z}$.)

9. Let $G = \left\{ \begin{bmatrix} a & b \\ [0] & [1] \end{bmatrix} \in GL(2, \boldsymbol{Z}_n) \mid a = [\pm 1]_n \right\}$. Show that $G \le GL(2, \boldsymbol{Z}_n)$ and that $G \cong D_n$.

6.2 BASIC RESULTS

In all the examples of group homomorphisms $\varphi : G \to B$ presented so far, $\varphi(e_G) = e_B$. This holds generally, and there are other properties of homomorphisms that will be useful to record as well.

THEOREM 6.2 Let $\varphi: G \to B$ be a homomorphism of the groups G and B. The following hold:

i) $\varphi(e_G) = e_B$
ii) for all $g \in G$, $\varphi(g^{-1}) = \varphi(g)^{-1}$
iii) for $g_1, \ldots, g_k \in G$, $\varphi(g_1 g_2 \cdots g_k) = \varphi(g_1)\varphi(g_2)\cdots\varphi(g_k)$
iv) $\varphi(G) \leq B$
v) if $g \in G$ with $o(g) = n$, then $o(\varphi(g)) \mid n$
vi) if $g \in G$ with $o(g) = n$ and φ is an isomorphism, then $o(\varphi(g)) = n$

Proof i) Since $e_G = e_G e_G$ and φ is a homomorphism, $\varphi(e_G) = \varphi(e_G e_G) = \varphi(e_G)\varphi(e_G)$ so $\varphi(e_G) = e_B$ by Theorem 2.2. For ii), use i) and $e_G = gg^{-1}$ to write $e_B = \varphi(e_G) = \varphi(gg^{-1}) = \varphi(g)\varphi(g^{-1})$. Similarly, $\varphi(g^{-1})\varphi(g) = e_B$ results from $e_G = g^{-1}g$. Thus $\varphi(g^{-1})$ is an inverse in B for $\varphi(g)$ so is the unique inverse, forcing $\varphi(g^{-1}) = \varphi(g)^{-1}$. Assume for $k \geq 1$ and $g_j \in G$ that $\varphi(g_1 g_2 \cdots g_k) = \varphi(g_1)\cdots\varphi(g_k)$. For $g_1, \ldots, g_{k+1} \in G$, $\varphi(g_1 \cdots g_{k+1}) = \varphi((g_1 \cdots g_k)g_{k+1}) = \varphi(g_1 \cdots g_k)\varphi(g_{k+1}) = \varphi(g_1)\cdots\varphi(g_k)\varphi(g_{k+1})$, and iii) holds by IND-I. Note that we have relied on Theorem 2.3 to write products without parentheses. For iv), let $\varphi(x), \varphi(y) \in \varphi(G)$ and use ii) to compute that $\varphi(x)\varphi(y)^{-1} = \varphi(x)\varphi(y^{-1}) = \varphi(xy^{-1}) \in \varphi(G)$, showing $\varphi(G) \leq B$ by Theorem 2.12. In v), since $o(g) = n$, $g^n = e_G$ so i) and iii) yield $e_B = \varphi(e_G) = \varphi(g^n) = \varphi(g)^n$ and $o(\varphi(g)) \mid n$ from Theorem 2.7. If $o(\varphi(g)) = k < n$, then $\varphi(e_G) = e_B = \varphi(g)^k = \varphi(g^k)$ so if φ is a bijection $g^k = e_G$. Hence when $o(g) = n$ and φ is an isomorphism, vi) holds. ∎

The image $\varphi(G)$ of G under φ is called a **homomorphic image** of G. It is important to realize that *every quotient group G/N of G is a homomorphic image* of G from Example 6.7: $\rho: G \to G/N$ given by $\rho(g) = Ng$ is a surjective homomorphism, so $G/N = \rho(G)$.

Let us illustrate the results in Theorem 6.2. If $A \in GL(n, \boldsymbol{R})$ we know $\det: GL(n, \boldsymbol{R}) \to (\boldsymbol{R}^*, \cdot)$ is a homomorphism and $\det(A^{-1}) = \det(A)^{-1}$. Also for $r \in (\boldsymbol{R}^+, \cdot)$ with $log: (\boldsymbol{R}^+, \cdot) \to (\boldsymbol{R}, +)$, $log(r^{-1}) = -log(r) = (log(r))^{-1} \in (\boldsymbol{R}, +)$. For the groups $(\boldsymbol{Z}_n, +)$, from Theorem 1.30 $\varphi([a])_n) = [a]_m$ is a function from $\boldsymbol{Z}_n$ to $\boldsymbol{Z}_m$ exactly when $m \mid n$. Now φ is also a group homomorphism via $\varphi([a]_n + [b]_n) = \varphi([a+b]_n) = [a+b]_m = [a]_m + [b]_m = \varphi([a]_n) + \varphi([b]_n)$. When $n = 12$ and $m = 4$, $\varphi([7]_{12}) = [7]_4 = [3]_4$ and $\varphi(-[7]_{12}) = \varphi([5]_{12}) = [5]_4 = [1]_4 = -[3]_4 = -\varphi([7]_{12})$. Looking at orders, $o([10]_{12}) = 12/(10, 2) = 6$ by Theorem 2.8 and $\varphi([10]_{12}) = [10]_4 = [2]_4$ has order 2 in $\boldsymbol{Z}_4$ and $2 \mid 6$. Now $o([6]_{12}) = 2$ and $\varphi([6]_{12}) = [6]_4 = [2]_4$ has order 2 also. Finally, $o([8]_{12}) = 3$ and $\varphi([8]_{12}) = [8]_4 = [0]_4$ has order 1, dividing three.

We can use Theorem 6.2 to show that some groups cannot be isomorphic by comparing orders of elements. For example, $|U_8| = 4 = |U_{10}|$ but these are not isomorphic since $o([3]_{10}) = 4$ in U_{10} but U_8 has no element or order 4. That is, if $\varphi: U_{10} \to U_8$ were an isomorphism then $o(\varphi([3]_{10})) = 4$ in U_8 by Theorem 6.2, a contradiction. Similarly, $G = U_8 \oplus U_8$ and U_{32} are Abelian of order 16 but not isomorphic: every $g \in G$ satisfies $x^2 = e_G$ by Theorem 3.20 but $[3]_{32} \in U_{32}$ with $o([3]_{32}) = 8$. For a non-Abelian example consider $D_4 \oplus D_4$ and D_{32}, each of order 64. Now $T_1 \in D_{32}$ has order 32, but $D_4 \oplus D_4$ has no element of order greater than 4, using Theorem 3.20 again, so these groups cannot be isomorphic.

If $\varphi: G \to B$ is a group isomorphism then Theorem 6.2 shows that φ must be a bijection between $\{g \in G \mid g^m = e_G\}$ and $\{b \in B \mid b^m = e_B\}$ (why?). Therefore $\mathbf{Z}_4 \oplus \mathbf{Z}_4$ cannot be isomorphic to $\mathbf{Z}_4 \oplus \mathbf{Z}_2 \oplus \mathbf{Z}_2$ since the first group has four elements satisfying $x^2 = e$ but the second has eight elements satisfying it (Theorem 3.20). In general it is difficult to determine the orders of elements in a group and how many elements of a given order a group has, so these approaches are not often useful. Finally, note that a non-Abelian group and an Abelian group cannot be isomorphic since $xy = yx$ for all $x, y \in G \Rightarrow \varphi(x)\varphi(y) = \varphi(xy) = \varphi(yx) = \varphi(y)\varphi(x)$ for all $\varphi(x), \varphi(y) \in \varphi(G)$.

The next result connects homomorphisms with compositions. Also, when φ is an isomorphism, then as a bijection of sets, φ has an inverse, but it is not completely clear that this inverse is a group homomorphism.

THEOREM 6.3 If $\varphi: G \to H$ and $\rho: H \to K$ are group homomorphisms then $\rho \circ \varphi: G \to K$ is a homomorphism. If φ is an isomorphism, $\varphi^{-1}: H \to G$ is also.

Proof To show $\rho \circ \varphi$ is a homomorphism take $x, y \in G$ and use that φ and ρ are homomorphisms to compute that $\rho \circ \varphi(xy) = \rho(\varphi(xy)) = \rho(\varphi(x)\varphi(y)) = \rho(\varphi(x))\rho(\varphi(y)) = (\rho \circ \varphi(x))(\rho \circ \varphi(y))$. An isomorphism φ is a bijection so has an inverse $\varphi^{-1}: H \to G$ by Theorem 0.11. Let $u, v \in H$. Since φ is a bijection, $u = \varphi(x)$ and $v = \varphi(y)$ for unique $x, y \in G$, and from the definition of inverse, $x = \varphi^{-1}(u)$ and $y = \varphi^{-1}(v)$. But φ is a homomorphism so $uv = \varphi(x)\varphi(y) = \varphi(xy)$, and applying φ^{-1} shows that $xy = \varphi^{-1}(uv)$. Hence $\varphi^{-1}(uv) = xy = \varphi^{-1}(u)\varphi^{-1}(v)$ and φ^{-1} is a homomorphism. By Theorem 0.13, φ^{-1} is a bijection, so φ^{-1} is indeed an isomorphism. ∎

COROLLARY On any set S of groups $\cong$ is an equivalence relation.

Proof For $G \in S$ the identity $I_G: G \to G$ is an isomorphism by Example 6.6 so $\cong$ is reflexive. If $G, H \in S$ with $G \cong H$ via an isomorphism $\varphi: G \to H$, then by the theorem, $\varphi^{-1}: H \to G$ is also an isomorphism so $H \cong G$ and $\cong$ is symmetric. When $G \cong H$ and $H \cong K$ in S there are isomorphisms $\varphi: G \to H$ and $\rho: H \to K$. The theorem shows $\rho \circ \varphi: G \to K$ is a homomorphism, and the composition $\rho \circ \varphi$ of bijections is also a bijection by Theorem 0.13. Thus $\rho \circ \varphi$ is an isomorphism, and $\cong$ is transitive, so $\cong$ is an equivalence relation on S. ∎

The first part of Theorem 6.2 shows that for any homomorphism $\varphi: G \to B$, $\varphi(e_G) = e_B$: the inverse image $\varphi^{-1}(\{e_B\}) = \{x \in G \mid \varphi(x) = e_B\} \neq \varnothing$. This set is extremely important and has a special name.

DEFINITION 6.2 If $\varphi: G \to B$ is a homomorphism of groups, the **kernel of** φ, denoted *ker* φ, is $\varphi^{-1}(\{e_B\}) = \{x \in G \mid \varphi(x) = e_B\}$.

The kernel leads to the first step in the development of a powerful approach used to analyze and better understand the structure of groups. It will enable us to use techniques of induction to relate the given group to simpler ones. Some important facts about *ker* φ come next.

THEOREM 6.4 If $\varphi : G \to B$ is a homomorphism of groups, then:

i) $ker\,\varphi \lhd G$.
ii) If $b = \varphi(g) \in \varphi(G)$, then $\varphi^{-1}(\{b\}) = \{x \in G \mid \varphi(x) = b\} = (ker\,\varphi)g$.
iii) φ is injective exactly when $ker\ \varphi = \langle e_G \rangle$.
iv) If φ is surjective, then it is an isomorphism exactly when $ker\,\varphi = \langle e_G \rangle$.

Proof For i), we need $ker\ \varphi \le G$ and use Theorem 2.12. We have seen that $e_G \in ker\ \varphi \ne \emptyset$. For $x, y \in ker\ \varphi$, $xy^{-1} \in ker\ \varphi$ requires $\varphi(xy^{-1}) = e_B$ from Definition 6.2. But by Theorem 6.2, $\varphi(xy^{-1}) = \varphi(x)\varphi(y^{-1}) = \varphi(x)\varphi(y)^{-1}$, and $x, y \in ker\,\varphi$ mean $\varphi(x) = \varphi(y) = e_B$, so $\varphi(xy^{-1}) = e_B$ follows and $ker\ \varphi \le G$. To see that $N = ker\ \varphi \lhd G$ we need $g^{-1}xg \in N$ for all $x \in N$ and $g \in G$. But $\varphi(g^{-1}xg) = \varphi(g^{-1})\varphi(x)\varphi(g) = \varphi(g^{-1})e_B\varphi(g) = \varphi(g^{-1})\varphi(g) = \varphi(g^{-1}g) = \varphi(e_G) = e_B$, so $g^{-1}xg \in N$ by the definition of N, and $N \lhd G$. Suppose next that $\varphi(g) = b$. If $x \in N$ then $\varphi(xg) = \varphi(x)\varphi(g) = e_B b = b$, proving $Ng \subseteq \varphi^{-1}(\{b\})$. When $h \in \varphi^{-1}(\{b\})$, $\varphi(h) = b = \varphi(g)$ implies $\varphi(h)\varphi(g)^{-1} = e_B$. By Theorem 6.2, $\varphi(hg^{-1}) = \varphi(h)\varphi(g)^{-1} = e_B$, so $hg^{-1} \in N$ and $h \in Ng$. This shows $\varphi^{-1}(\{b\}) \subseteq Ng$ and proves that ii) holds (why?). For iii), if $g \in G$, by part ii) $\varphi(Ng) = \{\varphi(g)\}$ so $\varphi(x) = \varphi(y)$ for any $x, y \in Ng$. Thus if φ is injective $|Ng| = 1$, so $|N| = 1$ and $N = \langle e_G \rangle$. When $N = \langle e_G \rangle$ and $\varphi(x) = \varphi(y)$ then by ii), $y \in Nx = \langle e_G \rangle x = \{x\}$, and thus $x = y$ and φ is injective. Isomorphisms are bijections so iii) $\Rightarrow$ iv). ∎

We apply the last result to Example 6.9 that shows $\det : GL(n, \boldsymbol{R}) \to \boldsymbol{R}^*$ is a surjective homomorphism. Now $A \in ker \det$ exactly when $\det A = 1$, so $ker \det = SL(n, \boldsymbol{R})$ (Definition 2.9) and $SL(n, \boldsymbol{R}) \lhd GL(n, \boldsymbol{R})$ directly from Theorem 6.4. If $r \in \boldsymbol{R}^*$ and $\det A = r$, then the theorem shows that the coset $SL(n, \boldsymbol{R})A$ is the set of all matrices having determinant r.

Given $N \lhd G$, Example 6.7 shows that $\rho(g) = Ng$ is a surjective homomorphism from G onto G/N. By definition, $ker\ \rho = \{x \in G \mid \rho(x) = e_{G/N}\} = \{x \in G \mid Nx = N\} = \{x \in G \mid x \in N\} = N \subseteq G$. Theorem 6.4 ii) now shows that for $x \in G$, $\rho(x) = Ng = \rho(g) \Leftrightarrow x \in Ng$. Since $\boldsymbol{Z}_n = \boldsymbol{Z}/n\boldsymbol{Z}$, a special case is $\rho : \boldsymbol{Z} \to \boldsymbol{Z}_n$ with $ker\ \rho = \{k \in \boldsymbol{Z} \mid \rho(k) = [k]_n = [0]_n\} = \{k \in \boldsymbol{Z} \mid n \mid k\} = n\boldsymbol{Z} \le \boldsymbol{Z}$. Thus $y \in \boldsymbol{Z}$ satisfies $\rho(y) = [a]_n = \rho(a) \Leftrightarrow y \in n\boldsymbol{Z} + a \subseteq \boldsymbol{Z}$.

We end this section by discussing homomorphisms and powers in a group and will see how to use the last parts of Theorem 6.4. For G any group, $g \in G$, and $n \in \boldsymbol{N}$, consider the power g^n (Definition 2.4) and then $F_n : G \to G$ given by $F_n(g) = g^n$. In general F_n is not a homomorphism. For example, in S_3, $F_2((1, 2))F_2((2, 3)) = (1, 2)^2(2, 3)^2 = I$ but $F_2((1, 2)(2, 3)) = F_2((1, 2, 3)) = (1, 2, 3)^2 = (1, 3, 2)$. However, F_6 is the trivial homomorphism on S_3 since $\theta^6 = I$ for all $\theta \in S_3$. For D_5 (Theorem 3.17) with $n = 3$ we see that $W_0T_1 = W_4$, $F_3(W_4) = W_4^3 = W_4$, but $F_3(W_0)F_3(T_1) = W_0^3T_1^3 = W_0T_3 = W_2$. Thus $F_3 : D_5 \to D_5$ is not a homomorphism. Note that neither S_3 nor D_5 is Abelian. Whenever $gh = hg$ in G then from Theorem 2.4, $(gh)^n = g^nh^n$, so F_n acting on an Abelian group is always a homomorphism. Restricting our attention to this case, we prove a result about powers in Abelian groups.

THEOREM 6.5 If G is an Abelian group, $m \in \boldsymbol{N}$, and $F : G \to G$ is given by $F(g) = g^m$ then F is a homomorphism and $ker\ F = \{x \in G \mid o(x) \mid m\}$, so F is injective when $\{x \in G \mid x^m = e\} = \langle e \rangle$. If G is finite, F is an isomorphism $\Leftrightarrow$ $(m, |G|) = 1$ and then $G = \{x^m \mid x \in G\}$.

Proof As we mentioned, by Theorem 2.4, $(gh)^m = g^m h^m$ when G is Abelian so F is a homomorphism. Since $ker\,F = \{x \in G \mid x^m = e\}$, Theorem 2.7 yields $ker\,F = \{x \in G \mid o(x) \mid m\}$. Now Theorem 6.4 shows that F is injective $\Leftrightarrow$ $ker\ F = \langle e\rangle$, so when $\{x \in G \mid x^m = e\} = \langle e\rangle$. The last statement in the theorem requires a bit of work. If G is finite and $g \in ker\ F$ then $o(g) \mid m$ and also $o(g) \mid |G|$ from Theorem 4.7 so $o(g) | (m, |G|)$ using Theorem 1.8 on greatest common divisors. Therefore, if $(m, |G|) = 1$ then $o(g) = 1$, forcing $g = e$ and $ker\ F = \langle e\rangle$. Thus $(m, |G|) = 1$ implies F is injective, so a bijection since G is finite, by applying Theorem 0.10: F is an isomorphism when $(m, |G|) = 1$. When F is an isomorphism but $(m, |G|) = k > 1$, let p be a prime divisor of k. Since $p \mid |G|$ there is $x \in G$ with $o(x) = p$ by Cauchy's Theorem (Theorem 5.9), and since $p \mid m$ we may write $m = ps$. But now $F(x) = x^m = (x^p)^s = e = F(e)$, so F a bijection forces $x = e$, contradicting $o(x) = p$. Consequently, if F is an isomorphism then $(m, |G|) = 1$, completing the proof. Note that when F is an isomorphism, $G = F(G) = \{x^m \mid x \in G\}$. ■

How does the theorem apply in the groups U_n (Definition 3.1)? When $n = 102$, $|U_{102}| = \varphi(102) = \varphi(2)\varphi(3)\varphi(17) = 1 \cdot 2 \cdot 16 = 32$. Therefore for any odd $m \in \boldsymbol{N}$, every $g \in U_{102}$ is an mth power, although for any given g it may be difficult to find $h \in U_{102}$ satisfying $g = h^m$. Specifically, in U_{102} every element is a fifth power, a thirty-fifth power, and a ninety-first power.

Now let $G = U_{19}$ so $|G| = 18$. Every $x \in G$ must be a fifth power by Theorem 6.5 since $(5,18) = 1$, although it is not easy to see which fifth power equals $[2]_{19}$. In fact $[2]_{19} = [15]_{19}^5$, and because taking fifth powers is an isomorphism, $[15]_{19}$ is the only element whose fifth power is $[2]_{19}$. This computation is not as mysterious as it may appear; it is not necessary to start computing all fifth powers in U_{19}. From Theorem 4.20 we know that U_{19} is a cyclic group, and it is not hard to verify directly that $U_{19} = \langle [2]_{19}\rangle$. Thus $\{g^5 \mid g \in U_{19}\} = \{([2]_{19}^k)^5 \mid k \in \boldsymbol{N}\} = \{[2^{5k}]_{19} \mid k \in \boldsymbol{N}\}$, so to find $g \in U_{19}$ with $[2]_{19} = g^5$ we must solve $[2]_{19} = [2^{5k}]_{19}$. This is equivalent to solving $5k \equiv_{18} 1$, by Theorem 2.9, and using $11 \cdot 5 \equiv_{18} 1$ we get $k = 11$. Computing $[2]_{19}^4 = [16]_{19} = [-3]_{19}$ shows that $[2]_{19}^{11} = [-3]_{19}[-3]_{19}[8]_{19} = [72]_{19} = [15]_{19}$, as claimed above. Now using $2^4 \equiv_{19} 16$, to write $[16]_{19} = g^5$ in U_{19} for $g = [2]_{19}^k$ we need $[2^4]_{19} = [2^{5k}]_{19}$. Again this is equivalent to solving $4 \equiv_{18} 5k$, which has solution $k = 8 \equiv_{18} 11 \cdot 4$. From our computation above, $[16]_{19} = [2^8]_{19}^5 = ([-3]_{19}[-3]_{19})^5 = [9]_{19}^5$. Similar computations work for other powers of 2.

EXERCISES

1. Show that the following are group homomorphisms and describe the kernel:

i) $\alpha : (8\boldsymbol{Z}, +) \to (2\boldsymbol{Z}, +)$ given by $\alpha(m) = m/2$

ii) $\beta : U_{21} \to G = (\{[3k]_{21} \in \boldsymbol{Z}_{21} \mid 1 \le k \le 6\}, \cdot) \subseteq \boldsymbol{Z}_{21}$ via $\beta([a]_{21}) = [15a]_{21}$

iii) $\varphi : (\boldsymbol{R}, +) \to (\boldsymbol{C}^*, \cdot)$ defined by $\varphi(r) = e^{ir} = \cos r + i \cdot \sin r$

iv) for $a \in \boldsymbol{R}$, $E_a : (\boldsymbol{R}[x], +) \to (\boldsymbol{R}, +)$ defined by $E_a(p(x)) = p(a)$

v) $\eta : (\boldsymbol{R}[x], +) \to (\boldsymbol{R}^*, \cdot)$ via $\eta(f(x)) = 3^{f(1)}\pi^{f(2)}$

vi) $\varphi : (\boldsymbol{R}[x], +) \to (\boldsymbol{R}, +)$ given by $\varphi(p(x)) = D(p(x))(1)$ (D is the derivative)

vii) $\beta : S_n \to \boldsymbol{Z}_2$ via $\beta(\theta) = [0]_2$ if $\theta \in A_n$ and $\beta(\theta) = [1]_2$ if $\theta \notin A_n$

viii) $\varphi : \boldsymbol{Z}_3 \oplus \boldsymbol{Z}_{12} \to \boldsymbol{Z}_3 \oplus \boldsymbol{Z}_{12}$ given by $\varphi([a]_3, [b]_{12}) = ([b]_3, [4a]_{12})$

ix) $\rho : \boldsymbol{Z}_3 \oplus S_4 \to \boldsymbol{Z}_3 \oplus S_4$ defined by $\rho([a]_3, \theta) = ([0]_3, (2, 3, 4)^a)$

x) $\alpha : \boldsymbol{Z}_{12} \to (\boldsymbol{C}^*, \cdot)$ via $\alpha([k]_{12}) = i^k$ for $i^2 = -1$ in $\boldsymbol{C}$.

2. Show that the following pairs of groups are *not* isomorphic:
   i) D_6 and $\mathbf{Z}_{12}$
   ii) U_{12} and U_{10}
   iii) A_4 and D_6
   iv) U_{16} and $\mathbf{Z}_8$
   v) U_{13} and U_{36}
   vi) D_{12} and S_4
   vii) $UT(2, \mathbf{Z})$ and $\mathbf{Z} \oplus \mathbf{Z}_2 \oplus \mathbf{Z}_2$
   viii) $U_{10} \oplus U_{10}$ and U_{32}

3. Let $n > 1$ and for each $1 \le i \le n$, set $G_i = \{\theta \in S_n \mid \theta(i) = i\}$. Show $G_i \le S_n$ and $G_i \cong S_{n-1}$. (Show first that $G_n \cong S_{n-1}$ and then that $(i, n)G_i(i, n) \cong G_n$.)

4. If $\varphi : G \to B$ is a surjective group homomorphism, then: a) $\varphi(Z(G)) \le Z(B)$; and b) $\varphi(G') \le B'$ (commutator subgroups in Definition 2.13).

5. If $G \cong H$ for groups G and H, show that: a) $Z(G) \cong Z(H)$; b) $G' \cong H'$ (see Exercise 4).

6. Show G is Abelian $\Leftrightarrow \varphi : G \to G$ via $\varphi(g) = g^{-1}$ is a group homomorphism.

7. If $\varphi : G \to B$ is a group homomorphism and $H \le G$, show that:
   a) $\varphi(H) \le B$
   b) if $H \lhd G$ and φ is surjective then $\varphi(H) \lhd B$
   c) if $K \le B$ then $\varphi^{-1}(K) = \{g \in G \mid \varphi(g) \in K\} \le G$
   d) if $K \lhd B$ then $\varphi^{-1}(K) \lhd G$

8. If $\varphi : G \to G$ is a group homomorphism, show $G^{\varphi} = \{g \in G \mid \varphi(g) = g\} \le G$.

9. Determine whether each $g \in G$ is an mth power:
   i) $G = U_{1000}$, $m = 3$
   ii) $G = U_{40}$, $m = 5$
   iii) $G = U_{61}$, $m = 35$
   iv) $G = D_{55}$, $m = 3$
   v) $G = D_{95}$, $m = 5$
   vi) $G = U_{72}$, $m = 35$
   vii) $G = S_4$, $m = 5$

10. For G any finite group and $m \in N$ with $(m, |G|) = 1$, show that $F : G \to G$ given by $F(g) = g^m$ is a bijection.

6.3 THE FIRST ISOMORPHISM THEOREM

Suppose $\varphi : G \to B$ is a surjective homomorphism of groups, so for $b \in B$ there is $g \in G$ with $\varphi(g) = b$. From Theorem 6.4, $\varphi^{-1}(\{b\}) = Ng$ for $N = ker\, \varphi$ and this gives a correspondence between B and $G/N = \{Ng\}$. This identification of $\varphi(g)$ with Ng is an isomorphism between B and G/N, which we prove next. Thus the image of G under any surjective homomorphism φ is abstractly the quotient group $G/ker\, \varphi$. Therefore studying homomorphic images $\varphi(G)$ is essentially the same as studying the quotient groups G/N.

THEOREM 6.6 (First Isomorphism Theorem) If $\varphi : G \to B$ is a surjective homomorphism of groups, then $B \cong G/ker\, \varphi$.

Proof The proof is outlined just above: use Theorem 6.4 to identify B with $G/ker\ \varphi$ and show that the identification is a homomorphism. Set $ker\,\varphi = N$. We claim that $\beta(Ng) = \varphi(g)$ is a function from G/N to B. The problem is to show that β does not depend on the representative for the coset Ng. That is, if $Ng = Nh$ is $\beta(Ng) = \beta(Nh)$? But $Ng = Nh \Rightarrow h \in Ng = \varphi^{-1}(\{\varphi(g)\})$ by Theorem 6.4, so $\varphi(h) = \varphi(g)$, $\beta(Ng) = \beta(Nh)$, and β is a function. Since φ is surjective, for each $b \in B$ there is $g \in G$ with $b = \varphi(g) = \beta(Ng)$, so β is surjective. To show β is a homomorphism, take Nx, $Ny \in G/N$ and use the definition of coset multiplication to see that $\beta(NxNy) = \beta(Nxy) = \varphi(xy) = \varphi(x)\varphi(y) = \beta(Nx)\beta(Ny)$. Finally, to see that β is injective, let $X = Nx \in ker\ \beta$ so $e_B = \beta(Nx) = \varphi(x)$, forcing $x \in N$. Thus $X = N = e_{G/N}$, $ker\ \beta = \langle e_{G/N}\rangle$, β is injective by Theorem 6.4, and hence is a bijection, proving the theorem. ■

The proof of the theorem, showing that the correspondence between the Ng and their images $\varphi(g) \in B$ preserves multiplication, is illustrated in Fig. 6.2.

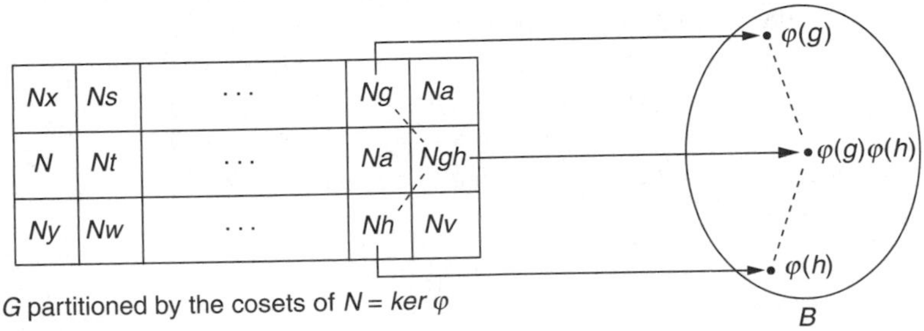

FIGURE 6.2

Theorem 6.6 enables us to identify various G/N with known groups merely by finding a surjective homomorphism from G onto B having kernel N. This avoids the problem of defining a map with the quotient as domain. We did this in the theorem, so we need not do it again in a special case.

Let us apply Theorem 6.6 to a number of examples. First we argue that $(\boldsymbol{R}^*, \cdot)/\langle -1\rangle \cong (\boldsymbol{R}^+, \cdot)$. By the theorem, we need a homomorphism $f : \boldsymbol{R}^* \to \boldsymbol{R}^+$ with $f(\pm 1) = 1$. Two choices are $f(r) = |r|$ or $g(r) = r^2$ and each is a surjective homomorphism with kernel $\langle -1\rangle$ (verify!). Now take $G = (\boldsymbol{C}^*, \cdot)$ and $m \in \boldsymbol{N}$. From Theorem 2.4, $B = \{c^m \in \boldsymbol{C}^* \mid c \in \boldsymbol{C}^*\} \le (\boldsymbol{C}^*, \cdot)$ (why?) and from Theorem 6.5, the mth-power map is a surjective homomorphism from $\boldsymbol{C}^*$ onto B with kernel $K = \{c \in \boldsymbol{C} \mid c^m = 1\}$. The polar form for elements of $\boldsymbol{C}$ shows that $K = \langle e^{2\pi i/m}\rangle$ where $e^{2\pi i/m} = \cos(2\pi/m) + i \cdot \sin(2\pi/m)$. We conclude that $(\boldsymbol{C}^*, \cdot)/\langle e^{2\pi i/m}\rangle \cong B$ by Theorem 6.6. However, every complex number has an mth root so is an mth power. Namely if $c = d(\cos(\theta) + i \cdot \sin(\theta))$ then $c = v^m$ for $v = d^{1/m}(\cos(\theta/m) + i \cdot \sin(\theta/m))$. It follows that $B = (\boldsymbol{C}^*, \cdot)$ so in fact $(\boldsymbol{C}^*, \cdot)/\langle e^{2\pi i/m}\rangle \cong (\boldsymbol{C}^*, \cdot)$! Thus $(\boldsymbol{C}^*, \cdot)/\langle -1\rangle \cong (\boldsymbol{C}^*, \cdot)$ and also $(\boldsymbol{C}^*, \cdot)/\langle i\rangle \cong (\boldsymbol{C}^*, \cdot)$. Strange things can happen when dealing with infinite groups.

Example 6.11 $GL(n, \boldsymbol{R})/SL(n, \boldsymbol{R}) \cong (\boldsymbol{R}^*, \cdot)$.

Recall that $SL(n, \boldsymbol{R}) = \{A \in GL(n, \boldsymbol{R}) \mid \det(A) = 1\}$. A natural choice for a homomorphism $\varphi : GL(n, \boldsymbol{R}) \to \boldsymbol{R}^*$ with $ker\ \varphi = SL(n, \boldsymbol{R})$ is $\varphi = \det$, which is a surjective homomorphism from Example 6.9. Applying Theorem 6.6, $GL(n, \mathbf{R})/SL(n, \mathbf{R}) \cong (\mathbf{R}^*, \cdot)$, giving a precise version of Example 5.12.

Example 6.12 If $G_1, \ldots, G_k$ are groups and $N_j = G_1 \oplus \cdots \oplus \langle e_j \rangle \oplus \cdots \oplus G_k$ for each $1 \le j \le k$ where e_j is the identity of G_j, then $(G_1 \oplus \cdots \oplus G_k)/N_j \cong G_j$.

By Example 6.10 the projections $\pi_j : G_1 \oplus \cdots \oplus G_k \to G_j$ are surjective homomorphisms. That $ker\,\pi_j = N_j = G_1 \oplus \cdots \oplus \langle e_j \rangle \oplus \cdots \oplus G_k$ is an exercise. Therefore the theorem shows that $G_j \cong (G_1 \oplus \cdots \oplus G_k)/N_j$. In particular, if A and B are groups then $A \cong (A \oplus B)/(\langle e_A \rangle \oplus B)$.

Example 6.13 For any $a \in \mathbf{Q}$, $(\mathbf{Q}[x], +)/(x-a)\mathbf{Q}[x] \cong (\mathbf{Q}, +)$.

Here $(x-a)\mathbf{Q}[x] = \{(x-a)p(x) \mid p(x) \in \mathbf{Q}[x]\}$. If $E_a : \mathbf{Q}[x] \to \mathbf{Q}$ is given by $E_a(p(x)) = p(a)$, then using the definition of the sum of polynomials shows E_a is a homomorphism of additive groups (exercise!). For any $c \in \mathbf{Q}$, $E_a(x^2 - ax + c) = c$ (or $c = E_a(c)$ for the constant polynomial $c \in \mathbf{Q}[x]$), so E_a is surjective. Because $0 = e_{\mathbf{Q}}$, $ker\ E_a = \{p(x) \in \mathbf{Q}[x] \mid p(a) = 0\}$. But $f(x) \in \mathbf{Q}[x]$ has a as a root $\Leftrightarrow f(x) \in (x-a)\mathbf{Q}[x]$. Hence $ker\ E_a = (x-a)\mathbf{Q}[x]$ and Theorem 6.6 shows the example holds as claimed.

Example 6.14 Let $n, m \in \mathbf{N}$ with $m \mid n$. Then $\mathbf{Z}_n/\langle [m]_n \rangle \cong \mathbf{Z}_m$.

By Theorem 1.30, $\varphi([a]_n) = [a]_m$ is a function from $\mathbf{Z}_n$ to $\mathbf{Z}_m$. Since $\varphi([a]_n + [b]_n) = \varphi([a+b]_n) = [a+b]_m = [a]_m + [b]_m = \varphi([a]_n) + \varphi([b]_n)$, we see (again!) that φ is a homomorphism, and it is clearly surjective. What is $ker\,\varphi$? By definition, $ker\,\varphi = \{[k]_n \in \mathbf{Z}_n \mid \varphi([k]_n) = [k]_m = [0]_m\}$, or, equivalently, $ker\,\varphi = \{[k]_n \in \mathbf{Z}_n \mid m|k\} = \{[mz]_n \in \mathbf{Z}_n \mid z \in \mathbf{Z}\} = \langle [m]_n \rangle$. Hence by Theorem 6.6, $\mathbf{Z}_m \cong \mathbf{Z}_n/ker\,\varphi = \mathbf{Z}_n/\langle [m]_n \rangle$.

Example 5.4 showing that $\mathbf{Z}_{12}/\langle [4]_{12} \rangle$ is cyclic of order 4 is a special case of the last example. From Example 6.14, $\mathbf{Z}_{12}/\langle [4]_{12} \rangle \cong \mathbf{Z}_4$ so we can now formally identify this quotient with another specific group but avoid an examination of the actual quotient construction.

It is natural to ask if $\varphi([a]_n) = [a]_m$ when $m \mid n$ in Example 6.14 can be used to consider quotients of $(U_n, \cdot)$. Although φ is a surjection onto $\mathbf{Z}_m$ it is *not* obvious that φ is a surjection from U_n onto U_m. For example, if $\varphi : \mathbf{Z}_{30} \to \mathbf{Z}_{10}$ then $[3]_{10} \in U_{10}$ and $\varphi([3]_{30}) = [3]_{10}$ but $[3]_{30} \notin U_{30}$. However, for $k = 1, 3, 7$, or 9, $[10+k]_{30} \in U_{30}$ and $\varphi([10+k]_{30}) = [k]_{10} \in U_{10}$. When m and n are small we can compute directly that $\varphi : U_n \to U_m$ is surjective. However, when $n = (10!)!$ then although $[3]_{10} = [3+10]_{10} = [3+20]_{10}$, all of 3, 13, 23, and many others, divide $(10!)!$ so cannot represent an element of U_n. Thus if $n = (10!)!$ it is not immediately clear that $[3]_{10} \in \varphi(U_n)$.

THEOREM 6.7 Let $n, m \in \mathbf{N} - \{1\}$, $m \mid n$, and $H(n, m) = \{[k]_n \in U_n \mid k \equiv_m 1\}$. Then $H(n, m) \le U_n$, $U_n/H(n, m) \cong U_m$, and U_m is a homomorphic image of U_n.

Proof We need not verify directly that $H(n, m) \leq U_n$ since this will follow from Theorem 6.4. Consider $\varphi([a]_n) = [a]_m$ used in Example 6.14. By Theorem 1.30, φ is a function from $\mathbf{Z}_n$ to $\mathbf{Z}_m$. We restrict the domain of φ to U_n and show first that $\varphi(U_n) \subseteq U_m$. If $[a]_n \in U_n$ then $(a, n) = 1$ from Definition 3.1 so $(a, m) = 1$ follows from $m \mid n$. That is, if $d = (a, m)$ then $d \mid a$ and $d \mid m$. But $m \mid n$ and the transitivity of division show that $d \mid n$, resulting in $d \mid (a, n)$ by Theorem 1.8. Hence $(a, m) = 1$ and we may consider $\varphi : U_n \to U_m$. Next $\varphi([a]_n[b]_n) = \varphi([ab]_n) = [ab]_m = [a]_m[b]_m = \varphi([a]_n)\varphi([b]_n)$, showing that φ is a homomorphism. *If* we can prove φ is surjective then by Theorem 6.6, $U_m \cong U_n/\ ker\ \varphi$. The definition of kernel and that $[1]_m$ is the identity of U_m show $ker\ \varphi = \{[k]_n \in U_n \mid [k]_m = [1]_m\} = \{[k]_n \in U_n \mid k \equiv_m 1\} = H(n, m)$, completing the proof. Thus we must show that φ is surjective.

Since $m \mid n$, by unique factorization in $\mathbf{N}$, $n = m't$ where for any prime p, $p \mid m \Leftrightarrow p \mid m'$, and $(m', t) = 1$. Let $1 \leq k < m$ with $(k, m) = 1$, suppose $t = 1$ or $t = q$ is a prime, and let $\eta : U_n \to U_m$ be given by $\eta([a]_n) = [a]_m$. If $(k, n) = 1$ then $[k]_n \in U_n$ and $\eta([k]_n) = [k]_m \in U_m$. When some prime v divides (k, n) then $v \mid k$, and $v \mid n$ forces $v \mid m'$ or $v = q$ by Theorem 1.14. But $v \mid m'$ implies $v \mid m$ so $v \mid (k, m)$, a contradiction. Hence $(k, n) = 1$ or $q \mid (k, n)$ and $q \mid k$. Should $(k + m, n) = 1$ then $\eta([k + m]_n) = [k + m]_m = [k]_m$ and otherwise $w \mid (k + m, n)$ for a prime w. As above, $w \mid n$ forces $w \mid m'$ or $w = q$. But $w \mid m'$ means $w \mid m$, and with $w \mid (k + m)$ yields $w \mid k$ (Theorem 1.6), so again $w \mid (k, m)$, a contradiction. Hence if $[k]_m \notin \{\eta([k]_n), \eta([k + m]_n)\}$ then $q \mid k$, and $q \mid (k + m)$. Again Theorem 1.6 forces $q \mid m$ so $q \mid (k, m)$, another contradiction. In conclusion, $\eta : U_n \to U_m$ is surjective when n and m have the same prime divisors, or if n has one prime divisor more, appearing once in n. Write $n = m'q_1^{a(1)} \cdots q_s^{a(s)}$ where for any prime p, $p \mid m \Leftrightarrow p \mid m'$, $q_1 < \cdots < q_s$ are primes, $a(j) \geq 1$, and $(m', q_1 \cdots q_s) = 1$. Since n and $n' = mq_1 \cdots q_s$ have the same prime divisors, our argument above shows $\mu : U_n \to U_{n'}$ given by $\mu([a]_n) = [a]_{n'}$ is surjective. If $n(0) = m$ and $n(j) = mq_1 \cdots q_j$ then $n(j + 1)$ has one more prime divisor than $n(j)$, appearing once in $n(j + 1)$, so $\mu_j : U_{n(j+1)} \to U_{n(j)}$ via $\mu_j([a]_{n(j+1)}) = [a]_{n(j)}$ is also surjective by the argument above. A composition of surjective homomorphism is surjective (verify!) so $\varphi : U_n \to U_m$ given by the composition $\varphi = \mu_{n(0)} \circ \cdots \circ \mu_{n(s-1)} \circ \mu$ is a surjective homomorphism and $\varphi([a]_n) = [a]_m$. ■

The surjectivity of φ in the theorem follows quickly from a difficult result of Dirichlet stating that for $a, b \in \mathbf{Z}$ with $(a, b) = 1$ the sequence $a, a + b, a + 2b, a + 3b, \ldots,$ contains *infinitely* many primes. In our case when $(k, m) = 1$ there is a prime $p = k + sm$ not dividing n, so both $[p]_n \in U_n$ and $\varphi([p]_n) = [k + sm]_m = [k]_m$. Looking back at Example 5.7, $U_{16}/\langle[9]_{16}\rangle$ is a noncyclic group of order 4 and $U_{16}/\langle[15]_{16}\rangle$ is a cyclic group of order 4. In the notation of the last theorem, $\langle[9]_{16}\rangle = H(16, 8)$, so $U_{16}/\langle[9]_{16}\rangle \cong U_8$. The quotient $U_{16}/\langle[15]_{16}\rangle$ is cyclic so not isomorphic to U_8, but is isomorphic to $\mathbf{Z}_4$ by Example 6.1. We also obtain this as a consequence of Theorem 6.6.

THEOREM 6.8 If $G = \langle g \rangle$ is infinite, then $G \cong (\mathbf{Z}, +)$, and if $|G| = n$, $G \cong \mathbf{Z}_n$.

Proof Let $\beta : (\mathbf{Z}, +) \to G$ be defined via $\beta(k) = g^k$. By the definition of cyclic group, β is surjective, and using Theorem 2.4, $\beta(k + t) = g^{k+t} = g^k g^t = \beta(k)\beta(t)$. Thus β is a homomorphism, and applying Theorem 6.6 yields $G \cong \mathbf{Z}/ker\,\beta$ where $ker\,\beta = \{k \in \mathbf{Z} \mid g^k = e_G\}$. When G is infinite, $o(g)$ is infinite by Theorem 2.16 so $g^k = e_G$ forces $k = 0$

and $ker\ \beta = \langle 0\rangle$. Hence $\mathbf{Z}/\langle 0\rangle \cong (\mathbf{Z}, +)$ by Example 6.7 so the transitivity of $\cong$ (corollary to Theorem 6.3) implies that $G \cong (\mathbf{Z}, +)$. If $|G| = n$, then $o(g) = n$ from Theorem 2.16, so $ker\ \beta = n\mathbf{Z}$ by Theorem 2.7, and $G \cong \mathbf{Z}/n\mathbf{Z} = \mathbf{Z}_n$ as in Example 5.3. ∎

The theorem formalizes the idea that $\mathbf{Z}_n$ is essentially the only cyclic group of order n. We showed in Example 5.5 that if $G = \langle g\rangle$ has order n and $m \mid n$ then $G/\langle g^m\rangle$ is cyclic of order m, so $G/\langle g^m\rangle \cong \mathbf{Z}_m$ by Theorem 6.8. Other examples are that $S_n/A_n \cong \mathbf{Z}_2$ since this quotient has two elements, and $H = \{I, (1,2)(3,4), (1,3)(2,4), (1,4)(2,3)\} \lhd A_4$ has index 3, so $A_4/H \cong \mathbf{Z}_3$. Similarly, $(\mathbf{Z}, +)$ is essentially the only infinite cyclic group. For $n\mathbf{Z} \le \mathbf{Z}$, we can see that $\mathbf{Z} \cong (n\mathbf{Z}, +)$ using $\varphi(k) = nk$. Other infinite cyclic groups like $(\mathbf{Z} \oplus \mathbf{Z})/\langle(2,3)\rangle$ in Example 5.9 may be more subtle to recognize. A less computational approach to a generalization of it comes next.

Example 6.15 If $a, b \in \mathbf{Z}^*$ with $(a, b) = 1$ then $(\mathbf{Z} \oplus \mathbf{Z})/\langle(a, b)\rangle \cong (\mathbf{Z}, +)$.

We want a homomorphism $\varphi : \mathbf{Z} \oplus \mathbf{Z} \to \mathbf{Z}$ with $ker\ \varphi = \langle(a, b)\rangle$. It is natural to let $\varphi((x, y)) = bx - ay$. Now $\varphi((x, y) + (x', y')) = \varphi((x + x', y + y')) = b(x + x') - a(y + y') = bx - ay + bx' - ay' = \varphi((x, y)) + \varphi((x', y'))$ so φ is a homomorphism. Since $(a, b) = 1$, by Theorem 1.8 $as + bt = 1$ for $s, t \in \mathbf{Z}$. It follows that $\varphi((tk, -sk)) = btk + ask = k$ for any $k \in \mathbf{Z}$, so φ is surjective and $(\mathbf{Z}, +) \cong (\mathbf{Z} \oplus \mathbf{Z})/ker\,\varphi$ from Theorem 6.6. Now $ker\ \varphi = \{(x, y) \in \mathbf{Z} \oplus \mathbf{Z} \mid bx - ay = 0\} = \{(x, y) \in \mathbf{Z} \oplus \mathbf{Z} \mid bx = ay\}$. For $(x, y) \in ker\ \varphi$, use $(a, b) = 1$, $a \mid bx$ and Theorem 1.12, to conclude that $a \mid x$. Write $x = ac$, so $bac = ay$, resulting in $bc = y$. Thus $(x, y) = (ac, bc)$ and $ker\ \varphi \subseteq \{(ak, bk) \in \mathbf{Z} \oplus \mathbf{Z} \mid k \in \mathbf{Z}\} = \{(a, b)^k \in \mathbf{Z} \oplus \mathbf{Z} \mid k \in \mathbf{Z}\} = \langle(a, b)\rangle$. Finally, observe $\langle(a, b)\rangle \subseteq ker\ \varphi$, so these are equal, completing the verification.

Next is another consequence of Theorem 6.6 that is useful to record; it generalizes Example 5.8 to more than two copies of $\mathbf{Z}$.

THEOREM 6.9 Let $G_1, \ldots, G_k$ be groups with $N_j \lhd G_j$ for each $1 \le j \le k$. Then $(G_1 \oplus \cdots \oplus G_k)/(N_1 \oplus \cdots \oplus N_k) \cong (G_1/N_1) \oplus \cdots \oplus (G_k/N_k)$.

Proof With Theorem 6.6 in mind, it is natural to let φ from $G_1 \oplus \cdots \oplus G_k$ to $(G_1/N_1) \oplus \cdots \oplus (G_k/N_k)$ be $\varphi((g_1, \ldots, g_k)) = (N_1g_1, \ldots, N_kg_k)$. For any $X \in (G_1/N_1) \oplus \cdots \oplus (G_k/N_k)$ we have $X = (N_1h_1, \ldots, N_kh_k)$ with $h_j \in G_j$ and it follows that $\varphi((h_1, \ldots, h_k)) = X$ so φ is surjective. Using the definition of multiplication in the direct sum, $\varphi((x_1, \ldots, x_k)(y_1, \ldots, y_k)) = \varphi((x_1y_1, \ldots, x_ky_k)) = (N_1x_1y_1, \ldots, N_kx_ky_k)$ and $(N_1x_1y_1, \ldots, N_kx_ky_k) = (N_1x_1, \ldots, N_kx_k)(N_1y_1, \ldots, N_ky_k) = \varphi((x_1, \ldots, x_k))\varphi((y_1, \ldots, y_k))$, so φ is a homomorphism. Thus by Theorem 6.6, $(G_1 \oplus \cdots \oplus G_k)/ker\,\varphi \cong (G_1/N_1) \oplus \cdots \oplus (G_k/N_k)$, and the definition of φ yields $ker\,\varphi = \{(g_1, \ldots, g_k) \in G_1 \oplus \cdots \oplus G_k \mid (N_1g_1, \ldots, N_kg_k) = (N_1, \ldots, N_k)\}$. This is equivalent to $g_j \in N_j$ for each j and implies $ker\ \varphi = N_1 \oplus \cdots \oplus N_k$ as claimed. ∎

COROLLARY Let $\{n_1, \ldots, n_k\} \subseteq \mathbf{N} - \{1\}$ be pairwise relatively prime. Then $\mathbf{Z}^k/(n_1\mathbf{Z} \oplus \cdots \oplus n_k\mathbf{Z}) \cong \mathbf{Z}_{n_1} \oplus \cdots \oplus \mathbf{Z}_{n_k} \cong \mathbf{Z}_n$ where $n = n_1 \cdots n_k$.

Proof Since for each $1 \leq i \leq k$, $n_i\mathbf{Z} \lhd (\mathbf{Z}, +)$ (why?), and since $\mathbf{Z}/n_i\mathbf{Z} = \mathbf{Z}_{n_i}$ the theorem shows that $\mathbf{Z}^k/(n_1\mathbf{Z} \oplus \cdots \oplus n_i\mathbf{Z}) \cong \mathbf{Z}_{n_1} \oplus \cdots \oplus \mathbf{Z}_{n_k}$. The assumption that $\{n_j\}$ is pairwise relatively prime means that Theorem 3.21 applies to prove that $\mathbf{Z}_{n_1} \oplus \cdots \oplus \mathbf{Z}_{n_k}$ is cyclic and of course has order $n = n_1 \cdots n_k$. By Theorem 6.8, $\mathbf{Z}_{n_1} \oplus \cdots \oplus \mathbf{Z}_{n_k} \cong \mathbf{Z}_n$. ∎

A special case of the corollary if $n > 1$ with prime factorization $n = p_1^{a_1} \cdots p_k^{a_k}$ is that $\mathbf{Z}_n \cong \mathbf{Z}_{p_1^{a_1}} \oplus \cdots \oplus \mathbf{Z}_{p_k^{a_k}}$. We use this to re-prove the multiplicative property of the Euler φ-function in Theorem 3.4, but without reference to the Chinese Remainder Theorem (Exercise 22 in §1.7). Since $\mathbf{Z}_n$ is cyclic of order n it has exactly $\varphi(n)$ generators by Theorem 3.5. Using Theorem 3.20, each $(g_1, \ldots, g_k) \in \mathbf{Z}_{p_1^{a_1}} \oplus \cdots \oplus \mathbf{Z}_{p_k^{a_k}}$ has order LCM$\{o(g_i)\}$ and of course $o(g_i) \mid p_i^{a_i}$ by Theorem 4.7. Thus $o((g_1, \ldots, g_k)) = n \Leftrightarrow o(g_i) = p_i^{a_i}$ for each $i \Leftrightarrow \mathbf{Z}_{p_i^{a_i}} = \langle g_i \rangle$ so by Theorem 3.5 there are exactly $\varphi(p_i^{a_i})$ choices for g_i. Thus $\varphi(n) = \varphi(p_1^{a_1}) \cdots \varphi(p_k^{a_k})$ by Theorem 0.3.

The last observation in this section is that Theorem 6.6 implies the Chinese Remainder Theorem for $\mathbf{Z}$. Our argument uses Theorem 6.6 to give another proof of the last corollary.

THEOREM 6.10 For $k > 1$, $\{n_j\} \subseteq N - \{1\}$ pairwise relatively prime, and $m_1, \ldots, m_k \in \mathbf{Z}$, the system $x \equiv m_1 (\bmod\, n_1), \ldots, x \equiv m_k (\bmod\, n_k)$ has a solution in $\mathbf{Z}$ and any two solutions are congruent modulo $n = n_1 \cdots n_k$.

Proof Define $\varphi : \mathbf{Z}_n \to \mathbf{Z}_{n_1} \oplus \cdots \oplus \mathbf{Z}_{n_k}$ by $\varphi([m]_n) = ([m]_{n_1}, \ldots, [m]_{n_k})$. It follows from Theorem 1.30 that φ is a function and it is straightforward to see that φ is group homomorphism (verify!). Now $ker\,\varphi = \{[m]_n \in \mathbf{Z}_n \mid n_j | m \text{ for all } j\}$. Use $\{n_j \mid 1 < j \leq k\}$ is pairwise relatively prime and Theorem 1.12 to get $ker\,\varphi = \{[m]_n \in \mathbf{Z}_n \mid n|m\} = \langle [0]_n \rangle$, so Theorem 6.4 shows φ is injective. Thus from Theorem 6.6, $Im\,\varphi \cong \mathbf{Z}_n/\langle [0]_n \rangle \cong \mathbf{Z}_n$ has n elements. But $|\mathbf{Z}_{n_1} \oplus \cdots \oplus \mathbf{Z}_{n_k}| = n$, forcing φ to be surjective, so an isomorphism. This has two consequences: $\mathbf{Z}_{n_1} \oplus \cdots \oplus \mathbf{Z}_{n_k} \cong \mathbf{Z}_n$; and for any $([m_1]_{n_1}, \ldots, [m_k]_{n_k}) \in \mathbf{Z}_{n_1} \oplus \cdots \oplus \mathbf{Z}_{n_k}$ there is a unique $[m]_n \in \mathbf{Z}_n$ with $\varphi([m]_n) = ([m_1]_{n_1}, \ldots, [m_k]_{n_k})$. Thus $m \in \mathbf{Z}$ is a solution of the congruences, and any other solution $d \in \mathbf{Z}$ satisfies $\varphi([d]_n) = \varphi([m]_n)$ (why?), so $[d]_n = [m]_n$ and $d \equiv_n m$ as claimed. ∎

Although the argument is a nice application of Theorem 6.6 to show the existence of a solution of a set of linear congruences for pairwise relatively prime moduli, it is not very helpful for actually finding a solution. To solve the congruences, it is best to refer to Exercise 22 in §1.7.

EXERCISES

1. Find a group G and a homomorphism φ of G so that:
   i) $\varphi(G) \cong \mathbf{Z}_n$, $ker\,\varphi \cong \mathbf{Z}_m$, and G cyclic
   ii) $\varphi(G) \cong \mathbf{Z}_2$, $ker\,\varphi \cong \mathbf{Z}_8$, and G not Abelian
   iii) $\varphi(G) \cong S_3$ and $ker\,\varphi \cong A_4$
   iv) $\varphi(G) \cong \mathbf{Z}_2 \oplus \mathbf{Z}_3$, $ker\,\varphi \cong \mathbf{Z}_7$, and G not Abelian.
2. Let G be an Abelian group and $k \in N$. If $|\{g^k \mid g \in G\}| = n$ and $|\{g \in G \mid g^k = e\}| = m$ with $n, m \in N$, show that $|G| = nm$.

3. Let $\varphi: G \to B$ be a group homomorphism and assume that G is finite.
   a) Show that $|G|/2 < |\varphi(G)|$ implies that φ is an injection.
   b) If $(|G|, |\varphi(G)|) = 1$, show that $ker\, \varphi = G$.

4. If $U = \{c \in \boldsymbol{C} \mid |c| = 1\} \leq (\boldsymbol{C}^*, \cdot)$ show:
   i) $(\boldsymbol{C}^*, \cdot)/U \cong (\boldsymbol{R}^+, \cdot)$
   ii) $(\boldsymbol{R}, +)/(\boldsymbol{Z}, +) \cong U$

5. Prove:
   i) $D_n/\langle T_1\rangle \cong \boldsymbol{Z}_2$
   ii) $D_{2n}/\langle T_2\rangle \cong \boldsymbol{Z}_2 \oplus \boldsymbol{Z}_2$
   iii) if $N \triangleleft D_n$ and D_n/N is cyclic, show that $N = \langle T_1\rangle$ or $N = D_n$
   iv) if $m > 2$ and $m \mid n$ then $D_n/\langle T_m\rangle \cong D_m$

6. If $n, m \in \boldsymbol{N}$ with $m \mid n$, show that $m\boldsymbol{Z}/n\boldsymbol{Z} \cong \boldsymbol{Z}_{n/m}$ as additive groups.

7. For $n \in \boldsymbol{N}$, let $n\boldsymbol{Z}[x] = \{a_m x^m + a_{m-1}x^{m-1} + \cdots + a_1 x + a_0 \in \boldsymbol{Z}[x] \mid \text{all } a_j \in n\boldsymbol{Z}\}$. Show that $(\boldsymbol{Z}[x], +)/n\boldsymbol{Z}[x] \cong (\boldsymbol{Z}_n[x], +)$.

8. In each case define a surjective homomorphism $\varphi: G \to B$ and find $ker\, \varphi$. Recall Definition 2.9 for $UT(n, \boldsymbol{R})$.
   a) For p a prime, let $G = UT(3, \boldsymbol{Z}_p)$ and $B = U_p^3 = U_p \oplus U_p \oplus U_p$.
   b) Let $G = UT(n, \boldsymbol{Z})$ and $B = \boldsymbol{Z}_2^n$.
   c) Let $G = UT(n, \boldsymbol{Z})$ and $B = \boldsymbol{Z}_2$.
   d) Let $G = GL(n, \boldsymbol{R})$ and $B = (\boldsymbol{R}^+, \cdot)$.
   e) Let $G = GL(n, \boldsymbol{Z}_7)$ and $B = \boldsymbol{Z}_3$.

9. **a)** If $K_2 = \{p/q \in \boldsymbol{Q}^+ \mid pq \text{ is odd}\}$, show $K_2 \leq (\boldsymbol{Q}^+, \cdot)$ and that $\boldsymbol{Q}^+/K_2 \cong (\boldsymbol{Z}, +)$.
   b) If $K_6 = \{p/q \in \boldsymbol{Q}^+ \mid (pq, 6) = 1\}$, show $K_6 \leq (\boldsymbol{Q}^+, \cdot)$ and $\boldsymbol{Q}^+/K_6 \cong \boldsymbol{Z}^2 = \boldsymbol{Z} \oplus \boldsymbol{Z}$.
   c) If $m \in \boldsymbol{N}$ with $m > 1$, let $K_m = \{p/q \in \boldsymbol{Q}^+ \mid (pq, m) = 1\}$. Show $K_m \leq (\boldsymbol{Q}^+, \cdot)$ and describe $\boldsymbol{Q}^+/K_m$ up to isomorphism using the prime factorization of m.

10. Let G be a finite group and $\varphi: G \to G$ a homomorphism. Show there is $m \in \boldsymbol{N}$ satisfying:
   i) $ker\, \varphi^m = ker\, \varphi^{m+j}$, all $j \geq 1$
   ii) $Im\, \varphi^m = Im\, \varphi^{m+j}$, all $j \geq 1$
   iii) $ker\, \varphi^m \bigcap Im\, \varphi^m = \langle e\rangle$
   iv) $ker\, \varphi^m \cdot Im\, \varphi^m = G$ (note that $\varphi^m(G) = \varphi^{2m}(G)$)

11. If $N_1, N_2, \ldots, N_k \triangleleft G$, let $\rho: G \to \bigoplus G/N_i$ be defined by $\rho(g) = (N_1 g, \ldots, N_k g)$.
   i) Show that ρ is a homomorphism.
   ii) Show that $G/(\bigcap N_i) \cong H$, some $H \leq \bigoplus G/N_i$.
   iii) If each G/N_i is Abelian, show that $G/(\bigcap N_i)$ is Abelian.
   iv) If $H, K \triangleleft G$, G/H and G/K are Abelian, and H and K are finite with $(|H|, |K|) = 1$, show that G is Abelian.

12. Let G be a group, A an Abelian group, $\varphi: G \to A$ a group homomorphism, and $\rho: G \to G/G'$ the homomorphism $\rho(g) = G'g$ for G' the commutator subgroup of G. Show there is a unique homomorphism $\eta: G/G' \to A$ so that $\eta \circ \rho = \varphi$.

13. Find an injective group homomorphism $\alpha: S_n \to A_{n+2}$ when $n \geq 2$.

14. Describe all possible group homomorphisms $\varphi: S_n \to S_k$ when $k < n$.

15. This exercise characterizes direct sums using homomorphisms. Let $G_1, G_2, \ldots, G_k$ be groups, let e_i be the identity of G_i, and set $G = \bigoplus G_i$. Define $\iota_j: G_j \to G$ by $\iota_j(g_j) = (e_1, \ldots, g_j, \ldots, e_k)$, so e_i is the ith entry of $\iota_j(g_j)$ if $i \neq j$.

a) Show the following: i) ι_j is an injective group homomorphism; ii) for $i \neq j$, $x \in \iota_i(G_i)$, and $y \in \iota_j(G_j)$, $xy = yx$; and iii) G is generated by $\{\iota_j(G_j)\}$.
b) If B is a group and for each j, $\varphi_j : G_j \to B$ is a homomorphism so that when $i \neq j$, $x \in \varphi_i(G_i)$ and $y \in \varphi_j(G_j) \Rightarrow xy = yx$, then prove there is a *unique* homomorphism $\varphi : G \to B$ so that $\varphi \circ \iota_j = \varphi_j$ for each $1 \leq j \leq k$.
c) Suppose a group H satisfies a) and b). That is, there are injective homomorphisms $\eta_j : G_j \to H$; for $i \neq j$, $x \in \eta_i(G_i)$ and $y \in \eta_j(G_j) \Rightarrow xy = yx$; H is generated by $\{\eta_j(G_j)\}$; and for any group C and homomorphisms $\rho_j : G_j \to C$ so that $u \in \rho_i(G_i)$ and $v \in \rho_j(G_j) \Rightarrow uv = vu$ when $i \neq j$, there is a unique homomorphism $\rho : H \to C$ so that $\rho \circ \eta_j = \rho_j$ for each j. Prove $H \cong G$.

16. Given groups $G_1, \ldots, G_k$ with e_i the identity element of G_i, let T be a proper nonempty subset of $\{1, 2, \ldots, k\}$ and $G_T = \{(g_1, \ldots, g_k) \in \bigoplus G_i \mid g_j = e_j \text{ if } j \in T\}$. Show that $(\bigoplus G_i)/G_T \cong \bigoplus G_j$ over all $j \in T$.

17. For any group G, let $G^* = \{\lambda : G \to (\mathbf{C}^*, \cdot) \mid \lambda \text{ is a group homomorphism}\}$.
a) If $\lambda, \mu \in G^*$ and $\lambda \cdot \mu(g) = \lambda(g)\mu(g)$ for all $g \in G$, show $(G^*, \cdot)$ is a group.
b) Show that $\mathbf{Z}_n^* \cong \mathbf{Z}_n$.
c) If $G = \mathbf{Z}_{n_1} \oplus \cdots \oplus \mathbf{Z}_{n_k}$ show that $G^* \cong G$.

18. In each case find a surjective group homomorphism from G onto B, or onto a group isomorphic to B, and identify the the kernel:
i) $G = \mathbf{Z}^2$, $B = \mathbf{Z} \oplus \mathbf{Z}_2$
ii) $G = \mathbf{Z}^k$, $B = \mathbf{Z}_3^k$
iii) $G = \mathbf{Z}^3$, $B = \mathbf{Z}_4 \oplus \mathbf{Z}_2$
iv) $G = \mathbf{Z}^k (k \geq 3)$, $B = \mathbf{Z}_4 \oplus \mathbf{Z}_6 \oplus \mathbf{Z}_{10}$
v) $G = \mathbf{Z}$, $B = \mathbf{Z}_3 \oplus \mathbf{Z}_{14}$
vi) $G = \mathbf{Z}^2$, $B = \mathbf{Z}_2 \oplus \mathbf{Z}_2 \oplus \mathbf{Z}_{15}$
vii) $G = \mathbf{Z} \oplus D_4$, $B = \mathbf{Z}_2 \oplus \mathbf{Z}_2$
viii) $G = A_4 \oplus D_4$, $B = \mathbf{Z}_6$
ix) $G = D_5 \oplus D_6$, $B = \mathbf{Z}_2 \oplus D_3$

19. For $m, n \in \mathbf{N}$ with $m \mid n$, set $H(n, m) = \{[k]_n \in U_n \mid k \equiv_m 1\} \leq U_n$.
a) Show that $|H(n, m)| = \varphi(n)/\varphi(m)$, where φ is the Euler phi-function.
b) Explicitly find: i) $H(48, 8)$; ii) $H(72, 9)$; iii) $H(72, 12)$.
c) Show that $U_8 \cong H(48, 8) \cong U_{48}/H(48/8)$.
d) Show that $\mathbf{Z}_6 \cong H(72, 12)$ and $U_8 \cong U_{72}/H(72/12)$.

20. Determine if the pairs of groups are isomorphic (use Exercise 5 in §6.1):
i) $\mathbf{Z}_{15} \oplus \mathbf{Z}_7$ and $\mathbf{Z}_5 \oplus \mathbf{Z}_{21}$
ii) $\mathbf{Z}_6 \oplus \mathbf{Z}_{10}$ and $\mathbf{Z}_{12} \oplus \mathbf{Z}_5$
iii) $\mathbf{Z}_6 \oplus \mathbf{Z}_{10} \oplus \mathbf{Z}_7$ and $\mathbf{Z}_{14} \oplus \mathbf{Z}_{30}$
iv) $\mathbf{Z}_{10} \oplus \mathbf{Z}_{10}$ and $\mathbf{Z}_2 \oplus \mathbf{Z}_2 \oplus \mathbf{Z}_{25}$

21. For any $a, b, c, \in \mathbf{Z}$ and $k \in \mathbf{N}$, show that there is $m \in \mathbf{N}$ satisfying $[m]_{2k+1} = [a]_{2k+1}$ in $\mathbf{Z}_{2k+1}$, $[m]_{2k+2} = [b]_{2k+2}$ in $\mathbf{Z}_{2k+2}$, and $[m]_{2k+3} = [c]_{2k+3}$ in $\mathbf{Z}_{2k+3}$.

22. Is there $m \in \mathbf{N}$ so that $m \equiv 104 \pmod{197}$, $m \equiv 197 \pmod{401}$, $m \equiv 104198 \pmod{197 \cdot 401 + 1}$, and $m \equiv 104199 \pmod{197 \cdot 401 + 2}$?

23. For all $n, m \in \mathbf{N}$ and any n distinct primes $q_1, \ldots, q_m$, use Theorem 6.10 to show that there are n consecutive integers $k + 1, \ldots, k + n$ so that $q_j^m \mid (k + j)$, e.g., $5^3 \mid 1375$, $2^3 \mid 1376$, and $3^3 \mid 1377$.

6.4 APPLICATIONS

To learn things about a group G from its quotients or its homomorphic images, G must have nonidentity normal subgroups. At times even isomorphic copies of G may be useful. For example, if $\varphi : G \to GL(n, \boldsymbol{R})$ is an injective homomorphism we may be able to use facts about matrices to help study G. A related idea is a result of Cayley that allows us to consider G as a subgroup of $\mathrm{Bij}(G)$ and opens up the possibility of using facts about symmetric groups to prove things about G.

THEOREM 6.11 If G is a group and $g \in G$, define $\tau_g : G \to G$ by $\tau_g(x) = gx$. Then $\tau_g \in \mathrm{Bij}(G)$, $B = \{\tau_g \in \mathrm{Bij}(G) \mid g \in G\} \le \mathrm{Bij}(G)$, and $\varphi : G \to B$ defined by $\varphi(g) = \tau_g$ is an isomorphism.

Proof Clearly τ_g is a function and if $g, h, c \in G$, $\tau_g(\tau_h(c)) = \tau_g(hc) = g(hc) = gh(c) = \tau_{gh}(c)$. Hence $\tau_a \circ \tau_b = \tau_{ab}$ and since $\tau_e = I_G$ it follows that $\tau_g^{-1} = \tau_{g^{-1}}$ (how?), which yields $\tau_g \in \mathrm{Bij}(G)$ by Theorem 0.11. These computations show $B \le \mathrm{Bij}(G)$ using Theorem 2.12. If $\varphi(g) = \tau_g$ then $\varphi : G \to B$ and $\varphi(hg) = \tau_{hg} = \tau_h \circ \tau_g = \varphi(h)\varphi(g)$ so φ is a homomorphism, surjective from the definition of B. Now $g \in ker\,\varphi \Leftrightarrow \varphi(g) = I_G \Leftrightarrow gx = \tau_g(x) = x$ for all $x \in G$, forcing $g = e$. Thus $ker\,\varphi = \langle e \rangle$ and φ is an isomorphism by Theorem 6.4. ■

This view of G as a group of bijections on the set G and the identification of $g \in G$ with the left multiplication by g is called the **left regular representation** of G. If $|G| = n$ and $G = \{g_1, \ldots, g_n\}$, by identifying g_j with $j \in I(n) = \{1, 2, \ldots, n\}$ we can see how the left multiplications correspond to elements in S_n. For a simple example, take $G = \langle g \rangle$ of order 4 and identify g^i with $i \in I(4)$ for $1 \le i \le 4$. Since $g \cdot g^i = g^{i+1}$, τ_g corresponds to $(1, 2, 3, 4) \in S_4$ and it follows that G is isomorphic to $\langle (1, 2, 3, 4) \rangle \le S_4$. By abusing notation a bit, it is sometimes convenient to write τ_g as a cycle reflecting how g multiplies elements of $G : \tau_g = (e, g, g^2, g^3)$, $\tau_{g^2} = (g, g^3)(e, g^2)$, and $\tau_{g^3} = (e, g^3, g^2, g)$. Replacing $g^i \in G$ with $i \in I(4)$ identifies τ_{g^2} with $(1, 3)(2, 4) \in S_4$ and τ_{g^3} with $(1, 4, 3, 2) \in S_4$.

For any finite G, we can write $\tau_g \in \mathrm{Bij}(G)$ as a product of cycles of elements of G. Specifically, if $G = D_4$ then we can use $\tau_{T_1}(T_j) = T_{j+1}$ and $\tau_{T_1}(W_j) = W_{j+1}$ to write $\tau_{T_1} = (I, T_1, T_2, T_3)(W_0, W_1, W_2, W_3)$ and similarly, $\tau_{T_2} = (I, T_2)(T_1, T_3)(W_0, W_2) \cdot (W_1, W_3)$; $\tau_{W_2} = (I, W_2)(T_1, W_1)(T_2, W_0)(T_3, W_3)$ since $W_2T_j = W_{2-j}$. Identifying T_i with $i + 1 \in I(8)$ for $0 \le i \le 3$ and W_j with $j + 5 \in I(8)$ for $0 \le j \le 3$, then τ_{T_1} is identified with $(1, 2, 3, 4)(5, 6, 7, 8) \in S_8$, τ_{T_2} with $(1, 3)(2, 4)(5, 7)(6, 8)$, and τ_{W_2} with $(1, 7)(2, 6)(3, 5)(4, 8)$.

The formal justification for viewing $\tau_g \in \mathrm{Bij}(G)$ as products of cycles follows from our next theorem, which identifies $\mathrm{Bij}(G)$ with $S_{|G|}$. The idea when $|G| = n$ is to identify $G = \{g_1, \ldots, g_n\}$ with $\{1, 2, \ldots, n\}$, then identify $\eta \in S_n$ with $\varphi \in \mathrm{Bij}(G)$ by defining $\varphi(g_i) = g_{\eta(i)} : \varphi$ acts on g_j in the same way that η acts on j. In effect this identification allows us to think of elements in $\mathrm{Bij}(G)$, specifically the τ_g, as in S_n but with the integers in the cycles expressing η replaced by the elements in G corresponding to them. This represents φ as a product of cycles in the elements of G.

THEOREM 6.12 If $|S| = |T| \ne 0$ then $(\mathrm{Bij}(S), \circ) \cong (\mathrm{Bij}(T), \circ)$.

Proof Since $|S|=|T|$ there is a bijection $\varphi : S \to T$ and by Theorem 0.13, its inverse $\varphi^{-1} : T \to S$ is also a bijection. To find an isomorphism between $\mathrm{Bij}(S)$ and $\mathrm{Bij}(T)$, we must first define a function between them. We have only φ and φ^{-1} to use to associate $\eta \in \mathrm{Bij}(S)$ with a function on T so we try $F(\eta)=\varphi \circ \eta \circ \varphi^{-1}$. The definition of composition shows $F(\eta) : T \to T$ and similarly for $\sigma \in \mathrm{Bij}(T)$, $H(\sigma)=\varphi^{-1} \circ \sigma \circ \varphi : S \to S$. Since $F(\eta)$ and $H(\sigma)$ are compositions of bijections, they are themselves bijections by Theorem 0.13. Therefore $F : \mathrm{Bij}(S) \to \mathrm{Bij}(T)$, $H : \mathrm{Bij}(T) \to \mathrm{Bij}(S)$, and we see that for $\eta \in \mathrm{Bij}(S)$, $H \circ F(\eta) = H(\varphi \circ \eta \circ \varphi^{-1}) = \varphi^{-1} \circ (\varphi \circ \eta \circ \varphi^{-1}) \circ \varphi = (\varphi^{-1} \circ \varphi) \circ \eta \circ (\varphi^{-1} \circ \varphi) = I_S \circ \eta \circ I_S = \eta$ so $H \circ F = I_{\mathrm{Bij}(S)}$. Similarly, $F \circ H = I_{\mathrm{Bij}(T)}$ and H and F are bijections by Theorem 0.11. Finally, $F(\eta \circ \theta) = \varphi \circ \eta \circ \theta \circ \varphi^{-1} = \varphi \circ \eta \circ \varphi^{-1} \circ \varphi \circ \theta \circ \varphi^{-1} = F(\eta) \circ F(\theta)$ so F is an isomorphism, completing the proof. ■

Viewing a finite group G as a subgroup of S_n means that if we know everything about subgroups of all S_n then we know everything about any finite group! Unfortunately, we are not is this position so as a practical matter the isomorphism in Theorem 6.11 is not often helpful. Sometimes, however, we can gain useful information. If G is finite and its image $\varphi(G)$ in $S_{|G|}$ contains both odd and even permutations then $\varphi(G) \cap A_{|G|} \lhd \varphi(G)$ and is not trivial. This yields some nontrivial $N \lhd G$ (exercise!). Using this argument, we present a result that may seem somewhat artificial but does show that certain groups cannot be simple.

THEOREM 6.13 Let G be a finite group with $|G|=n=2^k m$ with m odd. If $h \in G$ with $o(h)=2^k$, then $N=\{g \in G \mid o(g) \text{ is odd}\} \lhd G$ and $|N|=m$.

Proof Assume that $k \geq 1$ since otherwise G has no element of even order by Theorem 4.7, so $N=G$. By Theorem 6.11, $\varphi(g)=\tau_g$, left multiplication by g, is an injective homomorphism of G into $\mathrm{Bij}(G)$. If $\eta : G \to I(n)$ is a bijection then Theorem 6.12 shows that η induces an isomorphism υ between $\mathrm{Bij}(G)$ and S_n, which for τ_g is $\upsilon(\tau_g)(j) = \eta(\tau_g(\eta^{-1}(j))) = \eta(g \cdot \eta^{-1}(j))$. It follows that $\upsilon(\tau_g)^s(j) = \eta(\tau_g^s(\eta^{-1}(j)))$ (verify!), so the orbit of j under $\upsilon(\tau_g)$ has length $s \Leftrightarrow s$ is minimal with $\upsilon(\tau_g)^s(j)=j \Leftrightarrow s$ is minimal with $\eta^{-1}(j) = g^s \cdot \eta^{-1}(j) \Leftrightarrow s$ is minimal with $g^s = e_G$ (why?). Hence all orbits of $\upsilon(\tau_g)$ have length $o(g)$.

Composing φ with υ gives and injective homomorphism (why?) from G into S_n, so an isomorphism $F : G \to F(G)$ for $F(G) \leq S_n$ that consists of elements, each one of which has equal orbit lengths. Clearly $|F(G)|=n$ and F preserves orders of elements by Theorem 6.2. It is an exercise that $H \lhd G$ precisely when $F(H) \lhd F(G)$, so there is no loss of generality in assuming that $G \leq S_n$, and further, by our observation above, that each element of G has all its orbits of the same length. We are assuming that there is $h \in G$ with $o(h)=2^k$. By Theorem 3.11, all the orbit lengths of h must be 2^k, since they have the same length. Also, h has no fixed points, so h must have m orbits of length 2^k. Thus h is a product of an odd number of even length cycles, each an odd permutation, and it follows that $h \notin A_n$. Every $g \in N$ has odd order and so is a product of cycles of odd length $o(g)$, using Theorem 3.11 again, so $N \subseteq A_n$. For any $x, y \in G - A_n$, $xy^{-1} \in G \cap A_n$, resulting in $[G : G \cap A_n]=2$. Hence $|G \cap A_n| = 2^{k-1}m$ by Lagrange's Theorem, and clearly $h^2 \in G \cap A_n$ with $o(h^2)=2^{k-1}$ by Theorem 2.8. Using induction, we have that $N \lhd G \cap A_n$ and $|N|=m$. Finally, $N \leq G$, and N is normal in G since, in any group, $o(x)=o(g^{-1}xg)$ (Exercise 2 in §2.4). ■

Theorem 6.13 shows how to use Cayley's Theorem and S_n to study a group G. We record a consequence for simple groups.

Example 6.16 For G a group, $|G| = 2m > 2$ and m odd, G is not simple.

Since G has even order, it has an element of order 2 by Exercise 10 in §2.2. By Theorem 6.13 the set of elements in G of odd order is a normal subgroup N of G of order m, necessarily proper and not $\langle e\rangle$ since $m > 1$. Therefore G cannot be simple.

Another important application of the ideas in this chapter often enables us to find normal subgroups in a given non-Abelian group.

THEOREM 6.14 For G a finite group, $H \leq G$ and proper, $X = \{gH \mid g \in G\}$ and $g \in G$, set $\sigma(g)(yH) = gyH$. Then $\sigma(g) \in \text{Bij}(X)$, $\varphi : G \to \text{Bij}(X)$ given by $\varphi(g) = \sigma(g)$ is a homomorphism, $ker\,\varphi \leq H$, and if $N \leq H$ with $N \lhd G$ then $N \leq ker\,\varphi$. Furthermore: i) if $|G| \nmid [G : H]!$ then $ker\,\varphi \neq \langle e\rangle$; and ii) if $|G| > 2$ and $|G| \nmid ([G : H]!/2)$ then there is $\langle e\rangle \neq N \lhd G$ with $N \neq G$.

Proof First we show $\sigma(g)$ is a function. If $yH = xH$ then by definition of set multiplication, $gyH = \{g\}yH = \{g\}xH = gxH$. Thus $\sigma(g) : X \to X$ since $yH = xH \Rightarrow \sigma(g)yH = \sigma(g)xH$. It follows by a direct computation that $\sigma(g)^{-1} = \sigma(g^{-1})$ so Theorem 0.11 shows $\sigma(g)$ is a bijection and $\sigma : G \to \text{Bij}(X)$. Next, for $yH \in X, \sigma(gh)(yH) = (gh)yH = g(h(yH)) = \sigma(g)(\sigma(h)(yH)) = \sigma(g)\sigma(h)(yH)$; $\sigma(gh)$ and $\sigma(g)\sigma(h)$ act identically on X so are equal by Theorem 0.8, and σ is a homomorphism. Now $ker\,\sigma = \{g \in G \mid \sigma(g) = I_x\} = \{g \in G \mid gyH = yH, \text{ all } yH \in X\} \subseteq \{g \in G \mid gH = H\} = H$, and $ker\,\sigma \lhd G$ by Theorem 6.4. Let $a \in N \lhd G$ and $N \leq H$. For $y \in G$, $y^{-1}ay \in N \subseteq H$ so $y^{-1}ayH = H$ and $ayH = yH$ follows. Thus $\sigma(a) = I_x$ and $a \in ker\,\sigma$, so $N \leq ker\,\sigma$ as claimed.

For i), set $n = [G : H]$ and assume $|G| \nmid n!$. If $ker\,\sigma = \langle e\rangle$ then σ is an injection by Theorem 6.4 so $\sigma(G) \cong G$. It follows from Theorem 4.5 that $|\sigma(G)| \mid |\text{Bij}(X)|$, but $|\text{Bij}(X)| = |X|! = [G : H]! = n!$ using Theorem 0.14. This contradicts $|G| \nmid n!$ so $ker\,\sigma \neq \langle e\rangle$ and i) holds. For ii), if $N = ker\,\sigma \neq \langle e\rangle$ then $N \lhd G$ by Theorem 6.4 and $N \neq G$ since $ker\,\sigma \leq H$ and $H \neq G$ by assumption. When $ker\,\sigma = \langle e\rangle$ use $\text{Bij}(X) \cong S_n$ by Theorem 6.12 and compose σ with this isomorphism to get an isomorphism $\eta : G \to \eta(G) \leq S_n$, so $G \cong \eta(G)$. Now $\eta(G) \cap A_n \lhd \eta(G)$ by Theorem 5.4 and it is an exercise using Theorem 6.2 to see that $\eta^{-1}(\eta(G) \cap A_n) \lhd G$. Thus ii) holds unless $\eta^{-1}(\eta(G) \cap A_n)$ is $\langle e\rangle$ or G. In the first case, $\eta(G) \cap A_n = \langle I\rangle$ so all elements in $\eta(G) - \langle I\rangle$ are odd permutations, which means that any two such must have product in $\eta(G) \cap A_n = \langle I\rangle$. This forces $|G| = |\eta(G)| \leq 2$, a contradiction. When $\eta^{-1}(\eta(G) \cap A_n) = G$ then $\eta(G) \cap A_n = \eta(G) \leq A_n$, contradicting $|G| \nmid ([G : H]!/2)$. Therefore ii) holds, completing the proof. ■

Using Theorem 6.14 requires a subgroup of fairly large order. We will develop results giving this sort of information. For example, we will prove that in any group G of order 48 there is $H \leq G$ with $|H| = 16$. Since $[G : H] = 3$ and $48 \nmid 3!$, by Theorem 6.14 H contains some $\langle e\rangle \neq N \lhd G$; no group of order 48 is simple. Similarly, it turns out that any group G of order 112 must contain a subgroup K of order 16. Now $[G : K]! = 7! = 16 \cdot 315$ and so $|G| \mid [G:K]!$, but $|G| \nmid (7!/2)$ so G contains a nontrivial normal subgroup by Theorem 6.14. We proved in Theorem 5.11 that A_n is simple when $n \geq 5$. If $H \leq S_n$ had small index, it would contain a normal subgroup of S_n whose intersection with A_n would be normal in A_n. Thus S_n should not have subgroups of small index.

Example 6.17 If $n \geq 5$, $H \leq S_n$, and $[S_n : H] = k < n$, then $H = A_n$ or $H = S_n$.

Clearly $n! \nmid k!$ so by Theorem 6.14 there is $\langle I \rangle \neq N \lhd S_n$ with $N \leq H$. Now $A_n \cap N \lhd A_n$ (Theorem 5.4) so the simplicity of A_n from Theorem 5.11 forces $A_n \cap N$ to be A_n or $\langle I \rangle$. In the first case, $A_n \leq N \leq H$ and $S_n = A_n \cup A_n\theta$ for any $\theta \notin A_n$ so either $A_n = H$ or $H = S_n$. When $A_n \cap N = \langle I \rangle$, $\eta, \theta \in N - \langle I \rangle$ are odd so $\eta\theta \in A_n \cap N = \langle I \rangle$. It follows that $|N| = 2$ (why?), say $N = \{I, \theta\}$. Since $N \lhd S_n$ all $(i, j)\theta(i, j) \in N$ so $(i, j)\theta(i, j) = \theta$, forcing θ to commute with all transpositions. Now $o(\theta) = 2$ so θ is a product of pairwise disjoint transpositions by Theorem 3.11. If $\theta = (a, b)$ then for $c \notin \{a, b\}$, $(a, c)\theta \neq \theta(a, c)$, and if $\theta = (a, b)(c, d)(e, f)\eta$ (θ is odd!) with the symbols in η disjoint from $\{a, b, c, d, e, f\}$ then θ and (a, c) do not commute. Thus $A_n \cap N \neq \langle I \rangle$.

The rest of this section is somewhat of a digression. We see how all finite groups are in some sense constructed from simple groups. As we mentioned in Chapter 5, a finite group G always has a proper, maximal, normal subgroup N_1. Since no normal subgroup of G lies between G and N_1, as we shall see in the next chapter, G/N_1 is a simple group. When G is simple, $N_1 = \langle e \rangle$, and otherwise N_1 itself must have a proper, maximal, normal subgroup N_2, and again N_1/N_2 is a simple group. Note that N_2 need not be a normal subgroup of G. Since G is finite there is a chain $\langle e \rangle = N_k \leq N_{k-1} \leq \cdots \leq N_1 \leq N_0 = G$ with each $N_j \lhd N_{j-1}$ and each N_{j-1}/N_j a simple group.

In $\mathbf{Z}_{24}$ we may write the chains $\langle [0] \rangle \leq \langle [12] \rangle \leq \langle [4] \rangle \leq \langle [2] \rangle \leq \mathbf{Z}_{24}$, or $\langle [0] \rangle \leq \langle [12] \rangle \leq \langle [6] \rangle \leq \langle [3] \rangle \leq \mathbf{Z}_{24}$, or $\langle [0] \rangle \leq \langle [8] \rangle \leq \langle [4] \rangle \leq \langle [2] \rangle \leq \mathbf{Z}_{24}$. In each case, each subgroup is cyclic and so each quotient of consecutive groups is cyclic by Example 5.5. Since each of these quotients has prime order, each is a simple group by Theorem 4.9. Although the chains and lists of consecutive quotients differ, the quotients can be matched for each chain: three are isomorphic to $\mathbf{Z}_2$ and one to $\mathbf{Z}_3$.

A natural question is whether two such chains for an arbitrary finite group must have the same set of quotients. The answer is that they must. The proof of this is somewhat involved but will be a good illustration of how the various results we have developed are used, although we will not apply the result later. We use the notation $H \lhd N \lhd L \lhd G$ to mean $H \lhd N$, $N \lhd L$, and $L \lhd G$, without assuming that either H or N is a normal subgroup of G. We split the argument into two parts, the first of which is a technical result of Zassenhaus used to prove the second.

THEOREM 6.15 If $A \lhd B \leq G$ and $C \lhd D \leq G$, then $A(B \cap C) \lhd A(B \cap D)$, $C(D \cap A) \lhd C(D \cap B)$, and $A(B \cap D)/A(B \cap C) \cong C(D \cap B)/C(D \cap A)$.

Proof Now $A \lhd B$ and $B \cap C$, $B \cap D \leq B$ so $A(B \cap C)$, $A(B \cap D) \leq B$ by Theorem 5.4. It is an exercise that $A \cap D$, $B \cap C \lhd B \cap D$, and it follows from Theorem 5.4 that

$E = (A \cap D)(B \cap C) \lhd B \cap D$. We claim $A(B \cap C) \lhd A(B \cap D)$ and $A(B \cap D)/A(B \cap C) \cong (B \cap D)/E$. For $a \in A$ and $d \in B \cap D$, set $\varphi(ad) = Ed \in (B \cap D)/E$. Is φ a function from $A(B \cap D)$ to $(B \cap D)/E$? If $ad = a'd'$ for $a' \in A$ and $d' \in B \cap D$, then $d(d')^{-1} = a^{-1}a' \in A \cap D \leq E$, so by Theorem 4.2 $Ed = Ed'$, which demonstrates that $\varphi : A(B \cap D) \to (B \cap D)/E$, and is surjective (why?). To show φ is a homomorphism, use that $A \lhd B$ implies $Ab = bA$ for $b \in B$ by Theorem 5.2, so if $a, a' \in A$ and $d, d' \in B \cap D$ there is $a'' \in A$ with $ada'd' = aa''dd'$. But now $\varphi(ada'd') = \varphi(aa''dd') = Edd' = EdEd' = \varphi(ad)\varphi(a'd')$, proving that φ is a homomorphism. Theorem 6.6 applies to yield $(B \cap D)/E \cong A(B \cap D)/ker\,\varphi$ and $A(B \cap C) \subseteq ker\,\varphi$ from the definition of φ, using $(B \cap C) \subseteq E$. If $ad \in ker\,\varphi$ then $Ed = E$, hence $d \in (A \cap D)(B \cap C) \subseteq A(B \cap C)$, so $ad \in A(B \cap C)$. Thus $ker\,\varphi = A(B \cap C)$; this yields $A(B \cap C) \lhd A(B \cap D)$ from Theorem 6.4 and then shows that $A(B \cap D)/A(B \cap C) \cong (B \cap D)/E$. Arguing similarly, $C(D \cap A) \lhd C(D \cap B)$ and $C(D \cap B)/C(D \cap A) \cong (B \cap D)/E \cong A(B \cap D)/A(B \cap C)$. ∎

THEOREM 6.16 Let G be a group with chains $\langle e \rangle = N_0 \lhd N_1 \lhd \cdots \lhd N_k = G$, and $\langle e \rangle = H_0 \lhd H_1 \lhd \cdots \lhd H_s = G$ so that each N_i is a proper, maximal, normal subgroup in N_{i+1} and each H_j is a proper, maximal, normal subgroup in H_{j+1}. Then $k = s$, and there is $\sigma \in S_k$ so that $N_i/N_{i-1} \cong H_{\sigma(i)}/H_{\sigma(i)-1}$.

Proof For N_i with $i \geq 1$, consider $N_{i-1} = N_{i-1}(N_i \cap H_0) \leq N_{i-1}(N_i \cap H_1) \leq \cdots \leq N_{i-1}(N_i \cap H_s) = N_{i-1}N_i = N_i$, and set $N_{ij} = N_{i-1}(N_i \cap H_j)$. Note that $N_{is} = N_i$. Similarly, for $j \geq 1$, we have $H_{j-1} \leq H_{j-1}(H_j \cap N_1) \leq \cdots \leq H_{j-1}(H_j \cap N_k) = H_j$ and set $H_{ji} = H_{j-1}(H_j \cap N_i)$. From Theorem 6.15 we conclude that for all $1 \leq i \leq k$ and $1 \leq j \leq s$, $N_{i\,j-1} \lhd N_{ij}$, $H_{ji-1} \lhd H_{ji}$, and $N_{ij}/N_{i\,j-1} \cong H_{ji}/H_{j\,i-1}$. Since $N_{is} = N_i$, for $1 \leq i \leq k$ there is a minimal $j = j(i) \geq 1$ so that $N_{ij} = N_{i-1}(N_i \cap H_j) = N_i$. Using Theorem 6.15, $N_{i-1} \leq N_{i-1}(N_i \cap H_{j-1}) \lhd N_{i-1}(N_i \cap H_j) = N_i$ and because N_{i-1} is a maximal normal subgroup in N_i the choice of j forces $N_{i-1}(N_i \cap H_{j-1}) = N_{i-1}$. But now $N_{i-1} \leq N_{i-1}(N_i \cap H_v) \leq N_{i-1}(N_i \cap H_{j-1}) = N_{i-1}$ for all $v < j$. Thus for $1 \leq i \leq k$ there is exactly one quotient, $N_{ij(i)}/N_{i\,j(i)-1} \cong N_i/N_{i-1}$ and all other $N_{iv}/N_{i\,v-1}$ have order 1. Similarly, for $1 \leq j \leq s$, in the chain $H_{j-1} \leq H_{j-1}(H_j \cap N_1) \leq \cdots \leq H_{j-1}(H_j \cap N_k) = H_j$ each subgroup either is H_{j-1} or is H_j. As we mentioned, $N_{ij(i)}/N_{i\,j(i)-1} \cong H_{j(i)i}/H_{j(i)\,i-1}$. But then $H_{j(i)i}/H_{j(i)\,i-1}$ is not trivial so is isomorphic to $H_{j(i)}/H_{j(i)-1}$. If $j(i) = j(i')$ then both $H_{j(i)i}/H_{j(i)\,i-1} \cong N_{ij(i)}/N_{i\,j(i)-1} \cong N_i/N_{i-1}$ and $H_{j(i)i'}/H_{j(i)\,i'-1} \cong N_{i'j(i)}/N_{i'\,j(i)-1} \cong N_{i'}/N_{i'-1}$ are not trivial, which is impossible if $i \neq i'$ since each $H_{j(i)v}$ is $H_{j(i)}$ or $H_{j(i)-1}$ and $H_{j(i)v} \leq H_{j(i)\,v+1}$. Therefore the association of $j(i)$ to i, say $\sigma : \{1, \ldots, k\} \to \{1, \ldots, s\}$, is injective. But for each $1 \leq d \leq s$ there is one nontrivial quotient $H_d/H_{d-1} \cong H_{dm(d)}/H_{d\,m(d)-1} \cong N_{m(d)d}/N_{m(d)\,d-1}$ with $1 \leq m(d) \leq k$ so $d = \sigma(m(d))$ and σ is surjective. Thus σ is a bijection, $k = s$, and as $N_i/N_{i-1} \cong N_{i\sigma(i)}/N_{i\,\sigma(i)-1} \cong H_{\sigma(i)i}/H_{\sigma(i)\,i-1} \cong H_{\sigma(i)}/H_{\sigma(i)-1}$, the theorem is proved. ∎

The upshot of Theorem 6.16 is the interesting fact that for any chain of subgroups from $\langle e \rangle$ to G with each a maximal normal subgroup in the next, its set of quotients and the number of times each quotient appears are determined uniquely by G. The smallest nonidentity subgroup in the chain is a simple group and must be isomorphic to one of the quotients in any other such chain. A chain for S_4 would be $\langle I \rangle \lhd \langle (1,2)(3,4) \rangle \lhd H \lhd A_4 \lhd S_4$ where

$H = \{I, (1,2)(3,4), (1,3)(2,4), (1,4)(2,3)\}$, with quotients $\mathbf{Z}_2, \mathbf{Z}_3, \mathbf{Z}_2, \mathbf{Z}_2$ in order from S_4. As above, these are the same quotients as for $\mathbf{Z}_{24}$ so the quotients cannot determine G. Also it is not true that $|G|$ determines the simple quotients since the quotients for $\mathbf{Z}_{60}$ will all have prime order (why?) but the only chain for A_5 is $\langle I \rangle \lhd A_5$. The proper subgroups appearing in two chains of maximal normal subgroups can be completely different, and one example is for D_4. Two chains are $\langle I \rangle \lhd \langle T_2 \rangle \lhd \langle T_1 \rangle \lhd D_4$ and $\langle I \rangle \lhd \langle W_2 \rangle \lhd \langle \{W_2, T_2\} \rangle \lhd D_4$ with all three quotients isomorphic to $\mathbf{Z}_2$ for each chain.

EXERCISES

1. Use Theorem 6.11 and Theorem 6.12 to identify each $g \in G$ with $\tau_g \in \mathrm{Bij}(G) \cong S_{|G|}$ written as a product of disjoint cycles of elements of G for:
 i) $G = \mathbf{Z}_5$
 ii) $G = \mathbf{Z}_6$
 iii) $G = \mathbf{Z}_2 \oplus \mathbf{Z}_2$
 iv) $G = S_3$
 v) $G = D_6$
 vi) $G = Q$, the quaternions of Exercise 13 in §5.1

2. **a)** If $n \geq 5$, show that if $H \leq A_n$ and $[A_n : H] < n$ then $H = A_n$.
 b) Show that there is no injective homomorphism $\beta : S_n \to A_{n+1}$ when $n \geq 4$.

3. If G is a group with $|G| = 240$ and $H \leq G$ with $H \cong A_5$, show that $H \lhd G$.

4. Let G be a group with $|G| = 900$. If $H \leq G$ with $H \cong \mathbf{Z}_{100}$, show G contains a nonidentity, normal, cyclic subgroup.

5. If G is a finite non-Abelian group with the property that for any $m \in N$ with $m \mid |G|$ there is some $H \leq G$ with $|H| = m$, show that G is not a simple group.

6. For each group G below, find a chain $\langle e \rangle = N_1 \leq N_2 \leq \cdots \leq N_k = G$ so that each $N_{j-1} \lhd N_j$ and $N_j / N_{j-1} \cong \mathbf{Z}_p$ for some prime p:
 i) $\mathbf{Z}_{30}$
 ii) S_3
 iii) U_{33}
 iv) $\mathbf{Z}_6 \oplus \mathbf{Z}_{30}$
 v) $\left\{ \begin{bmatrix} a & b \\ 0 & c \end{bmatrix} \in UT(2, \mathbf{Z}_7) \mid a, c \in \{[1]_7, [-1]_7\} \right\}$
 vi) $UT(2, \mathbf{Z}_5)$
 vii) U_{63} (note that $o([5]_{63} = 6)$
 viii) $D_4 \oplus D_5$
 ix) $\mathbf{Z}_{12} \oplus D_6$

6.5 AUTOMORPHISMS

It is interesting to consider all isomorphisms from a group to itself. This set is never empty since it contains the identity map.

DEFINITION 6.3 For any group G, let $Aut(G) = \{\varphi : G \to G \mid \varphi \text{ is an isomorphism}\}$. Any $\varphi \in Aut(G)$ is called an **automorphism** of G.

The first observation about $Aut(G)$ is

THEOREM 6.17 $(Aut(G), \circ)$ is a group for $\circ$ the composition of functions.

Proof As in Example 6.6, the identity map $I_G \in Aut(G)$, so $Aut(G) \neq \varnothing$. Compositions of automorphisms are automorphisms by Theorem 6.3 and Theorem 0.13 so $Aut(G)$ is closed under composition. Composition is associative by Theorem 0.9, and if $\varphi \in Aut(G)$ then $\varphi^{-1} \in Aut(G)$ by Theorem 6.3. Therefore $(Aut(G), \circ)$ is a group by Definition 2.1. ∎

It is difficult in general to describe $Aut(G)$ from G and not easy to write down many nonidentity automorphisms for an arbitrary group. One type of automorphism is the **conjugation** T_x in Example 6.8 given by $T_x(g) = xgx^{-1}$. When G is an Abelian group, all $T_x = I_G$ so we do not get anything new. However these T_x are sufficiently important that we give them a name.

DEFINITION 6.4 For G a group and $x \in G$, the **inner automorphism determined by x** is the function T_x defined by $T_x(g) = xgx^{-1}$. The set of all inner automorphisms is $Inn(G) = \{T_x \in Aut(G) \mid x \in G\}$.

It is possible to define $T_x(g) = x^{-1}gx$ instead, but our definition will make the proofs of later results a little simpler. We can say something about $Inn(G)$ and also see how it is described as a group solely in terms of G. Recall that the center of G is $Z(G) = \{y \in G \mid yg = gy \text{ for all } g \in G\}$.

THEOREM 6.18 For any group G, $Inn(G) \lhd Aut(G)$ and $Inn(G) \cong G/Z(G)$.

Proof First we need $Inn(G) \leq Aut(G)$. Now $Inn(G) \neq \varnothing$ as $T_e = I_G$ and if $x, y, g \in G$, $T_x \circ T_y(g) = T_x(T_y(g)) = T_x(ygy^{-1}) = xygy^{-1}x^{-1} = (xy)g(xy)^{-1} = T_{xy}(g)$. Thus $T_x \circ T_y = T_xT_y = T_{xy} \in Inn(G)$ by Theorem 0.8 so $Inn(G)$ is closed under composition. Also $T_x \in Inn(G) \Rightarrow (T_x)^{-1} = T_{x^{-1}} \in Inn(G)$ from Example 6.8 or our last computation. Thus $Inn(G) \leq Aut(G)$ by Theorem 2.12. For $T_x \in Inn(G)$, $\varphi \in Aut(G)$, and $g \in G$, $\varphi^{-1}T_x\varphi(g) = \varphi^{-1}(x\varphi(g)x^{-1}) = \varphi^{-1}(x)g\varphi^{-1}(x^{-1}) = (\varphi^{-1}(x))g\varphi^{-1}(x)^{-1} = T_{\varphi^{-1}(x)}(g)$, using Theorem 6.2. Hence $\varphi^{-1}T_x\varphi = T_{\varphi^{-1}(x)} \in Inn(G)$ by Theorem 0.8 and $Inn(G) \lhd Aut(G)$. Finally, we use Theorem 6.6 to show that $Inn(G) \cong G/Z(G)$. Define $\eta : G \to Inn(G)$ by $\eta(g) = T_g$, so η is surjective by definition of $Inn(G)$. The computation above shows $\eta(xy) = T_{xy} = T_xT_y = \eta(x)\eta(y)$, proving η is a homomorphism, so $Inn(G) \cong G/ker\,\eta$ by Theorem 6.6. But $ker\,\eta = \{g \in G \mid T_g(y) = y, \text{ all } y \in G\} = \{g \in G \mid gyg^{-1} = y, \text{ all } y \in G\} = \{g \in G \mid gy = yg, \text{ all } y \in G\} = Z(G)$. ∎

We apply Theorem 6.18 to some non-Abelian groups. We showed in Theorem 3.18 that $Z(D_n) = \langle I \rangle$ when n is odd and $Z(D_n) = \langle T_{n/2} \rangle$ when n is even. Thus $Inn(D_{2k+1}) \cong D_{2k+1}$ by Theorem 6.18, so each $g \in D_{2k+1}$ gives rise to a different inner automorphism. Also $Inn(D_{2n}) \cong D_{2n}/\langle T_n \rangle \cong D_n$ by Exercise 5 in §6.3. From Exercise 4 in §3.3, $Z(S_n) = \langle I \rangle$ when $n > 2$ so then $Inn(S_n) \cong S_n$. If Q is the quaternion group of Exercise 13 in §5.1 it is an exercise to see that $Z(Q) = \langle -1 \rangle$, so $Inn(Q) \cong Q/\langle -1 \rangle \cong \mathbf{Z}_2 \oplus \mathbf{Z}_2$. This latter isomorphism follows from a direct computation with cosets or by applying Theorem 5.8 and then Example 6.2 (verify!).

We give a result that is helpful in determining the structure of certain groups. Given $N \lhd G$, we define a homomorphism from G to $Aut(N)$ and use this fact to show that certain elements in G must commute. Recall from Definition 2.11 that for $H \leq G$, the centralizer of H in G is $C_G(H) = \{g \in G \mid gh = hg \text{ for all } h \in H\}$, so $G = C_G(H)$ if and only if $H \leq Z(G)$.

THEOREM 6.19 For $N \triangleleft G$, $\varphi(g) = T_g|_N$ defines a homomorphism $\varphi : G \to Aut(N)$ and $\varphi(G) \cong G/C_G(N)$. If $H \leq G$, H is finite, N is finite, and $(|H|, |Aut(N)|) = 1$, then $H \leq C_G(N)$. Thus $(|G|, |Aut(N)|) = 1 \Rightarrow N \leq Z(G)$.

Proof Since $N \triangleleft G$, $xNx^{-1} = N$ for any $x \in G$. Hence $T_x|_N : N \to N$, has inverse $T_{x^{-1}}|_N$ so $T_x|_N \in \text{Bij}(N)$, and $T_x|_N$ is a homomorphism (T_x is) so $T_x|_N \in Aut(N)$. It is easy to verify that $T_x|_N T_y|_N = T_{xy}|_N$, showing φ is a homomorphism. By Theorem 6.6, $\varphi(G) \cong G/ker\,\varphi$ and $ker\,\varphi = \{g \in G \mid T_g|_N(y) = y, \text{ all } y \in N\} = \{g \in G \mid gyg^{-1} = y, \text{ all } y \in N\} = C_G(N)$, so $\varphi(G) \cong G/C_G(N)$. If $H \leq G$ then $\rho = \varphi|_H : H \to \varphi(H)$ is a surjective homomorphism. Now Theorem 6.6 shows $\varphi(H) \cong H/ker\,\rho$ and when H is finite $|\varphi(H)| \mid |H|$ by Lagrange's Theorem. Our assumptions force $(|\varphi(H)|, |Aut(N)|) = 1$, but since $\varphi(H) \leq Aut(N)$ (Exercise 7 in §6.2), $|\varphi(H)| \mid |Aut(N)|$ (Lagrange!). We must conclude that $|\varphi(H)| = 1$. Hence $\varphi(H) = \langle I \rangle$, so $H \leq ker\,\varphi = C_G(N)$, completing the proof. ∎

Using Theorem 6.19 we can determine the structure of any group of order pq for $p < q$ primes. We illustrate this for groups of order 35. The general case is more complicated and will be examined in a later chapter.

Example 6.18 If G is a group with $|G| = 35$ then $G \cong \mathbf{Z}_{35}$.

From Theorem 4.7, for $g \in G$, $o(g) \mid 35$, and if $o(y) = 35$ for $y \in G$ then $|\langle y \rangle| = 35$ by Theorem 2.16 and $G = \langle y \rangle \cong \mathbf{Z}_{35}$ using Theorem 6.8. Assume $x \in G$ with $o(x) = 7$. It follows from Theorem 4.12 (as in the proof of Theorem 6.1) that $\langle x \rangle$ is the only subgroup in G of order 7, so $\langle x \rangle \triangleleft G$ by Theorem 5.3. Let us find $|Aut(\langle x \rangle)|$. For $\varphi \in Aut(\langle x \rangle)$ Theorem 6.2 shows $o(\varphi(x)) = 7$ and $\varphi(x^k) = \varphi(x)^k$ for all $k \in N$, resulting in $\langle \varphi(x) \rangle = \varphi(\langle x \rangle) = \langle x \rangle$. Thus $\varphi(x)$ generates $\langle x \rangle$ and determines φ. Since $o(x) = 7$, $\{x, x^2, \ldots, x^6\}$ are the generators of $\langle x \rangle$ from Theorem 2.8 and it is an exercise that $\eta_j(x^k) = x^{kj}$ for $1 \leq j \leq 6$ defines $\eta_j \in Aut(\langle x \rangle)$. Therefore $|Aut(\langle x \rangle)| = 6$, and $(35, 6) = 1$ forces $\langle x \rangle \leq Z(G)$ by Theorem 6.19. But $\langle x \rangle$ is the only subgroup of G with order 7 so $g \in G - \langle x \rangle$ forces $o(g) = 5$ or $o(g) = 35$. As above, we are finished if $o(g) = 35$, and if $o(g) = 5$ then $xg = gx$ results in $o(xg) = 35$ by Theorem 4.16, again showing $G \cong \mathbf{Z}_{35}$. Finally, if $g \in G - \langle e \rangle \Rightarrow o(g) = 5$ then since different subgroups of order 5 intersect only in $\langle e \rangle$, as in the corollary to Theorem 4.6, there are $4k$ elements of order 5 in G. However, $35 \neq 4k + 1$ so there must be $g \in G$ of order 7 or 35, verifying the example.

As we have just seen, characterizing $Aut(\mathbf{Z}_n)$ is useful. We will do this but first find some automorphisms of Abelian groups. By Theorem 6.5 if G is an Abelian group and $m \in N$, then $F_m(g) = g^m$ is a homomorphism from G to G. When G is finite and $(|G|, m) = 1$ then F_m is an automorphism. Therefore $F_m \in Aut(\mathbf{Z}_p)$ for p a prime when $1 \leq m < p$. In $\mathbf{Z}_n^k$, the direct sum of $\mathbf{Z}_n$ with itself k times, any permutation of the coordinates is an automorphism (exercise!). Another example follows.

Example 6.19 For G a group and $f(g) = g^{-1}$, $f \in Aut(G) \Leftrightarrow G$ is Abelian.

Since $(g^{-1})^{-1} = g$ for all $g \in G$, $f^{-1} = f$ so f is bijective (Theorem 0.11). When G is Abelian and $x, y \in G$ then $f(xy) = (xy)^{-1} = y^{-1}x^{-1} = x^{-1}y^{-1} = f(x)f(y)$ and $f \in Aut(G)$ follows. Given $f \in Aut(G)$ then since $f(x)f(y) = f(xy)$ for all $x, y \in G$, we must have $x^{-1}y^{-1} = (xy)^{-1} = y^{-1}x^{-1}$. This holds also when x^{-1} replaces x and y^{-1} replaces y, yielding $xy = yx$. Therefore G is Abelian.

Our last results characterize $Aut(\mathbf{Z}_n)$ and $Aut(\mathbf{Z}_n^k)$ for $n, k > 1$. This will be crucial later for determining the structure of many groups.

THEOREM 6.20 If $n > 1$ then $Aut((\mathbf{Z}_n, +)) \cong U_n$. In particular, if $p > 2$ is a prime then $Aut(\mathbf{Z}_{p^n}) \cong \mathbf{Z}_{p^{n-1}(p-1)}$, and $Aut(\mathbf{Z}_p) \cong \mathbf{Z}_{p-1}$.

Proof We determine the homomorphisms of $(\mathbf{Z}_n, +)$ and then see which are bijections. Denote the elements of $\mathbf{Z}_n$ by $[t]$ and $a \equiv_n b$ by $a \equiv b$. For any homomorphism, $\varphi : \mathbf{Z}_n \to \mathbf{Z}_n$ let $\varphi([1]) = [s]$. If $m \in \mathbf{N}$, $\varphi([m]) = \varphi([1]^m) = \varphi([1])^m = [s]^m = [sm]$ using Theorem 6.2. Hence φ acts like multiplication by s. For $k \in \mathbf{Z}$ does $\mu_k([a]) = [ka]$ define a homomorphism of $\mathbf{Z}_n$? Note that $[a] = [b] \Rightarrow a \equiv b \Rightarrow ka \equiv kb \Rightarrow [ka] = [kb]$ by Theorem 1.24, and this shows that $\mu_k : \mathbf{Z}_n \to \mathbf{Z}_n$. But now $\mu_k([a] + [b]) = \mu_k([a + b]) = [k(a + b)] = [ka] + [kb] = \mu_k([a]) + \mu_k([b])$, proving that μ_k is a homomorphism of $\mathbf{Z}_n$. Also $\mu_k = \mu_d$ implies $[k] = \mu_k([1]) = \mu_d([1]) = [d]$ so $k \equiv d$. If $k \equiv d$ then $ka \equiv da$ so $\mu_k([a]) = [ka] = [da] = \mu_d([a])$ for all $[a] \in \mathbf{Z}_n$. Hence $\mu_k = \mu_d \Leftrightarrow k \equiv d$ and the set of homomorphisms from $\mathbf{Z}_n$ to itself is $\{\mu_k \mid 0 \leq k < n\}$.

We claim $\mu_k \in Aut(\mathbf{Z}_n) \leftrightarrow (k, n) = 1$. Since μ_k is just the kth power map on $\mathbf{Z}_n$, note that $(k, n) = 1 \Rightarrow \mu_k \in Aut(\mathbf{Z}_n)$ from Theorem 6.5. We will show directly that $\mu_k \in Aut(\mathbf{Z}_n) \Leftrightarrow (k, n) = 1$. Of course μ_0 is the trivial homomorphism and cannot be an automorphism so we assume $k \geq 1$. When $(k, n) = m > 1$ then $\mu_k([n/m]) = [kn/m] = [(k/m)n] = [0]$, so $ker\, \mu_k \neq \langle [0] \rangle$ and μ_k is not injective by Theorem 6.4. When $(k, n) = 1$ then $ka \equiv kb \Rightarrow a \equiv b$ by Theorem 1.24. This means that $\mu_k([a]) = \mu_k([b])$ forces $[a] = [b]$ so μ_k is injective by definition. Consequently, μ_k is injective $\Leftrightarrow (k, n) = 1$. But $\mathbf{Z}_n$ is finite so Theorem 0.10 shows that μ_k is injective if and only if $\mu_k \in Aut(\mathbf{Z}_n)$. Since $\mu_k = \mu_d \Leftrightarrow k \equiv d$ we have $Aut(\mathbf{Z}_n) = \{\mu_k \mid (k, n) = 1\} = \{\mu_k \mid 1 \leq k \leq n \text{ and } (k, n) = 1\}$. Hence there is a bijection $\eta : Aut(\mathbf{Z}_n) \to U_n$ given by $\eta(\mu_k) = [k]$. Note that η really is a function: if $\mu_k = \mu_d$ then $k \equiv d$ so $\eta(\mu_k) = [k] = [d] = \eta(\mu_d)$. Now η is a homomorphism since $\mu_k\mu_s([a]) = \mu_k(\mu_s([a])) = \mu_k([sa]) = [ksa] = \mu_{ks}([a])$. Thus $\mu_k\mu_s = \mu_{ks}$ and $\eta(\mu_k\mu_s) = \eta(\mu_{ks}) = [ks] = [k][s] = \eta(\mu_k)\eta(\mu_s)$ so η is an isomorphism. Finally, by Theorem 4.20, U_p is a cyclic group when p is a prime, and more generally, U_{p^n} is cyclic when p is odd by Theorem 4.22 so applying Theorem 6.8 completes the proof since $|U_{p^n}| = \varphi(p^n) = p^{n-1}(p - 1)$. ∎

An application of Theorem 6.20 and earlier results comes next.

THEOREM 6.21 Let G be a group with $|G| = p^n$ forp a prime.

a) If $H \lhd G$ and $|H| = p$ then $H \leq Z(G)$.
b) If $|G| = p^2$ then G is Abelian.

Proof For a), observe first that H is cyclic from Theorem 4.9 and then that $H \cong \mathbf{Z}_p$ by Theorem 6.8. By an exercise, $Aut(H) \cong Aut(\mathbf{Z}_p)$ so it follows from Theorem 6.20 that $|Aut(H)| = p - 1$. But now Theorem 6.19 yields $H \le Z(G)$ since $(p^n, p-1) = 1$. As for b), if $g \in G$ then $o(g) \mid p^2$ by Theorem 4.7 so $o(g) = 1$, p, or p^2. If $x \in G$ with $o(x) = p^2$ then $|\langle x \rangle| = p^2$ by Theorem 2.16 so $G = \langle x \rangle$ is cyclic and Abelian. If every $x \in G - \langle e \rangle$ has order p, then $H = \langle x \rangle$ has order p using Theorem 2.16 again, and $[G : H] = p$ by Lagrange's Theorem. In this case $|G| = p^2$ does not divide $p! = [G : H]!$. Applying Theorem 6.14 shows that $H \triangleleft G$, and part a) now yields $x \in H \le Z(G)$. But then every $x \in G$ is in $Z(G)$ and G is Abelian, proving b). ■

Our last result in the chapter generalizes Theorem 6.20 and follows the same basic approach. Rather than giving all the details, we will outline a complete argument. It is instructive to write out the details when $k = 2$.

THEOREM 6.22 If $n > 1$ and $k \in \mathbf{N}$, then $Aut(\mathbf{Z}_n^k) \cong GL(k, \mathbf{Z}_n)$.

Proof Note that for $k = 1$ the statement is $Aut(\mathbf{Z}_n) \cong GL(1, \mathbf{Z}_n)$, the invertible 1×1 matrices over $\mathbf{Z}_n$, or $Aut(\mathbf{Z}_n) \cong U_n$, which is true by Theorem 6.20. When $k > 1$ we identify a homomorphism φ of $\mathbf{Z}_n^k$ to itself as multiplication by some $A \in M_k(\mathbf{Z}_n)$. A typical element of $\mathbf{Z}_n^k$ is a k-tuple $([x_1], \ldots, [x_k])$ where each $[x_i] \in \mathbf{Z}_n$. Let $\varepsilon_j = ([0], \ldots, [1], \ldots, [0])$, the tuple with $[1]$ in the jth coordinate and $[0]$ elsewhere. Using that the group operation is addition in $\mathbf{Z}_n^k$ and that for $[m] \in \mathbf{Z}_n$ with $0 \le m < n$, $[m] = [1]^m$, we can write

$$\begin{aligned}([x_1], \ldots, [x_k]) &= ([x_1], [0], \ldots, [0]) + ([0], [x_2], \ldots, [0]) + \cdots + ([0], \ldots, [0], [x_k]) \\ &= \varepsilon_1^{x_1} + \cdots + \varepsilon_k^{x_k} = x_1\varepsilon_1 + x_2\varepsilon_2 + \cdots + x_k\varepsilon_k\end{aligned}$$

Therefore $\varphi(([x_1], \ldots, [x_k])) = x_1\varphi(\varepsilon_1) + x_2\varphi(\varepsilon_2) + \cdots + x_k\varphi(\varepsilon_k)$. If $\varphi(\varepsilon_j) = ([a_{1j}], [a_{2j}], \ldots, [a_{kj}])$ then

$$\begin{aligned}\varphi(([x_1], \ldots, [x_k])) &= x_1([a_{11}], [a_{21}], \ldots, [a_{k1}]) + \cdots + x_k([a_{1k}], [a_{2k}], \ldots, [a_{kk}]) \\ &= ([a_{11}x_1 + a_{12}x_2 + \cdots + a_{1k}x_k], \ldots, [a_{k1}x_1 + a_{k2}x_2 + \cdots + a_{kk}x_k])\end{aligned}$$

which means that $\varphi(([x_1], \ldots, [x_k])) = (A_\varphi([x_1], \ldots, [x_k])^T)^T$ where $A_\varphi = ([a_{ij}]) \in M_k(\mathbf{Z}_n)$ and $([x_1], \ldots, [x_k])^T$ is the transpose of $([x_1], \ldots, [x_k])$: the column with entry $[x_j]$ in the jth coordinate. We apply transpose a second time to change the column $A_\varphi([x_1], \ldots, [x_k])^T$ into an element of $\mathbf{Z}_n^k$, which is a row. Not only is φ given by multiplication by A_φ, but A_φ is uniquely defined by φ. If also $\varphi(([x_1], \ldots, [x_k])) = (B([x_1], \ldots, [x_k])^T)^T$ for $B \in M_k(\mathbf{Z}_n)$, then in particular, $(A_\varphi(\varepsilon_j)^T)^T = \varphi(\varepsilon_j) = (B(\varepsilon_j)^T)^T$. But $A_\varphi(\varepsilon_j)^T$ is just the jth column of A_φ, so A_φ and B would have the same columns and so be the same matrix. Observe that the identity map on $\mathbf{Z}_n^k$ is associated with I_k, the identity matrix in $M_k(\mathbf{Z}_n)$.

If η is another homomorphism identified with left multiplication by $A_\eta \in M_k(\mathbf{Z}_n)$ then we have $\varphi \circ \eta(([x_1], \ldots, [x_k])) = \varphi(\eta(([x_1], \ldots, [x_k]))) = \varphi((A_\eta([x_1], \ldots, [x_k])^T)^T) = (A_\varphi A_\eta([x_1], \ldots, [x_k])^T)^T$. The uniqueness of the matrix associated with $\varphi \circ \eta$ shows $A_{\varphi\circ\eta} = A_\varphi A_\eta$. Consequently, for $\varphi \in Aut(\mathbf{Z}_n^k)$ and $\eta = \varphi^{-1}$ it follows first that A_φ is invertible with inverse $A_\varphi^{-1} = A_{\varphi^{-1}}$, so $A_\varphi \in GL(k, \mathbf{Z}_n)$, and then that $F(\varphi) = A_\varphi$ defines a homomorphism from $Aut(\mathbf{Z}_n^k)$ to $GL(k, \mathbf{Z}_n)$. The uniqueness of A_φ associated to φ shows that

F is injective. It is readily checked using properties of matrix multiplication that for any $A \in M_k(\mathbf{Z}_n)$, $\gamma(([x_1], \ldots, [x_k])) = (A([x_1], \ldots, [x_k])^T)^T$ defines a homomorphism γ of $\mathbf{Z}_n^k$ and $A = A_\gamma$ by the uniqueness of the associated matrix. Furthermore if $A \in GL(k, \mathbf{Z}_n)$ then the γ it defines has inverse defined by $(A^{-1}([x_1], \ldots, [x_k])^T)^T$, so $\gamma \in Aut(\mathbf{Z}_n^k)$ and $F(\gamma) = A$, showing that F is surjective. This means that $F : Aut(\mathbf{Z}_n^k) \to GL(k, \mathbf{Z}_n)$ is an isomorphism, proving the theorem. ∎

EXERCISES

1. $N, K \leq G$ with $NK = G$, $N \cap K = \langle e \rangle$, and $N \lhd G$.
  i) For $k \in K$, show that $f_k(x) = kxk^{-1}$ defines $f_k \in Aut(N)$.
  ii) Show that the hypotheses of Theorem 3.22 hold for $\{f_k \in Aut(N) \mid k \in K\}$.
  iii) Referring to Theorem 3.22, show that $(N \times K, \cdot) \cong G$.

2. **a)** If G is any group and $\theta \in S_n$, show that $f_\theta(g_1, \ldots, g_n) = (g_{\theta(1)}, \ldots, g_{\theta(n)})$ defines $f_\theta \in Aut(G^n)$.
  b) If $G \neq \langle e \rangle$ is a group and $n \in \mathbf{N}$, show there is an injective homomorphism $\varphi : S_n \to Aut(G^n)$: that is, there is $H \leq Aut(G^n)$ with $S_n \cong H$.

3. For G a group, if $Aut(G)$ is cyclic, show that G is Abelian.

4. If G is a group and $Aut(G)$ is simple, show that G is Abelian or $G/Z(G)$ is a non-Abelian simple group.

5. For $H \leq G$ recall (Exercise 7 in §2.5) that $N_G(H) = \{g \in G \mid gHg^{-1} = H\} \leq G$ and that $C_G(H) = \{g \in G \mid gh = hg, \text{ all } h \in H\}$ (Definition 2.11). Show that $C_G(H) \lhd N_G(H)$ and prove that H finite and cyclic $\Rightarrow N_G(H)/C_G(H)$ is Abelian. (Think about $Aut(H)$.)

6. Using known groups, describe $Inn(G)$ for each of the following:
  i) $G = D_5$
  ii) $\mathbf{Z}_8 \oplus D_7$
  iii) $A_5 \oplus D_4$
  iv) $D_3 \oplus D_4 \oplus D_5$
  v) $D_4 \oplus D_4 \oplus D_4$
  vi) $UT(3, \mathbf{Z}_2)$
  vii) $UT(2, \mathbf{Z}_6)$
  viii) $SL(2, \mathbf{Z}_5) \cap UT(2, \mathbf{Z}_5)$ (Note Example 6.4.)

7. Show that $Aut((\mathbf{Z}, +)) \cong \mathbf{Z}_2$.

8. If $G \cong H$ are groups, show that: i) $Aut(G) \cong Aut(H)$; and ii) $Inn(G) \cong Inn(H)$.

9. If H and K are finite groups satisfying $(|H|, |K|) = 1$, show that $Aut(H \oplus K) \cong Aut(H) \oplus Aut(K)$.

10. If $n, m \in \mathbf{N}$ with $(n, m) = 1$, show that $U_{nm} \cong U_n \oplus U_m$ (use Exercises 8 and 9).

11. Use Exercise 10 to describe each of the following as a direct sum of cyclic groups: i) U_{36}; ii) U_{77}; iii) U_{105}; iv) U_{450}; v) U_{1000}; vi) U_{1485}.

12. **a)** Find $|Aut(D_n)|$. (Use $D_n = \langle T_1 \rangle \cup \langle T_1 \rangle W_j$ for any $0 \leq j \leq n-1$. Show any automorphism of D_n must send T_1 to an element of order n and W_0 to an element in $\{W_j\}$. Show that all such choices give rise to automorphisms.)
  b) For all $n \geq 3$, show that $Aut(D_n)$ has a subgroup isomorphic to D_n (see Example 6.4).
  c) For which $n > 2$ is $Aut(D_n) \cong D_n$?

13. Let G be a finite group, $|G| > 1$, p the smallest prime divisor of $|G|$, $x \in G$ with $o(x) = p$, and assume that $\langle x \rangle \lhd G$. Show that $x \in Z(G)$.

14. If G is a group, $|G| = 3 \cdot 5 \cdot 7^2 \cdot 13$, $N \lhd G$, and $|N| = 5$, show that $N \leq Z(G)$.

15. Suppose every element in the group G has finite order. If $m \in \mathbf{N}$ is the minimal order of any element in $G - \langle e \rangle$ and if $y \in G$ with $o(y) = m$ and $\langle y \rangle \lhd G$, show that $y \in \mathbf{Z}(G)$. Note that G may be infinite.

16. If G is a group, $|G| = pq$ for $p < q$ primes, $p \nmid (q - 1)$, and $o(y) = q$ for some $y \in G$, show G is cyclic. (Note that $\langle y \rangle \lhd G$ and that $o(x) = p$, some $x \in G$.)

17. Any $H \leq G$ is called **invariant** in G if $\varphi(H) \leq H$ for all $\varphi \in Aut(G)$, and is called **fully invariant** in G if $\varphi(H) \leq H$ for all homomorphisms $\varphi : G \to G$.

i) If H is invariant in G, show that $\varphi(H) = H$ for all $\varphi \in Aut(G)$.

ii) If H is invariant in G, show that $H \lhd G$.

iii) If $H = \langle h \rangle$ is invariant in G, then for any $K \leq H$, K is invariant in G.

iv) If $K \leq H$, K is invariant in H, and H is invariant in G, then show that K is invariant in G.

v) G', the commutator subgroup of G, is fully invariant in G.

vi) If $m \in \mathbf{N}$, $A = \{g \in G \mid o(g) | m^k \text{ for some } k \in \mathbf{N}\}$, and if $A \leq G$, then A is fully invariant in G.

vii) $Z(G)$ is invariant in G.

viii) Use $G = D_3 \oplus \mathbf{Z}_3$ to see that $Z(G)$ need not be fully invariant in G.

18. For $(G, +)$ an Abelian group, let $End(G) = \{\varphi : G \to G \mid \varphi \text{ is a homomorphism}\}$. If $\varphi, \eta \in End(G)$ define $\varphi + \eta : G \to G$ by $(\varphi + \eta)(g) = \varphi(g) + \eta(g)$, all $g \in G$.

a) Show that $\varphi + \eta \in End(G)$.

b) Show that $(End(G), +)$ is an Abelian group.

c) If G is a cyclic group, show that $G \cong End(G)$.

d) For p a prime, show that $End(\mathbf{Z}_p^2, +)$ is not Abelian.

e) Show that $End(\mathbf{Z}^n, +) \cong (M_n(\mathbf{Z}), +)$ (see Theorem 6.22).

CHAPTER 7

STRUCTURE THEOREMS

7.1 THE CORRESPONDENCE THEOREM

The quotient construction enables us to study finite groups by using induction on the smaller quotients. To be able to draw conclusions about the group G from facts about G/N, we study how subgroups of G and those of G/N are related. Since $G/N = \rho(G)$ for $\rho : G \to G/N$ given by $\rho(g) = Ng$, we concentrate on surjective homomorphisms of G.

For this section we assume $\varphi : G \to B$ is an **epimorphism** (surjective homomorphism) of groups with $ker\ \varphi = N$. Our first result relates groups in $Sub(B) = \{K \subseteq B \mid K \leq B\}$ to groups in $Sub(G, N) = \{H \leq G \mid N \leq H\}$. It is crucial to recall from Definition 0.15 that for $T \subseteq B$, the *inverse image of T under* φ is $\varphi^{-1}(T) = \{g \in G \mid \varphi(g) \in T\}$.

THEOREM 7.1 Let $\varphi : G \to B$ be an epimorphism of groups with $ker\ \varphi = N$.

i) If $H \leq G$ then $\varphi(H) \leq B$, and if $H \lhd G$ then $\varphi(H) \lhd B$.
ii) If $K \leq B$ then $\varphi^{-1}(K) \leq G$ with $N \leq \varphi^{-1}(K)$, and if $K \lhd B$ then $\varphi^{-1}(K) \lhd G$.

Proof We use Theorem 2.12 to show $\varphi(H) \leq B$ and $\varphi^{-1}(K) \leq G$. Suppose $H \leq G$. To see that $\varphi(H) \leq B$, for $\varphi(x), \varphi(y) \in \varphi(H)$ with $x, y \in H$, we need $\varphi(x)\varphi(y)^{-1} \in \varphi(H)$. Since $xy^{-1} \in H$ it follows from Theorem 6.2 that $\varphi(x)\varphi(y)^{-1} = \varphi(x)\varphi(y^{-1}) = \varphi(xy^{-1}) \in \varphi(H)$ so $\varphi(H) \leq B$. When $H \lhd G$, take $\varphi(x) \in \varphi(H)$ with $x \in H$ and let $b \in B$. The surjectivity of φ means that $b = \varphi(g)$ for some $g \in G$, so $b^{-1}\varphi(x)b = \varphi(g)^{-1}\varphi(x)\varphi(g) = \varphi(g^{-1}xg)$ (Theorem 6.2), and $g^{-1}xg \in H \Rightarrow b^{-1}\varphi(x)b \in \varphi(H)$, proving that $\varphi(H) \lhd B$. If $K \leq B$ and $x, y \in \varphi^{-1}(K)$, then $\varphi(x), \varphi(y) \in K$. Since $\varphi(xy^{-1}) = \varphi(x)\varphi(y^{-1}) = \varphi(x)\varphi(y)^{-1} \in K$, by Theorem 2.12 we have $xy^{-1} \in \varphi^{-1}(K)$, hence $\varphi^{-1}(K) \leq G$. But $\varphi(N) = \langle e_B \rangle \subseteq K$ so $N \leq \varphi^{-1}(K)$. Finally, when $K \lhd B$, $x \in \varphi^{-1}(K)$, and $g \in G$, $\varphi(g^{-1}xg) = \varphi(g)^{-1}\varphi(x)\varphi(g) \in K$ (why?), showing that $g^{-1}xg \in \varphi^{-1}(K)$ so $\varphi^{-1}(K) \lhd G$. ∎

Using the notation above, Theorem 7.1 shows that we can define $\alpha : Sub(G, N) \to Sub(B)$ and $\beta : Sub(B) \to Sub(G, N)$ by $\alpha(H) = \varphi(H)$ and $\beta(K) = \varphi^{-1}(K)$. It is useful to recall that by Theorem 6.4, $\varphi^{-1}(K)$ is just the union in G of $\{\varphi^{-1}(\{k\}) \mid k \in K\}$, which is a union of cosets of N in G.

Next is the **Correspondence Theorem,** the main result in this section. It gives a number of facts about $Sub(G, N)$ and $Sub(B)$, the first of which is that α and β are inverses, so bijections between $Sub(B)$ and $Sub(G, N)$. Although the theorem is somewhat lengthy, it is the essential ingredient in studying a group by means of its quotients. For this reason the theorem is certainly one of the most important results we will prove.

THEOREM 7.2 Let $\varphi : G \to B$ be an epimorphism of groups with $ker\ \varphi = N$. Then $\alpha : Sub(G, N) \to Sub(B)$ given by $\alpha(H) = \varphi(H)$ and $\beta : Sub(B) \to Sub(G, N)$ given by $\beta(K) = \varphi^{-1}(K)$ are inverses and inclusion-preserving bijections.

i) α and β each send normal subgroups to normal subgroups.
ii) $N \leq H \leq L \leq G \Rightarrow [L : H] = [\varphi(L) : \varphi(H)]$
iii) $K \leq V \leq B \Rightarrow [V : K] = [\varphi^{-1}(V) : \varphi^{-1}(K)]$
iv) if $H, L \in Sub(G, N)$, $H \lhd L \Leftrightarrow \varphi(H) \lhd \varphi(L)$, and then $L/H \cong \varphi(L)/\varphi(H)$
v) $N \leq L \leq G \Rightarrow \varphi(L) \cong L/N$
vi) $N \leq H \lhd G \Rightarrow G/H \cong B/\varphi(H)$
vii) N finite and $K \leq B$ finite $\Rightarrow |\varphi^{-1}(K)| = |N| \cdot |K|$

Proof From Theorem 7.1 α and β are functions with the given domains and codomains. If α and β are inverses then they are bijections (Theorem 0.11). Take $N \leq H \leq G$ and note that $h \in H \Rightarrow \varphi(h) \in \varphi(H)$, so $h \in \varphi^{-1}(\varphi(H))$, showing $H \subseteq \varphi^{-1}(\varphi(H))$. Now $x \in \varphi^{-1}(\varphi(H))$ means that $\varphi(x) \in \varphi(H)$. If $y \in H$ with $\varphi(x) = \varphi(y)$ then $x \in \varphi^{-1}(\{\varphi(y)\})$ and Theorem 6.4 forces $x \in Ny \subseteq H$. Hence $\varphi^{-1}(\varphi(H)) \subseteq H$, proving that $\varphi^{-1}(\varphi(H)) = H$ (why?). Consequently, $\beta(\alpha(H)) = H$ for all $H \in Sub(G, N)$. Now let $K \leq B$ and $x \in \varphi^{-1}(K)$. Then $\varphi(x) \in K$ shows that $\varphi(\varphi^{-1}(K)) \subseteq K$. Since φ is surjective, if $k \in K$ then $k = \varphi(g)$ for some $g \in G$. But now by definition of inverse image, $g \in \varphi^{-1}(K)$ so $k \in \varphi(\varphi^{-1}(K))$. Therefore $K \subseteq \varphi(\varphi^{-1}(K)) \subseteq K$, $K = \varphi(\varphi^{-1}(K))$, and $\alpha(\beta(K)) = K$. Thus α and β are inverses so each is a bijection.

Clearly $N \leq H \leq L \leq G \Rightarrow \varphi(H) \subseteq \varphi(L)$. If $\varphi(H) = \varphi(L)$ then $\alpha(H) = \alpha(L)$ and since $\beta = \alpha^{-1}$, $H = \beta(\alpha(H)) = \beta(\alpha(L)) = L$. Therefore α is inclusion preserving as claimed, and β is similarly. Let us move on to the other parts of the theorem. Part i) is just a restatement of Theorem 7.1.

If $N \leq H \leq L \leq G$ and $\{x_j\}_S$ is a transversal for the right cosets of H in L then $L = \bigcup_S Hx_j$ and $\{Hx_j\}_S$ is pairwise disjoint. From the definition of index, $[L : H] = |S|$, which we do not assume is finite. To prove $[\varphi(L) : \varphi(H)] = [L : H]$, it suffices to show that $\{\varphi(x_j)\}_S$ is a transversal for the right cosets of $\varphi(H)$ in $\varphi(L)$ and that $\varphi(x_i) = \varphi(x_j)$ forces $x_i = x_j$. Now $\varphi(L) = \varphi(\bigcup_S Hx_j) = \bigcup \varphi(H)\varphi(x_j)$ (exercise!) and if $\varphi(H)\varphi(x_j) = \varphi(H)\varphi(x_i)$, then by Theorem 4.2, $\varphi(x_j)\varphi(x_i)^{-1} \in \varphi(H)$. Thus $\varphi(x_j x_i^{-1}) \in \varphi(H)$, forcing $x_j x_i^{-1} \in \varphi^{-1}\varphi(H) = H$ from above. But now $Hx_j = Hx_i$, and since $\{x_j\}$ is a transversal, $x_j = x_i$ results. Therefore $\eta : \{x_j\}_S \to \{\varphi(x_j)\}_S$ given by $\eta(x_j) = \varphi(x_j)$ is a bijection and $\{\varphi(x_j)\}_S$ is a transversal for the cosets of $\varphi(H)$ in $\varphi(L)$: $[\varphi(L) : \varphi(H)] = |S| = [L : H]$. Using that α and β are inverses, when $K \leq V \leq B$ then $K = \varphi(\varphi^{-1}(K))$ and $V = \varphi(\varphi^{-1}(V))$. Also $\varphi^{-1}(K) \leq \varphi^{-1}(V)$ by definition of inverse image, so as we have just seen, $[\varphi^{-1}(V) : \varphi^{-1}(K)] = [\varphi(\varphi^{-1}(V)), \varphi(\varphi^{-1}(K))] = [V, K]$, proving iii).

For iv), if $N \le H \le L \le G$ then $H \lhd L \Leftrightarrow \varphi(H) \lhd \varphi(L)$ follows from Theorem 7.1 using the epimorphism $\varphi|_L : L \to \varphi(L)$ with kernel $N : \varphi(H) = \varphi|_L(H) \lhd \varphi|_L(L) = \varphi(L)$. As in Example 6.7 we see that $\rho : L \to \varphi(L)/\varphi(H)$ given by $\rho(x) = \varphi(H)\varphi(x)$ is an epimorphism, so the isomorphism theorem (Theorem 6.6) yields $\varphi(L)/\varphi(H) \cong L/ker\ \rho$. Now $ker\ \rho = \{x \in L \mid \varphi(H)\,\varphi(x) = \varphi(H)\} = \{x \in L \mid \varphi(x) \in \varphi(H)\} = \{x \in L \mid x \in \varphi^{-1}(\varphi(H)) = H\}$, using that α and β are inverses. Thus $ker\ \rho = H$, proving iv), and also v) by taking $H = N$, and vi) as well by taking $L = G$.

Finally, if $K \le B$ with both K and N finite, use Theorem 6.4 to write $\varphi^{-1}(K) = \varphi^{-1}(\bigcup_K k) = \bigcup_K \varphi^{-1}(k) = \bigcup_K N g_k$ where $g_k \in \varphi^{-1}(\{k\})$. Note that the middle equality follows from Exercise 3 in §0.6. Since φ is a function and $Ng_k = \varphi^{-1}(\{k\})$, Ng_k and $Ng_{k'}$ must be different if $k \neq k'$, so $\bigcup_K N g_k$ is a union of pairwise disjoint sets. By Theorem 4.4, $|Ng_j| = |N|$ and it follows from Theorem 0.3 that $|\varphi^{-1}(K)| = \sum_K |Ng_k| = |N| \cdot |K|$, completing the proof. ■

In the theorem, the correspondence between the subgroups of B and those of G containing $N = ker\ \varphi$ is just the taking of images or of inverse images under φ. By Theorem 6.4, $\varphi^{-1}(b)$ for $b \in B$ is some coset Ng, and so $\varphi^{-1}(K)$ for $K \le B$ is a union of cosets. Using Theorem 6.6 to think of B as G/N, any $H \le G$ containing N is identified in G/N with the cosets of N in H. Similarly, if $K \le B$ then those cosets in G/N composing K are associated in G to the set of elements that are in those cosets. The subgroups in G and in G/N that are identified are in a sense the same subset of G looked at in two ways: as a collection of cosets of N, with coset multiplication; and as the elements in G in those cosets, with multiplication in G. This point of view will help to see what subgroup in G corresponds to the image $\varphi(H)$ of an arbitrary $H \le G$, and also to see that every $K \le B$ is essentially a quotient.

THEOREM 7.3 Let $\varphi : G \to B$ be an epimorphism of groups with $ker\ \varphi = N$. If $H \le G$ and $K \le B$, then $\varphi^{-1}(\varphi(H)) = NH$, $\varphi(H) \cong NH/N$, and $K \cong \varphi^{-1}(K)/N$.

Proof If $x \in \varphi(H)$ then $x = \varphi(h)$ for $h \in H$, so the group in G corresponding to $\varphi(H)$, $\beta(H)$ in Theorem 7.2, is just the union of all $\varphi^{-1}(\varphi(h)) = Nh$ by Theorem 6.4. Thus $\varphi^{-1}(\varphi(H)) = NH \le G$. When $N \le H$ then $NH = H$, consistent with Theorem 7.2 since in this case H and $\varphi(H)$ correspond. Since $NH \in Sub(G, N)$ and $\varphi(H)$ correspond, by Theorem 7.2 $\varphi(H) \cong NH/N$. For $K \le B$, apply Theorem 7.2 to get $\varphi(\varphi^{-1}(K)) = K$ and then part iv) of the theorem with $H = N$ and $L = \varphi^{-1}(K)$ to see that $\varphi^{-1}(K)/N \cong \varphi(\varphi^{-1}(K))/\varphi(N) = K/\langle e_B\rangle$. But $K/\langle e_B\rangle \cong K$ by Example 6.7, so $K \cong \varphi^{-1}(K)/N$. ■

Let us consider the special case of Theorem 7.2 when $N \lhd G$ and $\rho : G \to G/N$ is given by $\rho(g) = Ng$ as in Example 6.7. Here Theorem 7.2 tells us that the set of subgroups of G/N and the various relations between them are essentially the same as for the collection of subgroups of G that contain N. That is, taking the quotient G/N isolates the subgroups containing N and retains a faithful picture of their relationships but loses all information about the other subgroups of G. In Fig. 7.1 we illustrate this preservation of subgroup structure for $24\mathbf{Z} \le H \le (\mathbf{Z}, +)$ and $K \le \mathbf{Z}_n = \mathbf{Z}/n\mathbf{Z}$ for some $n \mid 24$. These subgroups are all cyclic and easily found. Here $\rho_n : \mathbf{Z} \to \mathbf{Z}_n$ is given by $\rho_n(k) = n\mathbf{Z} + k = [k]_n$ and if $k \mid 24$ then $\rho_n(\langle k\rangle) = \langle [k]_n\rangle$. Lines indicate containment.

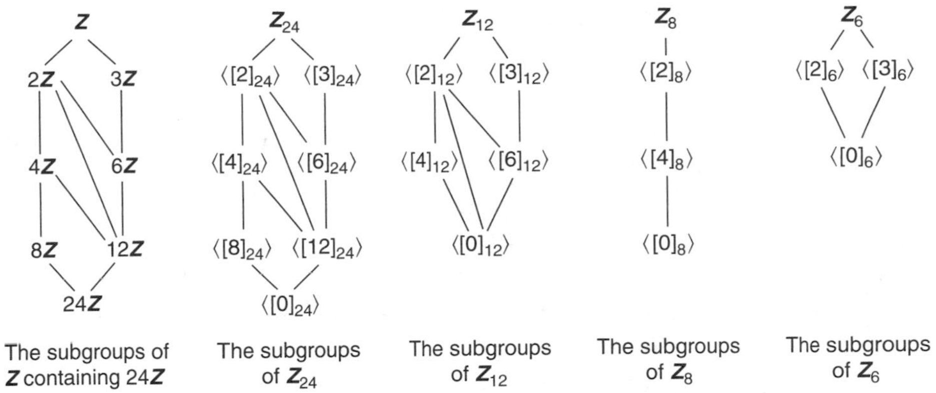

FIGURE 7.1

We can see clearly how all inclusions among subgroups of $\mathbf{Z}$ containing $24\mathbf{Z}$ are preserved in the quotients. For example, the diagram for $\mathbf{Z}_{12}$ is the same as that given for $\mathbf{Z}$ with $8\mathbf{Z}$ and $24\mathbf{Z}$ deleted, leaving only the subgroups containing $12\mathbf{Z}$, and the diagram for $\mathbf{Z}_8$ is the same as that part of the diagram for $\mathbf{Z}$ which deletes the subgroups not containing $8\mathbf{Z}: 3\mathbf{Z}, 6\mathbf{Z}, 12\mathbf{Z}$, and $24\mathbf{Z}$. If we start with $15\mathbf{Z} \leq \mathbf{Z}$ then using Theorem 7.2, $\rho_{24}(15\mathbf{Z})$ corresponds to $\rho_{24}^{-1}(\rho_{24}(15\mathbf{Z})) = 15\mathbf{Z} + 24\mathbf{Z} = (15, 24)\mathbf{Z} = 3\mathbf{Z}$ by Theorem 7.3 and Theorem 1.8 on GCDs. Hence $\rho_{24}(15\mathbf{Z}) = \rho_{24}(3\mathbf{Z}) = \langle [3]_{24}\rangle \leq \mathbf{Z}_{24}$. Note that since $(15, 8) = 1$, $\rho_8^{-1}\rho_8(15\mathbf{Z}) = 15\mathbf{Z} + 8\mathbf{Z} = \mathbf{Z}$ so $\rho_8(15\mathbf{Z}) = \rho_8(\mathbf{Z}) = \mathbf{Z}_8$.

The group D_6 is more interesting to consider since it has both normal and nonnormal subgroups. Using $D_6 = \{T_i, W_j \mid 0 \leq i, j \leq 5\}$ from Definition 3.10 we can compute all the subgroups of D_6 using Theorem 3.17. To do this observe that if $H \leq D_6$ with $W_i, W_j \in H$ then $T_{i-j} = W_i W_j \in H$. Furthermore, $H \cap \langle T_1\rangle \leq \langle T_1\rangle$ so must be either $\langle T_1\rangle$, $\langle T_2\rangle$, $\langle T_3\rangle$, or $\langle I\rangle$. The details of why the subgroups are those we describe are left as an exercise. Since $|D_6| = 12$, if $H \leq D_6$ then $|H| \mid 12$ by Lagrange's Theorem. The subgroups of order 6 are: $\langle T_1\rangle$; $\langle\{T_2, W_0\}\rangle = \{T_{2k}, W_{2j} \mid k, j \in \mathbf{Z}\}$; and $\langle\{T_2, W_1\}\rangle = \{T_{2k}, W_{2j+1} \mid k, j \in \mathbf{Z}\}$. If we call any of these N, then $[D_6 : N] = 2$ so $N \lhd D_6$ by Theorem 5.3. Any of these D_6/N has order 2 so is isomorphic to $\mathbf{Z}_2$ and has only itself and the identity $\langle N\rangle = \langle NT_0\rangle$ as subgroups.

The subgroups of order 4 in D_6 are: $\langle\{T_3, W_0\}\rangle = \{T_0, T_3, W_0, W_3\}$; $\langle\{T_3, W_1\}\rangle = \{T_0, T_3, W_1, W_4\}$; and $\langle\{T_3, W_2\}\rangle = \{T_0, T_3, W_2, W_5\}$, and none of these is normal since $T_1^{-1} W_j T_1 = W_{4+j}$. The only subgroup of order 3 is $\langle T_2\rangle$ and $\langle T_2\rangle \lhd D_6$ by Example 5.1. The subgroups of order 2 in D_6 are cyclic and generated by the elements of order 2, namely by T_3 and the six W_j. Now $\langle T_3\rangle = Z(D_6)$ so is normal in D_6 by Theorem 3.18, but none of the $\langle W_j\rangle$ is normal. We are interested in comparing the subgroups of D_6 with those of its quotients so we omit the $\langle W_j\rangle$ since they do not contain any other subgroup except $\langle I\rangle$, and none is normal. The remaining subgroups of D_6 are as shown in Fig. 7.2.

The quotient $D_6/\langle T_2\rangle$ has order $[D_6 : \langle T_2\rangle] = 12/3 = 4$ and to see its structure let us consider its subgroups, or, equivalently, those subgroups of D_6 containing $\langle T_2\rangle$. Observe that the three normal subgroups of order 6 in D_6 are precisely the subgroups containing $\langle T_2\rangle$ and if H is any of these then $[H : \langle T_2\rangle] = 6/3 = 2$. Thus $D_6/\langle T_2\rangle$ has order 4 and contains three

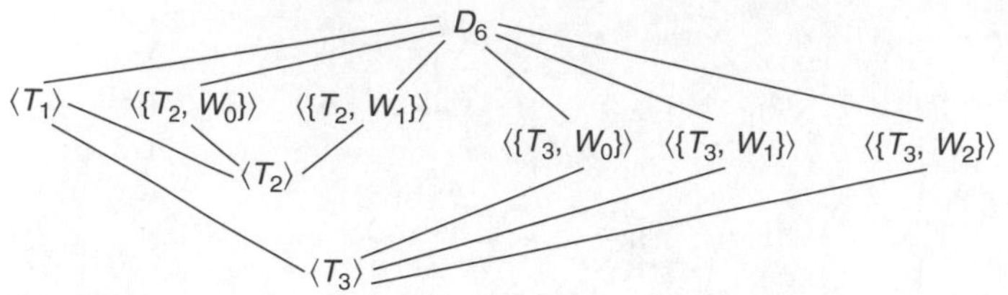

FIGURE 7.2

different subgroups of order 2, so $D_6/\langle T_2\rangle \cong \mathbf{Z}_2 \oplus \mathbf{Z}_2$ by Example 6.2 and Theorem 3.1 on cyclic groups. The diagram for the subgroups of D_6 containing $\langle T_2\rangle$ appears on the left in Fig. 7.3, and the diagram of subgroups for $D_6/\langle T_2\rangle$ appears on the right.

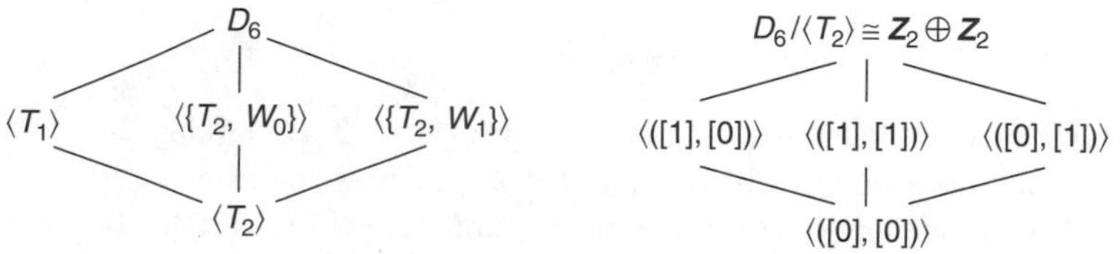

The subgroups of D_6 containing $\langle T_2\rangle$ The subgroups of $D_6/\langle T_2\rangle \cong \mathbf{Z}_2 \oplus \mathbf{Z}_2$

FIGURE 7.3

The diagram of subgroups of D_6 containing $\langle T_3\rangle$ is shown in Fig. 7.4.

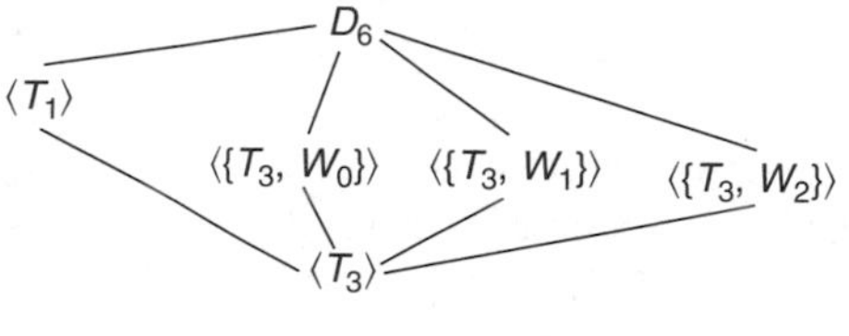

FIGURE 7.4

Here $\langle T_1\rangle$ is the only other normal subgroup and the remaining three subgroups have order 4 each. As usual, let $\rho : D_6 \to D_6/\langle T_3\rangle$ be $\rho(A) = \langle T_3\rangle A$. Each $\rho(\langle\{T_3, W_i\}\rangle) = \{\rho(I), \rho(W_i)\} = \langle\langle T_3\rangle W_i\rangle \leq D_6/\langle T_3\rangle$ has order 2. These will not be normal by Theorem 7.2 since no $\langle\{T_3, W_i\}\rangle$ is normal in D_6. Similarly, $\rho(\langle T_1\rangle) = \langle\langle T_3\rangle T_1\rangle \leq D_6/\langle T_3\rangle$ has order 3 and is normal in $D_6/\langle T_3\rangle$ since $\langle T_1\rangle \triangleleft D_6$. Now $D_6/\langle T_3\rangle = \{\rho(I), \rho(T_1), \rho(T_2), \rho(W_0), \rho(W_1), \rho(W_2)\}$ and the multiplication of these $\rho(T_i)$ and $\rho(W_j)$ mimics that of the T_i and W_j themselves, except that subscripts are reduced mod 3 rather than mod 6. For example $\rho(W_1)\rho(W_2) = \langle T_3\rangle W_1 \langle T_3\rangle W_2 = \langle T_3\rangle W_1 W_2 = \langle T_3\rangle T_5 = \langle T_3\rangle T_2 = \rho(T_2)$. The upshot is that by identifying $\rho(T_i)$ with $T_i \in D_3$ and $\rho(W_j)$ with $W_j \in D_3$ we see that $D_6/\langle T_3\rangle \cong D_3$. Observe that the only normal subgroup of D_3 is $\langle T_1\rangle$ and the other subgroups are $\langle W_0\rangle$, $\langle W_1\rangle$, and $\langle W_2\rangle$, each of order 2. These naturally correspond to the four proper subgroups of the diagram above, and the normal subgroups correspond as well.

The last paragraph shows in some detail how $D_6/\langle T_3\rangle$ is identified with D_3 by using the natural surjection of D_6 onto $D_6/\langle T_3\rangle$, Theorem 7.2, and the definition of coset multiplication in the quotient. There is a more elegant way of making this identification. From Theorem 7.2, $D_6/\langle T_3\rangle$ is a group of order 6 containing three subgroups of order 2. Since the quotient is not cyclic (Theorem 3.1) it is isomorphic to D_3 by Theorem 6.1.

We give a couple of typical applications of Theorem 7.2 showing how the theorem is used to transfer information from a quotient back to the group. We restrict our attention for now to Abelian groups since these will have easily found normal subgroups. The first result is the converse of Lagrange's Theorem for finite Abelian groups.

THEOREM 7.4 Let G be an Abelian group with $|G|=n$. For any $k \in \mathbf{N}$ with $k \mid n$, there is $H \leq G$ with $|H|=k$.

Proof Use induction on $|G|$ and note that there is nothing to prove if $|G|=1$. Assume $|G|>1$ and that the theorem holds for any Abelian group with order smaller than $|G|$. If $k=1$ take $H=\langle e\rangle$. When $k>1$ and $p \mid k$ for p a prime, Cauchy's Theorem (Theorem 5.9) yields $x \in G$ with $|\langle x\rangle|=N$ of order p and $N \lhd G$ by Theorem 5.1 since G is Abelian. From Example 6.7, $\rho: G \to G/N$ via $\rho(g)=Ng$ is an epimorphism with $ker\ \rho=N$. By Lagrange's Theorem, $|G/N|=n/p$, and G/N is Abelian by Theorem 5.7. The induction hypothesis guarantees some $K \leq G/N$ with $|K|=k/p$, since clearly $(k/p) \mid (n/p)$. Using Theorem 7.2, $\rho^{-1}(K) \leq G$ and $|\rho^{-1}(K)|=|N|\cdot|K|=p\cdot(k/p)=k$, so the theorem holds for G. Applying IND-II completes the argument. ■

The next result uses the preservation of containments between the subgroups of G and G/N. The previous theorem is a special case of it, but at this point the more examples that use Theorem 7.2, the better. The statement of the theorem refers to representing $n \in \mathbf{N}$ as *some* product of primes, for example $180=2\cdot2\cdot3\cdot3\cdot5$, or $3\cdot2\cdot3\cdot5\cdot2$, or $5\cdot2\cdot3\cdot2\cdot3$, etc.

THEOREM 7.5 Let G be an Abelian group, $|G|=n>1$, and $n=p_1p_2\cdots p_k$ for some primes p_j. There are $H_1 \leq H_2 \leq \cdots \leq H_k=G$ with $|H_j|=p_1p_2\cdots p_j$.

Proof There is $H_1 \leq G$ with $|H_1|=p_1$ by Theorem 5.9 (Cauchy), and G Abelian forces $H_1 \lhd G$ by Theorem 5.1. Now $|G/H_1|=p_2\cdots p_k$ by Lagrange's Theorem and G/H_1 is Abelian from Theorem 5.7 so by induction there are $L_2 \leq L_3 \leq \cdots L_k=G/H_1$ with $|L_j|=p_2\cdots p_j$. Apply the correspondence in Theorem 7.2: formally use the epimorphism $\rho: G \to G/H_1$ given by $\rho(g)=H_1g$. Since $ker\ \rho=H_1$, by Theorem 7.2 if $H_j=\rho^{-1}(L_j) \leq G$ then $H_1 \leq H_2 \leq \cdots \leq H_k=G$, since these must satisfy the same containment relations as $\{L_j\}$, and also $|H_j|=|H_1|\cdot|L_j|=p_1p_2\cdots p_j$ as required. ■

Subgroups of $\mathbf{Z}_n$ are easy to find so let us illustrate the theorem with $\mathbf{Z}_{60}$. Writing $60=2\cdot2\cdot3\cdot5$, the chain of subgroups is $\langle[30]_{60}\rangle \leq \langle[15]_{60}\rangle \leq \langle[5]_{60}\rangle \leq \mathbf{Z}_{60}$; for $60=3\cdot2\cdot5\cdot2$ the chain is $\langle[20]_{60}\rangle \leq \langle[10]_{60}\rangle \leq \langle[2]_{60}\rangle \leq \mathbf{Z}_{60}$.

Recall from Definition 5.3 that G *simple* means $K \lhd G \Rightarrow K=\langle e\rangle$ or $K=G$. If $N \lhd G$, when is G/N simple? By Theorem 7.2 the normal subgroups of G/N correspond to the normal subgroups of G containing N, so G/N is simple exactly when N is a proper, maximal, normal subgroup of G (*maximal* refers to containment). Thus Theorem 6.16 may be restated in terms of chains $N_1 \lhd N_2 \lhd \cdots \lhd N_k=G$ with N_{j+1}/N_j simple groups, as we indicated, and shows that the simple groups used to build G are uniquely determined by G and not the particular chain constructed.

Part of Theorem 5.10 on the commutator subgroup G' of G follows from Theorem 7.2, namely if $G' \leq H \leq G$ then $H \triangleleft G$. An easy computation with cosets in Theorem 5.10 shows G/G' is Abelian so by Theorem 5.1 any subgroup of G/G' is normal. But $H \leq G$ with $G' \leq H \leq G$ corresponds to a subgroup of G/G'. Thus Theorem 7.2 implies that $H \triangleleft G$.

One last observation using Theorem 7.2 is that if G is an infinite group having no infinite chain of different subgroups then there is no infinite chain of different subgroups in any quotient or homomorphic image of G (why?). As for the converse, if for every $\langle e\rangle \neq N \triangleleft G$ there is no infinite chain of distinct subgroups in G/N, then G has no infinite chain of distinct subgroups *containing a normal subgroup*. If $H_1 \leq H_2 \leq \cdots \leq H_k \leq \cdots \leq G$ with $H_j \triangleleft G$ then G/H_j contains the infinite chain $H_{j+1}/H_j \leq H_{j+2}/H_j \leq \cdots$. It is an exercise to see that S_∞ (Example 2.12) contains an infinite chain of subgroups but that its proper quotients are finite groups. Note that every proper quotient of $(\mathbf{Z}, +)$ is finite but $\mathbf{Z}$ has infinite descending chains of (normal) subgroups. One example would be $2\mathbf{Z} \geq 4\mathbf{Z} \geq \cdots \geq 2^n\mathbf{Z} \geq \cdots$.

EXERCISES

1. Let $\varphi: G \to B$ be a surjective homomorphism of groups, $H, K \leq G$ and $L, M \leq B$. Show that: i) $\varphi(HK) = \varphi(H)\varphi(K)$; and ii) $\varphi^{-1}(LM) = \varphi^{-1}(L)\varphi^{-1}(M)$.

2. Let $N \triangleleft G$, let $H, K \leq G$, and let $\rho: G \to G/N$ given by $\rho(g) = Ng$.
- **i)** If $\rho(H) \bigcap \rho(K) = e_{G/N}$, show that $H \bigcap K \leq N$.
- **ii)** If $H \bigcap K \leq N$, show by example that $\rho(H) \bigcap \rho(K) = e_{G/N}$ need not hold.

3. If $N, H, K \leq G$, $N \triangleleft G$, $N \leq H \bigcap K$, and $\rho: G \to G/N$ via $\rho(g) = Ng$, show that:
- **i)** $\rho(H) \bigcap \rho(K) = \rho(H \bigcap K)$
- **ii)** $\rho(H) \bigcap \rho(K) = e_{G/N} \Leftrightarrow H \bigcap K \leq N$
- **iii)** $\rho(H)\rho(K) = G/N \Leftrightarrow HK = G$

4. In each case, for $N \triangleleft G$ and $\rho: G \to G/N$ given by $\rho(g) = Ng$, identify each $K \leq G/N$, identify each $H \leq G$, and determine $\rho(H)$ for each $H \leq G$.
- **i)** $G = \mathbf{Z}_4 \oplus \mathbf{Z}_2$ and $N = \langle([2]_4, [1]_2)\rangle$
- **ii)** $G = D_5 \oplus \mathbf{Z}_2$ and $N = \langle T_1\rangle \oplus \langle[0]_2\rangle$

5. Find four different $H \leq G = \mathbf{Z}_2^3$, $|H| = 4$, and their images in G/N for:
- **i)** $N = \langle([1]_2, [0]_2, [0]_2)\rangle$
- **ii)** $N = \langle([1]_2, [1]_2, [0]_2)\rangle$

6. If $N \triangleleft G$ and $G/N \cong (\mathbf{Z}, +)$, show that whenever $H, K \leq G$ with $H \bigcap K = N$, then either $H = N$ or $K = N$.

7. In each case, if $N \triangleleft G$ find how many $H \leq G$ satisfy $N \leq H$, identify the H/N, and describe the different G/H when $H \triangleleft G$:
- **i)** $G/N \cong \mathbf{Z}_2 \oplus \mathbf{Z}_2$
- **ii)** $G/N \cong \mathbf{Z}_{30}$
- **iii)** $G/N \cong S_3$
- **iv)** $G/N \cong D_4$
- **v)** $G/N \cong S_3 \oplus \mathbf{Z}_5$
- **vi)** $G/N \cong D_{15}$
- **vii)** $G/N \cong D_6$

8. Let G be a finite group with $|G| = 90$, $N \lhd G$ with $|N| = 3$, and G/N Abelian.
 a) If $m \in \mathbf{N}$, $m \mid 90$, and $3 \mid m$, show that there is $H \leq G$ with $|H| = m$.
 b) If $m \in \mathbf{N}$ and $m \mid 90$, show that there is $K \leq G$ with $|K| = m$.

9. Given $N \lhd G$ find all $H \neq K$ with $N \leq H \leq K \leq G$ and $H, K \lhd G$. For all of these choices describe the K/H:
 i) $G/N \cong \mathbf{Z}_{45}$
 ii) $G/N \cong \mathbf{Z}_{46}$
 iii) $G/N \cong \mathbf{Z}_{49}$
 iv) $G/N \cong S_4$
 v) $G/N \cong S_5$
 vi) $G/N \cong D_4$
 vii) $G/N \cong D_5$
 viii) $G/N \cong D_6$
 ix) $G/N \cong GL(5, \mathbf{Z}_5)$
 x) $G/N \cong GL(3, \boldsymbol{R})$

10. Let G be a finite group with $A \lhd G$, A Abelian, and $G/A \cong A$. If $m \in \mathbf{N}$, $|A| \mid m$, and $m \mid |G|$, show that there is $H \leq G$ with $|H| = m$.

11. Let G be a group with $\langle x \rangle \lhd G$ and $o(x) = 7$. If $G/\langle x \rangle \cong D_5$, show that for each $m \in \mathbf{N}$ with $m \mid |G|$ there is $H \leq G$ with $|H| = m$. (See Example 6.18.)

12. Let G be a finite group so that every subgroup is normal.
 i) Show that for each prime divisor p of $|G|$, there is $x \in G$ with $o(x) = p$.
 ii) Show that for $m \in \mathbf{N}$ with $m \mid |G|$, there is $H_m \leq G$ with $|H_m| = m$.

13. Let a group G satisfy $(*)$ if there are $\langle e \rangle = H_0 \leq H_1 \leq \cdots \leq H_k = G$, each $H_i \lhd H_{i+1}$, and H_{i+1}/H_i is Abelian. Recall from Definition 2.13 that G' is the subgroup generated by all $x^{-1}y^{-1}xy$ for $x, y \in G$. Properties of G' are given in Theorem 5.10. Set $G^{(1)} = G'$, $G^{(2)} = (G')'$, and $G^{(s+1)} = (G^{(s)})'$.
 i) Prove that G satisfies $(*) \Leftrightarrow G^{(s)} = \langle e \rangle$ for some $s \geq 1$.
 ii) If G satisfies $(*)$ and $H \leq G$, show that H also satisfies $(*)$.
 iii) If G satisfies $(*)$ and $N \lhd G$, show G/N satisfies $(*)$.
 iv) If for some $N \lhd G$ both N and G/N satisfy $(*)$, show that G satisfies $(*)$.

14. Recall from Exercise 10 in §5.1 that $A_\infty \leq S_\infty$.
 i) Show that S_∞ has infinite chains of (different) subgroups.
 ii) Show that A_∞ is the only proper, normal, nonidentity subgroup of S_∞.
 iii) Show that S_∞/A_∞ has no infinite chain of (different) subgroups.

7.2 TWO ISOMORPHISM THEOREMS

Two standard consequences of Theorem 6.6 and Theorem 7.2 are the **Second** and **Third Isomorphism Theorems.** They are useful, mostly to prove other results. We state them next and then see how they can be used.

THEOREM 7.6 Let G be a group, $N, H, L \leq G$ with $N, L \lhd G$ and $N \leq L$. Then:

i) $NH/N \cong H/(H \cap N)$
ii) $(G/N)/(L/N) \cong G/L$

Proof Let $\rho : G \to G/N$ be the usual epimorphism $\rho(g) = Ng$. By Theorem 5.4, $NH \le G$ so $N \lhd NH$, and $H \cap N \lhd H$. Thus the quotients in i) make sense. From Theorem 7.3, $\rho^{-1}(\rho(H)) = NH$ and $\rho(H) = NH/N$. Clearly $\rho|_H : H \to \rho(H)(= NH/N)$ is an epimorphism and $ker\ \rho|_H = \{h \in H \mid Nh = N\} = \{h \in H \mid h \in N\} = H \cap N$, which shows again that $H \cap N \lhd H$ by Theorem 6.4. Thus $NH/N \cong H/(H \cap N)$ by Theorem 6.6. For ii), since $L \lhd G$, by Theorem 7.2 $\rho(L) = L/N \lhd G/N$ and $G/L \cong \rho(G)/\rho(L) = (G/N)/(L/N)$. ■

We use Theorem 7.6 to gain information about images of subgroups in quotients and about quotient groups of quotients. If $\rho : G \to G/N$ is the usual surjection with $\rho(g) = Ng$, then for $H \le G$, $\rho(H) = \{Nh \in G/N \mid h \in H\}$ and from Theorem 7.3, $\rho^{-1}(\rho(H)) = NH$. Also $NH/N = \rho(H)$ so part i) of Theorem 7.6 identifies $\rho(H) \le G/N$ as the quotient $H/(H \cap N)$ of H. We will see how this is used in arguments but a simple example is to take $G = \mathbf{Z}$ and $N = 6\mathbf{Z}$. Now in $\mathbf{Z}/6\mathbf{Z}$, $\rho(10\mathbf{Z}) = (10\mathbf{Z} + 6\mathbf{Z})/6\mathbf{Z} \cong 10\mathbf{Z}/(6\mathbf{Z} \cap 10\mathbf{Z}) = 10\mathbf{Z}/30\mathbf{Z} = \langle 10 + 30\mathbf{Z}\rangle$ in $\mathbf{Z}_{30}$, so $\rho(10\mathbf{Z}) \cong 10\mathbf{Z}/30\mathbf{Z} \cong \mathbf{Z}_3$.

Continuing with $\rho : G \to G/N$, when $N \le H$ then $\rho(H) = H/N$ so when $H \lhd G$ the theorem identifies the quotient of G/N by H/N as G/H. In Example 6.14 we saw that $\mathbf{Z}_n/\langle [m]_n\rangle \cong \mathbf{Z}_m$. Now $\mathbf{Z}_n = \mathbf{Z}/n\mathbf{Z}$ and when $m \mid n$, $\langle [m]_n\rangle = m\mathbf{Z}/n\mathbf{Z}$ so this observation is just $(\mathbf{Z}/n\mathbf{Z})/(m\mathbf{Z}/n\mathbf{Z}) \cong \mathbf{Z}/m\mathbf{Z} = \mathbf{Z}_m$. Also $\langle T_1\rangle, \langle T_3\rangle \lhd D_6$ by Example 5.1 so $\langle T_1\rangle/\langle T_3\rangle \lhd D_6/\langle T_3\rangle$ by Theorem 7.2. Thus Theorem 7.6 shows $(D_6/\langle T_3\rangle)/(\langle T_1\rangle/\langle T_3\rangle) \cong D_6/\langle T_1\rangle \cong \mathbf{Z}_2$ as $|D_6/\langle T_1\rangle| = 2$.

Another application of Theorem 7.6 uses $\mathbf{Z}^2/(2\mathbf{Z} \oplus 3\mathbf{Z}) \cong \mathbf{Z}_6$ from Example 5.8 and Theorem 6.8. The image of $\langle(1, 1)\rangle = \{(k, k) \in \mathbf{Z}^2 \mid k \in \mathbf{Z}\}$ in this quotient is $(\langle(1, 1)\rangle + (2\mathbf{Z} \oplus 3\mathbf{Z}))/(2\mathbf{Z} \oplus 3\mathbf{Z})$ by Theorem 7.3, and from Theorem 7.6 this is isomorphic to $\langle(1, 1)\rangle/(\langle(1, 1)\rangle \cap (2\mathbf{Z} \oplus 3\mathbf{Z}))$. Now $(x, y) \in \langle(1, 1)\rangle \cap (2\mathbf{Z} \oplus 3\mathbf{Z}) \Leftrightarrow x = y, 2 \mid x$, and $3 \mid y$. It follows from Theorem 1.12 that $(x, y) = (6k, 6k)$ for some $k \in \mathbf{Z}$, so $\langle(1, 1)\rangle \cap (2\mathbf{Z} \oplus 3\mathbf{Z}) = \langle(1, 1)^6\rangle = \{(6k, 6k) \in \mathbf{Z}^2 \mid k \in \mathbf{Z}\}$ (why?). It is an exercise to see that $\langle(1, 1)\rangle/\langle(1, 1)^6\rangle = \langle(1, 1) + \langle(1, 1)^6\rangle\rangle$ has order 6, so in fact $(\langle(1, 1)\rangle + (2\mathbf{Z} \oplus 3\mathbf{Z}))/(2\mathbf{Z} \oplus 3\mathbf{Z}) = \mathbf{Z}^2/(2\mathbf{Z} \oplus 3\mathbf{Z})$ since both groups have order 6 and one is a subgroup of the other. Thus $\mathbf{Z}^2 = \langle(1, 1)\rangle + (2\mathbf{Z} \oplus 3\mathbf{Z})$ follows from Theorem 7.2: both groups correspond to the same subgroup of $\mathbf{Z}^2/(2\mathbf{Z} \oplus 3\mathbf{Z})$. Next consider $G = (2\mathbf{Z} \oplus 3\mathbf{Z})/\langle(1, 1)^6\rangle$. To see what G is, we could find a transversal and examine the addition of its elements, but we instead use Theorem 7.6 again to avoid these computations. From above, $\langle(1, 1)^6\rangle = \langle(1, 1)\rangle \cap (2\mathbf{Z} \oplus 3\mathbf{Z})$ so $G = (2\mathbf{Z} \oplus 3\mathbf{Z})/(\langle(1, 1)\rangle \cap (2\mathbf{Z} \oplus 3\mathbf{Z})) \cong (\langle(1, 1)\rangle + (2\mathbf{Z} \oplus 3\mathbf{Z}))/\langle(1, 1)\rangle$ by Theorem 7.6 so $G \cong \mathbf{Z}^2/\langle(1, 1)\rangle \cong \mathbf{Z}$ by Example 6.15; G is an infinite cyclic group.

A couple of consequences of Theorem 7.6 will show how quotients are used to gain information about subgroups or about the group itself.

Example 7.1 If $H \le G$ and $G/Z(G)$ is Abelian then $H/Z(H)$ is Abelian.

From Theorem 5.1, $Z(G) \lhd G$ and $Z(G)H \le G$ using Theorem 4.11. Hence $Z(G)H/Z(G) \le G/Z(G)$ is Abelian by assumption. Theorem 7.6 shows that $Z(G)H/Z(G) \cong H/(H \cap Z(G))$ so $H/(H \cap Z(G))$ is Abelian. Note that $H \cap Z(G) \le Z(H)$ and that $H \cap Z(G), Z(H) \lhd H$ by Theorem 5.1, so $H/Z(H) \cong (H/(H \cap Z(G))/(Z(H)/(H \cap Z(G))$ using Theorem 7.6. Thus $H/Z(H)$ is isomorphic to a quotient of the Abelian group $H/(H \cap Z(G))$, so is Abelian by Theorem 5.7.

Before our next result we need a useful observation. Let $N, H \le G$ with $N \lhd G$ and $\rho : G \to G/N$ the usual epimorphism. If $H \cap N = \langle e \rangle$, then $\rho(H) = NH/N \cong H/(N \cap H) = H/\langle e \rangle \cong H$ (why?). Thus if G is finite, $H \le G$ with $(|H|, |N|) = 1$, and $|N||H| = |G|$, then $H \cap N = \langle e \rangle$ by Theorem 4.6 so $NH = G$ by Theorem 4.12, and $G/N \cong H$.

THEOREM 7.7 Let G be an Abelian group with $|G| = p^n$ for p a prime. If G/H is cyclic for every $\langle e \rangle \ne H \le G$ then either $|G| = p^2$ or G is cyclic.

Proof Observe that if $|G| = p$ then G is cyclic by Theorem 4.9. Also if $G = \mathbf{Z}_p \oplus \mathbf{Z}_p$ then $|G| = p^2$ and for each proper quotient, $|G/N|$ is 1 or p so is cyclic, but G is not. Hence we now assume $|G| = p^n$ for $n \ge 3$ and must show that G is cyclic. If $g \in G - \langle e \rangle$, $o(g) = p^m$ by Theorem 4.7 and then $o(g^{p^{m-1}}) = p$ using Theorem 2.8. Thus for $x = g^{p^{m-1}}$, $o(x) = p$, $|\langle x \rangle| = p$ (Theorem 2.16), and by assumption $G/\langle x \rangle$ is cyclic so $G/\langle x \rangle = \langle \langle x \rangle y \rangle$ for some $y \in G$. Since by Lagrange's Theorem $|G/\langle x \rangle| = p^{n-1}$, $o(\langle x \rangle y) = p^{n-1}$ in $G/\langle x \rangle$ by Theorem 2.16 and this implies $y^{p^{n-1}} \in \langle x \rangle$ and p^{n-1} is minimal in $\{k \in \mathbf{N} \mid y^k \in \langle x \rangle\}$. If $y^{p^{n-1}} \ne e$, then since $o(y) \mid p^n (= |G|)$, it follows that $o(y) = p^n$ and $G = \langle y \rangle$. Hence we assume $y^{p^{n-1}} = e$, and the minimality of p^{n-1} forces $o(y) = p^{n-1}$. For $1 \le t < p^{n-1}$, $y^t = x^j$ would contradict the minimality of $y^{p^{n-1}} \in \langle x \rangle$, which means that $\langle x \rangle \cap \langle y \rangle = \langle e \rangle$. Note that $\langle y^p \rangle \lhd G$ and $G/\langle y^p \rangle$ is cyclic since $n \ge 3$ means $y^p \ne e$. Let $\eta : G \to G/\langle y^p \rangle$ be the usual homomorphism and observe that $\eta(\langle y \rangle) = \langle y \rangle / \langle y^p \rangle \le G/\langle y^p \rangle$ of order p. Now using Theorem 7.3, $\eta(\langle x \rangle) = \langle y^p \rangle \langle x \rangle / \langle y^p \rangle$ so $\eta(\langle x \rangle) \cong \langle x \rangle / (\langle y^p \rangle \cap \langle x \rangle) \cong \langle x \rangle$ by both Theorem 7.6 and our observation that $\langle x \rangle \cap \langle y \rangle = \langle e \rangle$. Thus $\eta(\langle x \rangle)$ is another subgroup in $G/\langle y^p \rangle$ of order p, impossible in a cyclic group unless $\eta(\langle x \rangle) = \eta(\langle y \rangle)$, by Theorem 3.1. But in this case Theorem 7.3 and the correspondence of Theorem 7.2 show $\langle y^p \rangle \langle x \rangle = \eta^{-1}(\eta(\langle x \rangle)) = \eta^{-1}(\eta(\langle y \rangle)) = \langle y \rangle$, forcing $x \in \langle y \rangle$, a contradiction. Therefore $y^{p^{n-1}} = e$ is impossible, so indeed $y^{p^{n-1}} = x^j \ne e$, $o(y) = p^n$, and it follows that $G = \langle y \rangle$ is cyclic. ∎

EXERCISES

1. If $N \lhd M \le G$ and $H \le G$, show that:
   a) M/N Abelian $\Rightarrow (M \cap H)/(N \cap H)$ is Abelian
   b) M/N cyclic $\Rightarrow (M \cap H)/(N \cap H)$ is cyclic
2. Let $\varphi : G \to B$ be a homomorphism of groups with $\ker \varphi = N$. If $H \lhd G$ and $H \le N$, show that there is a surjective homomorphism $\theta : G/H \to \varphi(G)$.
3. If G is a group, $H \le Z(G)$, and G/H is cyclic, use Theorem 7.6 to show that G is Abelian.
4. If $H \le G$, G is finite, and $[G : H] = p$, the smallest prime $p \mid |G|$, then $H \lhd G$.
5. Suppose $N \lhd G$ with both N and G/N infinite cyclic groups. Show that for any $H \le G$ with $H \ne \langle e \rangle$, either H is infinite cyclic or for some $K \lhd H$ both K and H/K are infinite cyclic.
6. If $|G/Z(G)| = p^k$ for p a prime, show that for any $H \le G$, $|H/Z(H)| = p^s$.
7. Let a finite group G satisfy condition (#): $Z(G) \ne \langle e \rangle$ and $N \lhd G$ with $N \ne G$ implies $Z(G/N) \ne \langle e_{G/N} \rangle$.
   i) If $K \lhd G$ with $K \ne G$, show that G/K satisfies (#).
   ii) Find $\langle e \rangle = N_0 \le N_1 \le \cdots \le N_s = G$ with each $N_j \lhd G$, and $N_{i+1}/N_i = Z(G/N_i)$.

iii) If $\langle e\rangle \neq H \leq G$, show that $Z(H) \neq \langle e\rangle$.
iv) If $p \mid |G|$ for p a prime, then there is $x \in G$ with $o(x) = p$.
v) If $\langle e\rangle \neq M \lhd G$, show that $M \cap Z(G) \neq \langle e\rangle$. (Use induction on s in part ii).)

7.3 DIRECT SUM DECOMPOSITIONS

From Definition 3.11 the direct sum of the groups $G_1, \ldots, G_k$ is the Cartesian product $G_1 \times \cdots \times G_k$ with coordinatewise multiplication, denoted $G_1 \oplus \cdots \oplus G_k$ or $\bigoplus G_i$. This construction easily relates properties of $\bigoplus G_i$ to those of the G_i so it is desirable to view a given group G as a direct sum. Of course G is not likely to be the set of k-tuples of other groups, but we can hope that G is isomorphic to a direct sum. How can this happen? In $\bigoplus G_i$ if $H_j = \langle e_1\rangle \oplus \cdots \oplus G_j \oplus \cdots \oplus \langle e_k\rangle$, the k-tuples with arbitrary $g \in G_j$ in the jth coordinate and the identity $e_i \in G_i$ otherwise, then $H_j \lhd \bigoplus G_i$ (why?); $\{H_j\}$ generates $\bigoplus G_i$ as a group (exercise!), and $H_i \cap H_s = \langle$the identity of $\bigoplus G_i\rangle$ when $i \neq s$. If $G \cong G_1 \oplus \cdots \oplus G_k$ it follows from Theorem 7.2 that the subgroups in G corresponding to the H_j are normal in G, intersect pairwise in $\{e\}$, and generate G. These conditions on a collection of subgroups of G are a guide to the properties needed for G to look like their direct sum.

THEOREM 7.8 If $G = N_1 \cdots N_k$ for $N_i \lhd G$ and if $N_1N_2 \cdots N_{j-1} \cap N_j = \langle e\rangle$ when $2 \leq j$, then $G \cong N_1 \oplus \cdots \oplus N_k$. Conversely, if $G \cong G_1 \oplus \cdots \oplus G_k$, there are $N_1, \ldots, N_k \lhd G$, $N_i \cong G_i$, $G = N_1 \cdots N_k$, and $N_1N_2 \cdots N_{j-1} \cap N_j = \langle e\rangle$ for each $2 \leq j$.

Proof Suppose some $\{N_j \lhd G\}$ has the properties described. We must find an isomorphism $\varphi : G \to N_1 \oplus \cdots \oplus N_k$, so we need some way to identify $g \in G$ with some $x \in N_1 \times \cdots \times N_k$. By assumption $G = N_1 \cdots N_k$ so any $g \in G$ has the expression $g = x_1 \cdots x_k$ with $x_i \in N_i$. It is tempting to try the identification $\varphi(g) = (x_1, x_2, \ldots, x_k)$, but if also $g = y_1 \cdots y_k$ with $y_j \in N_j$ then φ may not be a function if some $x_i \neq y_i$. We prove that the assumption on intersections of $\{N_i\}$ forces $g = x_1 \cdots x_k$ for a unique $\{x_i\}$. Therefore φ will be a function, and it will be easy to see that φ is the required isomorphism.

For $i < j$, from $N_1 \cdots N_{j-1} \cap N_j = \langle e\rangle$ and $N_i \leq N_1 \cdots N_iN_{i+1} \cdots N_{j-1}$ we have $N_i \cap N_j = \langle e\rangle$. If $x \in N_i$, $y \in N_j$, and $h = x^{-1}y^{-1}xy$, then $N_i, N_j \lhd G$ imply $h = x^{-1}(y^{-1}xy) = (x^{-1}y^{-1}x)y \in N_i \cap N_j = \langle e\rangle$, forcing $xy = yx$, so elements in different N_i commute. Suppose $g \in G$ has two expressions $g = x_1 \cdots x_k = y_1 \cdots y_k$ with $x_i, y_i \in N_i$. We want $x_i = y_i$ for all i. If some $x_i \neq y_i$ let $m \geq 1$ be the largest such subscript, so $x_j = y_j$ for $m < j$. Now $g(x_{m+1} \cdots x_k)^{-1} = x_1 \cdots x_m = y_1 \cdots y_m$, (why?), and $x_m y_m^{-1} = (x_1 \cdots x_{m-1})^{-1}(y_1 \cdots y_{m-1}) = x_{m-1}^{-1} \cdots x_1^{-1} y_1 \cdots y_{m-1}$ follows. Since elements in different N_i commute, x_2^{-1} commutes with $(x_1^{-1}y_1)$, x_3^{-1} commutes with $(x_1^{-1}y_1)(x_2^{-1}y_2)$, etc., so $x_m y_m^{-1} = (x_1^{-1}y_1) \cdots (x_{m-1}^{-1}y_{m-1}) \in N_1N_2 \cdots N_{m-1} \cap N_m = \langle e\rangle$. Hence $x_m = y_m$, a contradiction. We are forced to conclude that $x_i = y_i$ for all $1 \leq i \leq k$, proving the uniqueness of the representation $g = x_1 \cdots x_k$ with $x_i \in N_i$.

Define $\varphi : G \to N_1 \oplus \cdots \oplus N_k$ by $\varphi(g) = (x_1, \ldots, x_k)$ when $g = x_1 \cdots x_k$ for $x_i \in N_i$. The expression $g = x_1 \cdots x_k$ with $x_i \in N_i$ is unique so φ is a function. For $(x_1, \ldots, x_k) \in \bigoplus N_i$, $(x_1, \ldots, x_k) = \varphi(x_1 \cdots x_k)$, and φ is surjective. If $\varphi(g) = \varphi(h)$ for $g, h \in G$, $g = x_1 \cdots x_k$, $h = y_1 \cdots y_k$, and $x_i, y_i \in N_i$, then $(x_1, \ldots, x_k) = \varphi(g) = \varphi(h) = (y_1, \ldots, y_k)$ forces $x_i = y_i$ for each i, so $g = h$ and φ is injective, so a bijection. To see that φ is a homomorphism, take $g, h \in G$ as above and note that $gh = (x_1 \cdots x_k)(y_1 \cdots y_k) = (x_1y_1) \cdots$

$(x_k y_k)$ with $x_j y_j \in N_j$, using the fact that elements in different N_i commute. This is the unique expression of gh as a product of elements in the N_i so in $(\bigoplus N_i, \cdot)$ $\varphi(g)\varphi(h) = (x_1, \ldots, x_k)(y_1, \ldots, y_k) = (x_1 y_1, \ldots, x_k y_k) = \varphi(gh)$ and φ is a homomorphism.

Finally, if $G \cong G_1 \oplus \cdots \oplus G_k$, let $H_i = \{(e_1, \ldots, e_{i-1}, g_i, e_{i+1}, \ldots, e_k) \in \bigoplus G_i \mid g_i \in G_i\}$ where $e_j \in G_j$ is the identity. It is an exercise that $H_j \lhd \bigoplus G_i$ and that $\eta_j : G_j \to H_j$ via $\eta_j(y_j) = (e_1, \ldots, e_{j-1}, y_j, e_{j+1}, \ldots, e_k)$ is a group isomorphism, so $G_j \cong H_j$. Let $\theta : \bigoplus G_i \to G$ be an isomorphism and set $\theta(H_i) = N_i$. By Theorem 7.1, $N_i \lhd G$, $G = \theta(\bigoplus G_i) = \theta(H_1 \cdots H_k) = N_1 \cdots N_k$, and similarly, $N_1 \cdots N_{j-1} \bigcap N_j = \langle e_G \rangle$ for each $2 \leq j \leq k$ ($\{H_i\}$ satisfies the same conditions). Now $G_i \cong N_i$ since $G_i \cong H_i$ via η_i, and $H_i \cong N_i$ via θ (restricted to H_i), proving the theorem. ■

The special case of Theorem 7.8 for two normal subgroups is sufficiently important to warrant its own statement.

THEOREM 7.9 If $H, K \lhd G$, $HK = G$, and $H \bigcap K = \langle e \rangle$, then $G \cong H \oplus K$.

When the hypothesis of Theorem 7.8 holds, we say G is **the direct sum of its normal subgroups** $\{N_i\}$, although G is only *isomorphic* to $\bigoplus N_i$. Some refer to this situation as G being the *internal* direct sum of the $\{N_i\}$. The important feature of the direct sum representation is what it might tell us about properties of G; whether or not the elements of G are k-tuples is not often important. Usually a given G is not a set of k-tuples, so saying that G is the direct sum of its subgroups $\{N_i\}$ should not cause any confusion.

We observed in Theorem 7.8 that if $N_1 \cdots N_{j-1} \bigcap N_j = \langle e \rangle$ when $j \geq 2$ then $N_i \bigcap N_j = \langle e \rangle$ for $i \neq j$. These statements are *not* equivalent when $k > 2$: the one in the theorem is required. A standard example is $G = \mathbf{Z}_2 \oplus \mathbf{Z}_2$ with normal subgroups $N_1 = \langle([1], [0])\rangle$, $N_2 = \langle([1], [1])\rangle$, and $N_3 = \langle([0], [1])\rangle$. Each subgroup is just $([0], [0])$ and its generator so $N_i \bigcap N_j = \langle e \rangle$ when $i \neq j$. It follows from Theorem 4.12 that $|N_i N_j| = 4$ so $G = N_i N_j$ when $i \neq j$ and by Theorem 7.9, $G \cong N_i \oplus N_j$. Now $G = N_1 N_2 N_3$ but even though $N_i \bigcap N_j = \langle e \rangle$ when $i \neq j$, $G \cong N_1 \oplus N_2 \oplus N_3$ is impossible since the latter group has eight elements. This does not contradict Theorem 7.8 since $N_1 N_2 \bigcap N_3 = N_3 \neq \langle e \rangle$.

Before looking at examples, we record some useful properties of direct sums, mostly for two groups but illustrating the general approach.

THEOREM 7.10 For A, B, C, D, E groups, $A \cong C$ and $B \cong D$ imply $A \oplus B \cong B \oplus A$, $A \oplus B \cong C \oplus D$, and $A \oplus B \oplus E \cong (A \oplus B) \oplus E \cong A \oplus (B \oplus E)$.

Proof For the first assertion, define $\varphi : A \oplus B \to B \oplus A$ by $\varphi((a, b)) = (b, a)$ so $\varphi((a, b)(a', b')) = \varphi((aa', bb')) = (bb', aa') = (b, a)(b', a') = \varphi((a, b))\varphi((a', b'))$, and φ is a group homomorphism. It is straightforward to see that φ is a bijection. Now let $f : A \to C$ and $g : B \to D$ be isomorphisms and define $\eta : A \oplus B \to C \oplus D$ by $\eta((a, b)) = (f(a), g(b))$. Again a simple computation (verify!) shows that η is an isomorphism. The maps $F((a, b, x)) = ((a, b), x)$ and $G(((a, b), x)) = (a, (b, x))$ are isomorphisms from $A \oplus B \oplus E$ to $(A \oplus B) \oplus E$ and from $(A \oplus B) \oplus E$ to $A \oplus (B \oplus E)$, respectively. The verifications are left as exercises. ■

The following examples illustrate Theorem 7.8 or Theorem 7.9, although they could be obtained differently. In particular, the first two examples are immediate consequences of the corollary to Theorem 6.9.

Example 7.2 $\mathbf{Z}_{12} \cong \mathbf{Z}_3 \oplus \mathbf{Z}_4$.

By direct calculation or Theorem 3.1, $H = \langle [4]_{12} \rangle \leq \mathbf{Z}_{12}$ is cyclic of order 3 and $K = \langle [3]_{12} \rangle \leq \mathbf{Z}_{12}$ has order 4. Hence by Theorem 4.6, $H \cap K = \langle [0]_{12} \rangle$ and then $H + K = \mathbf{Z}_{12}$ by Theorem 4.12. Both $H, K \lhd \mathbf{Z}_{12}$ since $\mathbf{Z}_{12}$ is Abelian, so $\mathbf{Z}_{12} \cong H \oplus K$ by Theorem 7.9. Since H and K are cyclic, from Theorem 6.8 $H \cong \mathbf{Z}_3$ and $K \cong \mathbf{Z}_4$, so $\mathbf{Z}_{12} \cong \mathbf{Z}_3 \oplus \mathbf{Z}_4$ by Theorem 7.10. To illustrate the second part of Theorem 7.8, $f(([a]_3, [b]_4)) = [4a + 3b]_{12}$ is an isomorphism from $\mathbf{Z}_3 \oplus \mathbf{Z}_4$ to $\mathbf{Z}_{12}$ (exercise!). Thus by Theorem 7.8, $\mathbf{Z}_{12}$ will be the direct sum of $f(\langle([1]_3[0]_4)\rangle) = \langle [4]_{12} \rangle$ and $f(\langle([0]_3[1]_4)\rangle) = \langle [3]_{12} \rangle$.

Example 7.3 $\mathbf{Z}_{30} \cong \mathbf{Z}_5 \oplus \mathbf{Z}_6$, $\mathbf{Z}_{30} \cong \mathbf{Z}_3 \oplus \mathbf{Z}_{10}$, and $\mathbf{Z}_{30} \cong \mathbf{Z}_2 \oplus \mathbf{Z}_3 \oplus \mathbf{Z}_5$.

If $S = \langle [6] \rangle$, $T = \langle [5] \rangle \leq \mathbf{Z}_{30}$, then $\mathbf{Z}_{30} \cong \mathbf{Z}_5 \oplus \mathbf{Z}_6$ by proceeding as in Example 7.2: $S \cong \mathbf{Z}_5$, $T \cong \mathbf{Z}_6$, $S \cap T = \langle [0] \rangle$, $S + T = \mathbf{Z}_{30}$, and $S, T \lhd \mathbf{Z}_{30}$. Similarly, in $\mathbf{Z}_{30}$ if $U = \langle [10] \rangle$ and $V = \langle [3] \rangle$ then we get $\mathbf{Z}_{30} \cong \mathbf{Z}_3 \oplus \mathbf{Z}_{10}$. For the last statement, observe that $\mathbf{Z}_6 \cong \mathbf{Z}_2 \oplus \mathbf{Z}_3$ as above by considering $\langle [3]_6 \rangle, \langle [2]_6 \rangle \leq \mathbf{Z}_6$, and then use Theorem 7.10 to get $\mathbf{Z}_{30} \cong \mathbf{Z}_5 \oplus \mathbf{Z}_6 \cong \mathbf{Z}_5 \oplus \mathbf{Z}_2 \oplus \mathbf{Z}_3$ and then to get $\mathbf{Z}_{30} \cong \mathbf{Z}_2 \oplus \mathbf{Z}_3 \oplus \mathbf{Z}_5$. Another approach takes $H = \langle [15] \rangle$, $K = \langle [10] \rangle$, $L = \langle [6] \rangle \leq \mathbf{Z}_{30}$. Direct calculation shows that $H = \{[0], [15]\}$ and $K = \{[0], [10], [20]\}$ intersect in $\langle [0] \rangle$ and then that $H + K = \langle [5] \rangle$ intersects L in $\langle [0] \rangle$. Furthermore, from above $\mathbf{Z}_{30} = \langle [5] \rangle + \langle [6] \rangle = H + K + L$, so by using Theorem 7.8 and Theorem 7.10, $\mathbf{Z}_{30} \cong H \oplus K \oplus L \cong \mathbf{Z}_2 \oplus \mathbf{Z}_3 \oplus \mathbf{Z}_5$.

In the last two examples we really need isomorphism since $\mathbf{Z}_n$ is *not* a Cartesian product of other $\mathbf{Z}_m$. Also note that a group may be isomorphic to different direct sums. Using Theorem 7.9 we restate some observations for non-Abelian groups from Chapter 3 that are now easier to obtain.

Example 7.4 $(\mathbf{R}^*, \cdot) \cong \mathbf{R}^+ \oplus \mathbf{Z}_2$.

Now $\mathbf{R}^+ \leq \mathbf{R}^*$, $\langle -1 \rangle \leq \mathbf{R}^*$, and it is clear that $\mathbf{R}^+ \cap \langle -1 \rangle = \langle 1 \rangle$. If $r \in \mathbf{R}^*$ then $r = |r|(r/|r|) \in \mathbf{R}^+ \cdot \langle -1 \rangle$ so $\mathbf{R}^* = \mathbf{R}^+ \cdot \langle -1 \rangle$. Since $\mathbf{R}^*$ is Abelian all of its subgroups are normal. Thus $\mathbf{R}^+ \oplus \langle -1 \rangle \cong \mathbf{R}^*$ by Theorem 7.9 and $\langle -1 \rangle \cong \mathbf{Z}_2$ from Theorem 6.8 so we can write $\mathbf{R}^* \cong \mathbf{R}^+ \oplus \mathbf{Z}_2$ using Theorem 7.10.

Example 7.5 $GL(3, \mathbf{R}) \cong SL(3, \mathbf{R}) \oplus \mathbf{R}^*$.

Using $\det(XY) = \det(X)\det(Y)$ and $\det(Diag(r_1, r_2, r_3)) = r_1 r_2 r_3$, if $A \in GL(3, \mathbf{R})$ then $A \cdot ((\det(A))^{-1/3} I_3) \in SL(3, \mathbf{R})$. Hence $SL(3, \mathbf{R})(\mathbf{R}^* \cdot I_3) = GL(3, \mathbf{R})$, and $rI_3 \in SL(3, \mathbf{R}) \cap \mathbf{R}^* \cdot I_3$ forces $r^3 = 1$, so $r = 1$. Thus $SL(3, \mathbf{R}) \cap \mathbf{R}^* \cdot I_3 = \langle I_3 \rangle$.

From Example 5.2, $SL(3, \boldsymbol{R}) \lhd GL(3, \boldsymbol{R})$ and from our discussion in Chapter 5, $\boldsymbol{R}^* \cdot I_3$ is central, so normal in $GL(3, \boldsymbol{R})$ by Theorem 5.1. Therefore applying Theorem 7.9 we get $GL(3, \boldsymbol{R}) \cong SL(3, \boldsymbol{R}) \oplus \boldsymbol{R}^* \cdot I_3$. It is apparent that $\boldsymbol{R}^* \cdot I_3 \cong (\boldsymbol{R}^*, \cdot)$ via $\varphi(r \cdot I_3) = r$, so Theorem 7.10 shows that $GL(3, \boldsymbol{R}) \cong SL(3, \boldsymbol{R}) \oplus \boldsymbol{R}^*$.

Example 7.6 $G = \{(a, b, c, d) \in \boldsymbol{Z}^4 \mid a + b + c + d = 0\} \cong \boldsymbol{Z}^3$.

It is straightforward to see that $G \leq \boldsymbol{Z}^4$, or observe that $G = ker\ \varphi$ for the homomorphism $\varphi : \boldsymbol{Z}^4 \to \boldsymbol{Z}$ given by $\varphi((a, b, c, d)) = a + b + c + d$. We claim $H = \{(a, -a, 0, 0) \in G \mid a \in \boldsymbol{Z}\}$, $K = \{(0, b, -b, 0) \in G \mid b \in \boldsymbol{Z}\}$, and $L = \{(0, 0, c, -c) \in G \mid c \in \boldsymbol{Z}\}$ are subgroups of G (Theorem 2.12!) satisfying the conditions in Theorem 7.8. Note that $H, K, L \lhd G$ by Theorem 5.1: G is Abelian. Observe that in G, $(a, b, c, d) = (a, -a, 0, 0) + (0, a + b, -a - b, 0) + (0, 0, -d, d)$, using that $-d = a + b + c$ from the definition of G. This shows $G = H + K + L$ (why?). The definitions of H and of K clearly force $H \cap K = \langle e_G \rangle$. A typical $\alpha \in H + K$ is $\alpha = (a, -a, 0, 0) + (0, b, -b, 0) = (a, b - a, -b, 0)$, and it follows that $(H + K) \cap L = \langle e_G \rangle$. Therefore Theorem 7.8 yields $G \cong H \oplus K \oplus L$. Finally, $H = \langle (1, -1, 0, 0) \rangle$, $K = \langle (0, 1, -1, 0) \rangle$, $L = \langle (0, 0, 1, -1) \rangle$, and each is infinite, forcing each to be isomorphic to $\boldsymbol{Z}$ by Theorem 6.8. Therefore $G \cong \boldsymbol{Z}^3$ using Theorem 7.10. Observe that G itself is not $\boldsymbol{Z}^3$ and that this representation of G is not immediately obvious from its definition.

It is desirable, but not always possible, to represent a group G as a direct sum of subgroups other than as $G \oplus \langle e \rangle$. If $G = A_4$, $H = \{I, (1, 2)(3, 4), (1, 3)(2, 4), (1, 4)(2, 3)\}$ and $K = \langle (1, 2, 3) \rangle$, then $H \cap K = \langle I \rangle$ and $HK = A_4$ using Theorem 4.12 on orders of products, but A_4 *cannot* be isomorphic to $H \oplus K$, an Abelian group. This does not conflict with Theorem 7.9 because K is not normal in A_4. Since $L \lhd A_4$ implies $L = \langle e \rangle$, H, or A_4, A_4 cannot be the direct sum of any two (or more) of its subgroups except trivially as $A_4 \cong A_4 \oplus \langle I \rangle$.

The corollary to Theorem 6.9 shows that if $\{n_1, \ldots, n_k\} \subseteq \boldsymbol{N}$ is pairwise relatively prime then $\boldsymbol{Z}_{n_1} \oplus \cdots \oplus \boldsymbol{Z}_{n_k} \cong \boldsymbol{Z}_n$ for $n = n_1 \cdots n_k$. Thus for any $n \geq 1$, if $n = n_1 \cdots n_k$ with $\{n_i\}$ pairwise relatively prime then $\boldsymbol{Z}_n \cong \boldsymbol{Z}_{n_1} \oplus \cdots \oplus \boldsymbol{Z}_{n_k}$. We use Theorem 7.8 to obtain this. Note that Theorem 3.21 shows that a direct sum of $\{\boldsymbol{Z}_{m_j}\}$ is cyclic $\Leftrightarrow \{m_j\}$ is pairwise relatively prime.

Example 7.7 If $n = n_1 \cdots n_k \in \boldsymbol{N}$ with $\{n_i\}$ pairwise relatively prime then $\boldsymbol{Z}_n \cong \boldsymbol{Z}_{n_1} \oplus \cdots \oplus \boldsymbol{Z}_{n_k}$.

By Theorem 3.1, $H_i = \langle [n/n_i]_n \rangle \leq \boldsymbol{Z}_n$ with $|H_i| = n_i$, and $H_i \cong \boldsymbol{Z}_{n_i}$ using Theorem 6.8. Since $\boldsymbol{Z}_n$ is Abelian, $H_1 H_2 \cdots H_i \leq \boldsymbol{Z}_n$ for each i (Theorem 4.11); hence Theorem 4.13 yields $|H_1 H_2 \cdots H_i| = n_1 \cdots n_i$. Now $(|H_1 H_2 \cdots H_i|, |H_{i+1}|) = 1$ follows from our assumption on $\{n_j\}$ (Theorem 1.11!), so Theorem 4.6 shows $H_1 H_2 \cdots H_i \cap H_{i+1} = \langle e \rangle$. Applying Theorem 7.8, we conclude that $\boldsymbol{Z}_n \cong H_1 \oplus \cdots \oplus H_k$, and then that $\boldsymbol{Z}_n \cong \boldsymbol{Z}_{n_1} \oplus \cdots \oplus \boldsymbol{Z}_{n_k}$ using Theorem 7.10.

An isomorphism for Example 7.7 can be written down explicitly. This has the advantage of also giving a direct sum decomposition for the groups U_n and provides another direct proof of Example 7.7.

THEOREM 7.11 If $n = n_1 \cdots n_k \in \boldsymbol{N}$ with $\{n_i\}$ pairwise relatively prime then $\rho : \boldsymbol{Z}_n \to \boldsymbol{Z}_{n_1} \oplus \cdots \oplus \boldsymbol{Z}_{n_k}$ via $\rho([a]_n) = ([a]_{n_1}, \ldots, [a]_{n_k})$ is an isomorphism that, when restricted to U_n, also shows that $U_n \cong U_{n_1} \oplus \cdots \oplus U_{n_k}$.

Proof Each $n_j \mid n$ so ρ is a function by Theorem 1.30. If $\rho([a]_n) = \rho([b]_n)$, then for all $1 \le i \le k$, $[a]_{n_i} = [b]_{n_i}$ and $n_i \mid (a - b)$. But $\{n_j\}$ is pairwise relatively prime so Theorem 1.12 forces $n \mid (a - b)$. Thus $[a]_n = [b]_n$ and ρ is injective. Hence $|\rho(\boldsymbol{Z}_n)| = n = |\boldsymbol{Z}_{n_1} \oplus \cdots \oplus \boldsymbol{Z}_{n_k}|$ by Theorem 0.3, so ρ is bijective. Now ρ is a homomorphism (verify!) so an isomorphism. We claim the restriction η of ρ to U_n is a bijection onto $U_{n_1} \oplus \cdots \oplus U_{n_k}$. If $(a, n) = 1$ then $[a]_n \in U_n$ and also $(a, n_j) = 1$ (why?) so $[a]_{n_j} \in U_{n_j}$. Thus $\eta : U_n \to U_{n_1} \oplus \cdots \oplus U_{n_k}$ and is injective since ρ is. But $|U_n| = \varphi(n)$ (Euler phi-function) so $\varphi(n) = |\eta(U_n)|$ and $|U_{n_1} \oplus \cdots \oplus U_{n_k}| = |U_{n_1}| \cdots |U_{n_k}| = \varphi(n_1) \cdots \varphi(n_k) = \varphi(n)$ from Theorem 3.4. This means that η is a bijection. Finally, using $[a]_m [b]_m = [ab]_m$ in U_m it is easy to see that $\eta : U_n \to U_{n_1} \oplus \cdots \oplus U_{n_k}$ is an isomorphism. ∎

Along with Theorem 7.10, Theorem 7.11 shows: $U_{30} \cong U_6 \oplus U_5 \cong \boldsymbol{Z}_2 \oplus \boldsymbol{Z}_4$, $U_{35} \cong U_5 \oplus U_7 \cong \boldsymbol{Z}_4 \oplus \boldsymbol{Z}_6$, and $U_{84} \cong U_3 \oplus U_4 \oplus U_7 \cong \boldsymbol{Z}_2 \oplus \boldsymbol{Z}_2 \oplus \boldsymbol{Z}_6$. The various $U_m \cong \boldsymbol{Z}_n$ follow by direct computation. Using U_m is cyclic of order $\varphi(m)$ when $m = p^k$ for $p > 2$ prime (Theorem 4.22), $U_{135} \cong U_{27} \oplus U_5 \cong \boldsymbol{Z}_{18} \oplus \boldsymbol{Z}_4 \cong \boldsymbol{Z}_2 \oplus \boldsymbol{Z}_4 \oplus \boldsymbol{Z}_9$ (why?), and $U_{2025} \cong U_{81} \oplus U_{25} \cong \boldsymbol{Z}_{54} \oplus \boldsymbol{Z}_{20}$. From Theorem 3.20 on orders of elements in direct sums, Theorem 6.2 showing that orders are preserved by isomorphisms, and direct sum representations, we can deduce facts about the U_n that are not otherwise obvious. For example, since $U_{35} \cong \boldsymbol{Z}_4 \oplus \boldsymbol{Z}_6$, U_{35} has three elements of order 2, four elements of order 4, six elements of order 6 (why?), and 8 elements of order 12, the maximal order of any element. Similarly, $U_{2025} \cong \boldsymbol{Z}_{54} \oplus \boldsymbol{Z}_{20}$ (why?) implies that the maximal order of any element in U_{2025} is $27 \cdot 20 = 540$ and that $\langle [0]_{54} \rangle \oplus \langle [4]_{20} \rangle$ is the only subgroup of order 5 in U_{2025}.

Direct sum representations can often determine isomorphisms. Using $U_{35} \cong \boldsymbol{Z}_4 \oplus \boldsymbol{Z}_6$ and $U_{84} \cong \boldsymbol{Z}_2 \oplus \boldsymbol{Z}_2 \oplus \boldsymbol{Z}_6$ from the last paragraph gives a rather painless way to see that these groups are not isomorphic. We observed that U_{35} has four elements satisfying $x^2 = e$ but from Theorem 3.20, U_{84} has eight such elements. Thus these groups cannot be isomorphic by Theorem 6.2. Now $|U_{72}| = \varphi(8 \cdot 9) = \varphi(8)\varphi(9) = 4 \cdot 6 = 24$ also and we claim $U_{72} \cong U_{84}$. Write $U_{72} \cong U_8 \oplus U_9 \cong (\boldsymbol{Z}_2 \oplus \boldsymbol{Z}_2) \oplus \boldsymbol{Z}_6$ so $U_{72} \cong U_{84}$ by the transitivity of isomorphism (corollary to Theorem 6.3). Next consider $U_{52} \cong U_4 \oplus U_{13} \cong \boldsymbol{Z}_2 \oplus \boldsymbol{Z}_{12}$. It might appear that U_{52} is isomorphic to neither U_{35} nor U_{84}, but from Theorem 7.11 and Theorem 7.10, $U_{52} \cong \boldsymbol{Z}_2 \oplus (\boldsymbol{Z}_3 \oplus \boldsymbol{Z}_4) \cong (\boldsymbol{Z}_2 \oplus \boldsymbol{Z}_3) \oplus \boldsymbol{Z}_4 \cong \boldsymbol{Z}_6 \oplus \boldsymbol{Z}_4 \cong \boldsymbol{Z}_4 \oplus \boldsymbol{Z}_6$ so indeed $U_{52} \cong U_{35}$.

Recall that for p a prime and G a group, $G_p = \{g \in G \mid o(g) = p^t$, some $t \ge 0\}$. If G is finite and Abelian Theorem 4.14 shows that $G_p \le G$ and $|G_p| = p^s$ when $|G| = p^s m$ for $(p, m) = 1$. In terms of direct sums, we have

THEOREM 7.12 Let G be an Abelian group with $|G| = p_1^{a_1} \cdots p_k^{a_k}$ for primes $p_1 < \cdots < p_k$. Then $|G_{p_i}| = p_i^{a_i}$ and $G \cong G_{p_1} \oplus \cdots \oplus G_{p_k}$.

Proof The statement of Theorem 4.14 shows that $\{G_{p_i}\}$ satisfies the hypothesis of Theorem 7.8 so applying that result proves the theorem. ■

When $G = \mathbf{Z}_n$ for $n = p_1^{a_1} \cdots p_k^{a_k}$, $G_{p_i} = \langle [n/p_i^{a_i}] \rangle$ by Theorem 3.1 so Example 7.7 follows from Theorem 7.12 (exercise!). We end this section with another example of how Theorem 7.8 can be used.

THEOREM 7.13 If G is an Abelian group, $|G| = p^n$ for p a prime, and $n \geq 1$, then the following are equivalent:

i) $g^p = e$ for all $g \in G$
ii) for $H \leq G$ there is $K \leq G$ with $HK = G$ and $H \cap K = \langle e \rangle$, so $G \cong H \oplus K$
iii) $G \cong \mathbf{Z}_p^n$

Proof Since G is Abelian the product of subgroups of G will be a subgroup by Theorem 4.11. Assume i), and let $H \leq G$. If $H = G$, ii) holds for $K = \langle e \rangle$. When $H \neq G$, let $x \notin H$, so $o(x) = p = o(\langle x \rangle)$ by Theorem 2.16. Now $H \cap \langle x \rangle \leq \langle x \rangle$, and Lagrange's Theorem gives $|H \cap \langle x \rangle| \mid p$ so either $H \cap \langle x \rangle = \langle e \rangle$ or else $|H \cap \langle x \rangle| = p$, forcing $H \cap \langle x \rangle = \langle x \rangle$ (why?). But $x \notin H$ so $H \cap \langle x \rangle = \langle e \rangle$ must occur. Let $K \leq G$ have maximal order among all subgroups with $H \cap K = \langle e \rangle$. Since G is Abelian, $HK \leq G$ and if $HK = G$ then ii) holds. On the other hand, if $HK \neq G$ then by the argument just given $HK \cap \langle y \rangle = \langle e \rangle$ if $y \notin HK$ and then $K\langle y \rangle \leq G$ is properly larger than K. However, $H \cap K\langle y \rangle = \langle e \rangle$ because if $h \in H \cap K\langle y \rangle$, say $h = ky^i$ for $k \in K$, then $hk^{-1} = y^i \in HK \cap \langle y \rangle = \langle e \rangle$. This forces $h = k \in H \cap K = \langle e \rangle$, showing that $H \cap K\langle y \rangle = \langle e \rangle$ and contradicting the maximality of $|K|$. Consequently, $HK = G$ and ii) holds.

Next we prove that ii) implies iii) by induction on n. If $n = 1$ then $G = \langle x \rangle \cong \mathbf{Z}_p$ for any $x \in G - \langle e \rangle$ using Theorem 4.9 and Theorem 6.8. Assume that if A is an Abelian group, $|A| = p^m$, and ii) holds for A, then $A \cong \mathbf{Z}_p^m$. Let $|G| = p^{m+1}$. By Theorem 4.7, if $g \in G$ then $o(g) \mid p^{m+1}$ so Theorem 2.8 gives some $x \in G$ with $o(x) = p$. Set $H = \langle x \rangle$ and use ii) to find $K \leq G$ with $G = HK$ and $H \cap K = \langle e \rangle$. From Theorem 4.12, $|G| = |H||K| = p|K|$ so $|K| = p^m$. We claim K satisfies ii). Let $L \leq K$ and use $L \leq G$ to find $M \leq G$ with $G = LM$ and $L \cap M = \langle e \rangle$. If $k \in K$ then $k = uv$ for $u \in L$ and $v \in M$, so $v = ku^{-1} \in K \cap M$. It follows that $K \leq L(K \cap M)$, and $L \leq K$ forces $K = L(K \cap M)$. Since $L \cap (K \cap M) \subseteq L \cap M = \langle e \rangle$, the subgroup $K \cap M$ of K shows that ii) holds for K. Now $|K| = p^m$ so the induction assumption applies and $K \cong \mathbf{Z}_p^m$. By Theorem 7.9, $G \cong H \oplus K$, then Theorem 7.10 implies that $G \cong \langle x \rangle \oplus \mathbf{Z}_p^m$, and Theorem 6.8 gives $G \cong \mathbf{Z}_p \oplus \mathbf{Z}_p^m \cong \mathbf{Z}_p^{m+1}$ by another application of Theorem 7.10. Therefore ii) implies iii) by IND-I.

Assuming $G \cong \mathbf{Z}_p^m$, it follows from Theorem 3.20 that any $x \in \mathbf{Z}_p^m$ satisfies $x^p = e$. Consequently, $g^p = e$ for all $g \in G$ by Theorem 6.2, proving that iii) implies i) and completing the proof of the theorem. ■

COROLLARY If $|G| = p^2$ for p a prime, then either $G \cong \mathbf{Z}_{p^2}$ or $G \cong \mathbf{Z}_p \oplus \mathbf{Z}_p$.

Proof G is Abelian from Theorem 6.21. Now $o(g) \mid p^2$ for all $g \in G$ by Theorem 4.7 so $g^p = e$ for all $g \in G$ and we can quote the theorem, or some element in G has order p^2, G is cyclic, and $G \cong \mathbf{Z}_{p^2}$ by Theorem 6.8. ■

EXERCISES

1. If $H, K \lhd G$ are simple groups, $G = HK$, and $H \cap K = \langle e \rangle$, show that either $H \cong K \cong \mathbf{Z}_p$ for p a prime, or $N \lhd G \Rightarrow N = \langle e \rangle$, H, K, or G.
2. For $n, m \in \mathbf{N}$ with $(n, m) = 1$, show that $F(([a]_n, [b]_m)) = [ma + nb]_{nm}$ defines a function and isomorphism from $\mathbf{Z}_n \oplus \mathbf{Z}_m$ to $\mathbf{Z}_{nm}$.
3. Let $N_1, \ldots, N_k \lhd G$ so that $N_1 N_2 \cdots N_j \cap N_{j+1} = \langle e \rangle$ for all $1 \le j < k$. If σ is any permutation in S_k, show that $N_{\sigma(1)} N_{\sigma(2)} \cdots N_{\sigma(j)} \cap N_{\sigma(j+1)} = \langle e \rangle$.
4. Let $N = \langle (1, 1, 1, 1) \rangle \le \mathbf{Z}^4$ and set $G = \mathbf{Z}^4/N$. Find $N \le H, K, L \le \mathbf{Z}^4$ properly between N and $\mathbf{Z}^4$ and satisfying $G \cong H/N \oplus K/N \oplus L/N$.
5. Find $H_j \le G$ with $G = \oplus H_j$ showing that:
   i) $G = \mathbf{Z}_{21} \cong \mathbf{Z}_3 \oplus \mathbf{Z}_7$
   ii) $G = \mathbf{Z}_6 \oplus \mathbf{Z}_{10} \cong \mathbf{Z}_2 \oplus \mathbf{Z}_2 \oplus \mathbf{Z}_{15}$
   iii) $G = U_{16} \cong \mathbf{Z}_4 \oplus \mathbf{Z}_2$
   iv) $G = U_{35} \cong \mathbf{Z}_2 \oplus \mathbf{Z}_3 \oplus \mathbf{Z}_4$
   v) $G = \mathbf{Z}_{630} \cong \mathbf{Z}_2 \oplus \mathbf{Z}_5 \oplus \mathbf{Z}_7 \oplus \mathbf{Z}_9$
   vi) $G = U_{60} \cong \mathbf{Z}_4 \oplus \mathbf{Z}_2 \oplus \mathbf{Z}_2$
6. Find all $m \in \mathbf{N}$ so that $U_m \cong U_n$ when $n =$: i) 30; ii) 24; iii) 100; iv) 70.
7. **a)** If $N \lhd D_n$, show that either $N \le \langle T_1 \rangle$, all $W_{2k} \in N$, or all $W_{2k+1} \in N$.
   b) If $n = 2(2m + 1)$, find proper $H, K \lhd D_n$ so that $D_n \cong H \oplus K$.
   c) If $n = 4m$ or $n = 2m + 1$, show that $D_n \ncong H \oplus K$ for proper $H, K \lhd D_n$.
8. If $G = \langle g \rangle$ with $|G| = p^n$ for p a prime, then $G \ncong H \oplus K$ for proper $H, K \le G$.
9. For G a finite Abelian group, p a prime, $G_p = \{g \in G \mid o(g) = p^k, k \ge 0\}$, and $H \le G$, show $H \cong (H \cap G_{p_1}) \oplus \cdots \oplus (H \cap G_{p_s})$ over all $p_i \mid |H|$.
10. Show that there is an epimorphism $\varphi : \mathbf{Z}_{m(1)} \oplus \cdots \oplus \mathbf{Z}_{m(s)} \to \mathbf{Z}_{n(1)} \oplus \cdots \oplus \mathbf{Z}_{n(k)}$ if $k \le s$ and $n(j) \mid m(j)$ for each $1 \le j \le k$.
11. For $n > 1$ and odd show that U_{n^2} contains a cyclic subgroup of order n. (Use Theorem 4.22.)

7.4 GROUPS OF SMALL ORDER

In this section we see how the ideas developed so far can be used to describe groups of small order in terms of known groups. Any group G of prime order is cyclic by Theorem 4.9, so Theorem 6.8 shows that G is isomorphic to $\mathbf{Z}_{|G|}$. Theorem 6.1 gives the possible structures for G of order 6, 10, or 14: each of these G has order $2p$ for an odd prime p so $G \cong \mathbf{Z}_{2p}$ or $G \cong D_p$. The remaining cases we consider are $|G|$ = 4, 8, 9, or 15. We could also classify groups of order 12 but this will be *much* easier to do later.

The corollary to Theorem 7.13 describes the possible structures of groups of order 4 or 9, but we prefer to give a more elementary argument for these. The task would be a bit easier using results developed in the next chapter, but our analysis here is good practice in using the material accumulated so far. It is useful and instructive to see how these results can be blended together to obtain definitive classifications. The approach illustrates a common theme in mathematical investigations: weave together simple results to obtain more complex ones. Let us specify the groups of orders 4, 8, 9, or 15 that we know. For order 4 there are $\mathbf{Z}_4$ and $\mathbf{Z}_2 \oplus \mathbf{Z}_2$. Abelian groups of order 8 include $\mathbf{Z}_8$, $\mathbf{Z}_4 \oplus \mathbf{Z}_2$ and $\mathbf{Z}_2 \oplus \mathbf{Z}_2 \oplus \mathbf{Z}_2$. Also of order 8 are D_4 and the quaternion group Q from Exercise 13 of §5.1. We know that $\mathbf{Z}_9$ and $\mathbf{Z}_3 \oplus \mathbf{Z}_3$ are Abelian of order 9. Finally, for order 15 we know only $\mathbf{Z}_{15}$. The question is whether any group of order 4, 8, 9, or 15 is isomorphic to one of these or whether there are other groups of these orders.

Groups of Order 4 or 9 These are special cases of groups of order p^2 for p a prime. The theorem is the corollary to Theorem 7.13, but the argument here is simpler.

THEOREM 7.14 If $|G| = p^2$ for p a prime, then $G \cong \mathbf{Z}_{p^2}$ or $G \cong \mathbf{Z}_p \oplus \mathbf{Z}_p$.

Proof Apply Theorem 4.7 to get $o(g) \mid p^2$ for any $g \in G$. By Theorem 2.16, G is cyclic $\Leftrightarrow$ there is $g \in G$ with $o(g) = p^2$, and then $G \cong \mathbf{Z}_{p^2}$ using Theorem 6.8. Hence we may assume $g^p = e$ for all $g \in G$. If $x \neq e$ and $y \in G - \langle x\rangle$ then $|\langle x\rangle| = |\langle y\rangle| = p$ using Theorem 2.16 again, and $\langle x\rangle \cap \langle y\rangle = \langle e\rangle$ by Theorem 4.6 as $\langle x\rangle \neq \langle y\rangle$. Thus $|\langle x\rangle \cap \langle y\rangle| = 1$ and $\langle x\rangle\langle y\rangle = G$ from Theorem 4.12. We also have $[G : \langle x\rangle] = p$ by Lagrange's Theorem, so $|G| = p^2$ does not divide $p! = [G : \langle x\rangle]!$ and Theorem 6.14 implies that $\langle x\rangle \lhd G$. Similarly, $\langle y\rangle \lhd G$. But now Theorem 7.9 shows $G \cong \langle x\rangle \oplus \langle y\rangle \cong \mathbf{Z}_p \oplus \mathbf{Z}_p$ by Theorem 6.8 and Theorem 7.10. This proves the theorem and shows that G is Abelian. ■

Groups of Order 8 If $|G| = 8$ then G is isomorphic to: D_4, Q, $\mathbf{Z}_8$, $\mathbf{Z}_4 \oplus \mathbf{Z}_2$, or $\mathbf{Z}_2 \oplus \mathbf{Z}_2 \oplus \mathbf{Z}_2$.

For any $g \in G$ $o(g) \mid 8$ by Theorem 4.7. If $o(g) = 8$ for some $g \in G$ then $G = \langle g\rangle \cong \mathbf{Z}_8$ from Theorem 2.16 and Theorem 6.8. Thus assume $g^4 = e$, for all $g \in G$. If $g^2 = e$ for all $g \in G$, G is Abelian by Theorem 2.10, and Theorem 7.13 gives $G \cong \mathbf{Z}_2^3$. We prefer to argue directly. Let $x \in G - \langle e\rangle$ and $y \in G - \langle x\rangle$. Since G is Abelian, $\langle x\rangle\langle y\rangle \leq G$ by Theorem 4.11 and $|\langle x\rangle\langle y\rangle| = 4$ from Theorem 4.12 so there is $z \in G - \langle x\rangle\langle y\rangle$. But $\langle z\rangle = \{e, z\}$, so these same results show $\langle x\rangle\langle y\rangle \cap \langle z\rangle = \langle e\rangle$ and $\langle x\rangle\langle y\rangle\langle z\rangle = G$. Hence Theorem 7.8 yields $G \cong \langle x\rangle \oplus \langle y\rangle \oplus \langle z\rangle \cong \mathbf{Z}_2^3$ by Theorem 6.8 and Theorem 7.10. We have reduced to: $x \in G$ with $o(x) = 4$. But $|\langle x\rangle| = 4$ (Theorem 2.16) so $[G : \langle x\rangle] = 2$ (Lagrange!) and $\langle x\rangle \lhd G$ by Theorem 5.3. If $y \notin \langle x\rangle$ with $o(y) = 2$, $\langle x\rangle \cap \langle y\rangle = \langle e\rangle$ (why?) so from Theorem 4.12, $G = \langle x\rangle\langle y\rangle$. Also since $y^{-1}x^iy = (y^{-1}xy)^i$ and $\langle x\rangle \lhd G$, $\langle x\rangle = y^{-1}\langle x\rangle y = \langle y^{-1}xy\rangle$. Thus $y^{-1}xy$ generates $\langle x\rangle$, of order 4, so either $y^{-1}xy = x$ or $y^{-1}xy = x^{-1}$. In the first case, $xy = yx$ so $G = \langle x\rangle\langle y\rangle$ is Abelian, $\langle y\rangle \lhd G$, and $G \cong \langle x\rangle \oplus \langle y\rangle \cong \mathbf{Z}_4 \oplus \mathbf{Z}_2$ using Theorem 7.9, Theorem 6.8, and also Theorem 7.10. When $y^{-1}xy = x^{-1}$ holds, then $G \cong D_4$ by Example 6.4.

Now we may assume $x \in G$ with $o(x) = 4$, $\langle x\rangle \lhd G$, and if $y \notin \langle x\rangle$ then $o(y) = 4$. Using Theorem 4.12, $|\langle x\rangle||\langle y\rangle|/|\langle x\rangle \cap \langle y\rangle| \leq |G| = 8$, forcing $\langle x^2\rangle = \langle x\rangle \cap \langle y\rangle = \langle y^2\rangle : x^2 = y^2$ is the only element of order 2. As above, $y^{-1}xy = x$ or $y^{-1}xy = x^{-1}$. If $y^{-1}xy = x$, then $xy = yx$ so $(xy^{-1})^2 = x^2y^{-2} = e$ and $xy^{-1} \notin \langle x\rangle$ force $o(xy^{-1}) = 2$, a contradiction since $xy^{-1} \neq x^2$, showing $y^{-1}xy = x^{-1}$. Note that $G = \langle x\rangle \cup \langle x\rangle y = \{e, x, x^2, x^3, y, xy, x^2y, x^3y\}$. Also $y^{-1}x^2y = (y^{-1}xy)^2 = (x^{-1})^2 = x^2$, so $x^2 = z \in Z(G)$, and the square of each element of order 4 is $x^2 = z$. Since $y^{-1}xy = x^{-1} = x^3$ we have $xy = yx^3 = zyx$. Further, $y \cdot xy = y \cdot zyx = zy^2x = x$; $xy \cdot x = zyx^2 = z^2y = y$; $yx = z^{-1}xy = zxy$; $xy \cdot y = zx$; and $x \cdot xy = zy$. This describes the **quaternion group** given in Exercise 13, §5.1. The usual notation is $Q = \{\pm 1, \pm i, \pm j, \pm k\}$ with identity 1; $-1 \in Z(Q)$ multiplies by "changing sign"; $i^2 = j^2 = k^2 = -1$, $ij = k$, $jk = i$, $ki = j$; and

$ab = -ba = (-1)ba$ for $a, b \in Q - \{\pm 1\}$ with $a \neq \pm b$. From the computations above, $G = \{e, x, z, zx, y, zy, xy, zxy\}$ and so $G \cong Q$ using the isomorphism $\varphi(z) = -1$, $\varphi(x) = i$, $\varphi(y) = j$ and $\varphi(xy) = k$. Another concrete version of Q is the subgroup of $GL(2, C)$ generated by the matrices "i" $= \begin{bmatrix} 0 & 1 \\ -1 & 0 \end{bmatrix}$ and "j" $= \begin{bmatrix} 0 & i \\ i & 0 \end{bmatrix}$, so "$k$" $= \begin{bmatrix} i & 0 \\ 0 & -i \end{bmatrix}$.

In conclusion, we have shown that if $|G| = 8$ then G is isomorphic to one of groups we already know: $\mathbf{Z}_8$, $\mathbf{Z}_4 \oplus \mathbf{Z}_2$, $\mathbf{Z}_2 \oplus \mathbf{Z}_2 \oplus \mathbf{Z}_2$, D_4, or Q. Up to isomorphism there are three Abelian groups of order 8, distinguished by the maximal order of their elements, or by the number of elements of order 2 they contain. There are two non-Abelian groups: D_4 has two elements of order 4 and five of order 2; Q has six elements of order 4 and one of order 2, and every subgroup of it is normal and cyclic.

Groups of Order 15 We consider groups of order pq for primes $p < q$.

THEOREM 7.15 If G is a group, $|G| = pq$ for primes $p < q$, and $p \nmid (q - 1)$, then $G \cong \mathbf{Z}_{pq}$.

Proof The most difficult part of the argument is to show that $g^p = e$ is not possible for all $g \in G$. Assuming that $g^p = e$ for all $g \in G$, set $H = \langle x \rangle$ for $x \neq e$. From Theorem 6.14, left multiplication by $y \in G$ on the set of left cosets of H defines a homomorphism $\varphi : G \to \mathrm{Bij}(\{gH \mid g \in G\})$. Now by Lagrange's Theorem (for left cosets!), or Exercise 4 in §4.1, $|\{gH\}| = [G : H] = q$. It follows from Theorem 6.2 that $o(\varphi(x)) \mid p$ so $\varphi(x) = I$ or $\varphi(x)$ is a permutation of order p and must have orbits of length p or 1 using Theorem 3.11 and Theorem 6.12. Since $\varphi(x)(H) = xH = H$, $\varphi(x)$ has a fixed point, and because $p \nmid (q - 1)$, the remaining $q - 1$ cosets cannot be partitioned by orbits of length p; that is, $\varphi(x)$ must have a fixed point other than H. Hence there is $v \notin \langle x \rangle$ with $vH = \varphi(x)(vH) = xvH$, forcing $v^{-1}xv \in H$. But $v^{-1}x^jv = (v^{-1}xv)^j \in H = \langle x \rangle$ so $v^{-1}Hv \leq H$ and $|H| = p$ implies that $v^{-1}Hv = H$, or, equivalently, $Hv = vH$. It follows that $H\langle v \rangle = \langle v \rangle H$ so $H\langle v \rangle \leq G$ using Theorem 4.10. But $o(v) = p$ so $|H\langle v \rangle| = p^2$ from Theorem 4.12, This contradicts Lagrange's Theorem since $p^2 \nmid pq$ and shows that it is not possible for every $g \in G$ to satisfy $g^p = e$.

From Theorem 4.8, for every $g \in G$, $o(g) \mid pq$ and we have seen $o(g) = p$ for all $g \neq e$ is impossible, so there is some $x \in G$ with $o(x) = q$ (verify!). Since $|\langle x \rangle| = q$ and $[G : \langle x \rangle] = p < q$ (Lagrange!) we conclude from Theorem 6.14 that $\langle x \rangle \triangleleft G$. This follows also by showing that $\langle x \rangle$ is the only subgroup of order q, using Theorem 4.12 and then Theorem 5.3, as in the discussion before Theorem 5.4. We give two separate arguments to finish the proof. The first is more sophisticated and observes that $|Aut(\langle x \rangle)| = q - 1$ by Theorem 6.20, using the exercise that $\langle x \rangle \cong \mathbf{Z}_q$ implies $Aut(\langle x \rangle) \cong Aut(\mathbf{Z}_q)$. Now, using $p \nmid (q - 1)$, apply Theorem 6.19 to see that $G \leq C_G(\langle x \rangle)$, forcing $\langle x \rangle \leq Z(G)$. Since $G/\langle x \rangle$ is of prime order it is cyclic, and G must be Abelian by Theorem 5.8 unless $\langle x \rangle \neq Z(G)$. But in this case, $G = Z(G)$ (why?) and again G is Abelian. Thus there is $y \in G$ with $o(y) = p$ from Theorem 5.9, and Theorem 4.16 implies that $o(xy) = pq$. Therefore $G = \langle xy \rangle$ is cyclic and $G \cong \mathbf{Z}_{pq}$ by Theorem 6.8.

For the second argument, observe that $G/\langle x \rangle$ is cyclic since it has prime order p, so $G/\langle x \rangle = (\langle x \rangle y)$ with $o(\langle x \rangle y) = p$, $y^p \in \langle x \rangle$, and $y^{pq} = e$ by Theorem 4.7. If $y^q = e$ then $(\langle x \rangle y)^q = \langle x \rangle y^q = \langle x \rangle$, so $p \mid q$ from Theorem 2.7, a contradiction. Thus $y^q \neq e$ and

$o(y^q) = p$. The upshot of this is that G has an element of order p, so assume $o(y) = p$. Now use $y^{-1}\langle x\rangle y = \langle x\rangle$ to define $T \in Aut(\langle x\rangle)$ by $T(x^j) = y^{-1}x^j y$ (Definition 6.4). Since $y^p = e$ it follows that $T^p = I_{\langle x\rangle}$ so $o(T) \mid p$ and, as above, $|Aut(\langle x\rangle)| = q - 1$. Our hypothesis forces $o(T) = 1$, which implies $xy = yx$. Hence $o(xy) = pq$ by Theorem 4.16 and $G \cong \mathbf{Z}_{pq}$ follows from Theorem 6.8. ■

In the theorem it was difficult to show that G has elements of order p and q. This will follow from the general version of Cauchy's Theorem in the next chapter. When $|G| = pq$ and $p \mid (q - 1)$, we will see that there is a unique non-Abelian group of this order that is isomorphic to G. Of course when $p = 2$ we have the non-Abelian groups D_q.

EXERCISES

1. a) Show that S_5 contains an isomorphic copy of all groups of order 6.
b) Show that S_6 contains an isomorphic copy of three groups of order 8.
c) For every group G with $|G| = 8$, show that $G \cong H \leq S_n$ exactly when $n \geq 8$.

2. Let $\mathbf{Z}_3 * \mathbf{Z}_4 = (\mathbf{Z}_3 \times \mathbf{Z}_4, \cdot)$ with $([a]_3, [b]_4) \cdot ([c]_3, [d]_4) = ([a + (-1)^b c]_3, [b + d]_4)$ (see Theorem 3.22). Verify:
i) $\mathbf{Z}_3 * \mathbf{Z}_4$ is a non-Abelian group of order 12
ii) $N = \langle([1]_3, [0]_4)\rangle \lhd \mathbf{Z}_3 * \mathbf{Z}_4$ of order 3
iii) $z = ([0]_3, [2]_4)$ has order 2 and $\langle z\rangle = Z(\mathbf{Z}_3 * \mathbf{Z}_4)$
iv) if $H = \langle([1]_3, [2]_4)\rangle$, $|H| = 6$ and $y \in G - H \Rightarrow o(y) = 4$

3. Describe the subgroup generated by the given set in terms of known groups:
i) $\langle\{(1, 2, 3, 4)(5, 6), (5, 6)(7, 8)\}\rangle \leq A_8$
ii) $\langle\{(1, 2, 3, 4)(5, 6, 7, 8), (2, 4)(6, 8)\}\rangle \leq A_8$
iii) $\langle\{(1, 2)(3, 4), (1, 2)(5, 6), (3, 4)(7, 8)\}\rangle \leq A_8$
iv) $\langle\{(2, 6)(3, 5), (1, 2, 3, 4, 5, 6)\}\rangle \leq S_6$
v) $\langle\{(1, 3, 2, 4)(5, 7, 6, 8), (1, 5, 2, 6)(3, 8, 4, 7)\}\rangle \leq S_8$
vi) $\langle\{(1, 2)(3, 4), (3, 4)(5, 6), (5, 6)(7, 8), (1, 2)(7, 8)\}\rangle \leq A_8$

4. Determine the possible structures of the following groups in terms of the usual known groups (results on automorphisms are needed also).
i) $|G| = 30$ with $2 \mid |Z(G)|$
ii) $|G| = 105$, $N \lhd G$ and $|N| = 3$
iii) $|G| = 245$, $N \lhd G$ and $|N| = 7$
iv) $|G| = 30$, $N \lhd G$ and $|N| = 15$

5. If G is a group and $|G| = 45$ use Theorem 6.14 as in Theorem 7.15 to show:
i) If $H \leq G$ and $|H|$ is prime, then $H \leq K \leq G$ with $H \neq K$ and $K \neq G$.
ii) If $g \in G \Rightarrow o(g) \mid 9$, then there is some $H \leq G$ with $|H| = 9$.
iii) If $H \leq G$ with $|H| = 9$, then not every $g \in G$ satisfies $o(g) \mid 9$.
iv) G contains elements of order 3 and of order 5.
v) There is $N \lhd G$ with $|N| = 3$ or 9.
vi) If $N \lhd G$ and $|N| = 3$, then $N \subseteq Z(G)$ and G is Abelian.
vii) If $N \lhd G$ and $|N| = 9$, then there is $H \leq G$ with $|H| = 15$.
viii) If $H \leq G$ with $|H| = 15$, then $H \subseteq Z(G)$ and G is Abelian.
Determine the possible structures of G.

7.5 THE FUNDAMENTAL THEOREM OF FINITE ABELIAN GROUPS

We prove two beautiful theorems showing that finite Abelian groups are isomorphic to direct sums of cyclic groups in two specific ways, each uniquely characterizing the group up to isomorphism. These representations yield a number of results about Abelian groups, including a description of all nonisomorphic Abelian groups of a given order.

We illustrate the approach, using $G = \mathbf{Z}_{12} \oplus \mathbf{Z}_6 \oplus \mathbf{Z}_6$. Recall that in G, $o(([a]_{12}, [b]_6, [c]_6)) = \mathrm{LCM}\{o([a]_{12}), o([b]_6), o([c]_6)\}$ by Theorem 3.20. We show that the orders of the summands of G, 12, 6, and 6, are intrinsically defined by elements of certain orders in G. The largest order of any $g \in G$ is 12, $x_1 = ([1]_{12}, [0]_6, [0]_6)$ has order 12, and $\langle x_1 \rangle = \mathbf{Z}_{12} \oplus \langle [0]_6 \rangle \oplus \langle [0]_6 \rangle$. By Theorem 6.9, $G/\langle x_1 \rangle \cong \mathbf{Z}_6 \oplus \mathbf{Z}_6$ so 6 is the maximal order of any $g \in G/\langle x_1 \rangle$. If $x_2 = ([2]_{12}, [1]_6, [0]_6) \in G$ then $o(x_2) = 6$ and $o(\langle x_1 \rangle + x_2) = 6$ in $G/\langle x_1 \rangle$. Now $\langle x_1 \rangle \cap \langle x_2 \rangle = \langle e \rangle$, $\langle x_1 \rangle + \langle x_2 \rangle = \mathbf{Z}_{12} \oplus \mathbf{Z}_6 \oplus \langle [0]_6 \rangle$ and $G/(\langle x_1 \rangle + \langle x_2 \rangle) \cong \mathbf{Z}_6$ by Theorem 6.9 again. Thus the maximal order of any $X \in G/(\langle x_1 \rangle + \langle x_2 \rangle)$ is 6 and for $x_3 = ([4]_{12}, [3]_6, [1]_6) \in G$ we have $o(x_3) = 6$, $o(\langle x_1 \rangle + \langle x_2 \rangle + x_3) = 6$ in $G/(\langle x_1 \rangle + \langle x_2 \rangle)$, $\langle x_1 \rangle + \langle x_2 \rangle + \langle x_3 \rangle = G$, and $(\langle x_1 \rangle + \langle x_2 \rangle) \cap \langle x_3 \rangle = \langle e \rangle$. From Theorem 7.8, $G \cong \langle x_1 \rangle \oplus \langle x_2 \rangle \oplus \langle x_3 \rangle$. Thus the summands of our G are determined by a sequence of maximal orders of elements in successive quotients, and we can find elements in G of these orders so that G is the direct sum of the cyclic subgroups they generate.

For $H = \mathbf{Z}_6 \oplus \mathbf{Z}_4 \oplus \mathbf{Z}_3$, the maximal order of any $h \in H$ is 12, $x = ([2]_6, [1]_4, [0]_3)$ is one such element, and $\langle x \rangle = \langle [2]_6 \rangle \oplus \mathbf{Z}_4 \oplus \langle [0]_3 \rangle$ by direct calculation (verify!). Thus by Theorem 6.9, $H/\langle x \rangle \cong \mathbf{Z}_2 \oplus \mathbf{Z}_3 \cong \mathbf{Z}_6$ by Theorem 7.11, and for $y = ([1]_6, [0]_4, [1]_3)$, $o(y) = 6$, $o(\langle\langle x \rangle y\rangle) = 6$, $\langle x \rangle \cap \langle y \rangle = \langle e \rangle$, and $\langle x \rangle \langle y \rangle = H$ (verify!) so $H \cong \langle x \rangle \oplus \langle y \rangle$ by Theorem 7.9. Therefore Theorem 6.8 and Theorem 7.10 show $H \cong \mathbf{Z}_{12} \oplus \mathbf{Z}_6$. Thus the sequence of maximal orders need not be the orders of the summands defining the group.

An essential part of our approach looks at elements of maximal order, so we want additional information about this order.

THEOREM 7.16 Let G be a finite Abelian group and $x \in G$ with $o(g) \leq o(x)$ for all $g \in G$. If $y \in G$ then $o(y) \mid o(x)$, and so $y^{o(x)} = e$ for all $y \in G$.

Proof Assume there is $y \in G$ with $o(y) \nmid o(x)$. By Theorem 1.18, or unique factorization in N, some prime q divides $o(y)$ more times than it divides $o(x)$. Thus $o(x) = q^s k$, $o(y) = q^t m$, $(km, q) = 1$, and $t > s \geq 0$. Using Theorem 2.8, $o(x^{q^s}) = k$ and $o(y^m) = q^t$. Now $(o(x^{q^s}), o(y^m)) = 1$ and G is Abelian so $o(x^{q^s} y^m) = q^t k$ by Theorem 4.16. This contradicts the maximality of $o(x)$ so we must conclude that $o(y) \mid o(x)$, and $y^{o(x)} = e$ follows from Theorem 2.7.

■

We show next that any finite Abelian group G is a direct sum of cyclic subgroups. As in the examples above, if $x, y \in G$ and for all $g \in G$, $o(g) \leq o(x)$ and $o(\langle x \rangle g) \leq o(\langle x \rangle y)$ in $G/\langle x \rangle$, can we ensure $\langle x \rangle \cap \langle y \rangle = \langle e \rangle$? Then, can we find $\langle x \rangle \langle y \rangle z$ of maximal order in $G/\langle x \rangle \langle y \rangle$ so $\langle x \rangle \langle y \rangle \cap \langle z \rangle = \langle e \rangle$, etc.? To do this it is important to recall from the definition of a quotient group that for $H \triangleleft G$, $Hx \in G/H$ has order $k > 1$ in G/H if and only if $x \notin H$ and $k \in N$ is minimal with $x^k \in H$. It follows that $H \cap \langle x \rangle = \langle x^k \rangle$ (verify!), an essential observation needed in our proof.

THEOREM 7.17 If G is a finite Abelian group, then $G \cong \mathbf{Z}_{n(1)} \oplus \cdots \oplus \mathbf{Z}_{n(k)}$ with $n(j) \mid n(j-1)$ for all $2 \leq j \leq k$.

Proof Since G is Abelian we use without further reference that any $H \leq G$ is normal so G/H exists, that $H, K \leq G \Rightarrow HK \leq G$ (Theorem 4.11), and that $(xy)^m = x^m y^m$ for any $x, y \in G$. Some $x_1 \in G$ has maximal order $n(1)$. If $G = \langle x_1 \rangle$ then $|G| = n(1)$ and $G \cong \mathbf{Z}_{n(1)}$ by Theorem 6.8. Otherwise set $H_1 = \langle x_1 \rangle$. For some $y \in G$, $o(H_1 y) = m$ is maximal in G/H_1. Since $G \neq H_1$, $m > 1$ so $y \notin H_1$, $y^m \in H_1$, and m is the minimal power of y in H_1. Note that $(H_1 y)^{o(y)} = H_1 y^{o(y)} = H_1$. Thus $m \mid o(y)$ by Theorem 2.7, and $o(y) \mid n(1)$ by Theorem 7.16, so $m \mid n(1)$ and we may write $n(1) = ms$. Now $y^m \in H_1$ so $y^m = x_1^k$. Use Theorem 1.1 to write $k = qm + j$ with $0 \leq j < m$, resulting in $y^m = x_1^{mq} x_1^j$. Therefore if $x_2 = x_1^{-q} y$, then $H_1 y = H_1 x_2$ so $o(H_1 x_2) = m$ in G/H_1, but now, using Theorem 7.16 again, $e = x_2^{n(1)} = x_2^{ms} = ((x_1^{-q} y)^m)^s = x_1^{js}$. Hence $n(1) \mid js$ (Theorem 2.7!), but $0 \leq j < m$ means that $0 \leq js < n(1)$. This forces $j = 0$ so $x_2^m = (x_1^{-q} y)^m = x_1^{-qm} y^m = e$. Since $m = o(H_1 x_2)$, $\langle x_1 \rangle \cap \langle x_2 \rangle = \langle x_2^m \rangle = \langle e \rangle$, as mentioned above. Furthermore, $o(x_2) = m$ since $o(x_2) \mid m$ from $x_2^m = e$ and Theorem 2.7, and $m \leq o(x_2)$ from $m = o(H_1 x_2)$. Set $H_2 = \langle x_2 \rangle$ and $n(2) = m$ so $n(2) \mid n(1)$ and $H_1 \cap H_2 = \langle e \rangle$. If $G = H_1 H_2$ then by Theorem 7.9, Theorem 6.8, and Theorem 7.10, $G \cong \langle x_1 \rangle \oplus \langle x_2 \rangle \cong \mathbf{Z}_{n(1)} \oplus \mathbf{Z}_{n(2)}$, finishing the proof.

When $G \neq H_1 H_2$ we continue with the same procedure. The general step is like the argument above but the initial conditions are a bit more complicated. Suppose we have a maximal set $\{x_1, x_2, \ldots, x_k\} \subseteq G$ so that the following hold: $o(x_i) = n(i)$; $n(i) \mid n(i-1)$ for $2 \leq i \leq k$; and, if $H_0 = \langle e \rangle$ and $H_j = \langle x_1 \rangle \cdots \langle x_j \rangle$ for $j \leq k$, then $o(H_{j-1} x_j) = n(j)$ is maximal among orders of all elements in G/H_{j-1}. When G is not cyclic, we constructed a set of two such elements above so $k \geq 2$. Of course k must be finite since G is. Since $o(x_j) = n(j) = o(H_{j-1} x_j)$, $H_{j-1} \cap \langle x_j \rangle = \langle x_j^{n(j)} \rangle = \langle e \rangle$, so $H_k \cong \oplus \langle x_i \rangle$ by Theorem 7.8. Consequently, if $G = H_k$, then $\langle x_i \rangle \cong \mathbf{Z}_{n(i)}$ and Theorem 7.10 finish the proof.

When $G \neq H_k$ we contradict the maximality of k by finding x_{k+1} satisfying $o(x_{k+1}) \mid n(k)$ and $o(H_k x_{k+1}) = o(x_{k+1})$ has maximal order in G/H_k. For some $y \in G$, $o(H_k y) = m$ is maximal in G/H_k. Now the definition of $n(k)$ and Theorem 7.16 show $y^{n(k)} \in H_{k-1} \leq H_k$ so $(H_k y)^{n(k)} = H_k$, and Theorem 2.7 forces $m \mid n(k)$. Although $y^m \in H_k$ it might be that $y^m \in H_j \leq H_k$ for $j < k$. Among all $x \in G$ satisfying $o(H_k x) = m$ in G/H_k there is one with $x^m \in H_j$ for j minimal (let $H_0 = \langle e \rangle$). That is, if $g \in G$ with $o(H_k g) = m$ and if $j > 0$ then $g^m \notin H_{j-1}$. Without loss of generality, we may assume that the initial choice of $y \in G$ satisfies this condition. If $j = 0$, then $y^m = e$. Since $(H_k y)^t \neq H_k$ for $1 \leq t < m$ it follows that $o(y) = m$, so $H_k \cap \langle y \rangle = \langle y^m \rangle = \langle e \rangle$. We have seen that $m \mid n(k)$ so using $x_{k+1} = y$ contradicts the maximality of k; thus $j \geq 1$.

By the choices of y and j, $y^m = x_1^{a(1)} x_2^{a(2)} \cdots x_j^{a(j)} = h x_j^{a(j)}$ with $h \in H_{j-1}$ and $a(j) > 0$. From above $m \mid n(k)$, and $n(k) \mid n(j)$ by assumption, so $n(j) = ms$ and the Division Theorem shows $a(j) = mq + v$ for $0 \leq v < m$. Now $y^m = h x_j^{mq} x_j^v$ so if $w = x_j^{-q} y$ then $w^m = h x_j^v$. Observe that $H_k w = H_k y$ so $o(H_k w) = m$ and also, since the order of any element in G/H_{j-1} divides $n(j)$, $w^{n(j)} \in H_{j-1}$. But $w^{n(j)} = w^{ms} = (h x_j^v)^s = h^s x_j^{vs} \in H_{j-1}$, forcing $x_j^{vs} \in H_{j-1}$. Thus $(H_{j-1} x_j)^{vs} = H_{j-1}$, $n(j) \mid vs$ by Theorem 2.7, but $0 \leq v < m$ forces $vs < ms = n(j)$, so $v = 0$. Consequently, $w^m = h \in H_{j-1}$, which contradicts the minimality of j. This shows that $G \neq H_k$ is not possible, proving the theorem. ∎

We illustrate the theorem when $|G|=72$. If $G=\oplus \mathbf{Z}_{n(i)}$ then $|G|=n(1)\cdots n(k)$ so we must list factors of 72, each a multiple of the next, with product equal to 72. There are six possibilities: 72; 36, 2; 24, 3; 18, 2, 2; 12, 6; or 6, 6, 2. These correspond to the direct sums $\mathbf{Z}_{72}$; $\mathbf{Z}_{36}\oplus\mathbf{Z}_2$; $\mathbf{Z}_{24}\oplus\mathbf{Z}_3$; $\mathbf{Z}_{18}\oplus\mathbf{Z}_2\oplus\mathbf{Z}_2$; $\mathbf{Z}_{12}\oplus\mathbf{Z}_6$; and $\mathbf{Z}_6\oplus\mathbf{Z}_6\oplus\mathbf{Z}_2$. By Theorem 7.17, G is isomorphic to one of these and in fact only to one since each of the six groups has the maximal order of its elements different from that in the other groups. For example, $\mathbf{Z}_{12}\oplus\mathbf{Z}_6$ is the only one of the six groups with 12 as the maximal order of its elements. Similarly, up to isomorphism the groups of order 16 are: $\mathbf{Z}_{16}$; $\mathbf{Z}_8\oplus\mathbf{Z}_2$; $\mathbf{Z}_4\oplus\mathbf{Z}_4$; $\mathbf{Z}_4\oplus\mathbf{Z}_2\oplus\mathbf{Z}_2$; and $\mathbf{Z}_2\oplus\mathbf{Z}_2\oplus\mathbf{Z}_2\oplus\mathbf{Z}_2$. Four is the maximal order of any element in $\mathbf{Z}_4\oplus\mathbf{Z}_4$ or $\mathbf{Z}_4\oplus\mathbf{Z}_2\oplus\mathbf{Z}_2$ but they have different numbers of elements of order 2 (how many in each?).

We saw in Example 7.3 that $\mathbf{Z}_{30}$ is a direct sum of cyclic groups in different ways but there is only one sum satisfying Theorem 7.17. Any $n(2)$ for $\mathbf{Z}_{30}$ must divide $n(1)$, but 30 had no repeated prime factor so $n(1)=30$. This same reasoning provides another proof of part of Theorem 4.17: G Abelian and $|G|=p_1\cdots p_k$ for distinct primes implies G is cyclic.

From Theorem 7.12, if G is Abelian, $|G|=p_1^{m(1)}\cdots p_k^{m(k)}$, and $G_i\le G$ with $|G_i|=p_i^{m(i)}$, then $G\cong G_1\oplus\cdots\oplus G_k$. Applying Theorem 7.17, we see that any G_j is the direct sum of cyclic subgroups of orders $p_j^{a(i)}$ for $a(1)\ge a(2)\ge\cdots\ge a(k_j)$. For example, if $|G_j|=3^6$ then three possibilities for the $\{a(i)\}$ are $a(1)=6$, $a(1)=3=a(2)$, and $a(1)=4$, $a(2)=a(3)=1$. Thus G has a representation as a direct sum of cyclic groups of prime power orders. We prove next that when G is Abelian with $|G|=p^n$ for p a prime, then the summands appearing in Theorem 7.17 are uniquely determined by the numbers of elements of order p in G and in its quotients. This result will lead to the uniqueness of the sequence $(n(1),\ldots,n(k))$ in Theorem 7.17.

THEOREM 7.18 Let G be an Abelian group with $|G|=p^n$ for p a prime. If $G\cong\mathbf{Z}_{p^{n(1)}}\oplus\cdots\oplus\mathbf{Z}_{p^{n(k)}}$ with $n(1)\ge\cdots\ge n(k)\ge 1$, and if $G\cong\mathbf{Z}_{p^{m(1)}}\oplus\cdots\oplus\mathbf{Z}_{p^{m(s)}}$ with $m(1)\ge\cdots\ge m(s)\ge 1$, then $k=s$ and each $n(j)=m(j)$.

Proof Let $A=\mathbf{Z}_{p^{n(1)}}\oplus\cdots\oplus\mathbf{Z}_{p^{n(k)}}$ and $B=\mathbf{Z}_{p^{m(1)}}\oplus\cdots\oplus\mathbf{Z}_{p^{m(s)}}$. We argue by induction on n. When $n=1$ certainly $k=s=1$ and $n(1)=m(1)=1$. Assume that when $|G|=p^t$ for $t<n$ there is only one representation for G, as in Theorem 7.17. Let $G(p)=\{g\in G\mid g^p=e\}$ and let $x,y\in G(p)$. Since G is Abelian, $(xy)^p=x^py^p=e$, so G finite and Theorem 2.12 show $G(p)\le G$, and $G(p)$ is uniquely determined by G. If $G(p)=G$ then $A(p)=A$ and $B(p)=B$ so all $n(i)=1=m(j)$ and $k=s=n$. When $G(p)\ne G$ let $a\ge 1$ be maximal with $n(a)>1$ and $b\ge 1$ maximal with $m(b)>1$. By Theorem 3.20, $g\in A(p)$ exactly when each of its coordinates g_j satisfies $g_j^p=[0]$. In $\mathbf{Z}_{p^{n(i)}}$ there is a unique $\langle[x_i]\rangle\le\mathbf{Z}_{p^{n(i)}}$ of order p (Theorem 3.1), $\langle[x_i]\rangle=\mathbf{Z}_{p^{n(i)}}(p)$ by Theorem 2.16, and $\mathbf{Z}_{p^{n(i)}}/\langle[x_i]\rangle$ is cyclic of order $p^{n(i)-1}$ by Theorem 5.7. Hence under the isomorphism between G and A, $G(p)\cong A(p)=\langle[x_1]\rangle\oplus\cdots\oplus\langle[x_k]\rangle$, so the correspondence in Theorem 7.2 yields $G/G(p)\cong A/\oplus\langle[x_i]\rangle$. Theorem 6.9, Theorem 6.8, and Theorem 7.10 show $G/G(p)\cong\mathbf{Z}_{p^{n(1)-1}}\oplus\cdots\oplus\mathbf{Z}_{p^{n(a)-1}}$. Also $|G(p)|=|A(p)|=p^k$. Similarly, $G/G(p)\cong B/B(p)\cong\mathbf{Z}_{p^{m(1)-1}}\oplus\cdots\oplus\mathbf{Z}_{p^{m(b)-1}}$ and $p^s=|B(p)|=|G(p)|=p^k$ forces $k=s$. But $|G/G(p)|=p^t$ for $t<n$, so by our induction assumption, $a=b$ and $n(j)=m(j)$ for $1\le j\le a$. Since $n(i)=m(i)=1$ for $a<i\le k$, the theorem holds by IND-II. ∎

The last two theorems show that for G an Abelian group of prime power order p^n, $G \cong \mathbf{Z}_{p^{m(1)}} \oplus \cdots \oplus \mathbf{Z}_{p^{m(s)}}$ with $m(1) \geq \cdots \geq m(s)$ and the sequence $(p^{m(1)}, \ldots, p^{m(s)})$ of **elementary divisors of** G is uniquely determined by G. Furthermore, for H an Abelian group, $G \cong H \Leftrightarrow G$ and H have the same elementary divisors. As $|G| = p^n$ is the product of the elementary divisors it follows that $n = m(1) + \cdots + m(s)$. Any nonincreasing sequence of positive integers adding to n is a **partition** of n so *the number of partitions of n is the number of nonisomorphic Abelian groups of order p^n*. We saw in Theorem 7.14 that G Abelian and $|G| = p^2 \Rightarrow G \cong \mathbf{Z}_p \oplus \mathbf{Z}_p$ or $G \cong \mathbf{Z}_{p^2}$ This follows from our observations here since the partitions of 2 are (1, 1) and (2).

Example 7.8 If G is an Abelian group and $|G| = p^5$ for p a prime, then G is isomorphic to exactly one of: $\mathbf{Z}_{p^5}$; $\mathbf{Z}_{p^4} \oplus \mathbf{Z}_p$; $\mathbf{Z}_{p^3} \oplus \mathbf{Z}_{p^2}$; $\mathbf{Z}_{p^3} \oplus \mathbf{Z}_p \oplus \mathbf{Z}_p$; $\mathbf{Z}_{p^2} \oplus \mathbf{Z}_{p^2} \oplus \mathbf{Z}_p$; $\mathbf{Z}_{p^2} \oplus \mathbf{Z}_p \oplus \mathbf{Z}_p \oplus \mathbf{Z}_p$; or $\mathbf{Z}_p \oplus \mathbf{Z}_p \oplus \mathbf{Z}_p \oplus \mathbf{Z}_p \oplus \mathbf{Z}_p$.

The groups in the list correspond to the partitions of 5: (5), (4, 1), (3, 2), (3, 1, 1), (2, 2, 1), (2, 1, 1, 1), and (1, 1, 1, 1, 1).

We can see from Theorem 7.17 why Theorem 7.13 holds. If G is Abelian and every nonidentity element has order p, then in its representation as a direct sum of cyclic groups every summand must be $\mathbf{Z}_p$, so $G \cong \mathbf{Z}_p^n$ when $|G| = p^n$. Using Theorem 7.17 and Theorem 7.18 enables us to answer other questions about the structure of Abelian groups of prime power order.

Example 7.9 How many nonisomorphic Abelian groups of order 256 have every element satisfy $x^4 = e$?

If $o(g) \mid 4$ for all $g \in G$ then every summand in Theorem 7.17 is $\mathbf{Z}_2$ or $\mathbf{Z}_4$. Since $256 = 2^8$, the groups we want correspond to the partitions of 8 using no integer greater than 2. These are (2, 2, 2, 2), (2, 2, 2, 1, 1), (2, 2, 1, 1, 1, 1), (2, 1, 1, 1, 1, 1, 1), and (1, 1, 1, 1, 1, 1, 1, 1), so there are exactly five nonisomorphic groups satisfying the condition given.

The *Fundamental Theorem of Finite Abelian Groups* restates Theorem 7.17 and proves the uniqueness of the summands given by it.

THEOREM 7.19 For G a finite Abelian group, $G \cong \mathbf{Z}_{n(1)} \oplus \cdots \oplus \mathbf{Z}_{n(k)}$ with $n(j) \mid n(j-1)$ for all $2 \leq j < k$, and $(n(1), \ldots, n(k))$ is uniquely determined by G. If $|G| = p_1^{a_1} \cdots p_t^{a_t}$ for $p_1 < \cdots < p_t$ primes, then $G \cong \mathbf{Z}_{p_1^{a(1,1)}} \oplus \cdots \oplus \mathbf{Z}_{p_1^{a(1,s(1))}} \oplus \cdots \oplus \mathbf{Z}_{p_t^{a(t,1)}} \oplus \cdots \oplus \mathbf{Z}_{p_t^{a(t,s(t))}}$ where for each $1 \leq i \leq t$, $a(i, 1) \geq \cdots \geq a(i, s(i))$ and $(a(i, 1), a(i, 2), \ldots, a(i, s(i)))$ is uniquely determined by G.

Proof By Theorem 7.17, $G \cong \mathbf{Z}_{n(1)} \oplus \cdots \oplus \mathbf{Z}_{n(k)}$ with $n(j) \mid n(j-1)$ for $2 \leq j$. Write $n(j) = p_1^{a(1,j)} p_2^{a(2,j)} \cdots p_t^{a(t,j)}$ with each $a(i, j) \geq 0$. Since $n(j) \mid n(j-1)$ for $j \geq 2$, by Theorem 1.18, $a(i, 1) \geq a(i, 2) \geq \cdots \geq a(i, k) \geq 0$ for all $1 \leq i \leq t$. Thus from Theorem 7.11, $\mathbf{Z}_{n(j)} \cong \mathbf{Z}_{p_1^{a(1,j)}} \oplus \cdots \oplus \mathbf{Z}_{p_t^{a(t,j)}}$, omitting any summand with $a(i, j) = 0$. Using Theorem 7.10

to rearrange the summands, it follows that

$$G \cong \mathbf{Z}_{p_1^{a(1,1)}} \oplus \cdots \oplus \mathbf{Z}_{p_1^{a(1,s(1))}} \oplus \cdots \oplus \mathbf{Z}_{p_t^{a(t,1)}} \oplus \cdots \oplus \mathbf{Z}_{p_t^{a(t,s(t))}}$$

where $s(i)$ is maximal, satisfying $a(i, s(i)) > 0$. Let A be this direct sum. Using Theorem 3.20 on orders of elements in direct sums, $x \in A_j = \{g \in A \mid o(g) = p_j^d$, some $d \geq 0\}$ exactly when the coordinates of x other than in $\{\mathbf{Z}_{p_j^{a(j,v)}}\}$ are the identity, so $\mathbf{Z}_{p_j^{a(j,1)}} \oplus \cdots \oplus \mathbf{Z}_{p_j^{a(j,s(j))}} \cong A_j \cong G_j = \{g \in G \mid o(g) = p_j^d$, some $d \geq 0\}$, uniquely determined by G. Thus $(p_j^{a(j,1)}, \ldots, p_j^{a(j,s(j))})$ is uniquely determined by G from Theorem 7.18, proving the second statement of the theorem. If also $G \cong \mathbf{Z}_{m(1)} \oplus \cdots \oplus \mathbf{Z}_{m(v)}$ with $m(j) \mid m(j-1)$ for $2 \leq j$, where $m(j) = p_1^{b(1,j)} \cdots p_t^{b(t,j)}$, we get a corresponding representation for G as a direct sum of cyclic groups of orders $\{p_i^{b(i,j)}\}$. But we saw above that there is only one such representation for G, so $a(i, j) = b(i, j)$ for all i and j, which means that $n(j) = m(j)$ for all j, proving the theorem. ∎

DEFINITION 7.1 If G is a finite Abelian group, the integers $n(1), \ldots, n(k)$ in Theorem 7.19 are the **invariant factors** of G and their prime power divisors $p_j^{a(j,i)}$, listed in descending order for each prime, are the **elementary divisors** of G.

Each finite Abelian group is determined up to isomorphism either by its invariant factors or by its elementary divisors, from Theorem 7.19. Given the elementary divisors, we can find the invariant factors: $n(1)$ is the product of the largest elementary divisor for each prime, $n(2)$ is the product of the second largest elementary divisor for each prime, if there is one, etc. Thus if the elementary divisors are 4, 4, 2, 2, 9, 9, 3, 5, 5, 5, 5, 5, then the invariant factors are 180, 180, 30, 10, 5.

Example 7.10 Describe all nonisomorphic Abelian groups of order 360.

The elementary divisors are easier to write systematically than the invariant factors. We consider each prime power divisor of $360 = 2^3 \cdot 3^2 \cdot 5$ and then choose one partition of the exponent for each prime. The partitions of 3 are (3), (2, 1), and (1, 1, 1), the partitions of 2 are (2), and (1, 1), and the only partition of 1 is (1). Since there are $3 \cdot 2 \cdot 1 = 6$ ways of choosing one partition for each prime, there are exactly six nonisomorphic Abelian groups of order 360. We list each of these, first as the direct sum corresponding to the elementary divisors, then as the direct sum defined by the invariant factors:

using elementary divisors	using invariant factors
$\mathbf{Z}_8 \oplus \mathbf{Z}_9 \oplus \mathbf{Z}_5$	$\mathbf{Z}_{360}$
$\mathbf{Z}_8 \oplus \mathbf{Z}_3 \oplus \mathbf{Z}_3 \oplus \mathbf{Z}_5$	$\mathbf{Z}_{120} \oplus \mathbf{Z}_3$
$\mathbf{Z}_4 \oplus \mathbf{Z}_2 \oplus \mathbf{Z}_9 \oplus \mathbf{Z}_5$	$\mathbf{Z}_{180} \oplus \mathbf{Z}_2$
$\mathbf{Z}_4 \oplus \mathbf{Z}_2 \oplus \mathbf{Z}_3 \oplus \mathbf{Z}_3 \oplus \mathbf{Z}_5$	$\mathbf{Z}_{60} \oplus \mathbf{Z}_6$
$\mathbf{Z}_2 \oplus \mathbf{Z}_2 \oplus \mathbf{Z}_2 \oplus \mathbf{Z}_9 \oplus \mathbf{Z}_5$	$\mathbf{Z}_{90} \oplus \mathbf{Z}_2 \oplus \mathbf{Z}_2$
$\mathbf{Z}_2 \oplus \mathbf{Z}_2 \oplus \mathbf{Z}_2 \oplus \mathbf{Z}_3 \oplus \mathbf{Z}_3 \oplus \mathbf{Z}_5$	$\mathbf{Z}_{30} \oplus \mathbf{Z}_6 \oplus \mathbf{Z}_2$

Using the idea in Example 7.10, if $n = p_1^2 p_2^2 \cdots p_k^2$ for $p_1 < \cdots < p_k$ primes then there are 2^k nonisomorphic Abelian groups of order n. The elementary divisors for each prime must be (p_i^2), or (p_i, p_i), so there are 2^k choices in all. For the choice $p_1, p_1, \ldots, p_k, p_k$, the corresponding invariant factors are $(p_1 p_2 \cdots p_k, p_1 p_2 \cdots p_k)$; for the choice $p_1^2, \ldots, p_k^2$, the invariant factors are just n, so the group is $\mathbf{Z}_n$; and when $k = 2s$, the choice $p_1^2, p_2, p_2, \ldots, p_{2s-1}^2, p_{2s}, p_{2s}$ yields the invariant factors $(p_1^2 p_2 p_3^2 p_4 \cdots p_{2s-1}^2 p_{2s}, p_2 p_4 \cdots p_{2s})$. For $m = p_1^4 \cdots p_k^4$ there are 5^k nonisomorphic Abelian groups of order m since for each prime there are five choices for the elementary divisors: (p^4), (p^3, p), (p^2, p^2), (p^2, p, p), and (p, p, p, p). For example, if $m = 3^4 \cdot 5^4$ and G has elementary divisors $(3^2, 3, 3, 5^2, 5^2)$, then the invariant factors are (225, 75, 3).

The **exponent** of a finite group G is $Exp(G) = \min\{m \in N \mid g^m = e \text{ for all } g \in G\}$ (see Exercises §4.5). By Theorem 4.7, $g^{|G|} = e$ for all $g \in G$, so $Exp(G)$ exists by Well Ordering. For example, $Exp(\mathbf{Z}_n) = n$ since $o([1]_n) = n$ and $Exp(S_3) = 6$ although each $\theta \in S_3$ has order 1, 2 or 3. It is an exercise to show that $Exp(D_5) = 10$, $Exp(D_6) = 6$, $Exp(A_4) = 12$, and $Exp(S_5) = 60$, so $Exp(G) = |G|$ can occur for non-Abelian groups. For Abelian groups, we have

THEOREM 7.20 A finite Abelian group G is cyclic $\Leftrightarrow |G| = Exp(G)$.

Proof If $G = \langle x \rangle$ then $o(x) = |G|$ by Theorem 2.16, so certainly $Exp(G) \geq |G|$. But $Exp(G) \leq |G|$ always holds, so $Exp(G) = |G|$. Now let G be Abelian with invariant factors $n(1), \ldots, n(k)$. By Theorem 7.19 and Theorem 3.20 on orders in direct sums, $g^{n(1)} = e$ for all $g \in G$, so $Exp(G) \leq n(1)$. Hence $|G| = Exp(G) \Rightarrow |G| = n(1)$, so $n(1)$ is the only invariant factor and $G \cong \mathbf{Z}_{|G|}$. ■

As another consequence of Theorem 7.19, every elementary divisor of a finite Abelian group G is a prime $\Leftrightarrow Exp(G) = p_1 \cdots p_k$ for $p_1 < \cdots < p_k$ primes. When $Exp(G) = p_1 \cdots p_k$ for $p_1 < \cdots < p_k$ primes then G contains no element of order p_j^m with $m > 1$, so every elementary divisor is a prime. If each elementary divisor is a prime, G is isomorphic to a direct sum of cyclic groups of prime order, so by Theorem 3.20, $Exp(G)$ is not divisible by any p^2.

A useful application of Theorem 7.19 concerns cyclic groups.

THEOREM 7.21 A finite Abelian group is cyclic $\Leftrightarrow$ for each prime divisor p of its order there are exactly p elements satisfying $x^p = e$.

Proof If $G = \langle x \rangle$ and $p \mid |G|$ for p a prime then G contains a unique subgroup H of order p by Theorem 3.1, and if $g \in G$ with $g^p = e$ then $g \in H$ by Theorem 2.16. Thus the p elements of H are precisely the elements in G satisfying $x^p = e$. Assume now that for any prime $p \mid |G|$, $\{x \in G \mid x^p = e\}$ has exactly p elements. If G has more than one invariant factor, then using Theorem 7.19, if a prime $q \mid n(2)$ we have $q \mid n(1)$ as well. Since $G \cong \mathbf{Z}_{n(1)} \oplus \mathbf{Z}_{n(2)} \oplus \cdots$, and since there are $\langle x \rangle \leq \mathbf{Z}_{n(1)}$ and $\langle y \rangle \leq \mathbf{Z}_{n(2)}$ with $|\langle x \rangle| = |\langle y \rangle| = q$, then all q^2 elements in G corresponding to $\{(x^i, y^j, [0], \ldots, [0]) \mid 0 \leq i, j \leq q\}$ satisfy $g^q = e$, contradicting our assumption. Therefore, G has only one invariant factor, so $G \cong \mathbf{Z}_{n(1)}$ is cyclic. ■

The proof of Theorem 7.21 shows again why any finite subgroup of the multiplicative group of a field must be cyclic (Theorem 4.20): more than one invariant factor gives rise to too many solutions of an equation $x^p = 1$.

EXERCISES

1. If G is an Abelian group with $|G| = p^n$ for p a prime, show G is cyclic $\Leftrightarrow$ there is no surjective homomorphism $\varphi : G \to \mathbf{Z}_p \oplus \mathbf{Z}_p$.

2. Let p be a prime and G an Abelian group.

a) How many subgroups does G have:

i) if G has elementary divisors (p, p)?

ii) if G has elementary divisors (p^2, p)?

b) How many cyclic subgroups does G have:

iii) if G has elementary divisors (p, p, p, p)?

iv) if G has elementary divisors (p^2, p^2)?

v) if G has elementary divisors (p^2, p^2, p)?

3. If G is Abelian, determine how many subgroups G has if:

i) $|G| = 77$

ii) $|G| = 45$ and $x^{15} = e$ for all $x \in G$

iii) $|G| = 100$ and $x^{10} = e$ for all $x \in G$

iv) $|G| = 2310$

v) $|G| = 120$, with invariant factors (60, 2)

4. For any Abelian group G with $|G| = n$, find the maximum possible number of elements in G of order m:

i) $n = 243$ $m = 9$

ii) $n = 36$ $m = 6$

iii) $n = 24$ $m = 6$

iv) $n = 1000$ $m = 10$

5. If G is a finite Abelian group, find all possibilities for the invariant factors of G and the corresponding elementary divisors when $|G|$ is:

i) 99

ii) 225

iii) 144

iv) 64

v) 385

vi) 900

vii) 1000

6. Let G be an Abelian group, $|G| = p^n$ for p a prime, and $G^p = \{g^p \in G \mid g \in G\} \leq G$. Show that $|G/G^p| = p^k \Leftrightarrow G$ has exactly k elementary divisors.

7. If G and H are finite Abelian groups and for a prime p, $G^p = \{g^p \mid g \in G\}$, does $|H^p| \leq |G^p|$ for all primes $p \Rightarrow H$ is a homomorphic image of G?

8. Let G be Abelian, $|G| = p^n$ for p a prime, and suppose G has exactly k elementary divisors. If $H \leq G$ and $H \cong H_1 \oplus \cdots \oplus H_s$, show that $s \leq k$.

9. Let G and H be finite Abelian groups and $G^p = \{g^p \mid g \in G\}$ for p a prime.

i) If $|G| \neq p^n$ for any prime p, and if $G^q \cong H^q$ for each $q \mid |G|$, show that $G \cong H$.

ii) In general, if for some prime p, $G^p \cong H^p$ and $G/G^p \cong H/H^p$, then show that $G \cong H$.

10. Let G be a finite group so that every subgroup is normal and has a complement (if $H \leq G$ then there is $K \leq G$ with $G = HK$ and $H \cap K = \langle e \rangle$).

i) Show that G is Abelian.

ii) For any $g \in G$, show $o(g)$ is square free (i.e., a product of distinct primes).

11. **a)** If A, B, and C are finite Abelian groups, show $A \oplus C \cong B \oplus C \Rightarrow A \cong B$.
b) Show that a) can fail when C is an infinite Abelian group.

12. Let G be Abelian with elementary divisors $(p^{n(1)}, \ldots, p^{n(k)})$, $|G| = p^n$, and $\langle x_j \rangle \leq G$ so that $o(x_j) = p^{n(j)}$ and $G \cong \langle x_1 \rangle \oplus \cdots \oplus \langle x_k \rangle$. Show that:
i) $G/\langle x_j \rangle$ has elementary divisors $(p^{n(1)}, \ldots, p^{n(j-1)}, p^{n(j+1)}, \ldots, p^{n(k)})$
ii) if $N = \{g \in G \mid g^p = e\}$, then either $G = N \cong \mathbf{Z}_p^n$ or there is $1 \leq t$ so that $n(r) = 1 \Leftrightarrow t < r$ and G/N has $(p^{n(1)-1}, \ldots, p^{n(t)-1})$ as elementary divisors

13. Let G be an Abelian group with $|G| = p^n$ for a prime p. If $(p^{n(1)}, \ldots, p^{n(k)})$ are the elementary divisors of G and if $H \leq G$ has $(p^{m(1)}, \ldots, p^{m(s)})$ as elementary divisors, show that $s \leq k$ and that for each $1 \leq j \leq s$, $m(j) \leq n(j)$. (One approach: Induction using Exercise 12 on G/N with $N = \{g \in G \mid g^p = e\}$.)

14. Let G be a finite Abelian group with invariant factors $(n(1), \ldots, n(k))$, and let $H \leq G$ have invariant factors $(m(1), \ldots, m(s))$. Show that $s \leq k$ and that for $1 \leq j \leq s$, $m(j) \mid n(j)$. (See Exercise 13.)

15. Let G be an Abelian group with invariant factors $(p^{n(1)}, \ldots, p^{n(k)})$ for p a prime, and let $H \leq G$. If the invariant factors of G/H are $(p^{m(1)}, \ldots, p^{m(s)})$, show that $s \leq k$ and for $1 \leq j \leq s$, $m(j) \leq n(j)$. (Hint: If $G = \langle x_1 \rangle \cdots \langle x_k \rangle$ gives G as a direct sum of subgroups of orders $p^{n(1)}, \ldots, p^{n(k)}$, respectively, then either $H \cap \langle x_i \rangle = \langle e \rangle$ for some $i \leq k$ and Exercise 12 i) and induction on $|G|$ can be used, or $g \in H$ for all $g \in G$ with $o(g) = p$ and Exercise 12 ii) can be used.)

16. Recall that $Exp(G) = \min\{n \in \mathbf{N} \mid g^n = e \text{ for all } g \in G\}$. If G is an Abelian group with $Exp(G)$ square free (distinct prime factors), then show that $Aut(G) \cong GL(n_1, \mathbf{Z}_{p_1}) \oplus \cdots \oplus GL(n_k, \mathbf{Z}_{p_k})$ for $p_1, \ldots, p_k$ the prime divisors of $|G|$ and some $n_i \in \mathbf{N}$.

17. Let G be a finite group. If for any $\varphi \in Aut(G)$, $\varphi(g) = g \Rightarrow g = e_G$ or $\varphi = I_G$:
i) Show that G is an Abelian group.
ii) Show that G is not cyclic of order p^k for p a prime and $k > 1$.
iii) Show that $G \cong \mathbf{Z}_p$ for p a prime.

CHAPTER

8

CONJUGATION

8.1 CONJUGATES

The idea behind Lagrange's Theorem is to partition a group G by cosets and compare the number of elements in the cosets with $|G|$. Here a different partition of G leads to a useful relation between the order of G and its center, which is helpful in studying non-Abelian groups. Our previous results and methods do not seem sufficient to answer some simple questions about non-Abelian groups G. For example, is there an element of order p for each prime $p \mid |G|$, and for which $m \mid |G|$ are there subgroups of order m? Our examination of groups of small order in the last chapter shows the need for additional results and techniques applicable to non-Abelian groups.

Much of this chapter is a special case of the notion of *groups acting on sets,* discussed in the next chapter. Here we illustrate the general idea concretely by studying *conjugation.* This leads to results about groups of order p^n for p a prime, and to the existence of subgroups of prime power order in finite groups. Other fundamental and useful information about such subgroups, known as the Sylow Theorems, requires the general notion of groups acting on sets, discussed in the next chapter. We begin with the definition of conjugate that we have seen and used before.

DEFINITION 8.1 For $x \in G$ a group, and $H \leq G$, a **conjugate of x** is $g^{-1}xg$ for any $g \in G$, and a **conjugate of H** is $g^{-1}Hg$ for any $g \in G$.

Conjugates are uninteresting when G is Abelian since $g^{-1}xg = x$ for all $x, g \in G$. In general $e^{-1}xe = x$, and for $H \leq G, e^{-1}He = H$, so every element and each subgroup of G is a conjugate of itself. Also $g^{-1}Hg \leq G$ when H is, and $N \lhd G$ exactly when the only conjugate of N is N itself (verify!). Since $g^{-1}eg = e$ the only conjugate of e is itself, and similarly, the only conjugate of $y \in Z(G)$ is itself. We call $g^{-1}xg$ the conjugate of x **by** g. Two examples in D_5 are $W_2T_3W_2 = W_2W_5 = T_{-3} = T_2$, and $T_3^{-1}W_2T_3 = T_3^{-1}W_{-1} = W_{-4} = W_1$. In S_5 we have $(1, 3, 5)^{-1}(2, 3, 4, 5)(1, 3, 5) = (1, 5, 3)(2, 3, 4, 5)(1, 3, 5) = (1, 4, 3, 2)$ and $((1, 2)(3, 4))^{-1}(1, 3, 5)(2, 4)((1, 2)(3, 4)) = (1, 3)(2, 4, 5)$.

Our first result is that conjugacy is an equivalence relation: its classes give the partition we need. As we observed above, conjugacy is the identity relation for Abelian groups.

THEOREM 8.1 For G a group and $x, y \in G$, set $x \sim_c y$ if $y = g^{-1}xg$ for some $g \in G$. Then $\sim_c$ is an equivalence relation on G.

Proof We must verify that $\sim_c$ satisfies the three properties of an equivalence relation. For reflexivity, we have seen that $e^{-1}xe = x$ for any $x \in G$ so $x \sim_c x$. Assume that $x \sim_c y$ with $y = g^{-1}xg$ for some $g \in G$. To show $y \sim_c x$ we need $x = h^{-1}yh$. But $y = g^{-1}xg$ implies $gyg^{-1} = x$, so taking $h = g^{-1}$ shows that $\sim_c$ is symmetric. Finally, if $x \sim_c y$ and $y \sim_c z$ then $y = g^{-1}xg$ and $z = h^{-1}yh$ for $g, h \in G$. Substitution yields $z = h^{-1}(g^{-1}xg)h = (gh)^{-1}xgh$ and proves that $\sim_c$ is transitive. Therefore $\sim_c$ is an equivalence relation. ∎

The relation $x \sim_c y$ is called **conjugacy** and, as for any equivalence relation, the set G is partitioned by its equivalence classes (Theorem 1.22).

DEFINITION 8.2 The equivalence classes of $\sim_c$ are called the **conjugacy classes of G** and the class of $x \in G$ is denoted $cls(x) = \{g^{-1}xg \mid g \in G\}$.

From Theorem 1.21, $cls(x) = cls(y) \Leftrightarrow x \sim_c y$. From the examples above, $cls(T_3) = cls(T_2)$ and $cls(W_2) = cls(W_1)$ in D_5, and $cls((2, 3, 4, 5)) = cls((1, 4, 3, 2))$ in S_5. Since $\{cls(x) \mid x \in G\}$ is a partition of G, when G is finite and $T \subseteq G$ is a transversal for the set of classes then $|G| = \sum_{t \in T} |cls(t)|$. We want some way to compute $|cls(x)|$, the number of conjugates of x. Conjugating x by different $g \in G$ will not always produce different conjugates, for example $g^{-1}xg = (xg)^{-1}x(xg)$. When $g^{-1}xg = h^{-1}xh$ then $hg^{-1}x = xhg^{-1}$ so $hg^{-1} \in C_G(\{x\}) = \{y \in G \mid xy = yx\} \leq G$, the centralizer of x in G (see Definition 2.11 and Theorem 2.17). We abbreviate $C_G(\{x\})$ to $C(x)$. Next we identify $cls(x)$ with the cosets $\{C(x)g \mid g \in G\}$, the number of which is given by Lagrange's Theorem.

THEOREM 8.2 For G a group and $x \in G$, $|cls(x)| = [G : C(x)]$.

Proof We need a bijection $\varphi : cls(x) \to \{C(x)g \mid g \in G\}$. A natural guess is $\varphi(g^{-1}xg) = C(x)g$, but it not clear that φ is a function since $g^{-1}xg = h^{-1}xh$ does not imply $g = h$. As above, $g^{-1}xg = h^{-1}xh \Leftrightarrow hg^{-1}x = xhg^{-1} \Leftrightarrow hg^{-1} \in C(x)$. Consequently, $g^{-1}xg = h^{-1}xh \Leftrightarrow C(x)h = C(x)g$ by Theorem 4.2. Therefore $\varphi : cls(x) \to \{C(x)g \mid g \in G\}$ and is injective; it is also surjective since for any $g \in G$, $\varphi(g^{-1}xg) = C(x)g$. Hence φ is a bijection, completing the proof. ∎

The theorem gives information about centralizers in D_n. Using $W_i T_j W_i = T_j^{-1}$ and that any T_j and T_k commute, $cls(T_j) = \{T_j, T_j^{-1}\}$. Thus when $n = 2m$ then $T_m^{-1} = T_m$ and $|cls(T_m)| = 1 = [D_n : C(T_m)]$. This shows again that $C(T_m) = D_n$; that is, $T_m \in Z(D_n)$. When n is odd and $T_j \neq I$, $|cls(T_j)| = 2 = [D_n : C(T_j)]$, so exactly n elements in D_n commute with T_j. Since $\langle T_1 \rangle \leq C(T_j)$ we have $C(T_j) = \langle T_1 \rangle$. In the next section we look in more detail at computations like these in S_n and describe $C(\theta)$ for various $\theta \in S_n$.

It is not easy to compute $[G : C(x)]$ so it is not yet clear how useful Theorem 8.2 is. There are special cases for which there is enough information about $[G : C(x)]$ to enable us to relate $|G|$ to $|Z(G)|$, but first we need a simple, crucial observation.

THEOREM 8.3 For a group G and $x \in G$, the following are equivalent:

i) $cls(x) = \{x\}$
ii) $x \in Z(G)$
iii) $C(x) = G$
iv) $[G{:}C(x)] = 1$

Proof Now $cls(x) = \{x\} \Leftrightarrow g^{-1}xg = x$ for all $g \in G \Leftrightarrow xg = gx$ for all $g \in G$. Thus i) and ii) are equivalent. The definitions of $C(x)$ and of index show that ii), iii), and iv) are equivalent statements, so the theorem is proved. ■

We have all the pieces needed to prove the **class equation:** the numerical consequence of G being partitioned by its conjugacy classes.

THEOREM 8.4 If G is a finite group then G is Abelian or there is some $T \subseteq G - Z(G)$ so that $|G| = |Z(G)| + \sum_{t \in T}[G : C(t)]$ and each $[G : C(t)] > 1$.

Proof By Theorem 8.1, $\sim_c$ is an equivalence relation on G so $\{cls(x) \mid x \in G\}$ is a partition of G by Theorem 1.22. Let $W \subseteq G$ be a transversal for the set of classes. Since there is a unique $w \in W$ in each conjugacy class it follows from Theorem 0.3 that $|G| = \sum_{w \in W} |cls(w)|$. We know from Theorem 8.2 that $|cls(w)| = [G : C(w)]$ and from Theorem 8.3 that $|cls(w)| = 1 \Leftrightarrow w \in Z(G)$. But if $z \in Z(G)$ then $cls(z) = \{z\}$, so the only choice of $t \in cls(z)$ is z itself, forcing $Z(G) \subseteq W$. Thus $|G| = \sum_{w \in Z(G)} |cls(w)| + \sum_{t \in T} |cls(t)|$ for $T = W - Z(G)$. Now $\sum_{w \in Z(G)} |cls(w)| = \sum_{z \in Z(G)} 1 = |Z(G)|$ so we have $|G| = |Z(G)| + \sum_{t \in T}[G : C(t)]$ as required, and each $[G : C(t)] > 1$ by Theorem 8.3 since $t \notin Z(G)$. ■

We will wait a bit to use Theorem 8.4. First we look at a couple of examples to illustrate this theorem. It is important in applying the theorem to realize that each $[G : C(t)] \mid |G|$ by Lagrange's Theorem.

Example 8.1 Conjugacy classes in D_5.

In D_5, $T_j^{-1} W_i T_j = T_{-j} W_i T_j = W_{i-2j}$, so $cls(W_i)$ contains W_i, $W_{i-2} = W_{i+3}$, $W_{i-4} = W_{i+1}$, $W_{i-6} = W_{i+4}$, and $W_{i-8} = W_{i+2}$, so all the W_k (why?). Since the other conjugates are $W_k W_i W_k = W_{2k-i}$ we have $cls(W_i) = \{W_k \mid 0 \le k \le 4\}$. These computations also show $|cls(W_i)| = 5 = [D_5 : C(W_i)] = 10/2$, forcing $C(W_i) = \langle W_i \rangle$. Of course $Z(D_5) = \{I = T_0\}$ from Theorem 3.18 and $cls(I) = \{I\}$. To find a conjugate of $T_j \ne I$, other than itself, we must conjugate by W_k. We get $W_k T_j W_k = T_{-j} = T_j^{-1} \ne T_j$ since $o(T_j) = 5$ for all $1 \le j \le 4$. Hence the conjugacy classes in D_5 are $cls(W_0)$, $cls(I)$, $cls(T_1) = \{T_1, T_4\}$, and $cls(T_2) = \{T_2, T_3\}$. For each of the last two classes, $2 = [D_5 : C(T_j)]$ so $C(T_j) = \langle T_1 \rangle$ has five elements and $|D_5| = 1 + 2 + 2 + 5$, as in the theorem with $T = \{T_1, T_2, W_0\}$.

Example 8.2 Conjugacy classes in A_4.

We could compute conjugates in A_4 but use the theorem instead. All $(a, b)(c, d) \in A_4$ of cycle structure (2, 2), together with I, form $H \lhd A_4$ by

Example 5.10, so the conjugates of these elements must have the same form. Now H is commutative so $H \subseteq C((a,b)(c,d)) = C$ and $|C| \geq 4$ but $C \neq A_4$ (why?). By Lagrange's Theorem, $3 = [A_4 : H] = [A_4 : C][C : H]$, $H = C$ follows, and by Theorem 8.2 $|cls((a,b)(c,d))| = 3$. Thus $cls((a,b)(c,d)) = H - \langle I \rangle$, and for all $\theta \in A_4 - H$, $\theta \notin C$. Now θ is a 3-cycle, so no 3-cycle commutes with any $(a,b)(c,d)$. Thus $Z(A_4) = \langle I \rangle$ and if $\rho = (a,b,c)$, $C(\rho)$ contains only 3-cycles and I. This forces $|C(\rho)|$ to be odd (Exercise 10 in §2.2) and Lagrange's Theorem shows $|C(\rho)| \mid 12$ so $|C(\rho)| = 3$. From Theorem 8.2, $|cls(\rho)| = [A_4 : C(\rho)] = 4$ so the eight 3-cycles in A_4 form two conjugacy classes. Computing $h^{-1}(1,2,3)h$ for $h \in H$ shows $cls((1,2,3)) = \{(1,2,3), (1,3,4), (1,4,2), (2,4,3)\}$ and then $cls((1,3,2)) = \{\eta^{-1} \mid \eta \in cls((1,2,3))\}$. One choice for T in Theorem 8.4 for A_4 is $T = \{(1,2)(3,4), (1,2,3), (1,3,2)\}$, and using $|cls(t)| = [A_4 : C(t)]$ we get $|A_4| = 12 = 1 + 3 + 4 + 4$.

EXERCISES

1. a) If $x \sim_c y$ in a group G, show that x and y have infinite order or $o(x) = o(y)$.
b) If $g, h \in G$, show that gh and hg have infinite order or $o(gh) = o(hg)$.

2. For G a group, let $S(G) = \{H \subseteq G \mid H \leq G\}$ and for $H, K \in S(G)$, set $H \# K$ when $K = g^{-1}Hg$ for some $g \in G$. a) Show that # is an equivalence relation on $S(G)$.
b) Using the argument in Theorem 8.2, show that the cardinality of each equivalence class of # on $S(G)$ is the index of some subgroup of G.

3. Describe all of the conjugacy classes in D_n for any $n \geq 3$.

4. For $H \leq G$, show that $H \lhd G \Leftrightarrow H$ is a union of conjugacy classes of G.

5. In any group G, show that $|cls(g)| = |cls(g^{-1})|$ for each $g \in G$.

6. If the finite group G has exactly two conjugacy classes, show that $|G| = 2$.

7. If G is finite with exactly three conjugacy classes, show $G \cong \mathbf{Z}_3$ or $G \cong D_3$.

8. If the commutator subgroup G' of G is finite, show that every $cls(g)$ is finite.

9. For G a group, set $\mathrm{FC}(G) = \{g \in G \mid cls(g) \text{ is finite}\}$. Show:
i) $\mathrm{FC}(G) \leq G$ and $\mathrm{FC}(G) \lhd G$
ii) $[G : Z(G)]$ finite $\Rightarrow \mathrm{FC}(G) = G$
iii) $\mathrm{FC}(G) = G$ and $G = \langle\{g_1, \ldots, g_k\}\rangle$ for $g_1, \ldots, g_k \in G \Rightarrow [G : Z(G)]$ is finite

8.2 CONJUGATES AND CENTRALIZERS IN S_n

The key to studying conjugates in S_n is the following observation that shows conjugates of k-cycles are again k-cycles.

THEOREM 8.5 For $\theta, \rho = (a_1, a_2, \ldots, a_k) \in S_n$, $\theta\rho\theta^{-1} = (\theta(a_1), \ldots, \theta(a_k))$.

Proof $\eta = (\theta(a_1), \ldots, \theta(a_k))$ acts on $I(n) = \{1, \ldots, n\}$ by $\eta(\theta(a_i)) = \theta(a_{i+1})$ where $a_{k+1} = a_1$, and if $j \in I(n) - \{\theta(a_1), \ldots, \theta(a_k)\}$ then $\eta(j) = j$. To see that $\theta\rho\theta^{-1} = \eta$ we must show $\theta\rho\theta^{-1}(x) = \eta(x)$ for all $x \in I(n)$. Since ρ is a cycle, $\theta\rho\theta^{-1}(\theta(a_i)) = \theta\rho(\theta^{-1}(\theta(a_i))) = \theta\rho(a_i) = \theta(a_{i+i})$. If $x \neq \theta(a_i)$ then $\theta^{-1}(x) \neq a_i$ for any i, so $\rho(\theta^{-1}(x)) = \theta^{-1}(x)$ and $\theta\rho\theta^{-1}(x) = x$. Thus $\theta\rho\theta^{-1} = (\theta(a_1), \ldots, \theta(a_k))$ from Theorem 0.8. ∎

By the theorem: $(1, 3, 4, 5)(2, 4, 5, 7)(1, 3, 4, 5)^{-1} = (2, 5, 1, 7)$ in S_7 since $(1, 3, 4, 5)$ fixes 2 and 7, sends 4 to 5 and 5 to 1; $(1, 3, 6)(2, 6)(3, 5)(1, 3, 6)^{-1} = (1, 2)(5, 6)$ since $(1, 3, 6)$ sends 6 to 1, 3 to 6 and fixes 2 and 5; and finally $((1, 2)(3, 4)(5, 6))(1, 2, 3, 4, 5, 6) \cdot ((1, 2)(3, 4)(5, 6)) = (2, 1, 4, 3, 6, 5)$.

Example 8.3 S_6 contains two nonconjugate cyclic subgroups of order 6.

Set $K = \langle(1, 2, 3, 4, 5, 6)\rangle$ and $L = \langle(1, 2, 3)(4, 5)\rangle$. Now K and L are both cyclic of order 6 but $\eta^{-1}K\eta \neq L$ for all $\eta \in S_6$ since $\eta^{-1}K\eta$ must contain $\eta^{-1}(1, 2, 3, 4, 5, 6)(\eta^{-1})^{-1}$, a 6-cycle by Theorem 8.5, and L contains no 6-cycle.

Theorem 8.5 implies the useful fact that the conjugates of any $\theta \in S_n$ are exactly the permutations with the same cycle structure as θ.

THEOREM 8.6 In S_n, $\eta \in cls(\theta) \Leftrightarrow \eta$ and θ have the same cycle structure.

Proof Using Theorem 3.10, $\theta = \rho_1\rho_2 \cdots \rho_t$ for $\{\rho_j\}$ pairwise disjoint cycles, one for each orbit of θ, where ρ_j is a k_j-cycle, $\sum k_i = n$, and the cycle structure of θ is $(k_1, \ldots, k_t)$. If $g \in S_n$, $g\theta g^{-1} = g\rho_1\rho_2 \cdots \rho_t g^{-1} = (g\rho_1 g^{-1}) \cdots (g\rho_t g^{-1})$ so for $\rho_j = (a_{j1}, a_{j2}, \ldots, a_{jk_j})$, $g\rho_j g^{-1} = (g(a_{j1}), g(a_{j2}), \ldots, g(a_{jk_j}))$ is a k_j-cycle by Theorem 8.5. Hence $g\theta g^{-1}$ is the product of cycles of the same lengths as in θ. The disjointness of $\{\rho_j\}$ shows there are no repetitions among the a_{ji}, $\{a_{ji}\} = I(n)$, and since $g \in S_n$, $\{g(a_{ji})\} = I(n)$ and there are no repetitions among the $g(a_{ji})$. Thus $\{g\rho_j g^{-1}\}$ is pairwise disjoint, so $g\theta g^{-1}$ has the same cycle structure as θ and proves that $cls(\theta) \subseteq \{\eta \in S_n \mid \eta$ has the same cycle structure as $\theta\}$.

Now let $\theta, \eta \in S_n$ both have cycle structure $(k_1, \ldots, k_t)$. As above, $\theta = (a_{11}, \ldots, a_{1k_1}) \cdot (a_{21}, \ldots, a_{2k_2}) \cdots (a_{t1}, \ldots, a_{tk_t})$, a product of pairwise disjoint cycles, one for each orbit of θ on $I(n) = \{1, \ldots, n\}$. Since η has this same cycle structure, we can write $\eta = (b_{11}, \ldots, b_{1k_1})(b_{21}, \ldots, b_{2k_2}) \cdots (b_{t1}, \ldots, b_{tk_t})$. We claim $\{a_{ji}\}$ and $\{b_{ji}\}$ can be matched. Every orbit of θ and of η is represented by a cycle, so $I(n) = \{a_{ij}\} = \{b_{uv}\}$, and the cycles appearing in each of θ and η are pairwise disjoint so no two a_{ij} are equal and no two b_{uv} are equal. The cycles in θ and η have matching lengths so the subscripts appearing in $\{a_{ij}\}$ and $\{b_{uv}\}$ are the same. Therefore $\sigma(a_{ji}) = b_{ji}$ defines $\sigma \in S_n$ and $\sigma\theta\sigma^{-1} = (\sigma(a_{11}, \ldots, a_{1k_1})\sigma^{-1}) \cdots (\sigma(a_{t1}, \ldots, a_{tk_t})\sigma^{-1}) = (b_{11}, \ldots, b_{1k_1})(b_{21}, \ldots, b_{2k_2}) \cdots (b_{t1}, \ldots, b_{tk_t}) = \eta$ using Theorem 8.5. Thus θ and η are conjugate so $\{\eta \in S_n \mid \eta$ has the same cycle structure as $\theta\} \subseteq cls(\theta)$, completing the proof of the theorem. ∎

The basic idea of the proof is quite simple and we look at a specific example to make it clear. Suppose $(1, 3, 4)(2, 7, 5), (2, 6, 5)(4, 7, 3) \in S_7$. As in the theorem we write $\alpha = (1, 3, 4)(2, 7, 5)(6)$, $\beta = (2, 6, 5)(4, 7, 3)(1)$, then get $\beta = \sigma\alpha\sigma^{-1}$ for $\sigma \in S_7$ that takes 1 to 2, 3 to 6, 4 to 5, 2 to 4, 7 to 7, 5 to 3, and 6 to 1. Thus $\sigma = (1, 2, 4, 5, 3, 6)$ but is not unique in S_n, satisfying $\sigma\alpha\sigma^{-1} = \beta$. For example, rewriting $\alpha = (3, 4, 1)(2, 7, 5)(6)$ and $\beta = (4, 7, 3)(2, 6, 5)(1)$ leads to $\sigma = (1, 3, 4, 7, 6)$, and there are many similar variations in the expressions for α and β that lead to different σ conjugating α into β.

From Chapter 3, cycle structures in S_n correspond to different ways of writing n as nonincreasing sums of positive integers: the **partitions** of n. By Theorem 8.6, the number of conjugacy classes in S_n is the number of partitions of n. Since there are seven partitions of five, namely (5), (4, 1), (3, 2), (3, 1, 1), (2, 2, 1), (2, 1, 1, 1), and (1, 1, 1, 1, 1), there are seven conjugacy classes in S_5: $cls((1,2,3,4,5))$, $cls((1,2,3,4))$, $cls((1,2,3)(4,5))$, $cls((1,2,3))$, $cls((1,2)(3,4))$, $cls((1,2))$, and $cls(I)$. In S_n all 3-cycles form one conjugacy class, as do all products of a 3-cycle with a disjoint transposition. However, Example 8.2 shows that the 3-cycles in A_4 form *two* conjugacy classes of four elements each. The difference comes about since in S_4 it is necessary to conjugate (1, 2, 3) by an odd permutation to get some 3-cycles, such as $(1,3,2)$.

Let us look at a few other consequences of Theorem 8.6 for conjugate subgroups and centralizers in S_n.

Example 8.4 S_8 contains five mutually nonconjugate cyclic subgroups of order 6, and any cyclic subgroup of order 6 is conjugate to one of them.

If $\theta \in S_8$ and $o(\theta) = 6$, by Theorem 3.11 the cycle structure of θ is (6, 2), (6, 1, 1), (3, 3, 2), (3, 2, 2, 1), or (3, 2, 1, 1, 1). Consequently, Theorem 8.6 shows $\eta^{-1}\theta\eta = \theta_i$ for $\theta_1 = (1, 2, 3, 4, 5, 6)(7, 8)$, $\theta_2 = (1, 2, 3, 4, 5, 6)$, $\theta_3 = (1, 2, 3)(4, 5, 6)(7, 8)$, $\theta_4 = (1,2,3)(4,5)(6,7)$, or $\theta_5 = (1, 2, 3)(4, 5)$, and we also have $\eta^{-1}\langle\theta\rangle\eta = \langle\eta^{-1}\theta\eta\rangle = \langle\theta_i\rangle$. If $\langle\theta_j\rangle = \rho^{-1}\langle\theta_i\rangle\rho = \langle\rho^{-1}\theta_i\rho\rangle$ then $\rho^{-1}\theta_i\rho$ generates $\langle\theta_j\rangle$ and so $o(\theta_j) = 6$ forces $\rho^{-1}\theta_i\rho = \theta_j$ or $\rho^{-1}\theta_i\rho = \theta_j^{-1}$. In either case, θ_i and θ_j have the same cycle structure by Theorem 8.6, so $i = j$.

Example 8.5 Any two cyclic subgroups of order 15 in S_8 are conjugate.

If $\langle\theta\rangle \leq S_8$ with $o(\theta) = 15$ then θ is a product of disjoint cycles of lengths 15, 5, and 3, so $\theta \in S_8$ forces θ to have cycle structure (5, 3). Should $\langle\eta\rangle \leq S_8$ with $o(\eta) = 15$ then θ and η have cycle structure $(5,3)$ so $\sigma\theta\sigma^{-1} = \eta$ for some $\sigma \in S_8$ by Theorem 8.6. Hence $\sigma\langle\theta\rangle\sigma^{-1} = \langle\sigma\theta\sigma^{-1}\rangle = \langle\eta\rangle$.

Another way to use Theorem 8.6 is to count the number of elements in $cls(\theta)$ and then try to characterize $C(\theta)$ using Theorem 8.2.

Example 8.6 If $\rho \in S_n$ is an n-cycle and $\rho\eta = \eta\rho$, then $\eta = \rho^j$ for $j \in \mathbf{N}$.

From Theorem 8.6, $cls(\rho)$ is the set of all n-cycles in S_n. How many are there? An n-cycle is $(a_1, a_2, \ldots, a_n)$ with $\{a_j\} = I(n)$ so there are n choices for a_1, any of the $n - 1$ choices in $I(n) - \{a_1\}$ for a_2, any of the $n - 2$ choices in $I(n) - \{a_1, a_2\}$ for a_3, etc. Thus there are $n \cdot (n-1) \cdots 2 \cdot 1 = n!$ ways of writing $(a_1, a_2, \ldots, a_n)$, but this permutation has n representations, one starting with any integer. That is, $(a_1, \ldots, a_n) = (a_2, \ldots, a_n, a_1) = \cdots = (a_n, a_1, \ldots, a_{n-1})$: there are $n!/n = (n-1)!$ different n-cycles and $|cls(\rho)| = (n-1)!$. By Theorem 8.2, $(n-1)! = [S_n : C(\rho)] = n!/|C(\rho)|$ using Lagrange's Theorem. Thus $|C(\rho)| = n$. But $\langle\rho\rangle \leq C(\rho)$ and $|\langle\rho\rangle| = o(\rho) = n$ by Theorem 3.8 and Theorem 2.16, so $\langle\rho\rangle = C(\rho)$. Hence if $\eta \in S_n$ and $\rho\eta = \eta\rho$ then $\eta \in \langle\rho\rangle$ so $\eta = \rho^k$ for some $k \in \mathbf{N}$.

Example 8.7 In S_n, $C((1,2)(3,4)) = \{(1,2)^i(3,4)^j((1,3)(2,4))^k\theta \mid \theta(v) = v \text{ if } 1 \leq v \leq 4\} = \{(1,2)^i(1,4,2,3)^j\theta \mid \theta(v) = v \text{ for } 1 \leq v \leq 4\} \cong D_4 \oplus S_{n-4}$.

We find $|C((1,2)(3,4))|$. Now $|cls((1,2)(3,4))| = [S_n : C((1,2)(3,4))]$ from Theorem 8.2, and by Lagrange's Theorem this is equivalent to $|C((1,2)(3,4))| = n!/|cls((1,2)(3,4))|$. Using Theorem 8.6, $cls((1,2)(3,4))$ is the set of permutations with cycle structure $(2,2,1,\ldots,1)$. To write $(a,b)(c,d)$ choose $a \in I(n)$, then any $b \in I(n) - \{a\}$, any $c \in I(n) - \{a,b\}$, and any $d \in I(n) - \{a,b,c\}$, resulting in $n(n-1)(n-2)(n-3)$ choices. But $(a,b) = (b,a)$, $(c,d) = (d,c)$, and $(a,b)(c,d) = (c,d)(a,b)$ so actually each $(a,b)(c,d)$ is counted eight times. For example, $(a,b)(c,d) = (b,a)(d,c) = (c,d)(b,a) = (d,c)(a,b)$. Hence $|cls((1,2)(3,4))| = n(n-1)(n-2)(n-3)/8$, so $|C((1,2)(3,4))| = 8(n-4)!$.

To see that $C((1,2)(3,4))$ is either of the sets described, it suffices to show that the sets are in $C((1,2)(3,4))$ and have $8(n-4)!$ elements. Observe that $(1,2), (3,4), (1,3)(2,4) \in C((1,2)(3,4))$. By Theorem 3.9, $B = \{\theta \in S_n \mid \theta(v) = v$ for $1 \leq v \leq 4\} \leq C((1,2)(3,4))$ since the cycles in $\theta \in B$ are disjoint from $(1,2)(3,4)$, and also $|B| = (n-4)!$ since $B \cong S_{n-4}$ (verify!). It is straightforward to see that $A = \{(1,2)^i(3,4)^j((1,3)(2,4))^k \mid i,j,k \in \mathbf{Z}\}$ has exactly eight elements, all of which centralize $(1,2)(3,4)$ and B. These also give different cosets of B by Theorem 4.2 since if $\alpha, \beta \in A$ with $\alpha \neq \beta$ then $\alpha^{-1}\beta \notin B$ since elements of B fix the first four integers. It follows that $C((1,2)(3,4))$ contains the eight cosets $\{(1,2)^i(3,4)^j((1,3)(2,4))^k B\}$, so their union with $8(n-4)!$ elements must be $C((1,2)(3,4))$, as claimed.

Next, note that $(3,4) \cdot (1,3)(2,4) = (1,4,2,3) \in C((1,2)(3,4))$, and from $(1,2)(1,4,2,3)(1,2) = (2,4,1,3) = (1,4,2,3)^{-1}$ we have $E = \{(1,2)^i(1,4,2,3)^j\}$ is closed under multiplication, so $E \leq C((1,2)(3,4))$ by Theorem 2.12. It is an exercise to see that $|E| = 8$ and $E = \{(1,2)^i(3,4)^j((1,3)(2,4))^k \mid 0 \leq i,j,k \leq 1\}$. Also $E \cong D_4$ by Example 6.4. The cycles representing any element in B are disjoint from those representing any element in E. Hence $EB = BE$ so $EB \leq C((1,2)(3,4))$ by Theorem 4.10. Since $E \cap B = \{I\}$, $|EB| = |E||B| = 8(n-4)!$ by Theorem 4.12. Thus $EB = \{(1,2)^i(1,4,2,3)^j\theta \mid \theta(v) = v \text{ for } 1 \leq v \leq 4\} = C((1,2)(3,4))$. Finally, since $E \leq C(B)$, $E, B \lhd EB$ so Theorem 7.9 shows that $C((1,2)(3,4)) = EB \cong E \oplus B \cong D_4 \oplus S_{n-4}$ using Theorem 7.10.

Example 8.7 applies to any $(a,b)(c,d)$. Note that $\rho((1,2)(3,4))\rho^{-1} = (a,b)(c,d)$ for $\rho \in S_n$ with $\rho(1) = a$, $\rho(2) = b$, $\rho(3) = c$, and $\rho(4) = d$. Now $C((a,b)(c,d)) = \rho C((1,2)(3,4))\rho^{-1}$ (verify!) and $T(\theta) = \rho\theta\rho^{-1}$ defines $T \in Aut(S_n)$ by Example 6.8, so $|C((a,b)(c,d))| = |C((1,2)(3,4))| = 8(n-4)!$ and $\rho C((1,2)(3,4))\rho^{-1} = \{(a,b)^i(c,d)^j \cdot ((a,c)(b,d))^k\theta \mid \theta \text{ fixes } a, b, c, \text{ and } d\}$.

EXERCISES

1. If $n > 2$, show the set of conjugacy classes in the group A_n is not the same as the set of conjugacy classes of $A_n \leq S_n$. (Some classes in S_n, contained in A_n, split into two classes in A_n. Consider cycles of maximal length in A_n.)

2. Describe those conjugacy classes of S_n whose union is A_n when:

i) $n=5$
ii) $n=6$
iii) $n=7$
iv) $n=8$
v) $n=9$

3. Find a maximal set of pairwise nonconjugate elements of:

i) order 2 in S_9
ii) order 6 in S_9
iii) order 8 in S_{16}
iv) order 12 in S_{16}
v) order 30 in S_{17}

4. In $\{H \leq S_{10}\}$ with the relation of conjugacy:

i) how many conjugacy classes of cyclic subgroups of order 4 does S_{10} have?
ii) how many conjugacy classes of cyclic subgroups of order 6 does S_{10} have?

5. If $\theta \in S_n$, show that $\theta \sim_c \theta^{-1}$ in S_n.

6. For $\theta \in S_n$, show that $o(\theta)$ is odd $\Leftrightarrow \theta \sim_c \theta^2$.

7. For $t, n \in N$, if $(t, n!) = 1$, show that for any $\theta \in S_n$, $\theta \sim_c \theta^t$.

8. **a)** If $n > 2$ and $\rho \in S_n$ is a k-cycle, which $\theta \in S_n$ commute with ρ?
b) If $k > n/2$, $\rho, \eta \in S_n$ are k-cycles, and $\rho\eta = \eta\rho$, show $\eta = \rho^s$ for $(s, k) = 1$.

9. Let $\theta = \rho\tau \in S_n$ where ρ is a k-cycle for k odd and τ is a transposition disjoint from ρ. Describe the centralizer $C(\rho\tau)$ in S_n.

10. If $\theta \in S_n$ has cycle structure $(3, 3, 1, \ldots, 1)$, find $|C(\theta)|$ and describe $\eta \in C(\theta)$.

11. **a)** Identify $S_3 = \{I, (1, 2), (1, 3), (2, 3), (1, 2, 3), (1, 3, 2)\} \leq S_n$ for $n \geq 4$ and describe $C(S_3) \leq S_n$. (Note that $C(S_3) = C((1, 2)) \cap C((1, 3))$.)
b) As in a), identify $S_k \leq S_n$ when $k \leq n$ and describe $C(S_k)$ in S_n.

8.3 *p*-GROUPS

The Class Equation in Theorem 8.4 relates $|G|$ to $|Z(G)|$ using certain $[G : C(t)]$. A special case leads to some nice and important results.

THEOREM 8.7 If G is a finite group, $d \in N$, and $d \mid [G : H]$ when $H \neq G$, then $d \mid |Z(G)|$.

Proof Since $d \mid [G : H]$ for every proper $H \leq G$, $d \mid [G : C(t)]$ for every $t \notin Z(G)$ by Theorem 8.3, so in Theorem 8.4, $d \mid \sum_{t \in T}[G : C(t)]$. Also $d \mid |G|$ using $|G| = [G : \langle e \rangle]$, so d divides $|G| - \sum_{t \in T}[G : C(t)]$, forcing $d \mid |Z(G)|$ as claimed. ∎

The main application of Theorem 8.7 is when $|G| = p^n$ and $d = p$ a prime since then Lagrange's Theorem gives $p \mid [G : H]$ when $H \neq G$.

DEFINITION 8.3 A group G is a ***p*-group** if p is a prime and $|G| = p^n$.

An Abelian p-group is a direct sum of various $\mathbf{Z}_{p^n}$ by Theorem 7.19. Examples of non-Abelian p-groups are found in $UT(n, \mathbf{Z}_p)$ (see Example 2.9).

Example 8.8 $H(n, p) = \{A \in UT(n, \mathbf{Z}_p) \mid \text{all } A_{ii} = [1]_p\} \le UT(n, \mathbf{Z}_p)$, if $n > 2$ then $H(n, p)$ is not Abelian, $|H(n, p)| = p^{(n^2-n)/2}$, and $|Z(H(n, p))| = p$.

$UT(n, \mathbf{Z}_p)$ is finite so $H(n, p) \le UT(n, \mathbf{Z}_p)$ by Theorem 2.12 if $H(n, p)$ is closed under multiplication. Given $A, B \in H(n, p)$, $A_{ij} = B_{ij} = 0$ for $i > j$ so AB satisfies $(AB)_{jj} = A_{jj}B_{jj} = [1]_p[1]_p = [1]_p$ from the definition of matrix multiplication. Thus $AB \in H(n, p)$. Now $H(n, p)$ contains all matrices with arbitrary entries from $\mathbf{Z}_p$ above the diagonal and fixed elements in the other positions. There are $(n^2 - n)/2$ positions above the diagonal and it follows that $|H(n, p)| = p^{(n^2-n)/2}$. Recall that $I_n \in M_n(\mathbf{Z}_p)$ is the identity matrix and $aE_{ij} \in M_n(\mathbf{Z}_p)$ is the matrix with $(aE_{ij})_{ij} = a \in \mathbf{Z}_p$, and [0] in all other entries. When $n > 2$, $(I_n + [1]E_{12})(I_n + [1]E_{23}) = I_n + [1]E_{12} + [1]E_{23} + [1]E_{13}$ and also $(I_n + [1]E_{23})(I_n + [1]E_{12}) = I_n + [1]E_{12} + [1]E_{23}$ so $H(n, p)$ is not Abelian. It is an exercise that $Z(H(n, p)) = \{I_n + aE_{1n} \mid a \in \mathbf{Z}_p\}$, which shows $|Z(H(n, p))| = p$.

Although $|H(n, p)|$ can be large, $|Z(H(n, p))| = p$ always. We see next that $Z(G) \ne \langle e\rangle$ for any p-group G. Since $Z(G)$ is normal, p-groups have nontrivial normal subgroups, so using quotients and the correspondence theorem (Theorem 7.2) becomes a powerful tool for studying these groups.

THEOREM 8.8 If G is a p-group, then $p \mid |Z(G)|$.

Proof If $H \le G$ is proper then by Theorem 4.5 (Lagrange), $[G : H] \mid |G|$ and $[G : H] \ne 1$. Since $|G| = p^n$ and p is a prime, $[G : H] = p^m$ for some $m \ge 1$. Thus $p \mid [G : H]$, forcing $p \mid |Z(G)|$ by Theorem 8.7. ■

Theorem 8.8 is the essential tool for proving things about p-groups. We record a few results showing how this theorem and Theorem 7.2 are used together. The first result re-proves Theorem 7.14.

THEOREM 8.9 If $|G| = p^2$ for p a prime, then $G \cong \mathbf{Z}_p \oplus \mathbf{Z}_p$ or $G \cong \mathbf{Z}_{p^2}$.

Proof If $x \in G$ with $o(x) = p^2$ then $G = \langle x\rangle \cong \mathbf{Z}_{p^2}$. Thus by Theorem 4.7 we may assume $g^p = e$ for all $g \in G$. If $G \ne Z(G)$, Theorem 8.8 gives $|Z(G)| = p$, so $Z(G) = \langle z\rangle$ using Theorem 4.9. For $x \notin \langle z\rangle$, $|\langle x\rangle| = p$ by Theorem 2.16, then $G = \langle z\rangle\langle x\rangle$ by Theorem 4.12. Thus $z \in Z(G)$ forces G to be Abelian, a contradiction. Hence G must be Abelian. The same argument using $y \ne e$ and $x \notin \langle y\rangle$ shows $G = \langle x\rangle\langle y\rangle$. Now $\langle x\rangle \cap \langle y\rangle = \langle e\rangle$, and $\langle x\rangle, \langle y\rangle \lhd G$ (G is Abelian!). Therefore Theorem 7.9 yields $G \cong \langle x\rangle \oplus \langle y\rangle \cong \mathbf{Z}_p \oplus \mathbf{Z}_p$. ■

If $|G| = p^3$ with $G \ne Z(G)$ then $|Z(G)| = p$ and $G/Z(G) \cong \mathbf{Z}_p \oplus \mathbf{Z}_p$; otherwise $G/Z(G)$ is cyclic using Theorem 8.9, and G is Abelian from Theorem 5.8. In $H(3, p)$, $\{I_n + aE_{12} + bE_{13} \mid a, b \in \mathbf{Z}_p\}$ and $\{I_n + aE_{23} + bE_{13} \mid a, b \in \mathbf{Z}_p\}$ correspond to the $\mathbf{Z}_p$-isomorphic summands in $H(3, p)/Z(H(3, p))$ (verify!). We see next that p-groups have long chains of normal subgroups.

THEOREM 8.10 If $|G| = p^n$ for p a prime, there are $\langle e\rangle = H_0 \le H_1 \le \cdots \le H_n = G$, all $H_k \lhd G$, and $|H_k| = p^k$.

Proof Use induction on n. When $n=1$, $H_0=\langle e\rangle$ and $H_1=G$. Assume the theorem is true for any group of order p^n, some $n\geq 1$, and let $|G|=p^{n+1}$. By Theorem 8.8 and Lagrange's Theorem $|Z(G)|=p^m$ for $m\geq 1$ so $y\in Z(G)-\langle e\rangle$ has $o(y)=p^s$ with $s\geq 1$ (Theorem 4.7). Thus Theorem 2.8 shows $o(y^{p^{s-1}})=p$ so there is $x\in Z(G)$ with $|\langle x\rangle|=p$. Now $H_1=\langle x\rangle \triangleleft G$ by Theorem 5.1 and G/H_1 exists. Applying the induction assumption to G/H_1 of order p^n gives $\langle H_1\rangle=K_1\leq K_2\leq\cdots\leq K_{n+1}=G/H_1$ with each $K_j\triangleleft G/H_1$ and $|K_j|=p^{j-1}$. Using Theorem 7.2 with $\rho:G\to G/H_1$ defined by $\rho(g)=H_1g$ shows the $H_j=\rho^{-1}(K_j)$ satisfy $H_1\leq H_2\leq\cdots\leq H_{n+1}=G$, each $H_j\triangleleft G$, and $|H_j|=|H_1||K_j|=p\cdot p^{j-1}=p^j$, proving the theorem when $|G|=p^{n+1}$. Applying IND-I completes the proof. ■

It is an exercise that in any p-group G, $\langle e\rangle\neq N\triangleleft G\Rightarrow N\cap Z(G)\neq\langle e\rangle$. We prove a special case of this, demonstrating again how Theorem 8.8, quotients, and the correspondence theorem are used. Note that the result is an immediate consequence of Theorem 6.19.

THEOREM 8.11 If G is a p-group, $N\triangleleft G$, and $|N|=p$, then $N\leq Z(G)$.

Proof If $|G|=p$ then $N=G=Z(G)$ (why?). When $n\geq 2$, use induction, assuming the theorem for any p-group B with $|B|<p^n$. Now $p\mid |Z(G)|$ from Theorem 8.8 so $|Z(G)|=p^t$ for $t\geq 1$ and $|G/Z(G)|=p^k$ for $k<n$ by Lagrange's Theorem. Let $\rho:G\to G/Z(G)$ be the usual epimorphism $\rho(g)=Z(G)g$. From Theorem 7.3 and Theorem 7.6, $\rho(N)\cong Z(G)N/Z(G)\cong N/(N\cap Z(G))$. Thus $|\rho(N)|\mid p$ yields $N\leq Z(G)$, or $N\cap Z(G)=\langle e\rangle$ and $|\rho(N)|=p$. Theorem 7.1 shows $\rho(N)\triangleleft G/Z(G)$, so $N\leq Z(G)$ or $\rho(N)\leq Z(G/Z(G))$ by induction. In the latter case, for $x\in N$, $Z(G)x\in Z(G/Z(G))$ and for all $g\in G$, $Z(G)xZ(G)g=Z(G)gZ(G)x$, forcing $xgx^{-1}g^{-1}\in Z(G)$. Now $N\triangleleft G$, so $gx^{-1}g^{-1}\in N$, implying that $xgx^{-1}g^{-1}\in N\cap Z(G)=\langle e\rangle$. Thus $xg=gx$, and $N\leq Z(G)$ follows, proving the theorem when $|G|=p^n$. Consequently, the theorem holds by IND-II. ■

For our last result here we need a definition from Exercise 7 in §2.5.

DEFINITION 8.4 If $H\leq G$, $N(H)=\{g\in G\mid g^{-1}Hg=H\}$ is the **normalizer** of H in G.

As in Exercise 7 in §2.5, $N(H)\leq G$ and $H\triangleleft N(G)$. Clearly $N(H)=G\Leftrightarrow H\triangleleft G$, and the centralizer of H, $C(H)\leq N(H)$, so $Z(G)\leq N(H)$. If $g\in G$ of order 2 then $N(\langle g\rangle)=C(g)$ since $x^{-1}\langle g\rangle x=\langle g\rangle=\{e,g\}\Rightarrow x^{-1}gx=g$. However, observe that $o((1,2,3))=3$ in S_4, $(2,3)\in N(\langle(1,2,3)\rangle)$ but $(2,3)\notin C((1,2,3))$. In fact in S_4, $N(\langle(1,2,3)\rangle)=\{\theta\in S_4\mid\theta(4)=4\}\cong S_3$: if $\beta\in S_4$ and $\beta(4)\neq 4$ then by Theorem 8.5, $\beta(1,2,3)\beta^{-1}=(\beta(1),\beta(2),\beta(3))\notin\langle(1,2,3)\rangle$.

We claim there are proper $H\leq D_n$ with $N(H)=H$. In D_n, $o(W_j)=2$ so by the last paragraph $N(\langle W_j\rangle)=C(W_j)$. But in D_{2k+1}, $T_iW_jT_{-i}=W_{j+2i}=W_j$ would force $(2k+1)\mid 2i$ so $T_i=T_0=I$. Also $W_iW_jW_i=W_{2i-j}=W_j$ would force $(2k+1)\mid 2(i-j)$ so $i=j$. Thus $N(\langle W_j\rangle)=\langle W_j\rangle$ in D_n when n is odd. However, in D_{2m}, $N(\langle W_j\rangle)=C(W_j)=\{I,W_j,W_{j+m},T_m\}$ (why?). Now if $C(W_0)=H=\langle W_0\rangle\langle T_m\rangle=\{I,W_0,T_m,W_m\}\leq D_{2m}$, then when m is odd we claim that $N(H)=H$. Our computations just above show that for $\alpha \in N(H)$, $\alpha W_0 \alpha^{-1} = W_k$ with $k \in 2\mathbf{Z}$. This forces $\alpha W_0 \alpha^{-1} = W_0$ and puts $\alpha \in C(W_0) = H$. The self-normalizing property of H cannot occur for subgroups of p-groups.

THEOREM 8.12 If $H \leq G$ is proper and G is a p-group, then $H \neq N(H)$.

Proof If $Z(G) \not\subseteq H$ then $H \neq HZ(G) \subseteq N(H)$, using Theorem 8.8. If $Z(G) \leq H$ then $H/Z(G) \neq N(H/Z(G))$ in $G/Z(G)$ by induction. Taking inverse images of $\varphi: G \to G/Z(G)$ shows that $H \neq \varphi^{-1}(N(H/Z(G))) \leq N(H)$ from Theorem 7.2. Specifically, if $x \in \varphi^{-1}(N(H/Z(G)))$ then $\varphi(x^{-1}Hx) = \varphi(H)$ (verify!) so Theorem 7.2 forces $x^{-1}Hx = H$, and $x \in N(H)$. ■

EXERCISES

1. If $H \leq G = Z(G)H$, G is finite, and H is a p-group, show that $p \mid |Z(G)|$.
2. Let $N \lhd G$ for G a p-group. Show that $N \cap Z(G) \neq \langle e \rangle$.
3. If G is a p-group and $H \leq G$ with $[G:H] = p$, show that $H \lhd G$.
4. A p-group G with $Z(G)$ cyclic is not a direct sum of two proper subgroups.
5. Show that the converse of Exercise 4 is false: there is a p-group G not the direct sum of two proper subgroups but $Z(G)$ is not cyclic. (Consider $G = (\mathbf{Z}_9 \times \mathbf{Z}_9, \cdot)$ where $([a],[b]) \cdot ([c],[d]) = ([a + 4^b c], [b + d])$. Show G is a group with $([a],[b])^3 = ([3a],[3b])$, $Z(G) \cong \mathbf{Z}_3 \oplus \mathbf{Z}_3$, and argue that $G = N_1 N_2$ with $N_i \lhd G$ and $N_1 \cap N_2 = \langle e \rangle$ is impossible unless some $N_i = G$.)
6. Show that any p-group is solvable. Recall that G is solvable if $G^{(k)} = \langle e \rangle$ for some $k \geq 1$ where $G^{(1)} = G'$ and $G^{(s+1)} = G^{(s)\prime}$. See Exercise 13 in §7.1.
7. If G is a p-group, show the following:
   i) if G has exactly four conjugacy classes, then $|G| = 4$
   ii) G cannot have exactly $p + 1$ conjugacy classes
   iii) if G has exactly $p + 2$ conjugacy classes, then $|G| = 4$
8. A proper homomorphic image of a group G is $\varphi(G)$ not isomorphic to G. If G is a p-group and every proper homomorphic image $\varphi(G)$ is cyclic, then: a) G is Abelian; b) G is cyclic or $G \cong \mathbf{Z}_p \oplus \mathbf{Z}_p$.

8.4 SYLOW SUBGROUPS

Information about subgroups is often crucial for studying the structure of a finite group. For arbitrary groups we have cyclic subgroups, subgroups generated by larger subsets, subgroups arising as centralizers, and the commutator subgroup. These and their orders are often hard to describe explicitly. Sylow provided basic and useful results about p-subgroups, and we discuss them in full in the next chapter. Here we introduce and illustrate the concept of Sylow subgroups, using the class equation to prove their existence. A consequence will be Cauchy's Theorem for arbitrary finite groups, in contrast to its restriction to Abelian groups as in Theorem 5.9.

DEFINITION 8.5 Let G be a finite group, $|G| = p^n m$ for p a prime, $n \geq 1$, and $(p, m) = 1$. A **p-Sylow subgroup** of G is any $H \leq G$ with $|H| = p^n$.

It is not obvious that p-Sylow subgroups must exist. If G is Abelian and the prime $p \mid |G|$ then $G_p = \{g \in G \mid o(g) = p^k$ for some $k \geq 0\}$ is a p-Sylow subgroup by Theorem 4.14. Note that these p-Sylow subgroups for Abelian groups are unique. For example, the 2-Sylow subgroup of $\mathbf{Z}_{18}$ is $\langle [17]_{18} \rangle$ and the 3-Sylow subgroup is $\langle [2]_{18} \rangle$. The 2-Sylow subgroup of U_{15} is U_{15} itself since $|U_{15}| = \varphi(15) = 8$, and for U_{33} of order $\varphi(33) = \varphi(3)\varphi(11) = 2 \cdot 10 = 2^2 \cdot 5$, the 2-Sylow subgroup is $\{[1], [10], [23], [32]\}$ and the 5-Sylow subgroup is $\langle [4] \rangle$ since $2^5 \equiv_{33} -1$ shows that $o([2]) = 10$.

Example 8.9 Sylow subgroups in D_6.

Since $|D_6| = 12$, a 2-Sylow subgroup has order 4 and a 3-Sylow subgroup has order 3. For any D_n, elements of odd order must be in $\langle T_1 \rangle$, so in D_6 $\langle T_2 \rangle$ is the unique 3-Sylow subgroup. As $Z(D_6) = \langle T_3 \rangle$ by Theorem 3.18, $\langle W_0 \rangle \langle T_3 \rangle = \{I, W_0, T_3, W_0 T_3 = W_3\} \leq D_6$ of order 4 (verify!). This is not the unique 2-Sylow subgroup since $\langle W_1 \rangle \langle T_3 \rangle$ and $\langle W_2 \rangle \langle T_3 \rangle$ also have order 4. Here a 2-Sylow subgroup does not contain all elements in D_6 with orders 2^s.

Example 8.10 Sylow subgroups in S_5.

$|S_5| = 5! = 2^3 \cdot 3 \cdot 5$. If $\rho \in S_5$ has order 5 then ρ is a 5-cycle and $\langle \rho \rangle$ is a 5-Sylow subgroup of S_5 by Theorem 2.16. There are twenty-four 5-cycles in S_5 and four appear in any subgroup of order 5, so these are distributed in six different subgroups of order 5 (why?). Thus S_5 has six different 5-Sylow subgroups, each of the form $\langle (a, b, c, d, e) \rangle$. Similarly, a subgroup of order 3 is cyclic, generated by a 3-cycle in S_5. There are $5 \cdot 4 \cdot 3/3 = 20$ different 3-cycles and two are in each subgroup of order 3, so S_5 has ten 3-Sylow subgroups, each some $\langle (a, b, c) \rangle$. A 2-Sylow subgroup in S_5 has order 8. We can identify $D_4 \cong H \leq S_5$ by the action of D_4 on the vertices of a square. Since $D_4 = \langle \{T_1, W_0\} \rangle$, take $H = \langle \{(1, 2, 3, 4), (1, 3)\} \rangle \leq S_5$. Note that since $(1, 3)(1, 2, 3, 4)(1, 3) = (1, 4, 3, 2) = (1, 2, 3, 4)^{-1}$, $H = \{(1, 3)^i (1, 2, 3, 4)^j \mid 0 \leq i \leq 1$ and $0 \leq j \leq 3\} \leq S_4$ (why?) and $H \cong D_4$ by Example 6.4. Other 2-Sylow subgroups are $\langle \{(1, 2, 4, 5), (1, 4)\} \rangle$ and $\langle \{(2, 3, 4, 5), (3, 5)\} \rangle$. Is every 2-Sylow subgroup of S_5 isomorphic to D_4? In the next chapter we will see that for any prime p, all p-Sylow subgroups in a given group are isomorphic, indeed conjugate.

The main theorem in this section is the first Sylow theorem: a finite group G always contains a p-Sylow subgroup for each prime divisor of $|G|$.

THEOREM 8.13 (Sylow) Let G be a group with $|G| = p^n m$ for p a prime, $n \geq 1$, and $(p, m) = 1$. For each $0 \leq k \leq n$ there is $H_k \leq G$ with $|H_k| = p^k$.

Proof If a p-Sylow subgroup exists then there are p-groups of all possible orders by Theorem 8.10. We use induction on $|G|$. If $|G| = p$ then G is the p-Sylow subgroup of G. Let G satisfy the hypothesis and assume the theorem holds for all groups B with $|B| < |G|$ and $p \mid |B|$. By Lagrange's Theorem, $|H| \mid |G|$ for any $H \leq G$ so $|H| = p^k s$ with $0 \leq k \leq n$ and $s \mid m$. If for some $H \leq G$, $|H| = p^n s$ and $s < m$ then by the induction assumption H

contains a p-Sylow subgroup P, necessarily of order p^n, so P is a p-Sylow subgroup of G. Thus we may assume that if $H \le G$ is proper then $p^n \nmid |H|$. But now Theorem 4.5 yields $p^n m = |G| = |H|[G:H]$, and $p^n \nmid |H|$ by assumption, so by unique factorization in $\mathbf{N}$ we are forced to conclude that $p \mid [G:H]$.

Using Theorem 8.7, $p \mid |Z(G)|$ so there is $x \in Z(G)$ with $o(\langle x\rangle) = p$ by Theorem 5.9. Now $\langle x\rangle \le Z(G)$ implies $\langle x\rangle \lhd G$ (Theorem 5.1), so $G/\langle x\rangle$ is defined and $|G/\langle x\rangle| = p^{n-1}m$ (why?). The induction assumption on $G/\langle x\rangle$ yields $K \le G/\langle x\rangle$ with $|K| = p^{n-1}$. To use the correspondence theorem, let $\rho : G \to G/\langle x\rangle$ be the epimorphism $\rho(g) = \langle x\rangle g$. Theorem 7.2 yields $\rho^{-1}(K) \le G$ with $|\rho^{-1}(K)| = |\langle x\rangle| \cdot |K| = p \cdot p^{n-1} = p^n$. Therefore $\rho^{-1}(K)$ is a p-Sylow subgroup of G, and applying IND-II completes the proof of the theorem. ∎

A special case of Theorem 8.13 extends Cauchy's Theorem from Abelian groups (Theorem 5.9) to arbitrary finite groups. In the next chapter we will re-prove Cauchy's Theorem without using the Abelian group case.

THEOREM 8.14 (Cauchy) If G is a finite group and $p \mid |G|$ for p a prime, then there is $x \in G$ with $o(x) = p$ and so $|\langle x\rangle| = p$.

We illustrate some standard techniques used with Sylow's Theorem to investigate the structure of finite groups, characterizing groups of order pq for $p < q$ primes. We begin with a technical result.

THEOREM 8.15 If $|G| = pq$ for $p < q$ primes, there is $\langle x\rangle \lhd G$ with $|\langle x\rangle| = q$.

Proof It is immediate from Theorem 8.14 that $|\langle x\rangle| = q$ for some $x \in G$. We claim that $\langle x\rangle$ is the only subgroup in G of order q. If $H \le G$ with $|H| = q$ then by Theorem 4.6 either $H = \langle x\rangle$ or $\langle x\rangle \cap H = \langle e\rangle$. In the second case, Theorem 4.12 implies that $|\langle x\rangle H| = |\langle x\rangle| \cdot |H| / |\langle x\rangle \cap H| = q^2$. But $\langle x\rangle H \subseteq G$ and $|G| = pq < q^2$ so $\langle x\rangle \cap H \ne \langle e\rangle$, forcing $H = \langle x\rangle$. Since $\langle x\rangle$ is the unique subgroup of order q in G, $\langle x\rangle \lhd G$ by Theorem 5.3. Another way to see that $\langle x\rangle \lhd G$ is to apply Theorem 6.14, noting that $pq \nmid p!$. ∎

From Theorem 7.15, if $|G| = pq$ for $p < q$ primes with $p \nmid (q-1)$ then $G \cong \mathbf{Z}_{pq}$. We obtain this more easily from Cauchy's Theorem. First, is there a non-Abelian group of order pq when $p \mid (q-1)$? We exhibit one next using the construction in Theorem 3.22.

Example 8.11 Let $p < q$ be primes with $p \mid (q-1)$. If $1 < c < q$ is minimal with $o([c]_q) = p$ in U_q then $(\mathbf{Z}_q \times \mathbf{Z}_p, \cdot)$ with $\cdot$ given by $([a]_q, [b]_p) \cdot ([x]_q, [y]_p) = ([a + c^b x]_q, [b+y]_p)$ for $a, b, x, y \ge 0$ is a non-Abelian group of order pq. For $X = ([1]_q, [0]_p)$ and $Y = ([0]_q, [1]_p)$, $\mathbf{Z}_q \cong \langle X\rangle \lhd (\mathbf{Z}_q \times \mathbf{Z}_p, \cdot)$, $YXY^{-1} = X^c$, and $(\mathbf{Z}_q \times \mathbf{Z}_p, \cdot) = \langle X\rangle\langle Y\rangle = \{X^j Y^i \mid 0 \le j < q \text{ and } 0 \le i < p\}$.

Certainly $|\mathbf{Z}_q \times \mathbf{Z}_p| = pq$ and there really is $[c]_q \in U_q$ with $o([c]_q) = p$. To see this, apply Theorem 8.14 using $|U_q| = q-1$ and $p \mid (q-1)$, or use Theorem 4.20 to see that U_q is cyclic and apply Theorem 3.1 on cyclic groups. Now if $[b]_p = [b']_p$ with $b, b' \ge 0$ then $c^b \equiv_q c^{b'}$ by Theorem 2.9 because $b \equiv_p b'$ and $o([c]_q) = p$, so $\cdot : (\mathbf{Z}_q \times \mathbf{Z}_p) \times (\mathbf{Z}_q \times \mathbf{Z}_p) \to \mathbf{Z}_q \times \mathbf{Z}_p$. It is easy to see that $([0]_q, [0]_p)$ is the identity for $(\mathbf{Z}_q \times \mathbf{Z}_p, \cdot)$. To find $([x]_q, [y]_p) = ([a]_q, [b]_p)^{-1}$ for $0 \le b < p$, we solve $a + c^b x \equiv_q 0$ and $b + y \equiv_p 0$ (why?). Thus $[y]_p = [-b]_p$, and using $c^p \equiv_q 1$ leads to $x \equiv_q -c^{p-b}a$ so we get $([a]_q, [b]_p)^{-1} = ([-c^{p-b}a]_q, [-b]_p)$, but does this "inverse" work on both sides? Computing

$$\begin{aligned}([-c^{p-b}a]_q, [p-b]_p) \cdot ([a]_q, [b]_p) &= ([-c^{p-b}a + c^{p-b}a]_q, [-b+b]_p) \\ &= ([0]_q, [0]_p)\end{aligned}$$

shows that $([-c^{p-b}a]_q, [-b]_p)$ is indeed the inverse of $([a]_q, [b]_p)$. To check that $\cdot$ is associative, take $([a]_q, [b]_p), ([u]_q, [v]_p), ([x]_q, [y]_p) \in \mathbf{Z}_q \times \mathbf{Z}_p$ and compute

$$\begin{aligned}&(([a]_q, [b]_p) \cdot ([u]_q, [v]_p)) \cdot ([x]_q, [y]_p) \\ &= ([a + c^b u]_q, [b+v]_p) \cdot ([x]_q, [y]_p) \\ &= ([a + c^b u + c^{b+v} x]_q, [b+v+y]_p) \\ &= ([a + c^b u + c^b c^v x]_q, [b+v+y]_p) \\ &= ([a]_q, [b]_p) \cdot ([u + c^v x]_q, [v+y]_p) \\ &= ([a]_q, [b]_p) \cdot (([u]_q, [v]_p) \cdot ([x]_q, [y]_p))\end{aligned}$$

Hence $\cdot$ is associative and $(\mathbf{Z}_q \times \mathbf{Z}_p, \cdot)$ is a group. Now $([1]_q, [0]_p) \cdot ([0]_q, [1]_p) = ([1]_q, [1]_p)$ but $([0]_q, [1]_p) \cdot ([1]_q, [0]_p) = ([c]_q, [1]_p) \ne ([1]_q, [1]_p)$ as $o([c]_q) = p > 1$ so $(\mathbf{Z}_q \times \mathbf{Z}_p, \cdot)$ is not Abelian. It is an exercise to see that $X = ([1]_q, [0]_p)$ has order q, and if $Y = ([0]_q, [1]_p)$, that $YXY^{-1} = X^c$ and $([a]_q, [b]_p) = X^a Y^b$. Hence $(\mathbf{Z}_q \times \mathbf{Z}_p, \cdot) = \langle X\rangle\langle Y\rangle = \{X^j Y^i \mid 0 \le j < q \text{ and } 0 \le i < p\}$ and we multiply these using the relation $YXY^{-1} = X^c$ so that $X^i Y^j X^u Y^v = X^{i + c^j u} Y^{j+v}$. With this representation of the elements of $(\mathbf{Z}_q \times \mathbf{Z}_p, \cdot)$, $\langle X\rangle \lhd (\mathbf{Z}_q \times \mathbf{Z}_p, \cdot)$ follows.

A more concrete representation of the group $(\mathbf{Z}_q \times \mathbf{Z}_p, \cdot)$ above is to let $G = \left\{\begin{bmatrix} a & b \\ [0] & [1]\end{bmatrix} \in GL\ (2,\ \mathbf{Z}_q) \mid a^p = [1]\right\}$. If $c \in U_q$ with $o(c) = p$ then here $X = \begin{bmatrix} c & [0] \\ [0] & [1]\end{bmatrix}$ and $Y = \begin{bmatrix} [1] & [1] \\ [0] & [1]\end{bmatrix}$.

DEFINITION 8.6 The group $(\mathbf{Z}_q \times \mathbf{Z}_p, \cdot)$ in Example 8.11 is denoted $\mathbf{Z}_q \diamond \mathbf{Z}_p$.

If $p > 2$ is a prime then the class $[c]_p$ of order 2 is just $[p-1]_p = [-1]_p$. In this case $\mathbf{Z}_p \diamond \mathbf{Z}_2$ has $2p$ elements, $\langle X\rangle = ([1]_p, [0]_2)$ has order p, and if $Y = ([0]_p, [1]_2)$ has order 2, then $YXY^{-1} = X^{-1}$ so $\mathbf{Z}_p \diamond \mathbf{Z}_2 \cong D_p$ by Example 6.4.

THEOREM 8.16 If G is a group and $|G| = pq$ for $p < q$ primes, then $G \cong \mathbf{Z}_{pq}$ or $p \mid (q-1)$ and $G \cong \mathbf{Z}_q \diamond \mathbf{Z}_p$.

Proof There is $x \in G$ with $o(x) = q$ and $\langle x\rangle \lhd G$ (Theorem 8.15) and $y \in G$ with $o(y) = p$ by Theorem 8.14. Now $y\langle x\rangle y^{-1} = \langle x\rangle$ so $yxy^{-1} = x^m$ for $1 \le m < q$. It follows that $y^2xy^{-2} = y(yxy^{-1})y^{-1} = yx^m y^{-1} = (yxy^{-1})^m = x^{m^2}$ and by induction $y^j x y^{-j} = x^{m^j}$ for $j > 1$. But $y^p = e$ implies $x = y^p x y^{-p} = x^{m^p}$ and $m^p \equiv_q 1$ follows from Theorem 2.9.

Thus in U_q, $o([m]_q) \mid p$ by Theorem 2.7 so $o([m]_q) = 1$ or $o([m]_q) = p$. In the first case $m = 1$, forcing $yx = xy$, and then $o(xy) = pq$ from Theorem 4.16. Thus $G = \langle xy \rangle \cong \mathbf{Z}_{pq}$ by Theorem 6.8.

When $o([m]_q) = p$, since $|U_q| = q - 1$, $o([m]_q) \mid (q - 1)$ by Theorem 4.7 so $p \mid (q - 1)$. As $\langle x \rangle \lhd G$, Theorem 5.4 shows $\langle x \rangle\langle y \rangle \leq G$. Also $|\langle x \rangle\langle y \rangle| = pq$ follows from Theorem 4.12, so $G = \langle x \rangle\langle y \rangle = \{x^j y^i \mid 0 \leq j < q \text{ and } 0 \leq i < p\}$. Since $x^j y^i = x^{j'} y^{i'}$ exactly when $j \equiv_q j'$ and $i \equiv_p i'$ (why?), there is a bijection $\varphi : G \to \mathbf{Z}_q \diamond \mathbf{Z}_p$ given by $\varphi(x^j y^i) = X^j Y^i$, using Example 8.11. If $0 < m < p$ is minimal representing a class in U_q of order p, then since $yxy^{-1} = x^m$ it follows from the computations in Example 8.11 that products of the $x^j y^i$ in G correspond to those of the $X^j Y^i$ in $\mathbf{Z}_q \diamond \mathbf{Z}_p$; that is, $G \cong \mathbf{Z}_q \diamond \mathbf{Z}_p$ via φ. In any case U_p is cyclic (Theorem 4.20) so $\langle [m]_q \rangle$ is the unique subgroup of U_q of order p, using Theorem 3.1 on cyclic groups. Thus if $o([c]_q) = p$ in U_q with $1 < c < q$ minimal then $\langle [c]_q \rangle = \langle [m]_q \rangle$ so $[c]_q = [m]_q^s$ for some $1 \leq s < p$. Hence $m^s \equiv_q c$ and Theorem 2.9 shows $y^s x y^{-s} = x^{m^s} = x^c$. Clearly $o(y^s) = p$ by Theorem 2.8, and Theorem 2.16 gives $\langle y \rangle = \langle y^s \rangle$. Replacing y with $z = y^s$, $G = \langle x \rangle\langle z \rangle$, but now with $zxz^{-1} = x^c$. As above, $\varphi(x^j z^i) = X^j Y^i$ shows $G \cong \mathbf{Z}_q \diamond \mathbf{Z}_p$. ∎

Another illustration of how Sylow's Theorem is typically combined with earlier results comes next. Here the special order of G forces the existence of normal subgroups; this is generally useful in determining the structure of G and for trying to do induction arguments in quotients.

THEOREM 8.17 Let p and q be primes, $p < q < 2p$, and G a group of order $p^n q$. Then there is $N \lhd G$ so that $[G : N] = p$ or $[G : N] = q$. Furthermore, if G has no normal subgroup of order p^n, then there is $M \lhd G$ with $|M| = p^{n-1}$.

Proof Use induction on n. If $n = 0$ then $N = \langle e \rangle \lhd G$, $|N| = p^0$, and $[G : N] = q$. When $n = 1$, Theorem 8.15 gives $N = \langle x \rangle \lhd G$ with $|N| = q$ so $[G : N] = p$ by Lagrange's Theorem. Assume $n > 1$, that the theorem is true for groups of order $p^k q$ when $k < n$, and that G contains no normal subgroup of order p^n. From Theorem 8.13 there is $H \leq G$ with $|H| = p^n$, so $[G : H] = q$. Now $q < 2p$ shows $p^2 \nmid q!$, so $n > 1$ implies that $|G| \nmid [G : H]!$. Thus Theorem 6.14 produces $\langle e \rangle \neq K \lhd G$ with $K \leq H$ and by Theorem 4.5, $|K| = p^m$ with $1 \leq m < n$ by our assumption. Now $|G/K| = p^{n-m} q$ with $n - m \geq 1$ so the induction assumption yields $L \lhd G/K$ with $[G/K : L] = p$ or $[G/K : L] = q$, and some $V \lhd G/K$ with order p^{n-m} or p^{n-m-1} as appropriate. Use the correspondence theorem on L or V to get the required subgroup in G via the epimorphism $\rho : G \to G/K$ given by $\rho(g) = Kg$. From Theorem 7.2, $N = \rho^{-1}(L) \lhd G$ and $[G/K : L] = [G : N] = p$ since $|N| \neq p^n$ by assumption. Similarly, $\rho^{-1}(V) \lhd G$ and $|\rho^{-1}(V)| \neq p^n$, so $|\rho^{-1}(V)| = |V| \cdot |K| = p^{n-m-1} p^m = p^{n-1}$. Therefore the theorem holds for G and applying IND-II completes the proof. ∎

The theorem applies when $|G| = 875 = 5^3 \cdot 7$: for some $N \lhd G$, $|N| = 125$ or else $|N| = 175$ and $|M| = 25$ for some $M \lhd G$. Also if $|G| = 48 = 2^4 \cdot 3$ then for some $N \lhd G$, $|N| = 16$ or $|N| = 24$ and $|M| = 8$ for some $M \lhd G$. When $G = A_4 \oplus \mathbf{Z}_4$ then $A_4 \oplus \langle [2] \rangle \lhd G$ of order 24 and $H \oplus \mathbf{Z}_4 \lhd G$ of order 16, for $H = \{g \in A_4 \mid g^2 = I\} \lhd A_4$ as in Example 5.10. If instead $G = S_3 \oplus \mathbf{Z}_8$ then G contains the normal subgroups $\langle I \rangle \oplus \mathbf{Z}_8$ of order 8, and $\langle (1, 2, 3) \rangle \oplus \mathbf{Z}_8$ of order 24, so index 2. The subgroups of order 16 look like $\langle (a, b) \rangle \oplus \mathbf{Z}_8$, so there are three of them, but none is normal since conjugation by $((1, 2, 3), [0])$ permutes them cyclically.

When $n = 2$ in Theorem 8.17 we get a stronger conclusion and also demonstrate a standard technique for investigating the structure of groups.

THEOREM 8.18 If G is a group and $|G| = p^2q \neq 12$ for $p < q$ primes with $q < 2p$, then G is Abelian and either $G \cong \mathbf{Z}_{|G|}$ or $G \cong \mathbf{Z}_{pq} \oplus \mathbf{Z}_p$.

Proof We use Theorem 6.19 to show that G is Abelian. There is $N \lhd G$ with $|N| = p^2$ or $|N| = pq$ from Theorem 8.17. Should $|N| = p^2$ then $N \cong \mathbf{Z}_{p^2}$ or $N \cong \mathbf{Z}_p \oplus \mathbf{Z}_p$ by Theorem 8.9, and Theorem 8.14 yields $y \in G$ with $o(y) = q$. If N is cyclic, $Aut(N) \cong U_{p^2}$ by Theorem 6.20, so $|Aut(N)| = p(p-1)$. Similarly, $N \cong \mathbf{Z}_p \oplus \mathbf{Z}_p$ and Theorem 6.22 show that $Aut(N) \cong GL(2, \mathbf{Z}_p)$ so $|Aut(N)| = (p^2 - 1)(p^2 - p)$. In either case $(|Aut(N)|, |\langle y \rangle|) = 1$, unless $q \mid (p + 1)$ and $|G| = 12$, so $\langle y \rangle \leq C_G(N)$ by Theorem 6.19. Since $(|N|, |\langle y \rangle|) = 1$, $|N\langle y \rangle| = p^2q$ results from Theorem 4.12, forcing $G = N\langle y \rangle$, and G must be Abelian.

When $|N| = pq$, then $p^2q \neq 12$ and $q < 2p$ imply $p \nmid (q - 1)$ (why?). Thus $N \cong \mathbf{Z}_{pq}$ by Theorem 8.16, and Theorem 6.20 gives $Aut(N) \cong U_{pq}$ so $|Aut(N)| = (p - 1)(q - 1)$. From Theorem 8.13 there is $P \leq G$ with $|P| = p^2$, and P is Abelian by Theorem 8.9. As in the last paragraph, since $(|Aut(N)|, |P|) = 1$, $P \leq C_G(N)$ by Theorem 6.19. But using Theorem 4.12 again gives $|NP| = |N| \cdot |P|/|N \cap P| = p^2q = |G|$ since $|N \cap P| \mid (|N|, |P|)(=p)$ by Theorem 4.6, and $|NP| \leq |G|$. Therefore $G = NP$ is Abelian ($P \leq C_G(N)$!).

Since G is Abelian the last statement of the theorem follows directly from Theorem 7.19. Alternatively, from Theorem 8.13 there is a p-Sylow subgroup P of G and a q-Sylow subgroup $\langle y \rangle$. If $P = \langle x \rangle$ then $o(xy) = p^2q$ from Theorem 4.16, and G is cyclic. When $g^p = e$ for all $g \in P$ then for $x \in P - \langle e \rangle$ again we have $o(xy) = pq$. If $z \in P - \langle x \rangle$ and $e \neq z^i = x^j y^j$ then $y^j = x^{-j}z^i$ is in $P \cap \langle y \rangle = \langle e \rangle$ by Theorem 4.6, resulting in $\langle z \rangle \cap \langle x \rangle \neq \langle e \rangle$, a contradiction (why?). Therefore $\langle xy \rangle \cap \langle z \rangle = \langle e \rangle$, so $G = \langle xy \rangle \langle z \rangle$ using Theorem 4.12 once again, and $G \cong \langle xy \rangle \oplus \langle z \rangle \cong \mathbf{Z}_{pq} \oplus \mathbf{Z}_p$ follows from Theorem 7.9. ■

Specific applications of the theorem show that any group of order $175 = 5^2 \cdot 7$, $539 = 7^2 \cdot 11$, $1573 = 11^2 \cdot 13$, $2057 = 11^2 \cdot 17$, or $2299 = 11^2 \cdot 19$ must be Abelian. These results are far from obvious from the definition of group and show how far we have come in developing methods for studying groups. We look at two more examples of groups but with orders involving three primes. Again we use Sylow's Theorem and the techniques above.

Example 8.12 Any group of order $255 = 3 \cdot 5 \cdot 17$ is cyclic.

By Theorem 8.14 there is $x \in G$ with $|\langle x \rangle| = 17$. Since $|G| \nmid [G : \langle x \rangle]!$, and since $H \leq \langle x \rangle \Rightarrow H = \langle e \rangle$ or $H = \langle x \rangle$, $\langle x \rangle \lhd G$ by Theorem 6.14. This also follows as in the proof of Theorem 8.15 by showing that $\langle x \rangle$ is the only subgroup in G of order 17, using Theorem 4.12 and then Theorem 5.3. Now $G/\langle x \rangle$ has order 15, so by Theorem 8.15 it contains a normal subgroup K of order 5. Apply Theorem 7.2 to the epimorphism $\rho : G \to G/\langle x \rangle$ given by $\rho(g) = \langle x \rangle g$ to see that $N = \rho^{-1}(K) \lhd G$ has index 3 and order 85. Note that N is cyclic by Theorem 8.16, so from Theorem 6.20, $Aut(N) \cong U_{85}$ and has order $4 \cdot 16 = 64$. Theorem 8.14 again shows there is

$P=\langle y\rangle \le G$ with $o(y)=3$. But $(|Aut(N)|, |P|)=1$ so Theorem 6.19 forces $P \le C_G(N)$. Finally, since $N \cap P=\langle e\rangle$ (why?), Theorem 4.12 shows that $|NP|=|G|$, so $G=NP$ and must be Abelian since both N and P are and $P \le C_G(N)$. Hence G is cyclic by Theorem 4.17.

Example 8.13 If G is a group and $|G|=130=2\cdot 5\cdot 13$ then either $G\cong \boldsymbol{Z}_{130}$, $G\cong D_{65}$, $G\cong \boldsymbol{Z}_5\oplus D_{13}$, or $G\cong \boldsymbol{Z}_{13}\oplus D_5$.

Let P be a 13-Sylow subgroup of G (Theorem 8.13) and observe as in Example 8.12 that $P \triangleleft G$ and $|G/P|=10$. By Theorem 8.15 there is $K \triangleleft G/P$ with $|K|=5$. Using the epimorphism $\rho: G\to G/P$ given by $\rho(g)=Pg$ and then Theorem 7.2, $N=\rho^{-1}(K)\triangleleft G$ with $|N|=5\cdot 13=65$. It follows from Theorem 8.16 that $N=\langle x\rangle$ so has unique subgroups of order 5 and of order 13 by Theorem 3.1, say $\langle a\rangle$ and $\langle b\rangle$, respectively, and $ab=ba$. Since $N\triangleleft G$, for all $h\in G$, $h\langle a\rangle h^{-1}\le N$ of order 5 so $h\langle a\rangle h^{-1}=\langle a\rangle$ and $\langle a\rangle \triangleleft G$. Similarly, $\langle b\rangle \triangleleft G$. Now $N=\langle ab\rangle$ by Theorem 4.16 and $N=\langle a\rangle\langle b\rangle$ by Theorem 4.12.

Using Theorem 8.14 there is $y\in G$ with $o(y)=2$ so $N\cap\langle y\rangle=\langle e\rangle$ (Theorem 4.6) and $|N\langle y\rangle|=|G|$ by Theorem 4.12. Thus $G=N\langle y\rangle=\langle a\rangle\langle b\rangle\langle y\rangle$. Since $\langle a\rangle\triangleleft G$, $y\langle a\rangle y^{-1}=\langle a\rangle$ so $T(x)=yxy^{-1}$ gives $T\in Aut(\langle a\rangle)$. Also $S(x)=yxy^{-1}$ defines $S\in Aut(\langle b\rangle)$. Using $y^2=e$, $o(T)\mid 2$ and $o(S)\mid 2$ follow. Both $Aut(\langle a\rangle)$ and $Aut(\langle b\rangle)$ are cyclic by Theorem 6.20, or here, by direct computation in U_5 and in U_{13}, so in each case $\varphi(x)=x^{-1}$ is its unique automorphism of order 2 by Theorem 3.1. If $o(T)=o(S)=1$ then y commutes with a and b, so with all $g\in\langle a\rangle\langle b\rangle$. But $ab=ba$ so $G=\langle a\rangle\langle b\rangle\langle y\rangle$ is Abelian and hence cyclic by Theorem 4.17. If $o(T)=1$ and $o(S)=2$ then $ya=ay$ and $yby^{-1}=b^{-1}$. Since $\langle b\rangle\triangleleft G$, $\langle b\rangle\langle y\rangle\le G$ (Theorem 5.4), $|\langle b\rangle\langle y\rangle|=26$ from Theorem 4.12, and $\langle b\rangle\langle y\rangle\cong D_{13}$ by Example 6.4. Also $\langle b\rangle\langle y\rangle\triangleleft G$ since its elements commute with those in $\langle a\rangle$. Now $(|\langle b\rangle\langle y\rangle|, |\langle a\rangle|)=1$ shows that $\langle e\rangle=\langle a\rangle\cap\langle b\rangle\langle y\rangle$, and Theorem 7.9 gives $G\cong\langle a\rangle\oplus\langle b\rangle\langle y\rangle\cong \boldsymbol{Z}_5\oplus D_{13}$ using Theorem 7.10. A similar argument shows $G\cong\boldsymbol{Z}_{13}\oplus D_5$ when $o(T)=2$ and $o(S)=1: yb=by$ and $yay^{-1}=a^{-1}$. Finally, when $o(T)=2=o(S)$, $yay^{-1}=a^{-1}$ and $yby^{-1}=b^{-1}$ imply $yaby^{-1}=yay^{-1}yby^{-1}=a^{-1}b^{-1}=(ab)^{-1}$ using that $ab=ba$. Now $N=\langle ab\rangle$ has order 65 so Example 6.4 yields $G\cong D_{65}$.

In general, when $|G|$ is a product of distinct primes the hope is to produce $\langle x\rangle=N\triangleleft G$ of prime *index* p in G and see how an element of order p acts on the q-Sylow subgroups of N for $q\ne p$. We outline the procedure when $|G|=3\cdot 5\cdot 19$ and leave the details to the exercises. As in the examples above, the 19-Sylow subgroup $P\triangleleft G$, G/P is cyclic, and we get $K\triangleleft G$ with $|K|=5\cdot 19$ by the correspondence theorem. Now Theorem 8.16 shows K is cyclic and $K=AB$ for normal subgroups A and B of K, with $|A|=5$ and $|B|=19$. Any $y\in G$ of order 3 acts by conjugation as an automorphism on each of A and B, and y must centralize A since $|Aut(A)|=4$; thus $G\cong A\oplus B\langle y\rangle$. Since $|B\langle y\rangle|=3\cdot 19$, by Theorem 8.16 either $B\langle y\rangle$ is cyclic and G is Abelian, so cyclic, or $B\langle y\rangle\cong\boldsymbol{Z}_{19}\diamond\boldsymbol{Z}_3$. Consequently, there are two possibilities for G, up to isomorphism: $G\cong\boldsymbol{Z}_{|G|}$, or $G\cong\boldsymbol{Z}_5\oplus\boldsymbol{Z}_{19}\diamond\boldsymbol{Z}_3$.

If $|G|=3\cdot 7\cdot 31$ there are four possibilities for G, as in Example 8.13. There is a normal cyclic subgroup of order $7\cdot 31$ that is a product of normal cyclic subgroups A of order 7 and B of order 31. Now $y \in G$ of order 3 acts by conjugation as an automorphism on each of A and B, and since $3 \mid 6$ and $3 \mid 30$, y can act nontrivially on A and on B. The four possibilities of its action being trivial or not on each of A and B give rise to four possible group structures for G via Theorem 8.16. These include $\mathbf{Z}_{651}$, $\mathbf{Z}_7 \oplus \mathbf{Z}_{31} \diamond \mathbf{Z}_3$, and $\mathbf{Z}_{31} \oplus \mathbf{Z}_7 \diamond \mathbf{Z}_3$. Finally, y can act nontrivially on A and on B, written $\mathbf{Z}_{217} \diamond \mathbf{Z}_3$.

EXERCISES

1. Prove Cauchy's Theorem directly from the class equation, using Theorem 5.9, and following the proof of Sylow's Theorem.

2. If G is a group, $|G| = 2^k m$ with $k \geq 1$ and m odd, and if a 2-Sylow subgroup of G is cyclic, show that either $|G| = 2$ or G is not simple. (Theorem 6.13)

3. For each prime $p \mid |G|$, describe a p-Sylow subgroup P of G and determine whether it is the unique p-Sylow subgroup of G:

i) $G = \mathbf{Z}_{360}$
ii) $G = A_4$
iii) $G = S_4$
iv) $G = D_{30}$
v) $G = D_{24}$
vi) $G = S_6$
vii) $G = S_7$
viii) $G = A_7$

4. Let $G = UT(3, \mathbf{Z}_7)$ (See Definition 2.9 and Example 2.9.)

a) Show that $|G| = 2^3 3^3 7^3$.
b) Find a p-Sylow subgroup of G for $p = 2$, $p = 3$, and $p = 7$.
c) Find $H \leq G$ with $|H| = 6^3$ and show that H is not normal in G.
d) Find a 7-Sylow subgroup in G and show that it is normal in G.

For Exercises 5–9, recall that $|GL(2, \mathbf{Z}_p)| = (p^2 - 1)(p^2 - p)$ for p a prime.

5. For p a prime, find a p-Sylow subgroup in $GL(2, \mathbf{Z}_p)$. Are there others?

6. i) Find a Sylow subgroup for each prime divisor of the order of $GL(2, \mathbf{Z}_3)$. $\left(\text{Consider the matrices } \begin{bmatrix} 0 & -1 \\ 1 & -1 \end{bmatrix}, A = \begin{bmatrix} 1 & -1 \\ 1 & 1 \end{bmatrix}, \text{ and } B = \begin{bmatrix} 0 & -1 \\ -1 & 0 \end{bmatrix}. \text{ Show that a 2-Sylow subgroup is } \{A^i B^j\} \text{ and describe the multiplication.}\right)$

ii) Show that a 2-Sylow subgroup of $SL(2, \mathbf{Z}_3)$ is generated by $\begin{bmatrix} 0 & 1 \\ -1 & 0 \end{bmatrix}$ and $\begin{bmatrix} -1 & 1 \\ 1 & 1 \end{bmatrix}$, and identify this group up to isomorphism.

7. Find a Sylow subgroup for each prime divisor of the order for:

i) $GL(2, \mathbf{Z}_5)$
ii) $SL(2, \mathbf{Z}_5)$ $\left(\text{Consider } \begin{bmatrix} 0 & -1 \\ 1 & -1 \end{bmatrix} \text{ and } Diag(2, \mathbf{Z}_5).\right)$

8. Find a Sylow subgroup for each prime divisor of the order of $GL(2, \mathbf{Z}_7)$. If $A = \begin{bmatrix} 0 & 1 \\ 1 & -3 \end{bmatrix}$ and $B = \begin{bmatrix} 0 & 1 \\ -1 & 0 \end{bmatrix}$, show that a 2-Sylow subgroup P_2 of $GL(2, \mathbf{Z}_7)$ is given by $P_2 = \{A^i B^j\}$, and describe the multiplication in P_2.

9. **i)** Use $\begin{bmatrix} 0 & -1 \\ 1 & -3 \end{bmatrix}$ to find a Sylow subgroup of $GL(2, \mathbf{Z}_{13})$ for each prime divisor of its order.

ii) Do the same for $SL(2, \mathbf{Z}_{13})$.

10. Let $p_1 < \cdots < p_k$ be primes with $p_j > p_1 p_2 \cdots p_{j-1}$ for all $1 < j \leq k$. If G is a group of order $p_1 p_2 \cdots p_k$, show that G contains a normal subgroup of index p_1.

11. **i)** Show that there is $H \leq S_9$ with $H \cong \mathbf{Z}_3^3$.

ii) Find a 3-Sylow subgroup of S_9.

12. Let G be a group with $3 \leq |G| = 2k + 1 \leq 59$.

i) Which of these orders force G to be cyclic?

ii) Which of the remaining orders force G to be Abelian?

13. If $|G| = 20$, show that G has only one 5-Sylow subgroup and that for any $k \in N$ with $k \mid 20$, G contains a subgroup of order k.

14. **a)** If $|G| = 28$, show that any 7-Sylow subgroup of G is normal and that $2 \mid |Z(G)|$.

b) If $|G| = 56$, assume G contains a normal Sylow subgroup and show that there is $N \triangleleft G$ with $|N| = 2^s$ for some $s \geq 1$.

c) If $|G| = 112 = 2^4 \cdot 7$ show there is $N \triangleleft G$ with $|N| = 2^s$ for some $s \geq 1$ or $[G : N] = 2$. (See Theorem 6.14.)

d) If $|G| = 2^k \cdot 7$ for $k \geq 5$, show that there is $N \triangleleft G$ with $|N| = 2^s$ for $s \geq k - 4$.

15. Let $|G| = mq^k$ for q prime, $k \geq 1$, and $m < q$. Show that the q-Sylow subgroup of G must be normal.

16. If G is a group with $21 \leq |G| \leq 28$, show that G must have prime order or contain a proper, nonidentity, normal subgroup. (Use Theorem 6.14.)

17. If G is a group with $42 \leq |G| \leq 50$, show that G must have prime order or contain a proper, nonidentity, normal subgroup. (Use Theorem 6.14.)

18. Find a concrete representation in S_q for $\mathbf{Z}_q \diamond \mathbf{Z}_p$ when:

i) $pq = 21$

ii) $pq = 39$

iii) $pq = 55$

iv) $pq = 57$

v) $pq = 11 \cdot 23 = 253$

19. Up to isomorphism determine the structure of groups of the following orders:

i) $3 \cdot 11 \cdot 41$

ii) $3 \cdot 17 \cdot 67$

iii) $5 \cdot 13 \cdot 67$

iv) $2 \cdot 11 \cdot 31$

v) $5 \cdot 11 \cdot 79$

vi) $7 \cdot 11 \cdot 83$

vii) $7 \cdot 13 \cdot 113$

20. Use Theorem 6.14 to show that any group of the following order must have a proper, nonidentity, normal subgroup:

i) $|G| = 80$

ii) $|G| = 189$

iii) $|G| = 1000$

iv) $|G| = 1750$

v) $|G| = 2^{25} \cdot 5^2$

vi) $|G| = 2 \cdot 3 \cdot 5 \cdot 7 \cdot 127^2$

21. Determine the possibilities for each group up to isomorphism (note Exercise 15 and recall that $Aut(\mathbf{Z}_p^2) \cong GL(2, \mathbf{Z}_p)$):

i) $|G| = 245$

ii) $|G| = 66$

iii) $|G| = 5 \cdot 7 \cdot 37^2$

iv) $|G| = 5 \cdot 11 \cdot 73^2$

v) $|G| = 5^2 \cdot 37^2$

vi) $|G| = 5 \cdot 7^2 \cdot 271$

vii) $|G| = 3 \cdot 5 \cdot 17 \cdot 257$

22. Let $|G| = 4225 = 5^2 \cdot 13^2$.

a) If G has a normal Sylow subgroup, show that G is Abelian and describe the possibilities for G up to isomorphism. (Use Theorem 6.19.)

b) Show that G contains a normal subgroup of order 13 or 169.

c) Show that $Z(G) \neq \langle e \rangle$. (Use Theorem 6.19.)

d) If $H \triangleleft G$ with $|H| = 13$ and P is a 5-Sylow subgroup of G, describe the possible structures of the product $HP \leq G$.

23. Suppose G is a group of order $2p_1 \cdots p_k$ for $3 \leq p_1 < \cdots < p_k$ primes and assume that G has a normal cyclic subgroup of index 2. Describe a collection S of 2^k pairwise nonisomorphic groups satisfying these conditions, all having order $|G|$. Argue that G must be isomorphic to some group in S.

CHAPTER 9

GROUP ACTIONS

In this chapter we look at a counting technique that leads to another proof of Cauchy's Theorem (Theorem 8.14) without using the special case for Abelian groups and enables us to answer questions like: how many necklaces can be made from 14 beads if two colors of beads are available? The answer is 687, and if three colors of beads are available then there are 173,088 possible necklaces! It is somewhat surprising that ideas in group theory can be used in sophisticated counting problems. The primary mathematical consequence of the counting technique introduced will be Sylow's Theorems, fundamental results in the theory of finite groups.

9.1 GROUP ACTIONS

Regarding group elements as bijections can be useful. Identifying $g \in G$ with $\sigma_g \in \mathrm{Bij}(G)$ by left multiplication gives Cayley's Theorem (Theorem 6.11); for $H \leq G$, identifying $g \in G$ with $\sigma(g) \in \mathrm{Bij}(X)$ for $X = \{xH \mid x \in G\}$ (Theorem 6.14) can produce normal subgroups in G; and identifying $g \in G$ with the inner automorphism $T_g \in \mathrm{Bij}(G)$ gives rise to the class equation (Theorem 8.4). In each case we think of group elements as functions acting on some set. Of course $S_n = \mathrm{Bij}(I(n))$, and each $A \in GL(2, F)$ acts as a function on $F^2 = \left\{\begin{bmatrix} a \\ b \end{bmatrix} \mid a, b \in F\right\}$ by matrix multiplication: $A\begin{bmatrix} a \\ b \end{bmatrix}$. The abstract formulation of these situations is called a *group action*.

DEFINITION 9.1 For a group G and nonempty set S, G **acts on** S if there is $\cdot : G \times S \to S$ so that when $g, h \in G$ and $s \in S$:

i) $gh \cdot s = g \cdot (h \cdot s)$
ii) $e_G \cdot s = s$

Condition i) in Definition 9.1 gives a relation between the structure of G and the way G is identified as a set of functions on S. The two conditions ensure that the function associated with each $g \in G$ is in $\mathrm{Bij}(S)$, a fact we prove next. Our result shows the equivalence of Definition 9.1 with the existence of a group homomorphism $\varphi : G \to \mathrm{Bij}(S)$.

THEOREM 9.1 G acts on $S \Leftrightarrow$ there is a homomorphism $\varphi: G \to \text{Bij}(S)$.

Proof An action of G on S produces a homomorphism from G to $\text{Bij}(S)$, and any such homomorphism gives a specific action of G on S. Let G act on S via $\cdot: G \times S \to S$ and for $g \in G$, define $f_g: S \to S$ by $f_g(s) = g \cdot s$. The first property of a group action implies that $f_{gh} = f_g \circ f_h$ and the second yields $f_e = I_S$. Combining these shows that $f_{g^{-1}} = (f_g)^{-1}$ so $f_g \in \text{Bij}(S)$ by Theorem 0.11. Now define $\varphi: G \to \text{Bij}(S)$ by $\varphi(g) = f_g$ and use $\varphi(gh) = f_{gh} = f_g \circ f_h = \varphi(g)\varphi(h)$ to see that φ is a homomorphism. Conversely, given a homomorphism $\varphi: G \to \text{Bij}(S)$ let $\cdot: G \times S \to S$ be $g \cdot s = \varphi(g)(s)$. Then $g \cdot (h \cdot s) = g \cdot (\varphi(h)(s)) = \varphi(g)(\varphi(h)(s)) = (\varphi(g)\varphi(h))(s) = \varphi(gh)(s)$ (why?) $= gh \cdot s$, and $\varphi(e) = I_S$ by Theorem 6.2, so if $s \in S$ then $e \cdot s = \varphi(e)(s) = s$, proving that G acts on S. ∎

Observe that φ in the theorem need not be injective since for any G and S let $g \cdot s = s$ for all $g \in G$ and $s \in S$. Then $\varphi(g) = I_S$ for all $g \in G$ and $ker\, \varphi = G$. If G acts on itself by conjugation, so $g \cdot x = T_g(x) = gxg^{-1}$, then the associated homomorphism $\varphi: G \to \text{Bij}(G)$ is $\varphi(g) = T_g$ and $ker(\varphi) = Z(G)$. If G is Abelian this action is $g \cdot x = x$ for all $g, x \in G$. When are different actions of G on S related? We will not explore this here other than to say when two actions $\cdot$ and # of G on S are considered to be equivalent. Let the homomorphisms of G into $\text{Bij}(S)$ corresponding to $\cdot$ and # be φ and η, respectively. Then $\cdot$ and # are *equivalent* if there is $\rho \in \text{Bij}(S)$ so that for all $g \in G$, $\rho^{-1}\varphi(g)\rho = \eta(g)$; that is, $\rho^{-1}(g \cdot \rho(s)) = g \# s$, or $g \cdot \rho(s) = \rho(g \# s)$.

We proceed to some general results. For simplicity, when G acts on S, we will usually write "gs" instead of $g \cdot s$.

DEFINITION 9.2 If G acts on S and $s, t \in S$, set $s \sim_G t$ if $t = gs$, some $g \in G$.

THEOREM 9.2 If G acts on S, then $\sim_G$ is an equivalence relation on S.

Proof By definition of action, for $s \in S$, $es = s$ so $s \sim_G s$ and $\sim_G$ is reflexive. Assume $s \sim_G t$ with $gs = t$ for some $g \in G$. Then $g^{-1}(gs) = g^{-1}t$ and it follows from Definition 9.1 that $g^{-1}t = (g^{-1}g)s = es = s$. Hence $t \sim_G s$ and $\sim_G$ is symmetric. If $s \sim_G t$ and $t \sim_G w$ then for $g, h \in G$, $t = gs$ and $w = ht = h(gs) = (hg)s$, so $s \sim_G w$. Therefore $\sim_G$ is transitive and so is an equivalence relation. ∎

The equivalence classes of $\sim_G$ partition S by Theorem 1.22. Just as for S_n, we call the equivalence class of $s \in S$ the *orbit* of s.

DEFINITION 9.3 If G acts on S, the equivalence class of $s \in S$ under $\sim_G$ is the **orbit** of s, denoted $orb(s) = Gs = \{gs \mid g \in G\}$. The **stabilizer** of s is $stab(s) = \{g \in G \mid gs = s\}$.

Thus $orb(s) \subseteq S$ contains all the elements in S obtained by applying all $g \in G$ to s, and $stab(s) \subseteq G$ consists of the group elements with s as a fixed point. Let us look at these notions in cases we have already considered. For S_n acting naturally on $I(n) = \{1, 2, \ldots, n\}$, there is only one orbit so $orb(j) = I(n)$ for all $j \in I(n)$, and $stab(j)$ are those $\theta \in S_n$ with j as a fixed point. When $G = \langle\theta\rangle \leq S_n$ then $orb(j) = [j]_\theta = \{\theta^i(j) \mid i \in \mathbf{Z}\}$ is just the orbit as defined in Chapter 3. Here $stab(j) = \{\theta^i \mid \theta^i(j) = j\}$ and depends on both j and θ. In S_5, when $\theta = (1,2,3)(4,5)$, $stab(1) = stab(2) = stab(3) = \langle \theta^3 \rangle$ and $stab(4) = stab(5) = \langle \theta^2 \rangle$, and when $\theta = (1,3,4,5)$ then $stab(2) = \langle \theta \rangle$ and for $j \neq 2$ $stab(j) = \langle I \rangle$.

If G acts on itself by conjugation, $g \cdot s = gsg^{-1}$, then $orb(x) = cls(x) = \{gxg^{-1} \mid g \in G\}$ and $stab(x) = \{g \in G \mid gxg^{-1} = x\} = C_G(x)$, the centralizer of x in G (see Definition 2.11). We generalize these ideas a bit and recall that for $H \leq G$ the normalizer of H is $N(H) = \{g \in G \mid gHg^{-1} = H\}$ (Definition 8.4).

Example 9.1 If G is a group then G acts on $S = \{H \leq G\}$ by conjugation. For $H \leq G$, $orb(H) = \{gHg^{-1} \mid g \in G\}$ and $stab(H) = N(H) \leq G$.

For $H \in S$ and $g \in G$, $g \cdot H = gHg^{-1}$. Clearly $g \cdot H \in S$ and $e \cdot H = H$ for all $H \in S$. To verify part i) of Definition 9.1 let $x, y \in G$, $H \in S$, and compute that $xy \cdot H = xyH(xy)^{-1} = xyHy^{-1}x^{-1} = x \cdot yHy^{-1} = x \cdot (y \cdot H)$. Therefore conjugation is an action of G on S. Now from Definition 9.3, for $H \in S$, $orb(H) = G \cdot H = \{gHg^{-1} \mid g \in G\}$ is the set of conjugates of H. Also $stab(H) = \{g \in G \mid g \cdot H = H\} = \{g \in G \mid gHg^{-1} = H\} = N(H)$.

Theorem 8.2, that $|cls(x)| = [G : C(x)]$ for $x \in G$, has a corresponding statement when G acts on S, replacing $cls(s)$ with $orb(s)$ and $C(s)$ with $stab(s)$.

THEOREM 9.3 If G acts on S, then the following hold for any $s \in S$:

i) $stab(s) \leq G$
ii) $|orb(s)| = [G : stab(s)]$
iii) $t \in orb(s) \Leftrightarrow orb(t) = orb(s)$
iv) If $t = xs$, then $x(stab(s))x^{-1} = stab(t)$, so $|stab(s)| = |stab(t)|$.

Proof If $g, h \in G$, Definition 9.1 shows the equivalence of: $gs = hs$, $g^{-1}(gs) = g^{-1}(hs)$, $g^{-1}gs = g^{-1}hs$, and $s = g^{-1}hs$. For i), if $x, y \in stab(s)$ then $xs = s = ys$ so, as we have just seen, $y^{-1}xs = s$. Therefore $y^{-1}x \in stab(s)$ and $stab(s) \leq G$ by Theorem 2.12. As above, $gs = hs \Leftrightarrow g^{-1}h \in stab(s) \Leftrightarrow gstab(s) = hstab(s)$ as left cosets. Since all elements in the coset $xstab(s)$ have the same action on s, $\varphi(xstab(s)) = xs$ defines $\varphi : \{xstab(s) \mid x \in G\} \to orb(s)$. If $gs \in orb(s)$, $gs = \varphi(gstab(s))$, so φ is surjective. We have seen that $gs = hs$ forces $gstab(s) = hstab(s)$, so φ is injective as well. Consequently, φ is a bijection, which implies that $|orb(s)| = |\{xstab(s) \mid x \in G\}| = [G : stab(s)]$, proving ii).

Now $\sim_G$ is an equivalence relation on S by Theorem 9.2, with classes $\{orb(s) \mid s \in S\}$. From Theorem 1.21, $orb(s) = orb(t) \Leftrightarrow t \in orb(s)$, proving iii). As for iv), note first that $t = xs$ implies $t \in orb(s)$ so if G is finite, then by ii), iii), and Lagrange's Theorem, $|stab(s)| = |G|/|orb(s)| = |G|/|orb(t)| = |stab(t)|$. In general, let $xs = t$ and $g \in stab(s)$. The definition of group action shows $gx^{-1}xs = s$ so acting by x yields $(xgx^{-1})xs = xs$, or $xgx^{-1}t = t$. Thus $t = xs$ implies $x(stab(s))x^{-1} \subseteq stab(t)$. But $t = xs$ implies $s = x^{-1}t$ so $x^{-1}(stab(t))x \subseteq stab(s)$, or $stab(t) \subseteq x(stab(s))x^{-1}$. Therefore $x(stab(s))x^{-1} = stab(t)$ as claimed in iv). This shows that $stab(t) = T_x(stab(s))$ for the inner automorphism T_x, and since an automorphism in injective, $|stab(s)| = |stab(t)|$. ■

The theorem has consequences for conjugates of subgroups, similar to those about conjugates of elements. By Example 9.1, if G acts on its subgroups by conjugation then for

$H \leq G$, $orb(H) = \{gHg^{-1} \mid g \in G\}$, so by Theorem 9.3, H has $[G : stab(H)] = [G : N(H)]$ conjugates. If G is finite this implies G cannot be the union of the conjugates of a fixed proper subgroup.

THEOREM 9.4 For G a finite group and $H \leq G$:

i) $|\{gHg^{-1} \mid g \in G\}| = [G : N(H)]$
ii) If $\bigcup_g gHg^{-1} = G$ then $H = G$.

Proof For $g \in G$, $T_g : G \to G$ via $T_g(x) = gxg^{-1}$ is a bijection by Example 6.8. Now $gHg^{-1} = T_g(H)$ so $T_g|_H : H \to gHg^{-1}$ is a bijection and $|H| = |gHg^{-1}|$. By Theorem 9.3 and Example 9.1, $|\{gHg^{-1} \mid g \in G\}| = [G : N(H)]$, and Lagrange's Theorem gives $[G : N(H)] = |G|/|N(H)| \leq |G|/|H| = [G : H]$. The intersection of two conjugates of H may contain more than $\langle e \rangle$, so using $|H| = |gHg^{-1}|$, when $G - \{e\} = \bigcup_g (gHg^{-1} - \{e\})$ then $|G| - 1 \leq (|H| - 1)[G : N(H)] \leq (|H| - 1)[G : H]$ (why?) $= |H|[G : H] - [G : H] = |G| - [G : H]$. Therefore $[G : H] \leq 1$, which forces $H = G$. ∎

In Chapter 8 the result corresponding to Theorem 9.3 gave the class equation and information on the $|Z(G)|$. Recall $x \in Z(G) \Leftrightarrow cls(x) = \{x\}$, or using our present notation, when $|orb(x)| = 1$. For group actions $|orb(s)| = 1 \Leftrightarrow stab(s) = G$ by Theorem 9.3 $\Leftrightarrow s$ is a fixed point of the action of all $g \in G$. For general group actions we get results for fixed points that correspond to those in Chapter 8 about centers. Much of the rest of this chapter is based on counting fixed points of group actions, so we give the appropriate definition and the theorem corresponding to the class equation.

DEFINITION 9.4 If G act on S and $\emptyset \neq A \subseteq G$, the **set of fixed points of A** is $Fix(A) = \{s \in S \mid gs = s \text{ for all } g \in A\}$; for $g \in G$, set $Fix(g) = Fix(\langle g \rangle)$.

Note that $Fix(G) = \{s \in S \mid orb(s) = \{s\}\}$. Clearly $\langle g \rangle$ acts on S and $gs = s \Leftrightarrow g^i s = s$ for all $i \in \mathbf{Z}$, using Definition 9.1 (verify!). The basic counting result corresponding to the class equation comes next.

THEOREM 9.5 When p is prime and the p-group G acts on the finite set S, then $|S| \equiv |Fix(G)| \pmod p$. Thus $(p, |S|) = 1 \Rightarrow |Fix(G)| \geq 1$, and $p \mid |S| \Rightarrow p \mid |Fix(G)|$.

Proof By Theorem 9.2, $\sim_G$ is an equivalence relation on S, so Theorem 1.22 shows that the equivalence classes $\{orb(s) \mid s \in S\}$ partition S. Let T be a transversal for $\{orb(s)\}$, so $|S| = \sum_T |orb(t)|$ by Theorem 0.3. Now $x \in Fix(G) \Leftrightarrow orb(x) = Gx = \{x\}$, so $Fix(G) \subseteq T$ and if $t \in Fix(G)$ then $|orb(t)| = 1$. Thus $|S| = |Fix(G)| + \sum_{T-Fix(G)} |orb(t)| = |Fix(G)| + \sum_{T-Fix(G)} [G : stab(t)]$, using Theorem 9.3. Should $[G : stab(t)] = 1$, then $|orb(t)| = 1$, forcing $t \in Fix(G)$. Hence from Lagrange's Theorem and G a p-group, whenever $t \in T - Fix(G)$ then $[G : stab(t)] = |G|/|stab(t)| = p^s$ for $s \geq 1$, and so $|S| \equiv |Fix(G)| \pmod p$. When $(p, |S|) = 1$ then $p \mid |Fix(G)|$ is impossible and we must have $|Fix(G)| \geq 1$. Finally, when $p \mid |S|$ then $|Fix(G)| \equiv 0 \pmod p$ and $p \mid |Fix(G)|$. ∎

Cauchy's Theorem now follows without using the Abelian version.

THEOREM 9.6 (Cauchy) If G is a finite group, p is a prime, and $p \mid |G|$, then there is $x \in G$ with $o(x) = p$ and so $H = |\langle x\rangle| \leq G$ with $|H| = p$.

Proof Let $T = G^p$, the Cartesian product of G with itself p times, and set $S = \{(g_1, \ldots, g_p) \in T \mid g_1 g_2 \cdots g_p = e\}$. Observe first that $(g_1, \ldots, g_p) \in S \Leftrightarrow g_p = (g_1 \cdots g_{p-1})^{-1}$, so $\sigma((g_1, \ldots, g_{p-1})) = (g_1, \ldots, g_{p-1}, (g_1 \cdots g_{p-1})^{-1})$ is a bijection from G^{p-1} to S. Since $p \mid |G|$ it follows that $p \mid |G^{p-1}|$, and so $p \mid |S|$. Next, if $(g_1, \ldots, g_p) \in S$ then $g_p g_1 g_2 \cdots g_{p-1} = g_p (g_1 g_2 \cdots g_p) g_p^{-1} = g_p e g_p^{-1} = e$, so the cyclic permutation $(g_p, g_1, \ldots, g_{p-1})$ of $(g_1, \ldots, g_p)$ is in S also. Clearly all the cyclic permutations $(g_j, \ldots, g_p, g_1, \ldots, g_{j-1}) \in S$. Another way to write these is $(g_{1+m}, g_{2+m}, \ldots, g_{p+m})$ for $m \in \mathbf{Z}$ where we define $g_{i+m} = g_j$ for $1 \leq j \leq p$ if $[i+m]_p = [j]_p \in \mathbf{Z}_p$: we identify the subscripts modulo p. Thus $(g_{p-1}, g_p, g_1, \ldots, g_{p-2})$ is the case $m = -2$ and $(g_j, g_{j+1}, \ldots, g_p, g_1, \ldots, g_{j-1})$ is the case $m = j - 1$. We define an action of $\mathbf{Z}_p$ on S by $[j]_p(g_1, \ldots, g_p) = (g_{1-j}, \ldots, g_{p-j})$. Is this really a function from $\mathbf{Z}_p \times S$ to S? If $[m]_p = [k]_p$ and $s = (g_1, \ldots, g_p) \in S$ then for all $1 \leq i \leq p$, $[i-m]_p = [i-k]_p$ and it follows that $g_{i-m} = g_{i-k}$ so by definition $[m]_p s = [k]_p s$. In particular, $[0]_p s = s$ for all $s \in S$. Also $([i]_p + [j]_p)s = [i+j]_p(g_1, \ldots, g_p) = (g_{1-i-j}, \ldots, g_{p-i-j}) = [i]_p(g_{1-j}, \ldots, g_{p-j}) = [i]_p([j]_p s)$, proving that we do have a group action of $\mathbf{Z}_p$ on S. Now $p \mid |Fix(\mathbf{Z}_p)|$ by Theorem 9.5 and $(e, \ldots, e) \in Fix(\mathbf{Z}_p)$ so $|Fix(\mathbf{Z}_p)| > 1$. Thus there is $(h_1, \ldots, h_p) \in Fix(\mathbf{Z}_p) - \{(e, \ldots, e)\}$. The definition of fixed point shows $(h_1, \ldots, h_p) = [1]_p(h_1, \ldots, h_p) = (h_p, h_1, \ldots, h_{p-1})$ so $h_p = h_1 = h_2 = \cdots = h_{p-1}$. That is, some $(h, h, \ldots, h) \in S$ with $h \neq e$, and the definition of S gives $h^p = e$. Consequently, $o(h) = p$ by Theorem 2.7, and then $|\langle h\rangle| = p$ by Theorem 2.16, completing the proof of the theorem. ■

The applications of Cauchy's Theorem in the previous chapters will not be repeated here, but we will need the result again when studying groups of various orders. A lot of effort was expended in Theorem 7.15 on groups of order pq to see that not every element could have order p, but this is immediate from Cauchy's Theorem. We end this section with another example that demonstrates how Cauchy's Theorem is instrumental in obtaining a result on the structure of groups with rather special orders.

Example 9.2 Let G be a group, $|G| = p_1 p_2 \cdots p_k$ for primes $p_1 < \cdots < p_k$, and $p_1 \cdots p_{j-1} < p_j$. There are $H_j \triangleleft G$ with $H_k \leq \cdots \leq H_2 \leq G$ and $|H_j| = p_j p_{j+1} \cdots p_k$.

By Cauchy's Theorem there is $H_k = \langle x\rangle$ with $|H_k| = p_k$. Thus Lagrange's Theorem shows H_k has no subgroups except $\langle e\rangle$ and itself. The hypothesis ensures that $|G|$ cannot divide $[G : H_k]! = (p_1 \cdots p_{k-1})!$ so apply Theorem 6.14 to conclude that $H_k \triangleleft G$. This same conclusion follows by using Theorem 4.12 to see that H_k is the only subgroup of G with order p_k. Consequently, G/H_k exists and also $|G/H_k| = p_1 \cdots p_{k-1}$ by Theorem 5.6. Now $|G/H_k|$ satisfies the same conditions as $|G|$, so by induction on k we may assume G/H_k contains normal subgroups $\{H_k\} = V_k \leq \cdots \leq V_2$ with $|V_i| = p_i p_{i+1} \cdots p_{k-1}$ when $i < k$. Using the usual surjection $\rho : G \to G/H_k$ and Theorem 7.2, it follows that for $H_j = \rho^{-1}(V_j)$, $H_k \leq H_{k-1} \leq \cdots \leq H_2 \leq G$, that each $H_j \triangleleft G$, and that for $i < k$, $|H_i| = |H_k||V_i| = p_i \cdots p_k$. Therefore the claim of the example holds by IND-I.

When $|G|=42=2\cdot 3\cdot 7$ the example yields normal subgroups $N\le M$ of G with $|N|=7$ and $|M|=21$. Specifically, if $G=D_{21}$ we could take $N=\langle T_3\rangle$ and $M=\langle T_1\rangle$. For $G=S_3\oplus \mathbf{Z}_7$, take $N=\langle I\rangle\oplus \mathbf{Z}_7$ and $M=\langle (1,2,3)\rangle\oplus \mathbf{Z}_7$.

EXERCISES

1. Let $\cdot : G\times S\to S$ be an action of the group G on the set S. If $H\le G$, show that $\cdot|_{H\times S}: H\times S\to S$ is a group action of H on S.

2. For G a group, determine which of the following are group actions:

i) $\#: G\times G\to G$ given by $g\#x=gx$

ii) $\sim : G\times G\to G$ given by $g\sim x=xg$

iii) $\approx : G\times (G\oplus G)\to (G\oplus G)$ given by $g\approx(x,y)=(gx,e)$

iv) $\cdot : G\times G\to G$ given by $g\cdot x=xg^{-1}$

v) $\wedge : (G\oplus G)\times G\to G$ given by $(g,h)\wedge x=gxh^{-1}$

vi) $* : (G\oplus G)\times G\to G$ given by $(g,h)*x=hxg^{-1}$

3. Show that each of the following operations is a group action.

i) $(g,h,k)\#s=gk\cdot s$ of G^3 on S for $\cdot : G\times S\to S$ a group action and G Abelian

ii) $(g,h,k)\wedge s=ghk\cdot s$ of G^3 on S for $\cdot : G\times S\to S$ a group action, G Abelian

iii) $(g,h,k)\sim s=ghk^{-1}\cdot s$ of G^3 on S for $\cdot : G\times S\to S$ a group action, G Abelian

iv) $\theta\cdot(a_1,\ldots,a_n)=(a_{\theta(1)},\ldots,a_{\theta(n)})$ of S_n on the set A^n

v) $([a_1]_2,\ldots,[a_n]_2)\#r=(-1)^{\Sigma a_i}r$ of $\mathbf{Z}_2^n$ on $\mathbf{R}$

vi) $c^*r=|c|r$ of $(C^*,\ \cdot)$ on $\mathbf{R}$

vii) $r\cdot a=ln(r)+a$ of $(\mathbf{R}^+,\ \cdot)$ on $\mathbf{R}$

viii) $r\#a=a^r$ of $(\mathbf{R}^*,\ \cdot)$ on $(\mathbf{R}^+,\ \cdot)$

4. Group actions $\cdot$, $\#: G\times S\to S$ with corresponding homomorphisms φ,η from G to $\mathrm{Bij}(S)$ are *equivalent* if for a $\rho\in \mathrm{Bij}(S)$ and all $g\in G$, $\eta(g)=\rho^{-1}\varphi(g)\rho$.

i) Show that the actions in i) and iv) of Exercise 2 are equivalent.

ii) Show that the actions in v) and vi) of Exercise 2 are equivalent.

iii) If G is a group, show that $(g,h)\cdot(x,y)=(hx,gy)$ is a group action of $G\oplus G$ on itself and is equivalent to the action of $G\oplus G$ on itself given by left multiplication, i.e., $(g,h)(x,y)=(gx,hy)$.

iv) If G is a group, $G^n=G\oplus\cdots\oplus G$, and $\theta\in S_n$, show $(g_{\theta(1)}x_1,\ldots,g_{\theta(n)}x_n)=(g_1,\ldots,g_n)\#_\theta(x_1,\ldots,x_n)$ is a group action of G^n on itself that is equivalent to the action of left multiplication of G^n on itself.

5. Let $H,K\le G$. Show that $(h,k)\cdot g=hgk^{-1}$ defines a group action of $H\oplus K$ on G. Describe $orb(g)$ for $g\in G$ (these are *double cosets* of H and K in G). Show that $stab(g)\le H\oplus K$ satisfies $stab(g)\cong K\bigcap xHx^{-1}$ for some $x\in G$.

6. If $e\in S\subseteq G$, a finite group, set $S_G=\{gSg^{-1}\subseteq G\mid g\in G\}$. Show that $|S_G|\mid |G|$.

7. Let G be a group, $H\le G$, and $X=\{Hg\mid g\in G\}$. Define $\cdot : G\times X\to X$ by $g\cdot(Hx)=Hxg^{-1}$. Show that this is an action of G on X and find $stab(Hx)$. If $\varphi: G\to \mathrm{Bij}(X)$ is given by $\varphi(g)(X)=g\cdot X$, show that $ker\ \varphi\le H$.

8. Let G be a finite group, $\langle e\rangle\ne H\le G$, $k\in N$ with $k\le|G|$, and $S=\{A\subseteq G\mid |A|=k\}$. Define $\cdot : H\times S\to S$ by $h\cdot\{x_1,\ldots,x_k\}=\{hx_1,\ldots,hx_k\}$.

i) Show that $\cdot$ is a group action of H on S.

ii) If $Fix(H)\ne\varnothing$, show that $|H|\mid k$ (think about cosets of H).

iii) If $|H|=p^i$ for p a prime and if $p^i\nmid k$, show that $p\mid |S|$.

9. Let G be any group and $H \leq G$. Recall for $x \in G$, $cls(x) = \{gxg^{-1} \mid g \in G\}$.
i) If G is finite and $H \neq G$, show that there is $x \in G$ so that $cls(x) \bigcap H = \varnothing$.
ii) Show G infinite and $[G : H] = n > 1 \Rightarrow cls(x) \bigcap H = \varnothing$ for some $x \in G$.

9.2 COUNTING ORBITS

The basic problem we consider is akin to the necklace problem mentioned above. How many different necklaces can be made using n beads of k possible colors? To analyze this problem mathematically, fix the positions of the beads as the vertices of the regular n-gon P_n. Each necklace corrresponds to specifying an allowable color for each vertex. We might guess there are k^n necklaces: choose any color for each of the n vertices. However, different colorings can give the same necklace. For example, if all the vertices but one are colored green and the remaining vertex is colored blue, these n different colorings all correspond to the one necklace with one blue bead and all other beads green. If instead two vertices are blue and the others are green then all colorings with adjacent blue vertices give the same necklace, but all colorings with the two blue vertices separated by a single green vertex give another necklace, assuming $n > 3$.

Each coloring corresponds to a necklace but different colorings may give the same necklace. Identifying all colorings giving the same necklace clearly partitions the set of all colorings into subsets corresponding to the different necklaces. How can we count the number of sets in the partition and how are group actions used? Suppose we are given two colorings and their corresponding necklaces. The necklaces are the same if we can rotate one or turn it over to match the other. These actions correspond to taking one coloring of P_n and applying elements of D_n to the colored vertices of P_n to see whether we can get the other coloring. Thus the number of necklaces, or subsets in the partition of all colorings, will be the number of orbits of D_n acting on the set of colorings by permutation of the colored vertices of P_n. This explanation will suffice for the applications we will consider, but we will discuss briefly how to make all of this mathematically precise.

Let G be a group (D_n above) acting on a set $S(P_n$ above), and let A be the set of colorings of S using k colors. We need a mathematical way to describe the colorings. Any coloring in A is an assignment of each $s \in S$ to an allowable color. Letting $I(k) = \{1, 2, \ldots, k\}$ represent the colors, each $a \in A$ is a function $a : S \to I(k)$, so $A = I(k)^s = \{f : S \to I(k)\}$. For $a, b \in A$, set $a \sim b$ if for some $g \in G$, $b(gs) = a(s)$ for all $s \in S$. This means that the coloring b is related to a when it is obtained from a by some permutation of the (colored!) elements of S. That is, $a \sim b$ when there is some $g \in G$ so that the coloring b is defined by giving gs the color $a(s)$ for all $s \in S$. It is an exercise to show that this is an equivalence relation on A. When $S = P_n$ as above, this relation identifies colorings corresponding to the same necklace since $a \sim b$ when for some $g \in D_n$ the color that b assigns to vertex gv_j is the same color that a assigns to vertex v_j. Thus if the necklace corresponding to the coloring a is rotated or turned over by applying $g \in D_n$, the result is the necklace corresponding to the coloring b. Consequently, the equivalence classes of $\sim$ correspond to those colorings in A that are essentially different.

How do we get G to act on A? Let $g \cdot a \in A$ be the coloring given by $g \cdot a(s) = a(g^{-1}s)$. It might seem more natural to use $g \cdot a(s) = a(gs)$ but because of our notation for functions, this will not quite give a group action. We use that G acts on S, take $a \in A$, and compute $g \cdot (h \cdot a)(s) = h \cdot a(g^{-1}s) = a(h^{-1}(g^{-1}s)) = a(h^{-1}g^{-1}s) = a((gh)^{-1}s) = gh \cdot a(s)$ for all

$s \in S$, so $g \cdot (h \cdot a) = gh \cdot a$, and also $e \cdot a(s) = a(es) = a(s)$ so $e \cdot a = a$. Thus G acts on A. Now G acts like a set of bijections of S by Theorem 9.1, so for $g \in G$, $S = gS$ and if $s = gt$ then $g \cdot a(gt) = a(t)$ for all $t \in S$. Hence $a \sim g \cdot a$ as above. When $a \sim b$ then for some $g \in G$ and all $s \in S$, $b(gs) = a(s) = a(g^{-1}gs) = g \cdot a(gs)$, and with $S = gS$ gives $b = g \cdot a$. Therefore the equivalence classes of $\sim$ on A are the same as the orbits of G acting on A, so the following are the same: the number of essentially different colorings, the number of equivalence classes of $\sim$ on A, and the number of orbits of G on A.

Clearly, counting the orbits for a group action will solve coloring problems like the one mentioned above about necklaces. It turns out that this number is the average number of fixed points per group element.

THEOREM 9.7 Let G be a finite group acting on the finite set S. The number of orbits for this action is $m = \frac{1}{|G|} \sum_{g \in G} |Fix(g)|$.

Proof Now $|Fix(g)|$ counts the elements of S fixed by $g \in G$, so $\sum_{g \in G} |Fix(g)|$ counts the total number of times elements of S are fixed by elements of G, summed over S. Since $s \in S$ is represented in the sum $|stab(s)|$ times, $\sum_{g \in G} |Fix(g)| = \sum_S |stab(s)|$. If $T = \{s_1, \ldots, s_m\}$ is a transversal for the orbits of G acting on S then by Theorem 9.3, $|stab(s)| = |stab(s_j)|$ when $s \in orb(s_j)$, and also $|orb(s_i)| = [G : stab(s_i)]$. Therefore $\sum_S |stab(s)| = \sum_i |stab(s_i)|[G : stab(s_i)] = \sum_i |G| = m|G|$, using Lagrange's Theorem, so $m = \frac{1}{|G|} \sum_{g \in G} |Fix(g)|$ as claimed. ■

How can we count fixed points for coloring problems? If G acts on S, and on the colorings A of S by permuting S, then for $g \in G$, $a \in Fix(g) \Leftrightarrow a(s) = g \cdot a(s) = a(g^{-1}s)$ so s and $g^{-1}s$ have the same color in a, for all $s \in S$. It follows that all $g^i s$ have the same color in a. Consider $\langle g \rangle$ acting on S, with orbits $orb_{\langle g \rangle}(s)$. Our observation means that all elements in $orb_{\langle g \rangle}(s)$ have the same color in a. Conversely, if for all $s \in S$, all $t \in orb_{\langle g \rangle}(s)$ have the same color under $a \in A$, then $a \in Fix(g)$. Hence $|Fix(g)| = k^{\lambda(g)}$ where k is the number of available colors and $\lambda(g)$ is the number of orbits of $\langle g \rangle$ acting on S. This observation and Theorem 9.7 give the method for computing different colorings. We state it as a theorem for reference.

THEOREM 9.8 Let the finite group G act on the finite set S and let A be the set of colorings of S using k colors. For $g \in G$, let $\lambda(g)$ be the number of orbits of S under the action of $\langle g \rangle$. If G acts on A by permutation of S then the number of orbits of G acting on A is $\frac{1}{|G|} \sum_{g \in G} k^{\lambda(g)}$.

Let us see how to use Theorem 9.8.

Example 9.3 A square made of nine equal smaller squares is painted on a piece of wood with each small square painted either red or white. How many paintings are possible?

By Theorem 9.8 the number of colors does not affect the method of computation, so suppose k colors are available. Number the small squares in each row from left to right, using 1, 2, and 3 for the squares in the first row, 4, 5, and 6 for the squares in the second row, and 7, 8, and 9 for the squares in the last row. How can a painting of the squares be moved to give another painting? The only possibilities are to rotate it about its center by multiples of $\pi/2$ radians. Reflections are not possible since then we would see only the unpainted back side of the wood. Thus the group that acts on the paintings and identifies two as giving the same final result is $\langle T\rangle \cong \mathbf{Z}_4$ for T the counter-clockwise rotation of $\pi/2$ radians about the center. Now T acting on the nine small squares has three orbits: the center square, the corner squares, and the "middle" squares, so T acts like the permutation $(5)(1,7,9,3)(2,4,8,6)$ of the small squares. Similarly, T^2 corresponds to $(5)(1,9)(7,3)(2,8)(4,6)$ and T^3 to $(5)(1,3,9,7)(2,6,8,4)$. Every small square is an orbit for $T^4 = I$. In the notation of Theorem 9.8, $\lambda(I) = 9$, $\lambda(T^2) = 5$, and $\lambda(T) = \lambda(T^3) = 3$, so the total number of paintings using k colors is $(k^9 + k^5 + 2k^3)/4$. When $k = 2$ there are 140 paintings, and when $k = 3$ there are 4995.

Example 9.4 If a square mobile is made from nine equal small squares of plastic, each either red or white, how many mobiles are possible?

The difference between this example and the last one is that since the small squares have the same color viewed from either side we can reflect the mobile about the horizontal, vertical, or either diagonal, besides rotating it. Hence the group here is D_4 acting on the square array of nine small squares. For $D_4 = \{T^i, W_j\}$ the action of the T^i is the same as in Example 9.3 and so $\lambda(I) = 9$, $\lambda(T^2) = 5$, and $\lambda(T) = \lambda(T^3) = 3$. It is easily checked that $\lambda(W_j) = 6$ for each j. For example, by numbering the squares as in Example 9.3, the reflection about the horizontal fixes the middle row of small squares and corresponds to $(1,7)(2,8)(3,9)(4)(5)(6)$. Consequently, by Theorem 9.8 the number of mobiles using k colors is $(k^9 + 4k^6 + k^5 + 2k^3)/8$. When $k = 2$ we get 102 mobiles, when $k = 3$ we get 2862 of them, and when $k = 10$ we get 125,512,750 mobiles.

Let us turn to the necklace problem mentioned at the beginning of the chapter.

Example 9.5 How many necklaces can be made from 14 beads using any of k colors?

As above, if A is the set of colorings of the vertices of the regular 14-gon P_{14}, then different colorings give the same necklace when they are in the same orbit of D_{14} acting on A by permutation of the vertices of P_{14}. By Theorem 9.8 we need to compute the number of orbits $\lambda(g)$ for each $g \in D_{14}$ acting on $V = \{\text{vertices of } P_{14}\}$. Now D_{14} consists of 14 rotations and 14 reflections. The rotations are generated by $T_1 = T$ whose action on V is represented by the 14-cycle $(1, 2, \ldots, 14)$. Either by direct computation or by Exercise 19 in §3.3, each T^i is a product of pairwise disjoint cycles all of the same length. Thus the generators of $\langle T\rangle$ are 14-cycles and these will be the six powers T^j for $(j, 14) = 1$. For each of these, $\lambda(T^j) = 1$; there is one orbit on V. Next $o(T^2) = 7$ and any $T^{2k} \neq I$ also has order 7 by Theorem 2.8. It is easy to compute that T^2 is represented by $(1, 3, \ldots, 11)(2, 4, \ldots, 14)$, and so has two orbits acting on V, and that the same is true for its six powers of the same order. This leaves. T^7 and I to consider. Clearly $\lambda(I) = 14$, and T^7 is represented by a product of seven transpositions, so $\lambda(T^7) = 7$. When n is odd, each reflection in D_n fixes one vertex and interchanges the others in pairs, but when n is even, matters are slightly more complicated. Of the 14 reflections $W_i \in D_{14}$, half are reflections through a diameter and fix two points. These are the $\{W_{2j}\}$ where W_{2j} fixes vertices v_j and v_{j+7}. Each of these interchanges the remaining 12 vertices in pairs so $\lambda(W_{2j}) = 8$. Each of the seven remaining reflections through the bisectors of the sides, $\{W_{2j+1}\}$, interchanges the vertices of P_{14} in pairs so $\lambda(W_{2j+1}) = 7$. Using these computations in Theorem 9.8 shows that the number of necklaces using k colors is $(k^{14} + 7k^8 + 8k^7 + 6k^2 + 6k)/28$. As given in the introduction, the number of necklaces using two colors is 687 and the number using three colors is 173,088. This can be verified from the formula fairly easily, even by hand calculation. It is also not hard to compute that with ten colors available, the number of necklaces is (a staggering!) 3,571,456,428,595.

We look at one more coloring problem when the underlying group is not so easy to identify.

Example 9.6 If a cube is painted so that each face is given one of k possible colors, how many different painted cubes are possible?

The underlying set S has the six faces of the cube as elements, and we need to determine the group G of allowable permutations. The cube can be rotated, but it cannot be reflected about planes since it is a solid object rather than a collection of points in $\boldsymbol{R}^3$. There are three rotations of order 4 about the axes of the cube, and four more of order 3 about the diameters (try it with a die!). To describe these, label the faces of the cube like a standard die. Looking at the cube straight on, label the visible face 1, the opposite face 6, the top face 3, the bottom face 4, the face to the right 2, and its opposite face (to the left) 5. The three rotations about the axes are represented by the 4-cycles $\alpha = (1, 2, 6, 5)$, $\beta = (1, 3, 6, 4)$, and $\gamma = (2, 3, 5, 4)$ satisfying: $\alpha\beta = \gamma\alpha = \beta\gamma$, $\alpha\gamma = \beta^3\alpha = \beta^{-1}\alpha$; $\gamma\beta\alpha = \beta$; $x^2y^2 = z^2$ for $\{x, y, z\} = \{\alpha, \beta, \gamma\}$; and $x^2yx^2 = y^3$ for $x, y \in \{\alpha, \beta, \gamma\}$ with $x \neq y$. Direct calculation shows that the four rotations about the diameters are $\alpha\beta$, $\beta\alpha$, $\alpha\gamma$, and $\gamma\beta$, so $G = \langle\{\alpha, \beta, \gamma\}\rangle$. Consider $T = \{\gamma^i\beta^j\alpha^k \mid 0 \le i, j, k \le 3\} \subseteq G$. Now $\gamma^i\beta^j\alpha^2 = \gamma^{i-2}(\gamma^2\beta^2)\beta^{j-2}\alpha^2 = \gamma^{i-2}\alpha^2\beta^{j-2}\alpha^2 = \gamma^{i-2}\beta^{2-j}$ and then $\gamma^i\beta^j\alpha^3 = \gamma^{i-2}\beta^{2-j}\alpha$ so we have $T = \{\gamma^i\beta^j\alpha^k \mid 0 \le i, j \le 3 \text{ and } 0 \le k \le 1\}$. Using $\gamma\beta\alpha = \beta$, it follows that $\gamma^i\beta\alpha = \gamma^{i-1}\beta$, and $\alpha\beta = \beta\gamma$ implies $\beta^3\alpha = \beta^{-1}\alpha = \gamma\beta^{-1} = \gamma\beta^3$ so $\gamma^i\beta^3\alpha = \gamma^{i+1}\beta^3$, and $T = \{\gamma^i\beta^j, \gamma^i\alpha, \gamma^i\beta^2\alpha \mid 0 \le i, j \le 3\}$ follows. These 24 elements are different, and $T = G$ *if* T is a group (why?), which holds if $\alpha T, \beta T, \gamma T \subseteq T$; clearly $\gamma T \subseteq T$. The other computations are exercises illustrated by $\beta\gamma^3\beta^j = \gamma\alpha\gamma^2\beta^j = \gamma^3\gamma^2\alpha\gamma^2\beta^j = \gamma^3\alpha^{-1}\beta^j = \gamma^3\alpha\alpha^2\beta^j\alpha^2\alpha^2 = \gamma^3\alpha\beta^{-j}\alpha^2 = \gamma^3\gamma^{-j}\alpha^3 \in T$. Calculating the cycle structures of the elements of $G = T$ from the representations of α, β, and γ, reveals that besides $I \in G$, six elements have cycle structure $(4, 1, 1)$, eight have cycle structure $(3, 3)$, six have cycle structure $(2, 2, 2)$, and three have cycle structure $(2, 2, 1, 1)$ (verify!). Applying Theorem 9.8 shows that $(k^6 + 3k^4 + 12k^3 + 8k^2)/24$ is the number of different paintings of the cube using k colors. If follows that there are 10 paintings using two colors, 57 using three colors, 800 using five colors, and 43,450 using ten colors.

EXERCISES

1. A child's toy is a collection of solid colored cubical blocks of the same size, with ten blocks of each color.

i) How many different rectangular columns of six blocks could be constructed using blocks of two given colors? How many using blocks of five given colors? How many using six colors?

ii) What are the answers in i) if the column is seven blocks tall?

2. A toy company makes pyramids, each resting on its square base and with a digit on each of its four congruent triangular sides. For $k \in N$, $G(k)$ is a set containing exactly one pyramid for every possible assignment of $\{1, \ldots, k\}$ to the four faces. Find:

i) $|G(3)|$

ii) $|G(4)|$

iii) $|G(5)|$

iv) $|G(6)|$

3. A wire manufacturer markets decorative mobiles in the shape of a regular pentagon, each side of which is a piece of wire in one of ten colors. How many different ornaments are possible?

4. How many bracelets of eleven beads can be made using two colors? How many are possible using ten colors?

5. How many bracelets are possible using twelve beads of two possible colors?

6. At a large dinner party, circular tables are set for ten people. Each table and the people at it are on a raised platform that rotates slowly so each person can see the views out of all the windows. How many arrangements of ten men and women around the table are possible? Note that there is only one possible arrangement of nine women and one man and only one arrangement of five women and five men if they sit alternately.

7. a) A piece of wood one foot on a side is divided into 16 smaller squares three inches on a side, and each small square is painted either red or white. How many different painted large squares are possible?

b) If a stained glass window one foot on a side is made from 16 smaller squares, each 3 inches on a side and colored blue or green, how many different windows are possible?

8. A design is made from eight equal squares, four of which are put together to form a larger square, and one remaining small square is attached at the center of each side of the larger square and extending out from it. The small squares are to be made from colored glass. Determine how many different finished designs there are if the possible number of colors used is:

i) 2

ii) 3

iii) 5

iv) 10

9. A piece of jewelry is made by putting together six pieces of colored glass to make a pendant in the shape of two concentric circles divided from their common center into three equal sections. Compute the number of different pendants that can be made if at least two colors are used and the number of available colors is:

i) 2

ii) 3

iii) 4
iv) 5
v) 10

10. A circular stained glass window is made in the form of four concentric circles divided from the common center into six equal sections, so it requires 24 suitable pieces of colored glass. The windows are assembled at a plant before installation. Find an expression that gives the number of different windows that the plant can produce if k different colors are available.

9.3 SYLOW'S THEOREMS

Recall from Chapter 8 that if G is a group with $|G| = p^n m$ for p a prime, $n \geq 1$, and $(p, m) = 1$, then $H \leq G$ is a ***p*-Sylow subgroup** if $|H| = p^n$. From Theorem 8.13, G must have a p-Sylow subgroup. Here we give a different proof that uses the notion of group actions. We also will obtain information about the set of p-Sylow subgroups and its cardinality: a key factor in analyzing the structure of groups of relatively small order.

The proof of Sylow's First Theorem on the existence of p-Sylow subgroups uses an action of G on those subsets of G having exactly p^n elements. This set of subsets has $\binom{p^n m}{p^n}$ elements and we need a fact about the powers of p dividing this binomial coefficient. By definition $\binom{p^n m}{p^n} = (p^n m)!/p^n!(p^n m - p^n)!$ is the product of $(p^n m - j)/(p^n - j)$ for $0 \leq j \leq p^n - 1$. Observe that when $j = 0$, $p^n m/p^n = m$ and $(p, m) = 1$, and for $j > 0$, $p^k \mid (p^n m - j)$ or $p^k \mid (p^n - j)$ is possible only for $k < n$. In this case$p^k \mid (p^n m - j)$ exactly when $p^k \mid j$, and this is equivalent to $p^k \mid (p^n - j)$. Therefore, all powers of p cancel from each quotient $(p^n m - j)/(p^n - j)$, forcing the conclusion that $\binom{p^n m}{p^n}$ is relatively prime to p.

THEOREM 9.9 (Sylow I) Let G be a group with $|G| = p^n m$ for p a prime, $n \geq 1$, and $(p, m) = 1$. Then there is $H \leq G$ with $|H| = p^n$.

Proof Fix p and use induction on $|G|$. If $|G| = p$ then G is a p-Sylow subgroup of G. Assume now that $|G| > p$ and that any group whose order is a multiple of p but less than $|G|$ has a p-Sylow subgroup. Set $S = \{A \subseteq G \mid |A| = p^n\}$; by our observation above, $(|S|, p) = 1$. We claim G acts on S by left multiplication: if $A = \{g_1, \ldots, g_{p^n}\} \in S$ and $x \in G$ then $x \cdot A = \{xg_1, \ldots, xg_{p^n}\}$. Clearly $x \cdot A \in S$, $e \cdot A = A$ for all $A \in S$, and $x \cdot (y \cdot A) = xy \cdot A$ for $x, y \in G$ since multiplication in G is associative, so G acts on S. If $p^n \nmid |stab(A)|$ for all $A \in S$, then $p \mid [G : stab(A)]$ by Lagrange's Theorem, and Theorem 9.3 shows $p \mid |orb(A)|$. But $\{orb(A) \mid A \in S\}$ partitions S so for some $\{A_j\} \subseteq S$, $|S| = |orb(A_1)| + \cdots + |orb(A_q)|$, forcing $p \mid |S|$, contradicting $(|S|, p) = 1$. Thus there must be $A \in S$ with $p^n \mid |stab(A)|$. If $stab(A) \neq G$ then our induction hypothesis yields $H \leq stab(A)$ with $|H| = p^n$ and since $H \leq G$ it is a p-Sylow subgroup of G. Finally, suppose that $stab(A) = G$. Then $gA = A$ for all $g \in G$. Thus $A = GA = G$ (why?) and $|G| = p^n$, so G itself is the required Sylow subgroup, and applying IND-II completes the proof of the theorem. ■

For examples of p-Sylow subgroups and how their existence is used in analyzing certain groups, see §8.4. The study of the set of p-Sylow subgroups for each prime $p \mid |G|$ is important so we give this set a name.

DEFINITION 9.5 For G a finite group and p a prime with $p \mid |G|$, let p-Syl(G) be the set of all p-Sylow subgroups of G.

The set p-Syl(G) can be large. For example S_7 has $6! = 720$ elements of order 7 so $720/6 = 120$ groups of order 7 (why?), and $|7\text{-Syl}(S_7)| = 120$. For $q > 2$ prime, $|q\text{-Syl}(D_q)| = 1$ but $|2\text{-Syl}(D_m)| = m$ for any odd $m > 1$. In each of these examples it is an exercise to verify that any two p-Sylow subgroups are conjugate. As we will see, this is no accident. It is important to observe that for any $P \in p\text{-Syl}(G)$ and $g \in G$, $gPg^{-1} \in p\text{-Syl}(G)$ also. That $gPg^{-1} \le G$ is straightforward and a fact we have seen before. Also $gPg^{-1} = T_g(P)$ for the inner automorphism T_g determined by g, so $T_g|_P : P \to gPg^{-1}$ is an isomorphism, showing that $|P| = |gPg^{-1}|$.

In a finite Abelian group A, the only conjugate of a subgroup is itself. This does not prove that the p-Sylow subgroup for each prime divisor of $|A|$ is unique, but in fact there is only one by Theorem 4.14. For a simple argument using Sylow's Theorem, if $P, Q \in p\text{-Syl}(A)$ with $|P| = p^n$, then p^n is the largest power of p dividing $|A|$, and since A is Abelian, $PQ \le A$ by Theorem 4.11. By Theorem 4.12, $|PQ| = p^n p^n / |P \cap Q| = p^s \le p^n$ so $|P \cap Q| = p^n$ forces $P = Q$. The same basic argument shows that there is a unique p-Sylow subgroup exactly when it is normal. This is an important fact and we prove it soon, but first we need a technical result required for the other Sylow Theorems.

THEOREM 9.10 Let G be a group, $|G| = p^n m$ for $(p, m) = 1$, $P \in p\text{-Syl}(G)$, and $A \le N(P)$, the normalizer of P in G. If $|A| = p^k$ then $A \le P$.

Proof Now $P \lhd N(P)$ from Definition 8.4, so when $A \le N(P)$, $AP \le G$ by Theorem 5.4. Using Theorem 4.12, $|AP| = |A||P|/|A \cap P| = p^k p^n / p^s$ where $|A \cap P| = p^s \le p^k$ (why?). But $AP \le G$ and Lagrange's Theorem show $|AP| \mid p^n$. Thus $n + k - s \le n$, forcing $k = s$ and so $A \cap P = A$, and $A \le P$ as claimed. ∎

In Chapter 8 we saw by example the benefit of having a normal Sylow subgroup when studying the structure of a group G. Theorem 9.10 is used to show that all $P \in p\text{-Syl}(G)$ are conjugate, a major step in determining when a p-Sylow subgroup is normal, and a special case of the second Sylow Theorem: every p-group in G is contained in a conjugate of any given p-Sylow subgroup. We combine this result with the third Sylow Theorem about $|p\text{-Syl}(G)|$, which is also used to see when $P \in p\text{-Syl}(G)$ is normal.

THEOREM 9.11 For a group G with $|G| = p^n m$ for p a prime and $(p, m) = 1$:

i) (Sylow-II) If $A \le G$, $|A| = p^k$, and $P \in p\text{-Syl}(G)$, then $A \le gPg^{-1}$ for some $g \in G$. In particular, if $Q \in p\text{-Syl}(G)$; then $Q = gPg^{-1}$ for some $g \in G$.
ii) (Sylow-III) $|p\text{-Syl}(G)| = [G : N(P)]$ for $P \in p\text{-Syl}(G)$, $|p\text{-Syl}(G)| \mid m$, and $|p\text{-Syl}(G)| \equiv 1 \pmod{p}$.

Consequently, for $P \in p\text{-Syl}(G)$, $P \lhd G \Leftrightarrow |p\text{-Syl}(G)| = 1$.

Proof From above, $gPg^{-1} \le G$ and $|gPg^{-1}| = |P|$, so $S = \{gPg^{-1} \mid g \in G\} \subseteq p\text{-Syl}(G)$. As in Example 9.1, G acts on S by conjugation, $S = orb(P)$, $stab(P) = N(P)$, and by Theorem 9.3, $|S| = [G : N(P)]$. Since $P \le N(P)$, $p^n \mid |N(P)|$ so $[G : N(P)] \mid m$ from Lagrange's Theorem; thus $(|S|, p) = 1$. Let A act on S by conjugation. Applying Theorem 9.5, $|Fix(A)| \ge 1$; for some $Q = gPg^{-1}$ in S and all $x \in A$, $xQx^{-1} = Q$, or, equivalently, $A \le N(Q)$. But now Theorem 9.10 shows that $A \le Q = gPg^{-1}$. When $k = n$, $A \in p\text{-Syl}(G)$, so $A = gPg^{-1}$ for some $g \in G$ since $A \le gPg^{-1}$ and both have the same order, proving i).

As above, $S = \{gPg^{-1} \mid g \in G\} = p\text{-Syl}(G)$, $|S| = [G : N(P)]$, and $|S| \mid m$, so the first two parts of ii) hold. Let P act on S by conjugation. Since $xPx^{-1} = P$ for all $x \in P$, P is a fixed point of this action. If $Q \in S - \{P\}$ is also fixed then $xQx^{-1} = Q$ for all $x \in P$, forcing $P \le N(Q)$ so $P \le Q$ by Theorem 9.10, and $Q = P$ (why?), contradicting $Q \ne P$. Thus if $Q \in S - \{P\}$, $1 < |orb(Q)| = [P : stab(Q)]$ by Theorem 9.3; from Lagrange's Theorem, $p \mid |orb(Q)|$. The orbits of P on S partition S, every orbit except $orb(P)$ has a multiple of p elements, and $|orb(P)| = 1$, so $|S| = 1 + pt$, or $|p\text{-Syl}(G)| \equiv 1 \pmod{p}$. Now $p\text{-Syl}(G) = \{gPg^{-1} \mid g \in G\}$ for any $P \in p\text{-Syl}(G)$, so we have $P \lhd G \Leftrightarrow |p\text{-Syl}(G)| = 1$. ■

For some groups of small order we can conclude immediately from Theorem 9.11 that certain p-Sylow subgroups are normal. If $|G| = 40$ then $|5\text{-Syl}(G)| \mid 8$ and $|5\text{-Syl}(G)| \equiv 1 \pmod{5}$. Neither 2, 4, nor 8 is congruent to 1 mod 5 so $|5\text{-Syl}(G)| = 1$ and the 5-Sylow subgroup of G is normal and unique. Of course if G is not a 2-group and $|G| = 2^n m$ with m odd then $|2\text{-Syl}(G)| \mid m$ so $|2\text{-Syl}(G)|$ is odd and $|2\text{-Syl}(G)| \equiv 1 \pmod{2}$ always holds.

Using Theorem 9.11 we get quicker proofs of Theorem 8.16 and Theorem 7.15 on groups G of order pq for $p < q$ primes. Such a G has a normal q-Sylow subgroup $Q = \langle y \rangle$ since the number of such is 1 or p and must be congruent to 1 modulo q. Furthermore, $|p\text{-Syl}(G)|$ is either 1 or q and if $q \equiv 1 \pmod{p}$ then $p \mid (q-1)$. Thus $p \nmid (q-1)$ forces $|p\text{-Syl}(G)| = 1$ so the p-Sylow subgroup $P = \langle x \rangle$ is normal. In this case, $G = \langle xy \rangle$ by one of several approaches. One proceeds as in Theorem 7.15 using Theorem 6.19, another uses the argument in Theorem 7.8 that $(|P|, |Q|) = 1$ forces elements in P and in Q to commute so $o(xy) = pq$ by Theorem 4.16, and still another observes that $P \cap Q = \langle e \rangle$ and $G = PQ$ so $G \cong P \oplus Q \cong \mathbf{Z}_{pq}$ (why?).

A useful observation with Theorem 9.11 is the following:

THEOREM 9.12 If G is finite, $N \lhd G$, $P \in p\text{-Syl}(N)$, and $P \lhd N$ then $P \lhd G$.

Proof For any $g \in G$, $gNg^{-1} = N$ so $gPg^{-1} \le N$. Since the automorphism T_g is a bijection, it follows that $|gPg^{-1}| = |P|$. Thus $gPg^{-1} \in p\text{-Syl}(N)$, so $P \lhd N$ forces $gPg^{-1} = P$ by Theorem 9.11, and $P \lhd G$ results. ■

We end this section with two examples that show how to describe the structure of some groups. The method in Chapter 8 using Theorem 6.14 is not applicable here because of the group orders.

Example 9.7 If G is a group with $|G| = 7 \cdot 11 \cdot 19$ then G is cyclic.

Since $|7\text{-Syl}(G)| \mid 11 \cdot 19$ it must be 1, 11, 19, or $11 \cdot 19$. The only one of these satisfying $x \equiv_7 1$ is 1, so if $P_7 \in 7\text{-Syl}(G)$ then $P_7 \triangleleft G$ by Theorem 9.11. It does not similarly follow that the 11 or 19 Sylow subgroups are normal since $7 \cdot 19 = 133 \equiv_{11} 1$ and $7 \cdot 11 = 77 \equiv_{19} 1$. However, $|G/P_7| = 11 \cdot 19$ and $M \in 19\text{-Syl}(G/P_7)$ is normal by Theorem 9.11, as above. Using Theorem 7.2 with the surjection $\rho : G \to G/P_7$ shows that G contains a normal subgroup $\rho^{-1}(M) = N$ of order $7 \cdot 19$ that is cyclic by Theorem 7.15. Now $N \cong \mathbf{Z}_{133}$ by Theorem 6.8 and we conclude from Theorem 6.20 that $|Aut(N)| = \varphi(7 \cdot 19) = \varphi(7)\varphi(19) = 6 \cdot 18$. Therefore $(|Aut(N)|, |G|) = 1$ and Theorem 6.19 forces $N \leq Z(G)$. Now for $P_{11} \in 11\text{-Syl}(G)$, $G = NP_{11}$ (why?), so G is Abelian, and then G is cyclic from Theorem 4.17.

Example 9.8 If G is any group with $|G| = 2 \cdot 5^2 \cdot 7^3$ then the 5-Sylow subgroup and the 7-Sylow subgroup of G are normal.

Theorem 9.11 does not imply that any Sylow subgroup is normal since $2 \cdot 5^2 \equiv_7 1$ and $2 \cdot 7^3 \equiv_5 2^4 \equiv_5 1$. Since $|G| = 2 \cdot 5^2 \cdot 7^3$, G has an element of order 2 by Cauchy's Theorem so Theorem 6.13 produces $N \triangleleft G$ with $|N| = 5^2 \cdot 7^3$. A direct application of Theorem 9.11 implies that the 5-Sylow subgroup and the 7-Sylow subgroup of N are normal in N. These are normal in G by Theorem 9.12, and by definition they are Sylow subgroups of G.

EXERCISES

1. Show that the group G must have a normal Sylow subgroup if $|G| =$:
   - **i)** $117 = 3^2 \cdot 13$
   - **ii)** 275
   - **iii)** $5^6 \cdot 7^3$
   - **iv)** $3 \cdot 5^2 \cdot 11^2$
   - **v)** $5 \cdot 13^4 \cdot 17^3$
   - **vi)** $3 \cdot 11 \cdot 17 \cdot 31$
   - **vii)** $11 \cdot 13 \cdot 43 \cdot 71$
   - **viii)** $2^9 \cdot 11$
   - **ix)** $2 \cdot 7 \cdot 13^2$
2. Show that G has a normal Sylow subgroup if $|G| = pq^2$, for p and q primes.
3. Show that there is $N \triangleleft G$ with $[G : N]$ prime if $|G| = p^2q^2$ for p and q odd primes.
4. If $2 < p < q$ are primes, show that any group G with $|G| = p^2q^k$ has a normal subgroup of prime index.
5. Show that any group of order p^nq for $p < q$ primes and $n = o([p]_q)$ in U_q has a normal Sylow subgroup (count elements of order q using Theorem 9.11).
6. If $|G| = p^3q$ for p and q odd primes, show that G has a normal Sylow subgroup. (Count elements of order q using Theorem 9.11).

7. Let G be a finite group, $N \lhd G$, and $P \in p\text{-Syl}(G)$.
 a) Show that $P \bigcap N \in p\text{-Syl}(N)$.
 b) Show that $NP/N \in p\text{-Syl}(G/N)$.
8. If G is a finite group, $P \in p\text{-Syl}(G)$, and $N_G(P) \leq K \leq G$, show that $N_G(K) = K$.
9. If G is a finite group, $K \lhd G$, and $P \in p\text{-Syl}(K)$, show that $G = K N_G(P)$.
10. For G a finite group, let $\Phi(G) = \bigcap M$ over all maximal $M \leq G$ with $M \neq G$.
 i) Show that $\Phi(G) \lhd G$.
 ii) If $H \leq G$ and $\Phi(G)H = G$ show that $H = G$.
 iii) Let p be a prime. If $P \in p\text{-Syl}(\Phi(G))$ show that $P \lhd G$. (See Exercise 9.)
11. If $\langle e \rangle \neq N \lhd G \Rightarrow N = G$ and if $|G| < 50$, show $G = \langle e \rangle$ or $G \cong \mathbf{Z}_p$, p a prime.
12. Let p be a prime, H and K groups with $|H| = p^n s$, $|K| = p^m t$, $n, m \geq 1$, $(s, t) = 1$, and both $|p\text{-Syl}(H)|, |p\text{-Syl}(K)| > 1$. Show that $|p\text{-Syl}(H \oplus K)| \geq 1 + 3p + 2p^2$.
13. Let $|G| = p^n s$ for p a prime, $n \geq 1$, and $(p, s) = 1$. If $O_p(G)$ is the intersection of all p-Sylow subgroups of G, show that:
 i) $O_p(G)$ is the unique maximal normal p-subgroup of G.
 ii) If $N \lhd G$ then $O_p(N) = O_p(G) \bigcap N$.
 iii) $O_p(G) \neq \langle e \rangle \Leftrightarrow$ there is $N \lhd G$ with $p \mid |N|$ and $\langle e \rangle \neq K \lhd N$, K a p-group.
14. **a)** Argue that if $|G| = 9p^2$ for $p > 3$ a prime, then Sylow's Theorems cannot be applied to show directly that G contains a normal 3-Sylow subgroup.
 b) Argue that if $|G| = 5^2 p_1^2 \cdots p_k^2$ for $5 < p_1 \cdots < p_k$ primes with $k \geq 2$, then Sylow's Theorem does not directly yield $P \lhd G$ with $|P| = 25$.
15. Suppose that G is a group with $|G| = p^n q^m$ for $p < q$ primes. If $n < o([p]_q)$ in U_q, show that there is a chain $\langle e \rangle = H_1 \leq H_2 \leq \cdots \leq H_{n+m} = G$ with $H_j \lhd H_{j+1}$ and H_{j+1}/H_j cyclic of prime order.
16. If G is a finite group so that for some prime p, $G/Z(G)$ has a unique p-Sylow subgroup, show that G has a unique p-Sylow subgroup.

9.4 APPLICATIONS OF SYLOW'S THEOREMS

We have seen that producing a nontrivial normal subgroup of G can be helpful in describing the structure of G. The following examples and results illustrate standard techniques that use Sylow's Theorems and earlier material to investigate the structure of groups. At times, Theorem 9.11 shows directly that a group G has some normal Sylow subgroup, but often some additional effort is needed to produce a normal subgroup. The next examples involve some counting arguments that use Theorem 9.11 and cannot be analyzed simply with the results available in Chapter 8.

Example 9.9 Any group G with $|G| = 56$ has a normal Sylow subgroup.

By Theorem 9.11, $|7\text{-Syl}(G)| = 1$ or $|7\text{-Syl}(G)| = 8$ and if $7\text{-Syl}(G) = \{P\}$ then $P \lhd G$. If $7\text{-Syl}(G) = \{P_1, \ldots, P_8\}$, each P_j has order 7 and so is cyclic by Theorem 4.9, and for $i \neq j$, $P_i \bigcap P_j = \langle e \rangle$ by Theorem 4.6. It follows that $P_1 \bigcup \cdots \bigcup P_8$ contains $8 \cdot (7 - 1) = 48$ elements of order 7. The leaves $56 - 48 = 8$ elements, none of order 7. But any 2-Sylow subgroup consists of eight elements, none of order 7, so there can be only one 2-Sylow subgroup, and Theorem 9.11 forces it to be normal.

Thus one of the Sylow subgroups in G is normal. Note that Theorem 6.14 is not applicable here since the index in G of any Sylow subgroup is too large.

Recall from Definition 8.6 and Theorem 8.16 that for $p < q$ primes with $p|(q-1)$, if G is a non-Abelian group with $|G| = pq$, then $G \cong \mathbf{Z}_q \diamond \mathbf{Z}_p$.

Example 9.10 G a group with $|G| = 105 \Rightarrow G \cong \mathbf{Z}_{105}$ or $G \cong \mathbf{Z}_5 \oplus \mathbf{Z}_7 \diamond \mathbf{Z}_3$.

Since $105 = 3 \cdot 5 \cdot 7$, $7 \equiv_3 1$, $21 \equiv_5 1$, and $15 \equiv_7 1$, we cannot conclude from Theorem 9.11 that any Sylow subgroup is normal. Also Theorem 6.14 is of no help. From Theorem 9.11, $|7\text{-Syl}(G)| = 1$ or 15 and if there are 15, then as in Example 9.9, any 7-Sylow subgroup is cyclic and the intersection of two different 7-Sylow subgroups is $\langle e \rangle$. Consequently, the fifteen 7-Sylow subgroups contain $15 \cdot 6 = 90$ elements of order 7, leaving 15 elements in G whose orders are not 7. Using Theorem 9.11 again, $|5\text{-Syl}(G)| = 1$ or 21, and if there are 21 then, as above, it follows that there are $21 \cdot 4 = 84$ elements of order 5. This computation shows that either $|7\text{-Syl}(G)| = 1$ or $|5\text{-Syl}(G)| = 1$, so one of these Sylow subgroup is normal by Theorem 9.11.

If the 5-Sylow subgroup P_5 is normal then $|G/P_5| = 21$ so there is $K \lhd G/P_5$ of order 7 by Theorem 8.15. Use Theorem 7.2 with the surjection $\rho : G \to G/P_5$ to get $\rho^{-1}(K) \lhd G$ of order 35. A similar argument when the 7-Sylow subgroup P_7 of G is normal also produces a normal subgroup of G of order 35. In either case there is $N \lhd G$, $|N| = 35$, and $N \cong \mathbf{Z}_{35}$ (Theorem 7.15). It follows by Theorem 3.1 that N contains subgroups of orders 5 and 7, necessarily normal in N since N is Abelian. Apply Theorem 9.12 to conclude that these subgroups are both normal in G, so $P_5, P_7 \lhd G$.

By Theorem 9.9 there is $\langle x \rangle = P_3 \le G$ of order 3. Also $P_5 \cong \mathbf{Z}_5$ and Theorem 6.20 show $|Aut(\mathbf{Z}_5)| = 4$. Since $(3, 4) = 1 = (7, 4)$, applying Theorem 6.19 results in $\langle x \rangle$ and P_7 centralizing P_5. Using $P_7 \lhd G$, $\langle x \rangle P_7 \le G$ from Theorem 5.4. Now Theorem 4.12 shows first that $|\langle x \rangle P_7| = 21$ and then that $P_5 \langle x \rangle P_7 = G$, so $\langle x \rangle P_7 \lhd G$ (P_5 centralizes it!), and $P_5 \cap \langle x \rangle P_7 = \langle e \rangle$ by considering orders. Hence Theorem 7.9 gives $G \cong P_5 \oplus \langle x \rangle P_7$. Finally, Theorem 8.16 tells us that $\langle x \rangle P_7 \cong \mathbf{Z}_{21}$ or $\langle x \rangle P_7 \cong \mathbf{Z}_7 \diamond \mathbf{Z}_3$, so Theorem 7.10 and Theorem 4.17 show $G \cong \mathbf{Z}_5 \oplus \mathbf{Z}_{21} \cong \mathbf{Z}_{105}$ or $G \cong \mathbf{Z}_5 \oplus \mathbf{Z}_7 \diamond \mathbf{Z}_3$.

The example shows how we can use the result on groups of order pq together with results about direct sums to obtain the structure of G. This approach depends on finding a normal Sylow subgroup. We can do this, just as in Example 9.10, for a more general situation.

THEOREM 9.13 If G is a group with $|G| = pqr$ for $p < q < r$ primes, then the r-Sylow subgroup of G is normal and G has a normal subgroup of index p.

Proof If neither the r-Sylow subgroup nor the q-Sylow subgroup of G is normal, then from Theorem 9.11 there is more than one of each. But $|r\text{-Syl}(G)| \equiv_r 1$ and divides pq, so must be

pq since both $p, q < r$. Any r-Sylow subgroup is cyclic by Theorem 4.9, so different ones intersect in $\langle e\rangle$ since r is prime. Hence G contains $pq(r-1)$ elements of order r. Similarly, there are at least r q-Sylow subgroups and these contain $r(q-1)$ elements of order q. We have counted $pq(r-1)+r(q-1)=pqr+r(q-1)-pq$ elements of G. That is, $|G|\geq |G|+r(q-1)-pq$. But $p<q<r$ implies that $p\leq q-1$, giving $r(q-1)-pq>0$, and this contradiction forces either the q-Sylow subgroup or the r-Sylow subgroup of G to be normal. If the r-Sylow subgroup $P_r \triangleleft G$ then G/P_r has order pq so there is $K \triangleleft G/P_r$ of order q by Theorem 8.15. Under the usual epimorphism $\rho: G \to G/P_r$, K corresponds to $\rho^{-1}(K)=N \triangleleft G$ with $|N|=qr$ by Theorem 7.2. From Lagrange's Theorem, $[G:N]=p$, proving the theorem. Now assume that the q-Sylow subgroup $P_q \triangleleft G$. By the argument just given, G/P_q contains a normal subgroup of order r and the Correspondence Theorem yields $M \triangleleft G$ with $|M|=qr$ and $[G:M]=p$. To finish the proof, we need only show that the r-Sylow subgroup of G is normal. Now M contains a normal subgroup H of order r by Theorem 8.15. Clearly H is an r-Sylow subgroup of both M and G so Theorem 9.12 shows that H is normal in G. ■

We illustrate again how the general approach in Example 9.10 can give the possible structures of some groups, essentially repeating the argument of Example 8.13 once we get normal Sylow subgroups.

Example 9.11 If G is a group with $|G|=30$ then G is isomorphic to exactly one of $\mathbf{Z}_{30}$, $\mathbf{Z}_3 \oplus D_5$, $D_3 \oplus \mathbf{Z}_5$, or D_{15}.

From Theorem 9.13 there is $N \triangleleft G$ of order 15, and N is cyclic by Theorem 7.15 or Theorem 8.16. Therefore N contains subgroups $\langle x\rangle$ of order 3 and $\langle y\rangle$ of order 5 by Theorem 7.1, necessarily normal in N since N is Abelian. Now $\langle x\rangle, \langle y\rangle \triangleleft G$ follow from Theorem 9.12, and $N=\langle x\rangle\langle y\rangle$ is a consequence of Theorem 4.12. If $\langle g\rangle \in 2\text{-Syl}(G)$ then $G=N\cup gN=\langle g\rangle N=\langle g\rangle\langle x\rangle\langle y\rangle$. Since $\langle x\rangle, \langle y\rangle \triangleleft G$, g acts on each of these by conjugation. The automorphism group of a cyclic group of prime order p is itself cyclic of order $p-1$ by Theorem 6.20 and so has a unique automorphism of order 2 when $p\geq 3$ (Theorem 3.1), which must be $\varphi(\alpha)=\alpha^{-1}$ by Example 6.19. Therefore $o(g)=2$ implies either $gxg^{-1}=x$ or $gxg^{-1}=x^{-1}$ and a similar dichotomy holds for gyg^{-1}. If g acts on each of $\langle x\rangle$ and $\langle y\rangle$ like the identity, then $gxg^{-1}=x$ and $gyg^{-1}=y$, so g commutes with x and y. Using $G=\langle g\rangle\langle x\rangle\langle y\rangle$ it follows that G is Abelian so $G\cong \mathbf{Z}_{30}$ by Theorem 4.17. If $gxg^{-1}=x^{-1}$ and $gyg^{-1}=y^{-1}$ then $g(xy)g^{-1}=gxg^{-1}gyg^{-1}=x^{-1}y^{-1}=(xy)^{-1}$ since $xy=yx$. Now Theorem 4.16 forces $o(xy)=15$ so $G\cong D_{15}$ by Example 6.4.

Next assume $gxg^{-1}=x$ but $gyg^{-1}=y^{-1}$. Note $\langle g\rangle\langle y\rangle \leq G$ since $\langle y\rangle \triangleleft G$ (Theorem 5.4), $|\langle g\rangle\langle y\rangle|=10$ from Theorem 4.12, and $\langle x\rangle \cap \langle g\rangle\langle y\rangle=\langle e\rangle$ by Theorem 4.6. Since $xg=gx$, $xy=yx$, and $G=\langle g\rangle\langle x\rangle\langle y\rangle=\langle x\rangle\langle g\rangle\langle y\rangle$, we have $\langle x\rangle \leq Z(G)$. Hence $\langle g\rangle\langle y\rangle, \langle x\rangle \triangleleft G$ and $G\cong\langle x\rangle\oplus\langle g\rangle\langle y\rangle$ using Theorem 7.9. Finally, $\langle x\rangle\cong\mathbf{Z}_3$ from Theorem 6.8 and $\langle g\rangle\langle y\rangle\cong D_{10}$ by Example 6.4, so Theorem 7.10 shows $G\cong\mathbf{Z}_3\oplus D_{10}$. Similarly, when $gxg^{-1}=x^{-1}$ and $gyg^{-1}=y$ we get $G\cong\langle g\rangle\langle x\rangle\oplus\langle y\rangle\cong D_3\oplus\mathbf{Z}_5$. Surely all four of these groups occur and no two are isomorphic since their centers have different orders.

We turn to examples for which we cannot easily determine the possible structures of groups of the given order but can as least find a normal subgroup by using Sylow's Theorems together with Theorem 6.14.

Example 9.12 If $|G| = 112$ there is $N \lhd G$ with $|N| = 2^k$ for $k \geq 1$.

Since $112 = 2^4 \cdot 7$, Theorem 9.9 gives $P_2 \in 2\text{-Syl}(G)$ with $|P| = 2^4$. From Lagrange's Theorem, $[G : P_2] = 7$ and so Theorem 6.14 produces some proper $\langle e\rangle \neq N \lhd G$ since $2^m \mid (7!/2)$ only for $m \leq 3$. If $|N| = 2^k$ we are finished, so assume $7 \mid |N|$. When $|N| = 7$, 14, or 28, Theorem 9.11 shows $7\text{-Syl}(N) = \{P_7\}$ so $P_7 \lhd N$, and $P_7 \lhd G$ by Theorem 9.12. When $|N| = 56$ then N has a normal Sylow subgroup by Example 9.9. This subgroup is normal in G using Theorem 9.12 again. Therefore we may assume that $P_7 \lhd G$.

Let G act on P_7 by conjugation. By Theorem 6.19 there is a homomorphism $\varphi : G \to Aut(P_7)$ given by $\varphi(g) = T'_g$, the restriction to P_7 of the inner automorphism T_g on G, and $G/C_G(P_7) \cong \varphi(G) \leq Aut(P_7)$, so $C_G(P_7) \lhd G$. Note that $Aut(P_7) \cong \mathbf{Z}_6$ using Theorem 6.20. It follows from $|G/C_G(P_7)| \mid 6$ that $C_G(P_7) = G$ or $|C_G(P_7)| = 56$ and in either case, let $K_2 \in 2\text{-Syl}(C_G(P_7))$. From Theorem 4.12 we get $C_G(P_7) = P_7K_2$, and since the elements in K_2 commute with the elements in P_7 it follows that $K_2 \lhd C_G(P_7)$. Applying Theorem 9.12 once again shows that $K_2 \lhd G$, completing the argument.

Theorem 9.11 can force some groups to be Abelian.

Example 9.13 If $|G| = 45$, G is Abelian and $G \cong \mathbf{Z}_{45}$ or $G \cong \mathbf{Z}_3 \oplus \mathbf{Z}_3 \oplus \mathbf{Z}_5$.

Since $45 = 3^2 \cdot 5$, Theorem 9.11 implies that the Sylow subgroups of G are normal : 1 is the only divisor of 5 congruent to 1 mod 3, and 1 is the only divisor of 9 congruent to 1 mod 5. Therefore $\{P_3\} = 3\text{-Syl}(G)$, $\{P_5\} = 5\text{-Syl}(G)$, $G = P_3P_5$ from Theorem 4.12, and $\langle e\rangle = P_3 \cap P_5$ by Theorem 4.6. Thus Theorem 7.9 shows $G \cong P_3 \oplus P_5$. But $P_5 \cong \mathbf{Z}_5$ by Theorem 6.8 and $P_3 \cong \mathbf{Z}_9$ or $P_3 \cong \mathbf{Z}_3 \oplus \mathbf{Z}_3$, by Theorem 7.14; Theorem 7.10 finishes the argument.

The next example follows from Theorem 6.13 and Theorem 9.12. Instead, we illustrate an interesting technique using Theorem 6.14 on the normalizer of a Sylow subgroup in a situation wherein it cannot be used on the Sylow subgroup itself.

Example 9.14 Any group G with $|G| = 90$ contains $\langle e\rangle \neq N \lhd G$ with $|N|$ odd.

Using $90 = 2 \cdot 3^2 \cdot 5$ and Theorem 9.11, if $P \in 5\text{-Syl}(G)$ then $P \lhd G$, or $[G : N(P)] = |5\text{-Syl}(G)| = 6$. Assume $[G : N(P)] = 6$ so $|N(P)| = 15$. Apply Theorem 6.14 to get a homomorphism $\varphi : G \to \text{Bij}(\{gN(P)\})$ with $ker\,\varphi \leq N(P)$. But $ker\,\varphi \lhd G$ is the required subgroup by Lagrange's Theorem, or $ker\,\varphi = \langle e\rangle$ and φ is an injection (Theorem 6.4). In the latter case, note first that $90 \mid [G : N(P)]!/2$ so Theorem 6.14 does not produce a normal subgroup of G. Using $\text{Bij}(\{gN(P)\}) \cong S_6$ from Theorem 6.12, the composition of φ with this isomorphism gives another homomorphism $\eta : G \to S_6$ with $ker\,\eta = \langle e \rangle$; hence η is injective and $|\eta(G)| = 90$. The product of any two odd permutations in $\eta(G)$ is in $\eta(G) \bigcap A_6$ and it follows that $[\eta(G) : \eta(G) \bigcap A_6] \leq 2$. If this index is two, $\eta(G) \bigcap A_6$ has order 45, is Abelian by Example 9.13, and then must contain an element of order 15 (why?), which is impossible in A_6. The possibility that $\eta(G) \leq A_6$ remains. But $|\eta(G)| = 90$ and $|A_6| = 6!/2 = 360$, so $[A_6 : \eta(G)] = 4$. Therefore Theorem 6.14 applied to $\eta(G) \leq A_6$ produces $\langle I \rangle \neq K \lhd A_6 \bigcap \eta(G)$, which is impossible since A_6 is a simple group by Theorem 5.11. This contradiction completes the verification of the example.

For a few special group orders we can show that all of the Sylow subgroups are normal, and this leads to a very pretty structure theorem for the group. The key observation is that the conditions of Theorem 7.8 on direct sums are satisfied when the Sylow subgroups are all normal.

THEOREM 9.14 Let G be a finite group, $p_1 < \cdots < p_k$ the prime divisors of $|G|$, and $P_j \in p_j\text{-Syl}(G)$. Then $G \cong P_1 \oplus \cdots \oplus P_k$ if and only if each $P_j \lhd G$.

Proof Assume $G \cong P_1 \oplus \cdots \oplus P_k$, and set $H_j = \langle e \rangle \oplus \cdots \oplus P_j \oplus \cdots \oplus \langle e \rangle$. From Theorem 7.8, the subgroup in G corresponding to H_j under the isomorphism is normal in G and has order $|P_j|$, so it is a normal p_j-Sylow subgroup of G. Thus $|p_j\text{-Syl}(G)| = 1$ and $P_j \lhd G$ by Theorem 9.11. On the other hand, if each $P_j \lhd G$ then Theorem 5.4 shows that for $1 \leq s \leq k$, $P_1 P_2 \cdots P_s \lhd G$, and $|P_1 P_2 \cdots P_s| = |P_1||P_2| \cdots |P_s|$ by Theorem 4.13. Hence using Theorem 4.6, $P_1 \cdots P_{s-1} \bigcap P_s = \langle e \rangle$ and the hypothesis of Theorem 7.8 is satisfied, allowing us to conclude that $G \cong P_1 \oplus \cdots \oplus P_k$, finishing the proof of the theorem. ∎

Using just Sylow's Theorems, we can show that all Sylow subgroups of G are normal only when the prime powers appearing in $|G|$ are nicely related. Note that Sylow's Theorems never enable us to conclude that the 2-Sylow subgroup is normal unless the group is a 2-group. Our first example using Theorem 9.14 can be handled without it, but not as easily.

Example 9.15 If G is a group of order 1001 then G is cyclic.

Since $1001 = 7 \cdot 11 \cdot 13$, Theorem 9.11 shows each Sylow subgroup is normal. Specifically, $|13\text{-Syl}(G)| = 1, 7, 11,$ or 77, and the only one of these satisfying $x \equiv_{13} 1$ is $x = 1$. Thus the 13-Sylow subgroup of G is normal by Theorem 9.11. Similarly, $13 \equiv_{11} 2$ and so $7 \cdot 13 \equiv_{11} 7 \cdot 2 \equiv_{11} 3$, forcing there to be a unique 11-Sylow subgroup. Finally, $11 \equiv_7 4$ and $13 \equiv_7 -1$, so $11 \cdot 13 \equiv_7 -4$. Hence the only divisor of $11 \cdot 13$ congruent to 1 modulo 7 is 1 and the 7-Sylow subgroup is normal. Since each Sylow subgroup is cyclic of prime order, each is isomorphic to the appropriate $\boldsymbol{Z}_p$, and Theorem 9.14 (together with Theorem 7.10) yields $G \cong \boldsymbol{Z}_7 \oplus \boldsymbol{Z}_{11} \oplus \boldsymbol{Z}_{13}$. Finally, $G \cong \boldsymbol{Z}_{1001}$ from Theorem 3.21, from the corollary to Theorem 6.9, or from Example 7.7.

Determining the structure of a group G is easiest when each prime divisor of $|G|$ appears at most twice. In this case the Sylow subgroups are Abelian, so if each is also normal then G is Abelian and we can describe the possible structures of G. When larger prime powers appear it is difficult to determine the structure of each Sylow subgroup, so the structure of G.

Example 9.16 Any group of order $1225 = 5^2 \cdot 7^2$ is Abelian and is isomorphic to exactly one of: $\mathbf{Z}_{1225}$; $\mathbf{Z}_{175} \oplus \mathbf{Z}_7$; $\mathbf{Z}_{245} \oplus \mathbf{Z}_5$; or $\mathbf{Z}_7 \oplus \mathbf{Z}_7 \oplus \mathbf{Z}_5 \oplus \mathbf{Z}_5$.

By Theorem 9.11, $n_7 = |7\text{-Syl}(G)| = 1, 5,$ or 25 and $n_7 \equiv_7 1$, so $n_7 = 1$. Hence if $P_7 \in 7\text{-Syl}(G)$ then $P_7 \triangleleft G$. Similarly, the divisors of 49 are 1, 7, and 49 and of these only 1 is congruent to 1 mod 5 so the 5-Sylow subgroup $P_5 \triangleleft G$. Therefore $G \cong P_5 \oplus P_7$ by Theorem 9.14. It follows from either Theorem 7.14 or Theorem 8.9 that $P_5 \cong \mathbf{Z}_{25}$ or $P_5 \cong \mathbf{Z}_5 \oplus \mathbf{Z}_5$ and that $P_7 \cong \mathbf{Z}_{49}$ or $P_7 \cong \mathbf{Z}_7 \oplus \mathbf{Z}_7$. Thus, as claimed, there are four possibilities for G using Theorem 7.10 together with the corollary to Theorem 6.9. For example, $\mathbf{Z}_{25} \oplus \mathbf{Z}_{49} \cong \mathbf{Z}_{1225}$ and $\mathbf{Z}_{25} \oplus \mathbf{Z}_7 \oplus \mathbf{Z}_7 \cong \mathbf{Z}_{175} \oplus \mathbf{Z}_7$ (see Theorem 7.11).

When $|G| = p_1^2 \cdots p_k^2$, Sylow's Theorems sometimes can be used to show all Sylow subgroups are normal, so G is Abelian and isomorphic to one of 2^k possible groups as in Example 9.16. This happens when $|G| = 7^2 \cdot 17^2 \cdot 31^2$ (why?), and G is isomorphic to one of eight Abelian groups. The details are left to an exercise. Occasionally, results from Chapter 6 can show that each Sylow subgroup is normal when Theorem 9.11 does not.

Example 9.17 Any group G of order $7^2 \cdot 11^2 \cdot 17^2$ is Abelian and is isomorphic to one of the eight groups obtained by taking the direct sum for each $p = 7, 11, 17$, of either $\mathbf{Z}_p \oplus \mathbf{Z}_p$ or $\mathbf{Z}_{p^2}$.

Since $11 \equiv_7 4$ and $17 \equiv_7 3$, $11 \cdot 17^2 \equiv_7 4 \cdot 9 \equiv_7 1$, so we cannot conclude from Theorem 9.11 that there is a normal 7-Sylow subgroup. The number of possible 11-Sylow subgroups is $1, 7, 7^2, 17, 7 \cdot 17, 7^2 \cdot 17, 17^2, 7 \cdot 17^2$, or $7^2 \cdot 17^2$. Using both that $17 \equiv_{11} 6$ and $7 \equiv_{11} -4$, it follows that the only one of these integers congruent to 1 mod 11 is 1, so the 11-Sylow subgroup P_{11} is normal by Theorem 9.11. A similar calculation shows that the 17-Sylow subgroup P_{17} is normal. If $P_q \in q\text{-Syl}(G)$ then $P_{11}, P_{17} \triangleleft G$ imply $P_7 P_{11} \triangleleft G$, and $P_7 P_{11} P_{17} \leq G$ by Theorem 5.4. But the orders of these Sylow subgroups are pairwise relatively prime so using Theorem 4.12 gives $P_7 P_{11} P_{17} = G$.

As $P_{11}, P_{17} \triangleleft G$, for $x \in P_{11}$ and $y \in P_{17}$, $x^{-1}y^{-1}xy \in P_{11} \cap P_{17} = \langle e \rangle$ (Theorem 4.6), forcing $xy = yx$. Each Sylow subgroup is Abelian and is isomorphic to $\mathbf{Z}_p \oplus \mathbf{Z}_p$ or $\mathbf{Z}_{p^2}$ by Theorem 8.9. Now Theorem 6.20 shows $|Aut(\mathbf{Z}_{121})| = 110$ and Theorem 6.22 shows $|Aut(\mathbf{Z}_{11} \oplus \mathbf{Z}_{11})| = 120 \cdot 110$ so Theorem 6.19 yields $P_7 \leq C_G(P_{11})$. Similarly, $P_7 \leq C_G(P_{17})$. But $G = P_7 P_{11} P_{17}$ and P_7 is Abelian, so G itself must be Abelian and $P_7 \triangleleft G$ by Theorem 5.1. Applying Theorem 9.14 yields $G \cong P_7 \oplus P_{11} \oplus P_{17}$, and just as in Example 9.16, this leads to the eight possibilities for G by Theorem 7.10.

We present another example for which only one Sylow subgroup is normal by Theorem 9.11. Since the example is very close in approach to the last one, we will be somewhat lax in citing all the appropriate results.

Example 9.18 Any group G of order $5^2 \cdot 7^2 \cdot 73^2$ is Abelian, and up to isomorphism there are eight possibilities for G.

If we show G is Abelian then its Sylow subgroups are normal and G is the direct sum of them by Theorem 9.14. The statement about the eight possibilities follows as in Example 9.17. After a little computation using Theorem 9.11 we see that the 73-Sylow subgroup $P_{73} \lhd G$ but do not know whether the other Sylow subgroups are normal since $7 \cdot 73 \equiv_5 2 \cdot 3 \equiv_5 1$, and $5 \cdot 73 \equiv_7 5 \cdot 3$ so $5 \cdot 73 \equiv_7 1$. Now $|G/P_{73}| = 5^2 \cdot 7^2$ and Theorem 9.11 shows that the 7-Sylow subgroup of this quotient is normal. Applying Theorem 7.2 produces $N \lhd G$ of order $7^2 \cdot 73^2$. But each of the Sylow subgroups of N is normal by Theorem 9.11, so each is normal in G by Theorem 9.12. Furthermore N is their product, $P_7 P_{73}$. If $P_5 \in 5\text{-Syl}(G)$ then, as in Example 9.17, $G = P_5 P_7 P_{73}$ and using Theorem 6.19 with G acting by conjugation on each of P_7 and P_{73} shows that both $P_5, P_{73} \le C_G(P_7)$ and $P_5, P_7 \le C_G(P_{73})$. Since each Sylow subgroup is Abelian and the elements in any one of them commute with the elements in any other, G must be Abelian, completing the argument.

It may be possible to show all Sylow subgroups of G are normal but not to be able to say more than Theorem 9.14 does about the structure of G. Even in these cases, since p-groups have nontrivial centers and (normal) subgroups of all possible orders, we can conclude the same about G (why?).

Example 9.19 If G is a group of order $11^a \cdot 13^b$ for $a \le 11$ and $b \le 9$, then $G \cong P_{11} \oplus P_{13}$ for $P_{11} \in 11\text{-Syl}(G)$ and $P_{13} \in 13\text{-Syl}(G)$.

Use Theorem 9.11. Observe that $13 \equiv_{11} 2$ and $o([2]_{11}) = 10$ in U_{11}, so $13^m \equiv_{11} 1$ for $0 \le m \le 9$ implies $m = 0$. Hence the 11-Sylow subgroup $P_{11} \lhd G$. Similarly, because $11 \equiv_{13} -2$ and $o([-2]_{13}) = 12$ in U_{13}, for $0 \le t \le 11$ if $11^t \equiv_{13} 1$ then $t = 0$. Thus the 13-Sylow subgroup $P_{13} \lhd G$, and by Theorem 9.14, $G \cong P_{11} \oplus P_{13}$. Unless $a \le 2$ or $b \le 2$ it is difficult to say more about the structure of G other than $Z(G) \ne \langle e \rangle$ and that G contains normal subgroups of all possible orders, using Theorem 8.8 and Theorem 8.10 (how?).

We have avoided groups in which $pq^2 \mid |G|$ and $p \mid (q-1)$ since these are more complicated to describe. One special group of order $2p^2$ is given next.

DEFINITION 9.6 For $p > 2$ prime, let $(\mathbf{Z}_p \oplus \mathbf{Z}_p) \diamond \mathbf{Z}_2 = ((\mathbf{Z}_p \times \mathbf{Z}_p) \times \mathbf{Z}_2, \cdot)$ with $(([a]_p, [b]_p), [c]_2) \cdot (([a']_p, [b']_p), [c']_2) = (([a + (-1)^c a']_p, [b + (-1)^c b']_p), [c + c']_2)$.

As in Example 8.10, $(\mathbf{Z}_p \oplus \mathbf{Z}_p) \diamond \mathbf{Z}_2$ is a non-Abelian group, $\mathbf{Z}_p \oplus \mathbf{Z}_p \cong (\mathbf{Z}_p \times \mathbf{Z}_p) \times \{[0]_2\} \lhd (\mathbf{Z}_p \oplus \mathbf{Z}_p) \diamond \mathbf{Z}_2$, $\mathbf{Z}_2 \cong \{([0]_p, [0]_p)\} \times \mathbf{Z}_2 \le (\mathbf{Z}_p \oplus \mathbf{Z}_p) \diamond \mathbf{Z}_2$, and conjugation by $(([0]_p,[0]_p),[1]_2)$ on the p-Sylow subgroup of $(\mathbf{Z}_p \oplus \mathbf{Z}_p) \diamond \mathbf{Z}_2$ sends each element to its inverse. It is an exercise to show that a group of order $2p^2$, having a normal p-Sylow subgroup $P \cong \mathbf{Z}_p \times \mathbf{Z}_p$ so that an element of order 2 conjugates each element of P to its inverse, must be isomorphic to $(\mathbf{Z}_p \oplus \mathbf{Z}_p) \diamond \mathbf{Z}_2$. A concrete example is $G = \left\{ \begin{bmatrix} a & b & c \\ [0] & [1] & [0] \\ [0] & [0] & [1] \end{bmatrix} \in GL(3,\mathbf{Z}_p) \mid a = [\pm 1] \in \mathbf{Z}_p \right\}$. Then $p\text{-Syl}(G) = \{P\}$ for $P = \{A \in G \mid A_{jj} = [1], \text{ all } j\}$, and conjugating P by $Diag([-1],[1],[1])$ sends each element of P to its inverse. It should be clear here that $G \cong (\mathbf{Z}_p \oplus \mathbf{Z}_p) \diamond \mathbf{Z}_2$. Now we can characterize all groups of order $2p^2$.

Example 9.20 Any group G of order $2p^2$ for $p > 2$ a prime is isomorphic to one of the groups: $\mathbf{Z}_{2p^2}$, D_{p^2}, $\mathbf{Z}_2 \oplus \mathbf{Z}_p \oplus \mathbf{Z}_p$, $D_p \oplus \mathbf{Z}_p$, or $(\mathbf{Z}_p \oplus \mathbf{Z}_p) \diamond \mathbf{Z}_2$.

We conclude from Theorem 9.11 that the p-Sylow subgroup $P \lhd G$. Since $|P| = p^2$ Theorem 8.9 shows P is cyclic or $P \cong \mathbf{Z}_p \oplus \mathbf{Z}_p$. In either case if $\langle h \rangle \in 2\text{-Syl}(G)$ then by Theorem 4.6, $\langle h \rangle \cap P = \langle e \rangle$ so $\langle h \rangle P = G$ follows from Theorem 4.12. Using $P \lhd G$ we have $hPh^{-1} = hPh = P$, so the inner automorphism T_h of G restricts to an automorphism $\varphi = T_h|_P : P \to P$ of P whose order divides $o(h) = 2$. Consider first when P is cyclic. If φ is the identity on P then $h \in C_G(P)$ so G is Abelian, $\langle h \rangle \lhd G$, and $G \cong \mathbf{Z}_{2p^2}$ by Theorem 4.17 (and Theorem 6.8). If $o(\varphi) = 2$ then since $Aut(\mathbf{Z}_{2p^2})$ is cyclic from Theorem 6.20, φ is the unique automorphism of order 2, using Theorem 3.1. Thus $\varphi(g) = g^{-1}$ for all $g \in P$ and $G \cong D_{p^2}$ by Example 6.4.

Next assume that $P \cong \mathbf{Z}_p \oplus \mathbf{Z}_p$. Now φ still acts on P and if $o(\varphi) = 1$ then, as above, $G = \langle h \rangle P$ is Abelian so $G \cong \mathbf{Z}_2 \oplus \mathbf{Z}_p \oplus \mathbf{Z}_p$ by Theorem 7.9. When $o(\varphi) = 2$, if $\varphi(x) \in \langle x \rangle$ for $x \in P$, then φ induces an automorphism on $\langle x \rangle$. Now $Aut(\langle x \rangle) \cong \mathbf{Z}_{p-1}$ by Theorem 6.20, so Theorem 3.1 shows that $\varphi(x) = x$ or $\varphi(x) = x^{-1}$. Should $\varphi(x) = x^{-1}$ for all $x \in P$ then $G \cong (\mathbf{Z}_p \oplus \mathbf{Z}_p) \diamond \mathbf{Z}_2$. Thus if $\varphi(v) \in \langle v \rangle$ for all $v \in P$, we may assume that $\varphi(x) = x$ and $\varphi(y) = y^{-1}$ for some $x \in P$ and $y \in P - \langle x \rangle$. Now $|\langle x \rangle| = |\langle y \rangle| = p$ implies $\langle x \rangle \cap \langle y \rangle = \langle e \rangle$, and then $\langle y \rangle \langle x \rangle = P$ and $G = \langle h \rangle \langle y \rangle \langle x \rangle$ follow from Theorem 4.12. But $hy = y^{-1}h$, so $\langle h \rangle \langle y \rangle = \langle y \rangle \langle h \rangle \le G$ by Theorem 4.10. Also $\langle x \rangle \le Z(G)$ (why?), which implies that $\langle h \rangle \langle y \rangle \lhd G$. Now $\langle h \rangle \langle y \rangle \cap \langle x \rangle = \langle e \rangle$ (verify!) and Theorem 7.9 show $G \cong \langle h \rangle \langle y \rangle \oplus \langle x \rangle$, and then from Example 6.4 we have $G \cong D_p \oplus \mathbf{Z}_p$. Finally, if for some $x \in P$, $\varphi(x) \notin \langle x \rangle$ then using P Abelian and $o(\varphi) = 2$, $\varphi(x\varphi(x)) = x\varphi(x)$ and $\varphi(x^{-1}\varphi(x)) = \varphi(x)^{-1}x = (x^{-1}\varphi(x))^{-1}$. It follows that $x^{-1}\varphi(x) \notin \langle x\varphi(x) \rangle$, so the argument just above shows again that $G \cong D_p \oplus \mathbf{Z}_p$, proving the theorem.

This characterization of groups of order $2p^2$ can be useful in describing other groups, especially when additional information is given.

Example 9.21 Let G be a group with $|G| = 350$. If $|Z(G)| = 7$ then either $G \cong D_{25} \oplus \mathbf{Z}_7$ or $G \cong (\mathbf{Z}_5 \oplus \mathbf{Z}_5) \diamond \mathbf{Z}_2 \oplus \mathbf{Z}_7$.

Since $350 = 2 \cdot 5^2 \cdot 7$, the 5-Sylow subgroup $P_5 \lhd G$ by Theorem 9.11, and, from our hypothesis, the 7-Sylow subgroup $P_7 = Z(G) \lhd G$. Thus the product $P_5P_7 \lhd G$, has order $5^2 \cdot 7$ (verify!), and $[G, P_5P_7] = 2$. It follows that if $\langle h \rangle \in 2\text{-Syl}(G)$ then

$G = \langle h\rangle P_5 P_7$. Now $P_5 \lhd G$ implies $\langle h\rangle P_5 \leq G$ by Theorem 5.4, and in fact $\langle h\rangle P_5 \lhd G$ since $P_7 = Z(G)$. From Theorem 4.6, $\langle h\rangle P_5 \cap P_7 = \langle e\rangle$ so $G \cong \langle h\rangle P_5 \oplus P_7$ results from Theorem 7.9. But $Z(G) \cong Z(\langle h\rangle P_5) \oplus Z(P_7)$ has order 7, forcing $Z(\langle h\rangle P_5) = \langle e\rangle$. Using Example 9.20, we see that the only possibilities for $\langle h\rangle P_5$ are D_{25} or $(\mathbf{Z}_5 \oplus \mathbf{Z}_5) \diamond \mathbf{Z}_2$ (verify!).

Our final example determines the structure of groups of order 12. Besides the groups we know, there is one more we need to mention. Let $\mathbf{Z}_3 \diamond \mathbf{Z}_4 = (\mathbf{Z}_3 \times \mathbf{Z}_4, \cdot)$ for $([a]_3, [b]_4) \cdot ([c]_3, [d]_4) = ([a + (-1)^b c]_3, [b + d]_4)$. It is straightforward to see that $\mathbf{Z}_3 \diamond \mathbf{Z}_4$ is a non-Abelian group of order 12, its 3-Sylow subgroup P_3 is normal, and its three 2-Sylow subgroups are cyclic with generators acting on P_3 by conjugating elements to their inverses. It is an exercise to show that any such group of order 12 is isomorphic to $\mathbf{Z}_3 \diamond \mathbf{Z}_4$.

Example 9.22 If G is a group of order 12 then G is isomorphic to $\mathbf{Z}_{12}$, to $\mathbf{Z}_6 \oplus \mathbf{Z}_2 \cong \mathbf{Z}_3 \oplus \mathbf{Z}_2 \oplus \mathbf{Z}_2$, to D_6, to A_4, or to $\mathbf{Z}_3 \diamond \mathbf{Z}_4$.

Let $P_2 \in$ 2-Syl(G) and $P_3 \in$ 3-Syl(G). If $P_2, P_3 \lhd G$ then $G \cong P_2 \oplus P_3$ by Theorem 9.14, so $G \cong \mathbf{Z}_4 \oplus \mathbf{Z}_3 \cong \mathbf{Z}_{12}$ or $G \cong \mathbf{Z}_2 \oplus \mathbf{Z}_2 \oplus \mathbf{Z}_3 \cong \mathbf{Z}_6 \oplus \mathbf{Z}_2$ from Theorem 8.9; G is Abelian. When G is not Abelian it still has a normal Sylow subgroup: if P_3 is not normal then $|$3-Syl$(G)| = 4$ by Theorem 9.11 and these contain eight elements of order 3 leaving room in G for only one 2-Sylow subgroup. Also note that $G = P_2 P_3$ from Theorem 4.12. If $\mathbf{Z}_4 \cong P_2 \lhd G$ then Theorem 6.20 shows $Aut(P_2) \cong \mathbf{Z}_2$, so $P_3 \leq C_G(P_2)$ by Theorem 6.19, forcing G to be Abelian, a contradiction. If $\mathbf{Z}_2 \oplus \mathbf{Z}_2 \cong P_2 \lhd G$ then $P_3 = \langle y\rangle$ acts nontrivially by conjugation on P_2 (G is not Abelian!), so conjugation by y permutes $P_2 - \langle e\rangle$ cyclicly. Consequently, for $x \in P_2 - \langle e\rangle$ and $z = yxy^{-1}$, $P_2 = \{e, x, z, xz\}$, $yzy^{-1} = xz$, and $yxzy^{-1} = x$. Hence $G = \{y^j, xy^j, zy^j, xzy^j \mid 0 \leq j \leq 2\}$, $yx = zy$, $yz = xzy$, and $yxz = xy$. It follows that $G \cong A_4$ using $\varphi : G \to A_4$ defined by $\varphi(x) = (1, 2)(3, 4)$, $\varphi(z) = (1, 4)(2, 3)$, and $\varphi(y) = (1, 2, 3)$.

Next assume $\langle y\rangle = P_3 \lhd G$ and $P_2 = \langle x\rangle \cong \mathbf{Z}_4$. By Theorem 6.20, $Aut(P_3) \cong \mathbf{Z}_2$, so G not Abelian forces $xyx^{-1} = y^{-1}$ and $G \cong \mathbf{Z}_3 \diamond \mathbf{Z}_4$. If instead $P_2 \cong \mathbf{Z}_2 \oplus \mathbf{Z}_2$ then for $x \in P_2 - \langle e\rangle$, $o(x) = 2$ and either $xyx^{-1} = y$ or $xyx^{-1} = y^{-1}$. In the first case, $xy = yx = g$ and $|\langle g\rangle| = 6$ by Theorem 4.16. When $xyx^{-1} = y^{-1}$ and also $zyz^{-1} = y^{-1}$ for $z \in P_2 - \langle x\rangle$, then $(xz)y = y(xz) = g$. Again $o(g) = 6$, $[G : \langle g\rangle] = 2$, and $\langle g\rangle \lhd G$ from Theorem 5.3. Since G is not Abelian, if $h \in P_2 - \langle g\rangle$, conjugation by h on $\langle g\rangle$ is an automorphism of order 2, using Theorem 6.20 again. Hence $hgh^{-1} = g^{-1}$ and Example 6.4 shows that $G \cong D_6$.

We end the chapter with some comments about finite groups that are the direct sum of their Sylow subgroups. Every p-group has a nontrivial center by Theorem 8.8 and a normal subgroup of any possible order by Theorem 8.10. Hence if G is the direct sum of its Sylow subgroups then $Z(G)$ is the direct sum of their centers by Theorem 3.20. Also, taking suitable direct sums of p-subgroups shows that G has a normal subgroup for every divisor of its order. Conversely, if G has a normal subgroup for each divisor of its order then its Sylow subgroups are normal and G is their direct sum by Theorem 9.14. Any such group is called **nilpotent.** The usual, and equivalent, definition of a nilpotent group uses a property of the centers of certain homomorphic images of G. In this spirit we prove a result relating centers of images to the normality of the Sylow subgroups.

THEOREM 9.15 A finite group $G \neq \langle e \rangle$ is the direct sum of its Sylow subgroups if and only if for every $N \triangleleft G$ with $N \neq G$, $|Z(G/N)| > 1$.

Proof Suppose $G \cong P_1 \oplus \cdots \oplus P_k$ where P_j is the p_j-Sylow subgroup of G, and let the isomorphism be φ. It follows from Theorem 7.2 that for $N \triangleleft G$, $\varphi(N) \triangleleft \varphi(G)$ and $G/N \cong \varphi(G)/\varphi(N)$, so to prove $|Z(G/N)| > 1$ we may assume $G = P_1 \oplus \cdots \oplus P_k$. Now P_j is not a subgroup of G but $G_j = \{(a_1, \ldots, a_k) \in G \mid a_i = e_i$, the identity of P_i, for all $i \neq j\} \cong P_j$ is the p_j-Sylow subgroup of G. Observe that $G_j \triangleleft G$ and that G_j consists of all elements in G with order a power of p_j. Let $N \triangleleft G$ with $N \neq G$. If $(|N|, p_j) = 1$ then Theorem 4.6 shows that $S_j = N \cap G_j = \langle e \rangle$. If $p_j \mid |N|$, let S_j be a p_j-Sylow subgroup of N. Every element in S_j has order p_j^m so $S_j \leq G_j$, and the definition of Sylow subgroup implies $S_j = N \cap G_j \triangleleft G$. It follows from Theorem 5.5 that $N = S_1 \cdots S_k = T_1 \oplus \cdots \oplus T_k$ where $T_j = \pi_j(S_j)$ for π_j the projection of G onto P_j, so $G/N \cong (P_1 \oplus \cdots \oplus P_k)/(T_1 \oplus \cdots \oplus T_k) \cong (P_1/T_1) \oplus \cdots \oplus (P_k/T_k)$ by Theorem 6.9. Since $N \neq G$, some$|P_j/T_j| = p_j^d > 1$. By Theorem 3.20, $Z(G/N) = \oplus Z(P_j/T_j)$ and, with Theorem 8.8, shows $|Z(G/N)| > 1$ proving half of the theorem.

Now assume the condition on $Z(G/N)$. Using Theorem 9.14 it suffices to prove that each Sylow subgroup of G is normal. When $N = \langle e \rangle$ our assumption yields $|Z(G)| > 1$. The Sylow subgroups of an Abelian group are normal so we may assume $Z(G) \neq G$. We claim $G/Z(G)$ satisfies the same hypothesis as G on the centers of quotient groups. If $K \triangleleft G/Z(G)$ with $K \neq G/Z(G)$ then by Theorem 7.3, $K = N/Z(G)$ for $Z(G) \leq N \leq G$, $N \neq G$, and $N \triangleleft G$ from Theorem 7.2. Now $(G/Z(G))/K = (G/Z(G))/(N/Z(G)) \cong G/N$ using Theorem 7.6 so $|Z(G/N)| > 1$ implies $|Z((G/Z(G))/K)| > 1$.

We show that the Sylow subgroups of G are normal by induction on $|G|$. Let p be a prime divisor of $|G|$ and $P \in p$-Syl(G). If $G = Z(G)$ then clearly $P \triangleleft G$. Otherwise, let $S \in p$-Syl$(Z(G))$ or set $S = \langle e \rangle$ if $(|Z(G)|, p) = 1$. Then Theorem 9.11 shows $S \leq gPg^{-1}$ for some $g \in G$, so $S = g^{-1}Sg \leq P$. It follows that $S = Z(G) \cap P$. Set $|P| = p^m$ and $|S| = p^k$ for $k \geq 0$. Since $|Z(G)| = p^k t$ with $(p, t) = 1$, we have $|G/Z(G)| = p^{m-k}d$ with $(p, d) = 1$. For the usual epimorphism $\rho: G \to G/Z(G)$ given by $\rho(g) = gZ(G)$, $\rho(P) = \{xZ(G) \mid x \in P\} = PZ(G)/Z(G) \cong P/(P \cap Z(G))$ by Theorem 7.6, so $|\rho(P)| = p^m/p^k = p^{m-k}$. Thus $\rho(P) \in p$-Syl$(G/Z(G))$. Now $|G/Z(G)| < |G|$ and $G/Z(G)$ satisfies the hypothesis on centers of quotients, as we showed in the last paragraph, so by the induction assumption, $\rho(P) \triangleleft G/Z(G)$. From Theorem 7.3 and Theorem 7.2, $\rho^{-1}(\rho(P)) = PZ(G) \triangleleft G$. Clearly $P \triangleleft PZ(G)$ and since $|P| = p^m$, P is a p-Sylow subgroup of both $PZ(G)$ and G. Therefore $P \triangleleft G$ by Theorem 9.12 and IND-II completes the claim that every Sylow subgroup of G is normal. ∎

EXERCISES

1. Use Theorem 6.14 to show that there is $\langle e \rangle \neq N \triangleleft G$ with $N \neq G$ whenever $|G| =$:

i) 80

ii) $2^{12} \cdot 13$

iii) $2^{15} \cdot 17$
iv) $5 \cdot 7 \cdot 17^3$
v) $3^4 \cdot 13^3$

2. Show that there is $N \lhd G$ with $[G:N]$ prime if $|G| =$:
i) 140
ii) $3^2 \cdot 7 \cdot 11 \cdot 13$
iii) $2^3 \cdot 5^3 \cdot 17^3$
iv) $3^4 \cdot 7^k \cdot 17$ (any $k \geq 0$)

3. If $|G| = 3^k \cdot 5 \cdot 7$ show that there is $N \lhd G$ with $|N| = 3^m$ for some $m \geq k - 2$.

4. If G is a group with $|G| = 90$, use Example 9.14 to show that the Sylow subgroups of odd order in G are normal.

5. If G is a group with $|G| = 72$, show that there is some $N \lhd G$ with $[G:N]$ prime.

6. If G is a group with $|G| = 120$, show that there is $\langle e \rangle \neq N \lhd G$ with $N \neq G$.

7. If G is a group with $|G| = 280$, show that there is some $N \lhd G$ with $[G:N]$ prime.

8. If $|G| = 5^5 \cdot 7^3$ with Abelian Sylow subgroups, show that G is Abelian. How many possibilities are there for the structure of G up to isomorphism?

9. If G is a group with $|G| = 300$, show that there is $N \lhd G$ with $5 \mid |N|$.

10. Show that a group of any of the following orders must be cyclic:
i) 255
ii) 805
iii) $3 \cdot 11 \cdot 17$
iv) $7 \cdot 11 \cdot 13$
v) $5 \cdot 17 \cdot 37 \cdot 47$
vi) $7 \cdot 17 \cdot 19 \cdot 23$
vii) $7 \cdot 17 \cdot 31 \cdot 59 \cdot 73$

11. Determine the possible structures for the group G if $|G| =$:
i) $5^2 \cdot 13$
ii) $7 \cdot 17^2$
iii) $5 \cdot 91$
iv) $7^2 \cdot 11^2$
v) $5 \cdot 7 \cdot 17$
vi) $5^2 \cdot 7 \cdot 37$
vii) $7 \cdot 11^2 \cdot 17$
viii) $7^2 \cdot 11 \cdot 17^2$
ix) $7^2 \cdot 17^2 \cdot 31^2$
x) $5^2 \cdot 7^2 \cdot 23^2$

12. Which group of order 12 in Example 9.22 is isomorphic to $D_3 \oplus \mathbf{Z}_2$?

13. Up to isomorphism describe the subgroup generated by the given set:
i) $\langle\{(1,3,5)(2,4,6), (1,2)(3,4), (1,2)(5,6), (3,4)(5,6)\}\rangle \leq A_6$
ii) $\langle\{(1,2)(6,7)(4,5,9,10), (1,2,3)(6,7,8)\}\rangle \leq S_{10}$
iii) $\langle\{(2,3)(4,6,5,7), (1,2,3)(4,5)(6,7)\}\rangle \leq A_7$

14. Up to isomorphism determine the possible structures the group G if $|G| =$:
i) 70
ii) 155
iii) 231
iv) $5 \cdot 7 \cdot 41$

v) $5 \cdot 7^2 \cdot 11$
vi) $5^2 \cdot 7 \cdot 29$
vii) $3 \cdot 5 \cdot 17 \cdot 23$
viii) $5 \cdot 13 \cdot 17 \cdot 19$

15. Up to isomorphism determine the possible structures of any group of order:
i) $3 \cdot 41 \cdot 61$
ii) $5^2 \cdot 13^2 \cdot 47$
iii) $2590 = 2 \cdot 5 \cdot 7 \cdot 37$
iv) $7 \cdot 11^2 \cdot 13 \cdot 17^2$
v) $5 \cdot 7 \cdot 17 \cdot 29$

16. Let G be a finite group with $A \lhd G$, A cyclic, and $G/A \cong A$. Assume each p-Sylow subgroup of G is cyclic. Let q be the largest prime divisor of $|G|$.
a) Show that $|q\text{-Syl}(G)| = 1$. (Hint: use Theorem 6.19.)
b) If $|G| = p^{2a}q^{2b}$ for $p < q$ primes, show that for $m \mid |G|$ and $m > 0$ there is $H \leq G$ with $|H| = m$.

17. For G a finite group, let $P_q \in q\text{-Syl}(G)$. Indicate what structures G can have if:
a) $|G| = 2 \cdot 5 \cdot 17^2$ and either any P_{17} is cyclic, or $P_2 \subseteq C_G(P_5)$, or $P_2 \subseteq C_G(P_{17})$.
b) $|G| = 3 \cdot 5 \cdot 7 \cdot 11$ and $P_3 \leq Z(G)$.
c) $|G| = 5 \cdot 7 \cdot 11 \cdot 29$ and $P_5 \leq Z(G)$.

18. If $G \neq \langle e \rangle$ is a finite group so that every $H \leq G$ is normal in G, show that $Z(G) \neq \langle e \rangle$ and that for each $m \mid |G|$ with $m > 0$, there is $K_m \leq G$ with $|K_m| = m$.

19. Indicate what $|Z(G)|$ could be if $|G| =$:
i) 105
ii) $5 \cdot 13^3$
iii) $5^3 \cdot 7^3$
iv) $7^2 \cdot 17^3$
v) $11^2 \cdot 13^4$
vi) $5 \cdot 13 \cdot 17^3$
vii) $5 \cdot 11 \cdot 13^3$
viii) $7 \cdot 17^3 \cdot 29$

20. If G is a group with $60 < |G| < 90$, show that either $|G|$ is a prime or there is $\langle e \rangle \neq N \lhd G$ with $N \neq G$.

21. If G is a finite group so that for all $g, h \in G$, $o(gh) \mid o(g)o(h)$, show that for any prime divisor p of $|G|$, $p \mid o(Z(G))$.

22. Let G be a finite group so that for all $x, y \in G$, $xyx^{-1} = y^j$ for some $j \in \mathbf{Z}$.
a) Show that G is the direct sum of its Sylow subgroups.
b) If $g \in G$ and $o(g) = p$, a prime, show that $g \in Z(G)$.
c) If $g \in G$ so that for all $n \in N$, $n^2 \mid o(g) \Rightarrow n = 1$, show that $g \in Z(G)$.

23. Let G be a finite group, $x \in Z(G)$ with $o(x) = p$, a prime, and suppose that $G/\langle x \rangle$ is Abelian. Show that $G \cong H \oplus K$ with H Abelian and K a p-group.

24. Let G be a finite group so that every subgroup H of G is a homomorphic image of G. Show that $p \mid |Z(G)|$ for each prime p dividing $|G|$.

CHAPTER 10

RINGS

Rings are additive groups with a multiplication. This extra operation constrains the structure of rings and makes it more difficult than for groups to obtain elementary structure theorems. In this chapter we introduce basic concepts concerning rings, give some standard examples and constructions of rings, and examine properties satisfied by certain natural collections of rings. These ideas will be developed and used in later chapters. The rings of primary interest are infinite, so many of the results about groups will not be helpful.

10.1 DEFINITIONS AND EXAMPLES

The properties defining a ring are those given in Theorem 1.28 for $\mathbf{Z}_n$. Here we use the definition of a group to shorten the list.

DEFINITION 10.1 A **ring** is a triple $(R, +, \cdot)$ for R a nonempty set with both $\cdot, + : R \times R \to R$ and satisfying the following conditions for all $a, b, c \in R$:

i) $(R, +)$ is an Abelian group with identity 0_R
ii) $(a \cdot b) \cdot c = a \cdot (b \cdot c)$
iii) $a \cdot (b + c) = (a \cdot b) + (a \cdot c)$, and $(a + b) \cdot c = (a \cdot c) + (b \cdot c)$

If all $a \cdot b = b \cdot a$ then R is **commutative,** and if there is $1_R \in R$ with $a \cdot 1_R = a = 1_R \cdot a$ for all $a \in R$, then R is called a **ring with identity** or a **ring with 1.**

Addition and multiplication in R are connected by iii), the distributive laws. By performing multiplications before additions in expressions without parentheses, we may write iii) as $a \cdot (b + c) = a \cdot b + a \cdot c$ and $(a + b) \cdot c = a \cdot c + b \cdot c$. We generally write multiplication as ab rather than as $a \cdot b$. Denote by $-a$ the unique inverse of $a \in (R, +)$, so $a - b = a + (-b)$. When the operations are understood or not relevant, we will write R for $(R, +, \cdot)$. Some take $1_R \in R$ as part of the definition of a ring, and often the rings we consider do have an identity. In these cases we write $1 \in R$.

Some easy examples of rings are $\boldsymbol{Z}$, $\boldsymbol{Q}$, $\boldsymbol{R}$, $\boldsymbol{C}$, and $\boldsymbol{Z}_n$. Recall that $\boldsymbol{C} = \{a + bi \mid a, b \in \boldsymbol{R}\}$ is the set of complex numbers, where $(a + bi) + (c + di) = (a + c) + (b + d)i$, and $(a + bi)(c + di) = (ac - bd) + (ad + bc)i$, so $i^2 = -1 = -1 + 0 \cdot i$. We leave as an exercise that these familiar examples satisfy Definition 10.1. Each of these rings is commutative with 1. A related example is $n\boldsymbol{Z}$ for $n \in \boldsymbol{N}$, a commutative ring without identity when $n > 1$.

We know from Theorem 2.1 that 0_R is the unique identity element for addition and will usually write 0 instead of 0_R. In rings such as $\boldsymbol{Q}$ and $\boldsymbol{Z}_n$ the product of any element with zero is again zero. Does this hold in any ring? Another question is whether computations with negatives work the same way in general rings as in the familiar examples.

THEOREM 10.1 Let R be a ring. For any $a, b \in R$ the following hold:

i) $a \cdot 0 = 0 = 0 \cdot a$
ii) $-(-a) = a$
iii) $-(a + b) = -a - b$ and $-(a - b) = -a + b$
iv) $-(ab) = (-a)b = a(-b)$ and $ab = (-a)(-b)$
v) If for $n \in \boldsymbol{N}$ we let $na = a + \cdots + a$ (n times), then $-(na) = n(-a)$.

When $1 \in R$, then for any $a, b \in R$:

vi) R has a unique element 1_R satisfying $1_R \cdot a = a = a \cdot 1_R$
vii) $-a = (-1_R)a$
viii) $(-1_R)(-1_R) = 1_R$

Proof The proofs of many parts use the distributive laws. For i), since 0 is the identity of $(R, +)$, $a0 = a(0 + 0) = a0 + a0$, so $a0 = 0$ by Theorem 2.2, or cancellation. Similarly, $0 = 0a$. Parts ii) and iii) follow from Theorem 2.1 on inverses in Abelian groups. To prove iv), use the uniqueness of inverses in a group. For example, $ab + (-a)b = (a + (-a))b = 0b = 0$ from i), so $(-a)b$ is the additive inverse, $-(ab)$, of ab. A corresponding computation proves $-(ab) = a(-b)$. Using this, $(-a)(-b) - (ab) = (-a)(-b) + (-a)b = (-a)((-b) + b) = (-a)0 = 0$ from i), proving that $(-a)(-b)$ is the inverse of $-(ab) \in (R, +)$, so $(-a)(-b) = -(-(ab)) = ab$ by ii). The definition of powers in a group (Definition 2.4), or iii) and induction, yields $-(na) = n(-a)$ and v) holds. If $u, v \in R$ are identity elements for multiplication then $uv = u$ since v is an identity and $uv = v$ since u is an identity, so $u = v$, proving vi). This unique element is denoted by either 1_R or just 1. Since $1a = a$, vii) and viii) are special cases of iv). ∎

Let us return to examples of rings. Any Abelian group $(G, +)$ can be made into a ring with **trivial multiplication:** $g \cdot h = e_G = 0$ for all $g, h \in G$. It is easy to verify Definition 10.1 and show that with this multiplication G is a commutative ring, without 1 if $|G| > 1$. The resulting ring is called a *trivial ring*. This is not a particularly interesting example but it does imply that any result that holds for *all* rings must also hold for all Abelian groups. Since hardly anything meaningful is true for all Abelian groups, to get reasonable results about rings we often need to put conditions on the ring to avoid trivial rings and others related to it. One such assumption is that the product of nonzero elements is not zero. Rings with this property are called **domains** and we will have more to say about them later.

Note that $\mathbf{Z}$, $\mathbf{Q}$, $n\mathbf{Z}$, and $\mathbf{Z}_p$ for p a prime are domains (Theorem 1.29), but $\mathbf{Z}_6$ is not since $[2]_6[3]_6 = [0]_6$.

Examples of noncommutative rings are somewhat more difficult to produce, and computation in them tends to be harder to carry out. One standard way to get such examples is to use matrices.

DEFINITION 10.2 For $n \in \mathbf{N}$ and R a ring, $M_n(R)$ denotes the ring of all $n \times n$ matrices over R.

The usual addition and multiplication in $M_n(\mathbf{Z})$ works in $M_n(R)$ as well, using the addition and multiplication in R. Because of the importance of $M_n(R)$ for examples, we review these operations again (see Chapter 2). Any $A \in M_n(R)$ is a square array of elements from R arranged in n rows and n columns. The i-j **entry** of A, denoted A_{ij}, is the element of R in the ith row and j^{th} column of A. Addition in $M_n(R)$ is by entry: if $A, B \in M_n(R)$, $A + B \in M_n(R)$ is defined by $(A + B)_{ij} = A_{ij} + B_{ij}$. Since $(R, +)$ is an Abelian group, it follows that $(M_n(R), +)$ is an Abelian group with identity $0_n \in M_n(R)$ having 0_R in each entry, and with the inverse $-A$ of A defined by $(-A)_{ij} = -A_{ij}$. Given $A, B \in M_n(R)$, the product $AB \in M_n(R)$ is defined by its entries as $(AB)_{ij} = A_{i1}B_{1j} + A_{i2}B_{2j} + \cdots + A_{in}B_{nj}$. With these operations $M_n(R)$ is a ring and not commutative unless R has trivial multiplication or is commutative and $n = 1$.

For $r \in R$ and $A \in M_n(R)$ there is a **scalar multiplication** $rA \in M_n(R)$ given by $(rA)_{ij} = rA_{ij} : rA$ multiplies each entry of A on its *left* by r. Let $rE_{ij} \in M_n(R)$ be the matrix with r in the i-j entry and 0_R elsewhere, and if $1_R \in R$, $E_{ij} = 1_R E_{ij}$. The notation rE_{ij} is ambiguous, depending on the context to determine the n for which $rE_{ij} \in M_n(R)$. Multiplication of the $\{r_{ij}E_{ij}\} \subseteq M_n(R)$ results in $aE_{ij}bE_{uv} = abE_{iv}$ when $j = u$ and in $aE_{ij}bE_{uv} = 0_n$ when $j \neq u$. Each $A \in M_n(R)$ can be expressed as $A = \sum A_{ij}E_{ij}$ and when $1 \in R$, the matrix $I_n = \sum E_{ii}$ is an identity for $M_n(R)$ under multiplication. When R fails to have an identity, so does $M_n(R)$. For example, every entry of $M_2(2\mathbf{Z})$ is even and $M_2(2\mathbf{Z})$ is a noncommutative ring without identity element.

Examples of E_{ij} in $M_3(\mathbf{Q})$ are

$$E_{12} = \begin{bmatrix} 0 & 1 & 0 \\ 0 & 0 & 0 \\ 0 & 0 & 0 \end{bmatrix}$$

and

$$E_{31} = \begin{bmatrix} 0 & 0 & 0 \\ 0 & 0 & 0 \\ 1 & 0 & 0 \end{bmatrix}$$

and

$$\begin{bmatrix} a & b \\ c & d \end{bmatrix} = aE_{11} + bE_{12} + cE_{21} + dE_{22} = \begin{bmatrix} a & 0 \\ 0 & 0 \end{bmatrix} + \begin{bmatrix} 0 & b \\ 0 & 0 \end{bmatrix} + \begin{bmatrix} 0 & 0 \\ c & 0 \end{bmatrix} + \begin{bmatrix} 0 & 0 \\ 0 & d \end{bmatrix}$$

in $M_2(\mathbf{Q})$. Some sample computations over $\mathbf{Q}$ are

$$3\begin{bmatrix} 2/3 & -1 \\ -1 & 1/3 \end{bmatrix} = \begin{bmatrix} 2 & -3 \\ -3 & 1 \end{bmatrix}$$

$$\begin{bmatrix} 1 & 1 \\ 1 & 1 \end{bmatrix} + \begin{bmatrix} -1 & 3 \\ -2 & -3 \end{bmatrix} = \begin{bmatrix} 0 & 4 \\ -1 & -2 \end{bmatrix}$$

$$\begin{bmatrix} 1 & 0 & 1 \\ 0 & 1 & 1 \\ 1 & 2 & 0 \end{bmatrix} - 2\begin{bmatrix} 2 & 0 & -1 \\ -1 & 1 & 1/2 \\ 0 & 3 & 1 \end{bmatrix} = \begin{bmatrix} -3 & 0 & 3 \\ 2 & -1 & 0 \\ 1 & -4 & -2 \end{bmatrix}$$

$$\begin{bmatrix} 1/2 & 1/2 \\ 1/2 & 1/2 \end{bmatrix}\begin{bmatrix} 1/2 & 1/2 \\ 1/2 & 1/2 \end{bmatrix} = \begin{bmatrix} 1/2 & 1/2 \\ 1/2 & 1/2 \end{bmatrix}; \quad \begin{bmatrix} 1 & -1 \\ 1 & -1 \end{bmatrix}\begin{bmatrix} 1 & 1 \\ 1 & 1 \end{bmatrix} = \begin{bmatrix} 0 & 0 \\ 0 & 0 \end{bmatrix}$$

$$\begin{bmatrix} 1 & 1 \\ 1 & 1 \end{bmatrix}\begin{bmatrix} 1 & -1 \\ 1 & -1 \end{bmatrix} = \begin{bmatrix} 2 & -2 \\ 2 & -2 \end{bmatrix}$$

and

$$\begin{bmatrix} -1 & 2 & 2 \\ 0 & 1 & -3 \\ 2 & 1 & 0 \end{bmatrix}\begin{bmatrix} 3 & 0 & -1 \\ -1 & 2 & -2 \\ 0 & 1 & 1 \end{bmatrix} = \begin{bmatrix} -5 & 6 & -1 \\ -1 & -1 & -5 \\ 5 & 2 & -4 \end{bmatrix}$$

Noncommutative rings like $M_n(R)$ lead to some *nonassociative rings*. These are triples as in Definition 10.1 but not satisfying associativity of multiplication. We will not dwell on this topic but mention an important example of such rings using $M_n(R)$ with $n > 1$ and the usual addition, but defining multiplication by $[A, B] = A \cdot B - B \cdot A$, where "$\cdot$" is the usual matrix multiplication. An example is the set of matrices in $M_n(\mathbf{Q})$ with trace zero, not a ring using matrix multiplication. The new product satisfies what is called the *Jacobi identity*: $[A, [B, C]] = [[A, B], C] + [B, [A, C]]$, and this nonassociative ring is called a *Lie ring*. Another nonassociative multiplication in $M_n(\mathbf{Q})$ is $A \cdot B = AB + BA$ and results in a *Jordan ring*.

The set of homomorphisms of an Abelian group to itself, hard to visualize concretely, yields other examples of noncommutative rings.

DEFINITION 10.3 If $(G, +)$ is an Abelian group, its **endomorphism ring** is $End(G) = \{f : G \to G \mid f \text{ is a group homomorphism}\}$.

We have called $End(G)$ a ring but have not said what the ring operations are; we use pointwise addition and *composition*.

THEOREM 10.2 For $(G, +)$ an Abelian group, $(End(G), +, \circ)$ is a ring using $(f + g)(x) = f(x) + g(x)$ for all $x \in G$, and composition for multiplication. $End(G)$ has I_G as an identity for multiplication.

Proof We verify Definition 10.1. It is important not to confuse addition in $End(G)$ with the addition in G. Given $f, g \in End(G)$, to see $f + g \in End(G)$, take $x, y \in G$, use the definition of addition in $End(G)$, and compute that $(f + g)(x + y) = f(x + y) + g(x + y) = f(x) + f(y) + g(x) + g(y) = f(x) + g(x) + f(y) + g(y) = (f + g)(x) + (f + g)(y)$, using that G is Abelian. Thus $f + g \in End(G)$ and $+ : (End(G) \times End(G)) \to End(G)$.

Next we show that $(End(G), +)$ is an Abelian group and observe first that it contains $I_G : G \to G$ and so is not empty.

The addition in $End(G)$ is Abelian since $(f+g)(x) = f(x)+g(x) = g(x)+f(x) = (g+f)(x)$ for all $x \in G$, using again that G is Abelian. The identity of $(End(G), +)$ is the trivial homomorphism $0_G(x) = e_G$ for all $x \in G$, since if $f \in End(G)$ then $(f+0_G)(x) = f(x)+e_G = f(x)$. The inverse of $f \in (End(G), +)$ is $-f$ defined by $(-f)(x) = -f(x) \in G$, so $(End(G), +)$ is an Abelian group. Now $f, g \in End(G)$ imply $f \circ g \in End(G)$ by Theorem 6.3, and "$\circ$" is associative from Theorem 0.9, so we need only prove the distributive laws. We verify one and leave the other as an exercise. For $f, g, h \in End(G)$, $f(g+h)(x) = f((g+h)(x)) = f(g(x)+h(x)) = f(g(x)) + f(h(x)) = (fg+fh)(x)$, so $f(g+h) = fg+fh$. It is clear that I_G is an identity element for composition. ■

In general, $End(G)$ is not a commutative ring. From Theorem 6.22, $GL(k, \mathbf{Z}_n) \cong Aut(\mathbf{Z}_n^k) \subseteq End(\mathbf{Z}_n^k)$ is not Abelian if $k > 1$. We also have:

Example 10.1 For p a prime, $R = End(\mathbf{Z}_p \oplus \mathbf{Z})$ is a noncommutative ring.

Define $\alpha, \beta : \mathbf{Z}_p \oplus \mathbf{Z} \to \mathbf{Z}_p \oplus \mathbf{Z}$ by $\alpha(([a]_p, m)) = ([a+m]_p, m)$ and $\beta(([a]_p, m)) = ([a]_p, 0)$. It is an exercise that $\alpha, \beta \in R$. Now $\alpha\beta(([1]_p, 1)) = \alpha(([1]_p, 0)) = ([1]_p, 0)$ but $\beta\alpha(([1]_p, 1)) = \beta(([2]_p, 1)) = ([2]_p, 0)$, so $\alpha\beta \neq \beta\alpha$.

We describe three other constructions of rings from given rings; they are noncommutative when the rings we start with are noncommutative.

Example 10.2 The (group) *direct sum* $R_1 \oplus \cdots \oplus R_k$ for rings R_j is a ring using coordinatewise multiplication. We write R^k when all $R_j = R$.

Recall that $\oplus R_j = R_1 \oplus \cdots \oplus R_k$, is the Cartesian product $R_1 \times \cdots \times R_k$ with group addition $(r_1, \ldots, r_k) + (s_1, \ldots, s_k) = (r_1 +_1 s_1, \ldots, r_k +_k s_k)$. Define multiplication similarly, using the ring multiplication in each coordinate: $(r_1, \ldots, r_k) \cdot (s_1, \ldots, s_k) = (r_1 s_1, \ldots, r_k s_k)$. The associative law holds by computing $((r_1, \ldots, r_k) \cdot (s_1, \ldots, s_k)) \cdot (u_1, \ldots, u_k) = (r_1 s_1, \ldots, r_k s_k) \cdot (u_1, \ldots, u_k) = ((r_1 s_1)u_1, \ldots, (r_k s_k)u_k)$ and using associativity of multiplication in each ring to obtain $((r_1 s_1)u_1, \ldots, (r_k s_k)u_k) = (r_1(s_1 u_1), \ldots, r_k(s_k u_k)) = ((r_1, \ldots, r_k) \cdot (s_1 u_1, \ldots, s_k u_k) = (r_1, \ldots r_k) \cdot ((s_1, \ldots, s_k) \cdot (u_1, \ldots, u_k))$. The distributive laws are verified similarly. Note that if some R_j is not commutative, neither is the direct sum.

The next example uses properties of both groups and rings and turns out to be an important object in advanced studies of noncommutative rings.

DEFINITION 10.4 For a group G and ring R with 1, the **group ring** $R[G] = \{\sum_G r_g g \mid r_g \in R$ and $r_g = 0_R$ for all but finitely many $g \in G\}$. Call r_g the **coefficient** of g in $\sum_G r_g g$.

The sums in $R[G]$ are formal and no relation is assumed between elements of R and of G. To avoid confusion it is sometimes better to write the elements of $R[G]$ as sums of

$r_g x_g$ with $x_g x_h = x_{gh}$. For example, if $G = \langle -1 \rangle$ in $(\boldsymbol{Q}^*, \cdot)$ then writing $\boldsymbol{Q}[G] = \{ax_1 + bx_{-1} \mid a, b \in \boldsymbol{Q}\}$ avoids the possibility of interpreting $a1 + b(-1) \in \boldsymbol{Q}[G] = \{c1 + d(-1) \mid c, d \in \boldsymbol{Q}\}$ as $a - b \in \boldsymbol{Q}$. We omit $r_g g$ when $r_g = 0$, write 0 if all $r_g = 0$, and write $1g = g$ if $1 \in R$.

Example 10.3 $(R[G], +, \cdot)$ is a ring with identity $1_R e_G$ using operations $\sum_G r_g g + \sum_G s_g g = \sum_G (r_g + s_g) g$ and $\sum_G r_g g \cdot \sum_G s_g g = \sum_{h \in G} \left(\sum_g r_{hg^{-1}} s_g\right) h$.

The sum and product of $\sum_G r_g g$ and $\sum_G s_g g$ make sense since only finitely many r_g and s_g are not zero. The coefficient of $h \in G$ in the product $\sum_G r_g g \cdot \sum_G s_g g$ is the sum of all $r_x s_y$ with $xy = h$. We would expect this, assuming the associative and distributive laws. The element $0 = \sum 0g$ is the identity for $(R[G], +)$, and the inverse of $\sum r_g g \in (R[G], +)$ is $\sum (-r_g) g$ (verify!). Using that $(R, +)$ is an Abelian group, it follows easily that $(R[G], +)$ is also. Given $\alpha = \sum a_g g$, $\beta = \sum b_g g$, and $\gamma = \sum c_g g$, $\alpha\beta = \sum_k \left(\sum_g a_{kg^{-1}} b_g\right) k$ and $(\alpha\beta)\gamma = \sum_k \left(\sum_y \left(\sum_g a_{ky^{-1}g^{-1}} b_g\right) c_y\right) k$: the coefficient in $(\alpha\beta)\gamma$ of $k \in G$ is $\sum_y \left(\sum_g a_{ky^{-1}g^{-1}} b_g\right) c_y = \sum_{y,g} a_{ky^{-1}g^{-1}} b_g c_y$. A similar computation shows that the coefficient of $k \in G$ in $\alpha(\beta\gamma)$ is $\sum_g a_{kg^{-1}} \left(\sum_y b_{gy^{-1}} c_y\right) = \sum_{y,g} a_{kg^{-1}} b_{gy^{-1}} c_y$. Since g and y vary over all elements of G independently, the latter sum is $\sum_y \left(\sum_g a_{kg^{-1}} b_{gy^{-1}} c_y\right) = \sum_y \left(\sum_{gy} a_{k(gy)^{-1}} b_{(gy)y^{-1}} c_y\right)$ since $\{gy \in G \mid g \in G\} = G$. Thus $\sum_{y,g} a_{kg^{-1}} b_{gy^{-1}} c_y = \sum_{y,g} a_{ky^{-1}g^{-1}} b_g c_y$ and the coefficients of each $k \in G$ in $(\alpha\beta)\gamma$ and $\alpha(\beta\gamma)$ are equal, forcing $(\alpha\beta)\gamma = \alpha(\beta\gamma)$: the multiplication in $R[G]$ is associative. The distributive laws are easier to verify and are left as exercises, as is the verification that $1_R e_G = e_G$ is the identity element for multiplication.

The multiplication in $R[G]$ is basically $ag \cdot bh = ab(gh)$ extended by the associative and distributive laws. Sample computations in $\boldsymbol{Q}[S_3]$ are: $(I + (1, 2))^2 = 2I + 2(1, 2)$; $((1, 2) + (1, 3))(I - 2(1, 2, 3)) = (1, 2) + (1, 3) - 2(2, 3) - 2(1, 2) = -(1, 2) + (1, 3) - 2(2, 3)$; and $(I - (1, 2, 3))(I + (1, 2, 3) + (1, 3, 2)) = I - (1, 2, 3) + (1, 2, 3) - (1, 3, 2) + (1, 3, 2) - I = 0$. If G is a finite group then $\left(\sum g\right) \cdot h = \sum g = h \cdot \left(\sum g\right)$ for all $h \in G$ since $Gh = hG = G$, so $\left(\sum g\right)^2 = \sum |G| g$. Thus in $\boldsymbol{Q}[G]$, $\left(\sum (1/|G|) g\right)^2 = \sum (1/|G|) g$. Similarly, if $x \in G$ with $o(x) = t$ then for each $x^i \in \langle x \rangle$, $\langle x \rangle x^i = x^i \langle x \rangle = \langle x \rangle$, so $\left(\sum_j (1/t) x^j\right)^2 = \sum_j (1/t) x^j$ in $\boldsymbol{Q}[G]$. Elements like these, satisfying $r^2 = r$, are called **idempotents.**

Let us move on to our last example in this section.

THEOREM 10.3 For $S \neq \varnothing$ and R any ring, the set $R^S = \{f : S \to R\}$ is a ring using $(f + g)(s) = f(s) + g(s)$ and $f \cdot g(s) = f(s)g(s)$, for $f, g \in R^S$ and $s \in S$.

Proof Since R is a ring, $f + g, f \cdot g \in R^S$ when $f, g \in R^S$. The addition in R^S is associative and commutative since $(R, +)$ is an Abelian group, as in the proof of Theorem 10.2. The identity element for addition in R^S is the function 0_S defined by $0_S(s) = 0_R$ for all $s \in S$, and the additive inverse of $f \in R^S$ is $-f$ given by $(-f)(s) = -(f(s)) \in (R, +)$ (verify!). Thus $(R^S, +)$ is an Abelian group. For $f, g, h \in R^S$, $f \cdot (g \cdot h)(s) = f(s)(g \cdot h(s)) = f(s)(g(s)h(s)) = (f(s)g(s))h(s) = f \cdot g(s)h(s) = (f \cdot g) \cdot h(s)$, using the associativity of multiplication in R, so "$\cdot$" in R^S is associative. The distributive laws for R^S follow from the distributive laws in R: $f \cdot (g + h)(s) = f(s)(g + h)(s) = f(s)(g(s) + h(s)) =$

$f(s)g(s) + f(s)h(s) = f \cdot g(s) + f \cdot h(s) = (f \cdot g + f \cdot h)(s)$; and $(f + g) \cdot h(s) = (f + g)(s)h(s) = (f(s) + g(s))h(s) = f(s)h(s) + g(s)h(s) = f \cdot h(s) + g \cdot h(s) = (f \cdot h + g \cdot h)(s)$. ■

It follows easily that if R is a commutative ring, so is R^S. On the other hand, if $C_a \in R^S$ is the constant function $C_a(s) = a$ then $C_a \cdot C_b(s) = ab$ and $C_b \cdot C_a(s) = ba$ so if R^S is commutative, R must be also. When $1 \in R$ the constant C_1 is an identity for R^S under multiplication. We will see in a later section how to use Theorem 10.3 to get a formal definition of polynomials by identifying the usual polynomial ring $R[x]$ with a part of R^N.

If $S = I(n) = \{1, 2, \ldots, n\}$ in Theorem 10.3 then each $f \in R^S$ can be identified with the n-tuple $(f(1), \ldots, f(n))$ and the definition of addition and multiplication in R^S corresponds to coordinate addition and multiplication of n-tuples. Thus $R^{I(n)}$ and $R^n = R \oplus \cdots \oplus R$ in Example 10.2 are essentially the same rings. In the next chapter we discuss the formal way of making this identification via the notion of isomorphism, just as for groups. When $S = N$, identify $f \in R^N$ with the sequence $(f(1), f(2), \ldots)$. The operations in R^N correspond to coordinatewise addition and multiplication of sequences, and we call R^N the **direct product** of R with itself a countable number of times (see Definition 3.13). The ring R^S in Theorem 10.3 is given this same name.

DEFINITION 10.5 For any nonempty set S and any ring R, the ring R^S is called the **direct product** of R with itself $|S|$ times.

The last observation in this section concerns an important subset of a ring R. If $1 \in R$ and if for some $a \in R$ there is $b \in R$ so that $ab = ba = 1$, then a is called a **unit** in R. Clearly b is also a unit. If $c, d \in R$ and $cd = dc = 1$ then $(ac)(db) = 1 = (db)(ac)$ (why?), so the set of units is closed under the associative multiplication in R, and it is not empty since 1_R is a unit. Therefore the set of units is a group using the multiplication in R.

DEFINITION 10.6 If R is a ring with $1 \in R$, then the **group of units in R** is $U(R) = \{a \in R \mid ab = ba = 1 \text{ for some } b \in R\}$.

Familiar examples are $U(\mathbf{Z}) = \{1, -1\}$, $U(\mathbf{Q}) = \mathbf{Q}^*$, $U(\mathbf{Z}_n) = U_n$, and $U(M_n(R)) = GL(n, R)$. Although we will discuss polynomials later, if as usual $\mathbf{Q}[x] = \{r_n x^n + \cdots + r_1 x + r_0 \mid r_i \in \mathbf{Q}\}$ then $U(\mathbf{Q}[x]) = \mathbf{Q}^*$ since $f(x)g(x) = 1$ forces $f(x) = r$ in $\mathbf{Q}^*$ and $g(x) = r^{-1}$. Similarly, $U(\mathbf{Z}[x]) = \{1, -1\}$. On the other hand, $[1]_8 + [2]_8 x \in U(\mathbf{Z}_8[x])$ has inverse $[1]_8 + [6]_8 x + [4]_8 x^2$. Interestingly, although we cannot prove it easily now, it turns out that $U(\mathbf{Z}_8[x]) = \{[a_0] + [a_1]x + \cdots + [a_k]x^k \in \mathbf{Z}_8[x] \mid a_0 \text{ is odd and } a_i \text{ is even when } i > 0\}$ but that $U(\mathbf{Z}_{15}[x]) = U_{15} \subseteq \mathbf{Z}_{15}[x]$. In general, $U(\mathbf{Z}_n[x])$ are those polynomials whose constant term is in U_n and all other coefficients have some power in $\mathbf{Z}_n$ equal to $[0]_n$. This result is neither obvious nor elementary. Thus in $\mathbf{Z}_{72}[x]$ the units are those $[a_0] + [a_1]x + \cdots + [a_k]x^k$ with $[a_0] \in U_{72}$ and all other $a_j \in 6\mathbf{Z}$.

EXERCISES

1. In each case is the set a ring? $(R, +, \cdot)$ denotes a ring with $R \neq \{0_R\}$.

i) $(R, +, \cdot)$ where $a \cdot b = a$

ii) $(R, +, *)$ with $a * b = ab + ba$

iii) $(\boldsymbol{Q}[x], +, \#)$ with $p(x)\#q(x) = p(0)q(0)$
iv) $(R, +, \cdot)$ where $a \cdot b = ba$
v) $(\boldsymbol{Q}[x], +, \sim)$ where $p(x) \sim q(x) = \lambda(p(x))\lambda(q(x))$ with

$$\lambda(a_0 + a_1x + \cdots + a_kx^k) = a_k \quad \text{when } a_k \neq 0 \text{ and } \lambda(0) = 0$$

vi) $(\boldsymbol{Q}[x], +, \circ)$ where $p(x) \circ q(x) = p(q(x))$
vii) $(R, +, *)$ with $a * b = a + b + ab$

2. **i)** Show that $U(\boldsymbol{C}) = \boldsymbol{C}^*$.
ii) Describe the functions in $U(\boldsymbol{R}^{\boldsymbol{R}})$.

3. For $R_1, \ldots, R_k$ rings with identity, show that $U(R_1 \oplus \cdots \oplus R_k) = U(R_1) \oplus \cdots \oplus U(R_k)$.

4. For $\{R_j\}_T$ a collection of rings, let $S = \bigcup_T R_j$ as sets. Show that $\prod_T R_j = \{f \in S^T \mid$ for each $j \in T$, $f(j) \in R_j\}$ is a ring using $(f + g)(i) = f(i) + g(i)$ and $f \cdot g(i) = f(i)g(i)$. This ring is called the **direct product** of $\{R_j\}_T$.

5. Given $\{R_j\}_T$ for rings R_j and $|T| > 1$, let $R = \prod_T R_j$ (Exercise 4). Show that:
i) R is commutative $\Leftrightarrow$ each R_j is.
ii) $1 \in \prod_T R_j \Leftrightarrow$ each R_j has as identity.

6. Show carefully that the following hold in any ring R:
i) $(a + b)(c + d) = ac + ad + bc + bd$
ii) $a(b - c) = ab - ac$ and $(a - b)c = ac - bc$
iii) $a(b_1 + b_2 + \cdots + b_n) = ab_1 + ab_2 + \cdots + ab_n$
iv) $(a_1 + \cdots + a_k)(b_1 + b_2 + \cdots + b_n) = \sum a_ib_j$
v) $-((a_1 + \cdots + a_k)b) = (-a_1 - \cdots - a_k)b = -a_1b - a_2b - \cdots - a_kb$

7. **i)** Find explicit $A, B \in M_2(\boldsymbol{Z})$ with $(A + B)(A - B) \neq A^2 - B^2$.
ii) Show that a ring R is commutative $\Leftrightarrow$ for all $a, b \in R$ $(a + b)(a - b) = a^2 - b^2$.

10.2 SUBRINGS

We consider subsets of a ring R that are themselves rings using the same operations as in R. Although useful, this idea is not as important in studying rings as the notion of subgroup is in studying groups.

DEFINITION 10.7 For R a ring and $\emptyset \neq S \subseteq R$, S is a **subring** of R if S is a ring using the restrictions to $S \times S$ of $\cdot, + : R \times R \to R$.

If $a, b \in S$ then $a + b, a \cdot b \in R$, so S a subring requires $a + b, a \cdot b \in S$. Formally, we want $+|_{S\times S}, \cdot|_{S\times S} : S \times S \to S$, or that S is **closed** under addition and multiplication, giving some $+$ and $\cdot$ defined **on S.** With this, verifying Definition 10.1 for S reduces to showing two simple facts akin to closure.

THEOREM 10.4 If R is a ring, $\emptyset \neq S \subseteq R$ is a subring $\Leftrightarrow a - b, ab \in S$ for all $a, b \in S$.

Proof If S is a subring then $(S, +)$ is an Abelian group so $a, b \in S \Rightarrow a - b \in S$ from Theorem 2.12. By Definition 10.7, $ab \in S$ as well. Now assume S is closed under subtraction and multiplication. We use the same symbols, $+$ and $\cdot$, for the restrictions to $S \times S$ of addition and multiplication in R. Since $a, b \in S \Rightarrow a - b \in S$, $(S, +) \leq (R, +)$ by Theorem 2.12, and $(S, +)$ is Abelian because $(R, +)$ is. The associativity of multiplication and the distributive laws in Definition 10.1 hold for any three elements in R and so for any three in S. Thus $(S, +, \cdot)$ is a ring and hence a subring of R. ∎

Note that $N \subseteq Z$ is closed under both addition and multiplication but is not a subring since $(N, +)$ is not a group. That is, for $a, b \in S$ we really do need $a - b \in S$, not just $a + b \in S$. Another observation is that in a subring S of R, $0_S = 0_R$ since the identity of $(S, +)$ and that of $(R, +)$ are the same by Theorem 2.11, and also the inverse of $s \in S$ is $-s \in (R, +)$. Two examples of subrings of an arbitrary ring R are R itself and $\{0\}$, using Theorem 10.1. It is easy to see that $n\boldsymbol{Z}$ is a subring of $\boldsymbol{Z}$, or of $\boldsymbol{Q}$, and that $\boldsymbol{Q}$ is a subring of both $\boldsymbol{R}$ and of $\boldsymbol{C}$. We look at other examples.

Example 10.4 $S = \{a + b\sqrt{5} \in \boldsymbol{R} \mid a, b \in \boldsymbol{Z}\}$ is a subring of $\boldsymbol{R}$.

Certainly S is not empty. If $\alpha = a + b\sqrt{5}$, $\beta = c + d\sqrt{5} \in S$, then $\alpha - \beta = (a - c) + (b - d)\sqrt{5}$ and $\alpha\beta = (ac + 5bd) + (ad + bc)\sqrt{5}$. Since $a, b, c, d \in \boldsymbol{Z}$, we have $\alpha - \beta, \alpha\beta \in S$, and S is a subring of $\boldsymbol{R}$ by Theorem 10.4.

Example 10.5 For $m > 1$, $S = \{a/b \in \boldsymbol{Q} \mid (b, m) = 1\}$ is a subring of $\boldsymbol{Q}$.

Assume for $a/b \in \boldsymbol{Q}$ that $b \in \boldsymbol{N}$ and a/b is reduced, so $(a, b) = 1$; otherwise $a/b = ma/mb$ and S would be empty. For $a/b, c/d \in S$, $(b, m) = 1 = (d, m)$. Reducing $a/b - c/d$ and $(a/b)(c/d)$ leaves denominators dividing bd and so relatively prime to m since $(bd, m) = 1$ (Theorem 1.11). Thus $a/b - c/d$, $(a/b)(c/d) \in S$ and S is a subring of $\boldsymbol{Q}$ by Theorem 10.4.

The next example uses the ring $R^S = \{f : S \to R\}$ from Theorem 10.3.

Example 10.6 If $S \neq \varnothing$ and R is a ring then $A = \{f \in R^S \mid f(s) = 0_R$ for all but finitely many $s \in S\}$ is a subring of R^S: the *direct sum* of $|S|$ copies of R.

The function 0_S defined by $0_S(x) = 0_R$ for all $x \in S$ is in A so A is not empty. Observe that for $\varphi \in R^S$, $\varphi \in A \Leftrightarrow B_\varphi = \{s \in S \mid \varphi(s) \neq 0_R\}$ is finite. For $f, g \in A$, $(f - g)(s) = f(s) - g(s)$ so $s \in B_{f-g} \Rightarrow s \in B_f \cup B_g$, forcing $B_{f-g} \subseteq B_f \cup B_g$, which is finite. Thus $f - g \in A$. Also $f \cdot g \in A$ since $f \cdot g(s) = f(s)g(s)$, so $B_{f \cdot g} \subseteq B_f \cap B_g$, a finite set. Hence A is a subring of R^S by Theorem 10.4.

Matrix rings give rise to many interesting subrings. It is straightforward to see that $M_n(m\boldsymbol{Z})$ is a subring of $M_n(\boldsymbol{Z})$, $M_n(\boldsymbol{Q})$, or $M_n(\boldsymbol{R})$.

Example 10.7 If R is a ring, $n > 1$, $L_k = \{A \in M_n(R) \mid A_{ij} = 0$ if $j = k\}$ and $T_k = \{A \in M_n(R) \mid A_{ij} = 0$ if $i = k\}$ then L_k and T_k are subrings of $M_n(R)$.

The result here is clear since L_k are those matrices with all zeros in a fixed column and T_k are those matrices with all zeros in a fixed row. More formally,

$A \in L_k \Leftrightarrow A_{ik} = 0$ for all $1 \le i \le n$ so $0_n \in L_k$ and $L_k \ne \varnothing$. For $A, B \in L_k$, $(A - B)_{ik} = A_{ik} - B_{ik} = 0$, and $(AB)_{ik} = A_{i1}B_{1k} + \cdots + A_{in}B_{nk} = 0$ since $B_{jk} = 0$ for all $1 \le j \le n$, so $A - B$, $AB \in L_k$, and L_k is a subring of $M_n(R)$ by Theorem 10.4. A similar argument shows that T_k is a subring of $M_n(R)$.

There is nothing special about using one row or column of zeros in Example 10.7. All matrices with all zero entries in any specified rows, or columns, will also be a subring (verify!)—for example, all matrices with their second and fifth rows (columns) zero in $M_6(R)$. We can also get subrings with other restrictions on entries. One specific and simple example is next.

Example 10.8 For $n, m \in \mathbf{Z}$, $S = \left\{\begin{bmatrix} a & b \\ c & d \end{bmatrix} \in M_2(\mathbf{Z}) \mid b \in n\mathbf{Z} \text{ and } c \in m\mathbf{Z}\right\}$ is a subring of $M_2(\mathbf{Z})$.

We use Theorem 10.4 again and note that $0_2 \in S$. If $A = \begin{bmatrix} a & b \\ c & d \end{bmatrix}$, $B = \begin{bmatrix} a' & b' \\ c' & d' \end{bmatrix} \in S$ then $A - B = \begin{bmatrix} a - a' & b - b' \\ c - c' & d - d' \end{bmatrix}$ and $AB = \begin{bmatrix} aa' + bc' & ab' + bd' \\ ca' + dc' & cb' + dd' \end{bmatrix}$. Clearly $b, b' \in n\mathbf{Z}$ force $b - b' \in n\mathbf{Z}$ and $c, c' \in m\mathbf{Z}$ force $c - c' \in m\mathbf{Z}$. Thus $A - B \in S$. Finally $b, b' \in n\mathbf{Z}$ imply $ab' + bd' \in n\mathbf{Z}$ and similarly $c, c' \in m\mathbf{Z}$ imply that $ca' + dc' \in m\mathbf{Z}$, so $AB \in S$, proving that S is a subring of $M_2(\mathbf{Z})$.

Example 10.8 gives the subring of diagonal matrices when $m = n = 0$ and gives the subring of upper triangular matrices when $m = 0$ and $n = 1$. It is straightforward using matrix addition and multiplication to see that the set of diagonal matrices in any $M_n(R)$ is a subring. It will be useful to have a notation for the (strictly) upper triangular matrices in general matrix rings.

DEFINITION 10.8 For $n \ge 2$ and R a ring, let $UT_n(R) = \{A \in M_n(R) \mid A_{ij} = 0 \text{ if } i > j\}$ and $StUT_n(R) = \{A \in UT_n(R) \mid A_{ii} = 0 \text{ all } 1 \le i \le n\}$.

Example 10.9 $UT_n(R)$ and $StUT_n(R)$ are subrings of $M_n(R)$. If $A_1, \ldots, A_n \in StUT_n(R)$, then $A_1A_2 \cdots A_n = (\cdots((A_1A_2)\cdots)A_n) = 0_n$.

The definition of matrix addition shows that when $A, B \in UT_n(R)$ so is $A - B$. Also, if $i > j$ then $(AB)_{ij} = A_{i1}B_{1j} + A_{i2}B_{2j} + \cdots + A_{in}B_{nj} = 0_R$ since $A_{i1} = \cdots = A_{ii-1} = 0$ and $B_{i+sj} = 0$ when $s \ge 0$. Therefore $(AB)_{ij} = 0$ if $i > j$ so $AB \in UT_n(R)$, and $UT_n(R)$ is a subring of $M_n(R)$ by Theorem 10.4. A similar computation proves that $StUT_n(R)$ is a subring of $M_n(R)$. For $A_i \in StUT_n(R)$, take $A_1A_2 \cdots A_n = (\cdots((A_1A_2)\cdots)A_n)$, although we will see that the particular way of placing parentheses does not matter. Since $A_1 \in StUT_n(R)$, $(A_1)_{ij} = 0$ when $i \ge j$. Suppose that for $1 \le k < n$, $(A_1A_2 \cdots A_k)_{ij} = 0$ if $i + k - 1 \ge j$. We claim that $(A_1A_2 \cdots A_{k+1})_{ij} = 0$ if $i + k \ge j$. Write $A_1A_2 \cdots A_k = B$ so $(A_1 \cdots A_{k+1})_{ij} = B_{i1}(A_{k+1})_{1j} + \cdots + B_{in}(A_{k+1})_{nj}$ and note that $i + k \ge j \Rightarrow i + k - 1 \ge j - 1 > j - 2 > \cdots > 1$, so $B_{i1} = \cdots = B_{ij-1} = 0$. But $(A_{k+1})_{jj} = \cdots = (A_{k+1})_{nj} = 0$ by definition of $StUT_n(R)$, so indeed $(A_1A_2\cdots A_{k+1})_{ij}=0$ if $i+k\ge j$. By finite induction (Theorem 0.7) we have $(A_1A_2\cdots A_n)_{ij}=0$ if $i+n-1\ge j$. This condition holds for all $1\le i, j\le n$ so it follows that $A_1A_2\cdots A_n=0_n$ as claimed.

A ring with the property that the product of a fixed number of elements is always zero is called **nilpotent.** Another example of this besides $StUT_n(R)$ is $S=\{[2n]\in \mathbf{Z}_{32}\mid n\in\mathbf{Z}\}$. It is an exercise to show that S is a subring of $\mathbf{Z}_{32}$. Now $[0]=[m]\in\mathbf{Z}_{32}\Leftrightarrow 32\mid m$, so the product of any five elements in S is $[0]$. For this same $S\subseteq\mathbf{Z}_{32}$, the condition on products remains true for the ring $M_n(S)$: any product of five matrices is 0_n (why?).

We claimed in Example 10.9 that any product of $A_1,\ldots,A_n$ written with increasing subscripts is equal to $(\cdots((A_1A_2)\cdots)A_n)$ regardless of how parentheses are placed. For example, $((A_1A_2)A_3)(A_4A_5)=(A_1A_2)((A_3A_4)A_5)=((A_1A_2)(A_3A_4))A_5$ (why?). This result is essentially Theorem 2.3: for any $g_1,\ldots,g_k\in G$, a group, there is a unique product $g_1g_2\cdots g_k$ with increasing subscripts, independent of placing parentheses. The proof used induction and associativity only, so it holds for products in a ring. Thus we need not put parentheses in $r_1r_2\cdots r_k$ for $r_i\in R$, a ring. When all $r_j=a$ we write this product unambiguously as a^k.

We turn to intersections and unions of subrings.

THEOREM 10.5 Let $\{A_i\}_I$ be subrings of a ring of R. Then $\bigcap_I A_i$ is a subring of R, and if $\{A_i\}_I$ is a chain under inclusion, $\bigcup_I A_i$ is a subring of R.

Proof Using $(A_i,+)\le(R,+)$, Theorem 2.14 gives $(\bigcap_I A_i,+)\le(R,+)$ and also $(\bigcup_I A_i,+)\le(R,+)$ when $\{A_i\}_I$ is a chain. Hence by Theorem 10.4 it suffices to show that each of these is closed under multiplication. If $a,b\in\bigcap_I A_i$ then $a,b\in A_i$ for all $i\in I$ (why?). Thus $ab\in A_i$ since it is a subring, forcing $ab\in\bigcap_I A_i$ and proving that $\bigcap_I A_i$ is a subring. If $\{A_i\}_I$ is a chain and $a,b\in\bigcup_I A_i$, take $a\in A_i$ and $b\in A_j$. Now $\{A_i\}$ a chain implies $A_i\subseteq A_j$ or $A_j\subseteq A_i$, so assume the former. Then $a,b\in A_j$ force $ab\in A_j$ since A_j is a subring, so $ab\in\bigcup_I A_i$ and $\bigcup_I A_i$ is a subring. ∎

Example 10.10 For R any ring, the matrices in $M_n(R)$ with all entries zero in any specified set of rows and in any specified set of columns is a subring.

As in Example 10.7, let T_i be the subring of all $A\in M_n(R)$ with $A_{ik}=0_R$ for all k, and let L_j be the subring of all $A\in M_n(R)$ with $A_{kj}=0_R$ for all k. By Theorem 10.5 the intersection of T_i over $i\in\{i_1,\ldots,i_s\}$ is a subring of $M_n(R)$ and consists of all matrices with zeros in rows $i_1,\ldots,i_s$. Similarly, the intersection of L_j over all $j\in\{j_1,\ldots,j_k\}$ is a subring of $M_n(R)$ consisting of all matrices with zeros in columns $j_1,\ldots,j_k$. The intersection of two such subrings is again a subring by Theorem 10.5 and consists of all matrices with zeros in rows $i_1,\ldots,i_s$ and columns $j_1,\ldots,j_k$. Specifically, the subring of $M_n(R)$ with all zero entries in the first and third rows and in the second, third and fourth columns is $T_1\cap T_3\cap L_2\cap L_3\cap L_4$, and $L_2\cap L_4\cap L_5$ are those matrices with all zero entries in the second, fourth, and fifth columns.

Chains of subrings do not often arise but we discuss one example as an illustration of Theorem 10.5. For $m \in \boldsymbol{N}$, let $\boldsymbol{Z}_{\{m\}} = \{a/m^i \in \boldsymbol{Q} \mid a, i \in \boldsymbol{Z}\}$, the fractions with a power of m as denominator. Here we need not assume $(a, m) = 1$. It is an exercise to show that $\boldsymbol{Z}_{\{m\}}$ is a subring of $\boldsymbol{Q}$ and when $m \mid m'$ that $\boldsymbol{Z}_{\{m\}} \subseteq \boldsymbol{Z}_{\{m'\}}$ (verify!). For $j \in \boldsymbol{N}$, set $S_j = \boldsymbol{Z}_{\{3 \cdot 5 \cdots (2j+1)\}}$ and observe that $\{S_j\}$ is a chain of subrings under inclusion, so $S = \bigcup S_j$ is a subring of $\boldsymbol{Q}$ by Theorem 10.5. We can identify this ring with the fractions having odd denominators, the ring given in Example 10.5 with $m = 2$ there.

For another example using intersections we give the ring theoretic version of Theorem 2.17.

DEFINITION 10.9 If $\varnothing \neq S \subseteq R$, a ring, the **centralizer of S in R** is $C_R(S) = \{x \in R \mid xs = sx \text{ for all } s \in S\}$. Set $C_R(R) = Z(R)$, called the **center** of R.

When R is commutative then $C_R(S) = R$ for all $\varnothing \neq S \subseteq R$. Note that $C_R(S)$ is not empty since $0_R \in C_R(S)$ by Theorem 10.1. In fact, $C_R(S)$ is a subring, proved next. When R is understood we write $C(S)$ for $C_R(S)$.

THEOREM 10.6 For any $\varnothing \neq S \subseteq R$, a ring, $C_R(S)$ is a subring of R.

Proof Observe that $C_R(S) = \bigcap_S C_R(\{s\})$, so from Theorem 10.5 it suffices to show that $C_R(\{s\})$ is a subring for any $s \in S$. If $x, y \in C(\{s\})$ then $xs = sx$ and $ys = sy$. Using Theorem 10.1 we have $(x - y)s = xs + (-y)s = sx + (-ys) = sx + (-sy) = sx + s(-y) = s(x - y)$. Therefore $x - y \in C(\{s\})$. Next, the associativity of multiplication yields $(xy)s = x(ys) = x(sy) = (xs)y = (sx)y = s(xy)$. Thus $xy \in C(\{s\})$ and $C(\{s\})$ is a subring of R by Theorem 10.4. ∎

We present an example of $Z(S)$ and recall from Example 10.7 that $L_k = \{A \in M_n(R) \mid A_{ij} = 0 \text{ if } j = k\}$, the matrices whose k^{th} column is zero, is a subring of $M_n(R)$. Assume $1 \in R$ and take $A \in Z(L_1)$. Now $E_{ij} \in L_1 \Leftrightarrow j \neq 1$, and then $E_{sj}A = AE_{sj}$. Thus for $k \neq s$, $0 = (AE_{sj})_{kj} = A_{ks}E_{kj}$, forcing $A_{ks} = 0$, so $A = A_{22}E_{22} + \cdots + A_{nn}E_{nn}$ ($A \in L_1 \Rightarrow A_{11} = 0$!). But now, if $j \neq 1$, $AE_{1j} = 0_n = E_{1j}A = A_{jj}E_{1j}$, and $A = 0_n$ follows. This shows that $Z(L_1) = \langle 0_n \rangle$. A similar approach shows that $Z(StUT_n(R)) = R \cdot E_{1n}$ when $1 \in R$. The details are left as an exercise. One useful observation regarding centers of matrix rings was mentioned in Chapter 2 and we prove a generalization of it next.

Example 10.11 If R is a ring with 1 then $Z(M_n(R)) = Z(R) \cdot I_n$, the scalar matrices over $Z(R)$.

Since $1 \in R$, $E_{ij} \in M_n(R)$ and commutes with any $A \in Z(M_n(R))$. When $n = 1$ the claim is clear so assume $n > 1$ and take $i \neq j$. The definition of matrix multiplication shows that $E_{1i}AE_{j1} = A_{ij}E_{11}$, so if $A \in Z(M_n(R))$ then $A_{ij}E_{11} = E_{1i}AE_{j1} = E_{1i}E_{j1}A = 0_nA = 0_n$. Hence $A_{ij} = 0$ if $i \neq j$ and $A = A_{11}E_{11} + \cdots + A_{nn}E_{nn}$. But now $A_{ii}E_{ij} = AE_{ij} = E_{ij}A = A_{jj}E_{ij}$, so $A_{ii} = A_{jj}$ and $A = aI_n$ for some $a \in R$. Finally, for $r \in R$, $rE_{11}A = A(rE_{11})$ implies $raE_{11} = arE_{11}$, forcing $ra = ar$, and so $a \in Z(R)$. Thus $Z(M_n(R)) \subseteq Z(R) \cdot I_n$. For the opposite inclusion, let $B \in M_n(R)$ and

$z \in Z(R)$. Then $(zI_n B)_{ij} = zB_{ij} = B_{ij}z = (B \cdot zI_n)_{ij}$ for all entries, so $zI_n B = B \cdot zI_n$ and $Z(R) \cdot I_n \subseteq Z(M_n(R))$, proving that $Z(M_n(R)) = Z(R) \cdot I_n$.

Given $\varnothing \neq S \subseteq R$, a ring, let A be the collection of all subrings in R containing S. Since $R \in A$, $A \neq \varnothing$, so $W(S) = \bigcap_{H \in A} H$ is a subring of R by Theorem 10.5, and $S \subseteq W(S)$ because $S \subseteq H$ for each $H \in A$. Since $W(S)$ is an intersection, $W(S) \subseteq H$ for all $H \in A$. Thus $W(S)$ is the smallest subring under inclusion that contains S. We give it a name and another notation.

DEFINITION 10.10 If $\varnothing \neq S \subseteq R$, a ring, and if A is the set of subrings of R containing S, then $\langle S \rangle = \bigcap_{H \in A} H$ is the **subring of *R* generated by *S*.**

In practice the context will make clear whether $\langle S \rangle$ represents the subring generated by S or the subgroup of $(R, +)$ generated by it. In the latter case we write $\langle S \rangle \leq (R, +)$. As for groups, the definition of $\langle S \rangle$ is not often useful, but there is a more concrete description. If H is a subring of R containing S then H contains all finite products of elements from S, say $x_1 x_2 \cdots x_k$ with all $x_i \in S$ and not necessarily distinct. Also H contains all sums of these products and their negatives, or inverses in $(R, +)$. These sums of products themselves form a subring containing S and so must be $\langle S \rangle$. The proof of this is much like that of Theorem 2.18, using Theorem 10.4, and is left as an exercise.

THEOREM 10.7 For $\varnothing \neq S \subseteq R$, a ring, $\langle S \rangle$ is the set of all finite sums of all products $x_1 x_2 \cdots x_k$ and their additive inverses $-x_1 x_2 \cdots x_k$ for $x_j \in S$ and $k \in \boldsymbol{N}$.

Clearly $\langle R \rangle = R$ and $\langle 0_R \rangle = \{0_R\}$. When $R = \langle S \rangle$ we can write elements of R in terms of S, which can be helpful when S is small. However, it is often impossible to find a small generating set for R.

Example 10.12 $\boldsymbol{Q} = \langle \{1/p \mid p \text{ is prime}\} \rangle$.

By unique factorization in $\boldsymbol{N}$, $0 < 1/n < 1$ is the product of appropriate $1/p$. Also $1/2 + 1/2 = 1/1$ so all $1/n \in \langle \{1/p \mid p \text{ is a prime}\} \rangle$. Adding $1/n$ to itself m times for $m \in \boldsymbol{N}$ gives m/n, and the additive inverse of this is $-m/n$. Thus $\boldsymbol{Q} = \langle \{1/p \mid p \text{ is a prime}\} \rangle$ as claimed. Note that $(\boldsymbol{Q}, +) \neq \langle \{1/p \mid p \text{ is a prime}\} \rangle$ as groups since addition alone will not produce denominators containing powers of primes.

For any $A \in M_n(R)$, $A = \sum A_{ij} E_{ij}$, so $M_n(\mathbf{Z}) = \langle \{E_{ij} \mid 1 \leq i, j \leq n\} \rangle$ follows easily. What simple set generates $M_n(R)$?

Example 10.13 When $1 \in R$, $M_n(R) = \langle \{rE_{11}, E_{j1}, E_{1j} \mid r \in R \text{ and } j \geq 2\} \rangle$.

Let $\{rE_{11}, E_{j1}, E_{1j} \mid r \in R \text{ and } j \geq 2\} = S$. Since for $A \in M_n(R)$, $A = \sum A_{ij} E_{ij}$, it suffices to show $rE_{ij} \in \langle S \rangle$ for any (i, j) and $r \in R$. But $rE_{11}E_{1j} = rE_{1j} \in \langle S \rangle$ and $rE_{j1} = E_{j1} rE_{11} \in \langle S \rangle$. When $i, j \neq 1$ then $rE_{ij} = rE_{i1}E_{1j} \in \langle S \rangle$.

When $S=\{a\}$, products of elements in S are just a^k for $k \in \boldsymbol{N}$. In rings, a^k always means the *product* of a with itself k times. *Adding* a to itself k times, which is the kth power of $a \in (R,+)$, will be written $k \cdot a$ or just ka when no confusion is possible. Since Theorem 10.1 shows that $-(ka)=k(-a)$, we will denote this element also by $(-k)a$ so that za is defined for any $z \in \boldsymbol{Z}$. Of course $0a$ is interpreted to mean 0_R, which we must be careful to distinguish from $0 \in \boldsymbol{Z}$. It will be convenient for purposes of reference to state this as a definition.

DEFINITION 10.11 For R a ring, $a \in R$, and $z \in \boldsymbol{Z}$, $z \cdot a = za$ is the zth power of $a \in (R,+)$, so $0 \cdot a = 0_R$, and for $n \in \boldsymbol{N}$, $n \cdot a = a + a + \cdots + a$ (n times), and $-n \cdot a = -(n \cdot a) = n \cdot (-a)$.

With this notation, $\boldsymbol{Z} \cdot a = \langle\{a\}\rangle \le (R,+)$, and the subring $\langle\langle\{a\}\rangle\rangle = \{n_k a^k + \cdots + n_1 a \mid k \ge 1, n_j \in \boldsymbol{Z}\}$. If $1 \in R$, the subring $\langle\langle\{1_R, a\}\rangle\rangle = \{n_k a^k + \cdots + n_1 a + n_0 1_R \mid k \ge 0$, all $n_j \in \boldsymbol{Z}\}$. When $R = \boldsymbol{C}$, since $\boldsymbol{Z} \subseteq \boldsymbol{C}$, $\boldsymbol{Z}a$ is just multiplication in $\boldsymbol{C}$ so, for example, $\langle\langle\{1,\pi\}\rangle\rangle = \{n_k \pi^k + \cdots + n_1 \pi + n_0 \mid n_j \in \boldsymbol{Z}\}$ looks like polynomials in π with integer coefficients. Although formally this is the same for any $c \in \boldsymbol{C}$, other choices can have a simpler expression.

Example 10.14 $\boldsymbol{Z}[i] = \langle\langle\{1,i\}\rangle\rangle = \langle\langle\{i\}\rangle\rangle = \{a+bi \mid a,b \in \boldsymbol{Z}\} \subseteq \boldsymbol{C}$ is a subring of $\boldsymbol{C}$ called the **Gaussian integers.**

In $\boldsymbol{C}$ the powers of i are just i, $i^2=-1$, $i^3=-i$, and $i^4=1$. Hence $1 = i^4 \in \langle\langle\{i\}\rangle\rangle$ and $\langle\langle\{1,i\}\rangle\rangle = \langle\langle\{i\}\rangle\rangle$. Clearly any sum or difference of products of i is just some $a+bi$ for $a,b \in \boldsymbol{Z}$ so $\boldsymbol{Z}[i] = \{a+bi \mid a,b \in \boldsymbol{Z}\}$ by Theorem 10.7.

Two related examples in $\boldsymbol{R}$ are $\langle\langle\{1,3^{1/2}\}\rangle\rangle = \{a+b3^{1/2} \mid a,b \in \boldsymbol{Z}\}$ and $\langle\langle\{1,2^{1/3}\}\rangle\rangle = \{a+b2^{1/3}+c2^{2/3} \mid a,b,c \in \boldsymbol{Z}\}$ (verify!).

EXERCISES

1. Let $\{R_j\}_T$ be rings.

i) Show that $\bigoplus_T R_j = \{f \in \prod_T R_j \mid f(i) = 0 \in R_i$ for all but finitely many $i \in T\}$ is a subring of $\prod_T R_j$ (see Exercise 4 in §10.1). This ring is called the **direct sum** of $\{R_j\}_T$.

ii) Show that $\bigoplus_T R_j$ is generated by $\{f_{ir} \mid i \in T, r \in R_i\}$ where $f_{ir}(i) = r$ and $f_{ir}(j) = 0 \in R_j$ for all $j \ne i$.

2. i) Find and describe all the subrings of $\boldsymbol{Z}$.

ii) Find and describe all the subrings of $\boldsymbol{Z}_n$.

3. If R is a ring, G is a group, S is a subring of R, and $H \le G$, show that $S[H] = \left\{\sum s_h h \in G[H] \mid s_h \in S \text{ and } h \in H\right\}$ is a subring of $R[G]$.

4. If R is a ring and G is a group, show that $A = \left\{\sum r_g g \in R[G] \mid \sum r_g = 0_R\right\}$ is a subring satisfying $\alpha\beta, \beta\alpha \in A$ for all $\alpha \in R[G]$ and $\beta \in A$.

5. a) For any subring S of $\boldsymbol{Q}$, show that $S \cap \boldsymbol{Z} = \langle 0 \rangle$ implies that $S = \{0\}$.

b) For any nonzero subrings $S_1, \ldots, S_k$ of $\boldsymbol{Q}$, show that $\bigcap S_j \ne \langle 0 \rangle$.

6. Let R be a ring, $a, b \in R$, and $n, m \in \mathbf{Z}$. Using Definition 10.11, show that:
  i) $n \cdot a + m \cdot a = (m + n) \cdot a$
  ii) $n \cdot (m \cdot a) = nm \cdot a$
  iii) $(n \cdot a)(m \cdot b) = nm \cdot ab = (m \cdot a)(n \cdot b)$

7. Prove Theorem 10.7.

8. For $n > 1$ let $W(\mathbf{Z}_n) = \{a \in \mathbf{Z}_n \mid a^2 = [0]_n\}$.
  i) Show that $W(\mathbf{Z}_n)$ is a subring of $\mathbf{Z}_n$ with trivial multiplication.
  ii) Find $W(\mathbf{Z}_n)$ using the prime factorization of n.

9. For R a commutative ring, set $N(R) = \{a \in R \mid a^k = 0_R \text{ for some } k(a) = k \geq 1\}$.
  i) Show that $N(R)$ is a subring of R (Binomial Theorem!).
  ii) If $R = \mathbf{Z}_n$ characterize $N(\mathbf{Z}_n)$ in terms of the prime factorization of n.

10. Show that the ring $\mathbf{R}$ is generated by the numbers in any interval (a, b).

11. If R is a commutative ring and $2r = 0_R$ for all $r \in R$, show that $\{e \in R \mid e^2 = e\}$ is a subring of R.

12. For R a ring with subring S, determine whether the following are subrings of R^N:
  i) $\{f \in R^N \mid f(j) \in S \text{ for all but finitely many } j \in N\}$
  ii) $\{f \in R^N \mid f(j) \in S \text{ for all } j \geq m, \text{ some } m \in N \text{ depending on } f\}$
  iii) $\{f \in R^N \mid f(i) = f(j) \in S \text{ for all } i, j \geq m, \text{ some } m \in N \text{ depending on } f\}$
  iv) $\{f \in R^N \mid f(i) = f(j) \text{ whenever } i \equiv_m j, \text{ some } m \in N \text{ depending on } f\}$

13. For $n > 1$ show that:
  a) $StUT_n(\mathbf{Z}) = \langle\{E_{12}, E_{23}, \ldots, E_{n-1n}\}\rangle$
  b) $M_n(\mathbf{Q}) = \langle\{(1/p)E_{ij} \mid p \text{ is prime and } i \neq j\}\rangle$

14. If R is a ring with identity, show that when $n > 1$, $M_n(R) = \langle S\rangle$ for $S =$:
  i) $\{A \in M_n(R) \mid A^2 = 0_n\}$
  ii) $\{rE_{11}, E_{12} + E_{23} + \cdots + E_{n-1n} + E_{n1} \mid r \in R\}$
  iii) $\{A \in M_n(R) \mid A^2 = A\}$ (note that $E_{11} + E_{12} \in S$)
  iv) $\{A \in M_n(R) \mid A^3 = 0_n\} - \{A \in M_n(R) \mid A^2 = 0_n\}$ if $n \geq 3$

15. If R is a ring with identity and $n > 1$, show that $Z(StUT_n(R)) = R \cdot E_{1n}$.

16. Let G be a finite group and let $C(1) = cls(x_1), \ldots, C(k) = cls(x_k)$ be the conjugacy classes of G. If R is a ring with 1, set $Y_i = \sum_{C(i)} g \in R[G]$ where we write $g = 1_R g$.
  a) Show that $Y_i \in Z(R[G])$.
  b) If $\alpha \in Z(R[G])$ show that $\alpha = a_1 Y_1, + \cdots + a_k Y_k$ where all $a_j \in Z(R)$.

17. Is the given subset of the ring a subring?
  i) $\{\alpha \in \mathbf{Z}_2^n = \mathbf{Z}_2 \oplus \cdots \oplus \mathbf{Z}_2 \mid \text{some } 2k \text{ coordinates of } \alpha \text{ are } [1]_2\} \subseteq \mathbf{Z}_2^n$
  ii) for $m \in \mathbf{N}$, $\{a/m^i \in \mathbf{Q} \mid i \in \mathbf{Z}\} \subseteq \mathbf{Q}$ (do not assume $(a, m) = 1$)
  iii) $\left\{\sum r_g g \in R[G] \mid r_g = r_h \text{ all } g, h \in G\right\} \subseteq R[G]$, for G a finite group
  iv) $\{A \in M_n(\mathbf{R}) \mid \det A = 0\} \subseteq M_n(\mathbf{R})$
  v) $\{A \in M_n(\mathbf{R}) \mid \text{for all } i, A_{i1} + A_{i2} + \cdots + A_{in} = 0\} \subseteq M_n(\mathbf{R})$
  vi) $\left\{\sum r_g g \in R[G] \mid \sum r_g = 0\right\} \subseteq R[G]$
  vii) $\{f \in \mathbf{R}^{\mathbf{R}} \mid f(r) = 0 \text{ for some } r \in \mathbf{R}\} \subseteq \mathbf{R}^{\mathbf{R}}$
  viii) If R is a ring and $T \subseteq R$ is not empty, $\{x \in R \mid tx = 0 \text{ for all } t \in T\} \subseteq R$.

10.3 POLYNOMIAL AND RELATED RINGS

We recall the notion of polynomial and look at some related rings, useful for examples. We assume that polynomials and computations with them are familiar. A formal construction of polynomials from Theorem 10.3 and Example 10.6 is given at the end of the chapter.

DEFINITION 10.12 For R a ring and x a formal symbol, or *indeterminate* over R, the set of **polynomials over** R is $R[x] = \{\sum_{i \geq 0} r_i x^i \mid \text{all } r_i \in R \text{ and all but finitely many } r_i = 0\}$. Each $p(x) = \sum r_i x^i$ is a **polynomial** and r_i is the **coefficient** of x^i in $p(x)$.

The symbols x^k are also formal and suggest multiplication of x with itself k times. Any formal symbol can be used in place of x. If $p(x) = \sum r_i x^i$, $q(x) = \sum s_i x^i \in R[x]$ then $p(x) = q(x) \Leftrightarrow r_i = s_i$ for all i. We usually omit any $0_R x^j$ but if $p(x) = \sum 0_R x^i$ we write $p(x) = 0$, the *zero polynomial.* With this convention $0 \neq p(x) \in R[x]$ can be written $p(x) = r_n x^n + \cdots + r_1 x + r_0 = 0_R x^m + \cdots + 0_R x^{n+1} + r_n x^n + \cdots + r_1 x + r_0$ for any $m > n$ when $n \geq 0$ and $r_n \neq 0_R$.

When $p(x) = r_n x^n + \cdots + r_1 x + r_0$ with $r_n \neq 0$, the **degree** of $p(x)$ is n, written $deg\ p(x) = n$, and r_n is the **leading coefficient** of $p(x)$. In particular, $p(x) = r_0$ has degree zero if $r_0 \neq 0$, but the zero polynomial has no degree. It is not uncommon to define $deg\, 0 = -\infty$. The **constant** of $p(x) = r_n x^n + \cdots + r_0$ is r_0, and $r_0 = 0_R$ if it does not appear explicitly in $p(x) : ax^2 + bx$ has constant term 0_R. Write $-rx^k$ for $(-r)x^k$ and when $1 \in R$, write x^k for $1x^k$ and write $-x^k$ for $(-1)x^k$. When the leading coefficient of $p(x)$ is 1, $p(x)$ is called **monic.** Note that $x \notin R[x]$ unless $1 \in R$. For example, $g(x) \in (2\mathbf{Z})[x]$ has every coefficient even so $x = 1x \notin (2\mathbf{Z})[x]$. We can always identify $R \subseteq R[x]$ as the constant polynomials. As an illustration of these notions for polynomials in $\mathbf{Z}[x] : x^5 - 7x$ is monic of degree 5 and has zero constant; and $-x^3 + 2x^2 + 3x - 4$ has degree 3, leading coefficient -1, and constant -4.

The sum of $p(x) = r_n x^n + \cdots + r_1 x + r_0$ and $q(x) = s_m x^m + \cdots + s_1 x + s_0$ in $R[x]$ just adds the coefficients of the corresponding x^i. Thus $p(x) + q(x) = \sum_{i=0}^{\max\{n,m\}} (r_i + s_i)x^i$. The product is $p(x)q(x) = \sum_{i=0}^{n+m} \left(\sum_{j=0}^{i} r_{i-j} s_j\right) x^i$ where $r_{i-j} = 0_R$ when $i - j > n$ and $s_j = 0$ when $j > m$; it is often abbreviated as $p(x)q(x) = \sum \left(\sum r_{i-j} s_j\right) x^i$ with the assumption that $r_k = 0_R$ if $k < 0$. This is just the usual multiplication of polynomials and amounts to assuming the associative and distributive laws, multiplying the terms and assuming $rx^j \cdot sx^i = (rs)x^{j+i}$, and adding coefficients of equal powers of x. Sample computations in $\mathbf{Z}[x]$ are:

$$(3x^4 + 2x^3 - x + 4) + (-2x^2 + x) = 3x^4 + 2x^3 - 2x^2 + 4$$
$$(2x^3 + x^2 - 3x + 1) + (-2x^3 + x^2 - 1) = 2x^2 - 3x$$
$$(x^2 + 2x - 1)(x + 2) = x^3 + 4x^2 + 3x - 2$$
$$(x^3 + 2x^2 + 1)(x^3 - 2x^2 + 1) = x^6 - 4x^4 + 2x^3 + 1$$

THEOREM 10.8 For any ring R, $R[x]$ together with polynomial addition and multiplication is a ring. When $1 \in R$ the constant polynomial $1 \in R[x]$ is an identity element for $R[x]$. Also $R[x]$ is commutative when R is.

Proof Verifying Definition 10.1 is straightforward, although showing that multiplication is associative is tedious and much like the computation for group rings in Example 10.3. It is quite easy to prove that $(R[x], +)$ is an Abelian group. We prove one distributive law and leave the other verifications as exercises. If $f(x) = \sum a_i x^i$, $g(x) = \sum b_i x^i$, $h(x) = \sum c_i x^i \in R[x]$, with all summations starting at $i = 0$, then $(f(x) + g(x))h(x) = \left(\sum (a_i + b_i)x^i\right)\left(\sum c_i x^i\right)$ has $d_k = \sum (a_{k-i} + b_{k-i})c_i$ as the coefficient of x^k, and $f(x)h(x) + g(x)h(x)$ has $e_k = \sum a_{k-i}c_i + \sum b_{k-i}c_i$ as the coefficient of x^k. Using the distributive law in R and the fact that $(R, +)$ is an Abelian group together show that $d_k = e_k$ for all k, so we must have $(f(x) + g(x))h(x) = f(x)h(x) + g(x)h(x)$ by our definition of equality of polynomials. When $1 \in R$, the definition of multiplication shows that the constant $1 \in R[x]$ is an identity. If R is commutative, then for $f(x)$ and $g(x)$, as above, the coefficient of x^k in $f(x)g(x)$ is

$$a_k b_0 + a_{k-1}b_1 + \cdots + a_0 b_k = b_0 a_k + b_1 a_{k-1} + \cdots + b_k a_0$$

and the latter is the coefficient of x^k in $g(x)f(x)$. Thus $R[x]$ is commutative. ■

In calculus and previous courses, $p(x) \in R[x]$ is viewed mainly as a function: x can be replaced by any $r \in R$, usually $\mathbf{R}$ or $\mathbf{Q}$, yielding $p(r) \in R$. Here we view $R[x]$ primarily as a ring of formal symbols. When $p(x) \in R[x]$ is used to represent a function by evaluation in R, we must be careful to distinguish the formal object from the function since equality in each case may be different. Any $p(x) \in \mathbf{Z}_2[x]$ looks like sums of powers of x when we omit $[0]_2 x^k$ and write $[1]_2 x^k$ as x^k. Now $x^5 + x^4 + x^3 + x$, $x^4 + x^2$, $x^3 + x$, $[0]_2 \in \mathbf{Z}_2[x]$ are four different polynomials but all give the zero function on $\mathbf{Z}_2$ by evaluation. In fact, considering all $p(x) \in \mathbf{Z}_2[x]$ results in only four different functions on $\mathbf{Z}_2$ by evaluation: $[0]_2$, $[1]_2$, x, and $x + [1]_2$ (why?).

We can define polynomial rings with many indeterminates, even infinitely many. *Indeterminates* are formal symbols having no relations among themselves or with R, except for commutativity. We will deal more precisely with this somewhat ambiguous and mysterious notion at the end of the chapter. Given a finite set of indeterminates $X = \{x_1, \ldots, x_n\}$ over R, a polynomial in $R[X]$ is a finite sum of terms $rx_1^{a_1}x_2^{a_2}\cdots x_n^{a_n}$ with $r \in R$ and $a_j \in \mathbf{N} \cup \{0\}$. Only one such term with any given $\{x_1^{a_1}, \ldots, x_n^{a_n}\}$ is allowed. Each $rx_1^{a_1}x_2^{a_2}\cdots x_n^{a_n}$ is a *monomial* with coefficient r. Any $p(X) \in R[X]$ is unchanged by adding monomials with 0_R as coefficient since two polynomials in $R[X]$ are equal when the coefficients of corresponding monomials are equal, as in $R[x]$. Addition of $f(X), g(X) \in R[X]$ is defined by taking the sum in R of the coefficients of the same monomials in $f(X)$ and $g(X)$, and multiplication is defined by using the distributive law, decreeing that the indeterminates commute with each other and with elements of R. For example, in $\mathbf{Z}[X]$ for $X = \{x, y, t\}$ and writing $1 \cdot m(X) = m(X)$ for any monomial $m(X)$, $(x^2y + yt) + (2xt - yt) = x^2y + 2xt$, and $(x^2y + 2yt)(xt + 3yt) = x^3yt + 3x^2y^2t + 2xyt^2 + 6y^2t^2$. Let us record this idea more formally.

DEFINITION 10.13 For R a ring, $X = \{x_1, \ldots, x_n\}$ indeterminates over R, and $\alpha = (a_1, \ldots, a_n) \in A = (\mathbf{N} \cup \{0\})^n \subseteq \mathbf{Z}^n$, let $X^\alpha = x_1^{a_1}x_2^{a_2}\cdots x_n^{a_n}$. The **set of polynomials over R in X** is $R[X] = \{\sum_A r_\alpha X^\alpha \mid r_\alpha \in R$ and all but finitely many $r_\alpha = 0_R\}$.

The *constant* of $f(X) \in R[X]$ is the term $r_z X^z$ where $z = (0, \ldots, 0) \in A$. Usually $r_z X^z$ is written just r_z. The set $R[X]$ has a ring structure, as we indicated just above, but the verification is left as an exercise. The statement of the next theorem uses coordinate addition in $\mathbf{Z}^n$ (see Example 10.2) to add $\beta, \gamma \in A \subseteq \mathbf{Z}^n$. For simplicity write $p(X) \in R[X]$ as $\sum r_\alpha X^\alpha$, omitting the A.

THEOREM 10.9 Let R be a ring and $X = \{x_1, \ldots, x_n\}$ indeterminates over R. Then $R[X] = R[x_1, \ldots, x_n]$ is a ring using $\sum r_\alpha X^\alpha + \sum s_\alpha X^\alpha = \sum (r_\alpha + s_\alpha) X^\alpha$ and

$\left(\sum r_\alpha X^\alpha\right)\left(\sum s_\alpha X^\alpha\right) = \sum\left(\sum r_\beta s_\gamma\right)X^\alpha$ where $\beta + \gamma = \alpha$ in $\mathbf{Z}^n$. When $1 \in R$ then writing $1 \cdot X^\alpha = X^\alpha$, $R[X]$ has identity element X^z where $z = (0, \ldots, 0)$. $R[X]$ is commutative when R is.

Addition and multiplication in $R[X]$ make sense because there are only finitely many nonzero r_β and s_γ, so only finitely many $\alpha = \beta + \gamma$ with $r_\beta s_\gamma \neq 0_R$. If $\alpha = (a_1, \ldots, a_n)$, rX^α is a *monomial of degree* $a_1 + \cdots + a_n$ when $r \neq 0$. The *degree* of $f(X) \in R[X]$ is the largest degree of any monomial appearing in $f(X)$. There may be several monomials of largest degree. As for $R[x]$, if $f(X) = \sum r_\alpha X^\alpha$ and all $r_\alpha = 0_R$ then $f(X) = 0$ has no degree and is called the **zero polynomial.** When all monomials with nonzero coefficients in $f(X)$ have the same degree k, we say that $f(X)$ is **homogeneous of degree k.** In $\mathbf{Z}[X]$ with $X = \{x, y\}$, a typical polynomial is $\sum c_{ij}x^i y^j$ for $i, j \geq 0$, all $c_{ij} \in \mathbf{Z}$, and only finitely many $c_{ij} \neq 0$. Specifically, $x^2y + 2xy^2 - 3x - 2$ has degree 3 and constant -2, and $x^2y + 2xy^2 - y^3$ is homogeneous of degree 3.

We can also define polynomial rings over R in which the indeterminates $\{x_1, \ldots, x_n\}$ do not commute with each other when $n > 1$. The usual notation is $R\{X\}$ rather than $R[X]$. We still assume the coefficients in R commute with the indeterminates. For example, in $\mathbf{Z}\{X\} = \mathbf{Z}\{x, y\}$, $xy \neq yx$, $(2x)(3y) = 6xy$, $xyx \neq x^2y = xxy$, $(x + y)^2 = x^2 + xy + yx + y^2 \neq x^2 + 2xy + y^2$ and $xyx \cdot xyx = xyx^2yx \neq x^4y^2$. We give these rings a special name.

DEFINITION 10.14 If R is a ring and $X = \{x_1, \ldots, x_n\}$ are indeterminates over R, then $R\{X\}$, the polynomial ring over R whose indeterminates do not commute with each other, is called the **free ring in X over R.**

Two other constructions related to polynomials are useful for examples, particularly in the commutative case. One takes all formal infinite sums of monomials $rx_1^{a_1}x_2^{a_2}\cdots x_n^{a_n}$ as its elements. These are formal sums since we cannot actually add infinitely many things. As in Definition 10.13, if $\alpha = (a_1, \ldots, a_n) \in A = (N \cup \{0\})^n$ let $X^\alpha = x_1^{a_1}x_2^{a_2}\cdots x_n^{a_n}$.

DEFINITION 10.15 For R a ring and $A = (N \cup \{0\})^n$, the **power series ring over R in the set of indeterminates** $X = \{x_1, \ldots, x_n\}$ is the ring $R[[X]] = \left\{\sum r_\alpha X^\alpha \mid \alpha \in A \text{ and } r_\alpha \in R\right\}$ with $\sum r_\alpha X^\alpha + \sum s_\alpha X^\alpha = \sum (r_\alpha + s_\alpha)X^\alpha$ for all $\alpha \in A$ and $\left(\sum r_\alpha X^\alpha\right)\left(\sum s_\alpha X^\alpha\right) = \sum\left(\sum r_\beta s_\gamma\right)X^\alpha$ over all $\beta + \gamma = \alpha$ in A.

The verification that $R[[X]]$ is a ring follows that for $R[X]$. Note that multiplication makes sense because for any $\alpha \in A$ there are only finitely many $\beta, \gamma \in A$ with $\beta + \gamma = \alpha$. As for $R[X]$, $1 \in R$ implies that $1 \cdot x_1^0 \cdots x_n^0$ is an identity for $R[[X]]$, and $R[[X]]$ is commutative when R is. We may write $R[[X]] = R[[x_1, \ldots, x_n]]$. The power series ring $R[[x]]$ in one indeterminate has elements that look like power series in calculus, at least when the coefficients are in the real numbers $\mathbf{R}$. For two indeterminates, each element in $R[[x, y]]$ can be considered to be in either $R[[x]][[y]]$ or $R[[y]][[x]]$, since, for example, $\sum r_{ij}x^i y^j = \sum\left(\sum r_{ij}y^j\right)x^i$. This identification of $R[[x, y]]$ with $R[[y]][[x]]$ should be justified and follows by defining ring isomorphisms (Chapter 11!). We do not pursue the matter here.

The second construction allows some negative powers of x in $R[[x]]$. Set $R\langle\langle x\rangle\rangle = \left\{\sum_{i \geq M} r_i x^i \mid r_i \in R \text{ and } M \in \mathbf{Z}\right\}$. This restriction to finitely many negative exponents of x implies that multiplication as in $R[[x]]$ makes sense here, and with pointwise addition as in

$R[[x]]$, it is an exercise to see that $R\langle\langle x\rangle\rangle$ is a ring. Just as for $R[[x]]$, $R\langle\langle x\rangle\rangle$ is commutative when R is, and when $1 \in R$, $R\langle\langle x\rangle\rangle$ has identity element $1_R x^0$.

DEFINITION 10.16 If R is a ring the **Laurent series ring over R** is $R\langle\langle x\rangle\rangle = \{\sum_{i\geq M} r_i x^i \mid r_i \in R \text{ and } M \in \mathbf{Z}\}$ with $\sum r_i x^i + \sum s_i x^i = \sum (r_i + s_i)x^i$ for all $i \in \mathbf{Z}$, and $(\sum r_i x^i)(\sum s_i x^i) = \sum (\sum r_j s_k)x^i$ for all $j, k \in \mathbf{Z}$ with $j + k = i$.

Can we define Laurent series rings in more indeterminates? One approach for two indeterminates takes formal infinite sums of $r_{ij}x^i y^j$ with i, j in $\mathbf{Z}$ and bounded below. This does not allow us to identify $R\langle\langle x, y\rangle\rangle$ with $R\langle\langle x\rangle\rangle\langle\langle y\rangle\rangle$, unlike the case for $R[[x, y]]$! Whereas our definition has lower bounds on the powers of x and y appearing in any element, $R\langle\langle x\rangle\rangle\langle\langle y\rangle\rangle$ does not. If $\alpha = \sum x^{-i} y^i$ over all $i \in \mathbf{N}$, then $\alpha \in R\langle\langle x\rangle\rangle\langle\langle y\rangle\rangle$ since each $x^{-i} \in R\langle\langle x\rangle\rangle$, but $\alpha \notin R\langle\langle x, y\rangle\rangle$. If instead we define $R\langle\langle x, y\rangle\rangle = R\langle\langle x\rangle\rangle\langle\langle y\rangle\rangle$, then $R\langle\langle x, y\rangle\rangle \neq R\langle\langle y\rangle\rangle\langle\langle x\rangle\rangle$: the latter has elements with no lower bound on the powers of y appearing, but it does have a lower bound for the powers of x appearing. If we attempt to define $R\langle\langle x, y\rangle\rangle$ as all formal sums of $r_{ij}x^i y^j$ for $i, j \in \mathbf{Z}$, then it is not clear how to define a multiplication like that in $R[[x, y]]$. Consequently, we will not give a definition of Laurent series for more than one indeterminate, but we may consider the iterated construction $R\langle\langle x\rangle\rangle\langle\langle y\rangle\rangle$.

Addition and multiplication in $R[x]$, $R[[x]]$, and $R\langle\langle x\rangle\rangle$ are essentially the same, so we may consider $R[x] \subseteq R[[x]] \subseteq R\langle\langle x\rangle\rangle$ as rings. Unlike the other two, $R[[x]]$ contains the "power series" in any $Y \in R[[x]]$ having constant term zero. That is, if $Y = r_1 x + r_2 x^2 + \cdots \in R[[x]]$ then we can make sense of the expression $s_1 Y + s_2 Y^2 + \cdots$. To define this element in $R[[x]]$ it suffices to specify its coefficients. The idea is that $s_j Y^j$ has no monomial of degree less than j so the coefficient of x^k in $\sum s_i Y^i$ is the uniquely determined coefficient of x^k in $s_1 Y + \cdots + s_k Y^k$. It is not feasible to give a general expression for the coefficients arising from an arbitrary $Y \in R[[x]]$. If $Y = x + x^3$, computing powers of Y shows that the sum of all positive powers is $\sum Y^i = x + x^2 + 2x^3 + 3x^4 + 4x^5 + 6x^6 + 9x^7 + 13x^8 + 19x^9 + \cdots$. If Y is sufficiently nice, a general expression for a simple series in it may be possible.

Example 10.15 If $Y = \sum_{i>0} x^i \in \mathbf{Z}[[x]]$ then $\sum_{i>0} Y^i = \sum_{i>0} 2^{i-1} x^i$.

For $k \geq 2$, x^k in Y^2 arises as $xx^{k-1} + x^2x^{k-2} + \cdots + x^{k-1}x = (k-1)x^k$. Thus $Y^2 = \sum_{i>1}(i-1)x^i$. Similarly, if $Y^s = \sum a_i x^i$ then for $t \geq s+1$ the coefficient of x^t in $Y^{s+1} = YY^s$ is $a_s + a_{s+1} + \cdots + a_{t-1}$. If in Y^s, $a_k = \binom{k-1}{s-1}$ then the coefficient of x^t in Y^{s+1} is $\binom{s-1}{s-1} + \cdots + \binom{t-2}{s-1} = \binom{t-1}{s}$, using Exercise 8 in §0.3. Thus the coefficient of x^t in $\sum_{i>0} Y^i$ is $\binom{t-1}{0} + \binom{t-1}{1} + \cdots + \binom{t-1}{t-1} = 2^{t-1}$ by induction and the Binomial Theorem.

Power series arise in the theory of counting, or combinatorics. In fact, Example 10.15 shows that the number of ways of writing $k \in N$ as a sum of positive integers, taking order into account, is 2^{k-1} (why?).

Using the existence of infinite series in any $Y \in R[[x]]$ without constant term, we can prove an important property of $R[[x]]$ that is needed in most examples it is used for. Recall from Definition 10.6 that when $1 \in R$, $U(R)$ is the group of invertible elements in R.

THEOREM 10.10 Let R be a ring with identity and $f(x) = \sum r_i x^i \in R[[x]]$.

i) $f(x) \in U(R[[x]]) \Leftrightarrow r_0 \in U(R)$
ii) $r_j = 0$ for $j < m$ and $r_m \in U(R) \Rightarrow f(x) = x^m v = w x^m$ for $v, w \in U(R[[x]])$
iii) $U(R) = R - \{0\} \Leftrightarrow U(R\langle\langle x\rangle\rangle) = R\langle\langle x\rangle\rangle - \{0\}$

Proof We identify $R \subseteq R[[x]]$ as those series with $r_j = 0$ for all $j > 0$, so we write $1_{R[[x]]} = 1_R = 1$. For $f(x)$, $g(x) \in R[[x]]$, the constant of $f(x)g(x)$ is the product of the constants of $f(x)$ and $g(x)$. Therefore, if for some $g(x) \in R[[x]]$ $f(x)g(x) = g(x)f(x) = 1$, then the constant s_0 of $g(x)$ satisfies $r_0 s_0 = 1 = s_0 r_0$ forcing $r_0 \in U(R)$. Suppose $r_0 \in U(R)$ and write $f(x) = r_0(h(x))$ for $h(x) = 1 + (r_0^{-1} r_1)x + \cdots + (r_0^{-1} r_j)x^j + \cdots = 1 - Y$. We claim that $h(x)$ has inverse $(1 - Y)^{-1} = 1 + Y + Y^2 + \cdots + Y^j + \cdots$. To prove this we show that for $k \geq 1$ the coefficient of x^k in $(1 - Y)(1 + Y + Y^2 + \cdots)$ is zero. Now $1 + Y + Y^2 + \cdots \in R[[x]]$, as we indicated above. Since for $t \leq k < j$, x^t does not appear in any Y^j, it follows that the coefficient of x^k in $(1 - Y)(1 + Y + Y^2 + \cdots)$ is the same as the coefficient of x^k in $(1 - Y)(1 + Y + \cdots + Y^k)$. Using that Y commutes with 1 and all Y^i, we have $(1 - Y)(1 + Y + \cdots + Y^k) = 1 - Y^{k+1}$, by induction on k, so the coefficient of x^k here is zero. The constant of $(1 - Y)(1 + Y + Y^2 + \cdots)$ is 1, so in fact this product is 1. Similarly, $(1 + Y + Y^2 + \cdots)(1 - Y) = 1$, so $h(x) = 1 - Y \in U(R[[x]])$. Since $r_0 \in U(R)$ we have $r_0 \in U(R[[x]])$ with the same inverse, so $h(x)^{-1} r_0^{-1} f(x) = 1$. Starting instead with $f(x) = g(x) \cdot r_0 = (1 - V)r_0$ leads to $f(x)q(x) = 1$ for some $q(x) \in R[[x]]$. The proof of the uniqueness of inverses in a group in Theorem 2.1 shows that these left and right inverses for $f(x)$ must be the same, so $f(x) \in U(R[[x]])$, proving i). Part ii) now follows since $f(x) = x^m h(x)$ with $h(x) = r_m - Y$ for some $Y \in R[[x]]$ without constant term. The constant of $h(x)$ is $r_m \in U(R)$ so $h(x) \in U(R[[x]])$ by i). Similarly, $f(x) = (r_m - V)x^m$ so again $(r_m - V) \in U(R[[x]])$.

For iii) we identify $R \subseteq R\langle\langle x\rangle\rangle$, note that $1_R \in R\langle\langle x\rangle\rangle$ is the identity of $R\langle\langle x\rangle\rangle$, and denote this element by 1. For any ring A let $A^* = A - \{0_A\}$. If $U(R\langle\langle x\rangle\rangle) = R\langle\langle x\rangle\rangle^*$ then for $r \in R^*$, $r \in U(R\langle\langle x\rangle\rangle)$ and has an inverse in $R\langle\langle x\rangle\rangle$. But if $1 = r\left(\sum a_j x^j\right) = \sum r a_j x^j$ then $r a_j = 0$ for $j \neq 0$, so $r^{-1} r a_j = 0$ forces $a_j = 0$ and $\sum a_j x^j = a_0 \in R$. Hence the inverse of r must be in R so $R^* = U(R)$. Now assume $R^* = U(R)$. Suppose that $f(x) = \sum_{i \geq m} r_i x^i \in R\langle\langle x\rangle\rangle$ with $r_m \neq 0$ and write $f(x) = r_m x^m (1 + r_m^{-1} r_{m+1} x + r_m^{-1} r_{m+2} x^2 + \cdots)$. By i), $1 + r_m^{-1} r_{m+1} x + r_m^{-1} r_{m+2} x^2 + \cdots \in U(R[[x]])$ and so has an inverse $g(x) \in R[[x]] \subseteq R\langle\langle x\rangle\rangle$. It follows that $g(x) r_m^{-1} x^{-m} \in R\langle\langle x\rangle\rangle$ and that $g(x) r_m^{-1} x^{-m} f(x) = 1$. Similarly, by writing $f(x) = (1 + r_{m+1} r_m^{-1} x + \cdots) r_m x^m$ we get $f(x)h(x) = 1$ for some $h(x) \in R\langle\langle x\rangle\rangle$. As above, using the proof of Theorem 2.1 we conclude that $f(x) \in U(R\langle\langle x\rangle\rangle)$, so $U(R\langle\langle x\rangle\rangle) = R\langle\langle x\rangle\rangle^*$, completing the proof of the theorem. ∎

For examples illustrating the last part of the theorem, take a ring R so that R^* is a group under multiplication. Using $\boldsymbol{Q}$, $\boldsymbol{R}$, $\boldsymbol{C}$, or $\boldsymbol{Z}_p$ for p a prime, we conclude that the nonzero elements in $\boldsymbol{Q}\langle\langle x\rangle\rangle$, $\boldsymbol{R}\langle\langle x\rangle\rangle$, $\boldsymbol{C}\langle\langle x\rangle\rangle$, or $\boldsymbol{Z}_p\langle\langle x\rangle\rangle$ are units. From this another example would be $\boldsymbol{Q}\langle\langle x\rangle\rangle\langle\langle y\rangle\rangle$. These are examples of **fields,** which we will define formally in the next section.

EXERCISES

1. Verify that the multiplication in $R[x]$ is associative.
2. Show carefully that the multiplication for $R\langle\langle x\rangle\rangle$ in Definition 10.16 makes sense: for $\alpha, \beta \in R\langle\langle x\rangle\rangle$, $\alpha\beta$ really is another element in $R\langle\langle x\rangle\rangle$.
3. Verify Theorem 10.9.
4. Let $p(x) = a_0 + a_1x + \cdots + a_kx^k$. Which of the following sets are subrings?
   i) $\{p(x) \in \mathbf{Q}[x] \mid a_j \in \mathbf{Z} \text{ if } j < 6\} \subseteq \mathbf{Q}[x]$
   ii) $\{p(x) \in \mathbf{Q}[x] \mid \sum a_j = 0\} \subseteq \mathbf{Q}[x]$
   iii) $\{p(x) \in \mathbf{Z}[x] \mid a_j \in 3\mathbf{Z} \text{ if } j \text{ is odd}\} \subseteq \mathbf{Z}[x]$
   iv) $\{p(x) \in \mathbf{Z}[x] \mid a_j \in 2^j\mathbf{Z}\} \subseteq \mathbf{Z}[x]$
   v) $\{p(x) \in \mathbf{Z}[x] \mid a_j \in j\mathbf{Z}\} \subseteq \mathbf{Z}[x]$
   vi) $\{p(x) \in \mathbf{Z}[x] \mid a_j \in j!\mathbf{Z}\} \subseteq \mathbf{Z}[x]$
5. For $i \in \mathbf{N}$, let $S_i = \{2x, 2x^2, \ldots, 2x^i\} \subseteq \mathbf{Z}[x]$. Describe $\langle S_i\rangle$ and $\bigcup_i \langle S_i\rangle$.
6. Is the given subset of the ring a subring?
   i) $\{f(x) \in \mathbf{Q}[[x]] \mid \text{there is } m \in \mathbf{Z} \text{ with } mf(x) \in \mathbf{Z}[[x]]\} \subseteq \mathbf{Q}[[x]]$
   ii) $\{0\} \cup \{f(x) \in \mathbf{Q}[[x]] \mid f(x) = q_kx^k + q_{k+1}x^{k+1} + \cdots$ with $q_k \in \mathbf{Z} - \langle 0\rangle\} \subseteq \mathbf{Q}[[x]]$
   iii) for $m \mid n$ in $\mathbf{N}$, $\{p(x) \in \mathbf{Z}_n[x] \mid a_j \in \langle [m]_n\rangle$, all coefficients of $f(x)\} \subseteq \mathbf{Z}_n[x]$
7. Show that $H \le (\mathbf{Z}[x], +)$ satisfying $Hx = \{hx \in \mathbf{Z}[x] \mid h \in H\} \subseteq H$ is a subring.
8. In $\mathbf{Z}[x]$ describe the subrings:
   i) $\langle\{x^3\}\rangle$
   ii) $\langle\{x^2, x^3\}\rangle$
   iii) $\langle\{x^6, x^8, x^{10}\}\rangle$
   iv) $\langle\{x^3, x^4\}\rangle$
   v) $\langle\{x^3, x^5\}\rangle$
   vi) $\langle\{x^4, x^7\}\rangle$
   vii) $\langle\{2x, 2x^2\}\rangle$
   viii) $\langle\{2x, 3x^2\}\rangle$
9. Describe explicitly the following subrings of $\mathbf{Z}[x, y]$:
   i) $\langle\{x, y\}\rangle$
   ii) $\langle\{x^2, y^2\}\rangle$
   iii) $\langle\{x^2, xy, y^2\}\rangle$
   iv) $\langle\{2x, y\}\rangle$
10. $R = End((\mathbf{Z}_2[x], +))$ is a ring by Theorem 10.2. Write $[1]_2x^i = x^i \in \mathbf{Z}_2[x]$.
    a) Show that $T = \{\varphi \in R \mid$ each x^j and also $[1]_2 \in \mathbf{Z}_2[x]$ appears in only finitely many $\varphi(x^k)\}$ is a subring of R. Thus, given $n \in \mathbf{N}$, if $\varphi \in T$ then for k large enough, $\varphi(x^k)$ contains no monomial of degree less than n.
    b) Show that $W = \{\varphi \in R \mid \varphi(x^i) = 0 \text{ for } i \ge m = m(\varphi) \in \mathbf{N}\}$ is a subring of R.
    c) Show that $W \subseteq T$.
    d) For any $t \in T$ and $w \in W$, show that $tw, wt \in W$.
    e) Show that the center $Z(W) = \{0\}$.
11. If R is a commutative ring, $a \in R$, and $f(x) = r_kx^k + \cdots + r_1x + r_0 \in R[x]$, let $f(a) = r_ka^k + \cdots + r_1a + r_0$.
    i) For $f, g \in R[x]$ show that $(f + g)(a) = f(a) + g(a)$, and $(f \cdot g)(a) = f(a)g(a)$.
    ii) Find $f, g \in M_2(\mathbf{Z})[x]$ and $A \in M_2(\mathbf{Z})$ such that $(f \cdot g)(A) \ne f(A)g(A)$.

12. Let R be a ring with identity, $a \in R$, and $f(x) \in R[x] - R$. Prove by induction on $deg\, f(x) \geq 1$ that there is $q(x) \in R[x]$ with $deg\, q(x) < deg\, f(x)$ and $f(x) = q(x)(x - a) + f(a)$ (see Exercise 11). Beware: $-R$ need not be commutative!

13. For R a ring with 1, let $R[x, x^{-1}] = \{\alpha = \sum_{i \geq M} r_i x^i \in R\langle\langle x \rangle\rangle \mid$ only finitely many $r_k \neq 0\}$.

a) Show that $R[x, x^{-1}]$ is a subring of $R\langle\langle x \rangle\rangle$.

b) Show that $U(R[x, x^{-1}]) = \{ux^i \mid u \in U(R)$ and $i \in \mathbf{Z}\}$.

c) Consider $R[x] \subseteq R[x, x^{-1}]$. If $\alpha \in R[x, x^{-1}]$ then $\alpha = uf(x)$ for $u \in U(R[x, x^{-1}])$ and $f(x) \in R[x]$.

d) Describe $\langle\{x^{-1}\}\rangle \subseteq R[x, x^{-1}]$.

14. Describe:

a) $U(\mathbf{Z}_n[[x]])$

b) $U(\mathbf{Z}_n\langle\langle x \rangle\rangle)$

10.4 ZERO DIVISORS AND DOMAINS

In $\mathbf{Z}$ or $\mathbf{Z}_p$ for p a prime, $ab = 0 \Rightarrow a = 0$ or $b = 0$, whereas in $\mathbf{Z}_6$, $[2][3] = [0]$, and in $M_3(\mathbf{Q})$, $E_{12}E_{13} = 0_3$. We want to have a name for nonzero elements with product zero. Recall again that in any ring R, $R^* = R - \{0\}$.

DEFINITION 10.17 In any ring R, if $a, b \in R^*$ with $ab = 0$, then a is a **left zero divisor** and b is a **right zero divisor.** If $a \in R^*$ is either a left or a right zero divisor it is called a **zero divisor,** and if it is both a left and a right zero divisor it is a **two-sided zero divisor.**

The notions of left, right, and two-sided zero divisor coincide in commutative rings. Noncommutative rings will not occupy much of our time but we do want to mention a couple of examples. In $M_3(\mathbf{Q})$, when $i \neq j$ each E_{ij} is a two-sided zero divisor, but note that $E_{32}E_{13} = 0_3$ whereas $E_{13}E_{32} = E_{12}$. It can happen that a right zero divisor is not a left zero divisor.

Example 10.16 The ring $R = (End(\mathbf{Q}[x], +), +, \circ)$ contains elements that are zero divisors on one side only.

Note that $0_R \in R$ is the function $0_R(p(x)) = 0 \in \mathbf{Q}[x]$ for all $p(x) \in \mathbf{Q}[x]$. Define $f, g \in R$ by $f(a_0 + a_1x + \cdots + a_kx^k) = a_1 + a_2x + \cdots + a_kx^{k-1}$, and $g(a_0 + a_1x + \cdots + a_kx^k) = a_0x + a_1x^2 + \cdots + a_kx^{k+1}$ (verify $f, g \in R$!). From Theorem 10.2 the multiplication in $End(\mathbf{Q}, +)$ is composition and it follows easily that $f \cdot g = 1_R$ is the identity map on $\mathbf{Q}[x]$. If $g \cdot h = 0_R$ for some $h \in R$ then $0_R = f \cdot 0_R = f \cdot (g \cdot h) = (f \cdot g) \cdot h = 1_R h = h$, so g is not a left zero divisor in R. For $\pi \in R$ given by $\pi(a_0 + a_1x + \cdots + a_kx^k) = a_0$, $\pi \cdot g = 0_R$, so g is a right zero divisor. Similarly, f is not a right zero divisor but $f \cdot \pi = 0_R$.

Let $p(x) = a_ix^i + \cdots + a_kx^k$, $q(x) = b_jx^j + \cdots + b_mx^m \in R[x]$ with $i \leq k$, $j \leq m$, and $p(x)q(x) = 0$. Then $p(x)q(x) = a_ib_jx^{i+j} + (a_{i+1}b_j + a_ib_{j+1})x^{i+j+1} + \cdots + a_kb_mx^{k+m}$, where the coefficients of the other powers of x are sums of products in R. It follows that $a_ib_j = 0 = a_kb_m$. This shows that if a polynomial is a left (right) zero divisor, then so are its leading coefficient and the coefficient of the smallest degree monomial in it. In particular, if R has no zero divisors neither does $R[x]$, and in this case $deg\ p(x)q(x) = deg\ p(x) + deg\ q(x)$ for nonzero $p(x), q(x) \in R[x]$. Let us state this as a theorem after giving a name to rings without zero divisors.

DEFINITION 10.18 A ring R is a **domain** if it has no zero divisors and if $|R| > 1$. A commutative domain with identity is called an **integral domain.**

The condition $|R| > 1$ eliminates calling $R = \{0_R\}$ a domain. The rings $n\boldsymbol{Z}$ for $n > 1$ are commutative domains but not integral domains since they have no identity. When $n \in \boldsymbol{N}$ with $n = ab$ for $a, b > 1$ then $\boldsymbol{Z}_n$ is not a domain since $[a][b] = [0]$. By Theorem 1.29, $\boldsymbol{Z}_p$ is an integral domain when p is a prime. Also $\boldsymbol{Z}, \boldsymbol{Q}$, and $\boldsymbol{R}$ are integral domains. It should be clear that if R is a domain then so is any nonzero subring of R. However, note that $M_2(\boldsymbol{Z})$ is not a domain, since $E_{11}E_{22} = 0_2$, but its subring $\boldsymbol{Z}E_{11} = \{mE_{11} \mid m \in \boldsymbol{Z}\}$ is an integral domain (why?). The argument of the last paragraph shows that if R is a domain, so is $R[x]$. We generalize this next.

THEOREM 10.11 If D is a domain, then so is:

i) $D[x]$, and $p(x), q(x) \in D[x]^* \Rightarrow deg\ p(x)q(x) = deg\ p(x) + deg\ q(x)$
ii) $D[X]$ for any nonempty set X of indeterminates over D
iii) $D[[Y]]$ for $Y = \{y_1, \ldots, y_n\}$ indeterminates over D
iv) $D\langle\langle x\rangle\rangle$

When D is an integral domain, so are $D[X]$, $D[[Y]]$, and $D\langle\langle x\rangle\rangle$.

Proof If $p(x) = a_0 + \cdots + a_kx^k, q(x) = b_0 + \cdots + b_mx^m \in D[x]^*$ with $a_k, b_m \neq 0$, then as above, the leading coefficient of $p(x)q(x)$ is $a_kb_m \neq 0$, proving i). For ii), if $X = \{x_1, \ldots, x_n\}$ then $p(X) \in D[X]$ is $p(X) = p_0 + p_1x_n + \cdots + p_mx_n^m$ with all $p_j \in D[x_1, \ldots, x_{n-1}]$ (verify!), so $D[X] = D[x_1, \ldots, x_{n-1}][x_n]$. Assuming that $D[x_1, \ldots, x_{n-1}]$ is a domain by induction, $D[x_1, \ldots, x_{n-1}][x_n] = D[X]$ is also from i). Thus X finite implies $D[X]$ is a domain by IND-I. If X is infinite and for some $f(X), g(X) \in D[X]$, $f(X)g(X) = 0$, then since elements of $D[X]$ are finite sums of monomials with coefficients from D, there is a finite $T \subseteq X$ so that $f(X), g(X) \in D[T]$. This forces $f(X) = 0$ or $g(X) = 0$, as we have just seen.

As for polynomials, $D[[Y]] = D[[y_1, \ldots, y_{n-1}]][[y_n]]$ (formally we need to use Chapter 11) and an induction argument like that given in ii) will prove iii), provided that $D[[y]]$ is a domain. Write each $f(y) \in D[[y]]^*$ as $f(y) = d_jy^j + d_{j+1}y^{j+1} + \cdots$, with $j \geq 0$ and $d_j \neq 0$. If $g(y) = c_iy^i + c_{i+1}y^{i+1} + \cdots$ with $c_i \neq 0$, then $f(y)g(y) = d_jc_iy^{i+j} + a_{i+j+1}y^{i+j+1} + \cdots \neq 0$ since D a domain implies $d_jc_i \neq 0$. The same argument shows $D\langle\langle x\rangle\rangle$ is a domain, simply by allowing $i, j \in \boldsymbol{Z}$ rather than $i, j \geq 0$. The last statement follows since we have seen that each of $D[X]$, $D[[Y]]$, and $D\langle\langle x\rangle\rangle$ is commutative with identity when D is. ■

By identifying $R \subseteq R[x] \subseteq R[[x]] \subseteq R\langle\langle x\rangle\rangle$, we see that if any of these rings is a domain, so is R, and then all are by Theorem 10.11. To get zero divisors in $R[x] - R$ when R has them, let $a, b \in R^*$ with $ab = 0$, and take $f(x), g(x) \in R[x]^*$ so that the coefficients of $f(x)$ are r_ia and the coefficients of $g(x)$ are bs_j for $r_i, s_j \in R$. Then $f(x)g(x) = 0$ since any product $r_i abs_j = 0$. Specifically, $ax \cdot bx^2 = 0$ and $(ax + rax^2)(bx^2 + bsx^3) = 0$. There are other ways to get zero divisors in $R[x]$, but all are related to our examples, as illustrated in the next theorem. The result we give about zero divisors in $R[X]$ is not obvious, even for $R[x]$. Our proof of this result shows how the degrees of monomials can be used to advantage. The statement of the theorem for a general set of indeterminates X adds some complexity to the proof, which is clearer in the special case when $X = \{x\}$. The idea of the argument is the same in all cases so it is useful to keep the special case of $R[x]$ in mind when reading the proof.

THEOREM 10.12 Let R be a commutative ring, X a set of indeterminates over R, $g(X)$, $f(X) \in R[X]^*$, and $g(X)f(X) = 0$. There is a $a \in R^*$ with $af(X) = 0$.

Proof Since $g(X)$ and $f(X)$ are sums of finitely many nonzero terms, both are polynomials in finitely many indeterminates from X, so we may assume $X = \{x_1, \ldots, x_n\}$. The monomials appearing in any $p(X) \in R[X]$ correspond to elements in $A = (N \cup \{0\})^n$, which is ordered lexicographically: for $\alpha = (a_1, \ldots, a_n)$, $\beta = (b_1, \ldots, b_n) \in A$, set $\alpha < \beta$ if either $a_1 < b_1$, or for some $j < n$, $a_i = b_i$ when $i \leq j$ but $a_{j+1} < b_{j+1}$ (Exercise 9 in §1.5). It is clear that for $\alpha \neq \beta$ in A, either $\alpha < \beta$ or $\beta < \alpha$. Also if $\alpha < \beta$ in A and $\gamma \in A$ then computing sums in $\mathbf{Z}^n$, $\alpha + \gamma < \beta + \gamma$ (why?). The monomials appearing in any $p(X) \in R[X]$ can be ordered in the same way: if we let $X^\alpha = x_1^{a_1} \cdots x_n^{a_n}$, set $X^\alpha < X^\beta$ when $\alpha < \beta$ and say that X^α has lower order than X^β. Note that $X^\alpha < X^\beta$ implies $X^\alpha X^\gamma < X^\beta X^\gamma$.

Write $f(X) = a(\beta_k)X^{\beta_k} + \cdots + a(\beta_1)X^{\beta_1}$ for $\beta_k > \cdots > \beta_1$ and all $a(\beta_i) \neq 0$, and $g(X) = r(\alpha_m)X^{\alpha_m} + \cdots + r(\alpha_1)X^{\alpha_1}$ for $\alpha_m > \cdots > \alpha_1$. Choose $g(X) \in R[X]$ of this form so that $g(X)f(X) = 0$ with m minimal and all $r(\alpha_j) \neq 0$: $g(X)$ has the fewest monomials appearing with nonzero coefficients. If $m = 1$ then $r(\alpha_1)X^{\alpha_1}f(X) = 0$ so all $r(\alpha_1)a(\beta_j) = 0$, resulting in $r(\alpha_1)f(X) = 0$ and we are finished. Otherwise $m > 1$, and using the ordering we have defined, since α_m is the largest α_i and β_k is the largest β_j, it follows that $r(\alpha_m)a(\beta_k)X^{\alpha_m+\beta_k}$ is the only term in the product with exponent $\alpha_m + \beta_k$ (verify!). Hence $r(\alpha_m)a(\beta_k) = 0 = a(\beta_k)r(\alpha_m)$ since R is commutative. Now $a(\beta_k)g(X)f(X) = 0$, and $a(\beta_k)g(X)$ has fewer terms than $g(X)$ (why?) so the choice of m forces $a(\beta_k)g(X) = 0$. If all $a(\beta_j)g(X) = 0$ then the product of any coefficient of $f(X)$ with any coefficient of $g(X)$ is zero, yielding $r(\alpha_i)f(X) = 0$. Therefore we may assume that some $a(\beta_j)g(X) \neq 0$ but that $a(\beta_i)g(X) = 0$ if $i > j$. The commutativity of R implies that for $i > j$, $g(X)a(\beta_i) = 0$ as well. But now $0 = g(X)f(X) = g(X)(a(\beta_k)X^{\beta_k} + \cdots + a(\beta_j)X^{\beta_j} + \text{lower-order terms}) = (r(\alpha_m)X^{\alpha_m} + \text{lower-order terms})(a(\beta_j)X^{\beta_j} + \text{lower-order terms})$. Once again, the ordering of the monomials shows that the only term in this product with exponent $\alpha_m + \beta_j$ is the product of $r(\alpha_m)X^{\alpha_m}$ in $g(X)$ with $a(\beta_j)X^{\beta_j}$ in $f(X)$ and we conclude that $r(\alpha_m)a(\beta_j) = 0$. Now $a(\beta_j)g(X)f(X) = 0$ and by the choice of j, $a(\beta_j)g(X) \neq 0$ but has fewer nonzero terms than $g(X)$ does because $a(\beta_j)r(\alpha_m) = 0$. This contradiction to the minimality of m when $m > 1$ forces $m = 1$ and completes the proof of the theorem. ∎

The result in Theorem 10.12 fails in general for $R[[x]]$.

Example 10.17 There is a commutative ring R and $f(x) \in R[[x]]^*$ that is a zero divisor, but $rf(x) \neq 0$ for all $r \in R^*$.

Let $Y = \{y_i\}_N$ be indeterminates over $\mathbf{Z}_2$, and $f_j \in End(\mathbf{Z}_2[Y], +)$ the formal partial derivative with respect to y_j. That is, if $p(Y) = p_0 + \cdots + p_m y_j^m$ where $p_i \in \mathbf{Z}_2[Y]$ does not contain y_j, then $f_j(p(Y)) = p_1 + 2p_2 y_j + \cdots + mp_m y_j^{m-1} = p_1 + p_3 y_j^2 + p_5 y_j^4 + \cdots$, since $2q(Y) = 0$ in $\mathbf{Z}_2[Y]$. Let $R = \langle\{f_i \mid i \in N\}\rangle$, the subring of $End(\mathbf{Z}_2[Y], +)$ generated by $\{f_j\}$. It is an exercise to check that R is a commutative ring, that $f_i^2 = 0$ for all $i \in N$ (the "derivative" of a polynomial in even powers of y_i is zero in $\mathbf{Z}_2[Y]$), and that the product of different f_j is not zero. For example, $f_1 f_2 \cdots f_k(y_1 \cdots y_k) = [1]_2 = 1 \in \mathbf{Z}_2[Y]$. Also since $2q(Y) = 0$ in $\mathbf{Z}_2[Y]$, $2r = 0$ in R for all $r \in R$. In $R[[x]]$, let $f(x) = \sum f_i x^i$ over *all* $i \in N$. We claim that $f(x)f(x) = 0$. The coefficient of x^{k+1} in this product is $f_1 f_k + f_2 f_{k-1} + \cdots + f_k f_1$. Using $f_i f_j = f_j f_i$ this coefficient is a sum of terms $2f_i f_{k-i+1} = 0$ and also $f_j^2 = 0$ when $k + 1 = 2m$ and $j = m$. Any $r \in R^*$ is a finite sum of products of *distinct* $f_k \in \{f_i\}$ by Theorem 10.7, the commutativity of R, and the fact that $f_j^2 = 0$. Thus for some $m \geq 1$, at most $\{f_1, \ldots, f_m\}$ appear in r. It is an exercise to see that either $r = f_1 \cdots f_m$ or some $f_{i_1} \cdots f_{i_k} r = f_1 \cdots f_m$. In either case, $r f_{m+1}(y_1 \cdots y_{m+1}) \neq [0]_2$ so $r f_{m+1} \neq 0$ and this implies that $r f(x) = \sum r f_j x^j \neq 0$.

The construction in Example 10.17 shows that in Theorem 10.12, $f(x)g(x) = 0$ in $R[x]$ does not force all products of coefficient of $f(x)$ with those of $g(x)$ to be zero. If $R = \langle\{f_i \mid i \in N\}\rangle \subseteq End(\mathbf{Z}_2[Y], +)$ as in the example, then $f(x) = f_1 + f_2 x \in R[x]$ satisfies $f(x)f(x) = f_1^2 + 2f_1 f_2 x + f_2^2 x^2 = 0$ using $f_1^2 = f_2^2 = 0$ and $2r = 0$ for all $r \in R$. However, $f_1 f_2 \neq 0$, and more generally, $h(x) = f_1 + f_2 x + \cdots + f_k x^{k+1} \in R[x]$ has square zero but for $i \neq j$, $f_i f_j \neq 0$. Note that $r = f_1 f_2 \cdots f_k \in R^*$ and $rh(x) = 0$, as required in Theorem 10.12.

Next we consider rings R for which $(R^*, \cdot)$ is an Abelian group.

DEFINITION 10.19 A ring R with 1 is a **field** if $(R^*, \cdot)$ is an Abelian group.

A field is a commutative ring with identity so that each $r \in R^*$ has an inverse under multiplication. Examples are $\boldsymbol{Q}$, $\boldsymbol{R}$, and $\mathbf{Z}_p$ for p a prime. We claim that any field F is an integral domain. By definition, F is commutative with identity. If $a \in F^*$ and $ab = 0$ then $0 = a^{-1} \cdot 0 = a^{-1}(ab) = (a^{-1}a) \cdot b = 1_F b = b$. Therefore F has no zero divisors and is a domain.

From Theorem 10.10, F a field implies that $F_1 = F\langle\langle x_1\rangle\rangle$, $F_2 = F_1\langle\langle x_2\rangle\rangle$, $F_3 = F_2\langle\langle x_3\rangle\rangle, \ldots$, are also. We describe some other fields.

Example 10.18 $\boldsymbol{Q}(\alpha) = \{a + b\alpha \in \boldsymbol{C} \mid \alpha \in \boldsymbol{C}, \alpha^2 \in \boldsymbol{Q}, a, b \in \boldsymbol{Q}\}$ is a field.

The operations in $\boldsymbol{C}$ show $(a + b\alpha) - (c + d\alpha) = (a - c) + (b - d)\alpha \in \boldsymbol{Q}(\alpha)$ and also $(a + b\alpha)(c + d\alpha) = (ac + bd\alpha^2) + (ad + bc)\alpha \in \boldsymbol{Q}(\alpha)$ since $\boldsymbol{Q}$ is a ring and $\alpha^2 \in \boldsymbol{Q}$. Therefore $\boldsymbol{Q}(\alpha)$ is a subring of $\boldsymbol{C}$ by Theorem 10.4 and contains $1 = 1 + 0 \cdot \alpha$, so $\boldsymbol{Q}(\alpha)$ is an integral domain. We must show that each $\beta \in \boldsymbol{Q}(\alpha)^*$ has an inverse under multiplication. If $\alpha \in \boldsymbol{Q}$ then $\boldsymbol{Q}(\alpha) = \boldsymbol{Q}$ a field; assume $\alpha \notin \boldsymbol{Q}$. For $\beta = a + b\alpha \in \boldsymbol{Q}(\alpha)$, let $N(\beta) = (a + b\alpha)(a - b\alpha) = a^2 - b^2\alpha^2$. If $N(\beta) = 0$ then $a = b = 0$ or $\alpha^2 = (a/b)^2$, forcing $\alpha = \pm a/b \in \boldsymbol{Q}$, a contradiction. Thus when $\beta \neq 0$

then $N(\beta) \in \boldsymbol{Q}^*$ and $\beta^{-1} \in \boldsymbol{Q}(\alpha)$ is given by $\beta^{-1} = (a/N(\beta)) - (b/N(\beta))\alpha$ (verify!). This shows that $\boldsymbol{Q}(\alpha)$ is a field, using Definition 10.19.

Special cases of Example 10.18 are $\boldsymbol{Q}(3^{1/2}) = \{a + b3^{1/2} \in \boldsymbol{R} \mid a, b \in \boldsymbol{Q}\}$ and $\boldsymbol{Q}(i) = \{a + bi \in \boldsymbol{C} \mid a, b \in \boldsymbol{Q}\}$ with $i^2 = -1$. Fields also arise as fractions of elements from a commutative domain. Indeed $\boldsymbol{Q} = \{a/b \mid a \in \boldsymbol{Z}, b \in \boldsymbol{N}\}$ is the set of all fractions of integers, and in $\boldsymbol{Q}$ we identify $z \in \boldsymbol{Z}$ with the element $z/1$. The mathematical construction of these "fractions" was discussed in §1.9 and we review it for an arbitrary commutative domain. In $\boldsymbol{Q} \subseteq \boldsymbol{R}$ some fractions of integers that are different, say 2/4 and 3/6, turn out to be equal. If we define fractions formally as "a/b" for $a, 0 \neq b \in D$, a commutative domain, it may not be clear when two such fractions should be identified as the same element. The construction of the fractions and the proper identifications use an equivalence relation on pairs of elements in D.

We will describe the construction briefly since many of the details are presented in §1.9. Let D be a commutative domain, $M = D \times D^*$, and define a relation on M by $(a, b) \sim (c, d)$ when $ad = bc$. It is an exercise to verify that $\sim$ is an equivalence relation on M. Set the equivalence class $[(a, b)]_\sim = a/b$. Let $QF(D)$ be the set of these classes and define an addition and multiplication on $QF(D)$ by $a/b + c/d = (ad + bc)/bd$ and $(a/b)(c/d) = ac/bd$. The first task is to show that these definitions are independent of the representatives of the classes, as explained in §1.9. Specifically, if $a/b = a'/b'$ and $c/d = c'/d'$ then $ab' = ba'$ and $cd' = dc'$ so $acb'd' = a'c'bd$, which shows that $ac/bd = a'c'/b'd'$. The verification for addition is similar. Thus "$\cdot$" and "+" really are functions from $QF(D) \times QF(D)$ to $QF(D)$. It is not hard to see that $QF(D)$ is a field using that for any $b, d \in D^* : a/b = ad/bd, 0/b = 0/d$ is the identity of $(QF(D), +)$, and $a/b = 0/d$ forces $a = 0$. For example, $a/b + c/d = (ad + bc)/bd = (cb + da)/db = c/d + a/b$ verifies commutativity of addition, and $0/b + c/d = (0d + bc)/bd = cb/db = c/d$, so $0/b = 0_{QF(D)}$ is the additive identity. The additive inverse of $a/b \in QF(D)$ is $(-a)/b$ and the multiplicative identity is $1_{QF(D)} = b/b$ for any $b \in D^*$. If $a/b \neq 0/d$, then $b/a = (a/b)^{-1}$ since $(a/b)(b/a) = ab/ab = b/b = 1_{QF(D)}$.

The field $QF(D)$ is the smallest we can construct starting with D, in the sense that it enlarges D only by allowing division by elements in D^*. Of course D is not contained in $QF(D)$ but $d \in D^*$ corresponds to $db/b \in D' = \{ab/b \in QF(D) \mid a \in D\}$. Note that for $c \in D^*$, $ab/b = ac/c$ since $abc = bac$. Should $ab/b = cb/b$, then $ab^2 = cb^2$ so $a = c$ (why?). Hence $\varphi(a) = ab/b$ is a bijection from D to D'. Now $ab/b + a'b/b = (a + a')b^2/b^2 = (a + a')b/b$ and $(ab/b)(a'b/b) = aa'b^2/b^2 = aa'b/b$ show that the ring structures of D and D' are abstractly the same: the bijection φ satisfies $\varphi(a + a') = \varphi(a) + \varphi(a')$ and $\varphi(aa') = \varphi(a)\varphi(a')$. Thus φ is an isomorphism between $(D, +)$ and $(D', +)$. Thinking of D' as a copy of D in $QF(D)$, for $c/d \in QF(D)$, $c/d = cb^2/db^2 = (cb/b)(b/db) = \varphi(c)\varphi(d)^{-1}$. In this way $QF(D)$ is the field obtained from D by inverting the elements in $(D')^*$.

The discussion above outlines the formal construction of the "field of fractions" of a commutative domain D, but it is convenient to view $QF(D)$ as $\{a/b\}$ subject to the usual identifications and operations on fractions.

THEOREM 10.13 The **quotient field** or field of fractions of a commutative domain D is $QF(D) = \{a/b \mid a \in D, b \in D^*\}$ with $a/b = c/d$ when $ad = bc$. $QF(D)$ is a field with

$a/b + c/d = (ad + bc)/bd$ and $(a/b)(c/d) = ac/bd$. The identity $1_{QF(D)} = b/b$ for any $b \in D^*$; $0/b = 0_{QF(D)}$; when $a/b \neq 0_{QF(D)}$ then $(a/b)^{-1} = b/a$; and D is identified in $QF(D)$ as $D' = \{ab/b \in QF(D) \mid a \in D\}$. Any $c/d \in QF(D)$ is a quotient of elements in D' via $c/d = (cb/b)(db/b)^{-1}$.

A familiar example is $QF(\mathbf{Z}) = \boldsymbol{Q}$. The identification of D in $QF(D)$ is easier when $1 \in D$, letting $d \in D$ correspond to $d/1 \in QF(D)$. Using this we see that for a field K, $QF(K) = K$, or, more accurately, $QF(K)$ is identified with K via $a/b = (a/1)(b/1)^{-1} = (a/1)(b^{-1}/1) = ab^{-1}/1$. The subring $2\mathbf{Z}$ of $\mathbf{Z}$ is a commutative domain and $QF(2\mathbf{Z})$ can be considered as a subring of $QF(\mathbf{Z}) = \boldsymbol{Q}$. But in $\boldsymbol{Q}$, $a/b = 2a/2b$, so $\boldsymbol{Q}$ can be identified with $QF(2\mathbf{Z})$. When $D = K[x]$ for K a field, $QF(K[x]) = K(x)$ is called the **rational function field** over K in one variable. The elements of $K(x)$ are taken to be the fractions $f(x)/g(x)$ with $g(x) \neq 0$. For $D = K[x_1, \ldots, x_n]$, $QF(D) = K(x_1, \ldots, x_n)$ is the rational function field over K in n variables, and its elements are $f(X)/g(X)$ for $f(X), g(X) \in D$ with $g(X) \neq 0$.

Some examples of fields result from the next theorem. A better approach to these examples uses the construction of quotient *rings* in the next chapter. However, the theorem is of some intrinsic interest.

THEOREM 10.14 If D is a finite domain then $1 \in D$ and $(D^*, \cdot)$ is a group. In particular, if D is a finite commutative domain then D is a field.

Proof If $D = \{d_1, \ldots, d_n\}$ then $n > 1$ by Definition 10.18, so let $d \in D^*$. We claim that $dD = \{dd_i \in D \mid d_i \in D\} = D$. Since $dD \subseteq D$ it suffices to show that $|dD| = n$. If $dd_i = dd_j$ then $d(d_i - d_j) = 0$, so $d \neq 0$ forces $d_i - d_j = 0$, and $d_i = d_j$. Thus $|dD| = n$ and $dD = D$. Hence $de = d$ for some $e \in D$ so for any $x \in D$, $dex = dx$. This yields $d(ex - x) = 0$, and $d \neq 0$ implies $ex = x$. But now $xex = x^2$ so $(xe - x)x = 0$, and $xe = x$ follows when $x \neq 0$. Also $0e = 0 = e0$ by Theorem 10.1, so $e = 1_D$. Using $dD = D$ for $d \in D^*$, there must be $b \in D$ with $db = e$. Since $b \neq 0$ (why?) we also have $bc = e$ for some $c \in D$. As in Theorem 2.1, $d = de = d(bc) = (db)c = ec = c$, so $b = d^{-1}$ and $(D^*, \cdot)$ is a group. When D is commutative, $(D^*, \cdot)$ is Abelian so D is a field by Definition 10.19. ■

We know that $\mathbf{Z}_p$ for p a prime is a finite field. With Theorem 10.14 in hand it follows easily that these are fields since they are domains: use the property of primes that $p \mid ab$ forces $p \mid a$ or $p \mid b$ (Theorem 1.14). It turns out that every finite domain *is* a field, but it is surprisingly difficult to prove this result, which is called Wedderburn's Theorem. Let us use Theorem 10.14 to produce some new finite fields that look like finite versions of the complex numbers.

Example 10.19 Let p be a prime, x an indeterminate over $\mathbf{Z}_p$, and $F(p^2) = \{a + bx \mid a, b \in \mathbf{Z}_p\}$. Setting $(a + bx) + (c + dx) = (a + c) + (b + d)x$ and $(a + bx) \cdot (c + dx) = (ac - bd) + (ad + bc)x$, $(F(p^2), +, \cdot)$ is a field with p^2 elements exactly when $[-1]_p$ is not a square in $\mathbf{Z}_p$.

Clearly $|F(p^2)| = p^2$, and it is an exercise to verify that $(F(p^2)), +, \cdot)$ is a commutative ring with $1 = [1] + [0]x$. By Theorem 10.14, $F(p^2)$ will be a field when it is a domain. The identity of $(F(p^2), +)$ is $0 = [0] + [0]x$. If $F(p^2)$ is not a domain, let $(a+bx), (c+dx) \in F(p^2)^*$ with $(a+bx)\cdot(c+dx) = 0$, so $ac - bd = [0] = ad + bc$ in $\boldsymbol{Z}_p$. If $b = [0]$ then either $a = [0]$ and $a + bx = 0$ or else $a \neq [0]$, forcing both $c = d = [0]$ and $c + dx = 0$. Thus we may assume $b \neq [0]$, and, similarly, that $a \neq [0]$. Now $b(c^2 + d^2) = c(ad + bc) - d(ac - bd) = [0]$ so $c^2 + d^2 = [0]$. Hence $c = [0]$ exactly when $d = [0]$, so neither is $[0]$ and there is $k \in \boldsymbol{Z}_p$ with $kd = [1]$. But now $(kc)^2 = -k^2d^2 = [-1]$, so if $F(p^2)$ is not a domain then $[-1]$ is a square in $\boldsymbol{Z}_p$. Finally, if $v^2 = [-1] \in \boldsymbol{Z}_p$ then $([1] + vx)([1] - vx) = [1] + v^2 + [0]x = 0$ and $F(p^2)$ is not a domain.

For which primes p is $[-1]_p$ not a square in $\boldsymbol{Z}_p$? Direct computation shows that $p = 3, 7, 11$, and 19 work but that $[2]_5^2 = [-1]_5$, $[5]_{13}^2 = [-1]_{13}$, and $[4]_{17}^2 = [-1]_{17}$. Thus by Example 10.19 there are fields with 9, 49, 121, or 361 elements. The pattern for small primes continues for all primes $p > 2$; $[-1]_p$ is not a square in $\boldsymbol{Z}_p \Leftrightarrow p \equiv_4 3$. We do not want to present a formal proof of this. However, that $p \equiv_4 1$ implies $[-1]_p = [a]_p^2$ is Exercise 27 in §1.7. For p any odd prime, if $[-1]_p = [a]_p^2$ in $\boldsymbol{Z}_p$ then $a^2 \equiv_p -1$. Theorem 4.8 (Euler!) shows $a^{p-1} \equiv_p 1$, and it follows that $1 \equiv (a^2)^{(p-1)/2} \equiv (-1)^{(p-1)/2}$ by Theorem 1.24. This forces $(p-1)/2$ to be even, or $p \equiv_4 1$.

Is there a field with p^2 elements when $[-1]_p$ is a square in $\boldsymbol{Z}_p$? A slight modification of Example 10.19 produces one. Fix some $e \in U_p$ and change the multiplication in the example to $(a+bx)\cdot(c+dx) = (ac + bde) + (ad + bc)x$. In Example 10.19, $e = [-1]_p$. As in the example, $F(p^2)$ is a field when $e \in U_p$ is not a square in $\boldsymbol{Z}_p$. There is always some such e when $p > 2$ since in U_p, $[a]_p^2 = [-a]_p^2 = [p-a]_p^2$, so only half the elements in U_p can be squares. Another approach uses that $U_p = \langle [c]_p \rangle$ (Theorem 4.20), so it follows from Theorem 2.8 on orders that $[c]_p$ cannot be a square. By a quick computation, $[2]_p \in \boldsymbol{Z}_p$ is not a square for $p = 5$ or $p = 13$ but is when $p = 17$, in which case $[3]_{17} \in \boldsymbol{Z}_{17}$ is not a square. Eventually we will see how to construct fields with p^m elements for any prime p and any $m \in \boldsymbol{N}$.

We end this section with one last notion about general rings. If R is any of $\boldsymbol{Z}$, $\boldsymbol{Z}[x]$, $\boldsymbol{Q}$, or $\boldsymbol{R}[[x]]$, then no $r \in (R, +) - \{0_R\}$ has finite order. Of course if R is finite then every $r \in (R, +)$ has finite order by Theorem 2.9. Observe that $\boldsymbol{Z}_n[x]$ is an infinite ring but all $p(x) \in (\boldsymbol{Z}_n[x], +)$ have finite order at most n (why?). In the ring $R = \boldsymbol{Z} \oplus \boldsymbol{Z}_{12}$, $o((0, [3]_{12})) = 4$ in $(R, +)$ but $o((3, [1]_{12}))$ is infinite: $(3, [1]_{12})^m = (3m, [m]_{12}) \neq (0, [0]_{12})$ for any $m \in \boldsymbol{N}$. There is a special term for elements with finite additive order in rings.

DEFINITION 10.20 For R any ring, $a \in (R, +) - \langle 0_R \rangle$ is an ***n*-torsion element** if $o(a) \mid n$ and is **torsion free** if $o(a)$ is infinite. If every $a \in R^*$ is an n-torsion element for a fixed $n \in \boldsymbol{N}$ with n minimal, then R **has characteristic *n*,** written *char* $R = n$. Call R **torsion free** if all $r \in R^*$ are torsion free, in which case R **has characteristic zero,** written *char* $R = 0$.

Recall that if $a \in (R, +)$ and $n \in \boldsymbol{N}$ then $na = a + \cdots + a$. When $na = 0$ then $o(a)$ in $(R, +)$ must divide n by Theorem 2.7, so a is an n-torsion element. By Definition 10.20, every $p(x) \in \boldsymbol{Z}_2[x]$ is a $2m$-torsion element for any $m \in \boldsymbol{N}$ since twice every coefficient of $p(x)$ is $[0]_2$. Now 2 is clearly minimal, so *char* $\mathbf{Z}_2[x] = 2$. Similarly, $\mathbf{Z}_n$, $\mathbf{Z}_n[X]$, or $\mathbf{Z}_n[[Y]]$ has characteristic n. The ring $R = \mathbf{Z}_6 \oplus \mathbf{Z}_8$ has 48 elements and *char* $R = 24$, since $24r = 0_R$ for all $r \in R$ and $o(([2]_6, [1]_8)) = 24$. Clearly *char* $\mathbf{Z}$ = *char* $\boldsymbol{Q}[x] = 0$ since these are torsion free. As above, $\mathbf{Z} \oplus \mathbf{Z}_{12}$ has both torsion and torsion-free elements so has no characteristic.

Even when each $r \in R$ is an n-torsion element for some n depending on r, R need not have a characteristic. For an example we use infinite direct sums of groups as in §3.5. If $G = \oplus \mathbf{Z}_n$ over $n \in \boldsymbol{N}$ then G is an Abelian group whose elements can be viewed as all sequences $(a_1, \ldots, a_n, \ldots)$ with $a_i \in \mathbf{Z}_i$, all but finitely many $a_i = [0]_i$, and addition is given coordinatewise. Every $\alpha \in G$ has finite order but G contains elements of arbitrarily large order (why?). To obtain a ring from G, take the trivial ring with all products zero, or define products coordinatewise. The details are left to the exercises.

If $1 \in R$ is an n-torsion element then $n \cdot 1 = 0$. By Theorem 10.1, for any $a \in R$, $0 = 0a = (n \cdot 1)a = (1 + \cdots + 1)a = a + \cdots + a = na$ by an extension of the distributive law (verify this!). Therefore every $a \in R$ is an n-torsion element and so R must have a characteristic (why?). Although an arbitrary ring need not have a characteristic, any domain must, and even more is true.

THEOREM 10.15 For any domain D, either D is torsion free so *char* $D = 0$, or *char* $D = p$ for some prime p.

Proof We may assume there is $a \in D^*$ with $na = 0$ for some integer $n > 1$. Let $d \in D$ and use the distributive law to see that $0 = (na)d = (a + \cdots + a)d = ad + \cdots + ad = a(d + \cdots + d) = a(nd)$. But D is a domain and $a \neq 0$ so $nd = 0 : nD = \{nd \in D \mid d \in D\} = 0$. By Well Ordering, $mD = 0$ for a minimal $m > 1$, and by our argument, $kd \neq 0$ for $0 < k < m$ and $d \in D^*$. If $m = uv$ for integers $u, v > 1$ then $vd \neq 0$ for $d \in D^*$ by the minimality of m. Now $mb = b + \cdots + b$ with m summands for $b \in D^*$, so mb is vb added to itself u times. Thus by associativity, $0 = mb = vb + \cdots + vb = u(vb)$. But $u < m$ and $vb \neq 0$ contradicts the choice of m. Consequently, $m = p$ is a prime and *char* $D = p$. ■

By Theorem 10.15 any field F has *char* $F = p$ or *char* $F = 0$. The fact that domains have a characteristic is sometimes helpful, particularly when taking powers in commutative rings. For example, it is easy to see that for p a prime, the binomial coefficient $\binom{p}{k} = pB_k$ for some $B_k \in \boldsymbol{N}$ when $2 \leq k \leq p - 1$, because p in the numerator is not canceled. If D is a commutative domain with *char* $D = p$ then in D, $(a + b)^p = a^p + b^p$ since the other terms in the expansion of $(a + b)^p$ are $\binom{p}{k} a^{p-k} b^k = pB_k a^{p-k} b^k = 0$. It follows by induction that $(a_1 + \cdots + a_m)^p = a_1^p + \cdots + a_m^p$, a pleasant computation! Another induction argument shows that the corresponding identity holds when p^n replaces p.

By Fermat's Theorem (see Theorem 4.8), for p a prime, $a^p = a$ in $\mathbf{Z}_p$, so for $f(x) = a_n x^n + \cdots + a_1 x + a_0 \in \mathbf{Z}_p[x]$, $f(x)^p = a_n x^{pn} + a_{n-1} x^{p(n-1)} + \cdots + a_1 x^p + a_0$ follows. This has implications for factoring in $\mathbf{Z}_p[x]$. For example, in $\mathbf{Z}_2[x]$, $x^4 + 1 = (x^2 + 1)^2 = (x + 1)^4$, and in $\mathbf{Z}_5[x]$, $2x^{15} - 3x^{10} + 2x^5 - 1 = (2x^3 - 3x^2 + 2x - 1)^5$. If $p(x) = x^6 + x^5 + x^4 + x^3 + x^2 + x + 1 \in \mathbf{Z}_7[x]$, then $(x - 1)p(x) = x^7 - 1 = (x - 1)^7$ and since

$\mathbf{Z}_7[x]$ is a domain by Theorem 10.11, it follows that $p(x) = (x-1)^6$ in $\mathbf{Z}_7[x]$, a somewhat unexpected result. Similarly, for any prime q, in $\mathbf{Z}_q[x]$, $x^{q-1} + x^{q-2} + \cdots + x + 1 = (x-1)^{q-1}$.

EXERCISES

1. For any ring R, show the equivalence of:
 i) R is a domain
 ii) $a, b, c \in R$, $ab = ac$ and $a \neq 0 \Rightarrow b = c$
 iii) $a, b, c \in R$, $ba = ca$ and $a \neq 0 \Rightarrow b = c$

2. Argue that for a field F, $QF(F[x])$ can be identified with a subfield of $F\langle\langle x\rangle\rangle$.

3. If F is a field and we regard $F[[x]] \subseteq F\langle\langle x\rangle\rangle$, show that $QF(F[[x]])$ can be identified with $F\langle\langle x\rangle\rangle$.

4. Let $m, n \in \mathbf{N}$ with $m \mid n$. In $\mathbf{Z}_n$, consider the subring $S(m) = \langle\{[m]_n\}\rangle$.
 i) Show that $S(m) = \langle [m]_n \rangle \leq (\mathbf{Z}_n, +)$.
 ii) If $(m, n/m) = 1$, show that $S(m)$ has an identity element.
 iii) If $n/m = p$ is prime and $(m, p) = 1$, show that $S(m)$ is a field.

5. If R is a commutative ring with identity, $f(x) = a_k x^k + \cdots + a_1 x + a_0 \in R[x]$, and the subring $\langle\{a_k, \ldots, a_0\}\rangle = R$, show that $f(x)$ is not a zero divisor in $R[x]$.

6. Let $m \in \mathbf{N}$, R a ring, and $R(m) = \{r \in R \mid m \cdot r = 0\}$.
 i) Show that $R(m)$ is a subring of R.
 ii) If $m, k \in \mathbf{N}$ with $(m, k) = 1$, show that $R(m) \cap R(k) = \langle 0 \rangle$.
 iii) If $char\ R = p_1^{a_1} \cdots p_d^{a_d}$ for $p_1 < \cdots < p_d$ primes, $x \in R(p_i^{a_i})$, and $y \in R(p_j^{a_j})$ for $i \neq j$, show that $xy = 0$.
 iv) Show $R \cong R(p_1^{a_1}) \oplus \cdots \oplus R(p_d^{a_d})$ as Abelian groups under addition.

7. Let $n \in \mathbf{N} - \{1\}$ and for any prime $p \mid n$, let $\mathbf{Z}_n(p) = \{[a] \in (\mathbf{Z}_n, +) \mid o([a]) = p^k$, some $k \geq 0\}$ be the p-Sylow subgroup of $(\mathbf{Z}_n, +)$. Show that:
 a) $\mathbf{Z}_n(p)$ is a subring of $\mathbf{Z}_n$.
 b) $\mathbf{Z}_n(p)[x]$ is a subring of $\mathbf{Z}_n[x]$.
 c) If $f(x) \in \mathbf{Z}_n(p)[x]$, $g(x) \in \mathbf{Z}_n(q)[x]$, and $p \neq q$, then $f(x)g(x) = 0$.
 d) As additive groups, $\mathbf{Z}_n[x] \cong \mathbf{Z}_n(p_1)[x] \oplus \cdots \oplus \mathbf{Z}_n(p_k)[x]$ for $n = p_1^{a(1)} \cdots p_k^{a(k)}$.

8. Consider $R = \oplus_i \mathbf{Z}_i$ as in Exercise 1 of §10.2. Show that any $r \in R$ is an n-torsion element for some $n \in \mathbf{N}$ but that R does not have a characteristic.

9. If R is a finite ring, show that R has a characteristic.

10. If F is a finite field, show that $|F| = p^n$ for p a prime.

11. If D is a domain with no proper nonzero subring, show that $|D| = p$, a prime. (Show that D is commutative and then that $1 \in D$ by considering aD for $a \in D$.)

12. If R is a ring and each proper nonzero subring of R is a domain, show that either: R is a domain; R has trivial multiplication and $|R| = p$, a prime; or R contains subrings A and B that are fields, $(R, +) = (A, +) \oplus (B, +)$, $ab = ba = 0$ for any $a \in A$ and $b \in B$, and $|A|$ and $|B|$ are primes. (Show that R has trivial multiplication $\Rightarrow |R| = p$. Then assume R neither is a domain nor has trivial multiplication and show that:
 i) $a \in R^*$ with $a^2 = 0 \Rightarrow a = 0$
 ii) $a, b \in R^*$ and $ab = 0$ imply $ba = 0$ and for any $r \in R$, $arb = bra = 0$
 iii) $ab = 0$ for $a, b \in R^* \Rightarrow R = \langle\{a, b\}\rangle = (\langle\{a\}\rangle, +) \oplus (\langle\{a\}\rangle, +)$.
 Now Apply Exercise 11.

13. For p a prime, set $q_p(x) = x^{p-1} + x^{p-2} + \cdots + x^2 + x + [1] \in \mathbf{Z}_p[x]$. Using that $q_p(x) = (x-1)^{p-1} \in \mathbf{Z}_p[x]$, show that $\binom{p-1}{k} \equiv (-1)^k \pmod p$.

14. If R is a commutative ring and $charR = 3$, is $\{a^3 \in R \mid a \in R\}$ a subring?

15. If F is a finite field, find $p(x) \in F[x]^*$ with $p(a) = 0$, all $a \in F$ (Exercise 11 in §10.3).

16. If R is a ring and $x^2 = x$ for all $x \in R$, show that $charR = 2$ and R is commutative.

17. Assume that R is a ring and $x^3 = x$ for all $x \in R$. Show the following:

a) if $r \in R$ and $r^2 = 0$ then $r = 0$

b) $x^4 = x^2$ for all $x \in R$

c) $x, y \in R \Rightarrow x^2 y = x^2 y x^2$, and then $x^2 \in Z(R)$

d) $x \in R \Rightarrow 2x, 3x \in Z(R)$ (consider $(x + x^2)^2$ and $(x + x^2)^3$)

e) R is commutative and $6x = 0$ for all $x \in R$

10.5 INDETERMINATES AS FUNCTIONS

We have avoided a formal definition of indeterminates, the question of whether they always exist, and a discussion of how they can be found. Also, when $1 \notin R$, "x" is not in $R[x]$, so does it really make sense to label it as an indeterminate and think of x^k as a power of x? Conceivably, any symbol we could think of is already in R and so is unavailable to be used as an indeterminate. Simply choosing an element not in R as an indeterminate over R may not work. An indeterminate over R must not have any relations with elements of R, but this rather imprecise notion may be difficult to verify. If we choose $5^{1/2} \in \mathbf{R} - \mathbf{Z}$ as an indeterminate over $\mathbf{Z}$ then $(5^{1/2})^2 = 5$ in $\mathbf{R}$ so it would seem that the polynomials $(5^{1/2})^2 - 5$ and 0 are equal in $\mathbf{Z}[5^{1/2}]$ (Example 10.4). This difficulty will not arise for $\pi \in \mathbf{R} - \mathbf{Z}$ but it is not easy to show that no polynomial expression in π over $\mathbf{Z}$ is zero.

We present a construction for polynomials and the related rings in §10.3 that does not depend on choosing indeterminates. Observe that we can think of $R[x]$ as functions. For $f(x) = r_n x^n + \cdots + r_1 x + r_0 \in R[x]$, associate the function $f : (\mathbf{N} \cup \{0\}) \to R$ given by $f(j) = r_j$, where $r_j = 0$ if $j > n$. We will define the *set* $R[X]$ in many indeterminates as functions and then define an addition and multiplication on these corresponding to the familiar "$+$" and "$\cdot$" for polynomials (see Theorem 10.9). The details will not be given.

Set $\mathbf{N} \cup \{0\} = \mathbf{N}+$. In identifying $R[x] \subseteq R^{\mathbf{N}+} = \{f : \mathbf{N}+ \to R\}$ as above, the monomials of $R[x]$ correspond to the domain $\mathbf{N}+$ of the functions. Similarly, to define polynomials in n variables over a ring R as functions contained in some R^A (Theorem 10.3), we want the domain A to correspond to all the monomials. We use $A = (\mathbf{N}+)^n \subseteq \mathbf{Z}^n$ for our domain. Now let $R[x_1, x_2, \ldots, x_n] = \{f \in R^A \mid f(\alpha) = 0_R$ for all but finitely many $\alpha \in A\}$. By Example 10.6 this is a subring of R^A but we take $R[x_1, \ldots, x_n]$ only as the *set* of functions, not the ring, although we will use the addition in R^A.

For any $\alpha \in A$ and $r \in R$, define $f_{r\alpha} \in R^A$ by $f_{r\alpha}(\beta) = 0$ for all $\beta \neq \alpha = (a_1, \ldots, a_n)$ and $f_{r\alpha}(\alpha) = r$. Denote $f_{r\alpha}$ by the *monomial* $r x_1^{a_1} x_2^{a_2} \cdots x_n^{a_n}$. Define addition in $R[x_1, \ldots, x_n]$ to be the addition in $R^A : (f + g)(\alpha) = f(\alpha) + g(\alpha)$ in R. It follows that for $f \in R[x_1, \ldots, x_n]$, $f(\alpha) \neq 0 \Leftrightarrow \alpha \in \{\alpha_1, \ldots, \alpha_k\}$ and if $f(\alpha_i) = r_i$ then $f = \sum f_{r_i \alpha_i}$ and so may be written familiarly as the sum of the monomials corresponding to the $f_{r_i \alpha_i}$. Multiplication is defined as in Theorem 10.9 as follows: for any $\beta, \gamma \in A$ we consider $\beta + \gamma \in \mathbf{Z}^n$ and set $f \cdot g(\alpha) = \sum f(\beta)g(\gamma)$ over all $\beta, \gamma \in A$ with $\beta + \gamma = \alpha$. This sum is finite since for $\alpha \in A$ there are only finitely many β and γ whose sum is α because all three of these have nonnegative entries. Also $f \cdot g(\alpha) = 0$ for all but finitely many $\alpha : f(\beta)g(\gamma) \neq 0_R$ for only finitely many $\beta, \gamma \in A$ so the set of their sums $\{\beta + \gamma\}$ is finite. It is an exercise to see that $R[x_1, \ldots, x_n]$ is a ring with these operations, is commutative exactly when R is, and has an identity exactly when $1 \in R$, namely the function $I(\beta) = 0$ for all $\beta \neq z = (0, \ldots, 0)$ and $I(z) = 1$. When $1 \in R$ we denote by x_j the function f_j satisfying $f_j(\varepsilon_j) = 1$, where ε_j is the n-tuple with 1 in the jth position and zeros elsewhere, and $f_j(\alpha) = 0$ otherwise. In this case $\{x_1, \ldots, x_n\}$ are n indeterminates over R.

Let us see explicitly that the multiplication we have just defined for $R[x_1, \ldots, x_n]$ corresponds to the usual polynomial multiplication. In $R[x]$ ax^i and bx^j denote the functions $f, g : N+ \to R$ satisfying $f(k) = a$ if $k = i$ and $f(k) = 0$ when $k \neq i$, and $g(k) = b$ when $k = j$ but $g(k) = 0$ otherwise. From our definition, $f \cdot g(k) = f(0)g(k) + f(1)g(k - 1) + \cdots + f(k)g(0)$, so $f \cdot g(k) = 0$ unless it contains the product $f(i)g(j)$. This happens exactly when $k = i + j$ in which case $f \cdot g(i + j) = f(i)g(j) = ab$. Thus $f \cdot g$ is represented by abx^{i+j}. In $R[x, y]$ (we let $x = x_1$ and $y = x_2$), if af_x and bf_y are the functions $af_x((1, 0)) = a$, $af_x(\alpha) = 0$ for all other $\alpha \in (N+)^2$, $bf_y((0, 1)) = b$, and $bf_y(\alpha) = 0$ for all other $\alpha \in (N+)^2$, then $g = af_x af_x bf_y(\alpha) = \sum af_x(\beta)af_x(\gamma)bf_y(\mu)$ over all $\beta + \gamma + \mu = \alpha$. But the only way this sum is not zero is for $\beta = \gamma = (1, 0)$ and $\mu = (0, 1)$, so $g(\alpha) = 0$ unless $\alpha = (2, 1)$ in which case $g(\alpha) = a^2 b$. Thus g is represented by the monomial a^2bx^2y, the usual product of ax, ax, and by. We mentioned above that each element in $R[x, y]$ is a finite sum $\sum f_{r(\alpha)\alpha} = \sum r(\alpha)X^\alpha$ of monomials. To find the term $f_{p \cdot q((3,2))(3,2)}$, or $p \cdot q((3, 2))x^3y^2$, in a product $p \cdot q$, we compute $p \cdot q((3, 2))$ as the sum of the twelve possible products $p((3, 2))q((0, 0)) + p((2, 2))q((1, 0)) + p((3, 1))q((0, 1)) + \cdots + p((0, 0))q((3, 2))$. This sum adds the product of the coefficient of x^3y^2 in p and the constant of q to the product of the coefficients of x^2y^2 in p and x in q to the product of the coefficients of x^3y in p and y in q, etc. This is the result of the usual multiplication of polynomials.

In $R[x_1, \ldots, x_n]$ there is a clear bijection between the symbols $x_1^{a_1}x_2^{a_2} \cdots x_n^{a_n}$ and the n-tuples of their exponents in $(N+)^n$. If we want to define polynomials in infinitely many variables X, the monomials will look essentially the same but there are infinitely many possibilities for the variables in any given monomial. Therefore, to define $R[X] \subseteq R^A$ we need A to be "infinite tuples" with only finitely many nonzero entries. For example, to get polynomials in $X = \{x_i\}_N$ we need tuples like $(1, 0, 1, 2, 0, 0, \ldots)$ corresponding to $x_1x_3x_4^2$ and $(3, 0, 0, 2, 1, 1, 0, 0, \ldots)$ corresponding to $x_1^3x_4^2x_5x_6$. Then $R[X]$ will be those functions from these tuples to R that have only finitely many nonzero values, and so those corresponding to finite sums of monomials. Even this will not work for $X = \{x_r\}_{\mathbf{R}}$ since we cannot write out sequences corresponding to all the elements in $\mathbf{R}$ (see Theorem 0.19).

The approach to defining $R[X]$ for X of arbitrary size is simply to extend Definition 10.13 and Theorem 10.9 from the case when X is finite.

DEFINITION 10.21 Let R be a ring, $\Lambda \neq \varnothing$, and $B = \{f \in (N+)^\Lambda \mid f(\lambda) = 0$ for all but finitely many $\lambda \in \Lambda\}$. The **set of polynomials over R** in $X = \{x_\lambda\}_\Lambda$ is $R[x] = R[\{x_\lambda\}_\Lambda] = \{f \in R^B \mid f(b) = 0_R$ for all but finitely many $b \in B\}$.

Each $b \in B$ assigns positive integers to each of finitely many $\lambda \in \Lambda$ and so naturally corresponds to a monomial in the variables $\{x_\lambda\}_\Lambda$, although the use of the symbols x_λ

is formal and unnecessary to represent the functions in $R[\{x_\lambda\}_\Lambda]$. For example, if $b \in B$ satisfies $b(\lambda)=2$, $b(\gamma)=1$, $b(\beta)=3$, and $b(\mu)=0$ for $\mu \in \Lambda - \{\lambda, \gamma, \beta\}$, then $f_b \in R^B$ defined by $f_b(b)=r$ and $f_b(b')=0_R$ for $b' \neq b$ corresponds to the monomial $rx_\lambda^2 x_\gamma x_\beta^3$. Alternatively, we can think of b itself as $x_\lambda^2 x_\gamma x_\beta$. Now $(\boldsymbol{N}+)^\Lambda \subseteq \mathbf{Z}^\Lambda$ and $\mathbf{Z}^\Lambda$ is a ring by Theorem 10.3, so the pointwise addition in $\mathbf{Z}^\Lambda$ allows us to add elements in B. For $b, b' \in B$, $(b+b')(\lambda)=b(\lambda)+b'(\lambda) \geq 0$, so $b+b' \in B$. Furthermore, for any $b \in B$, since $b(\lambda)=0$ for all but finitely many $\lambda \in \Lambda$, there are only finitely many ways of writing $b=b'+b''$ for $b', b'' \in B$ (why?).

THEOREM 10.16 $R[\{x_\lambda\}_\Lambda]$ is a ring with $(f+g)(b)=f(b)+g(b)$ for all $b \in B$ and $f \cdot g(b)=\sum f(c)g(d)$ over all $c, d \in B$ with $c+d=b$. If $1 \in R$ then $R[\{x_\lambda\}_\Lambda]$ has an identity element, and when R is commutative so is $R[\{x_\lambda\}_\Lambda]$.

Proof We leave most of the verifications as exercises. Observe that the operations make sense. That is, for $f, g \in R[\{x_\lambda\}_\Lambda]$, $f+g \in R[\{x_\lambda\}_\Lambda]$ by Example 10.6, and $f \cdot g \in R[\{x_\lambda\}_\Lambda]$ because for each $b \in B$ there are only finitely many $u, v \in B$ satisfying $u+v=b$. Also there are only finitely many $f(c), g(d) \neq 0$ and these c and d give rise to only finitely many sums $c+d$. The conditions in Definition 10.1 are fairly straightforward to verify. The hardest of these to prove is associativity of multiplication, which we write out. To see that $(f \cdot g) \cdot h = f \cdot (g \cdot h)$, we need these to have the same value on each $b \in B$. Using the definition of multiplication in $R[\{x_\lambda\}_\Lambda]$ and the associativity of multiplication in R,

$$\begin{aligned}(f \cdot g) \cdot h(b) &= \sum_{u+v=b} (f \cdot g)(u)h(v) = \sum_{u+v=b} \left(\sum_{a+c=u} f(a)g(c) \right) h(v) \\ &= \sum_{a+c+v=b} (f(a)g(c))h(v) = \sum_{a+c+v=b} f(a)(g(c)h(v)) \\ &= \sum_{a+q=b} f(a) \sum_{c+v=q} g(c)h(v) = \sum_{a+q=b} f(a)g \cdot h(q) = f \cdot (g \cdot h)(b)\end{aligned}$$

When R is commutative, $f \cdot g(b)=\sum f(c)g(d)=\sum g(d)f(c)=g \cdot f(b)$ so $R[\{x_\lambda\}_\Lambda]$ is commutative. If $1 \in R$, define $b_0 \in B$ by $b_0(\lambda)=0$ for all $\lambda \in \Lambda$, and let $f_0 \in R[\{x_\lambda\}_\Lambda]$ be given by $f_0(b)=0$ when $b \neq b_0$ and $f_0(b_0)=1$. Then $f_0 \cdot g(b)=\sum f_0(c)g(d)$ for $c+d=b$, but $f_0(c)=0$ unless $c=b_0$, and $b_0+b=b$ for all $b \in B$, so $f_0 \cdot g(b)=f_0(b_0)g(b)=g(b)$. Similarly, $g \cdot f_0=g$, showing that f_0 is an identity for $R[\{x_\lambda\}_\Lambda]$. ■

Why denote the functions in Definition 10.21 by $R[\{x_\lambda\}_\Lambda]$? Just as for $R[x_1, \ldots, x_n]$, it is usually easier to think about the elements in $R[\{x_\lambda\}_\Lambda]$ as polynomials, and there is a natural bijection between the functions in $R[\{x_\lambda\}_\Lambda]$ and sums of monomials. Let $b \in B$ and suppose that $b(\lambda)=0$ unless $\lambda \in \{\lambda_1, \ldots, \lambda_k\}$, and then $b(\lambda_i)=n_i$. Let $f_{rb}=rx_{\lambda_1}^{n_1} \cdots x_{\lambda_k}^{n_k} \in R[\{x_\lambda\}_\Lambda]$ denote the function that sends b to r and all other elements of B to zero. We are using two notations for this same function, one made to look like a monomial. Now any $f \in R[\{x_\lambda\}_\Lambda]$ is zero on all $b \in B$ except for some set $\{b_1, \ldots, b_m\}$ and $f(b_j)=r_j$. It follows from the definition of addition that $f=f_{r_1b_1}+\cdots+f_{r_mb_m}$ and hence is a finite sum of monomials. When $1 \in R$ we let $x_\lambda=f_{1b_\lambda}$ for $b_\lambda \in B$ defined by $b_\lambda(\lambda)=1$ and $b_\lambda(\mu)=0$ for $\mu \neq \lambda$. This shows how we can define indeterminates in $R[\{x_\lambda\}_\Lambda]$.

It is an exercise to see that the multiplication of the f_{rb} in $R[\{x_\lambda\}_\Lambda]$ looks like the usual multiplication of monomials.

We have not given much detail and generally have not provided proofs of our claims and constructions for $R[X] = R[\{x_\lambda\}_\Lambda]$. These are tedious but not particularly difficult and not central to the development of the important material. Our point is to indicate how it is possible to obtain polynomial rings formally without reliance on the ambiguous term "indeterminate." In practice, it is best to think of polynomials, in any set $\{x_\lambda\}_\Lambda$, finite or infinite, as a finite sum of monomials with coefficients from R. Our discussion provides a better mathematical basis for writing these formal expressions. The rings $R[X]$ will arise both as examples and as objects of further study.

EXERCISES

1. As in Definition 10.21, for R a ring with 1, and $\Lambda \neq \emptyset$, let $B = \{b \in (N, +)^\Lambda \mid b(\lambda) = 0$ for all but finitely many $\lambda\}$ and $R[\{x_\lambda\}_\Lambda] = \{f \in R^B \mid f(b) = 0_R$ for all but finitely many $b\}$. For $\lambda \in \Lambda$, define $b_\lambda \in B$ by $b_\lambda(\lambda) = 1$ and $b_\lambda(\mu) = 0$ if $\mu \in \Lambda - \{\lambda\}$, and $f_\lambda \in R[\{x_\lambda\}_\Lambda]$ by $f_\lambda(b_\lambda) = 1_R$ and $f_\lambda(b) = 0_R$ if $b \in B - \{b_\lambda\}$. For any $b \in B$ define $f_b \in R[\{x_\lambda\}_\Lambda]$ by $f_b(b) = 1_R$ and $f_b(b') = 0$ if $b' \neq b$.

i) If $b \in B$, suppose $b(\lambda) = 0$ unless $\lambda \in \{\lambda_1, \ldots, \lambda_k\}$ and $b(\lambda_j) = n_j > 0$. Show that the multiplication in $R[\{x_\lambda\}_\Lambda]$ forces $f_b = f_{\lambda_1}^{n_1} \cdots f_{\lambda_k}^{n_k}$.

ii) If $f \in R[\{x_\lambda\}_\Lambda]$ and $r \in R$, then $rf \in R[\{x_\lambda\}_\Lambda]$ via $rf(b) = r \cdot f(b)$. Show that there is no algebraic relation between R and $\{f_\lambda \in R[\{x_\lambda\}_\Lambda] \mid \lambda \in \Lambda\}$: for any distinct $b_1, \ldots, b_k \in B$, if $\sum r_i f_{b_i} = 0 \in R[\{x_\lambda\}_\Lambda]$ then all $r_i = 0_R$.

2. Show how to define $R\langle\langle x\rangle\rangle$ (Definition 10.16) as functions, as we did for $R[X]$ in Definition 10.21.

3. Show how to define $R[[x_1, \ldots, x_n]]$ (Definition 10.15) as functions, as we did for $R[X]$ in Definition 10.21.

4. How can the free ring $R\{X\}$ over R (Definition 10.14) be represented as functions, as we did in Definition 10.21?

5. For R a ring, use Definition 10.21 with $\Lambda = \{1, 2\}$ to see that $R[x_1, x_2]$ can be identified with $(R[x_1])[x_2]$, as follows. First identify $\{b : \Lambda \to N+\}$ with $(N+)^2$, then for $f \in R[x_1, x_2]$ and $(i, j) \in (N+)^2$, let $\varphi(f) \in (R[x_1])[x_2]$ be given by $(\varphi(f)(j))(i) = f(i, j)$. Show that for $f, g \in R[x_1, x_2]$, $\varphi(f + g) = \varphi(f) + \varphi(g)$ and $\varphi(f \cdot g) = \varphi(f) \cdot \varphi(g)$. For $F \in (R[x_1])[x_2]$ and $i, j \in N+$, define $\eta(F) \in R[x_1, x_2]$ by $\eta(F)(i, j) = (F(j))(i)$. Show that φ and η are inverses, so φ is a bijection.

6. Use the idea in Exercise 5 to see that $R[\{x_\lambda\}_\Lambda \cup \{x_\gamma\}_\Gamma]$ can be identified with $(R[\{x_\lambda\}_\Lambda])[\{x_\gamma\}_\Gamma]$, where $\Lambda, \Gamma \neq \emptyset$ and $\Lambda \cap \Gamma = \emptyset$.

CHAPTER 11

IDEALS, QUOTIENTS, AND HOMOMORPHISMS

Normal subgroups are vital for studying finite groups, using the quotient construction and induction on the group order. Homomorphic images of groups are identified with quotient groups via the kernel of the homomorphism, a normal subgroup. The role of normal subgroups in groups is played in rings by *ideals*. Our investigation of rings focuses on infinite rings so induction is not helpful. However, the notions of quotient and homomorphism are still important for rings, and their development here is closely related to the material in Chapter 5 and Chapter 6. Indeed, with modest additional verification or comment, many of the earlier constructions and arguments for groups extend to rings.

11.1 IDEALS

DEFINITION 11.1 For R a ring, $I \leq (R, +)$ is a **left ideal** of R, written $I \lhd_l R$, if $rx \in I$ for all $x \in I$ and $r \in R$; I is a **right ideal** of R, written $I \lhd_r R$, if $xr \in I$ for all $x \in I$ and $r \in R$; and I is an **ideal** of R, written $I \lhd R$, if $xr, rx \in I$ for all $x \in I$ and $r \in R$.

The notation $I \lhd R$ is consistent with the idea that ideals correspond to normal subgroups in groups, especially for the quotient construction in the next section. There will be no confusion in usage since when talking about rings we will never write $I \lhd R$ to mean merely that $I \lhd (R, +)$, but only that I is an ideal. Also $(R, +)$ is an Abelian group so all of its subgroups are normal.

It is immediate that any (left, right) ideal of R is a subring and that when R is commutative there is no difference between a left ideal, a right ideal, and an ideal. Sometimes we will refer to an ideal as a **two-sided ideal.** If $I \lhd R$ then for any $x \in I$ and $a, b \in R$, $axb \in I$ since $xb \in I$ and then $axb = a(xb) \in I$. At the extremes are $\langle 0 \rangle$, $R \lhd R$ in any ring, using Theorem 10.1 for $\langle 0 \rangle$. If R is a ring with trivial multiplication $ab = 0$, then any $A \leq (R, +)$ is an ideal, so the notion of ideal is not very interesting here. The subgroups of $(\mathbf{Z}, +)$ are just the various $n\mathbf{Z}$ (Theorem 2.13) and these are ideals of $\mathbf{Z}$ since from Exercise 6 in §10.2 $na \cdot b = n \cdot ab \in n\mathbf{Z}$ for any $b \in \mathbf{Z}$. Note that $\mathbf{Z}$ is a subring of $\mathbf{Q}$, but not an ideal since $(1/2) \cdot 3 \notin \mathbf{Z}$. There is an easy way to obtain one-sided ideals.

DEFINITION 11.2 Let R be a ring and $a \in R$. Set $aR = \{ar \in R \mid r \in R\}$, $Ra = \{ra \in R \mid r \in R\}$, $\mathbf{Z} \cdot a + aR = \{n \cdot a + ar \in R \mid n \in \mathbf{Z} \text{ and } r \in R\}$, and $\mathbf{Z} \cdot a + Ra = \{n \cdot a + ra \mid n \in \mathbf{Z} \text{ and } r \in R\}$.

When $1_R \in R$ then $na = n(1_R \cdot a) = (n \cdot 1_R)a$ and also $na = a(n \cdot 1_R)$ (why?) so $aR = \mathbf{Z} \cdot a + aR$ and $Ra = \mathbf{Z} \cdot a + Ra$. Otherwise we may have $a \notin aR$ (or Ra). For example, if $R = 2\mathbf{Z}$ then $4 \notin 4R = 8\mathbf{Z}$. The main difference between aR and $\mathbf{Z} \cdot a + aR$ is that $a \in \mathbf{Z} \cdot a + aR$ when R does not have an identity. We verify that the sets just defined are right or left ideals of R.

Example 11.1 If R is a ring and $a \in R$, then aR is the **principal right ideal generated by $a \in R$**, $\mathbf{Z} \cdot a + aR$ is the **right ideal generated by $a \in R$**, Ra is the **principal left ideal generated by $a \in R$**, and $\mathbf{Z} \cdot a + Ra$ is the **left ideal generated by $a \in R$**. When R is commutative we write $(a) = \mathbf{Z} \cdot a + aR$.

From Theorem 10.1, $a \cdot 0 = 0 = 0 \cdot a$, so none of the four sets is empty. If $ar, as \in aR$, then $ar - as = a(r - s) \in aR$ and $aR \leq (R, +)$ by Theorem 2.12. For all $x \in R$, $(ar)x = a(rx) \in aR$, so aR is a right ideal of R by Definition 11.1. Similarly, if $n \cdot a + ra, m \cdot a + sa \in \mathbf{Z} \cdot a + Ra$ and $x \in R$ then $(n \cdot a + ra) - (m \cdot a + sa) = n \cdot a - m \cdot a + ra - sa = (n - m) \cdot a + (r - s)a \in \mathbf{Z} \cdot a + Ra$ and $x(n \cdot a + ra) = x(n \cdot a) + x(ra) = (n \cdot x)a + (xr)a = (n \cdot x + xr)a \in Ra \subseteq \mathbf{Z} \cdot a + Ra$, using Exercise 6 in §10.2, so $\mathbf{Z} \cdot a + Ra$ is a left ideal of R. The other two verifications are left as exercises.

To see again that we must distinguish $\mathbf{Z} \cdot a + aR$ from aR, let $R = \{f(x, y) \in \mathbf{Z}[x, y] \mid f(0, 0) = 0\}$, a subring of $\mathbf{Z}[x, y]$ (exercise!). In R, xR is the set of those polynomials all of whose monomials contain x^2 or xy, and (x) adjoins to xR those polynomials having monomials in $\mathbf{Z}x$. Neither contains polynomials with any y^k as a monomial.

Generating ideals in noncommutative rings is more complicated, especially when $1 \notin R$. It is straightforward to verify that for any $a \in R$, $RaR = \{r_1as_1 + \cdots + r_kas_k \mid k \in \mathbf{N}$ and $r_j, s_j \in R\} \lhd R$. Also $aR + Ra + RaR = \{ar + sa + b \mid r, s \in R, b \in RaR\} \lhd R$, the **principal ideal generated by a.** The minimal ideal containing a when R does not have an identity is the **ideal generated by a,** $(a) = \mathbf{Z}a + aR + Ra + RaR$. We omit the verification that this is an ideal containing a and that $(a) = RaR$ when $1 \in R$. An example will illustrate the differences between left ideals, right ideals, and ideals.

Example 11.2 Let $R = M_n(\mathbf{Z})$ for $n > 1$, $L = \{A \in R \mid A_{i1} = 0, \text{ all } i\}$, $T = \{A \in R \mid A_{1j} = 0, \text{ all } j\}$, and $I = \{A \in R \mid A_{ij} \in 3\mathbf{Z} \text{ for all } i, j\} = M_n(3\mathbf{Z})$. Then $L \lhd_l R$ but is not a right ideal, $T \lhd_r R$ but is not a left ideal, and $I \lhd R$.

Now L and T are subrings by Example 10.7, so $L, T \leq (R, +)$. If $A \in L$ and $B \in R$ then $(BA)_{i1} = B_{i1}A_{11} + \cdots + B_{in}A_{n1} = 0$ since all $A_{j1} = 0$, so $BA \in L$ and $L \lhd_l R$. Because $E_{22} \in L$ but $E_{21} = E_{22}E_{21} \notin L$, L is not a right ideal. Similarly, for

$A \in T$ and $B \in R$, $(AB)_{1j} = A_{11}B_{1j} + \cdots + A_{1n}B_{nj} = 0$ since all $A_{1k} = 0$, so $T \lhd_r R$. Again $E_{22} \in T$ but $E_{12} = E_{12}E_{22} \notin T$ and T is not a left ideal. When $A, B \in I$, $(A - B)_{ij} = A_{ij} - B_{ij} \in 3\mathbf{Z}$, since $3\mathbf{Z} \leq (\mathbf{Z}, +)$. Finally, for any $U, V \in R$, $(UV)_{ij} = U_{i1}V_{ij} + \cdots + U_{in}V_{nj}$ so if all $U_{ij} \in 3\mathbf{Z}$, or if all $V_{ij} \in 3\mathbf{Z}$, then all $(UV)_{ij} \in 3\mathbf{Z}$ and it follows that $UV \in I$ if either $U \in I$ or $V \in I$, and thus $I \lhd R$.

Actually L and T of Example 11.2 are principal: $L = M_n(\mathbf{Z})(I_n - E_{11})$ and $T = (I_n - E_{11})M_n(\mathbf{Z})$. The ideals of $\mathbf{Z}$ are just the $n\mathbf{Z}$, as we mentioned above, and in the last example, $M_n(3\mathbf{Z}) \lhd M_n(\mathbf{Z})$. The same approach shows $M_n(k\mathbf{Z}) \lhd M_n(\mathbf{Z})$ also. Are these the only ideals of $M_n(\mathbf{Z})$? The answer, characterizing ideals in $M_n(R)$ when $1 \in R$, follows from our next theorem.

THEOREM 11.1 If R is a ring with 1, and $T \lhd M_n(R)$, then $T = M_n(I)$ for $I \lhd R$.

Proof Since $T \leq (M_n(R), +)$, $0_n \in T$ and $T = (0_n)$ implies $T = M_n((0))$. Let $0_n \neq A \in T$ and note that since $1 \in R$, all $E_{ij} \in M_n(R)$. Using the definition of ideal, $E_{1i}A \in T$ and then $E_{1i}AE_{j1} = A_{ij}E_{11} \in T$. Set $I = \{r \in R \mid r = A_{ij}$ for $A \in T$ and some $1 \leq i, j \leq n\}$. We just saw that $IE_{11} = \{rE_{11} \in M_n(R) \mid r \in I\} \subseteq T$. We claim that $I \lhd R$ and $T = M_n(I)$. If $a, b \in I$ then $aE_{11}, bE_{11} \in T$ (why?) so $C = (a - b)E_{11} = aE_{11} - bE_{11} \in T$. Thus $a - b = C_{11} \in I$ and $I \leq (R, +)$. For $r \in R$, $rE_{11} \in M_n(R)$, so both $raE_{11} = rE_{11}aE_{11} \in T$ and $arE_{11} = aE_{11}rE_{11} \in T$ follow from $T \lhd M_n(R)$. Therefore $ar, ra \in I$, showing $I \lhd R$. Using again that T is an ideal, that $IE_{11} \subseteq T$, and that $E_{ij} \in R$, we have $IE_{i1} = E_{i1}IE_{11} \subseteq T$ and then $IE_{ij} = IE_{i1}E_{1j} \subseteq T$. Adding these elements over all subscripts shows that $M_n(I) \subseteq T$. But for any $A \in T$, $A = \sum A_{ij}E_{ij} \in M_n(I)$, proving that $T = M_n(I)$. ∎

The theorem is false when $1 \notin R$: let $R = M_2(2\mathbf{Z})$ and $I = \{A \in R \mid A_{12} \in 4\mathbf{Z}\}$. After checking that $I \leq M_2(2\mathbf{Z})$, compute that for $A, B \in M_2(2\mathbf{Z})$, $AB \in M_2(4\mathbf{Z}) \subseteq I$, so $I \lhd R$. Now Theorem 11.1 is useful only to the extent that we can describe the ideals of a ring R. This is not usually easy to do, although it was for $\mathbf{Z}$. The ideals of fields are easy to describe.

THEOREM 11.2 A commutative ring R with identity is a field $\Leftrightarrow$ R and (0) are the only ideals of R. For F a field, $I \lhd M_n(F) \Leftrightarrow I = (0_n)$ or $I = M_n(F)$.

Proof Let $1 \in R$, R be a field, and $I \lhd R$. If $0 \neq a \in I$, then since R is a field, $a^{-1} \in R$ and $ra^{-1} \in R$ for any $r \in R$. But $I \lhd R$ so $r = r \cdot 1 = r(a^{-1}a) = (ra^{-1})a \in I$. Thus either $I = (0)$ or $I = R$. Now assume that (0) and R are the only ideals of R. We must show that every $0 \neq a \in R$ has an inverse under multiplication. For such an a, $aR \lhd R$ by Example 11.1 and $a = a \cdot 1 \in aR$, so $aR \neq (0)$, forcing $aR = R$. Hence $1 \in R = aR$ implies $1 = ab$ for some $b \in R$, $b = a^{-1}$, and R is a field. Now apply Theorem 11.1 to complete the proof. ∎

The proof shows that *a proper ideal cannot contain a unit of* R. By the theorem, the matrix rings $M_n(\mathbf{Z}_p)$ for p a prime, $M_n(\mathbf{Q})$, $M_n(\mathbf{R})$, $M_n(\mathbf{C})$, $M_n(\mathbf{Q}(x))$, and $M_n(\mathbf{Q}\langle\langle x\rangle\rangle)$ have no proper nonzero ideals. Any ring satisfying this property and not having trivial multiplication is called a **simple ring.** Thus, another description of a field is a simple commutative ring. An important type of ring for which we can describe all of the ideals is the ring of polynomials $F[x]$ for F a field.

THEOREM 11.3 If F is a field and $I \lhd F[x]$, then $I = (0)$ or $I = (f(x)) = f(x)F[x]$ for $f(x)$ the unique monic polynomial of least degree in I.

Proof Assume $I \neq (0)$. By the Well Ordering of $\mathbf{N}$, there is $g(x) = ax^k + \cdots + cx + d \in I$ with $a \neq 0$, of least degree k. Since $I \lhd F[x]$, $f(x) = a^{-1}g(x) \in I$ and $f(x) = x^k + \cdots + a^{-1}cx + a^{-1}d$ is monic. If $h(x) \in I$ is also monic of degree k, then $f(x) = h(x)$ or $f(x) - h(x) \in I$ has degree less than k. The minimality of k forces $f(x) = h(x)$, so $f(x)$ is the unique monic polynomial in I of smallest degree. Now $(f(x)) = f(x)F[x] = \{f(x)g(x) \in F[x] \mid g(x) \in F[x]\} \subseteq I$ by the definition of ideal, so it remains to show that $I \subseteq (f(x))$. If not, use Well Ordering again to find $p(x) = a_m x^m + \cdots + a_1 x + a_0 \in I - (f(x))$ of least degree $m \geq k$. Since $(f(x)) \subseteq I \lhd F[x]$, $q(x) = p(x) - a_m x^{m-k} f(x) \in I$ and either $q(x) = 0$ or $deg\ q(x) < m$, forcing $q(x) \in (f(x))$. In the first case, $p(x) = a_m x^{m-k} f(x)$, and in the second, $p(x) = a_m x^{m-k} f(x) + h(x)f(x)$, so $p(x) \in (f(x))$. This contradiction proves $I = (f(x))$. ■

The theorem shows that just as in $\mathbf{Z}$, every ideal in $F[x]$ is principal. This property fails for $D[x]$ when D is an integral domain but not a field. If $(0) \neq I \lhd D$ with $I \neq D$ (say $2\mathbf{Z} \lhd \mathbf{Z}$) then $K = \{f(x) \in D[x] \mid f(0) \in I\} \lhd D[x]$ (verify!) but cannot be principal. If $K = (p(x))$, then for $0 \neq a \in I \subseteq K$, $a = p(x)g(x)$ and by Theorem 10.11 on degrees of products, $p(x) = b \in I$ (why?). But $x \in K$ so $x = bh(x)$ forces b to be a unit and $K = (b) = D[x]$. This contradicts the choice of I. For F a field, $F[x, y] = F[x][y]$ and $K = \{f(x, y) \in F[x, y] \mid f(x, 0) \in xF[x]\} \neq (p(x, y))$ since $x, y \in K$ and both $x = p(x, y)g(x, y)$ and $y = p(x, y)h(x)$ are impossible in $F[x][y]$ (Theorem 10.11!).

If $(f(x)), (g(x))$ are any ideals in $F[x]$, then $(f(x)) \subseteq (g(x)) \Leftrightarrow f(x) = g(x)h(x)$ (verify!) $\Leftrightarrow g(x)$ is a factor of $f(x)$. Similarly, in $\mathbf{Z}$, $a\mathbf{Z} \subseteq b\mathbf{Z}$ exactly when $b \mid a$, assuming $b \neq 0$. Thus a maximal proper ideal $(f(x))$ in $F[x]$ or (a) in $\mathbf{Z}$ requires $f(x)$ or a to have no proper factors. In $\mathbf{Z}$ these are the primes, and in $F[x]$ they are called **irreducibles.** The corresponding ideals are called maximal; this notion is useful in any ring.

DEFINITION 11.3 A proper ideal M of R is **maximal** if for any ideal I of R with $M \subseteq I$, either $I = M$ or $I = R$. Replacing *ideal* with *left ideal* or with *right ideal* defines a **maximal left ideal** or a **maximal right ideal.**

As we mentioned, $(m) \lhd \mathbf{Z}$ is maximal $\Leftrightarrow m$ is a prime, and for F a field, $(f(x)) \lhd F[x]$ is maximal $\Leftrightarrow f(x)$ is irreducible. Some irreducibles are $x^2 + x + 1 \in \mathbf{R}[x]$, $x^3 - 2$, $x^2 - 3x + 1 \in \mathbf{Q}[x]$, $x^2 + x + 1 \in \mathbf{Z}_2[x]$, and $x^2 + 1 \in \mathbf{Z}_{11}[x]$. Note that $x^2 + 1 \in \mathbf{Z}_p[x]$ is irreducible $\Leftrightarrow p \equiv_4 3$ by Example 4.16. These examples use the fact that $p(x) \in F[x]$ of degree 2 or 3 is irreducible $\Leftrightarrow p(x)$ has no root in F (why?). Unfortunately, this equivalence fails for larger degrees. For example, $x^4 + x^2 + 1 \in \mathbf{R}[x]$ has no real root but factors as $x^4 + x^2 + 1 = (x^2 + x + 1)(x^2 - x + 1)$. We put off a further discussion of irreducibility until Chapter 12. In view of Theorem 11.1, the maximal ideals in $M_n(\mathbf{Z})$ are $M_n(p\mathbf{Z})$ for p a prime, and the maximal ideals in $M_n(F[x])$ for F a field are $M_n((f(x)))$ for $f(x)$ irreducible. The left and right ideals L and T in Example 11.2 are not maximal, but slightly larger versions of these are: $T + 2E_{11}M_n(\mathbf{Z})$ is a maximal right ideal in $M_n(\mathbf{Z})$ (verify!).

Now $([2]_{30}), ([3]_{30}), ([5]_{30}) \lhd \mathbf{Z}_{30}$ are maximal: they are maximal additive subgroups. Note that these are generated by the prime factors of 30, which also generate maximal ideals in $\mathbf{Z}$. For $[a]_{27} \notin ([3]_{27}) \lhd \mathbf{Z}_{27}$, $(a, 3) = 1$ so $[a]_{27} \in U_{27}$ and cannot be in any proper ideal, as we saw in Theorem 11.2. Thus $([3]_{27})$ is the *unique* maximal ideal of $\mathbf{Z}_{27}$. A similar example is $(x) \lhd F[[x]]$ for F a field, all series with zero constant terms. Any $f(x) \notin (x)$ has a nonzero constant and so is invertible by Theorem 10.10; hence it cannot be in any proper ideal. Thus (x) is the unique maximal ideal in $F[[x]]$. To describe the general situation recall that $U(R)$ is the group of invertible elements in R.

THEOREM 11.4 If R is a commutative ring with identity, then some proper $M \lhd R$ contains all proper ideals of R if and only if $M = R - U(R)$ is an ideal.

Proof Assume a proper $M \lhd R$ contains all proper ideals of R. Then for $x \in R - M$, $x \in xR \lhd R$ by Example 11.1, and since $x \notin M$, $xR \not\subset M$. Thus xR is not a proper ideal, forcing $xR = R$, so $xy = 1$ for some $y \in R$. Hence $x \in U(R)$ and $R - M \subseteq U(R)$. Since a unit cannot be in any proper ideal, as we saw in the proof of Theorem 11.2, $U(R) \cap M = \varnothing$, resulting in $R - M = U(R)$. Equivalently, $M = R - U(R)$. When $R - U(R) = M \lhd R$, M contains *all* proper $I \lhd R$ since $x \in I$ implies $x \notin U(R)$. ■

Ideals are preserved under sums, intersections, and some unions.

THEOREM 11.5 Let R be a ring, $I_1, \ldots, I_n$, and $\{I_j\}_J$ right ideals of R.

i) $\bigcap_J I_j \lhd_r R$
ii) If $\{I_j\}_J$ is a chain with respect to inclusion, then $\bigcup_J I_j \lhd_r R$.
iii) $\sum_J I_j = \{x_{j(1)} + \cdots + x_{j(k)} \in R \mid k \in \mathbf{N} \text{ and all } x_{j(i)} \in \bigcup_J I_j\} \lhd_r R$
iv) $I_1 I_2 \cdots I_n = \{\sum_{i=1}^{k} y_{i1} y_{i2} \cdots y_{in} \in R \mid k \in \mathbf{N} \text{ and } y_{is} \in I_s\} \lhd_r R$

When each $I_\lambda \lhd_l R (I_\lambda \lhd R)$, each of these four constructions is also.

Proof Each $(I_j, +)$ is an Abelian group so Theorem 2.14 shows that the intersection in i) and the union in ii) are Abelian groups. If $x \in \bigcap I_j$ and $r \in R$, then for all $j \in J$, $x \in I_j$ so $xr \in I_j$, using $I_j \lhd_r R$. Hence $xr \in \bigcap I_j$, and $\bigcap I_j \lhd_r R$. Similarly, in ii), if $x \in \bigcup I_j$ then for some $j \in J$, $x \in I_j$ and again $xr \in I_j$ so $xr \in \bigcup I_j : \bigcup I_j \lhd_r R$. In iii), for $x_1 + \cdots + x_k, y_1 + \cdots + y_s \in \sum I_j$, since $y_i \in I_j$ implies $-y_i \in I_j$ as well, we have $x_1 + \cdots + x_k - (y_1 + \cdots + y_s) = \sum x_i + \sum(-y_i) \in \sum I_j$, so $\sum I_j \leq (R, +)$ by Theorem 2.12. Also if $r \in R$ then $x_i \in I_j$ and $I_j \lhd_r R$ force $x_i r \in I_j$; hence $\left(\sum x_i\right) r = \sum(x_i r) \in \sum I_j$ and $\sum I_j \lhd_r R$. For iv), using $(-a)b = -ab$ in R from Theorem 10.1, if $y_1 y_2 \cdots y_n \in I_1 I_2 \cdots I_n$ then $-y_1 y_2 \cdots y_n = (-y_1) y_2 \cdots y_n \in I_1 I_2 \cdots I_n$. Thus $\sum_{i=1}^{k} x_{i1} x_{i2} \cdots x_{in}$, $\sum_{i=1}^{s} y_{i1} y_{i2} \cdots y_{in} \in I_1 I_2 \cdots I_n \Rightarrow \sum_{i=1}^{k} x_{i1} x_{i2} \cdots x_{in} - \sum_{i=1}^{s} y_{i1} y_{i2} \cdots y_{in} = \sum_{i=1}^{k} x_{i1} x_{i2} \cdots x_{in} + \sum_{i=1}^{s} (-y_{i1}) y_{i2} \cdots y_{in}$ so $(I_1 I_2 \cdots I_n, +) \leq (R, +)$. When $r \in R$, $(\sum_{i=1}^{k} x_{i1} x_{i2} \cdots x_{in}) r = \sum_{i=1}^{k} x_{i1} x_{i2} \cdots (x_{in} r) \in I_1 I_2 \cdots I_n$ since $I_n \lhd_r R$. Hence $I_1 I_2 \cdots I_n \lhd_r R$, proving iv). If instead all I_j are left ideals or ideals, the arguments are essentially the same as those given and are left as exercises. ■

The product of ideals takes all *sums* of products of elements from each ideal. This differs from the product $HK = \{hk \in G \mid h \in H, k \in K\}$ of $H, K \leq G$ in the group G.

The difference results from the fact that there are two operations in rings. To ensure that the product of ideals is even a subring, we must be able to add products of elements. Another useful fact about products of ideals is that for $I_1, \ldots, I_k \lhd R$, $I_1 I_2 \cdots I_k \subseteq \bigcap I_j \subseteq I_j$ (why?).

To illustrate the theorem, let $R = M_2(\mathbf{Z})$ with right ideals $E_{11}R$ and $E_{22}R$. Using computations in matrices it follows that $E_{11}R \bigcap E_{22}R = (0_2)$, $E_{11}R + E_{22}R = R$, and $E_{11}RE_{22}R = E_{11}R$. Taking $2\mathbf{Z}, 3\mathbf{Z}, 4\mathbf{Z}, 6\mathbf{Z} \lhd \mathbf{Z}$, $2\mathbf{Z} \cdot 4\mathbf{Z} = 8\mathbf{Z}$ but $2\mathbf{Z} \bigcap 4\mathbf{Z} = 4\mathbf{Z}$; $3\mathbf{Z} + 4\mathbf{Z} + 6\mathbf{Z} = \mathbf{Z}$ since $-3 + 4 = 1$; and $3\mathbf{Z} \cdot 4\mathbf{Z} \cdot 6\mathbf{Z} = 72\mathbf{Z}$ but $3\mathbf{Z} \bigcap 4\mathbf{Z} \bigcap 6\mathbf{Z} = 12\mathbf{Z}$ (see Example 2.11). Although in general the intersection of ideals and their product are different, they can be the same, aside from when one of the ideals is (0) or R. For example, $2\mathbf{Z} \bigcap 3\mathbf{Z} = 6\mathbf{Z} = 2\mathbf{Z} \cdot 3\mathbf{Z}$. A general version of this last example uses Theorem 11.5 but first requires some easy consequences of the definition of products of ideals.

THEOREM 11.6 Let R be a ring and $I, J, K \lhd R$.

i) If $1 \in R$ then $R^2 = RR = R$ and $RI = I = IR$.
ii) If R is commutative then $IJ = JI$.
iii) $(I + K)(J + K) \subseteq IJ + K$

Proof All products of elements in R are in R so $RR \subseteq R$. If $1 \in R$ then for any $r \in R$, $r = 1 \cdot r \in RR$ so $R \subseteq RR$. Similarly, when $1 \in R$ both $I \subseteq RI$ and $I \subseteq IR$. The opposite inclusions follow from $I \lhd R$, proving i). For ii), the commutativity of R immediately gives $IJ = JI$ from the definition of IJ in Theorem 11.5. To prove iii) let $a \in I$, $b \in J$, $x, y \in K$. The distributive law and definition of ideal show that $(a + x)(b + y) = ab + ay + xb + xy \in IJ + K$. ■

THEOREM 11.7 Let R be a commutative ring, $1 \in R$, and $I_1, \ldots, I_k \lhd R$. If $I_j + I_m = R$ when $j \neq m$, then:

i) for any $1 \leq j < k$, $I_1 I_2 \cdots I_j + I_{j+1} = R$
ii) $I_1 I_2 \cdots I_k = I_1 \bigcap \cdots \bigcap I_k$.

Therefore if $I, J \lhd R$, and $I + J = R$, then $IJ = I \bigcap J$.

Proof For i) use induction on $k \geq 2$. The case $k = 2$ is immediate. Assume i) for $k = n \geq 2$ ideals and let $I_1, \ldots, I_{n+1} \lhd R$ so that $I_j + I_m = R$ for $j \neq m$. From the induction assumption, $I_2 I_3 \cdots I_n + I_{n+1} = R$ so by Theorem 11.6, $R = R^2 = (I_1 + I_{n+1})(I_2 I_3 \cdots I_n + I_{n+1}) \subseteq I_1 I_2 \cdots I_n + I_{n+1}$. Since i) holds by the induction assumption when $j < n$, the $k = n + 1$ case is now proved and part i) is true by IND-I. We also use induction to prove ii). Start with $n = 2$ ideals, I and J. The definition of ideal shows that $IJ \subseteq I \bigcap J$. It follows from Theorem 11.5 that $I \bigcap J \lhd R$ and since $1 \in R$, by Theorem 11.6 we have $(I \bigcap J) = (I \bigcap J)R = (I \bigcap J)(I + J) \subseteq (I \bigcap J)I + (I \bigcap J)J$, using the definition of product of ideals. But since R is commutative, $(I \bigcap J)I, (I \bigcap J)J \subseteq JI = IJ$, so $I \bigcap J \subseteq IJ + IJ = IJ$, and the two inclusions prove $IJ = I \bigcap J$. Now assume that ii) holds for any $T_1, \ldots, T_n \lhd R$ satisfying the hypothesis and let $I_1, \ldots, I_{n+1} \lhd R$ satisfy $R = I_j + I_m$ when $j \neq m$. From the definitions of ideal and product, $I_1 I_2 \cdots I_{n+1} \subseteq I_1 \bigcap \cdots \bigcap I_{n+1}$, so we need the opposite containment. By i), $I_1 \cdots I_j + I_{j+1} = R$ for $j \leq n$. As in the $n = 2$ case, $I_1 \bigcap \cdots \bigcap I_{n+1} =$

$(I_1 \cap \cdots \cap I_{n+1})R = (I_1 \cap \cdots \cap I_{n+1})(I_{n+1} + I_1 I_2 \cdots I_n) \subseteq (I_1 \cap \cdots \cap I_{n+1})I_{n+1} + (I_1 \cap \cdots \cap I_{n+1})(I_1 I_2 \cdots I_n)$. By the induction assumption, $I_1 \cap \cdots \cap I_n = I_1 I_2 \cdots I_n$, so $I_1 \cap \cdots \cap I_{n+1} \subseteq I_1 I_2 \cdots I_n$, and certainly $I_1 \cap \cdots \cap I_{n+1} \subseteq I_{n+1}$. Hence $I_1 \cap \cdots \cap I_{n+1} \subseteq (I_1 I_2 \cdots I_n)I_{n+1} + I_{n+1}(I_1 I_2 \cdots I_n) = I_1 I_2 \cdots I_{n+1}$ follows using the commutativity of R again. This proves the $k = n + 1$ case of part ii), so ii) holds by IND-I. ■

When do ideals satisfy the condition in Theorem 11.7? For $n\mathbf{Z}, m\mathbf{Z}$ in $\mathbf{Z}$, $n\mathbf{Z} + m\mathbf{Z} = \mathbf{Z} \Leftrightarrow 1 = sn + tm$ for some $s, t \in \mathbf{Z} \Leftrightarrow (n, m) = 1$, by Theorem 1.10. Similarly, $n_1\mathbf{Z}, \ldots, n_k\mathbf{Z} \subseteq \mathbf{Z}$ satisfy the condition precisely when $\{n_i\}$ is pairwise relatively prime. Thus $6\mathbf{Z} \cdot 25\mathbf{Z} \cdot 7\mathbf{Z} = 6\mathbf{Z} \cap 25\mathbf{Z} \cap 7\mathbf{Z}$. If R is any commutative ring with 1, and $I_1, \ldots, I_k$ are different maximal ideals in R, then $I_j + I_m \lhd R$ by Theorem 11.5 and properly contains both I_j and I_m, so $I_j + I_m = R$, and $I_1 I_2 \cdots I_k = I_1 \cap \cdots \cap I_k$ by Theorem 11.7. Finally, consider $(x), (y) \lhd \mathbf{Q}[x, y]$. These are not maximal since $(x) + (y) = M = \{f(x, y) \in \mathbf{Q}[x, y] \mid f(0, 0) = 0\} \lhd \mathbf{Q}[x, y]$. However, $(x) \cap (y)$ are the polynomials that are multiples of both x and y, so $(x) \cap (y) = (xy) = (x)(y)$.

EXERCISES

1. If $R = \begin{bmatrix} \mathbf{Z} & 2\mathbf{Z} \\ 3\mathbf{Z} & \mathbf{Z} \end{bmatrix} = \{A \in M_2(\mathbf{Z}) \mid A_{12} \in 2\mathbf{Z} \text{ and } A_{21} \in 3\mathbf{Z}\}$ show that:

i) $I = \begin{bmatrix} \mathbf{Z} & 2\mathbf{Z} \\ 3\mathbf{Z} & 3\mathbf{Z} \end{bmatrix} \lhd R$

ii) $J = \begin{bmatrix} 2\mathbf{Z} & 2\mathbf{Z} \\ 3\mathbf{Z} & 2\mathbf{Z} \end{bmatrix} \lhd R$

iii) $K = \begin{bmatrix} 2\mathbf{Z} & 2\mathbf{Z} \\ 3\mathbf{Z} & 6\mathbf{Z} \end{bmatrix} \lhd R$

iv) Find a right ideal in R that is not a left ideal.

v) Find a left ideal in R that is not a right ideal.

2. If R is a ring, $s \in R$, $\varnothing \neq W \subseteq R$, $T \lhd_r R$, and $L \lhd_l R$, show that:

i) $r\text{-}ann(s) = \{x \in R \mid sx = 0\} \lhd_r R$

ii) $l\text{-}ann(s) = \{x \in R \mid xs = 0\} \lhd_l R$

iii) $r\text{-}ann(W) = \{x \in R \mid wx = 0 \text{ for all } w \in W\} \lhd_r R$

iv) $r\text{-}ann(T) \lhd R$

v) $l\text{-}ann(W) = \{x \in R \mid xw = 0 \text{ for all } w \in W\} \lhd_l R$

vi) $l\text{-}ann(L) \lhd R$

3. In each case, determine whether the subset of the given ring is an ideal:

i) $\{(r_1, \ldots, r_n) \in \mathbf{Q}^n \mid r_1 + r_2 + \cdots + r_n = 0\} \subseteq \mathbf{Q}^n$

ii) $\{a_0 + a_1x + \cdots + a_mx^m \in \mathbf{Z}[x] \mid a_j \in 3^j\mathbf{Z}\} \subseteq \mathbf{Z}[x]$

iii) $\{a_0 + a_1x + \cdots + a_mx^m \in \mathbf{Z}[x] \mid a_0 + a_1 + \cdots + a_m = 0\} \subseteq \mathbf{Z}[x]$

iv) given $\varnothing \neq S \subseteq \mathbf{Q}$, $\{k \in \mathbf{Z} \mid ks \in \mathbf{Z} \text{ for all } s \in S\} \subseteq \mathbf{Z}$

v) for $(G, +)$ an Abelian group, $\{\varphi \in End((G, +)) \mid \varphi(G) \text{ is finite}\} \subseteq End((G, +, \circ))$

vi) $\{[a]_n \in \mathbf{Z}_n \mid [a]_n^2 = [0]_n\} \subseteq \mathbf{Z}_n$

vii) $\{x \in R \mid xy = 0 \text{ for some } y \in R^*\} \subseteq R$, when R is commutative

viii) $\{x \in R \mid x^k = 0 \text{ for some } k(x) = k \in \mathbf{N}\} \subseteq R$, when R is commutative

ix) for a fixed $k \in N$, $k \cdot R = \{k \cdot x \in R \mid x \in R\} \subseteq R$, for any ring R

x) for a fixed $n \in N$, $\{r \in R \mid n \cdot r = 0_R\} \subseteq R$, for any ring R

4. Find ideals in $M_n(2\mathbf{Z})$ that are not of the form $M_n(k\mathbf{Z})$.

5. **i)** Find a maximal right ideal in $M_n(F)$ for F a field.

ii) Find a maximal left ideal in $M_n(F)$ for F a field.

iii) Find a maximal right ideal in $M_n(\mathbf{Z})$ (consider $(I_n - E_{11})M_n(\mathbf{Z})$).

iv) Find a maximal left ideal in $M_n(\mathbf{Z})$.

6. **i)** Show that every $I \lhd \mathbf{Z}_n$ is principal, and describe the maximal ideals.

ii) For which n is $\mathbf{Z}_n - U_n$ an ideal of $\mathbf{Z}_n$?

7. Find a maximal ideal in:

i) $\mathbf{Q}[[x]]$

ii) $\mathbf{Z}[x]$

iii) $\mathbf{Z}_n[x]$ (see Exercise 6)

8. **i)** Show that $\{([p]_n)\langle\langle x\rangle\rangle \mid p \text{ is prime and } p \mid n\}$ are maximal ideals in $\mathbf{Z}_n\langle\langle x\rangle\rangle$.

ii) For which $n > 1$ is $\mathbf{Z}_n\langle\langle x\rangle\rangle - U(\mathbf{Z}_n\langle\langle x\rangle\rangle$ an ideal of $\mathbf{Z}_n\langle\langle x\rangle\rangle$?

9. Let R be a commutative ring with 1.

i) If the set of all ideals of R is a chain, $(I, J \lhd R \Rightarrow I \subseteq J \text{ or } J \subseteq I)$, show that $R - U(R) \lhd R$.

ii) Show that the set of ideals in $\mathbf{Q}[[x]]$ is a chain.

iii) Show that $\mathbf{Q}[[x, y]] - U(\mathbf{Q}[[x, y]])$ is an ideal but that the set of ideals in $\mathbf{Q}[[x, y]]$ is not a chain with respect to inclusion.

10. **i)** Describe all principal ideals in $\mathbf{Q}[[x]]$.

ii) Show that each proper, nonzero, principal ideal in $\mathbf{Z}_n\langle\langle x\rangle\rangle$ has a generator with a nonzero constant, and describe what constants are possible.

11. Write each of the following ideals of $\mathbf{Q}[x]$ as $(f(x))$:

i) $(x^2 - 1)\mathbf{Q}[x] + (x^5 - 1)\mathbf{Q}[x]$

ii) $(x^2 - 1)\mathbf{Q}[x] + (x^3 + 1)\mathbf{Q}[x] + (x^4 + x^2 + 1)\mathbf{Q}[x]$

iii) $\sum_{j\geq 5}(x^j - 1)\mathbf{Q}[x]$

iv) $\bigcap_{j=1}^{10}(x - j)\mathbf{Q}[x]$

v) $\bigcap_{j>3}(x - j)\mathbf{Q}[x]$

vi) $(x^3 - x)\mathbf{Q}[x] \bigcap (x^4 - 1)\mathbf{Q}[x]$

vii) $(x^2 - 3x + 2)\mathbf{Q}[x] \bigcap (x^2 - 5x + 6)\mathbf{Q}[x] \bigcap (x^2 - 7x + 12)\mathbf{Q}[x]$

12. Let R be a ring, A and B subrings, $L \lhd_l R$, $T \lhd_r R$, $I \lhd R$, $AB = \{a_1b_1 + \cdots + a_kb_k \mid k \geq 1, a_j \in A, b_j \in B\}$, and $A + B = \{a + b \mid a \in A \text{ and } b \in B\}$. Show that:

a) AB is a subring of R if R is commutative

b) $LT \lhd R$

c) $LB \lhd_l R$

d) $T + RT \lhd R$

e) $A + LA$ is a subring of R

f) $A + I$ is a subring of R

g) $L + T + LT$ is a subring of R

h) $A \bigcap I \lhd A$

13. Let R be a ring and $I \lhd R$.

a) If $L \lhd_l I$, show that $IL \lhd_l R$ contained in L.

b) If $L \lhd_l I$, show that $RL^2 + L^2 \lhd_l R$ contained in $L (L^2 = LL)$.

c) If $J \lhd I$, show that $IJI \lhd R$ contained in J.

14. Let $I, J \lhd R$.

i) Show that $W(J, I) = \{x \in R \mid JxJ \subseteq I\} \lhd R$, where $JxJ = \{a_1xb_1 + \cdots + a_kxb_k \mid k \geq 1, a_j, b_j \in J\}$.

ii) What can you conclude about the structure of $W(J, I)$ if $I, J \lhd_l R$?

15. Let R be a ring so that when $I \lhd R$ and $I^2 = I \cdot I = (0)$ then $I = (0)$.

a) If $K \lhd R$ and $K^m = KK \cdots K = (0)$ for some $m \geq 1$, show that $K = (0)$.

b) If $T \lhd_r R$ and $T^k = TT \cdots T = (0)$ for some $k \geq 1$, show that $T = (0)$.

c) If $M \lhd R$, $J \lhd M$, and $J^2 = (0)$ then show that $J = (0)$.

d) If $M \lhd R$, $(0) \neq J \lhd M$, show that J contains a nonzero ideal of R (Exercise 13 c)).

16. Any $P \lhd R$ is called *prime* if $P \neq R$ and whenever $IJ \subseteq P$ for $I, J \lhd R$, then either $I \subseteq P$ or $J \subseteq P$. For the following, assume that $P \lhd R$ with $P \neq R$.

a) Show that P is prime exactly when for any $a, b \in R$, $aRb \subseteq P$ implies $a \in P$ or $b \in P$. Here $aRb = \{arb \in R \mid r \in R\}$. (See Exercise 14.)

b) When R is a commutative ring, show that P is prime $\Leftrightarrow$ for any $a, b \in R$, $ab \in P$ implies that $a \in P$ or $b \in P$.

c) If P is prime, $J \lhd R$, $a, b \in R$, and $aJb \subseteq P$ then $a \in P$, $b \in P$, or $J \subseteq P$.

d) If (0) is a prime ideal of R, show that (0) is a prime ideal of any $(0) \neq I \lhd R$.

e) If P is prime and $I \lhd R$ then either $I \bigcap P$ is a prime ideal of I or $I \subseteq P$.

f) If (0) is a prime ideal of R, show that *char* $R = 0$ or *char* $R = p$, a prime.

17. For any $m \geq 2$ show that there is $n \in \mathbf{N}$ such that $\mathbf{Z}_n$ contains m distinct nonzero ideals $\{I_j \mid 1 \leq j \leq m\}$ such that for any I_j, $I_j \bigcap \sum_{s \neq j} I_s = (0)$.

18. An ideal of I of a ring R is called **nil** if for each $x \in I$ there is $n = n(x) \geq 1$ such that $x^n = 0$. Show that R contains a nil ideal $N(R)$ (possibly (0)) that contains every nil ideal of R. (Consider $I + J$ when I and J are nil ideals.)

19. a) For R a commutative ring, show that $N(R) = \{r \in R \mid r^k = 0 \text{ for some } k \geq 1\} \lhd R$.

b) Explicitly describe $N(\mathbf{Z}_m)$ in terms of the prime factorization of m.

11.2 QUOTIENT RINGS

The construction of quotient rings is the same as for quotient groups. For a ring R and $A \leq (R, +)$, $(R, +)$ Abelian implies $A \lhd (R, +)$ so, as in Theorem 5.6, $(R, +)/A = \{r + A \mid r \in R\}$ with $(x + A) + (y + A) = (x + y) + A$. Also $(R, +)/A$ is an Abelian group by Theorem 5.7. Since R is a ring and has a multiplication, it is natural to ask whether the group $(R, +)/A$ has a corresponding ring structure using $(x + A) \cdot (y + A) = xy + A$ as the multiplication. First, is $\cdot : (R/A, +)^2 \to (R/A, +)$ really a function: is this "product" independent of the representatives for $x + A$ and $y + A$? If so, then for $x \in A$ and $r \in R$, since $0 + A = x + A$, $xr + A = (x + A) \cdot (r + A) = (0 + A) \cdot (r + A) = 0 + A = A$. Thus $xr + A = 0 + A$, forcing $xr \in A$, and this shows that $A \lhd_r R$. Similarly, $(r + A) \cdot (x + A) = (r + A) \cdot (0 + A)$ implies that $A \lhd_r R$. Consequently, if the "natural" coset multiplication on $(R, +)/(A, +)$ makes sense then $A \lhd R$.

We claim any $A \lhd R$ gives rise to a natural ring structure on $(R, +)/A$. As above, $(R, +)/(A, +) = \{r + A \mid r \in R\}$. For $x + a \in x + A$ and $y + b \in y + A$ with $a, b \in A$, $(x + a)(y + b) = xy + (xb + ay + ab) \in xy + A$ since A an ideal implies $xb, ay, ab \in A$. Because for any $\alpha \in x + A$ and any $\beta \in y + A$, $\alpha\beta \in xy + A$, it is natural to define the product of two cosets to be the coset containing the products of their elements. This observation about products of elements in cosets shows that when A is an ideal of R then the definition

$(x + A) \cdot (y + A) = xy + A$ is independent of the choice of coset representatives and so gives a multiplication $\cdot : (R/A, +)^2 \to (R/A, +)$ on $(R, +)/A$.

THEOREM 11.8 Let R be a ring, $I \triangleleft R$, and $R/I = (R, +)/(I, +) = \{r + I \mid r \in R\}$. Using the coset multiplication $(a + I) \cdot (b + I) = ab + I$, $(R/I, +, \cdot)$ is a ring that is commutative if R is and has an identity when R does.

Proof The verification is routine, as in Theorem 5.6. For associativity of multiplication, $((a + I) \cdot (b + I)) \cdot (c + I) = (ab + I) \cdot (c + I) = (ab)c + I = a(bc) + I = (a + I) \cdot (bc + I) = (a + I) \cdot ((b + I) \cdot (c + I))$, using the definition of coset multiplication and the associative law in R. The distributive laws are proved similarly, as is the fact that R/I is commutative when R is. If $1 \in R$ then it is easy to check that $1 + I = 1_{R/I}$. ■

DEFINITION 11.4 If $I \triangleleft R$, a ring, then $(R/I, +, \cdot)$ with coset addition and multiplication is the **quotient ring of R by I,** or the factor ring of R by I.

We look at a number of examples of the quotient construction. In some cases we see how to identify the quotient with known rings, and in others we see how new properties can arise. In the extreme case when $I = R$ then $R/R = \{0 + R\}$, so R/R is the ring of one element, namely $0_{R/R}$. If $I = (0)$, then $r + (0) = \{r\}$. Thus $R/(0) = \{\{r\} \mid r \in R\}$ consists of the singleton subsets of R, is clearly identified with R, and has essentially the same multiplication as R: $\{r\}\{s\} = \{rs\}$.

Example 11.3 The quotient of $R = E_{11}M_2(\mathbf{Q})$ by $I = \mathbf{Q} \cdot E_{12}$.

R is a subring of $M_2(\mathbf{Q})$ by Example 11.1, $I = \{qE_{12} \in R \mid q \in \mathbf{Q}\} \le (R, +)$, and for $A \in R$, $qE_{12}A = qE_{12}E_{11}A = 0_2$. Also $AqE_{12} = (A_{11}E_{11} + A_{12}E_{12})qE_{12} = (qA_{11})E_{12}$ so indeed $I \triangleleft R$. Since $IR = (0_2)$, R cannot have an identity, and $(0_2) \ne E_{11}I \subseteq RI$, so R is not commutative. For any $A = aE_{11} + bE_{12} \in R$, $A + I = aE_{11} + I \in R/I$, so it follows that $R/I = \{aE_{11} + I \mid a \in \mathbf{Q}\}$. Set $aE_{11} + I = [a]$ and observe that $[a] = [b]$ implies $a = b$. Now $[a] + [b] = (aE_{11} + I) + (bE_{11} + I) = (a + b)E_{11} + I = [a + b]$ and we also have $[a][b] = (aE_{11} + I)(bE_{11} + I) = aE_{11}bE_{11} + I = abE_{11} + I = [ab]$. Clearly $[a][b] = [ab] = [ba] = [b][a]$ so R/I is commutative, and $[1] = 1_{R/I}$. Identifying $aE_{11} + I = [a]$ with $a \in \mathbf{Q}$ shows that R/I is essentially $\mathbf{Q}$.

The example shows that R/I can be commutative or have an identity when R does not. Also, a nice set of representatives for R/I produced a natural bijection with $\mathbf{Q}$ so that addition and multiplication of corresponding elements correspond. Just as for groups, we say that the rings R/I and $\mathbf{Q}$ are *isomorphic;* the formal definition is in the next section. For now will just see whether there is a natural correspondence between our examples and known rings.

For $n\mathbf{Z} \triangleleft \mathbf{Z}$, $\mathbf{Z}/n\mathbf{Z} = \{i + n\mathbf{Z} \mid 0 \le i < n\}$, the equivalence classes of congruence modulo n. We know that coset addition and multiplication are just the addition and multiplication in $\mathbf{Z}_n$, so $\mathbf{Z}/n\mathbf{Z} = \mathbf{Z}_n$ as rings.

Example 11.4 For R a ring, the quotient of $R[x, y]$ by $I = \mathbf{Z}y + Ry = \{p(x, y) \in R[x, y] \mid \text{every monomial in } p(x, y) \text{ contains } y\}$ is essentially $R[x]$.

$I \lhd R[x, y]$ by Example 11.1, and $I = Ry$ if $1 \in R$. Write $f(x, y) \in R[x, y]$ as $f(x, y) = g(x) + h(x, y)$ for $g(x) \in R[x]$ and $h(x, y) \in I$, so $f(x, y) + I = g(x) + I$, and $R[x, y]/I = \{g(x) + I \mid g(x) \in R[x]\}$. Now $g(x) + I = q(x) + I$ implies $g(x) - q(x) \in I$, forcing $g(x) = q(x)$ (why?). Thus, associating $g(x) \in R[x]$ with $[g(x)] = g(x) + I \in R[x, y]/I$ gives a bijection. Observe that $[g(x)] + [h(x)] = [g(x) + h(x)]$ and $[g(x)][h(x)] = [g(x)h(x)]$ (verify!), so the elements of the quotient add and multiply like those they correspond to in $R[x]$.

Example 11.5 If R and S are rings, $(R \oplus S)/((0_R) \oplus S)$ is essentially R.

For $(0_R, a), (0_R, b) \in (0_R) \oplus S$ and $(x, y) \in R \oplus S$, $(0_R, a) - (0_R, b) = (0_R, a - b) \in (0_R) \oplus S$, $(0_R, a)(x, y) = (0_R, ay) \in (0_R) \oplus S$, and $(x, y)(0_R, a) = (0_R, ya) \in (0_R) \oplus S$, so $I = (0_R) \oplus S \lhd R \oplus S$. Now $(r, s) + I$ is represented by any $(r, s) + (0_R, s') \in (r, s) + I$, so taking $s' = -s$ gives the representative $(r, 0_S)$. Therefore $(R \oplus S)/I = \{(r, 0_S) + I \mid r \in R\}$ and $(r, 0_S) + I = (r', 0_S) + I \Leftrightarrow r = r'$ by Theorem 4.2. Hence identifying $r \in R$ with $[r] = r + I$ is a bijection between R and $(R \oplus S)/I$, and the operations in the quotient yield $[r] + [r'] = [r + r']$ and $[r][r'] = [rr']$, so essentially R and $(R \oplus S)/I$ are the same ring.

Example 11.6 The quotient ring $\mathbf{Z}_{mn}/([m]_{mn})$ is essentially the ring $\mathbf{Z}_m$.

Note that $I = ([m]_{mn}) = \{[km]_{mn} \in \mathbf{Z}_{mn} \mid k \in \mathbf{Z}\}$ (why?), and write $[a]$ for $[a]_{mn}$. From the division theorem, if $0 \le i < mn$ then $i = bm + r$ for $0 \le r < m$, so $[i] - [r] = [bm]$ and $[i] + ([m]) = [r] + ([m])$ by Theorem 4.2. Also, when $0 \le r \le r' < m$ and $[r] + ([m]) = [r'] + ([m])$ then $[r - r'] = [r] - [r'] = [km]$ since $[r] - [r'] \in ([m])$ so $r - r' - km = tmn$, forcing $m \mid (r - r')$ and so $r = r'$. Thus $\mathbf{Z}_{mn}/([m]) = \{[i] + ([m]) \mid 0 \le i < m\}$ has m distinct elements, and identifying $[i] + ([m]) = [[i]]$ with $[i]_m \in \mathbf{Z}_m$ for $0 \le i < m$ is a bijection. Do the ring operations in $\mathbf{Z}_{mn}/([m])$ and in $\mathbf{Z}_m$ give corresponding results for elements identified by this bijection? Take $0 \le i, j < m$, write $i + j = x + dm$ and $ij = y + em$ for appropriate $0 \le x, y < m$, and compute that $[[i]] + [[j]] = ([i] + ([m])) + ([j] + ([m])) = ([i] + [j]) + ([m]) = ([x] + [dm]) + ([m]) = [x] + ([m]) = [[x]]$. But in $\mathbf{Z}_m$, $[i]_m + [j]_m = [x]_m$, so the addition of corresponding elements corresponds. Similarly, in $\mathbf{Z}_{mn}/([m])$, $[[i]][[j]] = ([i] + ([m]))([j] + ([m])) = [i][j] + ([m]) = [ij] + ([m]) = [y + em] + ([m]) = [y] + [em] + ([m]) = [y] + ([m]) = [[y]]$ and in $\mathbf{Z}_m$ $[i]_m[j]_m = [y]_m$. Hence the multiplication in $\mathbf{Z}_{mn}/([m])$ mimics that of the corresponding elements in $\mathbf{Z}_m$, so $\mathbf{Z}_{mn}/([m]_{mn})$ is essentially $\mathbf{Z}_m$.

Let $([2]_8) = \{[0]_8, [2]_8, [4]_8, [6]_8\}$ in $\mathbf{Z}_8$ so $\mathbf{Z}_8/([2]_8) = \{[0]_8 + ([2]_8), [1]_8 + ([2]_8)\} = \{[[0]], [[1]]\}$, and $[[1]] + [[1]] = ([1]_8 + ([2]_8)) + ([1]_8 + ([2]_8)) = [2]_8 + ([2]_8) = [0]_8 + ([2]_8) = [[0]]$. This shows that $\mathbf{Z}_8/([2]_8)$ is essentially $\mathbf{Z}_2$ with $[0]_2$ corresponding to $[[0]]$ and $[1]_2$ to $[[1]]$. Although $([2]_8) \cong (\mathbf{Z}_4, +)$ as groups, $([2]_8)$ is not essentially $\mathbf{Z}_4$ *as a ring* since each of its elements has cube zero! For $([3]_{15}) = \{[0]_{15}, [3]_{15}, \ldots, [12]_{15}\} \subseteq \mathbf{Z}_{15}$, as in Example 11.6, $\mathbf{Z}_{15}/([3]_{15}) = \{[[0]] = [0]_{15} + ([3]_{15}), [[1]] = [1]_{15} + ([3]_{15}), [[2]] = [2]_{15} + ([3]_{15})\}$ is basically the ring $\mathbf{Z}_3$. For example, $[[1]] + [[2]] = ([1]_{15} + ([3]_{15})) + ([2]_{15} + ([3]_{15})) = [3]_{15} + ([3]_{15}) = [0]_{15} + ([3]_{15}) = [[0]]$, and $[[2]]^2 = ([2]_{15} + ([3]_{15}))^2 = [4]_{15} + ([3]_{15}) = [1]_{15} + [3]_{15} + ([3]_{15}) = [1]_{15} + ([3]_{15}) = [[1]]$.

These illustrations of Example 11.6 show that the quotient ring can be a domain when the original ring is not. Another instance of this is Example 11.5 when R is a domain and S is arbitrary. We have observed that $\mathbf{Z}/n\mathbf{Z} = \mathbf{Z}_n$ so some quotients of $\mathbf{Z}$ are domains (when n is prime) and some are not. A quotient R/I is a domain whenever $a + I, b + I \in R/I - 0_{R/I}$ imply $(a + I)(b + I) = ab + I \neq I = 0_{R/I}$. This is equivalent to $a, b \notin I \Rightarrow ab \notin I$, a situation sufficiently important to deserve a special name.

DEFINITION 11.5 If R is a commutative ring and $I \lhd R$, then I is a **prime ideal** if $I \neq R$ and if for any $a, b \in R$, $ab \in I$ implies that $a \in I$ or $b \in I$.

Since for $I \lhd R$, $(a + I)(b + I) = ab + I$ in R/I, when R is commutative I is a prime ideal $\Leftrightarrow$ R/I is a domain, so R is a domain $\Leftrightarrow$ (0) is a prime ideal. This equivalence remains true for noncommutative rings using Definition 11.5 but properties of commutative rings given by conditions on elements are often more useful for noncommutative rings when defined by the corresponding conditions for ideals. A more useful analogue of domain for noncommutative rings is the product of nonzero *ideals* not being zero. These rings are called *prime,* and it turns out (§11.5) that R/I is a prime ring $\Leftrightarrow$ I is a *prime ideal:* $J, K \lhd R$ and $JK \subseteq I \Rightarrow J \subseteq I$ or $K \subseteq I$. By Exercise 16 in §11.1, I a prime ideal is equivalent to: $aRb = \{arb \in R \mid r \in R\} \subseteq I$ for $a, b \in R \Rightarrow a \in I$ or $b \in I$. When R is commutative this is also equivalent to Definition 11.5. Now $M_n(\mathbf{Q})$ has no proper nonzero ideal by Theorem 11.1, so (0_n) is a prime ideal in the sense just given; $M_n(\mathbf{Q})$ is a prime ring but not a domain. Products of nonzero ideals can be zero, for example $([6]_{12})([4]_{12}) = ([0]_{12})$ in $\mathbf{Z}_{12}$, and using Theorem 11.1, $M_2(([3]_6))M_2(([2]_6)) = (0_2)$ in $M_2(\mathbf{Z}_6)$.

Let us turn to some examples that show how the quotient can satisfy other interesting properties not satisfied by the ring we start with.

Example 11.7 $\mathbf{Q}[x]/(x^2 - 2)$ is a field, essentially with a square root of 2.

Now $I = (x^2 - 2) \lhd \mathbf{Q}[x]$ and write $[f(x)] = f(x) + I \in \mathbf{Q}[x]/I$. The multiplication in $\mathbf{Q}[x]/I$ shows that for $q \in \mathbf{Q}$, $[qx^j] = qx^j + I = (q + I)(x^j + I) = (q + I)(x + I)^j = [q][x]^j$. Thus for $X = [x]$ we have $[a_n x^n + \cdots + a_1 x + a_0] = [a_n]X^n + \cdots + [a_1]X + [a_0]$. Now $X^2 = [x]^2 = [x^2] = [2]$ since $x^2 - 2 \in I$, so X is a square root of $[2]$ in $\mathbf{Q}[x]/I$. Also $X^{2n} = [2^n]$ and $X^{2n+1} = [2^n]X$, so $[a_n x^n + \cdots + a_1 x + a_0] = [a]X + [b]$. Thus $\mathbf{Q}[x]/I = \{[ax + b] \mid a, b \in \mathbf{Q}\}$ and $[ax + b] = [cx + d]$ precisely when $(a - c)x + (b - d) \in (x^2 - 2)$, forcing $ax + b = cx + d$ (why?). By Theorem 11.8, $\mathbf{Q}[x]/I$ is a commutative ring with identity and so will be a field when each nonzero $\alpha \in \mathbf{Q}[x]/I$ has an inverse under multiplication. Take $[a]X + [b] \neq [0]X + [0] = 0 \in \mathbf{Q}[x]/I$. Now $([a]X + [b])([a]X - [b]) = [a^2]X^2 - [b^2] = [2a^2 - b^2]$. If $2a^2 = b^2$ then $2 = (b/a)^2$ for $b/a \in \mathbf{Q}$, which is impossible by Theorem 1.17. Therefore $c = 2a^2 - b^2 \neq 0$ and it follows that $([a]X + [b])^{-1} = [a/c]X - [b/c]$, showing that $\boldsymbol{Q}[x]/(x^2 - 2)$ is a field.

The construction in Example 11.7 is quite important and will be generalized later in Theorem 11.18. Note that $K = \boldsymbol{Q}[x]/(x^2 - 2)$ contains a square root of $[2] = 2 + (x^2 - 2) \in K$, not a square root of $2 \in \boldsymbol{Q}$. We identify $\{[q] = q + (x^2 - 2) \in K \mid q \in \boldsymbol{Q}\}$ with $\boldsymbol{Q}$, *and so we think of* $\boldsymbol{Q} \subseteq K$ and $[2] \in K$ as being $2 \in \boldsymbol{Q}$. It is possible to replace K with a field K' actually containing $\boldsymbol{Q}$ and thereby to construct a square root of 2 in K'. The quotient construction was used, but disguised, in Example 10.19 to produce fields with p^2 elements. Using other kinds of ideals can lead to quite different properties.

Example 11.8 The quotient $\boldsymbol{Q}[x]/(x^2 - 3x + 2)$ is a commutative ring with identity, has zero divisors, and is esssentially the ring $\boldsymbol{Q} \oplus \boldsymbol{Q}$.

As in the last example, let $I = (p(x)) = (x^2 - 3x + 2) \triangleleft \boldsymbol{Q}[x]$ and write elements in $\boldsymbol{Q}[x]/I$ as $[a_n]X^n + \cdots + [a_1]X + [a_0]$ where $[a] = a + I$ for $a \in \boldsymbol{Q}$ and $X = x + I$. Since $x^2 - (3x - 2) = p(x) \in I$, here $X^2 = [x^2] = [3x - 2] = [3]X - [2]$. Thus $X^3 = X^2X = ([3]X - [2])X = [3]X^2 - [2]X = [3]([3]X - [2]) - [2]X = [7]X - [6]$. It follows by induction that $X^n = [2^n - 1]X - [2^n - 2]$, so every element in $\boldsymbol{Q}[x]/I$ is represented as $[a]X + [b]$ for $a, b \in \boldsymbol{Q}$. Also $[a]X + [b] = [c]X + [d]$ exactly when $a = c$ and $b = d$ (verify!). Now $(X - [2])(X - [1]) = X^2 - [3]X + [2] = [x^2 - 3x + 2] = [0]$, so $\boldsymbol{Q}[x]/I$ is not a domain. Of course $\boldsymbol{Q}[x]/I$ is commutative with 1 by Theorem 11.8.

We argue that $\boldsymbol{Q}[x]/I$ is essentially $\boldsymbol{Q} \oplus \boldsymbol{Q}$. If we identify $[a]X + [b]$ with $(a, b) \in \boldsymbol{Q}^2$ then products of corresponding elements will not correspond. For example, X would correspond to $(1, 0)$ but $X^2 = [3]X - [2]$ would not correspond to $(1, 0)^2 = (1, 0)$ in $\boldsymbol{Q}^2$. We proceed in a way that at this point is not very well motivated, but it works! Set $e = X - [1]$ and use $\boldsymbol{Q}[x]/I$ commutative to compute $e^2 = (X - [1])^2 = X^2 - [2]X + [1] = [3]X - [2] - [2]X + [1] = X - [1] = e$. Next, for $f = [1] - e$, $f^2 = ([1] - e)^2 = [1]^2 - 2e + e^2 = [1] - e = f$, since $e^2 = e$. Now $ef = fe = e([1] - e) = e - e^2 = [0]$, and $e + f = [1]$ (why?). For $\alpha \in \boldsymbol{Q}[x]/I$, $\alpha = \alpha[1] = \alpha e + \alpha f$, and $X^2 = [3]X - [2]$ implies that $([a]X + [b])e = [2a + b]e$ and $([a]X + [b])f = [a + b]f$. Thus $[a]X + [b] = [2a + b]e + [a + b]f$. Since for any $r, s \in \boldsymbol{Q}$ we can solve $2a + b = r$ and $a + b = q$ with $a, b \in \boldsymbol{Q}$, $\boldsymbol{Q}[x]/I = \{[r]e + [s]f \mid r, s \in \boldsymbol{Q}\}$. But $([r]e + [s]f)([r']e + [s']f) = [rr']e + [ss']f$ because $ef = fe = [0]$, and $([r]e + [s]f) + ([r']e + [s']f) = ([r] + [r'])e + ([s] + [s'])f = [r + r']e + [s + s']f$. Hence identifying $[r]e + [s]f$ with $(r, s) \in \boldsymbol{Q}^2$ shows that $\boldsymbol{Q}[x]/I$ is really the ring $\boldsymbol{Q} \oplus \boldsymbol{Q}$ in disguise.

Example 11.9 When F is a field and $n \in \boldsymbol{N}$, for any $f(x) \in F[x]$, $[f(x)] = f(x) + (x^n) \in F[x]/(x^n)$ is a unit when $f(0) \neq 0$ and otherwise $[f(x)]^n = [0]$.

When $f(0) = 0$ then $f(x) = xh(x)$ for $h(x) \in F[x]$ (why?), so $f(x)^n = x^n h(x)^n \in (x^n)$. Hence $[f(x)]^n = [f(x)^n] = [0]$. If $a = f(0) \neq 0$, then $f(x) = a + xh(x)$. Let $[xh(x)] = Y$, note that $Y^n = [0]$ as we just saw, and let

$a^{-1} = b \in F$. Then $([a] + Y)([b] - [b^2]Y + [b^3]Y^2 + \cdots + (-1)^s[b^{s+1}]Y^s) = [1] + (-1)^s[b^{s+1}]Y^{s+1}$ follows by induction on s. As $Y^n = [0]$, $([b] - [b^2]Y + \cdots + (-1)^{n-1}[b^n]Y^{n-1}) = [f(x)]^{-1}$ results, verifying the example.

Our last example in this section is an interesting variant of the last one and shows that all nonunits in a ring can have square zero but that there can be arbitrarily long products of nonunits that are not zero.

Example 11.10 There is a ring R with $y^2 = 0$ for all $y \in R - U(R)$, there is $\{y_i \in R - U(R) \mid i \in \mathbf{N}$ and $y_1 y_2 \cdots y_k \neq 0$ for any $k \in \mathbf{N}\}$, and $S = R - U(R)$ is an ideal of R and so contains all proper ideals of R.

Begin with $\mathbf{Z}_2[X]$ for $X = \{x_i \mid i \in \mathbf{N}\}$ indeterminates over $\mathbf{Z}_2$, let $I_j = (x_j^2) \triangleleft \mathbf{Z}_2[X]$, and set $I = \sum_N I_j$. Now $I \triangleleft \mathbf{Z}_2[X]$ by Theorem 11.5, and $f(X) \in I$ precisely when each monomial of $f(X)$ contains some x_j^2. As usual we omit writing monomials in $\mathbf{Z}_2[X]$ with coefficient $[0]_2$ and delete the coefficient $[1]_2$ in the others, so we write $x_1x_5^3 + x_3x_5x_6^2$ instead of $[1]_2x_1x_5^3 + [1]_2x_3x_5x_6^2$. Note that for any $f(X) \in \mathbf{Z}_2[X]$, $f(X) + f(X) = [2]_2 f(X) = 0 \in \mathbf{Z}_2[x]$. Since $x_j^2 \in I$ for all $x_j \in X$, a *monomial* $m(X) \notin I \Leftrightarrow m(X) = x_i x_j \cdots x_k$ for $i < j < \cdots < k$. Consequently, if $R = \mathbf{Z}_2[X]/I$ and $y_j = x_j + I \in R$, then every $f(X) + I \in R^*$ is a finite sum of monomials $m_1(Y) + \cdots + m_i(Y)$, each a product of *distinct* y_j, or such an expression plus $1_R = [1]_2 + I$. Now $y_j^2 = 0_R\,(x_j^2 \in I!)$ and $y_1 y_2 \cdots y_k = x_1 x_2 \cdots x_k + I \neq I$ for any $k \in \mathbf{N}$ because every monomial in I must contain some x_j^2. Thus $y_1 y_2 \cdots y_k \neq 0$. If $f(X) \in \mathbf{Z}_2[X]$ has constant zero, then $f(X) + I = g(Y) = m_1(Y) + \cdots + m_t(Y) \in R$ where $\{m_i(Y)\}$ are the monomials actually appearing in $g(Y)$. Since each $m_i(Y)^2 = 0$ (why?) it follows that $g(Y)^2 = \sum m_i(Y)^2 + \sum_{i \neq j} m_i(Y) m_j(Y) = \sum_{i<j} [2]_2 m_i(Y) m_j(Y) = 0$. If $f(X)$ has constant $[1]_2$ then $f(X) + I = 1_R + g(Y)$ with $g(Y)^2 = 0$. But now $(1_R + g(Y))^2 = 1_R + [2]_2 g(Y) + g(Y)^2 = 1_R$, so $f(X) + I = 1_R + g(Y) \in U(R)$. Therefore an element in R is a unit or it is $g(Y)$ with square zero. It follows that $S = R - U(R)$ is an ideal of R (verify!) in which every element has square zero, so Theorem 11.4 shows that S contains every proper ideal of R.

EXERCISES

1. Describe the quotient rings for the ideals in Exercise 1 of §11.1.
2. Let G be a group, R a ring with 1, and $I = \left\{\sum r_g g \in R[G] \mid \sum r_g = 0_R\right\}$ (Definition 10.4). Show that $I = \sum_G R[G](1 - g) \triangleleft R[G]$ and describe $R[G]/I$.
3. Show that the ring:
   i) $2\mathbf{Z}/6\mathbf{Z}$ is essentially $\mathbf{Z}_3$
   ii) $\mathbf{Z}[x]/(x)$ is essentially $\mathbf{Z}$
4. Show that $3\mathbf{Z}/9\mathbf{Z}$ is not essentially the ring $\mathbf{Z}_3$.
5. Find a minimal sum of principal ideals in $\mathbf{Z}[x]$, one generated by $a \in \mathbf{Z}$, equal to I. Describe R/I and determine whether I is maximal, or, if not, then prime, for $I =$:
   i) $(x - a), a \in \mathbf{N}$
   ii) $(x - 3) + (x - 7) + (x - 11)$

iii) $(x-3)+(x-8)$
iv) $(x-7)+(x-9)+(x-12)$
v) $\sum(x-p)$ over $\{p_j\}$ of odd primes

6. Describe the elements and multiplication in the following quotients (F is a field):
i) $Q[x]/(x^2-x)$
ii) $F[x]/(x^3)$
iii) $Q[x,y]/(xy)$
iv) $F[x,y,z]/I$ where $I=(x-y)+(y-z)$
v) R/I for $R=Q+x^3Q[x]\subseteq Q[x]$ and $I=x^4R$

7. If R is a commutative ring, show that $N(R/N(R))=0_{R/N(R)}$ (See Exercise 19, §11.1).

8. If R is a commutative ring and P is a prime ideal of R, show that $N(R)\subseteq P$.

9. Find the prime ideals in:
i) $\mathbf{Z}_{16}$
ii) $\mathbf{Z}_{24}$
iii) $\mathbf{Z}_{300}$
iv) $R=([2]_{60})\subseteq \mathbf{Z}_{60}$
v) $R=([2]_{64})\subseteq \mathbf{Z}_{64}$
vi) $Q[[x]]$
vii) $Q\langle\langle x\rangle\rangle$

10. For $m\in N$, which of $I_m=(m),(x)\lhd \mathbf{Z}[[x]]$ are prime? Describe $\mathbf{Z}[[x]]/I_m$.

11. If $I,J\lhd R$ and $I\subseteq J$, show that $J/I\lhd R/I$.

12. If $I\lhd R$ and $a\in U(R)$, show that $a+I\in U(R/I)$.

13. If R is a commutative ring with 1, P is a prime ideal of R, and R/P is finite, show that P is a maximal ideal of R.

14. For which $n\in N$ are there $([0])\neq I,J\lhd \mathbf{Z}_n$ $I,J\neq \mathbf{Z}_n$, I prime and J not prime?

15. **i)** Refer to Exercise 16 in §11.1 and show that (0_n) is a prime ideal in $M_n(\mathbf{Z})$.
ii) Find $(0_n)\neq I,J\lhd M_n(\mathbf{Z})$ and proper, with I prime and J not prime.

11.3 HOMOMORPHISMS OF RINGS

Rings come with an addition and a multiplication, so the functions of interest between rings are those whose action is consistent with these operations in each ring. The definition and results about these functions between rings are extensions of those between groups.

DEFINITION 11.6 Given rings R and S, $\varphi:R\to S$ is a **ring homomorphism** if for all $a,b\in R$, $\varphi(a+b)=\varphi(a)+\varphi(b)$ and $\varphi(ab)=\varphi(a)\varphi(b)$. If φ is also a bijection then it is an **isomorphism,** and we write $R\cong S$ and say that R and S are **isomorphic.** An isomorphism $\varphi:R\to R$ is called an **automorphism** of R. Set $End(R)=\{\varphi:R\to R\mid \varphi$ is a ring homomorphism$\}$.

If $\varphi:R\to S$ is a ring homomorphism then $\varphi:(R,+)\to(S,+)$ is a group homomorphism so the results we have for group homomorphisms hold. For example, $\varphi(0_R)=0_S$, $\varphi(-r)=-\varphi(r)$, and if $A\leq(R,+)$ then $\varphi(A)\leq(S,+)$ by Theorem 6.2. Observe that two additions and two multiplications appear in the definition of ring homomorphism, so in the condition $\varphi(ab)=\varphi(a)\varphi(b)$, the product ab is computed in R whereas $\varphi(a)\varphi(b)$ is computed in S. In general, when discussing rings we will refer to φ in Definition 11.6 as a homomorphism rather than as a ring homomorphism. By analogy with groups, we call $\varphi(R)$ a **homomorphic image** of R. Although $End((R,+))$ is a ring by Theorem 10.2 and $End(R) \subseteq End((R,+))$, $End(R)$ is essentially never closed under addition because usually $(\varphi+\eta)(ab) \neq (\varphi+\eta)(a)\cdot(\varphi+\eta)(b)$.

A homomorphism of rings must satisfy two conditions rather than the one condition for a group homomorphism, so there tend to be fewer homomorphisms between given rings than group homomorphisms between their additive subgroups. Let us look at a few examples.

Example 11.11 For any rings R and S, $I_R : R \to R$ is an automorphism and $0_S : R \to S$ given by $0_S(r) = 0_S \in S$ for all $r \in R$ is a homomorphism.

Each map is a homomorphism of additive groups (why?) and I_R is a bijection, so we need only check the multiplicative property. Now $I_R(rr') = rr' = I_R(r)I_R(r')$ and $0_S(rr') = 0_S = 0_S \cdot 0_S = 0_S(r)0_S(r')$.

Example 11.12 If R is a commutative ring and $a \in R$, then $E_a : R[x] \to R$ given by $E_a(f(x)) = f(a)$ is a homomorphism of rings.

The commutativity of R is essential here. The result is Exercise 11 of §10.3; it uses the definitions of addition and multiplication in $R[x]$ and shows that evaluation may not be a homomorphism if R is not commutative.

Example 11.13 Given rings $R_1, \ldots, R_k$, the projections $\pi_j : \bigoplus R_i \to R_j$ via $\pi_j((r_1, \ldots, r_k)) = r_j$ are surjective homomorphisms.

From Example 6.10, π_j is a surjective homomorphism of additive groups and the multiplicative property follows directly: if $\alpha = (r_1, \ldots, r_k)$, $\beta = (r_1', \ldots, r_k') \in \bigoplus R_i$ then from Example 10.2, $\alpha\beta = (r_1r_1', \ldots, r_kr_k')$. Therefore $\pi_j(\alpha\beta) = r_jr_j' = \pi_j(\alpha)\pi_j(\beta)$ and π_j is a surjective ring homomorphism.

Just as for groups, perhaps the most important example of a ring homomorphism uses the quotient construction.

Example 11.14 For any ring R and $I \lhd R$, $\rho : R \to R/I$ given by $\rho(r) = r + I$ is a surjective homomorphism of rings, and $\rho(r) = 0_{R/I} \Leftrightarrow r \in I$.

Since ρ is defined exactly the same as for additive groups, it is a surjective homomorphism from $(R,+)$ to $(R,+)/(I,+)$ by Example 6.7. The definition of coset multiplication shows that $\rho(ab) = ab + I = (a+I)(b+I) = \rho(a)\rho(b)$. Finally, $\rho(r) = r + I = 0_{R/I} = 0 + I \Leftrightarrow r \in I$ by Theorem 4.2 on cosets.

As a special case of the last example we have that $\rho : \mathbf{Z} \to \mathbf{Z}_n = \mathbf{Z}/n\mathbf{Z}$ given by $\rho(k) = [k]_n$ is a surjective ring homomorphism.

Example 11.15 For any ring $R, \sigma : R \to R[x]$ given by $\sigma(r) = r$, and when $1 \in R$ $\eta : R \to M_n(R)$ given by $\eta(r) = rI_n$ are injective ring homomorphisms.

The result for polynomials follows from the definitions of the operations in $R[x]$, and the result for matrix rings from the matrix computations (verify!) $rI_n + r'I_n = (r + r')I_n$ and $rI_n r'I_n = rr'I_n$. In particular, σ is an isomorphism from R to the subring of constant polynomials, and η is an isomorphism from R to the subring $R \cdot I_n$ of scalar matrices in $M_n(R)$.

We collect a few facts about homomorphisms, as in Theorem 6.2.

THEOREM 11.9 Let R and S be rings and $\varphi : R \to S$ a homomorphism.

i) For any $r_1, \ldots, r_k \in R$, $\varphi(r_1 r_2 \cdots r_k) = \varphi(r_1)\varphi(r_2) \cdots \varphi(r_k)$.
ii) If A is a subring of R then $\varphi(A)$ is a subring of S.
iii) If φ is surjective and I is a left (right, two-sided) ideal of R so is $\varphi(I) \subseteq S$.
If $1_R \in R$ then:
iv) $\varphi(1_R)$ is the identity for $\varphi(R)$
v) $\varphi(U(R)) \subseteq U(\varphi(R))$.

Proof We prove i) by induction on $k \geq 2$. When $k = 2$, $\varphi(r_1 r_2) = \varphi(r_1)\varphi(r_2)$ since φ is a homomorphism. If the statement is true for $n \geq 2$ elements of R and $x_1, \ldots, x_{n+1} \in R$ then $\varphi(x_1 x_2 \cdots x_{n+1}) = \varphi((x_1 x_2 \cdots x_n)x_{n+1}) = \varphi(x_1 x_2 \cdots x_n)\varphi(x_{n+1})$ since φ is a homomorphism. By the induction assumption, $\varphi(x_1 x_2 \cdots x_{n+1}) = \varphi(x_1) \cdots \varphi(x_n)\varphi(x_{n+1})$, proving i) by IND-I. If A is a subring of R and $a, b \in A$, it follows from Theorem 6.2 that $\varphi(a) - \varphi(b) = \varphi(a - b) \in \varphi(A)$, and φ a homomorphism shows that $\varphi(a)\varphi(b) = \varphi(ab) \in \varphi(A)$, so $\varphi(A)$ is a subring of S by Theorem 10.4. If $I \lhd_l R$, let $x \in I$ and $s \in S$. When φ is surjective, $s = \varphi(r)$ for some $r \in R$ and $s\varphi(x) = \varphi(r)\varphi(x) = \varphi(rx) \in \varphi(I)$, using $rx \in I$. The corresponding computation on the right shows that $\varphi(I) \lhd_r S$ when $I \lhd_r R$. Hence $\varphi(I)$ is an ideal of S when I is an ideal of R.

If $1_R \in R$ then for all $r \in R$, $1_R r = r = r1_R$ and applying φ yields $\varphi(1_R)\varphi(r) = \varphi(r) = \varphi(r)\varphi(1_R)$, so $\varphi(1_R)$ is the identity for $\varphi(R)$, a ring by ii). Finally, when $u \in U(R)$ with $u^{-1} = v$ then $uv = 1_R = vu$ and applying φ gives $\varphi(u)\varphi(v) = \varphi(1_R) = \varphi(v)\varphi(u)$. But $\varphi(1_R)$ is the identity of $\varphi(R)$, so $\varphi(u) \in U(\varphi(R))$ with inverse $\varphi(v)$. ∎

An important special case of the theorem is for $\rho : R \to R/I$. Here we get that $\rho(A) = \{a + I \mid a \in A\}$ is a subring of R/I when A is a subring of R, and $\{b + I \mid b \in B\}$ is a left, right, or two-sided ideal of R/I when B is in R.

We see next that $End(R)$ is smaller than $End((R, +))$ for $R = \mathbf{Z}$.

Example 11.16 As rings $End((\mathbf{Z}, +)) \cong \mathbf{Z}$, but $(End(\mathbf{Z}), \circ) \cong (\mathbf{Z}_2, +)$.

$End((\mathbf{Z}, +))$ is a ring using $(\varphi + \eta)(z) = \varphi(z) + \eta(z)$ and composition, by Theorem 10.2. For $k \in \mathbf{Z}$, the map $f_k(n) = kn$ is a group homomorphism of $(\mathbf{Z}, +)$, so $f_k \in End((\mathbf{Z}, +))$. Furthermore, every $\varphi \in End((\mathbf{Z}, +))$ is determined by $\varphi(1) = k$ since in $(\mathbf{Z}, +)$, $\varphi(z) = \varphi(z \cdot 1) = \varphi(1^z) = \varphi(1)^z = z\varphi(1)$, so $\varphi = f_{\varphi(1)}$. It follows that $\beta(k) = f_k$ is a bijection from $(\mathbf{Z}, +)$ to $End((\mathbf{Z}, +))$, and an exercise shows that β is a ring homomorphism (verify!). Thus β is an isomorphism of rings. For $\varphi \in End(\mathbf{Z})$, Theorem 11.9 shows that $\varphi(1) = m = 1_{\varphi(\mathbf{Z})}$, but the only $m \in \mathbf{Z}$ satisfying $m^2 = m$ is $m = 1$ or $m = 0$. These give $\varphi = I_{\mathbf{Z}}$ or $\varphi = f_0$ with $f_0(\mathbf{Z}) = 0$, respectively. It is straightforward to show that $(End(\mathbf{Z}), \circ) \cong (\mathbf{Z}_2, +)$. Note that $End(\mathbf{Z})$ has no natural addition since $I_{\mathbf{Z}} + I_{\mathbf{Z}} \notin (End(\mathbf{Z}), \circ)$.

For rings R and S with identity, and homomorphism $\varphi : R \to S$, it is not necessary that $\varphi(1_R) = 1_S$, unlike the case for groups. If $\varphi : \mathbf{Z} \to M_n(\mathbf{Z})$ is defined by $\varphi(m) = mE_{11}$ then $\varphi(nm) = nmE_{11} = nE_{11}mE_{11} = \varphi(n)\varphi(m)$ and $\varphi(n+m) = (n+m)E_{11} = nE_{11} + mE_{11} = \varphi(n) + \varphi(m)$. Thus $\varphi(1) \neq I_n$ but φ is a homomorphism. Next, by Example 11.14, $\rho : F[x] \to F[x]/(x^n)$ is a surjective homomorphism, $U(F[x]) = F^*$ by Theorem 10.10, and from Example 11.9, $\rho(g(x)) \in U(F[x]/(x^n))$ whenever $g(0) \neq 0$. Hence $\varphi(R)$ may have many more units than R. Finally, from Example 11.3, $\mathbf{Q}E_{12} = I \lhd R = E_{11}M_2(\mathbf{Q})$, R has no identity (why?), but $\rho(R) = R/I = \{qE_{11} + I \mid q \in \mathbf{Q}\}$ is a field with $E_{11} + I$ as an identity and $U(\rho(R)) = \rho(R)^*$.

We end this section with a version of Theorem 6.3 for rings.

THEOREM 11.10 Let $\varphi : R \to S$ and $\eta : S \to T$ be ring homomorphisms.

i) $\eta \circ \varphi : R \to T$ is a ring homomorphism.
ii) If φ is an isomorphism so is $\varphi^{-1} : S \to R$.
iii) $\cong$ is an equivalence relation on any set of rings.

Proof Theorem 6.3 shows that $\eta \circ \varphi$ is a homomorphism of additive groups, and its proof, but with multiplicative notation, shows that $\eta \circ \varphi$ is a ring homomorphism. Also, by Theorem 6.3, $\varphi^{-1} : S \to R$ is an isomorphism of additive groups when φ is an isomorphism. In this case $s, s' \in S$ and φ surjective yield $s = \varphi(r)$ and $s' = \varphi(r')$ for some $r, r' \in R$, so $\varphi^{-1}(s) = r$ and $\varphi^{-1}(s') = r'$. Now φ is a homomorphism so $\varphi(rr') = \varphi(r)\varphi(r') = ss'$ and φ bijective implies $\varphi^{-1}(ss') = rr' = \varphi^{-1}(s)\varphi^{-1}(s')$, proving that φ^{-1} is an isomorphism of rings. If A is a set of rings and $R \in A$, then I_R is an automorphism of R by Example 11.11, so $R \cong R$ and $\cong$ is reflexive. If $R, S \in A$ and $R \cong S$ there is an isomorphism $\varphi : R \to S$ and $S \cong R$ via φ^{-1} using ii), so $\cong$ is symmetric. Finally if $R, S, T \in A$ with $R \cong S$ and $S \cong T$ then there are isomorphisms $\varphi : R \to S$ and $\eta : S \to T$. By i), $\eta \circ \varphi : R \to T$ is a ring homomorphism and so an isomorphism by Theorem 0.13. ∎

EXERCISES

1. Is the map a ring homomorphism, and if so, is it injective or surjective? Let R denote a ring that does *not* have trivial multiplication $rs = 0$ for all $r, s \in R$.

i) $\varphi : R \oplus R \to R \oplus R$ given by $\varphi(a, b) = (b, a)$
ii) $f : R \to M_2(R)$ via $f(r) = rE_{12}$
iii) $g : R \to \mathbf{Z} \oplus R \oplus \mathbf{Q}$ given by $g(r) = (0, r, 0)$
iv) $h : R \to R \oplus R$ defined by $h(r) = (r - er, er)$ for $e \in Z(R)$ with $e^2 = e$
v) $\varphi : \mathbf{Z}_{30} \oplus \mathbf{Z}_{30} \to \mathbf{Z}_{30}$ via $\varphi([a]_{30}, [b]_{30}) = [6a + 10b]_{30}$
vi) $\varphi : \mathbf{Q}[x] \to M_2(\mathbf{Q}[x])$ via $\varphi(f(x)) = f(x)E_{11} + f(0)E_{22}$
vii) $T : M_n(R) \to R$ given by $T(A) = A_{11}$

viii) $g : M_n(R) \to M_n(R)$ defined by $g(A) = A^T$ (matrix transpose), so $g(A)_{ij} = A_{ji}$
ix) $f : M_n(\mathbf{C}) \to M_n(\mathbf{C})$ defined by $f(A)_{ij} = \overline{A_{ij}}$ (where $\overline{a+bi} = a - bi$).

2. Show that for a ring R, $Aut(R) = \{\varphi : R \to R \mid \varphi$ is an isomorphism$\}$ is a group using composition as the multiplication.

3. **a)** If R and S are rings with $R \cong S$, show that $(U(R), \cdot) \cong (U(S), \cdot)$.
b) If $R_1 \cong S_1, \ldots, R_k \cong S_k$ are ring isomorphisms, show that $\oplus R_i \cong \oplus S_i$ as rings and that $\oplus U(R_i) \cong \oplus U(S_i)$ as groups under multiplication.

4. Let $\varphi : R \to S$ be a ring homomorphism. Show that:
i) $\alpha\left(\sum r_i x^i\right) = \sum \varphi(r_i)x^i$ defines a ring homomorphism $\alpha : R[x] \to S[x]$
ii) $\beta((A_{ij})) = (\varphi(A_{ij}))$ defines a ring homomorphism $\beta : M_n(R) \to M_n(S)$.

5. If the rings R and S are isomorphic, prove the following ring isomorphisms:
i) $R[x] \cong S[x]$
ii) $R[[x]] \cong S[[x]]$
iii) $R[X] \cong S[X]$ for any nonempty set of commuting indeterminates X over R and S
iv) $M_n(R) \cong M_n(S)$

6. Let R be commutative with 1 and $X = \{x_i \mid i \in \mathbf{N}\}$ indeterminates over R.
i) For any $\{r_i \mid i \in \mathbf{N}\} \subseteq R$, set $\varphi(f(x_1, \ldots, x_m)) = f(r_1, \ldots, r_m)$. Show that $\varphi : R[X] \to R$ is a surjective ring homomorphism.
ii) For any $\{p_i(X) \in R[X] \mid i \in \mathbf{N}\}$, set $\eta(f(x_1, \ldots, x_m)) = f(p_1(X), \ldots, p_m(X))$. Show that $\eta : R[X] \to R[X]$ is a ring homomorphism.

7. If for each $1 \le i \le k$, $\varphi_i : R \to S_i$ is a ring homomorphism, show that $\varphi : R \to \oplus S_i$ defined by $\varphi(r) = (\varphi_1(r), \ldots, \varphi_k(r))$ is a ring homomorphism. Under what conditions will φ be injective?

8. If $\rho : \mathbf{Z}_n \to R$ is a ring isomorphism and $n > 1$ is not prime, show that R contains nonzero ideals I and J with $IJ = (0_R)$.

11.4 ISOMORPHISM THEOREMS

The results in this section mimic those in Chapter 6 and Chapter 7 for groups. The statements and arguments here are essentially the same as those for groups, usually requiring only a verification that the isomorphisms between groups are also ring homomorphisms. We begin with the kernel.

DEFINITION 11.7 For $\varphi : R \to S$ a ring homomorphism, the **kernel** of φ, denoted $ker\ \varphi$, is the inverse image $\varphi^{-1}(\{0_S\}) = \{r \in R \mid \varphi(r) = 0_S\}$.

Observe that $ker\ \varphi$ is precisely the kernel of $\varphi : (R, +) \to (S, +)$. In particular, from §6.2, $ker\ \varphi$ is an additive subgroup of $(R, +)$ containing 0_R.

THEOREM 11.11 If $\varphi : R \to S$ is a ring homomorphism, then:

i) $ker\ \varphi \lhd R$
ii) φ is injective $\Leftrightarrow ker\ \varphi = (0_R)$
iii) When φ is surjective then, it is an isomorphism $\Leftrightarrow ker\ \varphi = (0_R)$.

Proof By Theorem 6.4, $ker\ \varphi \le (R, +)$. Let $x \in ker\ \varphi$, $r \in R$, and use that φ is a homomorphism, the definition of kernel, and Theorem 10.1 to see that $\varphi(rx) = \varphi(r)\varphi(x) = \varphi(r)0_S = 0_S$, and $\varphi(xr) = \varphi(x)\varphi(r) = 0_S\varphi(r) = 0_S$. These computations prove $rx, xr \in ker\ \varphi$, which verifies that $ker\ \varphi \lhd R$. Clearly φ is injective or surjective precisely when it is as a homomorphism of additive groups, so ii) and iii) follow from Theorem 6.4. ∎

As in Example 11.14, when I is an ideal of R, $\rho(r) = r + I$ defines a surjective ring homomorphism $\rho : R \to R/I$ with $ker\ \rho = \{r \in R \mid \rho(r) = 0_{R/I}\} = \{r \in R \mid r + I = I\} = I$. Just as for groups, if $\varphi : R \to S$ is a surjective ring homomorphism then $S \cong R/ker\ \varphi$. The idea is exactly the same as for groups: the isomorphism from S to $R/ker\ \varphi$ identifies $s \in S$ with its inverse image $\varphi^{-1}(\{s\}) = r + ker\ \varphi$ for $r \in R$ with $\varphi(r) = s$. The three parts of the next theorem in order are the **First, Second,** and **Third Isomorphism Theorems.**

THEOREM 11.12 Let $\varphi : R \to S$ be a surjective ring homomorphism, A a subring of R, and $I \subseteq J \subseteq R$ with $I, J \lhd R$. Then the following hold:

i) $S \cong R/ker\ \varphi$
ii) $(A + I)/I \cong A/(A \cap I)$
iii) $(R/I)/(J/I) \cong R/J$

Proof If $\beta(r + ker\ \varphi) = \varphi(r)$, then $\beta : (R/ker\ \varphi, +) \to (S, +)$ is an isomorphism by Theorem 6.6, so to prove i) we need β to be a ring homomorphism. Coset multiplication gives $\beta((a + ker\ \varphi)(b + ker\ \varphi)) = \beta(ab + ker\ \varphi) = \varphi(ab) = \varphi(a)\varphi(b) = \beta(a + ker\ \varphi)\beta(b + ker\ \varphi)$, and β is a ring homomorphism.

For ii), note first that $(A + I)/I$ is defined. We could show that $(A + I, +) = \{a + x \in R \mid a \in A \text{ and } x \in I\} \leq R$ is a subring containing I as an ideal. Instead use the surjective homomorphism $\rho : R \to R/I$ in Example 11.14, where $\rho(r) = r + I$ and $ker\ \rho = I$. Then $\rho(A) = \{a + I \mid a \in A\}$, and $a + I = a + x + I$ for any $x \in I$ shows that $\rho(A) = \{(a + x) + I \mid a \in A, x \in I\} = (A + I)/I$ is a subring of $\rho(R) = R/I$ by Theorem 11.9. Define $\eta = \rho|_A : A \to (A + I)/I$ by $\eta(a) = \rho(a) = a + I$. Thus η is a surjective homomorphism and $ker\ \eta = \{a \in A \mid a \in ker\ \rho = I\} = A \cap I$. Applying part i) yields $(A + I)/I \cong A/ker\ \eta = A/(A \cap I)$, proving ii). Note that $A \cap I = ker\ \eta$ itself implies $A \cap I \lhd A$ by Theorem 11.11.

For iii), we need to make sense of $(R/I)/(J/I)$ and begin with the surjection $\rho : R \to R/I$ for $\rho(r) = r + I$. Now $I \lhd J$ since $I \lhd R$, so J/I is a ring. Next, Theorem 11.9 yields $\rho(J) = \{a + I \in R/I \mid a \in J\} = J/I \lhd \rho(R) = R/I$, and $(R/I)/(J/I)$ is defined using Theorem 11.8. Consider $\sigma : R \to (R/I)/(J/I)$ given by $\sigma(r) = (r + I) + J/I$. Observe that $\sigma(a + b) = ((a + b) + I) + J/I = ((a + I) + (b + I)) + J/I = ((a + I) + J/I) + ((b + I) + J/I) = \sigma(a) + \sigma(b)$, using the addition in R/I and in $(R/I)/(J/I)$. Similarly, $\sigma(ab) = (ab + I) + J/I = (a + I)(b + I) + J/I = ((a + I) + J/I)((b + I) + J/I) = \sigma(a)\sigma(b)$. Therefore σ is a ring homomorphism and is clearly surjective. Part i) shows $(R/I)/(J/I) \cong R/ker\ \sigma$. By Definition 11.7, $ker\ \sigma = \{r \in R \mid \sigma(r) = (r + I) + J/I = J/I\} = \{r \in R \mid r + I \in J/I\}$. Now $r + I \in J/I \Leftrightarrow r + I = x + I$ for some $x \in J \Leftrightarrow r \in J$, since $I \subseteq J$ implies that $r \in x + I \subseteq J$. Therefore $ker\ \sigma = J$ and $(R/I)/(J/I) \cong R/J$, completing the proof of the theorem. ∎

We give a number of examples to illustrate the theorem, particularly the first part. Ideals of rings are harder to find than normal subgroups in groups so our examples here are somewhat less varied than for groups. First, using the automorphism $I_R : R \to R$ yields

$R \cong R/ker\, I_R = R/(0)$. Just as for groups, this isomorphism identifies $r \in R$ with the singleton $r + (0) = \{r\}$. The analysis of some previous examples that identified quotient rings is easier using Theorem 11.12. In Example 11.4, if $\varphi(f(x, y)) = f(x, 0) \in R[x]$ then $\varphi : R[x, y] \to R[x]$ is a surjective ring homomorphism with $ker\, \varphi = (y) = yR[x, y]$, called I in the example (verify!). Now by Theorem 11.12, $R[x] \cong R[x, y]/I$ as rings. We consider some variants of this.

Example 11.17 If R is a ring with identity and $(x) = xR[x] \lhd R[x]$ is the ideal of polynomials with zero constant term, then $R[x]/(x) \cong R$.

The evaluation $\varphi : R[x] \to R$ given by $\varphi(f(x)) = f(0) \in R$ is a ring homomorphism and is surjective (exercise!). But $f(0) = 0 \Leftrightarrow f(x) \in xR[x]$, so $ker\, \varphi = (x)$ and then $R \cong R[x]/(x)$ by Theorem 11.12.

A slightly different but related set of examples is described next.

Example 11.18 If R is a ring and $I \lhd R$, then $R[X]/I[X] \cong (R/I)[X]$ for any nonempty set of indeterminates X, and $R[[x]]/I[[x]] \cong (R/I)[[x]]$.

Denote $r + I = [r] \in R/I$, take $X = \{x\}$ with one element, and define $\varphi : R[x] \to (R/I)[x]$ by $\varphi(r_n x^n + \cdots + r_1 x + r_0) = [r_n]x^n + \cdots + [r_1]x + [r_0]$. Using the addition and multiplication in $R[x]$, together with $\rho(r) = [r]$ a homomorphism from R to R/I it follows that φ is a ring homomorphism. For example, if $f(x) = \sum r_i x^i$ and $g(x) = \sum s_i x^i$ then the coefficient of x^k in $\varphi(f(x)g(x))$ is $[r_0 s_k + r_1 s_{k-1} + \cdots + r_k s_0]$. Since ρ is a homomorphism, this element is $[r_0][s_k] + \cdots + [r_k][s_0]$, the coefficient of x^k in $\varphi(f(x))\varphi(g(x))$. It follows that $\varphi(f(x)g(x)) = \varphi(f(x))\varphi(g(x))$ (why?). Clearly φ is surjective so $(R/I)[x] \cong R[x]/ker\, \varphi$. Now $f(x) \in ker\, \varphi$ exactly when $\varphi(f(x)) = 0$ in $(R/I)[x]$, so when $[r_j] = 0_{R/I}$ for every coefficient r_j of $f(x)$. This is equivalent to each $r_j \in I$, or to $f(x) \in I[x]$. Therefore $R[x]/I[x] \cong (R/I)[x]$ from Theorem 11.12. The same approach shows that $R[[x]]/I[[x]] \cong (R/I)[[x]]$ and, more tediously, that $R[X]/I[X] \cong (R/I)[X]$ for arbitrary X (verify!). In the latter case the argument works as well when the elements in X do not commute with each other.

Example 11.18 shows that $\mathbf{Z}[x]/(n\mathbf{Z})[x] \cong \mathbf{Z}_n[x]$. For another simple illustration consider $\mathbf{Q}[x, y] = \mathbf{Q}[x][y]$, let $(x) = x\mathbf{Q}[x] \lhd \mathbf{Q}[x]$, and apply Example 11.18 and Example 11.17 to write $\mathbf{Q}[x, y]/((x)[y]) \cong (\mathbf{Q}[x]/(x))[y] \cong \mathbf{Q}[y]$. This uses the exercise that $R \cong S$ implies $R[x] \cong S[x]$. If we interpret Example 11.18 to mean that ideals of polynomial rings behave well with respect to quotients, then the same is true for ideals of matrix rings.

Example 11.19 If R is a ring and $I \lhd R$ then $M_n(R)/M_n(I) \cong M_n(R/I)$.

Proceed as in Example 11.18: define $\varphi : M_n(R) \to M_n(R/I)$ by $\varphi(A)_{ij} = \rho(A_{ij}) = [A_{ij}]$ for $\rho(r) = r + I = [r]$, the surjection of R onto R/I. Thus φ sends each entry of $A \in M_n(R)$ to its class in R/I. For $A, B \in M_n(R)$, $\varphi(A + B)_{ij} = \rho((A + B)_{ij}) = \rho(A_{ij} + B_{ij}) = \rho(A_{ij}) + \rho(B_{ij}) = \varphi(A)_{ij} + \varphi(B)_{ij}$. Since the entries of $\varphi(A+B)$ and of $\varphi(A)+\varphi(B)$ are equal, these matrices are equal. Similarly, $\varphi(AB)_{ij} = \rho((AB)_{ij}) = \rho(A_{i1}B_{1j} + \cdots + A_{in}B_{nj}) = \rho(A_{i1})\rho(B_{1j}) + \cdots + \rho(A_{in})\rho(B_{nj}) = \varphi(A)_{i1}\varphi(B)_{1j} + \cdots + \varphi(A)_{in}\varphi(B)_{nj} = (\varphi(A)\varphi(B))_{ij}$ and $\varphi(AB) = \varphi(A)\varphi(B)$ so φ is a homomorphism. Given any $([r_{ij}]) \in M_n(R/I)$, we have $\varphi((r_{ij})) = ([r_{ij}])$ so φ is surjective. Finally, $ker\,\varphi$ consists of those $A \in M_n(R)$ such that every $\rho(A_{ij}) = [0]$, or, equivalently, every $A_{ij} \in I$. Thus $ker\,\varphi = M_n(I)$ and $M_n(R)/M_n(I) \cong M_n(R/I)$ by Theorem 11.12.

Special cases of the example are that $M_n(\mathbf{Z})/M_n(m\mathbf{Z}) \cong M_n(\mathbf{Z}_m)$ and that for R a ring with identity, $M_n(R[x])/M_n((x)) \cong M_n(R)$ using Example 11.17 and the exercise that $R \cong S$ implies $M_n(R) \cong M_n(S)$. From Example 11.18 we get results like $M_n(\mathbf{Z}[x])/M_n((m\mathbf{Z})[x]) \cong M_n(\mathbf{Z}_m[x])$ (How?). An unexpected consequence of Theorem 11.12 is presented next.

Example 11.20 If $X = \{x_i \mid i \in N\}$ are commuting indeterminates over $\mathbf{Z}$ then $\mathbf{Z}[X]/(x_1) \cong \mathbf{Z}[X]$.

Any $f(X) \in \mathbf{Z}[X]$ can be considered as $f(X) \in \mathbf{Z}[x_1, \ldots, x_t]$ for t large enough (why?). Define $\eta : \mathbf{Z}[X] \to \mathbf{Z}[X]$ by $\eta(f(x_1, \ldots, x_m)) = f(0, x_1, \ldots, x_{m-1})$, so if $f(x_1, \ldots, x_m) = \sum f_i(x_2, \ldots, x_m)x_1^i$, then $\eta(f(x_1, \ldots, x_m)) = f_0(x_1, \ldots, x_{m-1})$. It is not hard to verify that η is a ring homomorphism. Also η is surjective since $g(x_1, \ldots, x_m) = \eta(g(x_2, \ldots, x_{m+1}))$, and $ker\,\eta = (x_1) = x_1\mathbf{Z}[X]$. Therefore $\mathbf{Z}[X]/(x_1) \cong \mathbf{Z}[X]$ by Theorem 11.12.

For X as in Example 11.20, the quotient $\mathbf{Z}[X]/(T)$ for $\varnothing \neq T \subseteq X$ satisfying $X - T$ infinite, is also isomorphic to $\mathbf{Z}[X]$. By Theorem 11.5, $(T) = \sum_T (x_j)$ consists of all polynomials for which every monomial contains an indeterminate in T. That $\mathbf{Z}[X]/(T) \cong \mathbf{Z}[X]$ is left to the exercises.

The second and third parts of Theorem 11.12 have not been used yet. These arise more naturally in proving other results, but one application of interest is a different approach to Example 11.6.

Example 11.21 If $m, n \in N$ with $m \mid n$ then $\mathbf{Z}_n/([m]_n) \cong \mathbf{Z}_m$ as rings.

We know $\mathbf{Z}/n\mathbf{Z} = \mathbf{Z}_n$ and since $m \mid n, n\mathbf{Z} \subseteq m\mathbf{Z}$. By Theorem 11.12, $(\mathbf{Z}/n\mathbf{Z})/(m\mathbf{Z}/n\mathbf{Z}) \cong \mathbf{Z}/m\mathbf{Z} = \mathbf{Z}_m$. Now $m\mathbf{Z}/n\mathbf{Z} = \{km + n\mathbf{Z} \in \mathbf{Z}_n \mid k \in \mathbf{Z}\} = \{[km]_n \in \mathbf{Z}_n \mid k \in \mathbf{Z}\} = ([m]_n)$, so $\mathbf{Z}_m \cong (\mathbf{Z}/n\mathbf{Z})/([m]_n) = \mathbf{Z}_n/([m]_n)$ as required.

When $m \mid n$, since $n\mathbf{Z} \subseteq m\mathbf{Z}$ and $n\mathbf{Z} \triangleleft m\mathbf{Z}$, we can consider the ring $m\mathbf{Z}/n\mathbf{Z} = ([m]_n) \triangleleft \mathbf{Z}_n$. What is the structure of this ring? There is no elegant answer. The product of any four elements in $2\mathbf{Z}/16\mathbf{Z}$ is zero, so this four-element ring is not isomorphic to $\mathbf{Z}_4$. However, $5\mathbf{Z}/20\mathbf{Z} \cong \mathbf{Z}_4$ as rings using $\varphi([j]_4) = [5j] \in 5\mathbf{Z}/20\mathbf{Z}$ (verify by direct computation!). This forces $5\mathbf{Z}/20\mathbf{Z}$ to have an identity element! We can describe $m\mathbf{Z}/n\mathbf{Z}$ in a special case.

Example 11.22 If $m, n \in N$ with $(m, n) = 1$ then as rings $m\mathbf{Z}/nm\mathbf{Z} \cong \mathbf{Z}_n$.

One approach uses $1 = sm + tn$ and Theorem 11.12 i) with $\varphi(k) = [ksm]_{nm}$ for $k \in \mathbf{Z}$, but we prefer to illustrate the use of Theorem 11.12 ii). The facts we need are that $m\mathbf{Z} + n\mathbf{Z} = (m, n)\mathbf{Z} = \mathbf{Z}$ from Theorem 1.8 and that $m\mathbf{Z} \bigcap n\mathbf{Z} = \text{LCM}\{m, n\}\mathbf{Z}$ from Example 2.11. Using $(m, n) = 1$, the corollary to Theorem 1.19 shows that $\text{LCM}\{m, n\} = mn$, so from Theorem 11.12 we have $m\mathbf{Z}/nm\mathbf{Z} = m\mathbf{Z}/(m\mathbf{Z} \bigcap n\mathbf{Z}) \cong (m\mathbf{Z} + n\mathbf{Z})/n\mathbf{Z} = \mathbf{Z}/n\mathbf{Z} = \mathbf{Z}_n$.

The example lets us describe $\rho(m\mathbf{Z}) \lhd \mathbf{Z}_n$ for $\rho : \mathbf{Z} \to \mathbf{Z}_n$ defined by $\rho(k) = [k]_n$ when $n = p_1 \cdots p_s$ for $p_1 < \cdots < p_s$ primes. As in the proof of Example 11.21, using Theorem 11.12, $\rho(m\mathbf{Z}) = \{mk + n\mathbf{Z} \in \mathbf{Z}_n \mid k \in \mathbf{Z}\} = (m\mathbf{Z} + n\mathbf{Z})/n\mathbf{Z} \cong m\mathbf{Z}/(m\mathbf{Z} \bigcap n\mathbf{Z}) = m\mathbf{Z}/\text{LCM}\{m, n\}\mathbf{Z}$. Since n has no repeated primes in its factorization, $\text{LCM}\{m, n\} = ms$ where s is the product of the primes appearing in n but not in m. In particular, $(m, s) = 1$. Thus from Example 11.22, $\rho(m\mathbf{Z}) \cong \mathbf{Z}_s$ as rings. This implies that the ring $\rho(m\mathbf{Z}) = ([m]_n) \subseteq \mathbf{Z}_n$ has an identity element. For example, let $n = 30$ and $m = 14$. Since $\text{LCM}\{14, 30\} = 14 \cdot 15$, $([14]_{30}) \cong 14\mathbf{Z}/\text{LCM}\{14, 30\}\mathbf{Z} = 14\mathbf{Z}/14 \cdot 15\mathbf{Z} \cong \mathbf{Z}_{15}$. Calculation shows that the identity of $([14]_{30}) = ([2_{30}])$ is $[16]_{30}$ and one isomorphism is $\varphi([16k]_{30}) = [k]_{15} \in \mathbf{Z}_{15}$. Alternatively, observe $([14]_{30}) = ([2]_{30})$ and use $2\mathbf{Z}/\text{LCM}\{2, 30\}\mathbf{Z} = 2\mathbf{Z}/2 \cdot 15\mathbf{Z} \cong \mathbf{Z}_{15}$ by Example 11.22.

We can apply Theorem 11.12 to get Example 11.5. More generally, for the projection $\pi_j : \oplus R_i \to R_j$ in Example 11.13, $ker\, \pi_j = M_j = R_1 \oplus \cdots \oplus R_{j-1} \oplus (0) \oplus R_{j+1} \oplus \cdots \oplus R_k$ so $\oplus R_i / M_j \cong R_j$. It is useful to extend this idea to ideals.

THEOREM 11.13 If $R_1, \ldots, R_k$ are rings and $I_j \lhd R_j$, then $\oplus I_j = I_1 \oplus \cdots \oplus I_k$ is an ideal of $\oplus R_i$ and $\oplus R_j / \oplus I_j \cong \oplus (R_j/I_j)$.

Proof Let $\varphi : \oplus R_i \to \oplus (R_j/I_j)$ be defined by $\varphi((r_1, \ldots, r_k)) = (r_1 + I_1, \ldots, r_k + I_k)$. Take $(r_1, \ldots, r_k), (s_1, \ldots, s_k) \in \oplus R_i$ and use the definitions of addition and of multiplication in direct sums and in quotients to compute that

$$\begin{aligned}
\varphi((r_1, \ldots, r_k) + (s_1, \ldots, s_k)) &= \varphi((r_1 + s_1, \ldots, r_k + s_k)) \\
&= ((r_1 + s_1) + I_1, \ldots, (r_k + s_k) + I_k) \\
&= ((r_1 + I_1) + (s_1 + I_1), \ldots, (r_k + I_k) + (s_k + I_k)) \\
&= (r_1 + I_1, \ldots, r_k + I_k) + (s_1 + I_1, \ldots, s_k + I_k) \\
&= \varphi((r_1, \ldots, r_k)) + \varphi((s_1, \ldots, s_k)) \text{ and that} \\
\varphi((r_1, \ldots, r_k)(s_1, \ldots, s_k)) &= \varphi((r_1 s_1, \ldots, r_k s_k)) = (r_1 s_1 + I_1, \ldots, r_k s_k + I_k) \\
&= ((r_1 + I_1)(s_1 + I_1), \ldots, (r_k + I_k)(s_k + I_k)) \\
&= (r_1 + I_1, \ldots, r_k + I_k)(s_1 + I_1, \ldots, s_k + I_k) \\
&= \varphi((r_1, \ldots, r_k))\varphi((s_1, \ldots, s_k)).
\end{aligned}$$

Consequently, φ is a ring homomorphism, it is clearly surjective, and $ker\, \varphi = \{(r_1, \ldots, r_k) \in \oplus R_i \mid (r_1 + I_1, \ldots, r_k + I_k) = (I_1, \ldots, I_k) \in \oplus (R_j/I_j)\} = \{(r_1, \ldots, r_k) \in \oplus R_i \mid r_j \in I_j$ for all $j\} = \oplus I_i$. Thus $\oplus I_i \lhd \oplus R_i$ by Theorem 11.11, or directly, and Theorem 11.12 yields $\oplus (R_j/I_j) \cong \oplus R_i / ker\, \varphi = \oplus R_i / \oplus I_i$. ∎

We restate Theorem 7.11 for rings and use Theorem 11.12.

THEOREM 11.14 If $\{m(1), \ldots, m(k)\} \subseteq \boldsymbol{N}$ is pairwise relatively prime and $n = m(1) \cdots m(k)$, then $\mathbf{Z}_n \cong \mathbf{Z}_{m(1)} \oplus \mathbf{Z}_{m(2)} \oplus \cdots \oplus \mathbf{Z}_{m(k)}$ as rings. Also $U_n \cong U_{m(1)} \oplus \cdots \oplus U_{m(k)}$ as multiplicative groups.

Proof Define $\varphi : \mathbf{Z} \to \mathbf{Z}_{m(1)} \oplus \cdots \oplus \mathbf{Z}_{m(k)}$ by $\varphi(z) = ([z]_{m(1)}, \ldots, [z]_{m(k)})$. Take any t, $z \in \mathbf{Z}$ and observe that $\varphi(t+z) = ([t+z]_{m(1)}, \ldots, [t+z]_{m(k)}) = ([t]_{m(1)} + [z]_{m(1)}, \ldots, [t]_{m(k)} + [z]_{m(k)}) = ([t]_{m(1)}, \ldots, [t]_{m(k)}) + ([z]_{m(1)}, \ldots, [z]_{m(k)}) = \varphi(t) + \varphi(z)$ and also that $\varphi(tz) = ([tz]_{m(1)}, \ldots, [tz]_{m(k)}) = ([t]_{m(1)}[z]_{m(1)}, \ldots, [t]_{m(k)}[z]_{m(k)}) = ([t]_{m(1)}, \ldots, [t]_{m(k)}) \cdot ([z]_{m(1)}, \ldots, [z]_{m(k)}) = \varphi(t)\varphi(z)$. Thus φ is a homomorphism. By Definition 11.7, $ker\,\varphi = \{z \in \mathbf{Z} \mid ([z]_{m(1)}, \ldots, [z]_{m(k)}) = ([0]_{m(1)}, \ldots, [0]_{m(k)})\} = \{z \in \mathbf{Z} \mid z \in m(j)\mathbf{Z} \text{ for all } j\}$. Therefore $z \in ker\,\varphi \Leftrightarrow m(j) \mid z$ for all j. Since $\{m(j)\}$ is pairwise relatively prime, Theorem 1.12 (and induction!) forces $n \mid z$, so $ker\,\varphi = n\mathbf{Z}$. Now by Theorem 11.9, $\varphi(\mathbf{Z})$ is a subring of $\oplus\, \mathbf{Z}_{m(j)}$, and so Theorem 11.12 gives $\varphi(\mathbf{Z}) \cong \mathbf{Z}/n\mathbf{Z} = \mathbf{Z}_n$. But $|\oplus\, \mathbf{Z}_{m(j)}| = n$ (Theorem 0.3) and $|\varphi(\mathbf{Z})| = |\mathbf{Z}_n| = n$ also, so $\mathbf{Z}_n \cong \varphi(\mathbf{Z}) = \oplus\, \mathbf{Z}_{m(j)}$. Now $\alpha \in \oplus\, \mathbf{Z}_{m(j)}$ is a unit precisely if each coordinate is a unit (verify!). It follows that $U_n = U(\mathbf{Z}_n) \cong U(\oplus\, \mathbf{Z}_{m(j)}) = \oplus\, U(\mathbf{Z}_{m(j)}) = \oplus\, U_{m(j)}$ as groups. We have used the exercises: if $R \cong S$ as rings then $U(R) \cong U(S)$ as multiplicative groups, by restriction of the isomorphism between R and S; and that $U(A \oplus B) \cong U(A) \oplus U(B)$. ∎

Theorem 11.14 is also called the **Chinese Remainder Theorem.** The name comes from very old problems attributed to the Chinese and of the following form: find an integer whose remainder on division by n_1 is r_1, by n_2 is r_2, etc. A solution is $s \in \mathbf{Z}$ satisfying $s \equiv r_1 \pmod{n_1}$, $s \equiv r_2 \pmod{n_2}$, etc. The proof of Theorem 11.14 shows that there is a solution if $\{n_j\}$ is pairwise relatively prime, for then $([r_1]_{n_1}, \ldots, [r_k]_{n_k}) = \varphi(z) = ([z]_{n_1}, \ldots, [z]_{n_k})$ for some $z \in \mathbf{Z}$, so $z \equiv r_j \pmod{n_j}$ for each j. A method for computing a solution is given in Exercise 22 in §1.7. An amusing consequence of this (Exercise 24 in §1.7) is that for any $m \in \boldsymbol{N}$ there are m consecutive integers in $\boldsymbol{N}$ so that each is divisible by some cube larger than 1. For example, if p_i is the ith prime then there is a solution $n \in \boldsymbol{N}$ of $x \equiv -j \pmod{p_j^3}$ for $1 \leq j \leq m$, and so $p_j^3 \mid (n+j)$. The smallest choice for three consecutive positive integers divisible in order by 8, 27, and 125 is 21,248, 21,249 and 21,250.

Actually Theorem 11.14 is a special case of a result that holds in any ring with identity. For $n, m \in \mathbf{Z}$ the condition $(m, n) = 1$ is equivalent to $n\mathbf{Z} + m\mathbf{Z} = \mathbf{Z}$ and this is the formulation we need in general.

THEOREM 11.15 Let R be a ring wih 1 (or just $R^2 = R$), $I_1, \ldots, I_k \lhd R$ with $k \geq 2$, and $I_i + I_j = R$ if $i \neq j$. Then $R/\bigcap I_j \cong R/I_1 \oplus \cdots \oplus R/I_k$.

Proof Define $\varphi : R \to R/I_1 \oplus \cdots \oplus R/I_k$ by $\varphi(r) = (r + I_1, \ldots, r + I_k)$. As in the previous theorem, it is routine to verify that φ is a ring homomorphism. Thus $\varphi(R)$ is a subring of $\oplus\, R/I_j$ by Theorem 11.9, and $ker\,\varphi = \{r \in R \mid r \in I_j \text{ for all } j\} = \bigcap I_j$, so Theorem 11.12 forces $\varphi(R) \cong R/\bigcap I_j$. We need to show that φ is surjective: if $(r_1 + I_1, \ldots, r_k + I_k) \in \oplus\, R/I_j$ we must find $r \in R$ with $\varphi(r) = (r_1 + I_1, \ldots, r_k + I_k)$. Since $R^2 = R$ it follows that $R^{k-1} = RR\cdots R = R$, so for any $1 \le j \le k$, using $I_s \lhd R$ and applying Theorem 11.6 (and induction!)

$$R = (I_1 + I_j)\cdots(I_{j-1} + I_j)(I_{j+1} + I_j)\cdots(I_k + I_j) \subseteq I_1 \cdots I_{j-1}I_{j+1}\cdots I_k + I_j$$

Thus we may write $r_j = a_j + b_j$ with $b_j \in I_j$ and $a_j \in I_1 \cdots I_{j-1}I_{j+1}\cdots I_k$. Note that $r_j + I_j = a_j + I_j$, and for each $i \ne j, a_j \in I_i$. Set $r = a_1 + \cdots + a_k$ so $r + I_j = a_j + I_j$. Hence $r + I_j = r_j + I_j$ and we have $\varphi(r) = (r_1 + I_1, \ldots, r_k + I_k)$ as required. ■

The condition on the ideals in the theorem will be satisfied by any finite collection of different maximal ideals. Our last example in this section applies Theorem 11.15 to polynomial rings and identifies certain quotients.

Example 11.23 If F is a field, $a_1, \ldots, a_k \in F$ distinct, and $f(x) = \prod(x - a_i)$ in $F[x]$, then as rings $F[x]/(f(x)) \cong F^k = F \oplus \cdots \oplus F$. Thus $F[x]/(x - a) \cong F$.

If $I_j = (x - a_j)F[x] \lhd F[x]$ and $j \ne m$, then $a_m - a_j = x - a_j - (x - a_m) \in (I_j + I_m) \bigcap F^*$. But a proper ideal cannot contain a unit and $I_j + I_m \lhd F[x]$ by Theorem 11.5, so $I_j + I_m = F[x]$ when $j \ne m$. Hence Theorem 11.15 shows that $F[x]/\bigcap I_j \cong \bigoplus F[x]/I_j$ as rings. For $g(x) \in F[x]$ and $b \in F$, we need to know that $g(b) = 0 \Leftrightarrow g(x) = (x - b)h(x) \in (x - b)F[x]$. This standard fact was derived in the proof of Theorem 4.18. By Example 11.12, evaluation at $b \in F$, $E_b(g(x)) = g(b)$, is a surjective ring homomorphism of $F[x]$ onto F and $ker E_b = \{g(x) \in F[x] \mid g(b) = 0\} = (x - b)F[x]$. Thus Theorem 11.12 yields $F \cong F[x]/(x - b)$ for any $b \in F$, so, in particular, each $F[x]/I_j \cong F$. This proves that $\bigoplus F[x]/I_j \cong F \oplus \cdots \oplus F = F^k$. To complete the argument we need that $\bigcap I_j = (f(x))$. But Theorem 11.7 applies to show that $\bigcap I_j = I_1 I_2 \cdots I_k = (x - a_1)F[x]\cdots(x - a_k)F[x] = (x - a_1)\cdots(x - a_k)F[x] = (f(x))$.

The example holds for a commutative ring R with identity, in place of F, if the $\{a_i\}$ satisfy $a_i - a_j \in U(R)$ whenever $i \ne j$. This forces $R[x] = (x - a_i)R[x] + (x - a_j)R[x]$ and Theorem 11.15 is still applicable. The only tricky point is that $R[x]/(x - a_i)R[x] \cong R$ still holds by using evaluation as in Example 11.23: $g(a_i) = 0 \Leftrightarrow g(x) \in (x - a_i)R[x]$ because $x - a_i$ is monic of degree 1 (exercise!). For example, if $R = \boldsymbol{Q}[x]$ and $q_1, \ldots, q_k \in \boldsymbol{Q}$ are distinct, then $\boldsymbol{Q}[x, y]/\bigcap_{i=1}^k (y - q_i)\boldsymbol{Q}[x, y] \cong \boldsymbol{Q}[x]^k$ as rings.

EXERCISES

1. If $\varphi_1 : R_1 \to S_1, \ldots, \varphi_k : R_k \to S_k$ are ring homomorphisms, show that $\varphi : \bigoplus R_j \to \bigoplus S_j$ defined by $\varphi((r_1, \ldots, r_k)) = (\varphi_1(r_1), \ldots, \varphi_k(r_k))$ is a ring homomorphism and find $ker\,\varphi$ using $\{ker\,\varphi_j\}$.
2. **a)** For p a prime, show that $x^p - x = x(x - [1])\cdots(x - [p-1])$ in $\boldsymbol{Z}_p[x]$ (Fermat!).
   b) Show that $R = \boldsymbol{Z}_p[x]/(x^p - x)$ has exactly 2^p elements satisfying $r^2 = r$.
3. If $I \lhd R$, a ring, show that:
   i) $I[[x]] \lhd R[[x]]$ and $R[[x]]/I[[x]] \cong (R/I)[[x]]$
   ii) $I\langle\langle x\rangle\rangle \lhd R\langle\langle x\rangle\rangle$ and $R\langle\langle x\rangle\rangle/I\langle\langle x\rangle\rangle \cong (R/I)\langle\langle x\rangle\rangle$
   iii) $I[X] \lhd R[X]$ and $R[X]/I[X] \cong (R/I)[X]$ for any set X of indeterminates over R

4. Let R be a ring with 1, and let $I, J \lhd R$ so that $I + J = R$ and $I \bigcap J = (0)$.
   i) Show that there is $e^2 = e \in I \bigcap Z(R)$ with $eR = I$.
   ii) Show that $R \cong I \oplus J$ as rings.
   iii) If $R \cong A \oplus B$ as rings with 1, find $J, L \lhd R$, $J + L = R$ and $J \bigcap L = (0)$.

5. Let $R = \{(f(t), g(t)) \in \boldsymbol{Q}[t] \oplus \boldsymbol{Q}[t] \mid f(0) = g(0)\}$.
   i) Show that R is a subring of $\boldsymbol{Q}[t] \oplus \boldsymbol{Q}[t]$.
   ii) Show that $I = \{(0, th(t)) \in R \mid h(t) \in \boldsymbol{Q}[t]\} \lhd R$ and $R/I \cong \boldsymbol{Q}[t]$.
   iii) If $A, B \lhd R$ with $A \bigcap B = (0)$, show that $R \neq A + B$ (see Exercise 4).
   iv) Show that $R \cong \boldsymbol{Q}[x, y]/(xy)$ as rings.

6. If $UT_n(R)$ is the ring of upper triangular matrices over R (Example 10.9), find a surjective ring homomorphism $\varphi : UT_n(R) \to R^n$ and find $ker\,\varphi$.

7. For any ring R denote the upper triangular matrices $UT_2(R)$ by $S = \begin{bmatrix} R & R \\ 0 & R \end{bmatrix}$. If $I, J \lhd R$, show that $K = \begin{bmatrix} I & R \\ 0 & J \end{bmatrix} = \{A \in S \mid A_{11} \in I \text{ and } A_{22} \in J\} \lhd S$ and prove that $S/K \cong R/I \oplus R/J$.

8. For R a ring with 1, let $R[x, x^{-1}] = \{a_{-m}x^{-m} + \cdots + a_0 + a_1x + \cdots + a_nx^n \mid m, n \in \boldsymbol{N}$ and all $a_j \in R\} \subseteq R\langle\langle x \rangle\rangle$ be the subring $\langle R \cup \{x, x^{-1}\}\rangle$. Find a surjective ring homomorphism $\varphi : R[y, z] \to R[x, x^{-1}]$ and identify its kernel.

9. For $\{R_j\}_A$ rings, $\oplus_A R_j = \{f \in (\bigcup R_j)^A \mid f(j) \in R_j$ for all $j \in A$ and $f(i) = 0$ for all but finitely many $i\}$ is a ring by Exercise 1 in §10.2 using $f \cdot g(j) = f(j)g(j)$ and $(f + g)(j) = f(j) + g(j)$. If $\varnothing \neq S \subseteq A$ is proper, show that there is a subring $H(S)$ of $\oplus_A R_j$ with $H(S) \cong \oplus_S R_j$, $H(S) \lhd \oplus_A R_j$, and $\oplus_A R_j / H(S) \cong \oplus_{A-S} R_j$.

10. **a)** For R a ring and $I_1, \ldots, I_k \lhd R$, show that $R/(\bigcap I_j) \cong S \subseteq \oplus R/I_j$, S a subring.
   b) If $\{I_\lambda \lhd R\}_L$, show that $R/(\bigcap I_j) \cong S \subseteq \oplus R/I_\lambda$, a subring (Example 9).

11. Let $X = \{x_i\}_N$ be indeterminates over $\boldsymbol{Z}$, set $E = \{x_{2i} \in X \mid i \in \boldsymbol{N}\}$, and $I = \sum_E (x_{2i}) \lhd \boldsymbol{Z}[X]$.
   i) Show that $\boldsymbol{Z}[X] \cong \boldsymbol{Z}[X - E]$.
   ii) Show that $\boldsymbol{Z}[X]/I \cong \boldsymbol{Z}[X]$.
   iii) If $\varnothing \neq T \subseteq X$ and if $X - T$ is infinite, show that $\boldsymbol{Z}[X]/\sum_T (x_i) \cong \boldsymbol{Z}[X]$.

12. A ring R is a **subdirect sum** of rings $\{R_i\}_I$ if for each $i \in I$, some $\varphi_i : R \to R_i$ is a surjective ring homomorphism with $ker\,\varphi_i \neq (0)$, and if $\bigcap ker\,\varphi_i = (0)$.
   i) R a subdirect sum of $\{R_i\}_I$ with $R_i \cong S_i \Rightarrow R$ is a subdirect sum of $\{S_i\}_I$.
   ii) If $\{I_j\}_A$ are nonzero ideals of R and R is a subdirect sum of $\{R/I_j\}_A$ using the usual surjections $\rho_j : R \to R/I_j$, then find an injective homomorphism $\varphi : R \to \prod R/I_j$, overall all $j \in A$. (See §10.1 Exercise 4.)
   iii) Show that $\boldsymbol{Z}$ is a subdirect sum of $\{\boldsymbol{Z}_p \mid p$ is a prime$\}$.
   iv) Show that $\boldsymbol{Q}[x]$ is a subdirect sum of $\{R_i\}_N$ with each $R_i = \boldsymbol{Q}$.
   v) Show that $\boldsymbol{Q}[x]$ is a subdirect sum of $\{\boldsymbol{Q}[x]/(x^i) \mid i \in \boldsymbol{N}\}$.

13. Using Exercise 12:
   i) If R is a subdirect sum of commutative rings then R is commutative.
   ii) If R is a subdirect sum of domains, show that if $r \in R$ and $r^k = 0$ for some $k > 1$, then $r = 0$.
   iii) If R is a subdirect sum of $R_i = M_2(A_i)$ for A_i commutative rings, show that for every $r, s \in R$, $(rs - sr)^2 \in Z(R)$.

11.5 THE CORRESPONDENCE THEOREM

From Theorem 7.2, if $\varphi: R \to S$ is a surjective ring homomorphism then there is an inclusion-preserving bijection between the additive subgroups of R containing $ker\,\varphi$ and the additive subgroups of S. This bijection assigns any subgroup of $(R, +)$ containing $ker\,\varphi$ to its image under φ and assigns any subgroup of $(S, +)$ to its inverse image under φ. These maps preserve subrings, ideals, and quotients. This correspondence for rings is important, though not as crucial as it was for groups, and has interesting consequences.

THEOREM 11.16 If $\varphi: R \to S$ is a surjective ring homomorphism, $ker\,\varphi = I$, $A = \{H \leq (R, +) \mid I \leq H\}$, and $B = \{K \leq (S, +)\}$, then $\alpha: A \to B$ given by $\alpha(H) = \varphi(H)$, and $\beta: B \to A$ defined by $\beta(K) = \varphi^{-1}(K)$ are inverses, so each is an inclusion-preserving bijection. Furthermore:

i) W a subring of $R \Rightarrow \varphi(W)$ is a subring of S and $\varphi(W) \cong W/(W \cap I)$
ii) $V \subseteq S$ a subring $\Rightarrow \varphi^{-1}(V)$ is a subring of R and $V = \varphi(\varphi^{-1}(V))$
iii) if $I \subseteq T$ then $T \triangleleft_r R \Leftrightarrow \varphi(T) \triangleleft_r S$
iv) if $I \subseteq T$ then $T \triangleleft_l R \Leftrightarrow \varphi(T) \triangleleft_l S$
v) if $I \subseteq T$ then $T \triangleleft R \Leftrightarrow \varphi(T) \triangleleft S$
vi) if $I \subseteq K \subseteq M \subseteq R$ for $K, M \triangleleft R$ then as rings $M/K \cong \varphi(M)/\varphi(K)$, so $R/K \cong S/\varphi(K)$
vii) if $J \subseteq L \subseteq S$ for $J, L \triangleleft S$, then as rings $L/J \cong \varphi^{-1}(L)/\varphi^{-1}(J)$, so $L \cong \varphi^{-1}(L)/I$

Proof That α and β are inclusion-preserving inverses is just Theorem 7.2. For i), $\varphi(W) \subseteq S$ is a subring from Theorem 11.9 so define $\upsilon: W \to \varphi(W)$ by $\upsilon(w) = \varphi(w)$. Clearly υ is a ring homomorphism since φ is, and from its definition, υ is surjective. Now $ker\,\upsilon = \{w \in W \mid \varphi(w) = 0_S\} = W \cap I$ so Theorem 11.12 yields $\varphi(W) \cong W/(W \cap I)$. To prove ii), since $\varphi^{-1}(V) = \beta(V)$ is an additive subgroup of R and $V = \varphi(\varphi^{-1}(V))$ by Theorem 7.2, we need $\varphi^{-1}(V)$ to be closed under multiplication (Theorem 10.4). For $x, y \in \varphi^{-1}(V)$, $\varphi(x), \varphi(y) \in V$ so $\varphi(xy) = \varphi(x)\varphi(y) \in V$ and $xy \in \varphi^{-1}(V)$. The image under φ of any left, right, or two-sided ideal of R has the same structure in S by Theorem 11.9, proving one direction in iii), iv), and v). If $V \leq (S, +)$, then $I = \varphi^{-1}(\{0_S\}) \subseteq \varphi^{-1}(V) = T \in A$ and $V = \alpha(\beta(V)) = \varphi(T)$, so to prove iii), iv), or v), it suffices to show that when V is a right, left, or two-sided ideal in S then $\varphi^{-1}(V)$ is also in R. If $V \triangleleft_r S$, $x \in \varphi^{-1}(V)$ and $r \in R$, then $\varphi(xr) = \varphi(x)\varphi(r) \in V$ since $\varphi(x) \in V$ and $\varphi(r) \in S$. Thus $xr \in \varphi^{-1}(V)$ and $\varphi^{-1}(V) \triangleleft_r R$. Similarly, $V \triangleleft_l S$ implies $\varphi^{-1}(V) \triangleleft_l R$, and $V \triangleleft S$ implies $\varphi^{-1}(V) \triangleleft R$. For vi), $\varphi(K) \triangleleft S$ by v) so $\varphi(K) \triangleleft \varphi(M)$ and $\varphi(M)/\varphi(K)$ is defined. Then $\eta: M \to \varphi(M)/\varphi(K)$ via $\eta(x) = \varphi(x) + \varphi(K)$ is a surjective ring homomorphism (verify!). For example, $\eta(xy) = \varphi(xy) + \varphi(K) = (\varphi(x) + \varphi(K))(\varphi(y) + \varphi(K)) = \eta(x)\eta(y)$. Now $ker\,\eta = \{x \in M \mid \varphi(x) \in \varphi(K)\} = \{x \in M \mid x \in \varphi^{-1}(\varphi(K)) = \beta(\alpha(K)) = K\} = K$ and Theorem 11.12 show that $M/K \cong \varphi(M)/\varphi(K)$. Part vii) follows from vi) since $J = \alpha(\beta(J)) = \varphi(\varphi^{-1}(J))$, $L = \varphi(\varphi^{-1}(L))$, and $I \subseteq \varphi^{-1}(J) \subseteq \varphi^{-1}(L) \triangleleft R$. ■

The theorem shows that *every* subring V of S is a quotient since $\varphi^{-1}(V)$ is a subring of R from the theorem, so $V = \varphi(\varphi^{-1}(V)) \cong \varphi^{-1}(V)/(\varphi^{-1}(V) \cap ker\,\varphi)$. The most frequent application of Theorem 11.16 is when $S = R/I$ for I an ideal of R and $\rho: R \to R/I$ is the usual surjection $\rho(r) = r + I$. If A is a subring of R then by Theorem 11.12, $\rho(A) = (A + I)/I \cong A/(A \cap I)$. In particular, when $A \cap I = (0)$, then $\rho(A) \cong A$. This is an isomorphism of the appropriate structure of A: of subrings if A is a subring, of left ideals if A is a left ideal, etc. (verify!). Also, just as for groups, every subring or (left, right) ideal $K \subseteq R/I$ has the form $K = \rho(\rho^{-1}(K)) = (\rho^{-1}(K) + I)/I$ and $\rho^{-1}(K) \subseteq R$ has the same structure (subring, or (left, right) ideal) as K, from the theorem.

From Theorem 11.16, $M \lhd R$ is maximal exactly when R/M has no proper nonzero ideal, since ideals strictly between M and R correspond to nonzero proper ideals of R/M. We record this remark for commutative rings.

THEOREM 11.17 For R a commutative ring with identity, $M \lhd R$ is maximal $\Leftrightarrow$ R/M is a field.

Proof By Theorem 11.2, R/M is a field $\Leftrightarrow$ it has no proper nonzero ideal. But this is equivalent to M being a maximal ideal of R by Theorem 11.16. ■

The theorem gives a way to construct fields by finding maximal ideals in commutative rings. We can identify maximal ideals in $\mathbf{Z}$ and $F[x]$, using Theorem 11.13. For $\mathbf{Z}$, the maximal ideals are the $p\mathbf{Z}$ for p a prime, and the corresponding fields are the $\mathbf{Z}/p\mathbf{Z} = \mathbf{Z}_p$. In $F[x]$, for F a field, the maximal ideals are those $(f(x))$ with $f(x)$ irreducible: that is, $f(x) \notin F$ but $f(x) = g(x)h(x)$ forces $g(x) \in F$ or $h(x) \in F$. Example 10.19 and Example 11.7 are such quotients in disguise. The general description of these fields comes next.

THEOREM 11.18 For F a field and $f(x) = x^n + \cdots + a_1x + a_0 \in F[x] - F$ irreducible, $K = F[x]/(f(x))$ is a field. If $[g(x)] = g(x) + (f(x)) \in K$ then
$F' = \{[a] \in K \mid a \in F\} \cong F$ as rings and $X^n + \cdots + [a_1]X + [a_0] \in F'[X]$ has
$Y = [x] = x + (f(x))$ as a root. Each $k \in K^*$ is $k = [g(x)]$ for a unique $g(x) \in F[x]$ with
$deg\ g(x) < n$. In particular, when F is finite, $|K| = |F|^n$.

Proof K is a field by Theorem 11.17. This follows directly since if $[g(x)] \neq [0]$ then $g(x) \notin (f(x))$ so $(g(x)) + (f(x)) \neq (f(x))$, a maximal ideal, forcing $(g(x)) + (f(x)) = F[x]$. Thus $1 = h(x)g(x) + k(x)f(x)$ for $h(x), k(x) \in F[x]$ and $[h(x)] = [g(x)]^{-1}$. We claim $\varphi(a) = [a]$ gives a ring homomorphism $\varphi : F \to K$. Now $\varphi(ab) = [ab] = ab + (f(x)) = (a + (f(x)))(b + (f(x))) = [a][b] = \varphi(a)\varphi(b)$. Similarly, $\varphi(a + b) = \varphi(a) + \varphi(b)$ (exercise!). By definition $\varphi(F) = F'$, and $ker\ \varphi = (0_F)$ since $\varphi(F) \neq (0_K)$, $ker\ \varphi \lhd F$, and F has no nontrivial ideals by Theorem 11.2. Thus using Theorem 11.12, $F' \cong F/(0_F) \cong F$. From the definition of the coset operations, if $[g(x)] \in K$ and $g(x) = c_kx^k + \cdots + c_0$, then $[g(x)] = [c_k]Y^k + \cdots + [c_0]$. Since $f(x) \in (f(x))$, $Y^n + \cdots + [a_1]Y + [a_0] = [f(x)] = 0_K$. Y is a root of the polynomial in $F'[X]$ corresponding to $f(x)$, and $Y^n = [-a_{n-1}]Y^{n-1} + \cdots + [-a_1]Y + [-a_0]$. Thus $Y^{n+1} = [-a_{n-1}]Y^n + [-a_{n-2}]Y^{n-1} + \cdots + [-a_0]Y = [v_{n-1}]Y^{n-1} + \cdots + [v_1]Y + [v_0]$ (why?). An induction argument shows that for $k \geq n$, $Y^k = [d_0] + \cdots + [d_{n-1}]Y^{n-1}$ for some $[d_j] \in F'$. Hence any $[g(x)] \in K$ has this same form. We need to show the uniqueness of these expressions. If $[b_0] + \cdots + [b_{n-1}]Y^{n-1} = [d_0] + \cdots + [d_{n-1}]Y^{n-1}$, then $[b_0 - d_0] + \cdots + [b_{n-1} - d_{n-1}]Y^{n-1} = 0_K$, so $q(x) = (b_0 - d_0) + \cdots + (b_{n-1} - d_{n-1})x^{n-1} \in (f(x))$ (why?). But every $h(x) \in (f(x))$ is a multiple of $f(x)$, so is zero or has degree at least n. Consequently, $q(x) = 0 \in F[x]$, forcing $b_j = d_j$ for each j. Thus each $k \in K$ can be written as $k = [b_0] + \cdots + [b_{n-1}]Y^{n-1}$, uniquely. ■

We usually identify the field $F' \subseteq K$ in Theorem 11.18 *as* F and so write $k \in K = F[x]/(f(x))$ as a polynomial of degree less than n in Y (or X) with coefficients *in* F *itself.* It is possible to replace K formally with an isomorphic copy actually containing F, but is not often necessary to do so.

In general it is not easy to recognize when $f(x) \in F[x]$ is irreducible, especially if $deg\ f(x) \geq 3$. Now $f(x)$ is irreducible $\Leftrightarrow af(x)$ is irreducible for any $a \in F^*$, so we need only consider monic polynomials. Clearly linear polynomials $x - a$ are irreducible. In these cases $K = F[x]/(x - a) = F' \cong F$ using Example 11.23. If $f(x) \in F[x]$ with $2 \leq deg\ f \leq 3$ is not irreducible then $f(x)$ must have a linear factor and so a root. When $4 \leq deg\ f(x)$, $f(x)$ may not be irreducible even though $f(a) \neq 0$ for all $a \in F$. An earlier example was $f(x) = x^4 + x^2 + 1 \in \boldsymbol{R}[x]$. Clearly $f(r) > 0$ for all $r \in \boldsymbol{R}$, but $f(x)$ factors in $\boldsymbol{R}[x]$ as $f(x) = (x^2 + x + 1)(x^2 - x + 1)$. It is not clear at this point that there are many irreducibles in $\boldsymbol{Z}_p[x]$ for p a prime (there are!). To illustrate Theorem 11.18, let us construct a field of eight elements.

Example 11.24 $K = \boldsymbol{Z}_2[x]/(x^3 + x + 1)$ is a field with eight elements.

For $f(x) = x^3 + x + 1$, $f([0]_2) = f([1]_2) = [1]_2$, so $f(x)$ has no linear factor over $\boldsymbol{Z}_2$ and must be irreducible. The statement now follows from Theorem 11.18, but we want to look a little closer at this field. Write $Y = [1]_2 x + (f(x)) \in K$, $[1] = [1]_2 + (f(x))$, $[1]k = k$ for all $k \in K$, and $K = \{0, [1], Y, Y + [1], Y^2, Y^2 + [1], Y^2 + Y, Y^2 + Y + [1]\}$. If $k \in K$, $k + k = 2k = 0$ (why?), so, for example, $(Y^2 + [1]) + (Y^2 + Y + [1]) = Y$. How does multiplication in K work? Since $Y^3 + Y + [1] = 0$, $Y^3 = Y + [1]$ follows, using that $[1]_2 = -[1]_2 = [-1]_2$, a pleasant feature in characteristic two. Now $Y^4 = Y^2 + Y$, so we can compute any products. For example, $Y^5 = Y^3 + Y^2 = Y^2 + Y + [1]$, $(Y^2 + [1])^2 = Y^4 + [1] = Y^2 + Y + [1]$, and $(Y^2 + [1])(Y + [1]) = Y^3 + Y^2 + Y + [1] = Y^2$. Note that $|(K^*, \cdot)| = 7$ so $(K^*, \cdot) = \langle k \rangle$ for any $k \in K^*$. The consecutive powers of Y are: $Y, Y^2, Y + [1], Y^2 + Y, Y^2 + Y + [1], Y^2 + [1]$, and $[1]$. In particular, we have $Y(Y^2 + [1]) = [1] = Y^2(Y^2 + Y + [1]) = (Y + [1])(Y^2 + Y)$.

Now $f(x) = x^3 + x + 1 \in \boldsymbol{Z}_5[x]$ is irreducible, so the construction in the example gives a field of 125 elements with $Y^3 = -Y - [1] = [4]Y^2 + [4]$. Also $f(x) \in \boldsymbol{Z}_7[x]$ is irreducible, resulting in a field of 343 elements. However, $f(x) = (x - [1]_3)(x^2 + x - [1]_3) \in \boldsymbol{Z}_3[x]$, and in $\boldsymbol{Z}_{11}[x]$, $f(x) = (x - [2]_{11})(x^2 + [2]_{11}x + [5]_{11})$ so in these cases $\boldsymbol{Z}_p[x]/(x^3 + x + 1)$ is not a field.

Example 11.25 $\boldsymbol{Q}[x]/(x^2 - 3) \cong \boldsymbol{Q}(\sqrt{3}) = \{a + b\sqrt{3} \in \boldsymbol{R} \mid a, b \in \boldsymbol{Q}\}$.

By Example 11.12, $\varphi(f(x)) = f(\sqrt{3})$ defines a homomorphism $\varphi : \boldsymbol{Q}[x] \to \boldsymbol{R}$. Since $(\sqrt{3})^{2k} = 3^k$ and $(\sqrt{3})^{2k+1} = 3^k\sqrt{3}$, $\varphi(\boldsymbol{Q}[x]) = \boldsymbol{Q}(\sqrt{3})$ follows and shows that $\boldsymbol{Q}(\sqrt{3})$ is a subring of $\boldsymbol{R}$ by Theorem 11.9 (or by direct computation). Using Theorem 11.12, $\boldsymbol{Q}(\sqrt{3}) \cong \boldsymbol{Q}[x]/ker\ \varphi$ and $ker\ \varphi = \{g(x) \in \boldsymbol{Q}[x] \mid g(\sqrt{3}) = 0\}$. But $x^2 - 3 \in ker\ \varphi$ and $x^2 - 3$ is irreducible in $\boldsymbol{Q}[x]$ since it has no linear factor in $\boldsymbol{Q}[x]$, so $(x^2 - 3)$ is a maximal ideal. Now clearly $(x^2 - 3) \subseteq ker\ \varphi$ and $ker\ \varphi \neq F[x]$ (why?). We are forced to conclude that $ker\ \varphi = (x^2 - 3)$, so then $\boldsymbol{Q}(\sqrt{3}) \cong \boldsymbol{Q}[x]/(x^2 - 3)$.

It is possible to proceed directly as in Example 11.7 using the representation of $Q[x]/(x^2-3)=K$ in Theorem 11.18. Namely, if $Y=x+(x^2-3)$, so $Y^2=[3]$, then $K=\{[a]+[b]Y \mid a,b\in Q\}$ and the bijection $\beta([a]+[b]Y)=a+b\sqrt{3}$ between K and $Q(\sqrt{3})$ is an isomorphism by an easy verification using $([a]+[b]Y)([c]+[d]Y)=[ac+3bd]+[ad+bc]Y$ in K.

EXERCISES

Recall from Exercise 16 in §11.1 that a proper $P \triangleleft R$ is prime if for $I, J \triangleleft R$ with $IJ \subseteq P$, then $I \subseteq P$ or $J \subseteq P$.

1. **i)** Show that $P \triangleleft R$ is prime $\Leftrightarrow$ for $(0) \neq A, B \triangleleft R/P$, $AB \neq (0)$ in R/P.
ii) Find all the prime ideals in $\mathbf{Z}_{24}$.

2. For a ring R, let $N(R)$ be the maximal nil ideal of R (Exercise 18 in §11.1). Show that $N(R/N(R))=(0)$: the only nil ideal in $R/N(R)$ is $(0_{R/N(R)})$.

3. In $Q[x,y,z]$, let $I=(x^2)+(y^3)$ and set $R=Q[x,y,z]/I$. Characterize $N(R)=\{r\in R \mid r^k=0 \text{ for some } k\geq 1\}$ and show that it is a prime ideal.

4. Let F be a field, $n\in N$, $I=(x^n)\triangleleft F[x]$, $R=F[x]/I$, and $J=(x+I)R\triangleleft R$. Show that:
i) for all $r\in J$, $r^n=0$.
ii) Each ideal of R is principal.
iii) R/J is a field.
iv) If $W\triangleleft R$ with $W\neq J$ and $W\neq R$, then R/W is not a domain.

5. Let R be a ring, M a maximal ideal in R, and I an ideal of R.
i) Show that either $I\subseteq M$, or $I\cap M$ is a maximal ideal of I.
ii) If V is a maximal ideal of R/I, $\rho: R\to R/I$ is the usual homomorphism, and $L=\rho^{-1}(V)$, show that L is a maximal ideal of R.

6. Let $\varphi: R\to S$ be a surjective ring homomorphism.
i) If every right ideal of R is principal, show that every right ideal of S is also.
ii) If $J, K\triangleleft S$, are proper, and $J+K=S$, show that $\varphi^{-1}(J)+\varphi^{-1}(K)=R$.

7. **i)** Find a maximal ideal I of $(x)\triangleleft \mathbf{Z}[x]$ such that $(x)/I$ is a field.
ii) Find a maximal ideal J of $(2\mathbf{Z})[x]\triangleleft \mathbf{Z}[x]$ such that $(2\mathbf{Z})[x]/J$ is not a field.
iii) Find a maximal ideal M of $(2\mathbf{Z})[x]\triangleleft \mathbf{Z}[x]$ such that $(2\mathbf{Z})[x]/M$ is a field.

8. For $I\triangleleft R$ and $I\subseteq P\triangleleft R$, show that P is prime $\Leftrightarrow P/I$ is a prime ideal of R/I.

9. If R is a ring so that every prime ideal of R is an intersection of maximal ideals, show that the same is true in R/I for any $I\triangleleft R$ (see Exercise 8).

10. For R a commutative ring with maximal nil ideal $N(R)$ (Exercise 19, §11.1):
i) If $a_1,\ldots,a_k\in N(R)$ and $I=(a_1)+\cdots+(a_k)$, show that there is $m\in N$ such that $I^m=II\cdots I=(0)$
ii) $N(R[x])=N(R)[x]$
iii) $M_k(N(R))=N(M_k(R))$ (Exercise 18, §11.1)

11. Describe the following quotient rings R/I:
i) $R=\mathbf{Z}[x]/(x^6)$, $I=(x^2)/(x^6)$
ii) $R=\mathbf{Z}[x,y,z]/(x^2)$, $I=((x)+(y))/(x^2)$
iii) $R=Q[x,y]/(x^3-y^3)$, $I=(x-y)Q[x,y]/(x^3-y^3)$

iv) $R = \mathbf{Z}[x, y]/(x - y)$, $I = (3\mathbf{Z}[x, y] + (x^2 + 1) + (x - y))/(x - y)$

v) $R = \boldsymbol{Q}[x, y, z]/(y - z)$, $I = ((x - 2) + (x - y) + (y - z))/(y - z)$

12. If R is a subdirect sum of rings each having (0) as a prime ideal (§11.4 Exercise 12), then if $J \lhd R$ with $J^k = (0)$ for some $k \geq 1$, show that $J = (0)$.

11.6 CHAIN CONDITIONS

We close the chapter by introducing important conditions on ideals, essential in later studies of rings but not used here except peripherally.

DEFINITION 11.8 A set of right ideals $\{T_i\}_N$ of a ring R is an **ascending chain** if $T_i \subseteq T_{i+1}$ for all $i \in N$ and is a **descending chain** if $T_i \supseteq T_{i+1}$ for all i. The chains are **strict** if each inclusion is proper. R is **right Noetherian,** or satisfies ACC (the ascending chain condition) on right ideals, if it has no infinite, strictly ascending chain of right ideals. R is **right Artinian,** or satisfies DCC (the descending chain condition) on right ideals if it has no infinite strictly descending chain of right ideals.

In noncommutative rings these chain conditions can be defined just as well for left ideals or for ideals. Clearly any finite ring satisfies ACC and DCC for left or for right ideals. Fields satisfy both conditions for ideals: a field has only two ideals! It turns out that DCC restricts the possible structure of the ring more than ACC does. We record two familiar examples that are Noetherian but not Artinian.

Example 11.26 $\mathbf{Z}$ and $F[x]$ for F a field, satisfy ACC for ideals but not DCC for ideals.

The ideals in $\mathbf{Z}$ are $\{n\mathbf{Z} \mid n \geq 0\}$, and $n\mathbf{Z} \subseteq m\mathbf{Z} \Leftrightarrow m \mid n$. Thus an ideal strictly larger than $n\mathbf{Z}$ arises from a proper positive factor m of n. In this case $m < n$ so any strictly ascending chain of ideals must be finite, and $\mathbf{Z}$ satisfies ACC on ideals. Now $\{I_j = 2^j\mathbf{Z}\}$ is an infinite, strictly descending chain of ideals. Also $F[x]$ satisfies ACC on ideals since from Theorem 11.3 its ideals are principal, and strict inclusions are possible only by taking proper factors. But if $(f(x)) \subseteq (g(x))$ properly then $deg\ g(x) < deg\ f(x)$, so any strictly increasing chain of ideals must be finite. The set $\{(x^i) \mid i \in N\}$ is an infinite, strictly descending chain of ideals in $F[x]$.

Both chain conditions fail for infinite direct products (Theorem 10.3) or sums (Example 10.6) (verify!). A simpler example is described next.

Example 11.27 $R = x\boldsymbol{Q}[x] \subseteq \boldsymbol{Q}[x]$ has an infinite, strictly ascending chain of ideals and an infinite, strictly descending chain of ideals.

$R = (x)$ is the subring of all $f(x) \in \boldsymbol{Q}[x]$ with $f(0) = 0$. For the descending chain, take $T_j = \mathbf{Z}x^j + x^jR$ and note that $x^j \in T_j - T_{j+1}$. Now let $I_j = \{2^{-j}kx + x^2q(x) \in R \mid k \in \mathbf{Z}$ and $q(x) \in \boldsymbol{Q}[x]\}$. Observing that $RI_j \subseteq (x^2) \subseteq I_j$, it is straightforward to check that each $I_j \lhd R$ and that $\{I_j\}_N$ is an infinite, strictly ascending chain of ideals since $2^{-j}x \in I_j - I_{j-1}$.

To see that chain conditions put constraints on rings, we observe that a ring R with identity is right Noetherian $\Leftrightarrow$ every $T \lhd_r R$ is a sum of finitely many principal right ideals: $T = a_1R + \cdots + a_kR$ for some $a_i \in R$. A formal proof of this requires the Axiom of Choice in §0.5, or its equivalent, but the idea is not hard to see. If R is a right Noetherian ring with right ideal $T \neq a_1R$ for any $a_1 \in T$, then for $a_1 \in T$ and $a_2 \in T - a_1R$, $a_1R + a_2R \subseteq T$. If $a_3 \in T - (a_1R + a_2R)$, then $a_1R + a_2R + a_3R \subseteq T$. Continuing, $T = a_1R + \cdots + a_kR$ or we get a strictly ascending chain of right ideals $a_1R \subseteq a_1R + a_2R \subseteq a_1R + a_2R + a_3R \subseteq \cdots$, contradicting the Noetherian assumption for R. On the other hand, if every right ideal in R is a finite sum of principal right ideals and $(0) = T_0 \subseteq T_1 \subseteq T_2 \subseteq \cdots$ is an infinite, ascending chain of right ideals, then the right ideal (Theorem 11.5) $T = \sum T_j = b_1R + \cdots + b_kR$. Now $b_i \in T$ implies that b_i is in a finite sum of the $\{T_j\}$, so for some $m \in N$ $\{b_1, \ldots, b_k\} \subseteq T_1 + \cdots + T_m = T_m$. Thus $T_{m+i} \subseteq T = b_1R + \cdots + b_kR \subseteq T_m$, forcing all $T_{m+i} \subseteq T_m$, so the chain $\{T_j\}$ cannot be strictly increasing and R is right Noetherian.

Chain conditions can also affect properties of elements. There are rings in which $ab = 1$ but $ba \neq 1$. An example is $R = End(\mathbf{R}[x], +)$ (see Theorem 10.2) using formal differentiation and integration (verify!) or using the functions in Example 10.16. However, "one-sided" inverses are not possible in the presence of either ACC or DCC.

Example 11.28 If a ring R with identity satisfies either ACC or DCC for right (or left) ideals then for $a, b \in R$ with $ab = 1$, also $ba = 1$.

Assume $a, b \in R, ab = 1$, but $ba \neq 1$. We construct elements that generate strictly increasing and decreasing chains of right (or left) ideals. First note that for $j \in N, a^jb^j = 1$ (why?). For $i \in N$, set $f_i = b^ia^i$ and observe that $f_if_j = f_jf_i = f_k$ for $k = \max\{i, j\}$. If $i \leq j$ then $f_if_j = b^ia^ib^ja^j = b^ib^{j-i}a^j = b^ja^j = f_j$, and similarly, $f_jf_i = b^ja^jb^ia^i = b^ja^{j-i}a^i = f_j$. Thus if $i < j$ then $(f_i - f_{i+1})f_j = 0 = f_j(f_i - f_{i+1})$ and $(f_i - f_{i+1})^2 = (f_i - f_{i+1})f_i - (f_i - f_{i+1})f_{i+1} = (f_i - f_{i+1})$. If $e_j = f_j - f_{j+1}$ then $e_j^2 = e_j$ and also $e_je_k = 0$ if $j \neq k$ (verify!). Consider $\{T_i = e_1R + \cdots + e_iR\}$. If $e_{i+1} \in T_i$ then $e_{i+1} = e_1r_1 + \cdots + e_ir_i$, so $e_{i+1} = e_{i+1}^2 = e_{i+1}e_1r_1 + \cdots + e_{i+1}e_ir_i = 0$. Thus $\{T_i\}$ is a strictly ascending chain of right ideals. For $\{W_i = \sum_{j>i} e_jR\}$, if $e_i \in W_i$ then $e_i = e_{i+1}r_{i+1} + \cdots + e_{i+m}r_{i+m}$ for some $m \geq 1$ by Theorem 11.5. As above $e_i = e_i^2 = e_ie_{i+1}r_{i+1} + \cdots + e_ie_{i+m}r_{i+m} = 0$, a contradiction, so $\{W_i\}$ is a strictly descending chain of right ideals. Hence if R satisfies the ACC or the DCC on right ideals then $ab = 1$ implies $ba = 1$. Similar constructions of left ideals using $\{e_j\}$ show that ACC or DCC on left ideals and $ab = 1$ imply $ba = 1$.

We give a final example using DCC.

Example 11.29 If a ring R with identity is right Artinian then any $a \in R$ that is not a *left* zero divisor must be invertible.

By Definition 10.17, if $a \in R^*$ in not a left zero divisor then $ar = 0$ implies $r = 0$. Consider the descending chain of right ideals $aR \supseteq a^2R \supseteq \cdots$. By DCC, this chain has only finitely many strict inclusions, so for k large enough, $a^kR = a^{k+1}R$. Since $1 \in R$, $a^k = a^k \cdot 1 \in a^kR = a^{k+1}R$, so $a^k = a^{k+1}b$ for some $b \in R$. Thus $a^k(1 - ab) = 0$ forces $ab = 1$ (why?), so $ba = 1$ by Example 11.28, resulting in $a \in U(R)$.

It is curious in Example 11.29 that we assume R is *right* Artinian but $a \in R^*$ is not a *left* zero divisor. This difference in "sides" is necessary, as we will see at the end of the chapter. Next we illustrate how isomorphism theorems arise in exploring the structure of rings. The theorem is not hard but is a fundamental one in the study of Noetherian or Artinian rings.

THEOREM 11.19 If R is a right Noetherian ring then for any ideal I of R, R/I is right Noetherian and I contains no infinite, strictly ascending chain of right ideals of R. If some ideal I of R satisfies these two conditions then R is right Noetherian. The corresponding statements for ACC on left ideals, and DCC on right or on left ideals also hold.

Proof We prove the statement for ACC on right ideals and leave the others as exercises. If R is a right Noetherian ring then I cannot contain an infinite, strictly ascending chain of right ideals of R, by definition. Let $M_1 \subseteq M_2 \subseteq \cdots$ be an ascending chain of right ideals in R/I. By Theorem 11.16, if $\rho : R \to R/I$ is the surjection $\rho(r) = r + I$, then $\rho^{-1}(M_1) \subseteq \rho^{-1}(M_2) \subseteq \cdots$ is an ascending chain of right ideals of R, so R right Noetherian means that for some $k \geq 1$, $\rho^{-1}(M_k) = \rho^{-1}(M_{k+j})$ for all $j \geq 1$. By the correspondence of right ideals given in Theorem 11.16, this forces $M_k = M_{k+j}$ so the chain $\{M_j\}$ in R/I cannot be strictly increasing. Hence R/I is right Noetherian.

Now assume that some ideal I of R contains no infinite, strictly ascending chain of right ideals of R and that R/I is a right Noetherian ring. Let $T_1 \subseteq T_2 \subseteq \cdots$ be right ideals in R. To show R is right Noetherian, we prove $T_m = T_{m+i}$ for some $m \geq 1$ and all $i \geq 1$. By Theorem 11.5, $\{T_j \cap I\}$ is a set of right ideals of R contained in I and is clearly an ascending chain, so by assumption there is $m' \geq 1$ so that $T_{m'} \cap I = T_{m'+j} \cap I$ for all $j \geq 1$. Also by Theorem 11.5, $\{T_i + I\}$ is an ascending chain of right ideals of R containing I. Let $\rho : R \to R/I$ be the usual homomorphism $\rho(r) = r + I$. By Theorem 11.9, $\rho(T_i + I) = (T_i + I)/I$ is a right ideal of R/I. Now $\{(T_i + I)/I\}$ is an ascending chain of right ideals in R/I by Theorem 11.16, so our assumption gives $m'' \geq 1$ with $(T_{m''} + I)/I = (T_{m''+j} + I)/I$ for all $j \geq 1$. The correspondence in Theorem 11.16 shows $T_{m''} + I = T_{m''+j} + I$ for all $j \geq 1$. If $m = \max\{m', m''\}$ we claim that $T_m = T_{m+j}$ for all $j \geq 1$. For $x \in T_{m+j}$, $x \in T_{m+j} + I = T_m + I$ since $m \geq m''$, so $x = t + y$ for $t \in T_m$ and $y \in I$. It follows that $y = x - t \in T_{m+j} \cap I = T_m \cap I$ since $m \geq m'$. Thus $x = t + y \in T_m$, forcing $T_m = T_{m+j}$, so R is right Noetherian. ■

COROLLARY If R is a right Noetherian or a right Artinian ring, so is any homomorphic image S of R.

Proof A homomorphic image S of R has the form $S = \varphi(R)$ for $\varphi : R \to S$ a surjective ring homomorphism. But then $S \cong R/ker\,\varphi$ by Theorem 11.12 and so is right Noetherian or right Artinian by the theorem. ■

We end the chapter with some consequences of Theorem 11.19.

THEOREM 11.20 Each of the rings $R_1, \ldots, R_k$ is right Noetherian $\Leftrightarrow \oplus R_j$ is also. The same statement holds for Artinian replacing Noetherian, or for left ideals rather than right ideals.

Proof Assume first that each R_j is right Noetherian. We argue by induction on $k \geq 2$. Note that $M = R_1 \oplus (0) \lhd R_1 \oplus R_2$ and $(R_1 \oplus R_2)/M \cong R_2$ using Example 11.13 and Theorem 11.12, or by Theorem 11.13. Since R_2 is a right Noetherian ring, to prove that $R_1 \oplus R_2$ is right Noetherian it suffices by Theorem 11.19 to show that M contains no infinite, strictly ascending chain of right ideals of $R_1 \oplus R_2$. If $T \subseteq M$ with $T \lhd_r R_1 \oplus R_2$ then for $(t, 0) \in T$, $(t, 0)(r_1, r_2) = (tr_1, 0) \in T$, so $T = T_1 \oplus (0)$ for $T_1 \lhd_r R_1$ (why?). Consequently, since R_1 is right Noetherian, M cannot contain an infinite, strictly ascending chain of right ideals of $R_1 \oplus R_2$, proving the $k = 2$ case. Now assume that the direct sum of n rings with ACC on right ideals satisfies the same property and consider $R_1 \oplus \cdots \oplus R_{n+1}$. Let $J = (0) \oplus \cdots \oplus (0) \oplus R_{n+1}$, observe that $J \lhd R_1 \oplus \cdots \oplus R_{n+1}$, and use Theorem 11.12 to see that $(R_1 \oplus \cdots \oplus R_{n+1})/J \cong R_1 \oplus \cdots \oplus R_n$ via $\varphi((r_1, \ldots, r_{n+1})) = (r_1, \ldots, r_n)$. Hence the induction assumption shows that $(R_1 \oplus \cdots \oplus R_{n+1})/J$ is a right Noetherian ring, and repeating the argument for the case $k = 2$ shows that J does not contain an infinite, strictly ascending chain of right ideals of $R_1 \oplus \cdots \oplus R_{n+1}$. Applying Theorem 11.19 again proves that $R_1 \oplus \cdots \oplus R_{n+1}$ is right Noetherian, so one implication of the theorem holds by IND-I. On the other hand, if $\oplus R_j$ is a Noetherian ring then each R_j is also by the corollary to Theorem 11.19, since R_j is the image of $\oplus R_j$ under the projection $\pi_j((r_1, \ldots, r_k)) = r_j$. Obvious modifications of the arguments above work as well for DCC on right ideals or for either chain condition for left ideals. ∎

The previous two theorems can be used to show that certain rings satisfy ACC or DCC without doing a lot of computation.

Example 11.30 Let $R = \{A \in M_2(\mathbf{Q}) \mid A_{11} \in \mathbf{Z} \text{ and } A_{21} = 0\}$. Then R is a right Noetherian ring but is not left Noetherian.

We can represent R as $\begin{bmatrix} \mathbf{Z} & \mathbf{Q} \\ 0 & \mathbf{Q} \end{bmatrix}$. It is straightforward to see that R is a ring with identity, that $J = \begin{bmatrix} 0 & \mathbf{Q} \\ 0 & 0 \end{bmatrix} = \{A \in R \mid A_{11} = A_{22} = 0\} \lhd R$, and that the right ideals of R in J are either (0) or J. If $\varphi : R \to \mathbf{Z} \oplus \mathbf{Q}$ is given by $\varphi(A) = (A_{11}, A_{22})$, then φ is a surjective homomorphism of rings with $ker\, \varphi = J$. Hence $R/J \cong \mathbf{Z} \oplus \mathbf{Q}$, a right Noetherian ring by Theorem 11.20. Applying Theorem 11.19 shows that R is right Noetherian. However, if $L_i = \{A \in J \mid A_{12} \in 2^{-i}\mathbf{Z}\}$ then $\{L_j\}$ is an infinite, strictly ascending chain of left ideals of R. Note that R is neither right nor left Artinian by the corollary to Theorem 11.19. That is, since $\mathbf{Z}$ is the image of R under the homomorphism $\eta : R \to \mathbf{Z}$ given by $\eta(A) = A_{11}$, and since $\{n!\mathbf{Z}\}$ is an infinite, strictly descending chain of ideals in $\mathbf{Z}$, it follows that $\mathbf{Z}$, and hence R, cannot satisfy DCC.

Example 11.31 Let $Q(x)$ be the field of fractions of $Q[x]$, and set $R = \{A \in M_2(Q(x)) \mid A_{21} = 0 \text{ and } A_{11} \in Q\}$. Then R is a right Artinian ring that is not left Artinian.

As in the last example, represent R as $\begin{bmatrix} Q & Q(x) \\ 0 & Q(x) \end{bmatrix}$. Again R is a ring and $J = \begin{bmatrix} 0 & Q(x) \\ 0 & 0 \end{bmatrix} = \{A \in R \mid A_{11} = A_{22} = 0\} \triangleleft R$. Since $Q(x)$ is a field, it follows that the only right ideals of R contained in J are (0) and J itself. As in Example 11.30, we get $R/J \cong Q \oplus Q(x)$ via the surjective homomorphism $\varphi : R \to Q \oplus Q(x)$ defined by $\varphi(A) = (A_{11}, A_{22})$. Since both Q and $Q(x)$ are fields, and so Artinian rings, $Q \oplus Q(x)$ is a right Artinian ring by Theorem 11.20. Thus Theorem 11.19 shows that R is a right Artinian ring. Now $L_j = \{A \in J \mid A_{12} \in x^j Q[x]\}$ is a left ideal of R (verify!) and $\{L_j\}$ is an infinite, strictly descending chain, so R is not left Artinian. Note that R is also right Noetherian, by essentially the same argument given for the Artinian condition. The left ideals $V_i = \{A \in J \mid A_{12} \in Qx + \cdots + Qx^i\}$ of R form a strictly ascending chain, so R is not left Noetherian.

Finally, referring to Example 11.29, note that in Example 11.31 the element $E_{12} + xE_{22} \in R$ is not a right zero divisor but is not invertible since $(E_{12} + xE_{22})E_{11} = 0$, even though R is right Artinian.

EXERCISES

1. Let R be a ring, for any $n \in N$ set $R^n = R \oplus \cdots \oplus R$, and let $B = \{T \leq (R^n, +) \mid (t_1, \ldots, t_n) \in T \text{ implies that } (t_1 r, \ldots, t_n r) \in T \text{ for all } r \in R\}$. Use the arguments in Theorem 11.19 and Theorem 11.20 as a model to show that if R satisfies either ACC or DCC on right ideals, then chains in B satisfy the same condition. (Note that when $n = 2$ and $I = R \oplus (0)$, any additive subgroup $T \in B$ satisfies: $T + I$ is a right ideal in $(R \oplus R)/I \cong R$.)
2. Use Exercise 1 to show that if R is a right Noetherian (Artinian) ring, then so is $M_n(R)$.
3. Find an example of a ring that is left Noetherian but not right Noetherian.
4. Find an example of a ring that is left Artinian but not right Artinian.
5. Use Example 5.14 to find a commutative ring satisfying DCC on ideals but not ACC on ideals.
6. Let R be a ring and $\varphi : R \to R$ a ring homomorphism.
   i) If R is right Noetherian and φ is surjective, show that φ is an isomorphism (note that $ker\ \varphi \subseteq ker\ \varphi^2$).
   ii) If R is right Artinian and φ is injective, show that φ is an isomorphism (note that $\varphi(R) \supseteq \varphi^2(R)$).
7. If R is a commutative domain, $1 \in R$, and R contains only finitely many ideals, show that R is a field.
8. Let F be a field and $f(x) \in F[x] - F$. i) Show that $F[x]/(f(x))$ is a Noetherian ring. ii) Show that $F[x]/(f(x))$ is an Artinian ring.
9. Assuming that $\mathbf{Z}[x]$ is a Noetherian ring, show that $(x) = x\mathbf{Z}[x]$ is also.

CHAPTER 12

FACTORIZATION IN INTEGRAL DOMAINS

From Definition 10.18, an integral domain is a commutative domain with 1. The most familiar examples are the rings $\mathbf{Z}$ and $F[x]$, for F a field. In $\mathbf{Z}$ we have the notion of divisibility, and every positive integer has a unique representation as a product of prime powers. We take $\mathbf{Z}$ as a model and study these properties in arbitrary integral domains.

12.1 PRIMES AND IRREDUCIBLES

If D is an integral domain, we can define divisibility just as in $\mathbf{Z}$. Recall that $D^* = D - \{0\}$ and that $U(D) = \{a \in D \mid ab = 1 \text{ for some } b \in D\}$ is the group of units of D under the multiplication in D.

DEFINITION 12.1 For $a, b \in R$, a commutative ring, when $a \neq 0$ we say that ***a* divides *b***, which is written $a \mid b$, if $b = ac$ for some $c \in R$.

Clearly $a \mid b \Leftrightarrow a \neq 0$ and $b \in aR \lhd R$. The properties of division in $\mathbf{Z}$ from Theorem 1.5 are essentially the same in any integral domain D. We restate these elementary facts but leave the proofs as exercises.

THEOREM 12.1 If $a, b, c, d, e \in D$, an integral domain, then: $a \mid 0$ if $a \neq 0$; $1 \mid a$; $a \mid b \Rightarrow a \mid bc$; $ab \mid c \Rightarrow a \mid c$; if $c \in D^*$ then $a \mid b \Leftrightarrow ac \mid bc$; $a \mid b$ and $b \mid c \Rightarrow a \mid c$; $a \mid b$ and $a \mid c \Rightarrow a \mid (bd + ce)$; and $a \mid b$ and $c \mid d \Rightarrow ac \mid bd$.

Facts relating divisibility and negatives in $\mathbf{Z}$ involve $U(D)$ for general integral domains. Note that $U(\mathbf{Z}) = \{\pm 1\}$.

THEOREM 12.2 If D is an integral domain and $a, b \in D^*$, then:

i) $a \mid b$ and $b \mid a \Leftrightarrow a = ub$ for $u \in U(D)$
ii) for any $v \in U(D)$, $a \mid b \Leftrightarrow va \mid b \Leftrightarrow a \mid vb$

Proof For i), $a \mid b$ and $b \mid a$ yield $b = ax$ and $a = by$ for $x, y \in D$, so $b = byx$ and $1 = yx$ since D is a domain. Hence $a = yb$ for $y \in U(D)$. Now $a = ub$ for $u \in U(D)$ yields $b \mid a$ and then $a \mid b$ from $b = u^{-1}a$. Part ii) follows from $b = ac \Rightarrow b = av(v^{-1}c)$, $b = (va)y \Rightarrow vb = a(v^2 y)$, and $vb = ac \Rightarrow b = a(v^{-1}c)$. ■

Write $c \sim_a d$ when both $c \mid d$ and $d \mid c$. Using Theorem 12.1, it is easy to see that $\sim_a$ is an equivalence relation on D^*. By Theorem 12.2, $c \sim_a d$ if and only if $d = uc$ for $u \in U(D)$, so $\{cU(D) \mid c \in D\}$ is a partition of D into the equivalence classes of $\sim_a$ on D^* together with $0 \cdot U(D) = \{0\}$. Elements in the same class $cU(D)$ are called *associates*.

DEFINITION 12.2 In an integral domain D any $c, d \in D$ are called **associates,** written $c \sim_a d$, if $d = uc$ for some invertible $u \in U(D)$.

Now $U(\mathbf{Z}) = \{\pm 1\}$ so $a, b \in \mathbf{Z}$ are associates $\Leftrightarrow b = \pm a$. When F is a field, $U(F[x]) = F^*$, the nonzero constants, by Theorem 10.11 on degrees, so it is easy to tell whether $f(x) \sim_a g(x)$. It is not always easy to identify $U(D)$, or hence to recognize associates. We look at a collection of very important examples.

Example 12.1 For $\gamma \in \mathbf{C} - \mathbf{Z}$ with $\gamma^2 \in \mathbf{Z}$, $\mathbf{Z}[\gamma] = \{a + b\gamma \in \mathbf{C} \mid a, b \in \mathbf{Z}\}$ is a subring of $\mathbf{C}$. If $\alpha = a + b\gamma \in \mathbf{Z}[\gamma]$ then the **norm** of α is $N(\alpha) = (a + b\gamma)(a - b\gamma) = a^2 - b^2\gamma^2 \in \mathbf{Z}$, and for $\alpha, \beta \in \mathbf{Z}[\gamma], t \in \mathbf{Z}$, $N(\alpha\beta) = N(\alpha)N(\beta)$, and $N(t\alpha) = t^2 N(\alpha)$. Furthermore, $u \in U(\mathbf{Z}[\gamma]) \Leftrightarrow N(u) = \pm 1$.

It is easy to check that $\mathbf{Z}[\gamma]$ is a subring of $\mathbf{C}$ and so an integral domain since $1 = 1 + 0 \cdot \gamma \in \mathbf{Z}[\gamma]$. Straightforward computations show that $N(\alpha\beta) = N(\alpha)N(\beta)$ and $N(t\alpha) = t^2 N(\alpha)$ (verify!). If $u \in U(\mathbf{Z}[\gamma])$ then $1 = uu^{-1}$ implies $1 = N(1) = N(u)N(u^{-1})$, forcing $N(u) = \pm 1$. When $(a + b\gamma)(a - b\gamma) = N(a + b\gamma) = \pm 1$ then $\pm(a - b\gamma) = (a + b\gamma)^{-1}$ and $(a + b\gamma) \in U(\mathbf{Z}[\gamma])$.

The rings in Example 12.1 give some nontrivial examples of associates. When $\gamma = i$, so $\gamma^2 = -1$, $N(a + bi) = a^2 + b^2$ and $U(\mathbf{Z}[i]) = \{\pm 1, \pm i\}$. Thus $3 + 5i \sim_a 5 - 3i$ since $5 - 3i = -i(3 + 5i)$. In this ring $a + bi \sim_a x + yi$ implies $\{\pm a, \pm b\} = \{\pm x, \pm y\}$ so it is fairly easy to tell when elements are associates. Similarly, in $\mathbf{Z}[(-p)^{1/2}]$ with p a prime, $N(a + b(-p)^{1/2}) = a^2 + pb^2$ and so $U(\mathbf{Z}[(-p)^{1/2}]) = \{\pm 1\}$. In $\mathbf{Z}[p^{1/2}]$ there are units other than ± 1 (this is not easy to see!), so associates are not easily described. For example, in $\mathbf{Z}[5^{1/2}]$, $-11 + 5 \cdot 5^{1/2} = (9 - 4 \cdot 5^{1/2})(1 + 5^{1/2})$ and $N(9 - 4 \cdot 5^{1/2}) = 9^2 - 4^2 \cdot 5 = 1$ so $9 - 4 \cdot 5^{1/2} \in U(\mathbf{Z}[5^{1/2}])$ and $1 + 5^{1/2} \sim_a -11 + 5 \cdot 5^{1/2}$. In $\mathbf{Z}[2^{1/2}]$, $N(a + b2^{1/2}) = a^2 - 2b^2$ so $3 - 2 \cdot 2^{1/2}$ is a unit and $5 + 2^{1/2} \sim_a 11 - 7 \cdot 2^{1/2} = (3 - 2 \cdot 2^{1/2})(5 + 2^{1/2})$.

In $\mathbf{Z}$, $p > 1$ is prime when $p = ab$ for $a, b \in \mathbf{N} \Rightarrow p = a$ or $p = b$. Equivalently, $p > 1$ is prime if $p = ab$ for $a, b \in \mathbf{Z} \Rightarrow a \in U(\mathbf{Z})$ or $b \in U(\mathbf{Z})$. The requirement that $p > 1$ is simply a specific choice in $pU(\mathbf{Z}) = \{\pm p\}$. It is more difficult to make such a choice in a general integral domain D since $U(D)$ may be large or hard to describe. In $\mathbf{Z}$, a property equivalent to being prime is Euclid's Lemma (Theorem 1.14): if a prime divides a product then it must divide one of the factors. This property and that of not factoring into nonunits fail to be equivalent in some integral domains, as indicated in Example 1.7. In $\mathbf{Z}[(-19)^{1/2}]$, $N(a + b(-19)^{1/2}) = a^2 + 19b^2$ so for $\beta \in \mathbf{Z}[(-19)^{1/2}]$ $N(\beta) \neq 2, 5,$ or 10, and if $\alpha = 1 + (-19)^{1/2}$ then $N(\alpha) = 20$. Should $\alpha = \beta\gamma$, then $20 = N(\alpha) = N(\beta)N(\gamma)$, forcing $N(\beta) = 1$ or $N(\gamma) = 1$ (why?), so $\beta \in U(D)$ or $\gamma \in U(D)$ by Example 12.1 and α has no proper factorization into nonunits. Now $N(\alpha) = 20$ implies $\alpha \mid 4 \cdot 5$, but if $\alpha \mid 4$ then $4 = \alpha\eta$ so $16 = N(4) = 20N(\eta)$, impossible in $\mathbf{Z}$. Similarly, $\alpha \mid 5$ is impossible, and although α divides $4 \cdot 5$ and cannot be factored into nonunits, α divides neither 4 nor 5.

This last example shows that we need to consider separately the properties that "a fails to factor into nonunits" and "$a \mid bc \Rightarrow a \mid b$ or $a \mid c$."

DEFINITION 12.3 If D is an integral domain, $a \in D^* - U(D)$, and $b, c \in D$, then a is **irreducible** if $a = bc$ implies $b \in U(D)$ or $c \in U(D)$, and a is **prime** if $a \mid bc$ implies $a \mid b$ or $a \mid c$.

Using Theorem 12.2, it is an exercise that if $a \sim_a b$ then a is irreducible (or prime) $\Leftrightarrow$ b is also. The example above shows that unlike in $\mathbf{Z}$, an irreducible need not be a prime. For another example, in $\mathbf{Z}[x, y]$ let $D = \mathbf{Z} + xy\mathbf{Z}[x, y]$. Now D contains no polynomial of degree 1, so xy is irreducible in D, divides $(xy)^3 = (x^2y)(xy^2)$, but cannot divide either x^2y or xy^2 in D (why?). However, a prime is always irreducible.

THEOREM 12.3 Let D be an integral domain and $p \in D^*$. Then p is a prime implies p is irreducible, and p is a prime $\Leftrightarrow$ $D/(p)$ is a domain.

Proof Suppose that $p = ab$ for $a, b \in D$. If p is a prime then $p \mid ab$ forces $p \mid a$ or $p \mid b$. But $a = pc$ yields $p = pcb$ so $1 = cb$, and $b \in U(D)$. Similarly, if $p \mid b$ then $a \in U(D)$, so p is irreducible. From Definition 11.5, (p) is a prime ideal exactly when $ab \in (p)$ implies $a \in (p)$ or $b \in (p)$. This is clearly equivalent to p a prime: $p \mid ab$ implies $p \mid a$ or $p \mid b$. But $D/(p)$ is a domain $\Leftrightarrow$ $(ab + (p)) = (a + (p))(b + (p)) = 0_{D/(p)} = (p)$ implies $(a + (p)) = (p)$ or $(b + (p)) = (p)$, which is also equivalent to (p) being a prime ideal, and so is equivalent to p a prime in D. ∎

We can use Theorem 12.3 to find some primes in $D[x]$.

THEOREM 12.4 If D is an integral domain and $p \in D$ is a prime, then p is also a prime in $D[X]$, for any set X of commuting indeterminates over D.

Proof By Theorem 12.3, it suffices to show that $D[X]/(p)$ is a domain. In $D[X]$, $(p) = \{pg(X) \mid g(X) \in D[X]\}$, so from the multiplication in $D[X]$ it follows that $(p) = (p)_D[X]$ for $(p)_D = pD \triangleleft D$. Thus $D[X]/(p) = D[X]/(p)_D[X] \cong (D/(p)_D)[X]$ by Example 11.18. But $D/(p)_D$ is a domain by Theorem 12.3, so Theorem 10.11 shows that $D[X]/(p) \cong (D/(p)_D)[X]$ is a domain. ∎

Some consequences of Theorem 12.4 are not entirely obvious. For example, if for $f(x), g(x) \in \mathbf{Z}$ every coefficient of $f(x)g(x)$ is even, then every coefficient of $f(x)$, or of $g(x)$, is even (why?). Similarly, if $p(x, y), q(x, y) \in \mathbf{Z}[x, y]$ and $x \mid p(x, y)q(x, y)$, then $x \mid p(x, y)$ or $x \mid q(x, y)$, since Example 11.17 and Theorem 12.4 show $x \in \mathbf{Z}[x]$ is prime, then that $x \in \mathbf{Z}[x, y] = \mathbf{Z}[x][y]$ is prime.

EXERCISES

1. Consider the subring $D = \langle\{1, x^2, x^3\}\rangle \subseteq \mathbf{Z}[x]$.
  i) Describe those $p(x) \in D$.
  ii) Show that $x^2 \in D$ is irreducible but is not a prime.
  iii) Describe $D/(x^2)$.

2. Verify that $xy \in D = \mathbf{Z} + xy\mathbf{Z}[x, y]$ is irreducible but is not a prime.

3. If a and b are associates in the integral domain D, show that:
  i) a is an irreducible $\Leftrightarrow b$ is an irreducible
  ii) a is a prime $\Leftrightarrow b$ is a prime

4. Show a and b are associates in the integral domain $D \Leftrightarrow (a) = (b)$.

5. D an integral domain, $p \in D$ a prime, and $(p) \subseteq (q) \Rightarrow (p) = (q)$ or $(q) = D$.

6. Show that $x \in D = \mathbf{Z} + x\mathbf{Q}[x]$ is not irreducible in D.

7. Show that $x \in \mathbf{Z}[x, y, z]$ is irreducible and is a prime.

8. If $D = \{a/b \in \mathbf{Q} \mid b \text{ is odd}\}$, show that D is an integral domain with every irreducible and every prime an associate of $2 = 2/1 \in D$.

9. Show that $U(\mathbf{Z}[p^{1/2}])$ is infinite (consider $\langle\alpha\rangle$ for $\alpha \in U(\mathbf{Z}[p^{1/2}])$) when $p =$:
  i) 2
  ii) 3
  iii) 5
  iv) 7
  v) 11
  vi) 13
  vii) 17

10. For $d \in \mathbf{Z}$, show that each $\alpha \in \mathbf{Z}[d^{1/2}]$ can be written as some product of irreducibles. (Use induction on $|N(\alpha)|$.)

11. If D is an integral domain and $(a) \lhd D$ is maximal, then a is a prime.

12. If D is an integral domain and $S = \{I \lhd D\}$ satisfies $I, J \in S \Rightarrow I = (a)$ for some $a \in D$ and $I \subseteq J$ or $J \subseteq I$, then D is a field or p, q primes in $D \Rightarrow p \sim_a q$.

12.2 PIDs

We have seen that an irreducible need not be a prime. Are there identifiable collections of domains in which each irreducible is a prime?

DEFINITION 12.4 An integral domain D is a **principal ideal domain,** or a PID, if $I \lhd D \Rightarrow I = (a)$ for some $a \in D$.

Of course $\mathbf{Z}$ is a PID, as is $F[x]$ for F a field by Theorem 11.3. Other examples, left as exercises, are $F[[x]]$ and $F[x, x^{-1}]$ (Exercise 13 in §10.3). It turns out that irreducibles *are* primes in PIDs. To see this, we need to introduce the idea of greatest common divisor. Unlike $\mathbf{Z}$, a general PID D has no natural ordering so we need to make sense of "greatest."

DEFINITION 12.5 For any $\emptyset \neq S \subseteq D$, an integral domain, $c \in D$ is a **common divisor** for S if $c \mid s$ for all $s \in S$. A common divisor c for S is a **greatest common divisor,** denoted $c \in \text{GCD}(S)$, if $d \mid s$ for all $s \in S \Rightarrow d \mid c$.

If $c, d \in \text{GCD}(S)$, then $c \mid d$ since c is a common divisor, and $d \mid c$ since d is a common divisor, so $c \sim_a d$ by Theorem 12.2. The same result shows that if $d \in \text{GCD}(S)$ and $c \sim_a d$, then $c \in \text{GCD}(S)$ as well. *Thus when $d \in GCD(S)$ then $GCD(S) = dU(D)$, and we will write $GCD(S) = d$.* It is not obvious that $\text{GCD}(S) \neq \emptyset$. Although $1 \in D$ is a common divisor for S, why should any common divisor be a multiple of all the others?

THEOREM 12.5 If D is a PID and $\emptyset \neq S \subseteq D^* - U(D)$ there is $\text{GCD}(S) = d$, unique up to associate, and $d = c_1 s_1 + \cdots + c_k s_k$ for some $c_j \in D$ and $s_j \in S$.

Proof By Theorem 11.5, $\sum_{s \in S}(s) = I \lhd D$, so $I = (d)$ using that D is a PID. Thus $d = c_1 s_1 + \cdots + c_k s_k$ for some $c_j \in D$ and $s_j \in S$. If $s \in S$ then $1 \cdot s = s \in I$ so $s = xd$ and $d \mid s$, making d a common divisor for S. Let $a \in D$ be any common divisor for S. Then $a \mid s_j$ so $s_j = x_j a$ and $d = \sum c_j s_j = \sum c_j x_j a = \left(\sum c_j x_j\right) a$. Hence $a \mid d$, proving that $d = \text{GCD}(S)$. That $\text{GCD}(S) = dU(D)$ follows from our comments before the theorem. ■

By the Euclidean Algorithm in §1.2, given $n, m \in \mathbf{Z}$ we can find $s, t \in \mathbf{Z}$ so that $\text{GCD}(\{n, m\}) = sn + tm$. In $F[x]$, for F a field, the procedure is essentially the same, using the analogue of Theorem 1.1 on remainders.

THEOREM 12.6 For F a field let $f(x), g(x) \in F[x]$ with $g(x) \neq 0$. There are unique $q(x), r(x) \in F[x]$ satisfying $f(x) = q(x)g(x) + r(x)$ with $r(x) = 0$ or $deg\ r(x) < deg\ g(x)$.

Proof First we show that some appropriate $q(x)$ and $r(x)$ exist. If $f(x) = 0$ take $q(x) = 0 = r(x)$. Now let $f(x) = ax^n + \cdots + a_0$ and $g(x) = bx^m + \cdots + c$ with $ab \neq 0$. When $n < m$ then $q(x) = 0$ and $r(x) = f(x)$ work. Let $n \geq m$ and set $h(x) = f(x) - ab^{-1}x^{n-m}g(x)$. If $h(x) = 0$ then take $q(x) = ab^{-1}x^{n-m}$ and $r(x) = 0$. Otherwise, $deg\ h(x) < n$ so by induction, $h(x) = q'(x)g(x) + r'(x)$ with $r' = 0$ or $deg\ r' < m$. Hence $f(x) = (ab^{-1}x^{n-m} + q'(x))g(x) + r'(x)$, as required, and the existence of $q(x)$ and $r(x)$ follow from IND-II. Assume that $q(x)g(x) + r(x) = q'(x)g(x) + r'(x)$ satisfy the conditions. Then $(q'(x) - q(x))g(x) = r(x) - r'(x)$ and if $r(x) \neq r'(x)$ then $deg(r(x) - r'(x)) \geq deg\ g(x)$ by Theorem 10.11, contradicting the choices of $r(x)$ and $r'(x)$. Thus $r(x) = r'(x)$, so $q(x) = q'(x)$. ■

Just as in §1.2 we apply the Euclidean Algorithm: write out the equations (using long division of polynomials) $f(x) = q_1(x)g(x) + r_1(x)$, $g(x) = q_2(x)r_1(x) + r_2(x)$ if $r_1(x) \neq 0$, $r_1(x) = q_3(x)r_2(x) + r_3(x)$ if $r_2(x) \neq 0$, etc. The last nonzero remainder $r_m(x)$, which must divide $r_{m-1}(x)$, is $\text{GCD}(\{f(x), g(x)\})$ (why?) and using the equations we can write

$r_m(x) = s(x)f(x) + t(x)g(x)$. In $F[x]$, we can make a unique choice for GCD(S) by taking it to be monic.

Example 12.2 Find $s(x), t(x) \in \boldsymbol{Q}[x]$ so that $\text{GCD}(\{x^5 - 1, x^3 + x - 2\}) = s(x)(x^5 - 1) + t(x)(x^3 + x - 2)$.

Set $f(x) = x^5 - 1$ and $g(x) = x^3 + x - 2$. Divide $g(x)$ into $f(x)$ to get

$$x^5 - 1 = (x^3 + x - 2)(x^2 - 1) + (2x^2 + x - 3).$$

Now dividing $2x^2 + x - 3$ into $g(x)$ results in

$$x^3 + x - 2 = (2x^2 + x - 3)((1/2)x - 1/4) + ((11/4)x - 11/4),$$

and note that $((11/4)x - 11/4) = (11/4)(x - 1)$ and $(x - 1) \mid (2x^2 + x - 3)$, so $\text{GCD}(\{p(x), g(x)\}) = x - 1$. Expressing the remainders in each equation in terms of the other polynomials yields

$$\begin{aligned} x - 1 &= (4/11)g(x) - (4/11)(2x^2 + x - 3)((1/2)x - 1/4) \\ &= (4/11)g(x) - (4/11)(f(x) - g(x)(x^2 - 1))((1/2)x - 1/4) \\ &= -(1/11)(2x - 1)f(x) + (1/11)(2x^3 - x^2 - 2x + 5)g(x). \end{aligned}$$

THEOREM 12.7 If D is a PID and $x \in D$ is irreducible, then x is a prime.

Proof Let $x \mid ab$ in D and $\text{GCD}(\{a, x\}) = d$, using Theorem 12.5. By the definition of GCD, $x = dy$, so x irreducible implies that $d \in U(D)$ or $y \in U(D)$. In the latter case, $d = xy^{-1}$ divides a so $x \mid a$ from Theorem 12.2. When $d \in U(D)$, then from Theorem 12.5, $d = ca + ex$, leading to $b = (d^{-1}c)ba + (bd^{-1}e)x$. But $x \mid ab$ and $x \mid x$ so Theorem 12.1 yields $x \mid b$. Therefore, either $x \mid a$ or $x \mid b$, proving that x is a prime. ■

It is difficult to know whether a domain D is a PID unless we know precisely what the ideals in D are, and this is rare. In some cases there is another approach via factorization. Our next result shows that elements in a PID have factorizations into primes, just as in $\boldsymbol{Z}$.

THEOREM 12.8 If D is a PID and $r \in D^* - U(D)$ then $r = p_1 \cdots p_k$ with each p_i irreducible. If also $r = q_1 \cdots q_m$ with each q_j irreducible, then $k = m$ and for some permutation σ of $\{1, 2, \ldots, k\}$, $p_i \sim_a q_{\sigma(i)}$.

Proof To prove the existence of some factorization of r into irreducibles, we show first that any $r \in D^* - U(D)$ has an irreducible factor. The argument essentially proves that a PID satisfies ACC on ideals (§11.6). If r is not irreducible then $r = y_1x_1$ with $y_1, x_1 \notin U(D)$. Furthermore, $(r) \subseteq (x_1)$ properly since if $x_1 = cr$ then $r = y_1cr$, forcing $y_1 \in U(D)$, a contradiction. Similarly, if x_1 is not irreducible then $x_1 = y_2x_2$ with $y_2, x_2 \notin U(D)$ and $(x_1) \subseteq (x_2)$ properly. If r has no irreducible factor then for each $k \in \boldsymbol{N}$, induction yields $r = y_1 \cdots y_kx_k$ with $(x_1) \subseteq (x_2) \subseteq \cdots$ an infinite, strictly ascending chain. The union I of all these (x_j) is an ideal by Theorem 11.5, say $I = (d)$. But then $d \in (x_j)$ for some j, forcing the contradiction $(x_{j+1}) \subseteq I = (d) \subseteq (x_j)$. Hence r must have an irreducible factor, and we write $r = r \cdot 1$ if r is irreducible.

Let p_1 be irreducible and $r = b_1 p_1$. Unless $b_1 \in U(D)$, $b_1 = b_2 p_2$ with p_2 irreducible, and $r = b_2 p_1 p_2$. Now $(b_1) \subseteq (b_2)$ properly since $b_2 = ab_1$ implies $b_1 = ab_1 p_2$, forcing the contradiction $p_2 \in U(D)$. Continuing, either $r = (b_k p_1) p_2 \cdots p_k$ with b_k a unit so $b_k p_1$ and all other p_j are irreducible, or induction produces another infinite, strictly ascending chain $(b_1) \subseteq (b_2) \subseteq \cdots$. As above, this latter situation cannot occur, so r is a product of irreducibles.

Assume $r = p_1 \cdots p_k = q_1 \cdots q_m$ with all p_i and q_j irreducible. If $k = 1$ and $m = 1$ then clearly $p_1 = q_1$, completing the proof. If $k = 1$ but $m > 1$ then $p_1 = (q_1)(q_2 \cdots q_m)$ contradicting p_1 irreducible. When $k > 1$, since p_k is a prime by Theorem 12.7 and $p_k \mid r$, then $p_k \mid q_j$ for some j as in Theorem 1.14. By reordering $\{q_i\}$ we may assume $j = m$. Thus $q_m = ap_k$, and q_m irreducible forces $a \in U(D)$, so $p_k \sim_a q_m$ and $p_1 \cdots p_{k-1} = (aq_1) \cdots q_{m-1}$. By induction on k, we have $k - 1 = m - 1$ and there is a permutation σ of $\{1, 2, \ldots, k\}$ so that $\sigma(k) = k$ and $p_j \sim_a q_{\sigma(j)}$, proving the theorem by IND-I. ∎

By Theorem 12.8, every $p(x) \in F[x] - F$ is a product of irreducibles. Here we need to allow associates since factors can be multiplied by units. In $\boldsymbol{Q}[x]$, $x^3 - x = x(x-1)(x+1) = (x/6)(2x-2)(3x+3) = (x/5)(x-1)(5x+5)$.

For $d \in \mathbf{Z}$, each nonunit $\alpha \in \mathbf{Z}[d^{1/2}]$ factors into irreducibles (Exercise 9 in §12.1). If $\mathbf{Z}[d^{1/2}]$ is a PID this factorization of α is unique up to associates by Theorem 12.8; when uniqueness fails, $\mathbf{Z}[d^{1/2}]$ is not a PID.

Example 12.3 $\boldsymbol{Z}[(-19)^{1/2}]$ is not a PID.

From our discussion of this ring in the last section, $N(a + b(-19)^{1/2}) = a^2 + 19b^2$, so $U(\mathbf{Z}[(-19)^{1/2}]) = \{\pm 1\}$ and $\alpha \sim_a \beta \Leftrightarrow \beta = \pm\alpha$. We saw that $\alpha = 1 + (-19)^{1/2}$ is irreducible with $N(\alpha) = 20$. Similarly, $1 - (-19)^{1/2}$ is irreducible. But $(1 + (-19)^{1/2})(1 - (-19)^{1/2}) = N(\alpha) = 20 = 2 \cdot 2 \cdot 5$ and both 2 and 5 are irreducible: if either is a product $\gamma\eta$, then $N(\gamma)N(\eta) = 4$ (or 25), forcing γ or η to be a unit (why?). Thus 20 factors into two products of different numbers of irreducibles, so $\mathbf{Z}[(-19)^{1/2}]$ cannot be a PID by Theorem 12.8.

EXERCISES

1. For F a field, show that:
  i) $F[[x]]$ (Definition 10.15) is a PID.
  ii) $F[x, x^{-1}]$ is a PID. Recall that $F[x, x^{-1}] = \langle F[x], x^{-1} \rangle \subseteq QF(F[x])$ and see Exercise 13 in §10.3.

2. i) Is $\mathbf{Z}_{(p)} = \{a/b \in \boldsymbol{Q} \mid \text{GCD}(\{b, p\}) = 1\}$ a PID?
  ii) Is $\boldsymbol{Q} + x\boldsymbol{R}[x] \subseteq \boldsymbol{R}[x]$ a PID?

3. If D and H are integral domains and $\varphi : D \to H$ is a ring homomorphism, show that if D a PID then $\varphi(D)$ is a PID.

4. If F is a field and $p(x) \in F[x]$, show that $F[p(x)] = \{g(p(x)) \in F[x] \mid g(x) \in F[x]\}$ is a PID.

5. If $\alpha \in \mathbf{Z}[d^{1/2}]$, $d \in \mathbf{Z}$, and $N(\alpha) \in \mathbf{Z}$ is a prime, show that α is irreducible.

6. Show that $\mathbf{Z}[d^{1/2}] \subseteq \boldsymbol{C}$ is not a PID when:
  i) $d = -3$

ii) $d = -5$
iii) $d = -p$ for $p \geq 3$ a prime (Note Exercise 5.)

7. Show that $R = \mathbf{Z}[(-2)^{1/2}]$ is a PID. (For $(0) \neq I \lhd R$ let $\alpha \in I$ so $N(\alpha) > 0$ and minimal. If $\beta \in I$, let $\beta/\alpha = q_1 + q_2(-2)^{1/2} \in \mathbf{Q}((-2)^{1/2})$ (see Example 10.18). Write $q_1 = n + r$ and $q_2 = m + q$ with $n, m \in \mathbf{Z}$ and $|r|, |q| \leq 1/2$. Then $\gamma = \alpha(r + q(-2)^{1/2}) = \beta - \alpha(n + m(-2)^{1/2}) \in I$ and $N(\gamma) \leq N(\alpha)(3/4)$, so $\gamma = 0$.)

8. Find $s(x), t(x) \in F[x]$ so the monic GCD$(\{p(x), g(x)\}) = s(x)p(x) + t(x)g(x)$, for $p(x)$ and $g(x)$:
i) $x^5 - 1, x^3 + x - 2 \in \mathbf{Z}_{11}[x]$
ii) $x^5 + 1, x^2 + x + 1 \in F[x]$ (F any field)
iii) $x^4 - 1, x^4 + x^3 + x^2 + x + 1 \in \mathbf{Z}_5[x]$
iv) $x^6 + x^3 + 1, x^3 + x + 1 \in \mathbf{Z}_2[x]$
v) $x^4 + 5, x^4 + 4x^3 + 4x + 2 \in \mathbf{Z}_7[x]$

9. Find $p_i(x) \in \mathbf{Q}[x]$ with $1 = p_1(x)(x^3 + x^2 + x) + p_2(x)(x^2 - x) + p_3(x)(x^3 - 1)$.

10. If D is a PID and $I \lhd D$ show that any $K \lhd D/I$ is principal, i.e., $K = (d + I)$.

11. Show that any PID R satisfies ACC on ideals: any strictly ascending chain of ideals $I_1 \subseteq I_2 \subseteq \cdots \subseteq I_n \subseteq \cdots$ contains only finitely many different ideals.

12. If D is a PID and $(0) \neq I \lhd D$ is proper, show that D/I satisfies DCC on ideals: if $J_1 \supseteq J_2 \supseteq \cdots J_n \supseteq \cdots$ are ideals of D/I, then for some m, $J_m = J_{m+k}$ for all $k \geq 0$.

13. If $R[x]$ is a PID show that R must be a field.

14. For an integral domain D, $\varnothing \neq S \subseteq D$ is multiplicative if $0 \notin S$ and $ab \in S$ when $a, b \in S$. If $K = QF(D)$ (Theorem 10.13), let $D_S = \{d/m \in K \mid m \in S\}$. Show that:
i) D_S is an integral domain.
ii) If D a PID then D_S is a PID.

15. Let F be a field, $X = \{x_\lambda \mid \lambda \in \Lambda\}$ commuting indeterminates over F, and $F(X)$ the quotient field of $F[X]$ (Theorem 10.13). Let $x_\lambda \in X$.
i) If $p(X), q(X) \in F[X]$, show that $p(X)q(X) \in (x_\lambda) \Leftrightarrow p(X) \in (x_\lambda)$ or $q(X) \in (x_\lambda)$.

For ii) and iii) let $R = \{p(X)/q(X) \in F(X) \mid q(X) \notin (x_\rho)$ for any $x_\rho \in X\}$.
ii) Show that R is an integral domain.
iii) Show that R is a PID. (Each $r \in R$ is a monomial in $\{x_\lambda\}$ times a unit.)

12.3 UFDs

By Theorem 12.8, each nonunit in any PID factors into a product of irreducibles, essentially uniquely. In particular, there is such a factorization in $F[x]$ when F is a field. We saw that $\mathbf{Z}[(-19)^{1/2}]$ is not a PID because it does not have "unique" factorization, a notion we next define formally.

DEFINITION 12.6 An integral domain D is a **unique factorization domain,** or UFD, if for any $d \in D^* - U(D)$, $d = p_1 \cdots p_m$ with all p_j irreducible, and if $d = q_1 \cdots q_k$ with all q_i irreducible, then $m = k$, and $p_j \sim_a q_{\sigma(j)}$ for a permutation σ of $\{1, 2, \ldots, k\}$.

Any PID is a UFD by Theorem 12.8, but it is not easy to show directly that a domain is a UFD (or a PID). Our aim is to extend the fact that $F[x]$ is a UFD to $F[X]$ for any set of indeterminates X. It is easy to see that $F[X]$ is not a PID when $|X| > 1$. In any UFD R,

GCD(S) exists for $\emptyset \neq S \subseteq R$, but it is not usually in $\sum_S(s)$, as in Theorem 12.5 for PIDs. A key observation for us is the next result about divisors (see §1.4 for $R = \mathbf{Z}$).

THEOREM 12.9 If R is a UFD and $S = \{a_1, \ldots, a_n\} \subseteq R^* - U(R)$, then there are nonassociate irreducibles $p_1, \ldots, p_s \in R$ so that $a_i = u_i p_1^{k(i,1)} \cdots p_s^{k(i,s)}$ for $u_i \in U(R)$ and $k(i, j) \geq 0$. Furthermore, $a_i \mid a_j \Leftrightarrow k(i, t) \leq k(j, t)$ for all $1 \leq t \leq s$, and also $\text{GCD}(S) = p_1^{m(1)} \cdots p_s^{m(s)}$ for $m(t) = \min\{k(1, t), \ldots, k(n, t)\}$.

Proof Let the irreducible factors of a_j be $\{p_{ji}\}$ and let $p_1, \ldots, p_s$ be a maximal set of nonassociate irreducibles in $\bigcup_j\{p_{ji}\}$. Each $p_{ji} = u_{jiw} p_w$ for some p_w with $u_{jiw} \in U(R)$ so $a_j = u_j p_1^{k(j,1)} \cdots p_s^{k(j,s)}$ as claimed. Fix $i \neq j$ and assume $k(j, t) - k(i, t) = d(t) \geq 0$ for all $1 \leq t \leq s$. Then $a_j = u_j u_i^{-1} a_i p_1^{d(1)} \cdots p_s^{d(s)}$, so $a_i \mid a_j$. When $a_i \mid a_j$ write $a_j = a_i c$. By unique factorization, each irreducible factor of c is an associate of some p_t. Thus $c = v p_1^{c(1)} \cdots p_s^{c(s)}$ for $v \in U(R)$, and $u_j p_1^{k(j,1)} \cdots p_s^{k(j,s)} = v u_i p_1^{k(i,1)+c(1)} \cdots p_s^{k(i,s)+c(s)}$. Using that different p_t are not associates and that R is a UFD, it follows that $k(j, t) = k(i, t) + c(t)$ for all t, and $c(t) \geq 0$ forces $k(i, t) \leq k(j, t)$. If $c \in R$ divides all a_i, so $a_i = c y_i$, then as we have just argued, each irreducible factor of c must be an associate of some p_t so $c = u p_1^{c(1)} \cdots p_s^{c(s)}$, and $c \mid a_i$ implies that $c(t) \leq k(i, t)$ for all i and t. Consequently, $c(t) \leq m(t)$ so $c \mid p_1^{m(1)} \cdots p_s^{m(s)}$, and since $m(t) \leq k(j, t)$ for all j, $p_1^{m(1)} \cdots p_s^{m(s)} \mid a_j$. Thus $p_1^{m(1)} \cdots p_s^{m(s)} = \text{GCD}(S)$. ∎

To illustrate the theorem, let $p(x) = x^3 - x^2$, $q(x) = x^4 - x^3 - x^2 + x$, and $g(x) = x^5 - 2x^4 + x^3 \in \mathbf{Q}[x]$. Factor these into irreducibles: $p(x) = x^2(x - 1)$, $q(x) = (x + 1)(x/2 - 1/2)^2(4x)$, and $g(x) = x^3(x - 1)^2$. Here the irreducibles used are $\{x + 1, x - 1, x, 4x, (1/2)(x - 1)\}$, and $\{x + 1, x - 1, x\}$ is a maximal set of nonassociates. Then $p(x) = x^2(x + 1)^0(x - 1)$, $q(x) = x(x + 1)(x - 1)^2$, and $g(x) = x^3(x + 1)^0(x - 1)^2$, and their GCD is $x(x + 1)^0(x - 1) = x^2 - x$. Similar examples in $\mathbf{Z}$ are given in §1.4.

We digress briefly to see when a UFD is a PID. The result shows how unique factorization can be used, but it is not very helpful in practice.

THEOREM 12.10 If R is a UFD, then R is a PID exactly when $(p) \lhd R$ is maximal for each irreducible $p \in R$.

Proof First assume R is a PID, $p \in R$ irreducible, and $(p) \subseteq I \lhd R$. Since R is a PID, $I = (d)$, so $p \in (p) \subseteq (d)$ yields $p = dr$. Now p irreducible forces d or r to be a unit. Hence either $I = (d) = R$, or $p \sim_a d$ and $(p) = (d)$ (exercise!), proving that (p) is maximal.

Now assume R is a UFD and that (p) is maximal for any irreducible $p \in R$. If $(0) \neq I \lhd R$ is proper, there is $d \in I$ with $d = u p_1^{d(1)} \cdots p_n^{d(n)}$ for each p_j irreducible, p_i not an associate of p_j for $i \neq j$, $u \in U(R)$, and with $\sum d(j)$ minimal. Let $0 \neq x \in I$ and suppose that $p_1^{d(1)}$ does not divide x. Use unique factorization first to write $x = p_1^m y$ with $m < d(1)$ so that no irreducible factor of y is an associate of p_1, and then to see $y \neq r p_1$ for any $r \in R$. Thus $y \notin (p_1)$. But (p_1) is maximal so $(p_1) + (y) = R$ and it follows that $1 = a p_1 + c y$. If $z = u p_1^{d(1)-1} p_2^{d(2)} \cdots p_n^{d(n)}$ then $z = a z p_1 + (z y) c \in I$ since $z p_1 = d$, $x \mid z y$, and $x \in I$. This contradicts the minimality of $\sum d(j)$ and forces $p_1^{d(1)} \mid x$. Similarly, each $p_j^{d(j)} \mid x$. Write $x = v p_1^{x(1)} \cdots p_n^{x(n)} h$ with $v \in U(R)$, h having no irreducible factor an associate of any p_j, and each $0 \leq x(i)$. Since $p_k^{d(k)} \mid x$, Theorem 12.9 shows first that $d(k) \leq x(k)$ for each $1 \leq k \leq n$, and then that $d \mid x$, so $I = (d)$. ∎

Just as in PIDs, in UFDs any irreducible is a prime.

THEOREM 12.11 If R is a UFD and $p \in R$ is irreducible, then p is a prime.

Proof When $p \mid ab$ in R then $ab = cp$, so by the uniqueness of irreducible factorization in R, p is an associate of an irreducible q appearing in a or in b. Thus if $q \mid a$, then since $q = vp$ for a unit v, $p \mid a$ by Theorem 12.2; if $q \mid b$ then $p \mid b$. ∎

DEFINITION 12.7 If R is a UFD and $f(x) = r_n x^n + \cdots + r_1 x + r_0 \in R[x]^*$, the **content** of $f(x)$ is $c(f) = \text{GCD}(\{r_n, \ldots, r_0\})$. When $c(f) = 1$, $f(x)$ is **primitive.**

The content of $f(x)$ is unique only up to associate. When $f(x)$ is primitive its coefficients have no common irreducible divisor. Clearly, for $f(x) = 4x^2 - 6x + 12 \in \mathbf{Z}[x]$, $c(f) = 2$, and $9x^2 - 12x - 8 \in \mathbf{Z}[x]$ is primitive. Any $f(x) \in R[x]$ with some coefficient a unit is primitive, so any monic $f(x)$ is primitive, as is any $f(x) \in F[x]^*$ when F is a field.

Since a UFD R is a domain, it has a quotient field $K = QF(R)$ (Theorem 10.13) and we consider $R[x] \subseteq K[x]$ via $r = r/1 \in K$ for $r \in R$. We want to prove that $R[x]$ is also a UFD and will use Theorem 11.3 that $K[x]$ is a PID. Describing the irreducibles in $R[x]$ uses those in $K[x]$ and it is useful to have a standard form relating the polynomials in these two rings. The idea is that given $g(x) \in K[x]$, we can write its coefficients with a common denominator, factor this out, and then factor out the greatest common divisor of the resulting coefficients. This means that $g(x) = (a/b)f(x)$ for $f(x) \in R[x]$ and primitive. Some simple but useful observations come next.

THEOREM 12.12 Let R be a UFD with $QF(R) = K$.

i) If $f \in R[x]$ is primitive and $a \in R$, then $c(af) = a$.
ii) If $f, g \in R[x]$ are primitive and $af = bg$ for $a, b \in R^*$, then $b = ua$ and $f = ug$ for some $u \in U(R)$.
iii) If $f, g \in R[x]$ are primitive and $f \sim_a g$ in $K[x]$, then $f = ug$ for $u \in U(R)$.
iv) If $f \in R^*[x]$ with $f = kf_0$ for $k \in K$ and $f_0 \in R[x]$ primitive, then $k \in R$.
v) If $f \in K[x]^*$, then $f = kf_0$ for $k \in K$ and $f_0 \in R[x]$ primitive and unique up to associate.

Proof If $f = \sum r_j x^j \in R[x]$ then $af = \sum ar_j x^j$ and $a \in R$ divides all coefficients of af. When $d \in R$ with $d \mid ar_j$ for all j, then $ar_j = dy_j$. Using Theorem 12.9, write $a = up_1^{a(1)} \cdots p_n^{a(n)}$ and $d = vp_1^{d(1)} \cdots p_n^{d(n)}$ for p_j nonassociate irreducibles and $u, v \in U(R)$. Since $ar_i = dy_i$, by unique factorization either $d(j) \le a(j)$ or $p_j \mid r_i$ for all i. But if $c(f) = 1$ then p_j cannot divide all r_i, so all $d(j) \le a(j)$ and it follows from Theorem 12.9 that $d \mid a$. Thus $a = c(af)$ when $c(f) = 1$, proving i).

For ii), since f and g are primitive, i) shows that $a = c(af)$ and $b = c(bg)$. It follows that $a \sim_a b$ so $b = ua$ for $u \in U(R)$, and then $af = uag$ yields $f = ug$ as required. Note that $U(K[x]) = K^*$ by Theorem 10.11 on degrees, so in iii), $f \sim_a g$ in $K[x]$ means that $f = kg$ for $k \in K^*$. If $k = a/b$ for $a, b \in R$, then $bf = ag$ results and ii) shows that $f = ug$ for $u \in U(R)$. In iv), we may assume $k \ne 0$, so again $k = a/b$ and $bf = af_0$. By i), $c(af_0) = a$ and every coefficient of $af_0 = bf$ is divisible by b, forcing $b \mid a$. Thus $k \in R$. Finally, for v), uniqueness of f_0 follows from iii) since $f = kf_0 = k'g_0$ implies that $f_0 = k^{-1}k'g_0$. To show that k and f_0 exist let $f = \sum r_j x^j \in K[x]^*$ and write $r_j = a_j/b$ (how?), so $f = (1/b)\sum a_j x^j$. If $d = \text{GCD}(\{a_j\})$ and $dc_j = a_j$, then $f = (d/b)f_0$ for $f_0 = \sum c_j x^j$. Finally, we observe that $c(f_0) = 1$ since if $p \in R$ is irreducible and $p \mid c_j$ for all j, then $dp \mid a_j$ for all j, contradicting $d = \text{GCD}(\{a_j\})$ since dp cannot divide d (why?). ∎

Primitive polynomials are key in proving that $R[x]$ is a UFD when R is. An important observation known as *Gauss's Lemma* comes next.

THEOREM 12.13 For R a UFD and $f, g \in R[x]$, $c(fg) = 1 \Leftrightarrow c(f) = 1 = c(g)$.

Proof Assume first that $c(fg) = 1$. If f is not primitive there is an irreducible $p \in R$ with $p \mid c(f)$, so p divides each coefficient of f. It follows from the definition of multiplication in $R[x]$ and Theorem 12.1 that p must divide each coefficient of fg, contradicting $c(fg) = 1$. Thus $c(f) = 1$ and, similarly, $c(g) = 1$. Now suppose that $p \mid c(fg)$ for some irreducible $p \in R$. Hence $fg = ph$ for some $h \in R[x]$. Since p is irreducible, Theorem 12.11 shows that it is a prime in R, so p is a prime in $R[x]$ by Theorem 12.4. Consequently, using $ph = fg$, either $p \mid f$ or $p \mid g$. If $p \mid f$ then $f = pq$ for $q \in R[x]$, and p divides each coefficient of f; this implies that f is not primitive. Similarly, when $p \mid g$, then g is not primitive. Therefore $c(f) = 1 = c(g)$ forces $c(fg) = 1$. ∎

We now have the pieces needed to characterize the irreducibles in $R[x]$ and to show that $R[x]$ is a UFD.

THEOREM 12.14 If R is a UFD, then $R[x]$ is a UFD. For $f \in R[x]$, f is irreducible $\Leftrightarrow f \in R$ is irreducible or $c(f) = 1$ and f is irreducible in $QF(R)[x]$.

Proof Any irreducible $p \in R$ must be irreducible in $R[x]$ by considering degrees of its factors. Let $QF(R) = K$. If $f \in R[x]$ with $c(f) = 1$ is irreducible in $K[x]$ then $f \neq gh$ for $g, h \in R[x]$ with *deg* g, *deg* $h \geq 1$, so $c(f) = 1$ shows f is irreducible in $R[x]$. When $f \in R[x]$ is irreducible, Theorem 12.12 yields $f = rf_0$ for $r \in R$ and $f_0 \in R[x]$ primitive. Thus $r \in U(R)$ or $f_0 \in U(R)$, and in the latter case, $f \in R$ is irreducible. When r is a unit then by Theorem 12.12, f is primitive. If in $K[x]$, $f = gh$ with *deg* g, *deg* $h \geq 1$, then using Theorem 12.12 again, $g = ag_0$ and $h = bh_0$ for $a, b \in K$ and $g_0, h_0 \in R[x]$ primitive. Since g_0h_0 is primitive from Theorem 12.13, and $f = abg_0h_0$, Theorem 12.12 shows $ab \in R$, contradicting f irreducible in $R[x]$ and showing that f is irreducible in $K[x]$. We now show that $R[x]$ is a UFD.

Since R is a UFD, each $r \in R^* - U(R)$ factors into irreducibles in R, and so in $R[x]$ by the first paragraph. By Theorem 11.3 and Theorem 12.8, $K[x]$ is a UFD. Hence for any $f \in R[x] - R$, $f = p_1 \cdots p_n$ for irreducibles $p_j \in K[x]$. Apply Theorem 12.12 to get $p_j = a_j f_j$ for $a_j \in K$ and $f_j \in R[x]$ primitive. Of course each f_j is irreducible in $R[x]$ since p_j is in $K[x]$, and $f_1 \cdots f_n$ is primitive from Theorem 12.13 and induction. Now $f = a_1 \cdots a_n f_1 \cdots f_n$ so Theorem 12.12 yields $a_1 \cdots a_n = a \in R$. Since R is a UFD, by the first paragraph f can be written as a product of irreducibles in $R[x]$. Given $f = p_1 \cdots p_m f_1 \cdots f_n = q_1 \cdots q_s g_1 \cdots g_t$ for $p_j, q_i \in R$ irreducible and $f_j, g_i \in R[x] - R$ irreducible, so primitive, then $p_1 \cdots p_m = c(f) = q_1 \cdots q_s$, (why?). Thus $m = s$ and $p_j \sim_a q_{\sigma(j)}$ for some permutation σ of $\{1, 2, \ldots, m\}$. Also $f_1 \cdots f_n = g_1 \cdots g_t$ with f_i and g_j irreducible in $K[x]$ by the first paragraph, so $K[x]$ a UFD shows $n = t$, and with a suitable reordering,

f_j and g_j are associates in $K[x]$. By Theorem 12.12 these are associates in $R[x]$, proving that $R[x]$ is UFD. ∎

We extend the theorem to any $R[X]$.

THEOREM 12.15 If R is a UFD and X is a nonempty set of commuting indeterminates over R, then $R[X]$ is a UFD.

Proof If $f \in R[X]$ then $f \in R[\lambda_1, \ldots, \lambda_n]$ for some $\lambda_j \in X$. Now $R[\lambda_1, \ldots, \lambda_n] = D = (R[\lambda_1, \ldots, \lambda_{n-1}])[\lambda_n]$ is a UFD using Theorem 12.14 and induction on n (exercise!), so f factors into irreducibles in D. If $h \in D$ is irreducible then by Theorem 12.14, $h \in D[\lambda_{n+1}]$ is irreducible, and induction shows that $h \in D[\lambda_{n+1}, \ldots, \lambda_t]$ is irreducible. It follows that h is irreducible in $R[X]$ since otherwise its factors in $R[X]$, and h itself, are in some $E = D[\lambda_{n+1}, \ldots, \lambda_t]$. But as we just saw, h is irreducible in E and so must be irreducible in $R[X]$. Hence f factors into irreducibles in $R[X]$. If $f = f_1 \cdots f_k = g_1 \cdots g_s$ for $f_i, g_j \in R[X]$ and irreducible then all f_i and g_j are elements in some $D = R[\lambda_1, \ldots, \lambda_m]$. But D is a UFD so we may assume $k = s$ and $f_j = u_j g_j$ for $u_j \in U(D) = U(R) = U(R[X])$. Therefore $f_j \sim_a g_j$ in $R[X]$, proving that $R[X]$ is a UFD. ∎

From the theorem, $\mathbf{Z}[x]$, $\mathbf{Z}[x, y]$, and $F[X]$ for F a field and $|X| > 1$ are UFDs, but these are not PIDs (exercise!). Although $F[X]$ and $\mathbf{Z}[X]$ are UFDs, it is not at all clear how to recognize their irreducibles. Just before Example 11.24 we recalled that $p(x) \in F[x] - F$ with $deg\ p \leq 3$ is irreducible exactly when $p(x)$ has no root in F. Any $p(x) \in \mathbf{R}[x]$ of degree 3 has a root, using calculus, but this is not apparent algebraically. There is one result called *Eisenstein's Criterion* that is useful for determining irreducibility in $R[x]$ when R is a UFD, and we state it next.

THEOREM 12.16 For R a UFD, $f(x) = a_n x^n + \cdots + a_0 \in R[x]$ is irreducible in $QF(R)[x]$ if for some prime $p \in R$, $p \nmid a_n$, $p \mid a_j$ for $0 \leq j < n$, and $p^2 \nmid a_0$.

Proof From Theorem 12.12, $f = c(f) f_0$ for $f_0 \in R[x]$ primitive (why?), so it suffices to prove that f_0 is irreducible in $K[x]$, for $K = QF(R)$. Now $p \nmid a_n$ implies $p \nmid c(f)$, and it follows from the fact that p is a prime that the conditions on the a_j are satisfied by the coefficients of f_0. Replacing f with f_0, we may assume that $c(f) = 1$. If $g = b_m x^m + \cdots + b_0$, $h = c_s x^s + \cdots + c_0 \in R[x]$ and $gh = f$, then since $p \mid a_0$ we have $p \mid b_0 c_0$ so $p \mid b_0$ or $p \mid c_0$ since p is a prime. Assume $p \mid b_0$. If $p \mid b_m$ then $p \mid b_m c_s$, contradicting $p \nmid a_n$. Let $0 < k \leq m$ be minimal so that $p \nmid b_k$. Since p divides $a_k = b_k c_0 + b_{k-1} c_1 + b_{k-2} c_2 + \cdots$, and $p \mid b_j$ for $j < k$, we conclude from Theorem 12.1 that $p \mid b_k c_0$. Now p a prime and $p \nmid b_k$ force $p \mid c_0$. Thus, from $a_0 = b_0 c_0$, $p^2 \mid a_0$, contradicting our assumption. Similarly, if initially $p \mid c_0$ then again $p^2 \mid a_0$. Hence f is irreducible in $R[x]$ so in $K[x]$ by Theorem 12.14. ∎

An immediate consequence of Eisenstein's Criterion is that $x^m - p$ is irreducible in $\mathbf{Q}[x]$ for any $m \geq 1$ and any prime $p \in \mathbf{Z}$. We end this section with two less obvious applications of Theorem 12.16.

Example 12.4 $Q[x, y]/((x+y)^m - y)$ is an integral domain.

From Theorem 12.3 it suffices to show that $(x+y)^m - y \in Q[x, y]$ is a prime. But $Q[x, y]$ is a UFD by Theorem 12.15 so using Theorem 12.11, we need $(x+y)^m - y$ to be irreducible in $Q[x, y]$. Now $(x+y)^m - y = x^m + c_1 y x^{m-1} + \cdots + c_{m-1} y^{m-1} x + (y^m - y) \in (Q[y])[x]$, and $y \in Q[y]$ is irreducible, so prime by Theorem 12.11 (or $Q[y]/(y) \cong Q$ a domain, so y is a prime by Theorem 12.2). Clearly $y \nmid 1$, y divides the coefficient $c_i y^{m-i}$ of x^i in $(x+y)^m - y$ for $i \neq m$, and $y^2 \nmid y(y^{m-1} - 1)$ (verify!). Thus Theorem 12.16 applies to show that $(x+y)^m - y$ is irreducible in $QF(Q[y])[x]$, and so in $Q[y][x] = Q[x, y]$ by Theorem 12.14.

Example 12.5 If $p \in Z$ is prime, $x^{p-1} + x^{p-2} + \cdots + x + 1 \in Q[x]$ is irreducible.

Note first that $\varphi(f(x)) = f(x+1)$ defines a ring homomorphism from $Q[x]$ to itself (Exercise 6 in §11.3), which is an isomorphism with inverse $\eta(f(x)) = f(x-1)$. To see that $q(x) = x^{p-1} + x^{p-2} + \cdots + x + 1$ is irreducible in $Q[x]$, it suffices to prove that $\varphi(q(x))$ is irreducible since if $q(x) = g(x)h(x)$ in $Q[x]$, then $\varphi(q(x)) = g(x+1)h(x+1)$. Now $x^p - 1 = (x-1)q(x)$ so $(x+1)^p - 1 = \varphi(x^p - 1) = \varphi(x-1)\varphi(q(x)) = x\varphi(q(x))$. Using the Binomial Theorem yields $(x+1)^p - 1 = x^p + px^{p-1} + \cdots + B_j x^j + \cdots + px = x(x^{p-1} + \cdots + B_j x^{j-1} + \cdots + p)$ where $p \mid B_j$ for all j (why?). Thus Theorem 12.16 applies directly to show that $\varphi(q(x)) = x^{p-1} + \cdots + B_j x^j + \cdots + p \in Z[x]$ is irreducible in $Q[x]$, so $q(x)$ is also irreducible in $Q[x]$, as indicated above.

EXERCISES

In all the exercises below, D is a UFD with quotient field K containing D.

1. If $S = \{a_1, \ldots, a_k\} \subseteq D$ then $\mathrm{LCM}(S) = c$ if all $a_j \mid c$ and if, when $y \in D$ and all $a_j \mid y$, then $c \mid y$. Show that $\mathrm{LCM}(S)$ exists and is unique up to associate.
2. If $(0) \neq P \lhd D$ is a prime ideal then $(q) \subseteq P$ for some $(0) \neq (q)$ a prime ideal.
3. **i)** If for $p \in D$, $pU(D)$ are the only primes in D, show that for each $I \lhd D$ either $I = (0)$, $I = (p^m)$ some $m \in N$, or $I = D$.
   ii) If $P \lhd D$ is the only proper nonzero prime ideal of D, show that $I \lhd D \Rightarrow I = (0)$, $I = P^m$, or $I = D$.
4. If $d \in D$ and for $m \in N$, $d^{1/m} \in K$, show that $d = a^m$ for some $a \in D$.
5. If $p \in D$ is a prime, and $D_{(p)} = \{a/b \in K \mid \mathrm{GCD}(\{a, b\}) = 1 \text{ and } p \nmid b\}$, show that $D_{(p)}$ is a subring of K and is a PID, and so a UFD.
6. If $S \subseteq D$ is multiplicative ($0 \notin S$ and $a, b \in S \Rightarrow ab \in S$—see Exercise 14 in §12.2) and $D_S = \{a/b \in K \mid \mathrm{GCD}(\{a, b\}) = 1 \text{ and } b \in S\}$, show that D_S is a UFD.
7. Show that $Z[x, x^{-1}] = \langle\{1, x, x^{-1}\}\rangle \subseteq QF(Z[x])$ is a UFD.
8. Show that D satisfies ACC on principal ideals: if $(d_1) \subseteq (d_2) \subseteq \cdots \subseteq (d_i) \subseteq \cdots$ then $(d_k) = (d_{k+j})$ for some $k \geq 1$ and all $j \geq 1$.
9. Let $p(x) = a_n x^n + \cdots + a_1 x + a_0 \in D[x]$ with $a_n a_0 \neq 0$. If $c/d \in K$, $p(c/d) = 0$, and $\mathrm{GCD}(\{c, d\}) = 1$, show that $d \mid a_n$ and $c \mid a_0$.

10. For each $p(x) \in \mathbf{Z}[x]$ below, find a root in $\boldsymbol{Q}$ (use Exercise 9):
  i) $2x^3 - 5x^2 + 3x + 3$
  ii) $x^3 - 3x^2 - 3x - 4$
  iii) $2x^3 - 5x^2 + 7x - 6$
  iv) $4x^4 + 8x^3 + 21x^2 + 17x + 4$
  v) $x^5 + x^4 - 2x^3 + 2x^2 - 8$

11. Find a root in $QF(\boldsymbol{Q}[x])$ for $p(x) = xy^3 + x^2 + xy + 1 \in (\boldsymbol{Q}[x])[y]$.

12. Let $R = \mathbf{Z}[X]$ for X an infinite set of commuting indeterminates over $\mathbf{Z}$. If $f_i \in R$ with $(f_1) \subseteq (f_2) \subseteq \cdots \subseteq (f_j) \subseteq \cdots$, show that $f_k \sim_a f_{k+i}$ for some k and all $i \geq 1$.

13. If $\alpha \in K$ and $g(\alpha) = 0$ for $g(x) \in D[x]^*$ and *monic,* show that $\alpha \in D$.

14. Let $R = \mathbf{Z}[x^2, x^3] = \langle\{1, x^2, x^3\}\rangle \subseteq \mathbf{Z}[x]$.
  i) Show that R is not a UFD.
  ii) Show that $x \in QF(R) - R$ but $p(x) = 0$ for a monic $p(Y) \in R[Y]^*$.

15. Let $R = \mathbf{Z}[5^{1/2}]$.
  i) Show that $QF(R) = \boldsymbol{Q}[5^{1/2}]$.
  ii) Show that $(1 + 5^{1/2})/2 \in \boldsymbol{Q}[5^{1/2}] - R$ satisfies a monic $p(x) \in R[x]$ (compare with Exercise 13).

16. Show that $(1 + 21^{1/2})/2 \in \boldsymbol{Q}[21^{1/2}] - \mathbf{Z}[21^{1/2}]$ satisfies a monic $q \in (\mathbf{Z}[21^{1/2}])[x]$.

17. If $d \in \boldsymbol{N}$, $d^{1/2} \notin \boldsymbol{N}$, and $d \equiv_4 1$, show that $\mathbf{Z}[d^{1/2}]$ is not a UFD (Exercise 15!).

18. If $f \in K[x]$ is irreducible, show that for some $b \in K$, $bf \in D[x]$ is irreducible.

19. Show carefully that $D[x_1, \ldots, x_n]$ is a UFD for any $n > 1$.

20. If $P \lhd D$ is prime, must D/P be a UFD? (See Example 12.3.)

21. Determine whether the subring is a UFD:
  i) $\langle\{1, x^2, y^4\}\rangle \subseteq \mathbf{Z}[x, y]$
  ii) $\boldsymbol{Q}[x, 1/y] = \langle\{\boldsymbol{Q}, x, 1/y\}\rangle \subseteq QF(\boldsymbol{Q}[x, y])$
  iii) $\langle\{1, x^2, xy, y^2\}\rangle \subseteq \mathbf{Z}[x, y]$
  iv) $\langle\{1\} \cup \{x^i y^j \mid 0 \leq i < j\}\rangle \subseteq \mathbf{Z}[x, y]$
  v) $\langle\{1\} \cup \{x^i y^j \mid i = j\}\rangle \subseteq \mathbf{Z}[x, y]$
  vi) $\langle\{1\} \cup \{x^i y^j \mid i + j = 0\}\rangle \subseteq QF(\mathbf{Z}[x, y])$
  vii) $\langle\{1\} \cup \{x^i y^j z^k \mid i + j = k\}\rangle \subseteq \mathbf{Z}[x, y, z]$

22. For $m(1), \ldots, m(k), r(1), \ldots, r(k) \in \boldsymbol{N}$, show that $(x + 3^{m(1)})^{r(1)} \cdots (x + 3^{m(k)})^{r(k)} - 3$ in $\boldsymbol{Q}[x]$ is irreducible.

23. Recall that $\varphi(n)$ is the Euler phi-function (§3.2). If $p \in \boldsymbol{N}$ is a prime and if $p \equiv_4 3$, show that $x^p + \varphi(p-1)x^{p-1} + \cdots + \varphi(3)x^3 + \varphi(p)$ is irreducible in $\boldsymbol{Q}[x]$.

24. Show that $x^{2^n} + 1 \in \boldsymbol{Q}[x]$ is irreducible (see Exercise 10 in §0.3).

25. **i)** Is $(x + 2)(x^2 - 4) \cdots (x^n + (-2)^n) + 30$ irreducible in $\boldsymbol{Q}[x]$?
  ii) Is $(x - y)(x^2 + y^2) \cdots (x^n + (-y)^n) - y$ irreducible in $\mathbf{Z}[x, y]$?
  iii) Is $(x_1 + x_2)(x_1 + x_2 x_3) \cdots (x_1 + x_2 x_3 \cdots x_n) + x_2 \in \mathbf{Z}[x_1, \ldots, x_n]$ irreducible?

26. **i)** Show that $x^n + (y^{n+1} - z^{n+1})x^{n-1} + \cdots + (y^3 - z^3)x + (y^2 - z^2)$ is a prime in $\boldsymbol{Q}[x, y, z]$.
  ii) Is $x^n + (y^2 - z^2)^n x^{n-1} + \cdots + (y^2 - z^2)^2 x + (y^2 - z^2)$ a prime in $\boldsymbol{Q}[x, y, z]$?
  iii) Show that $(y - z)x^n + (y^2 - z^2)^n x^{n-1} + \cdots + (y^2 - z^2)^2 x + (y^2 - z^2)$ is a prime in $QF(\boldsymbol{Q}[y, z])[x]$.

27. Determine whether following are domains:
  i) $\mathbf{Z}[x, y, z]/(x^n + x^{n-1}y^n z + x^{n-2}y^{n-1}z^2 + \cdots + yz^n)$
  ii) $\mathbf{Z}[x, y]/(x^m + y^m - 12)$ for $m \geq 1$
  iii) $\mathbf{Z}[x_1, \ldots, x_n]/(x_1^k + x_2^k + \cdots + x_n^k - 180)$

28. For $n \geq 2$ set $q_n(x, y) = x^n + y^n - (x + y)^n$.
  i) Show $x \mid q_n(x, y)$ but $x^2 \nmid q_n(x, y)$.
  ii) Show that $(x + y)^n - (y + z)^n + z^n \in \boldsymbol{Q}[x, y, z]$ is irreducible.

29. Show that $x_1^{n-1} x_2^{n-2} \cdots x_{n-1} + x_n^{n-1} x_1^{n-2} \cdots x_{n-2} + x_{n-1}^{n-1} x_n^{n-2} \cdots x_{n-3} + \cdots + x_2^{n-1} x_3^{n-2} \cdots x_n$ is irreducible in $\boldsymbol{Z}[x_1, \ldots, x_n]$.

12.4 EUCLIDEAN DOMAINS

We have seen that PIDs are UFDs. Here we want to look at a special kind of PID that is of interest in its own right. Still using $\boldsymbol{Z}$ and $F[x]$ as models, we know that for each of these we can divide by a nonzero element and get a "smaller" remainder. The general condition is given next.

DEFINITION 12.8 The pair (D, υ) is a **Euclidean domain** if D is an integral domain and $\upsilon : D^* \to \boldsymbol{N} \cup \{0\}$ satisfies:

i) $\upsilon(a) \leq \upsilon(b)$ if $a \mid b$ in D^*
ii) for $d \in D$ and $x \in D^*$ there are $q, r \in D$ with $d = qx + r$ and $r = 0$ or $\upsilon(r) < \upsilon(x)$.

We may say that D is a Euclidean domain without specifying υ. Sometimes Euclidean domains are defined with $\upsilon(D) \subseteq \boldsymbol{N}$ and i) replaced by the stronger condition that $\upsilon(ab) = \upsilon(a)\upsilon(b)$. The familiar examples of Euclidean domains are $\boldsymbol{Z}$ with $\upsilon(n) = |n|$ and $F[x]$ with $\upsilon(p(x)) = deg\ p(x)$. The main result here is that Euclidean domains are PIDs and so are also UFDs.

THEOREM 12.17 If (D, υ) is a Euclidean domain, then D is a PID.

Proof Let $(0) \neq I \lhd D$. By the Well Ordering Axiom in $\boldsymbol{N} \cup \{0\}$, there is $x \in I$ with $\upsilon(x)$ minimal in $\{\upsilon(y) \mid y \in I - (0)\}$. Clearly $(x) \subseteq I$. For $d \in I$ use Definition 12.8 to write $d = qx + r$ for $r = 0$ or $\upsilon(r) < \upsilon(x)$. Should $r \neq 0$, then $r = d - qx \in I$ with $\upsilon(r) < \upsilon(x)$, contradicting the minimality of $\upsilon(x)$. Therefore $r = 0$, so $d \in (x)$, proving that $I = (x)$. ■

Nowhere in the proof of Theorem 12.17 did we use $1 \in D$, so if we define a Euclidean domain in Definition 12.8 for any commutative domain, we are forced to conclude that $1 \in D$ from $D = (x)$ (how?). Since any PID is a UFD and any Euclidean domain is a PID, and so a UFD, we can picture the relation between these properties on any set of UFDs as set inclusion (Fig. 12.1).

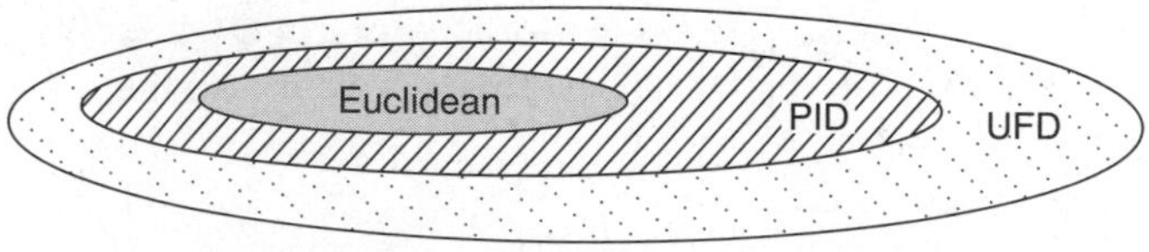

FIGURE 12.1

Aside from the familiar $\mathbf{Z}$ and $F[x]$ as examples of Euclidean domains, or PIDs, and the results from the last section to the effect that $D[X]$ is a UFD if D is, finding examples with these properties is difficult. To see that a domain is not a PID, and so is not Euclidean, we can try to show that it is not a UFD, as in Example 12.3 or some of the exercises in the last section. The next result provides another way to see that a domain is not Euclidean.

THEOREM 12.18 If (D, v) is a Euclidean domain, then:

i) $\alpha \in U(D)$ implies $v(\alpha) = v(1)$
ii) if $\alpha, \beta \in D^* - U(D)$ then $v(\alpha) < v(\alpha\beta)$
iii) if $v(\gamma)$ is minimal in $\{v(\alpha) \mid \alpha \in D^* - U(D)\}$ and if $\delta \in D$, then $\gamma|\delta$ or $\gamma|(\delta + u)$, some $u \in U(D)$.

Proof If α is a unit then $\alpha\alpha^{-1} = 1$ so $\alpha \mid 1$, and $v(\alpha) \le v(1)$ follows. But $1 \mid \alpha$ so $v(1) \le v(\alpha)$, and thus $v(1) = v(\alpha)$. Using the definition of Euclidean for ii), $\alpha = q\alpha\beta + r$ for $q, r \in D$ with $r = 0$ or $v(r) < v(\alpha\beta)$. If $q = 0$ then $r = \alpha$ and $v(\alpha) < v(\alpha\beta)$. When $q \ne 0$, $r = 0$ forces the contradiction $\beta \in U(D)$, and otherwise $r = \alpha(1 - q\beta)$, forcing $v(\alpha) \le v(r) < v(\alpha\beta)$. Now assume that $v(\gamma)$ is minimal in $\{v(\alpha) \mid \alpha \in D^* - U(D)\}$, assume that $\delta \in D$, and write $\delta = q\gamma + r$ with $r = 0$ or $v(r) < v(\gamma)$. Clearly $r = 0$ implies $\gamma \mid \delta$. If $r \ne 0$ then the minimality of $v(\gamma)$ forces $r \in U(D)$, so $\delta - r = q\gamma$ and $\gamma \mid (\delta + u)$ for $u = -r \in U(D)$. ■

Recall from Example 12.1 that in the domains $\mathbf{Z}[m^{1/2}]$ for $m \in \mathbf{Z}^*$ and not a square, $N(a + bm^{1/2}) = (a + bm^{1/2})(a - bm^{1/2}) = a^2 - mb^2$ is the *norm* of $a + bm^{1/2}$, $N(\alpha\beta) = N(\alpha)N(\beta)$, and $N(\alpha) = \pm 1 \Leftrightarrow \alpha$ is a unit. Using the norm, we saw that $\mathbf{Z}[(-19)^{1/2}]$ is not a UFD in Example 12.3. This relied on the observation that if $N(\alpha)$ is a prime then α must be irreducible, since from $\alpha = \beta\gamma$, $N(\alpha) = N(\beta)N(\gamma)$ forces $N(\beta) = \pm 1$ or $N(\gamma) = \pm 1$. It follows that $\mathbf{Z}[5^{1/2}]$ is not a UFD, and so is not Euclidean, since $|N(1 + 5^{1/2})| = 4 = 2 \cdot 2$. Here $1 \pm 5^{1/2}$ and 2 are irreducible: if either is $\beta\gamma$, then $|N(\beta)N(\gamma)| = 4$ and as $|a^2 - 5b^2| \ne 2$ (neither 2 nor 3 is a square modulo 5) we are forced to conclude that β or γ is a unit. Since 2 cannot be an associate of $1 \pm 5^{1/2}$ (why?), factorization into irreducibles in $\mathbf{Z}[5^{1/2}]$ is not unique: $(1 + 5^{1/2})(1 - 5^{1/2}) = 2 \cdot 2$. Rather than making many separate similar computations, we can deal with $\mathbf{Z}[m^{1/2}]$ for most $m < 0$ by using Theorem 12.18.

Example 12.6 $\mathbf{Z}[(-m)^{1/2}]$ is not a Euclidean domain when $m \ge 3$.

If $\alpha = a + b(-m)^{1/2}$ then $N(\alpha) = a^2 + mb^2$ so clearly $N(\alpha) = 1$ exactly when $\alpha = \pm 1$, and by Example 12.1, $U(\mathbf{Z}[(-m)^{1/2}]) = \{\pm 1\}$. Also, since $m \ge 3$, $N(\alpha) \ne 2$ and it follows that 2 must be irreducible: if $2 = \beta\gamma$ in $\mathbf{Z}[(-m)^{1/2}]$ then $4 = N(2) = N(\beta)N(\gamma)$ so $N(\beta) = 1$ or $N(\gamma) = 1$. When $m = 3$, $N(1 \pm (-3)^{1/2}) = 4$ and $1 \pm (-3)^{1/2}$ are irreducible in $\mathbf{Z}[(-3)^{1/2}]$ since $1 + (-3)^{1/2} = \beta\gamma$ implies $4 = N(\beta)N(\gamma)$, again forcing β or γ to be a unit (why?). Also $1 \pm (-3)^{1/2}$ and 2 are not associates and $(1 + (-3)^{1/2})(1 - (-3)^{1/2}) = 2 \cdot 2$ so $\mathbf{Z}[(-3)^{1/2}]$ is not a UFD. For $m = 4$ and $\alpha = a + 2bi \in \mathbf{Z}[(-4)^{1/2}]$, $N(\alpha) = a^2 + 4b^2$ so $N(2 + 2i) = 8 = 2 \cdot 2 \cdot 2$, and again $2 + 2i$ is irreducible in $\mathbf{Z}[(-4)^{1/2}]$, just as above, using that no element has norm 2. Since $2+2i \notin \{\pm 1\} \cdot 2$ we get different factorizations of $8 \in \mathbf{Z}[(-4)^{1/2}$ into irreducibles; this shows that $\mathbf{Z}[(-4)^{1/2}]$ cannot be Euclidean.

Now assume $m \geq 5$ and that $(\mathbf{Z}[(-m)^{1/2}], \upsilon)$ is Euclidean. There is some $\gamma \in \mathbf{Z}[(-m)^{1/2}] - \{0, \pm 1\}$ with $\upsilon(\gamma)$ minimal and by Theorem 12.18, $\gamma \mid 2$ or $\gamma \mid 3$. Should $\gamma \mid 2$ then since $2 \in \mathbf{Z}[(-m)^{1/2}]$ is irreducible, $\gamma = \pm 2$ follows. But now from Theorem 12.18, $2 \mid (1+(-m)^{1/2})$, $2 \mid (-m)^{1/2}$, or $2 \mid (2+(-m)^{1/2})$, all of which are impossible as $2(a+b(-m)^{1/2}) = 2a + 2b(-m)^{1/2}$. Consequently, $\gamma \mid 3$, implying that $N(\gamma) \mid 9$. Since $m \geq 5$, $N(\alpha) \neq 3$ for all $\alpha \in \mathbf{Z}[(-m)^{1/2}]$ so $N(\gamma) = 9$. If $\gamma = \pm 3$ then by Theorem 12.18, $3 \mid (1+(-m)^{1/2})$, $3 \mid (2+(-m)^{1/2})$, or $3 \mid (-m)^{1/2}$, all impossible as above. For $\gamma = a + b(-m)^{1/2}$, $N(\gamma) = 9$, and $\gamma \neq \pm 3$, we must have $m = 5$, 8, or 9. In these cases, if $N(\alpha) = 9$ then $\alpha \in \mathbf{Z}[(-m)^{1/2}]$ is irreducible as above, using that 3 is not a norm. But $N(2+(-5)^{1/2}) = 3 \cdot 3$, $N(1+(-8)^{1/2}) = 3 \cdot 3$, and $N((-9)^{1/2}) = 3 \cdot 3$, so when $m = 5$, 8, or 9, $(\mathbf{Z}[(-m)^{1/2}], \upsilon)$ cannot be Euclidean, completing the verification.

As we mentioned, it is difficult to show that domains are Euclidean, or are PIDs. A single approach shows that $(\mathbf{Z}[m^{1/2}], |N|)$ is Euclidean for a few different m. When $m \geq 2$ then $N(a+bm^{1/2}) = a^2 - mb^2$ so we need to take the absolute value of the norm. It is still true that $|N(\alpha\beta)| = |N(\alpha)||N(\beta)|$.

Example 12.7 $(\mathbf{Z}[m^{1/2}], |N|)$ is Euclidean when $m = -2, -1, 2, 3, 6$, or 7.

The particular m we consider here are not the only ones that give Euclidean domains, but our approach works for these. Now $\mathbf{Z}[m^{1/2}] \subseteq \mathbf{Q}[m^{1/2}]$, a field by Example 10.18, and $|N(a+bm^{1/2})| = |a^2 - mb^2|$ is defined for $\mathbf{Q}[m^{1/2}]$ as well as for $\mathbf{Z}[m^{1/2}]$. Note first that for the given m and for $\alpha, \beta \in \mathbf{Z}[m^{1/2}]$, $N(\alpha) = 0 \Leftrightarrow \alpha = 0$ and so when $\alpha \mid \beta$ then $N(\alpha) \leq N(\beta)$. For $\alpha = c + dm^{1/2} \in \mathbf{Z}[m^{1/2}]^*$, $\alpha^{-1} = (1/N(\alpha))(c - dm^{1/2}) \in \mathbf{Q}[m^{1/2}]$, so when $\beta \in \mathbf{Z}[m^{1/2}]$ and $\alpha \nmid \beta$ then $\beta\alpha^{-1} = q_1 + q_2 m^{1/2} \in \mathbf{Q}[m^{1/2}] - \mathbf{Z}[m^{1/2}]$. If for some $\gamma \in \mathbf{Z}[m^{1/2}]$, $\beta\alpha^{-1} - \gamma = \delta \neq 0$ (since $\alpha \nmid \beta$) with $|N(\delta)| < 1$, then $\beta = \gamma\alpha + \delta\alpha$, $\delta\alpha = \beta - \gamma\alpha \in \mathbf{Z}[m^{1/2}]$, and $|N(\delta\alpha)| = |N(\delta)||N(\alpha)| < |N(\alpha)|$, proving that $(\mathbf{Z}[m^{1/2}], |N|)$ is Euclidean. Therefore it suffices to show that for any $\rho = q_1 + q_2 m^{1/2} \in \mathbf{Q}[m^{1/2}] - \mathbf{Z}[m^{1/2}]$, there is $a + bm^{1/2} \in \mathbf{Z}[m^{1/2}]$ with $|(q_1 - a)^2 - m(q_2 - b)^2| = |N(\rho - (a + bm^{1/2}))| < 1$.

Write $q_1 = n_1 + r$ and $q_2 = n_2 + s$ for $n_1, n_2 \in \mathbf{Z}$ and $|r|, |s| \leq 1/2$. For $a, b \in \mathbf{Z}$, $|(q_1 - a)^2 - m(q_2 - b)^2| < 1 \Leftrightarrow |(r - (a - n_1))^2 - m(s - (b - n_2))^2| < 1$ so we may assume $|q_1|, |q_2| \leq 1/2$, and not both zero. For $m = -2, -1, 2$, or 3, take $a = b = 0$. If $m = -2, -1$, or 2, our inequality becomes $|q_1^2 - mq_2^2| \leq (1 + |m|)/4 < 1$, and for $m = 3$, $q_1^2 \in [0, 1/4]$ and $mq_2^2 \in [0, 3/4]$ so again $|q_1^2 - mq_2^2| < 1$. When $m = 6$ or $m = 7$, if $q_2 = 0$ then $a = b = 0$ gives $|q_1^2 - mq_2^2| = q_1^2 < 1$. If $q_2 \neq 0$ note that $mq_2^2 \neq 5/4$ since $q_2 = c/d$ with $\mathrm{GCD}(\{c, d\}) = 1$ and $4c^2 m = 5d^2$ would force $5 \mid c$, then $5 \mid d$, a contradiction. When $0 < mq_2^2 < 5/4$ take $a = \pm 1$ with $aq_1 \geq 0$ and $b = 0$. Since $1/4 \leq (q_1 - a)^2 \leq 1$, $|(q_1 - a)^2 - mq_2^2| < 1$. Finally, if $5/4 < mq_2^2 \leq 7/4$ then for $a = \pm 1$ with $aq_1 \leq 0$ and $b = 0$, $1 \leq (q_1 - a)^2 \leq 9/4$ and once again

$|(q_1 - a)^2 - mq_2^2| < 1$. Therefore in all cases $(\mathbf{Z}[m^{1/2}], |N|)$ is Euclidean as required if $m = -2, -1, 2, 3, 6$, or 7.

A ring of particular interest is $\mathbf{Z}[i] = \mathbf{Z}[(-1)^{1/2}]$, called the *Gaussian integers*. From Example 12.7 this domain is Euclidean, and hence is a PID and a UFD. The argument in Example 12.7 shows how to find a remainder when dividing $\alpha \in \mathbf{Z}[i]^*$ into $\beta \in \mathbf{Z}[i]$. For example, if $\alpha = 2 + i$ and $\beta = 1 + 2i$, then $\alpha^{-1} = (1/5)(2 - i)$ and $\beta\alpha^{-1} = (1/5)(4 + 3i) = (1 + i) + (1/5)(-1 - 2i)$, where we use $1 + i$ since 1 is the integer nearest to both 4/5 and 3/5. Hence $\beta = (1 + i)\alpha + (1/5)(-1 - 2i)(2 + i) = (1 + i)\alpha + (-i)$. Thus $-i \in U(\mathbf{Z}[i])$ is the remainder, and we can multiply by i to represent $1 = \text{GCD}(\{\alpha, \beta\})$ as a linear combination of α and β, namely $1 = i\beta + (1 - i)\alpha$. This method also shows when $\alpha \mid \beta$, for if $\beta\alpha^{-1} = \gamma \in \mathbf{Z}[i]$, then $\beta = \gamma\alpha$. As an illustration, let us try to find the remainder when we divide $6 + 7i$ by $1 + 2i$. We compute that $(6 + 7i)(1 + 2i)^{-1} = (6 + 7i)(1/5)(1 - 2i) = (1/5)(20 - 5i) = 4 - i$, so $(6 + 7i) = (4 - i)(1 + 2i)$. Of course we could also try to solve for $a, b \in \mathbf{Z}$ satisfying $(1 + 2i)(a + bi) = 6 + 7i$, giving the system $a - 2b = 6$ and $2a + b = 7$.

Mimicking the "Euclidean Algorithm" of successively dividing by remainders in $\mathbf{Z}$ or in $F[x]$, the GCD $(\{\alpha, \beta\})$ in $\mathbf{Z}[i]$ is the last nonzero remainder, and then we can represent it as a combination of α and β.

Example 12.8 Find GCD $(\{10 + 5i, 6 + 7i\}) = \alpha(10 + 5i) + \beta(6 + 7i)$.

We need a sequence of equations for division with remainder, and each remainder has to have smaller norm than the divisor. Now $N(10 + 5i) = 125$, $N(6 + 7i) = 85$, and clearly $N((10 + 5i) - (6 + 7i)) = N(4 - 2i) = 20$, so we can write $(10 + 5i) = (6 + 7i) + (4 - 2i)$ with $4 - 2i$ as the first remainder. Now divide $4 - 2i$ into $6 + 7i$ by the method above:

$$(6 + 7i)(4 - 2i)^{-1} = (6 + 7i)(1/20)(4 + 2i) = (1/20)(10 + 40i).$$

Thus $(6 + 7i)(4 - 2i)^{-1} = 1/2 + 2i = 2i + 1/2$ so $(6 + 7i) = 2i(4 - 2i) + (2 - i)$. Since clearly $(4 - 2i) = 2(2 - i)$, $2 - i = \text{GCD}(\{10 + 5i, 6 + 7i\}$ is the last nonzero remainder and we can find α and β by using the equations derived above. Specifically, the last division shows $2 - i = (6 + 7i) - 2i(4 - 2i)$, and now use the first "division" to write $2 - i = (6 + 7i) - 2i((10 + 5i) - (6 + 7i))$. Thus $2 - i = -2i(10 + 5i) + (1 + 2i)(6 + 7i)$.

In general it is hard to identify the primes in a PID or in a Euclidean domain, but we can do so in the Gaussian integers. We will need the nontrivial fact that for a prime $2 < p \in \mathbf{N}$, $p = a^2 + b^2$ in $\mathbf{N} \Leftrightarrow p \equiv 1 \pmod 4$. When $p = a^2 + b^2$ then one of a or b is odd so $(2k + 1)^2 \equiv 1 \pmod 4$ implies that $p \equiv 1 \pmod 4$. An argument that any prime $p \equiv 1 \pmod 4$ is a sum of two squares in $\mathbf{N}$ is outlined in Exercise 13 of §4.6 and depends on the fact that there is a solution in $\mathbf{Z}$ of $x^2 \equiv -1 \pmod p$ (Example 4.16). We also need to recall from Example 12.1 that $U(\mathbf{Z}[i]) = \langle i \rangle = \{1, -1, i, -i\}$.

THEOREM 12.19 For $\rho \in \mathbf{Z}[i]$, ρ is a prime exactly when either:

i) $\rho \in \{1+i, -1+i, 1-i, -1-i\} = (1+i)\cdot\langle i\rangle$; or
ii) $\rho \in p\cdot\langle i\rangle$ for $p \in \mathbf{N}$ a prime with $p \equiv 3 \pmod 4$; or
iii) $\rho = a+bi$ with $a^2+b^2 = p$ a prime in $\mathbf{N}$ and $p \equiv 1 \pmod 4$.

Proof If $\rho \in \mathbf{Z}[i]$ is prime then $\rho \mid N(\rho)$ and $N(\rho) > 1$ a product of primes in $\mathbf{N}$ force ρ to divide a prime factor of $N(\rho)$ in $\mathbf{N}$ (why?). If ρ divides primes $p, q \in \mathbf{Z}$ with $q \neq \pm p$, then $sp + tq = 1$ for some $s, t \in \mathbf{Z}$ by Theorem 1.10, and ρ divides $sp+tq$ by Theorem 12.1, forcing the contradiction $\rho \mid 1$. Thus ρ divides a unique prime $p \in \mathbf{N}$ so we can find the primes in $\mathbf{Z}[i]$ by factoring the primes in $\mathbf{N}$. Suppose $\gamma, \alpha, \beta \in \mathbf{Z}[i]$, $N(\gamma) \in \mathbf{N}$ is prime, and $\gamma = \alpha\beta$. Then $N(\gamma) = N(\alpha)N(\beta)$ so $N(\alpha) = 1$ or $N(\beta) = 1$. Thus γ is irreducible in $\mathbf{Z}[i]$ and hence is a prime by Example 12.7, Theorem 12.17, and Theorem 12.7. That is, $N(\gamma) \in \mathbf{N}$ a prime implies that γ itself is a prime.

Now in $\mathbf{Z}[i]$, $2 = (1+i)(1-i)$ with $N(1+i) = 2$, so $1+i \in \mathbf{Z}[i]$ is a prime, $1-i = (-i)(1+i)$, and any prime $\rho \in \mathbf{Z}[i]$ dividing 2 must be an associate of $1+i$ by unique factorization. When $p \in \mathbf{N}$ is a prime so that $p \equiv 1 \pmod 4$ then by our comments above, $p = a^2+b^2$ in $\mathbf{N}$. It follows that $p = (a+bi)(a-bi) = N(a \pm bi)$, so again $a \pm bi \in \mathbf{Z}[i]$ are primes and any prime divisor of p must be an associate of one of these. Finally, let $p \in \mathbf{N}$ be prime, $p \equiv 3 \pmod 4$, $p = \rho\sigma$ for $\rho = a+bi$, $\sigma = c+di \in \mathbf{Z}[i]$, and ρ a prime. Taking norms, $p^2 = (a^2+b^2)(c^2+d^2)$, so $a^2+b^2 = p$ or $a^2+b^2 = p^2$. But it is impossible that $a^2+b^2 = p$ since $p \equiv 3 \pmod 4$, so $a^2+b^2 = p^2$, σ is a unit, and $\rho \in p\cdot\langle i\rangle$. As an associate of ρ, p is a prime in $\mathbf{Z}[i]$, completing the proof of the theorem. ∎

We know that the prime factors of $\alpha \in \mathbf{Z}[i]$ must have norms dividing $N(\alpha)$ and therefore, from Theorem 12.19, we get the possibilities for these primes from the prime factors in $\mathbf{N}$ of $N(\alpha)$. To factor $\eta = -3+4i$, note that $N(\eta) = 25$ and that $5 \equiv_4 1$ so the prime factors of η must be associates of $1+2i$ or $1-2i$. If we try dividing η by $(1+2i)$ the result is $\eta(1+2i)^{-1} = (-3+4i)(1/5)(1-2i) = 1+2i$, so in fact $\eta = (1+2i)^2$. Another example follows.

Example 12.9 Factor $24+23i$ into primes in $\mathbf{Z}[i]$.

$N(24+23i) = 24^2+23^2 = 576+529 = 1105 = 5\cdot 221 = 5\cdot 13\cdot 17$, so the prime divisors of $24+23i$ are $a+bi$ with $a^2+b^2 = 5$, 13, or 17. Now $(24+23i)\cdot(1/5)(1+2i) \notin \mathbf{Z}[i]$ so $(1-2i) \nmid (24+23i)$. Hence $(1+2i) \mid (24+23i)$ and $(24+23i)(1/5)(1-2i) = 14-5i$ so $24+23i = (14-5i)(1+2i)$. Since $N(2+3i) = 13$ we see whether $2+3i$ divides $14-5i$: $(14-5i)(1/13)(2-3i) = (1/13)(13-52i)$. Thus $24+23i = (1+2i)(14-5i) = (1+2i)(2+3i)(1-4i)$.

There are PIDs that are not Euclidean. One such uses Exercise 3, which shows that $\mathbf{Z}[\gamma] = \{a+b\gamma \in \mathbf{C} \mid a, b \in \mathbf{Z}$ and $\gamma = (1+(-19)^{1/2})/2\}$ is not Euclidean. This ring is a PID but the computations needed to verify this are not easy and we do not pursue the matter any further.

There is a nice result about matrices over Euclidean domains that does not hold for every PID. It is essentially Theorem 5.12 that the group $SL(n, D)$ of $n \times n$ matrices with determinant 1 is generated by the transvections. Recall that for D an integral domain, I_n is the $n \times n$ identity matrix, E_{ij} is the matrix with i-j entry 1 and other entries zero, and for $i \neq j$ the transvection $T_{ij}(a) = I_n + aE_{ij}$. For $A \in SL(n, D)$ with $A_{jk} \neq 0$, using D Euclidean gives $A_{ik} = qA_{jk} + r$ with $r = 0$ or $\upsilon(r) < \upsilon(A_{jk})$. Thus for $B = T_{ij}(-q)A$, $B_{ik} = A_{ik} - qA_{jk} = r$ and $B_{ik} = 0$ or $\upsilon(B_{ik}) < \upsilon(A_{jk}) = \upsilon(B_{jk})$. This will be the crucial computation that uses D Euclidean.

THEOREM 12.20 If (D, υ) is a Euclidean domain and $n > 1$, then $SL(n, D) = \langle\{T_{ij}(a) \mid i \neq j \text{ and } a \in D\}\rangle \leq GL(n, D)$.

Proof For simplicity, set $SL(n, D) = SL(D)$. Each $T_{ij}(a) \in SL(D)$ so $G = \langle\{T_{ij}(a) \mid i \neq j$ and $a \in D\}\rangle \leq SL(D)$ and it suffices to prove that any $A \in SL(D)$ is in G. If we show that gAg' is diagonal for some $g, g' \in G$ then the proof of Theorem 5.12 works here to complete the argument since for $v \in U(D)$, $I_n + (v-1)E_{ii} + (v^{-1} - 1)E_{jj} \in G$ by the same computation as in Theorem 5.12.

Now $E(i, j) = I_n - E_{ii} - E_{jj} + E_{ij} - E_{ji} = T_{ij}(1)T_{ji}(-1)T_{ij}(1) \in G$ and for $A \in GL(n, D)$, $B = E(i, j)A$ interchanges the ith and j^{th} rows of A up to sign. Also $B' = AE(i, j)$ interchanges the ith and j^{th} columns of A up to sign. For $S = \{gAg' \mid g, g' \in G\}$ there is $B \in S$ so that $\upsilon(B_{ij})$ is minimal in $\upsilon(S) = \{\upsilon(s_{kt}) \mid s \in S \text{ and } s_{kt} \neq 0\}$. Multiply B by $E(i, j)$ as necessary to get $Y \in S$ with $\upsilon(Y_{11}) = \upsilon(B_{ij})$ minimal in $\upsilon(S)$ (how?—also see Exercise 1). For $j > 1$ the definition of Euclidean shows that $Y_{j1} = q_jY_{11} + r_j$ for $r_j = 0$ or $\upsilon(r_j) < \upsilon(Y_{11})$. Set $V = T_{n1}(-q_n) \cdots T_{21}(-q_2)Y$ in S. It follows that $V_{11} = Y_{11}$ and $V_{j1} = r_j$ for all $j > 1$. By the minimality of $\upsilon(Y_{11})$, all $r_j = 0 = V_{j1}$. Repeat this argument on the right by writing $V_{1k} = d_kV_{11} + t_k$ for $k > 1$ and $t_k = 0$ or $\upsilon(t_k) < \upsilon(V_{11})$. If $L = VT_{12}(-d_2) \cdots T_{1n}(-d_n)$ then $L_{11} = V_{11} = Y_{11}$, $L_{j1} = V_{j1} = 0$ if $j > 1$, and $L_{1k} = t_k = 0$ if $k > 1$ by the minimality of $\upsilon(V_{11})$.

Using that L is invertible, from $LL^{-1} = I_n$ we have $(LL^{-1})_{11} = 1$. Hence $1 = \sum L_{1j}L^{-1}_{j1}$, and since $L_{1j} = 0$ for $j > 1$, we see that $L_{11} \in U(D)$. Clearly L is diagonal if $n = 2$. If $n > 2$ then $L = \begin{bmatrix} u & 0 \\ 0 & C \end{bmatrix}$ for $u = L_{11} \in U(D)$ and $C \in GL(n-1, D)$. By induction on n, a diagonal matrix results by multiplying C on its left and right by some products of transvections in $M_{n-1}(D)$. Hence there are $g_1, g_2 \in G$, products of suitable $T_{ij}(a_{ij}) \in M_n(D)$ with $i, j \geq 2$ so that $g_1Lg_2 = gAg'$ is a diagonal matrix. When $A \in SL(D)$, as we indicated above, the proof is completed by applying the argument in Theorem 5.12. ∎

EXERCISES

1. Let (D, υ) be a Euclidean domain. Show that:
 i) if $u \in U(D)$ then $\upsilon(u) \leq \upsilon(d)$ for all $d \in D^*$
 ii) if $x, y \in D^*$ with $x \sim_a y$ then $\upsilon(x) = \upsilon(y)$
 iii) if $\upsilon(d) = 0$ then $d \in U(D)$. (Be careful! What if $\upsilon(d) = 0$ for all $d \in D^*$?)

2. Let $d \in \boldsymbol{Z}$, $d^{1/2} \notin \boldsymbol{Q}$, $d \equiv 1 \pmod 4$, $\boldsymbol{Z}[\delta] = \{a + b\delta \in \boldsymbol{C} \mid a, b \in \boldsymbol{Z}\}$ for $\delta = (1 + d^{1/2})/2 \in \boldsymbol{C}$, and $R(d) = \{a/2 + (b/2)d^{1/2} \mid a, b \in \boldsymbol{Z} \text{ and } a \equiv b \pmod 2\}$.
 i) Show that $R(d)$ is an integral domain using the arithmetic in $\boldsymbol{C}$.
 ii) Show that $\boldsymbol{Z}[\delta] = R(d)$.
 iii) If $\upsilon_d(a/2 + (b/2)d^{1/2}) = |a^2/4 - db^2/4|$, show that $\upsilon_d : R(d)^* \to \boldsymbol{N}$.

3. Use the approach in Example 12.6 to show that $(R(d), \upsilon_d)$ (Exercise 2) is not a Euclidean domain when:
 i) $d = -19$
 ii) $d = -23$
 iii) $d < -15$ and $d \equiv_4 1$

4. Use Exercise 2 and the approach in Example 12.7 to show that:
 i) $(R(-3), \upsilon_{-3})$ is a Euclidean domain.
 ii) $(R(5), \upsilon_5)$ is a Euclidean domain.

5. Let D be an integral domain and $\eta : D^* \to \boldsymbol{N}$ so that for $a, b \in D^*$ with $\eta(a) \le \eta(b)$, either $a \mid b$ or for some $x, y \in D^*$ $\eta(xa - yb) < \eta(a)$. Show that D is a PID.

6. Write $q \in \boldsymbol{Q}^*$ as $q = \pm p_1^{q(1)} \cdots p_n^{q(n)}$ for p_i the ith prime, n sufficiently large, and all $q(j) \in \boldsymbol{Z}$. Fix p_i and define $\upsilon_i : \boldsymbol{Q}^* \to \boldsymbol{Q}^+$ by $\upsilon_i(q) = p_i^{q(i)}$. For $s, t \in \boldsymbol{Q}^*$ show that:
 i) $\upsilon_i(st) = \upsilon_i(s)\upsilon_i(t)$, and if $s + t \neq 0$ then $\upsilon_i(s + t) \ge \min\{\upsilon_i(s), \upsilon_i(t)\}$
 ii) $R_i = \{q \in \boldsymbol{Q}^* \mid \upsilon_i(q) \ge 1\} \cup \{0\}$ is an integral domain
 iii) $U(R_i) = \{a/b \in \boldsymbol{Q}^* \mid \text{GCD}(\{a, b\}) = 1 = \text{GCD}(\{a, p_i\})\}$
 iv) if $x, y \in R_i^*$ then $x \mid y \Leftrightarrow \upsilon_i(x) \le \upsilon_i(y)$
 v) (R_i, η_i) satisfies i) and ii) in Definition 12.8 for η_i the restriction of υ_i to R_i
 vi) if $I \lhd R_i$ then $I = (0)$, $I = (p_i^m)$, or $I = R_i$
 vii) $\bigcap\{R_i \mid i \in \boldsymbol{N}\} = \boldsymbol{Z}$

7. As in Exercise 6, write $q \in \boldsymbol{Q}^*$ as $q = \pm p_1^{q(1)} \cdots p_n^{q(n)}$ for p_i the ith prime, fix j, and set $\omega_j(q) = q(j)$, so $\omega_j : \boldsymbol{Q}^* \to \boldsymbol{Z}$.
 i) Show that $A_j = \{q \in \boldsymbol{Q} \mid \omega_j(q) \ge 0\}$ is an integral domain.
 ii) Restricting ω_j to A_j, show that (A_j, ω_j) is Euclidean.

8. Can $deg : \boldsymbol{Q}[x, y] \to \boldsymbol{N} \cup \{0\}$ be altered to make $\boldsymbol{Q}[x, y]$ a Euclidean domain?

9. Since $\boldsymbol{Q}[x, y]$ is a UFD, write $\alpha \in QF(\boldsymbol{Q}[x, y])$ as $\alpha = x^i y^j \beta$ for $\beta = p(x, y)/q(x, y)$ with $\text{GCD}(\{xy, p(x, y)q(x, y)\}) = 1$. Show that $R = \{x^i y^j \beta \in QF(\boldsymbol{Q}[x, y])$ as above $\mid i, j \ge 0\}$ is a domain, and that (R, υ) is Euclidean for $\upsilon(x^i y^j \beta) = i + j$.

10. Show that any field is a Euclidean domain.

11. Find $\text{GCD}(\{\alpha, \beta\})$ for $\alpha, \beta \in (\boldsymbol{Z}[i], N)$ (consider norms!) when:
 i) $\alpha = 1 + 8i$, $\beta = 9 + 2i$
 ii) $\alpha = 9 - i$, $\beta = 11 + 7i$
 iii) $\alpha = 28 + 23i$, $\beta = 6 + 41i$

12. For $\alpha, \beta \in \boldsymbol{Z}[i]$, write $\text{GCD}(\{\alpha, \beta\}) = x\alpha + y\beta$ when:
 i) $\alpha = 5 + 4i$, $\beta = 7 + i$
 ii) $\alpha = -3 + 29i$, $\beta = -4 + 7i$
 iii) $\alpha = 6 + 4i$, $\beta = 3 + 5i$

13. Factor each of the following $\alpha \in \boldsymbol{Z}[i]$ into irreducibles:
 i) $3 + i$
 ii) $5 - 10i$
 iii) $9 - 7i$
 iv) $5 + 6i$
 v) $9 - 15i$
 vi) $45i$
 vii) $19 + i$
 viii) 210
 ix) $-10 + 49i$

14. If $\alpha = a + bi \in \mathbf{Z}[i]$ with $ab \neq 0$ and $(a, b) = 1$, and if $\beta \in \mathbf{Z}[i]^* - \langle i \rangle$ with $\beta \mid \alpha$, then $\beta = c + di$ with $cd \neq 0$ and $(c, d) = 1$.

15. Write $\begin{bmatrix} 1+i & 3+2i \\ 2+i & 6+i \end{bmatrix}$ as a product of transvections $T_{ij}(\alpha_{ij}) \in SL(2, \mathbf{Z}[i])$.

16. If $D = \{p(x, y) + q(x, y)i \in \boldsymbol{C}[x, y] \mid p(x, y), q(x, y) \in \mathbf{Z}[x, y]\}$ then using the addition and multiplication in $\boldsymbol{C}[x, y]$, show that D is a UFD. (Hint: Show that $D = (\mathbf{Z}[i])[x, y]$.)

CHAPTER 13

COMMUTATIVE RINGS

We introduce important notions and prove results essential for further studies of commutative rings, of algebraic number theory, or of algebraic geometry. The basic topics are maximal and prime ideals, localization, Noetherian rings and the Hilbert Basis Theorem, integrality, and an introduction to notions in algebraic geometry.

13.1 MAXIMAL AND PRIME IDEALS

We recall the notions of maximal ideal and prime ideal but first present *Zorn's Lemma,* an axiom that is very useful in algebra. We discuss its equivalence to the Axiom of Choice (see §0.5) at the end of this chapter.

From §1.5, a *partially ordered set,* or *poset,* is $(S, \leq')$ for a set $S \neq \emptyset$ and $\leq' : S \times S \to S$ satisfying the following properties for all $a, b, c \in S$:

i) $a \leq' a$
ii) $a \leq' b$ and $b \leq' a \Rightarrow a = b$
iii) $a \leq' b$ and $b \leq' c \Rightarrow a \leq' c$

Two examples are $(P(A), \subseteq)$, the set of subsets of a set A with $\leq'$ the usual set containment, and $(\boldsymbol{N}, |)$: "divides" on $\boldsymbol{N}$. Given a poset $(S, \leq')$, $L \subseteq S$ is *linearly ordered,* or a *chain,* if for all $x, y \in L$ either $x \leq' y$ or $y \leq' x$ (Definition 1.9). For a poset S, $\emptyset \neq T \subseteq S$ has an *upper bound* $d \in S$ if $t \leq' d$ for all $t \in T$, and d is the **least upper bound** for T, written $d = \text{LUB}(T)$, if d is an upper bound for T and $d \leq' e$ for any upper bound e of T. A *maximal element* of a poset S is any $m \in S$ so that for $s \in S$, $m \leq' s$ implies $m = s$.

Note that LUB(T) is unique if it exists (why?). The maximal element in any finite chain (there is one!) is the least upper bound for the chain. If we define $\leq'$ on $\boldsymbol{N} - \{1\}$ via $a \leq' b$ when $b \mid a$ then $B = (\boldsymbol{N} - \{1\}, \leq')$ is a poset. The primes in $\boldsymbol{N}$ are the maximal elements in B but are not upper bounds for B (why?). By unique factorization in $\boldsymbol{N}$, every chain in B has a least upper bound. Also, $\{6, 10, 15\} \subseteq B$ has no upper bound and $\{6\}$ has no least upper bound in B.

Example 13.1 The poset $S = (\{0\} \times [0, 1]) \cup (\{1\} \times (0, 1)) \cup ([2, 3) \times \boldsymbol{R}) \subseteq \boldsymbol{R}^2$ with $(a, b) \leq' (c, d)$ when $a = c$ and $b \leq d$ in $\boldsymbol{R}$ or when $2 \leq a < c$ in $\boldsymbol{R}$.

It is an exercise that $(S, \leq')$ is a poset. Any $\varnothing \neq A \subseteq \{0\} \times [0, 1]$ is a chain and has least upper bound $(0, m)$ for $m = \mathrm{LUB}\{a \in \boldsymbol{R} \mid (0, a) \in A\}$. Similarly, any $\varnothing \neq B \subseteq \{1\} \times (0, 1)$ is a chain, has no upper bound if $\pi(B) = \{b \in \boldsymbol{R} \mid (1, b) \in B\}$ has least upper bound 1 in $\boldsymbol{R}$, and has least upper bound $(1, d)$ if $\mathrm{LUB}(\pi(B)) = d < 1$. The chain $C = \{2\} \times \boldsymbol{R} \subseteq S$ has upper bound (c, d) for any $2 < c < 3$ but no least upper bound since a smaller upper bound is $(c, d - 1)$. The chain $D = \{(a, b) \in S \mid a \in [2, 5/2]$ and $b < 0\}$ in S has least upper bound $(5/2, 0)$. Finally, $E = \{(a, b) \in S \mid a \in [2, 3)$ and $b = 1/(3 - a)\}$ is a chain with no upper bound. Note that $(0, 1)$ is the only maximal element in S but that $(0, 1)$ is not an upper bound for S.

Example 13.2 The poset $(S, \subseteq)$ for $S = \{H \leq (\boldsymbol{Q}, +) \mid H \neq \boldsymbol{Q}\}$.

From Theorem 2.14, if $\{H_\lambda\}_\Lambda$ is a chain in S then $H = \bigcup H_\lambda \leq \boldsymbol{Q}$, and if $H \neq \boldsymbol{Q}$ then $H = \mathrm{LUB}\{H_\lambda\}$ in S. If $H = \boldsymbol{Q}$, the chain $\{H_\lambda\}$ has no upper bound in S. Each possibility can occur. Note that in $(\boldsymbol{Q}, +), q^m = mq$. If $A_j = \langle 1/j! \rangle \leq (\boldsymbol{Q}, +)$ then $A_j \subseteq A_{j+1}$ and for any $a/b \in \boldsymbol{Q}, a/b \in A_b$ so $\bigcup A_j = \boldsymbol{Q}$. For $B_i = \langle 1/2^i \rangle$, $\bigcup B_i = \{a/b \in \boldsymbol{Q} \mid b = 2^j$ some $j \geq 0\} \neq \boldsymbol{Q}$. Assume $M \in S$ is maximal with $q \in \boldsymbol{Q} - M$. Then $M + \langle q \rangle = (\boldsymbol{Q}, +)$ and it follows from Theorem 7.6 that $\boldsymbol{Q}/M \cong \langle q \rangle / (\langle q \rangle \cap M)$. If $a/b = q$ and if $m/n \in M$ with $b, n > 0$ then in $(\boldsymbol{Q}, +)$, $am\boldsymbol{Z} \leq \langle (a/b)^b \rangle \cap \langle (m/n)^n \rangle \subseteq \langle q \rangle \cap M$ so $\langle 0 \rangle \neq \langle q \rangle \cap M$. But every proper quotient of the cyclic group $\langle q \rangle$ is finite by Example 5.3. Thus $\boldsymbol{Q}/M$ is a finite group, say $|\boldsymbol{Q}/M| = k$, and Theorem 4.7 shows that for any $t \in \boldsymbol{Q}$, $M = (t + M)^k = kt + M$, so $kt \in M$. Thus $q = k(q/k) \in M$, a contradiction. This shows that S cannot contain a maximal element.

The examples show a variety of situations involving the existence of upper bounds for chains in a poset, or of maximal elements in a poset. Zorn's Lemma gives a relation between these.

ZORN'S LEMMA A nonempty poset S has a maximal element if every chain in S has an upper bound in S.

As we said, Zorn's Lemma is an axiom and we discuss its relation to the Axiom of Choice at the end of the chapter. A key result for some applications of Zorn's Lemma is Theorem 11.5: the union of any chain of ideals is itself an ideal. Clearly this union is the chain's least upper bound (using inclusion) in the set of all ideals. Thus an immediate and useful application of Zorn's Lemma is the following observation.

THEOREM 13.1 If S is any nonempty set of (right, left) ideals of the ring R, then the poset $(S, \subseteq)$ has maximal elements if $\bigcup I_\lambda \in S$ for any chain $\{I_\lambda\}$ in S.

Recall from Definition 11.3 that a *maximal* ideal of a ring R is just a maximal element in the poset of *proper* ideals of R, under set inclusion. Must every ring have a maximal ideal?

THEOREM 13.2 If R is a commutative ring with 1, and $I \lhd R$, then either $I = R$ or $I \subseteq M$, a maximal ideal of R. If $x \in R - U(R)$ then $x \in H$, a maximal ideal of R.

Proof If $I \neq R$ then $I \in S(I) = \{A \lhd R \mid I \subseteq A \neq R\}$, a poset using $\subseteq$. For $\{H_\lambda\}$ a chain in $S(I)$, $\bigcup H_\lambda = H \lhd R$ by Theorem 11.5, and $I \subseteq H_\lambda \subseteq H$. When $H = R$ then $1 \in H$ implies $1 \in H_\lambda$ for some λ so $H_\lambda = R$ (why?), contradicting $H_\lambda \in S(I)$. Thus $H \neq R$, $H \in S(I)$, and Theorem 13.1 yields a proper ideal M maximal with respect to containing I. Clearly M is maximal in R. For $x \in R$, $(x) \lhd R$ so if $(x) = R$ then $1 = rx$ and $x \in U(R)$. Hence if $x \notin U(R)$ then $x \in (x) \subseteq M$ for M maximal, by the first part of the theorem. ■

To see that $1 \in R$ is essential for the theorem, set $R = (x) \lhd \mathbf{Q}[x]$. For any proper $G \leq (\mathbf{Q}, +)$, $I = Gx + x^2\mathbf{Q}[x]$ is an ideal of R. If $I \subseteq M \subseteq R$ are ideals and if $K = \{k \in \mathbf{Q} \mid kx + x^2h(x) \in M\}$, then $G \leq K \leq (\mathbf{Q}, +)$, $M = Kx + x^2\mathbf{Q}[x]$ since $x^2\mathbf{Q}[x] \subseteq I$, and if $M = R$ then $K = \mathbf{Q}$. When $K \neq \mathbf{Q}$ then using the result in Example 13.2 that $(\mathbf{Q}, +)$ has no maximal subgroup, there is $L \leq (\mathbf{Q}, +)$ with $K \leq L$ properly and $L \neq \mathbf{Q}$. Thus $M \subseteq Lx + x^2\mathbf{Q}[x] \neq R$ properly, so R cannot have a maximal ideal containing I. At the other extreme, Theorem 11.4 and Theorem 10.10 show that (x) is the only maximal ideal in $F[[x]]$ (Definition 10.5) and contains all its ideals when F is a field. Next we characterize the intersection of the maximal ideals in terms of invertible elements.

THEOREM 13.3 If R is a commutative ring with 1, and $S = \{M \lhd R \mid M \text{ is maximal}\}$, then $J(R) = \bigcap\{M \mid M \in S\} = \{x \in R \mid 1 + rx \in U(R), \text{ all } r \in R\}$.

Proof Let $J(R)$ be the intersection of the maximal ideals. Now $J(R) \lhd R$ from Theorem 11.5 so $x \in J(R)$ implies $rx \in J(R)$ for all $r \in R$. If $1 + rx \notin U(R)$ then Theorem 13.2 shows that $1 + rx \in M_0$ for some maximal ideal M_0. Hence $1 = (1 + rx) + -r \cdot x \in M_0 + J(R) \subseteq M_0$, contradicting M_0 a maximal ideal. Thus all $1 + rx$ are units for $x \in J(R)$, proving one containment. Assume that for $x \in R$, $\{1 + rx \mid r \in R\} \subseteq U(R)$. If $x \notin H$ for some maximal ideal H, then $R = H + (x)$ forces $1 = h + yx$. Thus $h = 1 + -y \cdot x \in H \cap U(R)$, a contradiction. Therefore $x \in H$, so $x \in J(R)$, completing the proof. ■

As we observed just above, $J(\mathbf{Q}[[x]]) = (x)$, but $J(\mathbf{Q}[x]) = (0)$ since each $(x - q)$ is maximal by Example 11.23, and unique factorization in $\mathbf{Q}[x]$ (or Theorem 4.18) shows that $p(x) \neq 0$ cannot be in all of these maximal ideals. For finite rings we have $J(\mathbf{Z}_{30}) = ([0]_{30})$ since $([2]_{30})$, $([3]_{30})$, and $([5]_{30})$ are maximal ideals in $\mathbf{Z}_{30}$ (see p. 359) whose intersection is $([0]_{30})$ (verify!), but since $([3]_{27})$ is the only maximal ideal in $\mathbf{Z}_{27}$, then $J(\mathbf{Z}_{27}) = ([3]_{27})$.

Before moving on, we present a result using the characterization of $J(R)$ in Theorem 13.3. A more general version of the result is called *Nakayama's Lemma,* although the proof we give is essentially the same as for more general versions.

THEOREM 13.4 If R is a commutative ring with 1, $I = Ra_1 + \cdots + Ra_n \lhd R$, S is a nonzero subring of $J(R)$, and $SI = I$, then $I = (0)$.

Proof If $I \neq (0)$, let $\{b_1, \ldots, b_m\} \subseteq I$ be minimal with $I = \sum Rb_j$. It suffices to take $S = J(R)$ since $SI \subseteq J(R)I \subseteq I$ and $I = SI$ force $J(R)I = I$. Now $J(R)Rb_j = J(R)b_j$ (why?) so $b_m \in J(R)I$ implies that $b_m = x_1b_1 + \cdots + x_mb_m$ for $x_i \in J(R)$, and then $(1 - x_m)b_m = x_1b_1 + \cdots + x_{m-1}b_{m-1}$. By Theorem 13.3, $1 - x_m \in U(R)$. Thus $b_m = (1 - x_m)^{-1}x_1b_1 + \cdots + (1 - x_m)^{-1}x_{m-1}b_{m-1} \in Rb_1 + \cdots + Rb_{m-1}$, forcing $Rb_m \subseteq Rb_1 + \cdots + Rb_{m-1}$, so

$I = Rb_1 + \cdots + Rb_{m-1}$, contradicting the minimality of m. If $m = 1$ the argument shows $b_1 = 0$. Therefore $I = (0)$, proving the theorem. ∎

When R is commutative with 1, what condition forces $1 + rx \in U(R)$ for all $r \in R$? If $x \in R$ is *nilpotent*—that is $x^k = 0$—then $1 + x \in U(R)$ since $(1+x)^{-1} = 1 - x + \cdots + (-x)^{k-1}$ (verify!). Also $(rx)^k = 0$ for any $r \in R$, so all $1 + rx$ are units and $\{x \in R \mid x^k = 0\} \subseteq J(R)$. More directly, if $M \lhd R$ is maximal, R/M is a field by Theorem 11.17 so $(x + M)^k = 0_{R/M}$ forces $x + M = 0_{R/M}$, or $x \in M$. The crucial factor here is that M is a prime ideal.

From Definition 11.5, $P \lhd R$ is *prime* in a commutative ring R when P is proper and $ab \in P$ implies $a \in P$ or $b \in P$, or, equivalently, when R/P is a domain. Another way to view this is to note that $a, b \notin P$ imply that $ab \notin P : R - P$ is closed under multiplication. We look at this idea more formally and use it to construct prime ideals, essential in the study of commutative rings. Of course when $1 \in R$, each maximal ideal M is prime since R/M is a field.

DEFINITION 13.1 If R is a commutative ring then $\varnothing \neq T \subseteq R$ is called **multiplicative** if $0 \notin T$ and if $a, b \in T$ imply $ab \in T$.

For $R \neq \{0\}$, a commutative ring, it is easy to see that the following $T \subseteq R$ are multiplicative when $T \neq \varnothing$: $\{d \in R \mid dr = 0 \Rightarrow r = 0\}$; $R - P$ for P a prime ideal of R; $\{x^k \in R \mid k \in N\}$ if $x \in R$ is fixed and $x^k \neq 0$ for any $k \in N$; and $\{f(x) \in D[x] \mid f(0) \neq 0\}$ when D is a commutative domain.

THEOREM 13.5 If R is a commutative ring and $T \subseteq R$ is multiplicative, then there is a maximal $P \in \{I \lhd R \mid I \bigcap T = \varnothing\}$, and P is a prime ideal of R.

Proof Let $S = \{I \lhd R \mid I \bigcap T = \varnothing\}$ and note that $(0) \in S$ so $S \neq \varnothing$. If $\{H_\lambda\}$ is a chain in S with union H, then $H \lhd R$ by Theorem 11.5, and if $t \in H \bigcap T$ then $t \in H_\lambda$ for some λ, contradicting $H_\lambda \in S$. Therefore $H \in S$ and is an upper bound in S for $\{H_\lambda\}$ so there is a maximal $P \in S$ by Theorem 13.1. Should $a, b \in R - P$ with $ab \in P$, then P is *properly* contained in the ideals $I = (a) + P$ and $J = (b) + P$. The maximality of P in S forces $x \in I \bigcap T$ and $y \in J \bigcap T$, but $IJ \subseteq P$ (why?) shows that $xy \in P \bigcap T$, a contradiction. Hence $ab \in P$ implies $a \in P$ or $b \in P$, so P is prime. ∎

We use the theorem to characterize the intersection $N(R)$ of all the prime ideals in R. Now $N(R)$ must contain all nilpotent elements since if $x^k = 0$ then for any prime ideal P, $x^k \in P$ so $x \in P$. If R contains no prime ideal—for example, $R = ([2]_{16}) \lhd \mathbf{Z}_{16}$ then set $\bigcap\{P \lhd R \mid P \text{ is prime}\} = R$.

THEOREM 13.6 For any commutative ring R, $N(R) = \bigcap\{P \lhd R \mid P \text{ is prime}\} = \{x \in R \mid x^k = 0 \text{ for some } k \geq 1, \text{ depending on } x\}$.

Proof We just observed that every prime ideal of R contains every nilpotent element, so $\{x \in R \mid x^k = 0\} \subseteq N(R)$, the intersection of the prime ideals. When each $x \in R$ is nilpotent there are no prime ideals so $N(R) = R$. Otherwise, for any nonnilpotent $x \in R$, $\{x^k \in R \mid k \geq 1\}$ is multiplicative. By Theorem 13.5 there is a prime ideal P_x with $x \notin P_x$. Consequently, if $y \in R$ is in every prime ideal then $y^k = 0$ for some $k \geq 1$, proving the theorem. ∎

The theorem provides a proof of Exercise 18 in §11.1, that for R commutative, the set of all nilpotent elements in R is an ideal of R and is therefore the unique largest ideal consisting of nilpotent elements. Using Theorem 13.6 we now describe the units in $R[x]$.

THEOREM 13.7 If R is a commutative ring with 1, then $U(R)+N(R)(x)=\{a+b_1x+\cdots+b_kx^k \in R[x] \mid a \in U(R) \text{ and } b_j \in N(R)\}=U(R[x])$.

Proof Let $p(x)=a+b_1x+\cdots+b_mx^m$ with $a \in U(R)$ and all $b_j \in N(R)$. By Theorem 13.6, each $b_j^{k(j)}=0$ so there is $t \in \mathbf{N}$ with $b_j^t=0$ for all j. Set $Y=b_1x+\cdots+b_mx^m$. Each coefficient of Y^n is a sum of products of n of the b_j. Any product of the b_j not containing at least t occurrences of some b_i contains each b_i at most $t-1$ times and so has at most $m(t-1)$ factors in all. Hence for $n \geq tm-m+1$, some b_j must appear at least t times in any product of n of the b_i so such a product is zero. The upshot is that $Y^n=0$ for n large enough. Now $p(x)=a+Y$ and $(a+Y)(a^{-1}-a^{-2}Y+\cdots+(-1)^{n-1}a^{-n}Y^{n-1})=1$, showing $p(x) \in U(R[x])$. We need every unit in $R[x]$ to have the given form.

Let $p(x)=a_0+a_1x+\cdots+a_nx^n \in U(R[x])$ with $p(x)g(x)=1$. Then $a_0g(0)=1$ so $a_0 \in U(R)$. For $P \lhd R$ and prime, R/P is a domain so $(R/P)[x]$ is also. From Example 11.18, if for $r \in R$, $[r]=r+P \in R/P$, then for $f(x)=r_0+\cdots+r_mx^m \in R[x]$, $\rho(f(x))=[r_0]+\cdots+[r_m]x^m$ is a ring homomorphism $\rho: R[x] \rightarrow (R/P)[x]$ (verify!). Now $\rho(1)=[1]$ implies $[1]=\rho(p(x))\rho(g(x))$, and Theorem 10.11 forces $deg\ \rho(p(x))=0$. Thus $[a_j]=[0]$ when $j>0$, and this means that $a_j \in P$. Since P is arbitrary, we conclude that for $j \geq 1$, $a_j \in N(R)$ by Theorem 13.6, completing the proof. ∎

A consequence of the theorem is that $[a]_{16}+[b_1]_{16}x+\cdots+[b_n]_{16}x^n$ is a unit in $\mathbf{Z}_{16}[x]$ exactly when a is odd and all b_j are even.

EXERCISES

In the exercises below, R denotes a commutative ring.

1. Decide whether the following sets are multiplicative:
  i) $\{p(x) \in \mathbf{Z}[x] \mid p(0)=\pm 1\}$
  ii) $\{p(x) \in \mathbf{Z}[x] \mid p(0)=3^k\}$
  iii) $\{p(x) \in \mathbf{Z}[x] \mid p(x) \text{ is primitive}\}$
  iv) $\{p(x) \in \mathbf{Z}[x] \mid \text{each coefficient of } p(x) \text{ is some } 2^m\}$
  v) $1+I=\{1+x \mid x \in I\}$ for I a proper ideal of R, when $1 \in R$

2. If $T \subseteq R$ is multiplicative, show that $T \cup U(R) \cup \{tu \mid t \in T, u \in U(R)\}$ is also.

3. If $1 \in R$, show that R contains maximal multiplicative sets (with respect to $\subseteq$).

4. If $T \subseteq R$ is a maximal multiplicative set (with respect to $\subseteq$), show that when $t \in T$ and $t=xy$ for $x, y \in R$, then $x, y \in T$.

5. If $T \subseteq R$ is a maximal multiplicative set (using $\subseteq$), show that $R-T$ is a prime ideal of R and contains no other prime ideal properly.

6. Let $1 \in R$ and $\{P_\lambda\}$ a nonempty set of prime ideals of R.
  i) Show that $R-\bigcup P_\lambda$ is a multiplicative set in R.

ii) If $\{P_\alpha\}$ are all the prime ideals of R, and $\{M_\beta\}$ are all the maximal ideals, show that $R - \bigcup P_\alpha = R - \bigcup M_\beta$.

iii) For $R = \boldsymbol{Q}[X]$ and $P_\alpha = (x_\alpha) \triangleleft R$ for $x_\alpha \in X$, describe $f(X) \in R - \bigcup (x_\alpha)$.

7. If P is a minimal prime ideal of $R(Q \triangleleft R$, Q prime, and $Q \subseteq P \Rightarrow Q = P)$, show that every $x \in P$ is a zero divisor in R—that is, $xy = 0$ for some $y \in R^*$.

8. Let I, $P \triangleleft R$ with $I \subseteq P$. Show that P is prime in $R \Leftrightarrow P/I$ is prime in R/I.

9. Let $\varnothing \neq S$ be the set of all prime ideals of R and let $\{P_\lambda\}$ be a chain in S.

i) Show that $\bigcap P_\lambda$ and $\bigcup P_\lambda$ are also prime ideals of R.

ii) Show that S contains maximal elements with respect to $\subseteq$.

iii) Show that S contains minimal elements with respect to $\subseteq$.

10. If $I \triangleleft R$ is proper and $1 \in R$, show that there is $I \subseteq P \triangleleft R$ with P prime and minimal with this property ($I \subseteq Q \subseteq P$ and Q prime $\Rightarrow Q = P$). (Exercise 9!)

11. Call $I \triangleleft R$ finitely generated if $I = (a_1) + \cdots + (a_m)$ for some $a_j \in I$ and let $S = \{I \triangleleft R \mid I \text{ is } \textit{not} \text{ finitely generated}\}$. Show that $S = \varnothing$ or the poset $(S, \subseteq)$ has maximal elements. If $1 \in R$, a maximal $P \in S$ is prime, but this is not easy.

12. If $M \triangleleft R$ is maximal and $I \triangleleft R$, show that $I \subseteq M$ or $I \bigcap M$ is a maximal ideal of I.

13. If $P \triangleleft R$ is prime and $I \triangleleft R$, show that $I \subseteq P$ or $I \bigcap P$ is a prime ideal of I.

14. If $R = x\boldsymbol{Q}[x]$ and $M = \{p(x) \in R \mid p(1) = 0\}$, show that M is a maximal ideal in R.

15. Describe $U(R)$ for $R = \boldsymbol{Z}_n[x]$ for $n =$:

i) 18
ii) 20
iii) 24
iv) 30
v) 54
vi) 100
vii) 210
viii) 256

16. If R is finite and $1 \in R$, show that $N(R) = J(R)$. (See Theorem 10.14.)

17. Find $J(R)$ for $R =$:

i) $\boldsymbol{Z}_6[[x]]$
ii) $\boldsymbol{Z}_{12}[[x]]$
iii) $\boldsymbol{Z}_{12}[x]$
iv) $\boldsymbol{Z}_n[[x]]$
v) $\boldsymbol{Z}_n[x]$

18. If $1 \in R$, $I \triangleleft R$, and $I \neq R$, show that $R \neq I + J(R)$.

19. For any $I \triangleleft R$, show that $N(I) = I \bigcap N(R)$.

20. When $1 \in R$, show that $N(R[x]) = N(R)[x]$.

21. Let $I \triangleleft R$ and $\rho : R \to R/I$ be the usual homomorphism $\rho(r) = r + I$.

i) Show that $\rho(J(R)) \subseteq J(R/I)$ but equality need not hold.

ii) Show that $\rho(N(R)) \subseteq N(R/I)$ but equality need not hold.

22. Prove that:

i) $J(R/J(R)) = (0_{R/J(R)})$

ii) $N(R/N(R)) = (0_{R/N(R)})$

23. If A and B are commutative rings with 1, show that:

i) $J(A \oplus B) = J(A) \oplus J(B)$

ii) $N(A \oplus B) = N(A) \oplus N(B)$

24. a) If $1 \in R$, every ideal of R is principal (R may not be a domain), and $I \lhd R$, show that $\bigcap (J(R)^k)I = (0)$, when k varies over $\mathbf{N}$.

b) Let $K = QF(\mathbf{Q}[[x]])$ and $A = \begin{bmatrix} \mathbf{Q}[[x]] & K \\ 0 & K \end{bmatrix} \subseteq M_2(K)$. Show that A is a subring of $M_2(K)$, every ideal of A is principal,

$\{a \in A \mid 1_A + (a) \subseteq U(A)\} = J = \begin{bmatrix} (x) & K \\ 0 & 0 \end{bmatrix}$, and $\bigcap J^k \neq (0)$.

25. From §11.2, for a general ring K, $P \lhd K$ is prime if $I, L \lhd K$ and $IL \subseteq P$, then $I \subseteq P$ or $L \subseteq P$. If $x \in K$ with no $x^m = 0$, show that $x \notin P$ for some prime $P \lhd R$.

26. Let $1 \in R$, let $I, K \lhd R$ with $K \subseteq I$, $r \in R$, $x \in I$, and set $r \cdot (x + K) = rx + K$.

i) Verify that $\cdot$ really is a function—that is, $\cdot : R \times I/K \to I/K$.

ii) If $I = Ra_1 + \cdots + Ra_m$ and $J(R) \cdot (I/K) = I/K$, modify the proof of Theorem 13.4 to show that $I/K = (0_{I/K})$.

iii) If $I = Ra_1 + \cdots + Ra_m$ and $J(R)I + K = I$, show that $K = I$.

13.2 LOCALIZATION REVISITED

The process of forming fractions with specified denominators, called *localization,* was discussed in §1.9 for $\mathbf{Z}$ and used to obtain the quotient field of an integral domain in §10.4. We study this construction for an arbitrary commutative ring. As we observed in §1.9, the addition and the multiplication of fractions require the set of allowable denominators to be closed under multiplication. Since the elements appearing as denominators will become units, we do not want zero as a denominator so we restrict the set of denominators to be multiplicative. The question then is whether the construction of $QF(D)$ for a domain D, or of the localization $\mathbf{Z}_S$ in §1.9, works for a multiplicative subset (Definition 13.1) of an arbitrary commutative ring.

The construction of fractions for domains ($a/b = c/d$ when $ad = cd$) must be modified for a general commutative ring. For example, if $R = \mathbf{Z} \oplus \mathbf{Z}$ then $M = \{(a, 0) \in R \mid a > 0\} \cup \{(1, 1)\}$ is multiplicative. Forming fractions r/m as usual with $m \in M$ makes $m/(1, 1)$ invertible, so $(3, 0)/(1, 1)$ is invertible. Now $(0, 0)/(1, 1)$ is the zero fraction and $(3, 0)/(1, 1) \cdot (0, 1)/(1, 1) = (0, 0)/(1, 1)$, so $(3, 0)/(1, 1)$ a unit forces $(0, 1)/(1, 1) = (0, 0)/(1, 1)$. Cross multiplying yields $(0, 1) = (0, 0)$ in R! Thus we need some change in the construction that is used for domains. The new construction and verifications are quite similar to those in §1.9 and §10.4 so we will go through these rather briefly. Finally, it is convenient to *assume* $1 \in R$ *and that multiplicative sets contain* 1. It is not necessary to do so, and all that follows can be modified easily either when R does not have 1 or when 1 is not in the multiplicative set.

THEOREM 13.8 Let R be a commutative ring with 1, M a multiplicative set containing 1, and $S = R \times M$. If $(r, m) \sim (r', m')$ when $m''(rm' - r'm) = 0$ for some $m'' \in M$, then $\sim$ is an equivalence relation on S.

Proof Clearly $(r, m) \sim (r, m)$ for all $(r, m) \in S$ since $t(rm - rm) = 0$ for any $t \in M$, so $\sim$ is reflexive. Also $m''(rm' - r'm) = 0$ implies $m''(r'm - rm') = 0$ so $(r, m) \sim (r', m')$ implies $(r', m') \sim (r, m)$ and $\sim$ is symmetric. Finally, if $(a, s) \sim (b, t)$ and $(b, t) \sim (c, u)$ then $m(at - bs) = 0 = m'(bu - ct)$. Hence $m'u(mat) = m'umbs = ms(m'ct)$ and this yields

$mm't(au - cs) = 0$, forcing $(a, s) \sim (c, u)$ since M is multiplicative. Thus $\sim$ is transitive, and so it is an equivalence relation. ■

DEFINITION 13.2 If R is a commutative ring with 1, $1 \in M \subseteq R$, and M is multiplicative, then the **localization of** R **at** M is $R_M = \{r/m = [(r, m)]_\sim \mid (r, m) \in R \times M\}$, the equivalence classes of $\sim$ on $R \times M$, with $a/m \cdot b/m' = ab/mm'$ and $a/m + b/m' = (am' + bm)/mm'$ for $a/m, b/m' \in R_M$.

Another notation for R_M is RM^{-1}. We need to verify that the addition and multiplication defined for R_M are actually functions: that they are independent of the pairs (a, m) representing the classes a/m. Then we can see that R_M is a ring with the given operations. Recall from Theorem 1.21 that $a/s = b/t \Leftrightarrow m(at - bs) = 0$ for some $m \in M$.

THEOREM 13.9 R_M is a commutative ring with 1.

Proof The proof is essentially that carried out for domains in §1.9 and §10.4. Our definition of $\sim$ on $R \times M$ makes it a bit trickier to see that the operations are independent of the representatives. Suppose that $a/t = a'/t'$ and $b/s = b'/s'$. Then by Theorem 1.21, for some $m, m' \in M$, $m(at' - a't) = 0 = m'(bs' - b's)$, or $mat' = ma't$ and $m'bs' = m'b's$. Multiply these equalities to get $mm'abs't' = mm'a'b'st$, so $mm'(abs't' - a'b'st) = 0$, showing that $ab/st = a'b'/s't'$. Hence multiplication is independent of the representatives for the classes. Now $mat' = ma't \Rightarrow m'ss'mat' = m'ss'ma't$ and $m'bs' = m'b's \Rightarrow mtt'm'bs' = mtt'm'b's$ so adding shows that $mm'(ss'at' + tt'bs') = mm'(ss'a't + tt'b's)$, or $mm'(as + bt)s't' = mm'(a's' + b't')st$. Since $mm' \in M$, this shows that $a/t + b/s = a'/t' + b'/s'$ and addition is independent of the class representatives. The verification that $(R_M, +, \cdot)$ is a commutative ring is left as an exercise (see §1.9). It is important to observe that $0/m \in R_M$ is the zero element for any $m \in M$, that $-(a/m) = (-a)/m$, and that for any $m, m' \in M$, $a/m = am'/mm'$. Since $1 \in M$, the identity of R_M is 1/1. Without the assumption that $1 \in M$, or even $1 \in R$, the identity is $m/m = t/t$ for any $m, t \in M$. It follows that $1/m$ (or m/m^2) is the multiplicative inverse for $m/1$ (or m^2/m). ■

In our example $R = \mathbf{Z} \oplus \mathbf{Z}$ above, $(0, 1)/(1, 1) = (0, 0)/(1, 1)$ is now not a contradiction since the definition of $a/m = b/t$ is not cross multiplication. We can see that $(0, 1)/(1, 1) = (0, 0)/(1, 1)$ makes sense now because for $(1, 0) \in M$, $(1, 0)((0, 1)(1, 1) - (1, 1)(0, 0)) = (0, 0)$ in R. This same example shows that when we localize in rings R with zero divisors we cannot expect an isomorphic copy of R in R_M. In $(\mathbf{Z} \oplus \mathbf{Z})_M$, $(a, b)/(m, 0) = (a, 0)/(m, 0)$ (why?) and $(\mathbf{Z} \oplus \mathbf{Z})_M \cong \boldsymbol{Q}$ follows by using $\varphi((a, b)/(m, 0)) = a/m$ and $\varphi((a, b)/(1, 1)) = a/1$. Let us examine the relation between R and R_M.

THEOREM 13.10 If $\theta_M : R \to R_M$ is given by $\theta_M(r) = r/1$, then:

i) θ_M is a ring homomorphism
ii) $ker\,\theta_M = \{x \in R \mid xm = 0$ for some $m \in M\}$
iii) $\theta_M(M) \subseteq U(R_M)$ and $r/m = \theta_M(r)\theta_M(m)^{-1}$ for all $r/m \in R_M$
iv) if no $m \in M$ is a zero divisor in R then $\theta_M(R) \cong R$

Proof The proof of i) is an easy exercise. Now $\theta_M(x) = 0/1 \Leftrightarrow x/1 = 0/1 \Leftrightarrow m(x \cdot 1 - 0 \cdot 1) = 0$ for some $m \in M \Leftrightarrow mx = 0$, proving ii). When $m \in M$, $\theta_M(m)^{-1} = 1/m$

since $m/1 \cdot 1/m = m/m = 1/1$. Hence $r/m = r/1 \cdot 1/m = \theta_M(r)\theta_M(m)^{-1}$ and iii) holds. Finally, if no $m \in M$ is a zero divisor then by ii), $ker\,\theta_M = (0)$ so θ_M is an isomorphism from R to $\theta_M(R)$.

■

By the theorem, a domain R can be viewed as a subring of R_M via θ_M. In this case our comments in §1.9 and §10.4 show that each localization is (isomorphic to) a subring of the quotient field $QF(R)$.

Example 13.3 The localization of $\mathbf{Z}$ at $M = \{2^j \mid j \geq 0\}$.

Here $\mathbf{Z}_M = \{a/2^j \mid j \geq 0\}$ can be identified with those $q \in \mathbf{Q}$ having denominator a power of 2 since equality, sum, and product are the same for $\{a/2^j\}$ in either $\mathbf{Z}_M$ or $\mathbf{Q}$. Note that $(p/1) \lhd \mathbf{Z}_M$ is prime for $p > 2$ and prime.

Example 13.4 The localization of $\mathbf{Z}$ at $T = \{\text{odd positive integers}\}$.

As in Example 13.3, $\mathbf{Z}_T \cong \{q = a/b \in \mathbf{Q} \mid b \text{ is odd}\}$. Thus each $\alpha \in \mathbf{Z}_T$ can be expressed as $2^i u$ for $i \geq 0$ and $u \in U(\mathbf{Z}_T)$, so $\mathbf{Z}_T$ has a unique maximal ideal $(2/1)$ and every other nonzero, proper ideal is $(2^j/1)$ (verify!).

Example 13.5 The localization of $\mathbf{Z}[x, y]$ at $S = \mathbf{Z}[x, y] - ((x) \cup (y))$.

S is the set of polynomials divisible by neither x nor y, and S is multiplicative since (x) and (y) are primes by Example 11.17 or by §12.3. Here Theorem 13.10 shows that θ_S is an isomorphism so we can think of $\mathbf{Z}[x, y]_S$ as the fractions $p(x, y)/q(x, y) \in QF(\mathbf{Z}[x, y])$ with neither x nor y dividing $q(x, y)$. Unique factorization in $\mathbf{Z}[x, y]$ from Theorem 12.15, $\mathbf{Z}^* \subseteq S$, and each $s \in S$ a unit in $\mathbf{Z}[x, y]_S$ by Theorem 13.10, show that the only nonassociate nonunits in $\mathbf{Z}[x, y]_S$ are $\{x^i y^j \mid i \geq 0, j \geq 0, i + j \geq 1\}$. It follows that $\mathbf{Z}[x, y]_S = \{x^i y^j u \mid i \geq 0, j \geq 0, i + j \geq 0 \text{ and } u \in U(\mathbf{Z}[x, y]_S)\}$, so $\mathbf{Z}[x, y]_S$ is a PID with ideals $(x^i y^j/1)$ and maximal ideals $(x/1)$ and $(y/1)$ (verify!). Note that $\mathbf{Z}[x, y]$ is not a PID!

Example 13.6 For $\mathbf{C}$ the complex numbers and $\varnothing \neq S \subseteq \mathbf{C}^m$, the localization of $R = \mathbf{C}[x_1, \ldots, x_m]$ at $M(S) = \{p(x_1, \ldots, x_m) \mid p(s) \neq 0 \text{ for all } s \in S\}$.

It is clear that $1 \in M(S)$, $0 \notin M(S)$, and $M(S)$ is multiplicative. Now $R_{M(S)} \cong \{f(x_1, \ldots, x_m)/g(x_1, \ldots, x_m) \in QF(R) \mid g(s) \neq 0, \text{ all } s \in S\}$, the rational functions defined at $S \subseteq \mathbf{C}^m$. When $S = \mathbf{C}^m$ then $M(S) = \mathbf{C}^*$ (verify that any nonconstant $f \in R$ has some zero) and $R_{M(S)} \cong R$. When $S = \{(0, \ldots, 0)\}$ then $M(S) = \mathbf{C}^* + (x_1) + \cdots + (x_m)$ since $f(0, \ldots, 0) \neq 0 \Leftrightarrow f$ has a nonzero constant term, and for $S = \mathbf{C}^{m-1} \times \{0\}$, $M(S) = \mathbf{C}^* + (x_m)$ since, as we claimed just above, $f(x_1, \ldots, x_{m-1}, 0)$ has a zero in $\mathbf{C}^{m-1}$ unless it is a nonzero constant. In these last two cases is not clear what more we can say about $R_{M(S)}$.

Example 13.7 The localization of $R = \mathbf{Z}_6[x]$ at $M = \{[1], [2], [4]\} \subseteq \mathbf{Z}_6$.

Since in R, $[2]([3]) = ([0])$, by Theorem 13.10 if $p(x) \in ([3])$ then in R_M, $p(x)/m = [0]/[1]$. Now $[4]/[1] = [1]/[1] + [3]/[1] = [1]/[1]$, and similarly $[5]/[1] = [2]/[1]$. Thus $([2]/[1])^2 = [1]/[1]$ so $[2]/[1] = ([2]/[1])^{-1} = [1]/[2]$ and also $[4]/[1] = [1]/[1] = [1]/[4]$. Therefore in R_M, $p(x)/[m] = p(x)/[1] \cdot [m]/[1]$ so $p(x)/[m] = ([b_0] + [b_1]x + \cdots + [b_n]x^n)/[1] = [b_0]/[1] + \cdots + [b_n]x^n/[1]$ and we can assume all $0 \le b_j \le 2$. Since $[1]/[1] + [2]/[1] = [0]/[1]$, it follows that this representation of $p(x)/[m]$ is unique (verify!). Using $\varphi([b_0]/[1] + \cdots + [b_n]x^n/[1]) = [b_0]_3 + \cdots + [b_n]_3 x^n$ it follows that $R_M \cong \mathbf{Z}_3[x]$. If we used $M' = \{[1], [3]\}$ instead of M, the corresponding analysis would give $R_{M'} \cong \mathbf{Z}_2[x]$.

What is the relation between the set of ideals in R and the set of ideals in R_M? Recall from Theorem 1.21 that in R_M, $a/s = b/t \Leftrightarrow m(at - bs) = 0$ for some $m \in M$, and recall from Theorem 13.10 that $\theta_M(r) = r/1$.

THEOREM 13.11 Let R be a commutative ring, $1 \in M \subseteq R$ multiplicative, $I \lhd R$, $J \lhd R_M$, $H = \{a/m \in R_M \mid a \in I\}$, and $J^c = \{r \in R \mid \text{some } r/m \in J\}$. Then $H = \theta_M(I)R_M = \sum_I \theta(a)R_M \lhd R_M$, $H = R_M \Leftrightarrow I \cap M \ne \varnothing$, $J^c \lhd R$, and $\theta_M(J^c)R_M = J$.

Proof It is easy to show that $H \lhd R_M$. If $r/m \in R_M$ and $a \in I$, then $a/1 \cdot r/m = ar/m$ and $ar \in I$ so $\theta_M(a)R_M \subseteq H$. Thus $\sum_I \theta_M(a)R_M \subseteq H$ since H is closed under addition (verify!). If $x/m \in H$, then $x/m = x/1 \cdot 1/m \in \theta_M(x)R_M : \sum_I \theta_M(a)R_M = H$. When $m \in M \cap I$, $1/1 = m/1 \cdot 1/m \in H$ forces $H = R_M$. If $H = R_M$ then $1/1 = a/m$ for $a \in I$, and $m'(m - a) = 0$ for some $m' \in M$ so $m'a = mm' \in I \cap M$. Now let $a, b \in J^c$ with $a/s, b/t \in J$. Using $J \lhd R_M$, J contains $a/s \cdot r/m = ar/sm$ for any $r \in R$ and also contains $a/s \cdot s/1 - b/t \cdot t/1 = as/s + (-bt)/t = (a - b)st/st = (a - b)/1$. The definition of J^c yields $a - b, ar \in J^c$, proving $J^c \lhd R$. If $a/m \in J$ then $a \in J^c$ so $a/m = a/1 \cdot 1/m \in \theta_M(J^c)R_M$ and $J \subseteq \theta_M(J^c)R_M$. Finally, for $b \in J^c$ some $b/m \in J \lhd R_M$, so for any $m' \in M$ and $r \in R$, $b/1 \cdot r/m' = br/m' = b/m \cdot rm/m' \in J$. This shows that $\theta_M(J^c)R_M = J$ and completes the proof. ∎

It will be convenient to give a name to the ideals in Theorem 13.11.

DEFINITION 13.3 For R a commutative ring with 1 and $1 \in M \subseteq R$ multiplicative, the **extension** of $I \lhd R$ to R_M is $I^e = \theta_M(I)R_M \lhd R_M$ and the **contraction** of $J \lhd R_M$ to R is $J^c = \{a \in R \mid \text{some } a/m \in J\} \lhd R$.

Perhaps the most important example of R_M is for $M = R - P$ when P is a prime ideal of R. This localization is usually written as R_P. No confusion results from this notation since $0 \in P$ means that P cannot be multiplicative. In this case Theorem 13.11 shows that for $I \lhd R$, unless $I \subseteq P$, we have $I^e = R_P$. Therefore in R_P we lose all information about the ideals of R not contained in P. This is the reverse of the situation for quotients. In R/P we lose all information about ideals contained in P, but the relations among those ideals containing P are preserved. By analogy, we might hope that we can study the ideals contained in P by considering R_P. Unfortunately, this is not the case. The localization in

Example 13.4 of $\mathbf{Z}$ at the odd integers is just $\mathbf{Z}_{(2)}$ for the prime ideal $(2) \lhd \mathbf{Z}$. In $\mathbf{Z}$, $(30) \subseteq (6) \subseteq (2)$ and these inclusions are proper. From Theorem 13.11, $(30)^e \subseteq (6)^e \subseteq (2)^e$ but $2k/1 = 30k/1 \cdot 1/15 \in (30)^e$ and this implies that $(30)^e = (2)^e$.

The importance of the localization R_P is that this ring has P^e as its unique maximal ideal. To see this, take $a/m \in R_P - P^e$ and note that $a \notin P$ from Theorem 13.11. Consequently, $(a/m)^{-1} = m/a \in R_P$ and so $U(R_P) = R_P - P^e$. Now Theorem 11.4 shows that P^e is the unique maximal ideal of R_P. Thus the term localization comes from the following definition.

DEFINITION 13.4 A commutative ring with 1 is **local** if it contains a unique maximal ideal.

Some use the term R *local* to mean that R has a unique maximal ideal and satisfies ACC on ideals (§11.6). We will not pursue this idea here so the definition is not crucial. Local rings are important in more advanced studies of commutative rings since they have some nice properties. Note that if R is local then $J(R)$ is its unique maximal ideal. We have seen that R_P is local, that for F a field $F[[x]]$ is local with maximal ideal (x) (Theorem 11.4 and Theorem 10.10), and that $F[x]/(x^n)$ is local using Example 11.9. When R has only finitely many maximal ideals, as in Example 13.5, it is **semi-local.**

From Theorem 13.11, if $K \lhd R_M$ then $K^c \lhd R$ and $K = (K^c)^e$ so K is the extension of an ideal of R. However, not every $I \lhd R$ is the contraction of an ideal of R_M. If $(6) = K^c$ for $K \lhd \mathbf{Z}_{(2)}$ then all $6k/m \in K$ and so $2k/m = 6k/m \cdot 1/3 \in K$, forcing $(2) \subseteq K^c = (6)$, a contradiction. Properties of extended ideals come next. Similar results for contracted ideals are left as exercises.

THEOREM 13.12 If R is a commutative ring, $1 \in M \subseteq R$ is multiplicative, $P, I, J \lhd R$, and P is a prime ideal, then:

i) $I^{ec} = \{r \in R \mid mr \in I, \text{ some } m \in M\}$
ii) $P = P^{ec}$ if $P \cap M = \varnothing$
iii) $I \subseteq J \Rightarrow I^e \subseteq J^e$
iv) $(I \cap J)^e = I^e \cap J^e$
v) $(IJ)^e = I^e J^e$
vi) $(I + J)^e = I^e + J^e$

Proof If $r \in I^{ec}$ some $r/m \in I^e$ so $r/m = x/s$ for $x \in I$ (Theorem 13.11). Thus $m'sr = m'mx \in I$ for some $m', s \in M$ and $r \in \{y \in R \mid ty \in I, \text{ some } t \in M\}$. If $mr = x \in I$ for some $m \in M$, then for any $s \in M$, $r/s = mr/ms = x/ms \in I^e$ so $r \in I^{ec}$, proving i). If $P \cap M = \varnothing$, $m \in M$, and $r \in R$, then $rm \in P$ forces $r \in P$, and i) shows $P^{ec} = P$. The definition of extension proves iii). Now $I \cap J \subseteq I, J$, so iii) implies that $(I \cap J)^e \subseteq I^e \cap J^e$. For $\alpha \in I^e \cap J^e$ let $\alpha = a/s = b/t$ with $a \in I$ and $b \in J$. For some $m \in M$, $mta = mbs \in I \cap J$ so $\alpha = mta/mts \in (I \cap J)^e$, proving iv). When $a_i \in I, b_i \in J, r \in R$, and $m \in M$, $(\sum a_i b_i)/1 \cdot r/m = \sum a_i/m \cdot b_i r/1 \in I^e J^e$, showing $(IJ)^e \subseteq I^e J^e$. If $a/s \in I^e$ and $b/t \in J^e$ with $a \in I$ and $b \in J$ then $a/s \cdot b/t = ab/st \in (IJ)^e$. Thus v) holds. Using that $I, J \subseteq I + J$ and iii), $I^e + J^e \subseteq (I + J)^e$ follows, and for $a \in I, b \in J, r \in R$, and $m \in M$, $(a + b)/1 \cdot r/m = ar/m + br/m \in I^e + J^e$, proving vi) and so the theorem. ■

We can say more about prime ideals in R and in R_M.

THEOREM 13.13 Let R be a commutative ring and $1 \in M \subseteq R$ multiplicative. Then $Q \to Q^c$ and $P \to P^e$ are inclusion-preserving bijections between $\{Q \lhd R_M \mid Q \text{ is prime}\}$ and $\{P \lhd R \mid P \text{ is prime and } P \cap M = \varnothing\}$.

Proof By Theorem 13.11, for any proper $Q \lhd R_M$, $Q^{ce} = Q$ so $Q^c \cap M = \varnothing$. If $a, b \in R$ with $ab \in Q^c$ then $a/1 \cdot b/1 = ab/1 \in Q^{ce} = Q$, and Q prime implies $a/1 \in Q$ or $b/1 \in Q$, forcing $a \in Q^c$ or $b \in Q^c$. Thus Q^c is a prime ideal of R with $Q^c \cap M = \varnothing$. If $P \lhd R$ and $P \cap M = \varnothing$, then $P^e \neq R_M$ by Theorem 13.11. Assume P is prime and $x/s \cdot y/t \in P^e$. Therefore $xy/st = a/m$ with $a \in P$ and there is $m' \in M$ so that $m'mxy = m'sta \in P$. Since $m'm \in M$, P is prime, and $P \cap M = \varnothing$, we conclude that $x \in P$ or $y \in P$. Hence $x/s \in P^e$ or $y/t \in P^e$, proving that P^e is a prime ideal in R_M. Using $Q = Q^{ce}$ for any (prime) ideal in R_M and Theorem 13.12 to see that $P = P^{ec}$ for any prime ideal of R satisfying $P \cap M = \varnothing$, it follows that $Q \to Q^c$ and $P \to P^e$ are inverses and so bijections. If $P \subseteq P_1$ are prime ideals of R with $P_1 \cap M = \varnothing$ but $P^e = P_1^e$, then $P = P^{ec} = P_1^{ec} = P_1$. Similarly, if $Q \subseteq Q_1$ are prime ideals in R_M with $Q^c = Q_1^c$ then $Q = Q_1$, completing the proof. ∎

The correspondence in Theorem 13.13 when $M = R - P$ for P a prime ideal of R shows that the prime ideals in R_P correspond to the prime ideals of R contained in P and that all containments are preserved.

Example 13.8 R_P for $R = \mathbf{Q}[x, y, z]$ and $P = (x) + (y) + (z)$.

Since R is a UFD by Theorem 12.15, if $f \in R$ is irreducible then Theorem 12.11 implies that $(f) \lhd R$ is prime. Now $P = (x) + (y) + (z)$ is a maximal ideal of R since $R/P \cong \mathbf{Q}$, so P is a prime ideal. Also $(x) + (y)$ is prime since $R/((x) + (y)) \cong \mathbf{Q}[z]$ (why?). In R_P, $(1 + x)^e = R_P = (1 + x + y - z)^e$ since $1 + x, 1 + x + y - z \notin P$. Proper containments of prime ideals in P, such as $(0) \subseteq (x) \subseteq (x) + (y) \subseteq P$, $(x - y) \subseteq (x - y) + (y) = (x) + (y)$, and $(x - y) \subseteq (x - y) + (z)$, are maintained under extension. Noncomparable prime ideals in P have noncomparable extensions in R_P; some examples are $(x + y + z)$, $(x^2 - z)$, and $(x - y) + (y - z)$. If $C = \{p(x, y, z) \in R \mid p(x, 0, 0) \neq 0\}$, then C is multiplicative (verify!) and the extensions of the primes (y), (z), and $(y) + (z)$ to R_C maintain their relative containments. Since $x/1$ is a unit in R_C, $(xy)^e = (y)^e$, and also $(x + y)^e = R_C = (x^2 + y + z^2)^e$ (why?).

We turn to a result relating localization and quotients.

THEOREM 13.14 Let R be a commutative ring, $1 \in M \subseteq R$ multiplicative, $I \lhd R$ with $I \cap M = \varnothing$ and $\rho : R \to R/I$ the surjection $\rho(r) = r + I$. Then $\rho(M)$ is a multiplicative set in R/I, $1_{R/I} \in \rho(M)$, and $(R/I)_{\rho(M)} \cong R_M/I^e$.

Proof By Example 11.14, ρ is a ring homomorphism so $\rho(M)$ is closed under multiplication and $1_{R/I} = \rho(1) \in \rho(M)$. If for $m \in M$, $\rho(m) = 0_{R/I} = I$, then $m \in ker\,\rho$ so $m \in I \cap M$, a contradiction. Hence $\rho(M)$ is multiplicative. For $r/m \in R_M$, set $\beta(r/m) = \rho(r)/\rho(m)$. Is this a function? If $r/m = r'/m'$ then for some $m'' \in M$, $m''m'r = m''mr'$

so $\rho(m'')\rho(m')\rho(r) = \rho(m''m'r) = \rho(m''mr') = \rho(m'')\rho(m)\rho(r')$. Therefore in $(R/I)_{\rho(M)}$, $\beta(r/m) = \rho(r)/\rho(m) = \rho(r')/\rho(m') = \beta(r'/m')$ and $\beta : R_M \to (R/I)_{\rho(M)}$. It is routine to show that β is a surjective ring homomorphism so $(R/I)_{\rho(M)} \cong R_M/ker\,\beta$ by Theorem 11.12. Now $r/m \in ker\,\beta \Leftrightarrow \rho(r)/\rho(m) = \rho(0)/\rho(1) \Leftrightarrow$ for some $m' \in M$, $\rho(m')\rho(1)\rho(r) = \rho(0) = I \Leftrightarrow \rho(m'r) = m'r + I = I \Leftrightarrow m'r \in I$. Thus $r/m = m'r/m'm \in I^{e}$, so $ker\,\beta \subseteq I^{e}$. Now $I^{e} \subseteq ker\,\beta$ follows from Theorem 13.11 and $I = ker\,\rho$, proving the theorem. ■

It is easy to see that R_M is a domain if R is. When $P \lhd R$ is prime then R/P is a domain so for $\rho : R \to R/P$, Theorem 13.14 shows that $R_P/P^{e} \cong (R/P)_{\rho(R-P)}$ is a domain. This proves again that P^{e} is prime in R_P and is in fact maximal since $(R/P)_{\rho(R-P)} \cong QF(R/P)$ (why?). We illustrate next how Theorem 13.14 can be used, although a direct argument is simple.

THEOREM 13.15 For any commutative ring R with 1 and multiplicative subset M containing 1, $N(R_M) = \{\alpha \in R_M \mid \alpha^k = 0, \text{ some } k \geq 1\} = N(R)^{e}$.

Proof If $x \in N(R)$ then $x^k = 0$ for some $k \geq 1$, so for $r/m \in R_M$, $(x/1 \cdot r/m)^k = x^k r^k/m^k = 0/1$ in R_M. Thus $N(R)^{e} \subseteq N(R_M)$. Observe that $N(R) \bigcap M = \varnothing$. Setting $I = N(R)$ and $\rho : R \to R/N(R)$ in Theorem 13.14 we have $(R/N(R))_{\rho(M)} \cong R_M/N(R)^{e}$. Now $N(R/N(R)) = (0_{R/N(R)})$ since $(y + N(R))^j = N(R)$ forces $y^j \in N(R)$, then $y \in N(R)$ so $y + N(R) = 0_{R/N(R)}$. We claim that when D is commutative and $N(D) = (0)$ then any $N(D_T) = (0)$ also. Should $(d/t)^k = 0/1$ then $d^k/t^k = 0/1$ so $sd^k = 0$ in D for some $s \in T$. Since $(sd)^k = 0$, $sd \in N(D) = (0)$ follows and shows that $d/t = 0/1$ in D_T (how?). Therefore $N(R_M/N(R)^{e}) \cong N((R/N(R))_{\rho(M)}) = (0)$ but must contain $N(R_M)/N(R)^{e}$. This forces $N(R_M) \subseteq N(R)^{e}$, so these are equal, proving the theorem. ■

The statement of Theorem 13.15 with $J(R)$ replacing $N(R)$ is false. If $R = \boldsymbol{Q}[[x]]$ and $M = \{x^i \mid i \geq 0\}$ then $(x) \bigcap M \neq \varnothing$, $J(R) = (x)$, $J(R)^{e} = R_M$, but $J(R_M) = (0)$ since $R_M \cong \boldsymbol{Q}\langle\langle x \rangle\rangle$ is a field by Theorem 10.10. For $1 \in R$ commutative, if $H \lhd R$ is prime then $J(R_H) = H^{e}$ since it is the unique maximal ideal of R_H. Thus for $R = \boldsymbol{Q}[x]$ and $H = (x)$, $J(R) = (0)$ so in R_H, $J(R)^{e} = (0/1)$ but $J(R_H) = (x)^{e}$.

EXERCISES

For the following, R is a commutative ring, $M \subseteq R$ is multiplicative, and $1 \in M$.

1. **a)** For $I \lhd R$, show that $I^{ece} = I^{e}$.
   b) For $H \lhd R_M$, show that $H^{cec} = H^{c}$.
   c) Prove that there is a bijection between $\{J \lhd R \mid J = L^{c} \text{ for } L \lhd R_M\}$ and $\{H \lhd R_M\}$.
2. If $H, L \lhd R_M$ show that:
   i) $H \subseteq L \Rightarrow H^{c} \subseteq L^{c}$
   ii) $H^{c} \bigcap L^{c} = (H \bigcap L)^{c}$
   iii) $H^{c}L^{c} \subseteq (HL)^{c}$
   iv) $H^{c} + L^{c} \subseteq (H + L)^{c}$
   v) Use $\boldsymbol{Q}[x, y]_M$ to show that equality can fail in iv).
3. Show carefully that R a domain implies that R_M is a domain.
4. Show that $U(R_M) = \{r/m \in R_M \mid sr \in M \text{ for some } s \in R\}$.

5. If R is a PID, show that R_M is a PID (this is Exercise 14 in §12.2).
6. If R is a UFD, show that R_M is a UFD (this is Exercise 6 in §12.3).
7. For F a field, $X=\{x_\lambda\}_\Lambda$ indeterminates over F, and $R=F[X]$, show that $M=R-\bigcup(x_\lambda)$ is multiplicative and that R_M is a PID.
8. If $M\subseteq U(R)$, show that $\theta_M:R\to R_M$ via $\theta_M(r)=r/1$ is an isomorphism.
9. If $I\lhd R$ and proper, and if $T=\{1+x\mid x\in I\}$, show that T is multiplicative and that $I^{\mathrm{e}}\subseteq J(R_T)$.
10. For any $1\in M\subseteq R$ multiplicative, let $M^{\#}=\{r\in R\mid rs\in M,\text{ some }s\in R\}$. Show that $M^{\#}$ is multiplicative and that $R_M\cong R_{M^{\#}}$.
11. $M=M^{\#}$ (Exercise 10) and R_M local $\Rightarrow M=R-P$ for P a prime ideal of R.
12. **a)** If $V\lhd R$ is maximal, then either $V^{\mathrm{e}}=R_M$ or $V^{\mathrm{e}}\lhd R_M$ is maximal.
b) If $Q\lhd R_M$ is maximal, then $Q^{\mathrm{c}}\lhd R$ is prime but may not be maximal.
13. Given $k\in\mathbf{N}$, find $M(k)\subseteq\mathbf{N}\subseteq\mathbf{Z}$ multiplicative and containing 1 so that $\mathbf{Z}_{M(k)}$ has exactly k maximal ideals.
14. For any prime $P\lhd R$, show that R_P/P^{e} is a field.
15. If for all maximal ideals W of R, $N(R_W)=(0/1)$, show that $N(R)=(0)$. (For $x\in N(R)$ consider $ann(x)=\{r\in R\mid rx=0\}\lhd R$ and then use Theorem 13.2.)
16. If $P\lhd R$ is a minimal prime ($Q\lhd R$ prime and $Q\subseteq P\Rightarrow Q=P$) and $N(R)=(0)$ show that R_P is a field.
17. If R is a domain and $1\in M\subseteq M^\wedge\subseteq R$ are both multiplicative, show that there is an injective homomorphism $\eta:R_M\to R_{M^\wedge}$.
18. Let $R=\mathbf{Z}\oplus\mathbf{Z}\oplus\mathbf{Z}$, $M=\{(1,1,1)\}\bigcup\{(0,a,b)\mid a,b\in\mathbf{N}\}$, and $M^\wedge=M\bigcup\{(0,0,c)\mid c\in\mathbf{N}\}$. Show that $M\subseteq M^\wedge$ are both multiplicativc, that $R_M\cong\mathbf{Q}\oplus\mathbf{Q}$, and that $R_{M^\wedge}\cong\mathbf{Q}$ (compare with Exercise 17).
19. Let $\varphi:R\to S$ be a ring homomorphism, $1\in T\subseteq R$ a multiplicative set, $1_S=\varphi(1_R)$, and $T\bigcap ker\,\varphi=\varnothing$.
i) Show that $\varphi(T)\subseteq S$ is multiplicative.
ii) If $\varphi_T:R_T\to S_{\varphi(T)}$ is $\varphi_T(r/t)=\varphi(r)/\varphi(t)$, show that φ_T is a function and a ring homomorphism.
iii) If φ is injective, show that φ_T is injective.
20. Let $1\in R\subseteq S$ be commutative rings and $1\in T\subseteq R$ multiplicative.
Call $A\leq(S,+)$ an *R-submodule*, written ${}_RA$ if $ra\in A$ for all $r\in R$ and $a\in A$.
i) Given ${}_RA$, show that $A_T=\{a/t\in S_T\mid a\in A,t\in T\}$ is an R_T-submodule of S_T.
For ${}_RA$, ${}_RB\subseteq S$, $\varphi:{}_RA\to{}_RB$ is an *R-homomorphism* if $\varphi:(A,+)\to(B,+)$ is a group homomorphism and $\varphi(ra)=r\varphi(a)$ for all $r\in R$ and $a\in A$.
ii) For an R-homomorphism $\varphi:{}_RA\to{}_RB$, show that $\varphi_T(a/t)=\varphi(a)/t$ gives a function $\varphi_T:A_T\to B_T$ and that φ_T is an R_T-homomorphism.
iii) Show that $\varphi:{}_RA\to{}_RB$ is an injective *R-homomorphism* $\Leftrightarrow\varphi_T:A_T\to B_T$ is an injective R_T-homomorphism for all $T=R-V$ with $V\lhd R$ and maximal.
iv) Show that $\varphi:{}_RA\to{}_RB$ is a surjective *R-homomorphism* $\Leftrightarrow\varphi_T:A_T\to B_T$ is a surjective R_T-homomorphism for all $T=R-V$ with $V\lhd R$ and maximal. (For $b\in B$, consider $I_b=\{r\in R\mid rb\in\varphi(A)\}$ and use Theorem 13.2.)

13.3 NOETHERIAN RINGS

Before we begin to discuss Noetherian rings, defined in §11.6, we introduce a useful concept that generalizes the notion of an ideal.

DEFINITION 13.5 The Abelian group $(M, +)$ is a **left R-module** for the ring R, written $_RM$, if there is $\cdot : R \times M \to M$ so that for all $r, s \in R$, $g, h, \in M$:

i) $r \cdot (g + h) = r \cdot g + r \cdot h$
ii) $(rs) \cdot g = r \cdot (s \cdot g)$
iii) $(r + s) \cdot g = r \cdot g + s \cdot g$
iv) if $1 \in R$ then $1 \cdot g = g$

Henceforth we will call $_RM$ simply an R-module. In $_RM$ we usually write $r \cdot m$ as rm. The main example for us is $_RR$ with $\cdot$ the multiplication in R. The associative and distributive properties in R show that R *is* a (left) R-module (verify!). In $_RR$, R-modules are just (left) ideals of R so any result for modules holds for (left) ideals. Modules arising naturally are $_R(R/I)$ when $I \lhd R$ where $r \cdot (a + I) = ra + I$. More generally, if $I \subseteq J$ are ideals of R then $_R(J/I)$ with the same action: $r \cdot (j + I) = rj + I$. Also if $R \subseteq S$ are rings then for any subring T of S containing R, $_RT$ using the ring operations in S. Thus if $R[x^2] = \{p(x) \in R[x] \mid p(x) = q(x^2) \text{ for } q(x) \in R[x]\}$ is the ring of polynomials in even powers of x, then $_RR[x^2]$, and $_WR[x^2]$ for $W = R[x^4]$.

Other common examples of modules are: $_{\mathbf{Z}}(G, +)$ for any Abelian group using $m \cdot g = g^m$; $_R(R[x], +)$ using the multiplication in $R[x]$; $_R(M_n(R), +)$ using scalar multiplication; and for F a field and $A \in M_n(F)$, $_{F[x]}(F^n, +)$ using $p(x) \cdot (a_1, \ldots, a_n) = (p(A)(a_1, \ldots, a_n)^{\mathrm{T}})^{\mathrm{T}}$ (the notation $(\ )^{\mathrm{T}}$ means transpose).

For a group we defined subgroups, sums of subgroups, and quotient groups. Corresponding objects for modules are also useful.

DEFINITION 13.6 If $_RM$ then $A \leq (M, +)$ is a **submodule,** written $A \leq_R M$, if $Ra = \{ra \in M \mid r \in R\} \subseteq A$ for all $a \in A$. The **quotient** of M by $A \leq_R M$ is $(M/A, +)$. The **sum** of $\{A_j \leq_R M\}$ is $\sum A_j = \{b_1 + \cdots + b_n \in M \mid n \geq 1 \text{ and each } b_i \in A_j \text{ for some } j\}$. Finally, for $m \in M$, $\mathbf{Z}m = \langle m \rangle \leq (M, +)$.

THEOREM 13.16 If $_RM$, $m \in M$, and A, $A_j \leq_R M$, then $\sum A_j \leq_R M$, $Rm \leq_R M$, $Rm + \mathbf{Z}m \leq_R M$, and M/A is an R-module using $r \cdot (m + A) = rm + A$. When $1 \in R$, $Rm + \mathbf{Z}m = Rm$.

Proof The arguments are left as exercises. Note that the action of R on the quotient group M/A is defined since if $m + A = m' + A$ then $m - m' \in A$, and using $A \leq_R M$ shows $rm - rm' = r(m - m') \in A$, so $rm + A = rm' + A$. ∎

DEFINITION 13.7 An R-module M is **finitely generated** if $M = Rm_1 + \mathbf{Z}m_1 + \cdots + Rm_n + \mathbf{Z}m_n = \{r_1m_1 + z_1m_1 + \cdots + r_nm_n + z_nm_n \in M \mid r_i \in R, z_i \in \mathbf{Z}\}$ for some $m_1, \ldots, m_n \in M$, or when $1 \in R$ if $M = Rm_1 + \cdots + Rm_n$.

For any left ideal L in a ring R, $L \leq_R R$, and when $_RL$ is finitely generated we say that *L is a finitely generated left ideal.* If R is commutative and $L \lhd R$ then *L finitely generated* means $L = \sum (a_i)$ for $1 \leq i \leq n$, and if $1 \in R$ then $L = Ra_1 + \cdots + Ra_n$, which is often written $L = (a_1, \ldots, a_n)$. In a PID, every ideal is finitely generated by a single element.

The notion of a left Noetherian ring in §11.6 is equivalent to all $L \leq_R R$ finitely generated. To see this we first generalize Definition 11.8 to modules.

DEFINITION 13.8 An R-module ${}_RM$ is **Noetherian** if there is no infinite, strictly ascending chain of submodules $M_1 \subseteq M_2 \subseteq \cdots \subseteq M_j \subseteq \cdots$ in M. R is **left Noetherian** if ${}_RR$ is Noetherian.

When R is commutative we say that R is Noetherian rather than left Noetherian. At this point we do not have many examples of Noetherian modules or rings. Of course any finite module for a ring R is Noetherian since it can have only finitely many submodules. Examples are ${}_{\mathbf{Z}}G$ for any finite Abelian group and $M_n(R)$ when R is a finite ring, either as an R module via scalar multiplication or as a module over itself. Also $\mathbf{Z}$ is a Noetherian ring. Any ideal in $\mathbf{Z}$ is (n) for $n \geq 0$, and $(0) \neq (n) \subseteq (m) \Leftrightarrow m \mid n$ so unique factorization shows that any strictly ascending chain must terminate in finitely many steps (why?). Indeed, any PID is a Noetherian ring since for any chain $I_1 \subseteq I_2 \subseteq \cdots$ of ideals, the union of these is an ideal (Theorem 11.5) whose single generator must lie in some I_j. This same argument applies if every ideal of R, or submodule of ${}_RM$, is finitely generated.

THEOREM 13.17 If R is a ring and M is an R-module, then the following conditions are equivalent:

i) M is Noetherian.
ii) Any nonempty set of submodules of M contains a maximal element with respect to inclusion.
iii) Every submodule of M is finitely generated.

Proof Assume that ${}_RM$ is Noetherian and that S is a nonempty collection of submodules of M. We use Zorn's Lemma to show that the poset $(S, \subseteq)$ has a maximal element. Let $T = \{(A_1, \ldots, A_m) \mid \{A_j\} \subseteq S, m \geq 1$ and $A_1 \subseteq \cdots \subseteq A_m$ is a strictly increasing chain$\}$. Define $\leq'$ on T by $(A_1, \ldots, A_m) \leq' (B_1, \ldots, B_k)$ when $m \leq k$ and $A_1 = B_1, \ldots, A_m = B_m$. It is easy to see that $(T, \leq')$ is a poset. Let $\{T_\lambda\}$ be a chain in T with $T_\lambda = (A_{\lambda 1}, \ldots, A_{\lambda j(\lambda)})$. Our definition of $\leq'$ in T shows that if $T_\lambda \leq' T_\mu$ and if each has at least k submodules from S, then $A_{\lambda k} = A_{\mu k}$. Thus, if some T_λ has at least k submodules, setting $M_k = A_{\lambda k}$ is independent of λ. Now $\bigcup\{A_{\lambda j}\} = \{M_k\}$ is a strictly ascending chain of submodules of M and so is finite since ${}_RM$ is Noetherian. Hence there is a maximal $M_k = A_{\upsilon k}$ and $T_\upsilon = (M_1, \ldots, M_k)$ is an upper bound for the chain $\{T_\lambda\}$ in T. Applying Zorn's Lemma to $(T, \leq')$ yields a maximal $(B_1, \ldots, B_m) \in T$. Now B_m must be maximal in S since otherwise $B_m \subseteq V$ properly for $V \in S$, so $(B_1, \ldots, B_m) \leq' (B_1, \ldots, B_m, V)$ in $(T, \leq')$ and $(B_1, \ldots, B_m)$ is not maximal in T. This contradiction proves that S contains a maximal element.

Assume ii), let $A \leq_R M$, and let $S = \{V \leq_R M \mid V \subseteq A$, and V is finitely generated$\}$. By Theorem 13.16, $Rm + \mathbf{Z}m \in S$ for any $m \in A$, so $S \neq \emptyset$ and there is a maximal $B = Ra_1 + \mathbf{Z}a_1 + \cdots + Ra_m + \mathbf{Z}a_m \in S$, for some $a_j \in A$. Should $A \neq B$, then for $a \in A - B$, $B \subseteq B + Ra + \mathbf{Z}a$ properly (why?) and Theorem 13.16 implies $B + Ra + \mathbf{Z}a \in S$. This contradiction to the maximality of B shows $A = B$ is finitely generated. Finally, if every $A \leq_R M$ is finitely generated and $M_1 \subseteq \cdots \subseteq M_j \subseteq \cdots$ is a strictly ascending chain of submodules in M, let $\bigcup M_i = H$. We know that $H \leq (M, +)$ by Theorem 2.14 and for $r \in R$ and $h \in H$, since $h \in M_j$ for some j and $M_j \leq_R M$, it follows that $rh \in M_j$ so $rh \in H$.

By Definition 13.6, $H \leq_R M$ so by iii) there are $h_j \in H$ with $H = Rh_1 + \mathbf{Z}h_1 + \cdots + Rh_n + \mathbf{Z}h_n$. Now each $h_j \in M_k$ for some k, and $\{M_i\}$ a chain force all $h_j \in M_w$ for some w. Thus $H \subseteq M_w$ and the chain $\{M_i\}$ must be finite with M_w as its maximal element. Therefore M is a Noetherian R-module, completing the proof. ∎

We will see how the Noetherian condition can be used to study the structure of commutative rings. First we use Theorem 13.17 to expand our collection of Noetherian rings by proving the famous *Hilbert Basis Theorem.*

THEOREM 13.18 (Hilbert Basis Theorem) If R is a commutative Noetherian ring with 1, so is $R[x]$. Hence $R[x_1, \ldots, x_n]$ is also a Noetherian ring.

Proof From Theorem 13.17 it suffices to prove that each $I \lhd R[x]$ is finitely generated. Let $(0) \neq I \lhd R[x]$, $C_n = \{r \in R^* \mid rx^n + \cdots + b_1x + b_0 \in I\}$, the set of leading coefficients of $p(x) \in I$ with degree n, and $A_n = C_n \cup \{0\} \lhd R$ (why?). Now $x \in R[x]$ implies $A_n \subseteq A_{n+1}$ and R is Noetherian so the ascending chain $\{A_j\}$ has a maximal element, say A_m. Theorem 13.17 applied to R shows that each of $A_0, \ldots, A_m$ is finitely generated. If $(0) \neq A_j = Ra_{j1} + \cdots + Ra_{jk(j)}$, let $p_{ji}(x) \in I$ of degree j have leading coefficient a_{ji}. We claim that I is generated by $\{p_{ji}\}$. Let $f \in I$ have least degree $d \leq m$. If $f = rx^d + \cdots + r_0$ then $r \in A_d$ so $r = r_1a_{d1} + \cdots + r_{k(d)}a_{dk(d)}$ and $h = f - \sum r_i p_{di} \in I$. This forces $f = \sum r_i p_{di}$ since otherwise $deg\ h < deg\ f$ (why?) contradicts the choice of f having least degree in I. Assume that all $f \in I$ with $d \leq deg\ f \leq n$ satisfy $f \in \sum(p_{ji})$ and let $q \in I$ with $deg\, q = n + 1$. The leading coefficient of q is $b = b_1a_{e1} + \cdots + b_{k(e)}a_{ek(e)}$ where $e = n + 1$ if $n + 1 \leq m$ and $e = m$ if $n + 1 > m$. Now $q - \sum b_i p_{ei} \in I$ and is zero or has smaller degree than q. In either case our induction assumption implies $q \in \sum(p_{ji})$, proving by IND-II that I is finitely generated. Hence $R[x]$ is Noetherian. Another easy induction using $R[x_1, \ldots, x_n] = R[x_1, \ldots, x_{n-1}][x_n]$ proves this ring is Noetherian also. ∎

The proof of Theorem 13.18 works as well when R is a left Noetherian ring with 1. The theorem shows that *any* ideal in $\mathbf{Q}[x_1, \ldots, x_n]$ must be finitely generated. We record some other consequences.

THEOREM 13.19

a) $T = (x_1) + \cdots + (x_n) \lhd \mathbf{Z}[x_1, \ldots, x_n]$ is a Noetherian ring.
b) If R is a commutative Noetherian ring with 1, $I \lhd R$, and $A \lhd R[x_1, \ldots, x_n]$, then R/I and $R[x_1, \ldots, x_n]/A$ are Noetherian rings.
c) Any finitely generated commutative ring $S = \langle\{s_1, \ldots, s_m\}\rangle$ is Noetherian.

Proof For a), note that any ideal of T is an additive subgroup so it is an ideal of $\mathbf{Z}[x_1, \ldots, x_n]$, Noetherian by Theorem 13.18. Therefore T contains no infinite, strictly ascending chain of ideals and so is Noetherian. In b), $R[x_1, \ldots, x_n]$ is Noetherian by Theorem 13.18, so we need only show that R/I is Noetherian. This is Theorem 11.19 and follows from Theorem 11.16: chains of ideals in R/I correspond to chains of ideals of R containing I. Now $T = (x_1) + \cdots + (x_m) \lhd \mathbf{Z}[x_1, \ldots, x_m]$ in a) is a Noetherian ring, and using Theorem 10.7 to describe $S = \langle\{s_1, \ldots, s_m\}\rangle$ as sums of products of the s_i, $\varphi : T \to S$ defined by $\varphi(p(x_1, \ldots, x_m)) = p(s_1, \ldots, s_m)$ is a surjective homomorphism (verify!).

Since Theorem 11.12 shows $S \cong T/I$ for $ker\,\varphi = I \lhd T$, as in b) the correspondence of ideals in T/I and in T yields T/I, so S itself, is Noetherian. ■

By the theorem, if $T = \{x^i\} \subseteq \mathbf{Z}[x]$ is any *finite* set of powers of x then $R = \langle\{1\} \cup T\rangle \subseteq \mathbf{Z}[x]$ is a Noetherian integral domain. It is easy to see that $\langle\{1, x^2, x^3\}\rangle = \{a_0 + a_2x^2 + \cdots + a_nx^n \in \mathbf{Z}[x]\}$, and it is not very difficult to see that $\langle\{1, x^3, x^5\}\rangle = \{\sum a_ix^i \in \mathbf{Z}[x] \mid a_1 = a_2 = a_4 = a_7 = 0\}$, but it is harder to characterize $\langle\{1, x^{10}, x^{17}, x^{23}, x^{28}\}\rangle$. We can use Theorem 13.19 to construct Noetherian rings with zero divisors. Recall that $Ra_1 + \cdots + Ra_n = (a_1, \ldots, a_n)$.

Example 13.9 $R = \mathbf{Q}[x, y, z]/(x^2 - z, y^2 - z)$ is Noetherian with zero divisors.
If $I = (x^2 - z, y^2 - z)$ and $R = \{[f(x, y, z)] = f(x, y, z) + I \mid f \in \mathbf{Q}[x, y, z]\}$ then R is a Noetherian ring by Theorem 13.18 and Theorem 13.19. In R, $[x - y][x + y] = [x^2 - y^2] = [x^2 - z - (y^2 - z)] = [0]$ since $x^2 - z - (y^2 - z) \in I$, but $x - y$, $x + y \notin I$. For example, if $x - y = g(x, y, z)(x^2 - z) + h(x, y, z)(y^2 - z)$ then for $z = 0$, $x - y = g_1(x, y)x^2 + h_1(x, y)y^2$, a contradiction since $g_1(x, y)x^2 + h_1(x, y)y^2$ contains no monomial of degree 1.

We turn to some structural results for commutative Noetherian rings. The first makes use of Theorem 13.17 to show that $N(R)$ is *nilpotent*. That is, for some $m \geq 1$ and all $a_1, \ldots, a_m \in N(R)$, $a_1 \cdots a_m = 0$.

THEOREM 13.20 If R is a commutative Noetherian ring, then there is some $m \geq 1$ with $N(R)^m = (0)$.

Proof $N(R)$ is an ideal of R (Theorem 13.6) and hence is an R-submodule. Applying Theorem 13.17 shows $N(R) = (x_1) + \cdots + (x_k)$ for some $x_j \in N(R)$ and $x_i^{m(i)} = 0$ with $m(i) \geq 1$. Let $m = m(1) + \cdots + m(k)$, and for $1 \leq s \leq m$, let $a_s = r_{s1}x_1 + z_{s1}x_1 + \cdots + r_{sk}x_k + z_{sk}x_k \in N(R)$ with $r_{si} \in R$ and $z_{si} \in \mathbf{Z}$. Every term in the expansion of $a_1 \cdots a_m$ is $\alpha = rx_1^{t(1)} \cdots x_k^{t(k)}$ or $\alpha = zx_1^{t(1)} \cdots x_k^{t(k)}$ with $t(j) \geq 0$ and some $t(i) \geq m(i)$ since $\sum t(i) = m$, so $\alpha = 0$. Hence $N(R)^m = (0)$ as required. ■

The theorem shows again why the equivalence of the Noetherian condition and the finite generation of ideals is useful. Next we see how the existence of maximal ideals in any set of ideals is used to see that $N(R)$ is the intersection of *finitely* many prime ideals of R (note Theorem 13.6). This leads to other results about R. First we need a definition.

DEFINITION 13.9 If R is a commutative ring and $I \lhd R$, then the **radical** of I is $\mathrm{rad}(I) = \{x \in R \mid x^k \in I\}$.

It is an exercise that $I \subseteq \mathrm{rad}\,(I)$, $\mathrm{rad}((0)) = N(R)$, and $y^k \in \mathrm{rad}(I) \Leftrightarrow y \in \mathrm{rad}(I)$. A direct computation using the Binomial Theorem shows that $\mathrm{rad}(I) \lhd R$, but it may be easier to observe that for $\rho : R \to R/I$ the homomorphism $\rho(r) = r + I$, $\mathrm{rad}(I) = \rho^{-1}(N(R/I)) \lhd R$ from Theorem 11.16, so $\mathrm{rad}(I)/I = N(R/I)$. If $L = (x_1^2x_2^3) \lhd \mathbf{C}[x_1, \ldots, x_n]$, then $\mathrm{rad}(L) = (x_1x_2) = (x_1) \cap (x_2)$ (verify!). In general, for $I \lhd R$ commutative, Theorem 13.6

shows that $\text{rad}(I)/I = N(R/I)$ is an intersection of prime ideals of R/I. Thus Exercise 8 in §13.1 implies that $\text{rad}(I)$ is the intersection of the corresponding prime ideals of R, but it is not clear that $\text{rad}(I)$ must be a finite intersection of primes in R.

THEOREM 13.21 If R is a commutative Noetherian ring, $1 \in R$, and $I \lhd R$ is proper, then $\text{rad}(I)$ is the intersection of finitely many prime ideals of R.

Proof Let $S = \{I \lhd R \mid I \neq R$ and $\text{rad}(I)$ is *not* an intersection of finitely many prime ideals of $R\}$. Since $1 \in R$, it follows that $\text{rad}(I) = R \Leftrightarrow I = R$, so $I \in S$ implies $\text{rad}(I) \neq R$. The theorem holds when $S = \varnothing$. If $S \neq \varnothing$ there is a maximal $I \in S$ by Theorem 13.17, and so $\text{rad}(I) \neq R$ cannot be a prime ideal. Thus there are $x, y \in R - \text{rad}(I)$ with $xy \in \text{rad}(I)$. Now $A = \text{rad}(I) + (x)$ and $B = \text{rad}(I) + (y)$ are ideals of R properly larger than $\text{rad}(I)$ and $AB \subseteq \text{rad}(I)$. Should $A = R$, then $y \in B = AB \subseteq \text{rad}(I)$, a contradiction, and similarly, $B \neq R$. Thus $\text{rad}(A)$ and $\text{rad}(B)$ are finite intersections of prime ideals of R by the maximality of I in S, and $\text{rad}(I) \subseteq \text{rad}(A) \bigcap \text{rad}(B)$, a finite intersection of primes (why?). For $r \in \text{rad}(A) \bigcap \text{rad}(B)$, $r^d \in A$ and $r^e \in B$, so $r^{d+e} = r^d r^e \in AB \subseteq \text{rad}(I)$. Hence $r \in \text{rad}(I)$ and $\text{rad}(I) = \text{rad}(A) \bigcap \text{rad}(B)$, contradicting $I \in S$. Therefore $S = \varnothing$, proving the theorem. ■

The special case of $I = (0)$ in the theorem shows that when $1 \in R$, $N(R)$ is a finite intersection of primes. This enables us to view $R/N(R)$, or R itself when $N(R) = (0)$, as a subring of a finite direct sum of Noetherian integral domains. If $N(R) = P_1 \bigcap \cdots \bigcap P_n$ with each P_j a prime ideal, then $\varphi : R \to R/P_1 \oplus \cdots \oplus R/P_n$ given by $\varphi(r) = (r + P_1, \ldots, r + P_n)$ is a ring homomorphism with $ker\,\varphi = N(R)$ (verify!), so by Theorem 11.12 $R/N(R) \cong \varphi(R)$, a subring of $\oplus R/P_j$. Clearly each R/P_j is an integral domain and is Noetherian by Theorem 13.19. Any subring of a direct sum of domains has no nonzero nilpotent element but need not be Noetherian if each domain is. An example with one summand is $S = \boldsymbol{Z} + \boldsymbol{Q}[x]x \subseteq \boldsymbol{Q}[x]$ containing the infinite, strictly ascending chain of ideals $\{((1/2^i)x) \mid i \in \boldsymbol{N}\}$.

We present a few illustrations of the representation of $R/N(R)$ mentioned above, using Theorem 13.18, and leave most of the verifications as exercises. For R a commutative ring and $a_1, \ldots, a_m \in R$, we use the notation $(a_1, \ldots, a_m) = (a_1) + \cdots + (a_m)$, where (a_i) is the ideal of R generated by a_i. For $r \in R$ and $I \lhd R$, write $[r] = r + I$ in R/I. Then if $I \subseteq J \lhd R$, by Theorem 11.12 $(R/I)/[J] \cong R/J$.

Example 13.10 $R = F[x, y]/(xy)$ with F a field.

In $F[x, y]$, x and y are primes so $\text{rad}((xy)) = (xy)$ (why?). Thus $N(R) = (xy)/(xy) = (0_R) = ([x]) \bigcap ([y])$. Here $\varphi(r) = (r + ([x]), r + ([y]))$ shows $R \cong \varphi(R) \subseteq R/([x]) \oplus R/([y]) \cong F[x, y]/(x) \oplus F[x, y]/(y) \cong F[z] \oplus F[z]$ (why?).

Example 13.11 $R = F[x, y]/(x^2, y^2)$ with F a field.

In R, $[x]^2 = [y]^2 = [0]$, so if $[p(x, y)] = [a + bx + cy + dxy] \in R$ then $[bx + cy + dxy]^3 = 0$, and $[p(x, y)]$ is a unit $\Leftrightarrow [a] \neq [0]$ (see Example 11.9). Thus $N(R) = ([x]) + ([y])$ is maximal and $R/N(R) \cong F$.

Example 13.12 $R = F[x, y, z]/(xy^2z^3)$ with F a field.

Since $x, y, z \in F[x, y, z] = T$ are primes, if $p(x, y, z)^m \in (xy^2z^3)$ then $p(x, y, z) \in (xyz)$. Thus $\text{rad}((xy^2z^3)) = (xyz) = (x) \bigcap (y) \bigcap (z)$. Now $T/(x) \cong T/(y) \cong T/(z) \cong F[s, t]$, the polynomial ring in two indeterminates, and $\varphi(t) = (t + (x), t + (y), t + (z))$ is a ring homomorphism from T to $T/(x) \oplus T/(y) \oplus T/(z)$ with $ker\,\varphi = (xyz)$. Thus by Theorem 11.12, $R/N(R) \cong T/(xyz) \cong S \subseteq F[s, t] \oplus F[s, t] \oplus F[s, t]$ (why?).

Example 13.13 $R = \mathbf{Z}[x, y]/(4x, xy^2)$.

In $T = \mathbf{Z}[x, y]$, (x) and $(2, y)$ are prime ideals (why?) and $N(R) = [(2x, xy)] = [(x)] \bigcap [(2, y)]$ since this intersection of prime ideals consists of nilpotent elements in R (all have cube zero). Now $T/(x) \cong \mathbf{Z}[y]$ and $T/(2, y) \cong \mathbf{Z}_2[x]$, so as in Example 13.12, by Theorem 11.12, $R/N(R) \cong T/(2x, xy) \cong S \subseteq \mathbf{Z}[t] \oplus \mathbf{Z}_2[z]$ (why?).

We end this section with two results relating the set of zero divisors in a commutative Noetherian ring to annihilator ideals in the ring.

DEFINITION 13.10 For R any commutative ring and $\varnothing \neq S \subseteq R$, the **annihilator** of S is $\text{ann}(S) = \{r \in R \mid sr = 0 \text{ for all } s \in S\}$, and $ZD(R) = \{x \in R \mid xy = 0 \text{ for some } y \in R^*\}$ is the **set of zero divisors** in R.

THEOREM 13.22 If R is a commutative ring with 1 and $\varnothing \neq S \subseteq R$, then $\text{ann}(S) \lhd R$. If $I = \text{ann}(S)$ is maximal in $\{\text{ann}(T) \mid T \neq \{0\}\}$, then I is a prime ideal of R and $I = \text{ann}(\{a\})$ for some $a \in R$.

Proof It is an exercise that $\text{ann}(S)$ is an ideal of R. Using $1 \in R$, $\text{ann}(S) = R \Leftrightarrow S = \{0\}$, for $1 \cdot s = 0$ requires $s = 0$. Assume that $I = \text{ann}(S)$ is maximal in the poset $(\{\text{ann}(T) \mid T \neq \{0\}\}, \subseteq)$. Let $x, y \in R$ with $xy \in I$, so $\{xys \mid s \in S\} = xyS = \{0\}$. If $yS = \{0\}$ then $y \in \text{ann}(S) = I$, and if $yS \neq \{0\}$ then $x \in \text{ann}(yS)$. Now $yst = 0$ for any $s \in S$ and all $t \in I$, so $I \subseteq \text{ann}(yS)$. The maximality of I forces $I = \text{ann}(yS)$, thus $x \in I$ and I is a prime ideal of R as claimed. For $a \in S$, $aI = \{0\}$ so $I \subseteq \text{ann}(\{a\})$ and again the choice of I forces $I = \text{ann}(\{a\})$. ∎

Now $ZD(R) = \bigcup \text{ann}(T)$ with $T \neq (0)$, since if $xy = 0$ for $x, y \in R^*$ then $x \in \text{ann}(\{y\})$ and $y \in \text{ann}(\{x\})$. For Noetherian rings we can say more about the relation between $ZD(R)$ and the annihilator ideals and can also prove that for any subring W consisting of zero divisors, $aW = \{0\}$ for some $a \in R$.

THEOREM 13.23 Let R be a commutative Noetherian ring with 1. Then $\{\text{ann}(S) \lhd R \mid S \neq \{0\}\}$ has only finitely many maximal elements $P_1, \ldots, P_k$, each $P_j = \text{ann}(\{a_j\})$ is a prime ideal, and $ZD(R) = P_1 \cup \cdots \cup P_k$. If $T \subseteq ZD(R)$ is a subring, then $T \subseteq P_j$ for some j, so $a_jT = \{0\}$.

Proof From Theorem 13.17, $\{\mathrm{ann}(S) \mid S \neq \{0\}\}$ has maximal elements $\{P_\lambda\}$. Each such P_λ is a prime ideal and $P_\lambda = \mathrm{ann}(\{a_\lambda\})$ by Theorem 13.22. Note that each $\mathrm{ann}(S) \subseteq P_\lambda$ for some P_λ using Theorem 13.17 on $\{\mathrm{ann}(A) \mid \mathrm{ann}(S) \subseteq \mathrm{ann}(A),\ A \neq 0\}$. Consider all the finite sums $\{(a_{\lambda(1)}, \ldots, a_{\lambda(j)}) = a_{\lambda(1)}R + \cdots + a_{\lambda(j)}R\}$. Theorem 13.17 now produces a maximal such sum, say $(a_1, \ldots, a_m)$, and $P_1P_2 \cdots P_m \subseteq P_1 \cap \cdots \cap P_m \subseteq \mathrm{ann}((a_1, \ldots, a_m))$. For any maximal annihilator $P = \mathrm{ann}(\{a\})$, $(a_1, \ldots, a_m) = (a_1, \ldots, a_m, a)$ using that $(a_1, \ldots, a_m)$ is maximal, so $a \in (a_1, \ldots, a_m)$. Therefore $P_1P_2 \cdots P_m \subseteq \mathrm{ann}(\{a\}) = P$, and P prime forces some $P_i \subseteq P$, so $P_i = P$ (why?). Thus $P_1, \ldots, P_m$ are the only maximal annihilator ideals. Now $a_jP_j = \{0\}$, so $P_j \subseteq ZD(R)$. If $x, y \in R^*$ with $xy = 0$ then $x \in \mathrm{ann}(\{y\}) \subseteq P_j$ for some j, so $ZD(R) = P_1 \cup \cdots \cup P_m$.

Finally, suppose that $T \subseteq R$ is a subring of zero divisors. By rearranging the $\{P_j\}$ if necessary, for some $k \geq 1$, $T \subseteq P_1 \cup \cdots \cup P_k$ and T is not contained in any union of fewer P_i. Thus when $k \geq 2$ there are $t_j \in (T \cap P_j)$ with $t_j \notin P_i$ if $j \neq i$. Since $t = t_1 \cdots t_{m-1} + t_m \in T$, we must have $t \in P_j$ for some j. If $j = m$ then $t_1 \cdots t_{m-1} = t - t_m \in P_m$ so $t_i \in P_m$ for some $i < m$ using that P_m is a prime ideal. This contradicts the choice of the t_i so $t \in P_j$ for $j < m$. Now $t_j \in P_j$ so $t_m = t - t_1 \cdots t_{m-1} \in P_j$, contradicting the choice of t_m. Therefore $k = 1$, $T \subseteq P_i = \mathrm{ann}(\{a_i\})$ for some i, and so $a_iT = \{0\}$. ∎

For any commutative domain R, $ZD(R) = \{0\}$, a prime ideal. In Example 13.10 for $R = F[x, y]/(xy)$, $ZD(R) = ([x]) \cup ([y])$ with $([x]) = \mathrm{ann}(\{[y]\})$ and $([y]) = \mathrm{ann}(\{[x]\})$ prime ideals: $R/([x]) \cong F[x, y]/(x) \cong F[y]$. Two more examples follow and show that a prime annihilator ideal need not be a maximal ideal or a maximal annihilator.

Example 13.14 $ZD(R)$ for $R = F[x, y, z]/(xy, xz)$, F a field.

In $F[x, y, z]$, (x) and (y, z) are prime ideals (why?). It follows that if $f(x, y, z) \cdot g(x, y, z) \in (xy, xz)$ then one of these factors is in (x) and one, maybe the same, is in (y, z). Using the same symbols for elements in $F[x, y, z]$ and in R, we have $ZD(R) = (x) \cup (y, z)$. In R, $(x) = \mathrm{ann}(\{y\}) = \mathrm{ann}(\{z\}) = \mathrm{ann}(\{yz\})$ and $(y, z) = \mathrm{ann}(\{x\})$ are maximal annihilator ideals, and each is properly contained in $(x) + (y) + (z)$.

Example 13.15 $R = F[x, y]/(x^2y, xy^2)$.

Once again, since $x, y \in F[x, y]$ are primes, $\mathrm{rad}((x^2y, xy^2)) = (xy) = (x) \cap (y)$ and $(x), (y) \lhd F[x, y]$ are prime. Identifying $g \in F[x, y]$ with its image in R, $N(R) = (xy) = (x) \cap (y)$ with $(x) = \mathrm{ann}(\{y^2\})$ and $(y) = \mathrm{ann}(\{x^2\})$, both prime ideals. Thus $(x) \cup (y) \subseteq ZD(R)$, but $(x, y) = (x) + (y) = \mathrm{ann}(\{xy\})$ is a maximal ideal so $(x, y) = ZD(R)$, and neither (x) nor (y) is a maximal annihilator.

EXERCISES

1. Let R be a left Noetherian ring, ${}_RM$, and $m \in M$. Show that Rm is a left Noetherian R-module.
2. Given ${}_RM$ and $A \leq_R M$, show that there is an inclusion-preserving bijection between the sets of R-modules $\{H \leq_R M/A\}$ and $\{L \leq_R M \mid A \leq_R L\}$.

3. If ${}_RM$ is Noetherian and $A \leq_R M$, show that ${}_R(M/A)$ is Noetherian (Exercise 2!).
4. If ${}_RM = Rm_1 + \cdots + Rm_k$ and ${}_RR$ is Noetherian, show that ${}_RM$ is Noetherian.
5. Let R be a left Noetherian ring, $I \triangleleft R$, and $R/I \cong R$. Prove that $I = (0)$.
6. Show that any surjective ring homomorphism $\varphi : R \to R$ of a left Noetherian ring R must be an isomorphism.
7. If R is a commutative Noetherian ring with 1, and $S \subseteq R$ is multiplicative, show that R_S is also Noetherian.
8. If R is a commutative Noetherian ring with 1, let $S = \{A \subseteq R \mid A = \{e_j \in R \mid e_j \neq 0, e_j^2 = e_j, \text{ and } e_je_i = 0 \text{ when } i \neq j\}\}$. Show that:
i) Each $A \in S$ is finite.
ii) There is $m \in \boldsymbol{N}$ so each $A \in S$ has at most m elements.
9. Show that the following are Noetherian domains (F is a field):
i) $\boldsymbol{Z}[x, y, z]/(x + y + z)$
ii) $F[x, y, z]/(x^2y^3 - x^3y^2z^2 + xyz^3 - z)$
iii) $F[x, y]/(xy - 1)$
iv) $F[x, y]/(x^m + y^m - 1)$, any $m \geq 1$
v) $\boldsymbol{Z}[x, y, z]/(x^2 + 1, y + z)$
10. For $n, k \in \boldsymbol{N}$ and F a field, set $S(k, n) = \{f(x_1, \ldots, x_n) \in F[x_1, \ldots, x_n] \mid$ every monomial in f has degree in $k\boldsymbol{Z}\}$, so $S(2, n)$ are the polynomials with every monomial of even degree. Show that the following are Noetherian:
i) $S(2, 2)$
ii) $S(2, 3)$
iii) $S(2, n)$
iv) $S(3, 2)$
v) $S(3, 3)$
vi) $S(3, n)$
vii) $S(k, n)$
11. Let $X = \{x_i \mid i \in \boldsymbol{N}\}$ be indeterminates over $\boldsymbol{Q}$ and $R = \boldsymbol{Q}[X]$. If $A_1, \ldots, A_k \in M_n(R)$, show that the subring generated by the A_j, $S = \langle\{A_i\}\rangle \subseteq M_n(R)$, is a subring of $M_n(D)$ for $D \subseteq R$ a Noetherian domain.
12. **a)** If R is a commutative Noetherian ring with 1, argue that $R[[x]]$ is a Noetherian ring.
b) Show that $R[[x_1, \ldots, x_n]]$ is a Noetherian ring.
13. If $M \leq_{\boldsymbol{Z}} \boldsymbol{Q}$, show that ${}_{\boldsymbol{Z}}M$ is Noetherian $\Leftrightarrow$ for some $a \in \boldsymbol{Z}^*$ $am \in \boldsymbol{Z}$ for all $m \in M$.
14. Let $R = \boldsymbol{Z}[x_1, \ldots, x_n]$, $K = QF(R)$, and $A \leq_R K$. Show that ${}_RA$ is Noetherian $\Leftrightarrow$ there is $r \in R$, $r \neq 0$, so that $ry \in R$ for all $y \in A$.
15. If $n \in \boldsymbol{N}$ show that there is $m \in \boldsymbol{N}$ so that for $k > m$ there exist $p_{kj} \in \boldsymbol{Z}[x_1, \ldots, x_n]$ with $x_1^k + \cdots + x_n^k = (x_1 + \cdots + x_n)p_{k1} + \cdots + (x_1^m + \cdots + x_n^m)p_{km}$.
16. Let $\{p_i \mid i \in \boldsymbol{N}\} \subseteq R = \boldsymbol{Q}[x_1, \ldots, x_n]$. Assume that for any $k > 0$ there is $\alpha_k = (c_1, \ldots, c_n) \in \boldsymbol{C}^n$ so that $f_i(\alpha_k) = 0$ if $i \leq k$. Show that $p_j(\beta) = 0$ for some $\beta \in \boldsymbol{C}^n$ and all $j \in \boldsymbol{N}$.
17. Let R be any ring with 1 and $A, B, C \in M_n(R)$ so that every permutation of ABC is I_n. Show that $S = \langle\{A, B, C\}\rangle \subseteq M_n(R)$ is a (left) Noetherian ring.
18. If R is a commutative Noetherian ring with 1, show that each nonzero, nonunit $r \in R$ is a product of irreducibles.
19. If $1 \in R$, a commutative Noetherian ring, $I \triangleleft R$, and $I^2 = I \neq (0)$, show that $I = Re$ for $e^2 = e$. (Let $M = \{1 + x \mid x \in I\}$, note $I^e \subseteq J(R_M)$ by Exercise 9 in §13.2, and use

Theorem 13.4 and I finitely generated to get $0 \neq h^2 = h \in I$. Now $I = Rh + (I \cap R(1-h))$.)

20. If R is a commutative Noetherian ring with 1, $P_1, \ldots, P_k$ are prime ideals of R, and $P_1 \cap \cdots \cap P_k \subseteq N(R)$, then $P_1 \cap \cdots \cap P_k = N(R)$.

21. If R is a commutative Noetherian ring with 1 and $R \neq I \lhd R$, show that there are prime ideals $P_1, \ldots, P_k$ of R, not necessarily distinct, with $I \subseteq P_j$ for each P_j, and $P_1 P_2 \cdots P_k \subseteq I$.

22. Let R be a commutative Noetherian ring with 1 and $N(R) = (0)$. If $P = \text{ann}(S)$ is a prime ideal of R, show that P is a maximal annihilator ideal of R.

23. For $n > 1$, show that $ZD(\boldsymbol{Z}_n) = \boldsymbol{Z}_n - U_n$, that each nonzero maximal ideal in $\boldsymbol{Z}_n$ is an annihilator ideal, and that their union is $ZD(\boldsymbol{Z}_n)$.

24. Let $R = \boldsymbol{Q}[x, y, z]/(xy, yz, xz)$.

i) Find $N(R)$ and prime ideals P_j of R so that $N(R) = P_1 \cap \cdots \cap P_k$.

ii) Find maximal annihilator ideals $A_1, \ldots, A_m$ in R.

25. For $n \geq 3$, in $\boldsymbol{Q}[x_1, \ldots, x_n]/(x_1 \cdots x_{n-1}, x_1 \cdots x_{n-2}x_n, \ldots, x_2 \cdots x_n)$ describe the maximal annihilator ideals.

26. Let $X = \{x_i \mid i \in \boldsymbol{N}\}$ be indeterminates over the field F, and in $F[X]$, let I be the ideal generated by all $x_i x_j$ for $i \neq j$, so $I = (\{x_i x_j \mid i \neq j\})$. Describe prime ideals of $F[X]/I$ whose intersection is $N(F[X]/I)$, and describe maximal annihilator ideals whose union is $ZD(F[X]/I)$.

13.4 INTEGRALITY

Another useful idea for studying commutative rings is to look at elements that satisfy a monic polynomial over a fixed subring. *When $R \subseteq S$ are rings, we assume that R is a subring of S.*

DEFINITION 13.11 If $R \subseteq S$ are commutative rings with the same identity element, then $\alpha \in S$ is **integral over** R if there is a monic $p(x) \in R[x]$ with $p(\alpha) = 0$. If all $\alpha \in S$ are integral over R then S is an **integral extension** of R. When R is a field, any $\alpha \in S$ integral over R is called **algebraic over** R.

We often say S is integral over R when it is an integral extension of R. Any R with 1 is integral over itself since $r \in R$ satisfies $x - r \in R[x]$. When $R \subseteq S$, any $\alpha \in N(S)$ is integral over R since α satisfies some x^m. Also, $s \in S$ integral over R implies rs is integral over R for any $r \in R$ (why?). For $n, m \in \boldsymbol{N}$, $n^{1/m} \in \boldsymbol{R}$ is integral over $\boldsymbol{Z}$ and algebraic over $\boldsymbol{Q}$ since $p(n^{1/m}) = 0$ for $p(x) = x^m - n \in \boldsymbol{Z}[x]$. Let us consider a more substantial example.

Example 13.16 The elements integral over $\boldsymbol{Z}$ in $\boldsymbol{Q}(5^{1/2})$.

If $a = s/t \in \boldsymbol{Q}$ is integral over $\boldsymbol{Z}$, $\text{GCD}\{s, t\} = 1$, and $f(x) \in \boldsymbol{Z}[x]$ with $f(a) = a^n + z_{n-1}a^{n-1} + \cdots + z_0 = 0$, then $t^n f(a) = s^n + td = 0$ for some $d \in \boldsymbol{Z}$. Any prime p with $p \mid t$ must divide $s^n = -td$, so $p \mid s$, forcing $p \mid \text{GCD}\{s, t\}$. Thus $t = \pm 1$ and $a \in \boldsymbol{Z}$. For $\alpha = a + b5^{1/2} \in \boldsymbol{Q}(5^{1/2})$ and $m(x) = x^2 - 2ax + (a^2 - 5b^2) \in \boldsymbol{Q}[x]$, $m(\alpha) = 0$ and if $b \neq 0$ then $m(x)$ has least degree in $I(\alpha) = \{g(x) \in \boldsymbol{Q}[x] \mid g(\alpha) = 0\} \lhd \boldsymbol{Q}[x]$ (why?). Thus $I(\alpha) = (m(x))$ by Theorem 11.3. Assume $f(x) =$

$m(x)q(x) \in \mathbf{Z}[x] \cap (m(x))$ is monic. By Theorem 12.12, $m(x) = r_1 m_1(x)$ and $q(x) = r_2 q_1(x)$ with m_1 and q_1 primitive in $\mathbf{Z}[x]$, so $m_1 q_1$ is primitive by Theorem 12.13. From $f(x) = r_1 r_2 m_1(x) q_1(x)$ and Theorem 12.12 again, $r_1 r_2 \in \mathbf{Z}$, and f monic forces $r_1 r_2 = \pm 1$ and $\pm m_1(x)$ to be monic as well. Thus if $\alpha \in \mathbf{Q}(5^{1/2}) - \mathbf{Q}$ is integral over $\mathbf{Z}$ then $m(x) \in \mathbf{Z}[x]$ and both $2a, a^2 - 5b^2 = k \in \mathbf{Z}$. It follows that $(2a)^2 - 5(2b)^2 = 4k$ and $5(2b)^2 \in \mathbf{Z}$. Unique factorization in $\mathbf{Z}$ shows that $(2b)^2 \in \mathbf{Z}$, then that $2b \in \mathbf{Z}$ (Theorem 1.17). Since $(2a)^2 - (2b)^2 \equiv_4 (2a)^2 - 5(2b)^2 \equiv_4 0$, $2a \equiv_2 2b$ follows (why?). Finally, if $a = u/2$ and $b = v/2$ with $u \equiv_2 v$, then both $2a, a^2 - 5b^2 \in \mathbf{Z}$. In conclusion, $\alpha \in \mathbf{Q}(5^{1/2})$ is integral over $\mathbf{Z}$ exactly when $\alpha = u/2 + (v/2) \cdot 5^{1/2}$ for $u, v \in \mathbf{Z}$ with $u \equiv_2 v$. Note that our argument shows that $\mathbf{Q}(5^{1/2})$ is algebraic over $\mathbf{Q}$.

We cannot often identify the polynomials satisfied by integral elements as we did in the example. It would be convenient to have other methods to use to study integrality. First we establish some notation.

DEFINITION 13.12 If $R \subseteq S$ are commutative rings with the same 1 and $T \subseteq S$ then $R[T] = \langle R \cup T \rangle$ is the subring of S generated by R and T. When $T = \{s_1, \ldots, s_m\}$, we write $R[s_1, \ldots, s_m]$ for $R[\{s_1, \ldots s_m\}]$.

The notation $R[s]$ suggests polynomials, and with S commutative, Theorem 10.7 shows that $R[s] = \{r_n s^n + \cdots + r_1 s + r_0 \in S \mid r_j \in R\}$. It is important to recognize that $R[s]$ contains R and is an R-module.

THEOREM 13.24 If $R \subseteq S$ are commutative rings with the same 1, then the following are equivalent:

i) $\alpha \in S$ is integral over R.
ii) $R[\alpha]$ is a finitely generated R-module.
iii) There is a ring T with $R \cup \{\alpha\} \subseteq T \subseteq S$ so that T is a finitely generated R-module.

Proof From i), $\alpha^n = r_{n-1}\alpha^{n-1} + \cdots + r_1\alpha + r_0$ with $n \geq 1$ and all $r_j \in R$. If $n \leq k$ and $\alpha^j \in R + R\alpha + \cdots + R\alpha^{n-1}$ for all $j \leq k$, then also $\alpha^{k+1} = \alpha^{k-n+1}\alpha^n = r_{n-1}\alpha^k + \cdots + r_0\alpha^{k-n+1} \in R + R\alpha + \cdots + R\alpha^{n-1}$. By induction (IND-II), all $\alpha^m \in R + R\alpha + \cdots + R\alpha^{n-1}$, so $R[\alpha] = R \cdot 1 + R\alpha + \cdots + R\alpha^{n-1}$ and ii) holds. Taking $T = R[\alpha]$ shows that ii) implies iii). To see that iii) implies i), we use the fact that the product of $A \in M_n(R)$ and its adjoint is $\det(A) \cdot I_n$. In iii), let $T = Rt_1 + \cdots + Rt_m$. Since $\alpha \in T$, $\alpha t_j = r_{1j}t_1 + \cdots + r_{mj}t_m$ and so $(t_1, \ldots, t_m) \in T^m$ satisfies each linear equation $\sum_j (\delta_{ji}\alpha - r_{ji})x_j = 0$ for $\delta_{ji} = 0$ if $i \neq j$ and $\delta_{ii} = 1$. Thus if $A \in M_m(T)$ is given by $A_{ji} = \delta_{ji}\alpha - r_{ji}$, then $(t_1, \ldots, t_m)A = 0_m \in T^m$, and $(t_1, \ldots, t_m)A \cdot \mathrm{adj}(A) = (\det(A)t_1, \ldots, \det(A)t_m) = 0_m$ follows. Since $R \subseteq T$, $\det(A) \in T$ and satisfies $\det(A)t_j = 0$, so $\det(A)T = 0$ and $1 \in R \subseteq T$ forces $\det(A) = 0$. Expanding (this is just the characteristic polynomial of (r_{ji}) evaluated at α) shows that $\alpha^m = r_{m-1}\alpha^{m-1} + \cdots + r_1\alpha + r_0$. Therefore α is integral over R, proving i) and so the theorem. ■

It is convenient to make an observation before using Theorem 13.24.

THEOREM 13.25 If $R \subseteq S$ are commutative rings with the same 1, and if V and W are subrings of S and finitely generated R-modules, then $VW = \left\{\sum v_i w_i \mid v_i \in V, w_i \in W\right\}$ is a subring of S, finitely generated as an R-module.

Proof Write $V = Rv_1 + \cdots + Rv_n$ and $W = Rw_1 + \cdots + Rw_m$. Since $v_i v_j = \sum r_{ijk} v_k$ and $w_i w_j = \sum s_{ijk} w_k$, for $r_{ijk}, s_{ijk} \in R$, $VW = \sum Rv_i w_j$ for $1 \le i \le n$ and $1 \le j \le m$ is a subring using Theorem 10.4 as in Theorem 11.5. ■

We can now see that the elements in S integral over R form a ring and that integrality is "transitive."

THEOREM 13.26 If $R \subseteq S$ are commutative rings with the same 1 and $W \subseteq S$ consists of elements integral over R, then $R[W]$ is integral over R. Thus the set of all integral elements in S is a subring. Also, if $R \subseteq T \subseteq S$ with T a ring integral over R, then any $\alpha \in S$ integral over T is integral over R.

Proof To show that $R[W]$ is integral over R it suffices to show that for $\alpha, \beta \in S$ integral over R, both $\alpha\beta$ and $\alpha + \beta$ are integral over R, using Theorem 10.7 and that $-\beta$ is integral over R. From Theorem 13.24, $R[\alpha]$ and $R[\beta]$ are finitely generated R-modules. Thus Theorem 13.25 and $1 \in R$ show that $R[\alpha, \beta] = R[\alpha]R[\beta]$ is a finitely generated R-module and subring of S containing R, α, and β. Since $\alpha\beta, \alpha + \beta \in R[\alpha, \beta]$, these are integral over R by Theorem 13.24. Now assume T is integral over R and $\alpha \in S$ is integral over $T \subseteq S$ with $\alpha^m = t_{m-1}\alpha^{m-1} + \cdots + t_1\alpha + t_0$ for $t_j \in T$. From the integrality of each t_j over R and Theorem 13.24, $t_j \in T_j$, a subring of S containing R with ${}_R T_j$ finitely generated. Hence Theorem 13.25 and induction show that $W = T_0 \cdots T_{m-1} = \sum Rw_i$ is a finitely generated R-module and is a subring of S containing R and all t_j. By the proof of Theorem 13.24, $V = W + W\alpha + \cdots + W\alpha^{m-1} = \sum_{i,j} Rw_i\alpha^j$ for $0 \le j < m$ is a subring of S containing R and α. A final application of Theorem 13.24 yields α integral over R, completing the proof. ■

DEFINITION 13.13 For $R \subseteq S$ integral domains with the same 1, the **integral closure of *R* in *S*** is $\text{Int}_R(S) = \{\alpha \in S \mid \alpha \text{ is integral over } R\}$. R is **integrally closed in *S*** if $R = \text{Int}_R(S)$, and R is **integrally closed** if $R = \text{Int}_R(QF(R))$, with R identified in $QF(R)$ as $\{r/1 \mid r \in R\}$.

When $R \subseteq S$ are integral domains with the same 1, Theorem 13.26 shows that $\text{Int}_R(S) = R[\text{Int}_R(S)]$ *is a ring and is integrally closed in S*. Example 13.16 shows that $\text{Int}_{\mathbf{Z}}(\mathbf{Q}(5^{1/2})) = \{a/2 + b/2 \cdot 5^{1/2} \mid a, b \in \mathbf{Z} \text{ and } a \equiv_2 b\}$ and also proves that $\mathbf{Z}$ is integrally closed. More generally, we have

THEOREM 13.27 Any UFD R is integrally closed.

Proof Let $K = QF(R)$ and $a/b \in K$ with $\text{GCD}\{a, b\} = 1$. If $a/b \in \text{Int}_R(K)$ then for some $m \ge 1$, $(a/b)^m + \cdots + r_1(a/b) + r_0 = 0$ with all $r_j = r_j/1 \in R$. Multiply by b^m to get $a^m + bd = 0$ for $d \in R$. If $p \in R$ is a prime and $p \mid b$ then $p \mid a^m$, so $p \mid a$ and $\text{GCD}\{a, b\} \ne 1$. Thus $b \in U(R)$, forcing $a/b \in R$ and proving that $\text{Int}_R(K) = R$. Thus R is integrally closed. ■

Examples of UFDs are given in §12.3 and §12.4 and include polynomial rings over $\boldsymbol{Z}$ or over a field F. We give next a standard example of an integral domain that is not integrally closed.

Example 13.17 If $R = \boldsymbol{Q}[x^2, x^3] \subseteq \boldsymbol{Q}[x]$, then $R \neq \text{Int}_R(QF(R))$.

Now $x/1 \notin R \subseteq K = QF(R)$ but is essentially $x^3/x^2 \in K - R$ and satisfies $Y^2 - x^2 \in R[Y]$. Thus $x^3/x^2 \in \text{Int}_R(K) - R$ (verify!).

Prime ideals and chains of prime ideals are important in advanced studies of commutative rings. We can say something about the relation between prime ideals in a ring R and prime ideals in an integral extension S of R. A consequence of our next result is that in any integral extension $R \subseteq S$, a chain of primes in one ring corresponds to a chain of primes in the other. As we will see, this result has nice implications when R or S is a field.

THEOREM 13.28 Let $R \subseteq S$ with S integral over R. If $P_1 \subseteq \cdots \subseteq P_n$ are prime ideals of R, then:

i) There is $Q_1 \lhd S$, prime with $Q_1 \cap R = P_1$.
ii) If for $j < n$ there are $Q_j \lhd S$, prime with $Q_1 \subseteq Q_2 \subseteq \cdots \subseteq Q_{n-1}$ and $Q_j \cap R = P_j$, then there is $Q_n \lhd S$, prime with $Q_{n-1} \subseteq Q_n$ and $Q_n \cap S = P_n$.
iii) If $W \subseteq V$ are ideals of S with W prime and $W \cap R = V \cap R$, then $W = V$.

Proof Now $Y = \{I \lhd S \mid I \cap R \subseteq P = P_1\} \neq \varnothing$ since $(0) \in Y$. We use Zorn's Lemma to obtain a maximal element in Y. If $\{I_\lambda\}$ is a chain in Y then $A = \bigcup I_\lambda \lhd S$ by Theorem 11.5, and any $s \in A \cap R$ is in some $I_\lambda \cap R \subseteq P$ using $I_\lambda \in Y$. Hence $A \in Y$, and Theorem 13.1 shows that some $Q \in Y$ is maximal. To see that Q is a prime ideal of S, let $a_1, a_2 \in S - Q$ with $a_1a_2 \in Q$. Clearly $Q \subseteq (Q + Sa_i) \lhd S$ properly and $(Q + Sa_1)(Q + Sa_2) \subseteq Q$. The maximality of Q in Y yields $r_i \in ((Q + Sa_i) \cap R) - P$ and $r_1r_2 \in (Q + Sa_1)(Q + Sa_2) \cap R \subseteq Q \cap R$ so $r_1r_2 \in P$. Since P is a prime ideal, either $r_1 \in P$ or $r_2 \in P$, a contradiction. Therefore $a_1a_2 \in Q$ implies $a_1 \in Q$ or $a_2 \in Q$, so Q is a prime ideal in S.

If $Q \cap R \neq P$ let $a \in P - (Q \cap R)$. Then $Q \subseteq Q + Sa \lhd S$ properly since $a \in Q$ implies $a \in Q \cap R$. There are $q \in Q$ and $s \in S$ with $y = q + sa \in ((Q + Sa) \cap R) - P$ by the maximality of Q, and S integral over R gives $s^m + r_{m-1}s^{m-1} + \cdots + r_0 = 0$. Thus $(sa)^m + r_{m-1}a(sa)^{m-1} + \cdots + a^m r_0 = 0$, so $sa = y - q$ shows $(y - q)^m + b_{m-1}(y - q)^{m-1} + \cdots + b_0 = 0$ with all $b_j \in P$ since $a \in P$. Expanding and using $y \in R$ gives $y^m + b_{m-1}y^{m-1} + \cdots + b_0 = dq \in Q \cap R \subseteq P$. Since all $b_j \in P$, we have $y^m \in P$, and P a prime ideal forces the contradiction $y \in P$. Therefore we cannot find $a \in P - (Q \cap R)$ so $Q \cap R = P$, proving i). It suffices by induction to prove ii) for $n = 2$. Repeat the argument in i) for $\{I \lhd S \mid Q_1 \subseteq I$ and $I \cap R \subseteq P_2\}$ to obtain a maximal element and prime ideal Q_2 with $Q_2 \cap R = P_2$ and $Q_1 \subseteq Q_2$. For iii), let $W \subseteq V$, $W \cap R = V \cap R$, and $s \in V - W$. Because s is integral over R, there is $p(s) = 0$ for $p(x) \in R[x]$ and monic, so there is n minimal with $s^n + \cdots + r_1s + r_0 = m(s) \in W$ and all r_j in R. Since $s \in V$, $r_0 = -s(s^{n-1} + \cdots + r_1) + m(s) \in V \cap R = W \cap R$, and it follows that $s(s^{n-1} + \cdots + r_1) = -r_0 + m(s) \in W$. Using that W is a prime ideal in S forces the contraction $s \in W$ or $s^{n-1} + \cdots + r_1 \in W$, proving $V - W = \varnothing$. Hence $W = V$, completing the proof of iii) and so of the theorem. ∎

When the ring S is an integral extension of the ring R then for any prime ideal Q of S, $P = Q \cap R$ is a prime ideal of R (exercise!). It follows from Theorem 13.28 that for any prime ideals $Q_1 \subseteq \cdots \subseteq Q_m$ in S with all inclusions proper, $(Q_1 \cap R) \subseteq \cdots \subseteq (Q_m \cap R)$ are prime ideals in R with all inclusions proper. Thus if either R or S has a chain of prime ideals of a given length, then so does the other ring. Also by Theorem 13.28, if $P \lhd R$ and $Q \lhd S$ are prime with $Q \cap R = P$, then P is maximal $\Leftrightarrow$ Q is maximal. From Theorem 11.17, R is a field if and only if $(0) \lhd R$ is maximal, so R is a field $\Leftrightarrow$ S is a field. In general when $R \subseteq S$, $\mathrm{Int}_R(S)$ is an integral domain by Theorem 13.26, $\mathrm{Int}_R(S)$ is integral over R by definition, and $\mathrm{Int}_R(S) = S$ when S is integral over R, so our comments apply to prove the following theorem.

THEOREM 13.29 Let $R \subseteq S$ be integral domains with the same 1. Then R is a field $\Leftrightarrow$ $\mathrm{Int}_R(S)$ is a field. In particular, if R is a field and S is algebraic over R then S is a field, and if S is a field integral over R then R is a field.

The theorem shows that $\mathrm{Int}_{\boldsymbol{Q}}(\boldsymbol{C})$, all $\alpha \in \boldsymbol{C}$ algebraic over $\boldsymbol{Q}$, is a field and that if $\beta \in \boldsymbol{C} - \mathrm{Int}_{\boldsymbol{Q}}(\boldsymbol{C})$ then β is not algebraic over $\boldsymbol{Q}$, or even over the field $\mathrm{Int}_{\boldsymbol{Q}}(\boldsymbol{C})$, by Theorem 13.26. For $\alpha \in \boldsymbol{C}$, define the ring homomorphism $\varphi_\alpha : \boldsymbol{Q}[x] \to \boldsymbol{C}$ by $\varphi_\alpha(f(x)) = f(\alpha)$. Then $ker\, \varphi_\alpha = (m_\alpha(x)) \lhd \boldsymbol{Q}[x]$ for $m_\alpha(x)$ the monic of least degree in $ker\, \varphi_\alpha$, or else $ker\, \varphi_\alpha = (0)$ (see Theorem 11.3 and verify!). In the first case, $\alpha \in \mathrm{Int}_{\boldsymbol{Q}}(\boldsymbol{C})$, and in the second case, $\alpha \notin \mathrm{Int}_{\boldsymbol{Q}}(\boldsymbol{C})$ and is called *transcendental* over $\boldsymbol{Q}$. Any $\beta \in \boldsymbol{C}$ transcendental over $\boldsymbol{Q}$ behaves like an indeterminate over $\boldsymbol{Q}$ since using φ_β shows that $\boldsymbol{Q}[x] \cong \varphi_\beta(\boldsymbol{Q}[x]) = \boldsymbol{Q}[\beta] \subseteq \boldsymbol{C}$. Whenever $F \subseteq L$ are fields there is a maximal subfield $F \subseteq A \subseteq L$ with A algebraic over F, namely $A = \mathrm{Int}_F(L)$. This A is called the *algebraic closure* of F in L. Once again Theorem 13.26 shows $\beta \in L - A$ is not algebraic over F or over A, and β is called transcendental over F.

We show next that $\mathrm{Int}_{\boldsymbol{Q}}(\boldsymbol{C})$ is not finitely generated over $\boldsymbol{Q}$.

Example 13.18 $\mathrm{Int}_{\boldsymbol{Q}}(\boldsymbol{C})$ is not finitely generated over $\boldsymbol{Q}$.

Assume $\mathrm{Int}_{\boldsymbol{Q}}(\boldsymbol{C}) = \boldsymbol{Q}\alpha_1 + \cdots + \boldsymbol{Q}\alpha_k$ and let $\beta \in \mathrm{Int}_{\boldsymbol{Q}}(\boldsymbol{C})$. The last part of the proof of Theorem 13.24 shows $f(\beta) = 0$ for some $f(x) \in \boldsymbol{Q}[x]$ monic with $deg\, f \leq k$. There is a prime $p > k$, $2^{1/p}$ satisfies $x^p - 2 \in \boldsymbol{Q}[x]$ so $2^{1/p} \in \mathrm{Int}_{\boldsymbol{Q}}(\boldsymbol{C})$, and $x^p - 2$ is irreducible by Eisenstein's criterion (Theorem 12.16). Thus if $g(2^{1/p}) = 0$ for $g(x) \in \boldsymbol{Q}[x]^*$ then $deg\, g \geq p$, so $2^{1/p} \notin \mathrm{Int}_{\boldsymbol{Q}}(\boldsymbol{C})$. This contradiction shows that R cannot be a finitely generated $\boldsymbol{Q}$ module.

We remark that a module over a field is called a *vector space* and is *finite dimensional* when it is finitely generated. We explore extensions of fields a bit further. For $M \lhd F[x_1, \ldots, x_n]$ and maximal, $L = F[x_1, \ldots, x_n]/M$ is a field by Theorem 11.17 and if $\rho : F[x_1, \ldots, x_n] \to L$ is the usual surjection then $F \cong \rho(F) = \{a + M \mid a \in F\}$, and $L = \rho(F)[\rho(x_1), \ldots, \rho(x_n)]$. When $n = 1$, $M = (p(x))$ with $p(x)$ irreducible, and Theorem 11.18 shows that $\rho(x)$ is algebraic over $\rho(F)$ so L is algebraic over $\rho(F)$ by Theorem 13.26. In general, if for each x_i some $g_i(x_i) \in M$ then $\rho(x_i)$ is algebraic over $\rho(F)$ (why?) and then L is algebraic over $\rho(F)$, again by Theorem 13.26. Is L always algebraic over $\rho(F)$, or, equivalently, must M always contain a polynomial in each indeterminate? If $H \lhd \mathbf{Z}[x_1, \ldots, x_n]$ is maximal then the quotient K is a field generated by the images of $\mathbf{Z}$ and the x_j. Must H contain some $h_i(x_i)$ for each x_i? Is K algebraic (over what is not clear)?

We prove a result of Zariski that leads to answers of these questions. The theorems that follow are crucial for the material in the next section, which introduces basic ideas from algebraic geometry. For the proof of Zariski's Theorem we need to know that if F is a field then $F[x]$ contains infinitely many nonassociate irreducibles. This is easy when F is infinite because $x - a$ is irreducible for any $a \in F$. We give an argument independent of the cardinality of F. To be clear, in the next result, F is any field.

THEOREM 13.30 $F[x]$ contains infinitely many nonassociate irreducibles.

Proof We claim that if $2 \leq n < m$ with $(n, m) = 1$ then $\mathrm{GCD}\{x^n - 1, x^m - 1\} = x - 1$. This implies that $\{f(x) \in F[x] \mid f(x)$ irreducible and $f(x) \mid (x^p - 1)$, some prime $p\}$ contains infinitely many nonassociates. Suppose $h(x) \in F[x]$ is irreducible, $h(x) \notin F \cdot (x - 1)$, and $h(x) \mid \mathrm{GCD}\{x^n - 1, x^m - 1\}$ for $2 \leq n < m$, $(n, m) = 1$, and $n + m$ minimal with these properties (Well Ordering!). Since $F[x]$ is a UFD by Theorem 11.3 and Theorem 12.8, $x^n - 1 = h(x)(x - 1)p(x)$, $x^m - 1 = h(x)(x - 1)q(x)$, and subtracting yields $(x^m - 1) - (x^n - 1) = x^n(x^{m-n} - 1) = h(x)(x - 1)(q(x) - p(x))$. Since $h(x) \mid (x^m - 1)$, $h(x) \notin F \cdot x$ so we must have $h(x) \mid (x^{m-n} - 1)$ (why?); thus $m - n \geq 2$. This shows that $h(x) \mid \mathrm{GCD}\{x^{m-n} - 1, x^n - 1\}$. Clearly $(m - n, n) = 1$ since $(m, n) = 1$ so $n + (m - n) < n + m$ contradicts the minimality of $n + m$. Therefore, when $2 \leq n < m$ and $(n, m) = 1$, $\mathrm{GCD}\{x^n - 1, x^m - 1\} = x - 1$ as claimed. ■

THEOREM 13.31 (Zariski) If $F \subseteq L$ are fields and $L = F[a_1, \ldots, a_n]$ for some $a_1, \ldots, a_n \in L$, then L is algebraic over F.

Proof If each a_j is algebraic over F then L is also by Theorem 13.26. We may as well assume that a_1 is not algebraic over F. As we have seen, this implies that $F[x] \cong F[a_1]$ by the map $\varphi(f(x)) = f(a_1)$. Extending φ to $QF(F[x]) = F(x)$ via $\varphi(f(x)/g(x)) = f(a_1)g(a_1)^{-1} \in L$ shows (verify!) that $\varphi(F(x)) \cong F(a_1) = \{st^{-1} \mid s, t \in F[a_1] \text{ with } t \neq 0\}$ is a subfield of L containing F. Now $L = F(a_1)[a_2, \ldots, a_n]$ so by induction on n, L is algebraic over $F(a_1)$. Hence for each a_j with $j \geq 2$, $a_j^{m(j)} + s_{m(j)-1}t(j)^{-1}a_j^{m(j)-1} + \cdots + s_0t(j)^{-1} = 0$, where $t(j)$ and all s_i are in $F[a_1]$ and $t(j)$ is a common denominator for the coefficients in $F(a_1)$. Multiply by $t(j)^{m(j)}$ to see that $t(j)a_j$ is integral over $F[a_1]$. Then $t(2) \cdots t(n)a_j$ is integral over $F[a_1]$ by Theorem 13.26 and the fact that $F[a_1]$ is integral over itself. Therefore for $p(a_1) = t(2) \cdots t(n)$, all $p(a_1)a_j$ are integral over $F[a_1]$. Any $\alpha \in L$ is a polynomial expression in the $\{a_j\}$ over F, and if the largest degree of any monomial in α is k, then $p(a_1)^k\alpha$ is a sum of terms, each a product of some of the $p(a_1)a_j$ and $h(a_1) \in F[a_1]$. Consequently, $p(a_1)^k\alpha$ is integral over $F[a_1]$, again by Theorem 13.26. In particular, for any $g(a_1) \in F[a_1]^*$ there is some $k \geq 0$ with $p(a_1)^k(g(a_1))^{-1}$ integral over $F[a_1]$. Using that $F[x] \cong F[a_1]$ is integrally closed by Theorem 13.27, we have $p(a_1)^k(g(a_1))^{-1} \in F[a_1]$ so $p(a_1)^k = g(a_1)f(a_1)$ for some $f(a_1) \in F[a_1]$. From Theorem 12.7, any irreducible $g(a_1) \in F[a_1]$ is a prime, and as we have just seen, it must divide $p(a_1)$. Unique factorization in $F[a_1] \cong F[x]$ shows that $p(a_1)$ has finitely many nonassociate irreducible divisors, but $F[a_1]$ contains infinitely many nonassociate irreducibles $g(a_1)$ by Theorem 13.30. This contradiction forces all a_j to be algebraic over F, so L is algebraic over F. ■

The answer to our questions above about maximal ideals in polynomial rings over fields and about $\boldsymbol{Z}[x_1, \ldots, x_n]/M$ follow from Zariski's result. We will see shortly afterwards that a precise description of maximal ideals in $F[x_1, \ldots, x_n]$ exists for some special fields, and in particular for maximal ideals in $\boldsymbol{C}[x_1, \ldots, x_n]$.

THEOREM 13.32 If F is a field and $M \lhd R = F[x_1, \ldots, x_n]$ is maximal, then R/M is an algebraic extension of $\{a+M \mid a \in F\}$ and for each x_i there is some irreducible $g_i(x_i) \in M$.

Proof Let $\rho : R \to R/M$ be the usual surjective homomorphism $\rho(g) = g + M$, so $F \cong \rho(F) = \{a + M \mid a \in F\}$. If $y_i = x_i + M$ then $R/M = \rho(F)[y_1, \ldots, y_n]$. Now R/M is a field by Theorem 11.17, so Theorem 13.31 shows that R/M is algebraic over $\rho(F)$. In particular, each y_j satisfies some polynomial over $\rho(F)$, which means that $h_j(x_j) \in M$ for some $h_j(x) \in F[x]^*$. Since R/M is a field, and hence a domain, M is a prime ideal. It follows that some irreducible factor of $h_j(x_j)$ must be in M as well. ■

Not every choice of an irreducible for each indeterminate defines a maximal ideal in $F[x_1, \ldots, x_n]$. If $g(x_1, x_2) \in M = (x_1^2 + 1, x_2^2 + 1) \lhd \boldsymbol{Q}[x_1, x_2]$ then $g(\pm i, \pm i) = 0$ in $\boldsymbol{C}$, so M contains no polynomial of degree 1. Since $x_1^2 - x_2^2 = (x_1 - x_2)(x_1 + x_2) \in M$, M cannot be prime and so is not maximal. We can describe the maximal $M \lhd F[x_1, \ldots, x_n]$ explicitly only for certain fields.

DEFINITION 13.14 A field F is **algebraically closed** if the irreducibles in $F[x]$ all have degree 1.

Equivalently, F is algebraically closed if it has no algebraic extension other than itself. This is true for the complex numbers $\boldsymbol{C}$ and is known as the *Fundamental Theorem of Algebra*. We will assume this hard-to-prove fact. It is also true, but difficult to prove, that for any field F there is an algebraic extension A of F with A algebraically closed (see Theorem 14.19). We can describe the maximal ideals in $F[x_1, \ldots, x_n]$ when F is algebraically closed.

THEOREM 13.33 If F is an algebraically closed field, then $M \lhd F[x_1, \ldots, x_n]$ is maximal exactly when $M = (x_1 - c_1, \ldots, x_n - c_n)$ for some $c_j \in F$. This association is a bijection between F^n and $\{M \lhd F[x_1, \ldots, x_n] \mid M \text{ is maximal}\}$.

Proof If $M \lhd F[x_1, \ldots, x_n] = R$ is maximal then Theorem 13.32 shows that for each x_i there is an irreducible $g_i(x_i) \in M$. Since F is algebraically closed we can assume $g_i(x_i) = x_i - c_i$, so the ideal $I = (x_1 - c_1, \ldots, x_n - c_n) \subseteq M$. In R/I, $f(x_1, \ldots, x_n) + I = f(c_1, \ldots, c_n) + I$ (verify!) so for $f \in M$, $f + I = a + I$ for some $a \in F$ and so $a \in f + I \subseteq M \cap F$. Therefore $a = 0$ and $f \in I$, proving that $M = I = (x_1 - c_1, \ldots, x_n - c_n)$. If some $A = (x_1 - a_1, \ldots, x_n - a_n)$ is not maximal then by Theorem 13.2 some maximal $M = (x_1 - c_1, \ldots, x_n - c_n) \supseteq A$. In this case $c_i - a_i = (x_i - a_i) - (x_i - c_i) \in M \cap F = (0)$, so in fact $A = M$ is maximal. This same computation shows that the association of $(c_1, \ldots, c_n) \in F^n$ with $M = (x_1 - c_1, \ldots, x_n - c_n)$ is a bijection. ■

Associating points in $\boldsymbol{C}^n$ with ideals in $\boldsymbol{C}[x_1, \ldots, x_n]$ provides a way to the study of the geometry of $\boldsymbol{C}^n$ by using the ring structure of $\boldsymbol{C}[x_1, \ldots, x_n]$. The beginnings of this are presented in the next section. Our final theorem here shows that maximal ideals in $\mathbf{Z}[x_1, \ldots, x_n]$ are large.

THEOREM 13.34 If $M \lhd \mathbf{Z}[x_1, \ldots, x_n] = R$ is maximal, R/M is a finite field.

Proof As we have seen, $F = R/M$ is a field so $\rho(\mathbf{Z}) = (\mathbf{Z} + M)/M$ is a domain, where $\rho(r) = r + M$ is the usual surjection. Now $\rho(\mathbf{Z}) \cong \mathbf{Z}/(\mathbf{Z} \cap M)$ from Theorem 11.12, so $\mathbf{Z} \cap M = p\mathbf{Z}$ for a prime p, or $\mathbf{Z} \cap M = (0)$. In the first case $\rho(\mathbf{Z}) \cong \mathbf{Z}_p$ and $F = \rho(\mathbf{Z})[\rho(x_1), \ldots, \rho(x_n)]$ is algebraic over $\rho(\mathbf{Z})$ by Theorem 13.31. Hence $\rho(\mathbf{Z})[\rho(x_j)]$ is a finite ring by Theorem 13.24, and then $F = \rho(\mathbf{Z})[\rho(x_1)] \cdots \rho(\mathbf{Z})[\rho(x_n)]$ is finite also using Theorem 13.25. When $\mathbf{Z} \cap M = (0)$ then $\rho(\mathbf{Z}) \cong \mathbf{Z}$ and the inverses of its nonzero elements are in F so $H = \{\rho(a)\rho(b)^{-1} \mid a, b \in \mathbf{Z}, b \neq 0\} \cong \mathbf{Q}$ is a subfield of F, and $F = H[\rho(x_1), \ldots, \rho(x_n)]$ is algebraic over H using Theorem 13.31 again. The proof of that theorem shows that each $\rho(x_j)$ satisfies $\rho(x_j)^{m(j)} + \rho(a_{m(j)-1})\rho(b_j)^{-1}\rho(x_j)^{m(j)-1} + \cdots + \rho(a_0)\rho(b_j)^{-1} = 0$ and $\rho(b_j)\rho(x_j)$ is integral over $\rho(\mathbf{Z})$. Let $\rho(b) = \rho(b_1) \cdots \rho(b_n)$. Since any $\alpha \in F$ is a polynomial in $\{\rho(x_j)\}$, for a large enough $k \geq 0$, $\rho(b)^k\alpha$ is a polynomial in the $\rho(b)\rho(x_j)$ and $\rho(b)$ so is integral over $\rho(\mathbf{Z})$ by Theorem 13.26. In particular, for $q \in \mathbf{N}$ a prime, some $\rho(b)^k\rho(q)^{-1}$ is integral over $\rho(\mathbf{Z}) \cong \mathbf{Z}$, which is integrally closed by Theorem 13.27. Hence $\rho(b)^k = \rho(q)\rho(m)$ for $m \in \mathbf{Z}$. Since ρ is an isomorphism from $\mathbf{Z}$ to $\rho(\mathbf{Z})$, $\rho(q)$ is prime in $\rho(\mathbf{Z})$ and must divide $\rho(b)$, for every prime $q \in \mathbf{N}$. This is impossible by unique factorization, so $\mathbf{Z} \cap M \neq (0)$ and, as we saw above, F is a finite field. ■

The last proof shows that any maximal $M \lhd \mathbf{Z}[x_1, \ldots, x_n]$ must contain a prime $p \in \mathbf{N}$ and for each x_i some $\sum_j a_{ij}x_i^j \in M$ with $\sum_j [a_{ij}]_p x^j \in \mathbf{Z}_p[x]$ irreducible.

EXERCISES

1. Let $R \subseteq T \subseteq S$ be commutative rings with the same 1 and ${}_RT$ finitely generated. Show that:
   i) $T \subseteq \text{Int}_R(S)$
   ii) if $S = \text{Int}_T(S)$ then $S = \text{Int}_R(S)$
2. If $\alpha \in \mathbf{C}$ with $\alpha^k \in \mathbf{Q}$, show that $\mathbf{Q}[\alpha]$ is algebraic over $\mathbf{Q}$.
3. Argue that $2^{1/2} + 3^{1/3} \cdot 5^{1/5} - 7^{1/7} \cdot 11^{1/11} \cdot 13^{1/13} \in \mathbf{R}$ satisfies a monic $p(x) \in \mathbf{Z}[x]$.
4. Let $\alpha, \beta \in \mathbf{C}$ satisfy $\alpha^6 - 6\alpha^4 + 4\alpha^2 + 2\alpha - 18 = 0$ and $\beta^8 - 9\beta^4 - 3 = 0$. Show that $\alpha^5 + \alpha\beta + \beta^3$ satisfies a monic $p(x) \in \mathbf{Z}[x]$.
5. For $\alpha, \beta \in \mathbf{C}$, decide whether α is integral over $\mathbf{Z} \Leftrightarrow \beta$ is integral over $\mathbf{Z}$ when:
   i) $\alpha + \beta = 17$
   ii) $\alpha\beta = 1$
   iii) $\alpha/\beta = 4$
   iv) $\alpha^4 + \beta = 2$
   v) $\alpha^2 + \beta^2 = 1$
6. If $R \subseteq S$ are commutative rings with the same 1, then if $s \in \text{Int}_R(S)$ and $s^{1/k} \in S$ for some $k \in \mathbf{N}$ (there is $t \in S$ with $t^k = s$) then $s + s^{1/k} \in \text{Int}_R(S)$.
7. If $R, S \subseteq D$ are integral domains with the same 1, then show that R and S integrally closed imply that $R \cap S$ is integrally closed.

8. Let R be a UFD, $QF(R) = K \subseteq L$ a field, $\alpha \in \text{Int}_R(L)$ and $m(x) \in K[x]$ the monic of least degree with $m(\alpha) = 0$. Show that $m(x) \in R[x]$ (Example 13.16!).

9. Show the following:

i) $\text{Int}_{\boldsymbol{Z}}(\boldsymbol{Q}(i)) = \boldsymbol{Z}[i]$

ii) $\text{Int}_{\boldsymbol{Z}}(\boldsymbol{Q}(3^{1/2})) = \boldsymbol{Z}[3^{1/2}]$

iii) $\text{Int}_{\boldsymbol{Z}}(\boldsymbol{Q}((-3)^{1/2})) = \{a/2 + b/2 \cdot (-3)^{1/2} \in \boldsymbol{C} \mid a, b \in \boldsymbol{Z} \text{ and } a \equiv_2 b\}$

10. If R is an integrally closed domain and $M \subseteq R$ is a multiplicative set, show that R_M is integrally closed.

11. If R is an integral domain, $K = QF(R)$, $K \subseteq L$ for L a field, and L is a finitely generated K-module, show that, $L = K\alpha_1 + \cdots + K\alpha_m$ for all $\alpha_j \in \text{Int}_R(L)$.

12. If $R \subseteq S$ are integral domains with the same 1, show that there can be prime ideals $W \subseteq V$ in S with $W \bigcap R = V \bigcap R$ but $W \neq V$. (Hint: use $\boldsymbol{Z} \subseteq \boldsymbol{Z}[x]$.)

13. Find $Q_1, Q_2 \lhd \boldsymbol{Z}[i]$, prime and incomparable, with $Q_1 \bigcap \boldsymbol{Z} = Q_2 \bigcap \boldsymbol{Z}$.

14. For S an integral extension of R, show that:

i) $S[x]$ is integral over $R[x]$

ii) for $M \subseteq R$ multiplicative, show that S_M is integral over R_M.

15. If S is integral over R, $P \lhd R$, $Q \lhd S$, and $Q \bigcap R = P$, show that:

i) P is maximal $\Leftrightarrow Q$ is

ii) When S is a domain then P is a minimal nonzero prime $\Rightarrow Q$ is also

16. For S an integral extension of R, show that all nonzero prime ideals in R are maximal if and only if all nonzero prime ideals in S are maximal.

17. For $c_j \in \boldsymbol{C}$, let $M = (x_1 - c_1, \ldots, x_n - c_n) \lhd \boldsymbol{C}[x_1, \ldots, x_n]$. Show that $f(x_1, \ldots, x_n) \in M$ if and only if $f(c_1, \ldots, c_n) = 0$.

18. If $I \lhd \boldsymbol{C}[x_1, \ldots, x_n]$ is proper, show that there is $\alpha \in \boldsymbol{C}^n$ with $f(\alpha) = 0$ for all $f \in I$.

19. a) If $M \lhd \boldsymbol{Q}[x_1, \ldots, x_n]$ is maximal, show that $S = \{H \lhd \boldsymbol{C}[x_1, \ldots, x_n] \mid H$ is maximal and $M \subseteq H\}$ is finite. (See Exercise 18.)

b) If $P \lhd \boldsymbol{C}[x_1, \ldots, x_n]$ is prime and $M \subseteq P$, show that $P \in S$.

20. If $M \lhd \boldsymbol{Z}[x]$ is maximal, show that $M = (p, g(x))$ for $p \in \boldsymbol{N}$ a prime and $\rho(g(x))$ irreducible in $\rho(\boldsymbol{Z})[x]$, where $\rho : \boldsymbol{Z}[x] \to \boldsymbol{Z}[x]/M$ is $\rho(r) = r + M$.

21. If F is any field, $n > 1$, and $g \in F[x_1, \ldots, x_n]$, show that (g) cannot be maximal.

13.5 ALGEBRAIC GEOMETRY

Algebraic geometry is the study of the relation between subsets of F^n and ideals in $F[x_1, \ldots, x_n]$ for F a field. The goal is to use both algebraic and geometric results on problems of either type. Here we introduce the basic ideas and develop a few simple consequences of them. The general idea relates $I \lhd F[x_1, \ldots, x_n]$ with the zeros in F^n common to all $f \in I$. Note that $(c_1, \ldots, c_n) \in \boldsymbol{C}^n$ is the only common zero of all $f \in M = (x_1 - c_1, \ldots, x_n - c_n)$. Now $x^2 + 1, x^2y^2 - 2 \in \boldsymbol{Q}[x, y]$ have no roots at all in $\boldsymbol{Q}$. Thus to find roots, we need to consider extensions of F rather than just F itself. The usual context for doing this is to take fields $F \subseteq K$ with K algebraically closed (Definition 13.14) and then to consider roots in K^n of $p \in F[x_1, \ldots, x_n]$. We will use $\boldsymbol{C}$, the complex numbers, as the algebraically closed field for our examples. Thus there is no harm in replacing the general algebraically closed field K with $\boldsymbol{C}$ in the definitions and statements of results that follow.

Analytic geometry shows that the set of zeros of a polynomial may be interesting. If $a^2 + b^2 \neq 0$ then the set of zeros in $\boldsymbol{R}^2$ of $ax + by + c \in \boldsymbol{R}[x, y]$ is a straight line, and each conic section is the zero set of a quadratic in $\boldsymbol{R}[x, y]$. The set of zeros in $\boldsymbol{R}^3$: of $ax + by + cz + d \in \boldsymbol{R}[x, y, z]$ is a plane, of xyz is the three coordinate planes, and of $x^2 + y^2 + z^2 - 1$ is the unit sphere.

Throughout this section, K will be an algebraically closed field (that can be taken to be $\boldsymbol{C}$), and $F \subseteq K$ is some subfield.

DEFINITION 13.15 For $I \lhd R = F[x_1, \ldots, x_n]$, $\mathcal{V}(I) = \{\alpha \in K^n \mid f(\alpha) = 0 \text{ for all } f \in I\}$ is the **variety** of I. For $S \subseteq K^n$, $\mathcal{I}(S) = \{g \in R \mid g(s) = 0 \text{ for all } s \in S\}$ is the **ideal** of S.

It is an exercise that $\mathcal{I}(S)$ *is* an ideal of $R = F[x_1, \ldots, x_n]$. Note that $\mathcal{V}(R) = \varnothing$ since $1 \in R$, $\mathcal{V}((0)) = K^n$, and $\mathcal{I}(\varnothing) = R$ since for $f \in R$, $f(\alpha) = 0$ *whenever* $\alpha \in \varnothing$! Two other useful facts require some justification.

Example 13.19 $f \in F[x_1, \ldots, x_n]^* \Rightarrow K^n - \mathcal{V}((f))$ is infinite, so $\mathcal{I}(K^n) = (0)$.

By Theorem 13.30, $F[x]$ has infinitely many nonassociate irreducibles, and if $p, q \in F[x]$ are two such, then $fp + gq = 1$ for some $f, g \in F[x]$, since $F[x]$ is a PID. Both p and q have linear factors in $K[x]$ and so have roots in K. Since $p, q \in K[x]$ can have no common zero, K is infinite. If $f = f(x_1)$ then by Theorem 4.18, $\{a \in K \mid f(a) = 0\}$ is finite. Thus $f(c) \neq 0$ for infinitely many $c \in K$ so $f(\gamma) \neq 0$ for infinitely many $\gamma \in K^n$. Now write $f = p_0 + p_1 x_n + \cdots + p_m x_n^m$ with all $p_j \in F[x_1, \ldots, x_{n-1}]$. Some $p_i \neq 0$ since $f \neq 0$ so by induction on n, $p_i(\gamma) \neq 0$ for infinitely many $\gamma \in K^{n-1}$. Hence $g_\gamma(x_n) = \sum p_j(\gamma) x_n^j \neq 0$ so $g_\gamma(c) \neq 0$ for infinitely many $c \in K$, and it follows that $f((\gamma, c)) \neq 0$ for all of these. Thus $K^n - \mathcal{V}((f))$ is infinite by induction, and $\mathcal{I}(K^n) = (0)$ results.

Example 13.20 If $I \lhd F[x_1, \ldots, x_n]$ is proper then $\mathcal{V}(I) \neq \varnothing$.

It is an exercise that $KI = \left\{\sum k_j g_j \mid k_j \in K, g_j \in I\right\} \lhd K[x_1, \ldots, x_n]$. If KI is proper it is contained in a maximal ideal M by Theorem 13.2, and Theorem 13.33 shows that $M = \sum (x_i - c_i)$. Thus $(c_1, \ldots, c_n) \in \mathcal{V}(I) \neq \varnothing$ since $I \subseteq M$. If $KI = K[x_1, \ldots, x_n]$ then $1 = a_1 g_1 + \cdots + a_t g_t$ for $a_j \in K$ and $g_j \in I$. Multiplying this out gives a polynomial whose coefficients are linear expressions in the coefficients of the $\{g_j\}$: *e.g.* $a(b_0 + b_1 x + b_2 xy + b_3 y^2) + c(d_0 + d_2 xy + d_3 y^2) = (ab_0 + cd_0) + ab_1 x + (ab_2 + cd_2)xy + (ab_3 + cd_3)y^2$. Thus $(a_1, \ldots, a_t) \in K^t$ is a solution of this consistent system of linear equations over F in t unknowns, so some solution $(b_1, \ldots, b_t)$, say determined by row reducing the coefficient matrix, is in F^t. Therefore $1 = b_1 g_1 + \cdots + b_t g_t \in I$, contradicting I a proper ideal. Consequently, $KI \lhd K[x_1, \ldots, x_n]$ is proper.

These examples show that every proper $I \lhd F[x_1, \ldots, x_n]$ does have some common zero—that is, $\mathcal{V}(I) \neq \varnothing$—but $K^n - \mathcal{V}(I)$ is infinite, unless $I = (0)$. We see next how $I \to \mathcal{V}(I)$ and $S \to \mathcal{I}(S)$ behave with respect to containments.

THEOREM 13.35 If $I, J \lhd R = F[x_1, \ldots, x_n]$ and $S, T \subseteq K^n$, then:

i) $I \subseteq J \Rightarrow \mathcal{V}(I) \supseteq \mathcal{V}(J)$
ii) $S \subseteq T \Rightarrow \mathcal{I}(S) \supseteq \mathcal{I}(T)$
iii) $\mathcal{I}(\mathcal{V}(I)) \supseteq I$
iv) $\mathcal{V}(\mathcal{I}(S)) \supseteq S$
v) $\mathcal{I}(\mathcal{V}(I)) = I \Leftrightarrow I = \mathcal{I}(S)$ for some S
vi) $\mathcal{V}(\mathcal{I}(S)) = S \Leftrightarrow S = \mathcal{V}(I)$ for some I
vii) $I \to \mathcal{V}(I)$ and $S \to \mathcal{I}(S)$ are inverses, and so bijections, between $\{\mathcal{V}(I) \subseteq K^n \mid I \lhd R\}$ and $\{\mathcal{I}(S) \lhd R \mid S \subseteq K^n\}$

Proof Parts i) and ii) follow from Definition 13.15. If $f \in I$ then $f(\alpha) = 0$ for all $\alpha \in \mathcal{V}(I)$, so $f \in \mathcal{I}(\mathcal{V}(I))$, proving iii). Similarly, if $\alpha \in S$ and $f \in \mathcal{I}(S)$ then $f(\alpha) = 0$ so $\alpha \in \mathcal{V}(\mathcal{I}(S))$ and iv) holds. When $I = \mathcal{I}(S)$ then $\mathcal{V}(I) = \mathcal{V}(\mathcal{I}(S)) \supseteq S$ by iv), and now ii) shows that $\mathcal{I}(\mathcal{V}(I)) \subseteq \mathcal{I}(S) = I$. That $I \subseteq \mathcal{I}(\mathcal{V}(I))$ is just iii) so $I = \mathcal{I}(S) \Rightarrow I = \mathcal{I}(\mathcal{V}(I))$. If $I = \mathcal{I}(\mathcal{V}(I))$ then take $S = \mathcal{V}(I)$ to finish the proof of v). The argument for vi) is similar and is omitted. The statements of v) and vi) imply that $I \to \mathcal{V}(I)$ and $S \to \mathcal{I}(S)$ are inverses between the set of varieties and the set of all $\mathcal{I}(S)$ for $S \subseteq K^n$, proving vii). ■

From the theorem, $I \subseteq \mathcal{I}(\mathcal{V}(I))$ for any $I \lhd F[x_1, \ldots, x_n]$ and equality holds when $I = \mathcal{I}(S)$. It is not often easy to determine whether I is the ideal of some $S \subseteq K^n$. One approach is to characterize $\mathcal{I}(\mathcal{V}(I))$ algebraically in terms of I and use Theorem 13.35 v). This is the important *Hilbert Nullstellensatz*. Recall from Definition 13.9 that if R is any commutative ring and $I \lhd R$ then $\mathrm{rad}(I) = \{x \in R \mid x^k \in I \text{ for some } k > 0\}$. We observed in §13.3 that $N(R/I) = \mathrm{rad}(I)/I$, and it is clear that when I is a prime ideal, $I = \mathrm{rad}(I)$.

THEOREM 13.36 (Hilbert) If $I \lhd R = F[x_1, \ldots, x_n]$, then $\mathcal{I}(\mathcal{V}(I)) = \mathrm{rad}(I)$.

Proof Let $f \in R$, $f^k \in I$, and $\alpha \in \mathcal{V}(I)$. Then $0 = f^k(\alpha) = f(\alpha)^k$ so $f(\alpha) = 0$ and $f \in \mathcal{I}(\mathcal{V}(I))$, proving $\mathrm{rad}(I) \subseteq \mathcal{I}(\mathcal{V}(I))$. We need $\mathcal{I}(\mathcal{V}(I)) \subseteq \mathrm{rad}(I)$. If $I = (0)$ then R a domain implies $\mathrm{rad}(I) = (0) = \mathcal{I}(K^n) = \mathcal{I}(\mathcal{V}(I))$, by Example 13.19. When $I \neq (0)$, $\mathcal{I}(\mathcal{V}(I)) \neq (0)$ by Theorem 13.35. Let $f \in \mathcal{I}(\mathcal{V}(I))^*$. The Hilbert Basis Theorem shows R is Noetherian so from Theorem 13.17 $I = (f_1, \ldots, f_s)$. Set $H = (f_1, \ldots, f_s, 1 - x_{n+1}f) \lhd F[x_1, \ldots, x_n, x_{n+1}] = R[x_{n+1}]$. When $H \neq R[x_{n+1}]$, apply Example 13.20: there is $\beta = (a_1, \ldots, a_{n+1}) \in \mathcal{V}(H)$, so that for all j, $f_j((a_1, \ldots, a_n)) = 0$. Thus $\alpha = (a_1, \ldots, a_n) \in \mathcal{V}(I)$. Since $f \in \mathcal{I}(\mathcal{V}(I))$, it follows that $f(\alpha) = 0$ also. Using $\beta \in \mathcal{V}(H)$ we have $0 = 1 - a_{n+1}f(\alpha) = 1$, a contradiction. Consequently, $H = R[x_{n+1}]$ and $1 = g_1f_1 + \cdots + g_sf_s + g \cdot (1 - x_{n+1}f)$ where $g_j, g \in R[x_{n+1}]$. Define $\varphi : R[x_{n+1}] \to QF(R)$ by $\varphi(a) = a$ for $a \in F$, $\varphi(x_j) = x_j$ if $j \leq n$, and $\varphi(x_{n+1}) = 1/f$. Now φ is a ring homomorphism (verify!) so $1 = g_1(x_1, \ldots, x_n, 1/f)f_1 + \cdots + g_s(x_1, \ldots, x_n, 1/f)f_s$, since $\varphi(1 - x_{n+1}f) = 0$. If $m > 0$ is large, all $f^m g_i(x_1, \ldots, x_n, 1/f) \in R$, forcing $f^m = \sum f^m g_i(x_1, \ldots, x_n, 1/f)f_i \in I$. This proves $f \in \mathrm{rad}(I)$ and so $\mathcal{I}(\mathcal{V}(I)) \subseteq \mathrm{rad}(I)$ as required. ■

THEOREM 13.37 Let $I \lhd R = F[x_1, \ldots, x_n]$ be proper. Then $\mathcal{I}(\mathcal{V}(I)) = R \bigcap (\bigcap \{M \lhd K[x_1, \ldots, x_n] \mid I \subseteq M \text{ and } M \text{ is maximal}\})$, and if I is a prime ideal, $I = \mathcal{I}(\mathcal{V}(I))$.

Proof Let $S = \{M \lhd K[x_1, \ldots, x_n] \mid M \text{ is maximal and } I \subseteq M\}$. Now $M \in S$ implies $M = \sum (x_i - a_i)$ with $a_j \in K$ by Theorem 13.33 so $(c_1, \ldots, c_n) \in \mathcal{V}(I) \Leftrightarrow \sum (x_i - c_i) \in S$ using

Exercise 17 in §13.4. Hence for $f \in R$, $f \in \bigcap\{M \in S\} \Leftrightarrow f((c_1, \ldots, c_n)) = 0$ for all $M = (x_1 - c_1, \ldots, x_n - c_n) \in S \Leftrightarrow f \in \mathcal{I}(\mathcal{V}(I))$. Thus $\mathcal{I}(\mathcal{V}(I)) = R \bigcap (\bigcap \{M \in S\})$, proving the first statement. For the second we merely observe that when I is prime, $I = \text{rad}(I) = \mathcal{I}(\mathcal{V}(I))$ by Theorem 13.36. ■

Example 13.21 If $P_1, \ldots, P_k \lhd F[x_1, \ldots, x_n]$ are prime then $\mathcal{I}(\mathcal{V}(\bigcap P_j)) = \bigcap P_j$.

If $P_1, \ldots, P_n \lhd F[x_1, \ldots, x_n] = R$ are prime, $g \in R$, and $g^k \in \bigcap P_j$, then $g^k \in P_j$ for each j, so P_j prime forces $g \in P_j$. Thus $\text{rad}(\bigcap P_j) = \bigcap P_j$, and Theorem 13.36 shows that $\bigcap P_j = \mathcal{I}(\mathcal{V}(\bigcap P_j))$.

Example 13.22 $I = (x^2 + 1, xy - 1) \lhd \mathbf{C}[x, y]$.

If $(a, b) \in \mathcal{V}(I)$ then $a^2 + 1 = 0$ so $a = \pm i$ and $ab = 1$ forces $b = -a$. Thus $\mathcal{V}(I) = \{(i, -i), (-i, i)\}$ and $\mathcal{I}(\mathcal{V}(I)) = (x - i, y + i) \bigcap (x + i, y - i)$ by Theorem 13.37. We claim that this intersection is I. Since $xy - 1 = x(y + i) - i(x - i)$ and $y + i = iy(x - i) - i(xy - 1)$, we have $(x - i, xy - 1) = (x - i, y + i)$, and similarly, $(x + i, xy - 1) = (x + i, y - i)$. Now $I = (x - i, xy - 1) \bigcap (x + i, xy - 1)$ (why?), so directly or by Example 13.21, $I = \mathcal{I}(\mathcal{V}(I)) = \text{rad}(I)$. If we consider $(x^2 + 1, xy - 1) \lhd F[x, y]$ for $F \subseteq \mathbf{C}$, then by Theorem 13.37, $\mathcal{I}(\mathcal{V}(I)) = F[x, y] \bigcap (x - i, y + i) \bigcap (x + i, y - i) = I$.

Example 13.23 $I = ((x^2 + y^2 + z^2 - 3)^2, (x^2 - y^2)^2, (y^2 - z^2)^3) \lhd \mathbf{C}[x, y, z]$.

For $(a, b, c) \in \mathcal{V}(I)$, $(a^2 - b^2)^2 = 0 = (b^2 - c^2)^3$ force $a^2 = b^2 = c^2$. Thus $3 = a^2 + b^2 + c^2 = 3a^2$ shows that $a, b, c \in \{\pm 1\}$ and by Theorem 13.37, $\mathcal{I}(\mathcal{V}(I))$ is the intersection of the eight maximal ideals $(x - c_1, y - c_2, z - c_3)$ with all $c_j \in \{\pm 1\}$. Here $I \neq \mathcal{I}(\mathcal{V}(I))$ since $x^2 - y^2, y^2 - z^2 \in \text{rad}(I) - I$. Note that by Theorem 13.36, $(xyz)^2 - 1 \in \mathcal{I}(\mathcal{V}(I)) = \text{rad}(I)$, so some $((xyz)^2 - 1)^k \in I$.

We record some other general properties of varieties and ideals.

THEOREM 13.38 Let $I_\lambda, I, J \lhd F[x_1, \ldots, x_n]$ and $\{S_\mu\}$ a collection of subsets of K^n. Then:

i) $\mathcal{I}(\bigcup S_\mu) = \bigcap \mathcal{I}(S_\mu)$
ii) $\mathcal{V}(\sum I_\lambda) = \bigcap \mathcal{V}(I_\lambda)$
iii) $\mathcal{V}(I \bigcap J) = \mathcal{V}(IJ) = \mathcal{V}(I) \bigcup \mathcal{V}(J)$
iv) $\{\mathcal{V}(I_\lambda)\}$ has a minimal element.

Proof The verifications for i) and ii) are straightforward from the definitions. For iii), $IJ \subseteq I \bigcap J \subseteq I, J$ yields $\mathcal{V}(I) \bigcup \mathcal{V}(J) \subseteq \mathcal{V}(I \bigcap J) \subseteq \mathcal{V}(IJ)$ using Theorem 13.35. Hence we need $\mathcal{V}(IJ) \subseteq \mathcal{V}(I) \bigcup \mathcal{V}(J)$. If $\alpha \in K^n$ but $\alpha \notin \mathcal{V}(I) \bigcup \mathcal{V}(J)$ there are $f \in I$ and $g \in J$ with $f(\alpha) \neq 0$ and $g(\alpha) \neq 0$. It follows that $fg(\alpha) = f(\alpha)g(\alpha) \neq 0$ and since $fg \in IJ$, $\alpha \notin \mathcal{V}(IJ)$. This shows that $\mathcal{V}(IJ) \subseteq \mathcal{V}(I) \bigcup \mathcal{V}(J)$ and finishes the proof of iii). Finally, $\{\mathcal{I}(\mathcal{V}(I_\lambda))\}$ has a maximal element by Theorem 13.17, using that $F[x_1, \ldots, x_n]$ is a Noetherian ring from Theorem 13.18. If $\mathcal{V}(I_\lambda) \supseteq \mathcal{V}(I_\nu)$ properly then Theorem 13.35 implies

$\mathcal{I}(\mathcal{V}(I_\lambda)) \subseteq \mathcal{I}(\mathcal{V}(I_\nu))$ properly. Thus the maximality of any $\mathcal{I}(\mathcal{V}(I_\gamma))$ implies the minimality of $\mathcal{V}(I_\gamma)$ (verify!). ■

The theorem shows that the set of varieties is closed under intersection and under finite union, and that a variety can be the union of other varieties. Since any variety $V = \mathcal{V}(I)$ for some ideal I, any set of varieties must have a minimal element by Theorem 13.38. These ideas lead to a nice represention for varieties in terms of special ones.

DEFINITION 13.16 A variety V is **irreducible** if whenever $V = V_1 \cup V_2$ for varieties V_1 and V_2, then $V_1 = V$ or $V_2 = V$.

THEOREM 13.39 A variety V is irreducible $\Leftrightarrow \mathcal{I}(V)$ is a prime ideal.

Proof If $\mathcal{I}(V)$ is prime and $V = V_1 \cup V_2$ for varieties V_1 and V_2, then from Theorem 13.38, $\mathcal{I}(V) = \mathcal{I}(V_1 \cup V_2) = \mathcal{I}(V_1) \cap \mathcal{I}(V_2) \supseteq \mathcal{I}(V_1)\mathcal{I}(V_2)$. Since $\mathcal{I}(V)$ is prime, either $\mathcal{I}(V) \supseteq \mathcal{I}(V_1)$ and Theorem 13.35 shows $V = \mathcal{V}(\mathcal{I}(V)) \subseteq \mathcal{V}(\mathcal{I}(V_1)) = V_1$, or $\mathcal{I}(V) \supseteq \mathcal{I}(V_2)$ and $V \subseteq V_2$. Therefore V is irreducible. Assume V is irreducible and let $f, g \in F[x_1, \ldots, x_n]$ with $fg \in \mathcal{I}(V)$. Set $I = \mathcal{I}(V) + (f)$ and $J = \mathcal{I}(V) + (g)$. If $\alpha \in V$ then $0 = fg(\alpha) = f(\alpha)g(\alpha)$, so $f(\alpha) = 0$ or $g(\alpha) = 0$ and $\alpha \in \mathcal{V}(I) \cup \mathcal{V}(J)$. Thus $V \subseteq \mathcal{V}(I) \cup \mathcal{V}(J)$, and from $I, J \supseteq \mathcal{I}(V)$ Theorem 13.35 implies that $\mathcal{V}(I) \cup \mathcal{V}(J) \subseteq \mathcal{V}(\mathcal{I}(V)) = V$. These containments yield $\mathcal{V}(I) \cup \mathcal{V}(J) = V$ and the irreducibility of V forces $V = \mathcal{V}(I)$ or $V = \mathcal{V}(J)$. If $V = \mathcal{V}(I)$ Theorem 13.35 gives $\mathcal{I}(V) = \mathcal{I}(\mathcal{V}(I)) \supseteq I$, resulting in $f \in \mathcal{I}(V)$. When $V = \mathcal{V}(J)$ then $g \in \mathcal{I}(V)$. Hence $\mathcal{I}(V)$ is prime. ■

THEOREM 13.40 Let $\emptyset \neq V = \mathcal{V}(I) \subseteq K^n$ for $I \lhd F[x_1, \ldots, x_n] = R$. Then $V = V_1 \cup \cdots \cup V_k$ for a unique minimal set $\{V_j\}$ of irreducible varieties. If $V_i = \mathcal{V}(P_i)$ then $P_i \lhd R$ is prime, $\mathcal{I}(V) = \bigcap P_i$, and $\{P_i\}$ are minimal in the set of primes containing I.

Proof If V is not a union of finitely many irreducible varieties, then by Theorem 13.38 there is a minimal such Y (why?). Since Y is not irreducible, it properly contains two varieties V_1 and V_2 with $Y = V_1 \cup V_2$. By the choice of Y, each of V_1 and V_2 is a finite union of irreducible varieties, so Y itself is. This contradiction implies that every variety is a finite union of irreducible varieties. Let $V = V_1 \cup \cdots \cup V_k = W_1 \cup \cdots \cup W_s$ for minimal sets of irreducible varieties $\{V_j\}$ and $\{W_i\}$. Now $W_i = W_i \cap V = W_i \cap (\bigcup V_j) = (W_i \cap V_1) \cup \cdots \cup (W_i \cap V_k)$. Each $W_i \cap V_j$ is a variety, as are unions of these, by Theorem 13.38, so W_i irreducible implies that $W_i = W_i \cap V_j$ for some j, and $W_i \subseteq V_j$ results. Similarly, $V_j \subseteq W_t$ for some W_t so $W_i \subseteq W_t$ and the minimality of $\{W_i\}$ forces $W_i = W_t = V_j$. Thus $\{W_i\} \subseteq \{V_j\}$, and by a similar argument $\{V_j\} \subseteq \{W_i\}$, proving that $\{V_j\}$ is unique. From Theorem 13.38, $\mathcal{I}(V) = \mathcal{I}(\bigcup V_j) = \bigcap \mathcal{I}(V_j) = \bigcap P_j$ and each P_j is prime by Theorem 13.39. Using Theorem 13.35, $I \subseteq \mathcal{I}(V)$ so $I \subseteq P_j$ for each j. If $P \lhd R$ is prime with $I \subseteq P$, $V = \mathcal{V}(I) \supseteq \mathcal{V}(P)$ so $\mathcal{I}(V) \subseteq \mathcal{I}(\mathcal{V}(P)) = P$ (Theorem 13.37), implying that $\bigcap P_j \subseteq P$ and thus $P_1 \cdots P_k \subseteq P$. Since P is prime, some $P_j \subseteq P$, proving that each P_j is minimal in the set of primes containing I (if $P_i \subseteq P_j$ then $V_j \subseteq V_i$). ■

Example 13.24 $\mathcal{V}(I)$ for $I = (x_1^2 \cdots x_{n-1}^2, x_1^2 \cdots x_{n-2}^2 x_n^2, \ldots, x_2^2 \cdots x_n^2)$.

For $I \subseteq F[x_1, \ldots, x_n] = R$, F a subfield of C, it is easy to see that $(a_1, \ldots, a_n) \in \mathcal{V}(I) \Leftrightarrow$ for some $i \neq j$, $a_i = 0 = a_j$ (verify!). Now $P_{ij} = Rx_i + Rx_j$ is a prime ideal of R (consider R/P_{ij}), $\mathcal{V}(I) = \bigcup \mathcal{V}(P_{ij})$ over all $i \neq j$, $\mathcal{V}(P_{ij})$ is irreducible, and every $\mathcal{V}(P_{ij})$ is required in the union to get $\mathcal{V}(I)$ (why?). Thus Theorem 13.40 and Theorem 13.36 show that $\text{rad}(I) = \bigcap P_{ij}$ so that $\text{rad}(I) = (x_1 \cdots x_{n-1}, x_1 \cdots x_{n-2}x_n, \ldots, x_2 \cdots x_n) \neq I$.

Since $\text{rad}(I) = \mathcal{I}(\mathcal{V}(I))$ by Hilbert's Theorem, taking $V = \mathcal{V}(I)$ in Theorem 13.40 shows that $\mathcal{I}(V) = \text{rad}(I)$ is the intersection of finitely many primes. This is a geometric proof of Theorem 13.21 when $R = F[x_1, \ldots, x_n]$.

We mentioned that Theorem 13.38 shows that the set of varieties is closed under intersection and under finite union. For $R = F[x_1, \ldots, x_n]$ we also have $\emptyset = \mathcal{V}(R)$ and $K^n = \mathcal{V}((0))$. These properties make the set of varieties into the *closed sets of a topology*.

DEFINITION 13.17 A **topology** on a nonempty set S is any collection Γ of subsets satisfying: $\emptyset, S \in \Gamma$; $T_1, \ldots, T_m \in \Gamma \Rightarrow T_1 \cap \cdots \cap T_m \in \Gamma$; and for any $T_\lambda \in \Gamma$, $\bigcup T_\lambda \in \Gamma$. The elements of Γ are the **open sets** of the topology, and their complements are the **closed sets.**

Any collection of subsets containing $\emptyset$ and the whole set and closed under finite unions and arbitrary intersections defines a topology. The given sets are closed and their complements are open. For $\mathbf{R}$, the usual topology is the collection of all unions of open intervals. We saw above that the varieties in K^n are the closed sets of a topology, called the *Zariski topology*. In this topology the following are equivalent: $\emptyset \neq U_1$ and $\emptyset \neq U_2$, both open $\Rightarrow U_1 \cap U_2 \neq \emptyset$; and $\emptyset \neq U \subseteq \mathcal{V}(I)$ with U open $\Rightarrow \mathcal{V}(I) = K^n$. It is an exercise that these two implications are equivalent. They are true since for U_i open, $U_1 \cap U_2 = \emptyset \Leftrightarrow (K^n - U_1) \cup (K^n - U_2) = K^n$, and Theorem 13.39 shows that $K^n = \mathcal{V}((0))$ is irreducible so some $(K^n - U_i) = K^n$, forcing $U_i = \emptyset$.

The Zariski topology can be useful in studying $M_n(K)$. We demonstrate this by proving a result about characteristic polynomials. Recall from linear algebra that the characteristic polynomial for $A \in M_n(K)$ is $\chi_A(x) = \det(xI_n - A) \in K[x]$, monic of degree n.

Example 13.25 For $A, B \in M_n(K)$, $\chi_{AB}(x) = \chi_{BA}(x)$.

Write $\chi_{AB}(x) = p_0(A, B) + \cdots + p_{n-1}(A, B)x^{n-1} + x^n$ where each $p_i(A, B)$ is a fixed polynomial in the $2n^2$ entries of A and B. That is, $p_i(A, B)$ is the evaluation of some $p_i \in K[x_{11}, \ldots, x_{nn}, y_{11}, \ldots, y_{nn}] = R$ at the entries of A and B where the $\{x_{ij}\} \cup \{y_{ij}\}$ are $2n^2$ commuting indeterminates over K representing the entries of two matrices in $M_n(K)$. Identify any pair of such matrices (A, B) with the $2n^2$-tuple of their entries (in a fixed order) in K^{2n^2}. Let $W = \{(A, B) \in K^{2n^2} \mid \chi_{AB}(x) = \chi_{BA}(x)\} = \{(A, B) \mid p_j(A, B) - p_j(B, A) = 0 \text{ for all } 0 \leq j \leq n-1\}$. Equivalently, W is the intersection over $0 \leq j \leq n-1$ of $\mathcal{V}(p_j(x_{11}, \ldots, x_{nn}, y_{11}, \ldots, y_{nn}) - p_j(y_{11}, \ldots, y_{nn}, x_{11}, \ldots, x_{nn}))$ in K^{2n^2}, so W itself is a variety by Theorem 13.38. If $\det(A) \neq 0$ then $AB = A(BA)A^{-1}$ and since conjugates have the same characteristic polynomial, $\chi_{AB}(x) = \chi_{BA}(x)$. This also holds if $\det(B) \neq 0$. Now $\det(A)$ is the evaluation of a polynomial $D(x_{11}, \ldots, x_{nn}) \in R$ at the entries of A so $U = \{(A, B) \mid \det(A) \neq 0\} \cup \{(A, B) \mid \det(B) \neq 0\} = (K^{2n^2} - \mathcal{V}(D(x_{ij}))) \cup (K^{2n^2} - \mathcal{V}(D(y_{ij})))$.

Therefore $\emptyset \neq U$ is open in the Zariski topology and $U \subseteq W$. Our comments above force W to contain all pairs (A, B), verifying the claim.

EXERCISES

In the exercises below, $R = \mathbf{C}[x_1, \ldots, x_n]$, $n \geq 2$, and $F \subseteq \mathbf{C}$ is a subfield of $\mathbf{C}$.

1. Find $\mathcal{V}(I) \subseteq \mathbf{C}^n$ for the following I:
   i) $(xy - y, yz - z, xz - x) \subseteq \mathbf{C}[x, y, z]$
   ii) $(xy^2 - xy, xz^3) \subseteq \mathbf{C}[x, y, z]$
   iii) $(x^2 + 1, (x + y)^2 + 1) \subseteq \mathbf{C}[x, y]$
   iv) $(x_1^2 - 1, (x_1 + x_2)^2 - 1, (x_1 + x_2 + x_3)^2 - 1) \subseteq \mathbf{C}[x_1, x_2, x_3]$

2. If $f \in R$ is irreducible and $g \in R$, then $g \in (f)$ if and only if for all $\alpha \in \mathbf{C}^n$, $f(\alpha) = 0 \Rightarrow g(\alpha) = 0$.

3. For $I \lhd R$, show that R/I has no nonzero nilpotent element $\Leftrightarrow I = \mathcal{I}(\mathcal{V}(I))$.

4. If $f, g \in R$, then either $f - g \in \mathbf{C}$ or there is $\alpha \in \mathbf{C}^n$ with $f(\alpha) = g(\alpha)$.

5. **i)** Any prime ideal P of R is an intersection of maximal ideals.
   ii) A prime ideal P of R is maximal $\Leftrightarrow \mathcal{V}(P)$ is finite.

6. For $f_1, \ldots, f_m \in R$, either $R = (f_1, \ldots, f_m)$ or there is $\alpha \in \mathbf{C}^n$ with all $f_j(\alpha) = 0$.

7. If $f \in R$ is irreducible and $g, h \in R$, then $g(\alpha) = h(\alpha)$ for all $\alpha \in \mathcal{V}((f)) \Leftrightarrow g + (f) = h + (f)$ in $R/(f)$.

8. If $f \in R$ with $f((a_1, \ldots, a_n)) = 0 \Leftrightarrow$ some $a_j = 1$, then $f \mid (\prod(x_i - 1))^k$, some k.

9. Let $I \lhd R$. Show that:
   i) There is $m > 0$ with $(\mathrm{rad}(I))^m \subseteq I$.
   ii) R/I is algebraic over $(\mathbf{C} + I)/I \cong \mathbf{C} \Leftrightarrow \mathcal{V}(I)$ is finite.

10. Show that there is $m > 0$ so that for any $f(x, y, z, t) \in \mathbf{C}[x, y, z, t]$, $(f(x, y, z, t) - f(0, 0, 0, 0))^m \in (x^{13}y^{15}z^{17}t^{19}, x^{21} + y^{23}, y^{25} + z^{27}, z^{27} + t^{29})$.

11. In $\mathbf{C}[x, y]$, show that:
   i) Some power of $(x + y)(x^2 + y^4 - 2)$ is in $(x^3 + y^2, y^3 + xy)$.
   ii) Some power of $x^3y^3 - xy$ is in $(x^3 + xy^2 + 2x, x^4y + x^2y^3 + y^3 - y)$.

12. In R with $n \geq 3$, show that $(x_i^s - x_i)^m \in (x_1 \cdots x_{n-1} - x_n, x_1 \cdots x_{n-2}x_n - x_{n-1}, \ldots, x_2 \cdots x_n - x_1)$ for some fixed $s, m > 0$ and all $1 \leq i \leq n$.

13. For $1 \leq i \leq k$, let $f_i(x, y) = a_i x^2 + b_i xy + c_i y^2 \in \mathbf{C}[x, y]^*$. Show that there is $(d, e) \in \mathbf{C}^2$ so that $d^2 + e^2 = 1$ and all $f_i(d, e) \neq 0$.

14. For $1 \leq i \leq n$, let $g_i = c_i + a_{i1}x_1 + \cdots + a_{in}x_n \in R$ and then set $A = (a_{ij}) \in M_n(\mathbf{C})$. Show that $(g_1, \ldots, g_n) \lhd R$ is maximal $\Leftrightarrow A$ is invertible.

15. For $I, J \lhd R$, if for all $\alpha \in \mathbf{C}^n$, $f(\alpha) = 0$ for all $f \in I \Leftrightarrow g(\alpha) = 0$ for all $g \in J$, then show that $(I + J)/J$ is a nil (and nilpotent!) ideal of R/J.

16. Let $g \in R^*$ and $A = \{f/g^j \in QF(R) \mid f \in R\}$, a subring of $QF(R)$. If $I \lhd R$, show that $IA = A \Leftrightarrow \mathcal{V}(I) \subseteq \mathcal{V}((g))$. Show that A has infinitely many maximal ideals.

17. If $I \lhd F[x_1, \ldots, x_n]$ is proper, show carefully that $\mathbf{C}I \lhd R$ is proper.

18. If $M \lhd F[x_1, \ldots, x_n]$ is maximal, show that $\mathcal{V}(M)$ is finite and not empty.

19. If $\{P_\lambda\}$ are prime ideals in $F[x_1, \ldots, x_n]$, show that $\bigcap P_\lambda = \mathcal{I}(\mathcal{V}(\bigcap P_\lambda))$.

20. If $I = (x_1 - 1, x_2^2 - 1, \ldots, x_n^n - 1) \lhd R$, describe the maximal ideals containing I and specify their number.

21. For $\{V_\lambda\}$ varieties in $\mathbf{C}^n$, show that $\bigcap V_\lambda = V_{\lambda(1)} \cap \cdots \cap V_{\lambda(k)}$, some $\lambda(1), \ldots, \lambda(k)$.

22. Let $I = (x_1 \cdots x_{n-1}(x_n - 1), x_1 \cdots x_{n-2}(x_{n-1} - 1)x_n, \ldots, (x_1 - 1)x_2 \cdots x_n) \lhd R$. Describe the prime ideals $\{P_\mu\}$ of R so that $\mathcal{V}(I) = \bigcup \mathcal{V}(P_\mu)$.

23. Let $I = \sum Rx_i x_j$ over all $i \neq j$. Describe $\mathcal{V}(I)$ geometrically and find the prime ideals whose varieties have union $\mathcal{V}(I)$.

24. Consider varieties in $M_n(\boldsymbol{C})$ by identifying each $A \in M_n(\boldsymbol{C})$ with the n^2-tuple of its entries, $(A_{11}, A_{12}, \ldots, A_{nn})$. So for $f(x_{11}, \ldots, x_{nn}) \in \boldsymbol{C}[x_{11}, \ldots, x_{nn}]$ we define $f(A) = f(A_{11}, A_{12}, \ldots, A_{nn})$. Using the Zariski topology:
   i) Show that $GL(n, \boldsymbol{C})$ is an open set.
   ii) Show that every nonempty open set in $M_n(\boldsymbol{C})$ contains an invertible matrix.
   iii) Show that the set of symmetric matrices $(A^T = A)$ is an irreducible variety.
   iv) Is $\{A \in M_n(\boldsymbol{C}) \mid A^2 = A\}$ a variety?

25. If $A \in M_n(\boldsymbol{C}) - \boldsymbol{C} \cdot I_n$, show that there is $P \in GL(n, \boldsymbol{C})$ so that every $(P^{-1}AP)_{ij}$ is not zero. (Show that for any (i, j), $(P_{ij}^{-1}AP_{ij})_{ij} \neq 0$ for some P_{ij}; then use Exercise 24.)

13.6 ZORN'S LEMMA AND CARDINALITY

We prove the equivalence of Zorn's Lemma, the Axiom of Choice in §0.5, and the existence of a *well-ordering* on any set (defined shortly) and then apply Zorn's Lemma to obtain results on cardinality. Recall that the Axiom of Choice states that for any set $S \neq \emptyset$ there is a *choice function* $H : (P(S) - \{\emptyset\}) \to S$ with $H(A) \in A : H$ chooses an element from each nonempty subset. Next is the notion of a well-ordering (see §1.5).

DEFINITION 13.18 If $(S, \leq')$ is a partially ordered set (poset) then a **chain** in S is any $\emptyset \neq A \subseteq S$ such that $s, t \in A \Rightarrow s \leq' t$ or $t \leq' s$. A chain $A \subseteq S$ is **well-ordered** if every $\emptyset \neq B \subseteq A$ contains a minimal element; that is, for some $m \in B$, $m \leq' b$ for all $b \in B$. If $s, t \in S$ write $s < t$ when $s \leq' t$ and $s \neq t$. When $A \subseteq S$ is a chain and $x \in A$ then the **initial segment of x** in A is $I_A(x) = \{a \in A \mid a < x\}$.

When $(S, \leq')$ is itself well-ordered we call $\leq'$ a *well-ordering* on S. We have seen that $\boldsymbol{Z}$ or $\boldsymbol{R}$ with the usual $\leq$ is a chain but is not well-ordered and that $\boldsymbol{N} \subseteq \boldsymbol{Z}$ is well-ordered in $(\boldsymbol{Z}, \leq)$. By Exercise 9 in §1.5, $(V = \boldsymbol{N}^2, \leq')$ is well-ordered for $(a, b) \leq' (c, d)$ when $a < c$ or when $a = c$ and $b \leq d$ in $(\boldsymbol{Z}, \leq)$. In $(V, \leq')$, $I_V((1,1)) = \emptyset$, $I_V((1, 5)) = \{(1, 1), (1, 2), (1, 3), (1, 4)\}$, $I_V((2, 1)) = \{(1, m) \mid m \in \boldsymbol{N}\}$, and $I_V((3, 2)) = \{(a, m) \mid m \in \boldsymbol{N}$ and $1 \leq a \leq 2\} \cup \{(3, 1)\}$.

In any chain $(S, \leq')$, if $a \in I_S(x)$ then $I_S(a) \subseteq I_S(x)$ (why?). We claim that if $(S, \leq')$ is well-ordered, $A \subseteq (S, \leq')$ properly, and $a \in A$ implies $I_S(a) \subseteq A$, then $A = I_S(x)$ for some $x \in S$. For $y \in S - A$, the assumption on A, and S a chain, force $a < y$ for all $a \in A$. Since S is well-ordered, $A = I_S(m)$ for m the minimal element in $S - A$ (verify!).

THEOREM 13.41 The Axiom of Choice implies Zorn's Lemma.

Proof To prove Zorn's Lemma we take a poset $(S, \leq')$, assume that every chain in S has an upper bound, and must prove that S contains a maximal element. *We assume S has no maximal element, so cannot be a chain*, and use the Axiom of Choice to derive a contradiction. If C is a chain in S with upper bound d, since d is not maximal in S there is $x \in S$ with $d < x$. If $x \in C$ then $x \leq' d$, so with $d \leq' x$, we get the contradiction $x = d$.

Hence $x \notin C$ and since both $s \leq' d$ for all $s \in C$ and $d < x$, we get $s < x$: that is, x is an upper bound for C. Thus $\{x \in S - C \mid c < x \text{ for all } c \in C\} \neq \emptyset$. By the Axiom of Choice there is $H : (P(S) - \{\emptyset\}) \to S$ with all $H(A) \in A$. If W is the set of chains in S, define $h : W \to S$ by $h(C) = H(\{x \in S - C \mid c < x, \text{ all } c \in C\})$. Observe that $h(C) \notin C$ and $c < h(C)$ for all $c \in C$.

For $H(S) = a \in S$ let T be the set of all chains C in $(S, \leq')$ satisfying the following:

i) $(C, \leq')$ is well-ordered.
ii) a is the minimal element of C.
iii) If $x \in C - \{a\}$ then $h(I_C(x)) = x$.

Now $T \neq \emptyset$ since $\{a\} \in T$, $\{a, h(\{a\})\} \in T$, $\{a, h(\{a\}), h(\{a, h(\{a\})\})\} \in T$, etc. Just as for these examples, if $A, B \in T$ we claim $A = B$ or one is an initial segment of the other. By the defnition of T, $a \in D = \{z \in A \cap B \mid I_A(z) = I_B(z)\}$ so $D \neq \emptyset$. For $d \in D$, if $e \in I_A(d) = I_B(d) \subseteq A \cap B$ then $I_A(e) \subseteq I_A(d) \subseteq B$ so $I_A(e) \subseteq I_B(e)$. Similarly, $I_B(e) \subseteq I_A(e)$ so $e \in D$. This proves $I_A(d) \subseteq D$ if $d \in D$. Since A is well-ordered, our comment before the theorem shows $D = I_A(d_1)$ for $d_1 \in A$ and minimal in $A - D$, unless $D = A$. By the same argument, $D = I_B(d_2)$ for $d_2 \in B$ and minimal in $B - D$, unless $D = B$. If $A \neq D$ and $B \neq D$ then using condition iii) yields $d_1 = h(I_A(d_1)) = h(D) = h(I_B(d_2)) = d_2$. Thus $d_1 = d_2 \in D$, contradicting the choice of d_1. Thus either $A = B$, $A = D = I_B(d_2)$, or $B = D = I_A(d_1)$, verifying our claim.

Since for $A, B \in T$, either $A = B$ or one is an initial segment of the other, $(T, \subseteq)$ is a chain. We show that $V = \bigcup\{C \mid C \in T\} \in T$. The definition of T shows that $a \in V$ is minimal. If $x, y \in V$ then for some $C_i \in T$, $x \in C_1$ and $y \in C_2$. Since T is a chain assume $C_1 \subseteq C_2$. Then $x, y \in C_2$, a chain, so $x \leq' y$ or $y \leq' x$ and V is a chain. If $\emptyset \neq Y \subseteq V$ then $Y \cap A \neq \emptyset$ for some $A \in T$, and set $m = min\,\{Y \cap A\}$ using the well-ordering of A. Should $y \in Y - A$ with $y < m$, then $y \in B$ for some $B \in T$. Now $y \notin A$ implies $A \subseteq B$ properly, so $A = I_B(b)$ and $m < b$ forces $y < b$, giving the contradiction $y \in A \cap Y$. Hence $m = min\,Y$, proving that V is well-ordered. To verify iii) for V, let $v \in V - \{a\}$. Then $v \in A$ for some $A \in T$ and $I_V(v) = I_A(v)$, for if $w \in I_V(v)$ and $w \in B - A$ for $B \in T$, then again $A = I_B(b')$, and $w < v < b'$ forces $w \in A$, a contradiction. Thus $I_V(v) \subseteq I_A(v)$ and since $I_A(v) \subseteq I_V(v)$ from the definition of V we get $I_V(v) = I_A(v)$. By assumption, $v = h(I_A(v)) = h(I_V(v))$, showing that $V \in T$. Clearly V must be the maximal element in T. Since V is a chain, $V \neq V \cup \{h(V)\} \in T$. This contradiction arises by assuming that S has no maximal element. Thus S has a maximal element, proving the theorem. ■

THEOREM 13.42 The following statements are equivalent:

i) The Axiom of Choice
ii) Zorn's Lemma
iii) Any nonempty set can be well-ordered.

Proof That i) implies ii) is just Theorem 13.41. Assume Zorn's Lemma holds, let S be a nonempty set, and let $T = \{(A, \leq') \mid \emptyset \neq A \subseteq S \text{ and } \leq' \text{ is a well-ordering of } A\}$ (verify that T is a set!). For each singleton $\{s\} \subseteq S$, $(\{s\}, =) \in T$ so $T \neq \emptyset$. Define a partial order # on T by $(A, \leq')\,\#\,(B, \leq'')$ when $A = B$ and $\leq' = \leq''$, or when $A \subseteq B$ and $(A, \leq') = (I_B(b), \leq'')$. It is an exercise that $(T, \#)$ is a poset. Let $C = \{(A_\lambda, \leq'_\lambda)\}$ be a chain in T and $A = \bigcup A_\lambda$.

Since either $A_\lambda \subseteq A_\mu$ or $A_\mu \subseteq A_\lambda$, for $x, y \in A$ there is β with $x, y \in A_\beta$. Define a partial order $\leq_A$ on A as follows: if $x, y \in A$ with $x, y \in A_\lambda$, then $x \leq_A y \Leftrightarrow x \leq'_\lambda y$. When $x, y \in A_\eta$ also, then using that C is a chain and the definition of #, $x \leq'_\eta y \Leftrightarrow x \leq'_\lambda y$, so $x \leq_A y$ is independent of the λ with $x, y \in A_\lambda$. Each A_λ is a chain so $(A, \leq_A)$ is as well. If $\varnothing \neq Y \subseteq A$ and $Y \cap A_\lambda \neq \varnothing$ then $(A_\lambda, \leq'_\lambda)$ well-ordered yields a minimal $m \in Y \cap A_\lambda$. Should $y \in Y - A_\lambda$ with $y <_A m$, then if $y, m \in A_\beta$, $A_\lambda = I(b) \subseteq A_\beta$ follows so $y <'_\beta m <'_\beta b$ forces $y \in A_\lambda$, a contradiction. Therefore in A, $m = min\, Y$ so $\leq_A$ is a well-ordering of A and $(A, \leq_A) \in T$. For $x \in A_\lambda$, $I_{A_\lambda}(x) = I_A(x)$, essentially by the argument just above (verify!), so $I_A(x) \subseteq A_\lambda$. Hence by our earlier comment $A_\lambda = I_A(y)$ for $y \in A$ and the definition of $\leq_A$ and that of # show that $(A, \leq_A)$ is an upper bound in T for C. Zorn's Lemma yields a maximal $(W, \leq_W) \in T$. If $s \in S - W$, define $\leq''$ on $W'' = W \cup \{s\}$ by $w_1 \leq'' w_2 \Leftrightarrow w_1 \leq_W w_2$ for $w_i \in W$, and $y \leq'' s$ for all $y \in W''$. Then $\leq''$ is a well-ordering (why?) so $(W'', \leq'') \in T$ and since $W = I(s)$ in W'', $(W, \leq_W)\,\#\,(W'', \leq'')$ properly in T, contradicting the maximality of $(W, \leq_W)$. Consequently, $W = S$ and $\leq_W$ is a well-ordering of S. This proves that ii) implies iii). To see that iii) implies i), let $S \neq \varnothing$ and let $\leq'$ be a well-ordering of S. For $\varnothing \neq A \subseteq S$ the assignment $H(A) = min\, A$ in $(S, \leq')$ is a choice function for S. ∎

The Axiom of Choice allows us to choose an element from each nonempty subset of S. In §0.5 we explained why this is not entirely obvious, although it seems quite reasonable to assume the truth of this axiom, and we will. By Theorem 13.42, something nontrivial results from this assumption: every nonempty set can be well-ordered! It is not at all obvious what a well-ordering of $\boldsymbol{R}$ might look like. The advantage of well-ordering a set is to allow inductive arguments on the elements of the set (see Exercise 19).

Zorn's Lemma leads to some useful results about cardinality, similar to those in §0.7 for countable sets. By Definition 0.20, if there is a bijection $\varphi : S \to T$ we write $|S| = |T|$ and say that S and T have the same *cardinality*, or the same *cardinal number* $|S|$. When $g : S \to T$ is an *injection* we write $|S| \leq |T|$. The Schroeder-Bernstein Theorem (Theorem 0.16) shows that $|S| \leq |T|$ and $|T| \leq |S|$ force $|S| = |T|$. It turns out that any set of sets with different cardinal numbers is a chain using $\leq$.

THEOREM 13.43 For nonempty sets A and B, either $|A| \leq |B|$ or $|B| \leq |A|$.

Proof Let $W = \{(L, f) \mid \varnothing \neq L \subseteq A$ and $f : L \to B$ is injective$\}$. If $a \in A, b \in B$, then $(\{a\}, f) \in W$ for $f(a) = b$, so $W \neq \varnothing$. Define $(L, f) \leq' (K, g)$ in W if $L \subseteq K$ and $g(x) = f(x)$ for all $x \in L$. It is an exercise that $(W, \leq')$ is partially ordered. For a chain $\{(L_\lambda, f_\lambda)\}$ in W, set $V = \bigcup L_\lambda$ and define $g : V \to B$ by $g(v) = f_\lambda(v)$ when $v \in L_\lambda$. Note that if $v \in L_\lambda \cap L_\mu$ then using that $\{(L_\lambda, f_\lambda)\}$ is a chain, either $L_\lambda \subseteq L_\mu$ or $L_\mu \subseteq L_\lambda$ so by the definition of $\leq'$, $f_\lambda(v) = f_\mu(v)$ and g *is* a function: $g(v)$ is independent of the L_λ containing v. Suppose that $v, w \in V$ and $g(v) = g(w)$. Now $v \in L_\lambda$, $w \in L_\mu$, and $\{(L_\lambda, f_\lambda)\}$ a chain imply that $v, w \in L_\beta$ for $\beta = \lambda$ or $\beta = \mu$, so $f_\beta(v) = g(v) = g(w) = f_\beta(w)$. Hence $v = w$ since f_β is injective, and g injective follows. Thus $(V, g) \in W$ and clearly is an upper bound for $\{(L_\lambda, f_\lambda)\}$. Applying Zorn's Lemma yields a maximal $(M, h) \in W$. If $x \in A - M$ and $y \in B - h(M)$ then $h' : (M \cup \{x\}) \to B$ via $h'(m) = h(m)$ for $m \in M$ and $h'(x) = y$ is injective, so $(M \cup \{x\}, h') \in W$ is properly larger than the maximal (M, h), a contradiction. Therefore either $M = A$ and $|A| \leq |B|$, or $h(M) = B$ and $h : M \to B$ is a bijection, so by Theorem 0.11, $h^{-1} : B \to M$ is injective, forcing $|B| \leq |A|$ and completing the proof. ∎

It is important to single out the last observation in the proof: if $g : S \to T$ is surjective then $|T| \le |S|$ since the right inverse $h : T \to S$ for g is injective by Theorem 0.11. Other useful observations, whose proofs are left as exercises, assume that $|A| = |B|$ and $|C| = |D|$. Then $|A| \le |C| \Leftrightarrow |B| \le |D|$, $|A \times C| = |B \times D|$, and if $A \cap C = \emptyset = B \cap D$ then $|A \cup C| = |B \cup D|$.

THEOREM 13.44 If S is an infinite set, then there is a partition $\{A_\lambda\}$ of S so that $|\mathbf{N}| = |A_\lambda|$ for all A_λ. In particular, $|\mathbf{N}| \le |S|$.

Proof Is there any $A \subseteq S$ with $|A| = |\mathbf{N}|$? Let H be a choice function on S and for $n \in \mathbf{N}$ define $f : \mathbf{N} \to S$ by $f(1) = H(S)$, $f(2) = H(S - \{f(1)\})$, and, in general, $f(n+1) = H(S - \{f(1), \ldots, f(n)\})$. This is possible since S is infinite. Now f is injective by its definition so $|f(\mathbf{N})| = |\mathbf{N}|$ and $f(\mathbf{N}) \subseteq S$. Let $W = \{\{A_\lambda\} \mid A_\lambda \subseteq S, |A_\lambda| = |\mathbf{N}|$, and $A_\lambda \cap A_\mu = \emptyset$ if $\lambda \ne \mu\}$. Clearly $(W, \subseteq)$ is partially ordered and $W \ne \emptyset$ since $\{f(\mathbf{N})\} \in W$. Let $\{C_\lambda\}$ be a chain in W with $C_\lambda = \{A_{\lambda\beta}\}$ and set $C = \{A_{\lambda\beta}\}$ over all λ and β. If $A_{\lambda\beta}, A_{\mu\delta} \in C$ then since $C_\lambda \subseteq C_\mu$ or $C_\mu \subseteq C_\lambda$, $A_{\lambda\beta}, A_{\mu\delta} \in C_\lambda$ (or C_μ) and are identical or disjoint by definition of W. Thus C is an upper bound for $\{C_\lambda\}$ in W, and Zorn's Lemma yields a maximal $\{S_\mu\} \in W$. We want $\bigcup S_\mu = S$. If $S - \bigcup S_\mu$ is infinite then by the first part of the proof, for some $B \subseteq S - \bigcup S_\mu$, $|B| = |\mathbf{N}|$ so $\{S_\mu\} \cup \{B\} \in W$ is properly larger than $\{S_\mu\}$, contradicting the maximality of $\{S_\mu\}$. If $S - \bigcup S_\mu = \{s_1, \ldots, s_m\}$ then $|S_\beta| = |\mathbf{N}|$ implies $|S_\beta \cup \{s_1, \ldots, s_m\}| = |\mathbf{N}|$ also (why?). For any S_β it follows that $\{S_\mu\}_{\mu \ne \beta} \cup \{S_\beta \cup \{s_1, \ldots, s_m\}\}$ is a partition of S as required, completing the proof. ■

We state some useful consequences of Theorem 13.44.

THEOREM 13.45 If S is an infinite set, then the following hold:

i) For $m \in \mathbf{N}$, $S = A_1 \cup \cdots \cup A_m$, $A_i \cap A_j = \emptyset$ if $i \ne j$, and $|A_j| = |S|$.
ii) If T is a set with $T \cap S = \emptyset$ then $|S \cup T| = max\{|S|, |T|\}$.
iii) If T is any set then $|S \cup T| = max\{|S|, |T|\}$.

Proof By Theorem 13.44, S is partitioned by $\{B_\lambda\}$ with all $|B_\lambda| = |\mathbf{N}|$. Now $\mathbf{N}$ itself is partitioned by $\{m\mathbf{N} + i \mid 1 \le i \le m\}$ and $|m\mathbf{N} + i| = |\mathbf{N}|$. Thus if $f_\lambda : \mathbf{N} \to B_\lambda$ is a bijection then $\{B_{\lambda i} = f_\lambda(m\mathbf{N} + i)\}$ partitions B_λ so $\{B_{\lambda i}\}$ over all λ and i partitions S. For $A_j = \bigcup_\lambda B_{\lambda j}$, $\{A_1, \ldots, A_m\}$ is a partition of S, and each $|B_{\lambda j}| = |B_\lambda|$ so $|A_j| = |S|$ (verify!) and i) holds. For ii), assume $|T| \le |S|$ and use i) to write $S = A \cup B$ with $A \cap B = \emptyset$ and $|A| = |S| = |B|$ via bijections $g_A : S \to A$ and $g_B : S \to B$. If $f : T \to S$ is injective then $g_A \circ f : T \to A$ is also so defining $h : (S \cup T) \to S$ by $h(s) = g_B(s)$ and $h(t) = g_A \circ f(t)$ (note that $T \cap S = \emptyset$) gives an injection as well. Thus $|S \cup T| \le |S|$. Clearly $|S| \le |S \cup T|$ using $p(s) = s \in S \cup T$, and now, by Theorem 0.16, $|S| = |S \cup T|$. By Theorem 13.43 either $|T| \le |S|$ or $|S| \le |T|$, and in the latter case an analogous argument shows that $|S \cup T| = |T|$, proving ii).

For iii), once again by Theorem 13.43 we may assume $|T| \le |S|$ since the argument when $|S| \le |T|$ is essentially the same. As in ii), $f(s) = s \in S \cup T$ shows $|S| \le |S \cup T|$ and from Theorem 0.16 we need only show $|S \cup T| \le |S|$. If $S' = S \cup \{1\}$, $T' = T \cup \{2\} \subseteq (S \cup T) \times \{1, 2\}$ then $S' \cap T' = \emptyset$, $|S'| = |S|$, and $|T'| = |T|$. From ii), $|S' \cup T'| = |S'| = |S|$, and $\pi : (S' \cup T') \to (S \cup T)$ given by $\pi((a, b)) = a$ is surjective hence $|S \cup T| \le |S' \cup T'| = |S|$, finishing the proof. ■

If we think of $|A|$ as the number of elements in A then it is reasonable to define $|A|+|B|=|A' \cup B'|$ where $|A'|=|A|$, $|B'|=|B|$, and $A' \cap B'=\emptyset$. By Theorem 13.45 this is the larger of $|A|$ and $|B|$ when at least one set is infinite! The usual definition for products is $|A||B|=|A \times B|$ and this turns out again to be the larger of $|A|$ and $|B|$ when at least one is infinite. This result follows from the special case when $A=B$ (see Exercise 8).

THEOREM 13.46 If S is an infinite set, then $|S \times S|=|S|$.

Proof Let $W=\{(A, f) \mid A \subseteq S, A \text{ is infinite, and } f: A \to A \times A \text{ is a bijection}\}$. Now $W \neq \emptyset$ since Theorem 13.44 shows there is $B \subseteq S$ with $|B|=|N|$ and by Theorem 0.18, $|B|=|B \times B|$. In W, define $(A, f) \leq' (B, g)$ if $A \subseteq B$ and $g(a)=f(a)$ for all $a \in A$. It is straightforward to see that $\leq'$ is a partial order on W. As in the proof of Theorem 13.43, any chain $\{(A_\lambda, f_\lambda)\}$ in W has an upper bound (V, g) for $V=\bigcup A_\lambda$, and $g(v)=f_\lambda(v)$ when $v \in A_\lambda$. The only difference here is to see that g is also surjective. If $(v, w) \in V \times V$ then $v \in A_\lambda$, $w \in A_\mu$, and one of these, say A_μ, is larger using that $\{(A_\lambda, f_\lambda)\}$ is a chain. Therefore $(v, w) \in A_\mu \times A_\mu$, the surjectivity of f_μ shows that $f_\mu(x)=(v, w)$ for some $x \in A_\mu \subseteq V$, and $g(x)=f_\mu(x)=(v, w)$ follows. This proves g is surjective and implies that $(V, g) \in W$ and is an upper bound for $\{(A_\lambda, f_\lambda)\}$. Now Zorn's Lemma provides a maximal $(M, h) \in W$, and in particular $|M|=|M \times M|$. If $|S-M| \leq |M|$ then $|S|=|(S-M) \cup M|=|M|$ by Theorem 13.45, and $|S|=|S \times S|$ results. Otherwise $|M| \leq |S-M|$ by Theorem 13.43 so there is $L \subseteq S-M$ with $|L|=|M|$. Since $|M|=|M \times M|$, it follows (how?) that $|L \times M|=|M \times L|=|L \times L|=|M \times M|=|M|$. From Theorem 13.45 there is a partition $\{L_1, L_2, L_3\}$ of L with all $|L_j|=|L|$, so by composition there are bijections $g_1: L_1 \to L \times M$, $g_2: L_2 \to M \times L$, and $g_3: L_3 \to L \times L$. Using that $M \cap L=\emptyset$ and that

$$(M \cup L) \times (M \cup L)=(M \times M) \cup (L \times M) \cup (M \times L) \cup (L \times L),$$

we can extend h to a bijection $f:(M \cup L) \to (M \cup L) \times (M \cup L)$ defined by $f(m)=h(m)$ if $m \in M$ and $f(a_j)=g_j(a_j)$ for $a_j \in L_j$. Therefore $(M \cup L, f) \in W$ is properly larger than the maximal (M, h). This contradiction shows that we must have $|S-M| \leq |M|$, and so $|S|=|S \times S|$ as required. ■

Some final applications of our previous results will be useful later when we discuss fields and are of independent interest.

THEOREM 13.47 Let S be an infinite set and T a set with $|T| \leq |S|$:

i) If $\{A_t\}_T$ are sets with $|A_t| \leq |S|$, then $|\bigcup A_t| \leq |S|$.
ii) If $\text{Fin}(S)=\{B \subseteq S \mid B \text{ is finite}\}$, then $|\text{Fin}(S)|=|S|$.
iii) If S is a ring, then $|S[x]|=|S|$.

Proof In i), for each $t \in T$ there is an injection from A_t to S whose left inverse $g_t: S \to A_t$ is a surjection (Theorem 0.11). Then $g:(T \times S) \to \bigcup A_t$ given by $g((t, s))=g_t(s)$ is surjective, so by an exercise $|\bigcup A_t| \leq |T \times S| \leq |S \times S|=|S|$, using Theorem 13.46. This theorem and induction show that $|S^m|=|S|$ for any $m \in N$, and since $|N| \leq |S|$ by Theorem 13.44, i) implies that $|\bigcup S^m|=|S|$. Define $h: \bigcup S^m \to \text{Fin}(S)$ by $h((s_1, \ldots, s_k))=\{s_1, \ldots, s_k\}$. Clearly h is surjective so $|\text{Fin}(S)| \leq |\bigcup S^m|=|S|$. To finish the proof of ii), observe that $|S| \leq |\text{Fin}(S)|$ via the injection $s \to \{s\} \in \text{Fin}(S)$ and apply Theorem 0.16.

For iii), it is clear that $|S| \le |S[x]|$ so we need only show that $|S[x]| \le |S|$ by Theorem 0.16 again. Define $f : \bigcup S^m \to S[x]$ by $f((s_1, \ldots, s_k)) = s_1 + s_2x + \cdots + s_kx^{k-1}$. Again, f is certainly a surjection so $|S[x]| \le |\bigcup S^m| = |S|$, completing the argument. ∎

EXERCISES

For all exercises below, A, B, C, D, S, and T are nonempty sets.

1. Show that $|A| \le |B| \Leftrightarrow$ there is a surjection $g : B \to A$.

2. Prove from the definitions that $|S| = |N|$ and T finite $\Rightarrow |S \cup T| = |N|$.

3. If $|A| = |B|$ and $|C| = |D|$, show that:
- **i)** $|A \times C| = |B \times D|$
- **ii)** if $|A| \le |C|$ then $|B| \le |D|$
- **iii)** $|A \cup C| = |B \cup D|$ if $A \cap B = \emptyset = C \cap D$

4. If $|A| \le |B|$ and $|C| \le |D|$, show that $|A \times C| \le |B \times D|$.

5. Let $\{A_t\}_T$ and $\{B_t\}_T$ be collections of sets with $A_t \cap A_s = \emptyset$ and $B_t \cap B_s = \emptyset$ if $t \ne s$.
- **i)** If $|A_t| = |B_t|$ for all $t \in T$, show that $|\bigcup A_t| = |\bigcup B_t|$.
- **ii)** If $|A_t| \le |B_t|$ for all $t \in T$, show that $|\bigcup A_t| \le |\bigcup B_t|$.

6. If $T \subseteq S$, and $A \subseteq S$ is maximal with $|A| = |T|$, show that T is finite or $A = S$.

7. If S is infinite, $A \subseteq S$, and $|A| \ne |S|$, show that $|S - A| = |S|$.

8. If S or T is infinite, show that $|S \times T| = max\{|S|, |T|\}$.

9. For S infinite and $T \subseteq S$, show there is a partition $\{A_\lambda\}$ of S so all $|A_\lambda| = |T|$.

10. For S infinite, $T \subseteq S$, $|T| \ne |S|$, and $\{A_\lambda\}_\Lambda$ a partition of S with all $|A_\lambda| = |T|$, show that $|\Lambda| = |S|$. (See Exercise 8.)

11. Describe explicitly a well-ordering of $\boldsymbol{Q}$.

12. If $(\boldsymbol{R}, \le')$ is well-ordered, describe a well-ordering of $\boldsymbol{R}[x]$.

13. Show that the following are well-orderings for $\boldsymbol{N}$($\le$ is the usual "is less than"):
- **i)** $a \le' b \Leftrightarrow 2 \le a \le b$ or $a \in \boldsymbol{N}$ and $b = 1$.
- **ii)** $a \le'' b \Leftrightarrow 3 \le a \le b$, or $a \ne 2$ and $b = 1$, or $a \in \boldsymbol{N}$ and $b = 2$
- **iii)** $a \# b \Leftrightarrow a$ is odd and b is even, or $a \equiv_2 b$ and $a \le b$
- **iv)** $a \#' b \Leftrightarrow a \in 3\boldsymbol{N} + i$ and $b \in 3\boldsymbol{N} + j$ for $0 \le i \le j \le 2$, or $a \equiv_3 b$ and $a \le b$

14. Describe a well-ordering $\le'$ of $\boldsymbol{N}$ with infinitely many $m \in \boldsymbol{N}$ having no immediate predecessor (if $a <' m$ then also $b <' m$ with $a <' b$).

15. Describe an explicit well-ordering of $\boldsymbol{N}^{\boldsymbol{N}} = \{g : \boldsymbol{N} \to \boldsymbol{N}\}$.

16. Well-ordered sets $(A, \le)$ and $(B, \#)$ are **order isomorphic,** $(A, \le) \cong_o (B, \#)$, if there is a bijection $g : A \to B$ with $a_1 \le a_2 \Leftrightarrow g(a_1) \# g(a_2)$. Show that $\cong_o$ is an equivalence relation on any set of well-ordered sets.

17. Show that no two different orderings (including $(\boldsymbol{N}, \le)$) in Exercise 13 are order isomorphic.

18. If $(A, \le')$ is well-ordered, show that $(A, \le')$ is *not* order isomorphic to $I_A(x)$ for any $x \in A$.

19. For any group G show there is a maximal Abelian $A(G) \le G$.

20. Show that any ring R with 1 has a proper ideal I maximal with the property that $I^2 = I$.

21. Let R be a commutative ring and $(R, \le)$ a well-ordering of R with 0 minimal in $(R, \le)$. Define $I_0 = (0)$ and for $0 \ne b \in R$ assume that $I_a \lhd R$ has been defined for all

$a < b$. If $\bigcup I_a = N(R)$, set $I_b = N(R)$, and otherwise set $I_b = \bigcup I_a + yR$ for y minimal in: $(R, \leq) \bigcap \{r \in R - \bigcup I_a \mid r^2 \in \bigcup I_a\}$. Show that $\{I_b \mid b \in R\}$ is a chain (using $\subseteq$), $I_b \subseteq N(R)$, and $N(R) = \bigcup I_r$, over all $r \in R$.

For Exercises 22–24, if $(A, \leq)$ and $(B, \#)$ are well-ordered and $A \bigcap B = \emptyset$, define $(A, \leq) * (B, \#) = (A \bigcup B, \leq'')$ using $x \leq'' y$ if $x, y \in A$ and $x \leq y$, or $x, y \in B$ and $x \# y$, or $x \in A$ and y in B.

22. If $(A, \leq')$ and $(B, \#)$ are well-ordered and $A \bigcap B = \emptyset$, show that:

i) $(A, \leq') * (B, \#)$ is well-ordered.

ii) $(A, \leq') * (B, \#)$ and $(B, \#) * (A, \leq')$ need *not* be order isomorphic (Exercise 16).

iii) $*$ is associative on any three well-ordered, mutually disjoint sets.

23. If $(A, \leq')$ is well-ordered and has a maximal element m, show that there is $x \in A$ so that $(A, \leq') \cong_o (I_A(x), \leq) * (\{a_1, \ldots, a_k\}, \leq')$ (Exercise 16), $I_A(x)$ has no maximal element, and $a_j \in A$ satisfy $x = a_1 \leq' a_2 \leq' \cdots \leq' a_k = m$.

24. Let $(A, \leq')$, $(B, \leq'')$, and $(C, \#)$ be well-ordered and mutually disjoint.

i) If $(A, \leq') * (B, \leq'') \cong_o (A, \leq') * (C, \#)$ (Exercise 16), show that $(B, \leq'') \cong_o (C, \#)$.

ii) Does $(A, \leq') * (C, \#) \cong_o (B, \leq'') * (C, \#) \Rightarrow (A, \leq') \cong_o (B, \leq'')$?

CHAPTER 14

FIELDS

We study fields $F \subseteq L$, and particularly those elements in L that are algebraic over F. The *relative dimension* of L over F is an important idea and plays a key role in understanding classical geometric constructions. First we give a brief review of vector spaces and basic notions regarding them but generalized a bit using Zorn's Lemma and §13.6.

14.1 VECTOR SPACES

A vector space is simply a module over a field. For the sake of clarity we restate the properties from Definition 13.5.

DEFINITION 14.1 An Abelian group $(V, +)$ is a **vector space over the field** F if there is $\cdot : F \times V \to V$ satisfying the following for all $c, d \in F$ and $\alpha, \beta \in V$:

i) $(c+d)\cdot\alpha = c\cdot\alpha + d\cdot\alpha$
ii) $(cd)\cdot\alpha = c\cdot(d\cdot\alpha)$
iii) $c\cdot(\alpha+\beta) = c\cdot\alpha + c\cdot\beta$
iv) $1\cdot\alpha = \alpha$

Write V_F to indicate that V is a vector space over F.

We will denote elements of the vector space V by Greek letters and elements of the field F by Roman letters. The product $c\cdot\alpha$ is *scalar multiplication* and we write it as juxtaposition $c\alpha$. We record some simple computational consequences of the definition (see Theorem 10.1).

THEOREM 14.1 If V_F, $c \in F$, and $\alpha \in V$, then:

i) $0_F\alpha = 0_V$
ii) $c0_V = 0_V$
iii) $(-1)\alpha = -\alpha$
iv) $c\alpha = 0_V \Rightarrow c = 0_F$ or $\alpha = 0_V$

Proof Using the properties in Definition 14.1, $0_F\alpha = (0_F + 0_F)\alpha = 0_F\alpha + 0_F\alpha$ so adding the inverse $-(0_F\alpha) \in V$ to both sides results in $0_V = 0_F\alpha$, proving i). A similar argument verifies ii). Now $0_V = 0_F\alpha = (1 + (-1))\alpha = 1\alpha + (-1)\alpha = \alpha + (-1)\alpha$, and $(-1)\alpha = -\alpha$ follows. Finally, when $c\alpha = 0_V$ but $c \neq 0_F$, $c^{-1}(c\alpha) = c^{-1}0_V = 0_V$ by ii), and then $0_V = (c^{-1}c)\alpha = 1\alpha = \alpha$. ∎

The theorem shows that $0_V = 0$ acts like a zero element under multiplication. Thus the trivial group $V = \{0\}$ with $c \cdot 0 = 0$ is a vector space over any field. The associative and distributive properties in any field F make F a vector space over itself. Similarly, if F is a *subfield* of L—that is, $F \subseteq L$ are fields and both the addition and multiplication in F are the restrictions of those in L—then L is a vector space over F.

Standard examples of vector spaces are $F[X]_F$ with the usual multiplication, $M_n(F)_F$ using matrix scalar multiplication, and $(F, +)^n_F$ using the multiplication $c(a_1, \ldots, a_n) = (ca_1, \ldots, ca_n)$. An important example for us is the vector space $(F[x]/(p(x)))_F$ using $c(f(x) + (p(x))) = cf(x) + (p(x))$. Useful examples arise from mappings between vector spaces. As for groups or rings, the maps of interest should preserve the vector space structure.

DEFINITION 14.2 Given V_F and W_F, $T : V \to W$ is a **linear transformation** if $T(\alpha + \beta) = T(\alpha) + T(\beta)$ and $T(c\alpha) = cT(\alpha)$ for all $\alpha, \beta \in V$ and all $c \in F$.

Some examples of linear transformations are the identity $I_V : V \to V$, scalar multiplication $m_c(\alpha) = c\alpha$ from V to V, various projections π from F^n to F^m with $m \leq n$, such as $\pi(a_1, \ldots, a_n) = (a_2, a_n)$, and formal differentiation $D : F[x] \to F[x]$. Any $m \times n$ matrix A over F defines a linear transformation $T_A((a_1, \ldots, a_n)) = (A(a_1, \ldots, a_n)^T)^T$ from F^n to F^m where $(\ )^T$ is matrix transpose.

Collections of linear transformations lead to interesting examples of vector spaces (and rings!).

Example 14.1 Given V_F and W_F, $Hom_F(V, W) = \{T : V \to W \mid T \text{ is a linear transformation}\}$ is a vector space over F using $(S + T)(\alpha) = S(\alpha) + T(\alpha)$ and $(cT)(\alpha) = c(T(\alpha))$ for $\alpha \in V$ and $c \in F$.

The verification is straightforward. For example, if $T \in Hom_F(V, W)$ and $c \in F$ then for $\alpha, \beta \in V$ and $d \in F$, $(cT)(\alpha + \beta) = c(T(\alpha + \beta)) = c(T(\alpha) + T(\beta)) = cT(\alpha) + cT(\beta) = (cT)(\alpha) + (cT)(\beta)$, and $(cT)(d\alpha) = c(T(d\alpha)) = c(dT(\alpha)) = (cd)T(\alpha) = d(cT(\alpha)) = d((cT)(\alpha))$. Thus $cT \in Hom_F(V, W)$. That $Hom_F(V, W)$ is an Abelian group follows easily using $0(\alpha) = 0_W$ as the identity element and $(-T)(\alpha) = -(T(\alpha))$. The other properties needed for a vector space are routine calculations.

Next comes the notion of a subspace of V_F.

DEFINITION 14.3 Given a vector space V over F, $W \leq (V, +)$ is a **subspace** of V if $\cdot(F \times W) \subseteq W$ and W_F using $\cdot$. In this case we write $W \leq_F V$.

With this notation, $\langle 0_F \rangle \leq_F F_F$ and $F \leq_F F_F$. Also $\boldsymbol{Q} \leq_{\boldsymbol{Q}} \boldsymbol{R}_{\boldsymbol{Q}}$ but it is not the case that $\boldsymbol{Q} \leq_{\boldsymbol{R}} \boldsymbol{R}_{\boldsymbol{R}}$ since, for example, $\pi(1/2) \notin \boldsymbol{Q}$. There is an easy test to determine whether $S \subseteq V_F$ is a subspace.

THEOREM 14.2 Given $\emptyset \neq S \subseteq V_F$, $S \leq_F V$ exactly when $c\alpha, \alpha - \beta \in S$ for all $c \in F$ and all $\alpha, \beta \in S$.

Proof If $S \leq_F V$, $c \in F$, and $\alpha, \beta \in S$ then from Definition 14.3, $S \leq (V, +)$ so $\alpha - \beta \in S$, and $F \cdot S \subseteq S$ so $c\alpha \in S$. When all appropriate $c\alpha, \alpha - \beta \in S$ then clearly $S \leq (V, +)$ and $F \cdot S \subseteq S$. That S_F follows from Definition 14.1 and the fact that V_F satisfies the properties in this definition. ∎

Our goal is to find effective ways of representing elements of V_F in terms of "small" subsets of V. Clearly every $p(x) \in F[x]$ is a sum (uniquely!) of multiples of elements in $\{x^j \mid j \geq 0\} \subseteq F[x]$, and $\alpha = (a_1, \ldots, a_n) \in F^n$ can be written as $\alpha = a_1\varepsilon_1 + \cdots + \alpha_n\varepsilon_n$ for $\varepsilon_j = (0, \ldots, 0, 1, 0, \ldots, 0)$, with a 1 in the jth position. It is routine to see that the power series ring $F[[x]]$ is a vector space over F using ring multiplication, but it is not at all clear if every $\alpha \in F[[x]]$ is a sum of multiples of elements from a fixed "small" subset of $F[[x]]$. Representing $\alpha \in V$ as a sum of multiples of elements from a nice subset of V makes computation in V easier and provides a more concrete way to visualize V. Ideally we want to identify V with something like F^n.

DEFINITION 14.4 A **linear combination** of $\alpha_1, \ldots, \alpha_m \in V_F$ is any $c_1\alpha_1 + \cdots + c_m\alpha_m$ with all $c_j \in F$. For $\emptyset \neq S \subseteq V_F$, the **span** of S, written *span*(S), is the set of all linear combinations of all finite subsets of S.

Clearly $span(V_F) = V$, but this is no help in representing elements of V. An important special case is $span(\{\alpha\}) = F\alpha = \{c\alpha \mid c \in F\}$, which is a subspace of V (verify!). Other examples are $\boldsymbol{C}_{\boldsymbol{R}} = span(\{1, i\})$ and for any $d \in \boldsymbol{C} - \boldsymbol{Q}$ with $d^2 \in \boldsymbol{Q}$, $\boldsymbol{Q}[d]_{\boldsymbol{Q}} = \{a + bd \in \boldsymbol{C} \mid a, b \in \boldsymbol{Q}\} = span(\{1, d\})$. In these cases we can identify $\boldsymbol{C}_{\boldsymbol{R}}$ with $\boldsymbol{R}^2$ via $a + bi \to (a, b)$ and $\boldsymbol{Q}[d]_{\boldsymbol{Q}}$ with $\boldsymbol{Q}^2$ via $a + bd \to (a, b)$. Observe that $\boldsymbol{C}_{\boldsymbol{R}} = span(\{1, i, 1 + i\})$ and also that $\boldsymbol{C}_{\boldsymbol{R}} = span(\{1 - i, 1 + i\})$, using $a + bi = ((a - b)/2)(1 - i) + ((a + b)/2)(1 + i)$. In $F[x]$, $span(\{x^{3k} \mid k \geq 0\}) = F[x^3]$ and an exercise shows that $span(\{x^m - 1\}) = \{p(x) \in F[x] \mid p(1) = 0\}$. It is always true that *span*(S) is a subspace.

THEOREM 14.3 If $\emptyset \neq S \subseteq V_F$, then $span(S) \leq_F V$.

Proof For $\alpha_j, \beta_i \in S$ and $c, c_j, d_i \in F$ simply note that $c\sum d_i\beta_i = \sum(cd_i)\beta_i \in span(S)$, $\sum c_j\alpha_j - \sum d_i\beta_i = \sum c_j\alpha_j + \sum(-d_i)\beta_i \in span(S)$, and use Theorem 14.2. ∎

For the purpose of representing elements of V economically we need a condition to help identify a minimal set S with $span(S) = V$.

DEFINITION 14.5 $S \subseteq V_F$ is ***F*-independent** if for distinct $\alpha_i \in S$ and any $c_i \in F$, $c_1\alpha_1 + \cdots + c_m\alpha_m = 0$ forces all $c_j = 0_F$. Otherwise S is ***F*-dependent.**

We often refer to an F-independent set simply as independent. If $0_V \in S$ then S must be dependent since $1 \cdot 0_V = 0_V$ but $1 \neq 0_F$. For $\alpha \in V_F$, Theorem 14.1 shows that $\{\alpha\} \subseteq V$ is independent $\Leftrightarrow \alpha \neq 0$. A useful observation is that $\{\alpha, \beta\} \subseteq V$ is dependent $\Leftrightarrow \alpha = c\beta$ or $\beta = c\alpha$ for some $c \in F$. Certainly if $\alpha = c\beta$ then $1\alpha + (-c)\beta = 0$ and $1 \neq 0_F$ so $\{\alpha, \beta\}$ is dependent. If $\{\alpha, \beta\}$ is dependent then $c\alpha + d\beta = 0$ with $c \neq 0_F$ or $d \neq 0_F$. Should $c \neq 0_F$ then $\alpha = (-c^{-1}d)\beta$ and if $d \neq 0_F$ then $\beta = (-d^{-1}c)\alpha$.

Example 14.2 If $S = \{g_i(x) \in F[x] \mid deg\ g_i = i\}$ then S is F-independent.

Suppose that $c_1 g_1 + \cdots + c_m g_m = 0$ for $c_j \in F$ (why is it sufficient to consider this special case?). If $g_m(x) = a_m x^m + \cdots + a_0$ then the coefficient of x^m in $c_1 g_1 + \cdots + c_m g_m$ is $c_m a_m = 0$ and $c_m = 0$ since $deg\ g_m = m$ means that $a_m \neq 0$. It follows by induction that all $c_j = 0$, so S is F-independent.

Consequences of the example are that the sets $\{x^m + x \mid m \geq 2\}$, $\{x^m - m \mid m \geq 1\}$, and $\{x^m + x^{m-1} + \cdots + x + 1 \mid m \geq 1\}$ are F-independent.

The next important step in our development uses Zorn's Lemma.

THEOREM 14.4 Let $V_F \neq \{0\}$ and $B \subseteq V$ be empty or F-independent. Then V contains maximal F-independent subsets $M \neq \varnothing$ with $B \subseteq M$.

Proof By maximal we mean with respect to inclusion. Let $T = \{\varnothing \neq S \subseteq V \mid S$ is F-independent and $B \subseteq S\}$. If $B \neq \varnothing$ then $B \in T$ and when $B = \varnothing$, $\{\alpha\} \subseteq V$ is F-independent whenever $\alpha \neq 0$, and $\varnothing \subseteq \{\alpha\} \in T$, so $T \neq \varnothing$. Clearly $(T, \subseteq)$ is a poset. Let $\{S_\lambda\}$ be a chain in T and set $A = \bigcup S_\lambda$. If $c_1\alpha_1 + \cdots + c_m\alpha_m = 0$ with $\alpha_j \in A$, then since $\{S_\lambda\}$ is a chain it follows that there is λ so that all these $\alpha_j \in S_\lambda$. The F-independence of S_λ forces all $c_j = 0$, so A must be F-independent also. Now $B \subseteq S_\lambda \subseteq A$, so A is an upper bound in T for $\{S_\lambda\}$ and Zorn's Lemma shows that $(T, \subseteq)$ contains maximal elements. ■

DEFINITION 14.6 A **basis** of V_F is any maximal F-independent $B \subseteq V$.

It is immediate from Theorem 14.4 that every $V_F \neq \{0\}$ has a basis and that if $B \subseteq V$ is F-independent then it is contained in some basis. We give some conditions equivalent to a set being a basis of V.

THEOREM 14.5 If $V \neq \{0\}$ is a vector space over F and $B \subseteq V$, then the following statements are equivalent:

i) B is a basis.
ii) B is F-independent and $span(B) = V$.
iii) $span(B) = V$ and B is minimal with this property.
iv) If $\gamma \in V - \{0\}$ there are unique $\beta_i \in B$ and $c_i \in F^*$ with $\gamma = c_1\beta_1 + \cdots + c_m\beta_m$.

Proof If B is a basis of V then B is F-independent by definition. If some $\gamma \in V - span(B)$, we claim $B \cup \{\gamma\}$ is independent. Assume that for $\beta_j \in B$, $b_1\beta_1 + \cdots + b_m\beta_m + c\gamma = 0$ for $b_j, c \in F$. If $c = 0$ then B independent forces all $b_j = 0$, and when $c \neq 0$, $\gamma = \sum(-c^{-1}b_j)\beta_j \in span(B)$, a contradiction. Hence $B \cup \{\gamma\}$ is F-independent, contradicting the maximality of B. Therefore i) implies ii). If ii) holds but $span(A) = V$

for $A \subseteq B$ properly, then $\beta \in B - A \subseteq span(A)$ means $\beta = \sum a_i\alpha_i$ for some $\alpha_i \in A$. Thus $(-1)\beta + \Sigma a_i\alpha_i = 0$, and $-1 \neq 0$ shows that B is dependent. Hence $span(A) \neq V$, B is minimal with $span(B) = V$, and iii) follows from ii). Assume that $span(B) = V$ and B is minimal with this property. If $c_i, d_j \in F$ and $\alpha_i, \beta_j \in B$ so that $\sum c_i\alpha_i = \sum d_j\beta_j$ are different representations, except for order, then we may write $\sum a_i\gamma_i = \sum b_i\gamma_i$ for $\{\gamma_i\} = \{\alpha_i\} \cup \{\beta_i\}$ and some $a_t \neq b_t$. Now $(a_t - b_t)\gamma_t = \sum_{i\neq t}(b_i - a_i)\gamma_i$ forces $\gamma_t \in span(B - \{\gamma_t\})$. It follows that $V = span(B - \{\gamma_t\})$ (verify!), contradicting the minimality of B and proving iv). Finally, if iv) is true then $0 \notin B$ since for any $0 \neq \alpha \in B$, $1 \cdot \alpha = 1\alpha + 1 \cdot 0$. Should $c_1\beta_1 + \cdots + c_m\beta_m = 0$ for $\beta_j \in B$ distinct and all $c_j \neq 0$, then $m = 1$ would force $c_1 = 0$, and if $m > 1$ then we have $0 \neq \beta_1 = (-c_1^{-1}c_2)\beta_2 + \cdots + (-c_1^{-1}c_m)\beta_m$, contradicting iv). This shows that B is F-independent. For $\gamma \in V - (B \cup \{0\})$ there are $\beta_j \in B$ and $c_j \in F$ with $\gamma + (-c_1)\beta_1 + \cdots + (-c_m)\beta_m = 0$ and it follows that $B \cup \{\gamma\}$ cannot be F-independent. Therefore B is a maximal F-independent set, completing the proof of the theorem. ∎

One way to show that $B \subseteq V_F$ is a basis is to prove that $span(B) = V$ and that B is F-independent. These conditions are clearly satisfied for $\{x^m \in F[x] \mid m \geq 0\}$ in $F[x]$, for $\{E_{ij} \in M_n(F) \mid 1 \leq i, j \leq n\} \subseteq M_n(F)$, and for $\{\varepsilon_i\} \subseteq F^n$ where, as above, $\varepsilon_j = (0, \ldots, 0, 1, 0, \ldots, 0)$ with the 1 in position j. It is not at all clear how to specify a basis for $\boldsymbol{R}_{\boldsymbol{Q}}$, $\boldsymbol{Q}[[x]]_{\boldsymbol{Q}}$, or $Hom_F(V, W)$. Any $\alpha \in F^*$ gives a basis of F_F since $\beta \in F$ can be expressed as $\beta = (\beta\alpha^{-1})\alpha$. Using some linear algebra it follows that a finite set $B \subseteq F^n$ is a basis exactly when the matrix with the elements of B as its rows is invertible (why?).

It is an exercise that $\{x^m + x^{m+1} \mid m \geq 0\} \cup \{1\}$ is a basis of $F[x]_F$, that $\{1, i\}$ and $\{3 + 2i, 1 - i\}$ are bases of $\boldsymbol{C}_{\boldsymbol{R}}$, and that $\{\varepsilon_1 + \cdots + \varepsilon_k \in F^n \mid 1 \leq k \leq n\}$ is a basis of F^n_F. When V_F and W_F have *finite* bases, say $\{\alpha_1, \ldots, \alpha_n\}$ for V and $\{\beta_1, \ldots, \beta_m\}$ for W, then a basis for $Hom_F(V, W)$ is the set of all T_{ij} defined by $T_{ij}(a_1\alpha_1 + \cdots + a_n\alpha_n) = a_i\beta_j$ (exercise!). It is natural to ask whether all bases of V have the same number of elements.

THEOREM 14.6 If A and B are bases of V_F, then $|A| = |B|$.

Proof If $V = \{0\}$ then $\varnothing$ is its only possible basis. When $V \neq \{0\}$ suppose $A = \{\alpha_1, \ldots, \alpha_n\}$ has n elements and $\beta_1, \ldots, \beta_{n+1} \in B$ are distinct. Then by Theorem 14.5, $\beta_j = \sum_i c_{ji}\alpha_i$ for $c_{ji} \in F$ since $span(A) = V$. If $\{d_j\} \subseteq F$ then $\gamma = \sum d_j\beta_j = \sum_i(\sum_j d_jc_{ji})\alpha_i$. Now $\gamma = 0$ when $(d_1, \ldots, d_{n+1})$ is a solution of the n linear equations $c_{1i}x_1 + \cdots + c_{n+1\,i}x_{n+1} = 0$, for $1 \leq i \leq n$. Since n homogeneous linear equations over F in $n + 1$ unknowns must have a nonzero solution, there is $\{d_j\} \subseteq F$ with $\gamma = 0$ but some $d_k \neq 0$. This contradicts the F-independence of B and proves that B has at most n elements. Thus $|B| \leq |A|$ and now, similarly, $|A| \leq |B|$. Hence if one basis is finite, both are, and they have the same number of elements. Assume now that A and B are infinite. By Theorem 13.47, if $\text{Fin}(B) = \{T \subseteq B \mid T \text{ is finite}\}$ then $|B| = |\text{Fin}(B)|$. For $T \in \text{Fin}(B)$ let $A_T = A \cap span(T)$. By Theorem 14.3, $span(T) \leq_F V$ and has basis T from Theorem 14.5. We have just seen that any basis of $span(T)$ has exactly $|T|$ elements, so by Theorem 14.4, $|A_T| = |A \cap span(T)| \leq |T|$, forcing A_T to be finite. Any $\alpha \in A$ is a linear combination of finitely many elements from B, so $\alpha \in span(T)$ for some $T \in \text{Fin}(B)$ and $\alpha \in A_T$. Consequently, $A = \bigcup A_T$ and since both $|A_T| < |B|$ and $|\text{Fin}(B)| = |B|$, Theorem 13.47 shows that $|A| = |\bigcup A_T| \leq |B|$. Interchanging A and B in the argument shows that $|B| \leq |A|$, and therefore Theorem 0.16 gives $|A| = |B|$ as required. ∎

DEFINITION 14.7 If V is a vector space over F then the **dimension of V** is $dim_F(V) = |B|$ for any basis B of V.

The dimension of V_F is unique by Theorem 14.6. Whenever we write $dim_F(V) = n$ we mean that $n \in N$, so V has a finite basis. In this case V is called *finite dimensional*. When $\{\alpha_1, \ldots, \alpha_n\}$ is a basis of V then we can identify V with F^n via $\sum a_j\alpha_j \to (a_1, \ldots, a_n)$.

EXERCISES

1. If p is a prime and $(G, +)$ is an Abelian group with $o(g) = p$ for all $g \neq 0_G$, show that G is a vector space over $\mathbf{Z}_p$ using $[j]_p \cdot g = g^j = g + \cdots + g$ in G for $0 \leq j < p$.

2. Let $V = \mathbf{R}^+ = \{r \in \mathbf{R} \mid r > 0\}$. Show that $(V, \#)$ is an Abelian group for $v \# w = vw$. Show that $(V, \#)$ is a vector space over $\mathbf{R}$ using $c \cdot \alpha = \alpha^c$.

3. For $S \neq \varnothing$, F a field, and $V = \{g : S \to F\}$, define $(g + h)(s) = g(s) + h(s)$ and $(cg)(s) = c(g(s))$ for $g, h \in V$ and $c \in F$. Show that V_F.

4. Show that $\{a_0 + \cdots + a_n x^n \in \mathbf{R}[x] \mid \sum a_j \in \mathbf{Q}\} \leq_{\mathbf{Q}} \mathbf{R}[x]$.

5. Show that $\{\alpha \in \mathbf{Q}[[x]] \mid m\alpha \in \mathbf{Z}[[x]]$ for some $m \in \mathbf{N}\} \leq_{\mathbf{Q}} \mathbf{Q}[[x]]$.

6. If $T \in Hom_F(V, W)$ and $U \leq_F V$, show that:

i) $T(U) \leq_F W$

ii) $\{\alpha \in V \mid T(\alpha) = 0_W\} \leq_F V$

7. Let $\varnothing \neq S, T \subseteq V_F$. Show that:

i) $span(S) = span(T) \Leftrightarrow \alpha \in span(T)$ for all $\alpha \in S$ and $\beta \in span(S)$ for all $\beta \in T$

ii) $span(S \cup T) = span(span(S) \cup span(T))$

iii) for $\gamma \in V$, $span(S \cup \{\gamma\}) = span(S) \Leftrightarrow \gamma \in span(S)$

8. If $span(A) = V_F \neq (0)$ and if $dim_F(V)$ is finite, show that V has a basis $B \subseteq A$.

9. Let $U, W \leq_F V_F$, $dim_F(U) = n$, $dim_F(W) = m$, and set $U + W = span(U \cup W)$. Show that $dim_F(U + W) \leq n + m$.

10. Show that $\{T \in Hom_F(V, W) \mid dim_F(T(V))$ is finite$\} \leq_F Hom_F(V, W)$. (Exercise 9)

11. If $g_j(x) \in F[x]$ so that $deg\ g_1 < deg\ g_2 < \cdots deg\ g_m < \cdots$, show that $\{g_j(x)\} \subseteq F[x]_F$ is an F-independent subset.

12. Show that if $\{\alpha_1, \ldots, \alpha_m\}$ is a basis of V_F, so are:

i) $\{c_j\alpha_j\}$ when all $c_j \neq 0$

ii) $\{\alpha_1, d_2\alpha_1 + \alpha_2, \ldots, d_m\alpha_1 + \alpha_m\}$

iii) $\{\alpha_1, \alpha_1 + \alpha_2, \ldots, \alpha_1 + \alpha_2 + \cdots + \alpha_m\}$

13. Let $W = \{A \in M_2(\mathbf{C}) \mid A_{11} + A_{22} \in \mathbf{R}$ and $A_{12} + A_{21} \in \mathbf{R} \cdot i\}$. Show that $W \leq_{\mathbf{R}} M_2(\mathbf{C})$ and find a basis for $W_{\mathbf{R}}$.

14. For $V = M_n(\mathbf{R})$ let $S = \{A \in V \mid A^T = A\}$ and $K = \{A \in V \mid A^T + A = 0_n\}$ where $(\ )^T$ is transpose. Show that $S, K \leq_{\mathbf{R}} V$ and find their dimensions.

15. If $p(x) \in F[x] - F$ show that $R = F[x]/(p(x))$ is a vector space over F using $c(f(x) + (p(x))) = cf(x) + (p(x))$. Show that $dim_F(R) = deg\ p(x)$.

16. Prove that the following are bases of $(x - 1)F[x] \leq_F F[x]$:

i) $\{x^{m+1} - 1 \mid m \geq 0\}$

ii) $\{x^{m+1} - x^m \mid m \geq 0\}$

iii) $\{(x - 1)^m\} \mid m \geq 1\}$

17. Show that any $T \in Hom_F(V, W)$ is uniquely determined by its values on any basis.

18. If $B=\{\alpha_\lambda\}$ is a basis of V_F and $\{\gamma_\lambda\}\subseteq W_F$ show that $f:B\to W$ given by $f(\alpha_\lambda)=\gamma_\lambda$ can be extended uniquely to some $T_f\in Hom_F(V,W)$ (*i.e.* $T_f(\alpha_\lambda)=\gamma_\lambda$).

19. Let $B=\{\alpha_\lambda\}$ be a basis of V_F and $\{\beta_\mu\}$ be a basis of W_F. For $\alpha\in V$ write $\alpha=a_\lambda\alpha_\lambda+\rho_\lambda(\alpha)$ for $\rho_\lambda(\alpha)\in span(B-\{\alpha_\lambda\})$.
- **i)** Show that $T_{\lambda\mu}(\alpha)=a_\lambda\beta_\mu$ defines $T_{\lambda\mu}\in Hom_F(V,W)$.
- **ii)** Show that $\{T_{\lambda\mu}\}\subseteq Hom_F(V,W)$ is independent.
- **iii)** When $dim_F(V)$ is finite, show that $\{T_{\lambda\mu}\}$ is a basis of $Hom_F(V,W)$.
- **iv)** When $dim_F(V)$ is infinite, show that $span(\{T_{\lambda\mu}\})\neq Hom_F(V,W)$.
- **v)** If $dim_F(V)=n\in N$, determine $dim_F(Hom_F(V,W))$ when $dim_F(W)=m$.
- **vi)** If $dim_F(V)=n$ determine $dim_F(Hom_F(V,W))$ when $dim_F(W)$ is infinite.

20. Show that $dim_F(Hom_F(F[x],F))$ is uncountable.

21. Show that:
- **i)** $dim_Q(\boldsymbol{R})$ is uncountable
- **ii)** $dim_F(F[[x]])$ is uncountable for $F=\boldsymbol{Z}_p$

14.2 SUBFIELDS

For fields F and L we write $F\subseteq L$ to mean that F is a subring of L, called a *subfield* of L. As we saw in §14.1, L is naturally a vector space over F and we set $dim_F(L)=[L:F]$. For any $\emptyset\neq S\subseteq L$ let $F[S]=\langle F\cup S\rangle\subseteq L$ be the subring of L generated by F and S, so by Theorem 10.7 any $\alpha\in F[S]$ looks like a polynomial in the elements of S with coefficients from F. In the case when $S=\{\alpha_1,\ldots,\alpha_n\}$ we write $F[S]=F[\alpha_1,\ldots,\alpha_n]$.

If $F\subseteq L$ are fields and $\alpha\in L$, $\varphi_\alpha(p(x))=p(\alpha)$ defines a surjective ring homomorphism $\varphi_\alpha:F[x]\to F[\alpha]$ (verify!) so by Theorem 11.12, $F[\alpha]\cong F[x]/ker\,\varphi_\alpha$. Since $ker\,\varphi_\alpha\lhd F[x]$, either $ker\,\varphi_\alpha=(0)$ or $ker\,\varphi_\alpha=(m_\alpha(x))$ for m_α monic of least degree with $m_\alpha(\alpha)=0$, using Theorem 11.3. In the latter case $m_\alpha(x)$ is irreducible since $F[\alpha]\subseteq L$ is a domain (verify!).

DEFINITION 14.8 If $F\subseteq L$ are fields and $\alpha\in L$ then α is **transcendental over F** if $p(\alpha)\neq 0$ for all $p(x)\in F[x]^*$ and α is **algebraic over F** if $p(\alpha)=0$ for some $p(x)\in F[x]^*$. If α is algebraic over F its **minimal polynomial over F** is $m_\alpha(x)\in F[x]$, the monic polynomial of least degree with $m_\alpha(\alpha)=0$. When every $\alpha\in L$ is algebraic over F we say that L is **algebraic over F.**

We next describe $F[\alpha]$ when α is either transcendental or algebraic over F. The situation in the latter case is given explicitly in Theorem 11.18.

THEOREM 14.7 Let $F\subseteq L$ be fields and $\alpha\in L^*$. If α is transcendental over F, then $F[\alpha]\cong F[x]$. If α is algebraic over F with minimal polynomial $m_\alpha(x)$ of degree k, then $F[\alpha]$ is a field and a vector space over F, with basis $\{1,\alpha,\ldots,\alpha^{k-1}\}$ over F.

Proof From above, $F[\alpha]\cong F[x]/(m_\alpha(x))$ so if α is transcendental then $m_\alpha=0$ and $F[\alpha]\cong F[x]$. If α is algebraic over F with minimal polynomial of degree k, say $m_\alpha(x)=x^k+\cdots+a_1x+a_0$, then $\alpha^k\in F+\cdots+F\alpha^{k-1}$ so $\alpha^{k+1}\in F\alpha+\cdots+F\alpha^k\subseteq F+\cdots+F\alpha^{k-1}$. As in Theorem 13.24, by induction $\alpha^m\in F+\cdots+F\alpha^{k-1}$ for all $m\in N$, showing $F[\alpha]=F+\cdots+F\alpha^{k-1}$. Hence $span(\{1,\alpha,\ldots,\alpha^{k-1}\})=F[\alpha]$ and this set is a basis.

Otherwise $c_0 \cdot 1 + \cdots + c_{k-1}\alpha^{k-1} = 0$ with some $c_j \neq 0$, and α satisfies a polynomial of degree $d < k$, a contradiction. We observed above that L a field forces m_α to be irreducible. Therefore, if $\beta \in F[\alpha]^*$ with $\beta = g(\alpha)$ for $g(x) \in F[x]$ and *deg* $g < k$, then $\text{GCD}\{m_\alpha(x), g(x)\} = 1 = f(x)g(x) + h(x)m_\alpha(x)$ for some $f(x), h(x) \in F[x]$, using that $F[x]$ satisfies a Euclidean algorithm (see §12.2). Consequently, $f(\alpha)g(\alpha) = 1$ and $f(\alpha) = \beta^{-1}$ in $F[\alpha]$ so $F[\alpha]$ is a field, completing the proof. ■

To illustrate the theorem, $2^{1/3} \in \boldsymbol{R}$ satisfies the irreducible $x^3 - 2 \in \boldsymbol{Q}[x]$ (Theorem 12.16) so $\boldsymbol{Q}[2^{1/3}]$ is field with $[\boldsymbol{Q}[2^{1/3}] : \boldsymbol{Q}] = 3$ and basis $\{1, 2^{1/3}, 2^{2/3}\}$. If $1 + 2^{1/3} = \beta \in \boldsymbol{Q}[2^{1/3}]$ then $\beta = g(2^{1/3})$ for $g(x) = 1 + x$. Now $x^3 - 2 = (x+1)(x^2 - x + 1) - 3$, so $1 = (x+1)(x^2/3 - x/3 + 1/3) - (1/3)(x^3 - 2)$ and $(1 + 2^{1/3})^{-1} = 1/3 - (1/3)2^{1/3} + (1/3)2^{2/3}$.

THEOREM 14.8 Let $F \subseteq L \subseteq K$ be fields. Then $[K : F]$ is finite exactly when both $[K : L]$ and $[L : F]$ are finite, and then $[K : F] = [K : L][L : F]$.

Proof If $[K : F]$ is finite and B is a basis of K over F then $span_L(B) = K$ and some $A \subseteq B$ must be minimal with this property. By Theorem 14.5, A is a basis of K over L so $[K : L]$ is finite. Any F-independent $E \subseteq L$ is contained in some basis of K over F by Theorem 14.4. Thus by Theorem 14.6, $[L : F]$ is finite. Now let $\{\alpha_1, \ldots, \alpha_n\}$ be a basis of L over F and let $\{\beta_1, \ldots, \beta_m\}$ be a basis of K over L. To complete the proof of the theorem it suffices to show that $\{\alpha_i\beta_j\}$ is a basis of K over F. If $0 = \sum c_{ij}\alpha_i\beta_j = \sum_j(\sum_i c_{ij}\alpha_i)\beta_j$ then the independence of $\{\beta_j\}$ over L forces all $\sum_i c_{ij}\alpha_i = 0$, and then the independence of $\{\alpha_i\}$ over F yields all $c_{ij} = 0$, so $\{\alpha_i\beta_j\}$ is F-independent. If $\gamma \in K$ then $\gamma = \sum_j d_j\beta_j$ for $d_j \in L$, and $d_j = \sum_i b_{ij}\alpha_i$ for $b_{ij} \in F$. Therefore $\gamma = \sum b_{ij}\alpha_i\beta_j$ so $span_F(\{\alpha_i\beta_j\}) = K$, proving that $\{\alpha_i\beta_j\}$ is a basis of K over F by Theorem 14.5. ■

The last two results yield easy but important observations.

THEOREM 14.9 Let $F \subseteq L$ be fields. If $[L : F]$ is finite, then L is algebraic over F. If $\alpha_1, \ldots, \alpha_m \in L$ are algebraic over F, then $F[\alpha_1, \ldots, \alpha_m]$ is a field and is finite dimensional over F and hence is algebraic over F.

Proof If $\beta \in L$ is transcendental over F then $F[\beta] \cong F[x]$ by Theorem 14.7 and so is infinite dimensional over F. When $[L : F]$ is finite L cannot contain any infinite F-independent set by Theorem 14.6 and Theorem 14.4; hence L is algebraic over F. Let $\alpha \in L$ be algebraic over F. Theorem 14.7 shows that $[F[\alpha] : F]$ is finite and that $F[\alpha]$ is a field, so $F[\alpha]$ is algebraic over F. Proceed by induction when $\alpha_1, \ldots, \alpha_m \in L$ are algebraic over F. Write $F[\alpha_1, \ldots, \alpha_m] = K[\alpha_m]$ for $K = F[\alpha_1, \ldots, \alpha_{m-1}]$. The induction assumption shows that K is a field, algebraic over F with $[K : F]$ finite. Clearly α_m is algebraic over K since its minimal polynomial over F is in $K[x]$, so $K[\alpha_m] = F[\alpha_1, \ldots, \alpha_m]$ is a field and $[K[\alpha_m] : K]$ is finite. Apply Theorem 14.8 to get that $[K[\alpha_m] : F] = [K[\alpha_m] : K][K : F]$ is finite, completing the proof. ■

In Theorem 14.9, how does $[F[\alpha_1, \ldots, \alpha_m] : F]$ compare with all the $[F[\alpha_j] : F]$? We look at some examples.

Example 14.3 $[Q[2^{1/3}, 3^{1/2}] : Q] = 6$.

By Theorems 12.16 and 14.7, $x^3 - 2, x^2 - 3 \in Q[x]$ are irreducible, $[Q[2^{1/3}] : Q] = 3$, and $[Q[3^{1/2}] : Q] = 2$. Now $Q \subseteq Q[2^{1/3}], Q[3^{1/2}] \subseteq L = Q[2^{1/3}, 3^{1/2}]$ so Theorem 14.8 implies $2 \mid [Q[2^{1/3}, 3^{1/2}] : Q]$ and $3 \mid [Q[2^{1/3}, 3^{1/2}] : Q]$, forcing $6 \mid [Q[2^{1/3}, 3^{1/2}] : Q]$. The minimal polynomial of $3^{1/2}$ over $Q[2^{1/3}]$ has degree at most 2, so $[Q[2^{1/3}, 3^{1/2}] : Q] = [Q[2^{1/3}, 3^{1/2}] : Q[2^{1/3}]][Q[2^{1/3}] : Q] \leq 2 \cdot 3$ by Theorem 14.8 and $[Q[2^{1/3}, 3^{1/2}] : Q] = 6$ results. A consequence is that $[Q[2^{1/3}, 3^{1/2}] : Q[2^{1/3}]] = 2$ so $3^{1/2} \notin Q[2^{1/3}]$. Similarly, $[Q[2^{1/3}, 3^{1/2}] : Q[3^{1/2}]] = 3$ (why?) so $x^3 - 2$ remains the minimal polynomial of $2^{1/3}$ over $Q[3^{1/2}]$.

Example 14.4 For $m \in N$, $2^{1/m} \in R$, and $i \in C$, $[Q[2^{1/m}, i] : Q] = 2m$.

The minimal polynomial over Q of $2^{1/m}$ is $x^m - 2$ (Theorem 12.16) and of $i \in C$ is $x^2 + 1$. From Theorem 14.7, $[Q[2^{1/m}] : Q] = m$ and $[Q[i] : Q] = 2$, so Theorem 14.8 shows that $[Q[2^{1/m}, i] : Q] = [Q[2^{1/m}, i] : Q[2^{1/m}]][Q[2^{1/m}] : Q]$. Now $i \notin Q[2^{1/m}] \subseteq R$, so $[Q[2^{1/m}, i] : Q[2^{1/m}]] = 2$ (why?) and $[Q[2^{1/m}, i] : Q] = 2m$. Clearly, the example still holds if any prime p replaces 2.

Example 14.5 If $\rho = e^{\pi i/3} \in C$ then $[Q[\rho, i] : Q] = 4$.

Since $\rho = 1/2 + 3^{1/2}/2 \cdot i$ it follows that $Q[\rho, i] = Q[3^{1/2}, i]$ (verify!) so from Example 14.4, $[Q[\rho, i] : Q] = 4$. Note that ρ satisfies $x^3 + 1 \in Q[x]$ so its minimal polynomial over Q is $x^2 - x + 1$ (why?).

Example 14.6 If $\omega = e^{2\pi i/3} \in C$ and $2^{1/3} \in R$, then $[Q[2^{1/3}, 2^{1/3}\omega] : Q] = 6$.

Since $\omega^3 = 1$, $2^{1/3}$ and $2^{1/3}\omega$ have minimal polynomial $x^3 - 2$ over Q (why?) so $[Q[2^{1/3}] : Q] = 3 = [Q[2^{1/3}\omega] : Q]$. Clearly $[Q[2^{1/3}, 2^{1/3}\omega] = Q[2^{1/3}, \omega]$ and the minimal polynomial of ω over Q is $x^2 + x + 1$ so $[Q[\omega] : Q] = 2$. Using Theorem 14.8, $[Q[2^{1/3}, \omega] : Q] = [Q[2^{1/3}, \omega] : Q[2^{1/3}]][Q[2^{1/3}] : Q]$. Since $\omega \notin Q[2^{1/3}] \subseteq R$ it follows that $[Q[2^{1/3}, \omega] : Q] = 2 \cdot 3 = 6$.

Example 14.7 If $S = \{\alpha \in R \mid \alpha$ is algebraic over $Q\}$ then $Q[S]$ is a field algebraic over Q with $[Q[S] : Q]$ infinite.

Any $0 \neq \beta \in Q[S]$ is a polynomial expression in $s_1, \ldots, s_m \in S$ so $\beta \in Q[s_1, \ldots, s_m]$, a field finite dimensional over Q by Theorem 14.9. The same result shows that β is algebraic over Q, and since $\beta^{-1} \in Q[s_1, \ldots, s_m]$, $Q[S]$ is a field algebraic over Q. If $[Q[S] : Q] = n \in N$ then $[Q[\beta] : Q] \leq n$ using Theorem 14.8. This fails for $\beta = 2^{1/m} \in Q[S]$ when $m > n$ since $x^m - 2$ is irreducible in $Q[x]$ by Theorem 12.16. Therefore $[Q[2^{1/m}] : Q] = m$ by Theorem 14.7 and this shows that $[Q[S] : Q]$ must be infinite.

Example 14.8 Let $F \subseteq \mathbf{C}$, let $\alpha_1, \ldots, \alpha_n \in \mathbf{C}$ be algebraic over F, and let $K = F[\alpha_1, \ldots, \alpha_n]$. If the minimal polynomial $m_j(x) \in F[x]$ of α_j has degree 2 for all j, then $[K : F] = 2^s$ for $s \leq n$, and the minimal polynomial over F of any $\beta \in K - F$ has degree 2^t for some $t \geq 1$.

Use induction on n. If $n = 1$ then $[F[\alpha_1] : F] = 2$ by Theorem 14.7. The induction assumption shows $[L : F] = 2^d$ for $L = F[\alpha_1, \ldots, \alpha_{n-1}]$. Now $m_n(x) \in L[x]$ and remains irreducible, or $\alpha_n \in L$. Therefore Theorem 14.8 shows $[K : F] = [F[\alpha_1, \ldots, \alpha_{n-1}][\alpha_n] : L][L : F] = [L[\alpha_n] : L] \cdot 2^d = 2^{d+1}$ if $\alpha_n \notin L$, and $[K : F] = 2^d$ if $\alpha_n \in L$. For $\beta \in K - F$, $F[\beta] \subseteq K$ are fields (Theorem 14.7) so using Theorem 14.8 again yields $[F[\beta] : F] \mid 2^s$, forcing $[F[\beta] : F] = 2^k > 1$, with 2^k the degree of the minimal polynomial for β over F by Theorem 14.7.

Example 14.6 shows that the 2 in Example 14.8 cannot be replaced with 3, and it also shows that for α and β algebraic over F, $[F[\alpha, \beta] : F]$ is not always the product of $[F[\alpha] : F]$ and $[F[\beta] : F]$. What *can* be said about how these dimensions are related? It will be helpful to introduce some notation.

DEFINITION 14.9 If H and L are subfields of K containing F then their **join,** denoted $H \vee L$, is the intersection of all fields $M \subseteq K$ with $H, L \subseteq M$.

Note that $H \vee L$ does exist since K itself is a subfield of K containing H and L, and the intersection of fields containing F is a field (verify!). Thus $H \vee L$ is a field containing H and L and is the smallest such subfield of K. That is, $H \vee L = F(H \cup L)$ is the subfield of K generated by H and L. If H and L are algebraic over F there is a nicer description of $H \vee L$.

THEOREM 14.10 Let $F, H, L \subseteq K$ be fields with $F \subseteq H \cap L$.

i) If H and L are algebraic over F, then $H \vee L = F[H \cup L]$ is algebraic over F.
ii) $[H \vee L : F]$ is finite $\Leftrightarrow$ $[H : F]$ and $[L : F]$ are finite.
iii) If $[H : F] = n$ and $[L : F] = m$, then n and m divide $[H \vee L : F]$ and $[H \vee L : F] \leq nm$.
iv) If in iii) $(n, m) = 1$, then $H \cap L = F$, $[H \vee L : F] = nm$, and $[H \vee L : H] = [L : F]$.

Proof If M is a field with $H, L \subseteq M \subseteq K$ then M contains $F[H \cup L]$, so for i), it suffices to prove that $T = F[H \cup L]$ is a field, algebraic over F. Any $\beta \in T$ is a polynomial in $s_1, \ldots, s_m \in H \cup L$ so $\beta \in F[s_1, \ldots, s_m] \subseteq T$. Theorem 14.9 shows that $F[s_1, \ldots, s_m]$ is a field, algebraic over F, and both $\beta^{-1} \in T$ and β algebraic over F follow. Thus $F[H \cup L] = H \vee L$ is algebraic over F when H and L are. Since $H, L \subseteq H \vee L$, Theorem 14.8 shows that $[H : F]$ and $[L : F]$ are finite when $[H \vee L : F]$ is and then both divide $[H \vee L : F]$. When $[H : F] = n$ and $[L : F] = m$, with respective bases $\{\alpha_i\}$ and $\{\beta_j\}$ over F, then $H \vee L = F[H \cup L] = HL = \left\{\sum h_i l_i \mid h_i \in H, l_i \in L\right\} = \sum F\alpha_i\beta_j$. Hence $span_F(\{\alpha_i\beta_j\}) = H \vee L$. There is a minimal $B \subseteq \{\alpha_i\beta_j\}$ with $span(B) = H \vee L$, so B is a basis by Theorem 14.5. Thus $[H \vee L : F] \leq nm$, verifying ii) and iii). For iv), when $(n, m) = 1$, iii) shows that $nm \mid [H \vee L : F]$, forcing $[H \vee L : F] = nm$. Also, Theorem 14.8, $H \cap L \subseteq H$, and $H \cap L \subseteq L$ show that $[H \cap L : F]$ divides n and m and so is 1. By this same theorem,

$nm = [H \vee L : F] = [H \vee L : H][H : F] = [H \vee L : H] \cdot n$. Therefore $[H \vee L : H] = m = [L : F]$, completing the proof. ∎

The theorem applies directly to Example 14.3 and Example 14.6. A standard visual representation of the last part of the theorem when $[H : F] = n$, $[L : F] = m$, and $(n, m) = 1$ is shown in Figure 14.1.

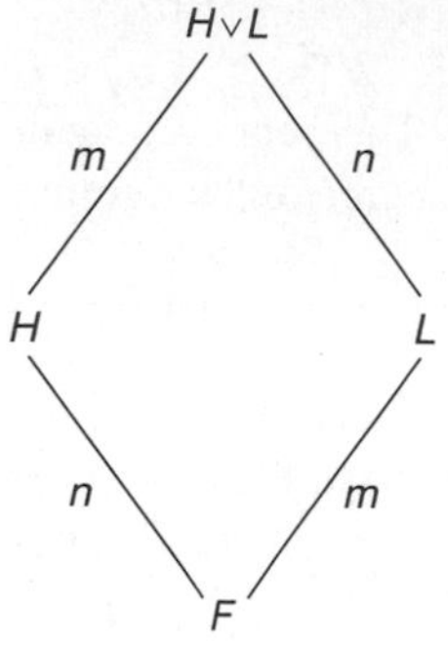

FIGURE 14.1

EXERCISES

In the following exercises, F, H, K, L, and M are fields.

1. Find the degree of the minimal polynomial for:
   i) $1 + 3^{1/5} \in \boldsymbol{R}$ over $\boldsymbol{Q}$
   ii) $1 + 2^{1/6} \in \boldsymbol{R}$ over $\boldsymbol{Q}$ and over $\boldsymbol{Q}[2^{1/2}]$
   iii) $e^{\pi i/4} = (2^{1/2}/2)(1 + i) \in \boldsymbol{C}$ over $\boldsymbol{Q}$
   iv) $2^{1/2} + 3^{1/2} \in \boldsymbol{R}$ over $\boldsymbol{Q}$
2. If $F \subseteq L$ with $[L : F] = p$ a prime, show that $F \subseteq M \subseteq L \Rightarrow M = F$ or $M = L$.
3. If $\boldsymbol{Q} \subseteq L$ with $[L : \boldsymbol{Q}] = 2$, show that $L = \boldsymbol{Q}[d^{1/2}]$ for some $d \in \boldsymbol{Q}$. Can $\boldsymbol{Q}$ be replaced with $\boldsymbol{Z}_p$ or with any F so that *char* $F \neq 2$ (see Theorem 10.15)?
4. If $F \subseteq L \subseteq K$ and $\beta \in K$ is algebraic over F, show that $[L[\beta] : L] \leq [F[\beta] : F]$.
5. For $F \subseteq L$ and $\{\alpha_1, \ldots, \alpha_m\}$ a basis of L over F, show that $L = F[\alpha_1, \ldots, \alpha_m]$.
6. Let $F \subseteq L$ and $\varnothing \neq S \subseteq L$ a set of elements algebraic over F. Prove that $F[S]$ is a field algebraic over F.
7. For $F \subseteq L \subseteq M$, if M is algebraic over L and L is algebraic over F, show that M is algebraic over F.
8. Let $S = \{\alpha \in \boldsymbol{C} \mid \alpha^k = 1$ for some $k \in \boldsymbol{N}\}$ and $L = \boldsymbol{Q}[S]$. Show that L is a field algebraic over $\boldsymbol{Q}$ and contains $\sin(2\pi/n)$ and $\cos(2\pi/n)$ for all $n \in \boldsymbol{N}$.
9. If $\boldsymbol{Z}_p \subseteq L$, $|L| = p^n$, and $\beta \in L$ with $m_\beta[x] \in \boldsymbol{Z}_p[x]$ what can *deg* m_β be?
10. Let $\boldsymbol{Z}_2 \subseteq L$ with $|L| = 4096$. If $F \subseteq L$ what can $|F|$ be?
11. Let $F \subseteq L$, $\alpha \in L$ with minimal polynomial $m_\alpha(x) \in F[x]$, and $m_\alpha(x) = p(x^n)$ for $p(x) \in F[x]$ irreducible of degree k. If $s \in \boldsymbol{N}$ and $s \mid n$, show that $[F[\alpha] : F[\alpha^s]] = s$ and find the degree of the minimal polynomial of α^s over F.

12. If $F \subseteq H, L \subseteq K$ with H and L algebraic over F, show that $H \vee L = H[L] = L[H]$.
13. Let $F \subseteq H, L \subseteq K$. If $[H \vee L : F]$ is finite, show that:
 i) $[H \vee L : H] = [L : F] \Leftrightarrow [H \vee L : L] = [H : F]$
 ii) $([H : H \cap L], [L : H \cap L]) = 1 \Rightarrow [H \vee L : L] = [H : H \cap L]$
14. Let $k_1, \ldots, k_m \in \boldsymbol{N}$ so that $(k_i, k_j) = 1$ for $i \neq j$ and no k_j is a square in $\boldsymbol{N}$. Prove that $[\boldsymbol{Q}[k_1^{1/2}, \ldots, k_m^{1/2}] : \boldsymbol{Q}] = 2^m$. (Observe first that $\boldsymbol{Q}[k_1^{1/2}, \ldots, k_m^{1/2}] = \boldsymbol{Q}[k_1^{1/2}, (k_1k_2)^{1/2}, \ldots, (k_1k_2 \cdots k_m)^{1/2}]$ and use induction on m.)
15. If $T = \{p^{1/2} \in \boldsymbol{R} \mid p \in \boldsymbol{N} \text{ is a prime}\}$, show that $\boldsymbol{Q}[T]$ is an infinite dimensional algebraic extension of $\boldsymbol{Q}$ and contains $n^{1/2}$ for all $n \in \boldsymbol{N}$. If $\beta \in T$ what can $[\boldsymbol{Q}[\beta] : \boldsymbol{Q}]$ be? (See Exercise 14.)

14.3 GEOMETRIC CONSTRUCTIONS

The results of the last section imply the impossibility of certain classical geometric constructions that use a straightedge and compass alone. A straightedge can be considered to be a ruler without marks, so it enables us to draw the straight line (segment!) passing through two distinct points in the plane. A compass allows the construction of a circle with a given point as center and with radius defined by that point and any other. The constructions asked for classically are: construct a square with area that of a given circle (*square the circle*); construct the edge of a cube having twice the volume of a given cube (*duplicate the cube*); and trisect a given angle. We will see that in general each of these is impossible, although special cases may be constructible. To be clear, *the allowable constructions in the plane are the lines and circles obtained from constructed points, together with the points of intersection of such straight lines and circles.*

Recall from plane geometry that by using a straightedge and compass we can construct a line perpendicular to a given straight line and passing through a given point. This implies that we can construct a line parallel to a given line and through a given point (how?). In particular, we can construct two perpendicular lines, the axes, and use these as a coördinate system for the plane. The intersection of the axes is O, the origin. Take any other point on either axis and define the distance between it and O to be the *unit distance,* constructing $1 \in \boldsymbol{N}$. The circle with unit radius and center O yields four points on the axes, each a unit distance from O. Constructing a circle from each of these points with unit radius gives points two units from O. Continuing allows the construction of points at any distance $n \in \boldsymbol{N}$ from O. By specifying the direction from O on either axis, we can consider that $\boldsymbol{Z}$ is constructed as well. The points constructed so far, called *integer points,* are assigned coordinates of $(n, 0)$ or $(0, n)$, as usual.

DEFINITION 14.10 A point is **constructible** if it is an integer point or if it is an intersection of lines and circles constructed by straightedge and compass using constructible points already obtained. A real number $r \in \boldsymbol{R}$ is **constructible** if it is a coordinate of a constructible point.

To determine what numbers are constructible, we first need to see that the segment joining any two constructible points can be copied to any nonperpendicular line L using any point on L as one endpoint. A picture of this when the segment is not on L appears in Figure 14.2, followed by an explanation.

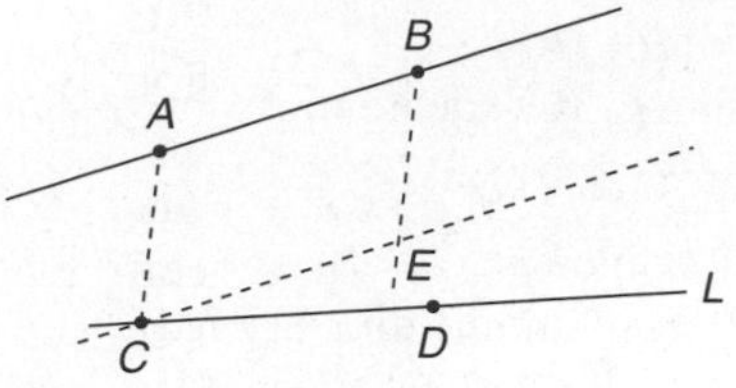

FIGURE 14.2

Given constructible points A and B, and a constructible point C on the line L, construct a line through C parallel to the line defined by A and B. Now construct a line through B parallel to the line through A and C, intersecting the first line constructed at the point E. The circle with center C and radius the segment from C to E intersects L at D, and the distance from A to B is the same as the distance from C to D. In this construction, if d is the distance from A to B, L is an axis, and $C = \mathrm{O}$, then D has coordinates $(d, 0)$: d is a constructible number. Note also that we could now copy the segment from C to D to a different place on the line through A and B.

If P is a constructible point, we can construct lines through P parallel to the axes and so construct its coordinates (or their negatives) as distances from O on the axes. For any set of constructible numbers we claim that the field generated by these consists of constructible numbers. To show this it suffices to represent sums, products, and reciprocals of positive constructible numbers as constructed distances. Since we can move segments from one line to another, it is clear how to construct sums and differences. To construct products and reciprocals we use similar triangles. The diagram shown in Figure 14.3 is the crucial one.

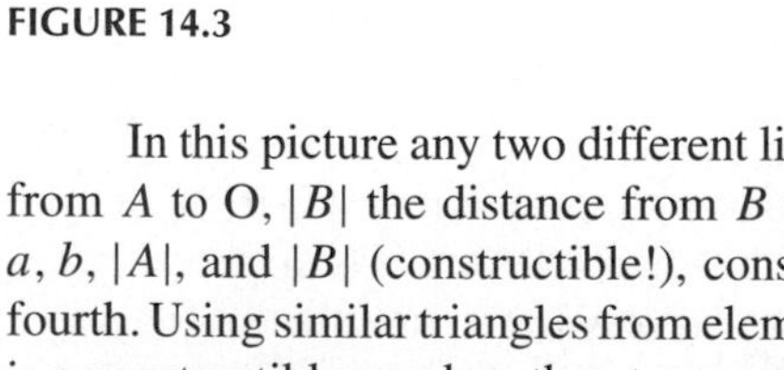

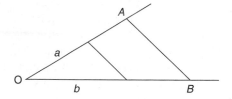

FIGURE 14.3

In this picture any two different lines through O may be used. Let $|A|$ be the distance from A to O, $|B|$ the distance from B to O, $|A| > a > 0$, and $|B| > b > 0$. Given three of $a, b, |A|$, and $|B|$ (constructible!), constructing a parallel line as in Figure 14.3 gives the fourth. Using similar triangles from elementary geometry shows that $a/b = |A|/|B|$. If $d > 0$ is a constructible number, then to construct $1/d$ when $d < 1$, let $a = d$, $b = 1$, and $|A| = 1$. Then $|B| = |A|b/a = 1/d$. When $d > 1$, set $|A| = d$ and $|B| = a = 1$, so $b = a|B|/|A| = 1/d$. To construct products, suppose that $x, y > 0$ are constructible numbers and $y > 1$. Then for $a = x, b = 1$, and $|B| = y$, $|A| = a|B|/b = xy$. If $0 < x, y < 1$ then we can construct $1/x \cdot 1/y$ and then its reciprocal xy.

THEOREM 14.11 The constructible numbers form a field $Con(\boldsymbol{Q})$. If $\beta \in Con(\boldsymbol{Q})$, then $[\boldsymbol{Q}[\beta]:\boldsymbol{Q}] = 2^m$.

Proof The discussion above shows that $Con(\boldsymbol{Q})$ is a field containing $\boldsymbol{Q}$. Let $F \subseteq Con(\boldsymbol{Q})$ be a field containing $\boldsymbol{Q}$, and $P = (d, e)$ the intersection of two straight lines, of one such with a circle, or of two circles, all constructed from points with coordinates in F. We claim $F[d] \subseteq Con(\boldsymbol{Q})$ is a field with $[F[d] : F] \le 2$, and similarly for $F[e]$. If U and V are points with coordinates in F, the straight line through them has equation with coefficients in F, and the circle with center U and radius the distance $r \in F$ from U to V has equation with its coefficients in F. The intersection (if any!) of two equations of these types has each coordinate in F, or $F[t^{1/2}]$ for some $t \in F$ (why?). When $d = u + vt^{1/2}$ with $u, v \in F$ then $t^{1/2} = (d - u)/v$ is constructible. Theorem 14.7 shows that $F[t^{1/2}]$ is a field with $[F[t^{1/2}]: F] \le 2$, and any $\alpha = a + bt^{1/2} \in F[t^{1/2}]$ is constructible as well (why?). This verifies our claim about $F[d]$.

Any $\beta \in Con(\boldsymbol{Q})$ is a coordinate of a constructible point P_n obtained as an intersection and requiring the construction of points $P_1, \ldots, P_n$ where P_1 is constructed from the integer points, and P_i is constructed from $P_1, \ldots, P_{i-1}$ and the integer points. Let $P_j = (d_{2j-1}, d_{2j})$, observe that the rational numbers are constructible, and set $F = \boldsymbol{Q}[d_1, \ldots, d_{2n}]$. From the last paragraph, both $[\boldsymbol{Q}[d_1] : \boldsymbol{Q}] \le 2$ and $[\boldsymbol{Q}[d_1, \ldots, d_{j+1}] : \boldsymbol{Q}[d_1, \ldots, d_j]] \le 2$ so Theorem 14.8 yields first that $[F : \boldsymbol{Q}] = 2^s$ and then that $[F : \boldsymbol{Q}] = [F : \boldsymbol{Q}[\beta]][\boldsymbol{Q}[\beta] : \boldsymbol{Q}]$. Therefore $[\boldsymbol{Q}[\beta] : \boldsymbol{Q}] = 2^m$ as required. ■

The reason for the impossibility of the three classical constructions mentioned earlier is essentially Theorem 14.11: any constructible number satisfies an irreducible in $\boldsymbol{Q}[x]$ of degree 2^m. A constructible square with area π, the area of the unit circle, has sides of length $\pi^{1/2}$. Thus $\pi^{1/2}$, and hence π, must be constructible. It is well known that π is transcendental over $\boldsymbol{Q}$ and so is not constructible by Theorem 14.11.

As for duplicating the cube, given a constructible $s > 0$, the cube with edge length s has volume s^3, so duplicating this cube requires constructing $2^{1/3}s$. When $s = 1$ we need to construct $2^{1/3}$. By Eisenstein's criterion, $x^3 - 2 \in \boldsymbol{Q}[x]$ is irreducible so $[\boldsymbol{Q}[2^{1/3}] : \boldsymbol{Q}] = 3$, and $2^{1/3}$ is not constructible. Therefore in general, cubes cannot be duplicated.

Given $0 < \theta < \pi/2$, the intersection of the unit circle with an angle of radian measure θ and vertex at O leads to the construction of $\cos(\theta)$. It is possible to construct an equilateral triangle with one vertex at O and base on the horizontal axis (how?). Thus if every angle can be trisected using straightedge and compass, then trisecting the angle $\pi/3$ in an equilateral triangle leads to the construction of $2\cos(\pi/9)$. Trigonometric identities yield

$$\begin{aligned}1/2 = \cos(\pi/3) &= \cos(2\pi/9)\cos(\pi/9) - \sin(2\pi/9)\sin(\pi/9)\\ &= \cos^3(\pi/9) - \sin^2(\pi/9)\cos(\pi/9) - 2\sin^2(\pi/9)\cos(\pi/9)\\ &= \cos^3(\pi/9) - 3(1 - \cos^2(\pi/9))\cos(\pi/9)\end{aligned}$$

and $2\cos(\pi/9)$ satisfies $x^3 - 3x - 1 \in \boldsymbol{Q}[x]$. This polynomial is irreducible over $\boldsymbol{Q}$ since it has no root in $\boldsymbol{Q}$ (verify!). Hence $[\boldsymbol{Q}[2\cos(\pi/9)] : \boldsymbol{Q}] = 3$, and $2\cos(\pi/9)$ is not constructible by Theorem 14.11, so an angle of radian measure $\pi/3$ cannot be trisected using straightedge and compass.

EXERCISES

1. Show carefully that an equilateral triangle is constructible.
2. Given a square, show how to construct a square of twice its area.
3. Describe how to bisect a given angle using a straightedge and compass.
4. If $k \in N$, show that $\cos(\pi/2^k)$ and $\sin(\pi/2^k)$ are constructible (Exercise 3).
5. Using straightedge and compass only, show how to "essentially" trisect an angle. That is, show how to construct an angle with measure as close as desired to one-third that of the given angle. (See Exercise 3.)
6. If $a, b \in \boldsymbol{R}$ are constructible, show that $(a^2+b^2)^{1/2}$ is also.
7. If $q \in \boldsymbol{Q}$ with $q > 0$, show that $q^{1/2}$ is constructible. (Use Exercise 6 to show first that for any $n \in \boldsymbol{N}$, $n^{1/2}$ is constructible.)
8. Given $m \in \boldsymbol{N}$ and a constructible circle C, show that a circle of area m times that of C is constructible.
9. If $m \in \boldsymbol{R}$ is constructible and P is a constructible point, show that the straight line passing through P and with slope m is constructible.
10. Let F be the field of all constructible real numbers. If $\beta \in F$ and $\beta > 0$ show that $\beta^{1/2} \in F$. (If $\gamma \in F$ with $0 < \gamma < 1$, construct an appropriate right triangle to see that $(1-\gamma^2)^{1/2}$ is constructible. Using that $1+\gamma$ is constructible and Exercise 6, show that $(1+\gamma)^{1/2}$ is constructible.)

14.4 SPLITTING FIELDS

Given a field F and irreducible $p(x) \in F[x]$, we want to construct a field L containing F with $p(x) = a(x-\alpha_1)\cdots(x-\alpha_n)$ in $L[x]$. By Theorem 11.3, $F[x]$ is a PID and by Theorem 12.10, $(p(x)) \lhd F[x]$ is a maximal when $p(x)$ is irreducible. Hence Theorem 11.17 shows that $F[x]/(p(x))$ is a field when $p(x)$ is irreducible. A precise description of $K = F[x]/(p(x))$ is given in Theorem 11.18 and we recall its content now. For $g(x) \in F[x]$ write $[g(x)] = g(x) + (p(x)) \in K$ and let $deg\ p(x) = n$. Then $[g(x)] = [r(x)]$ for $r(x)$ the remainder when $p(x)$ divides $g(x)$ so $K = \{[a_0 + a_1x + \cdots + a_{n-1}x^{n-1}] \mid a_j \in F\}$ and any $\beta \in K$ is represented by a *unique* polynomial of degree at most $n-1$. If $[g(x)] \neq 0$ then since $p(x)$ is irreducible and $F[x]$ is a PID, for some $h(x), q(x) \in F[x]$, $1 = g(x)h(x) + p(x)q(x)$ and $[g(x)]^{-1} = [h(x)]$. Finally, if $Y = [x]$ and if $p(x) = x^n + b_{n-1}x^{n-1} + \cdots + b_0$, then in K, $[0] = [p(x)] = Y^n + [b_{n-1}]Y^{n-1} + \cdots + [b_0]$. Thus Y in K is, in effect, a root of $p(x)$.

The defect in the construction above is that F is not a subset of K, so $Y \in K$ is not really a root of $p(x)$. One way to deal with this is to identify F with $\{[a] \in K \mid a \in F\}$. Specifically, $\varphi : F \to K$ given by $\varphi(a) = [a]$ is a ring isomorphism from F to $\{[a] \in K \mid a \in F\} = \varphi(F)$ (verify!). Also, φ extends naturally to an isomorphism $\eta : F[x] \to \varphi(F)[x]$ via $\eta(a_mx^m + \cdots + a_1x + a_0) = [a_m]x^m + \cdots + [a_1]x + [a_0]$. Thus if we identify F with $\varphi(F) \subseteq K$, and $g(x) \in F[x]$ with $\eta(g(x)) \in K[x]$, then since $(\eta(p(x)))(Y) = 0_K$, we more or less have constructed a field containing F and a root $Y = [x]$ of $p(x)$.

We can use the construction above of $K = F[x]/(p(x))$ to find a field L that *actually contains F and a root of $p(x)$* by replacing $\varphi(F)$ with F. Start with the partition $\{\varphi(F), K - \varphi(F)\}$ of K, let $L = F \cup (K - \varphi(F))$, and note that $(K - \varphi(F)) \cap F = \emptyset$. (Alternatively, take $L = (F \times \{1\}) \cup ((K - \varphi(F)) \times \{2\})$ in $(K \cup F) \times \{1, 2\}$.)

Clearly $\sigma : K \to L$ defined by $\sigma(k) = k$ for $k \in K - \varphi(F)$, and $\sigma(\varphi(a)) = a$ for $a \in F$ is a bijection of sets. Define a ring $(L, +_L, \cdot_L)$ by mimicking the operations of the corresponding elements in K. For example, if $u, v \in L$ let $u +_L v = \sigma(\varphi(u+v))$ if $u, v \in F$, let $u +_L v = \sigma(u + \varphi(v))$ if $v \in F$ and $u \in K - \varphi(F)$, and let $u +_L v = \sigma(u + v)$ if $u, v \in K - \varphi(F)$. For multiplication, $u \cdot_L v = \sigma(uv)$ if $u, v \in K - \varphi(F)$, $u \cdot_L v = \sigma(u\varphi(v))$ if $v \in F$ and $u \in K - \varphi(F)$, and $u \cdot_L v = \sigma(\varphi(uv))$ if $u, v \in F$. With these operations it follows that $L \cong K$ as fields. Therefore $F \subseteq L = F[\sigma(Y)]$ and $p(\sigma(Y)) = 0$.

Using the ideas described in the last paragraph, we can assume that given an irreducible $p(x) \in F[x]$, there is a field L containing F and of the form $L = F[\alpha]$ with $p(\alpha) = 0$. This leads to the following result.

THEOREM 14.12 If F is a field and $p(x) \in F[x] - F$ is monic, then there is a field $L = F[\alpha_1, \ldots, \alpha_m]$ with $p(x) = (x - \alpha_1)\cdots(x - \alpha_m)$ in $L[x]$.

Proof Use induction on $deg\ p = m$. If $m = 1$, take $L = F$. Assume the theorem holds for any field K and monic $g(x) \in K[x]$ with $deg\ g < m$. Since $F[x]$ is a PID, $p(x)$ has a monic irreducible factor $h(x)$. As we have seen, there is a field $L_1 = F[\alpha_1] \supseteq F$ with $h(\alpha_1) = 0$. In $L_1[x]$, $p(x) = (x - \alpha_1)g(x)$ for $g(x) \in L_1[x]$ with $deg\ g = m - 1$. By the induction assumption there is a field L containing L_1 so that $L = L_1[\alpha_2, \ldots, \alpha_m]$ and $g(x) = (x - \alpha_2) \cdots (x - \alpha_m)$. Hence $F \subseteq L = F[\alpha_1, \ldots, \alpha_m]$ and $p(x) = (x - \alpha_1)g(x) = (x - \alpha_1) \cdots (x - \alpha_m)$ as required. ■

There is a special name for the field L in Theorem 14.12.

DEFINITION 14.11 For $F \subseteq L \subseteq M$ fields and $p(x) \in F[x] - F$, L is a **splitting field for** $p(x)$ **over** F if $p(x) = a(x - \alpha_1) \cdots (x - \alpha_m) \in L[x]$ and $L = F[\alpha_1, \ldots, \alpha_m]$. In this case we say that $p(x)$ **splits over** M.

By Theorem 14.12 there is a splitting field over F for any $p(x) \in F[x] - F$. Often we cannot explicitly describe roots of $p(x)$, even for $F = \boldsymbol{Q}$. For example, no root of $g(x) = x^5 - 2x^3 + 6x^2 - 2 \in \boldsymbol{Q}[x]$ in $\boldsymbol{C}$ is apparent. Note that $g(x)$ is irreducible in $\boldsymbol{Q}[x]$ by Theorem 12.16. We consider some examples that can be described rather completely.

Example 14.9 A splitting field of $p(x) = x^4 - 2$ over $\boldsymbol{Q}$.

By Theorem 12.16, $p(x)$ is irreducible and its set of roots in $\boldsymbol{C}$ is $S = \{2^{1/4}, -2^{1/4}, i2^{1/4}, -i2^{1/4}\}$. If $L = \boldsymbol{Q}[S] \subseteq \boldsymbol{C}$ then $L = \boldsymbol{Q}[2^{1/4}, i]$ (why?), $[L : \boldsymbol{Q}] = 8$ by Example 14.4, and $B = \{(2^{1/4})^j i^k\}$ is a basis of L over $\boldsymbol{Q}$.

Example 14.10 A splitting field of $f(x) = x^3 - 2$ over $\boldsymbol{Q}[2^{1/2}]$.

In $\boldsymbol{C}$ the roots of $f(x)$ are $2^{1/3}$, $2^{1/3}\omega$, and $2^{1/3}\omega^2$ for $\omega = e^{2\pi i/3} (\omega^3 = 1)$. By Theorem 12.16, all $x^m - 2 \in \boldsymbol{Q}[x]$ are irreducible so $[\boldsymbol{Q}[2^{1/2}] : \boldsymbol{Q}] = 2$. Since $\boldsymbol{Q}[2^{1/2}, 2^{1/3}] = \boldsymbol{Q}[2^{1/6}]$ (why?) we get $[\boldsymbol{Q}[2^{1/2}, 2^{1/3}] : \boldsymbol{Q}] = 6$. From Theorem 14.8, $[\boldsymbol{Q}[2^{1/2}, 2^{1/3}] : \boldsymbol{Q}[2^{1/2}]] = 3$ so $f(x) \in \boldsymbol{Q}[2^{1/2}]$ is irreducible using Theorem 14.7. Now $L = \boldsymbol{Q}[2^{1/2}][2^{1/3}, 2^{1/3}\omega, 2^{1/3}\omega^2] = \boldsymbol{Q}[2^{1/2}][2^{1/3}, \omega] = \boldsymbol{Q}[2^{1/6}, \omega]$ is a splitting field for

$f(x)$ over $\mathbf{Q}[2^{1/2}]$. Since $\omega \notin \mathbf{Q}[2^{1/6}] \subseteq \mathbf{R}$ and $[L:\mathbf{Q}] = [L:\mathbf{Q}[2^{1/6}]][\mathbf{Q}[2^{1/6}]:\mathbf{Q}]$, Theorem 14.8 shows first that $[L:\mathbf{Q}] = 12$ and then that $[L:\mathbf{Q}[2^{1/2}]] = 6$.

Example 14.11 A splitting field of $X^2 - x \in F(x)[X]$ over $F(x)$.

Recall that $F(x) = QF(F[x])$ is the field of fractions of $F[x]$. By Theorem 14.12 there is a field $F(x) \subseteq L = F(x)[y, -y] = F(x)[y]$ with $y^2 = x$. For $\beta \in L$, $\beta = f(x)/g(x) + (p(x)/q(x))y$ and $x = y^2$, so $\beta = u(y)/v(y) \in F(y)$. Thus $L = F(y)$ and it follows that $L \cong F(x)$ (how?)!

Example 14.12 A splitting field of $x^n - 1$ over $\mathbf{Q}$.

If $\rho = e^{2\pi i/n} \in \mathbf{C}$ then $\rho^n = 1$ and $\rho, \rho^2, \ldots, \rho^n$ are n distinct roots of $x^n - 1$ so $L = \mathbf{Q}[\rho, \rho^2, \ldots, \rho^n] = \mathbf{Q}[\rho] \subseteq \mathbf{C}$ is a splitting field of it over $\mathbf{Q}$. At this point it is not clear what $[L:\mathbf{Q}]$ is. Using Example 12.5 we have that $[L:\mathbf{Q}] = n - 1$ when n is a prime. This last fact fails for $x^n - 1 \in \mathbf{Z}_p[x]$. If $K = \mathbf{Z}_2[\alpha]$ splits $x^7 - 1 = (x-1)(x^3 + x + 1)(x^3 + x^2 + 1) \in \mathbf{Z}_2[x]$, then $[K:\mathbf{Z}_2] = 3$.

Our few examples shed little light on the relation between *deg p* for $p(x) \in F[x]$ and $[L:F]$ for L a splitting field for $p(x)$ over F. If $p(x)$ is irreducible then $(\deg p) \mid [L:F]$ by Theorem 14.7 and Theorem 14.8.

THEOREM 14.13 If F is a field, $p(x) \in F[x]$ with $\deg p = n > 0$, and L is a splitting field for $p(x)$ over F, then $[L:F] \mid n!$.

Proof If $n = 1$ then $L = F$ and there is nothing to prove. Assume the result holds for $g(x) \in K[x] - K$, K a field, and $\deg g < n$. Suppose that $p(x)$ is irreducible. By Theorem 14.7, $[F[\alpha_1]:F] = n$ if $\alpha_1 \in L$ is a root of $p(x)$. Let $p(x) = (x - \alpha_1)\cdots(x - \alpha_n) \in L[x]$, so $p(x) = (x - \alpha_1)g(x)$ in $F[\alpha_1][x]$ and $L = F[\alpha_1, \ldots, \alpha_n] = F[\alpha_1][\alpha_2, \ldots, \alpha_n]$ is a splitting field for $g(x)$ over $F[\alpha_1]$. Since $\deg g = n - 1$, $[L:F[\alpha_1]] \mid (n-1)!$ by the induction assumption. Applying Theorem 14.8 shows that $[L:F] \mid n(n-1)!$, proving the theorem when $p(x)$ is irreducible. When $p(x)$ is not irreducible then $p(x) = g(x)h(x)$ with $\deg g, \deg h < n$. By unique factorization in $L[x]$, $g(x)$ and $h(x)$ split over L with sets of roots S_g and S_h, respectively. The roots of $f(x)$ are all $\alpha \in S_g \cup S_h$. Now $F[S_g]$ is a splitting field of $g(x)$ over F, and $L = F[S_g \cup S_h] = F[S_g][S_h]$ is a splitting field for $h(x)$ over $F[S_g]$. By induction $[F[S_g]:F] \mid (\deg g)!$ and $[L:F[S_g]] \mid (\deg h)!$. Using Theorem 14.8 we have $[L:F] \mid (\deg g)!(\deg h)!$. The definition of the binomial coefficients and the fact that $n = \deg g + \deg h$ show that $(\deg g)!(\deg h)! \mid n!$, completing the proof. ■

A splitting field for $p(x) \in F[x]$ over F adjoins the roots of $p(x)$ to F. When can $p(x)$ have fewer distinct roots than its degree? Clearly $(x^3 - 1)^3 \in \mathbf{Q}[x]$ has degree 9 but only three roots in $\mathbf{C}$. Not all examples are so obvious. Consider $g(x) = x^{10} - [3] \in \mathbf{Z}_5[x]$. Since $3^5 \equiv_5 3$ by Fermat's Theorem, $x^{10} - [3] = (x^2 - [3])^5 \in \mathbf{Z}_5[x]$ (verify!). Thus a splitting field L over $\mathbf{Z}_5$ for $g(x)$ is $L = \mathbf{Z}_5[\alpha]$ with $\alpha^2 = [3]$, $[L:\mathbf{Z}_5] = 2$, and $\pm\alpha$ are the only roots of $g(x)$. Can an irreducible $p(x)$ have repeated roots? Theorem 12.16 shows that $X^5 - x \in QF(\mathbf{Z}_5[x])$ is irreducible, but if $\alpha \in L$ is a root, then $X^5 - x = (X-\alpha)^5 \in L[X]$.

The problem of whether a polynomial has distinct roots is related to the *characteristic* of the field, *char* F, discussed in §10.4. Briefly, if F is a field then $\varphi: \mathbf{Z} \to F$ given by $\varphi(k) = k \cdot 1_F$ is a ring homomorphism. Since F is a field, $\varphi(\mathbf{Z}) \cong \mathbf{Z}/(ker\,\varphi)$ is a domain. If $ker\,\varphi = (0)$ then $char\,F = 0: na = 0$ for $a \in F$ and $n \in \mathbf{Z}$ forces $a = 0_F$ or $n = 0_{\mathbf{Z}}$. When $ker\,\varphi = (p)$ for p a prime, then $char\,F = p: pa = 0_F$ for all $a \in F$. The connection between characteristic and roots requires looking at derivatives. For a field F of any characteristic we can define the formal derivative $D: F[x] \to F[x]$ just as in calculus, namely $D\left(\sum a_i x^i\right) = \sum i a_i x^{i-1}$. The usual computations still hold: for $g, h \in F[x]$ and $a \in F$, $D(ag+h) = aD(g) + D(h)$ and $D(gh) = gD(h) + D(g)h$.

THEOREM 14.14 Let F be a field, $g(x) \in F[x] - F$, and L a splitting field for $g(x)$ over F. In $L[x]$, $g(x) = (x-a)^2 h(x) \Leftrightarrow (x-a)$ divides both $g(x)$ and $D(g(x))$. If $g(x) \in F[x]$ is irreducible, then $g(x) = (x-a)^2 h(x) \in L[x] \Leftrightarrow char\,F = p > 0$ and $g(x) = v(x^p)$ for $v(x) \in F[x]$.

Proof If $g(x) = (x-a)h(x)$ in $L[x]$ then $D(g(x)) = h(x) + (x-a)D(h(x))$. Since $L[x]$ is a UFD, $(x-a) \mid D(g(x)) \Leftrightarrow (x-a) \mid h(x) \Leftrightarrow (x-a)^2 \mid g(x)$. When $g(x) \in F[x]$ is irreducible and $D(g(x)) \neq 0$, then $deg\ D(g(x)) < deg\ g(x)$ and $(g(x))$ maximal force $1 = q(x)g(x) + s(x)D(g(x))$. Thus $g(a) = 0 = D(g)(a)$ in L is impossible so $g(x)$ has no multiple root by the first statement of the theorem. Now $0 = D(g(x)) \Leftrightarrow char\,F = p > 0$ and $g(x) = v(x^p)$ for $v(x) \in F[x]$ (verify!), so in this case $(x-a) \mid g(x)$ implies $(x-a) \mid D(g(x))$ and again $(x-a)^2 \mid g(x)$. ■

We use Theorem 14.14 and the existence of splitting fields from Theorem 14.12 to show that there are finite fields of all allowable cardinalities. If F is a finite field then certainly $char F = p > 0$, and it follows that every $0 \neq \alpha \in (F, +)$ has order p, so $|F| = p^n$ by Cauchy's Theorem.

THEOREM 14.15 For $p, n \in \mathbf{N}$ with p prime let L be a splitting field of $g(x) = x^{p^n} - x \in \mathbf{Z}_p[x]$ over $\mathbf{Z}_p$. Then L is the set of p^n distinct roots of $g(x)$ in L so $|L| = p^n$ and $[L : \mathbf{Z}_p] = n$. Furthermore $L = \mathbf{Z}_p[\alpha]$ for some $\alpha \in L$.

Proof Now $D(g(x)) = -1 \neq 0$ so by Theorem 14.14, $g(x)$ has p^n distinct roots in L. Let the set of these be $S \subseteq L$ and note that $\mathbf{Z}_p \subseteq S$ by Fermat's Theorem and induction. Since $char\,L = p$, it follows that for $\alpha, \beta \in S$, $(\alpha - \beta)^p = \alpha^p - \beta^p$, so $\alpha^{p^n} = \alpha$ and $\beta^{p^n} = \beta$ imply both $(\alpha - \beta)^{p^n} = \alpha - \beta$ and $(\alpha\beta)^{p^n} = \alpha\beta$. Consequently S is a ring. In fact S is a field by Theorem 10.14, or notice that for $0 \neq \alpha \in S$, $\alpha^{p^n - 1} = 1$ so $\alpha^{-1} = \alpha^k \in S$ for $k = p^n - 2$ (verify!). Hence $S = \mathbf{Z}_p[S] = L$ and $|L| = p^n$. Since L is a vector space over $\mathbf{Z}_p$, $[L : \mathbf{Z}_p] = n$ follows (why?). Finally, $(L^*, \cdot) = \langle \alpha \rangle$ is a cyclic group by Theorem 4.20, so $L = \mathbf{Z}_p[\alpha]$. ■

Since any splitting field of $x^{p^n} - x \in \mathbf{Z}_p$ over $\mathbf{Z}_p$ is the set of roots of this polynomial, it is natural to suspect that any two such fields are isomorphic. There may be many different splitting fields over F for a fixed $p(x) \in F[x]$. For example, $\mathbf{R}[x]/(x^2+1)$, $\{aI_2 + b(E_{12} - E_{21}) \in M_2(\mathbf{R}) \mid a, b \in \mathbf{R}\}$, and $\mathbf{C} = \mathbf{R}[i]$ are all different splitting fields for

x^2+1 over $\boldsymbol{R}$ but are isomorphic. Are any two splitting fields for $p(x) \in F[x]$ over F isomorphic? We will show that this is the case by looking at extensions of isomorphisms.

Given a ring isomorphism $\varphi: F_1 \to F_2$ between fields F_i, we also use φ to represent the isomorphism from $F_1[x]$ to $F_2[x]$ (verify!) defined by $\varphi(\sum a_j x^j) = \sum \varphi(a_j)x^j$. *Observe that $p(x) \in F_1[x]$ is irreducible if and only if $\varphi(p(x)) \in F_2[x]$ is irreducible.* It is routine to see that the ideals $\varphi((p(x))) = (\varphi(p(x)))$ so the Correspondence Theorem (Theorem 11.16) shows that $F_1[x]/(p(x)) \cong F_2[x]/(\varphi(p(x)))$. Recall our observation in §14.2 that if $F \subseteq L$, a field, and if $\alpha \in L$ is a root of the irreducible $p(x) \in F[x]$, then $F[\alpha] \cong F[x]/(p(x)) = K$ via $\eta_\alpha([f(x)]) = f(\alpha)$, where $[f(x)] = f(x) + (p(x)) \in K$. Under this isomorphism, $[x]$ gets sent to α, so $\eta_\alpha^{-1}(\alpha) = [x]$. These observations are used to prove an important result on isomorphisms.

THEOREM 14.16 Let $F_1 \subseteq L_1$ and $F_2 \subseteq L_2$ be fields, $p(x) \in F_1[x]$ irreducible, $\varphi: F_1 \to F_2$ an isomorphism, $\alpha \in L_1$ a root of $p(x)$, and $\beta \in L_2$ a root of $\varphi(p(x))$. There is an isomorphism $\eta: F_1[\alpha] \to F_2[\beta]$ extending φ, that is $\eta(a) = \varphi(a)$ for all $a \in F_1$, and $\eta(\alpha) = \beta$.

Proof From the comments at the beginning of §14.2 and our remarks above, there are isomorphisms $\eta_\alpha: F_1[x]/(p(x)) \to F_1[\alpha]$ and $\eta_\beta: F_2[x]/(\varphi(p(x))) \to F_2[\beta]$ with $\eta_\alpha(g(x) + (p(x))) = g(\alpha)$ and $\eta_\beta(h(x) + (\varphi(p(x)))) = h(\beta)$. The ideals $\varphi((p(x))) = (\varphi(p(x)))$, so by Theorem 11.16, $F_1[x]/(p(x)) \cong F_2[x]/(\varphi(p(x)))$ via ρ defined by $\rho(g(x) + (p(x))) = \varphi(g(x)) + (\varphi(p(x)))$. The isomorphism $\eta = \eta_\beta \circ \rho \circ \eta_\alpha^{-1}: F_1[\alpha] \to F_2[\beta]$ is the required one. That η is an isomorphism follows from Theorem 11.10 and Theorem 0.14. As for the other properties, note that if $a \in F_1$, then $\eta_\beta \circ \rho \circ \eta_\alpha^{-1}(a) = \eta_\beta \circ \rho(a + (p(x))) = \eta_\beta(\varphi(a) + (\varphi(p(x)))) = \varphi(a)$ and that

$\eta_\beta \circ \rho \circ \eta_\alpha^{-1}(\alpha) = \eta_\beta \circ \rho(x + (p(x))) = \eta_\beta(x + (\varphi(p(x)))) = \beta$. ■

Example 14.13 $\eta: \boldsymbol{Q}[2^{1/4}] \to \boldsymbol{Q}[2^{1/4}i]$ with $\eta(2^{1/4}) = 2^{1/4}i$.

As in Example 14.9, $x^4 - 2 \in \boldsymbol{Q}[x]$ is irreducible with $2^{1/4}, 2^{1/4}i \in \boldsymbol{C}$ as roots. By Theorem 14.16 there is an isomorphism $\eta: \boldsymbol{Q}[2^{1/4}] \to \boldsymbol{Q}[2^{1/4}i]$ extending $I_{\boldsymbol{Q}}$ with $\eta(2^{1/4}) = 2^{1/4}i$. It follows that $\eta(a + b2^{1/4} + c2^{1/2} + d2^{3/4}) = a + b2^{1/4}i - c2^{1/2} - d2^{3/4}i$.

Theorem 14.16 is a crucial step in proving that splitting fields over F for a given $p(x) \in F[x] - F$ must be isomorphic.

THEOREM 14.17 If F_1 and F_2 are fields, $\varphi: F_1 \to F_2$ is an isomorphism, $p(x) \in F_1[x] - F_1$, L_1 is a splitting field for $p(x)$ over F_1, and L_2 is a splitting field for $\varphi(p(x))$ over F_2, then there is an isomorphism $\eta: L_1 \to L_2$ extending φ. In particular, if $F_1 = F_2 = F$ and $\varphi = I_F$ then $L_1 \cong L_2$ via η extending I_F.

Proof Use induction on $deg\ p = n$. If $n = 1$ then $L_1 = F_1$, $L_2 = F_2$, and $\eta = \varphi$ is the required isomorphism. Assume the theorem holds for polynomials of degree less than n. If $g(x) \in F_1[x]$ is an irreducible factor of $p(x) \in F_1[x]$ with $g(\alpha) = 0$ for $\alpha \in L_1$, then $\varphi(g(x)) \in F_2[x]$ is an irreducible factor of $\varphi(p(x))$ in $F_2[x]$ and has a root $\beta \in L_2$. By Theorem 14.16 there is an isomorphism $\mu : F_1[\alpha] \to F_2[\beta]$ extending φ with $\mu(\alpha) = \beta$. Let S be the set of roots of $p(x)$ in L_1 and write $p(x) = (x - \alpha)^k h(x)$ for $h(x) \in F_1[\alpha][x]$ not divisible by $(x - \alpha)$. Thus $L_1 = F_1[S] = F_1[\alpha][S - \{\alpha\}]$ is a splitting field for $h(x)$ over $F_1[\alpha]$. Now $\mu(p(x)) = (x - \beta)^k \mu(h(x))$, β is a root in L_2 of $\mu(p(x))$, and L_2 is a splitting field for $\mu(h(x))$ over $F_2[\beta]$ (why?). Since $deg\, h = n - k$, the induction assumption shows that there is an isomorphism $\eta : L_1 \to L_2$ extending μ, so extending φ, and hence completing the proof. ■

The theorem shows that any two splitting fields for $p(x) \in F[x]$ over F must be isomorphic, so splitting fields are essentially unique. Another very important consequence of the theorem is the ability to define specific automorphisms of a splitting field of an irreducible, as we see next.

THEOREM 14.18 If F is a field, $p(x), q(x) \in F[x]$, L is a splitting field for $p(x)$ over F, and $\alpha, \beta \in L$ are roots of the irreducible $q(x)$, then there is an automorphism $\eta : L \to L$ extending I_F with $\eta(\alpha) = \beta$.

Proof By Theorem 14.16 there is an isomorphism $\mu : F[\alpha] \to F[\beta]$ extending I_F with $\mu(\alpha) = \beta$. Now L is a splitting field for $p(x)$ over both $F[\alpha]$ and $F[\beta]$ (verify!) so Theorem 14.17 shows that there is an isomorphism $\eta : L \to L$ extending μ. Hence η is an automorphism of L, η extends I_F, and $\eta(\alpha) = \beta$. ■

The last three results are extremely useful for studying individual automorphisms and the group of automorphisms of a splitting field.

DEFINITION 14.12 For fields $F \subseteq L$, the ***F*-automorphisms of *L***, denoted $Aut(L/F)$, is $\{\varphi : L \to L \mid \varphi$ is an automorphism of L and $\varphi(a) = a$ for all $a \in F\}$.

It is an exercise that $(Aut(L/F), \circ)$ is a group. If $L = F[\alpha_1, \ldots, \alpha_m]$ is the splitting field over F of the irreducible $p(x) \in F[x]$ with $S = \{\alpha_j\}$, the roots of $p(x)$ in L, then each $\varphi \in Aut(L/F)$ is certainly uniquely determined by $\varphi(\alpha_1), \ldots, \varphi(\alpha_m)$, and also all $\varphi(\alpha_j) \in S$. To see why, note that $p(\alpha_j) = 0$ implies $0 = \varphi(p(\alpha_j)) = p(\varphi(\alpha_j))$ since φ fixes each coefficient of $p(x)$. *Thus φ acts as a permutation on S.* We use these comments and the previous results to illustrate how $Aut(L/F)$ can be determined in a specific case.

Example 14.14 $Aut(L/\mathbf{Q})$ for L a splitting field for $x^4 - 2 \in \mathbf{Q}[x]$ over $\mathbf{Q}$.

By Example 14.9, $x^4 - 2 \in \mathbf{Q}[x]$ is irreducible with $S = \{\pm 2^{1/4}, \pm 2^{1/4} i\}$ in $\mathbf{C}$ as roots, and $L = \mathbf{Q}[S] = \mathbf{Q}[2^{1/4}, i] = \mathbf{Q}[2^{1/4} i, i]$ with $[L : \mathbf{Q}] = 8$. In L, $\pm i$ are the roots of the irreducible $x^2 + 1 \in \mathbf{Q}[x]$, so our comments above show that $\varphi \in G = Aut(L/\mathbf{Q})$ is uniquely determined by $\varphi(i) = \pm i$ and $\varphi(2^{1/4}) \in S$. It follows that $|G| \leq 8$. Let $\rho : \mathbf{Q}[i] \to \mathbf{Q}[i]$ be complex conjugation ($\rho(a + bi) = a - bi$), an automorphism. Now $L = \mathbf{Q}[i][2^{1/4}]$, $[L : \mathbf{Q}] = 8$, and $[\mathbf{Q}[i] : \mathbf{Q}] = 2$ imply that $[L : \mathbf{Q}[i]] = 4$ by Theorem 14.8, so $x^4 - 2 \in \mathbf{Q}[i]$ is irreducible (Theorem 14.7). Applying Theorem 14.16 yields $\mu_1, \ldots, \mu_4 \in G$ extending ρ on $\mathbf{Q}[i]$ with $\mu_1(2^{1/4}) = 2^{1/4}$, $\mu_2(2^{1/4}) = -2^{1/4}$, $\mu_3(2^{1/4}) = 2^{1/4} i$, and $\mu_4(2^{1/4}) = -2^{1/4} i$. Similarly, using $I_{\mathbf{Q}[i]} : \mathbf{Q}[i] \to \mathbf{Q}[i]$ there are $\upsilon_1, \ldots, \upsilon_4 \in G$ extending $I_{\mathbf{Q}[i]}$ with $\upsilon_j(2^{1/4}) = \mu_j(2^{1/4})$. We show that these eight automorphisms correspond to distinct permutations of S and so constitute G. Since $\mu_1(i)=-i$ and $\mu_1(2^{1/4})=2^{1/4}$, it follows that $\mu_1(-2^{1/4})=-2^{1/4}$ (why?), $\mu_1(2^{1/4}i)=\mu_1(2^{1/4})\mu_1(i)=-2^{1/4}i$, and $\mu_1(-2^{1/4}i)=\mu_1(-2^{1/4})\mu_1(i)=-2^{1/4}(-i)=2^{1/4}i$, so μ_1 corresponds to the 2-cycle $(2^{1/4}i, -2^{1/4}i)$. Similarly, $\upsilon_4(i)=i$ and $\upsilon_4(2^{1/4})=-2^{1/4}i$ lead to $\upsilon_4(-2^{1/4}i)=\upsilon_4(-2^{1/4})\upsilon_4(i)=(2^{1/4}i)i=-2^{1/4}$, and $\upsilon_4(-2^{1/4})=2^{1/4}i$. Therefore υ_4 corresponds to the 4-cycle $(2^{1/4}, -2^{1/4}i, -2^{1/4}, 2^{1/4}i)$. Each μ_j and υ_j can be identified in the same way with a permutation of S, and these can be used to find products in G. Observe that $\mu_1^{-1}\upsilon_4\mu_1=\mu_1\upsilon_4\mu_1$ corresponds to $(2^{1/4}i, -2^{1/4}i)(2^{1/4}, -2^{1/4}i, -2^{1/4}, 2^{1/4}i)(2^{1/4}i, -2^{1/4}i)=(2^{1/4}, 2^{1/4}i, -2^{1/4}, -2^{1/4}i)$, which corresponds to υ_4^{-1}. It follows directly from Example 6.4 that G is isomorphic to the dihedral group D_4.

EXERCISES

In all the exercises below, E, F, K, H, L, and M are fields.

1. For each $p(x)\in \boldsymbol{Q}[x]$ that follows, find its roots $\{\alpha_i\}\subseteq \boldsymbol{C}$, at most two β_j in $\boldsymbol{C}$ so that $L=\boldsymbol{Q}[\{\alpha_i\}]=\boldsymbol{Q}[\{\beta_j\}]$, and $[L:\boldsymbol{Q}]$:
  i) x^4-5x^2+6
  ii) x^6-3x^3+2
  iii) x^4-4
  iv) $x^{12}-1$ (if $\omega\in \boldsymbol{C}-\boldsymbol{Q}$ with $\omega^3=1$, then $(\omega i)^{12}=1$)
  v) x^6-2
  vi) x^5-2 (see Example 12.5)

2. If $F\subseteq L$ and $[L:F]=2$, show that L is a splitting field over F for $p(x)\in F[x]$.

3. If $F\subseteq M\subseteq L$ and L is a splitting field over F for $p(x)\in F[x]$ then L is splitting field over M for $p(x)\in M[x]$.

4. Show that the splitting fields in $\boldsymbol{C}$ for x^8-1 and x^4-4 over $\boldsymbol{Q}$ are the same.

5. If $H, M\subseteq L$ and if H and M are splitting fields over F for $p_H, p_M\in F[x]$, respectively, show that $H\vee M$ is a splitting field over F for some $p(x)\in F[x]$.

6. If $[L:F]=n$, show that there is $p(x)\in F[x]$ and a splitting field M for $p(x)$ over F so that $L\subseteq M$.

7. **a)** If E and F are finite with $|E|=|F|$, show that $E\cong F$.
  b) If $\boldsymbol{Z}_p, E, F\subseteq L$ and $|E|=p^n=|F|$, show that $E=F$.

8. Let $d, n, p\in \boldsymbol{N}$ with $d\mid n$ and p a prime, and let $\boldsymbol{Z}_p\subseteq L$, a field with $|L|=p^n$.
  i) Use Example 0.18 to show that $(x^d-1)\mid(x^n-1)$.
  ii) Show that $(x^{p^d}-x)\mid(x^{p^n}-x)$.
  iii) Show that there is a field $L_d\subseteq L$ with $|L_d|=p^d$.
  iv) Characterize n so that $\boldsymbol{Z}_p$ and L are the only subfields of L.

9. For $\boldsymbol{Q}\subseteq L\subseteq \boldsymbol{C}$, show that $Aut(L/\boldsymbol{Q})=Aut(L)$.

10. Describe:
  i) $Aut(\boldsymbol{C}/\boldsymbol{R})$
  ii) $Aut(\boldsymbol{Q}[3^{1/m}]/\boldsymbol{Q})$ for $m\in \boldsymbol{N}$

11. Show that $Aut(\boldsymbol{Q}[2^{1/4}]/\boldsymbol{Q})\cong \boldsymbol{Z}_2\cong Aut(\boldsymbol{Q}[2^{1/4}]/\boldsymbol{Q}[2^{1/2}])$.

12. Show that $\boldsymbol{Q}[2^{1/4}]$ is not a splitting field over $\boldsymbol{Q}$ for any $p(x)\in \boldsymbol{Q}[x]$.

13. Show that $Aut(\boldsymbol{R}/\boldsymbol{Q})=\{I_{\boldsymbol{R}}\}$. (If $\varphi\in Aut(\boldsymbol{R}/\boldsymbol{Q})$ and $r>0$ then $\varphi(r)>0$.)

14. If $F \subseteq L \subseteq M$, L is a splitting field for $p(x) \in F[x]$ over F, and $\varphi \in Aut(M/F)$, then $\varphi(L) = L$.

15. If $F \subseteq L \subseteq M$, M is a splitting field for $p(x) \in F[x]$ over F, and $\varphi \in Aut(L/F)$, show that there is $\mu \in Aut(M/F)$ so that $\mu|_L = \varphi$(that is, $\mu(a) = \varphi(a)$ for all $a \in L$).

16. Let $p \in \boldsymbol{N}$ be prime, $g(x) \in \boldsymbol{Z}_p[x]$ irreducible, and $deg\ g = n$.
   i) If L is a splitting field for $g(x)$ over $\boldsymbol{Z}_p$ and if $g(\alpha) = 0$ for $\alpha \in L$, show that $L = \boldsymbol{Z}_p[\alpha]$. (If $g(\beta) = 0$ for $\beta \in L$, show that $\boldsymbol{Z}_p[\alpha] = \boldsymbol{Z}_p[\beta]$—see Exercise 7.)
   ii) Show that $g(x) \mid (x^{p^n} - x)$ in $\boldsymbol{Z}_p[x]$.
   iii) If $h(x) \in \boldsymbol{Z}_p[x]$ is irreducible with $deg\ h \mid n$, show that $h(x)$ splits over L. (See Exercise 8.)

17. Let $\boldsymbol{Z}_p \subseteq F$ with $|F| = p^n$.
   i) Show that $Aut(F/\boldsymbol{Z}_p) = Aut(F)$.
   ii) If $\gamma : F \to F$ is given by $\gamma(a) = a^p$, show that $\gamma \in Aut(F)$.
   iii) Show that $o(\gamma) = n$ in $(Aut(F), \circ)$.
   iv) Show that $Aut(F) = \langle \gamma \rangle$. (If $F = \boldsymbol{Z}_p[\alpha]$ then α has minimal polynomial $m_\alpha(x) \in \boldsymbol{Z}_p[x]$ of degree n. Show that $\{\gamma^j(\alpha) \mid 0 \le j \le n-1\}$ are n distinct roots of $m_\alpha(x)$ in F.)

18. If $L \subseteq \boldsymbol{C}$ is a splitting field over $\boldsymbol{Q}$ for $x^8 - 1 \in \boldsymbol{Q}[x]$, show that $Aut(L/\boldsymbol{Q}) \cong U_8$.

19. Let $L \subseteq \boldsymbol{C}$ be a splitting field for $x^3 - 2 \in \boldsymbol{Q}[x]$ over $\boldsymbol{Q}$. Describe the elements of $Aut(L/\boldsymbol{Q})$ explicitly and determine the structure of $(Aut(L/\boldsymbol{Q}), \circ)$.

20. Let $L \subseteq \boldsymbol{C}$ be a splitting field for $x^4 - 5 \in \boldsymbol{Q}[x]$ over $\boldsymbol{Q}[i]$. Describe the elements of $Aut(L/\boldsymbol{Q}[i])$ explicitly and the structure of $(Aut(L/\boldsymbol{Q}[i]), \circ)$.

21. Let $p_1 < \cdots < p_k$ be primes in $\boldsymbol{N}$ and let L be the splitting field over $\boldsymbol{Q}$ for $(x^2 - p_1)\cdots(x^2 - p_k) \in \boldsymbol{Q}[x]$. Find the order of $Aut(L/\boldsymbol{Q})$ and describe its structure. (Use Exercise 14 in §14.2 and then induction on k.)

22. Let $p > 2$ be prime, $\rho = e^{2\pi i/p} \in \boldsymbol{C}$, and $L = \boldsymbol{Q}[\rho] \subseteq \boldsymbol{C}$. From Example 12.5 the minimal polynomial in $\boldsymbol{Q}[x]$ for ρ is $x^{p-1} + x^{p-2} + \cdots + x + 1$.
   i) For any integer $1 \le k \le p-1$, show that $\varphi : L \to L$ defined by $\varphi(a_0 + a_1\rho + \cdots + a_{p-1}\rho^{p-1}) = a_0 + a_1\rho^k + \cdots + a_{p-1}\rho^{k(p-1)}$ gives $\varphi \in Aut(L/\boldsymbol{Q})$.
   ii) Show that $Aut(L/\boldsymbol{Q}) \cong U_p = (\boldsymbol{Z}_p^*, \cdot)$.
   iii) If $\alpha \in L$ and $\varphi(\alpha) = \alpha$ for all $\varphi \in Aut(L/\boldsymbol{Q})$, show that $\alpha \in \boldsymbol{Q}$. (Let $\alpha = a_0 + a_1\rho + \cdots + a_{p-1}\rho^{p-1}$ and use $U_p = \langle [k]_p \rangle$ for some k to show that $a_1 = a_2 = \cdots = a_{p-1}$.)

14.5 ALGEBRAIC CLOSURES

A splitting field for $f(x) \in F[x]$ over F is algebraic over F. Is there a maximal algebraic extension L of F? If so, then L itself can have no algebraic extension M since then M would be algebraic over F from Theorem 13.26 or Exercise 7 in §14.2. Therefore any maximal algebraic extension of F must be algebraically closed. Recall from Definition 13.14 that a field L is algebraically closed if every irreducible in $L[x]$ has degree 1.

DEFINITION 14.13 An **algebraic closure** of the field F is a field *extension* L of $F (F \subseteq L)$ that is algebraic over F and is algebraically closed.

If an algebraic closure L exists for a field F, then the minimal polynomial in $F[x]$ for any $\alpha \in L$ must split over L. Thus it is natural to consider a splitting field for all irreducible polynomials in $F[x]$.

DEFINITION 14.14 Let F be a field and $\emptyset \neq S \subseteq F[x] - F$. A field L is a **splitting field for S over F** if every $p(x) \in S$ splits over L and if $L = F[T]$ for T the set of all roots in L of all $p(x) \in S$.

The construction of a splitting field for $S \subseteq F[x]$ and the proof of its uniqueness use Zorn's Lemma. This proceeds by constructing algebraic extensions over F inside a fixed set A. Since we need to construct ever larger algebraic extensions within A, we need to ensure that A is large enough to do this. It suffices to observe that there is an A with $|A| > |M|$ for any algebraic extension M of F since by Theorem 13.45 if $M \subseteq A$ with $|M| < |A|$ then $|A - M| = |A| > |M|$. If F is an infinite field then Theorem 13.47 shows that $|F[x]| = |F|$. Each $p(x) \in F[x] - F$ has finitely many roots in any splitting field, and any $\alpha \in L$, an algebraic extension of F, is a root of some $q(x) \in F[x]$. Thus L is contained in the union of F and $|F| = |F[x]|$ finite subsets so by Theorem 13.47, $|L| = |F|$. When F is finite, the results in §0.7 show similarly that any algebraic extension of F is countable. Consequently there is a set A with $|A| > |M|$ for any algebraic extension M of F. Specifically, when F is infinite its power set $P(F)$ satisfies $|F| < |P(F)|$ by Theorem 0.15, and when F is finite take $A = P(\mathbf{N})$.

THEOREM 14.19 Any field F has an algebraic closure. For $F \subseteq L$ fields, L is an algebraic closure of $F \Leftrightarrow L$ is a splitting field for $F[x] - F$ over F.

Proof Let S be a set with $|F| < |S|$ and $|\mathbf{N}| < |S|$. Consider $T = \{L \subseteq F \cup S \mid L$ is a field and an algebraic extension of $F\}$. Now $(T, \subseteq)$ is a partially ordered set using Theorem 13.26, and $F \in T$ so $T \neq \emptyset$. If $\{L_\lambda\}$ is a chain in T then $F \subseteq M = \bigcup L_\lambda$ is a field and an algebraic extension of F (verify). Thus M is an upper bound for $\{L_\lambda\}$ in $(T, \subseteq)$, so Zorn's Lemma shows that T has a maximal element L. Let $p(x) \in L[x]$ be irreducible of degree $n > 1$, and let K be a splitting field for $p(x)$ over L. Using Theorem 13.26 again, K is algebraic over F so $|L|, |K - L| \leq |F|$ (or $|\mathbf{N}|$) as we observed above. Since by Theorem 13.45 $|S - L| = |S|$ there is $A \subseteq S - L$ with $|A| = |K - L|$ (Theorem 13.43). Therefore there is a bijection $\varphi : K \to (L \cup A)$ and, as in the discussion before Theorem 14.12, we can use φ to define a ring structure on $L \cup A$ so that $K \cong L \cup A$. This gives an algebraic extension of F contained in S and larger than L, contradicting the maximality of L. Therefore $p(x) \in L[x]$ is irreducible precisely when $deg\ p = 1$ so L is an algebraic closure of F.

For any algebraic extension L of F each $\alpha \in L$ is a root of some $f(x) \in F[x]^*$ and $L = F[L]$. If L is algebraically closed, its irreducibles have degree 1 so the irreducible factors of $p(x) \in F[x] - F$ have degree 1 in $L[x]$ and L is a splitting field for $F[x] - F$. When $F \subseteq L$ and L is a splitting field for $F[x] - F$, let $q(x) \in L[x]$ be irreducible. If $L \subseteq M$ is a splitting field for $q(x)$ over L with $\alpha \in M$ a root of $q(x)$, then $L[\alpha] \subseteq M$ is algebraic over L and so is algebraic over F by Theorem 13.26. Hence $g(\alpha) = 0$ for some $g(x) \in F[x] - F$, so $g(x)$ splits over L, forcing $\alpha \in L$. Therefore $deg\ q = 1$. This shows that L is an algebraic closure of F. ∎

To see that the algebraic closure of a field F is unique up to isomorphism, we first explore the existence and uniqueness of splitting fields for arbitrary sets of polynomials.

THEOREM 14.20 If F is any field and $\emptyset \neq S \subseteq F[x] - F$, then there is a splitting field for S over F.

Proof From Theorem 14.19 there is an algebraic closure L of F, and L is a splitting field for $F[x] - F$. Thus each $p(x) \in S$ splits in L. If $T \subseteq L$ is the set of all roots in L of all $p(x) \in S$ then $F[T]$ is a splitting field for S over F. ∎

THEOREM 14.21 Let $\varphi : F_1 \to F_2$ be an isomorphism of fields, $\emptyset \neq S_1 \subseteq F_1[x] - F_1$, $S_2 = \{\varphi(p(x)) \in F_2[x] \mid p(x) \in S_1\}$, L_1 a splitting field for S_1 over F_1, and L_2 a splitting field for S_2 over F_2. Then there is an isomorphism $\eta : L_1 \to L_2$ extending φ.

Proof Let $S = \{(A, B, \eta) \mid A$ and B are fields, $F_1 \subseteq A \subseteq L_1$, $F_2 \subseteq B \subseteq L_2$, and $\eta : A \to B$ is an isomorphism extending $\varphi\}$. Note that $(F_1, F_2, \varphi) \in S$. In S, define $(A, B, \eta) \leq' (C, D, \mu)$ when $A \subseteq C$, $B \subseteq D$, and μ extends η. It is routine to verify that $(S, \leq')$ is a partially ordered set. If $\{(A_\lambda, B_\lambda, \eta_\lambda)\}$ is a chain in S, let $A = \bigcup A_\lambda$ and $B = \bigcup B_\lambda$. It follows that A and B are fields, that $F_1 \subseteq A \subseteq L_1$, $F_2 \subseteq B \subseteq L_2$, and that $\eta : A \to B$ defined by $\eta(a) = \eta_\lambda(a)$ when $a \in A_\lambda$ is an isomorphism extending φ (verify!). Observe that η is a function since if $a \in A_\lambda \subseteq A_\beta$ then $\eta_\lambda(a) = \eta_\beta(a)$ using that η_β extends η_λ. The definition of $\leq'$ shows that (A, B, η) is an upper bound in $(S, \leq')$ for the chain $\{(A_\lambda, B_\lambda, \eta_\lambda)\}$ so Zorn's Lemma yields a maximal $(M_1, M_2, \upsilon) \in (S, \leq')$. If $M_1 \neq L_1$ then M_1 is not a splitting field over F_1 for S_1 so some $p(x) \in S_1$ does not split over M_1. If T_1 are the roots of $p(x)$ in L_1 then $M_1 \neq M_1[T_1] \subseteq L_1$ is a splitting field for $p(x)$ over M_1. Using that υ is a isomorphism extending φ, $\upsilon(p(x))$ does not split over M_2 and if T_2 are its roots in L_2 then $M_2[T_2] \subseteq L_2$ is a splitting field for $\upsilon(p(x))$ over M_2. Hence by Theorem 14.17 there is an isomorphism $\rho : M_1[T_1] \to M_2[T_2]$ extending υ so $(M_1[T_1], M_2[T_2], \rho) \in (S, \leq')$ contradicting the maximality of (M_1, M_2, υ). This proves that $M_1 = L_1$. Since each $p(x) \in S_1$ splits over L_1, each $\varphi(p(x)) = \upsilon(p(x))$ must split over $\upsilon(L_1) = M_2$. Therefore $M_2 = L_2$ and υ is the required isomorphism, completing the proof. ∎

An algebraic closure of F is a splitting field by Theorem 14.19, so the following is immediate from Theorem 14.21.

THEOREM 14.22 Any two algebraic closures of a field F are isomorphic.

EXERCISES

In the exercises below, F, H, K, L, and M are fields.

1. Let $F \subseteq L \subseteq M$ with L an algebraic extension of F and M an algebraic closure of L. Show that M is an algebraic closure of F.
2. Let $F \subseteq H \subseteq L$ with L an algebraic closure of F. Show that:
   i) L is an algebraic closure of H.
   ii) If $\varphi \in Aut(H/F)$ then there is $\eta \in Aut(L/F)$ extending φ.

3. If L is an algebraic closure of $\mathbf{Q}$, show that there is $\mathbf{Q} \subseteq H \subseteq L$ with $[H : \mathbf{Q}]$ infinite and $[L : H]$ infinite. (Let H be a splitting field for some $S \subseteq \mathbf{Q}[x]$.)

4. For p a prime and L an algebraic closure of $\mathbf{Z}_p$, show that:

i) L is the union of a countable set of finite subfields.

ii) There is $\mathbf{Z}_p \subseteq K \subseteq L$ with both K and $[L : K]$ infinite.

iii) If $\mathbf{Z}_p \subseteq H \subseteq L$ then H is a splitting field over $\mathbf{Z}_p$ for some $S \subseteq \mathbf{Z}_p[x]$ (see Exercise 16 in §14.4).

iv) If $\mathbf{Z}_p \subseteq H \subseteq L$ and $\eta \in Aut(L/\mathbf{Z}_p)$ then $\eta(H) = H$.

5. If $\mathbf{Q} \subseteq L \subseteq \mathbf{C}$ and if L is an algebraic closure of $\mathbf{Q}$, show that $L = (L \cap \mathbf{R})[i]$.

14.6 TRANSCENDENTAL EXTENSIONS

We have studied fields $F \subseteq L$ when L is algebraic over F, and we now look at the situation when L contains transcendental elements over F. From Theorem 14.7, if α is transcendental over F then $F[\alpha] \cong F[x]$ so $QF(F[x]) \cong F(\alpha) = \{\beta\gamma^{-1} \in L \mid \beta, \gamma \in F[\alpha] \text{ with } \gamma \neq 0\}$. We generalize the transcendental condition that $g(\alpha) \neq 0$ for any $g(x) \in F[x]^*$ to sets larger than $\{\alpha\}$.

DEFINITION 14.15 If $F \subseteq L$ are fields, $T \subseteq L$ is **algebraically independent over** F if $T = \varnothing$ or if whenever $f(x_1, \ldots, x_n) \in F[x_1, \ldots, x_n]$ and $f(\alpha_1, \ldots, \alpha_n) = 0$ for $\alpha_i \in T$ and distinct, then $f = 0$.

In particular, when $T = \{\alpha\} \subseteq L$ is algebraically independent over F then α is transcendental over F. Clearly, any subset of an algebraically independent set is algebraically independent and consists of transcendental elements that must act like indeterminates over F in the following sense.

THEOREM 14.23 Let $F \subseteq L$ be fields, $T \subseteq L$ algebraically independent over F, $X = \{x_\lambda\}$ a set of indeterminates over F, and $|X| = |T| > 0$. Then $F[X] \cong F[T] \subseteq L$ and $QF(F[X]) \cong F(T) = \{pq^{-1} \in L \mid p, q \in F[T] \text{ with } q \neq 0\}$.

Proof If $\varphi : X \to T$ is a bijection, set $\varphi(x_\lambda) = t_\lambda \in T$. Then $\rho : F[X] \to F[T]$ defined by $\rho(f(x_{\lambda(1)}, \ldots, x_{\lambda(n)})) = f(t_{\lambda(1)}, \ldots, t_{\lambda(n)})$ is a ring homomorphism (verify!) that is clearly surjective, and $\ker \rho = (0)$ since $f(t_{\lambda(1)}, \ldots, t_{\lambda(n)}) = 0$ and Definition 14.15 force $f(x_{\lambda(1)}, \ldots, x_{\lambda(n)}) = 0$. It is routine to see that the isomorphism ρ extends to an isomorphism of $QF(F[X]) = \{p(X)/q(X) \mid p(X), q(X) \in F[X] \text{ and } q(X) \neq 0\}$ onto $F(T)$ via $\rho(p(X)/q(X)) = \rho(p(X))\rho(q(X))^{-1}$. ■

In general when $F \subseteq L$ are fields and $T \subseteq L$, *we denote the quotient field of the ring* $F[T]$ *by* $F(T)$. For example, $\mathbf{Q}(x, y) = \{p(x, y)/q(x, y) \mid p(x, y), q(x, y) \in \mathbf{Q}[x, y] \text{ with } q(x, y) \neq 0\}$. Whether elements in L are algebraically independent depends on the field F. Now $\{x, y\}, \{x^2, y\} \subseteq \mathbf{Q}(x, y)$ are algebraically independent sets over $\mathbf{Q}$, but if $F = \mathbf{Q}(x^2)$ then $\{x, y\}$ is not algebraically independent over F since x satisfies $Y^2 - x^2 \in F[Y]$. Since $x^2 - y^2$ is transcendental over $\mathbf{Q}$ (verify!), $\{x^2 - y^2\} \subseteq \mathbf{Q}(x, y)$ is algebraically independent over $\mathbf{Q}$. If $M = \mathbf{Q}(x^2 - y^2)$ then $\{x, y\} \subseteq \mathbf{Q}(x, y)$ is not algebraically independent over M since for $p(X_1, X_2) = X_1^2 - X_2^2 - (x^2 - y^2) \in M[X_1, X_2]^*$, $p(x, y) = 0$.

Similarly, $\{x+y, x-y\} \subseteq \boldsymbol{Q}(x, y)$ is algebraically independent over $\boldsymbol{Q}$ (why?) but not over M: for $q(X_1, X_2) = X_1X_2 - (x^2 - y^2) \in M[X_1, X_2]$, $q(x+y, x-y) = 0$.

Example 14.15 If F is a field, x an indeterminate over F, and $F(x) = QF(F[x])$, then any $\alpha \in F(x) - F$ is transcendental over F, $F(x)$ is algebraic over $F(\alpha)$, and $\{\alpha, \beta\} \subseteq F(x)^*$ is not algebraically independent unless $\alpha = \beta$.

Write $\alpha = h(x)/g(x) \notin F$ with $\mathrm{GCD}\{h, g\} = 1$ and observe that $f(Y) = g(Y)\alpha - h(Y) \in F(\alpha)[Y]^*$ satisfies $f(x) = 0$. Thus x is algebraic over $F(\alpha)$ so $F(x) = F(\alpha)[x]$ is finite dimensional, and thus algebraic, over $F(\alpha)$ by Theorem 14.9. Now α must be transcendental over F since otherwise $F(\alpha)$ is algebraic over F, forcing $F(x)$ to be algebraic over F by Theorem 13.26. Using x algebraic over $F(\alpha)$ shows that any $\beta \in F(x)$ is algebraic over $F(\alpha)$ so $\beta^m + p_{m-1}(\alpha)q_{m-1}(\alpha)^{-1}\beta^{m-1} + \cdots + p_0(\alpha)q_0(\alpha)^{-1} = 0$ where $p_j(\alpha), q_j(\alpha) \in F(\alpha)$. Clearing denominators yields $g_m(\alpha)\beta^m + \cdots + g_0(\alpha) = 0$. If $\alpha \neq \beta$ then we have $f(\alpha, \beta) = 0$ for $f(X, Y) = g_m(X)Y^m + \cdots + g_0(X) \in F[X, Y]^*$, since $g_m(\alpha) \neq 0$. Therefore $\{\alpha, \beta\}$ cannot be algebraically independent over F when $\beta \neq \alpha$.

The example shows that over F any maximal algebraically independent subset of $F(x)$ has one element, although $F(x)$ has infinitely many transcendental elements. Maximal algebraically independent subsets play a role akin to bases in vector spaces.

DEFINITION 14.16 Given fields $F \subseteq L$, a **transcendence basis** of L over F is any maximal algebraically independent subset $T \subseteq L$ over F.

THEOREM 14.24 If $F \subseteq L$ are fields, then:

i) There is a transcendence basis of L over F.
ii) $\varnothing \neq T \subseteq L$ is a transcendence basis of L over $F \Leftrightarrow T \neq \varnothing$ is algebraically independent over F and L is algebraic over $F(T)$.

Proof Now $\alpha \in L$ transcendental over F is equivalent to $\{\alpha\}$ algebraically independent over F, so L is algebraic over F precisely when $\varnothing$ is a transcendence basis of L over F. Assume L is not algebraic over F. Then $S = \{\varnothing \neq Y \subseteq L \mid Y \text{ is algebraically independent over } F\} \neq \varnothing$ and clearly $(S, \subseteq)$ is partially ordered. If $\{Y_\beta\}$ is a chain in S, let $Y = \bigcup Y_\beta$ and $W \subseteq Y$ finite. There is $Y_\beta \supseteq W$, so W is algebraically independent over F, showing that Y is also (why?). Consequently $Y \in S$ is an upper bound for $\bigcup Y_\beta$, and by Zorn's Lemma S contains a maximal element T. By Definition 14.16, T is a transcendence basis of L over F, proving i).

If $T \subseteq L$ is a nonempty transcendence basis of L over F then T is algebraically independent over F by definition. The maximality of T shows that for any $\alpha \in L - T$, $T \cup \{\alpha\}$ is not algebraically independent over F. Thus there is a nonzero $f(x_1, \ldots, x_{n+1}) \in F[x_1, \ldots, x_{n+1}]$ with $f(t_1, \ldots, t_n, \alpha) = 0$ for $\{t_j\} \subseteq T$. Writing $f \in F[x_1, \ldots, x_n][x_{n+1}]$ as $g_m(x_1, \ldots, x_n)x_{n+1}^m + \cdots + g_0(x_1, \ldots, x_n)$ and letting $\gamma_j = g_j(t_1, \ldots, t_n)$ yields $\gamma_m\alpha^m + \cdots + \gamma_1\alpha + \gamma_0 = 0$. Now T is algebraically independent over F and $f \neq 0$ so some $\gamma_j \neq 0$ with $j > 0$. It follows that α is algebraic over $F(T)$, and α arbitrary shows that L is also algebraic over $F(T)$.

Assume that $T \subseteq L$ is algebraically independent over F, T is not empty, and that L is algebraic over $F(T)$. If $\beta \in L - T$ then $\beta^n + \upsilon_{n-1}\beta^{n-1} + \cdots + \upsilon_0 = 0$ for $\upsilon_j = p_j(T)/q_j(T) \in F(T)$ and clearing denominators yields $h_n(t_1, \ldots, t_k)\beta^n + \cdots + h_0(t_1, \ldots, t_k) = 0$. This implies that for a suitable $f(x_1, \ldots, x_{k+1}) \in F[x_1, \ldots, x_{k+1}]^*$, $f[t_1, \ldots, t_k, \beta) = 0$ so $T \cup \{\beta\}$ is not algebraically independent over F. Hence $T \subseteq L$ is a maximal algebraically independent set over F, and so a transcendence basis of L over F, completing the proof of the theorem. ■

From Example 14.15, any $\alpha \in F(x) - F$ is transcendental over F and $F(x)$ is algebraic over $F(\alpha)$ so $\{\alpha\}$ is a transcendence basis of $F(x)$ over F.

Example 14.16 When $n > 1$, in $QF(\mathbf{Q}[x_1, \ldots, x_n]/(x_1^2 + \cdots + x_n^2 - 1))$, $\{x_1, \ldots, x_{n-1}\}$ is a transcendence basis over $\mathbf{Q}$.

The field $\mathbf{Q}$ plays no role here. By Theorem 12.15, $\mathbf{Q}[x_1, \ldots, x_n] = R$ is a UFD, so Theorem 12.16 (Eisenstein), $n > 1$, and induction show that $p(X) = x_1^2 + \cdots + x_n^2 - 1$ is irreducible. Thus $\mathbf{Q}[x_1, \ldots, x_n]/(p(X)) = D$ is a domain and so has a quotient field L. Using $p(X) = 0$ in D, we can write the elements of D as $p(x_1, \ldots, x_{n-1}) + q(x_1, \ldots x_{n-1})x_n$ with multiplication defined via $x_n^2 = 1 - x_1^2 - \cdots - x_{n-1}^2$. Let x_j also denote the image in D and L of $x_j \in R$. Since $p(X) \in R$ is irreducible and R is a UFD, $(p(X)) \cap \mathbf{Q}[x_1, \ldots, x_{n-1}] = (0)$. Thus $T = \{x_1, \ldots, x_{n-1}\} \subseteq L$ is algebraically independent over $\mathbf{Q}$. Clearly x_n is algebraic over $F = \mathbf{Q}(T)$ so $L = F[x_n]$ is algebraic over F, and T is a transcendence basis for L over $\mathbf{Q}$ by Theorem 14.24.

We will show that different transcendence bases of L over F have the same number of elements, but first we need a preliminary result.

THEOREM 14.25 Let $F \subseteq L$ be fields, $\{x_1, \ldots, x_n\}, \{y_1, \ldots, y_m\} \subseteq L$, $\{y_j\}$ algebraically independent over F, and L algebraic over $F(\{x_i\})$. After re-ordering $\{x_i\}$, there is $T = \{y_1, \ldots, y_m, x_{m+1}, \ldots, x_n\}$ with L algebraic over $F(T)$. Thus $m \le n$. If $\{y_j\}$ is a transcendence basis of L over F, then for $S \subseteq L$ algebraically independent over F, $|S| \le m$, and when $|S| = m$, S is a transcendence basis of L over F.

Proof If $m = 0$ take $T = \{x_i\}$. For $0 \le k < m$ assume L is algebraic over $F(S)$ for some $S = \{y_1, \ldots, y_k, x_{k+1}, \ldots, x_n\}$. Thus $y_{k+1}^d + \beta_{d-1}y_{k+1}^{d-1} + \cdots + \beta_0 = 0$ with all $\beta_j = p_j(S)/q_j(S) \in F(S)$. After clearing denominators there is some nonzero $f(X_1, \ldots, X_{n+1}) \in F[X_1, \ldots, X_{n+1}]$ of least degree with $f(y_1, \ldots, y_{k+1}, x_{k+1}, \ldots, x_n) = 0$. Since $\{y_j\}$ is algebraically independent, some X_i with $k+2 \le i \le n+1$ must appear in f, and by re-ordering we can assume it to be X_{k+2}. Writing f as a polynomial in X_{k+2}, it follows from the choice of f that the leading coefficient is not zero when evaluated at $S \cup \{y_{k+1}\}$, so x_{k+1} is algebraic over $F(S')$ for $S' = \{y_1, \ldots, y_{k+1}, x_{k+2}, \ldots, x_n\}$. Now L is algebraic over the field $F(S \cup \{y_{k+1}\}) = F(S')[x_{k+1}]$, which is algebraic over $F(S')$, so L is algebraic over $F(S')$ by Theorem 13.26. By induction on k (Theorem 0.7) there is $T = \{y_1, \ldots, y_m, x_{m+1}, \ldots, x_n\}$ with L algebraic over $F(T)$, and this forces $m \le n$.

When $\{y_j\}$ with m elements is a transcendence basis of L over F then L is algebraic over $F(\{y_j\})$ so by the first part of the theorem any algebraically independent subset of L over F has at most m elements. Finally, if $S=\{\alpha_1,\ldots,\alpha_m\}\subseteq L$ with m elements is algebraically independent, then the first paragraph shows that $T=S$ and L is algebraic over $F(S)$, so $\{\alpha_i\}$ is a transcendence basis for L over F. ■

In Example 14.15, that $F(x)$ has no set of two or more algebraically independent elements follows from Theorem 14.25. The theorem also shows in Example 14.16 that $QF(Q[x_1,\ldots,x_n]/(x_1^2+\cdots+x_n^2-1))$ cannot have a subset of n algebraically independent elements.

THEOREM 14.26 If $F\subseteq L$ are fields with $S, T\subseteq L$ transcendence bases of L over F, then $|S|=|T|$.

Proof If either S or T is empty then L is algebraic over F so both are empty. Now assume neither S nor T is empty. Since T is algebraically independent over F and L is algebraic over $F(S)$ by Theorem 14.24, if S is finite then Theorem 14.25 shows that $|T|\leq|S|$. Reversing the roles of S and T, we get $|S|\leq|T|$ so $|S|=|T|$. Hence we may assume that $|S|$ and $|T|$ are infinite. The argument now is much like that in Theorem 14.6. For any finite $A\subseteq S$, let $B_A=\{t\in T\mid t \text{ is algebraic over } F(A)\}$ so $F(A)[B_A]$ is algebraic over $F(A)$ by Theorem 13.26. Since $B_A\subseteq T$ it is algebraically independent over F so Theorem 14.25 forces $|B_A|\leq|A|$. Using that L is algebraic over $F(S)$, any $t\in T$ must be algebraic over $F(A)$ for $A\subseteq S$ finite. Therefore T is the union of all B_A and we have $|T|=|\bigcup B_A|\leq|S|$ by Theorem 13.47 (verify!). A similar argument starting with finite subsets $B\subseteq T$ and looking at elements in S algebraic over $F(B)$ gives $|S|\leq|T|$. Hence $|S|=|T|$ by Theorem 0.16 when both are infinite, finishing the proof. ■

The theorem proves that the cardinality of all transcendence bases of L over F is unique; it is called the *transcendence dimension.*

DEFINITION 14.17 If $F\subseteq L$ are fields and $T\subseteq L$ is a transcendence basis of L over F, then $|T|=trdim_F L$, the **transcendence dimension of L over F.**

It is difficult to use Definition 14.16 to produce transcendence bases. We next provide some examples of transcendence bases of *symmetric functions* in commuting indeterminates $\{x_1,\ldots,x_n\}$ over a field F.

DEFINITION 14.18 Let $X=\{x_1,\ldots,x_n\}$ be commuting indeterminates over the field F. For each $1\leq i\leq n$ the **i^{th} elementary symmetric function in X** is the sum of all products of i different elements from X, denoted $\sigma_i(X)$.

Specifically, $\sigma_1(X)=x_1+\cdots+x_n$, $\sigma_n(X)=x_1x_2\cdots x_n$, and in general, $\sigma_k(X)=\sum_{i_1<\cdots<i_k}x_{i_1}x_{i_2}\cdots x_{i_k}$. Thus $\sigma_2(\{x_1,x_2,x_3\})=x_1x_2+x_1x_3+x_2x_3$, and $\sigma_3(\{x_1,x_2,x_3,x_4\})=x_1x_2x_3+x_1x_2x_4+x_1x_3x_4+x_2x_3x_4$. Let the monic $p(x)\in F[x]$ split over L, so $p(x)=x^n+a_{n-1}x^{n-1}+\cdots+a_1x+a_0=(x-\beta_1)\cdots(x-\beta_n)$ in $L[x]$. It follows that $a_j=(-1)^{n-j}\sigma_{n-j}(\beta_1,\ldots,\beta_n)$ (verify!) so the coefficients of $p(x)$, up to sign, are the

evaluations of the $\sigma_i(X)$ at the roots of $p(x)$ in any splitting field L. This observation is the key to showing that the elementary symmetric functions are algebraically independent.

THEOREM 14.27 For $X = \{x_1, \ldots, x_n\}$ the set $\{\sigma_1(X), \ldots, \sigma_n(X)\}$ of elementary symmetric functions is a transcendence basis of $F(X)$ over F.

Proof It is immediate from Theorem 14.24 that X is a transcendence basis of $F(X)$ over F, so any transcendence basis must have exactly n elements by Theorem 14.26. As we saw just above, $(Y - x_1) \cdots (Y - x_n) \in F(X)[Y]$ has coefficients that are $\pm\sigma_j(X)$ (and 1) so each $x_j \in X$ is algebraic over $F(\{\sigma_j(X)\}) = K$. It follows from Theorem 14.9 that $F(X) = K[X]$ is algebraic over K. An exercise shows that if $F \subseteq L$ are fields and $A = \{\alpha_1, \ldots, \alpha_m\} \subseteq L$ so that L is algebraic over $F(A)$, then a subset of A must be a transcendence basis of L over F. Therefore a subset of $\{\sigma_j(X)\}$ is a transcendence basis for $F(X)$ over F and must have n elements by our first remark. Hence $\{\sigma_j(X)\}$ itself must be a transcendence basis for $F(X)$ over F, as claimed. ∎

We can use Theorem 14.27 to find another interesting transcendence basis for $\mathbf{Q}(X)$ over $\mathbf{Q}$ when $X = \{x_1, \ldots, x_n\}$. If $T_j = x_1^j + \cdots + x_n^j$ for $1 \leq j \leq n$ then it is not clear by direct calculation whether $\{T_j\}$ is a transcendence basis for $\mathbf{Q}(X)$. Suppose that each $\sigma_j(X) \in \mathbf{Q}[T_1, \ldots, T_n]$. Since each x_j is integral over $\mathbf{Q}[\sigma_1(X), \ldots, \sigma_n(X)]$ from the proof of Theorem 14.27, it would follow that each x_j is integral over $\mathbf{Q}[T_1, \ldots, T_n]$. Hence $\mathbf{Q}(X)$ would be algebraic over $\mathbf{Q}(T_1, \ldots, T_n)$. Now any transcendence basis of $\mathbf{Q}(X)$ over $\mathbf{Q}$ has n elements by Theorem 14.26, so $\{T_1, \ldots, T_n\}$ would be a transcendence basis for $\mathbf{Q}(X)$ over $\mathbf{Q}$ (why?). This result cannot hold for any field F, because if *char* $F = p \leq n$ then $T_1^p = T_p$, and $\{T_1, \ldots, T_n\}$ cannot be algebraically independent over F. For the next example it suffices to take *char* $F > n$ rather than $F = \mathbf{Q}$.

Example 14.17 For $X = \{x_1, \ldots, x_n\}$ and $T_j = x_1^j + \cdots + x_n^j$ for $1 \leq j \leq n$, each $\sigma_j(X) \in \mathbf{Q}[T_1, \ldots, T_n]$, so $\{T_j\}$ is a transcendence basis for $\mathbf{Q}(X)$ over $\mathbf{Q}$.

The argument here is not difficult but requires some complicated notation. For any $1 \leq j \leq n$, a partition of j is $(n_1, \ldots, n_k) \in \mathbf{N}^k$ with $\sum n_i = j$, and let $[n_1, \ldots, n_k]$ represent the partition obtained by re-ordering the $\{n_i\}$ to make them decreasing. Thus $[2, 3, 1, 3]$ is the partition $(3, 3, 2, 1)$ of 9. Let $S = \{(j, [n_1, \ldots, n_k]) \mid 1 \leq j \leq n$ and $[n_1, \ldots, n_k]$ is a decreasing partition of $j\}$. Order S as follows: $(j, [n_1, \ldots, n_k]) < (j', [m_1, \ldots, m_p])$ if either $j < j'$, or $j = j'$ and $k < p$, or $j = j', k = p$, some $n_a < m_a$, and $n_c = m_c$ for all $c < a$. The ordering begins $(1, [1]) < (2, [2]) < (2, [1, 1]) < (3, [3]) < (3, [2, 1]) < (3, [1, 1, 1]) < (4, [4]) < \cdots$. For $(j, [n_1, \ldots, n_k]) \in S$ define $\Delta(j, [n_1, \ldots, n_k]) = \sum x_{i_1}^{n_1} x_{i_2}^{n_2} \cdots x_{i_k}^{n_k}$ with the sum taken over all k-elements subsets $\{i_1, \ldots, i_k\} \subseteq \{1, 2, \ldots, n\}$ but without repetition of monomials. For example, if $n = 4$ then $\Delta(4, [2, 2]) = x_1^2x_2^2 + x_1^2x_3^2 + x_1^2x_4^2 + x_2^2x_3^2 + x_2^2x_4^2 + x_3^2x_4^2$, and $\Delta(3, [2, 1]) = x_1^2(x_2 + x_3 + x_4) + x_2^2(x_1 + x_3 + x_4) + x_3^2(x_1 + x_2 + x_4) + x_4^2(x_1 + x_2 + x_3)$.

We claim that all $\Delta(j, [n_1, \ldots, n_k]) \in \mathbf{Q}[T_1, \ldots, T_n]$. Note that $\Delta(j, [j]) = T_j$ and $\Delta(j, [1, 1, \ldots, 1]) = \sigma_j(X)$, so verifying this claim completes the example. Some other easy computations are: $\Delta(2, [1, 1]) = 1/2 \cdot T_1^2 - 1/2 \cdot T_2$, $\Delta(3, [2, 1]) = T_2T_1 - T_3$, $\Delta(3, [1, 1, 1]) = 1/6 \cdot T_1^3 + 1/3 \cdot T_3 - 1/2 \cdot T_2T_1$, and $\Delta(4, [3, 1]) = T_3T_1 - T_4$. Assume that $(j, [n_1, \ldots, n_k]) \in S$ and that for $s \in S$ with $s < (j, [n_1, \ldots, n_k])$,

$\Delta s \in \boldsymbol{Q}[T_1, \ldots, T_n]$. If $n_1 = j$ then $\Delta(j, [j]) = T_j$ and if $k \geq 2$ consider $T_{n_1}\Delta(j - n_1, [n_2, \ldots, n_k])$. The monomials in $\Delta(j, [n_1, \ldots, n_k])$ occur in this product, each arising q times for $q \in N$ the number of $n_d = n_1$. The other monomials in $T_{n_1}\Delta(j - n_1, [n_2, \ldots, n_k])$, in $k - 1$ variables, each arise only once since $n_1 \geq n_i$ for $i > 1$. Thus $q\Delta(j, [n_1, \ldots, n_k]) = T_{n_1}\Delta(j - n_1, [n_2, \ldots, n_k]) - \Delta(j, [n_1 + n_2, n_3, \ldots, n_k]) - \cdots - \Delta(j, [n_2, n_3, \ldots, n_1 + n_k])$ where the other terms Δs appear without repetition. Thus $4\Delta(8, [2, 2, 2, 2]) = T_2\Delta(6, [2, 2, 2]) - \Delta(8, [4, 2, 2])$ and $2\Delta(5, [2, 2, 1]) = T_2\Delta(3, [2, 1]) - \Delta(5, [4, 1]) - \Delta(5, [3, 2])$. All Δs appearing in the representation of $q\Delta(j, [n_1, \ldots, n_k])$ above satisfy $s < (j, [n_1, \ldots, n_k])$, so our inductive assumption shows that $\Delta(j, [n_1, \ldots, n_k]) \in \boldsymbol{Q}[T_1, \ldots, T_n]$. Proceeding through all the $s \in S$ in order shows that all $\Delta s \in \boldsymbol{Q}[T_1, \ldots, T_n]$, as required.

THEOREM 14.28 If $F \subseteq L \subseteq M$ are fields, $trdim_F M = trdim_L M + trdim_F L$.

Proof Let S be a transcendence basis of M over L, and let T be a transcendence basis of L over F. If $\alpha \in S \cap T$ then $Y - \alpha \in L[Y]$ has α as a root so S cannot be algebraically independent over L, contradicting Theorem 14.24. Hence $S \cap T = \varnothing$. We claim $S \cup T$ is algebraically independent over F. Let $f(x_1, \ldots, x_n) \in F[x_1, \ldots, x_n]^*$ with $f(s_1, \ldots, t_n) = 0$ for some $s_i, t_j \in S \cup T$. If some $s_1 \in S$ actually appears, then S is not algebraically independent over L (why?). If only $t_j \in T$ appear, then T is not algebraically independent over F. Hence $S \cup T$ is algebraically independent over F. By Theorem 14.24, M is algebraic over $L(S)$ and L is algebraic over $F(T)$. If $A = \{\alpha \in M \mid \alpha$ is algebraic over $F(S \cup T)\}$ then A is a subfield of M algebraic over $F(S \cup T)$ by Theorem 13.26 or Exercise 6 in §14.2. Clearly $L, S \subseteq A$, so $L(S) \subseteq A$ is algebraic over $F(S \cup T)$. This, together with M algebraic over $L(S)$ and Theorem 13.26 or Exercise 7 in §14.2, yields M algebraic over $F(S \cup T)$. Thus Theorem 14.24 proves that $S \cup T$ is a transcendence basis of M over F. The proof is complete since $|S| + |T| = |S \cup T|$: in the case of infinite cardinalities, see Theorem 13.45 and the comment after it. ■

When $\{\alpha_1, \ldots, \alpha_n\} \subseteq F(x_1, \ldots, x_m)$ it is difficult to say much about the subfield $F(\alpha_1, \ldots, \alpha_n)$, even in the simplest case of $m = 1$. Example 14.15 shows that any $\alpha \in F(x) - F$ is transcendental over F, so $F(\alpha) \cong F(x)$, and $F(x)$ is algebraic over $F(\alpha)$ (also by Theorem 14.28). It is easy to characterize the elements in $F(x^2)$ but not easy to describe those in the rather simple example $F(x^2 + 1/x)$. If $\{\alpha_1, \ldots, \alpha_n\} \subseteq F(x) - F$ then by Example 14.15, $F(\{\alpha_1, \ldots, \alpha_n\})$ is algebraic over $F(\alpha_1)$. Clearly $F(x^4, x^2) = F(x^2)$ and also $F(x^2, x^3) = F(x)$ since $x = x^3 \cdot (x^2)^{-1} \in F(x^2, x^3)$, but it is not clear whether $F((x^2 + 1)/x, (x^3 + 1)/x^2))$ is $F(x)$, is $F((x^2 + 1)/x)$, or is something else.

Example 14.18 $L = F((x^2 + 1)/x, (x^3 + 1)/x^2)) = F(x)$.

Now $(x^2 + 1)/x = x + 1/x$, $(x^3 + 1)/x^2 = x + 1/x^2$, and $(x + 1/x)^2 = x^2 + 2 + 1/x^2$, so $x^2 + 1/x^2 \in L$. Using $x + 1/x^2 \in L$ we get $x^2 - x \in L$. Also $(x + 1/x) - (x + 1/x^2) = 1/x - 1/x^2 = (x - 1)/x^2 \in L$ and $(x^2 - x)x^2/(x - 1) = x^3 \in L$ follows. Thus $x^3 + 1 \in L$ so $(x^3 + 1)^{-1}(x^3 + 1)/x^2 \in L$ implies $1/x^2 \in L$, then $x \in L$.

The difficulty of identifying even simply given examples may show why our final result, known as Lüroth's Theorem, is a bit surprising.

THEOREM 14.29 (Lüroth) If $F \subseteq L \subseteq F(x)$ are fields, then $L = F(\beta)$.

Proof If $L = F$ then $L = F(1)$, and if $\alpha \in L - F$ then Example 14.15 shows that x is algebraic over $F(\alpha)$, so over L. Let $m(Y) = Y^n + \alpha_{n-1}Y^{n-1} + \cdots + \alpha_0$ be the minimal polynomial in $L[Y]$ for x. Since x is transcendental over F, some $\alpha_i \in L - F$. Write each $\alpha_j = h_j(x)/g_j(x) \in F(x)$ for $h_j(x), g_j(x) \in F[x]$ with $\text{GCD}(\{h_j(x), g_j(x)\}) = 1$. If $g(x)$ is the least common multiple of $\{g_j(x)\}$ then $g(x)m(Y) = g(x)Y^n + v_{n-1}(x)Y^{n-1} + \cdots + v_1(x)Y + v_0(x)$ with $g(x), v_j(x) \in F[x]$ *and note that* $\text{GCD}(\{g(x), v_{n-1}(x), \ldots, v_0(x)\}) = 1$ (see Theorem 1.19, and verify!). Hence $g(x)m(Y) = f(Y) \in F[x][Y]$ is primitive. Next observe that using x as the indeterminate, $deg_x\, f(Y) = \max\{deg_x\, g, deg_x\, v_0, \ldots, deg_x\, v_{n-1}\}$, and this implies that $deg_x\, f(Y) \geq deg_x\, g_j$ and $deg_x\, f(Y) \geq deg_x\, h_j$ for each j. Assume that $\alpha_i = h_i(x)/g_i(x) \notin F$. If $d(Y) = g_i(Y)\alpha_i - h_i(Y) \in L[Y]$ then $d(x) = 0$ forces $d(Y) = u(Y)m(Y)$ in $L[Y]$. Thus $g_i(x)d(Y) = g_i(x)u(Y)(1/g(x))f(Y)) = w(Y)f(Y)$ with $w(Y) \in F(x)[Y]$. Since $F[x]$ is a UFD, from Theorem 12.12 we can write $w(Y) = s(x)t(Y)$ for some $s(x) \in F(x)$ and a primitive $t(Y) \in F[x][Y]$. Substituting yields $s(x)(t(Y)f(Y)) = g_i(x)d(Y) = g_i(Y)h_i(x) - h_i(Y)g_i(x) \in F[x][Y]$. Now Theorem 12.13 shows that $t(Y)f(Y) \in F[x][Y]$ is primitive so $s(x) \in F[x]$ by Theorem 12.12. Therefore $g_i(Y)h_i(x) - h_i(Y)g_i(x) = w(Y)f(Y)$ with $w(Y) \in F[x][Y]$. We have observed that $deg_x\, f(Y) \geq deg_x\, h_i, deg_x\, g_i$, and this forces $deg_x\, w(Y) = 0$ because $g_i(Y), h_i(Y) \in F[Y]$.

We claim $w(Y) \in F$. If not, let M be a splitting field for $w(Y)$ over $F(x)$ with $\beta \in M$ a root of $w(Y)$. It follows that $0 = w(\beta)f(\beta) = g_i(\beta)h_i(x) - h_i(\beta)g_i(x)$. Since $\text{GCD}(\{g_i(x), h_i(x)\}) = 1$, $g_i(\beta) = h_i(\beta) = 0$ is impossible since otherwise each would be divisible by the minimal polynomial for β over F. If one of these were zero then x would be algebraic over F, a contradiction. Hence $h_i(\beta)/g_i(\beta) = h_i(x)/g_i(x) = \alpha_i \in F[\beta]$ is algebraic over F, contradicting the choice of α_i. Therefore $w(Y) \in F$ as claimed. We conclude first that $deg_Y(g_i(Y)h_i(x) - h_i(Y)g_i(x)) = deg_Y\, f(Y)$ and then that $deg_Y\, m(Y) = deg_Y\, f(Y) = deg_Y(g_i(Y)\alpha_i - h_i(Y))$. Consequently, $[F(x) : L] = deg_Y\, m(Y) \geq [F(x) : F(\alpha_i)]$ since x satisfies $g_i(Y)\alpha_i - h_i(Y) \in F(\alpha_i)[Y]$. Finally, $F \subseteq F(\alpha_i) \subseteq L \subseteq F(x)$ and Theorem 14.8 force $L = F(\alpha_i)$, proving the theorem. ∎

EXERCISES

1. Show that $\{x + y, x - y\} \subseteq \boldsymbol{Q}(x, y)$ is algebraically independent over $\boldsymbol{Q}$.
2. If $F \subseteq L$ are fields, show that: $S \subseteq L$ is not algebraically independent over F if and only if some $\alpha_j \in S$ is algebraic over $F(S - \{\alpha_j\})$.
3. If $F \subseteq L$ are fields, $\{\alpha_1, \ldots, \alpha_n\} \subseteq L$, and $\{\alpha_j\}$ is algebraically independent over F, show that for any $k(i) \in \boldsymbol{Z} - \{0\}$, $\{\alpha_1^{k(1)}, \ldots, \alpha_n^{k(n)}\}$ is also algebraically independent over F.
4. If $F \subseteq L$ are fields, $\{\alpha_1, \ldots, \alpha_n\} \subseteq L$, and L is algebraic over $F(\{\alpha_j\})$, show that some subset of $\{\alpha_j\}$ is a transcendence basis of L over F. (See Exercise 2.)
5. Show that $S = \{x^2, y^2, xy\} \subseteq \boldsymbol{Q}(x, y)$ is not algebraically independent over $\boldsymbol{Q}$. Show that $\boldsymbol{Q}(S) = \boldsymbol{Q}(x^2 + xy, y^2 + xy)$.

6. If $F \subseteq L$ are fields, $g(x) = x^n + a_{n-1}x^{n-1} + \cdots + a_0 \in F[x]$, and if in $L[x]$, $g(x) = (x - \beta_1) \cdots (x - \beta_n)$, verify that $a_j = (-1)^{n-j}\sigma_{n-j}(\beta_1, \ldots, \beta_n)$ (symmetric function!).

7. Can $\{x + y, x + y^2, x^2 + y^2\} \subseteq \boldsymbol{Q}(x, y, z)$ be algebraically independent over $\boldsymbol{Q}$?

8. Is S a transcendence basis of $\boldsymbol{Q}(x, y, z)$ over $\boldsymbol{Q}$:

i) if $S = \{x^m + y^m, y^m + z^m, z^m + x^m\} (m \in \boldsymbol{N})$?

ii) if $S = \{x^u + y^v + z^w, x^{2u} + y^{2v} + z^{2w}, x^{3u} + y^{3v} + z^{3w}\}$, $u, v, w \in \boldsymbol{Z}^*$?

9. If F is a field and $\alpha = (ax + b)/(cx + d)) \in F(x) - F$, show that $F(\alpha) = F(x)$.

10. If $p(x), q(x) \in F[x]$ are quadratic and $p(x)/q(x) \in S = \{\alpha_1, \ldots, \alpha_n\} \subseteq F(x) - F$, show that $F(S) = F(x)$ or $F(S) = F(p(x)/q(x))$.

11. Find $\beta \in \boldsymbol{Q}(x)$ so that $\boldsymbol{Q}(S) = \boldsymbol{Q}(\beta)$ when $S =$:

i) $\{(x^4 + 1)/x^2, (x^6 + 1)/x^4\}$

ii) $\{x^2 + x^3, x^4 - x^6\}$

iii) $\{x^{12}, x^{40} + 1\}$

iv) $\{x^5(x + 1)^5 - 1, (x + 1)^2 - x\}$

CHAPTER 15

GALOIS THEORY

The quadratic formula gives the roots of a quadratic equation in terms of its coefficients. It is well known that any polynomial $p(x) \in \boldsymbol{R}[x]$ of degree 3 or 4 has some root in $\boldsymbol{C}$ that is given by an algebraic expression in its coefficients. It is true but not obvious that when $k \geq 5$ there is no algebraic expression in $\{x_k, \ldots, x_0\}$ giving a root of $a_k x^k + \cdots + a_1 x + a_0$ for all choices of $\{a_j\}$ when $a_k \neq 0$ and when a_i replaces x_i. We will prove this fact using Galois Theory, an approach that studies the relation between the subfields of a field L and the subgroups of its automorphism group $Aut(L)$. The interplay between these, using results from group theory and results about fields, is a powerful tool for understanding field extensions $F \subseteq L$ and the structure of the intermediate subfields $F \subseteq M \subseteq L$.

15.1 THE GALOIS CORRESPONDENCE

For fields $F \subseteq L$ the automorphisms of L fixing each element in F form the group $Aut(L/F)$ (Definition 14.12). Our interest is primarily when $[L:F]$ is finite and, more specifically, when L is a splitting field over F. Thus the results in §14.4 are crucial here. We begin by renaming $Aut(L/F)$.

DEFINITION 15.1 Given fields $F \subseteq L$, the **Galois group of L over F** is $Aut(L/F)$, denoted $Gal(L/F)$.

As we remarked in §14.4, $(Gal(L/F), \circ)$ is a group. Since for $\varphi \in Gal(L/F)$ and $a \in F$, $\varphi(a) = a$, it follows that if $p(x) \in F[x]$ and $\beta \in L$ with $p(\beta) = 0$, then $0 = \varphi(p(\beta)) = p(\varphi(\beta))$. Hence φ permutes the roots in L of any $p(x) \in F[x]$. When $L = F[S]$ for S the set of roots of $p(x) \in F[x]$, it follows that $Gal(L/F)$ is identified with a subgroup of permutations of S. Further, by Theorem 14.18, if $p(x) \in F[x]$ is irreducible and $\alpha, \beta \in S$ then there is some $\rho \in Gal(L/F)$ with $\rho(\alpha) = \beta$.

Example 15.1 $Gal(\boldsymbol{Q}[2^{1/3}]/\boldsymbol{Q}) = \langle I_{\boldsymbol{Q}[2^{1/3}]}\rangle$ where $\boldsymbol{Q}[2^{1/3}] \subseteq \boldsymbol{C}$.

If $\varphi \in Gal(\boldsymbol{Q}[2^{1/3}]/\boldsymbol{Q})$ then φ is uniquely determined by $\varphi(2^{1/3})$ (why?) and this must be another root of $x^3 - 2 \in \boldsymbol{Q}[x]$. The roots of $x^3 - 2$ in $\boldsymbol{C}$ are $2^{1/3}$, $2^{1/3}\omega$, and $2^{1/3}\omega^2 \in \boldsymbol{C}$ for $\omega = -1/2 + 3^{1/2}/2 \cdot i$, and $2^{1/3}$ is the only root in $\boldsymbol{Q}[2^{1/3}] \subseteq \boldsymbol{R}$. This forces $\varphi(2^{1/3}) = 2^{1/3}$, so $\varphi = I_{\boldsymbol{Q}[2^{1/3}]}$.

Example 15.2 $Gal(\boldsymbol{Q}[2^{1/4}]/\boldsymbol{Q}) \cong \boldsymbol{Z}_2$ where $\boldsymbol{Q}[2^{1/4}] \subseteq \boldsymbol{C}$.

Now $x^4 - 2 \in \boldsymbol{Q}[x]$ has roots $\pm 2^{1/4}$ and $\pm 2^{1/4} i$ in $\boldsymbol{C}$ and is irreducible by Eisenstein's criterion. Since $\boldsymbol{Q}[2^{1/4}] \subseteq \boldsymbol{R}$, if $\varphi \in Gal(\boldsymbol{Q}[2^{1/4}]/\boldsymbol{Q})$ then either $\varphi(2^{1/4}) = 2^{1/4}$ and $\varphi = I$ or $\varphi(2^{1/4}) = -2^{1/4}$. From Theorem 14.16 there is $\varphi \in Gal(\boldsymbol{Q}[2^{1/4}]/\boldsymbol{Q})$ with $\varphi(2^{1/4}) = -2^{1/4}$, so $Gal(\boldsymbol{Q}[2^{1/4}]/\boldsymbol{Q}) = \{I, \varphi\} \cong \boldsymbol{Z}_2$.

Example 14.14 shows that a splitting field for $x^4 - 2 \in \boldsymbol{Q}[x]$ over $\boldsymbol{Q}$ is $L = \boldsymbol{Q}[2^{1/4}, i] \subseteq \boldsymbol{C}$ and that $Gal(L/\boldsymbol{Q}) \cong D_4$. We revisit this example for the simpler case of $x^3 - 2 \in \boldsymbol{Q}[x]$. Recall that $\omega = -1/2 + 3^{1/2}/2 \cdot i \in \boldsymbol{C}$ satisfies $\omega^3 = 1$ and has minimal polynomial $x^2 + x + 1 \in \boldsymbol{Q}[x]$.

Example 15.3 A splitting field for $x^3 - 2 \in \boldsymbol{Q}[x]$ is $L = \boldsymbol{Q}[2^{1/3}, \omega] \subseteq \boldsymbol{C}$, $[L : \boldsymbol{Q}] = 6$ and $Gal(L/\boldsymbol{Q}) \cong S_3$.

The roots in $\boldsymbol{C}$ of $x^3 - 2 \in \boldsymbol{Q}[x]$ are $S = \{2^{1/3}, 2^{1/3}\omega, 2^{1/3}\omega^2\}$ so $L = \boldsymbol{Q}[S] = \boldsymbol{Q}[2^{1/3}, \omega]$ is a splitting field for $x^3 - 2$ over $\boldsymbol{Q}$. That $[L : \boldsymbol{Q}] = 6$ and $[L : \boldsymbol{Q}[\omega]] = 3$ follow from Example 14.6 and Theorem 14.8. Thus $x^3 - 2 \in \boldsymbol{Q}[\omega][x]$ is irreducible by Theorem 14.7, and of course L is still a splitting field of $x^3 - 2$ over $\boldsymbol{Q}[\omega]$. Clearly any $\varphi \in Gal(L/\boldsymbol{Q})$ is uniquely determined by $\varphi(2^{1/3}) \in S$ and $\varphi(\omega) \in \{\omega, \omega^2\}$ so $|Gal(L/\boldsymbol{Q})| \leq 6$. Using Theorem 14.16 there are $\sigma_0, \sigma_1, \sigma_2 \in Gal(L/\boldsymbol{Q})$ that extend $I_{\boldsymbol{Q}[\omega]}$ and satisfy $\sigma_i(2^{1/3}) = 2^{1/3}\omega^i$. Note that $\sigma_0 = I_L$, that $\sigma_1^2(2^{1/3}) = \sigma_1(2^{1/3}\omega) = \sigma_1(2^{1/3})\sigma_1(\omega) = 2^{1/3}\omega \cdot \omega = 2^{1/3}\omega^2$, so $\sigma_1^2 = \sigma_2$ (why?), and that $\sigma_1^3 = I_L$. Complex conjugation is an automorphism of $\boldsymbol{Q}[\omega]$ sending ω to ω^2 and extends to $\rho_0, \rho_1, \rho_2 \in Gal(L/\boldsymbol{Q})$, with $\rho_i(2^{1/3}) = 2^{1/3}\omega^i$, again by Theorem 14.16. Since $\rho_0(\omega) = \omega^2$ and $\rho_0(2^{1/3}) = 2^{1/3}$ we have $\rho_0(2^{1/3}\omega) = 2^{1/3}\omega^2$ and $\rho_0(2^{1/3}\omega^2) = 2^{1/3}\omega$, so ρ_0 fixes one root of $x^3 - 2$ and interchanges the other two. Similarly, ρ_1 and ρ_2 are the other transpositions of the roots. It follows that $\{\sigma_i, \rho_i\}$ has six elements and so is $Gal(L/\boldsymbol{Q})$, and as there are only six possible permutations of the roots, $Gal(L/\boldsymbol{Q}) \cong S_3$. This holds also by Example 6.4 since $\rho_0\sigma_1\rho_0 = \sigma_1^{-1}$.

In the last example $Gal(L/\boldsymbol{Q})$ has four proper subgroups, three of order 2 and one of order 3. What are the fields between $\boldsymbol{Q}$ and L? The only obvious proper ones are $\boldsymbol{Q}[2^{1/3}]$, $\boldsymbol{Q}[2^{1/3}\omega]$, $\boldsymbol{Q}[2^{1/3}\omega^2]$, and $\boldsymbol{Q}[\omega]$, three of dimension 3 over $\boldsymbol{Q}$ and one of dimension 2. It is not obvious that these are the only proper subfields between L and $\boldsymbol{Q}$ (they are!). To study intermediate fields we introduce the basic Galois correspondence between the subfields of L and the subgroups of $Gal(L/F)$.

DEFINITION 15.2 If $F \subseteq E \subseteq L$ are fields and $H \leq G = Gal(L/F)$ then the **fixed field of H** is $H' = \{\alpha \in L \mid h(\alpha) = \alpha$ for all $h \in H\}$, and $E' = Gal(L/E)$.

The maps in Definition 15.2 between the set of *intermediate fields* E and the set of subgroups of G are called the *priming* operations. The compositions of these are the *double primes* $H'' = (H')'$ and $E'' = (E')'$. It is easy to see that $L' = \langle I_L \rangle$, $F' = G$, and $\langle I_L \rangle' = L$. Example 15.1 shows that it can happen that $G' \neq F$. A way to visualize these maps and part of the next result about inclusions is shown in Figure 15.1.

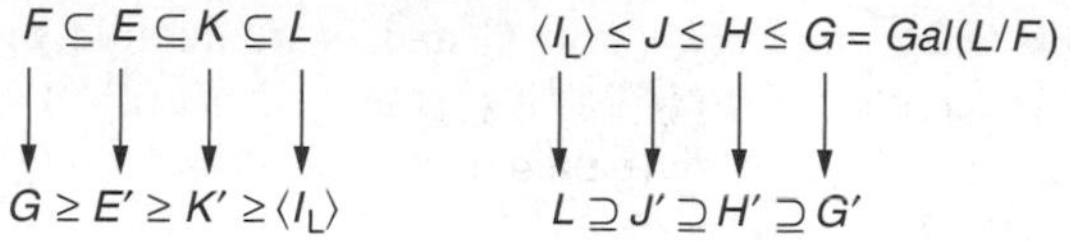

FIGURE 15.1

THEOREM 15.1 Let $F \subseteq E \subseteq K \subseteq L$ be fields and $J \leq H \leq G = Gal(L/F)$. The following hold:

i) H' is a subfield of L containing F
ii) $K' \leq E'$
iii) $H' \subseteq J'$
iv) $H \leq H''$
v) $E \subseteq E''$
vi) $H' = H''' = (H'')'$
vii) $E' = E''' = (E'')'$

Hence taking primes is an inclusion reversing bijection between $\{Gal(L/E) \mid E$ is a subfield of L containing $F\}$ and $\{F \subseteq H' \subseteq L \mid H \leq G\}$.

Proof i), ii), and iii) are routine verifications using the definitions. If $\varphi \in H$ and $\alpha \in H'$ then $\varphi(\alpha) = \alpha$, so $\varphi \in (H')'$. Thus iv) follows, and the proof of v) is similar. From iv), $H \leq H''$, applying iii) yields $H' \supseteq H'''$, and using v) results in $H' \subseteq (H')'' = H'''$, proving vi). A similar argument gives vii). The last statement follows from ii), iii), vi), and vii) since the sets described are the "primed" objects. ■

Example 15.3 illustrates the priming correspondence well since $[\boldsymbol{Q}[2^{1/3}, \omega] : \boldsymbol{Q}] = 6$ implies that any proper intermediate field must have prime dimension over $\boldsymbol{Q}$. Recall that $S = \{2^{1/3}, 2^{1/3}\omega, 2^{1/3}\omega^2\}$ is the set of all roots of $x^3 - 2$ in $L = \boldsymbol{Q}[2^{1/3}, \omega]$ and that $Gal(L/\boldsymbol{Q}) = G$ is essentially the set of six permutations of S. From Example 15.3, $\langle \sigma_1 \rangle$ is the only subgroup of order 3 in G and since σ_1 extends $I_{\boldsymbol{Q}[\omega]}$, $\boldsymbol{Q}[\omega] \subseteq \langle \sigma_1 \rangle'$. Now $2^{1/3} \notin \langle \sigma_1 \rangle'$ so $\langle \sigma_1 \rangle' \neq L$ and Theorem 14.8 force $\langle \sigma_1 \rangle' = \boldsymbol{Q}[\omega]$. Also since $\rho_j(\omega) = \omega^2$, $\boldsymbol{Q}[\omega]' \neq G$, and Theorem 15.1 shows that $\langle \sigma_1 \rangle \leq \langle \sigma_1 \rangle'' = \boldsymbol{Q}[\omega]' \neq G$, resulting in $\langle \sigma_1 \rangle = \langle \sigma_1 \rangle'' = \boldsymbol{Q}[\omega]'$, so $\boldsymbol{Q}[\omega]'' = \langle \sigma_1 \rangle' = \boldsymbol{Q}[\omega]$. Similarly, $L \neq \langle \rho_0 \rangle' \supseteq \boldsymbol{Q}[2^{1/3}]$ implies $\langle \rho_0 \rangle' = \boldsymbol{Q}[2^{1/3}]$, then $\langle \rho_0 \rangle = \langle \rho_0 \rangle'' = \boldsymbol{Q}[2^{1/3}]'$, and $\boldsymbol{Q}[2^{1/3}] = \boldsymbol{Q}[2^{1/3}]''$. In the same way we get $\langle \rho_1 \rangle' = \boldsymbol{Q}[2^{1/3}\omega^2]$, $\langle \rho_1 \rangle = \boldsymbol{Q}[2^{1/3}\omega^2]'$, $\langle \rho_2 \rangle' = \boldsymbol{Q}[2^{1/3}\omega]$, and $\langle \rho_2 \rangle = \boldsymbol{Q}[2^{1/3}\omega]'$. If $\alpha \in G'$ then $\alpha \in \langle \sigma_1 \rangle' \cap \langle \rho_0 \rangle' = \boldsymbol{Q}[\omega] \cap \boldsymbol{Q}[2^{1/3}] = \boldsymbol{Q}$, so $G' = \boldsymbol{Q}$. For *any* field $\boldsymbol{Q} \subseteq E \subseteq L$, we have

$\langle I_L \rangle \leq E' \leq G$ and so $\mathbf{Q} \subseteq E \subseteq E'' \subseteq L$. Using $[L : \mathbf{Q}] = 6$, a dimension argument shows $E = E''$ is a primed field we have already identified, or else $E' = \langle I_L \rangle$.

We want a condition equivalent to $E = E'' = Gal(L/E)'$ since a direct computation of this is often difficult, as in our discussion above of Example 15.3. Note that $Gal(L/E)' = E$ means that for $\alpha \in L - E$, $\varphi(\alpha) \neq \alpha$ for some $\varphi \in Gal(L/E)$. It turns out that we need to consider polynomials with distinct roots in a splitting field.

DEFINITION 15.3 Let $F \subseteq L$ be fields, $\alpha \in L$, $p(x) \in F[x]$ irreducible, and $deg\ p(x) = n$. Then $p(x)$ is **separable** if it has n distinct roots in a splitting field over F, α is **separable over** $\boldsymbol{F}$ if $p(\alpha) = 0$ and $p(x)$ is separable, and L is **separable over** $\boldsymbol{F}$ if every $\alpha \in L$ is separable over F.

If the irreducible $p(x) \in F[x]$ has distinct roots in some splitting field L_1 then it has distinct roots in any splitting field L by Theorem 14.17: $L_1 \cong L$ so the factorization of $p(x) \in L_1[x]$ into distinct linear factors yields a similar factorization in $L[x]$. Using Theorem 14.14 we see that an irreducible $g(x)$ is separable unless $char\ F = p > 0$ and $g(x) = f(x^p)$. In particular, every irreducible $g(x) \in \mathbf{Q}[x]$ is separable, and since in $\mathbf{Z}_p[x]$, $f(x^p) = (f(x))^p$, any irreducible $g(x) \in \mathbf{Z}_p[x]$ is separable also.

Most of our examples are splitting fields for a separable $p(x) \in F[x]$ but it is not clear that such extensions are separable. They are, and a more general statement holds that is equivalent to $Gal(L/F)' = F$.

THEOREM 15.2 Let $F \subseteq L$ be fields, $[L : F]$ finite, and $Gal(L/F) = G$. Then G is finite and the following are equivalent:

i) $F'' = G' = F$.
ii) The minimal polynomial $m_\beta(x) \in F[x]$ of any $\beta \in L$ is separable and splits over L.
iii) L is a splitting field over F and is separable over F.
iv) L is a splitting field over F of a polynomial whose irreducible factors over F are separable.

Proof If $\{\alpha_1, \ldots, \alpha_k\}$ is a basis of L over F then since $[L : F]$ is finite each α_j satisfies an irreducible $p_j(x) \in F[x]$ by Theorem 14.9. Consider $T = \{\gamma \in L \mid p_j(\gamma) = 0 \text{ for some } j\}$. Now $\{\alpha_j\} \subseteq T$ implies $L = F[T]$, and clearly each $\varphi \in G$ is uniquely determined by its action on the elements of T. As we have observed, φ is a permutation of the roots in L of each $p_j(x)$, so φ permutes the elements in T and $|G| \leq |T|!$ follows. Assuming i), let $\beta \in L$ with minimal polynomial $m_\beta(x) \in F[x]$. For $\varphi \in G$, $0 = \varphi(m_\beta(\beta)) = m_\beta(\varphi(\beta))$, so $\{\varphi(\beta) \in L \mid \varphi \in G\} = \{\beta_1, \ldots, \beta_n\}$. Set $g(x) = (x - \beta_1) \cdots (x - \beta_n) = \sum a_j x^j \in L[x]$ and note that $g(\beta) = 0$. Since $\varphi \in G$ permutes $\{\beta_i\}$, $g(x) = (x - \varphi(\beta_1)) \cdots (x - \varphi(\beta_n)) = \varphi(g(x)) = \sum \varphi(a_j) x^j$. Thus i) forces all $a_j \in F$, and $g(x) \in F[x]$. Since $m_\beta(x)$ must divide $g(x)$, and since $g(x)$ splits over L into distinct linear factors, $m_\beta(x)$ itself splits over L into distinct linear factors, which proves ii).

By ii), L is separable over F and splits any irreducible in $F[x]$ that has a root in L. If $\{\alpha_1, \ldots, \alpha_k\}$ is a basis of L over F and if $m_j(x) \in F[x]$ is the minimal polynomial for α_j, then L is a splitting field for $m_1(x) \cdots m_k(x)$ over F and iii) holds. Assuming iii), since $[L : F]$ is finite we may assume that L is the splitting field over F of $\{p_1(x), \ldots, p_m(x)\} \subseteq F[x]$, all irreducible (why?). Now each $p_i(x)$ is separable by assumption and L is a splitting field over F of $f(x) = p_1(x) \cdots p_m(x)$ as required for iv).

If $[L:F] = 1$ then clearly iv) implies i). If iv) holds but i) fails, let $K \subseteq M$ be fields with $[M:K] > 1$ and minimal so that: M is a splitting field over K of a polynomial $f(x) \in K[x]$ whose irreducible factors are separable, and there is $\beta \in M - K$ satisfying $\varphi(\beta) = \beta$ for all $\varphi \in Gal(M/K)$. Let $g(x) \in K[x]$ be an irreducible factor of $f(x)$ with $deg\, g(x) = t > 1$ and $\alpha \in M$ a root of $g(x)$. If $\beta \notin K[\alpha]$, since $\eta \in Gal(M/K[\alpha])$ implies $\eta \in Gal(M/K)$, we have $\eta(\beta) = \beta$. Clearly M is a splitting field for $f(x)$ over $K[\alpha]$ and the irreducible factors of $f(x)$ over $K[\alpha]$ are divisors of the irreducible factors of $f(x)$ over K and so are separable. Since $[M:K[\alpha]] < [M:K]$ by Theorem 14.8, we get a contradiction to the minimality of $[M:K]$, forcing $\beta \in K[\alpha]$. By assumption, M contains t distinct roots of $g(x)$, say $\alpha = \alpha_1, \ldots, \alpha_t$. Using $\beta \in K[\alpha]$ and $deg\, g(x) = t$, Theorem 14.7 shows that $\beta = k_0 + k_1\alpha + \cdots + k_{t-1}\alpha^{t-1}$ for $k_j \in K$. From Theorem 14.18, for each i there is $\eta_i \in Gal(M/K)$ with $\eta_i(\alpha) = \alpha_i$. It follows that $k_0 + k_1\eta_i(\alpha) + \cdots + k_{t-1}\eta_i(\alpha^{t-1}) = \eta_i(\beta) = \beta = k_0 + k_1\alpha_i + \cdots + k_{t-1}\alpha_i^{t-1}$ so if $h(x) = k_{t-1}x^{t-1} + \cdots + k_1x + (k_0 - \beta) \in K[\alpha][x]$ then the t different α_j are roots of $h(x)$ in M. This is impossible for a polynomial of degree at most $t-1$ and forces $h(x) = 0$, so $\beta = k_0 \in K$, a contradiction. This shows that iv) must imply i) and finishes the proof. ∎

When $F \subseteq E \subseteq L$ are fields, $[L:F] = n$, $G = Gal(L/F)$, and $G' = F$, then the theorem shows that L is a separable splitting field over F. It is an exercise that L is separable over E and that L is a splitting field over E, so by the theorem, $E'' = Gal(L/E)' = E$. Thus if $E, M \subseteq L$ are subfields containing F and $E' = M'$, then $E = E'' = M'' = M$. This proves the next result.

THEOREM 15.3 If $F \subseteq L$ are fields, $[L:F] = n$, $G = Gal(L/F)$, and $G' = F$, then $E'' = E$ for all fields $F \subseteq E \subseteq L$. Therefore the priming operation is an injection from the set of the intermediate fields to the set of subgroups of G.

Since any irreducible in $\boldsymbol{Q}[x]$ or in $\boldsymbol{Z}_p[x]$ is separable, any splitting field L over $F = \boldsymbol{Q}$ or $\boldsymbol{Z}_p$ with $[L:F] = n$ satisfies $Gal(L/F)' = F$. In this case, if we can find as many subfields of L containing F as there are subgroups of G then by Theorem 15.3 these are all the intermediate subfields. The four proper subfields of $\boldsymbol{Q}[2^{1/3}, \omega]$ containing $\boldsymbol{Q}$ in Example 15.3 are all the proper intermediate fields by Theorem 15.3 since there are exactly four $\langle e \rangle \neq H \leq S_3$ with $H \neq S_3$. The next example is analyzed similarly and uses Theorem 15.2. Recall that $\omega = e^{2\pi i/3} \in \boldsymbol{C}$ satisfies $\omega^3 = 1$ but $\omega \neq 1$.

Example 15.4 The subfields of a splitting field of $x^6 - 2$ over $\boldsymbol{Q}[\omega]$.

Since $\omega \in (\boldsymbol{C}^*, \cdot)$ has order 3, $-\omega$ has order 6, and clearly $\boldsymbol{Q}[-\omega] = \boldsymbol{Q}[\omega]$. By Eisenstein's criterion (Theorem 12.16), $x^6 - 2 \in \boldsymbol{Q}[x]$ is irreducible and its roots in $\boldsymbol{C}$ are $\beta_i = 2^{1/6}(-\omega)^i$ for $0 \leq i \leq 5$. Thus $L = \boldsymbol{Q}[2^{1/6}, \omega]$ is a splitting field for $x^6 - 2$ over $\boldsymbol{Q}$ (why?) so over $\boldsymbol{Q}[\omega]$, and $Gal(L/\boldsymbol{Q}[\omega])' = \boldsymbol{Q}[\omega]$ from Theorem 15.2. Now $[\boldsymbol{Q}[\omega]:\boldsymbol{Q}] = 2$ and $\omega \notin \boldsymbol{Q}[2^{1/6}]$ so $[L:\boldsymbol{Q}] = [L:\boldsymbol{Q}[2^{1/6}]][\boldsymbol{Q}[2^{1/6}]:\boldsymbol{Q}] = 2 \cdot 6 = 12$, using Theorem 14.7 and Theorem 14.8. The latter result now shows that

$[L:\boldsymbol{Q}[\omega]]=6$. Hence $x^6-2\in\boldsymbol{Q}[\omega][x]$ is irreducible using Theorem 14.7 again. Since any $\varphi\in G=Gal(L/\boldsymbol{Q}[\omega])$ is uniquely determined by $\varphi(2^{1/6})\in\{\beta_j\}$, it follows that $|G|\le 6$. Extending $I_{\boldsymbol{Q}[\omega]}$ to L via Theorem 14.16 yields $\rho_j\in G$ with $\rho_j(2^{1/6})=2^{1/6}(-\omega)^j$ and $\rho_j(\omega)=\omega$. Consequently, $G=\langle\rho_1\rangle$ and its subgroups are G, $\langle\rho_1^2\rangle=\langle\rho_2\rangle$, $\langle\rho_1^3\rangle=\langle\rho_3\rangle$, and $\langle I_L\rangle$. As we have seen, $G'=\boldsymbol{Q}[\omega]$ and of course $\langle I_L\rangle'=L$. Now $\rho_2(2^{1/6})=2^{1/6}\omega^2$ and $\rho_2^2(2^{1/6})=2^{1/6}\omega^4=2^{1/6}\omega$. Since $o(\rho_2)=3$ it follows that ρ_2 fixes $2^{1/6}\rho_2(2^{1/6})\rho_2^2(2^{1/6})=2^{1/2}$. Thus $\boldsymbol{Q}[2^{1/2},\omega]\subseteq\langle\rho_2\rangle'$, and $[\boldsymbol{Q}[2^{1/2},\omega]:\boldsymbol{Q}[\omega]]=2$ (why?) forces $[L:\boldsymbol{Q}[2^{1/2},\omega]]=3$ by Theorem 14.8. This same theorem shows that $\langle\rho_2\rangle'=\boldsymbol{Q}[2^{1/2},\omega]$ since $\langle\rho_2\rangle'\ne L$. Similarly, $\rho_3(2^{1/6})=2^{1/6}(-\omega^3)=-2^{1/6}$ leads to $\langle\rho_3\rangle'=\boldsymbol{Q}[2^{1/3},\omega]$. Thus $\boldsymbol{Q}[\omega]$, $\boldsymbol{Q}[2^{1/2},\omega]$, $\boldsymbol{Q}[2^{1/3},\omega]$, and L are the only subfields of L containing $\boldsymbol{Q}[\omega]$ by Theorem 15.3.

In the examples of splitting fields we have looked at, when $H\le G=Gal(L/F)$ then $|H|=[L:H']$ and for $F\subseteq E\subseteq L$, $[E:F]=[G:E']$. These relations are generally true and are part of a further study of the subgroups of G and the subfields of L in the next section. Our final example here shows that the separability condition in Theorem 15.2 is crucial.

Example 15.5 A splitting field over $\mathbf{Z}_p(x)$ for $Y^p-x\in\mathbf{Z}_p(x)[Y]$.

Note first that $f(x)=Y^p-x\in\mathbf{Z}_p(x)[Y]$ is irreducible by Eisenstein's Criterion but is not separable by Theorem 14.14. Let L be a splitting field for $f(x)$ over $\mathbf{Z}_p(x)$ and $\beta\in L$ a root. In $L[Y]$, $(Y-\beta)^p=Y^p-\beta^p=Y^p-x$. Thus $L=\mathbf{Z}_p(x)[\beta]$ and β is the only root of Y^p-x in L. For any $\varphi\in Gal(L/\mathbf{Z}_p(x))$, $\varphi(\beta)=\beta$, so $\varphi=I_L$, $G=Gal(L/\mathbf{Z}_p(x))=\langle I_L\rangle$, and $\mathbf{Z}_p(x)\ne G'=L$.

EXERCISES

1. For $n\in N$ describe $Gal(\boldsymbol{Q}[2^{1/n}]/\boldsymbol{Q})$.
2. If $p_1<\cdots<p_m$ are primes, show that $Gal(\boldsymbol{Q}[p_1^{1/2},\ldots,p_m^{1/2}]/\boldsymbol{Q})$ is Abelian.
3. Let $F\subseteq L$ be fields with $G=Gal(L/F)$.
   i) If $E, M\subseteq L$ are subfields containing F, show that $(E\vee M)'=E'\cap M'$.
   ii) If $J, H\le G$, show that $\langle\{J\cup H\}\rangle'=J'\cap H'$.
4. Let $F\subseteq E\subseteq L$ be fields, $G=Gal(L/F)$, $\varphi\in G$, and $H\le G$. Show that:
   i) $\varphi(E)'=\varphi E'\varphi^{-1}$
   ii) $\varphi(H')=(\varphi H\varphi^{-1})'$
5. If $f(x)\in F[x]$ with $deg\ f=n>0$ and L is a splitting field for $f(x)$ over F, show that:
   i) $Gal(L/F)\cong H\le S_n$
   ii) if $n=5$ then $|Gal(L/F)|\ne 15$
6. Let $F\subseteq E\subseteq L$ be fields with E a splitting field of $S\subseteq F[x]$ over F.
   i) If $\varphi\in Gal(L/F)$, show that $\varphi(E)=E$
   ii) Show that $\eta: Gal(L/F)\to Gal(E/F)$ given by $\eta(\varphi)=\varphi|_E$ is a group homomorphism.

7. Let $F \subseteq L$ be fields and $H = \{\varphi_1, \ldots, \varphi_m\} \le G = Gal(L/F)$. For any $\beta \in L$, show that $Tr(\beta) = \varphi_1(\beta) + \cdots + \varphi_m(\beta) \in H'$ and $N(\beta) = \varphi_1(\beta)\varphi_2(\beta)\cdots\varphi_m(\beta) \in H'$.

8. For F a field and $\alpha \in F$, show that $\varphi_\alpha(p(x)/q(x)) = p(x+\alpha)/q(x+\alpha)$ defines $\varphi_\alpha \in Gal(F(x)/F)$. When F is infinite, show that $Gal(F(x)/F)' = F$. (If $s(x)/t(x) \in Gal(F(x)/F)'$ then reduce to $s(x)$ and $t(x)$ monic and relatively prime. Use $\varphi_\alpha(s(x)/t(x)) = s(x)/t(x)$ for all $\alpha \in F$ to show $s(x+y)t(x) = s(x)t(x+y)$ in $F[x, y]$. Since $s(x+y)$ and $t(x+y) \in F[x][y]$ are primitive, $s(x)$ and $t(x)$ are associates in $F[x]$ forcing $s(x)/t(x) \in F$.)

9. If $F \subseteq E \subseteq L$ are fields with L separable over F, show that L is separable over E and that E is separable over F.

10. Let $L = \boldsymbol{Q}[2^{1/2}, 3^{1/2}]$ and $G = Gal(L/\boldsymbol{Q})$. Describe the structure of G, find all $H \le G$, and find the corresponding H'. Are there other subfields of L?

11. If $\rho = (2^{1/2}/2)(1+i) \in \boldsymbol{C}$, show that $L = \boldsymbol{Q}[\rho]$ is a splitting field for $x^8 - 1$ over $\boldsymbol{Q}$. Show that $L = \boldsymbol{Q}[2^{1/2}, i]$, determine $G = Gal(L/\boldsymbol{Q})$, find all $H \le G$, find all fields $E \subseteq L$, and describe all H' and all E'.

12. If $\emptyset \neq S \subseteq F[x]$ are separable polynomials and if L is a splitting field of S over F, show that L is separable over F. Note that S may be infinite.

13. If $F \subseteq L$ are fields and $\alpha_1, \ldots, \alpha_m \in L$ are separable over F, show that $F[\alpha_1, \ldots, \alpha_m]$ is separable over F.

14. If $F \subseteq L$ are fields, show that $Sep_F(L) = \{\beta \in L \mid \beta \text{ is separable over } F\}$ is a subfield of L containing F. (See Exercise 13.)

15. Let $F \subseteq L \subseteq M$ be fields with L separable over F and with M separable over L. Show that M is separable over F. (If $\beta \in M - Sep_F(M)$ (Exercise 14), show that for a prime p and some $v = p^k$, $\beta^v \in Sep_F(M)$. If $g(x)$ is the minimal polynomial of β over $Sep_F(M)$, then $g(x)$ is separable over $Sep_F(M)$ and $g(x)$ divides $x^v - \beta^v \in Sep_F(M)[x]$. For γ any root of $g(x)$ in a splitting field over M, $0 = \gamma^v - \beta^v = (\gamma - \beta)^v$, forcing $\gamma = \beta$ and $g(x) = x - \beta$.)

16. Let $F \subseteq L$ be fields, $[L : F] = n$, and L a splitting field over F. If $\beta \in L$ with irreducible $m(x) \in F[x]$, show that $m(x)$ splits in L. (For $L \subseteq M$ a splitting field of $m(x)$, if $m(\gamma) = 0$ for $\gamma \in M - L$, extend $\eta : F[\beta] \to F[\gamma]$ to M; see Exercise 6.)

17. If L is a splitting field over F of $\{p_1(x), \ldots, p_m(x)\} \subseteq F[x]$ where each $p_j(x)$ has all its irreducible factors separable over F, then L is the splitting field over F of $f(x) \in F[x]$ with all its irreducible factors separable over F.

18. Let p be a prime and L a field with $\boldsymbol{Z}_p \subseteq L$ and $[L : \boldsymbol{Z}_p] = n$. Show that L is separable over $\boldsymbol{Z}_p$ using Theorem 15.2 and the observation that for $\alpha \in L$, $\varphi(\alpha) = \alpha^p$ defines $\varphi \in Gal(L/\boldsymbol{Z}_p)$. Use $L = \boldsymbol{Z}_p[\beta]$ (Theorem 4.20) to see that for any $n \in \boldsymbol{N}$ there is an irreducible $p_n(x) \in \boldsymbol{Z}_p[x]$ of degree n.

19. For $L \subseteq \boldsymbol{C}$, a splitting field of $x^m - 1 \in \boldsymbol{Q}[x]$, show that $Gal(L/\boldsymbol{Q})$ is Abelian.

20. If $S = \{\rho \in \boldsymbol{C} \mid \rho^n = 1, \text{ some } n \in \boldsymbol{N}\}$, set $F = \boldsymbol{Q}[S] \subseteq \boldsymbol{C}$. For $p(1) < \cdots < p(k)$ primes and $\alpha_1, \ldots, \alpha_k \in \boldsymbol{Q}$, let $L = F[\alpha_1^{1/p(1)}, \ldots, \alpha_k^{1/p(k)}]$ and $G = Gal(L/F)$.

i) Show that $G' = F$.

ii) Show that G is cyclic.

21. Let $L = \boldsymbol{Q}[2^{1/2}, 3^{1/2}, 5^{1/2}] \subseteq \boldsymbol{R}$ and $G = Gal(L/\boldsymbol{Q})$.

i) Show that there are $\varphi_p \in G$ for $p \in \{2, 3, 5\}$ with $\varphi_p(p^{1/2}) = -p^{1/2}$ and $\varphi_p(q^{1/2}) = q^{1/2}$ for $q \in \{2, 3, 5\} - \{p\}$.

ii) Show that $G = \langle\{\varphi_2, \varphi_3, \varphi_5\}\rangle$, $|G| = 8$, and $\varphi^2 = I_L$ for all $\varphi \in G$.
iii) For $p, q \in \{2, 3, 5\}$, find H' as $\boldsymbol{Q}[\alpha, \beta]$ or $\boldsymbol{Q}[\gamma]$ for $H =$:
  a) $\langle \varphi_p \rangle$
  b) $\langle \varphi_p \varphi_q \rangle$ when $p \neq q$
  c) $\langle \varphi_2 \varphi_3 \varphi_5 \rangle$
  d) $\langle \{\varphi_p, \varphi_q\} \rangle$ with $p \neq q$

22. Let $\omega \in \boldsymbol{C} - \boldsymbol{Q}$ be a root of $x^3 - 1$, and set $F = \boldsymbol{Q}[\omega]$. For each field $F \subseteq L$ below find $[L : F]$, describe the order and structure of $G = Gal(L/F)$, show $G' = F$, and find all the subfields of L containing F:
  i) $L = F[2^{1/2}, 3^{1/3}]$
  ii) $L = F[2^{1/3}, 3^{1/3}]$
  iii) $L = F[2^{1/2}, 3^{1/2}, 5^{1/3}]$

15.2 THE FUNDAMENTAL THEOREM

When $F \subseteq L$ are fields with $[L : F] = n$ then $Gal(L/F)' = F$ exactly when L is a separable splitting field over F by Theorem 15.2. In this case the priming operations are bijections between the subgroups of G and the fields betweeen F and L. To prove this and other important relations between these objects requires more work. First we give a name to this situation.

DEFINITION 15.4 The field L is **Galois over the field** F if $F \subseteq L$, $[L : F]$ is finite, and $Gal(L/F)' = F$.

We note that some texts do not include the condition that $[L : F]$ is finite in the definition of L Galois over F. By Theorem 15.2, L Galois over F is equivalent to either L is a separable splitting field over F, or L is the splitting field over F of $f(x) \in F[x]$ whose irreducible factors are separable. The next result gives another important equivalence.

THEOREM 15.4 If $F \subseteq L$ are fields with $[L : F] = n$, then L is Galois over F if and only if $|Gal(L/F)| = n$.

Proof The theorem is clearly true if $L = F$, so we may take $n > 1$. Assume L is Galois over F, $G = Gal(L/F)$, and $\alpha \in L - F$ has minimal polynomial $m(x) \in F[x]$ of degree $k > 1$. By Theorem 15.2, $m(x)$ splits over L and has k distinct roots $\alpha = \alpha_1, \ldots, \alpha_k \in L$. This same result and Theorem 14.18 show that there are $\eta_j \in G$ with $\eta_j(\alpha) = \alpha_j$. For any $\varphi \in G$, $m(\varphi(\alpha)) = \varphi(m(\alpha)) = 0$ so $\varphi(\alpha) = \eta_j(\alpha)$ for some $j \leq k$. Thus $\eta_j^{-1}\varphi \in F[\alpha]'$ and since $\eta_j^{-1}\eta_i \in F[\alpha]' \Leftrightarrow i = j$ (why?) it follows that $[G : F[\alpha]'] = k$ with $\{\eta_1, \ldots, \eta_k\}$ as coset representatives. If $F[\alpha] = L$ then $k = n$ by Theorem 14.7, and $F[\alpha]' = \langle I_L \rangle$ shows $n = [G : \langle I_L \rangle] = |G|$ as well. When $F[\alpha] \neq L$ then since G is finite by Theorem 15.2, $|G| = [G : F[\alpha]'] |F[\alpha]'| = k|F[\alpha]'| = [F[\alpha] : F] |F[\alpha]'|$. Now L Galois over F forces L Galois over $F[\alpha]$ (Theorem 15.3), so induction on n yields $|F[\alpha]'| = |Gal(L/F[\alpha])| = [L : F[\alpha]]$. Therefore $|G| = [F[\alpha] : F] |F[\alpha]'| = [F[\alpha] : F][L : F[\alpha]] = [L : F] = n$ as required. Now assume that $[L : F] = |G|$. Since $(Gal(L/G'))' = (G'')' = G'$ from Theorem 15.1, L is Galois over G'. By what we just proved above, $[L : G'] = |Gal(L/G')| = |G''| = |G| = [L : F]$ ($G = G''$ always!), forcing $G' = F$. Hence L is Galois over F, completing the proof. ∎

It is worthwhile to point out that the proof of the theorem shows that L *is always Galois over* G'. When $F \subseteq L$ are fields and $\alpha \in L$, then $F(\alpha)$ is called a *simple extension* of F. We need to know that every Galois extension is simple and this follows from an interesting, more general result.

THEOREM 15.5 If $F \subseteq L$ are fields with $[L:F]=n$, then $L=F[\alpha]$ for some $\alpha \in L$ if and only if $\{M \subseteq L \mid M$ is a field containing $F\}$ is finite. Thus $L=F[\alpha]$ for some $\alpha \in L$ when L is Galois over F.

Proof When F is finite then $|L|=|F|^n$ and $(L^*,\ \cdot)=\langle\alpha\rangle$ for some $\alpha \in L$ by Theorem 4.20 so clearly $L=F[\alpha]$. Thus we may assume that F is infinite. If $L=F[\alpha]$, let E be a field with $F \subseteq E \subseteq L$, and let $m_E(x) \in E[x]$ be the minimal polynomial for α over E. Note that $m_E(x)$ is a factor of $m_F(x)$ in $L[x]$ and that $m_F(x)$ has only finitely many different monic factors in $L[x]$ by unique factorization. If $m_E(x)=x^k+b_{k-1}x^{k-1}+\cdots+b_0$ then $m_E(x)$ is irreducible over $F[b_{k-1},\ldots,b_0] \subseteq E$ so Theorem 14.7 yields $[L:E]=k=[L:F[b_{k-1},\ldots,b_0]]$, forcing $E=F[b_{k-1},\ldots,b_0]$ by Theorem 14.8 (why?). Thus every intermediate field E is determined by $m_E(x)$, and as we have observed, there are only finitely many such monic factors of $m_F(x)$, so there are only finitely many intermediate fields. When this set of fields is finite there is some $F[\beta] \subseteq L$ with $[F[\beta]:F]$ maximal. When $F[\beta] \neq L$, let $\gamma \in L - F[\beta]$ and for $c \in F$, set $F_c = F[\beta + c\gamma]$. Since F is infinite our assumption forces $F_c = F_d$ for some $c \neq d$ in F. Now $\beta + c\gamma, \beta + d\gamma \in F_c$ imply that $(c-d)\gamma \in F_c$ and it follows that $\gamma \in F_c$ and then $\beta \in F_c$. That $F[\beta] \subseteq F_c = F[\beta + c\gamma]$ properly, by the choice of γ, contradicts the maximality of $[F[\beta]:F]$. Therefore no such γ can exist, which means that $L=F[\beta]$ as required.

Finally, when L is Galois over F then $G=Gal(L/F)$ is finite by either Theorem 15.4 or Theorem 15.2, so Theorem 15.3 shows that there are only finitely many subfields of L containing F. Thus $L=F[\alpha]$ as above. ∎

It is not always easy to find $\alpha \in L$ with $L=F[\alpha]$ when L is Galois over F. In the simple case of $L=\boldsymbol{Q}[2^{1/2},3^{1/2}]$ Galois over $\boldsymbol{Q}$ (why?) it is an exercise that $|Gal(L/\boldsymbol{Q})|=4$, so $[L:\boldsymbol{Q}]=4$. If $\alpha=2^{1/2}+3^{1/2}$ then $\alpha^2=2\cdot 6^{1/2}+5$ so $(\alpha^2-5)^2=24$ and α satisfies $m(x)=x^4-10x^2+1$. An easy computation shows that $m(x)$ has no monic irreducible factors in $\boldsymbol{Q}[x]$ and so is irreducible. Therefore Theorem 14.7 shows that $L=\boldsymbol{Q}[\alpha]$. Alternatively, we could note that $\alpha \notin \boldsymbol{Q}[2^{1/2}] \cup \boldsymbol{Q}[3^{1/2}]$, then argue that $\alpha \notin \boldsymbol{Q}[6^{1/2}]$, and observe that $\boldsymbol{Q}[2^{1/2}]$, $\boldsymbol{Q}[3^{1/2}]$, and $\boldsymbol{Q}[6^{1/2}]$ are the only subfields $E \subseteq L$ with $[L:E]=2$ by Theorem 15.3. Again we get $L=\boldsymbol{Q}[\beta]$.

From Theorem 15.5 or Theorem 4.20, every finite field has the form $\boldsymbol{Z}_p[\alpha]$. Hence, in view of Theorem 14.15 and Theorem 14.7, we see that for any $n \in \boldsymbol{Z}$ and any prime p there is an irreducible $g(x) \in \boldsymbol{Z}_p[x]$ of degree n. An important consequence of Theorem 15.5 is that the priming operations are bijections when L is Galois over F.

THEOREM 15.6 If L is a Galois extension of F, then the priming operations are bijections between $\{H \leq Gal(L/F)\}$ and $\{E \subseteq L \mid E$ is a field and $F \subseteq E\}$.

Proof From Theorem 15.1 it suffices to see that every intermediate field and every subgroup of $G=Gal(L/F)$ are primed objects. Now Theorem 15.3 shows that for any intermediate field E, $E''=E$, so $E=(E')'$. Therefore we need only prove that $H''=H$ for all

$H \leq G$. Using Theorem 15.5, $L = F[\alpha]$, so $L = H'[\alpha]$. If $S = \{\eta_1(\alpha), \ldots, \eta_t(\alpha)\} = \{\eta(\alpha) \in L \mid \eta \in H\}$ has t elements, let $p(x) = (x - \eta_1(\alpha)) \cdots (x - \eta_t(\alpha))$. Observe that H a group implies $\eta(S) = S$ for any $\eta \in H$, so $\eta(p(x)) = p(x)$ forces $p(x) \in H'[x]$. Since $p(\alpha) = 0$, the minimal polynomial for α over H' is a factor of $p(x)$, so by Theorem 14.7, $[L : H'] \leq t \leq |H|$. From L Galois over F we have L Galois over H' by Theorem 15.3, and Theorem 15.4 yields $|H''| = |Gal(L/H')| = [L : H'] \leq |H|$. Since $H \leq H''$ from Theorem 15.1, $H = H''$ as required. ■

Example 15.6 The splitting field of $x^5 - 1$ over $\boldsymbol{Q}$.

If $\rho = e^{2\pi i/5} = \cos(2\pi/5) + i \cdot \sin(2\pi/5) \in \boldsymbol{C}$ then $\rho, \rho^2, \rho^3, \rho^4$, and $\rho^5 = 1$ are five roots of $x^5 - 1$ in $\boldsymbol{C}$ so $L = \boldsymbol{Q}[\rho]$ is a splitting field for $x^5 - 1$ over $\boldsymbol{Q}$. By Example 12.5, $x^4 + x^3 + x^2 + x + 1$ is the minimal polynomial for ρ over $\boldsymbol{Q}$ so $[L : \boldsymbol{Q}] = 4 = |Gal(L/\boldsymbol{Q})|$ by Theorems 14.7, 15.2, and 15.4. From Theorem 14.16 there is $\varphi \in G = Gal(L/\boldsymbol{Q})$ so that $\varphi(\rho) = \rho^2$ (why?). It follows first that $G = \langle\varphi\rangle$, then from Theorem 15.6 that L contains exactly three subfields. Clearly $G' = \boldsymbol{Q}$ and $\langle I_G\rangle' = L$ are two of these; $\boldsymbol{Q}' = G$ and $L' = \langle I_G\rangle$ also. What is the other field? Note that $\varphi^2(\rho) = \rho^4 = \rho^{-1}$ is the complex conjugate of ρ and that $\rho + \rho^4 = 2\cos(2\pi/5) \in (L - \boldsymbol{Q}) \bigcap \boldsymbol{R}$ since it is a root of $x^2 + x - 1 \in \boldsymbol{Q}[x]$. Hence $E = \boldsymbol{Q}[\rho + \rho^4]$ must be the third subfield and $E = \langle\varphi^2\rangle' = L \bigcap \boldsymbol{R}$, which can be verified independently.

In Example 15.6, $G = Gal(L/\boldsymbol{Q})$ cyclic and Theorem 15.6 tell us the number of subfields of L. When we can find enough subfields, Theorem 15.6 helps to determine the structure of G. Using $L = \boldsymbol{Q}[2^{1/3}, \omega]$ in Example 15.3, we know by Theorem 15.4 that $[L : \boldsymbol{Q}] = 6$ forces $|Gal(L/\boldsymbol{Q})| = 6$ and that $\boldsymbol{Q}[2^{1/3}]$, $\boldsymbol{Q}[2^{1/3}\omega]$, $\boldsymbol{Q}[2^{1/3}\omega^2]$, and $\boldsymbol{Q}[\omega]$ are intermediate fields. Consequently, Theorem 15.6 shows that $Gal(L/\boldsymbol{Q})$ cannot be cyclic, since then it would have only two proper $\langle e\rangle \neq H \leq G$, so $Gal(L/\boldsymbol{Q}) \cong S_3$, as in Example 15.3.

The main result about Galois extensions comes next and is often called the *Fundamental Theorem of Galois Theory.*

THEOREM 15.7 Let the field L be Galois over the field F, $G = Gal(L/F)$, $Sub(G) = \{H \leq G\}$, and $Fld(L/F) = \{E \subseteq L | E \text{ is a field with } F \subseteq E\}$.
For $H, J \in Sub(G)$ and $E, M \in Fld(L/F)$, the following hold:

i) $[L : F] = |G|$
ii) $H \to H'$ and $E \to E'$ are inclusion-reversing inverses and hence are bijections between $Sub(G)$ and $Fld(L/F)$
iii) if $E \subseteq M$ then $[M : E] = [E' : M']$, so $[L : E] = |E'|$ and $[E : F] = [G : E']$
iv) if $H \leq J$ then $[J : H] = [H' : J']$, so $|H| = [L : H']$ and $[G : H] = [H' : F]$
v) for all $\varphi \in G$, $\varphi(E)' = \varphi E' \varphi^{-1}$
vi) $\varphi(E) = E$ for all $\varphi \in G \Leftrightarrow E$ is Galois over F
vii) E is Galois over $F \Leftrightarrow E' \triangleleft G$, in which case $Gal(E/F) \cong G/E'$

Proof i) is Theorem 15.4 and ii) is Theorem 15.6. For iii), it follows from Theorem 15.2 that L is Galois over E and over M so $[L : E] = |E'|$ and $[L : M] = |M'|$ by i).

Hence $|E'| = [L:E] = [L:M][M:E] = |M'|[M:E]$, and by Lagrange's theorem, $|E'| = [E':M']|M'|$, so we get $[M:E] = [E':M']$. Using ii) and Theorem 15.1, $H' = H''' = (H'')'$ forces $H = H''$, so by iii), $[H':J'] = [J'':H''] = [J:H]$, proving iv). To see that v) holds, note that $\eta \in \varphi(E)' \Leftrightarrow \eta(\varphi(a)) = \varphi(a)$ for all $a \in E \Leftrightarrow \varphi^{-1}\eta\varphi(a) = a$ for all $a \in E \Leftrightarrow \varphi^{-1}\eta\varphi \in E' \Leftrightarrow \eta \in \varphi E'\varphi^{-1}$. If $\varphi(E) = E$ for $E \neq F$ and all $\varphi \in G$ then the restriction $\varphi|_E$ of φ to E is in $Gal(E/F)$. By assumption, for $\alpha \in E - F$ there is $\varphi \in G$ with $\varphi(\alpha) \neq \alpha$, so $\varphi|_E(\alpha) \neq \alpha$ and $Gal(E/F)' = F$, showing that E is Galois over F. When E is Galois over F then by Theorem 15.2 the minimal polynomial $m_\beta(x)$ over F for any $\beta \in E$ splits over E. Thus if $\varphi \in G$ then $0 = \varphi(m_\beta(\beta)) = m_\beta(\varphi(\beta))$ forces $\varphi(\beta) \in E$ and proves that $\varphi(E) \subseteq E$, then that $\varphi(E) = E$ (why?). Assuming E is Galois over F, vi) gives $\varphi(E) = E$ for all $\varphi \in G$, and then v) yields $E' \lhd G$. Also $\varphi(E) = E$ implies that the restriction $\eta(\varphi) = \varphi|_E$ defines a group homomorphism $\eta: G \to Gal(E/F)$. Clearly $ker\ \eta = E'$ so $G/E' \cong \eta(G)$. Now L Galois over F yields $L = F[\alpha]$ by Theorem 15.5. Using Theorem 15.2, L is Galois over E, so L is a splitting field for the minimal polynomial of α over E. Thus Theorem 14.17 shows that any $\upsilon \in Gal(E/F)$ extends to $\varphi_\upsilon \in G$. Therefore $\eta(\varphi_\upsilon) = \upsilon$ and $\eta(G) = Gal(E/F)$, so $G/E' \cong Gal(E/F)$. Finally, when $E' \lhd G$, v) shows that $\varphi(E)' = E'$ for all $\varphi \in G$ and then ii) results in $E = E'' = \varphi(E)'' = \varphi(E)$. Hence $\varphi(E) = E$ and by vi), E is Galois over F, completing the proof of the theorem. ∎

We look at a few examples that illustrate parts of Theorem 15.7 and show how to get information about $Gal(L/F)$ and about subfields of L.

Example 15.7 $Gal(L/\boldsymbol{Q})$ for L a splitting field of $g(x) = x^3 + x + 1 \in \boldsymbol{Q}[x]$.

A quick check shows that $g(x)$ has no rational root and so is irreducible. If $\alpha \in L \subseteq \boldsymbol{C}$ is a root of $g(x)$ then $\boldsymbol{Q}[\alpha] \subseteq L$ and $[\boldsymbol{Q}[\alpha]:\boldsymbol{Q}] = 3$ from Theorem 14.7; hence Theorem 14.8 yields $3 \mid [L:\boldsymbol{Q}]$. Since $g'(x) > 0$ it follows from calculus that $g(x)$ is an increasing function on $\boldsymbol{R}$. Thus $g(x)$ has two (conjugate!) roots in $\boldsymbol{C} - \boldsymbol{R}$, and complex conjugation acts as a nonidentity element $\eta \in G = Gal(L/\boldsymbol{Q})$ of order 2, so $2 \mid |G|$, forcing $6 \mid |G|$. Since each $\varphi \in G$ is determined by a permutation of the three roots of $g(x)$ in L, $G \cong H \leq S_3$, so $G \cong S_3$. Now Theorem 15.7 shows that $[L:\boldsymbol{Q}] = 6$ and that L contains four proper subfields $E_j \neq \boldsymbol{Q}$: three generated by the roots of $g(x)$ with $[E_j:\boldsymbol{Q}] = 3$, and one, say E_4, is Galois over $\boldsymbol{Q}$ with $[E_4:\boldsymbol{Q}] = 2$.

Example 15.7 generalizes to the splitting field of an irreducible $p(x) \in \boldsymbol{Q}[x]$ of prime degree.

Example 15.8 If $q(x) \in \boldsymbol{Q}[x]$ is irreducible of prime degree $p > 2$ and has exactly two roots in $\boldsymbol{C} - \boldsymbol{R}$ then if $L \subseteq \boldsymbol{C}$ is a splitting field of $q(x)$ over $\boldsymbol{Q}$, $G = Gal(L/\boldsymbol{Q}) \cong S_p$, the full symmetric group on p symbols.

Let $\alpha \in L \subseteq \boldsymbol{C}$ be a root of $q(x)$. Using $\boldsymbol{Q}[\alpha] \subseteq L$, Theorem 14.7 and Theorem 14.8 show that $p \mid |L:\boldsymbol{Q}]$, so $p \mid |G|$ by Theorem 15.7. The assumption on $q(x)$ implies that complex conjugation acts as $\eta \in G - \langle I_L \rangle$. Each $\varphi \in G$ is uniquely determined by how it permutes the p roots of $q(x)$ in L so we may consider $G \cong \upsilon(G) \leq S_p$. Since $p \mid |G|$, G has an element of order p by Theorem 9.6 (Cauchy), forcing $\upsilon(G)$

to contain a p-cycle. Clearly η interchanges the two complex roots of $p(x)$ and fixes the others. Thus $\upsilon(G)$ contains a 2-cycle and a p-cycle, and hence $\upsilon(G) = S_p$ (see Exercise 12 in §3.3).

Are there polynomials satisfying the conditions in Example 15.8?

Example 15.9 For a prime $p \geq 5$, $g_p(x) = x^2(x-2)(x-4)\cdots(x-2(p-2)) - 2$ is irreducible over $\boldsymbol{Q}$ and has exactly two roots in $\boldsymbol{C} - \boldsymbol{R}$.

Let $h(x) = x^2(x-2)(x-4)\cdots(x-2(p-2))$ and observe that $deg\, h = p$ and that $h(x)$ has $p-1$ roots in $\boldsymbol{R}$. From calculus we see that $h(x)$ has $(p-1)/2$ local maxima $M_1 = 0, M_2 \in (2,4), \ldots, M_{(p-1)/2} \in (2(p-4), 2(p-3))$ and $(p-1)/2$ local minima $m_1 \in (0,2), m_2 \in (4,6), \ldots, m_{(p-1)/2} \in (2(p-3), 2(p-2))$. Since these are $p-1$ distinct roots of $h'(x)$ it follows that $h(x)$ must be either increasing or decreasing between consecutive local extrema, increasing on $(-\infty, 0)$ and on $(2(p-2), \infty)$. Observe that for $k \in \boldsymbol{N}$ and odd, $|h(k)| > 2$. Thus $h(x) - 2 = g_p(x)$ has two real roots on each interval $(2(2i-1), 2(2i))$ and one real root greater than $2(p-2)$. Hence $g_p(x)$ has $p-2$ real roots and so has exactly two roots in $\boldsymbol{C} - \boldsymbol{R}$. Finally, $g_p(x)$ is irreducible by Theorem 12.16 since it is monic with constant -2 and all its other coefficients are even (verify!).

Example 15.10 $Gal(L/\boldsymbol{Q})$ for L a splitting field of $g(x) = x^6 - 2 \in \boldsymbol{Q}[x]$.

By Eisenstein's Criterion, $g(x)$ is irreducible. By Example 15.4, $L = \boldsymbol{Q}[2^{1/6}, \omega]$ for $\omega \neq 1$ in $\boldsymbol{C}$ with $\omega^3 = 1$ is a splitting field for $g(x)$ over $\boldsymbol{Q}$. Since $\omega \notin \boldsymbol{Q}[2^{1/6}] \subseteq \boldsymbol{R}$, Theorem 14.7 and Theorem 14.8 give $[L : \boldsymbol{Q}] = 12$, so by Theorem 15.7, $12 = |G|$ for $G = Gal(L/\boldsymbol{Q})$. In Example 15.4 we showed that $Gal(L/\boldsymbol{Q}[\omega]) = \langle\sigma\rangle$ for $\sigma(2^{1/6}) = 2^{1/6}(-\omega)$ and $\sigma(\omega) = \omega$. Thus $[G : \langle\sigma\rangle] = 2$ and $\langle\sigma\rangle \triangleleft G$ by Theorem 5.3. Complex conjugation acts as $\eta \in G$, $\eta(\omega) = \omega^2$ so $\eta \notin \langle\sigma\rangle$, and $G = \langle\eta\rangle\langle\sigma\rangle$ results. A direct computation using $\eta(2^{1/6}) = 2^{1/6}$ shows that $\eta\sigma\eta = \sigma^{-1}$ and Example 6.4 proves that $G \cong D_6$.

Example 15.11 $Gal(K/\boldsymbol{Q})$ for K a splitting field of $h(x) = x^5 - 2 \in \boldsymbol{Q}[x]$.

By Example 15.6 if $\rho = e^{2\pi i/5} \in \boldsymbol{C}$ then $x^5 - 1 \in \boldsymbol{Q}[x]$ splits over $\boldsymbol{Q}[\rho]$, $[\boldsymbol{Q}[\rho] : \boldsymbol{Q}] = 4$, and $Gal(\boldsymbol{Q}[\rho]/\boldsymbol{Q}) = \langle\varphi\rangle$ of order 4. Also, $\boldsymbol{Q}[2^{1/5}, \rho] = K$ is a splitting field for $h(x)$ over $\boldsymbol{Q}$ (verify!). Since $h(x)$ is irreducible over $\boldsymbol{Q}$ by Eisenstein's Criterion, Theorem 14.7 shows that $[\boldsymbol{Q}[2^{1/5}] : \boldsymbol{Q}] = 5$. Therefore $[K : \boldsymbol{Q}] = 20$ by Theorem 14.10, observing that $K = \boldsymbol{Q}[2^{1/5}] \vee \boldsymbol{Q}[\rho]$. By Theorem 15.2, K and $\boldsymbol{Q}[\rho]$ are Galois over $\boldsymbol{Q}$, so Theorem 15.7 shows that if $G = Gal(K/\boldsymbol{Q})$ then $|G| = 20$, and also $\boldsymbol{Q}[\rho]' \triangleleft G$ with $|\boldsymbol{Q}[\rho]'| = 5$. Indeed $\boldsymbol{Q}[\rho]' = \langle\eta\rangle$ for η the extension to K of $I_{\boldsymbol{Q}[\rho]}$ with $\eta(2^{1/5}) = 2^{1/5}\rho$, using Theorem 14.18: note that $x^5 - 2 \in \boldsymbol{Q}[\rho][x]$ is irreducible by Theorem 14.7 since $[K : \boldsymbol{Q}[\rho]] = 5$ from Theorem 14.8. Similarly, φ extends to K with $\varphi(2^{1/5}) = 2^{1/5}$ so $\langle\varphi\rangle = \boldsymbol{Q}[2^{1/5}]'$ since this group has order 4 by

Theorem 15.7. It follows that $G = \langle\eta\rangle\langle\varphi\rangle$ (why?). Computing the actions of η and φ on $2^{1/5}$ shows that $\varphi\eta\varphi^{-1} = \eta^2$ and this defines the multiplication in G. Note that Theorem 15.7 itself implies that G is not Abelian since G has five 2-Sylow subgroups, namely the $\boldsymbol{Q}[2^{1/5}\rho^i]'$. Also, since $G/\boldsymbol{Q}[\rho]' \cong Gal(\boldsymbol{Q}[\rho]/\boldsymbol{Q})$ is cyclic of order 4, G has a unique subgroup of order 10, $H = \langle\eta\rangle\langle\varphi^2\rangle \cong D_5$, and $\boldsymbol{Q}[\rho] \cap \boldsymbol{R} = H' \subseteq \boldsymbol{Q}[\rho]$ is the only subfield of K with $[H' : \boldsymbol{Q}] = 2$.

Example 15.12 $Gal(F/\boldsymbol{Q})$ for F splitting $g(x) = (x^3 - 2)(x^3 - 3) \in \boldsymbol{Q}[x]$.

From Example 15.3, $F = \boldsymbol{Q}[2^{1/3}, 3^{1/3}, \omega] \subseteq \boldsymbol{C}$ splits $g(x)$ over $\boldsymbol{Q}$, with $\omega \neq 1$ and $\omega^3 = 1$, and $[\boldsymbol{Q}[2^{1/3}, \omega] : \boldsymbol{Q}] = 6$. Using Theorem 14.8 we conclude that $[F : \boldsymbol{Q}] = [F : \boldsymbol{Q}[2^{1/3}, \omega]][\boldsymbol{Q}[2^{1/3}, \omega] : \boldsymbol{Q}] = 18$, or some $3^{1/3}\omega^i \in \boldsymbol{Q}[2^{1/3}, \omega] = L$ (why?), and this forces $3^{1/3} \in L$. By considering dimensions over $\boldsymbol{Q}$, $\boldsymbol{Q}[2^{1/3}] = L \cap \boldsymbol{R} = \boldsymbol{Q}[3^{1/3}]$. As in Example 15.3, there is $\beta \in Gal(L/\boldsymbol{Q})$ so that $\beta(2^{1/3}) = 2^{1/3}\omega$ and $\beta(\omega) = \omega$. If $3^{1/3} = a + b2^{1/3} + c2^{2/3}$ with $a, b, c \in \boldsymbol{Q}$ then $\beta(3^{1/3}) = a + b2^{1/3}\omega + c2^{2/3}\omega^2$ and also $\beta(3^{1/3}) = 3^{1/3}\omega^i$ a root of $x^3 - 3$. If $\beta(3^{1/3}) = 3^{1/3} = a + b2^{1/3} + c2^{2/3}$ then $b2^{1/3}(\omega - 1) + c2^{2/3}(\omega^2 - 1) = 0$. This in turn implies $\omega + 1 \in \boldsymbol{R}$ or else $b = c = 0$. Neither possibility can occur, and a similar contradiction arises when $\beta(3^{1/3}) = 3^{1/3}\omega$ or $\beta(3^{1/3}) = 3^{1/3}\omega^2$ (verify!). Hence $3^{1/3} \notin \boldsymbol{Q}[2^{1/3}]$, forcing $[F : \boldsymbol{Q}] = 18$. If $G = Gal(F/\boldsymbol{Q})$ then Theorem 15.7 shows that $|G| = 18$. What is the structure of G? Using $G \cong H \leq S_6$ (why?), we see that G cannot have an element of order 9 so a 3-Sylow subgroup P_3 of G must be isomorphic to $\boldsymbol{Z}_3 \oplus \boldsymbol{Z}_3$. Now $P_3 \lhd G$ since $[G : P_3] = 2$ (Theorem 5.3) and we can take a 2-Sylow subgroup of G to be $\langle\eta\rangle$ for η complex conjugation. Thus $G = \langle\eta\rangle P_3$. For $0 \leq i, j \leq 2$, set $\boldsymbol{Q}[2^{1/3}\omega^i, 3^{1/3}\omega^j] = F_{ij}$ and observe that $[F_{ij} : \boldsymbol{Q}] = [F_{ij} : \boldsymbol{Q}[2^{1/3}\omega^i]][\boldsymbol{Q}[2^{1/3}\omega^i] : \boldsymbol{Q}] \leq 9$, $F = F_{ij}[\omega]$, and $[F : F_{ij}] \leq 2$. Since $18 = [F : \boldsymbol{Q}] = [F : F_{ij}][F_{ij} : \boldsymbol{Q}]$ we have $[F_{ij} : \boldsymbol{Q}] = 9$. If $F_{ij} = F_{st}$ then $i \neq s$ implies that $\omega \in F_{ij}$, giving the contradiction $F = F_{ij}$. Similarly, $j \neq t$ cannot hold so $\{F_{ij}\}$ contains nine different subfields. Using Theorem 15.7 again we conclude that $\{F_{ij}'\}$ are nine different 2-Sylow subgroups of G, all of which are conjugates of $\langle\eta\rangle$. By Theorem 9.11 the normalizer of $\langle\eta\rangle$ is itself and it follows that no element of $P_3 - \langle e\rangle$ can commute with η so if $\varphi \in P_3 - \langle e\rangle$ then $\eta\varphi\eta \neq \varphi$. Let $\eta\varphi\eta = \upsilon$, so $\eta\upsilon\eta = \varphi$ (why?), and $\eta\varphi\upsilon\eta = \eta\varphi\eta\eta\upsilon\rho = \upsilon\varphi = \varphi\upsilon$. Hence $\upsilon = \varphi^{-1}$ and using Definition 9.5 and the comments after it, $G \cong (\boldsymbol{Z}_3 \times \boldsymbol{Z}_3) \diamondsuit \boldsymbol{Z}_2$.

EXERCISES

In the exercises below, F, E, M, and L denote fields and $G = Gal(L/F)$.

1. If $F \subseteq E \subseteq L$ and L is Galois over F, show that L is Galois over E.
2. If $p(x) \in F[x]$ is irreducible of degree n and if L is a splitting field of $p(x)$ over F, show carefully that $G \cong H \leq S_n$. If $p(x)$ is also separable, show that $n \mid |G|$.
3. If $F \subseteq L$ and $[L : F] = n$, show that $G = Gal(L/F'')$ and that L is Galois over F''.
4. Let $char\ F = p > 0$ and $E = F(x^p, y^p) \subseteq L = F(x, y)$. Show that $[L : E] = p^2$ but that $L \neq E[\alpha]$ for any $\alpha \in L$. (Use Theorem 15.5.)
5. If $[L : F] = n$ and L is separable over F, show that $L \subseteq M$ with M Galois over F.

6. Let $F \subseteq E \subseteq L$.
  i) If L is Galois over F, show that $E = F[\alpha]$ for some $\alpha \in E$.
  ii) Show that $[L : F] = n$ and L separable over $F \Rightarrow L = F[\alpha]$. (See Exercise 5.)
  iii) If $\boldsymbol{Q} \subseteq L$ and $[L : \boldsymbol{Q}] = n$, show that $L = \boldsymbol{Q}[\beta]$ for some $\beta \in L$.

7. If $\boldsymbol{Q} \subseteq L$ is algebraic and if, for a fixed $n \in \boldsymbol{N}$, $[\boldsymbol{Q}[\alpha] : \boldsymbol{Q}] \leq n$ for all $\alpha \in L$, then show that $[L : \boldsymbol{Q}] \leq n$. (See Exercise 6.)

8. If $F \subseteq E \subseteq L$, L is Galois over F, and G is Abelian, show that E is Galois over F and that $Gal(E/F)$ is Abelian.

9. Let $F \subseteq E \subseteq L$, $[L : F] = n$, E Galois over F, and L a splitting field over F.
  i) If $\varphi \in Gal(E/F)$, then φ extends to L ($\varphi = \eta|_E$ for $\eta \in G$).
  ii) $Gal(E/F) \cong G/E'$

10. If L is Galois over F, G is Abelian and $\beta \in L$ with minimal polynomial $m(x) \in F[x]$, then $m(x)$ splits over $F[\beta]$.

11. If L is Galois over F there is $F \subseteq E \subseteq L$ with $E \neq F$, E Galois over F, and $Gal(E/F)$ a simple group ($N \lhd Gal(E/F) \Rightarrow N = \langle e \rangle$ or $N = Gal(E/F)$).

12. If $\boldsymbol{Q} \subseteq L \subseteq \boldsymbol{C}$, L is Galois over $\boldsymbol{Q}$ with $[L : \boldsymbol{Q}] = n$, and $L \not\subseteq \boldsymbol{R}$, then $n = 2m$ and $[L \cap \boldsymbol{R} : \boldsymbol{Q}] = m$.

13. Let $L \subseteq \boldsymbol{C}$ be a splitting field of $x^4 + 2 \in \boldsymbol{Q}[x]$ over $\boldsymbol{Q}$.
  i) Show that $x^4 + 2$ is irreducible in $\boldsymbol{Q}[x]$ and find $[L : \boldsymbol{Q}]$.
  ii) Describe the structure of $Gal(L/\boldsymbol{Q})$.

14. Redo Example 15.12 (find $Gal(F/\boldsymbol{Q})$ for $F \subseteq \boldsymbol{C}$ a splitting field over $\boldsymbol{Q}$ for $(x^3 - 2)(x^3 - 3)$) by explicitly finding elements in $Gal(F/\boldsymbol{Q})$.

15. For each $\alpha \in \boldsymbol{C}$ below, find its minimal polynomial $m_\alpha(x) \in \boldsymbol{Q}[x]$, and for $L \subseteq \boldsymbol{C}$ a splitting field for $m_\alpha(x)$ over $\boldsymbol{Q}$, find $[L : \boldsymbol{Q}]$ and $Gal(L/\boldsymbol{Q})$:
  i) $\alpha = 3^{1/2} + 2i$
  ii) $\alpha = 1/2 + 3^{1/4}$
  iii) $\alpha = 2^{1/2}(1 + i)$
  iv) $\alpha = 2^{1/2} + 3^{1/2}i$

16. Let L be Galois over F. Determine how many subfields of L contain F and their dimensions over F, the number and dimensions of those $E \subseteq L$ Galois over F and $Gal(E/F)$ when $G \cong$:
  i) D_5
  ii) A_4
  iii) $\boldsymbol{Z}_{60}$
  iv) $D_5 \oplus \boldsymbol{Z}_3$
  v) D_{15}

17. If $L \subseteq \boldsymbol{C}$ is a splitting field over $\boldsymbol{Q}$ of $(x^3 - 2)(x^2 + 1)$:
  i) Find $|Gal(L/\boldsymbol{Q})|$.
  ii) Show that $Gal(L/\boldsymbol{Q})$ has a normal subgroup of order 2.
  iii) Show that the 3-Sylow subgroup of $Gal(L/\boldsymbol{Q})$ is normal.

18. Let $f(x) = x^6 - 6x^2 + 2 \in \boldsymbol{Q}[x]$ and $L \subseteq \boldsymbol{C}$ a splitting field for $f(x)$ over $\boldsymbol{Q}$. Show that $f(x)$ is irreducible over $\boldsymbol{Q}$ and has exactly two roots in $\boldsymbol{C} - \boldsymbol{R}$ (use calculus!) but $Gal(L/\boldsymbol{Q})$ is not isomorphic to either S_6 or A_6. (Some $\alpha \in L$ satisfies $x^3 - 6x + 2 \in \boldsymbol{Q}[x]$ irreducible. Consider $[Gal(L/\boldsymbol{Q}) : \boldsymbol{Q}[\alpha]']$.)

19. Find $Gal(L/\boldsymbol{Q})$ for L a splitting field over $\boldsymbol{Q}$ of:
  i) $x^5 - 10x + 2$
  ii) $x^3(x^2 - 2)(x^2 - 4) + 2$

20. For each $p(x) \in F[x]$ with splitting field L over F, find $|G|$ and describe the group structure of G. Determine the number of $F \subseteq E \subseteq L$, their dimensions over F, those E Galois over F, and $Gal(E/F)$:
i) $x^{10} - 1 \in \boldsymbol{Q}[x]$ (use Example 15.6)
ii) $x^{20} - 1 \in \boldsymbol{Q}[x]$ (Example 15.6!)
iii) $x^{15} - 1 \in \boldsymbol{Q}[x]$
iv) $x^{12} - 1 \in \boldsymbol{Q}[x]$ (note that $\omega, i \in L$)
v) $x^4 - 2 \in \boldsymbol{Q}[i][x]$
vi) $x^4 - 2 \in \boldsymbol{Q}[2^{1/2}][x]$
vii) $(x^4 - 2)(x^4 - 3) \in \boldsymbol{Q}[i][x]$ (find various $\varphi \in G$)

21. Let $S = \{\rho \in \boldsymbol{C} \mid \rho^k = 1, \text{ some } k > 0\}$ and $F = \boldsymbol{Q}[S]$. If $p_1 < \cdots < p_k$ are primes, let $L = F[p_1^{1/p_1}, \ldots, p_k^{1/p_k}] \subseteq \boldsymbol{C}$. Show that:
i) L is Galois over F
ii) G is cyclic
iii) if $F \subseteq E \subseteq L$ then $E = F[T]$ for $T \subseteq \{p_1^{1/p_1}, \ldots, p_k^{1/p_k}\}$

22. Let L be Galois over F. If $E_1, \ldots, E_t$ are all of the subfields of L containing F and if $[E_i : F] = [E_j : F] \Rightarrow i = j$, show that each E_j is Galois over F and that t is the number of divisors in $\boldsymbol{N}$ of $[L : F]$. (What must G be?)

23. If L is Galois over F and $[L : F] = n > 1$ then there is a unique subfield E, $F \subseteq E \subseteq L$, so that E is maximal satisfying E is Galois over F and $Gal(E/F)$ is Abelian.

24. Let L be Galois over F, $f(x) \in F[x]$ irreducible, and $f(x) = g_1(x) \cdots g_k(x) \in L[x]$ with each $g_j(x)$ irreducible. Show $deg\ g_1 = deg\ g_2 = \cdots = deg\ g_k$. (Consider $L \subseteq L[\alpha]$ with $g_1(\alpha) = 0$ and show that $f(x)$ divides $\prod_G \varphi(g_1(x))$.)

25. Let L be Galois over F so that for each $m \in \boldsymbol{N}$ with $m \mid [L : F]$ there is a unique $F \subseteq E_m \subseteq L$ with $[E_m : F] = m$. Show that E_m is Galois over F and that $Gal(E_m/F)$ is cyclic.

26. Let L be Galois over F with $[L : F] = n = \prod p_i^{a(i)}$ (prime factorization).
i) Show that for each i there is $F \subseteq E_i \subseteq L$ with $[L : E_i] = p_i^{a(i)}$.
ii) If for each i there is $M_i \subseteq L$ Galois over F and $[M_i : F] = p_i^{a(i)}$, show that if $m|n$ with $m > 0$ there is $F_m \subseteq L$ Galois over F, and $[F_m : F] = m$.

27. Let $\boldsymbol{Q} \subseteq E \subseteq L$, L Galois over $\boldsymbol{Q}$, and $[L : \boldsymbol{Q}] = p^2 q$ for $p < q$ primes. If $\boldsymbol{Q}$, E, and L are the only intermediate fields Galois over $\boldsymbol{Q}$, and if $[E : \boldsymbol{Q}] = q$, determine $Gal(L/\boldsymbol{Q})$ up to isomorphism. (See Example 9.22.)

28. Let $F \subseteq L$ with $char\ F \neq 2$. Show that $L = F[\alpha_1^{1/2}, \ldots, \alpha_k^{1/2}]$ for $\alpha_j \in F \Leftrightarrow L$ is Galois over F and G is Abelian with $\varphi^2 = I_L$ for all $\varphi \in G$.

29. If L is Galois over F then $|G| = p^k$ for p a prime $\Leftrightarrow$ for any $\alpha, \beta \in L$ the degree of the minimal polynomial over F of one of these divides the degree of the minimal polynomial of the other.

30. Let $f(x) \in \boldsymbol{Q}[x]$ be irreducible of degree $p > 3$ a prime, L a splitting field for $f(x)$ over $\boldsymbol{Q}$, $G = Gal(L/\boldsymbol{Q}) \cong S_p$ (the symmetric group), and let $E \subseteq L$ with $2 < [E : \boldsymbol{Q}] < [L : \boldsymbol{Q}]$.
i) Show that there is $M \subseteq L$, $M \neq E$, and $[M : \boldsymbol{Q}] = [E : \boldsymbol{Q}]$.
ii) If $[E : \boldsymbol{Q}] = (p - 1)!$ and M is as in i), show that $\varphi(E) = M$ for some $\varphi \in G$.

31. Assume that L is Galois over F with $G \cong S_n$, the symmetric group.
i) If $n \geq 2$, show that there is $F \subseteq E \subseteq L$ with $[L : E] = n$.

ii) If H is any group with $|H| = n$, show that there is $F \subseteq E \subseteq L$ with $Gal(L/E) \cong H$.

iii) If $n > 2$, show that there is $T \subseteq L$ with T Galois over F and $T \neq F, L$. What is $Gal(L/T)$?

32. If $E, M \subseteq L$ with E, M, and L Galois over F, and $E \cap M = F$, show that:

i) $E = F[\alpha]$ is a splitting field over F of the irreducible $m(x) \in F[x]$ for α.

ii) $\varphi \in Gal(M/F)$ extends to $\eta \in G \cap E'$ (write $M = F[\beta] \subseteq F[\beta][\alpha] \subseteq L$).

33. Let $E, M \subseteq L$ with E, M Galois over F, $E \cap M = F$, and $E \vee M = L$. Show that:

i) L is Galois over F

ii) $Gal(L/E) \cong Gal(M/F)$ (Exercise 32)

iii) $[L : F] = [L : E][L : M]$

iv) $Gal(L/F) \cong Gal(L/M) \oplus Gal(L/E) \cong Gal(E/F) \oplus Gal(M/F)$

34. Let $F, L \subseteq \boldsymbol{C}$ with F and L Galois over $\boldsymbol{Q}$, $[F : \boldsymbol{Q}]$ and $[L : \boldsymbol{Q}]$ relatively prime, and set $F \vee L = K$. Show that K is Galois over $\boldsymbol{Q}$. If $\varphi \in Gal(F/\boldsymbol{Q})$ and $\eta \in Gal(L/\boldsymbol{Q})$, show that there is $\nu \in Gal(K/\boldsymbol{Q})$ with $\nu|_F = \varphi$ and $\nu|_L = \eta$.

15.3 APPLICATIONS

We apply the results above to finite fields, describing their subfields and Galois groups, and to the polynomial rings $F[x_1, \ldots, x_n]$, characterizing those polynomials invariant under all permutations of $\{x_j\}$.

Recall that for L a finite field, *char* $L = p > 0$, and so the subring, actually subfield (why?), $P = \langle 1_L \rangle$ of L generated by $1_L \in L$ is isomorphic to $\boldsymbol{Z}_p$. Thus $|L| = p^n \Leftrightarrow [L : P] = n$. Also Theorem 4.20 shows that $(L^*, \cdot) = \langle \alpha \rangle$ for some $\alpha \in L$ of order $p^n - 1$. Since every $\beta \in L^*$ is $\beta = \alpha^k$, the p^n elements of L are all roots of $x^{p^n} - x \in P[x]$ so L is a splitting field over P of this polynomial. From Theorem 14.15 and Theorem 14.17 L is isomorphic to a splitting field of $x^{p^n} - x \in \boldsymbol{Z}_p[x]$. The upshot is that to describe the subfields of L, or the Galois groups of subfields of L, we may assume that $\boldsymbol{Z}_p \subseteq L$. The main result on finite fields includes some things we already know.

THEOREM 15.8 For p a prime and $n \in \boldsymbol{N}$, there is a field L with p^n elements, and any two fields with p^n elements are isomorphic. If $F \subseteq L$ are fields, $|F| = p^k$, and $|L| = p^n$, then $k \mid n$, L is a Galois over F, and $Gal(L/F) = \langle \varphi \rangle$ of order n/k where φ takes $\beta \in L$ to its p^k-th power. There is a unique subfield $F_m \subseteq L$ for each divisor m of n, $|F_m| = p^m$, F_m is Galois over $P = \langle 1_L \rangle$, and $Gal(F_m/P) = \langle \eta_m \rangle$ where $\eta_m(\alpha) = \alpha^p$ in F_m.

Proof From Theorem 14.15 there is a field of p^n elements, and our comments above show that any two such fields are isomorphic to a splitting field of $x^{p^n} - x \in \boldsymbol{Z}_p[x]$ and hence are isomorphic to each other. Thus without loss of generality, L is a splitting field of $x^{p^n} - x \in \boldsymbol{Z}_p[x]$ over $\boldsymbol{Z}_p$. If $F \subseteq L$ with $[L : F] = s$ then $|L| = |F|^s$. Thus, when $|F| = p^k$ it follows that $ks = n$ and $k \mid n$. Since L is a splitting field over $\boldsymbol{Z}_p$ of a polynomial with distinct roots in L, the same is true for L over F, so L is Galois over F by Theorem 15.2. Using *char* $L = p$ it is routine to verify that $\eta(\alpha) = \alpha^p$ is a ring homomorphism of L to itself. Clearly η is injective, and so is a bijection since L is finite, and $\eta(\alpha) = \alpha \Leftrightarrow \alpha$ is a root of $x^p - x \in \boldsymbol{Z}_p[x] \Leftrightarrow \alpha \in \boldsymbol{Z}_p$. Thus $\eta \in Gal(L/\boldsymbol{Z}_p)$. Now $|Gal(L/\boldsymbol{Z}_p)| = [L : \boldsymbol{Z}_p] = n$ by Theorem 15.4, and if $\eta^s = I_L$ for $s < n$ then all $\beta \in L$ are roots of $x^{p^s} - x$, a contradiction since a nonzero polynomial over a field cannot have more roots than its degree. Hence $Gal(L/\mathbf{Z}_p) = \langle\eta\rangle$ and so has exactly one subgroup for each divisor m of n. From Theorem 15.6 we conclude that the subfields of L are precisely those $F_m = \langle\eta^m\rangle'$. Now L is Galois over F_m with $Gal(L/F_m) = \langle\eta^m\rangle$ by Theorem 15.7. Note that $o(\eta^m) = n/m$, and $\eta^m(\alpha) = \alpha^{p^m}$. Therefore $[L : F_m] = |\langle\eta^m\rangle| = n/m$, so $[F_m : \mathbf{Z}_p] = m$ and $|F_m| = p^m$, F_m is Galois over $\mathbf{Z}_p$ since $\langle\eta^m\rangle' \lhd \langle\eta\rangle$, and $Gal(F_m/\mathbf{Z}_p) \cong \langle\eta\rangle/\langle\eta^m\rangle$ (Theorem 15.7!). Since $\eta(\alpha) = \alpha^p$, η acts on F_m by restriction, say $\eta_m \in Gal(F_m/\mathbf{Z}_p)$. By what we have seen for L, $Gal(F_m/\mathbf{Z}_p) = \langle\eta_m\rangle$: $\eta_m(\beta) = \beta^p$ for $\beta \in F_m$ and $o(\eta_m) = m$ since $\langle\eta_m\rangle \cong \langle\eta\rangle/\langle\eta^m\rangle$. ■

Let $F \subseteq L$ be fields, $|F| = p^k$, $|L| = p^n$, and $G = Gal(L/F) = \langle\eta\rangle$ for η the p^k-power map. If $\beta \in L$ with minimal polynomial $g(x) \in F[x]$, then by the theorem, $F[\beta]$ is Galois over F, so Theorem 15.2 shows that $g(x)$ splits over $F[\beta]$. The roots of $g(x)$ in L are $\{\eta^m(\beta) = \beta^{p^{mk}} \mid 1 \le m \le deg\, g\}$, using Theorem 14.18. Also observe that $deg\ g \mid [L : F]$ (why?) and $[L : F] \mid n$ so $deg\ g \mid n$, and furthermore $\beta \in L$ satisfies $x^{p^n} - x$, forcing $g(x) \mid x^{p^n} - x$. Any $h(x) \in F[x]$ irreducible with $deg\ h \mid [L : F]$ satisfies the same divisibility property. The theorem shows $F = F_k \subseteq F_{kdeg\,h} \subseteq L$. If M is a splitting field for $h(x)$ over F and $F \subseteq F[\theta] \subseteq M$ with $h(\theta) = 0$, then again from the theorem, $F[\theta] \cong F_{kdeg\,h}$ via υ fixing F (both are splitting fields over F and have the same number of elements). It follows from above that $h(x) \mid (x^{p^n} - x)$.

Example 15.13 A field $L \supseteq \mathbf{Z}_p$ with $|L| = p^3$ has only two subfields and splits any irreducible $f(x) \in \mathbf{Z}_p$ of degree 3.

Now $[L : \mathbf{Z}_p] = 3$ and if $\mathbf{Z}_p \subseteq F \subseteq L$ for a field F then Theorem 15.8 shows $|F| = p$ and $F = \mathbf{Z}_p$ or $|F| = p^3$ and $F = L$. The splitting field for any irreducible cubic $f(x)$ over $\mathbf{Z}_p$ is a field $\mathbf{Z}_p[\beta]$ of p^3 elements (why?) and so is isomorphic to L by Theorem 15.8. Thus L splits $f(x)$.

Example 15.14 For any prime p, $x^4 + 1 \in \mathbf{Z}_p[x]$ is not irreducible.

If $p = 2$ then $x^4 + 1 = (x+1)^4$ in $\mathbf{Z}_2[x]$, so assume that $p > 2$. By Theorem 1.3, $8 \mid (p^2 - 1)$ and we know that there is $\mathbf{Z}_p \subseteq L$ with L a field of p^2 elements. Thus $(L^*, \cdot)$ is a cyclic group (Theorem 4.20) with order divisible by 8 so there is $\beta \in L$ with $\beta^8 = 1$ but $\beta^4 \neq 1$. Since $x^8 - 1 = (x^4 - 1)(x^4 + 1)$, β must be a root of $x^4 + 1$. If $x^4 + 1$ were irreducible over $\mathbf{Z}_p$ then using Theorem 14.7, $[\mathbf{Z}_p[\beta] : \mathbf{Z}_p] = 4$, but $\mathbf{Z}_p[\beta] \subseteq L$ and $[L : \mathbf{Z}_p] = 2$. Consequently $x^4 + 1 \in \mathbf{Z}_p[x]$ cannot be irreducible.

In Example 15.14, when $p > 2$ is a prime a little experimentation shows that $x^4 + 1 = (x^2 + x - 1)(x^2 - x - 1) \in \mathbf{Z}_3[x]$, $x^4 + 1 = (x^2 + 2)(x^2 + 3) \in \mathbf{Z}_5[x]$, and $x^4 + 1 = (x^2 - 4)(x^2 + 4) = (x - 2)(x - 8)(x - 9)(x - 15) \in \mathbf{Z}_{17}[x]$.

Example 15.15 The factorization of $x^{16} + x \in \mathbf{Z}_2[x]$.

A splitting field L for this polynomial over $\mathbf{Z}_2$ has 16 elements by Theorem 14.15, so $[L : \mathbf{Z}_2] = 4$ and $x^{16} + x$ is divisible by all irreducibles in $\mathbf{Z}_2[x]$ of degree 4 (see above). The linear polynomials in $\mathbf{Z}_2[x]$ are x and $x+1$, and the only irreducible quadratic is x^2+x+1. Since L must contain a subfield of dimension 2 over $\mathbf{Z}_2$, L splits x^2+x+1. Factoring gives first

$$x^{16}+x=x(x^{15}+1)=x(x^3+1)(x^{12}+x^9+x^6+x^3+1)$$

and $x^3+1=(x+1)(x^2+x+1)\in\mathbf{Z}_2[x]$. Any irreducible of degree 4 in $\mathbf{Z}_2[x]$ cannot be divisible by $x, x+1$, or x^2+x+1, and a quick check shows that the irreducibles of degree 4 are $p_1(x)=x^4+x+1$, $p_2(x)=x^4+x^3+1$, and $p_3(x)=x^4+x^3+x^2+x+1$. Thus the required factorization in the UFD $\mathbf{Z}_2[x]$ is $x^{16}+x=x(x+1)(x^2+x+1)p_1(x)p_2(x)p_3(x)$.

We turn now to what are often called *symmetric polynomials*. For any $n>1$, let S_n be the symmetric group on n symbols. If F is a field, $\theta\in S_n$, and $p,q\in F[x_1,\ldots,x_n]$, set

$$T_\theta(p(x_1,\ldots,x_n)/q(x_1,\ldots,x_n))=p\big(x_{\theta(1)},\ldots,x_{\theta(n)}\big)/q\big(x_{\theta(1)},\ldots,x_{\theta(n)}\big).$$

It is an exercise to verify that $T_\theta\in Gal(F(x_1,\ldots,x_n)/F)$, that $\theta\to T_\theta$ is a group homomorphism, and that $S_n\cong\{T_\theta\mid\theta\in S_n\}\leq Gal(F(x_1,\ldots,x_n)/F)$. We are interested in those $g(x_1,\ldots,x_n)\in F[x_1,\ldots,x_n]$ so that $T_\theta(g)=g$ for all $\theta\in S_n$. Examples of such polynomials are the **elementary symmetric functions** σ_j that are defined in §14.6: $\sigma_k(X)=\sum_{i_1<\cdots<i_k}x_{i_1}x_{i_2}\cdots x_{i_k}$. We observed in §14.6 that

$$Y^n-\sigma_1(X)Y^{n-1}+\cdots+(-1)^j\sigma_j(X)Y^{n-j}+\cdots+(-1)^n\sigma_n(X)\in F[\sigma_i(X),\ldots,\sigma_n(X)][Y]$$

factors as $(Y-x_1)(Y-x_2)\cdots(Y-x_n)\in F[x_1,\ldots,x_n][Y]$. Thus, setting $\sigma_j=\sigma_j(X)$, $F(x_1,\ldots,x_n)=M$ is a splitting field over $F(\sigma_1,\ldots,\sigma_n)=L$ for this polynomial with distinct roots so Theorem 15.2 shows that M is Galois over L. From Theorem 14.13, $[M:L]\mid n!$, and Theorem 15.4 shows that $[M:L]=|Gal(M/L)|$. Now $T_\theta(\sigma_j)=\sigma_j$ for each $\theta\in S_n$ and each σ_j so $T_\theta\in Gal(M/L)$. Hence, using $S_n\cong\{T_\theta\}\leq Gal(M/L)$ forces $[M:L]=|Gal(M/L)|=n!$ and $Gal(M/L)=\{T_\theta\mid\theta\in S_n\}\cong S_n$. These observations yield the following result.

THEOREM 15.9 If G is any group of order n and F is any field, then there is a field $E(G)\subseteq F(x_1,\ldots,x_n)=M$ with $Gal(M/E(G))\cong G$.

Proof Our comments above show that for $L=F(\sigma_1,\ldots,\sigma_n)$ for $\sigma_j=\sigma_j(X)$, M is Galois over L with $Gal(M/L)\cong S_n$. By Cayley's Theorem (Theorem 6.11) the left regular representation of G on itself that identifies $g\in G$ with the bijection $\tau_g(x)=gx$ of G is an injective group homomorphism from G to the group of bijections of G. Therefore by Theorem 6.12, $G\cong H\leq S_n$. Now there is $V\leq Gal(M/L)$ with $V\cong G$, and from Theorem 15.7, $E(G)=V'$ satisfies $Gal(M/E(G))=E(G)'=V''=V\cong G$. ∎

The theorem shows that up to isomorphism every finite group is a Galois group. An interesting and unsolved problem is to characterize the finite groups G so that there is a field $\mathbf{Q}\subseteq E$ with $Gal(E/\mathbf{Q})\cong G$. From Example 15.8 and Example 15.9, for p an odd prime, each S_p is such a Galois group. We will see in the next section that every finite cyclic group is also a Galois group of an extension of $\boldsymbol{Q}$.

Using $\sigma_j = \sigma_j(X)$, we saw above that $Gal(F(x_1, \ldots, x_n)/F(\sigma_1, \ldots \sigma_n)) \cong S_n$ identified as the permutations $\{T_\theta\}$ of $\{x_j\}$. Since $F(x_1, \ldots, x_n)$ is Galois over $F(\sigma_1, \ldots, \sigma_n)$, if $\beta \in F(x_1, \ldots, x_n)$ satifies $T_\theta(\beta) = \beta$ for all $\theta \in S_n$ then $\beta \in F(\sigma_1, \ldots, \sigma_n)$ so $\beta = f(\sigma_1, \ldots, \sigma_n)/g(\sigma_1, \ldots, \sigma_n)$ for $f, g \in F[\sigma_1, \ldots, \sigma_n]$. Using results from §14.6 we see next that when $\beta \in F[x_1, \ldots, x_n]$ then $g \in F$, and β is a polynomial in $\{\sigma_j\}$. We continue with setting $\sigma_j = \sigma_j(X)$.

THEOREM 15.10 For $n \geq 1$ and F a field, let $F[X] = F[x_1, \ldots, x_n]$, $F[\sigma] = F[\sigma_1, \ldots, \sigma_n]$, and M and L the quotient fields of these, respectively. If $\beta \in M$ with $T_\theta(\beta) = \beta$ for all $\theta \in S_n$ then $\beta \in L$. If $p(x_1, \ldots, x_n) \in F[X]$ with $T_\theta(p(X)) = p(X)$ for all $\theta \in S_n$, then $p(X) \in F[\sigma]$.

Proof The first statement is our observation just above. If $p(X) \in F[X]$ with $T_\theta(p(X)) = p(X)$ for all $\theta \in S_n$, then to see that $p(X) \in F[\sigma]$ we need to recall some facts from §14.6. First, Theorem 14.27 shows that $\{\sigma_j\}$ is a transcendence basis for $F(X)$ over F, so $\{\sigma_j\}$ is algebraically independent. It follows from Theorem 14.23 that $F[\sigma]$ is the polynomial ring over F in the indeterminates $\{\sigma_j\}$. Therefore $F[\sigma]$ is a UFD by Theorem 12.15. Also in §14.6 we saw that each $x_j \in F[X]$ is a root of the polynomial

$$(Y - x_1)(Y - x_2) \cdots (Y - x_n) = Y^n + \cdots + (-1)^j \sigma_j Y^{n-j} + \cdots + (-1)^n \sigma_n \in F[\sigma].$$

Consequently, each $x_i \in F[X]$ is integral over $F[\sigma]$ and so Theorem 13.26 shows that $F[X]$ is integral over $F[\sigma]$. Since $T_\theta(p(X)) = p(X)$ for all $\theta \in S_n$, it follows from the first statement of the theorem that $p(X) \in L$. We just observed that $p(X) \in F[X]$ must be integral over $F[\sigma]$, and since the UFD $F[\sigma]$ is integrally closed by Theorem 13.27, $p(X) \in F[\sigma]$ as required. ∎

EXERCISES

For the following exercises let F, E, L, and M be fields, and p a prime.

1. If $F \subseteq L$ are finite with $[L : F] = n$, indicate how many subfields of L contain F and what is $\{Gal(E/F) \mid E \subseteq L \text{ with } E \text{ Galois over } F\}$ if $n =$:
   i) 24
   ii) 60
   iii) p^t for p a prime
   iv) $p^2 q$ for $p < q$ primes
   v) $p_1 \cdots p_k$ for primes $p_1 < \cdots < p_k$

2. Let $F \subseteq L$ be finite, $[L : F] = n$, and $Gal(L/F) = \langle \varphi \rangle$.
   i) If for $a_i \in L$, $\sum a_i \varphi^i(\beta) = 0$ for all $\beta \in L$ then all $a_i = 0$. (Take a minimal set $\{i(1), \ldots, i(k)\}$ so that $\sum a_{i(j)} \varphi^{i(j)}(\beta) = 0$ with all $a_{i(j)} \neq 0$, multiply by $\gamma \in L$, and compare to what results by replacing β with $\gamma\beta$.)
   ii) Show that $T_{L/F}(\beta) = \beta + \varphi(\beta) + \cdots + \varphi^{n-1}(\beta)$ is not the zero map on L.
   iii) Show that $T_{L/F}$ is an F-linear transformation from L onto F.

3. If $\mathbf{Z}_{11} \subseteq L$, $|L| = 11^3$, and $\alpha \in L - \mathbf{Z}_{11}$ with minimal polynomial $m(x) \in \mathbf{Z}_{11}[x]$, show that $m(x) = (x - \alpha)(x - \alpha^{11})(x - \alpha^{121}) \in L[x]$.

4. Let $|L| = p^n$, $m \in N$ with $m \mid n$, and set $A = \{\alpha^{p^m} + \alpha^{p^{2m}} + \cdots + \alpha^{p^n} \mid \alpha \in L\}$. Show that $A \subseteq F \subseteq L$ with $|F| = p^m$. Show that $A = F$ (see Exercise 2).

5. If $\varphi \in Aut(L)$ has *finite order*, $n \in N$, and $\varphi(\alpha) = \alpha^n$ for all $\alpha \in L$, show that:

i) L is finite

ii) if $char\ L = p$ there are $m, k \in N$ with $p^k \equiv n \pmod{p^m - 1}$

6. Let F be finite and R a subring of $M_n(F)$ with $F \cdot I_n \subseteq R$. If R is a field, show that $[R : F \cdot I_n] \leq n$. Show that there is such a subfield R with $[R : F \cdot I_n] = n$.

7. Let $\mathbf{Z}_p \subseteq F \subseteq L$ be finite fields with $|F| = p^n$. If $\alpha \in L$ then $\alpha^{(p^n-1)/(p-1)} \in \mathbf{Z}_p$ exactly when $\alpha \in F$.

8. If $F \subseteq L$ are finite, $f(x), g(x) \in F[x]$ are irreducible of the same degree, and $\alpha, \beta \in L$ with $f(\alpha) = 0 = g(\beta)$, then $\beta = h(\alpha)$ for some $h(x) \in F[x]$.

9. If F is finite, $f(x) \in F[x]$ is irreducible, and L is a splitting field of $f(x)$ over F, prove that $|Gal(L/F)| = deg\ f$.

10. Indicate how may subfields L has when $|L|$ is:

i) 32

ii) 64

iii) 256

iv) 512

11. For F finite containing subfields with cardinalities 2, 4, 8, 16, ..., 2^{10}, what is the least $|F|$ can be, and then how many subfields does F contain?

12. If $n \geq 2$ and $p > 2$ is prime, show that $g(x) = x^{p^n} - x - 1 \in \mathbf{Z}_p[x]$ is not irreducible. (Let L be a splitting field over $\mathbf{Z}_p$ of $g(x)$ and $\alpha \in L$ a root of $g(x)$, so $\alpha^{p^n} = \alpha + 1$. Take a succession of p^n-th powers.)

13. If $|F| = p^n$ and $f(x) \in F[x]$ is irreducible, show that $deg\ f \mid m \Leftrightarrow f(x) \mid (x^{p^{nm}} - x)$.

14. Let $F \subseteq L$ be finite so that for any intermediate fields $E, M \subseteq L$, either $E \subseteq M$ or $M \subseteq E$. What are the possibilities for $[L : F]$?

15. Let $F \subseteq L$ be finite, $|F| = p^m$, $\alpha \in L - F$, $m(x) \in F[x]$ the minimal polynomial for α over F, and $deg\ m = k$. Show carefully that the roots of $m(x)$ in L are precisely $\{\alpha, \alpha^{p^m}, \alpha^{p^{2m}}, \ldots, \alpha^{p^{m(k-1)}}\}$.

16. Let F be finite and $S_F = \{\alpha^2 \in F \mid \alpha \in F\}$. Note that $S_F^* \leq (F^*, \cdot)$.

i) If $char\ F = 2$, show that $S_F = F$.

ii) If $char\ F > 2$, show that $|S_F - \{0\}| = (|F| - 1)/2$.

iii) Show that $F = S_F + S_F = \{\alpha^2 + \beta^2 \in F \mid \alpha, \beta \in F\}$. (Consider when $S_F = S_F + S_F$ or not.)

17. Let $F \subseteq L$ be finite.

i) If $|L| = |F|^2$, show that for any $\alpha \in F$, $\alpha = \beta^2$ for $\beta \in L$.

ii) If $|L| = |F|^3$, show that for any $\alpha \in F$, $\alpha = \beta^3$ for $\beta \in L$.

18. Let $F \subseteq L$ be finite, $|L| = |F|^5$, and $g(x) \in F[x]$ with $deg\ g = 5$. Must $g(x)$ split over L? What if $g(x)$ is irreducible?

19. For $n > 1$, show that there is $\alpha \in F(x_1, \ldots, x_n) - F$ so that its minimal polynomial $m(Y) \in F(\sigma_1(X), \ldots, \sigma_n(X))[Y]$ has degree $n!$ and that $\{T_\theta(\alpha) \mid \theta \in S_n\}$ are the $n!$ distinct roots of $m(Y)$ in $F(X)$.

20. Show that $\alpha = x_1(1 + x_2)(1 + x_2x_3) \cdots (1 + x_2x_3 \cdots x_n) \in F(x_1, \ldots, x_n)$ has minimal polynomial of degree $n!$ over $F(\sigma_1(X), \ldots, \sigma_n(X))$.

21. Let $F[X] = F[x_1, \ldots, x_n]$ and $F[\sigma] = F[\sigma_1(X), \ldots, \sigma_n(X)]$. If $f(X), g(X) \in F[\sigma]$, $h(X) \in F[X]$, and $f(X)h(X) = g(X)$, show that $h(X) \in F[\sigma]$.

22. Let $F[X]=F[x_1,\ldots,x_n]$ and $F[\sigma]=F[\sigma_1(X),\ldots,\sigma_n(X)]$.

i) If $f(X)\in F[X]$ is irreducible and $\{T_\theta(f(X))\mid\theta\in S_n\}=\{f_1(X),\ldots,f_k(X)\}$ has k distinct elements with $f(X)=f_1(X)$ and $f_j(X)\notin F\cdot f(X)$ for all $j>1$, show that $f_1(X)f_2(X)\cdots f_k(X)\in F[\sigma]$ is irreducible in $F[\sigma]$.

ii) Show that $f_j(X)\notin F\cdot f(X)$ for all $j>1$ is necessary by considering $f(X)=x_1+\omega x_2+\omega^2 x_3\in \mathbf{C}[x_1,x_2,x_3]$.

15.4 CYCLOTOMIC EXTENSIONS

An important application of Galois Theory describes the splitting fields L of polynomials of the form $x^n-1\in F[x]$, and $Gal(L/F)$. These are called *cyclotomic extensions*, and the roots in L of x^n-1 are *roots of unity*.

DEFINITION 15.5 For F a field, a **cyclotomic extension** of F is any splitting field L over F of $x^n-1\in F[x]$ for some $n\geq 1$. The roots of x^n-1 in L are called ***n*-th roots of unity** and an n-th root of unity $\rho\in L$ is a **primitive *n*-th root of unity** if $o(\rho)=n$ in $(L^*,\cdot)$.

Any $\rho\in L$ is a primitive n-th root of unity $\Leftrightarrow n$ is the minimal positive integer satisfying $\rho^n=1$ (verify!). When $char F\neq 2$, then $-1\in F$ is a primitive second root of unity. For any $n\geq 1$ there *is* a primitive n-th root of unity in $\mathbf{C}$, namely $e^{2\pi i/n}=\cos(2\pi/n)+i\cdot\sin(2\pi/n)$. Since splitting fields of $x^n-1\in F[x]$ are unique up to isomorphism by Theorem 14.17, if some splitting field L of $x^n-1\in F[x]$ contains a primitive n-th root of unity then all do. If $\rho\in L$ is a primitive n-th root of unity, $\{1,\rho,\rho^2,\ldots,\rho^{n-1}\}=\langle\rho\rangle\leq(L^*,\cdot)$ are the n distinct n-th roots of unity in L (Theorem 2.16). It follows from Theorem 2.8 that ρ^j is a primitive n-th root of unity precisely when $(j,n)=1$, so there are exactly $\varphi(n)$ (Euler phi-function in §3.2!) different primitive n-th roots of unity in L. For example, $i\in C$ is a primitive fourth root of unity and the $\varphi(4)=2$ such are i and $i^3=-i$. If $\rho_{15}=e^{2\pi i/15}\in\mathbf{C}$, then the $\varphi(15)=8$ primitive 15-th roots of unity in $\mathbf{C}$ are $\rho_{15},\rho_{15}^2,\rho_{15}^4,\rho_{15}^7,\rho_{15}^8,\rho_{15}^{11},\rho_{15}^{13}$, and ρ_{15}^{14}.

Note that $x^{14}-1=(x^7-1)^2\in\mathbf{Z}_2[x]$ and $x^{45}-1=(x^5-1)^9\in\mathbf{Z}_3[x]$ so primitive roots of unity do not always exist. The relation between *char F* and the existence of primitive roots is described next.

THEOREM 15.11 If $n\geq 1$ and L is a splitting field over F of $x^n-1\in F[x]$, then L contains a primitive n-th root of unity exactly when either $char F=0$ or $char\ F$ does not divide n.

Proof If L contains a primitive n-th root of unity ρ then $\{\rho^j\}$ contains n distinct n-th roots of unity and x^n-1 has n distinct linear factors in $L[x]$. By Theorem 14.14, x^n-1 and its derivative nx^{n-1} have no common linear divisor, so $n\neq 0$ in F and either $char F=0$ or $char F=p>0$ but does not divide n. When these conditions hold, x^n-1 has n distinct roots in L. If $\alpha^n=1$ in $(L^*,\cdot)$ then $o(\alpha)=d$ for $d\mid n$, and $\langle\alpha\rangle$ has $\varphi(d)$ generators of order d by Theorem 2.8. In the field L there are at most d solutions of x^d-1 so these must be $\langle\alpha\rangle$, forcing $(L^*,\cdot)$ to contain only $\varphi(d)$ elements of order d. Therefore $|\{\beta\in L\mid\beta^n=1\}|=n\leq\sum\varphi(d)$, summing over *all* the divisors of n. In any cyclic group $G=\langle g\rangle$ of order n, each element has order some $d\mid n$, and for any $d\mid n$ there is a unique $\langle h\rangle\leq G$ of order d by Theorem 3.1, so G has exactly $\varphi(d)$ elements of order d and $n=\sum\varphi(d)$ (Exercise 4 in

§3.2). Thus $\varphi(n)$ roots in L of $x^n - 1$ have order n in $(L^*, \cdot)$ so L has primitive n-th roots of unity. ∎

Note that $x^{16} - 1 = (x - 1)^{16} \in \mathbf{Z}_2[x]$. Here $1 \in \mathbf{Z}_2$ is the only 16-th root of unity and has order 1. The same situation holds for any 2^n-th root of unity when $char F = 2$ and, more generally, for any p^n-th root of unity when $char F = p > 0$. Similarly, $(x^{15} - 1) = (x^3 - 1)^5 \in \mathbf{Z}_5[x]$ so any 15-th root of unity in characteristic five is actually a third root of unity.

The nature of a cyclotomic extension when $char F = p > 0$ and $p \nmid n$ is somewhat easier to deal with than when $char F = 0$ so we consider it first. In this case, by Theorem 15.11 there is a primitive n-th root of unity over $\mathbf{Z}_p$ and we determine the degree of its minimal polynomial over $\mathbf{Z}_p$.

THEOREM 15.12 Let p be a prime, $n > 1$ not divisible by p, and L a splitting field of $x^n - [1] \in \mathbf{Z}_p[x]$ over $\mathbf{Z}_p$. Then $[L : \mathbf{Z}_p] = o([p]_n)$ in U_n.

Proof By Theorem 15.11, $(L^*, \cdot)$ contains a primitive n-th root of unity ρ so $o(\rho) = n$ in $(L^*, \cdot)$ and $L = \mathbf{Z}_p[\rho]$. If $|L| = p^k$ then $(L^*, \cdot)$ is a cyclic group of order $p^k - 1$ by Theorem 4.20, so Theorem 3.1 shows that $(L^*, \cdot)$ has an element ρ of order n exactly when $n \mid (p^k - 1)$. Now $[L : \mathbf{Z}_p] = k$ is the degree of the minimal polynomial $m(x) \in \mathbf{Z}_p[x]$ for ρ, so k must be minimal satisfying $p^k \equiv 1 \pmod{n}$. Since p does not divide n, $[p]_n \in U_n$ so $k = o([p]_n)$ in U_n. ∎

To illustrate the theorem, consider $x^9 - [1] \in \mathbf{Z}_p[x]$ and a splitting field L over $\mathbf{Z}_p$ for this polynomial. Now $U_9 = \{[1], [2], [4], [5], [7], [8]\} = \langle [2] \rangle$ so for $p \neq 3$, $[p] \in U_9$ has order dividing 6. Since $o([2]) = 6 = o([5])$ in U_9, when $p = 2$, $[L : \mathbf{Z}_2] = 6$. Some other cases with $[L : \mathbf{Z}_p] = 6$ are $p = 5, 11, 23$, and 41. The elements of order 3 in U_9 are [4] and [7] so $[L : \mathbf{Z}_p] = 3$, for example, when $p = 7$, 13, or 43. Only $[8] \in U_9$ has order 2 so $[L : \mathbf{Z}_p] = 2$ if $p = 17, 53$, or 71. Finally, $L = \mathbf{Z}_p$ if $[p] = [1]$ in U_9, so when $9 \mid (p - 1)$. Examples are $p = 19, 37, 73$, and 109.

If $x^n - 1 \in L[x]$ and if L contains a primitive n-th root of unity ρ then each of the $\varphi(n)$ primitive n-th roots of unity is also in L. The polynomial over L having these as roots has a special name.

DEFINITION 15.6 Let $x^n - 1$ split over L, let $\rho \in L$ be a primitive n-th root of unity, and let $\{\rho_1, \ldots, \rho_{\varphi(n)}\} \subseteq L$ be the $\varphi(n)$ primitive n-th roots of unity in L. The ***n*-th cyclotomic polynomial** is $\Phi_n = (x - \rho_1) \cdots (x - \rho_{\varphi(n)}) \in L[x]$.

Let $\varepsilon \in L$ be a primitive n-th root of unity. We have seen that $P = QF(\langle 1_L \rangle) \subseteq L$ is isomorphic to $\mathbf{Q}$ when $char\, L = 0$ and to $\mathbf{Z}_p$ when $char\, L = p > 0$. Thus to study a splitting field of $x^n - 1 \in P[x]$, we may assume $P = \mathbf{Q}$ or $P = \mathbf{Z}_p$, by Theorem 14.17. When $P = \mathbf{Z}_p$, $L = \mathbf{Z}_p[\varepsilon]$ so the degree of the minimal polynomial $m_1(x) \in \mathbf{Z}_p[x]$ of ε is $[L : \mathbf{Z}_p] = o([p]_n)$ in U_n by Theorem 14.7 and Theorem 15.12. For any primitive n-th root of unity $\varepsilon^k \in L$, $P[\varepsilon] = P[\varepsilon^k]$ so $m_k(x) = m_{\varepsilon^k}(x) \in \mathbf{Z}_p[x]$ has $deg\, m_k = deg\, m_1$. Now $m_k(x) \mid x^n - 1$ but $m_k(x) \nmid x^d - 1$ if $d < n$ so $m_k(x) \mid \Phi_n$ forcing Φ_n to be a product of some $m_j(x)$. Unique factorization in $\mathbf{Z}_p[x]$ proves the next theorem.

THEOREM 15.13 If p is a prime then $\Phi_n \in \boldsymbol{Z}_p[x]$ depends only on p and factors into irreducibles all having degree $o([p]_n)$ in U_n.

We look at the factorization of Φ_n for small n and p. If $n=3$ then $x^3-1=(x-1)\Phi_3$ so $\Phi_3=x^2+x+1\in \boldsymbol{Z}_p[x]$ is irreducible unless $[p]=[1]\in U_3$ or, equivalently, $p\equiv_3 1$. Hence Φ_3 is irreducible in $\boldsymbol{Z}_p[x]$ when $p=2, 5, 11, 17, 23$, etc., and $\Phi_3=(x-2)(x-4)\in \boldsymbol{Z}_7[x]$, $\Phi_3=(x-3)(x-9)\in \boldsymbol{Z}_{13}[x]$, and $\Phi_3=(x-7)(x-11)\in \boldsymbol{Z}_{19}[x]$. Now let $n=5$. Any fifth root of unity is primitive or 1 so $x^5-1=(x-1)\Phi_5$ and the irreducible factors of Φ_5 in $\boldsymbol{Z}_p[x]$ all have degree 4, 2, or 1. As above, Φ_5 splits over $\boldsymbol{Z}_p$ exactly when $5\mid(p-1)$ so, for example, when p is 11 or 31: here the irreducible factors are $(x-3)(x-4)(x-5)(x-9)\in \boldsymbol{Z}_{11}[x]$ and $(x-2)(x-4)(x-8)(x-16)\in \boldsymbol{Z}_{31}[x]$. When $o([p])=4$ in U_5 then $\Phi_5\in \boldsymbol{Z}_p[x]$ is irreducible. Since $o([2])=4=o([3])$ in U_5 we need $p\equiv_5 2$ or $p\equiv_5 3$ to get $\Phi_5=x^4+x^3+x^2+x+1\in \boldsymbol{Z}_p[x]$ irreducible. Such primes are 2, 7, 13, 17, 23, and 37. Finally, when $p\equiv_5 4$ then $o([p])=2$ in U_5 and Φ_5 factors as a product of two irreducible quadratics. For example, in $\boldsymbol{Z}_{19}[x]$ we have $\Phi_5=(x^2+5x+1)(x^2+15x+1)$ and in $\boldsymbol{Z}_{29}[x]$, $\Phi_5=(x^2+6x+1)(x^2+24x+1)$.

When *char* $L=0$ then $\Phi_n\in \boldsymbol{Z}[x]$, is irreducible over $\boldsymbol{Q}$ and is independent of the splitting field L. To prove this we will use §12.3.

THEOREM 15.14 If $x^n-1\in \boldsymbol{Q}[x]$, then $\Phi_n\in \boldsymbol{Z}[x]$ and is independent of L.

Proof We use induction on n. Clearly $\Phi_1=x-1\in \boldsymbol{Z}[x]$ and since $-1\in \boldsymbol{Q}$ is the only second root of unity, $\Phi_2=x+1\in \boldsymbol{Z}[x]$. As $\boldsymbol{Q}$ contains no cube root of unity, $x^3-1=(x-1)(x^2+x+1)$ and $\Phi_3=(x^2+x+1)\in \boldsymbol{Z}[x]$ is irreducible. Assume that $n>1$ and for all $m<n$, $\Phi_m\in \boldsymbol{Z}[x]$ is monic and is independent of a splitting field L for x^n-1 over $\boldsymbol{Q}$. Let $\varepsilon\in L$ be a primitive n-th root of unity, so the roots of x^n-1 in L are the n distinct powers of ε. If $o(\varepsilon^j)=k$ dividing n, then ε^j is a primitive k-th root of unity (why?) and so a root of Φ_k. It follows that $x^n-1=\prod\Phi_m\in L[x]$ over all divisors m of n. By our induction assumption, the product $H(x)$ of the Φ_m for m a *proper* divisor of n is in $\boldsymbol{Z}[x]$ so $x^n-1=H(x)\Phi_n\in L[x]$, and $H(x)$ is monic and independent of L. Each primitive n-th root of unity in L satisfies some minimal polynomial in $\boldsymbol{Q}[x]$ not dividing $H(x)$, and so Φ_n is a product of some of these, forcing $\Phi_n\in \boldsymbol{Q}[x]$. Write $\Phi_n=qg(x)$ for $q\in \boldsymbol{Q}$ and a primitive $g(x)\in \boldsymbol{Z}[x]$, using Theorem 12.12. Thus $x^n-1=qH(x)g(x)$ for $H(x)g(x)\in \boldsymbol{Z}[x]$ and primitive (Theorem 12.13). Now x^n-1 monic and Theorem 12.12 show $q\in \boldsymbol{Z}$, so $q=1$ and $\Phi_n=g(x)\in \boldsymbol{Z}[x]$. Since $x^n-1=H(x)\Phi_n\in \boldsymbol{Z}[x]$ and $H(x)$ is independent of L, by unique factorization in $\boldsymbol{Z}[x]$ (Theorem 12.14), Φ_n must be independent of L as well. ∎

THEOREM 15.15 $\Phi_n\in \boldsymbol{Z}[x]$ is irreducible in $\boldsymbol{Q}[x]$.

Proof Let $\varepsilon\in \boldsymbol{C}$ be a primitive n-th root of unity, let $m(x)\in \boldsymbol{Q}[x]$ be its minimal polynomial, and set $L=\boldsymbol{Q}[\varepsilon]\subseteq \boldsymbol{C}$. Now $\Phi_n=m(x)g(x)\in \boldsymbol{Q}[x]$ and by Theorem 12.12, $m(x)g(x)=q_1m_1(x)q_2g_1(x)$ for $q_i\in \boldsymbol{Q}$ and primitive $m_1, g_1\in \boldsymbol{Z}[x]$. Since $m_1(x)g_1(x)$ is primitive by Theorem 12.13, we conclude from Theorem 12.12 that $q_1q_2\in \boldsymbol{Z}$ so Φ_n monic forces $\Phi_n=\pm m_1(x)g_1(x)$ and $m_1(x)$ is irreducible in $\boldsymbol{Q}[x]$ by Theorem 12.14. Hence we may assume that $m(x), g(x)\in \boldsymbol{Z}[x]$ are monic and so are primitive. If $H(x)$ is the product of the Φ_m for m the proper divisors of n, then, as in the proof of Theorem 15.14, $x^n-1=H(x)\Phi_n=H(x)m(x)g(x)$ in $\boldsymbol{Z}[x]$. Suppose that p is a prime, $p\nmid n$, and $m(\varepsilon^p)\neq 0$.

Then ε^p is still a primitive n-th root of unity by Theorem 2.8 and hence is a root of Φ_n, forcing $g(\varepsilon^p)=0$. Thus ε is a root of $g(x^p)$, and $g(x^p)=m(x)f(x)\in \boldsymbol{Q}[x]$. Using that $g(x^p), m(x)\in \boldsymbol{Z}[x]$ are monic and Theorem 12.12 as above, we have $f(x)\in \boldsymbol{Z}[x]$. Now $\eta(\sum a_j x^j)=\sum [a_j]_p x^j$ defines a ring homomorphism $\eta: \boldsymbol{Z}[x]\to \boldsymbol{Z}_p[x]$ (verify!). Therefore in $\boldsymbol{Z}_p[x]$, $x^n-[1]_p=\eta(x^n-1)=\eta(H(x))\eta(m(x))\eta(g(x))$ and also using $[a]^p=[a]$ for $[a]\in \boldsymbol{Z}_p$, $\eta(g(x))^p=\eta(g(x^p))=\eta(m(x))\eta(f(x))$. By unique factorization in $\boldsymbol{Z}_p[x]$ if $s(x)\in \boldsymbol{Z}_p[x]$ is irreducible and $\eta(m(x))=s(x)t(x)$ then $\eta(g(x))=s(x)v(x)$. Hence $x^n-[1]_p=\eta(H(x))s^2(x)t(x)v(x)\in \boldsymbol{Z}_p[x]$ and $x^n-[1]_p$ cannot have n distinct roots in a splitting field F over $\boldsymbol{Z}_p$, so there is no primitive n-th root of unity in F, contradicting Theorem 15.11. This proves that $m(\varepsilon^p)=0$ when $p\nmid n$. If ε^j is any primitive n-th root of unity then $(j,n)=1$ by Theorem 2.8 and $j=p_1p_2\cdots p_k$ where each p_s is a prime not dividing n. If $\varepsilon_i=\varepsilon^{p_1\cdots p_i}$ then the argument just given shows that all ε_i are roots of $m(x)$ so, in particular, $\varepsilon^j=\varepsilon_k$ is: all primitive n-th roots of unity are roots of $m(x)$. Since $m(x)$ is a factor in $\boldsymbol{Z}[x]$ of Φ_n we must have that $\Phi_n=m(x)\in \boldsymbol{Z}[x]$ is irreducible in $\boldsymbol{Q}[x]$. ■

With Theorem 15.15 we can find $\Phi_n\in \boldsymbol{Z}[x]$ for small n. First we observe that we have another approach to Example 12.5.

Example 15.16 If p is prime $\Phi_p=x^{p-1}+\cdots+x+1\in \boldsymbol{Q}[x]$ is irreducible.

Since $x^p-1=(x-1)(x^{p-1}+\cdots+x+1)$ and $deg\,\Phi_p=\varphi(p)=p-1$, it follows that $\Phi_p=x^{p-1}+\cdots+x+1\in \boldsymbol{Q}[x]$ is irreducible by Theorem 15.15.

Example 15.17 In $\boldsymbol{Z}[x]$, $\Phi_2=x+1$, $\Phi_4=x^2+1$, $\Phi_6=x^2-x+1$, $\Phi_8=x^4+1$, $\Phi_9=x^6+x^3+1$, $\Phi_{10}=x^4-x^3+x^2-x+1$, and $\Phi_{12}=x^4-x^2+1$.

Since $x^2-1=(x-1)(x+1)$ and $x^4-1=(x^2-1)(x^2+1)$, it is clear that $\Phi_2=x+1$, and $\Phi_4=x^2+1$. Similarly, $x^8-1=(x^4-1)(x^4+1)$ implies that $\Phi_8=x^4+1$ since no primitive 8-th root of unity is a root of x^4-1. Now $x^6-1=(x^3-1)\cdot(x^3+1)=(x^3-1)(x+1)(x^2-x+1)$ so the primitive 6-th roots of unity satisfy $x^2-x+1=\Phi_6$. Using $\varphi(9)=6$ and the factorization of $x^9-1=(x^3)^3-1=(x^3-1)(x^6+x^3+1)$, $\Phi_9=x^6+x^3+1$. The expressions for Φ_{10} and Φ_{12} follow from $\varphi(10)=4=\varphi(12)$ and the two factorizations: $x^{10}-1=(x^5-1)(x^5+1)=(x^5-1)(x+1)(x^4-x^3+x^2-x+1)$, and also $x^{12}-1=(x^6-1)(x^6+1)=(x^6-1)(x^2+1)(x^4-x^2+1)$.

It remains to describe the Galois group of a cyclotomic extension of the rational numbers.

THEOREM 15.16 If L is a splitting field of $x^n-1\in \boldsymbol{Q}[x]$, $Gal(L/\boldsymbol{Q})\cong U_n$.

Proof A splitting field over $\boldsymbol{Q}$ of Φ_n is $L=\boldsymbol{Q}[\varepsilon]$ for ε a primitive n-th root of unity. Since Φ_n is irreducible by Theorem 15.15, $[L:\boldsymbol{Q}]=deg\,\Phi_n=\varphi(n)$, using Theorem 14.7. Thus $|Gal(L/\boldsymbol{Q})|=\varphi(n)$ by Theorem 15.4. Also note that the roots of Φ_n in L are the primitive

n-th roots of unity $\{\varepsilon^j \mid (j,n)=1\}$, and for any of these $L=\mathbf{Q}[\varepsilon^j]$. It follows from Theorem 14.16 that for any $1 \le j \le n$ with $(j,n)=1$ there is an isomorphism η_j of L extending the identity on $\mathbf{Q}$ and with $\eta_j(\varepsilon)=\varepsilon^j$. Thus $\eta_j \in Gal(L/\mathbf{Q})=G$ and $\eta_j \neq \eta_i$ for $j \neq i$ so $G=\{\eta_j \mid 1 \le j \le n \text{ and } (j,n)=1\}$. We claim that $\beta: U_n \to G$ given by $\beta([j])=\eta_j$, when $1 \le j \le n$, is a group isomorphism. Now β is a function since each $[a] \in U_n$ has a unique representative in $\{j \in \mathbf{N} \mid 1 \le j \le n \text{ with } (j,n)=1\}$ (why?). For $[j],[k] \in U_n$ let $[j][k]=[m]$ for $m \equiv jk \pmod n$, $1 \le m \le n$, and $(m,n)=1$. Thus $\beta([j])\beta([k])=\eta_j\eta_k$ and $\eta_j\eta_k(\varepsilon)=\eta_j(\varepsilon^k)=(\eta_j(\varepsilon))^k=\varepsilon^{jk}=\varepsilon^m$ since $o(\varepsilon)=n$ in $(L^*,\cdot)$ and $m \equiv jk \pmod n$ (Theorem 2.9). Any $\alpha \in G$ is uniquely defined by $\alpha(\varepsilon)$, so $\eta_j\eta_k=\eta_m=\beta([m])=\beta([j][k])$ and β is a group homomorphism. Clearly β is surjective, and since $|G|=\varphi(n)=|U_n|$, β must be an isomorphism. ∎

One consequence of Theorem 15.16 is that the Galois group of a cyclotomic extension of $\mathbf{Q}$ is always Abelian and so by Theorem 15.17 every intermediate field is Galois over $\mathbf{Q}$. Using results from §4.6, U_n is a cyclic group exactly when $n=1, 2, 4, p^k$, or $2p^k$ for p an odd prime, so these are the n for which the splitting fields of $x^n-1 \in \mathbf{Q}[x]$ have a cyclic Galois group. The fact that the splitting field of x^p-1 has a cyclic Galois group enables us to show that every cyclic group arises as $Gal(L/\mathbf{Q})$, although we also use a sophisticated result of Dirichlet.

THEOREM 15.17 If G is a finite cyclic group, then there is a field L Galois over $\mathbf{Q}$ with $Gal(L/\mathbf{Q}) \cong G$.

Proof By a famous result of Dirichlet, for any $n>1$ the arithmetic sequence $\{1+tn \mid t \in \mathbf{N}\}$ contains infinitely many primes. Given $n>1$, let $p=1+tn$ be a prime and let E be a splitting field over $\mathbf{Q}$ of x^p-1. Then E is Galois over $\mathbf{Q}$ with $Gal(E/\mathbf{Q}) \cong U_p$, cyclic of order $p-1=tn$. From Theorem 3.1 there is $H \le Gal(E/\mathbf{Q})$ with H cyclic of order t. Since $Gal(E/\mathbf{Q})$ is Abelian, $H \lhd Gal(E/\mathbf{Q})$ so by Theorem 15.7, $L=H'$ is Galois over $\mathbf{Q}$ with $Gal(L/\mathbf{Q}) \cong Gal(E/\mathbf{Q})/H \cong \mathbf{Z}_n$. ∎

A special case of Theorem 15.17 is worth doing separately.

Example 15.18 There is a field L Galois over $\mathbf{Q}$ with $[L:\mathbf{Q}]=3$.

Let $E=\mathbf{Q}[\varepsilon] \subseteq \mathbf{C}$ be a splitting field of $x^7-1 \in \mathbf{Q}[x]$ where $\varepsilon=\cos(2\pi/7)+i\cdot\sin(2\pi/7)$. From Theorem 15.16, $[E:\mathbf{Q}]=\varphi(7)=6$ and $G=Gal(E/\mathbf{Q})=\langle\eta\rangle \cong U_7=\langle[3]_7\rangle$ with $\eta(\varepsilon)=\varepsilon^3$. Complex conjugation on $\mathbf{C}$ restricts to E and is $\eta^3 \in G: \eta^3(\varepsilon^i)=\varepsilon^{27i}=\varepsilon^{-i}=\varepsilon^{7-i}$. Since the subgroups of G are $\langle e\rangle$, $\langle\eta^3\rangle$, $\langle\eta^2\rangle$, and G, by Theorem 15.7 the subfields of E are $E=\langle e\rangle'$, $L=\langle\eta^3\rangle'$, $K=\langle\eta^2\rangle'$, and $\mathbf{Q}=G'$. Also L and K are Galois over $\mathbf{Q}$, $[L:\mathbf{Q}]=3$, and $[K:\mathbf{Q}]=2$ (why?). What are L and K? Since η^3 is complex conjugation, $L=\mathbf{R}\cap E$ and $\varepsilon+\varepsilon^6=\varepsilon+\eta^3(\varepsilon)=2\cos(2\pi/7) \in L-\mathbf{Q}$. Since $[L:\mathbf{Q}]=3$, Theorem 14.8 forces $L=\mathbf{Q}[\cos(2\pi/7)]$. Clearly $\alpha=\varepsilon+\eta^2(\varepsilon)+\eta^4(\varepsilon)=\varepsilon+\varepsilon^9+\varepsilon^{81}=\varepsilon+\varepsilon^2+\varepsilon^4 \in K$ and $\eta^3(\alpha)=\varepsilon^{-1}+\varepsilon^{-2}+\varepsilon^{-4}=\varepsilon^3+\varepsilon^5+\varepsilon^6$; hence $0=\Phi_7(\varepsilon)=1+\alpha+\eta^3(\alpha)$. By a direct computation, $\alpha^2=\alpha+2\eta^3(\alpha)=\alpha+2(-\alpha-1)=-\alpha-2$ and α satisfies the irreducible $x^2+x+2 \in \mathbf{Q}[x]$. The quadratic formula now shows $K=\mathbf{Q}[(-7)^{1/2}]$.

Example 15.19 The splitting field of $x^{10}-1 \in \boldsymbol{Q}$.

Let $L=\boldsymbol{Q}[\varepsilon] \subseteq \boldsymbol{C}$ split $x^{10}-1$ over $\boldsymbol{Q}$ for ε a primitive 10-th root of unity. Then $\varphi(10)=4$ and $U_{10}=\langle[3]\rangle$ (why?) so $[L:\boldsymbol{Q}]=4$, $Gal(L/\boldsymbol{Q})=G=\langle\eta\rangle$ for $\eta(\varepsilon)=\varepsilon^3$, and L has three subfields: $\boldsymbol{Q}$, E, and L, with $[E:\boldsymbol{Q}]=2$. All three fields are Galois over $\boldsymbol{Q}$ since $G=Gal(L/\boldsymbol{Q})$ is cyclic and so all of its subgroups are normal. Now $\eta^2(\varepsilon)=\varepsilon^{-1}$ so η^2 is the restriction of complex conjugation to L and has order 2, so $E=\langle\eta^2\rangle'=L\cap\boldsymbol{R}$. As in Example 15.18, $E=\boldsymbol{Q}[\cos(2\pi/10)]$. Using that the minimal polynomial of ε over $\boldsymbol{Q}$ is $x^4-x^3+x^2-x+1$ (why?) and that $\varepsilon^5=-1$, if $\alpha=\varepsilon+\varepsilon^{-1}=\varepsilon-\varepsilon^4$ then α satisfies $x^2-x-1\in\boldsymbol{Q}[x]$ and it follows that $E=\boldsymbol{Q}(5^{1/2})$ also.

Example 15.20 The splitting field of $x^{12}-1 \in \boldsymbol{Q}[x]$.

Let $L\subseteq\boldsymbol{C}$ split $x^{12}-1$ over $\boldsymbol{Q}$. Here $\varphi(12)=4$ but U_{12} is not cyclic: every $[a]\in U_{12}$ satisfies $[a]^2=[1]$. If $G=Gal(L/\boldsymbol{Q})$ and $\varepsilon=e^{2\pi i/12}$ is a primitive twelfth root of unity in L then $G=\{I,\eta_5,\eta_7,\eta_{11}\}$ with $\eta_j(\varepsilon)=\varepsilon^j$. Since G has exactly three subgroups of order 2, the $\langle\eta_j\rangle$, L contains exactly three subfields not $\boldsymbol{Q}$ or L and each is a quadratic, and hence a Galois extension of $\boldsymbol{Q}$. Since L contains $\varepsilon^4=\omega$ a 3-rd root of unity, and $\varepsilon^3=i$, two of the subfields are $\boldsymbol{Q}[\varepsilon^4]$ and $\boldsymbol{Q}[\varepsilon^3]$. Now $\eta_5(\varepsilon^3)=\varepsilon^{15}=\varepsilon^3$ so $\boldsymbol{Q}[\varepsilon^3]=\langle\eta_5\rangle'$, and $\eta_7(\varepsilon^4)=\varepsilon^{28}=\varepsilon^4$ so $\boldsymbol{Q}[\varepsilon^4]=\langle\eta_7\rangle'$. To identify the third field $\langle\eta_{11}\rangle'$, note that $\varepsilon+\varepsilon^{-1}=\varepsilon+\eta_{11}(\varepsilon)=2\cos(\pi/6)=3^{1/2}\in\langle\eta_{11}\rangle'$. Thus $\langle\eta_{11}\rangle'=\boldsymbol{Q}[3^{1/2}]=L\cap\boldsymbol{R}$.

EXERCISES

For the exercises below, let p be a prime and L a field.

1. If $[L:\boldsymbol{Z}_p]=2$, show that $x^8-1\in\boldsymbol{Z}_p$ splits over L.
2. If L is a splitting field of $x^n-1\in\boldsymbol{Z}_p$ over $\boldsymbol{Z}_p$, find $[L:\boldsymbol{Z}_p]$ and the number of subfields of L when:
   i) $n=7$, $p=5$
   ii) $n=7$, $p=23$
   iii) $n=15$, $p=11$
   iv) $n=98$, $p=7$
   v) $n=98$, $p=2$
   vi) $n=75$, $p=3$
   vii) $n=11$, $p=43$
3. Show that over $\boldsymbol{Q}$, $(x^k-1)\Phi_p(x^k)=x^{pk}-1$. When is $\Phi_p(x^k)$ irreducible?
4. If $p>2$, show that $\Phi_{2p}=x^{p-1}-x^{p-2}+\cdots+(-1)^{i+1}x^{p-i}+\cdots-x+1\in\boldsymbol{Q}[x]$.
5. Show $\Phi_{2^{k+1}}=x^{2^k}+1\in\boldsymbol{Q}[x]$ and factor $x^{2^n}-1\in\boldsymbol{Q}[x]$ into irreducibles.
6. Show that $\Phi_{p^n}(x)=\Phi_{p^d}(x^{p^{n-d}})\in\boldsymbol{Q}[x]$ for any $1\le d\le n$. Write out as polynomials in $\boldsymbol{Q}[x]$, Φ_{25}, Φ_{49}, Φ_{125}, and Φ_{81}.

7. Let $L \subseteq \boldsymbol{C}$ be a splitting field over $\boldsymbol{Q}$ of $x^p - 2 \in \boldsymbol{Q}[x]$ with $p > 2$.
  i) Show that $|Gal(L/\boldsymbol{Q})| = p(p-1)$ and for some $N \triangleleft Gal(L/\boldsymbol{Q})$, $Gal(L/\boldsymbol{Q})/N \cong U_p$.
  ii) Explicitly describe the elements in $Gal(L/\boldsymbol{Q})$. Is $Gal(L/\boldsymbol{Q})$ Abelian?
  iii) Show that $Gal(L/\boldsymbol{Q}) \cong \left\{ \begin{bmatrix} a & b \\ 0 & 1 \end{bmatrix} \in GL(2, \boldsymbol{Z}_p) \right\} \leq GL(2, \boldsymbol{Z}_p)$.

8. Let $L \subseteq \boldsymbol{C}$ be a splitting field of $x^p - p \in \boldsymbol{Q}[x]$ over $\boldsymbol{Q}$ and $G = Gal(L/\boldsymbol{Q})$. How many subfields E of L are Galois over $\boldsymbol{Q}$ (see Exercise 7.)? For these E find $Gal(E/\boldsymbol{Q})$ when $p =$:
  i) 5
  ii) 7
  iii) 13
  iv) 31
  v) 37
  vi) 61

9. Let $p > 2$ and $\varepsilon = e^{2\pi i/p} \in \boldsymbol{C}$. Show that $L = \boldsymbol{Q}[\varepsilon]$ contains a unique subfield E_d with $[E_d : \boldsymbol{Q}] = d$ for each positive divisor d of $p - 1$. For each E_d describe $Gal(L/E_d)$ and $Gal(E_d/\boldsymbol{Q})$.

10. Let $L = \boldsymbol{Q}[e^{2\pi i/p}]$ and $S = \{E \subseteq L \mid E \text{ is a field}\}$. If $|S|$ is odd, characterize p and find five specific such $p > 2$.

11. Let $\varepsilon, \rho \in \boldsymbol{C}$ with ε a primitive n-th root of unity and ρ a primitive m-th root of unity. Show that $\boldsymbol{Q}[\varepsilon, \rho]$ is a cyclotomic extension of $\boldsymbol{Q}$ and find $[\boldsymbol{Q}[\varepsilon, \rho] : \boldsymbol{Q}]$. (Consider first when $(n, m) = 1$ and look at Theorem 4.16.)

12. If $p > 2$, L is a splitting field over $\boldsymbol{Q}$ of $x^p - 1 \in \boldsymbol{Q}[x]$, and for any subfield $E \subseteq L$, $[E : \boldsymbol{Q}]$ is square-free ($m^2 \mid [E : \boldsymbol{Q}] \Rightarrow m = \pm 1$), show that $p = 1 + p_1 \cdots p_k$ for primes $2 = p_1 < p_2 < \cdots < p_k$.

13. Let $n > 2$ and $\varepsilon = e^{2\pi i/n} \in \boldsymbol{C}$. Show that:
  i) $[\boldsymbol{Q}[\varepsilon] \cap \boldsymbol{R} : \boldsymbol{Q}] = \varphi(n)/2$
  ii) $\boldsymbol{Q}[\varepsilon] = \boldsymbol{Q}[\varepsilon + \varepsilon^{-1}][i \cdot \sin(2\pi/n)]$
  iii) $[\boldsymbol{Q}[\varepsilon] : \boldsymbol{Q}[\varepsilon + \varepsilon^{-1}]] = 2$
  iv) $\boldsymbol{Q}[\varepsilon + \varepsilon^{-1}] = \boldsymbol{R} \cap \boldsymbol{Q}[\varepsilon] = \boldsymbol{Q}[\cos(2\pi/n)]$
  v) What is the degree of the minimal polynomial over $\boldsymbol{Q}$ for $\cos(2\pi/n)$?

14. If $\varepsilon = e^{2\pi i/n} \in \boldsymbol{C}$ explicitly find all the subfields of $\boldsymbol{Q}[\varepsilon]$ (use Example 13) when $n =$:
  i) 9
  ii) 27
  iii) 81
  iv) 243

15. If L is a splitting field over $\boldsymbol{Q}$ of $x^{98} - 1 \in \boldsymbol{Q}[x]$, find $[L : \boldsymbol{Q}]$ and describe $Gal(L/\boldsymbol{Q})$. Determine the number of subfields E of L, explain which of these E are Galois over $\boldsymbol{Q}$, and find $Gal(E/\boldsymbol{Q})$.

16. Let L be a splitting field over $\boldsymbol{Q}$ of $(x^m - 1)(x^n - 1) \in \boldsymbol{Q}[x]$. Find $[L : \boldsymbol{Q}]$ and the explicit structure of $Gal(L/\boldsymbol{Q})$ (See Exercise 11.) when:
  i) $m = 5, n = 7$
  ii) $m = 6, n = 9$
  iii) $m = 12, n = 18$
  iv) $m = 24, n = 45$
  v) $m = 20, n = 28$

17. a) If $n = p_1 p_2 \cdots p_k$ for $p_1 < \cdots < p_k$ primes, show that $\mathbf{Z}_n \cong Gal(L/\mathbf{Q})$ for some $\mathbf{Q} \subseteq L$. Do not use Theorem 15.17 but instead find $m \in \mathbf{N}$ with $n \mid \varphi(m)$.

b) Use the same idea to show that for any odd n, $\mathbf{Z}_n \cong Gal(F/\mathbf{Q})$ for some $\mathbf{Q} \subseteq F$.

18. Let $p > 2$, $\varepsilon = e^{2\pi i/p} \in \mathbf{C}$, $L = \mathbf{Q}[\varepsilon]$, and $U_p = \langle [t]_p \rangle$.

i) Show that $Gal(L/\mathbf{Q}) = \langle \eta \rangle$ where $\eta(\varepsilon) = \varepsilon^t$.

ii) For $\theta = \varepsilon^{t^2} + \varepsilon^{t^4} + \cdots + \varepsilon^{t^{p-1}} \in L$, $[\mathbf{Q}[\theta] : \mathbf{Q}] = 2$. ($\theta$ is integral over $\mathbf{Z}$!)

iii) Show that $\prod_{i=1}^{p-1}(1 - \varepsilon^i) = p$.

iv) Show that $\prod_{i=1}^{p-1}(1 - \varepsilon^i) = (-1)^{(p-1)/2}\alpha^2 \prod_{i=1}^{(p-1)/2} \varepsilon^{-i}$ for $\alpha = \prod_{i=1}^{(p-1)/2}(1 - \varepsilon^i)$ and conclude, using iii), that $p^{1/2} \in L$ if $p \equiv_4 1$ and $(-p)^{1/2} \in L$ if $p \equiv_4 3$. (Note that $(\varepsilon^i)^{-1} = \varepsilon^{p-i}$ and that $\varepsilon = (\varepsilon^{(p+1)/2})^2$.)

15.5 SOLVABLE GROUPS

We study *solvable* groups, mentioned in §5.3. These groups are crucial for the seminal result of Galois Theory in §15.6 that relates the existence of formulas for roots of polynomials to the Galois group of a splitting field of the polynomial. Recall from §5.3 that the *commutator subgroup* G' of a group G is the subgroup generated by $\{x^{-1}y^{-1}xy \mid x, y \in G\}$. It follows from Theorem 2.19 that $G' \lhd G$ and, from Theorem 5.10, both that G/G' is Abelian and for $N \lhd G$, G/N is Abelian exactly when $G' \leq N$.

DEFINITION 15.7 For G any group, set $G^{(0)} = G$, $G^{(1)} = G'$ and for $k \geq 1$ define the $k+1$-st **commutator subgroup** of G to be $G^{(k+1)} = (G^{(k)})'$. Then G is **solvable of length k** if $G^{(k)} = \langle e \rangle$ with $k \geq 1$ and minimal.

Of course $G' = \langle e \rangle$ exactly when G is Abelian since $x^{-1}y^{-1}xy = e \Leftrightarrow xy = yx$. Thus the notion of a solvable group generalizes that of an Abelian group and offers the possibility of proof by induction on the solvable length of the group. We will not pursue this idea here, but a hint of how it can be used is given at the end of §5.3.

Examples of solvable, non-Abelian groups are the dihedral groups D_n, solvable of length 2 by Example 5.15, and the groups of upper triangular matrices $UT(n, F)$ over the field F, by the discussion at the end of §2.6 (see the exercises below). For $UT(n, F)$ the solvable length increases as n increases. The symmetric groups S_n are not solvable when $n \geq 5$ by Example 5.16 since then $S_n' = A_n$, the alternating group, and $A_n' = A_n$, so $S_n^{(k)} = A_n$ for all $k \geq 2$. It will be useful to record results about higher commutators of subgroups and of quotients.

THEOREM 15.18 Let G be a group, $H \leq G$, $N \lhd G$, and $k \geq 1$. Then:

i) $H^{(k)} \leq G^{(k)}$
ii) $H^{(s+t)} = (H^{(s)})^{(t)}$
iii) $N^{(k)} \lhd G$
iv) $(NH/N)^{(k)} = (NH^{(k)})/N$

Proof The definition of commutator subgroup shows that $H' \leq G'$, and easy inductions on k yield i) and ii). Any $\alpha \in N'$ has the form $\alpha = c_1 \cdots c_m$ for some

$c_j \in \{x^{-1}y^{-1}xy \mid x, y \in N\}$. If $g \in G$ then $g\alpha g^{-1} = T_g(\alpha)$ for the automorphism T_g, conjugation by g. Thus $g\alpha g^{-1} = T_g(c_1)\cdots T_g(c_m)$ and $N' \lhd G$ if each $T_g(c_j) = v^{-1}w^{-1}vw$ for some $v, w \in N$. Using $c_j = x^{-1}y^{-1}xy$, it follows that $T_g(c_j) = T_g(x)^{-1}T_g(y)^{-1}T_g(x)T_g(y)$ (Theorem 6.2!) with $T_g(x), T_g(y) \in N$ since $N \lhd G$. Hence $N' \lhd G$. If $N^{(k)} \lhd G$, our computation now shows that $N^{(k+1)} \lhd G$ so iii) holds by induction on k. For $x, y \in H$, $(Nx)^{-1}(Ny)^{-1}NxNy = Nx^{-1}y^{-1}xy$ in G/N and for $ch \in NH, Nch = Nh \in NH/N$, so $(NH/N)' = NH'/N$ (verify!). Using this, $(NH/N)^{(k)} = (N \cdot H^{(k)})/N$ implies $(NH/N)^{(k+1)} = ((NH/N)^{(k)})' = ((NH^{(k)}/N))' = N(H^{(k)})'/N = NH^{(k+1)}/N$, proving iv) by induction. ∎

We apply Theorem 15.18 to relate the solvability of a group to the solvability of its subgroups and quotient groups. Let G be solvable of length k. By part i) of the theorem, for any $H \leq G$, $H^{(k)} \leq G^{(k)} = \langle e \rangle$, so H is solvable of length at most k. When $N \lhd G$ then by part iv), $(G/N)^{(k)} = NG^{(k)}/N = N/N$, so G/N is solvable of length at most k. It is perhaps more interesting and useful to ask whether G must be solvable when it has a solvable normal subgroup with a solvable quotient.

THEOREM 15.19 Let G be a group, $H \leq G$, and $N \lhd G$. When G is solvable of length k, then H and G/N are solvable of length at most k. If N is solvable of length t and G/N is solvable of length s, then G is solvable of length at most $t + s$.

Proof Our comments above prove the first statement so assume that $N \lhd G$ with $N^{(t)} = \langle e \rangle$ and $(G/N)^{(s)} = \langle Ne \rangle$. From Theorem 15.18, $NG^{(s)}/N = (G/N)^{(s)} = \langle Ne \rangle$, forcing $G^{(s)} \leq N$. It follows that $G^{(s+t)} = (G^{(s)})^{(t)} \leq N^{(t)} = \langle e \rangle$, again by Theorem 15.18. Thus G is solvable of length at most $s + t$, as claimed. ∎

We illustrate Theorem 15.19 with a couple of examples. These could be attacked directly, but we prefer to see how the theorem is used.

Example 15.21 Any group G with $|G| = pq$ for p and q primes is solvable of length at most 2.

When $p = q$ then $|G| = p^2$ and G is Abelian by Theorem 7.14 or Theorem 8.9, so $G' = \langle e \rangle$. If $p < q$ then there is $\langle x \rangle = N \lhd G$ with $|N| = q$ from Sylow's Theorem or Theorem 8.15. Thus G/N is cyclic of order p using Theorem 4.9. Therefore both N and G/N are solvable of length 1 and Theorem 15.19 forces G to be solvable of length at most 2.

Example 15.22 Every p-group is solvable.

Recall from Definition 8.3 that G is a p-group when p is a prime and $|G| = p^n$. If $n \leq 2$ then G is Abelian by Theorem 8.9, so G is solvable. Assume that $n \geq 2$ and that any p-group of order p^m is solvable if $m < n$. We know $\langle e \rangle \neq Z(G) \leq G$ from Theorem 8.8 so $|G/Z(G)| = p^m$ for some $0 \leq m < n$. By the induction assumption $G/Z(G)$ is solvable, and $Z(G)$ is Abelian so solvable. Thus G must be solvable by Theorem 15.19.

If G is a group and is solvable of length $k \geq 1$ then each $G^{(j)} \lhd G$ by Theorem 15.18 and we have $\langle e \rangle = G^{(k)} \leq G^{(k-1)} \leq \cdots \leq G' \leq G$, a chain of normal subgroups of G with each quotient $G^{(j)}/G^{(j+1)}$ an Abelian group. This latter fact follows from $(G^{(j)}/G^{(j+1)})' = G^{(j+1)}(G^{(j)})'/G^{(j+1)} = G^{(j+1)}/G^{(j+1)}$, once again using Theorem 15.18. For finite groups, solvability is equivalent to two useful conditions on chains of subgroups.

THEOREM 15.20 If G is a finite group the following are equivalent:

i) G is solvable
ii) There are $\langle e \rangle = H_0 \leq H_1 \leq \cdots \leq H_m = G$ so that for $k \geq 1$, $H_{k-1} \lhd H_k$ and H_k/H_{k-1} is Abelian
iii) There are $\langle e \rangle = L_0 \leq L_1 \leq \cdots \leq L_m = G$ so that for $k \geq 1$, $L_{k-1} \lhd L_k$ and L_k/L_{k-1} is cyclic of prime order

Proof If G is solvable of length k then our comments above show that $H_j = G^{(k-j)}$ defines the required chain of subgroups, so i) implies ii). Assume that ii) holds. Since each H_k/H_{k-1} is a finite Abelian group, applying Theorem 7.5 yields a chain $\langle H_{k-1}e \rangle = B_0 \leq \cdots \leq B_s = H_k/H_{k-1}$ with each B_t/B_{t-1} of prime order. Now $B_i \lhd H_k/H_{k-1}$ since this quotient group is Abelian, so by the Correspondence Theorem (Theorem 7.3) each B_i corresponds to some $H_{k-1} \leq E_i \lhd H_k$, $B_i \cong E_i/H_{k-1}$, and $H_{k-1} = E_0 \leq E_1 \leq \cdots \leq E_s = H_k$. Also $E_{t-1} \lhd E_t$ and $E_t/E_{t-1} \cong B_t/B_{t-1}$ by either Theorem 7.3 or Theorem 7.6: this quotient is cyclic of prime order. Replace each inclusion $H_{k-1} \leq H_k$ with a chain $H_{k-1} \leq E_1 \leq \cdots \leq E_{s-1} \leq H_k$ to get the chain required in iii). Finally, if iii) holds, proceed by induction on the length m of the chain of subgroups. When $m = 1$, G is Abelian and so is solvable. Assume that when a group has a chain as in iii) of length $k < m$ then the group is solvable. Now $\langle e \rangle \leq L_1 \leq \cdots \leq L_{m-1}$ and $m - 1 < m$, so L_{m-1} is solvable by our induction assumption. Also $L_{m-1} \lhd L_m = G$ and G/L_{m-1} has prime order and so is cyclic and thus solvable. Using Theorem 15.19 we see that G must be solvable, completing the proof of the theorem. ■

The chain of subgroups in Theorem 5.20 need not be unique, as shown in §7.1. Different chains for $\mathbf{Z}_{18}$ are $\langle[1]\rangle \leq \langle[9]\rangle \leq \langle[3]\rangle \leq \mathbf{Z}_{18}$ and $\langle[1]\rangle \leq \langle[6]\rangle \leq \langle[2]\rangle \leq \mathbf{Z}_{18}$. In the non-Abelian dihedral group D_{15} the subgroups of the subgroup $\langle T_1 \rangle$ of rotations are all normal from Example 5.1 so a chain with prime order quotients is $\langle e \rangle \leq \langle T_5 \rangle \leq \langle T_1 \rangle \leq D_{15}$.

Example 15.23 A chain with prime order quotients in $G = UT(2, \mathbf{Z}_3)$.

The considerations here hold for any prime replacing 3. It is an exercise that $G' = \{A \in G \mid A_{11} = A_{22} = [1]\}$ and $H = \{A \in G \mid A_{11} \in \langle[2]\rangle, A_{22} = [1]\} \lhd G$. Now $\langle I_7 \rangle \leq G' \leq H \leq G$ satisfied $G' \cong \mathbf{Z}_3$, $H/G' \cong \mathbf{Z}_2$, and $G/H \cong \mathbf{Z}_2$).

Example 15.24 A chain for A_4 with prime order quotients but with not all subgroups normal.

Set $H = \{I, (1,2)(3,4), (1,3)(2,4), (1,4)(2,3)\} \lhd A_4$ and observe that $\langle I \rangle \leq \langle (1,2)(3,4) \rangle \leq H \leq A_4$, $\langle (1,2)(3,4) \rangle$ is not normal in A_4, each subgroup is normal in the next, $\langle (1,2)(3,4) \rangle \cong H/\langle (1,2)(3,4) \rangle \cong \mathbf{Z}_2$, and $A_4/H \cong \mathbf{Z}_3$.

Given fields $F \subseteq E \subseteq L$, since $Gal(L/E) \leq Gal(L/F)$, when $Gal(L/F)$ is solvable so is $Gal(L/E)$. Must $Gal(E/F)$ be solvable as well? We answer this when L is a splitting field over F, but first we need a general result about splitting fields.

THEOREM 15.21 Let $F \subseteq E \subseteq L$ be field with L a splitting field of $p(x) \in F[x]$ over F. Then:

i) Any $\varphi \in Gal(E/F)$ extends to some $\mu \in Gal(L/F)$.
ii) The minimal polynomial over F of $\gamma \in L$ splits over L.

Proof Certainly L is a splitting field over E of $p(x) \in E[x]$, and L is also a splitting field over E of $\varphi(p(x)) = p(x) \in E[x]$ for $\varphi \in Gal(E/F)$. Hence φ extends to some $\mu \in Gal(L/F)$ by Theorem 14.17. For ii), let $\gamma \in L$ with minimal polynomial $m(x) \in F[x]$, let $M \supseteq L$ be a splitting field for $m(x)$ over L, and let $\beta \in M$ with $m(\beta) = 0$. By Theorem 14.16 there is an isomorphism $\eta : F[\gamma] \to F[\beta]$ extending I_F and with $\eta(\gamma) = \beta$. Clearly L is a splitting field of $p(x)$ over $F[\gamma]$ since $\gamma \in L$, and $L[\beta]$ is a splitting field over $F[\beta]$ of $\eta(p(x)) = p(x) \in F[\beta][x]$. Therefore from Theorem 14.17 there is an isomorphism $\sigma : L \to L[\beta]$ extending η and a consequence is that $[L[\beta] : F] = [L : F]$. Using Theorem 14.8, $[L[\beta] : L][L : F] = [L : F]$ and we must conclude that $[L[\beta] : L] = 1$. Hence $\beta \in L$ and $m(x)$ splits over L. ∎

THEOREM 15.22 Let $F \subseteq E \subseteq L$ be fields with L a splitting field of $p(x) \in F[x]$ over F. Then $H = \{\mu \in Gal(L/F) \mid \mu(E) = E\} \leq Gal(L/F)\}$ and $H/E' \cong Gal(E/F)$, so $Gal(L/F)$ solvable implies $Gal(E/F)$ is solvable. When E is a splitting field over F then $H = Gal(L/F)$ so $Gal(E/F) \cong Gal(L/F)/E'$, and $Gal(L/F)$ is solvable if $E' = Gal(L/E)$ and $Gal(E/F)$ are both solvable.

Proof It is an exercise that $H \leq Gal(L/F)$ and each $\mu \in H$ restricts to an automorphism of E over F. In fact $\beta(\mu) = \mu|_E$ gives a group homomorphism $\beta : H \to Gal(E/F)$, and β is surjective by Theorem 15.21. Hence $Gal(E/F) \cong H/ker\ \beta$ by Theorem 6.6. Now $ker\ \beta = \{\mu \in H \mid \mu(\alpha) = \alpha$ for all $\alpha \in E\} = E' = Gal(L/E)$. Thus $Gal(E/F) \cong H/E'$, a quotient of a subgroup of $Gal(L/F)$, and so is solvable by Theorem 15.19 when $Gal(L/F)$ is.

If E is a splitting field over F, $\beta \in E$, and $\varphi \in Gal(L/F)$, then $\varphi(\beta)$ is a root of the minimal polynomial for β over F so Theorem 15.21 shows that $\varphi(\beta) \in E$. Therefore $Gal(L/F) = H$ follows, and the last paragraph shows that $Gal(E/F) \cong Gal(L/F)/E'$. From Theorem 15.19, $Gal(L/F)$ is solvable when both $Gal(E/F)$ and E' are, completing the proof. ∎

We show how the theorem is used and preview the content of the next section by proving a result on solvability in an important special case.

THEOREM 15.23 If L is a splitting field over F of $x^n - \alpha \in F[x]$, then $Gal(L/F)$ is solvable of length at most 2.

Proof If $\alpha = 0$ then $L = F$ and we are finished. When $\alpha \neq 0$ we claim that L is a splitting field of $g(x) = (x^n - \alpha)(x^n - 1) \in F[x]$ over F. Let $M \supseteq L$ be a splitting field of $g(x)$ over L and $\{\beta_1, \ldots, \beta_k\}$ the roots of $x^n - \alpha$ in L. Since $x^n - \alpha$ splits in L, each of its linear factors in L is some $x - \beta_j$, so $\{\beta_j\}$ are the only roots of $x^n - \alpha$ in M. For any $\varepsilon \in M$ with

$\varepsilon^n = 1$, $(\varepsilon\beta_1)^n = \varepsilon^n\beta_1^n = \alpha$. This shows that $\varepsilon\beta_1 \in M$ is a root of $x^n - \alpha$, forcing $\varepsilon\beta_1 = \beta_j$ for some j and so $\varepsilon = \beta_j/\beta_1 \in L$. Hence $M = L$ as claimed. Let $E \subseteq L$ be a splitting field over F of $x^n - 1$. By Theorem 15.22, $Gal(E/F) \cong Gal(L/F)/E'$ and applying Theorem 15.19 will finish the proof if both $Gal(E/F)$ and E' are Abelian. If $char\ F = p > 0$ write $n = p^t m$ with $p \nmid m$, and if $char\ F = 0$ let $m = n$ and $t = 0$. Using $x^n - 1 = (x^m - 1)^{p^t} \in F[x]$, E contains a primitive m-th root of unity ρ by Theorem 15.11 and $E = F[\rho]$. Any $\varphi \in Gal(E/F)$ is determined by $\varphi(\rho) = \rho^i$ for $(i, m) = 1$ and it follows that $Gal(E/F)$ is Abelian (verify!). Now $L = E[\{\beta_j\}]$ and since $(\beta_j/\beta_1)^n = 1$, $\{\beta_j\} = \{\rho^i\beta_1 \mid 1 \le i < m\}$ so $L = E[\beta_1]$. Therefore each $\eta \in Gal(L/E) = E'$ is uniquely determined by $\eta(\beta_1) = \rho^i\beta_1$. Since $\eta(\rho) = \rho$, an easy computation shows that $Gal(L/E)$ is Abelian. ■

EXERCISES

In all the exercises below, G and H are groups.

1. If $G \ne \langle e\rangle$ is a finite solvable group, use $G' \ne G$ to show that there is $H \lhd G$ with $[G : H] = p$, a prime.
2. Show that $UT(2, \mathbf{Z})$ (Example 2.6) is solvable.
3. Find a chain of subgroups $\langle e\rangle = H_0 \le \cdots \le H_m = G$ with $H_{i-1} \lhd H_i$ and $[H_i : H_{i-1}]$ a prime when $G =$:
   i) $\mathbf{Z}_{28}$
   ii) $\mathbf{Z}_{72}$
   iii) D_9
   iv) D_{24}
   v) S_4
   vi) $\mathbf{Z}_4 \oplus \mathbf{Z}_6$
   vii) $\mathbf{Z}_6 \oplus D_6$
   viii) $UT(2, \mathbf{Z}_{10})$
   ix) $UT(2, \mathbf{Z}_{12})$
   x) $UT(3, \mathbf{Z}_5)$
   xi) $UT(4, \mathbf{Z}_2)$
4. **a)** Show that G and H are solvable $\Leftrightarrow G \oplus H$ is solvable.
   b) Show that $G_1, \ldots, G_m$ are solvable groups $\Leftrightarrow G_1 \oplus \cdots \oplus G_m$ is a solvable group.
5. If $p < q < r$ are primes, show that any G of the following order is solvable:
   i) pqr
   ii) p^2q
   iii) p^2q^2
   iv) p^iq^j when $i < o([p]_q)$ in U_q
6. If $|G| = p^3$ for p a prime, show that $G^{(2)} = \langle e\rangle$.
7. If G is solvable and $\langle e\rangle \ne H \le G$ is a simple group, show that $H \cong \mathbf{Z}_p$, p a prime.
8. Show that G is solvable $\Leftrightarrow Inn(G)$ is solvable (see §6.5).
9. If $G = UT(n, F)$ for F a field, show that $G^{(n)} = \langle e\rangle$. (Show that $G' \subseteq \{A \in G \mid A_{jj} = 1 \text{ for all } j\}$ and note that $A \in G' \Rightarrow A = I_n + X$ with $X^n = 0_n$, so $A^{-1} = I_n - X + X^2 - \cdots$.)
10. If G is finite and not solvable, show that there is $\langle e\rangle \ne N \lhd G$ with $N' = N$.

11. For G finite, show that G is solvable if each of its Sylow subgroups is normal.
12. If G is finite and solvable then any maximal $N \lhd G$ has prime index in G.
13. If $H, K \leq G$, $K \lhd G$, and both H and K are solvable, show that HK is solvable.
14. Any finite G must contain a unique maximal $M \lhd G$ with M solvable. (Use Exercise 13.)
15. Let L be a Galois extension of F. Show that there is $E \subseteq L$ with E Galois over F, $Gal(L/E)$ is solvable, and $E \subseteq K$ whenever $K \subseteq L$, K is Galois over F, and $Gal(L/K)$ is solvable. (Exercise 14!)
16. If $H, K \lhd G$ so that G/H and G/K are solvable, show that $G/(H \cap K)$ is solvable.
17. If G is finite it has a unique minimal normal subgroup N with G/N solvable. (See Exercise 16.)
18. Let L be a Galois extenion of F. Show that there is $E \subseteq L$, Galois over F so that $Gal(E/F)$ is solvable, and E contains any other subfield $K \subseteq L$ that is Galois over F with $Gal(K/F)$ solvable. (Use Exercise 17.)
19. If $F \subseteq L$ are fields, p is a prime, and $[L : F] = p^m$, show that $Gal(L/F)$ is solvable. (Note that $|Gal(L/F)| = [L : F'']$.)
20. If G is finite and solvable, show that any nonidentity, minimal normal subgroup (with respect to inclusion) is isomorphic to $\boldsymbol{Z}_p \oplus \cdots \oplus \boldsymbol{Z}_p$ for p a prime.
21. Show that $GL(n, F)$ is not solvable if $n \geq 3$ or $|F| \geq 4$ (see §5.4).
22. If $K \leq G$ is maximal and proper, and if G is finite and solvable, show that $[G : K] = p^m$ for p a prime. (Use induction on the solvable length of G. If $H = G^{(n-1)} \neq \langle e \rangle$, $G^{(n)} = \langle e \rangle$, and H is not in K, consider KP_i for each Sylow subgroup P_i of H.)

15.6 RADICAL EXTENSIONS

We come to the question that motivated the development of Galois Theory: when is some root of an irreducible $p(x) \in F[x]$ an algebraic expression in the coefficients of $p(x)$? An "algebraic expression" is one that involves only the usual operations of addition, subtraction, multiplication, and division, together with taking roots. Any such expression is built from a finite number of these basic operations and so can be viewed as an element in an extension $L = F[\alpha_1, \ldots, \alpha_m]$ where for integers $k(j) \in \boldsymbol{N}$ $\alpha_1^{k(1)} \in F$, $\alpha_2^{k(2)} \in F[\alpha_1]$, and so on. For example, $\gamma = (2^{1/3} + 3^{2/5})^{1/7}/(4 - 5^{4/9})^{1/2} \in \boldsymbol{C}$ is an algebraic expression over, or starting with, $\boldsymbol{Q}$ and $\gamma \in L$ for

$$L = \boldsymbol{Q}\left[2^{1/3}, 9^{1/5}, \left(2^{1/3} + 3^{2/5}\right)^{1/7}, 625^{1/9}, \left(4 - 5^{4/9}\right)^{1/2}\right].$$

Note that $(2^{1/3})^3 = 2 \in \boldsymbol{Q}$, $(9^{1/5})^5 \in \boldsymbol{Q}[2^{1/3}]$, $((2^{1/3} + 3^{2/5})^{1/7})^7 \in \boldsymbol{Q}[2^{1/3}, 9^{1/5}]$, $(625^{1/9})^9 \in \boldsymbol{Q} \subseteq \boldsymbol{Q}[2^{1/3}, 9^{1/5}, (2^{1/3} + 3^{2/5})^{1/7}]$, and $((4 - 5^{4/9})^{1/2})^2 \in \boldsymbol{Q}[5^{4/9}] \subseteq \boldsymbol{Q}[2^{1/3}, 9^{1/5}, (2^{1/3} + 3^{2/5})^{1/7}, 625^{1/9}]$. Since L is an algebraic extension of $\boldsymbol{Q}$, γ is the root of some polynomial over $\boldsymbol{Q}$.

DEFINITION 15.8 For $F \subseteq L$ fields, L is a **radical extension** of F if $L = F[\alpha_1, \ldots, \alpha_m]$ and there are $k(j) \in \boldsymbol{N}$ for $1 \leq j \leq m$, so that $\alpha_1^{k(1)} \in F$ and for $1 < j$, $\alpha_j^{k(j)} \in F[\alpha_1, \ldots, \alpha_{j-1}]$.

From our comments above, if $p(x) \in F[x]$, L is a splitting field for $p(x)$ over F, and a root γ of $p(x)$ in L is an algebraic expression in the coefficients of $p(x)$, then γ is contained in a radical extension of F. We will study conditions necessary for a root of $p(x)$ to lie in a radical extension of F.

Some useful observations about radical extensions are immediate from the definition. Clearly any radical extension of F is algebraic over F. If $F \subseteq E \subseteq L$, E is a radical extension of F, and L is a radical extension of E, then L is a radical extension of F. To see this, use Definition 15.8 to write $E = F[\alpha_1, \ldots, \alpha_m]$ and $L = E[\beta_1, \ldots, \beta_s]$, so $L = F[\alpha_1, \ldots, \alpha_m, \beta_1, \ldots, \beta_s]$ shows that L is a radical extension of F. Finally, we may take the integers $k(j)$ in Definition 15.8 to be primes. For example if $L = F[\alpha]$ with $\alpha^n \in F$ then write $n = p_1 \cdots p_t$ for primes p_j. If follows that $L = F[\alpha^{p_1 \cdots p_{t-1}}, \alpha^{p_1 \cdots p_{t-2}}, \ldots, \alpha]$ also expresses L as a radical extension of F, and all "powers" $k(j)$ are primes.

Our first goal is to show that if L is a radical extension of F then *Gal*(L/F) must be solvable. To this end we need a result that will enable us to assume that L is also a splitting field over F.

THEOREM 15.24 If L is a radical extension of F, then there is $M \supseteq L$ so that M is a splitting field over F and a radical extension of F.

Proof Let $L = F[\alpha_1, \ldots, \alpha_m]$ express L as a radical extension of F, let $p_j(x) \in F[x]$ be the minimal polynomial for α_j over F, and let M be a splitting field over L of $f(x) = p_1(x) \cdots p_m(x) \in L[x]$. Since $\{\alpha_j\} \subseteq L$, M is also a splitting field over F. If $\beta_j \in M$ is any root of $p_j(x)$ then Theorem 14.18 shows that there is $\varphi \in Gal(M/F)$ with $\varphi(\alpha_j) = \beta_j$. Thus $\beta_j \in \varphi(L) = F[\varphi(\alpha_1), \ldots, \varphi(\alpha_m)]$ and it follows that M is generated by the subfields $\{\varphi(L) \mid \varphi \in Gal(M/F)\}$. It is an exercise that each $\varphi(L)$ is a radical extension of F and by our remarks above, the join of these (Theorem 14.10) is also a radical extension of F. That is, if $Gal(M/F) = \{\varphi_1, \ldots, \varphi_t\}$ then by Theorem 14.10,

$$\varphi_1(L) \vee \varphi_2(L) = F[\varphi_1(\alpha_1), \ldots, \varphi_1(\alpha_m), \varphi_2(\alpha_1), \ldots, \varphi_2(\alpha_m)]$$

is clearly a radical extension of F. In this way, adjoining one $\varphi_j(L)$ at a time leads to M, a radical extension of F as required. ∎

THEOREM 15.25 If L is a radical extension of F, then *Gal*(L/F) is solvable.

Proof By Theorem 15.24 there is a splitting field M over F that is a radical extension of F and $F \subseteq L \subseteq M$. If $Gal(M/F)$ is solvable then Theorem 15.22 shows that $Gal(L/F)$ is solvable, so we may assume that L is a splitting field over F. Write $L = F[\alpha_1, \ldots, \alpha_m]$ where $\alpha_1^k \in F$ and $\alpha_j^{k(j)} \in F[\alpha_1, \ldots, \alpha_{j-1}]$ for $j > 1$. We use induction on m and let $V \supseteq L$ be a splitting field of $x^k - \alpha_1^k \in F[x]$ over $L : V = L[S]$ for S the set of all roots of $x^k - \alpha_1^k$ in V. Clearly V is a radical extension of L, and so of F by our earlier comments. If $E = F[S]$ then E is a splitting field over F of $x^k - \alpha_1^k$ and by Theorem 15.23, $Gal(E/F)$ is solvable. When $m = 1$ then $L \subseteq E$ and Theorem 15.22 shows $Gal(L/F)$ is solvable. If $m > 1$ then since $\alpha_1 \in E$, $V = E[\alpha_2, \ldots, \alpha_m]$ and is a radical extension of E and a splitting field over F and so over E. To see that V is a splitting field over F, when L is a splitting field of $g(x) \in F[x]$ over F then V is a splitting field of $g(x)(x^k - \alpha_1^k) \in F[x]$ over F, so both E and V are splitting fields over F. Our inductive assumption now implies that $Gal(V/E)$ is solvable. Since $F \subseteq E \subseteq V$, Theorem 15.22 shows first that $Gal(V/F)$ is solvable and then that $Gal(L/F)$ is solvable, proving the theorem by induction. ∎

Next we see how to use Theorem 15.25 to shed light on the question of whether $f(x) \in F[x]$ has a root in some radical extension of F, at least when $f(x)$ is irreducible.

THEOREM 15.26 If $p(x) \in F[x]$ is irreducible with a root in some radical extension of F, and L is a splitting field of $p(x)$ over F, then $Gal(L/F)$ is solvable.

Proof Let E be a radical extension of F and $\alpha \in E$ a root of $p(x)$. By Theorem 15.24 there is a field $M \supseteq E$ that is a radical extension of F and a splitting field over F. Now Theorem 15.25 shows that $Gal(M/F)$ is solvable, and by Theorem 15.21, M contains a splitting field K of $p(x)$ over F since $\alpha \in M$. Also $Gal(K/F)$ is solvable using Theorem 15.22 again. Finally, $L \cong K$ from Theorem 14.17, so $Gal(L/F) \cong Gal(K/F)$ is solvable (exercise!). ■

We can use Theorem 15.26 negatively. That is, if $f(x) \in F[x]$ is irreducible, L is a splitting field for $f(x)$ over F, and $Gal(L/F)$ is not solvable, then no root of $f(x)$ can lie in a radical extension of F. Thus no algebraic expression in the coefficients of $f(x)$ is a root. When $F = \boldsymbol{Q}$ we will see that the converse of Theorem 15.26 holds: $Gal(L/\boldsymbol{Q})$ solvable implies that all roots of $f(x)$ lie in some radical extension of $\boldsymbol{Q}$.

Whether a root of a particular $f(x) \in F[x]$ lies in a radical extension of F depends on F and on $f(x)$. When $f(x)$ is not irreducible it may well have linear factors and hence roots in F itself. If F_q is a finite field with q elements then any finite field extension $F_q \subseteq L$ is a radical extension, namely $L = F_q[\theta]$ for $(L^*, \cdot) = \langle\theta\rangle$ (Theorem 4.20) with $\theta^{|L|-1} = 1 \in F_q$. Thus any root of any $f(x) \in F_q[x]$ lies in a radical extension of F_q. It is usually difficult to determine the Galois group G of a splitting field of a given polynomial and hence to tell whether G is solvable. We have already done this for some irreducible polynomials of prime degree over $\boldsymbol{Q}$ in Example 15.9.

Example 15.25 If $p \geq 5$ is prime there is no fixed algebraic expression $A(q_{p-1}, \ldots, q_0)$ in the coefficients of an arbitrary $x^p + q_{p-1}x^{p-1} + \cdots + q_1x + q_0 \in \boldsymbol{Q}[x]$ that gives a root of every monic $f(x) \in \boldsymbol{Q}[x]$ of degree p.

As we just observed, it suffices to find some $g(x) \in \boldsymbol{Q}[x]$ of degree p with splitting field L over $\boldsymbol{Q}$ so that $Gal(L/\boldsymbol{Q})$ is not solvable. As we mentioned in §15.5, the symmetric groups S_n are not solvable when $n \geq 5$. Hence by Example 15.8 and Example 15.9 there is $g_p(x) \in \boldsymbol{Q}[x]$, irreducible of degree p, so that if L is a splitting field of $g_p(x)$ over $\boldsymbol{Q}$ then $Gal(L/\boldsymbol{Q}) \cong S_p$.

The result in Example 15.25 holds for any $n \geq 5$, and most of these cases follow easily using this example. The more difficult cases to verify are $n = 6$, 8, and 9, and we omit these.

Example 15.26 If $n \geq 10$ there is no fixed algebraic expression, say $h(q_{n-1}, \ldots, q_0)$, in the coefficients of an arbitrary $x^n + q_{n-1}x^{n-1} + \cdots + q_0 \in \boldsymbol{Q}[x]$ that gives a root of every monic $f(x) \in \boldsymbol{Q}[x]$ of degree n.

For each irreducible polynomial $g_p(x) \in \boldsymbol{Q}[x]$ from Example 15.25, and splitting field L over $\boldsymbol{Q}$ for it, $Gal(L/\boldsymbol{Q}) \cong S_p$ is not solvable, so $g_p(x)$ has no root in a radical extension of $\mathbf{Q}$ by Theorem 15.26. If $g_{10}(x)=g_5(x)^2$ then any root of $g_{10}(x)$ is a root of $g_5(x)$. Thus no root of $g_{10}(x)$ lies in a radical extension of $\mathbf{Q}$; there is no fixed algebraic expression in the coefficients of an arbitrary monic polynomial of degree 10 over $\mathbf{Q}[x]$ giving a root of the polynomial. A similar analysis shows that $g_{12}(x)=g_5(x)g_7(x)$ and $g_{14}(x)=g_7(x)^2$ have no roots in a radical extension of $\mathbf{Q}$, so there is no algebraic expression giving a root for monic polynomials of degree 12 or of degree 14. The same holds for degree 11 or degree 13 by Example 15.25. We now have $g_{10}(x), g_{11}(x), \ldots, g_{14}(x) \in \mathbf{Q}[x]$ with no roots in a radical extension of $\mathbf{Q}$. For $n \geq 15$ and not a prime, consider $g_n(x)=g_{n-5}(x)g_5(x)$. Any root of $g_n(x)$ is a root of either $g_5(x)$ or of $g_{n-5}(x)$, so by induction, $g_n(x)$ cannot have a root in a radical extension of $\mathbf{Q}$, verifying the example.

Example 15.18 shows that there is a Galois extension L of $\mathbf{Q}$ with $[L:\mathbf{Q}]=3$. It follows that $Gal(L/\mathbf{Q}) \cong \mathbf{Z}_3$ and so is solvable, but L is not a radical extension of $\mathbf{Q}$. Otherwise $L=\mathbf{Q}[\alpha]$ with $\alpha^3 \in \mathbf{Q}$, and L Galois over $\mathbf{Q}$ would force $g(x)=x^3-\alpha^3 \in \mathbf{Q}[x]$ to split over L by Theorem 15.2. If $\beta \in L$ is a second root of $g(x)$ then $\varepsilon=\alpha/\beta \in L$ is a primitive 3-rd root of unity. This is impossible since $[\mathbf{Q}[\varepsilon]:\mathbf{Q}]=2$ and $\mathbf{Q}[\varepsilon] \subseteq \mathbf{Q}[\alpha]$. Although L is not a radical extension of $\mathbf{Q}$, from Example 15.18, $L \subseteq \mathbf{Q}[\rho]$ for ρ a 7-th root of unity, so L embeds in a radical extension of $\mathbf{Q}$.

Could every field extension $F \subseteq L$ with $Gal(L/F)$ solvable embed in a radical extension of F? This would be a partial converse of Theorem 15.25. The answer is actually no, but we need a result first for *char* $F=p$.

THEOREM 15.27 Let $F \supseteq \mathbf{Z}_p$ be a field containing all roots of unity over $\mathbf{Z}_p$. If K is a radical extension of F, then $p \nmid |Gal(K/F)|$.

Proof Using Theorem 15.24 we may assume that K is a splitting field over F since Theorem 15.22 shows that $|Gal(K/F)| \mid |Gal(M/F)|$ for any splitting field M over F containing K. Let $K=F[\beta_1,\ldots,\beta_m]$ express K as a radical extension of F. We may assume all $\beta_j \notin F$, $\beta_1^q \in F$ and for $j>1$, $\beta_j^{k(j)} \in F[\beta_1,\ldots,\beta_{j-1}]$ with q and all $k(j)$ primes. If $q=p$ then $Y^p-\beta_1^p=(Y-\beta_1)^p \in K[Y]$ so β_1 is the only root of $Y^p=\beta_1^p$ in K, and $\varphi(\beta_1)=\beta_1$ for any $\varphi \in Gal(K/F)$ follows. Thus $Gal(K/F)=Gal(K/F[\beta_1])$, and by induction on m, $p \nmid |Gal(K/F)|$. When $q \neq p$ and $\varepsilon \in F$ is a primitive q-th root of unity, then $\varepsilon^i\beta_1$ are q distinct roots of $Y^q-\beta_1^q$ in K so $F[\beta_1]$ is a splitting field over F. If $m(x) \in F[x]$ is the minimal polynomial for β_1 over F then $m(x) \mid (Y^q-\beta_1^q)$, and if $\varepsilon^i\beta_1 \neq \beta_1$ is another root of $m(x)$ then Theorem 14.18 yields $\eta \in Gal(F[\beta_1]/F)$ with $\eta(\beta_1)=\varepsilon^i\beta_1$. Since q is a prime it follows that $o(\eta)=q$, so $m(x)$ has q distinct roots and $Gal(F[\beta_1]/F)=\langle\eta\rangle \cong \mathbf{Z}_q$ (why?). Induction on m shows that $p \nmid |Gal(K/F[\beta_1])|$. Since K and $F[\beta_1]$ are splitting fields over F, Theorem 15.22 gives the desired conclusion, $p \nmid |Gal(K/F)|$. Thus the theorem holds by induction. ∎

There are fields F as described in Theorem 15.27. We can take an algebraic closure of $\mathbf{Z}_p$ or, equivalently, a splitting field over $\mathbf{Z}_p$ of $\{x^n-1 \in \mathbf{Z}_p[x] \mid n \geq 1\}$. Theorem 15.27 yields an interesting example.

Example 15.27 There is a Galois extension $E \subseteq L$ with $Gal(L/E)$ solvable, but L is not contained in any radical extension of E.

Let $F \supseteq \mathbf{Z}_p$ be a field containing all roots of unity over $\mathbf{Z}_p$ and set $E = F(x)$, the quotient field of $F[x]$. Now E contains all roots of unity over $\mathbf{Z}_p$ so Theorem 15.27 applies to E. Let L be a splitting field over E of $g(Y) = Y^p - Y - x \in E[Y]$ and $\alpha \in L$ a root of $g(Y)$. It is an exercise that $\alpha \notin E$ and that $\alpha, \alpha+1, \ldots, \alpha+p-1$ are p distinct roots of $g(Y)$ in L, so $L = E[\alpha]$ and L is Galois over E by Theorem 15.2. If α has minimal polynomial $m(Y) \in E[Y]$ over E, then $m(Y) \mid g(Y)$, and if $\alpha + i$ is another root of $m(x)$, then there is $\varphi \in Gal(L/E)$ with $\varphi(\alpha) = \alpha + i$ by Theorem 14.18. Since $\varphi^m(\alpha) = \alpha + im$ and $1 \leq i \leq p-1$, it follows that $o(\varphi) = p$. Using Theorem 15.4, $|Gal(L/E)| = [L : E] \leq p$ and we conclude that $\mathbf{Z}_p \cong \langle \varphi \rangle = Gal(L/E)$. If L embeds in a radical extension V of E we may assume that V is a splitting field over E by Theorem 15.24. Now Theorem 15.22 shows that $|Gal(L/E)| \mid |Gal(V/E)|$, a contradiction since $p = |Gal(L/E)|$ but $p \nmid |Gal(V/E)|$ by Theorem 15.27.

Let $F \subseteq L$ be fields with $[L : F]$ finite. Example 15.27 shows that to embed L in a radical extension of F we must avoid $[L : F] = p = char\ F$. Also, to show that L embeds in a radical extension of F when $Gal(L/F)$ is solvable we need to assume that L is a splitting field over F. Let us see what can happen when L is not a splitting field over F. From Example 15.8 and Example 15.9 there is an irreducible $g(x) \in \mathbf{Q}[x]$ with $deg\ g = 5$ and $Gal(L/\mathbf{Q}) \cong S_5$ for L a splitting field of $g(x)$ over $\mathbf{Q}$. If $\alpha \in L$ is a root of $g(x)$ then $[\mathbf{Q}[\alpha] : \mathbf{Q}] = 5$ so $Gal(\mathbf{Q}[\alpha]/\mathbf{Q})$ is solvable, actually trivial since $\mathbf{Q}[\alpha]$ can contain no root of $g(x)$ other than α: otherwise $[L : \mathbf{Q}] < 5!$. However $\mathbf{Q}[\alpha]$ does not embed in a radical extension of $\mathbf{Q}$ by Theorem 15.26. Thus we need the restrictions in the statement of our final theorem.

THEOREM 15.28 Let L be a splitting field of $f(x) \in F[x]$ over F. If $Gal(L/F)$ is solvable, and $char\ F = 0$ or $char\ F = p > 0$ and $p \nmid [L : F]$, then L embeds in a radical extension of F.

Proof There is nothing to prove if $L = F$, and otherwise $|Gal(L/F)| > 1$ by Theorem 14.18. We claim that L is a Galois extension of F. Let $\gamma \in L$ with minimal polynomial $m_\gamma(x) \in F[x]$. Since $[F[\gamma] : F] \mid [L : F]$ and $deg\ m_\gamma = [F[\gamma] : F]$, when $char\ F = p > 0$, $p \nmid deg\ m_\gamma$ so $m_\gamma(x)$ is separable over F and splits over L by Theorem 15.21. Thus L is Galois over F by Theorem 15.2, and since now $|Gal(L/F)| = [L : F]$ from Theorem 15.4, $p \nmid |Gal(L/F)|$ when $char\ F = p > 0$.

If $G = Gal(L/F)$, then $G \neq \langle I_F \rangle$ and G solvable imply that G/G' is a nontrivial Abelian group. If q is a prime dividing $|G/G'|$ then G/G' contains a (normal) subgroup K of index q by Theorem 7.5. Note that $q \neq char\ F$. The Correspondence Theorem now yields $H \lhd G$ of index q. Let $H' = E \subseteq L$. From Theorem 15.7, E is Galois over F, $[E : F] = q \neq char\ F$, and $Gal(E/F) \cong G/Gal(L/E)$. Since $char\ F = 0$ or $char\ F = p \neq q$, there is a primitive q-th root of unity ε so that $L[\varepsilon]$ is a splitting field of $x^q - 1$ over L. Now L is a splitting field of $f(x) \in F[x]$ over F so $L[\varepsilon]$ is a splitting field of $f(x)(x^q - 1) \in F[x]$

over F and this polynomial has distinct roots in $L[\varepsilon]$ so $L[\varepsilon]$ is Galois over F, using Theorem 15.2 again. Similarly, since E is Galois over F, $E[\varepsilon]$ is Galois over F and so over $F[\varepsilon]$ (Theorem 15.3). By Theorem 15.22 and our assumption that $Gal(L/F)$ is solvable, $Gal(L[\varepsilon]/F)$ is solvable since $Gal(L[\varepsilon]/L)$ is solvable: it is Abelian since each of its automorphisms θ is determined by $\theta(\varepsilon)=\varepsilon^j$. Clearly $L[\varepsilon]$ is a radical extension of L so it suffices to show that $L[\varepsilon]$ embeds in a radical extension of F.

Let $m_\varepsilon(x) \in L[x]$ be the minimal polynomial for ε over L and $w_\varepsilon(x)$ its minimal polynomial over E. since $E[\varepsilon]$ is Galois over F, $w_\varepsilon(x)$ splits in $E[\varepsilon]$ and divides x^q-1 over E (why?), so all its roots are primitive q-th roots of unity and each generates $L[\varepsilon]$ over L. Thus $[L[\varepsilon]:L]$ is the degree of each irreducible factor of $w_\varepsilon(x)$ in $L[x]$, so $[L[\varepsilon]:L]=deg\, m_\varepsilon$ and also $[L[\varepsilon]:L]=deg\, m_\varepsilon$ divides $deg\, w_\varepsilon=[E[\varepsilon]:E]$. From Theorem 14.8, $[L[\varepsilon]:L][L:E]=[L[\varepsilon]:E[\varepsilon]][E[\varepsilon]:E]$ and we conclude that $[L[\varepsilon]:E[\varepsilon]] \mid [L:E]$. Now $[L:E] \mid [L:F]$ so $p \nmid [L[\varepsilon]:E[\varepsilon]]$ when $char\ F=p$. Also $[L[\varepsilon]:E[\varepsilon]] \le [L:E] < [L:F]$, and $Gal(L[\varepsilon]/E[\varepsilon]) \le Gal(L[\varepsilon]/F)$, which is solvable from above. Applying induction on $[L:F]$ shows that $L[\varepsilon]$ embeds in a radical extension V of $E[\varepsilon]$. To finish the proof we need only show that $E[\varepsilon]$ is a radical extension of F since then V is a radical extension of F.

Clearly $F[\varepsilon]$ is a radical extension of F, so to show that $E[\varepsilon]$ is a radical extension of F, it suffices to show that $E[\varepsilon]$ is a radical extension of $F[\varepsilon]$. The computation of the last paragraph for $E \subseteq L$ works as well for $F \subseteq E$ to show that $[E[\varepsilon]:F[\varepsilon]] \mid [E:F]$. Now $[E:F]=q$ so either $E[\varepsilon]=F[\varepsilon]$ and $E[\varepsilon]$ is trivially a radical extension of $F[\varepsilon]$, or $[E[\varepsilon]:F[\varepsilon]]=q$, which we may assume. Since $E[\varepsilon]$ is Galois over $F[\varepsilon]$, $Gal(E[\varepsilon]/F[\varepsilon])=\langle\varphi\rangle$ of order q from Theorem 15.4.

If there are $\alpha_0,\ldots,\alpha_{q-1} \in E[\varepsilon]$, not all zero, so that $\alpha_0\gamma+\alpha_1\varphi(\gamma)+\cdots+\alpha_{q-1}\varphi^{q-1}(\gamma)=0$ for all $\gamma \in E[\varepsilon]$, choose such an expression $J(\gamma)$ with a minimal number of $\alpha_i \ne 0$. Clearly in $J(\gamma)$ there are at least two nonzero α_i. If $\alpha_j \ne 0$ then $\varphi^{q-j}(J(\gamma))=\varphi^{q-j}(\alpha_j)\gamma+\varphi^{q-j}(\alpha_{j+1})\varphi(\gamma)+\cdots$, using $\varphi^q=I_{E[\varepsilon]}$. Thus we may assume that $J(\gamma)=\alpha_0\gamma+\alpha_1\varphi(\gamma)+\cdots+\alpha_{q-1}\varphi^{q-1}(\gamma)=0$ with α_0 and another α_j not zero. For any γ, $\upsilon \in E[\varepsilon]$,

$$0=\gamma J(\upsilon)-J(\gamma\upsilon)=\alpha_1(\gamma-\varphi(\gamma))\varphi(\upsilon)+\cdots+\alpha_{q-1}(\gamma-\varphi^{q-1}(\gamma))\varphi^{q-1}(\upsilon).$$

Since some $\alpha_j \ne 0$ with $j>0$, $\langle\varphi\rangle=\langle\varphi^j\rangle$, and $E[\varepsilon]$ is Galois over $F[\varepsilon]$, there is $\mu \in E[\varepsilon]$ so that $\varphi^j(\mu) \ne \mu$. This forces $\alpha_j(\mu-\varphi^j(\mu)) \ne 0$ but for all $\upsilon \in E[\varepsilon]$, $\mu J(\upsilon)-J(\mu\upsilon)=0$, and this contradicts the choice of $J(\upsilon)$ as having a minimal number of nonzero coefficients. Hence whenever $\alpha_0\gamma+\alpha_1\varphi(\gamma)+\cdots+\alpha_{q-1}\varphi^{q-1}(\gamma)=0$ for all $\gamma \in E[\varepsilon]$, with all $\alpha_i \in E[\varepsilon]$, we must have all $\alpha_i=0$. In particular, there is some $\gamma \in E[\varepsilon]$ with $\beta=\varepsilon\gamma+\varepsilon^2\varphi(\gamma)+\cdots+\varepsilon^{q-1}\varphi^{q-2}(\gamma)+\varphi^{q-1}(\gamma) \ne 0$. Note that $\varphi(\beta)=\varepsilon\varphi(\gamma)+\varepsilon^2\varphi^2(\gamma)+\cdots+\varepsilon^{q-1}\varphi^{q-1}(\gamma)+\gamma=\varepsilon^{-1}\beta$. Thus $\beta \notin F[\varepsilon]=\langle\varphi\rangle'$ so $E[\varepsilon]=F[\varepsilon][\beta]$, using $[E[\varepsilon]:F[\varepsilon]]=q$ is prime. Also $\varphi(\beta^q)=(\varphi(\beta))^q=(\varepsilon^{-1})^q\beta^q=\beta^q$. This means that $\beta^q \in \langle\varphi\rangle'=F[\varepsilon]$, and $E[\varepsilon]$ is a radical extension of $F[\varepsilon]$, finally completing the proof. ■

Let $f(x) \in F[x]$ have degree n and let L be a splitting field for $f(x)$ over F. Recall that every $\varphi \in Gal(L/F)$ permutes the roots of $f(x)$ and is determined by this permutation, so $Gal(L/F) \cong H \le S_n$, the symmetric group. By Example 5.16, S_n is not solvable when $n \ge 5$, but otherwise S_n, so $Gal(L/F)$, is solvable. Specifically, $S_2 \cong \mathbf{Z}_2$ is Abelian, and

$\langle(1,2,3)\rangle \lhd S_3$ is Abelian with $S_3/\langle(1,2,3)\rangle \cong \mathbf{Z}_2$ so $S_3^{(2)} = \langle I\rangle$. It is useful to record the following fact about S_4.

Example 15.28 $S_4^{(3)} = \langle I\rangle$, so S_4 is solvable.

It is an exercise that $S_4' = A_4$, and as we have seen before, $H = \{I, (1,2)(3,4), (1,3)(2,4), (1,4)(2,3)\} \lhd A_4$ and is Abelian with $A_4/H \cong \mathbf{Z}_3$. Hence $A_4' \le H$, and since no proper subgroup of H of order 2 is normal in A_4, $S_4^{(2)} = A_4' = H$. Now H Abelian forces $S_4^{(3)} = \langle I\rangle$.

As a consequence of these observations, we can show that certain splitting fields do lie in radial extensions.

Example 15.29 Let $p(x) = x^4 + 4x^3 - 12x + 2$, let $q(x) = x^4 + 9x + 3 \in \mathbf{Q}[x]$, and let L be a splitting field for $p(x)q(x)$ over $\mathbf{Q}$. Then L embeds in a radical extension of $\mathbf{Q}$.

Using Theorem 15.28, it suffices to show that $Gal(L/\mathbf{Q})$ is solvable. Let $E \subseteq L$ be a splitting field of $p(x)$ over $\mathbf{Q}$. Then E and L are Galois over $\mathbf{Q}$ so $Gal(E/\mathbf{Q}) \cong Gal(L/\mathbf{Q})/Gal(L/E)$ (why?). Since E is a splitting field of $p(x)$ over $\mathbf{Q}$, L is a splitting field of $q(x)$ over E, and up to isomorphism $Gal(E/\mathbf{Q}), Gal(L/E) \le S_4$, each of these is solvable by Example 15.28. Thus $Gal(L/\mathbf{Q})$ is solvable by Theorem 15.22.

EXERCISES

1. If $\mathbf{Q} \subseteq L$ is a Galois extension and $Gal(L/\mathbf{Q})$ is Abelian, show that L embeds in a radical extension of $\mathbf{Q}$.
2. Let L be a Galois extension of F with $[L:F] = 2^n$.
   i) Show that there are fields $F = E_0 \subseteq E_1 \subseteq \cdots \subseteq E_n = L$ with all E_j Galois over F and $[E_{j+1} : E_j] = 2$.
   ii) If *char* $F \neq 2$, show that L is a radical extension of F.
3. Let L be a splitting field over F with $[L:F] = 3^k$. If *char* $F \neq 3$, show that L embeds in a radical extension of F. Show that L need not be a radical extension of F.
4. Let L be a Galois extension of $\mathbf{Q}$ with $[L:\mathbf{Q}] = p$, an odd prime. Show that L cannot be a radical extension of $\mathbf{Q}$.
5. If $k > 1$, $p \in \mathbf{N}$ is prime, and $E = \mathbf{Q}[p^{1/k}] \in \mathbf{C}$, show that either $k = 2$ or E is not a Galois extension of $\mathbf{Q}$.
6. If $f(x) \in \mathbf{Q}[x]$ is irreducible and has some root in a radical extension of $\mathbf{Q}$, show that any root of $f(x)$ in some $F \supseteq \mathbf{Q}$ lies in a radical extension of $\mathbf{Q}$.
7. Let L be a splitting field over $\mathbf{Q}$ of $(x^8 + p_1x^4 + p_1)\cdots(x^8 + p_mx^4 + p_m)$, where each p_j is a prime. Show that $Gal(L/\mathbf{Q})$ is solvable.

8. Let $f(x) \in F[x]$ have all its irreducible factors in $F[x]$ of degree at most 4. If L is a splitting field of $f(x)$ over F, and if *char*$F \neq 2, 3$, show that L embeds in a radical extension of F.

9. For $k \geq 1$, set $p_i(x) = x^{2k} + a_i x^k + b_i \in \boldsymbol{Q}[x]$ for $1 \leq i \leq t$. If L is a splitting field over $\boldsymbol{Q}$ of $p_1(x) \cdots p_t(x)$ that show that $Gal(L/\boldsymbol{Q})$ is solvable.

10. Let F be a field, *char* $F \neq 2, 3$, and $p_1(x), \ldots, p_n(x) \in F[x]$ with *deg* $p_j \leq 4$ for all $j \leq n$. Show that any root of the composition $g_n(x) = p_1(p_2(\cdots(p_n(x))\cdots))$ lies in a radical extension of F when:

i) $n = 2$
ii) $n = 3$
iii) $n \in \boldsymbol{N}$

HINTS FOR SELECTED ODD-NUMBERED PROBLEMS

Chapter 0

§0.1 13iii). If $A \cap B \neq \emptyset$ and $a \in A - B$, consider $\{a\} \cup (A \cap B)$.

§0.2 3. Use Definition 0.8 or Theorem 0.2.
17. Consider palindromes with a fixed number of digits.

§0.3 3b). Show $2n + 1 < (n - 1)!$. 3c). Show $2n + 1 < 2^{n-1}$.
7a). Pascal's identity. 11. Define $f(n)$ for n odd and for n even.
15. Example 0.15 and Exercise 1c.

§0.4 11a). See Example 0.18 and Example 0.27. 13. Example 0.18.

§0.5 3. Define $g \circ f$ as ordered pairs.

§0.6 1v). Hyperbolas! 7. Any $f : S \to T$ is $f \subseteq S \times T$.
13. Theorem 0.11 and Theorem 0.13. 21. Find f^{-1} and use Theorem 0.11.
23. Pigeonhole! 27ii). $g \circ f(k) = g(k)$ for all large $k \in N$.

§0.7 3. Use the bijections between A and C and between B and D.

Chapter 1

§1.1 5. No for b) and e). 7. Theorem 1.2 i).
11. Follow the argument in Theorem 1.3. 13. Theorem 1.3.

§1.2 1. Theorem 1.4. 5. Theorem 1.6. 7. Theorem 1.8.
9a). Theorem 1.8.

§1.3 1. Theorem 1.11. 3. Use $k\binom{n}{k} = n\binom{n-1}{k-1}$. 5. Theorem 1.12 iii).
11. Consider $\{a^j \mid 0 \leq j \leq n\}$ and use Theorem 1.2.

§1.4 1. Theorem 1.8. 7. Use prime factorization. 15. Theorem 1.20.
17. Each switch is flipped once for each of its divisors.

§1.5 7. Visualize the ordering on the lattice points in the first quadrant of the plane.
9. The ordering mimics the way words are ordered in a dictionary. For iii), project the nonempty set onto the first coordinate. The elements with minimal first coordinate are smaller than the others. Now project these onto their second coordinates, etc. (use induction on n).

§1.6 1ii). Careful: t depends on f and g. vi). Note that $A - C \subseteq (A - B) \cup (B - C)$.
5i). See Example 0.12. iv). What are the possibilities for $a - b$?
vi). Use prime factorization.
§1.7 7. Consider n even, then n odd. 9. See Example 1.19.
11. Example 0.15 and Exercise 1 in §0.3.
17. Note that $x^3 - 1 = (x - 1)(x^2 + x + 1)$ and $10^{3^{k+1}} = (10^{3^k})^3$. Use Theorem 1.26.
21. Theorem 1.12.
§1.8 11i). Show that $[a]_{nm} = [b]_{nm} \Rightarrow g([a]_{nm}) = g([b]_{nm})$. iii). Theorem 0.10.
§1.9 3i). $T = \boldsymbol{Z} - (\{0\} \cup \{p^k \mid k \geq 1\})$ and $\boldsymbol{Z}_T = \boldsymbol{Q}$. ii). Same as i).
iii). One choice: $T = \boldsymbol{Z} - (\{0\} \cup \{2^i 3^j \mid 0 \leq i < j\})$ and $\boldsymbol{Z}_T = \boldsymbol{Q}$.

Chapter 2

§2.2 5. Consider $\{g^k \in G \mid k \in N\} \subseteq G$.
7. Consider separately $|G| = 2, 3, 4$, and 5. If $G = \{e, a, b\}$, note that $ab \neq a$ and $ab \neq b$. When $G = \{e, a, b, c\}$ what are a^2 and a^3?
9. Think about inverses.
§2.3 9. Show $H \in E(S)$: S is the union of circles with centers $(0, 0)$.
§2.4 3. Theorem 2.9. 7. Theorem 2.8. 9c). Theorem 0.10.
11. Show F is injective (see Theorem 1.28) and use Theorem 0.10.
§2.5 3c). Note that $g(h(r)) - r = g(h(r)) - h(r) + h(r) - r$, and use the triangle inequality.
9a). Make sure to use that G is Abelian.
11. Use G is Abelian, and use Theorem 2.7.
§2.6 3a). Otherwise there is $a \in L - H$ and $b \in L - K$. Use L is a subgroup.
3b). Use part a).
7b). Either there is a maximal order for elements in H or there is not. For any element g of order p^t, $\langle g \rangle$ contains all elements of order p^s for $s \leq t$.
9d, g, h, i). Composition, not multiplication!
15a). The set contains all the reflections.
17b). What can the diagonals of the generators be?
17c). Consider denominators of the entries of the generators.
21. As in Example 2.14, $n > 6$ and $(n, 3) = 1$. Follow the argument in the example: when n is even, using the example, $n > 6$ and $(n, 3) = 1$; when n is odd the result can fail when $3 \mid n$. If $3 \mid n$ then as n-tuples of ± 1, $s_3 s_6 \cdots s_n$ and the product of all s_j both give $(-1, -1, \ldots, -1)$.

Chapter 3

§3.1 3. If G is infinite note that $\langle g \rangle = \langle g^2 \rangle$ for $g \in G$.
7a). Use Theorem 3.1.
9ii). Theorem 3.1!
§3.2 3. If $x^4 \equiv 1 \pmod{56}$ then $x^4 \equiv 1 \pmod 7$ and $x^4 \equiv 1 \pmod 8$.
§3.3 3. Consider cycle structures. For example, if θ has cycle structure $(4, \ldots,)$ then $o(\theta) \leq 12$.

5. Let $i = \theta(x_j)$; then let $i \in \{1, \ldots, n\} - \{\theta(x_j)\}$.
9. If ρ is a k-cycle then ρ^2 is a k-cycle $\Leftrightarrow k$ is odd.
11. Use Theorem 2.8.
19i). First look at ρ^t for $k = 6, 8, 9$, and 10 to find the pattern. If $\rho = (a_1, \ldots, a_k)$ cosider $\rho(a_j) = a_{j+1}$ with subscripts computed in $\{1, \ldots, k\}$ but mod k.
ii). View the action of ρ as in the second statement of 19i) above.
27. How do the transpositions in θ and in θ^{-1} compare?
29. The restrictions of $\alpha, \beta \in A_\infty$ to $I(k) = \{1, \ldots, k\}$ are both in A_k for k large.

§3.5 11ii). Use prime factorization to characterize $(a, b) \in H$.
iii). Reduce to ii): note $sx + ty = 0 \Leftrightarrow ksx + kty = 0$ when $k \neq 0$.
17. If $2m = a + b$ then a and b are both even or both odd.

Chapter 4

§4.1 3. Redo Theorem 4.1.
5d). Show that the cosets determined by the elements in H are all different and that every element in G is in one of these four cosets.
11. Use Theorem 3.17 to compute products in D_{2m}.
13. Identify $f \in S_\infty$ with its restrictions $\theta \in S_n$ for any n suitably large. See Theorem 3.15.
17. Consider$\langle(1, [1]_3)\rangle$ and $\langle(1, [2]_3)\rangle$ and then see what is missing from $\mathbf{Z} \oplus \mathbf{Z}_3$.

§4.2 3. $H = \langle h \rangle$ for $h = g^k$.
5. Theorem 4.3 and induction.
7. Look at coset representatives of $H \bigcap K$ in H.
11ii). For any $g \in G$ and $a \in H^\#$, $a = (gx)^{-1}h_g(gx)$ for $h_g \in H$, so $xax^{-1} \in g^{-1}Hg$.
13. Identify $D_4 \subseteq S_4$ by its action on vertices.
15. Since G is Abelian, H and K are more than just subsets.
23. See Theorem 1.7.
25. What result in this section gives this kind of conclusion?

§4.3 9. $K_iK_j \leq D_{2m} \Leftrightarrow i \equiv j \pmod m \Leftrightarrow K_i = K_j$.
11. Theorem 4.10, Theorem 4.12, and Lagrange.

§4.4 3. If $x, y \in G_q$ use Theorem 4.12 on $\langle x \rangle$ and $\langle y \rangle$.

§4.5 3ii). $G_2 = U_{17} \oplus \langle [-1]_{19} \rangle \oplus \langle [0]_5 \rangle$ is not cyclic (why?), so neither is G.
vii). $G_2 = \langle [-1]_{47} \rangle \oplus \langle [0]_{81} \rangle \oplus \langle [25]_{400} \rangle$, not cyclic.
5. If $o(x)$ does not divide $o(g)$ then for some prime p, $p^k \mid o(x)$ but not $o(g)$. Find $y \in \langle x \rangle$ of order p^k.

§4.6 3. What has order 2 in U_{23}? 9. Use that U_p is cyclic.

Chapter 5

§5.1 5. Do not assume that G is finite! If $N = \langle x \rangle$ and $g \in G$, what is $g^{-1}xg$? What does H look like?
9. If $f \in S_\infty$ and $f(k) = k$ for $k \geq n_0$, consider $g(\{1, 2, \ldots, n_0\})$ for $g \in \mathrm{Bij}(\mathbf{N})$.
11. Note that $(1, n) \in S_n$.
17a). If $gH = g'H$ then is $xgH = xg'H$?

§5.2 3. What is $(Ng)^{o(g)}$ in G/N? 5. Use Exercise 3 or Exercise 4.
7. Use Theorem 2.7. 13. Note that $Z(G) \leq G$ and G/N is a group.
15. K is a transversal for the cosets of N in G.

§5.3 3. Remember what elements in quotients look like.
7. Theorem 5.10.
13a). Compute $y^{-k}xy^k$ and note that $j^p \equiv 1 \pmod{p}$ forces $j \equiv 1 \pmod{p}$.
b). Part a) applies if $k = 2$; use induction on G' to see that $k \leq 2$.

§5.4 13iii). Use the prime factorization of n and Theorem 3.1.

Chapter 6

§6.3 1. What is a non-Abelian group with order $2m$?
5ii). What are the possibilities for the quotient? iii). See Theorem 5.10.
iv). T_j and W_j are the same symbols in any D_k.
9a). $x \in \mathbf{Q}^+$ can be written $x = p_1^{a(1)} \cdots p_k^{a(k)}$ for $p_1 < \cdots < p_k$ primes and $a(j) \in \mathbf{Z}$.
17b). $\lambda \in \mathbf{Z}_n^*$ is determined by $\lambda([1]_n)$. c). See the hint for b).

§6.4 3. Use Theorem 5.11 and Theorem 6.14.
5. Note that $|G|$ is not prime and has a smallest prime divisor.

§6.5 3. See Theorem 5.8.
5. Define a homomorphism from $N_G(H)$ to $Aut(H)$.
9. If $\varphi \in Aut(G)$ and $L \leq G$, how are $|L|$ and $|\varphi(L)|$ related?
13. Apply Theorem 6.20 or Theorem 6.19. 15. See Exercise 13.

Chapter 7

§7.1 3i). If $X \in \rho(H) \cap \rho(K)$ then $X = Nh = Nk$. 7. Use Theorem 7.2.
11. Show that any group of order 14 has an element of order 2 (see Exercise 10 in §2.2) and that any group of order 35 has an element of order 5 (Theorem 4.12). Use Theorem 6.19 to see that G has $H \lhd G$ with $|H| = 5$.
13i). Theorem 5.10 and induction on k. ii) and iii). Use part i).

§7.2 1. Observe that $N(M \cap H)/N \leq M/N$. 3. Use Theorem 5.8 also.
5. When $H \not\subset N$ consider its image HN/N in G/N.
7iii). Use part ii). $H \cap N_1 = H \cap Z(G) \leq Z(H)$. If $H \cap N_1 = \langle e \rangle$ then there is a minimal m with $H \cap N_m \neq \langle e \rangle$ and note $N_m/N_{m-1} = Z(G/N_{m-1})$.
iv). Use ii), induction on s, and Cauchy's Theorem. v). As in iii), let t be minimal with $M \cap N_t \neq \langle e \rangle$. If $g \in G$ and $x \in M \cap N_t$ show $g^{-1}x^{-1}gx = e$.

§7.3 1. Use Theorem 7.9. Note for $A, B \lhd G$ that $A \cap B \lhd A$ and if $A \cap B = \langle e \rangle$ then elements in A commute with elements in B.
3. From Theorem 7.8, if $x \in N_i$, $y \in N_j$, and $i \neq j$ then $xy = yx$.
7bc). If $H \lhd D_n$ contains $\{W_{2i}\}$ or $\{W_{2i+1}\}$ then which T_j must H contain? What must be the order of K?

§7.5 1. Theorem 7.18. 7. What is G^p for $G = \mathbf{Z}_p^n$?
9. What is G^p for $G = \mathbf{Z}_m$ when $m = q^k$ for $q \neq p$ primes? Consider the group homomorphism(!) $[x] \to [x]^p$.

17ii). From Theorem 6.20, $[x] \to [ax]$ is an automorphism of $\mathbf{Z}_n$ if $(a, n) = 1$. Note that if $k > 1$ then $p^{k-1} + 1 < p^k$.

iii). Define automorphisms on G componentwise.

Chapter 8

§8.1 1a). Theorem 2.7.

7. By Theorem 8.4, $|G| = 1 + a + b$ and a, b divide $|G|$. Now $(a, b) > 1$ is impossible so either $a = b = 1$ and $|G| = 3$ or $(a, b) = 1$ and $a < b$. Thus $|G| = abm = 1 + a + b$ resulting in $a = 2$ and $b = 3$.

9i). Theorem 4.3. ii). Conjugates must come from a transveral of $Z(G)$ in G.

§8.2 7. Theorem 8.6. 9. Count conjugates.

§8.3 3. Consider when $Z(G) \leq H$ or not.

7. Theorem 8.4 and Theorem 8.8.

§8.4 5. What is the order of a p-Sylow subgroup? (See p. 98).

11. Use i) to do ii).

13. Use Theorem 4.12 and note that the 5-Sylow subgroup is normal.

17. Use Theorem 8.13 and perhaps Theorem 4.12 or Theorem 6.14.

19. Follow the procedure on p. 290.

Chapter 9

§9.1 5. Note that $hgk^{-1} = g \Leftrightarrow k = g^{-1}hg \in K \cap g^{-1}Hg$.

9. Theorem 9.4. For ii) see Theorem 4.3 and Exercise 11 in §4.2.

§9.2 7a). Z_4 acts by rotation: the generator has four orbits.

b). D_4 acts and two reflections have eight orbits and two have ten orbits.

9. D_3 acts: the rotations have two orbits and the reflections have four orbits.

§9.3 1ix). First apply Theorem 6.13.

7a). $Q \in p\text{-Syl}(N)$ implies $Q = S \cap N$ for $S \in p\text{-Syl}(G)$. Show $g^{-1}(P \cap N)g = g^{-1}Pg \cap N$.

b). Theorem 7.6.

9. For $g \in G$, $g^{-1}Pg \in p\text{-Syl}(K)$.

§9.4 1. Use Theorem 6.14.

3. If a 3-Sylow subgroup P_3 is not normal then $|3\text{-Syl}(G)| = 7$, so $|N(P_3)| = 3^k \cdot 5$. If $k > 2$, H contains a nontrivial normal subgroup L of G by Theorem 6.14. If $|L| = 3^t \cdot 5$ then the 3-Sylow subgroup J of L is normal in G. Consider G/J.

5. If $P_3 \in 3\text{-Syl}(G)$ is not normal then $[G : N(P_3)] = 4$ so $N(P_3)$ contains a normal subgroup H of G. Look at G/H.

7. If G has no normal Sylow subgroup, count the elements of orders 5 and 7.

9. If $P_5 \in 5\text{-Syl}(G)$ is not normal then $[G : N(P_5)] = 6$ so $N(P_5)$ contains a normal subgroup H of G. By Theorem 9.12 assume $|H| = 2$. Apply Theorem 6.13 to G/H.

11. The Sylow subgroups are all normal.

15. Either each Sylow subgroup is normal or some $\mathbf{Z}_q \diamond \mathbf{Z}_p$ appear.

Chapter 10

§10.1 7i). Find A and B that do not commute.
§10.2 17. Verify Theorem 10.4 or give a counter-example to one of its conditions.
i). Note that $0 = 2 \cdot 0$. vii). r depends on f!
§10.3 5. If $j > i$ then $2x^j \notin \langle S_i \rangle$.
11i). Use the definition of addition and multiplication in $R[x]$.
ii). Find A, f, and g with A not commuting with coefficients of f and g.
§10.4 3. See the proof of Theorem 10.10.
7c). $(p^s, q^t) = 1$. d). See Theorem 7.11 or Theorem 9.14.
11. $\langle a \rangle$ is a subring for any $a \in D$.
17c). $(x^2y - x^2yx^2)x^2 = 0$; compute $(x^2y - x^2yx^2)^2$. e). Consider $(2x)^3$.

Chapter 11

§11.1 3vi). Consider the prime factorization of a and of n.
vii). Consider $\mathbf{Z} \oplus \mathbf{Z}$. viii). Use the Binomial Theorem.
15b). Let $K = RT$ and use a). c). Consider $MJM \subseteq J$.
17. Prime factorization!
19. Use the Binomial Theorem.
§11.2 1. Find transversals for the cosets. 5ii). Note that $4 \in \mathbf{Z}$ is in the ideal.
13. See §10.4. 15ii). If $K = M_n(k\mathbf{Z})$ then $M_n(\mathbf{Z})/K$ is essentially $M_n(\mathbf{Z}_k)$.
§11.4 3. Theorem 11.12. 5ii). Project on the first coordinate.
§11.5 3. Note that $N(R)$ contains $(x) + (y)$.
5i). Theorem 11.12. ii). Correspondence Theorem.
7i). Start with $x^2 - x \in I$. iii). Find an ideal containing $x(2\mathbf{Z})[x] = (2\mathbf{Z})(x)$.
11. Start with Theorem 11.12.

Chapter 12

§12.1 7. What is $\mathbf{Z}[x, y, z]/(x)$? See Example 11.17.
9. Find $\alpha = a + bp^{1/2} \in U(\mathbf{Z}[p^{1/2}])$ with $a, b \in \mathbf{N}$ then look at α^n.
11. Theorem 11.2.
§12.2 3. Correspondence Theorem in Chapter 11. 11. Theorem 11.5.
13. If R has a proper nonzero ideal I, consider all $p(x)$ with $p(0) \in I$.
15i). Theorem 12.3.
§12.3 3ii). See Theorem 12.3.
7. Write $\alpha \in \mathbf{Z}[x, x^{-1}]$ as a unit times $p(x) \in \mathbf{Z}[x]$.
9. Clear denominators and use unique factorization.
13. See Exercise 9. 15ii). Norms!
21. Find an isomorphism with a UFD.
iv). Use $x^j y^{j+1}$ irreducible to find a nonunique factorization.
27. Eisenstein (one indeterminate at a time!).
§12.4 5. By definition of η, $xa - yb \neq 0$.
9. If $\alpha, \beta \in R$ then $\alpha = x^i y^j u$ for $u \in U(R)$ and $i, j \geq 0$. When neither $\alpha \mid \beta$ nor $\upsilon(\beta) < \upsilon(\alpha)$, consider $\beta - \alpha$ and use Definition 12.8.

Chapter 13

§13.1 5. Theorem 13.5.

7. If $x \in P$ is not a zero divisor, consider the multiplicative set generated by x and $R - P$.

9ii). Use i). iii). Use i) on the poset $(S, \leq)$ where $\leq$ reverses $\subseteq$.

13. An ideal P of R is prime $\Leftrightarrow R/P$ is a domain; see Theorem 11.12.

21. Use $\mathbf{Z}$ for examples.

§13.2 7. Note that every element in R_M is a monomial in X times a unit in R_M. If I is an ideal in R_M take an element in I using the fewest elements from X.

9. Theorem 13.3 or Theorem 13.11.

11. If the maximal ideal of R_M is I then $P = I^c \subseteq R - M$ is a prime ideal of R. If $x \in R - P$, show that $\theta_M(x) \notin I$ so is invertible. See Exercise 4.

13. Start with k different primes.

17. Use the natural map of "fractions" and verify that this is a function.

§13.3 5. There is $I \lhd R$ maximal with $R/I \cong R$. What corresponds to I in R/I?

9i)–iv). Each case is $R/(a)$. Show that $a \in R$ is irreducible.

11. Note that S need not be commutative. 13. Denominators!

15. Theorem 13.18. 21. Theorem 13.21 and Theorem 13.20.

§13.4 3. Use Theorem 13.26.

11. If $\beta \in L$ then β is integral over K so for some $y \in R$, $y\beta$ is integral over R.

13. Find nonassociate primes with the same norm; see Theorem 12.19.

17. If $f \notin M$ consider $(f) + M$. 19. Use Theorem 13.32.

§13.5 3. What is $\text{rad}(I)$ considered in R/I?

5ii). Any $\alpha \in \mathcal{V}(I)$ defines a maximal ideal of R containing I; the product of ideals is contained in their intersection.

7. An irreducible $f \in R$ generates a prime ideal in R, so what is $\text{rad}((f))$?

9i). See Theorem 13.20. ii). A finite intersection of maximal ideals in R contains a polynomial in each variable.

11. Explicitly find the variety of the given ideal.

13. Otherwise $\mathcal{V}((x^2 + y^2 - 1)) \subseteq \mathcal{V}((f_1 \cdots f_k))$. Use $f_1 \cdots f_k$ homogeneous and consider degrees.

21. Theorem 13.38.

§13.6 9. Mimic the proof of Theorem 13.44.

15. Extend the lexicographic ordering on N^m.

17. Look at whether there are maximal elements and how many elements have immediate predecessors.

Chapter 14

§14.1 9. See Theorem 14.5.

19ii). Evaluate a linear combination of $T_{\lambda\mu}$ at various $\alpha_\gamma \in B$.

iv). Any $T \in span(\{T_{\lambda\mu}\})$ satisfies $T(\alpha) = 0$ for almost all $\alpha \in B$.

vi). Use iii) and Theorem 13.45.

21. Theorem 13.47 and Theorem 13.46.

§14.2 1iv). $2^{1/2} + 3^{1/2}$ satisfies $f(x) \in \boldsymbol{Q}[x]$ of degree 4. Is there one of smaller degree?
3. Complete squares. 7. See Theorem 13.26.
11. Use Theorem 14.7 and Theorem 14.8.

§14.4 1i), ii). Quadratic formula! 7. Theorem 14.15.
9. If $\varphi \in Aut(L/\boldsymbol{Q})$ then $\varphi(1) = 1$.
11. Consider roots of polynomials.
13. If $\varphi(r) \neq r$, consider rationals between them.
15. Apply Theorem 14.17.
19. Note that $L = \boldsymbol{Q}[2^{1/3}, \omega]$ for $\omega \in \boldsymbol{C}$, with $\omega^3 = 1$. Describe $\varphi \in Aut(L/\boldsymbol{Q})$ by its action on $2^{1/3}$ and on ω.

§14.5 3. Let S be all the quadratic polynomials in $\boldsymbol{Q}[x]$.
5. If $a + bi \in L$, show that $a, b \in L \cap \boldsymbol{R}$ (use complex conjugation).

§14.6 1. Note that $\boldsymbol{Q}(x, y)$ is algebraic over $\boldsymbol{Q}(x - y, x + y)$.
7. It is easier to use the theorems than the definition.
9. The 2×2 matrix with rows (a, b) and (c, d) has an inverse. Use this to see that some $(a'\alpha + b')/(c'\alpha + d') = x$ with $a', b', c', d' \in \boldsymbol{Q}$.

Chapter 15

§15.1 1. Think about roots of polynomials.
9. The minimal polynomial for $\alpha \in L$ over F is in $E[x]$.
11. Using Theorem 14.16, define automorphisms by their action on $2^{1/2}$ and i.
13. Consider a suitable extension of $F[\alpha_1, \ldots, \alpha_m]$ and use Theorem 15.2.
21. Compute that $[L : \boldsymbol{Q}] = 8$ (Exercise 14 in §14.2) and use Theorem 14.18.

§15.2 5. Splitting fields! 9ii). If $\varphi \in Gal(L/F)$ then $\varphi(E) \subseteq E$.
11. Every finite group contains some maximal normal subgroup.
15. Complex conjugation!
19. Use calculus to find the number of real roots.
23. What must $Gal(L/E)$ be?
25. What can one say about the subgroups of $Gal(L/F)$, and what does this say about the structure of $Gal(L/F)$?
29. Does the condition on degrees of elements say anything about dimensions of intermediate fields? Use Theorem 15.7.

§15.3 5. Compute powers of φ acting on α.
7. What is the exponent in terms of powers of p?
13. See Exercise 9. 19. See p. 505.

§15.4 1. If $p > 2$ then $p^2 \equiv 1 \pmod 8$.
3. How many primitive pk-roots of unity are there?
11. Show that $\boldsymbol{Q}[\varepsilon, \rho] = \boldsymbol{Q}[\mu]$ for μ a primitive LCM($\{n, m\}$)-root of unity.
13. Complex conjugation is in $Gal(\boldsymbol{Q}[\varepsilon]/\boldsymbol{Q})$.

§15.5 5. Find a normal Sylow subgroup. 11. See Theorem 9.14.
13. Theorem 15.18 and Theorem 15.19.

§15.6 3. See p. 523.
7. What do the roots of the degree-8 factors look like?
9. What do the roots of the $p_i(x)$ look like?

INDEX